Calculus

Applications and Technology

THIRD EDITION

Life Sciences

Physical Sciences

Social Sciences

Calculus

Applications and Technology

THIRD EDITION

Edmond C. Tomastik
University of Connecticut

With Interactive Illustrations by
Hu Hohn, Massachusetts School of Art

Australia • Canada • Mexico • Singapore • Spain
United Kingdom • United States

Publisher: Curt Hinrichs
Development Editor: Cheryll Linthicum
Assistant Editor: Ann Day
Editorial Assistant: Katherine Brayton
Technology Project Manager: Earl Perry
Marketing Manager: Tom Ziolkowski
Marketing Assistant: Jessica Bothwell
Advertising Project Manager: Nathaniel Bergson-Michelson
Senior Project Manager, Editorial Production: Janet Hill
Print/Media Buyer: Barbara Britton
Production Service: Hearthside Publication Services/Anne Seitz
Text Designer: John Edeen
Art Editor: Ann Seitz
Photo Researcher: Gretchen Miller
Copy Editor: Barbara Willette
Illustrator: Hearthside Publishing Services/Jade Myers
Cover Designer: Ron Stanton
Cover Image: Gary Holscher
Interior Printer: Quebecor/Taunton
Cover Printer: Phoenix Color Corp
Compositor: Techsetters, Inc.

Printed in the United States of America

1 2 3 4 5 6 7 08 07 06 05 04

For more information about our products, contact us at:
Thomson Learning Academic Resource Center
1-800-423-0563
For permission to use material from this text or product, submit a request online at
http://www.thomsonrights.com.
Any additional questions about permissions can be submitted by email to **thomsonrights@thomson.com**

Library of Congress Control Number: 2004104623

Student Edition: ISBN 0-534-46496-3

Instructor's Edition: ISBN 0-534-46498-X

Thomson Brooks/Cole
10 Davis Drive
Belmont, CA 94002
USA

Asia
Thomson Learning
5 Shenton Way #01-01
UIC Building
Singapore 068808

Australia/New Zealand
Thomson Learning
102 Dodds Street
Southbank, Victoria 3006
Australia

Canada
Nelson
1120 Birchmount Road
Toronto, Ontario M1K 5G4
Canada

Europe/Middle East/Africa
Thomson Learning
High Holborn House
50/51 Bedford Row
London WC1R 4LR
United Kingdom

Latin America
Thomson Learning
Seneca, 53
Colonia Polanco
11560 Mexico D.F.
Mexico

Spain/Portugal
Paraninfo
Calle Magallanes, 25
28015 Madrid, Spain

An Overview of Third Edition Changes

1. In this new edition we have followed a general philosophy of dividing the material into **smaller, more manageable sections**. This has resulted in an increase in the number of sections. We think this makes it easier for the instructor and the student, gives more flexibility, and creates a better flow of material.
2. To add to the flexibility, many sections now have **enrichment subsections**. Material in such enrichment subsections is not needed in the subsequent text (except possibly in later enrichment subsections). Now instructors can easily tailor the material in the text to teach a course at different levels.
3. The third edition has even more **referenced real-life examples**. It is important to realize that the mathematical models presented in these referenced examples are models created by the experts in their fields and published in refereed journals. So not only is the data in these referenced examples real data, but the mathematical models based on this real data have been created by experts in their fields (and not by us).
4. **Mathematical modeling** is stressed in this edition. Mathematical modeling is an attempt to describe some part of the real world in mathematical terms. Already at the beginning of Section 1.2 we describe the three steps in mathematical modeling: formulation, mathematical manipulation, and evaluation. We return to this theme often. For example, in Section 5.6 on optimization and modeling we give a six-step procedure for mathematical modeling specifically useful in optimization. Essentially every section has examples and exercises in mathematical modeling.
5. This edition also includes many more opportunities to **model by curve fitting**. In this kind of modeling we have a set of data connecting two variables, x and y, and graphed in the xy-plane. We then try to find a function $y = f(x)$ whose graph comes as close as possible to the data. This material is found in a new Chapter 2 and can be skipped without any loss of continuity in the remainder of the text. Curve-fitting exercises are clearly marked as such.
6. The text is now **technology independent**. Graphing calculators or computers work just as well with the text.
7. A disk with **interactive illustrations** is now included with each text. These interactive illustrations provide the student and instructor with wonderful demonstrations of many of the important ideas in the calculus. They appear in every chapter. These demonstrations and explorations are highlighted in the text at appropriate times. They provide an extraordinary means of obtaining deep and clear insights into the important concepts. We are extremely excited to present these in this format.

Chapter 1. Functions. This chapter now contains five sections: 1.1, Functions; 1.2, Mathematical Models; 1.3, Exponential Models; 1.4, Combinations of Functions; and 1.5, Logarithms. The material that covered modeling with least squares has all been moved to a new Chapter 2. Most of the material in the sections on quadratics and special functions has been moved to the Review Appendix. A geometric definition of continuity now appears in the first section.

Chapter 2. Modeling with Least Squares. This is a new chapter and places all the material on least squares that was originally in Chapter 1 into this new chapter. Instructors who wish can ignore the material in this chapter.

Chapter 3. Limits and the Derivative. This chapter has been substantially revised. The material on the limit definition of continuity is now an "enrichment" subsection of the first section on limits and is not needed in the remainder of the text. The material on limits at infinity has been moved to a later chapter. The section on rates of change now has more examples of average rates of change. More emphasis is put on interpretations of rates of change and on units. The old section on derivatives has been made into two sections, the first on derivatives and the second on local linearity. The new section on derivatives has more emphasis on graphing the derivative given the function and also on interpretations. The section on local linearity now includes marginal analysis and the economic interpretation of the derivatives of cost, revenue, and profits. This latter material was formerly in a later chapter.

Chapter 4. Rules for Derivatives. This chapter now includes more "intuitive," that is, geometrical and numerical, sketches of a number of proofs, the formal proofs being given in enrichment subsections. Thus, a geometrical sketch of the proof for the derivative of a constant times a function is given, and numerical evidence for the proof for the derivative of the sum of two functions is given. The formal proofs of these, together with the proof of the derivative of the product, are in optional subsections. More geometrical insight has been added to the chain rule, and more emphasis is put on determining units. The more difficult proofs in the exponential and logarithm section have been placed in an enrichment subsection. Elasticity of demand now has it's own section. The introductory material on elasticity has been rewritten to make the topic more transparent. The last section on applications on renewable resources has been updated with timely new material.

Chapter 5. Curve Sketching and Optimization. This chapter has been extensively reorganized. The second section on the second derivative now contains only material specific to concavity and the second derivative test and is much shorter and much more manageable. The material on additional curve sketching that was previously in this section has been given its own section, Section 5.4. Limits at infinity are now discussed in Section 3, having been moved from an earlier chapter. It is in this chapter that this material is actually used, so it seems appropriate that it be located here. The old section on optimization has been split into two sections, the first on absolute extrema and the second on optimization and mathematical modeling. A new section on the logistic curve has been created from material found scattered in various sections. With its own section, new material has been added to give this important model its proper due (although instructors can omit this material without effecting the flow of the text).

Chapter 6. Integration. The section on substitution has been refocused to have a more intuitive as opposed to formal approach and is now more easily accessible. To the third section, on distance traveled, more examples of Riemann sums have been added, and taking the limit as $n \to \infty$ is postponed until the next section. The section on the definite integral now contains some properties of integrals that were not found in the last edition. The section on the fundamental theorem of calculus has been extensively rewritten, with a different proof of the fundamental theorem given. We first show that the derivative of $\int_a^x f(t)\,dt$ is $f(x)$ using a geometric argument using the new properties of integrals that were included in the previous section and then proceed to prove $\int_a^b f(t)\,dt = F(b) - F(a)$, where F is an antiderivative. The more formal proof is given in an enrichment subsection. Finally, a new Section 6.7 has been created to include the various applications of the integral that had been scattered in previous sections.

Chapter 7. Additional Topics in Integration. The interactive illustrations in the numerical integration section yield considerable insight into the subject. Students can move from one method to another and choose any n and see the graphs and the numerical answers immediately.

Chapter 8. Functions of Several Variables. Graphing in several variables and visualizing the geometric interpretation of partial derivatives is always difficult. There are several interactive illustrations in this chapter that are extremely helpful in this regard.

Chapter 9. The Trigonometric Functions. This chapter covers an introduction to the trigonometric functions, including differentiation and integration.

Chapter 10. Taylor Polynomials and Infinite Series. This chapter covers Taylor polynomials and infinite series. Sections 10.1, 10.2, and 10.7 constitute a subchapter on Taylor polynomials. Section 10.7 is written so that the reader can go from Section 10.2 directly to Section 10.7.

Chapter 11. Probability and Calculus. This chapter is on probability. Section 11.1 is a brief review of discrete probability. Section 11.2 considers continuous probability density functions and Section 11.3 presents the expected value and variance of these functions. Section 11.4 covers the normal distribution.

Chapter 12. Differential Equations. This chapter is a brief introduction to differential equations and includes the technique of separation of variables, approximate solutions using Euler's method, some qualitative analysis, and mathematical problems involving the harvesting of a renewable natural resource.

Preface

Calculus: Applications and Technology is designed to be used in a one- or two-semester calculus course aimed at students majoring in business, management, economics, or the life or social sciences. The text is written for a student with two years of high school algebra. A wide range of topics is included, giving the instructor considerable flexibility in designing a course.

Since the text uses technology as a major tool, the reader is required to use a computer or a graphing calculator. The Student's Suite CD with the text, gives all the details, in user friendly terms, needed to use the technology in conjunction with the text. This text, together with the accompanying Student's Suite CD, constitutes a completely organized, self-contained, user-friendly set of material, even for students without any knowledge of computers or graphing calculators.

Philosophy

The writing of this text has been guided by four basic principles, all of which are consistent with the call by national mathematics organizations for reform in calculus teaching and learning.

1. **The Rule of Four:** Where appropriate, every topic should be presented graphically, numerically, algebraically, and verbally.
2. **Technology:** Incorporate technology into the calculus instruction.
3. **The Way of Archimedes:** Formal definitions and procedures should evolve from the investigation of practical problems.
4. **Teaching Method:** Teach calculus using the investigative, exploratory approach.

The Rule of Four. By always bringing graphical and numerical, as well as algebraic, viewpoints to bear on each topic, the text presents a conceptual understanding of the calculus that is deep and useful in accommodating diverse applications. Sometimes a problem is done algebraically, then *supported* numerically and/or graphically (with a grapher). Sometimes a problem is done numerically and/or graphically (with a grapher), then *confirmed* algebraically. Other times a problem is done numerically or graphically because the algebra is too time-consuming or impossible.

Technology. Technology permits more time to be spent on concepts, problem solving, and applications. The technology is used to assist the student to think about the geometric and numerical meaning of the calculus, without undermining the algebraic aspects. In this process, a balanced approach is presented. I point out clearly that the computer or graphing calculator might not give the whole story, motivating the need to learn the calculus. On the other hand, I also stress common situations in which exact solutions are impossible, requiring an approximation technique using the technology. Thus, I stress that the graphers are just another needed tool, along with the calculus, if we are to solve a variety of problems in the applications.

Applications and the Way of Archimedes. The text is written for *users* of mathematics. Thus, applications play a central role and are woven into the development of the material. Practical problems are always investigated first, then used to motivate, to maintain interest, and to use as a basis for developing definitions and procedures. Here too, technology plays a natural role, allowing the forbidding and time-consuming difficulties associated with real applications to be overcome.

The Investigative, Exploratory Approach. The text also emphasizes an investigative and exploratory approach to teaching. Whenever practical, the text gives students the opportunity to explore and discover for themselves the basic calculus concepts. Again, technology plays an important role. For example, using their graphers, students discover for themselves the derivatives of x^2, x^3, and x^4 and then generalize to x^n. They also discover the derivatives of $\ln x$ and e^x. None of this is realistically possible without technology.

Student response in the classroom has been exciting. My students enjoy using their computers or graphing calculators and feel engaged and part of the learning process. I find students much more receptive to answering questions about their observations and more ready to ask questions.

A particularly effective technique is to take 15 or 20 minutes of class time and have students work in small groups to do an exploration or make a discovery. By walking around the classroom and talking with each group, the instructor can elicit lively discussions, even from students who do not normally speak. After such a minilab the whole class is ready to discuss the insights that were gained.

Fully in sync with current goals in teaching and learning mathematics, every section in the text includes a more challenging exercise set that encourages exploration, investigation, critical thinking, writing, and verbalization.

Interactive Illustrations. The Student's Suite CD with **interactive illustrations** is now included with each text. These interactive illustrations provide the student and instructor with wonderful demonstrations of many of the important ideas in the calculus. These demonstrations and explorations are highlighted in the text at appropriate times. They provide an extraordinary means of obtaining deep and clear insights into the important concepts. We are extremely excited to present these in this format. They are one more important example of the use of technology and fit perfectly into the investigative and exploratory approach.

Content Overview

Chapter 1. Section 1.0 contains some examples that clearly indicate instances when the technology fails to tell the whole story and therefore motivates the need to learn the calculus. This failure of the technology to give adequate information is complemented elsewhere by examples in which our current mathematical knowledge is inadequate to find the exact values of critical points, requiring us to use some approximation technique on our computers or graphing calculators. This theme of needing both mathematical analysis and technology to solve important problems continues throughout the text. Section 1.1 begins with functions; the second section contains applications of linear and nonlinear functions in business and economics, including an introduction to the theory of the firm. Next is a section on exponential functions, followed by the algebra of functions, and finally logarithmic functions.

Chapter 2. This chapter consists entirely of fitting curves to data using least squares. It includes linear, quadratic, cubic, quartic, power, exponential, logarithmic, and logistic regression.

Chapter 3. Chapter 3 begins the study of calculus. Section 3.1 introduces limits intuitively, lending support with many geometric and numerical examples. Section 3.2 covers average and instantaneous rates of change. Section 3.3 is on the derivative. In this section, technology is used to find the derivative of $f(x) = \ln x$. From the limit definition of derivative we know that for h small, $f'(x) \approx \dfrac{f(x+h) - f(x)}{h}$. We then take $h = 0.001$ and graph the function $g(x) = \dfrac{\ln(x + 0.001) - \ln x}{0.001}$. We see on our grapher that $g(x) \approx 1/x$. Since $f'(x) \approx g(x)$, we then have strong evidence that $f'(x) = 1/x$. This is confirmed algebraically in Chapter 4. Section 3.4 covers local linearity and introduces marginal analysis.

Chapter 4. Section 4.1 begins the chapter with some rules for derivatives. In this section we also discover the derivatives of a number of functions using technology. Just as we found the derivative of $\ln x$ in the preceding chapter, we graph $g(x) = \dfrac{f(x + 0.001) - f(x)}{0.001}$ for the functions $f(x) = x^2$, x^3, and x^4 and then discover from our graphers what particular function $g(x)$ is in each case. Since $f'(x) \approx g(x)$, we then discover $f'(x)$. We then generalize to x^n. In the same way we find the derivative of $f(x) = e^x$. This is an exciting and innovative way for students to find these derivatives. Now that the derivatives of $\ln x$ and e^x are known, these functions can be used in conjunction with the product and quotient rules found in Section 4.2, making this material more interesting and compelling. Section 4.3 covers the chain rule, and Section 4.4 derives the derivatives of the exponential and logarithmic functions in the standard fashion. Section 4.5 is on elasticity of demand, and Section 4.6 is on the management of renewable natural resources.

Chapter 5. Graphing and curve sketching are begun in this chapter. Section 5.1 describes the importance of the first derivative in graphing. We show clearly that the technology can fail to give a complete picture of the graph of a function, demonstrating the need for the calculus. We also consider examples in which the exact values of the critical points cannot be determined and thus need to resort to using an approximation technique on our computers or graphing calculators. Section 5.2 presents the second derivative, its connection with concavity, and its use in graphing. Section 5.3 covers limits at infinity, Section 5.4 covers additional curve sketching, and Section 5.5 covers absolute extrema. Section 5.6 includes optimization and modeling. Section 5.7 covers the logistic model. Section 5.8 covers implicit differentiation and related rates. Extensive applications are given, including Laffer curves used in tax policy, population growth, radioactive decay, and the logistic equation with derived estimates of the limiting human population of the earth.

Chapter 6. Sections 6.1 and 6.2 present antiderivatives and substitution, respectively. Section 6.3 lays the groundwork for the definite integral by considering left- and right-hand Riemann sums. Here again technology plays a vital role. Students can easily graph the rectangles associated with these Riemann sums and see graphically and numerically what happens as $n \to \infty$. Sections 6.4, 6.5, and 6.6 cover the definite integral, the fundamental theorem of calculus, and area between two curves, respectively. Section 6.7 presents a number of additional applications of the integral, including average value, density, consumer's and producer's surplus, Lorentz's curves, and money flow.

Chapter 7. This chapter contains material on integration by parts, integration using tables, numerical integration, and improper integrals.

Chapter 8. Section 8.1 presents an introduction to functions of several variables, including cost and revenue curves, Cobb-Douglas production functions, and level curves. Section 8.2 then introduces partial derivatives with applications that include competitive and complementary demand relations. Section 8.3 gives the second derivative test for functions of several variables and applied application on optimization. Section 8.4 is on Lagrange multipliers and carefully avoids algebraic complications. The tangent plane approximations is presented in Section 8.5. Section 8.6, on double integrals, covers double integrals over general domains, Riemann sums, and applications to average value and density. A program is given for the graphing calculator to compute Riemann sums over rectangular regions.

Chapter 9. This chapter covers an introduction to the trigonometric functions. Section 9.1 starts with angles, and Sections 9.2, 9.3, and 9.4 cover the sine and cosine functions, including differentiation and integration. Section 9.5 covers the remaining trigonometric functions. Notice that these sections include extensive business applications, including models by Samuelson and Phillips. Notice in Section 9.3 that the derivatives of $\sin x$ and $\cos x$ are found by using technology and that technology is used throughout this chapter.

Chapter 10. This chapter covers Taylor polynomials and infinite series. Sections 10.1, 10.2, and 10.7 constitute a subchapter on Taylor polynomials. Section 10.7 is written so that the reader can go from Section 10.2 directly to Section 10.7. Section 10.1 introduces Taylor polynomials, and Section 10.2 considers the errors in Taylor polynomial approximation. The graphers are used extensively to compare the Taylor polynomial with the approximated function. Section 10.7 looks at Taylor series, in which the interval of convergence is found analytically in the simpler cases while graphing experiments cover the more difficult cases. Section 10.3 introduces infinite sequences, and Sections 10.4, 10.5, and 10.6 are on infinite series and includes a variety of test for convergence and divergence.

Chapter 11. This chapter is on probability. Section 11.1 is a brief review of discrete probability. Section 11.2 then considers continuous probability density functions, and Section 11.3 presents the expected value and variance of these functions. Section 11.4 covers the normal distribution, arguably the most important probability density function.

Chapter 12. This chapter is a brief introduction to differential equations and includes the technique of separation of variables, approximate solutions using Euler's method, some qualitative analysis, and mathematical problems involving the harvesting of a renewable natural resource. The graphing calculator is used to graph approximate solutions and to do some experimentation.

Important Features

Style. The text is designed to implement the philosophy stated earlier. Every chapter and section opens by posing an interesting and relevant applied problem using familiar vocabulary; this problem is solved later in the chapter or section after the appropriate mathematics has been developed. Concepts are always introduced intuitively, evolve gradually from the investigation of practical problems or particular cases, and culminate in a definition or result. Students are given the opportunity to investigate and discover concepts for themselves by using the technology, including the interactive illustrations, or by doing the explorations. Topics are presented graphically, numerically, and algebraically to give the reader a deep and conceptual understanding. Scattered throughout the text are historical and anecdotal comments. The historical comments not only are interesting in themselves, but also indicate that mathematics is a continually developing subject. The anecdotal comments relate the material to contemporary real-life situations.

Applications. The text includes many meaningful applications drawn from a variety of fields, including over 500 referenced examples extracted from current journals. Applications are given for all the mathematics that is presented and are used to motivate the students. See the Applications Index.

Explorations. These explorations are designed to make the student an active partner in the learning process. Some of these explorations can be done in class, and some can be done outside class as group or individual projects. Not all of these explorations use technology; some ask students to solve a problem or make a discovery using pencil and paper.

Interactive Illustrations. The interactive illustrations provide the student and instructor with wonderful demonstrations of many of the important ideas in the calculus. These demonstrations and explorations are highlighted in the text at appropriate times. They provide an extraordinary means of obtaining deep and clear insights into the important concepts. They are one more important example of the use of technology and fit perfectly into the investigative and exploratory approach.

Worked Examples. Over 400 worked examples, including warm up examples mentioned below, have been carefully selected to take the reader progressively from the simplest idea to the most complex. All the steps that are needed for the complete solutions are included.

Connections. These are short articles about current events that connect with the material being presented. This makes the material more relevant and interesting.

Screens. About 100 computer or graphing calculator screens are shown in the text. In almost all cases, they represent opportunities for the instructor to have the students reproduce these on their graphers at the point in the lecture when they are needed. This makes the student an active partner in the learning process, emphasizes the point being made, and makes the classroom more exciting.

Enrichment Subsections. Many sections in the text have an enrichment subsection at the end. Sometimes this subsection will include proofs that not all instructors might wish to present. Sometimes this subsection will include material that goes beyond what every instructor might wish to cover for the particular topic. It seems likely that most instructors will use some of the enrichment subsections, but very few will use all of them. This feature gives added flexibility to the text.

Warm Up Exercises. Immediately preceding each exercise set is a set of warm up exercises. These exercises have been very carefully selected to bridge the gap between the exposition in the section and the regular exercise set. By doing these exercises and checking the complete solutions provided, students will be able to test or check their comprehension of the material. This, in turn, will better prepare them to do the exercises in the regular exercise set.

Exercises. The book contains over 2600 exercises. The exercises in each set gradually increase in difficulty, concluding with the more challenging exercises mentioned below. The exercise sets also include an extensive array of realistic applications from diverse disciplines, including numerous referenced examples extracted from current journals.

More Challenging Exercises. Every section in the text includes a more challenging exercise set that encourages exploration, investigation, critical thinking, writing, and verbalization.

Mathematical Modeling Exercises. Every section in the text has exercises that provide the opportunity for mathematical modeling. The discipline of taking a problem and translating it into a mathematical equation or construct may well be more important than learning the actual material.

Modeling Exercises by Curve Fitting. Some instructors are interested in curve fitting using least squares. Ample exercises are provided throughout the text that use curve fitting as part of the problem.

End-of-Chapter Cases. These cases, found at the end of each chapter, are especially good for group assignments. They are interesting and will serve to motivate the mathematics student.

Learning Aids.

- **Boldface** is used when new terms are defined.
- **Boxes** are used to highlight definitions, theorems, results, and procedures.
- **Remarks** are used to draw attention to important points that might otherwise be overlooked.
- **Warnings** alert students against making common mistakes.
- **Titles** for worked examples help to identify the subject.
- **Chapter summary outlines** at the end of each chapter conveniently summarize all the definitions, theorems, and procedures in one place.
- **Review exercises** are found at the end of each chapter.
- **Answers** to selected exercises and to all the review exercises are provided in an appendix.

Instructor Aids

- The **Instructor's Suite CD** contains electronic versions of the Instructor's Solutions Manual, Test Bank, and a Microsoft® Power-Point® presentation tool.
- The **Instructor's Solutions Manual** provides completely worked solutions to all the exercises and to all the Explorations.
- The **Student Solutions Manual** contains the completely worked solutions to selected exercises and to all chapter review exercises. Between the two manuals all exercises are covered.
- The Graphing Calculator Manual and Microsoft® Excel Manual, available electronically, have all the details, in user friendly terms, on how to carry out any of the graphing calculator operations and Excel operations used in the text. The Graphing Calculator Manual includes the standard calculators and computer algebra systems.
- The **Test Bank** written by James Ball (University of Indiana) includes a combination of multiple-choice and free-response test questions organized by section.
- A **BCA/iLrn Instructor Version** allows instructors to quickly create, edit, and print tests or different versions of tests from the set of test questions accompanying the text. It is available in IBM or Mac versions.

Custom Publishing. Courses in business calculus are structured in various ways, differing in length, content, and organization. To cater to these differences, Brooks/Cole Publishing is offering Applied Calculus with Technology and Applications in a custom-publishing format. Instructors can rearrange, add, or cut chapters to produce a text that best meets their needs. Chapters on differential equations, trigonometric functions, Taylor polynomials and infinite series, and probability are also available.

Thomson Brooks/Cole is working hard to provide the highest-quality service and product for your courses. If you have any questions about custom publishing, please contact your local Brooks/Cole sales representatives.

Acknowledgments. I owe a considerable debt of gratitude to Curt Hinrichs, Publisher, for his support in initiating this project, for his insightful suggestions in preparing the manuscript, and for obtaining the services of Hu Hohn, mentioned below, who created the interactive illustrations that are in this text. These interactive illustrations are wonderful enhancements to the text.

I wish to thank Hu Hohn for his considerable work and creativity in developing all of the Interactive Illustrations that are in this text.

I wish also to thank the other editorial, production, and marketing staff of Brooks/Cole: Katherine Brayton, Ann Day, Janet Hill, Hal Humphrey, Cheryll Linthicum, Earl Perry, Jessica Perry, Barbara Willette, Joseph Rogove, and Marlene Veach. I wish to thank David Gross and Julie Killingbeck for doing an excellent job ensuring the accuracy and readability of this edition.

I would like to thank Anne Seitz for an outstanding job as Production Editor and to thank Jade Myers for the art and Barbara Willette for the copyediting.

I would like to thank the Mathematics Department at the University of Connecticut for their collective support, with particular thanks to Professors Jeffrey Tollefson, Charles Vinsonhaler, and Vince Giambalvo, and to our computer manager Kevin Marinelli.

I would especially like to thank my wife Nancy, since without her support, this project would not have been possible.

Many thanks to all the reviewers listed below.

Bruce Atkinson, Samford University; Robert D. Brown, University of Kansas; Thomas R. Caplinger, University of Memphis; Janice Epstein, Texas A&M University; Tim Hagopian, Worcester State College. Fred Hoffman, Florida Atlantic University; Miles Hubbard, Saint Cloud State University; Kevin Iga, Pepperdine University; David L. Parker, Salisbury University; Georgia Pyrros, University of Delaware; Geetha Ramanchandra, California State University, Sacramento; Jennifer Stevens, University of Tennessee; Robin G. Symonds, Indiana University, Kokomo; Stuart Thomas, University of Oregon at Eugene; Jennifer Whitfield, Texas A&M University; and Richard Witt, University of Wisconsin, Eau Claire.

Table of Contents

Calculus

Applications and Technology

THIRD EDITION

1 Functions

This chapter covers functions, which form the basis of calculus. We introduce the notion of a function, introduce a variety of functions, and then explore the properties and graphs of these functions.

CASE STUDY

We consider here certain data found in a recent detailed study by Cotterill and Haller[1] of the costs and pricing for a number of brands of breakfast cereals. The data shown in Table 1.1 were in support of Cotterill's testimony as an expert economic witness for the state of New York in State of New York v. Kraft General Foods et al. It is the first, and probably only, full-scale attempt to present in a federal district court analysis of a merger's impact using scanner-generated brand-level data and econometric techniques to estimate brand- and category-level responses of demand to pricing. Keep in mind that the data in this study were obtained from Kraft by court order as part of New York's challenge of the acquisition of Nabisco Shredded Wheat by Kraft General Foods. Otherwise, such data would be extremely difficult, and most likely impossible, to obtain.

Table 1.1

Item	$/lb	$/ton
Manufacturing cost:		
Grain	0.16	320
Other ingredients	0.20	400
Packaging	0.28	560
Labor	0.15	300
Plant costs	0.23	460
Total manufacturing costs	1.02	2040
Marketing expenses:		
Advertising	0.31	620
Consumer promo (mfr. coupons)	0.35	700
Trade promo (retail in-store)	0.24	480
Total marketing costs	0.90	1800
Total costs per unit	1.92	3840

[1] Ronald W. Cotterill and Lawrence E. Haller. 1997. An economic analysis of the demand for RTE cereal: product market definition and unilateral market power effects. Research Report No. 35. Food Marketing Policy Center. University of Connecticut.

The manufacturer obtains a price of \$2.40 a pound, or \$4800 a ton. Nevo[2] estimated the costs of construction of a typical plant to be \$300 million. We want to find the cost, revenue and profit equations.

Let x be the number of tons of cereal manufactured and sold, and let p be the price of a ton sold. Notice that, according to Table 1.1, the cost to manufacture each ton of cereal is \$3840. So the cost of manufacturing x tons is $3840x$ dollars. To obtain (total) cost, we need to add to this the cost of the plant itself, which was \$300 million. To simplify the cost equation, let total cost C be given in thousands of dollars. Then the total cost C, in thousands of dollars, for manufacturing x tons of cereal is given by $C = 300{,}000 + 3.84x$. This is graphed in Figure 1.1.

A ton of cereal sold for \$4800. So selling x tons of cereal returned revenue of $4800x$ dollars. If we let revenue R be given in thousands of dollars, then the revenue from selling x tons of cereal is $R = 4.8x$. This is shown in Figure 1.1.

[2] Aviv Nevo. 2001. Measuring market power in the ready-to-eat cereal industry. *Econometrica* 69(2):307–342.

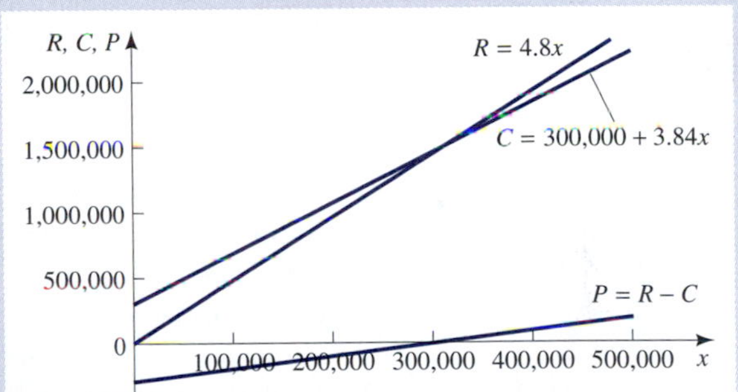

Figure 1.1

Profits are always just revenue less costs. So if P is profits in thousands of dollars, then

$$\begin{aligned} P = R - C &= (4.8x) - (3.840x + 300{,}000) \\ &= 0.96x - 300{,}000 \end{aligned}$$

This equation is also graphed in Figure 1.1.

We might further ask how many tons of cereal we need to manufacture and sell before we break even. The answer can be found in Example 1 of Section 1.2 on page 28.

1.0 Graphers Versus Calculus

We (informally) call a graph *complete* if the portion of the graph that we see in the viewing window suggests all the important features of the graph. For example, if some interesting feature occurs beyond the viewing window, then the graph is not complete. If the graph has some important wiggle that does not show in the viewing window because the scale of the graph is too large, then again the graph is not complete. Unfortunately, no matter how large or small the scale of the graph, we can never be certain that some interesting behavior might not be occurring outside the viewing window or some interesting wiggles aren't hidden within the curve that we see. Thus, if we use only a graphing utility on a graphing calculator or computer, we might overlook important discoveries. This is one reason why we need to carefully do a *mathematical* analysis.

If you do not know how to use your graphing calculator or computer, consult the Technology Resource Manual that accompanies this text. Any time a term or operation is introduced in this text, the Technology Resource Manual clearly explains the term or operation and gives all the necessary keystrokes. Therefore, you can read the text and the manual together.

EXAMPLE 1 **Complete Graphs**

Graph $y = x^4 - 12x^3 + x^2 - 2$ in a window with dimensions $[-10, 10]$ by $[-10, 10]$ using your grapher. If this is not satisfactory, find a better window.

Solution

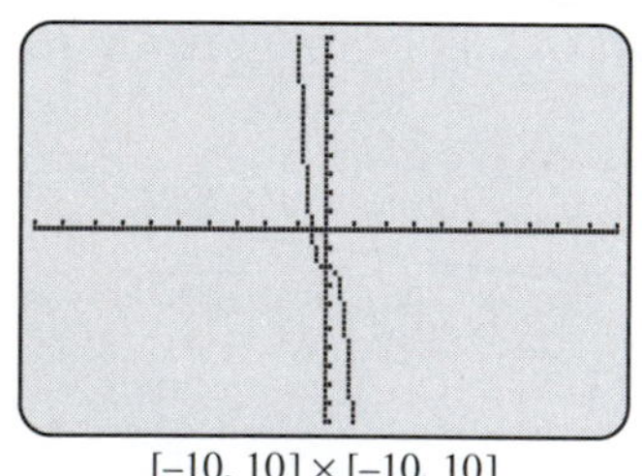

$[-10, 10] \times [-10, 10]$

Screen 1.1
A graph of $y = x^4 - 12x^3 + x^2 - 2$ in a standard window.

The graph is shown in Screen 1.1. You might reflect whether this is a complete graph. Suppose, for example, that x is huge, say, a billion. Then the first term x^4 can be written as $x \cdot x^3$, or one billion times x^3. The second term $-12x^3$ can be thought of as -12 times x^3. Since -12 is insignificant compared to one billion, the term $-12x^3$ is insignificant compared to x^4. The other terms x^2 and 10 are even less significant. So the polynomial for huge x should be approximately equal to the leading term x^4. But x^4 is a huge positive number when x is huge. This is not reflected in the graph found in Screen 1.1. Therefore, we should take a screen with larger dimensions. If we set the dimensions of our viewing window to $[-5, 14]$ by $[-2500, 1000]$, we obtain Screen 1.2. Notice the missing behavior we have now discovered. ■

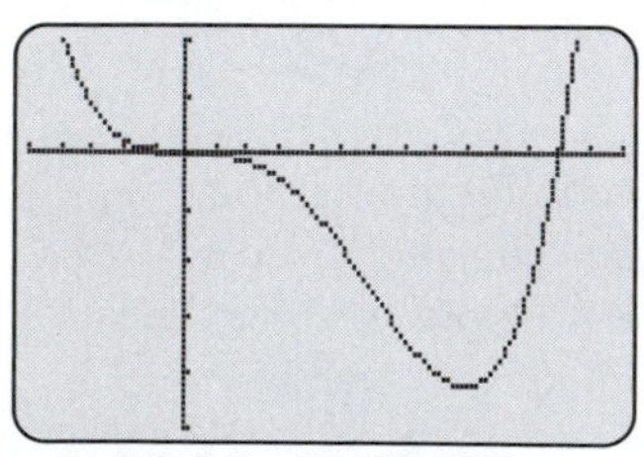

$[-5, 14] \times [-2500, 1000]$

Screen 1.2
A graph of $y = x^4 - 12x^3 + x^2 - 2$ showing some hidden behavior.

EXAMPLE 2 **Complete Graphs**

Graph $y = x^4 - 2x^3 + x^2 + 10$ using a window with dimensions $[-10, 10]$ by $[-10, 10]$ on your grapher. If this is not satisfactory, find a better window.

Solution

If we graph using the given viewing window, we see nothing! Try it. Where is the graph? We must examine the function more carefully to see which window to use. Notice that when $x = 0$, $y = 10$. We then might think to center our screen on the point $(0, 10)$. So take a screen with dimensions $[-10, 10]$ by $[0, 20]$ and obtain Screen 1.3. Now we see something! But are we seeing everything? Either using the ZOOM feature of your grapher to ZOOM about $(0, 10)$ or setting the screen dimensions to

$[-2.5, 2.5]$ by $[7.5, 12.5]$, obtain Screen 1.4. Notice the missing behavior, in the form of a wiggle, that we have now discovered.

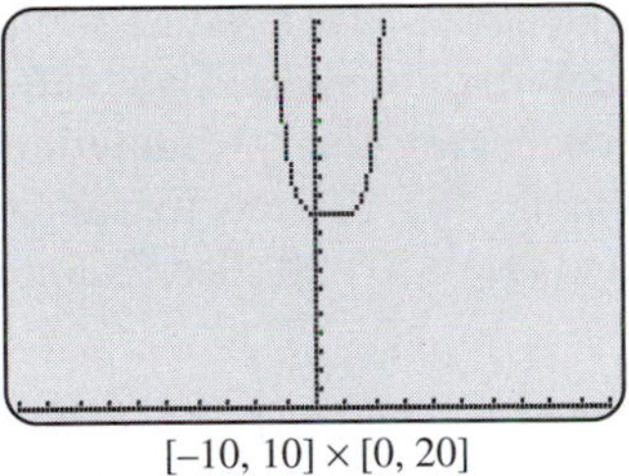

$[-10, 10] \times [0, 20]$

Screen 1.3
A graph of $y = x^4 - 2x^3 + x^2 + 10$.

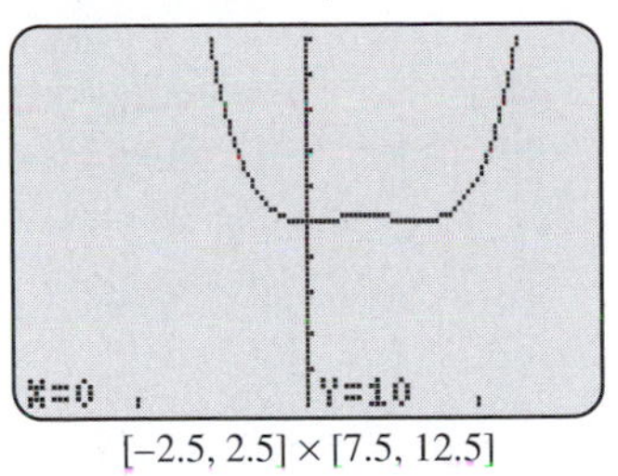

$[-2.5, 2.5] \times [7.5, 12.5]$

Screen 1.4
A graph of $y = x^4 - 2x^3 + x^2 + 10$ showing some hidden behavior.

■

The previous two examples indicate the shortfalls of using a graphing utility on a graphing calculator or computer. Determining the dimensions of the viewing screen can represent a major difficulty. We can never know whether some interesting behavior is taking place just outside the viewing screen, no matter how large it is. Also, if we use only a graphing utility, how can we ever know whether there are some hidden wiggles somewhere in the graph? We cannot ZOOM everywhere and forever!

We will be able to determine complete graphs by expanding our knowledge of mathematics, and, in particular, by using calculus. In Chapter 5 we will use calculus to find all the wiggles and hidden behavior of a graph.

1.1 Functions

Definition of Function

Graphs of Functions

Increasing, Decreasing, Concavity, and Continuity

Applications and Mathematical Modeling

The Bettmann Archive/Hutton

Lejeune Dirichlet, 1805–1859

Dirichlet was one of the mathematical giants of the 19th century. He formulated the notion of function that is still used today and is also known for the Dirichlet series, the Dirichlet function, the Dirichlet principle, and the Dirichlet problem. The Dirichlet problem is fundamental to the study of thermodynamics and electrodynamics. Although described as noble, sincere, humane, and possessing a modest disposition, Dirichlet was known as a dreadful teacher. He was also a failure as a family correspondent. When his first child was born, he neglected to inform his father-in-law, who, when he found out about the event, commented that Dirichlet might have at least written a note saying "$2 + 1 = 3$."

APPLICATION **State Income Tax**

The following instructions are given on the Connecticut state income tax form to determine your income tax.

If your taxable income is less or equal to \$16,000, multiply by 0.03. If it is more that \$16,000, multiply the excess over \$16,000 by 0.045 and add \$480.

Let x be your taxable income. Now write a formula that gives your state income tax for any value of x. Use this formula to find your taxes if your taxable income is \$15,000 and also \$20,000. See Example 12 on page 16 for the answer.

Definition of Function

We are all familiar with the correspondence between an element in one set and an element in another set. For example, to each house there corresponds a house number, to each automobile there corresponds a license number, and to each individual there corresponds a name.

Table 1.2 lists eight countries and the capital city of each. The table indicates that to each country there corresponds a capital city. Notice that there is one and only one capital city for each country. Table 1.3 gives the gross domestic product (GDP) for the United States in trillions of (current) dollars for each of 12 recent years.[3] Again, there is one and only one GDP associated with each year.

Table 1.2

Country	Capital City
Afghanistan	Kabul
Albania	Tirana
Algeria	Algiers
Angola	Luanda
Argentina	Buenos Aires
Armenia	Yerevan
Australia	Canberra
Austria	Vienna

Table 1.3

Year	U.S. GDP (trillions)	Year	U.S. GDP (trillions)
1990	\$5.8	1996	\$7.8
1991	6.0	1997	8.3
1992	6.3	1998	8.8
1993	6.6	1999	9.3
1994	7.1	2000	10.0
1995	7.4	2001	10.2

We call any rule that assigns or corresponds to each element in one set precisely one element in another set a **function**. Thus, the correspondences indicated in Tables 1.2 and 1.3 are functions.

As we have seen, a table can represent a function. Functions can also be represented by formulas. For example, suppose you are going a steady 40 miles per hour in a car. In one hour you will travel 40 miles; in two hours you will travel 80 miles; and so on. The distance you travel depends on (corresponds to) the time. Indeed, the equation relating distance (d), velocity (v), and time (t), is $d = v \cdot t$. In our example, we have $d = 40 \cdot t$. We can view this as a correspondence or rule: Given the time t in hours, the rule gives a distance d in miles according to $d = 40 \cdot t$. Thus, given $t = 3$, $d = 40 \cdot 3 = 120$. Notice carefully how this rule is *unambiguous*. That is,

[3] *Statistical Abstract of the United States*, 2002.

given any time t, the rule specifies one and only one distance d. This rule is therefore a function; the correspondence is between time and distance.

Often the letter f is used to denote a function. Thus, using the previous example, we can write $d = f(t) = 40 \cdot t$. The symbol $f(t)$ is read "f of t." One can think of t as the "input" and the value of $d = f(t)$ as the "output." For example, an input of $t = 4$ results in an output of $d = f(4) = 40 \cdot 4 = 160$.

The following gives a general definition of function.

DEFINITION **Function**

Let D and R be two nonempty sets. A **function** f from D to R is a rule that assigns to each element x in D one and only one element $y = f(x)$ in R. (See Figure 1.2.)

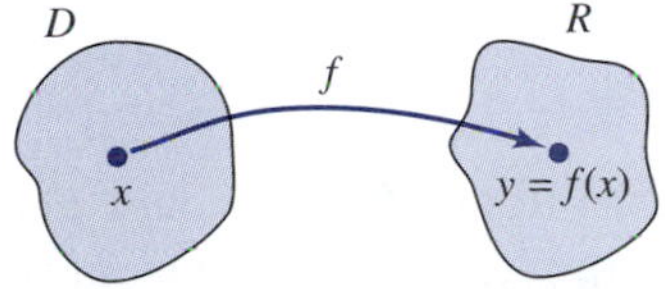

Figure 1.2

The set D in the definition is called the **domain** of f. We might think of the domain as the set of inputs. We then can think of the values $f(x)$ as outputs.

Another helpful way to think of a function is shown in Figure 1.3. Here the function f accepts the input x from the conveyor belt, operates on x, and outputs (assigns) the new value $f(x)$.

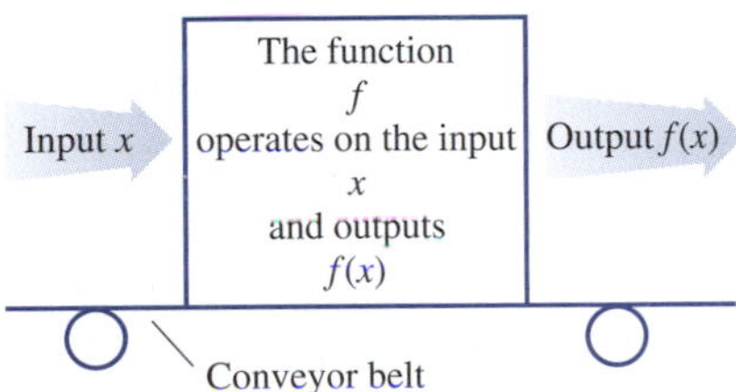

Figure 1.3

The set of all possible outputs is called the **range** of f. The letter representing elements in the domain is called the **independent variable**, and the letter representing the elements in the range is called the **dependent variable**. Thus, if $y = f(x)$, x is the independent variable, and y is the dependent variable, since y *depends* on x. In the equation $d = 40t$, we can write $d = f(t) = 40t$ with t as the independent variable. We are free to set the independent variable t equal to any number of values. The dependent variable is d. Notice that d *depends* on the particular value of t that is used.

EXAMPLE 1 **Determining When Rules Are Functions**

Which of the rules in Figure 1.4 are functions? Find the range of each function.

Solution

The rule indicated in Figure 1.4a is a function from $A = \{a, b, c\}$ to $B = \{p, q, r, s\}$, since each element in A is assigned a unique element in B. The domain of this function is A, and the range is $\{p, r\}$. The rule indicated in Figure 1.4b is not a function, since the element a has been assigned more than one element in B. The rule indicated in Figure 1.4c is a function with domain A and with range the single point $\{q\}$.

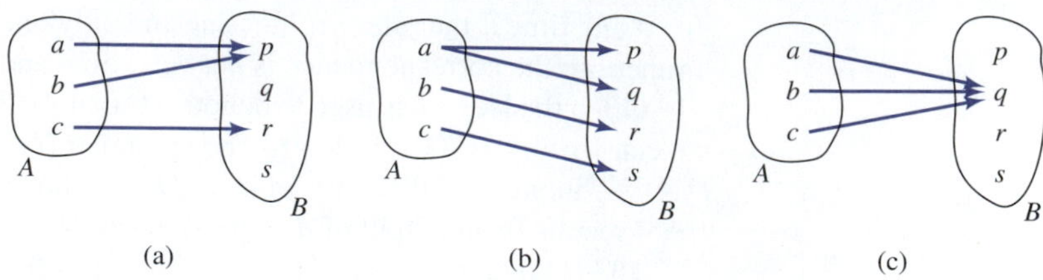

Figure 1.4
Which correspondences are functions?

■

Notice that it is possible for the same output to result from different inputs. For example, inputting a and b into the function in part (a) results in the same output p. In part (c), the output is always the same regardless of the input.

Functions are often given by equations, as we saw for the function $d = f(t) = 40 \cdot t$ at the beginning of this section. As another example, let both the sets D and R be the set of real numbers, and let f be the function from D to R defined by $f(x) = 3x^2 + 1$. In words, the output is obtained by taking three times the square of the input and adding one to the result. Thus, the formula

$$f(x) = 3x^2 + 1$$

can also be viewed as

$$f(\text{input}) = 3(\text{input})^2 + 1$$

If we want $f(2)$, then we replace x (or the input) with 2 and obtain

$$f(2) = 3(2)^2 + 1 = 13$$

If we want $f(-4)$, then we replace x (or the input) with -4 and obtain

$$f(-4) = 3(-4)^2 + 1 = 49$$

If we want $f(s + 1)$, then we replace x (or the input) with $s + 1$ and obtain

$$f(s + 1) = 3(s + 1)^2 + 1 = 3(s^2 + 2s + 1) + 1 = 3s^2 + 6s + 4$$

If we want $f(x + 2)$, then we replace x (or the input) with $x + 2$ and obtain

$$f(x + 2) = 3(x + 2)^2 + 1 = 3(x^2 + 4x + 4) + 1 = 3x^2 + 12x + 13$$

Convention

When the domain of a function f is not stated explicitly, we shall agree to let the domain be all values x for which $f(x)$ makes sense and is real.

For example, if $f(x) = 1/x$, the domain is assumed to be all real numbers other than zero.

EXAMPLE 2 **Finding the Domain of a Function**

Let $f(x) = \sqrt{2x - 4}$. Find the domain of f. Evaluate $f(2)$, $f(4)$, $f(2t + 2)$.

Solution

Since $f(x)$ must be a real number, we must have $2x - 4 \geq 0$ or $x \geq 2$. Thus, the domain is $[2, \infty)$.

$$f(2) = \sqrt{2(2) - 4} = \sqrt{0} = 0$$
$$f(4) = \sqrt{2(4) - 4} = \sqrt{4} = 2$$
$$f(2t + 2) = \sqrt{2(2t + 2) - 4} = \sqrt{4t} = 2\sqrt{t}, \ t \geq 0$$

■

We often encounter functions with domains divided into two or more parts with a different rule applied to each part. We call such functions **piecewise-defined functions**. The absolute value function, $|x|$, is such an example.

DEFINITION **Absolute Value Function, |*x*|**

$$|x| = \begin{cases} x & \text{if } x \geq 0 \\ -x & \text{if } x < 0 \end{cases}$$

For example, since $-5 < 0$, $|-5| = -(-5) = 5$. Do not make the mistake of thinking that the absolute value of an *algebraic* expression can be obtained by "dropping the sign." (See Exercise 96.)

Graphs of Functions

When the domain and range of a function are sets of real numbers, the function can be graphed.

DEFINITION **Graph of a Function**

The **graph** of a function f consists of all points (x, y) such that x is in the domain of f and $y = f(x)$.

EXAMPLE 3 **Graphing a Function**

Construct a graph of the function represented in Table 1.3.

Solution

Let us label the x-axis in years from 1990, and let $y = f(x)$, with the y-axis in trillions of dollars. Thus, for example, the year 1995 corresponds to 5 on the x-axis. Corresponding to the year 1995, we see a GDP of \$7.4 trillion. Thus, the point $(5, 7.4)$

is on the graph, and $f(5) = 7.4$. In a similar fashion, the points $(6, 7.8)$ and $(7, 8.3)$ are on the graph with $f(6) = 7.8$ and $f(7) = 8.3$. A graph is shown in Figure 1.5.

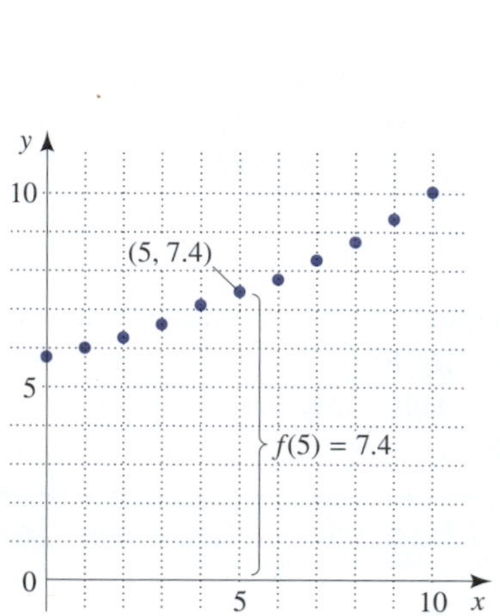

Figure 1.5

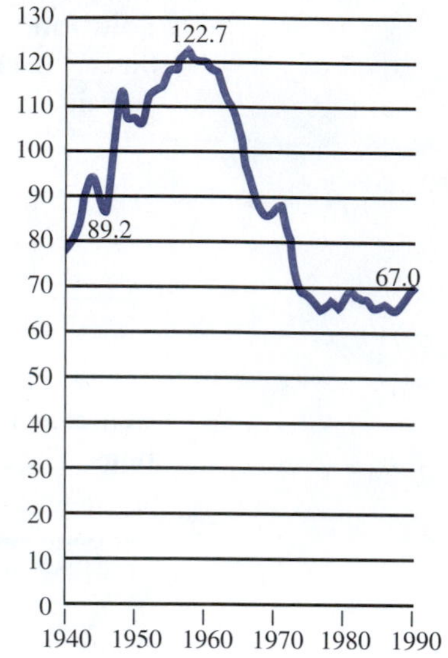

Figure 1.6
Baby boom and bust, 1940–90 (Births per 1000 U.S. women aged 15–44).

Notice that the graph in Figure 1.5 gives a *picture* of the function that is represented in Table 1.3.

We have already seen that a function can be represented by a table or formula. A function can also be represented by a graph. Figure 1.6 represents the function that gives the births per 1000 women[4] aged 15–44 in the United States. We see that the birthrate peaked in 1960 at 122.7. If the name of the function represented by the graph in Figure 1.6 is f, then this means that $f(1960) = 122.7$.

Exploration 1

A Function Given by a Graph

Graph the sine function, $y = \sin x$, on your grapher using a window with dimensions $[-4.17, 4.17]$ by $[-1, 1]$. (Be sure to be in radian mode.) This graph determines a function. Let us call the function f, so $f(x) = \sin x$. Now estimate $f(0)$, $f(1)$, $f(3.1)$, $f(-1)$, $f(-3.1)$.

[4] John W. Write. 1993. *The Universal Almanac*. New York: Andrews and McMeel.

We now determine the graphs of several important functions.

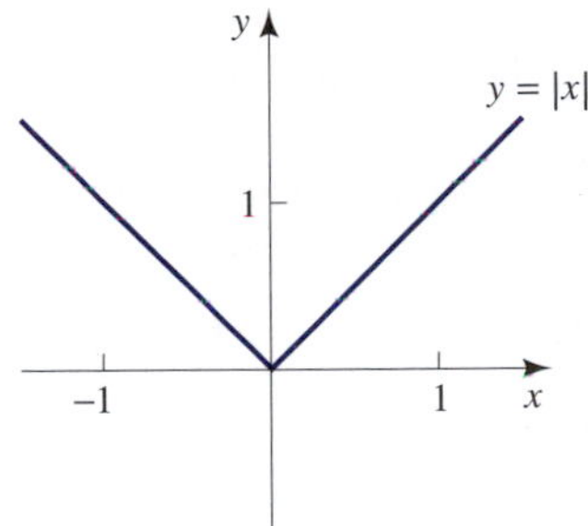

Figure 1.7
A graph of $y = |x|$.

EXAMPLE 4 **The Graph of the Absolute Value Function**

Graph the function $y = f(x) = |x|$.

Solution

If $x \geq 0$, then $|x| = x$. This is a line through the origin with slope 1. This line is graphed in the first quadrant in Figure 1.7. Also if $x < 0$, then $|x| = -x$, which is a line through the origin with slope -1. This is shown in Figure 1.7. Since we can take the absolute value of any number, the domain is $(-\infty, \infty)$. ■

For convenient reference and review, Figures 1.8, 1.9, 1.10, and 1.11 show the graphs of the important functions $y = f(x) = x^2$, $y = f(x) = x^3$, $y = f(x) = \sqrt{x}$, and $y = f(x) = \sqrt[3]{x}$. If these are not familiar to you, refer to Review Appendix A, Sections A.7 and A.8. In these review sections you will also find review material on quadratic, power, polynomial, and rational functions and further graphing techniques.

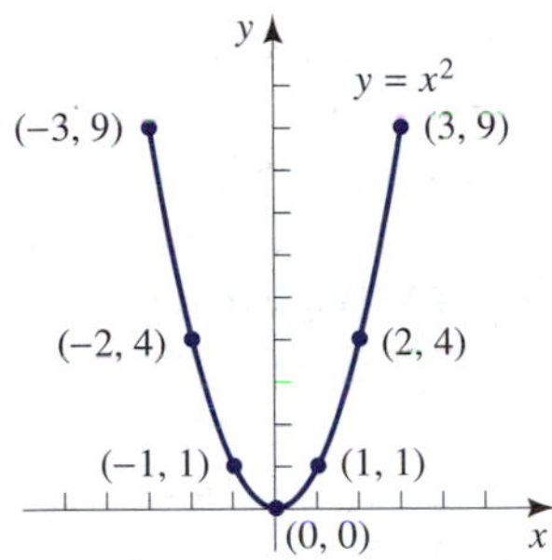

Figure 1.8
A graph of $y = x^2$.

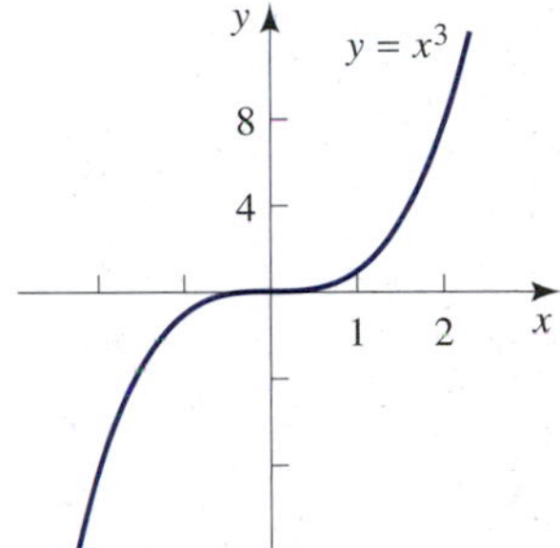

Figure 1.9
A graph of $y = x^3$.

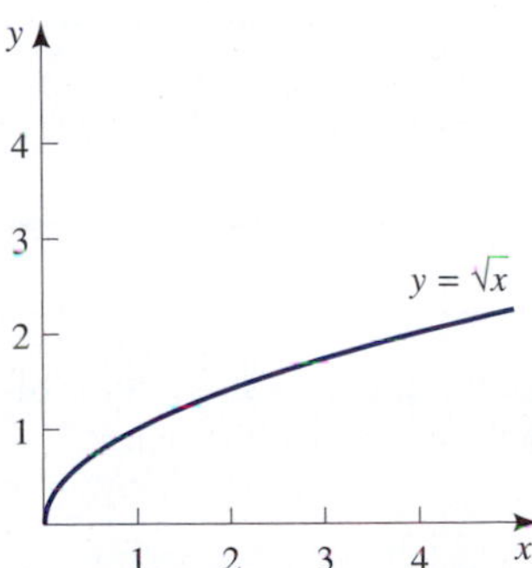

Figure 1.10
A graph of $y = \sqrt{x}$.

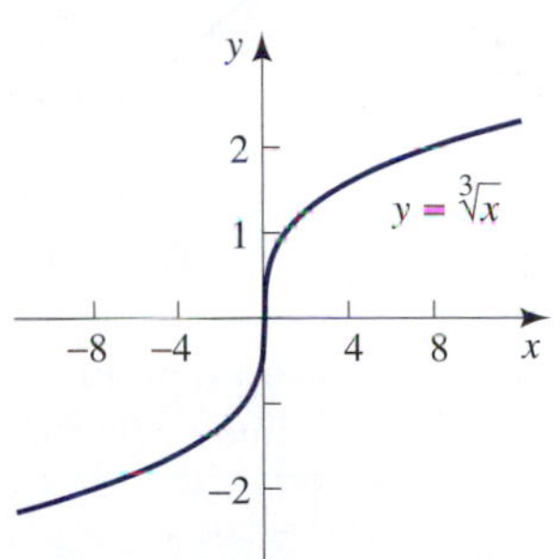

Figure 1.11
A graph of $y = \sqrt[3]{x}$.

Since we can take the square, cube, or cube root of any number, the domains of x^2, x^3, and $\sqrt[3]{x}$ are $(-\infty, \infty)$. Since we can take the square root only of nonnegative numbers, the domain of $\sqrt{x}$ is $[0, \infty)$.

Exploration 2

Finding a Domain Graphically

Estimate the domain of $f(x) = \sqrt{6-3x}$ by graphing $y = f(x)$ on your grapher. Your grapher will automatically graph only on the domain of f. Confirm your answer algebraically.

We now see how to determine whether a graph is the graph of a function.

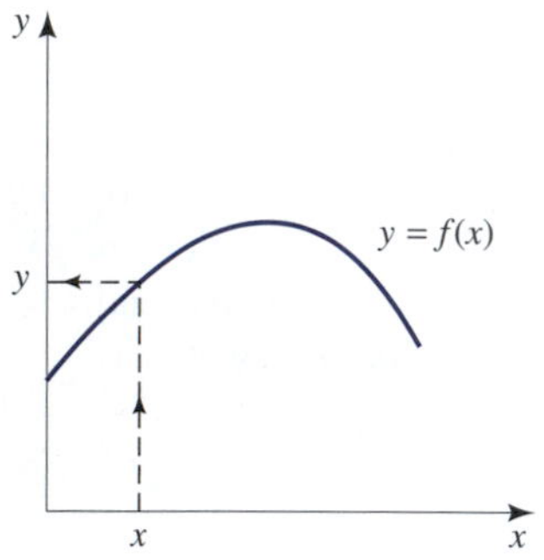

Figure 1.12
The function $y = f(x)$ takes x to y.

EXAMPLE 5 Determining Whether a Graph Is the Graph of a Function

A graph is shown in Figure 1.12. Does this represent the graph of a function of x?

Solution

Figure 1.12 indicates that, given any number x, there is one and only one value of y. For any value of x, moving vertically until you strike the graph and then moving horizontally until you strike the y-axis will never result in more than one value of y. This process assigns unambiguously one single value y to any value x in the domain. Thus, this graph describes a function. ■

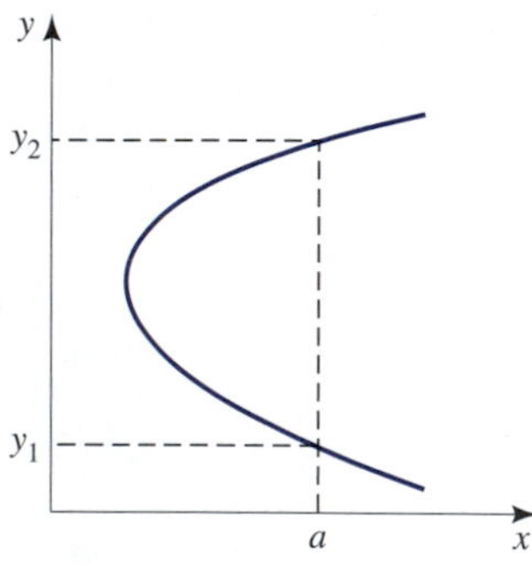

Figure 1.13
Not a function, since a vertical line strikes the graph in two places.

EXAMPLE 6 Determining Whether a Graph Is the Graph of a Function

A graph is shown in Figure 1.13. Does this represent the graph of a function of x?

Solution

Notice that for the value a indicated in Figure 1.13, a vertical line drawn from a strikes the graph in *two* places, resulting in assigning two values to x when $x = a$. This cannot then be the graph of a function of x. ■

The process used in Example 6 works in general.

Vertical Line Test

A graph in the xy-plane represents a function of x, if and only if, every vertical line intersects the graph in at most one place.

EXAMPLE 7 A Graph That Fails to Be the Graph of a Function

Since $\sqrt{x^2 + y^2}$ is the distance from the point (x, y) to the origin, the graph of the equation $x^2 + y^2 = 1$ is a circle centered at the origin with radius 1. See Figure 1.14a. Does this represent a function of x?

Solution

As Figure 1.14b indicates, the vertical line test fails, so this cannot be the graph of a function of x.

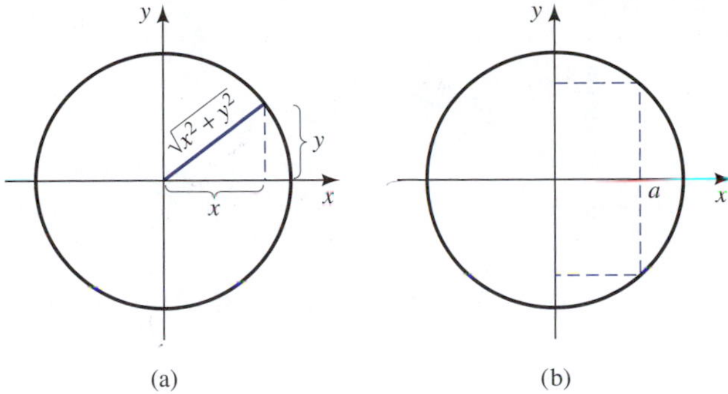

Figure 1.14
Not a function, since a vertical line strikes the graph in two places.

Remark In the previous example, if we try to solve for y, we obtain $y = \pm\sqrt{1 - x^2}$. Thus, for any value of x in the interval $(-1, 1)$ there are *two* corresponding values of y. We are not told which of these two values to take. Thus, the rule is ambiguous and not a function.

Increasing, Decreasing, Concavity, and Continuity

In Figure 1.15 we see that the population of the United States[5] has steadily *increased* over the time period shown, whereas in Figure 1.16, we see that the percentage of the population that are farmers[6] has steadily *decreased*. In Figure 1.17 we see that the median age of the U.S. population[7] increased from 1910 to 1950, decreased from 1950 to 1970, and then increased again from 1970 to 1990.

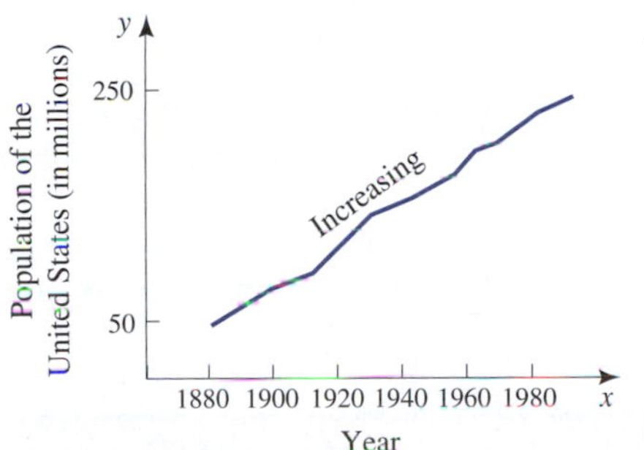

Figure 1.15
Population of the United States (in millions).

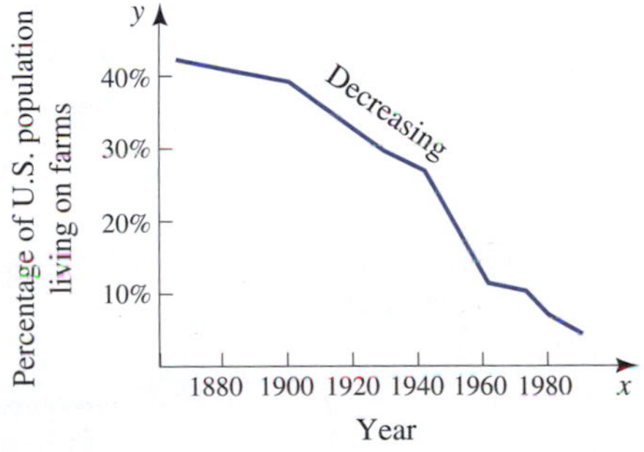

Figure 1.16
Percentage of U.S. population living on farms.

In this subsection we will examine the concepts of *increasing*, *decreasing*, and *concavity*. First we give the definitions of increasing and decreasing.

---

[5] U.S. Bureau of the Census.
[6] John W. Write. 1993. *The Universal Almanac*. New York: Andrews and McMeel.
[7] Ibid.

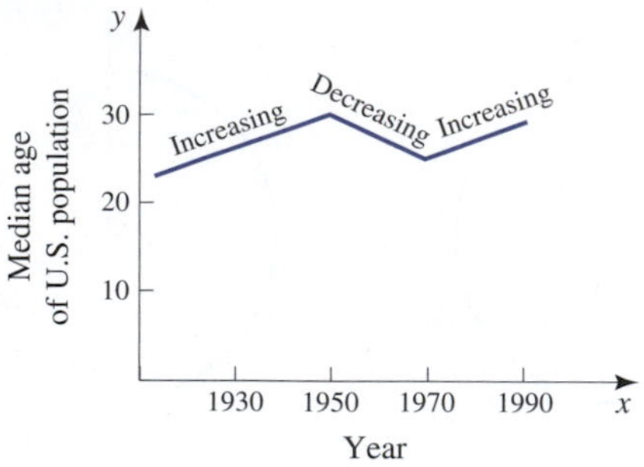

Figure 1.17
Median age of U.S. population.

DEFINITION **Increasing and Decreasing Functions**

A function $y = f(x)$ is said to be **increasing** (denoted by $\nearrow$) on the interval (a, b) if the graph of the function rises while moving left to right or, equivalently, if

$$f(x_1) < f(x_2) \qquad \text{when} \qquad a < x_1 < x_2 < b$$

A function $y = f(x)$ is said to be **decreasing** (denoted by $\searrow$) on the interval (a, b) if the graph of the function falls while moving left to right or, equivalently, if

$$f(x_1) > f(x_2) \qquad \text{when} \qquad a < x_1 < x_2 < b$$

EXAMPLE 8 **Increasing and Decreasing**

Determine the x-intervals on which each of the functions in Figure 1.7 and Figure 1.8 is increasing and the intervals on which each is decreasing.

Solution

Each function is decreasing on $(-\infty, 0)$ and increasing on $(0, \infty)$. ■

We continue with the definition of concave up and concave down.

DEFINITION **Concave Up and Concave Down**

If the graph of a function bends upward ($\smile$), we say that the function is **concave up**. If the graph bends downward ($\frown$), we say that the function is **concave down**. A straight line is neither concave up nor concave down.

EXAMPLE 9 **Concavity**

Determine the concavity of the functions graphed in Figures 1.8 and 1.12.

Solution

The function $y = x^2$ graphed in Figure 1.8 is concave up. The function graphed in Figure 1.12 is concave down. ■

We now introduce the concept of *continuity* from a graphical point of view. The definition that we give here is not precise. A precise definition will be given in Section 3.1, where continuity is discussed in greater depth.

DEFINITION **Continuity**

A function is said to be **continuous** on an interval if the graph of the function does not have any breaks, gaps, or holes in that interval. A function is said to be **discontinuous** at a point c if the graph has a break or gap at the point $(c, f(c))$ or if $f(c)$ is not defined, in which case the graph has a hole where $x = c$.

EXAMPLE 10 **Continuity**

A graph of a function is shown in Figure 1.18. Find the points of discontinuity.

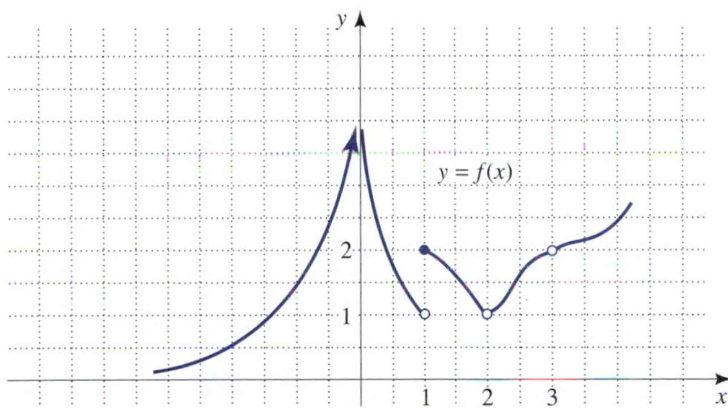

Figure 1.18

Solution

The function is discontinuous at $x = 0$ since the graph has a break here. The function is discontinuous at $x = 1$, since the graph has a jump here. The function is discontinuous at $x = 2$ and $x = 3$, since the graph has a hole at each of these points. ■

Almost every function we deal with will be continuous. Except possibly for piecewise-defined functions, the functions that we encounter in this text are continuous on their domains. For example, polynomials, such as $y = 2x^4 - 3x^3 + x^2 - 5x + 7$, are continuous everywhere, while rational functions, such as $\dfrac{x^3 - 2x^2 - 4}{x^2 - 1}$, are continuous everywhere except where the denominator is zero. Functions that we consider later in this chapter, such as exponential and logarithmic functions, are continuous everywhere on their domains.

Applications and Mathematical Modeling

To solve any applied problem, we must first take the problem and translate it into mathematics. This requires us to create equations and functions. This is called *mathematical modeling*. We already did this at the beginning of the section when we related distance traveled with velocity and time. We also have created other mathematical models using (linear) equations in Appendix A, Section A.6. In the next example we will model the application using a *nonlinear* function that gives the volume of a certain package. Later in the text we will use this function to find such important information as the dimensions that yield the maximum volume. For now we will just find the function.

EXAMPLE 11 **Finding a Function**

From all four corners of a 10-inch by 20-inch rectangular piece of cardboard, squares are cut with dimensions x by x. The sides of the remaining piece of cardboard are turned up to form an open box. See Figure 1.19. Find the volume V of the box as a function of x.

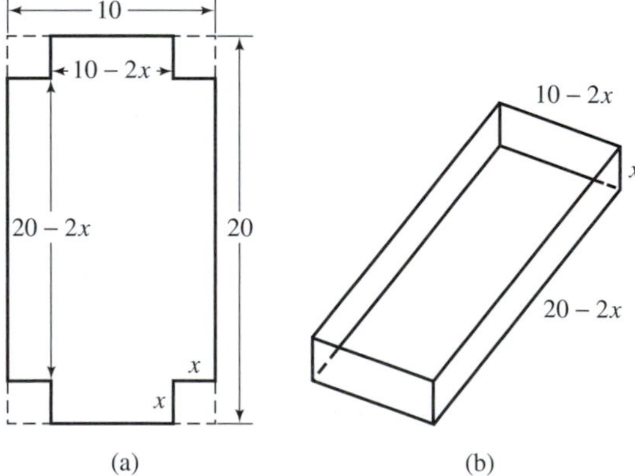

Figure 1.19
Fold the cut cardboard at the left to obtain the box without a top at the right.

Solution

The volume of a box is given by the product of the width (w), the length (l), and the height (h), that is, $V = w \cdot l \cdot h$. The width of the box is $w = 10 - 2x$, the length is $20 - 2x$, and the height is x. Thus, the volume is

$$V = V(x) = (10 - 2x)(20 - 2x)x$$

■

EXAMPLE 12 **A Piecewise-Defined Function**

The following instructions are given on the Connecticut state income tax form to determine your income tax.

If your taxable income is less or equal to \$16,000, multiply by 0.03. If it is more that \$16,000, multiply the excess over \$16,000 by 0.045 and add \$480.

Let x be your taxable income. Then write a formula that gives your state income tax for any value of x. Find the taxes on \$15,000 and the taxes on \$20,000. Graph the function. Is this function continuous?

Solution

If $x \le 16{,}000$, then $T(x) = 0.03x$. Let $x > 16{,}000$; then the excess over 16,000 is $(x - 16{,}000)$. Then $T(x) = 480 + 0.045(x - 16{,}000)$. We can write this as the piecewise-defined function

$$T(x) = \begin{cases} 0.03x & \text{if } x \le 16{,}000 \\ 480 + 0.045(x - 16{,}000) & \text{if } x > 16{,}000 \end{cases}$$

When $x = 15{,}000$, $x \le 16{,}000$, so we use the formula found in the first line of the function, that is, $T(x) = 0.03x$. So $T(15{,}000) = 0.03(15{,}000) = 450$, or \$450. When $x = 20{,}000$, $x > 16{,}000$, so we must use the formula found in the second line of the function, namely, $T(x) = 480 + 0.045(x - 16{,}000)$. So

$$T(20{,}000) = 480 + 0.045(20{,}000 - 16{,}000) = 480 + 180 = 660$$

Now consider the graph. When $x \le 16{,}000$, we have $T(x) = 0.03x$, a line through the origin with slope 0.03. We note that $T(16{,}000) = 480$. We graph this in Figure 1.20a and note that we graph this function only for $0 \le x \le 16{,}000$. Let us now consider the function on the second line, $T(x) = 480 + 0.045(x - 16{,}000)$. This is a line with slope 0.045 and goes through the point (16,000, 480). However, we use only this function when $x > 16{,}000$. So we graph only that part of this function for which $x > 16{,}000$. This is shown in Figure 1.20b. We then put these two graphs together to obtain the graph of the piecewise-defined function found in Figure 1.20c.

We do not see any breaks, gaps, or holes in this graph, so we conclude that $T(x)$ is continuous on $(0, \infty)$.

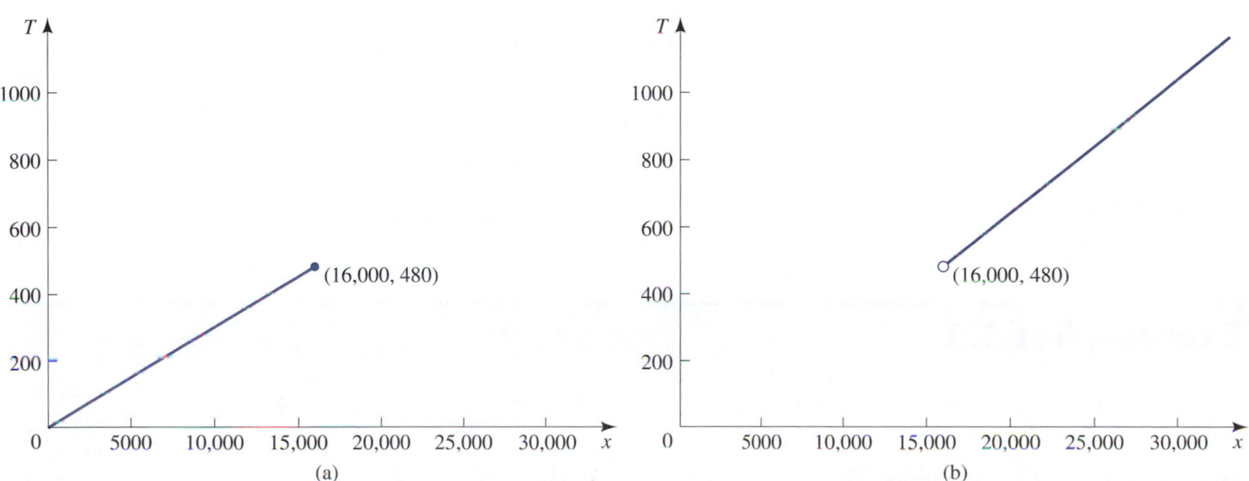

Figure 1.20

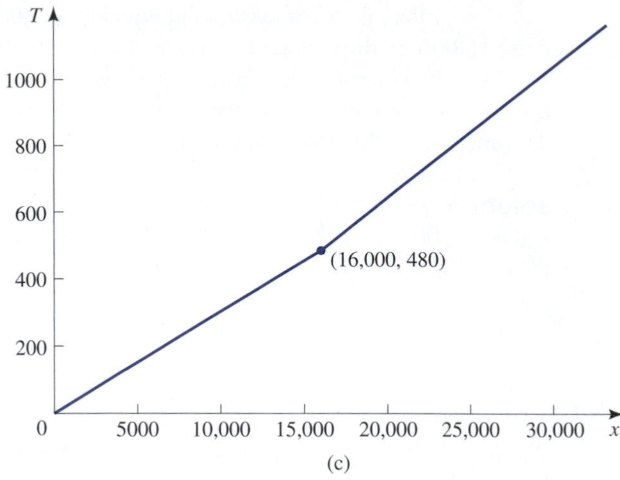

Figure 1.20
Continued

■

Exploration 3

Graphing Piecewise-Defined Functions

Graph $y = f(x)$ on your grapher, where

$$f(x) = \begin{cases} x^2 + 1, & x \leq 0 \\ 3, & 0 < x < 3 \\ x - 3, & x \geq 3 \end{cases}$$

You might wish to consult the Technology Resource Manual.

Warm Up Exercise Set 1.1

Which of the graphs in the figures represents the graph of a function?

1. Find the domain of $f(x) = \dfrac{\sqrt{x+1}}{x}$.
2. For the function in Exercise 2, evaluate (a) $f(-1)$, (b) $f(3)$, (c) $f(a+2)$, (d) $f(x+2)$.
3. A certain telephone company charges $0.10 for the first minute of a long-distance call and $0.07 for each additional minute or fraction thereof. If C is the cost of a call and x is the length of the call in minutes, find the cost as a function of x, and sketch a graph. Assume that $x \leq 3$. Find any points of discontinuity. Would these points affect your behavior?

Exercise Set 1.1

For Exercises 1 through 6, determine which of the rules in the figures represent functions.

1.

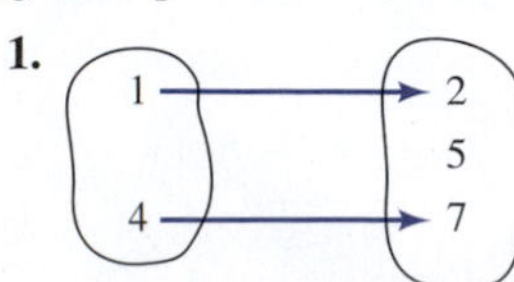

2.

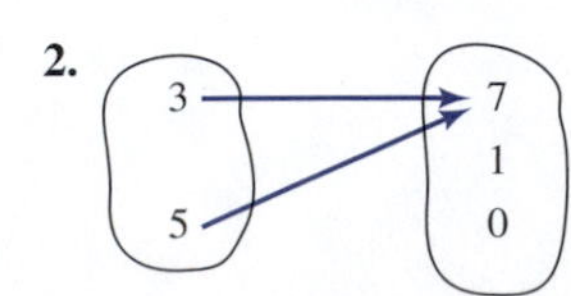

3.

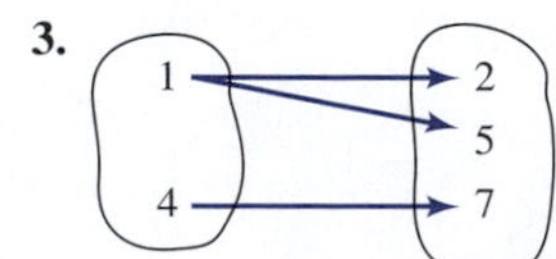

4.

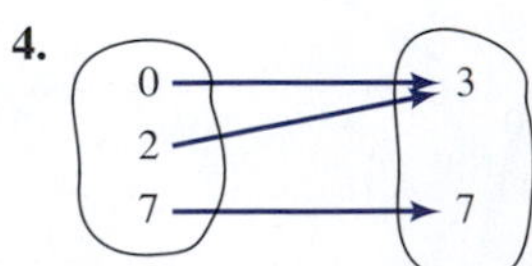

5.

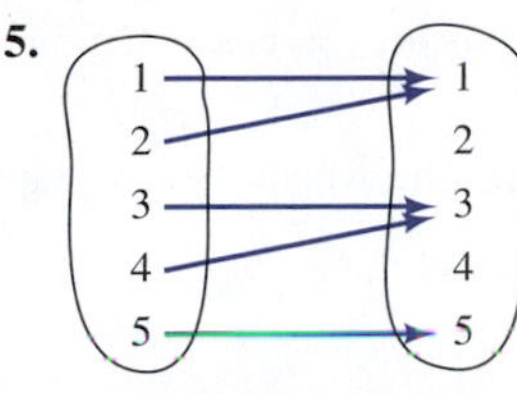

6.

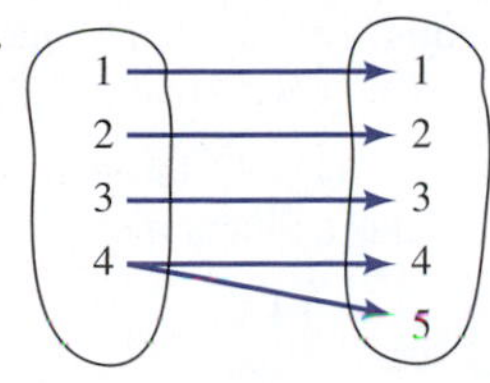

7. Consider the function graphed in the figure. In each part, find all values of x satisfying the given conditions.

a. $f(x) = 0$ **b.** $f(x) = 3$

c. $f(x) \geq 0$ **d.** $f(x) \leq 0$

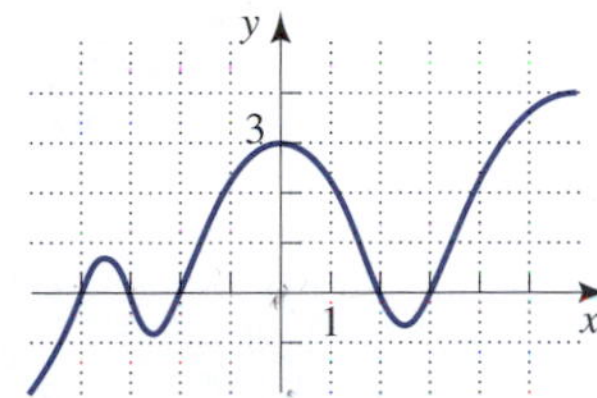

8. Specify the domain and range for each rule that represents a function in Exercises 1 through 6.

9. Match each of the following stories with one of the graphs. Write a story for the remaining graph.

a. Profits at our company have grown steadily.

b. At our company profits grew at first and then held steady.

c. At our company profits rose dramatically, and then things took a turn for the worse, resulting in recent losses.

d. Profits dropped dramatically at our company, and so we hired a new CEO. She turned the company around, and we are now making profits again.

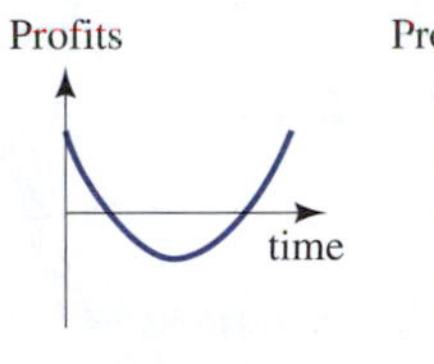

(i)

Profits

time

(ii)

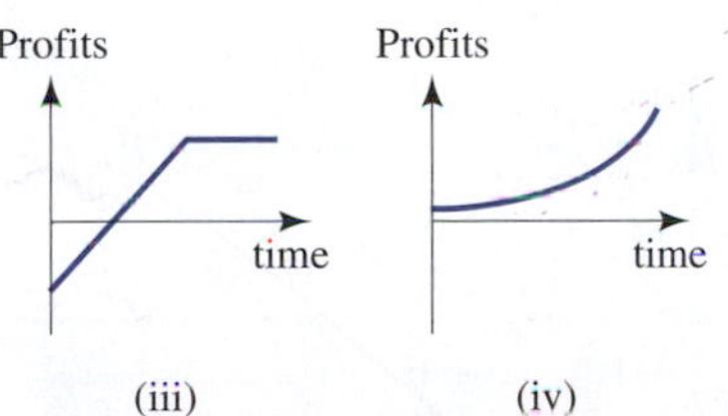

(iii) (iv)

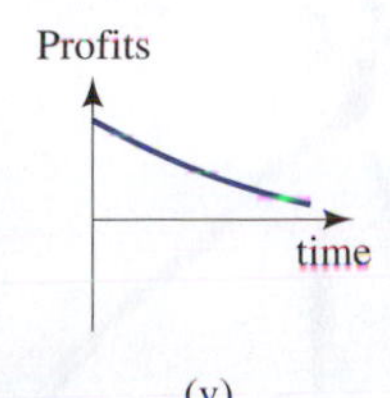

(v)

10. Match each of the following stories with one of the graphs. Write a story for the remaining graph.

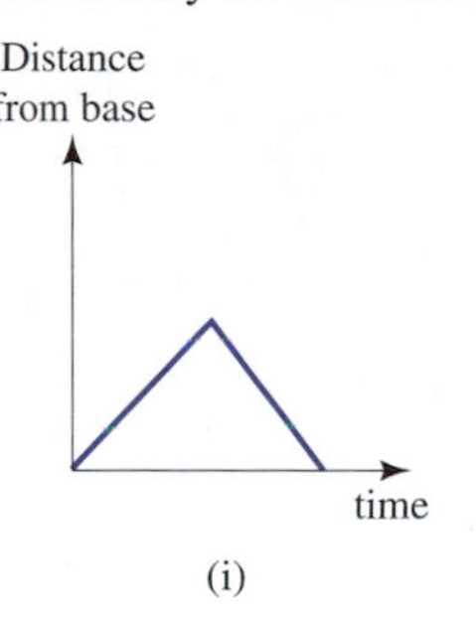

(i)

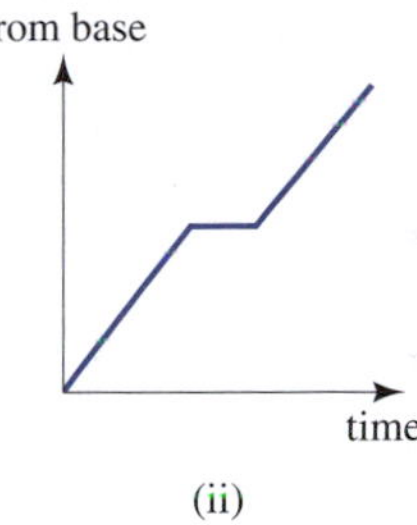

(ii)

Distance from base

time

(iii)

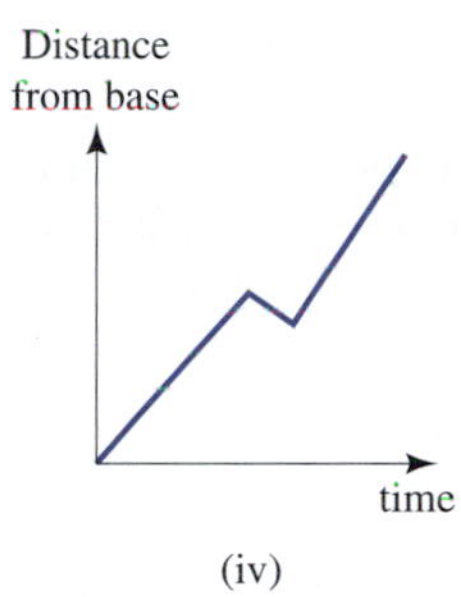

(iv)

a. We steadily climbed the mountain, got exhausted, and had to return to base camp.

b. We steadily climbed the mountain, rested for a while, and then resumed our steady climb to the top.

c. We climbed straight to the top of the mountain, going fast at first but slowing down later.

For Exercises 11 through 24, find the domain of each function.

11. $\dfrac{x}{x-1}$ **12.** $\dfrac{x-1}{x+2}$ **13.** $\sqrt[4]{x}$

14. $\sqrt{x-1}$ **15.** $\left|\dfrac{1}{\sqrt{x-1}}\right|$ **16.** $\sqrt{2-x}$

17. $\dfrac{x-1}{x^2-5x+6}$ **18.** $\dfrac{x}{x^2-1}$

19. $x^7 - 2x^4 - 3$ **20.** $x^{-1/5}$

21. $\dfrac{\sqrt{x+2}}{x}$ **22.** $\dfrac{x}{\sqrt{x-1}}$

23. $\sqrt[5]{x}$ **24.** $\sqrt{x^2-1}$

For Exercises 25 through 30, find the domain of each function.

25.

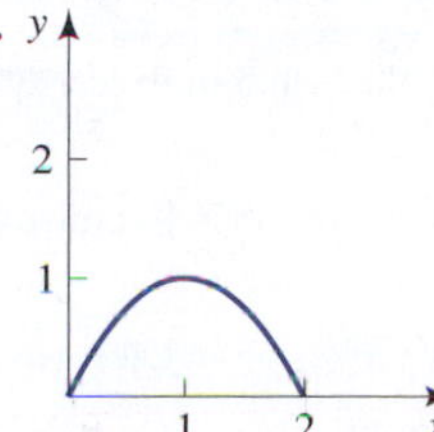

26.

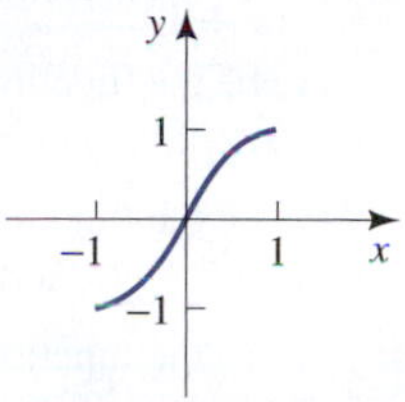

27.

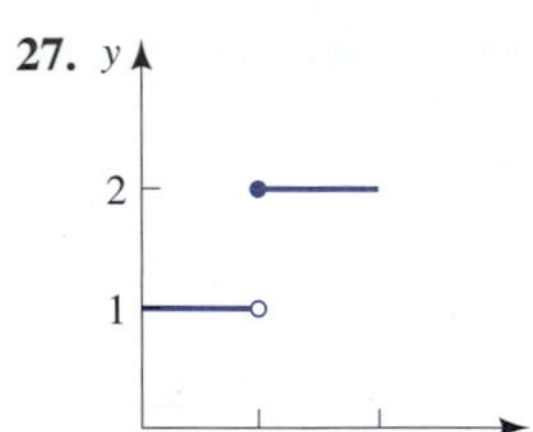

28.

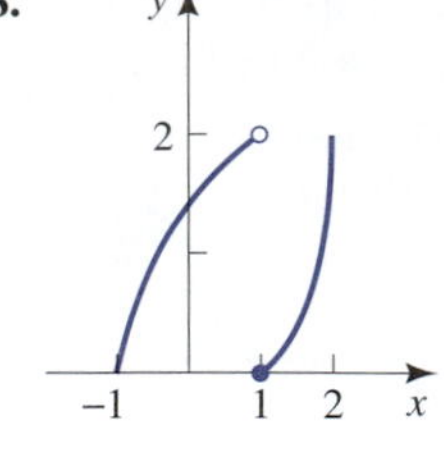

29.

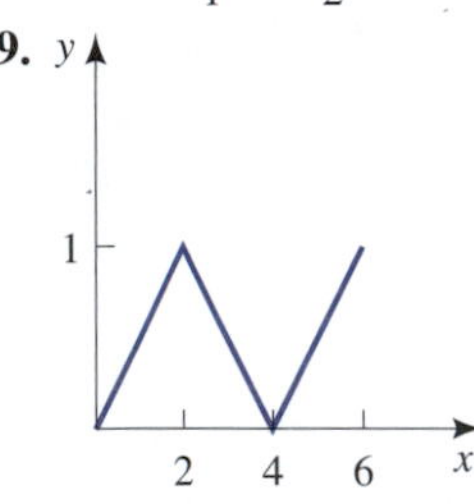

30.

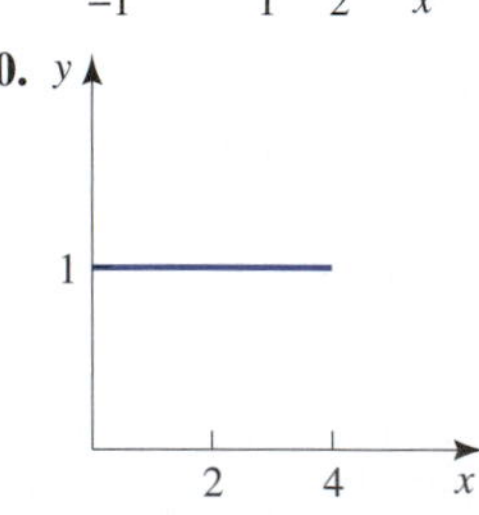

31. Consider the functions whose graphs are shown in Exercises 25, 26, and 28. In each case, find the interval(s) on which the function is increasing and the interval(s) on which the function is decreasing. In each case, find the interval(s) on which the function is concave up and the interval(s) on which the function is concave down. Also in each case, find the interval(s) on which the function is continuous.

In Exercises 32 through 38, evaluate the given function at the given values.

32. $f(x) = 2x^4 - 1$, $f(0)$, $f(2)$, $f(-\frac{1}{2})$, $f(-\sqrt{2})$, $f(x^2)$, $f(\sqrt{x})$

33. $f(x) = \dfrac{1}{x+1}$, $f(1)$, $f\left(\frac{1}{2}\right)$, $f(-2)$, $f(\sqrt{2}-1)$, $f(x+2)$, $f\left(\dfrac{1}{x+1}\right)$

34. $f(x) = \dfrac{1}{x^2+1}$, $f(2)$, $f\left(\frac{1}{2}\right)$, $f(-3)$, $f(2\sqrt{2})$, $f(\sqrt{x})$, $f(x-1)$

35. $f(x) = \sqrt{2x-1}$, $f(5)$, $f\left(\frac{1}{2}\right)$, $f\left(x^2 - \frac{1}{2}\right)$, $f\left(\dfrac{1}{x}\right)$, $f(-x)$, $-f(x)$

36. $f(x) = \sqrt{2-3x}$, $f(-1)$, $f(0)$, $f(x+1)$, $f\left(\dfrac{1}{x}\right)$, $f(-x)$, $-f(x)$

37. $f(x) = |x+1|$, $f(-2)$, $-f(-2)$, $f(x-1)$, $f(x^2-1)$

38. $f(x) = \dfrac{x-1}{x+1}$, $f(1)$, $f(-1)$, $f(x+1)$, $f(x^2)$

39. Let $f(x)$ be the function given by the graph in Exercise 27. Find $f(0)$, $f(1)$, $f(1.5)$, and $f(2)$.

40. Let $f(x)$ be the function given by the graph in Exercise 28. Find $f(-1)$, $f(1)$, and $f(2)$.

41. Let $f(x)$ be the function given by the graph in Exercise 29. Find $f(0)$, $f(2)$, $f(4)$, and $f(6)$.

42. Graph the *cosine* function, $y = \cos x$, on your grapher. (Be sure to be in radian mode.) This graph determines a function. Let us call the function f, so $f(x) = \cos x$. Estimate $f(0)$, $f(1.57)$, and $f(3.14)$.

For Exercises 43 through 50, determine whether each graph represents a function of x.

43.

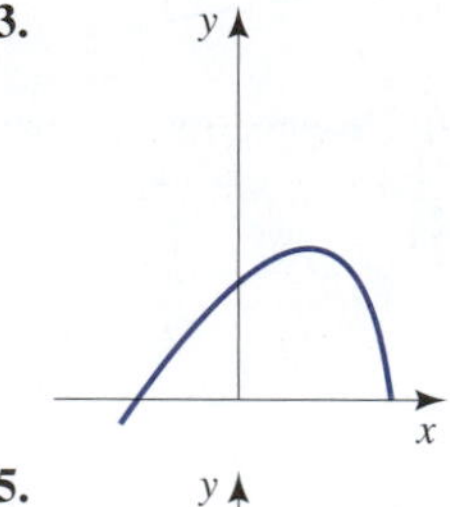

44.

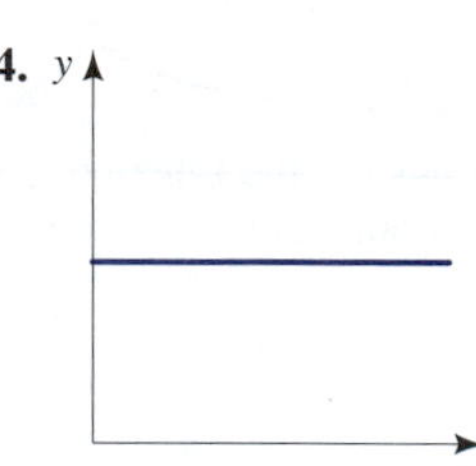

45.

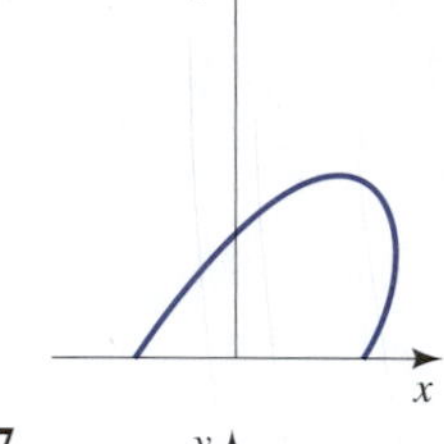

46.

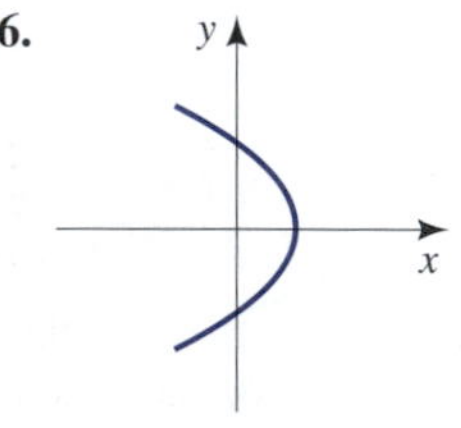

47.

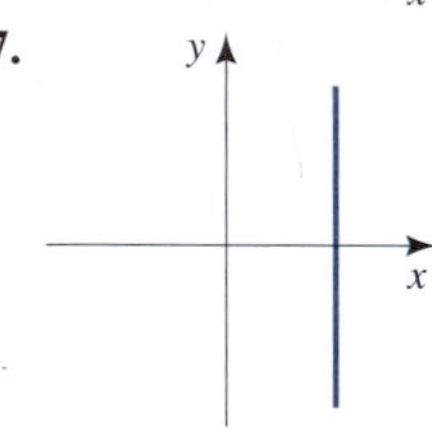

48.

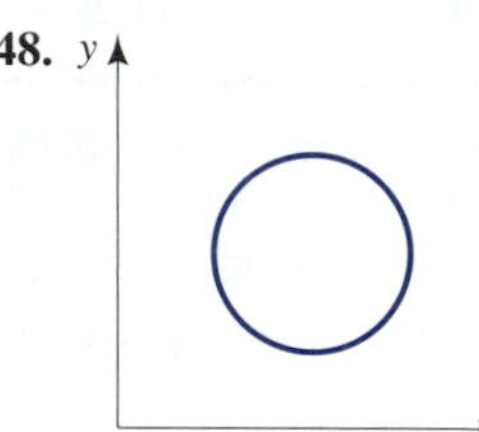

49.

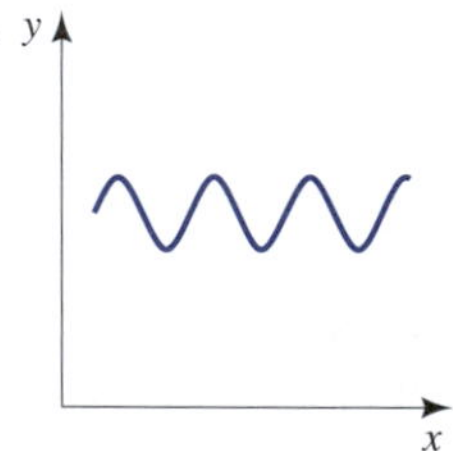

50.

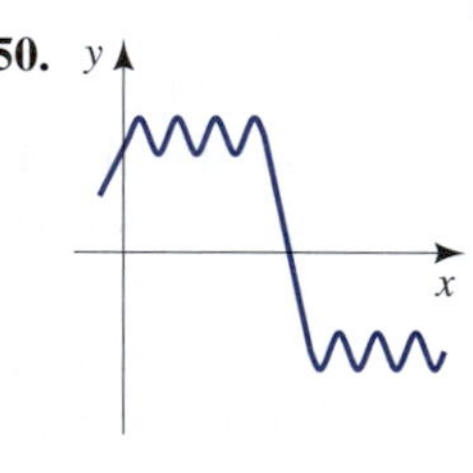

For Exercises 51 through 58, graph the indicated function. Find the interval(s) on which each function is continuous.

51. $f(x) = \begin{cases} 3x - 1 & \text{if } x \le -1 \\ x & \text{if } x > -1 \end{cases}$

52. $f(x) = \begin{cases} x & \text{if } x \le 1 \\ -x & \text{if } x > 1 \end{cases}$

53. $f(x) = \begin{cases} x^2 & \text{if } x \le 0 \\ x & \text{if } x > 0 \end{cases}$

54. $f(x) = \begin{cases} -1 & \text{if } x < 0 \\ 1 & \text{if } x \ge 0 \end{cases}$

55. $f(x) = \begin{cases} x^2 & \text{if } x \le 0 \\ x & \text{if } 0 < x \le 1 \\ 2x & \text{if } x > 1 \end{cases}$

56. $f(x) = \begin{cases} 1 & \text{if } x < 0 \\ 0 & \text{if } 0 \le x \le 1 \\ -1 & \text{if } x > 1 \end{cases}$

57. $f(x) = \begin{cases} 1 & \text{if } 0 \leq x < 1 \\ 2 & \text{if } 1 \leq x < 2 \\ 3 & \text{if } 2 \leq x \leq 3 \end{cases}$

58. $f(x) = \begin{cases} -x & \text{if } x \leq -1 \\ 1 & \text{if } -1 < x < 1 \\ x & \text{if } x \geq 1 \end{cases}$

In Exercises 59 through 64, graph, using your grapher, and estimate the domain of each function. Confirm algebraically.

59. $f(x) = |-x^2 + 8|$

60. $f(x) = \sqrt{x^2 + 4}$

61. $f(x) = \sqrt{9 - x^2}$

62. $f(x) = \sqrt{x^2 - 9}$

63. $f(x) = \dfrac{\sqrt{x}}{x - 5}$

64. $f(x) = \begin{cases} \dfrac{\sqrt{1 - x}}{x + 4}, & x \leq -2 \\ \sqrt{4 - x}, & x > -2 \end{cases}$

65. Find all points where the function given in the following graph is not continuous.

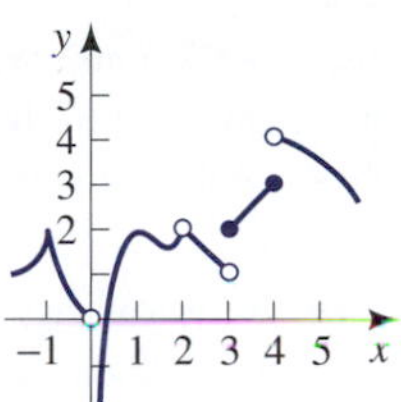

66. Find all points where the function given in the following graph is not continuous.

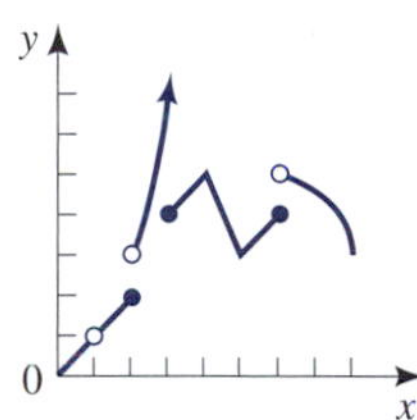

Applications and Mathematical Modeling

67. Packaging A box has a square base with each side of length x and height equal to $3x$. Find the volume V as a function of x.

68. Packaging Find the surface area S of the box in Exercise 67 as a function of x.

69. Velocity A car travels at a steady 60 miles per hour. Write the distance d in miles that the car travels as a function of time t in hours.

70. Navigation Two ships leave port at the same time. The first ship heads due north at 5 miles per hour while the second heads due west at 3 miles per hour. Let d be the distance between the ships in miles and let t be the time in hours since they left port. Find d as a function of t.

71. Revenue A company sells a certain style of shoe for \$60. If x is the number of shoes sold, what is the revenue R as a function of x?

72. Alfalfa Yields Generally, alfalfa yields are highest in the second year of production and then decline as the stands thin out and grow less vigorously. This yield decline was estimated by Knapp[8] as

$$y = \begin{cases} 4.93, & \text{if } t = 1 \\ 7.47 - 0.584(t - 2), & \text{if } t \geq 2 \end{cases}$$

where y gives the yield in tons per acre per year and t is an integer and denotes the age of the crop in years. Determine the yield in each of the first three years.

73. Sales Commission A salesman receives a commission of \$1 per square yard for the first 500 yards of carpeting sold in a month and \$2 per square yard for any additional carpet sold during the same month. If x is the number of yards of carpet sold and C is the commission, find C as a function of x and graph this function. Is this function continuous?

74. Taxes A certain state has a tax on electricity of 1% of the monthly electricity bill, the first \$50 of the bill being exempt from tax. Let $P(x)$ be the percent one pays in taxes, and let x be the amount of the bill in dollars. Graph this function. Is this function continuous?

75. Taxes In Exercise 74, let $T(x)$ be the total amount in dollars paid on the tax, where x is the amount of the monthly bill. Graph $T(x)$. Is this function continuous?

76. Production A production function that often appears in the literature (see Kim and Mohtadi[9]) is

$$y = \begin{cases} 0 & \text{if } x < M \\ x - M & \text{if } x \geq M \end{cases}$$

where y is production (output) and x is total labor. Graph this function. Is this function continuous?

77. Postage Rates In 2003 the rate for a first-class letter weighing one ounce or less mailed in the United States was 37 cents. For a letter weighing more than 1 ounce but less than or equal to 2 ounces, the postage was 60 cents. For a letter weighing more than 2 ounces but less than or equal to

[8] Keith Knapp. 1987. Dynamic equilibrium in markets for perennial crops. *Amer. Agr. J. Econ.* 69:97–105.

[9] Sunwoong Kim. 1992. Labor specialization and endogenous growth. *Amer. Econ. Rev.* 82:404–408.

3 ounces, the postage was 83 cents. Write the postage $P(x)$ as a piecewise-defined function of the weight x in ounces for $0 < x \leq 3$. Graph. Where is this function discontinuous?

78. Unemployment and Inflation Atkeson and Ohanian[10] gave the graph shown in the figure that relates the unemployment rate x as a percentage of the labor force and the change i in the inflation as a percentage for the years 1960–1983. Explain what this graph is saying.

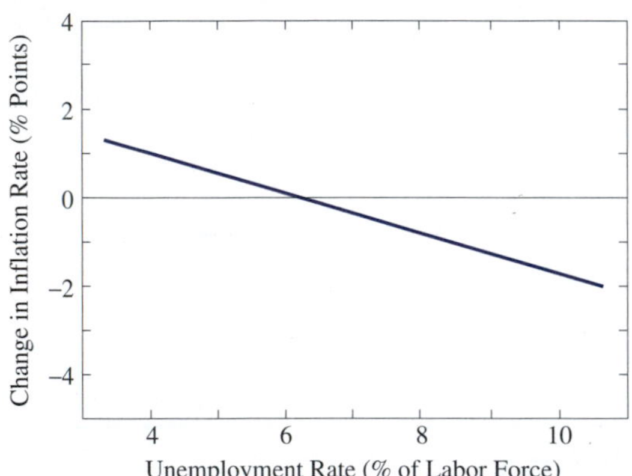

79. Steelhead Trout Wu and colleagues[11] developed a mathematical model that related stream temperature to trout density. The figure shows a graph of the function that describes this relationship, where T is temperature in degrees Celsius and A measures trout abundance given by the number of trout in a cubic meter of volume. Find $A(10)$, $A(13)$, and $A(19)$, and explain what they mean. Explain in a sentence what this graph is saying.

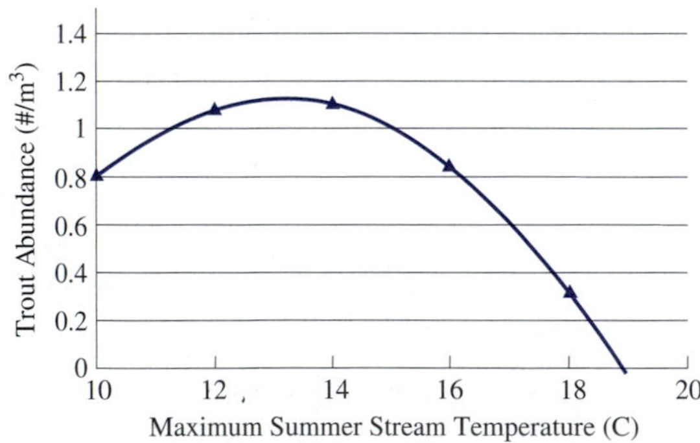

80. Parking Tax Voith[12] gave the following graph representing the effect of a parking tax x in dollars on the number of people who work in the central business district and commute by automobile. Explain what the graph of this function is saying.

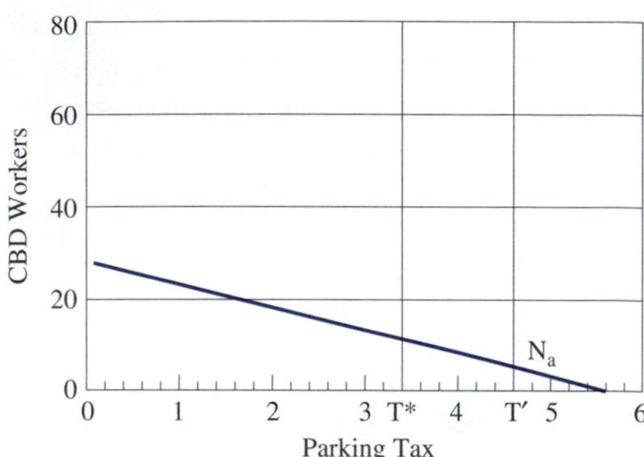

81. Bathymetric Distribution of Fish The graph was developed by Moore and Bronte[13] and gives the relationship between the depth x in meters and the number N of siscowet lake trout in Lake Superior. Find $N(0)$, $N(120)$, and $N(260)$, and explain what they mean. Explain in a sentence what this graph is saying.

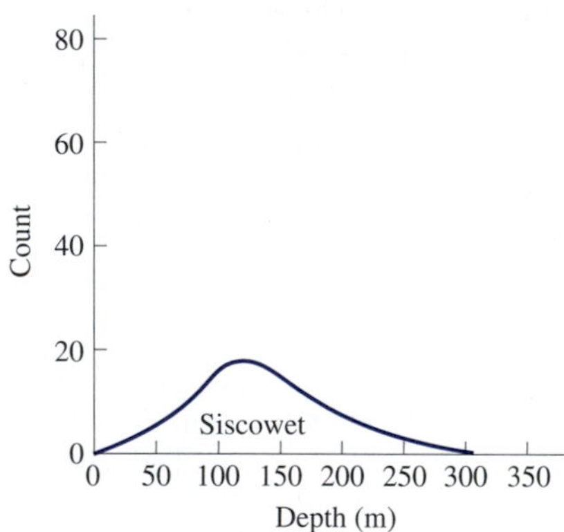

82. Interest Rates The figure[14] shows a graph of the prime rate (the rate that banks charge to their best customers) given in percent over a recent period of time. Let $p(x)$ be this function, where x is the year. What is $p(1990.3)$? $p(1993.5)$?

[10] Andrew Atkeson and Lee E. Ohanian. 2001. Are Phillips curves useful for forecasting inflation? *Quarterly Review Fed. Reserve Bank Minneapolis* Winter 2001:2–10.

[11] JunJie Wu, Richard M. Adams, and William G. Boggess. 2000. Cumulative effects and optimal targeting of conservation efforts: steelhead trout habitat enhancement in Oregon. *Amer. J. Agr. Econ.* 82:400–413.

[12] Richard Voith. 1998. The downtown parking syndrome: does curing the illness kill the patient? *Business Review Fed. Reserve Bank Philadelphia* Jan./Feb. 1998:3–14.

[13] Seith A. Moore and Charles R. Bronte. 2001. Delineation of sympatric morphotypes of lake trout in Lake Superior. *Trans. Amer. Fisheries Soc.* 130(6):1233–1240.

[14] *The Wall Street Journal*, August 17, 1994.

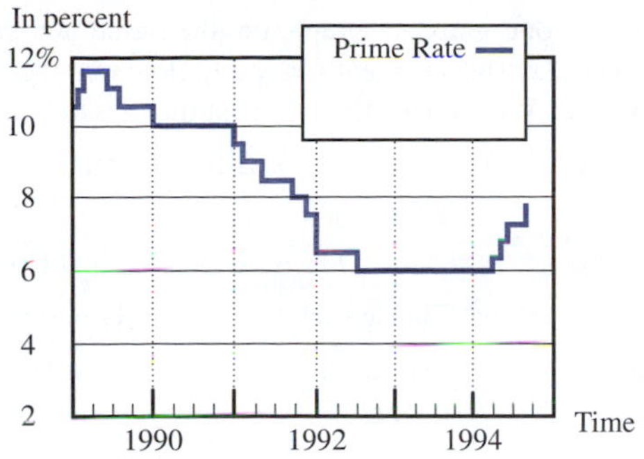

83. Biology In an experiment Senanayake and coworkers[15] showed that the yield y of potato flea beetles treated with an insecticide, given as a percentage of untreated beetles, was approximated by the piecewise linear function $y = 100 + 0.24x$ when $x \leq 59$ and by $y = 114.3 - 0.219x$ when $x > 59$, where x is the peak number of potato flea beetles per plant. What was the yield when $x = 40$? 200? Graph on $[0, 500]$. Is this function continuous?

84. Biology Hardmann[16] found that the survival rates S of the summer eggs of the European red mite were approximated by the piecewise linear function given by $S = 0.998 - 0.007(18 - T)$ if $T \geq 18$ and by $S = 0.998$ if $T \geq 18$, where T is temperature in degrees Centigrade. Determine the survival rate if $T = 15$. If $T = 20$. Graph.

85. Federal Income Tax The 2001 federal income tax rate schedule for a single person was given in the 2001 1040 Forms and Instruction publication as follows.

If the amount on line 39, is over	But not over	Enter on line 40	of the amount over
\$0	\$27,050	15%	\$0
27,050	65,550	\$4057.50 + 27.5%	27,050
65,550	136,750	14,645.00 + 30.5%	65,550
136,750	297,350	36,361.00 + 35.5%	136,750
297,350	...	93,374.00 + 39.1%	297,350

The first column is taxable income and the third column is taxes. Translate the information in the table into a piecewise-defined function.

86. Hydrology Heyman and Kjerfve[17] developed a mathematical model of the runoff for the Rio Grande dam in Belize given by $Q(H) = 26.5165(H - 0.4863)^{1.165}$, where Q is discharge in cubic meters per second and H is river stage height in meters. They report that their model is valid when $0.49 < H < 10$. Find $Q(1)$, and explain what this means. Graph this function.

87. Diversity of Species Usher and coworkers[18] developed a mathematical model for certain farm woodlands, described by the equation $S(A) = 1.81A^{0.284}$, where S is the number of species of plants and A is the area of woodland in square meters. Graph. Explain what $S(118)$ means. What happens to the number of species if the area is reduced by one half?

88. Forestry McNeil and associates[19] showed that for small loblolly pine trees $V(H) = 0.0000837H^{3.191}$, where V is the volume in cubic meters and H is the tree height in meters. Find $V(10)$. Explain what this means. Find $V(1)$, and explain what this means. What happens to V when H doubles? Graph $V = V(H)$.

[15] D. G. Senanayake, S. F. Pernal, and N. J. Holliday. 1993. Yield response of potatoes to defoliation by the potato flea beetle in Manitoba. *J. Econ. Entomol.* 86(5):1527–1533.

[16] J. M. Hardmann. 1989. Model simulating the use of miticides to control European red mite in Nova Scotia apple orchards. *J. Econ. Entomol.* 82:1411–1422.

[17] William D. Heyman and Bjorn Kjerfve. 1999. Hydrological and oceanographic considerations for integrated coastal zone management in southern Belize. *Environmental Management* 24(2):229–245.

[18] M. B. Usher, A. C. Brown, S. E. Bedford. 1992. Plant species richness in farm woodlands. *Forestry* 65:1–13.

[19] Robert C. McNeil, Russ Lea, Russell Ballard, and H. Lee Allen. 1988. Predicting fertilizer response of loblolly pine using foliar and needle-fall nutrients sampled in different seasons. *Forest Sci.* 34:698–707.

More Challenging Exercises

In Exercises 89 through 92, find $\frac{f(x+h)-f(x)}{h}$ for the indicated functions.

89. $f(x) = 3x + 1$

90. $f(x) = x^2 - 1$

91. $f(x) = 3x^2 + 1$

92. $f(x) = |x|$

93. Express the function $f(x) = 2x + |2 - x|$ in piecewise form without using absolute values and sketch its graph.

94. Find a formula for the function graphed in the figure.

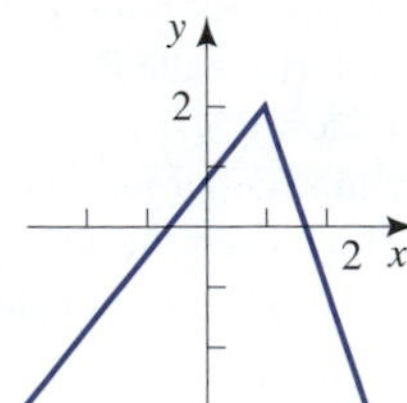

95. Using your grapher, graph on the same screen $y_1 = f(x) = |x|$ and $y_2 = g(x) = \sqrt{x^2}$. How many graphs do you see? What does this say about the two functions?

96. Is it always true that $|-a| = a$ for any number a? Explain why or why not.

97. Let $f(x) = mx$. Show $f(a + b) = f(a) + f(b)$.

98. Let $f(x) = x^2$. Find x such that $f(x + 1) = f(x + 2)$.

99. Let $f(x) = 3^x$. Show that $3f(x) = f(x + 1)$ and that $f(a + b) = f(a) \cdot f(b)$.

100. Let $f(x) = x^2 + 1$. Find $f[f(1)]$ and $f[f(x)]$.

Solutions to WARM UP EXERCISE SET 1.1

1. The function $f(x) = \frac{\sqrt{x+1}}{x}$ is defined if $x + 1 \geq 0$, or $x \geq -1$, and if $x \neq 0$. Thus, the domain is all points on $[-1, 0)$ together with all points on $(0, \infty)$.

2. For the function in the previous exercise

a. $f(-1) = \dfrac{\sqrt{-1+1}}{-1} = 0.$

b. $f(3) = \dfrac{\sqrt{3+1}}{3} = \frac{2}{3}.$

c. $f(a+2) = \dfrac{\sqrt{a+2+1}}{a+2} = \dfrac{\sqrt{a+3}}{a+2}.$

d. $f(x+2) = \dfrac{\sqrt{x+2+1}}{x+2} = \dfrac{\sqrt{x+3}}{x+2}.$

3. For the first minute the cost is a constant \$0.10. During the second minute the cost is a constant \$0.10 + \$0.07 = \$0.17. During the third minute the cost is \$0.24. We can write the function as

$$C(x) = \begin{cases} 0.10 & \text{if } 0 < x \leq 1 \\ 0.17 & \text{if } 1 < x \leq 2 \\ 0.24 & \text{if } 2 < x \leq 3 \end{cases}$$

The graph is shown in the figure. Notice that the graph has breaks at $x = 1$ and $x = 2$. These are points of discontinuity. These points might affect your behavior. Notice that a 59.9-second call costs \$0.10 but a 60.1-second call costs \$0.17. You might be inclined to wrap up your conversation in just under 1 minute rather than just over 1 minute.

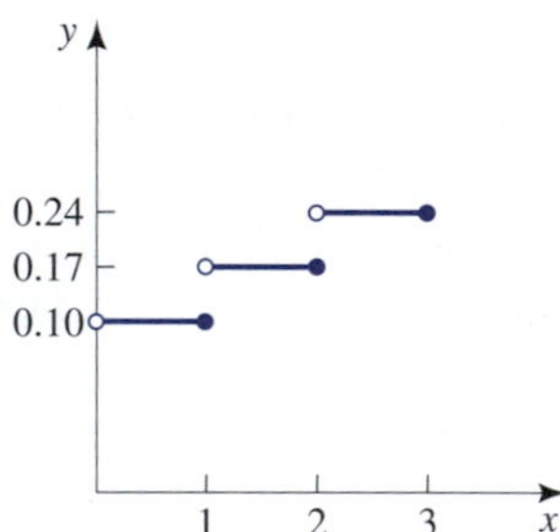

1.2 Mathematical Models

Mathematical Modeling

Mathematical Models of Cost, Revenue, and Profits

Mathematical Models of Supply and Demand

Quadratic Mathematical Models

Enrichment: Supply and Demand Dynamics

Stock Montage, Inc.

Augustin Cournot, 1801–1877

The first significant work dealing with the application of mathematics to economics was Cournot's "Researches into the Mathematical Principles of the Theory of Wealth," published in 1836. It was Cournot who originated the supply and demand curves that are discussed in this section. Irving Fisher, a prominent economics professor at Yale University and one of the first exponents of mathematical economics in the United States, wrote that Cournot's book "seemed a failure when first published. It was far in advance of the times. Its methods were too strange, its reasoning too intricate for the crude and confident notions of political economy then current."

APPLICATION **Cost, Revenue, and Profit Models**

Cotterill[20] estimated the costs and prices in the cereal-manufacturing industry. The table summarizes the costs in both pounds and tons in the manufacture of a typical cereal. The manufacturer obtained a price of \$2.40 a pound, or \$4800 a ton. Let x be the number of tons of cereal manufactured and sold, and let p be the price of a ton sold. Nevo[21] estimated fixed costs for a typical plant to be \$300 million. What are the cost, revenue and profit equations? See Example 1 on page 28 for the answer.

Item	\$/lb	\$/ton
Manufacturing cost:		
Grain	0.16	320
Other ingredients	0.20	400
Packaging	0.28	560
Labor	0.15	300
Plant costs	0.23	460
Total manufacturing costs	1.02	2040
Marketing expenses:		
Advertising	0.31	620
Consumer promo (mfr. coupons)	0.35	700
Trade promo (retail in-store)	0.24	480
Total marketing costs	0.90	1800
Total variable costs	1.92	3840

[20] Ronald W. Cotterill and Lawrence E. Haller. 1997. An economic analysis of the demand for RTE cereal: product market definition and unilateral market power effects. Research Report No. 35. Food Marketing Policy Center, University of Connecticut.

[21] Aviv Nevo. 2001. Measuring market power in the ready-to-eat cereal industry. *Econometrica* 69(2):307–342.

Mathematical Modeling

As we saw in the preceding section, mathematical modeling is an attempt to describe some part of the real world in mathematical terms. There are three steps in mathematical modeling: formulation, mathematical manipulation, and evaluation.

Formulation

First, on the basis of observations, we must state a question or formulate a hypothesis. If the question or hypothesis is too vague, we need to make it precise. If it is too ambitious, we need to restrict it or subdivide it into manageable parts. Second, we need to identify important factors. We must decide which quantities and relationships are important to answer the question and which can be ignored. We then need to formulate a *mathematical* description. For example, each important quantity should be represented by a variable. Each relationship should be represented by an equation, inequality, or other mathematical construct. If we obtain a function, say, $y = f(x)$, we must carefully identify the input variable x and the output variable y and the units for each. We should also indicate the interval of values of the input variable for which the model is justified.

Mathematical Manipulation

After the mathematical formulation, we then need to do some mathematical manipulation to obtain the answer to our original question. We might need to do a calculation, solve an equation, or prove a theorem. Sometimes the mathematical formulation gives us a mathematical problem that is impossible to solve. In such a case, we will need to reformulate the question in a less ambitious manner.

Evaluation

Naturally, we need to check the answers given by the model with real data. We normally expect the mathematical model to describe only a very limited aspect of the world and to give only approximate answers. If the answers are wrong or not accurate enough for our purposes, then we will need to identify the sources of the model's shortcomings. Perhaps we need to change the model entirely, or perhaps we just need to make some refinements. In any case, this requires a new mathematical manipulation and evaluation. Thus, modeling often involves repeating the three steps of formulation, mathematical manipulation, and evaluation.

We will create mathematical equations that relate cost, revenue, and profits of a manufacturing firm to the number of units produced and sold. We now consider linear models. Later in this section we will consider quadratic models.

CONNECTION

What Are Costs?

Isn't it obvious what the costs to a firm are? Apparently not. On July 15, 2002, Coca-Cola Company announced that it would begin treating stock-option compensation as a cost, thereby lowering earnings. If all companies in the Standard and Poors 500 stock index were to do the same, the earnings for this index[22] would drop by 23%.

[22] *The Wall Street Journal*, July 16, 2002.

Mathematical Models of Cost, Revenue, and Profits

Any manufacturing firm has two types of costs: fixed and variable. **Fixed costs** are those that do not depend on the amount of production. These costs include real estate taxes, interest on loans, some management salaries, certain minimal maintenance, and protection of plant and equipment. **Variable costs** depend on the amount of production. They include the cost of material and labor. Total cost, or simply **cost**, is the sum of fixed and variable costs:

$$\text{cost} = (\text{variable cost}) + (\text{fixed cost}).$$

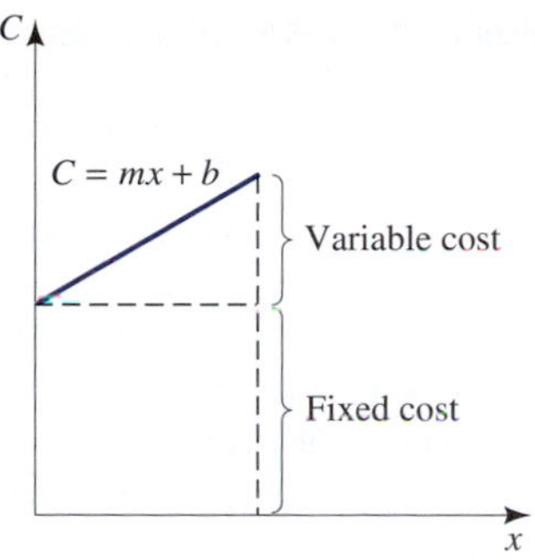

Figure 1.21
Cost is fixed cost plus variable cost.

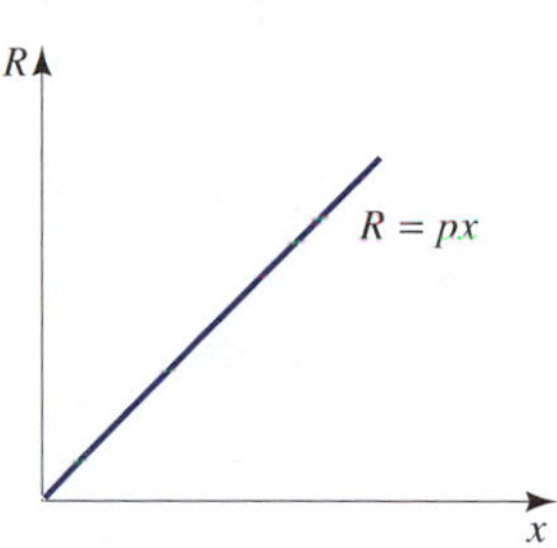

Figure 1.22
Revenue is the price of each unit times the number of units sold.

Let x denote the number of units of a given product or commodity produced by a firm. (Notice that we must have $x \geq 0$.) The units could be bales of cotton, tons of fertilizer, or number of automobiles. In the **linear cost model** we assume that the cost m of manufacturing one unit is the same no matter how many units are produced. Thus, the variable cost is the number of units produced times the cost of each unit:

$$\begin{aligned} \text{variable cost} &= (\text{cost per unit}) \times (\text{number of units produced}) \\ &= mx \end{aligned}$$

If b is the fixed cost and $C(x)$ is the cost, then we have the following:

$$\begin{aligned} C(x) &= \text{cost} \\ &= (\text{variable cost}) + (\text{fixed cost}) \\ &= mx + b \end{aligned}$$

Notice that we must have $C(x) \geq 0$. In the graph shown in Figure 1.21, note that the y-intercept is the fixed cost and the slope is the cost per item.

In the **linear revenue model** we assume that the price p of a unit sold by a firm is the same no matter how many units are sold. (This is a reasonable assumption if the number of units sold by the firm is small in comparison to the total number sold by the entire industry.) Thus, the revenue is the price per unit times the number of units sold. Let x be the number of units sold. (For convenience, we always assume that *the number of units sold equals the number of units produced.*) Then, if we denote the revenue by $R(x)$,

$$\begin{aligned} R(x) &= \text{revenue} \\ &= (\text{price per unit}) \times (\text{number sold}) \\ &= px \end{aligned}$$

Notice that we must have $R(x) \geq 0$. Notice in Figure 1.22 that the straight line goes through (0, 0) because nothing sold results in no revenue. The slope is the price per unit.

CONNECTION
What Are Revenues?

The accounting practices of many telecommunications companies, such as Cisco and Lucent, have been criticized for what the companies consider revenues. In particular, these companies have loaned money to other companies, which then use the proceeds of the loan to buy telecommunications equipment from Cisco and Lucent. Cisco and Lucent then book these sales as "revenue." But is this revenue?

Regardless of whether our models of cost and revenue are linear or not, **profit** P is always revenue less cost. Thus

$$\begin{aligned} P &= \text{profit} \\ &= (\text{revenue}) - (\text{cost}) \\ &= R - C \end{aligned}$$

Recall that both cost $C(x)$ and revenue $R(x)$ must be nonnegative functions. However, profit $P(x)$ can be positive or negative. Negative profits are called *losses*.

Let's now determine the cost, revenue, and profit equations for the cereal-manufacturing industry.

EXAMPLE 1 Cost, Revenue, and Profit Equations in the Cereal-Manufacturing Industry

Cotterill[23] estimated the costs and prices in the cereal-manufacturing industry. Table 1.4 summarizes the costs in both pounds and tons in the manufacture of a typical cereal. The manufacturer obtained a price of \$2.40 a pound, or \$4800 a ton. Let x be the number of tons of cereal manufactured and sold, and let p be the price of a ton sold. Nevo[24] estimated fixed costs for a typical plant to be \$300 million. Find the cost, revenue, and profit equations.

Table 1.4

Item	\$/lb	\$/ton
Manufacturing cost:		
Grain	0.16	320
Other ingredients	0.20	400
Packaging	0.28	560
Labor	0.15	300
Plant costs	0.23	460
Total manufacturing costs	1.02	2040
Marketing expenses:		
Advertising	0.31	620
Consumer promo (mfr. coupons)	0.35	700
Trade promo (retail in-store)	0.24	480
Total marketing costs	0.90	1800
Total variable costs	1.92	3840

Solution

Let the cost, revenue, and profits be given in thousands of dollars. Then

(a)
$$\begin{aligned} C(x) &= (\text{variable cost}) + (\text{fixed cost}) \\ &= mx + b \\ &= 3.84x + 300{,}000 \end{aligned}$$

See Figure 1.23a. Notice that $x \geq 0$ and $C(x) \geq 0$.

[23] Ronald W. Cotterill and Lawrence E. Haller. 1997. An economic analysis of the demand for RTE cereal: product market definition and unilateral market power effects. Research Report No. 35. Food Marketing Policy Center. University of Connecticut.

[24] Aviv Nevo. 2001. Measuring market power in the ready-to-eat cereal industry. *Econometrica* 69(2):307–342.

(b)

$$\begin{aligned} R(x) &= (\text{price per ton}) \times (\text{number tons sold}) \\ &= px \\ &= 4.8x \end{aligned}$$

See Figure 1.23b. Notice that $x \geq 0$ and $R(x) \geq 0$.

(c)

$$\begin{aligned} P(x) &= (\text{revenue}) - (\text{cost}) = R - C \\ &= (4.8x) - (3.84x + 300{,}000) \\ &= 0.96x - 300{,}000 \end{aligned}$$

See Figure 1.23c.

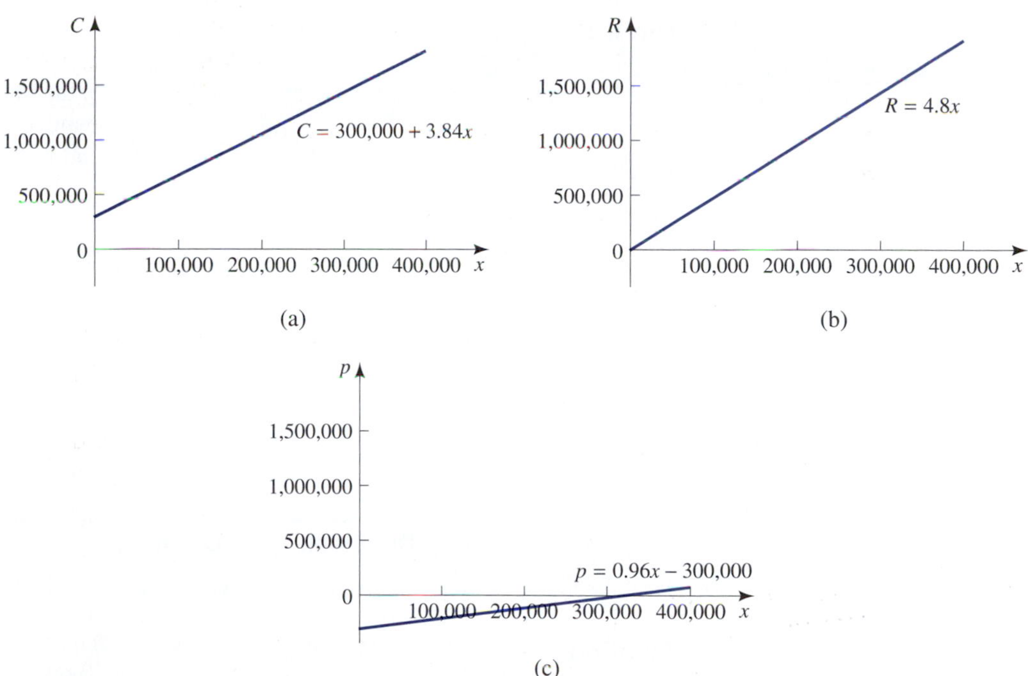

Figure 1.23

To be specific, if $x = 200{,}000$, then

$$\begin{aligned} C(200{,}000) &= 3.84(200{,}000) + 300{,}000 = 1{,}068{,}000 \\ R(200{,}000) &= 4.8(200{,}000) = 960{,}000 \\ P(200{,}000) &= 960{,}000 - 1{,}068{,}000 = -108{,}000 \end{aligned}$$

Thus, if 200,000 tons are produced and sold, the cost is 1,068,000 thousand dollars or \$1068 million. The revenue is \$960 million, and there is a *loss* of \$108 million.

Doing the same for some other values of x, we have the results shown in Table 1.5. Notice in Figure 1.23c that profits can be negative.

Table 1.5

Number Made and Sold	200,000	300,000	400,000
Cost in millions of dollars	1068	1452	1836
Revenue in millions of dollars	960	1440	1920
Profit (or loss) in millions of dollars	−108	−12	84

■

We can see in Figure 1.23c or in Table 1.5, that for smaller values of x, $P(x)$ is *negative*; that is, the plant has losses. For larger values of x, $P(x)$ turns positive and the plant has (positive) profits. The value of x at which the profit is zero is called the **break-even quantity**. Figure 1.24 shows that the break-even quantity is the value of x at which cost equals revenue. This is true, since $P = R - C$, and $P = 0$ if and only if $R = C$.

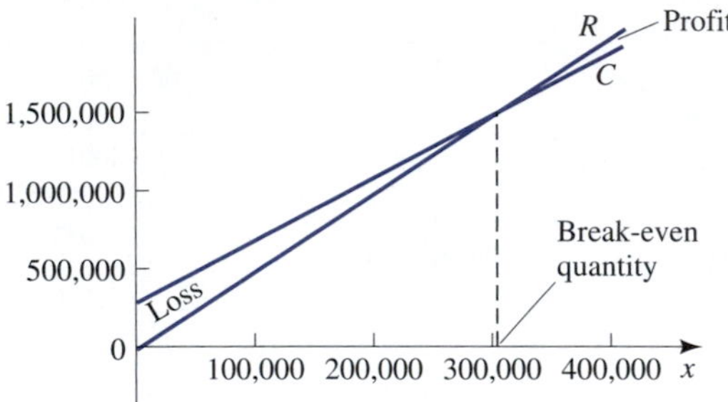

Figure 1.24

EXAMPLE 2 **Finding the Break-Even Quantity**

Find the break-even quantity in the previous example.

Solution

We can solve this problem graphically or algebraically. We leave the graphical solution to Exploration 1 and find the algebraic solution here.

To find the break-even quantity, set $P(x) = 0$ and solve for x.

$$\begin{aligned} 0 &= P(x) \\ &= 0.96x - 300{,}000 \\ 0.96x &= 300{,}000 \\ x &= 312{,}500 \end{aligned}$$

Thus, the plant needs to produce and sell 312,500 tons of cereal to break even (i.e., for profits to be zero). ■

Remark Since $P = 0$ if and only if $R = C$, we could have found the break-even quantity in Example 2 by setting $R = C$, obtaining $4800x = 3840x + 300{,}000$. Then solve for x.

Exploration 1

Finding the Break-Even Quantity Graphically

Find the approximate break-even quantity in Example 2 on your graphing calculator or computer by finding where $P = 0$ and also by finding where $C = R$.

In 1997 Fuller and coworkers[25] at Texas A & M University estimated the operating costs of cotton gin plants of various sizes. For the largest plants, the total cost in thousands of dollars was given approximately by $C(x) = 22.8073x + 377.3865$, where x is the annual quantity of bales in thousands produced. Plant capacity was 50,000 bales. Revenue was estimated at \$63.25 per bale. So $R(x) = 63.25x$. The following interactive illustration allows you to explore this example.

Interactive Illustration 1.1

Revenue, Cost, and Profit

This interactive illustration demonstrates how to graphically find where profits are positive, negative, and zero (break-even point) for the cotton gin plant mentioned in the previous paragraph. You will find a slider under the top graph. As you move the slider, you will notice a vertical line connecting the revenue and cost curves. When the revenue curve is above the cost curve, profits are positive, and the line is green. When the revenue curve is below the cost curve, profits are negative (losses), and the line is red.

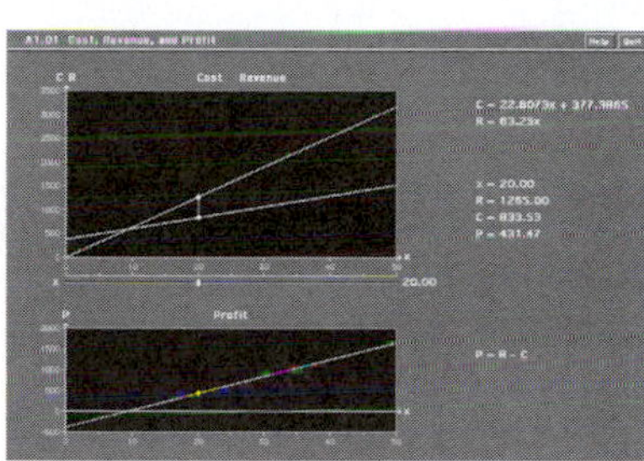

1. Estimate the interval where profits are positive.
2. Estimate the interval where profits are negative.
3. What seems to happen to profits as x increases?
4. Estimate the break-even point.

Mathematical Models of Supply and Demand

In the previous discussion we assumed that the number of units produced and sold by the given firm was small in comparison to the number sold by the industry. Under this assumption it was reasonable to conclude that the price was constant and did not vary with the number x sold. But if the number of units sold by the firm represented a *large* percentage of the number sold by the entire industry, then trying to sell significantly more units could only be accomplished by *lowering* the price of each unit. Thus, under these assumptions the price p of each unit would depend on the number sold and be a decreasing function. If, in addition, the price is assumed to be linear, then the graph of this equation is a straight line that must slope downward as shown in Figure 1.25.

[25] S. Fuller, M. Gillis, C. Parnell, and R. Childers. 1997. Effect of air pollution abatement on financial performance of Texas cotton gins. *Agribusiness* 13(5):521–532.

CONNECTION
Demand for Apartments

The figure shows[26] that during the minor recession of 2001, vacancy rates for apartments rose, that is, the demand for apartments decreased. Also notice from the figure that as demand for apartments *decreased*, rents also *decreased*. For example, in San Francisco's South Beach area, a two-bedroom apartment that had rented for \$3000 a month two years before saw the rent drop to \$2100 a month.

Down and Out
Quarterly percentage change in rent, and vacancy rates, for the top 50 markets

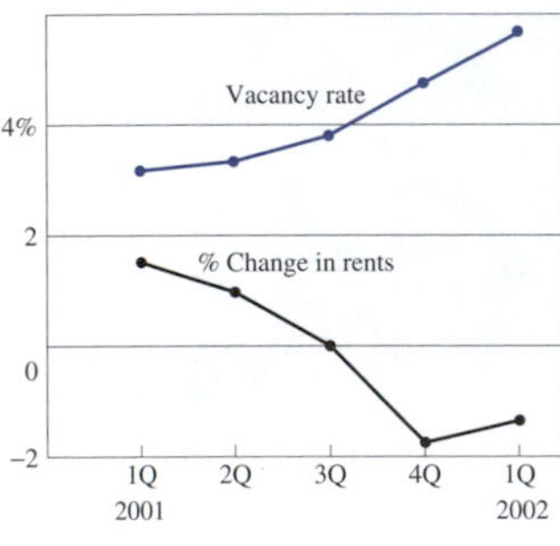

[26] *The Wall Street Journal*, April 11, 2002.

We assume that x is the number of units produced and sold by the entire industry during a given time period and that $p = -cx + d$, $c > 0$, is the price of one unit if x units are sold; that is, $p = -cx + d$ is the price of the xth unit sold. We call $p = -cx + d$ the **demand equation** and the graph the **demand curve**.

Estimating the demand equation is a fundamental problem for the management of any company or business.

The **supply equation** $p = p(x)$ gives the price p necessary for suppliers to make available x units to the market. The graph of this equation is called the **supply curve**. Figure 1.26 shows a *linear* supply curve. A reasonable supply curve rises, moving from left to right, because the suppliers of any product naturally want to sell more if the price is higher. (See Shea,[28] who looked at a large number of industries and determined that the supply curve does indeed slope upward.)

The best-known law of economics is the law of supply and demand. Figure 1.27 shows a demand equation and a supply equation that intersect. The point of intersection, or the point at which supply equals demand, is called the **equilibrium point**. The x-coordinate of the equilibrium point is called the **equilibrium quantity** and the p-coordinate is called the **equilibrium price**.

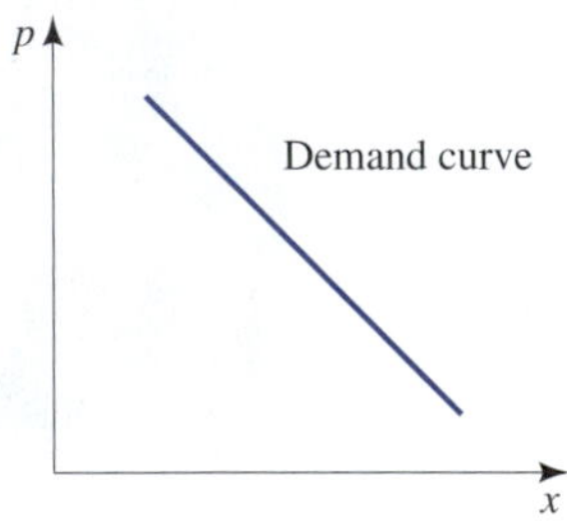

Figure 1.25
A typical demand curve slopes downward.

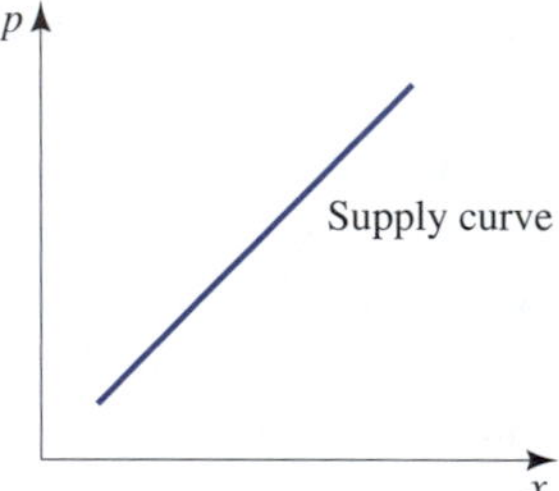

Figure 1.26
A typical supply curve slopes upward.

CONNECTION
Demand for Wheat

On June 7, 2002, without warning, China canceled a 100,000 metric ton soft red winter-wheat purchase for the 2002–2003 marketing year. The price of wheat immediately dropped 5.5 cents to \$2.75 a bushel.[27]

[27] *The Wall Street Journal*, June 1, 2002.

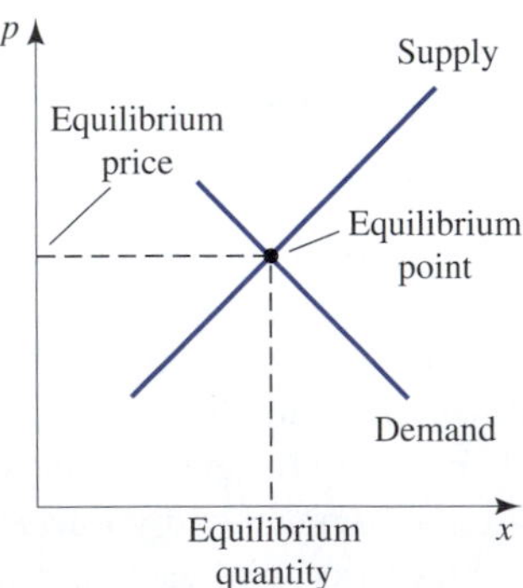

Figure 1.27
The equilibrium point is the point of intersection of the supply and demand curves.

[28] John Shea. 1993. Do supply curves slope up? *Quart. J. Econ.* cviii:1–32.

CONNECTION
Supply of Cotton

On May 2, 2002, the U.S. House of Representatives passed a farm bill that promises billions of dollars in subsidies to cotton farmers. With the prospect of a greater supply of cotton, cotton prices dropped 1.36 cents to 33.76 cents per pound.[29]

[29] *The Wall Street Journal*, May 3, 2002.

CONNECTION
Demand for Steel Outpaces Supply

In early 2002 President George W. Bush imposed steep tariffs on imported steel to protect domestic steel producers. As a result millions of tons of imported steel were locked out of the country. Domestic steelmakers announced on March 27, 2002, that they had been forced to ration steel to their customers and boost prices because demand has outpaced supply.[30]

[30] *The Wall Street Journal*, March 28, 2002.

EXAMPLE 3 **Finding the Equilibrium Point**

Tauer[31] determined demand and supply curves for milk in this country. If x is billions of pounds of milk and p is in dollars per hundred pounds, he found that the demand function for milk was $p = 55.9867 - 0.2882x$ and the supply function was $p = 0.0865x$. Graph the demand and supply equations. Find the equilibrium point.

Solution

We can solve this problem graphically or algebraically. We leave the graphical solution to Exploration 2 and find the algebraic solution here.

The demand equation $p = 55.9867 - 0.2882x$ is a line with negative slope -0.2882 and y-intercept 55.9867 and is graphed in Figure 1.28. The supply equation $p = 0.0865x$ is a line with positive slope 0.0865 with y-intercept 0. This is also graphed in Figure 1.28.

To find the point of intersection of the demand curve and the supply curve, set $55.9867 - 0.2882x = p = 0.0865x$ and solve:

$$\begin{aligned} 0.0865x &= 55.9867 - 0.2882x \\ (0.0865 + 0.2882)x &= 55.9867 \\ 0.3747x &= 55.9867 \\ x &\approx 149.42 \end{aligned}$$

rounded. Then since $p(x) = 0.0865x$,

$$p(149.42) = 0.0865(149.42) = 12.92$$

approximately. We then see that the equilibrium point is approximately $(x, p) = (149.42, 12.92)$.

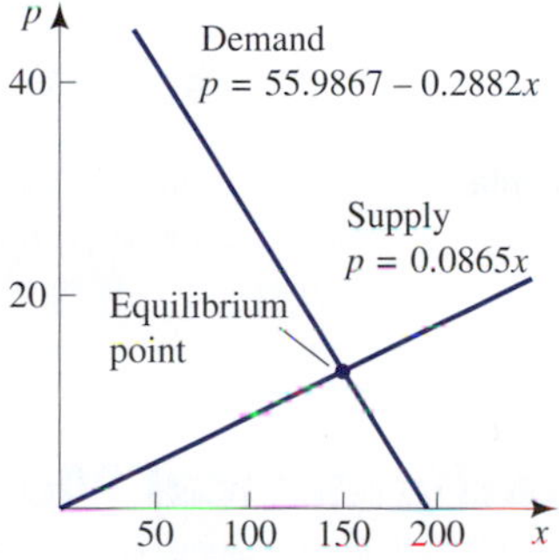

Figure 1.28
The equilibrium of supply and demand for milk.

■

[31] Loren W. Tauer. 1994. The value of segmenting the milk market into bST-produced and non-bST-produced milk. *Agribusiness* 10(1):3–12.

Exploration 2

Finding the Equilibrium Point Graphically

Find the approximate equilibrium point in Example 3 on your graphing calculator or computer.

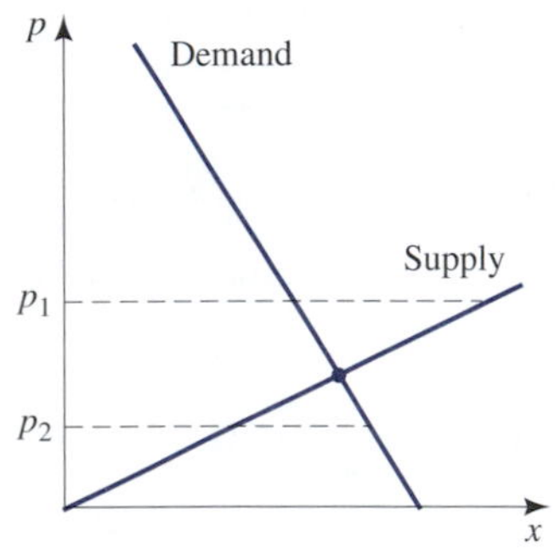

Figure 1.29
What will consumers and suppliers do at price p_1? At price p_2?

EXAMPLE 4 **Supply and Demand**

Refer to Example 3. What will consumers and suppliers do if the price is p_1 shown in Figure 1.29. What if the price is p_2?

If the price is at p_1 shown in Figure 1.29, then referring to Figure 1.30, we see that consumers will consume x_1 billion pounds of milk, but suppliers will produce x_2 billion pounds. There will be a surplus of $x_2 - x_1$ billion pounds; to work off the surplus, the price should fall toward the equilibrium price.

If the price is at p_2 shown in Figure 1.29, then referring to Figure 1.31 we see that consumers will want to buy x_4 billion pounds, but suppliers will produce only x_3 billion pounds. There will be a shortage of $x_4 - x_3$ billion pounds, and the price should rise toward the equilibrium price.

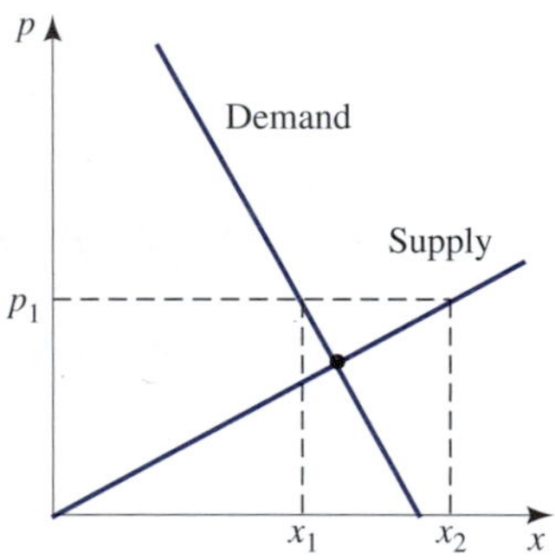

Figure 1.30
At price p_1, consumers will consume x_1 units, but suppliers will produce x_2 units.

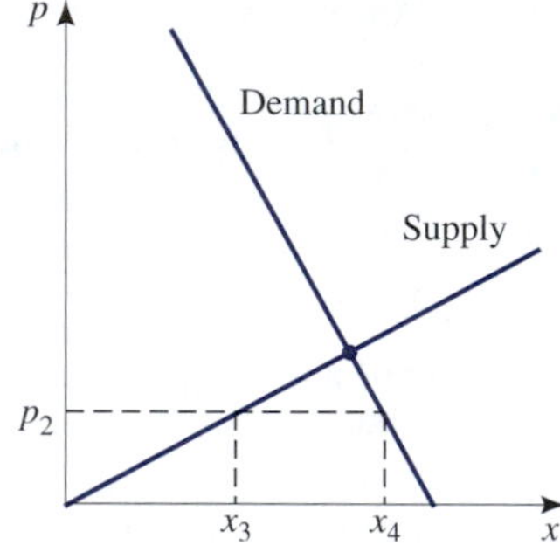

Figure 1.31
At price p_2, consumers will want to buy x_4 units but suppliers will produce x_3 units.

■

Quadratic Mathematical Models

A quadratic is a polynomial function of order 2, $q(x) = ax^2 + bx + c$, $a \neq 0$. Recall (see the review in Appendix A.7) that every quadratic equation $ax^2 + bx + c$ can be placed in standard form $a(x - h)^2 + k$. The point (h, k) is called the vertex and can be found by using the formulas

$$h = -\frac{b}{2a} \qquad k = c - \frac{b^2}{4a}$$

Figure 1.32 shows the two possible cases. If $a > 0$, then the graph is concave up, and $q(x)$ assumes a minimum of k when $x = h$. If $a < 0$, then the graph is concave down, and $q(x)$ assumes a maximum of k when $x = h$.

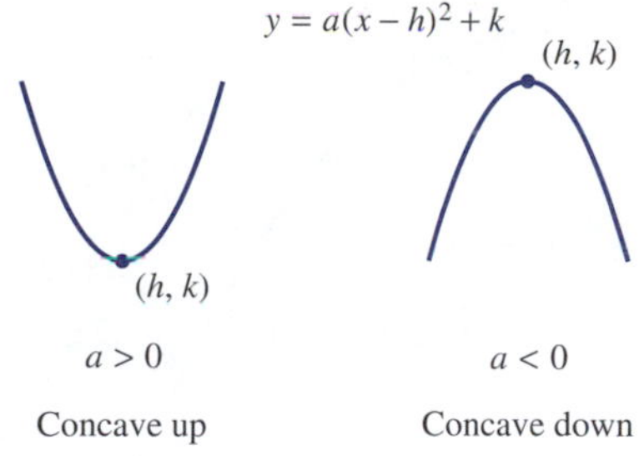

Figure 1.32
The two types of quadratics.

The following interactive illustration is given here as a convenient review. If you are already familiar with quadratic functions, you can skip it.

Interactive Illustration 1.2

The Graph of $y = a(x - h)^2 + k$

This interactive illustration allows you to explore how changing the values of a, h, and k changes the graph of $y = a(x - h)^2 + k$. There are three sliders, one for a, one for h, and one for k. Click or drag the mouse on each of the sliders and see how the graph changes.

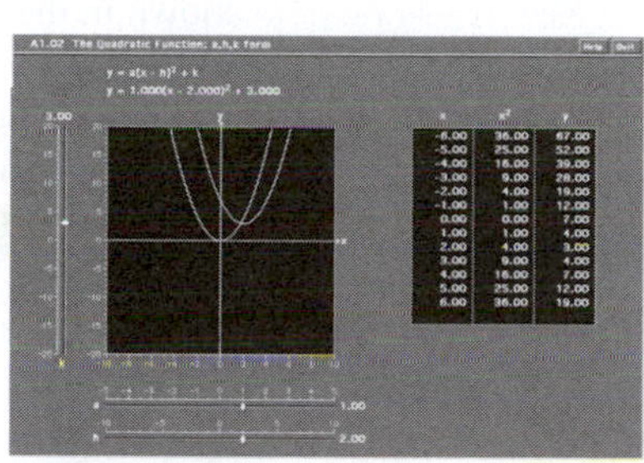

1. Make h larger. What happened to the graph?
2. Make h smaller. What happened to the graph?
3. Make k larger. What happened to the graph?
4. Make k smaller. What happened to the graph?
5. Make a larger. What happened to the graph?
6. Make a smaller. What happened to the graph?
7. Make a negative. What happened to the graph?

We will now consider quadratic mathematical models of revenue and profit. Why do this? We commented earlier that if the number of units of a product sold by an industry is to increase significantly, then the price per unit will most likely need to be decreased. Therefore, we decided that if the demand equation for an entire industry is linear, say, $p = -cx + d$, then $c > 0$. (So the demand curve slopes downward.) In this circumstance the revenue, which is always the number x sold times the price p of each, is

$$R = px = (-cx + b)x = -cx^2 + bx$$

This is a *quadratic*. Since $-c < 0$, the quadratic is concave down. If no units are sold, then we have no revenue, that is, $R(0) = 0$. Then the graph of a typical revenue equation of this form will look like that shown in Figure 1.33.

Compare this graph to the graph of a linear revenue function shown in Figure 1.22. Notice from Figure 1.22 that the linear model of revenue implies that we can increase our revenue to any level we want just by selling more units. This is hardly realistic, since flooding the market with an excessive number of units will require that the price

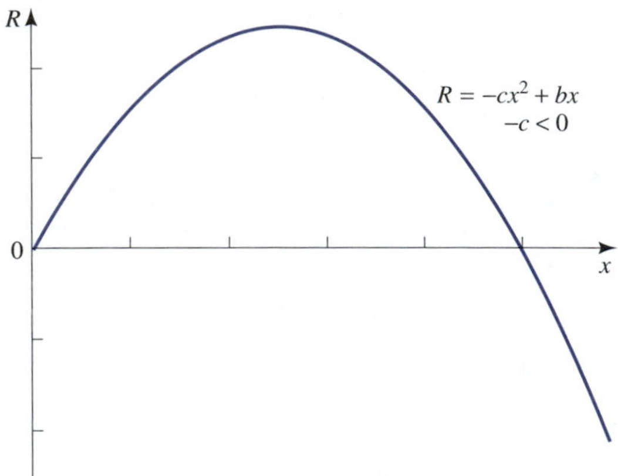

Figure 1.33

CONNECTION
Demand in the Airline Industry

Leo Mullin, CEO and Chairman of Delta Air Lines, gave an interview on CNBC on October 3, 2002, stating that after September 11, 2001, air passenger traffic dropped off. In response, the airline industry lowered prices to fill the available seats. (To increase demand x, they lowered the price p.) The result was more tickets sold but *lower* revenue. (Notice in Figure 1.33 that the curve turns *downward* after a certain point, so larger x translates into smaller R.) After some months the airline industry began to mothball some airplanes. As a result of the decreased number of available seats, they could increase prices again.

per unit be lowered. In fact, the price per unit might need to be lowered so much that the revenue will become lower (even though more units are being sold). This is shown in the graph of the quadratic revenue model shown in Figure 1.33, where the revenue first increases but after a certain point begins to decrease.

Notice how this relates to the comments made at the beginning of this section on mathematical modeling. For certain purposes the *evaluation* of the linear model led us to realize that this model was not accurate enough. We then *reformulated* the model as a quadratic one. Thus, we find ourselves repeating the three steps of formulation, mathematical manipulation, and evaluation.

We will also find that quadratic models can be used in other applications, in which they are often more realistic than linear models.

EXAMPLE 5 **A Quadratic Revenue Function**

In 1973 Braley and Nelson[32] studied the effect of price increases on school lunch participation in the city of Pittsburgh. They estimated the demand equation for school lunches to be $p = -0.00381x + 62.476$, where x is the number of lunches purchased and p is the price in cents. Find the value of x for which the revenue will be maximum, and find the maximum revenue. For this value of x, find the price.

Solution

We can solve this problem graphically or algebraically. We leave the graphical solution to Interactive Illustration 1.3 and find the algebraic solution here.

If $R(x)$ is the revenue in cents, then

$$R(x) = xp = x(-0.00381x + 62.476) = -0.00381x^2 + 62.476x$$

Notice that the coefficient of x^2 is negative, so the graph is concave down. To find

[32] George A. Braley and P. E. Nelson, Jr. Effect of a controlled price increase on school lunch participation: Pittsburgh. *Amer. J. Agri. Econ.* February 1975:90–96.

the vertex, we can use the formulas given earlier:

$$h = -\frac{b}{2a} = -\frac{62.476}{2(-0.00381)} \approx 8199$$

$$k = c - \frac{b^2}{4a} = 0 - \frac{(62.476)^2}{4(-0.00381)} \approx 256{,}119$$

Thus, the revenue is maximized at about 256,119 cents, or \$2561.19, when 8199 lunches are sold. According to the demand equation, if 8199 lunches are sold, then the price of each lunch is

$$p = -0.00381(8199) + 62.476 \approx 31$$

cents.

A graph is shown in Figure 1.34.

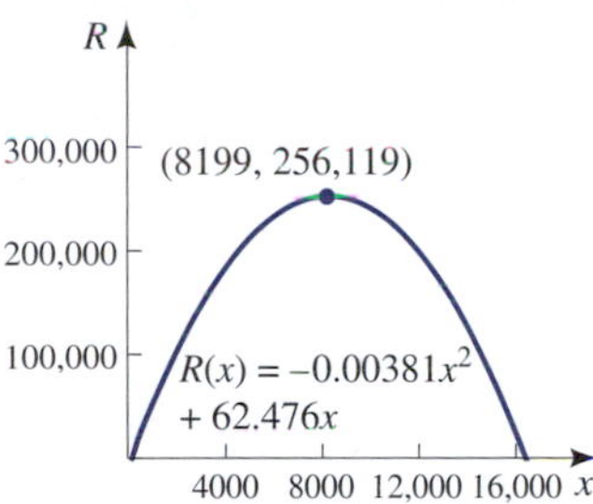

Figure 1.34
This quadratic has vertex (8199, 256,119) and opens downward, since $a = -0.00381 < 0$.

Interactive Illustration 1.3

Revenue

This interactive illustration allows you to experiment with the graphs found in Example 5. Notice that in the first window there is a graph of the demand equation and a colored rectangle formed by the two axes and the point (x, p) on the demand line. The *area of this rectangle* is xp, which is the *revenue*. As you move the cursor, the point (x, p) moves, and the rectangle changes. In the bottom window is plotted the graph of the revenue.

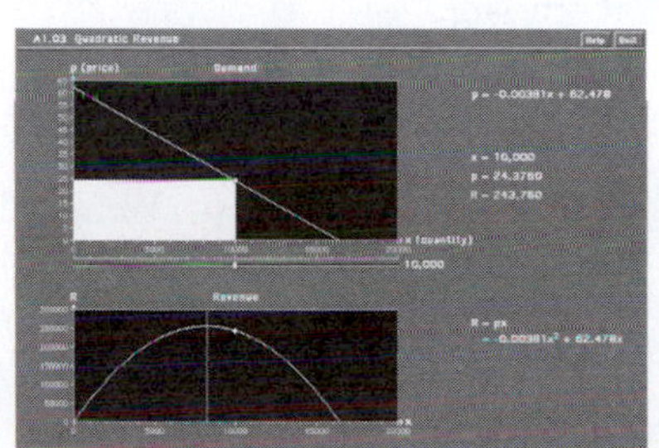

1. Take demand to be 1000, 2000, 3000, and so on, and make a table of the corresponding revenues. Describe what is happening to revenue. Is it always increasing? Does it decrease at some point?
2. Locate values for demand for which revenue increases with increases in demand.
3. Locate values for demand for which revenue decreases with increases in demand.

The graph of the revenue function shown in Figure 1.34 is typical. To sell an unusually large number of an item to consumers, the price needs to be lowered to such an extent as to begin lowering the (total) revenue.

Enrichment: Supply and Demand Dynamics

Let us consider some commodity—corn, to be specific. For this discussion we will find it convenient to have demand as a function of price. Thus, for this discussion the price p will be on the horizontal axis, and the demand x will be on the vertical axis. Let us now suppose that the supply equation is $x = p - 1$ and the demand equation is $x = 11 - 2p$, where x is bushels and p is in dollars per bushel.

The graphs are shown in Figure 1.35. Notice from the supply curve that as the price decreases, the supply does also until a point is reached at which there will be no supply. Normally, there is a price at which suppliers refuse to produce anything (sometimes called the "choke" price). We can readily find the equilibrium price for our model:

$$\begin{aligned} p - 1 &= 11 - 2p \\ 3p &= 12 \\ p &= 4 \end{aligned}$$

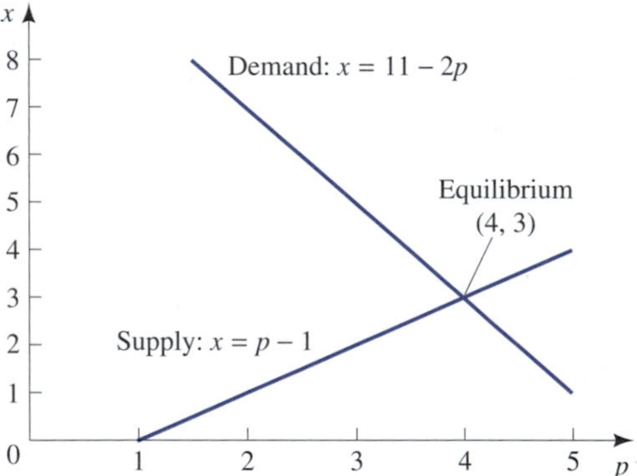

Figure 1.35

We noted that in equilibrium, demand must equal supply, which corresponds to the point where the demand curve and the supply curve intersect. This rarely happens. What actually happens is the farmer bases production on the price p_n of corn that prevails at planting time. The supply equation will then determine the supply of corn. Owing to the time lag in the growing process, the resulting supply of corn is not available until the next period, that is, the fall. At that time the price will be determined by the demand equation. The price will adjust so that all the corn is sold. The farmer notes the new price p_{n+1} at planting time, and a new cycle begins.

The supply s_{n+1} is determined by the equation $s_{n+1} = p_n - 1$, and the demand is determined by the equation $d_{n+1} = 11 - 2p_{n+1}$ according to the discussion in the previous paragraph. We impose the realistic condition that the price (at the time the corn is brought to market) is adjusted so that demand equals supply, that is, so that all the corn is sold. Imposing the condition that supply must equal demand gives the equation

$$p_n - 1 = s_{n+1} = d_{n+1} = 11 - 2p_{n+1}$$

This gives

$$p_{n+1} = 6 - 0.5p_n$$

That is, the price next year, p_{n+1}, will be 6 less one half times this year's price.

Now let's suppose that the price starts out at $p_0 = 2$. Then rounding off to the nearest cent, we have

$$\begin{aligned}
p_1 &= 6 - 0.5(2) \\
&= 5 \\
p_2 &= 6 - 0.5(5) \\
&= 3.5 \\
p_3 &= 6 - 0.5(3.5) \\
&= 4.25 \\
p_4 &= 6 - 0.5(4.25) \\
&= 3.88 \\
p_5 &= 6 - 0.5(3.88) \\
&= 4.06 \\
p_6 &= 6 - 0.5(4.06) \\
&= 3.97 \\
p_7 &= 6 - 0.5(3.97) \\
&= 4.02 \\
p_8 &= 6 - 0.5(4.02) \\
&= 3.99 \\
p_9 &= 6 - 0.5(3.99) \\
&= 4.01 \\
p_{10} &= 6 - 0.5(4.01) \\
&= 4.00
\end{aligned}$$

So the price in this case moves steadily toward the equilibrium price.

There is a dramatic graphical way of determining and seeing the price movements. This graphical technique is called a **cobweb**. The cobweb is shown in Figure 1.36. We start at the point labeled A. The price is \$2. With such a low price, far below the equilibrium, farmers plant a small crop and bring the crop to market in the fall of that year. Fortunately for the farmer, the small supply results from the demand equation with a high price of \$5.00 (point B). This is obtained by drawing the horizontal line seen in Figure 1.36. Now proceed to the spring of the next year by drawing a vertical line until we hit the supply curve, at point C. Now the farmer sees a high price of \$5 for corn, so the supply curve determines the supply to be produced, which is relatively large. When the corn is brought to market in the fall (follow the horizontal to the demand curve), the price is set by the demand curve, at point D. The large amount of corn results in a poor price (for the farmer) of \$3.50. So the farmer produces little corn, resulting in a higher price of \$4.25, point E, and so on. As the cobweb indicates, the price gets closer and closer to the equilibrium price.

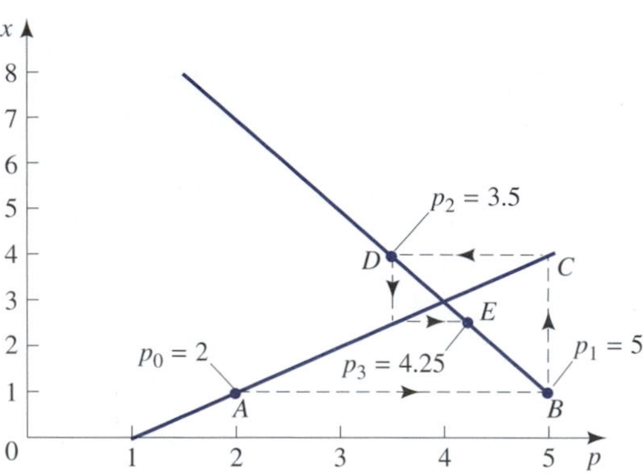

Figure 1.36

The next interactive illustration reproduces the above cobweb, allowing you to explore a more general set of supply and demand equations and to explore behavior different from that given above.

Interactive Illustration 1.4

Supply and Demand Dynamics

This interactive illustration allows you to produce the cobweb diagrams for a variety of supply and demand curves. There are four sliders to allow the four parameters in the supply and demand equations to be changed.

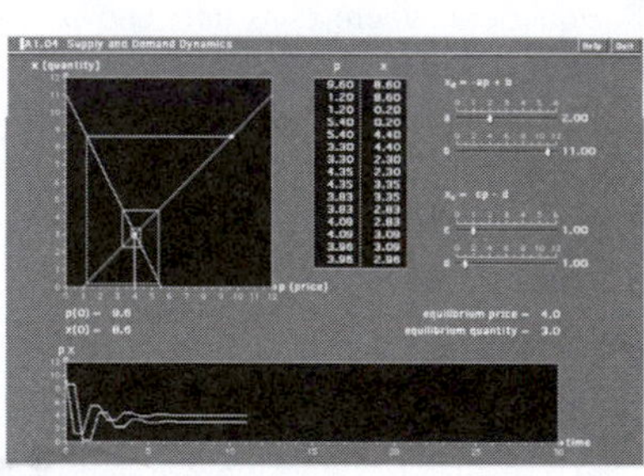

1. Locate at least one set of parameters for the supply and demand equations that give a cobweb diagram like the one in Figure 1.36.
2. Sometimes the price, instead of moving closer to equilibrium can move *away* from the equilibrium. Locate at least one set of parameters for the supply and demand equations that give a cobweb diagram for which the cobweb spirals *farther away* from equilibrium.
3. Sometimes the price can stay constant. Locate at least one set of parameters for the supply and demand equations that give a cobweb diagram for which the cobweb moves in a repeated cycle, that is, does not move closer to equilibrium or farther from equilibrium. (Pick a point other than the equilibrium point.)

Warm Up Exercise Set 1.2

1. Rogers and Akridge[33] of Purdue University studied fertilizer plants in Indiana. For a typical medium-sized plant they estimated fixed costs at \$366,821 and estimated that it cost \$208.52 to produce each ton of fertilizer. The plant sells its fertilizer output at \$266.67 per ton. Find the cost, revenue, and profit equations. Determine the cost, revenue, and profits when the number of tons produced and sold is 5000, 6000, and 7000. Find the break-even point.

2. The excess supply and demand curves for wheat worldwide were estimated by Schmitz and coworkers[34] to be

$$\text{Supply:} \quad x = 57.4 + 0.143p$$
$$\text{Demand:} \quad x = 142.8 - 0.284p$$

where p is price per metric ton and x is in millions of metric tons. Excess demand refers to the excess of wheat that producer countries have over their own consumption. Graph and find the equilibrium price and equilibrium quantity.

3. Suppose that the demand equation for a certain commodity is given by $p = -2x + 12$, where x is given in thousands per week and p is the price of one item. Suppose the cost function, in thousands of dollars per week, is given by $C(x) = 4x + 6$. Find the value of x for which profits will be maximized. Find any break-even quantities.

4. A bus company charges \$70 per person for a sightseeing trip if 40 people travel in a group. If for each person above 40 the company reduces the charge per person by \$1, how many people will maximize the total revenue for the bus company? What is the maximum revenue?

[33] Duane S. Rogers and Jay T. Akridge. 1996. Economic impact of storage and handling regulations on retail fertilizer and pesticide plants. *Agribusiness* 12(4):327–337.

[34] Andrew Schmits, Dale Sigurdson, and Otto Doering. 1986. Domestic farm policy and the gains from trade. *Amer. J. Agr. Econ.* 68:820–827.

Exercise Set 1.2

In Exercises 1 and 2 you are given the cost per item and the fixed costs over a certain time period. Assuming a linear cost model, find the cost equation, where C is cost and x is the number produced.

1. Cost per item = \$3, fixed cost = \$10,000
2. Cost per item = \$6, fixed cost = \$14,000

In Exercises 3 and 4 you are given the price of each item, which is assumed to be independent of the number of items sold. Find the revenue equation, where R is revenue and x is the number sold, assuming that the revenue is linear in x.

3. Price per item = \$5
4. Price per item = \$0.1
5. Using the cost equation found in Exercise 1 and the revenue equation found in Exercise 3, find the profit equation for P, assuming that the number produced equals the number sold.
6. Using the cost equation found in Exercise 2 and the revenue equation found in Exercise 4, find the profit equation for P, assuming that the number produced equals the number sold.

Exercises 7 through 10 show linear cost and revenue equations. Find the break-even quantity.

7. $C = 2x + 4$, $R = 4x$
8. $C = 3x + 10$, $R = 6x$
9. $C = 0.1x + 2$, $R = 0.2x$
10. $C = 0.03x + 1$, $R = 0.04x$

In Exercises 11 through 14 you are given a demand equation and a supply equation. Sketch the demand and supply curves, and find the equilibrium point.

11. Demand: $p = -x + 6$, supply: $p = x + 3$
12. Demand: $p = -3x + 12$, supply: $p = 2x + 5$
13. Demand: $p = -10x + 25$, supply: $p = 5x + 10$
14. Demand: $p = -0.1x + 2$, supply: $p = 0.2x + 1$

In Exercises 15 through 18, find where the revenue function is maximized using the given linear demand function.

15. $p = -x + 10$
16. $p = -2x + 3$
17. $p = -5x + 15$
18. $p = -0.5x + 20$

In Exercises 19 through 22, find the profit function using the given revenue and cost functions. Find the value of x that maximizes the profit. Find the break-even quantities (if they exist); that is, find the value of x for which the profit is zero. Graph the solution.

19. $R(x) = -2x^2 + 30x$, $C(x) = 10x + 42$
20. $R(x) = -3x^2 + 20x$, $C(x) = 2x + 15$
21. $R(x) = -x^2 + 30x$, $C(x) = 15x + 36$
22. $R(x) = -2x^2 + 15x$, $C(x) = 5x + 8$

In Exercises 23 through 26, use your grapher to find the break-even quantities for the given profit functions and the value of x that maximizes the profit.

23. $P(x) = -x^2 + 3.1x - 1.98$

24. $P(x) = -x^2 + 5.9x - 7.92$

25. $P(x) = -4x^2 + 38.4x - 59.36$

26. $P(x) = -2.4x^2 + 14.64x - 14.112$

Applications Using Linear Mathematical Models

27. Straight-Line Depreciation Many assets, such as machines or buildings, have a *finite* useful life and furthermore *depreciate* in value from year to year. For purposes of determining profits and taxes, various methods of depreciation can be used. In **straight-line depreciation** we assume that the value V of the asset is given by a *linear* equation in time t, say, $V = mt + b$. The slope m must be *negative*, since the value of the asset *decreases* over time. Consider a new machine that costs $50,000 and has a useful life of nine years and a scrap value of $5000. Using straight-line depreciation, find the equation for the value V in terms of t, where t is in years. Find the value after one year and after five years.

28. Straight-Line Depreciation A new building that costs $1,100,000 has a useful life of 50 years and a scrap value of $100,000. Using straight-line depreciation (refer to Exercise 27), find the equation for the value V in terms of t, where t is in years. Find the value after one year, after two years, and after 40 years.

29. Demand Suppose that 500 units of a certain item are sold per day by the entire industry at a price of $20 per item and that 1500 units can be sold per day by the same industry at a price of $15 per unit. Find the demand equation for p, assuming the demand curve to be a straight line.

30. Demand Suppose that 10,000 units of a certain item are sold per day by the entire industry at a price of $150 per item and that 8000 units can be sold per day by the same industry at a price of $200 per item. Find the demand equation for p, assuming the demand curve to be a straight line.

31. Costs of Manufacturing Fenders Saur and colleagues[35] did a careful study of the cost of manufacturing automobile fenders using five different materials: steel, aluminum, and three injection-molded polymer blends: rubber-modified polypropylene (RMP), nylon-polyphenyle-neoxide (NPN), and polycarbonate-polybutylene terephthalate (PPT). The following table gives the fixed and variable costs of manufacturing each pair of fenders.

Variable and Fixed Costs of Pairs of Fenders

Costs	Steel	Aluminum	RMP	NPN	PPT
Variable	$5.26	$12.67	$13.19	$9.53	$12.55
Fixed	$260,000	$385,000	$95,000	$95,000	$95,000

Write down the cost function associated with each of the materials.

32. Comparing Costs of Materials In the previous example notice that using steel requires the least variable costs but not the least fixed costs. Suppose you try to decide whether to use steel or RMP and your only consideration is cost.

a. If you were going to manufacture 15,000 pairs of fenders, which material would you pick? Why?

b. If you were going to manufacture 25,000 pairs of fenders, which material would you pick? Why?

c. How many pairs of fenders are required for the cost of the steel ones to equal the cost of the RMP ones?

33. Break-even Analysis In the Saur study of fenders mentioned in Exercises #31 and 32, the amount of energy consumed by each type of fender was also analyzed. The total energy was the sum of the energy needed for production plus the energy consumed by the vehicle used in carrying the fenders. If x is the miles traveled, then the total energy consumption equations for steel and RMP were as follows:

$$\text{Steel:} \quad E = 225 + 0.012x$$
$$\text{RPM:} \quad E = 285 + 0.007x$$

Setting $x = 0$ gives the energy used in production, and we note that steel uses less energy to produce these fenders than does RPM. However, since steel is heavier than RPM, carrying steel fenders requires more energy. Graph these equations, and find the number of pairs of fenders for which the total energy consumed is the same for both fenders.

34. Oil Production Technology D'Unger and coworkers[36] studied the economics of conversion to saltwater injection for inactive wells in Texas. (By injecting saltwater into the

[35] Konrad Saur, James A. Fava, and Sabrina Spatari. 2000. Life cycle engineering case study: automobile fender designs. *Environ. Progress* 19(2):72–82.

[36] Claude D'Unger, Duane Chapman, and R. Scott Carr. 1996. Discharge of oil field-produced water in Nueces Bay, Texas: a case study. *Environ. Management* 20(1):143–150.

wells, pressure is applied to the oil field, and oil and gas are forced out to be recovered.) The expense of a typical well conversion was estimated to be $31,750. The monthly revenue as a result of the conversion was estimated to be $2700. If x is the number of months the well operates after conversion, determine a revenue function as a function of x. How many months of operation would it take to recover the initial cost of conversion?

35. Costs, Revenues, and Profits on Kansas Beef Cow Farms Featherstone and coauthors[37] studied 195 Kansas beef cow farms. The average fixed and variable costs are found in the following table.

Variable and Fixed Costs	
Costs per cow	
Feed costs	$261
Labor costs	$82
Utilities and fuel costs	$19
Veterinary expenses costs	$13
Miscellaneous costs	$18
Total variable costs	$393
Total fixed costs	$13,386

The farm can sell each cow for $470. Find the cost, revenue, and profit functions for an average farm. The average farm had 97 cows. What was the profit for 97 cows? Can you give a possible explanation for your answer?

36. Cost, Revenue, and Profit Equations In 1996 Rogers and Akridge[38] of Purdue University studied fertilizer plants in Indiana. For a typical small-sized plant they estimated fixed costs at $235,487 and estimated that it cost $206.68 to produce each ton of fertilizer. The plant sells its fertilizer output at $266.67 per ton. Find the cost, revenue, and profit equations.

37. Cost, Revenue, and Profit Equations In 1996 Rogers and Akridge[39] of Purdue University studied fertilizer plants in Indiana. For a typical large-sized plant they estimated fixed costs at $447,917 and estimated that it cost $209.03 to produce each ton of fertilizer. The plant sells its fertilizer output at $266.67 per ton. Find the cost, revenue, and profit equations.

38. Ecotourism Velazquez and colleagues[40] studied the economics of ecotourism. A grant of $100,000 was given to a certain locality to use to develop an ecotourism alternative to destroying forests and the consequent biodiversity. The community found that each visitor spent $40 on average. If x is the number of visitors, find a revenue function. How many visitors are needed to reach the initial $100,000 invested? (This community was experiencing about 2500 visits per year.)

39. Make or Buy Decision A company includes a manual with each piece of software it sells and is trying to decide whether to contract with an outside supplier to produce the manual or to produce it in-house. The lowest bid of any outside supplier is $0.75 per manual. The company estimates that producing the manuals in-house will require fixed costs of $10,000 and variable costs of $0.50 per manual. Which alternative has the lower total cost if demand is 20,000 manuals?

40. Make or Buy Decision Repeat the previous exercise for a demand of 50,000 manuals.

41. Rail Freight In a report of the Federal Trade Commission (FTC)[41] an example is given in which the Portland, Oregon, mill price of 50,000 board square feet of plywood is $3525 and the rail freight is $0.3056 per mile.

a. If a customer is located x rail miles from this mill, write an equation that gives the total freight f charged to this customer in terms of x for delivery of 50,000 board square feet of plywood.

b. Write a (linear) equation that gives the total c charged to a customer x rail miles from the mill for delivery of 50,000 board square feet of plywood. Graph this equation.

c. In the FTC report, a delivery of 50,000 board square feet of plywood from this mill is made to New Orleans, Louisiana, 2500 miles from the mill. What is the total charge?

42. Supply and Demand In Example 3 we considered demand and supply equations for milk given by Tauer.[42] In the same paper he estimated demand and supply equations for bovine somatotropin-produced milk. The demand equation is $p = 55.9867 - 0.3249x$, and the supply equation is $p = 0.07958$, where again p is the price in dollars per hundred pounds and x is the amount of milk measured in billions of pounds.

[37] Allen M. Featherstone, Michael R. Langemeier, and Mohammad Ismet. 1997. A nonparametric analysis of efficiency for a sample of Kansas beef cow farms. *J. Agric. Appl. Econ.* July 1997:175–184.

[38] Duane S. Rogers and Jay T. Akridge. 1996. Economic impact of storage and handling regulations on retail fertilizer and pesticide plants. *Agribusiness* 12(4):327–337.

[39] Ibid.

[40] Alejandro Velazquez, Gerardo Bocco, Alejandro Torres. 2001. Turning scientific approaches into practical conservation actions: the case of Comunidad Indegena de Nuevo San Juan Parangaricutiro. *Environ. Management* 27(5):655–665.

[41] Thomas Gilligan. 1992. Imperfect competition and basing-point pricing. *Amer. Econ. Rev.* 82:1106–1119.

[42] Loren W. Tauer. 1994. The value of segmenting the milk market into bST-produced and non-bST-produced produced milk. *Agribusiness* 10(1):3–12.

a. Find the equilibrium point.
b. What should happen if the price is \$12.00 per hundred pounds? \$10.00?

43. **Break-even Analysis** At the University of Connecticut there are two ways to pay for copying. You can pay 10 cents a copy, or you can buy a plastic card for 50 cents and then pay 7 cents a copy. Let x be the number of copies you make. Write an equation for your costs for each way of paying. How many copies do you need to make before buying the plastic card is cheaper?

44. **Demand for Rice** Suzuki and Kaiser[43] estimated the demand equation for rice in Japan to be $p = 1,195,789 - 0.1084753x$, where x is in tons of rice and p is in yen per ton. Graph this equation. In 1995, the quantity of rice consumed in Japan was 8,258,000 tons. According to the demand equation, what was the price in yen per ton?

45. **Demand for Rice** Using the demand equation in the previous exercise, what happens to the price of a ton of rice when the demand increases by 1 ton. What has this number to do with the demand equation?

[43] Nobuhiro Suzuki and Harry M. Kaiser. 1998. Market impacts of Japanese rice policies with and without supply control. *Agribusiness* 14(5):355–362.

Applications using Quadratic and Other Mathematical Models

46. **Cost** A farmer wishes to enclose a rectangular field of an area of 200 square feet using an existing wall as one of the sides. The cost of the fencing for the other three sides is \$1 per foot. Find the dimensions of the rectangular field that minimizes the cost of the fence.

47. **Revenue** An apple orchard produces annual revenue of \$60 per tree when planted with 100 trees. Because of overcrowding, the annual revenue per tree is reduced by \$0.50 for each additional tree planted. How many trees should be planted to maximize the revenue from the orchard?

48. **Profit** If, in the previous problem the cost of maintaining each tree is \$5 per year, how many trees should be planted to maximize the profit from the orchard?

49. **Costs, Revenue, and Profits in Texas Cotton Gins** In 1997, Fuller and coworkers[44] at Texas A & M University estimated the operating costs of cotton gin plants of various sizes. For the next-to-largest plant, the total cost in thousands of dollars was given approximately by $C(x) = 0.059396x^2 + 22.7491x + 224.664$, where x is the annual quantity of bales in thousands produced. Plant capacity was 30,000 bales. Revenue was estimated at \$63.25 per bale. Using this quadratic model, find the break-even quantity, and determine the production level that will maximize profit.

50. **Maximizing Revenue** Chakravorty and Roumasset[45] showed that the revenue R in dollars for cotton in California is approximated by the function $R(w) = -0.2224 + 1.0944w - 0.5984w^2$, where w is the amount of irrigation water in appropriate units paid for and used. What happens to the revenue if only a small amount of water is paid for and used? A large amount? What is the optimal amount of water to use?

51. **Advertising** Let x be a measure (in percent) of the degree of concentration in an industry. Sutton[46] noted that the advertising intensity y, defined as the advertising/sales ratio (in percent), will rise to a peak at intermediate levels of concentration x and decline again for the most concentrated sectors. One economist noted that in a sample of consumer industries, $y = -3.1545 + 0.1914x - 0.0015x^2$ approximately modeled this situation. Sketch a graph, find the value of the concentration ratio for which the advertising intensity is largest, and find the maximum value of this intensity. Confirm graphically.

52. **Physics** If an object is initially at a height above the ground of s_0 feet and is thrown straight upward with an initial velocity of v_0 feet per second, then from physics it can

[44] S. Fuller, M. Gillis, C. Parnell, and R. Childers. 1997. Effect of air pollution abatement on financial performance of Texas cotton gins. *Agribusiness* 13(5):521–532.

[45] Ujjayant Chakravorty and James Roumasset. 1994. Incorporating economic analysis in irrigation design and management. *J. Water Res. Plan. Man.* 120:819–835.

[46] C. J. Sutton. 1974. Advertising, concentration and competition. *Econ. J.* 84:56–69.

be shown the height in feet above the ground is given by $s(t) = -16t^2 + v_0t + s_0$, where t is in seconds. Find how long it takes for the object to reach maximum height. Find when the object hits the ground.

53. Production Function Spencer[47] found that the relative yield y of vegetables on farms in Alabama was related to the number x of hundreds of pounds of phosphate (fertilizer) per acre by $y = 0.15 + 0.63x - 0.11x^2$. How much phosphate per acre, to the nearest pound, would you use to fertilize? Why?

54. Irrigation Caswell and coauthors[48] indicated that the cotton yield y in pounds per acre in the San Joaquin Valley in California was given approximately by $y = -1589 + 3211x - 462x^2$, where x is the annual acre-feet of water application. Determine the annual acre-feet of water application that maximizes the yield and determine the maximum yield.

55. Political Action Committees (PACs) PACs are formed by corporations to funnel political contributions. Grier and collaborators[49] showed that the percentage P of firms with PACs within a manufacturing industry was represented approximately by $P = -23.21 + 0.014x - 0.0000006x^2$, where x is the average industry sales in millions of dollars. Determine the sales that result in the maximum percentage of firms with PACs and find this maximum.

56. Political Action Committees (PACs) In the same study as was mentioned in the previous exercise, P was also given approximately as $P = 74.43 + 0.659x - 0.008x^2$, where this time x represents the concentration ratio of the industry. (This ratio is defined as the proportion of total sales of the four largest firms to total industry sales.) Find the concentration ratio that maximizes the percentage of PACs and find this maximum.

57. Biology Kaakeh and colleagues[50] showed that the amount of photosynthesis in apple leaves in appropriate units was approximated by $y = 19.8 + 0.28x - 0.004x^2$, where x is the number of days from an aphid infestation. Determine the number of days after the infestation until photosynthesis peaked.

58. Biology Hardman[51] showed the survival rate S of European red mite eggs in an apple orchard after insect predation was approximated by

$$y = \begin{cases} 1, & t \le 0 \\ 1 - 0.01t - 0.001t^2, & t > 0 \end{cases}$$

where t is the number of days after June 1. Determine the predation rate on May 15. On June 15.

59. Entomology Dowdy[52] found that the percentage y of the lesser grain borer beetle initiating flight was approximated by the equation $y = -240.03 + 17.83T - 0.29T^2$ where T is the temperature in degrees Celsius.

a. Find the temperature at which the highest percentage of beetles would fly.

b. Find the minimum temperature at which this beetle initiates flight.

c. Find the maximum temperature at which this beetle initiates flight.

60. Ecological Entomology Elliott[53] studied the temperature affects on the alder fly. He showed that in a certain study conducted in the laboratory, the number of pupae that successfully completed pupation y was approximately related to temperature t in degrees Celsius by $y = -0.718t^2 + 21.34t - 112.42$.

a. Determine the temperature at which this model predicts the maximum number of successful pupations.

b. Determine the two temperatures at which this model predicts that there will be no successful pupation.

---

[47] Milton H. Spencer. 1968. *Managerial Economics.* Homewood, Ill.: Richard D. Irwin.

[48] Margriet Caswell, Erik Lichtenberg, and David Zilberman. 1990. The effects of pricing policies on water conservation and drainage. *Amer. J. Agr. Econ.* 72:883–890.

[49] Kevin B. Grier, Michael C. Munger, and Brian E. Roberts. 1991. The industrial organization of corporate political participation. *Southern Econ.* 57:727–738.

[50] W. Kaakeh, D. G. Pfeiffer, and R. P. Marini. 1992. Combined effects of spirea aphid nitrogen fertilization on net photosynthesis, total chlorophyll content, and greenness of apple leaves. *J. Econ. Entomol.* 85:939–946.

[51] J. M. Hardman. 1989. Model simulating the use of miticides to control European red mite in Nova Scotia apple orchards. *J. Econ. Entomol.* 82:1411–1422.

[52] Alan K. Dowdy. 1994. Flight initiation of lesser grain borer as influenced by temperature, humidity, and light. *J. Econ. Entomol.* 87(6):1714–1422.

[53] J. M. Elliott. 1996. Temperature-related fluctuations in the timing of emergence and pupation of Windermere alder-flies over 30 years. *Ecolog. Entomol.* 21:241–247.

More Challenging Exercises For Linear Models

61. Assume that the linear cost model applies and fixed costs are \$1000. If the *total* cost of producing 800 items is \$5000, find the cost equation.

62. Assume that the linear revenue model applies. If the *total* revenue from producing 1000 items is \$8000, find the revenue equation.

63. Assume that the linear cost model applies. If the *total* cost of producing 1000 items at \$3 each is \$5000, find the cost equation.

64. Assume that the linear cost and revenue models applies. An item that costs \$3 to make sells for \$6. If profits of \$5000 are made when 1000 items are made and sold, find the cost equation.

65. Assume the linear cost and revenue models applies. An item costs \$3 to make. If fixed costs are \$1000 and profits are \$7000 when 1000 items are made and sold, find the revenue equation.

66. Assume that the linear cost and revenue model applies. An item sells for \$10. If fixed costs are \$2000 and profits are \$9000 when 1000 items are made and sold, find the cost equation.

67. Process Selection and Capacity A machine shop needs to drill holes in a certain plate. An inexpensive manual drill press could be purchased that will require large labor costs to operate, or an expensive automatic press can be purchased that will require small labor costs to operate. The following table summarizes the options.

Machine	Annual Fixed Costs	Variable Labor Costs	Production Rate
Manual	\$1000	\$16.00/hr	10 plates/hr
Automatic	\$8000	\$2.00/hr	100 plates/hr

If these are the only fixed and variable costs, determine which machine should be purchased by filling in the costs on the following table if the number produced is

a. 1000. **b.** 10,000.

Volume	Total Costs Manual	Automatic
1000		
10,000		

68. Process Selection and Capacity If x is the number of plates produced in Exercise 67 and C is the total cost using the manual drill press, find an equation that x and C must satisfy. Repeat if C is the total cost using the automatic drill press. Graph both equations on your grapher. Find the number of plates produced per hour for which the manual and automatic drill presses will cost the same.

69. Costs per Unit If C is the total cost to produce x items, then the cost per unit is $\bar{C} = C/x$. Fill in the table with the costs per unit for the indicated volumes for each of the machines in the Exercise 67.

Volume	Costs per Unit Manual	Automatic
1000		
10,000		
100,000		

Explain in words what happens to the costs per unit as the number of items made becomes larger. Find the equation that $\bar{C}$ and x must satisfy and graph this equation, using your grapher. What happens to the cost per unit as x becomes larger?

70. Cost, Revenue, and Profit Equations. Assuming a linear cost and revenue model, explain in complete sentences where you expect the y-intercepts to be for the cost, revenue, and profit equations. Give reasons for your answers.

More Challenging Exercises for Quadratic and Other Models

71. Let $q(x) = x^2 - 4\alpha x + \beta$. Determine α and β so that the graph of the quadratic has a vertex at $(4, -8)$.

72. Suppose the demand equation is given by $p(x) = -x + b$. Determine b so that the revenue is maximized when $x = 10$.

73. **Profit Function** If you use the quadratic function $P(x) = ax^2 + bx + c$ to model profits on a very large interval, what sign should the coefficient a have? Explain carefully.

74. **Cost Function** If you use the quadratic function $C(x) = ax^2 + bx + c$ to model costs on a very large interval, what sign should the coefficient a have? Explain carefully.

75. **Revenue Function** Suppose we assume that the demand equation for a commodity is given by $p = mx + e$, where x is the number sold and p is the price. Explain carefully why the resulting revenue function is of the form $R(x) = ax^2 + bx$ with the sign of a negative and the sign of b positive.

76. **Peak Profit** Suppose the demand equation for a commodity is of the form $p = mx + b$, where $m < 0$ and $b > 0$. Suppose the cost function is of the form $C = dx + e$, where $d > 0$ and $e < 0$. Show that profit peaks before revenue peaks.

77. **Demand for Cereal** Cotterill and Haller[54] recently estimated a demand curve for Kellogg's Special K cereal to be approximated by $x = Ap^{-2.0598}$, where p is the price of a unit of cereal, x is the quantity sold, and A is a constant. Find the percentage decrease in quantity sold if the price increases by 1%. By 2%.

78. **Demand for Water Clarity** Boyle and colleagues[55] estimated the demand for water clarity in freshwater lakes in Maine. Shown in the figure is the graph of their demand curve. The horizontal axis is given in meters of visibility as provided by the Maine Department of Environmental Protection. (The average clarity of all Maine lakes is 3.78 m.) The vertical axis is the price decrease of lakeside housing and is given in thousands of dollars per house. Explain what you think is happening when the visibility gets below 3 m.

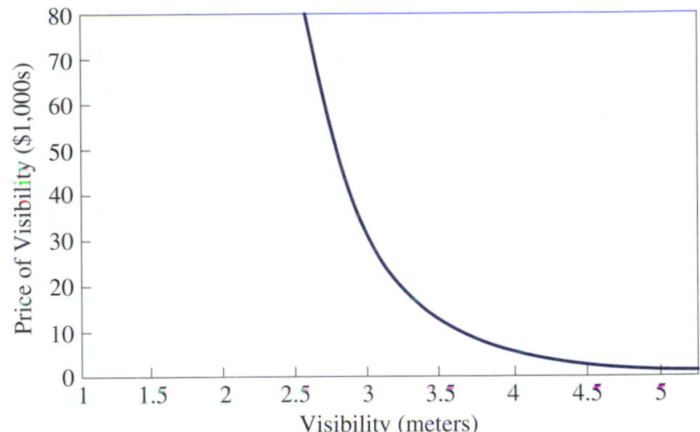

79. **Demand for Recreational Fishing Boats** The graph shows a demand curve estimated by Li.[56] The horizontal axis is labeled by q, the number of boat trips. The owners of these boats take out parties of recreational anglers on a day trip for snapper and other game fish. On the vertical axis is p the cost the owner charges the party for using the boat on the day trip. Explain the significance to the fact that the demand curve strikes the p-axis (at $p = b$).

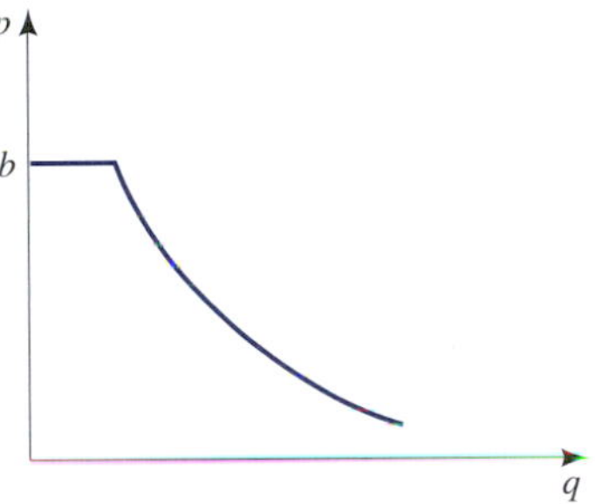

80. **Bull Trout Length and Weight** Donald[57] formulated a mathematical model relating the length of bull trout to their weight. He used the equation $W = 4.8 \cdot 10^{-6} L^{3.15}$, where L is length in millimeters and W is weight in grams. Plot this graph for $L \leq 500$. What happens to weight when length is doubled?

[54] Ronald W. Cotterill and Lawrence E. Haller. 1997. An economic analysis of the demand for RTE cereal product market definition and unilateral market power effects. Research Report No. 35. Food Marketing Policy Center, University of Connecticut.

[55] Kevin J. Boyle, P. Joan Poor, and Laura O. Taylor. 1999. Estimating the demand for protecting freshwater lakes from eutrophication. *Amer. J. Agr. Econ.* 81(5):1118–1122.

[56] Eric A. L. Li. 1999. A travel cost demand model for recreational snapper angling in Port Phillip Bay, Australia. *Trans. Amer. Fisheries Soc.* 128:639–647.

[57] David B. Donald. 1999. Seasonal food habits of bull trout from small alpine lakes in the Canadian Rocky Mountains. *Trans. Amer. Fisheries Soc.* 128:1176–1192.

81. Demand Function for Recreational Fishing Shafer and colleagues[58] formulated a demand curve for outdoor recreation in Pennsylvania. The equation they used was $N(E) = 19.8E^{-0.468}$, where N is visitor days and E was dollars spent traveling. Graph this equation. How many visitor days can be expected if cost is near zero (say \$1)? For any value of E, what happens to the number of visitor days if the cost E doubles?

82. Sewage Treatment Cost Khan[59] developed a mathematical model based on various sewage treatment plants in Saudi Arabia. The sewage treatment cost equation he gave was $C(X) = 0.62 \cdot X^{1.143}$, where C is cost in millions of dollars and X is sewage treated in millions of cubic meters per year. Graph this equation. What is the cost of treating 1 million m^3 of sewage in a year? What percentage increase in costs will incur if the amount of sewage treated doubles?

[58] Elwood L. Shafer, Robert Carline, Richard W. Gulden, and H. Ken Cordell. 1993. Economic amenity values of wildlife: six studies in Pennsylvania. *Environ. Man.* 17(5):669–682.

[59] Mohammed J. Abdulrazzak and Muhammad Z. A. Khan. 1990. Domestic water conservation potential in Saudi Arabia. *Environ. Man.* 14(2):167–178.

83. Species Richness Weller[60] formulated a mathematical model of species richness as related wetland size and gave the equation $N(A) = 7.1 \cdot A^{0.22}$, where N is the number of species of wildlife and A is the area in hectares of the wetland. Graph this function. How many species do you expect if the area is one hectare? What percentage increase in the number of species are expected in these wetlands if the area is ten times greater?

84. Demand for Northern Cod Crafton[61] created a mathematical model of demand for northern cod and formulated the demand equation

$$p(x) = \frac{173213 + 0.2x}{138570 + x}$$

where p is the price in dollars and x is in kilograms. Graph this equation. Does the graph have the characteristics of a demand equation? Explain. Find $p(0)$, and explain what the significance of this is.

[60] Milton W. Weller. 1988. Issues and approaches in assessing cumulative impacts on waterbird habitat in wetlands. *Environ. Man.* 12(5):695–701.

[61] R. Quentin Grafton, Leif K. Sandal, and Stein Ivar Steinhamn. 2000. How to improve the management of renewable resources: the case of Canada's northern cod fishery. *Amer. J. Agr. Econ.* 82:570–580.

Solutions to WARM UP EXERCISE SET 1.2

1. Let x be the number of tons of fertilizer produced and sold. Then the cost, revenue, and profit equations are

$$\begin{aligned} C(x) &= (\text{variable cost}) + (\text{fixed cost}) \\ &= mx + b \\ &= 208.52x + 366{,}821 \end{aligned}$$

$$\begin{aligned} R(x) &= (\text{price per ton}) \times (\text{number tons sold}) \\ &= px \\ &= 266.67x \end{aligned}$$

$$\begin{aligned} P(x) &= (\text{revenue}) - (\text{cost}) = R - C \\ &= (266.67x) - (208.52x + 366{,}821) \\ &= 58.15x - 366{,}821 \end{aligned}$$

If $x = 5000$, then

$$\begin{aligned} C(5000) &= 208.52(5000) + 366{,}821 = 1{,}409{,}421 \\ R(5000) &= 266.67(5000) = 1{,}333{,}350 \\ P(5000) &= 1{,}333{,}350 - 1{,}409{,}421 = -76{,}071 \end{aligned}$$

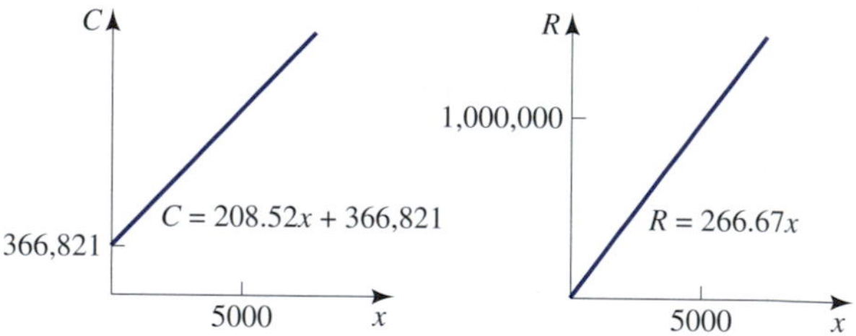

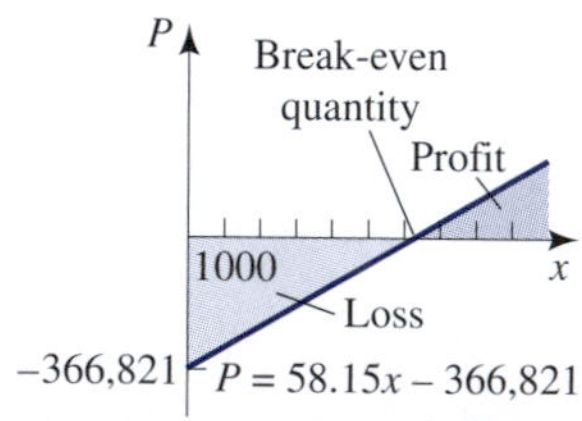

A linear profit function shows losses if only a small number of items are sold but shows profits if a large number of items are sold.

Thus, if 5000 tons are produced and sold, the cost is \$1,409,421, the revenue is \$1,333,350, and there is a *loss* of \$76,071.

Doing the same for some other values of x, we have the results shown in the following table.

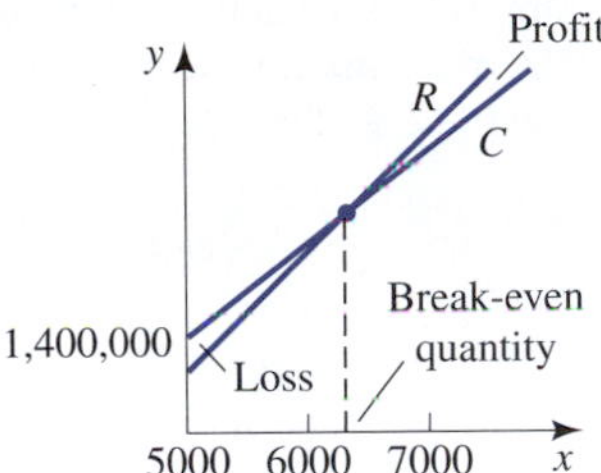

Profit is revenue less cost.

	Number Made and Sold		
	5000	6000	7000
Cost	1,409,421	1,617,941	1,826,461
Revenue	1,333,350	1,600,020	1,866,690
Profit (or loss)	−76,071	−17,921	40,229

The cost, revenue, and profit equations are graphed in the figure.

To find the break-even quantity set $P(x) = 0$ and solve for x:

$$\begin{aligned} 0 &= P(x) \\ &= 58.15x - 366{,}821 \\ 58.15x &= 366{,}821 \\ x &= 6308 \end{aligned}$$

rounded to the nearest ton. Thus, the plant needs to produce and sell 6308 tons of fertilizer to break even (i.e., for profits to be zero).

2. The graphs are shown in the figure. Note that the horizontal axis is p, not x. To find the equilibrium price, set $57.4 + 0.143p = x = 142.8 - 0.284p$ and obtain

$$\begin{aligned} 57.4 + 0.143p &= 142.8 - 0.284p \\ 0.143p + 0.284p &= 142.8 - 57.4 \\ 0.427p &= 85.4 \\ p &= 200 \end{aligned}$$

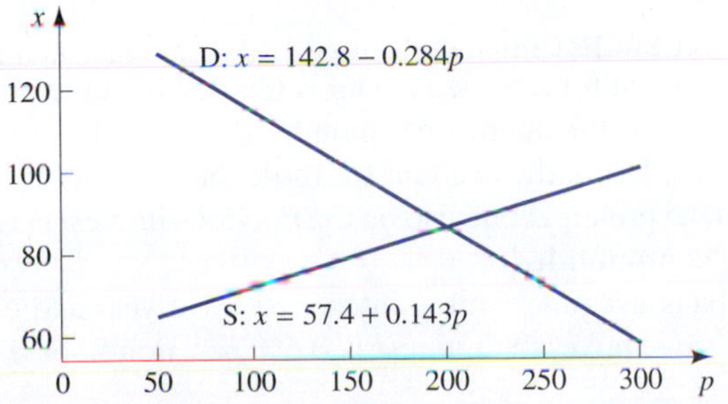

With $p = 200$, $x = 57.4 + 0.143(200) = 86$. The equilibrium price is \$200 per metric ton, and the equilibrium quantity is 86 million metric tons.

3. The revenue is $R(x) = xp = -2x^2 + 12x$, and the profit is

$$\begin{aligned} P(x) &= R(x) - C(x) \\ &= [-2x^2 + 12x] - [4x + 6] \\ &= -2x^2 + 8x - 6 \end{aligned}$$

Using the formulas for finding the vertex, we have

$$h = -\frac{b}{2a} = -\frac{8}{2(-2)} = 2$$

$$k = c - \frac{b^2}{4a} = -6 - \frac{8^2}{4(-2)} = 2$$

So profits will be maximized when 2000 items per week are made and sold, and the maximum profit occurs when the price of each item is \$2.

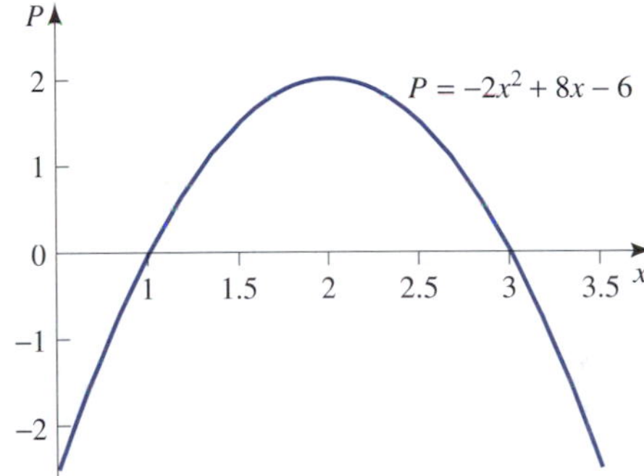

To find the break-even quantities, we must set $P = 0$. This gives

$$\begin{aligned} 0 = P(x) &= -2x^2 + 8x - 6 \\ &= -2(x^2 - 4x + 3) = -2(x-1)(x-3) \end{aligned}$$

From this factorization it is apparent that $x = 1$ and $x = 3$ are the break-even quantities. See the figure, which is typical of profit functions. As more items are produced and sold, initial losses turn into profits. But if too many items are made and sold, this flooding of the market will result in a downturn in profits.

4. The revenue R for the bus company is the number n of people who take the trip times the price p each person pays, that is, $R = n \cdot p$. Let x be the number of people in excess of 40 who take the trip. Then the total number who take the trip is $n = 40 + x$. For each person in excess of 40 who takes the trip, the company lowers the price by \$1. Thus the price per person is $p = 70 - x$. Revenue is then

$$R = n \cdot p = (40 + x)(70 - x) = 2800 + 30x - x^2$$

This is a quadratic equation. Completing the square yields

$$2800 + 30x - x^2 = -(x^2 - 30x - 2800)$$
$$= -(x^2 - 30x + 225 - 225 - 2800) \quad \text{since } \left(\tfrac{-30}{2}\right)^2 = 225$$
$$= -(x^2 - 30x + 225 - 3025)$$
$$= -(x - 15)^2 + 3025$$

The quadratic equation is now in standard form. The maximum occurs at $x = 15$, and the maximum is 3025. Thus, the bus company will have maximum revenue if the number who take the trip is $40 + x = 40 + 15 = 55$. The maximum revenue will be \$3025.

1.3 Exponential Models

A Mathematical Model of Compound Interest
Effective Yield
Present Value
The Base e
Population Growth
Exponential Decay
Solving an Equation with an Exponent
Enrichment: Radioactive Decay

APPLICATION Calculating Interest over A Long Period of Time

Suppose that after the Canarsie Indians sold Manhattan for \$24 in 1626, the money was deposited into a Dutch guilder account that yielded an annual rate of 6% compounded quarterly. How much money would be in this account in 2003? See Example 3 on page 54 for the answer.

A Mathematical Model Of Compound Interest

We first consider how to calculate the most common type of interest, **compound interest**.

If the principal is invested for a period of time at an interest rate of i, where i is given as a decimal, then the amount at the end of the first period is

$$\text{principal} + \text{interest} = P + Pi = P(1 + i)$$

The interest i is called the **interest per period**.

If, for example, the annual interest for a bank account is 6% and the time period is one month, then $i = \frac{0.06}{12} = 0.005$ is the interest per month.

If the interest and principal are left in the account for more than one period and interest is calculated not only on the principal but also on the previous interest earned, we say that the interest is being **compounded**.

If \$1000 is deposited in a bank account earning interest at 6% a year and compounding monthly, then, as we saw above, the interest is 0.5% per month, or 0.005

as a decimal. The amount in the account at the end of any month is then always $1 + 0.005 = 1.005$ times the amount at the beginning of the month.

$$\text{amount at end of month} = [\text{amount at beginning of month}] \times (1.005)$$

If we let the amount at the beginning of any month be inside square brackets, then we have the following.

$$\begin{aligned}
\text{amount at end of month 1} &= [\text{amount at beginning of month 1}] \times (1.005)\\
&= \$1000(1.005)\\
\text{amount at end of month 2} &= [\text{amount at beginning of month 2}] \times (1.005)\\
&= [\$1000(1.005)](1.005)\\
&= \$1000(1.005)^2\\
\text{amount at end of month 3} &= [\text{amount at beginning of month 3}] \times (1.005)\\
&= [\$1000(1.005)^2](1.005)\\
&= \$1000(1.005)^3
\end{aligned}$$

If we continue in this manner, the future amount of money F in the account at the end of n months will be

$$F = \$1000(1.005)^n$$

In the same way, if a principal P earns interest at the rate per period of i, expressed as a decimal, and interest is compounded, then the amount F after n periods is

$$F = P(1 + i)^n$$

Exploration 1

Importance of Interest Rate

It is important to understand the relationship between the amount in a compounding account and the interest rate per period. Suppose you invest \$1000 and leave your investment for 20 years. Write an equation that gives the amount in the account in terms of the interest rate x, where x is a decimal. Graph on your computer or graphing calculator, using a window with dimensions $[0, 0.15]$ by $[0, 20{,}000]$. What do you observe?

EXAMPLE 1 **Finding Compound Interest**

Suppose \$1000 is deposited into an account with an annual yield of 8% compounded quarterly. Find the amount in the account at the end of five years, 10 years, 20 years, 30 years, and 40 years.

Solution

We have $P = \$1000$, and the interest per quarter is $i = \frac{0.08}{4} = 0.02$. Thus, $F = \$1000(1 + 0.02)^n$. We obtain the following result.

Years (t)	Periods ($n = 4t$)	Future Value
5	20	$\$1000(1+0.02)^{20} = \$1485.95$
10	40	$\$1000(1+0.02)^{40} = \$2208.04$
20	80	$\$1000(1+0.02)^{80} = \$4875.44$
30	120	$\$1000(1+0.02)^{120} = \$10{,}765.16$
40	160	$\$1000(1+0.02)^{160} = \$23{,}769.91$

■

Notice that during the first 10 years the account grows by about \$1200, but during the last 10 years, it grows by about \$13,000. In fact, each year the account grows by more than in the previous year. Refer to Figure 1.37.

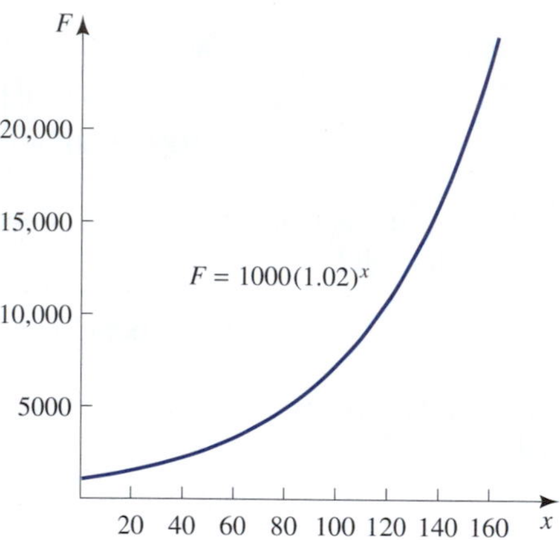

Figure 1.37
This account is growing exponentially.

Exploration 2

Graph of Exponential Function

Use your grapher to explore the function $y = (1.02)^x$ for negative x. What happens as x becomes more negative?

The graph of $y = 1000(1.02)^x$ shown in Figure 1.37 is typical of the graphs of functions of the form $y = ba^x$ for $b > 0$ and $a > 1$. Figure 1.38 shows the graph of $y = ba^x$ for $b > 0$ and $a > 1$ and indicates some properties for this function.

Remark We will use the letter i to designate the interest rate for any period, whether annual or not. But we reserve the letter r to always designate an annual rate.

Suppose r is the annual interest rate expressed as a decimal and interest is compounded m times a year. Then the interest rate per time period is $i = r/m$. If the compounding goes on for t years, then there are $n = mt$ time periods, and the amount

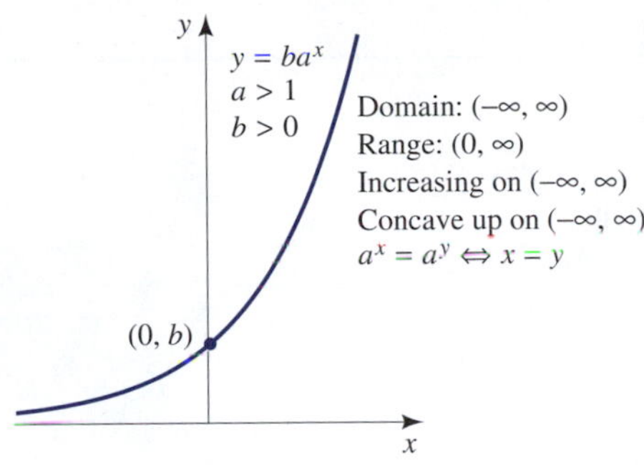

Figure 1.38

F after t years is

$$F = P(1+i)^n = P\left(1+\frac{r}{m}\right)^{mt}$$

Compound Interest

Suppose a principal P earns interest at the annual rate of r, expressed as a decimal, and interest is compounded m times a year. Then the amount F after t years is

$$F = P(1+i)^n = P\left(1+\frac{r}{m}\right)^{mt}$$

where $n = mt$ is the number of time periods and $i = \frac{r}{m}$ is the interest per period.

Table 1.6

Compounding (t)	m
Annually	1
Semiannually	2
Quarterly	4
Monthly	12
Daily	365

Table 1.6 relates the type of compounding with m.

EXAMPLE 2 **Finding Compound Interest**

Suppose \$1000 is deposited into an account that yields 9% annually. Find the amount in the account at the end of the fifth year if the compounding is (a) quarterly, (b) monthly.

Solution

(a) Here $P = \$1000$, $r = 0.09$, $m = 4$, and $t = 5$:

$$F = P\left(1+\frac{r}{m}\right)^{mt} = \$1000\left(1+\frac{0.09}{4}\right)^{4\cdot5} = \$1560.51$$

(b) Here $P = \$1000$, $r = 0.09$, $m = 12$, and $t = 5$:

$$F = P\left(1+\frac{r}{m}\right)^{mt} = \$1000\left(1+\frac{0.09}{12}\right)^{12\cdot5} = \$1565.68$$ ■

Remark Notice that the amount at the end of the time period is larger if the compounding is done more often.

Interactive Illustration 1.5

Compounding

This interactive illustration allows you to experiment with the formula $F = P\left(1 + \frac{r}{m}\right)^{mt}$. Sliders are provided for P and r, and you can select a number of values for m. Let t be given in years. In the formula t has been set equal to 30, so the graph shows the amount after 30 years.

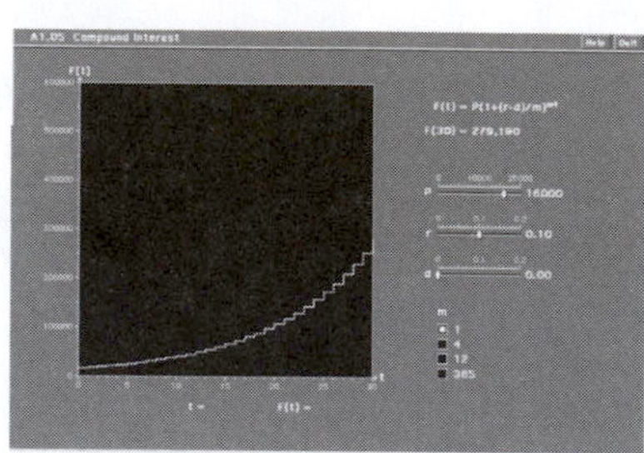

1. Set $P = 10{,}000$ and r to 10%. Now change m and notice what happens to the amount in the account after 30 years. Specifically, what happens when m goes from 1 to 12? From 12 to 365? Which was the bigger change?
2. Now fix m and change r, and observe the result. Specifically, set $m = 1$, and double r from 5% to 10%. How has the amount changed? Has it doubled? More than doubled?
3. Change P, and observe the result. Specifically, change P from 10,000 to 20,000. Has the amount at the end of the period doubled?
4. You can also observe the effect of inflation on your investment. A slider is given for d that reduces your investment by a rate of inflation given by d. Thus, your ending amount will be given in "constant" dollars. Set the rate to 6% and the inflation rate to 3%. What is the difference between your ending amount with $d = 0$ and your "real" amount that includes the effects of inflation? Repeat with just a "little" inflation of 1%.

EXAMPLE 3 **Calculating Interest over a Very Long Period of Time**

Suppose that after the Canarsie Indians sold Manhattan for \$24 in 1626, the money was deposited into a Dutch guilder account that yielded an annual rate of 6% compounded quarterly. How much money would be in this account in 2003?

Solution

Since $2003 - 1626 = 377$, we have

$$\begin{aligned} F &= P\left(1 + \frac{r}{m}\right)^{mt} \\ &= \$24\left(1 + \frac{0.06}{4}\right)^{377\cdot 4} \\ &\approx \$135 \text{ billion} \end{aligned}$$

■

Table 1.7 indicates what the value of this account would have been at some intermediate times.

Remark It is, of course, very unlikely that any investment could have survived through the upheavals of wars and financial crises that occurred during this 377-year span of time. Nonetheless, it is examples such as this one that inspire some to use the phrase "the wonders of compounding."

Table 1.7

Year			Future Value
1626			\$24
1650	$\$24\left(1+\frac{0.06}{4}\right)^{4\cdot 24}$	$\approx$	\$100
1700	$\$24\left(1+\frac{0.06}{4}\right)^{4\cdot 74}$	$\approx$	\$2,000
1750	$\$24\left(1+\frac{0.06}{4}\right)^{4\cdot 124}$	$\approx$	\$39,000
1800	$\$24\left(1+\frac{0.06}{4}\right)^{4\cdot 174}$	$\approx$	\$760,000
1850	$\$24\left(1+\frac{0.06}{4}\right)^{4\cdot 224}$	$\approx$	\$15,000,000
1900	$\$24\left(1+\frac{0.06}{4}\right)^{4\cdot 274}$	$\approx$	\$293,000,000
1950	$\$24\left(1+\frac{0.06}{4}\right)^{4\cdot 324}$	$\approx$	\$6,000,000,000
2003	$\$24\left(1+\frac{0.06}{4}\right)^{4\cdot 377}$	$\approx$	\$135,000,000,000

Interactive Illustration 1.6

Compound Interest over Very Long Periods

This interactive illustration allows you to experiment with using different interest rates over long periods of time. We revisit Example 3. Move the cursor to any time and see how much money was in the account at that time. A slider is provided so you can change the interest rate. You are offered a number of possible compounding values.

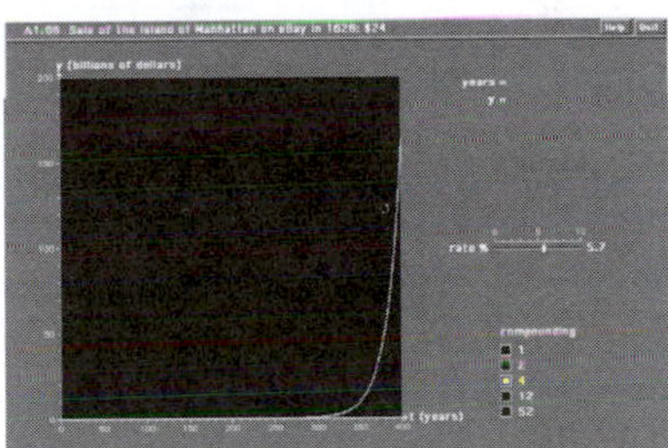

1. Change the rate a small amount and see the effect over a long period. For example, change the rate from 5.5% to 6%. What is the difference in the amount at the end of the 400 year period?
2. Change the compounding, and observe the effect. For example, set the rate at 6% and change the compounding from once a year to monthly. What is the difference in the amount at the end of the period?
3. Now change the compounding to weekly. What is the difference in the amount at the end of the period for weekly compounding compared to monthly?

Effective Yield

If \$1000 is invested at an annual rate of 9% compounded monthly, then at the end of a year there is

$$F = \$1000\left(1+\frac{0.09}{12}\right)^{12} = \$1093.81$$

in the account. This is the same amount that is obtainable if the same principal of \$1000 is invested for one year at an *annual* rate of 9.381% (or 0.09381 expressed as a decimal). We call the rate 9.381% the **effective annual yield**. The 9% annual rate is often referred to as the **nominal rate**.

Suppose r is the annual interest rate expressed as a decimal and interest is compounded m times a year. If the compounding goes on for one year, then the amount F after one year is

$$F = P\left(1+\frac{r}{m}\right)^{m}$$

If we let r_{eff} be the effective annual yield, then r_{eff} must satisfy

$$P\left(1+\frac{r}{m}\right)^m = P\left(1+r_{eff}\right)$$

Solving for r_{eff}, we obtain

$$r_{eff} = \left(1+\frac{r}{m}\right)^m - 1$$

Effective Yield

Suppose a sum of money is invested at an annual rate of r expressed as a decimal and is compounded m times a year. The effective yield r_{eff} is

$$r_{eff} = \left(1+\frac{r}{m}\right)^m - 1$$

EXAMPLE 4 **Comparing Investments**

One bank advertises a nominal rate of 7.1% compounded semiannually. A second bank advertises a nominal rate of 7% compounded daily. What are the effective yields? In which bank would you deposit your money?

Solution

For the first bank $r = 0.071$ and $m = 2$. Then

$$r_{eff} = \left(1+\frac{0.071}{2}\right)^2 - 1 = 0.0723$$

or, as a percent, 7.23%.

For the second bank $r = 0.07$ and $m = 365$. Then

$$r_{eff} = \left(1+\frac{0.07}{365}\right)^{365} - 1 = 0.0725$$

or, as a percent, 7.25%.

Despite the higher nominal rate given by the first bank, the effective yield for the second bank is higher than that for the first. Thus, money deposited in the second bank will grow faster than money deposited in the first bank. ■

Present Value

If we have an account initially with P earning interest at an annual rate of r expressed as a decimal and interest is compounded m times a year, then the amount F in the account after t years is

$$F = P\left(1+\frac{r}{m}\right)^{mt}$$

If we wish to know how many dollars P to set aside now in this account so that we will have a future amount of F dollars after t years, we simply solve the above expression for P. Thus,

$$P = \frac{F}{\left(1 + \frac{r}{m}\right)^{mt}}$$

This is called the **present value**.

Present Value

Suppose an account earns an annual rate of r expressed as a decimal and compounds m times a year. Then the amount P, called the **present value**, needed currently in this account so that a future amount of F will be attained in t years is given by

$$P = \frac{F}{\left(1 + \frac{r}{m}\right)^{mt}}$$

EXAMPLE 5 **Finding the Present Value of a Future Balance**

How much money must grandparents set aside at the birth of their grandchild if they wish to have \$20,000 when the grandchild reaches his or her eighteenth birthday. They can earn 9% compounded quarterly.

Solution

Here $r = 0.09$, $m = 4$, $t = 18$, and $F = \$20,000$. Thus,

$$\begin{aligned} P &= \frac{F}{\left(1 + \frac{r}{m}\right)^{mt}} \\ &= \frac{\$20{,}000}{\left(1 + \frac{0.09}{4}\right)^{4(18)}} \approx \$4029.69 \end{aligned}$$

Thus, their investment of \$4029.69 will become \$20,000 in 18 years. ■

The Base e

As we shall see in later chapters, the most important base a is the irrational number 2.7182818.... This number is so important that a special letter e has been set aside for it. Thus, to seven decimal places, $e = 2.7182818$.

We noted in Example 2 that for an account earning compound interest, the more we compound each year, the larger is the amount at the end of each year. In Example 2 we found that \$1000 deposited into an account that yields 9% annually becomes \$1560.51 in five years when compounded quarterly and the larger amount of \$1565.68 when compounded monthly. Suppose we compound monthly, then weekly, then daily, then by the hour, then by the minute, then by the second, and so on indefinitely. What will happen to the amount in the account at the end of a year? We explore this in the next example.

EXAMPLE 6 **Compounding Ever More Often**

Suppose \$1000 is deposited into an account that yields 6% annually. Find the principal at the end of the first year if it is compounded (a) annually, (b) monthly, (c) weekly, (d) daily, (e) hourly, (f) by the minute.

Solution

In all cases $P = \$1000$, $r = 0.06$, and $t = 1$.

(a) Here $m = 1$. We have

$$F = P(1 + r) = \$1000(1.06) = \$1060.00$$

(b) Here $m = 12$. Thus,

$$F = P\left(1 + \frac{r}{m}\right)^m = \$1000\left(1 + \frac{0.06}{12}\right)^{12} = \$1061.68$$

(c) Here $m = 52$. Thus,

$$F = P\left(1 + \frac{r}{m}\right)^m = \$1000\left(1 + \frac{0.06}{52}\right)^{52} = \$1061.80$$

(d) Here $m = 365$. Thus,

$$F = P\left(1 + \frac{r}{m}\right)^m = \$1000\left(1 + \frac{0.06}{365}\right)^{365} = \$1061.83$$

(e) Here $m = 8760$. Thus,

$$F = P\left(1 + \frac{r}{m}\right)^m = \$1000\left(1 + \frac{0.06}{8760}\right)^{8760} = \$1061.84$$

(f) Here $m = 525600$. Thus,

$$F = P\left(1 + \frac{r}{m}\right)^m = \$1000\left(1 + \frac{0.06}{525600}\right)^{525600} = \$1061.84$$ ■

Thus, as the number of compoundings per year increases, the amount of money in the account increases but seems to be limited by some finite amount, as shown in Table 1.8. In the above example the limiting amount, or upper bound, was \$1061.84, rounded to the nearest cent. The effective yield based on this upper bound is 6.184%, rounded to three decimal places. When the effective yield is set at this upper bound, we say that interest is **compounding continuously**. We say *continuously* because the compounding is done not just every day, or every hour, or every minute, or every second, but *continuously*.

This number can also be obtained using the base e. In fact $e^{0.06} = 1.06184$, rounded to five decimal places. Thus, $1000e^{0.06} = 1061.84$ to the nearest cent. A similar analysis (done in Chapter 5) leads us to the following result.

Table 1.8

Compounded	Amount at End of 1 Year
Yearly	\$1060.00
Monthly	1061.68
Weekly	1061.80
Daily	1061.83
Hourly	1061.84
By the minute	1061.84

DEFINITION **Continuous Compounding**

If a principal P earns interest at the annual rate of r, expressed as a decimal, and interest is compounded continuously, then the amount F after t years is $F = Pe^{rt}$.

Interactive Illustration 1.7

The Exponent e

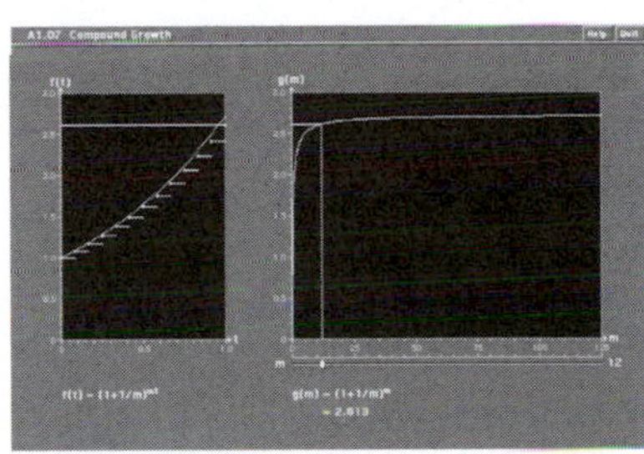

This interactive illustration allows you to observe what happens as the number of compoundings increases. A slider is provided for the number of compoundings. Increase this number, and observe what happens to $g = \left(1 + \frac{1}{m}\right)^m$.

1. Does it appear that g is bounded by some number?
2. Increase m from 1 to 12. What is the change in g?
3. Now increase m from 12 to 125. What is the change in g?
4. How does the change in (3) compare with the change in (2)?
5. What does your answer in (4) mean in terms of the importance of compounding and how often?

EXAMPLE 7 **Finding Amounts in a Continuously Compounded Account**

Suppose \$1000 is invested at an annual rate of 9% compounded continuously. How much is in the account after (a) 1 year, (b) 3 years?

Solution

(a) Here $P = \$1000$, $r = 0.09$, and $n = 1$. Thus,

$$F = Pe^{rt} = \$1000e^{0.09} = \$1094.17$$

(b) Here $P = \$1000$, $r = 0.09$, and $n = 3$. Thus

$$F = Pe^{rt} = \$1000e^{(0.09)(3)} = \$1309.96$$

We might wish to know when the amount in the account in Example 7 will become \$1500. To do this, we need to solve the equation $1500 = 1000e^{0.09t}$ for t. So we need to solve $1.5 = e^{0.09t}$ for t. But t is an *exponent*. At this time we do not know how to solve, since solving analytically requires logarithms. We will see how to solve such equations in Section 5. However, we can find an approximate solution on a computer or graphing calculator as indicated in the following exploration. ■

Exploration 3

Solving for Time

Using your computer or graphing calculator, find the time for the amount in the account in Example 7 to reach \$1500. Graph on the same screen $y_1 = 1000e^{0.09x}$ and $y_2 = 1500$, and notice that the two graphs do intersect. Use available operations to find the point of intersection.

Since $e > 1$, we know that e^x is an increasing function. Furthermore, since $2 < e < 3$, the graph of e^x lies between the graphs of 2^x and 3^x.

When an account is compounding continuously, we have the future value, F, after t years given by $F = Pe^{rt}$, where r is the annual rate and P is the initial amount. If we want to know the present value, we simply solve for P and obtain $P = Fe^{-rt}$. As was done before, to find the effective yield r_{eff}, we solve $e^r = 1 + r_{eff}$ and obtain $r_{eff} = e^r - 1$. We summarize this as follows.

DEFINITION Present Value and Effective Yield For Continuous Compounding

Suppose an account earns an annual rate of r expressed as a decimal and compounds continuously. Then the amount P, called the **present value**, needed presently in this account so that a future amount of F will be attained in t years is given by

$$P = Fe^{-rt}$$

The **effective yield**, r_{eff}, is given by

$$r_{eff} = e^r - 1$$

Population Growth

Exponential functions have a wide variety of applications. We give one now for population growth.

EXAMPLE 8 Population Growth

Table 1.9 gives the population in thousands of the United States for some selected years.[62] On the basis of data given for the years 1790–1794 find a mathematical model that approximately describes the annual population. What does the model predict the population will be in 1804?

[62] John W. Wright. 1993. *The Universal Almanac*. New York: Andrews and McMeel.

Table 1.9

Year	1790	1791	1792	1793	1794	1804	1805
Population	3926	4056	4194	4332	4469	6065	6258

Solution

We first check the annual increases in population. We obtain

$$\frac{\text{population in 1791}}{\text{population in 1790}} = \frac{4056}{3926} \approx 1.033$$

$$\frac{\text{population in 1792}}{\text{population in 1791}} = \frac{4194}{4056} \approx 1.034$$

$$\frac{\text{population in 1793}}{\text{population in 1792}} = \frac{4332}{4194} \approx 1.033$$

$$\frac{\text{population in 1794}}{\text{population in 1793}} = \frac{4469}{4332} \approx 1.032$$

Thus, for these early years the population is increasing at about 3.3% per year. Taking 1790 to be our initial year, we therefore expect the population in thousands n years after 1790 to be $P(n) = 3926(1.033)^n$.

On the basis of this model we would predict the population in 1804 to be

$$P(14) = 3926(1.033)^{10} \approx 6{,}185{,}000$$

As we can see from Table 1.9 this is close. Thus, we can say that during these early years the population of the United States was growing exponentially. See Figure 1.39 for a graph of $y = P(t) = 3926(1.033)^t$.

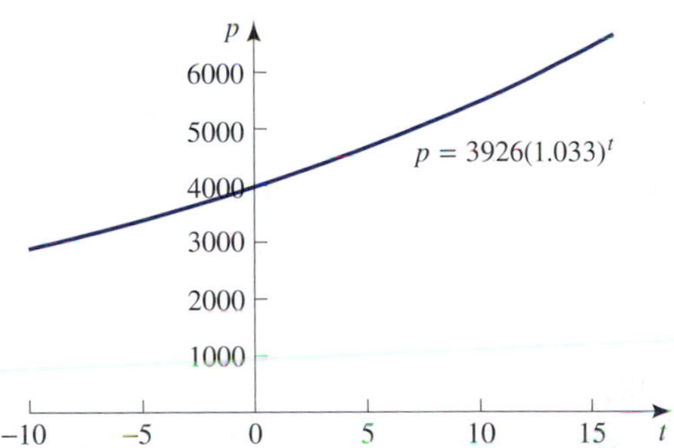

Figure 1.39

■

The population of the United States has *not* been increasing exponentially in recent years. Thus, the population model given above is adequate only for the earlier years. Later in this text we will use a more complex mathematical model, called a logistic curve, that yields rather accurate estimates of the population over the entire last 200 years.

Exponential Decay

We now consider an important case in which money can *decrease* in value. We also refer to this as decay. Suppose, for example, that you placed \$1000 under your mattress and left it there. Suppose further that inflation continued at an annual rate of 3%. What would this \$1000 be worth in real terms, that is, in real purchasing power, in future years? Each year your cash would have 3% less purchasing power. So at the end of any year your purchasing power (real value) would be $1 - 0.03 = 0.97$ times the purchasing power (real value) at the beginning of the year.

If we let the real value at the beginning of any year be inside square brackets, then we have the following:

$$\begin{aligned}\text{real value at end of year 1} &= [\text{real value at beginning of year 1}] \times (0.97)\\ &= \$1000(0.97)\end{aligned}$$

$$\begin{aligned}\text{real value at end of year 2} &= [\text{real value at beginning of year 2}] \times (0.97)\\ &= [\$1000(0.97)](0.97)\\ &= \$1000(0.97)^2\\ \text{real value at end of year 3} &= [\text{real value at beginning of year 3}] \times (0.97)\\ &= [\$1000(0.97)^2](0.97)\\ &= \$1000(0.97)^3\end{aligned}$$

If we continue in this manner, the future real value of the money F in the account at the end of n years will be

$$F = \$1000(0.97)^n$$

In the same way, if a principal P loses value at the rate per period of i, expressed as a decimal, and this loss is compounded, then the amount F after n periods is

$$F = P(1 - i)^n$$

This is called **exponential decay**.

EXAMPLE 9

Suppose that you placed \$1000 under your mattress and left it there. Suppose further that inflation continued at an annual rate of 3%. What would this \$1000 be worth in real terms, that is, in real purchasing power, in five years? 10 years?

Solution

We have $P = 1000$ and $i = 0.03$ so when $n = 5$,

$$F = 1000(1 - 0.03)^5 = 858.73$$

or \$858.73. If $n = 10$,

$$F = 1000(1 - 0.03)^{10} = 737.42$$

or \$737.42. See Figure 1.40.

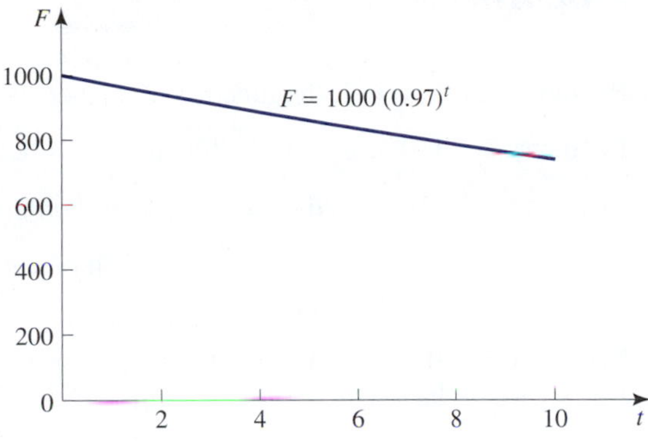

Figure 1.40

■

Exploration 4

Graph of Exponential Function

Use your grapher to explore the function $y = (0.97)^x$ for negative x. What happens as x becomes more negative?

Exploration 5

Solving for Time

Using your computer or graphing calculator, find the time for the real value of the money in Example 9 to be reduced by half. Graph on the same screen $y_1 = 1000(0.97)^t$ and $y_2 = 500$, and notice that the two graphs do intersect. Use available operations to find the point of intersection.

The graph of $y = (0.97)^t$ shown in Figure 1.40 is typical of the graphs of functions of the form $y = Pa^t$ for $a > 1$ and $P > 0$. Figure 1.41 shows the graph of $y = Pa^x$ for $0 < a < 1$ and $P > 0$ and indicates some properties for this function.

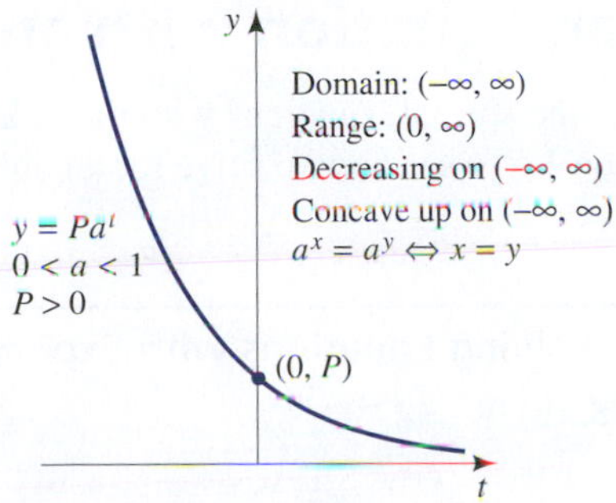

Figure 1.41

Properties of a^x

Property 1. If $a > 1$, the function a^x is increasing and concave up.

Property 2. If $0 < a < 1$, the function a^x is decreasing and concave up.

Property 3. If $a \neq 1$, then $a^x = a^y$ if and only if $x = y$.

Property 4. If $0 < a < 1$, then the graph of $y = a^x$ approaches the x-axis as x becomes large.

Property 5. If $a > 1$, then the graph of $y = a^x$ approaches the x-axis as x becomes negatively large.

The following definitions and properties are listed here as a convenient review.

DEFINITION Some Definitions Involving Exponentials

$a^0 = 1$, $a^{-1} = 1/a$, and, in general, if $a \neq 0$, $a^{-x} = 1/a^x$

$a^{1/2} = \sqrt{a}$, $a^{1/3} = \sqrt[3]{a}$, and, in general, $a^{1/n} = \sqrt[n]{a}$

$a^{m/n} = \sqrt[n]{a^m} = (\sqrt[n]{a})^m$

Further Properties of Exponentials

	Property	Example
Property 6.	$a^x \cdot a^y = a^{x+y}$	$2^3 \cdot 2^4 = 2^7$
Property 7.	$\dfrac{a^x}{a^y} = a^{x-y}$	$\dfrac{3^6}{3^2} = 3^4$
Property 8.	$(a^x)^y = a^{xy}$	$(4^2)^3 = 4^6$

Solving an Equation with an Exponent

We now solve some special equations with the unknown as the exponent. We must wait until Section 1.5 to see how to solve for an unknown exponent in general.

EXAMPLE 10 Solving Equations with Exponents

Solve (a) $2^x = 8$, (b) $9^x = 27^{4x-10}$.

Solution

(a) First rewrite $2^x = 8$ as $2^x = 2^3$. Then by Property 3, $x = 3$.

(b) Since $9 = 3^2$ and $27 = 3^3$, we have

$$
\begin{aligned}
9^x &= 27^{4x-10} \\
(3^2)^x &= (3^3)^{4x-10} \\
3^{2x} &= 3^{12x-30} && \text{Property 8} \\
2x &= 12x - 30 && \text{Property 3} \\
x &= 3
\end{aligned}
$$

■

Enrichment: Radioactive Decay

We now turn to another case of $y = a^x$ when $0 < a < 1$, radioactive decay. Some elements exhibit radioactive decay. For these elements, on occasion one nucleus spontaneously divides into two or more other nuclei. Thus, over time the amount of the original substance will decrease (decay). The rate at which decay takes place depends on the substance. In the next example we see a way in which this property can be exploited.

EXAMPLE 11 **Radioactive Decay**

Iodine-131 is a radioactive substance that is used as a tracer for medical diagnosis. Table 1.10 gives the amount of iodine-131 that is observed in a laboratory over a number of days. On the basis of this data, determine a mathematical model that gives the amount of iodine-131 at any time. Then use this model and your computer or graphing calculator to find how long it will take until this substance decays to half its original amount.

Table 1.10

Day	0	1	2	3	4
Number of Units	20.00	18.34	16.82	15.42	14.14

Solution

Taking ratios of the amount of substance at the end of a day with the amount at the beginning, we obtain

$$\frac{18.34}{20} \approx 0.917$$

$$\frac{16.82}{18.34} \approx 0.917$$

$$\frac{15.42}{16.82} \approx 0.917$$

$$\frac{14.14}{15.42} \approx 0.917$$

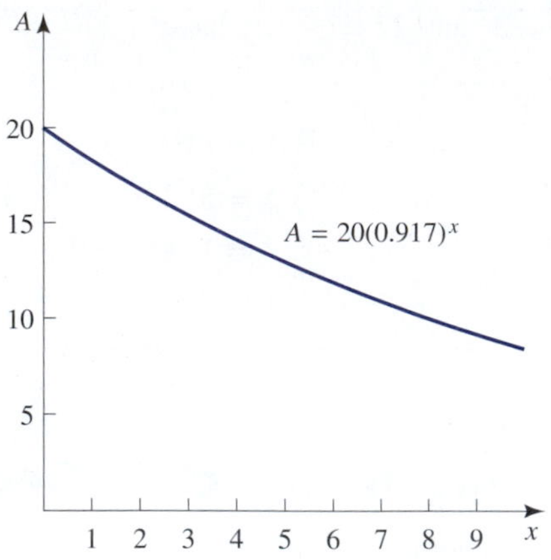

Figure 1.42
A graph of an exponential decay model.

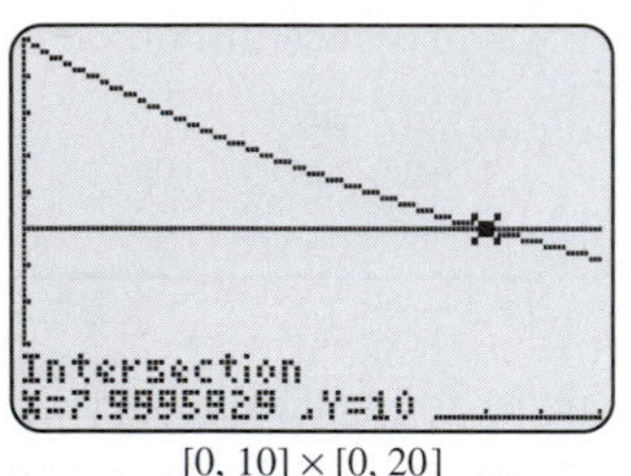

[0, 10] × [0, 20]

Screen 1.5
We see that the intersection of the graphs of $y_1 = 20(0.917)^x$ and $y_2 = 10$ occurs when x is approximately 8.

We see we have a constant daily decay of $1 - 0.917 = 0.083$. In exact analogy with the analysis given just before Example 9, we have the amount after the nth day is $A(n) = 20(0.917)^n$. (See Figure 1.42.)

To find the time until the substance decays to half its original amount, we graph the two functions $y_1 = 20(0.917)^x$ and $y_2 = 10$ in our grapher and obtain Screen 1.5. Now using the operations available to us on our computers or graphing calculators, we find that the x-coordinate of the point of intersection is 8. Thus, it will take eight days until the 20 units of iodine-131 decays to 10 units. ■

For any quantity that exhibits exponential decay, such as radioactive decay, let $A(t)$ be the amount at time t. Then there is a positive constant k, called the **decay constant**, such that

$$A(t) = A_0(1 - k)^t$$

Since $A(0) = A_0(1 - k)^0 = A_0$, A_0 is the initial amount, or amount at time $t = 0$.

In an analogous manner for any quantity that exhibits exponential growth, such as money growing at a compounding rate, let $A(t)$ be the amount at time t. Then there is a positive constant c, called the **growth constant**, such that

$$A(t) = A_0(1 + c)^t$$

Since $A(0) = A_0(1 + c)^0 = A_0$, A_0 is the initial amount, or amount at time $t = 0$.

Warm Up Exercise Set 1.3

1. Solve $3^{2x+1} = 1/3^x$.
2. An account with \$1000 earns interest at an annual rate of 8%. Find the amount in this account after 10 years if the compounding is (a) monthly, (b) continuous.
3. Find the effective yield if the annual rate is 8% and the compounding is weekly.
4. How much money should be deposited in a bank account earning an annual interest rate of 8% compounded quarterly so that there will be \$10,000 in the account at the end of 10 years?

Exercise Set 1.3

In Exercises 1 through 8, sketch a graph of each of the functions without using your grapher. Then support your answer with your grapher.

1. $y = 5^x$
2. $y = 7^x$
3. $y = 3^{x^2}$
4. $y = 3^{-x^2}$
5. $y = (1/2)^x$
6. $y = (0.1)^x$
7. $y = 10^x$
8. $y = 5^{-x}$

In Exercises 9 through 16, solve for x.

9. $3^x = 9$
10. $4^x = 8$
11. $16^x = 8$
12. $\left(\frac{1}{2}\right)^x = \frac{1}{16}$
13. $\left(\frac{1}{16}\right)^x = \frac{1}{8}$
14. $\left(\frac{1}{4}\right)^x = 8$
15. $4^{2x} = 8^{9x+15}$
16. $5^{3x} = 125^{4x-4}$

In Exercises 17 through 24, graph each of the functions without using a grapher. Then support your answer with a grapher.

17. $y = 2^{3x}$
18. $y = 3^{-2x}$
19. $y = 3^{x^3}$
20. $y = 5^{-x^2}$
21. $y = 2 - 3^{-x}$
22. $y = 2 + 3^{-x}$
23. $y = \frac{1}{2}(3^x + 3^{-x})$
24. $y = \frac{1}{2}(2^x - 2^{-x})$

In Exercises 25 through 38, solve for x.

25. $3^x = \frac{1}{3}$
26. $5^x = \frac{1}{25}$
27. $2^x = \frac{1}{8}$
28. $2^{5x} = 2^{x+8}$
29. $(3^{x+1})^2 = 3$
30. $5^{2x-1}5^x = \frac{1}{5^x}$
31. $3^{x+1} = \frac{1}{3^x}$
32. $7^{x^2+2x} = 7^{-x}$
33. $3^{-x^2+2x} = 3$
34. $x2^x = 2^x$
35. $3^{-x}(x+1) = 0$
36. $x^2 5^{2x} = 5^{2x}$
37. $x5^{-x} = x^2 5^{-x}$
38. $(x^2 + x - 2)2^x = 0$

Applications and Mathematical Modeling

In Exercises 39 through 42, determine how much is in each account on the basis of the indicated compounding after the specified years have passed; P is the initial principal, and r is the annual rate given as a percent.

39. After one year, where $P = \$1000$ and $r = 8\%$, compounded (a) annually, (b) quarterly, (c) monthly, (d) weekly, (e) daily, (f) continuously.

40. After 40 years, where $P = \$1000$ and $r = 8\%$, compounded (a) annually, (b) quarterly, (c) monthly, (d) weekly, (e) daily, (f) continuously.

41. After 40 years, where $P = \$1000$, compounded annually, and r is (a) 3%, (b) 5%, (c) 7%, (d) 9%, (e) 12%, (f) 15%.

42. $P = \$1000$ and $r = 9\%$, compounded annually, after (a) 5, (b) 10, (c) 15, (d) 30 years.

In Exercises 43 and 44, find the effective yield given the annual rate r and the indicated compounding.

43. $r = 8\%$, compounded (a) semiannually, (b) quarterly, (c) monthly, (d) weekly, (e) daily, (f) continuously.

44. $r = 10\%$, compounded (a) semiannually, (b) quarterly, (c) monthly, (d) weekly, (e) daily, (f) continuously.

In Exercises 45 and 46, find the present value of the given amounts F with the indicated annual rate of return r, the number of years t, and the indicated compounding.

45. $F = \$10{,}000$, $r = 9\%$, $t = 20$, compounded (a) annually, (b) monthly, (c) weekly, (d) continuously.

46. $F = \$10{,}000$, $r = 10\%$, $t = 20$, compounded (a) annually, (b) quarterly, (c) daily, (d) continuously.

47. Your rich uncle has just given you a high school graduation present of \$1 million. The present, however, is in the form of a 40-year bond with an annual interest rate of 9% compounded annually. The bond says that it will be worth \$1 million in 40 years. What is this million-dollar gift worth at the present time?

48. Redo Exercise 47 if the annual interest rate is 6%.

49. Your rich aunt gives you a high school graduation present of \$2 million. Her present is in the form of a 50-year bond with an annual interest rate of 9%. The bond says that it will be worth \$2 million in 50 years. What is this gift worth at the present time? Compare your answer to that in Exercise 47.

50. Redo Exercise 49 if the annual interest rate is 6%.

51. Using Example 3 in this section, find the amount in 2003 if the annual interest was 7% compounded quarterly. Compare your answer to that of Example 3.

52. In Example 3 in this section, find the amount in 2003 if the annual interest was 5% compounded quarterly. Compare your answer to that of Example 3 in the text.

53. Real Estate Appreciation The United States paid about 4 cents an acre for the Louisiana Purchase in 1803. Suppose the value of this property grew at an annual rate of 5.5% compounded annually. What would an acre have been worth in 1994? Does this seem realistic?

54. Real Estate Appreciation Redo Exercise 53 using a rate of 6% instead of 5.5%. Compare your answer with the answer to Exercise 53.

55. Comparing Rates at Banks One bank advertises a nominal rate of 6.5% compounded quarterly. A second bank advertises a nominal rate of 6.6% compounded daily. What are the effective yields? In which bank would you deposit your money?

56. Comparing Rates at Banks One bank advertises a nominal rate of 8.1% compounded semiannually. A second bank advertises a nominal rate of 8% compounded weekly. What are the effective yields? In which bank would you deposit your money?

57. Saving for Machinery How much money should a company deposit in an account with a nominal rate of 8% compounded quarterly to have $100,000 for a certain piece of machinery in five years?

58. Saving for Machinery Repeat Exercise 57 with the annual rate at 7% and compounding monthly.

59. Doubling Time Suppose $1000 is invested at an annual rate of 6% compounded monthly. Using your computer or graphing calculator, determine (to two decimal places) the time it takes for this account to double.

60. Comparing Two Accounts One account with an initial amount of $1 grows at 100% a year compounded quarterly, and a second account with an initial amount of $2 grows at 50% a year compounded quarterly. Graph two appropriate functions on the same viewing screen, and using available operations, find (to two decimal places) the time at which the two accounts will have equal amounts.

61. Finding the Rate An account earns an annual rate of r, expressed as a decimal, and is compounded quarterly. The account initially has $1000 and five years later has $1500. What is r?

62. Saving Suppose an account earns an annual rate of 9% and is compounded continuously. Determine the amount of money grandparents must set aside at the birth of their grandchild if they wish to have $20,000 by the grandchild's eighteenth birthday.

63. Effective Yield One bank has an account that pays an annual interest rate of 6.1% compounded annually, and a second bank pays an annual interest rate of 6% compounded continuously. In which bank would you earn more money? Why?

64. Inflation Inflation, as measured by the U.S consumer price index,[63] increased by 2.8% in the year 2001. If this rate were to continue for the next 10 years, what would be the real value of $1000 after 10 years?

65. Inflation Inflation, as measured by the U.S consumer price index,[64] increased by 2.8% in the year 2001. If this rate were to continue for the next 10 years, use your computer or graphing calculator to determine how long before the value of a dollar would be reduced to 90 cents.

66. Deflation Inflation, as measured by Japan's consumer price index,[65] *decreased* (thus the word deflation) by 0.7% in the year 2001. If this rate were to continue for the next 10 years, use your computer or graphing calculator to determine how long before the value of a typical item would be reduced to 95% of its value in 2001.

67. Real Yield Suppose you were one of the people who in 1955 purchased a 40-year Treasury bond with a coupon rate of 3% per year. Presumably, you thought that inflation over the next 40 years would be low. (The inflation rate for the 30 years before 1955 was just a little over 1% a year.) In fact, the inflation rate for the 40-year period of this bond was 4.4% a year. Thus, in real terms you would have lost 1.4% a year. If you bought $1000 of such a bond at issuance and held it to maturity, how much would the $1000 bond have been worth in real terms?

68. Population of the World The human population of the world was about 6 billion in the year 2000 and increasing at the rate of 1.3% a year[66] Assume that this population will continue to grow exponentially at this rate, and use your computer or graphing calculator to determine the year in which the population of the world will reach 7 billion.

69. Population of the World The human population of the world was about 6 billion in the year 2000 and increasing at the rate of 1.3% a year[67] Assume that this population will continue to grow exponentially at this rate, and determine the population of the world in the year 2010.

70. Population of Latvia The population of Latvia was decreasing at the rate of 0.52% a year[68] from 1995 to 2000. Assuming that the population continued to decline at this annual rate, use your computer or graphing calculator to determine how long until the population of Latvia will become 95% of what is was in the year 2000.

71. Population of Bulgaria The population of Bulgaria was decreasing at the rate of 0.47% a year[69] from 1995 to 2000. Assuming that the population continued to decline at this annual rate, determine the total percentage drop in population over the subsequent 10 years.

---

63 U.S. Bureau of Labor Statistics.

64 Ibid.

65 Japan Information Network. http://www.jinjapan.org/stat/stats/05ECN32.html.

66 U.N. report. The world at six billion. http:www.un.org/esa/population/publications/sixbillion.htm.

67 Ibid.

68 Ibid.

69 Ibid.

More Challenging Exercises

72. Radioactive Decay Plutonium-239 is a product of nuclear reactors with a half-life of about 24,000 years. What percentage of a given sample will remain after 10,000 years?

73. Medicine As we noted in the text, radioactive tracers are used for medical diagnosis. Suppose 20 units of iodine-131 are shipped and take two days to arrive. Determine how much of the substance arrives.

74. Medicine Suppose that 20 units of iodine-131 are needed in two days. Determine how much should be ordered.

75. On the same screen of dimension $[-2, 2]$ by $[0, 5]$, graph 2^x, 3^x, and 5^x. Determine the interval for which $2^x < 3^x < 5^x$ and the interval for which $2^x > 3^x > 5^x$.

For each function graphed in Exercises 76 and 77, give a possible formula of the form Pa^x.

76. **77.**

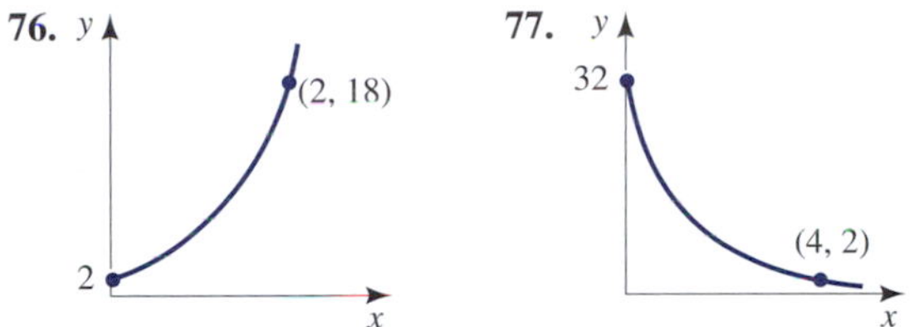

78. The graphs of two functions are shown. Could either one be the graph of an exponential function? Why or why not?

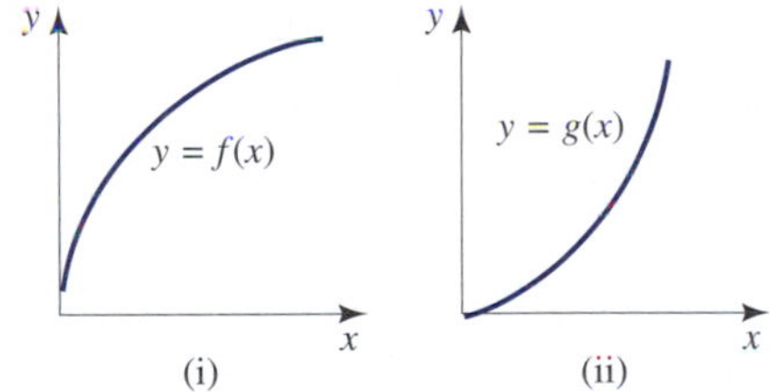

79. Growth Rates of Graysby Groupers Potts and Manooch[70] studied the growth habits of graysby groupers. These groupers are important components of the commercial fishery in the Caribbean. The mathematical model that they created was given by the equation $L(t) = 446(1 - e^{-0.13[t+1.51]})$, where t is age in years and L is length in millimeters. Graph this equation. What seems to be happening to the length as the graysby become older? Potts and Manooch also created a mathematical model that connected length with weight and was given by the equation $W(L) = 8.81 \times 10^{-6} \cdot L^{3.12}$, where L is length in millimeters and W is weight in grams. Find the length of a 10-year old graysby. Find the weight of a 10-year old graysby.

80. Growth Rates of Coney Groupers Potts and Manooch[71] studied the growth habits of coney groupers. These groupers are important components of the commercial fishery in the Caribbean. The mathematical model that they created was given by the equation $L(t) = 385(1 - e^{-0.32[t-0.49]})$, where t is age in years and L is length in millimeters. Graph this equation. What seems to be happening to the length as the coneys become older? Potts and Manooch also created a mathematical model that connected length with weight and was given by the equation $W(L) = 2.59 \times 10^{-5} \cdot L^{2.94}$, where L is length in millimeters and W is weight in grams. Find the length of a 10-year old coney. Find the weight of a 10-year old coney.

81. Vessel Fuel Use Mistiaen and Strand[72] needed to develop a vessel's fuel consumption per mile to study the cost structure of commercial fishing. They developed the equation $F(L) = 0.21e^{0.33L}$, where L is the length of the vessel in feet and F is the fuel used per mile in gallons per mile. Determine graphically the value of L such that a vessel twice the length of L will use twice the fuel per mile.

82. Weight of Trees Tritton and coworkers[73] needed estimates of the weights of large red oak trees in Connecticut. By using samples of felled trees, they developed the mathematical model given by the equation $W(D) = 0.1918 \cdot 10^{2.3454D}$, where D is the diameter of the tree at breast height in centimeters and W is the weight of the tree in kilograms. Using this formula, estimate the weight of a red oak tree in Connecticut with a diameter of 2 feet. (1 centimeter equals 0.3937 inch.) Do you think there is something wrong with their formula? Why or why not?

---

[70] Jennifer C. Potts and Charles S. Manooch III. 1999. Observations on the age and growth of graysby and coney from the southeastern United States. *Trans Amer. Fisheries Soc.* 128:751–757.

[71] Ibid.

[72] Johan A. Mistiaen and Ivar E. Strand. 2000. Location choice of commercial fisherman with heterogeneous risk preferences. *Amer. J. Agr. Econ.* 82(5):1184–1190.

[73] Louise M. Tritton, C. Wayne Martin, and James W. Hornbeck. 1987. Biomass and nutrient removals from commercial thinning and whole-tree clearcutting of central hardwoods. *Environ. Man.* 11(5):659–666.

Solutions to WARM UP EXERCISE SET 1.3

1. Write the equation as $3^{2x+1} = 3^{-x}$. Then from Property 3 we must have $2x + 1 = -x$ and $x = -1/3$.

2. We have $P = \$1000$ and $r = 0.08$.
   a. Here $n = 10$, $m = 12$. Thus,

$$F = P\left(1 + \frac{r}{m}\right)^{mt} = 1000\left(1 + \frac{0.08}{12}\right)^{12\cdot 10} = 2219.64$$

   b. For continuous compounding we have

$$F = Pe^{rt} = 1000e^{0.08(10)} = 2225.54$$

   Notice that this is greater than the amount in part (a).

3. If the annual rate is 8%, the effective yield is given by

$$r_e = \left(1 + \frac{r}{m}\right)^m - 1 = \left(1 + \frac{0.08}{52}\right)^{52} - 1 \approx 0.0832$$

4. The present value of \$10,000, if the annual interest rate is 8% compounded quarterly for 10 years is

$$P = \frac{F}{\left(1 + \frac{r}{m}\right)^{mt}} = \frac{10{,}000}{\left(1 + \frac{0.08}{4}\right)^{4\cdot 10}} = 4528.90$$

   Thus, a person must deposit \$4528.90 in an account earning 8% compounded quarterly so that there will be \$10,000 in the account after 10 years.

1.4 Combinations of Functions

The Algebra of Functions
The Composition of Functions

APPLICATION Area of a Forest Fire

Suppose a forest fire is spreading in a circular manner with the radius of the burnt area increasing at a steady rate of three feet per minute. What is the area of the circular burnt region as a function of the time? See Example 6 on page 75 for the answer.

The Algebra of Functions

A function such as $h(x) = x^4 + \sqrt{x+1}$ can be viewed as the sum of two simpler functions, $f(x) = x^4$ and $g(x) = \sqrt{x+1}$. So $h(x) = x^4 + \sqrt{x+1} = f(x) + g(x)$. We have always written profit as revenue less costs, that is, $P(x) = R(x) - C(x)$. The function $h(x) = xe^x$ can be written as the product of the two functions $f(x) = x$ and $g(x) = e^x$. So $h(x) = x \cdot e^x = f(x) \cdot g(x)$. A function such as

$$r(x) = \frac{1 + 10x^3}{1 + x^2}$$

can be viewed as the quotient of the two polynomials $p(x) = 1 + 10x^3$ and $q(x) = 1 + x^2$. So

$$r(x) = \frac{1 + 10x^3}{1 + x^2} = \frac{p(x)}{q(x)}$$

The two polynomials are much simpler than the original rational function. Breaking a complicated function into less complicated parts can yield insight into the behavior of the original function.

As we noted above, functions can be added, subtracted, multiplied, and divided. We state now the formal definitions of these concepts.

DEFINITION The Algebra of Functions

Let f and g be two functions and define

$$D = \{x \mid x \in (\text{domain of } f) \text{ and } x \in (\text{domain of } g)\}$$

Then for all $x \in D$ we define

the sum $f + g$ by	$(f + g)(x) = f(x) + g(x)$
the difference $f - g$ by	$(f - g)(x) = f(x) - g(x)$
the product $f \cdot g$ by	$(f \cdot g)(x) = f(x) \cdot g(x)$
the quotient $\dfrac{f}{g}$ by	$\left(\dfrac{f}{g}\right)(x) = \dfrac{f(x)}{g(x)}$ for x with $g(x) \neq 0$

EXAMPLE 1 Examples of Combining Functions

Let $f(x) = \sqrt{x}$ and $g(x) = 1 - x$. Find $(f + g)(x)$, $(f - g)(x)$, $(f \cdot g)(x)$, $(f/g)(x)$; also find the domain of each of them.

Solution

The domain of f is $[0, \infty)$, and the domain of g is all real numbers. Thus, the common domain of the first three functions is $D = [0, \infty)$.

$$(f + g)(x) = f(x) + g(x) = \sqrt{x} + (1 - x) = \sqrt{x} + 1 - x$$
$$(f - g)(x) = f(x) - g(x) = \sqrt{x} - (1 - x) = \sqrt{x} - 1 + x$$
$$(f \cdot g)(x) = f(x) \cdot g(x) = \sqrt{x}\,(1 - x)$$
$$\left(\frac{f}{g}\right)(x) = \frac{f(x)}{g(x)} = \frac{\sqrt{x}}{1 - x}$$

The domain of f/g is all nonnegative numbers other than 1. ■

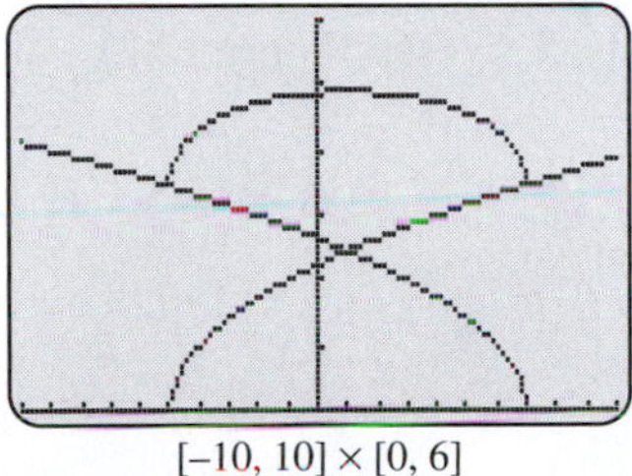

$[-10, 10] \times [0, 6]$

Screen 1.6
Graphs of $y_1 = \sqrt{x + 5}$, $y_2 = \sqrt{7 - x}$, and $y_3 = \sqrt{x + 5} + \sqrt{7 - x}$. From these graphs we can estimate the domains.

EXAMPLE 2 Combining Functions on a Grapher

On the same screen, graph $y_1 = f(x) = \sqrt{x + 5}$, $y_2 = g(x) = \sqrt{7 - x}$, $y_3 = y_1 + y_2$. From this obtain an indication of the domain of $f + g$. Verify analytically.

Solution

The graphs are shown in Screen 1.6. Notice how the graph gives us an indication of the domain of $f + g$. The domain appears to be $[-5, 7]$. The domain of f is $[-5, \infty)$, and the domain of g is $(-\infty, 7]$. Thus, the domain of $f + g$ is $[-5, 7]$. ■

Exploration 1

Profit

We have that $P(x) = R(x) - C(x)$, where $P(x)$ is profit, $R(x)$ is revenue, and $C(x)$ is cost. Graph on the same screen $y_1 = R(x) = 12x - x^2$, $y_2 = C(x) = 5x + 4$, $y_3 = y_1 - y_2$, and $y_4 = P(x) = -x^2 + 7x - 4$. What happened to the graph of the fourth function? Does this indicate that $y_3 = y_4$?

The Composition of Functions

Consider the function h given by the equation

$$h(x) = \sqrt{25 - x^2}$$

To evaluate $h(3)$, for example, we would first take $25 - (3)^2 = 16$ and then take the square root of 16 to obtain $h(3) = 4$. Thus, we can view this function as the combination of two simpler functions as follows. The first is $g(x) = 25 - x^2$, and the second is the square root function. Let

$$u = g(x) = 25 - x^2$$
$$y = f(u) = \sqrt{u}$$

then

$$y = \sqrt{u} = \sqrt{25 - x^2}$$

In terms of the two functions f and g,

$$h(x) = \sqrt{25 - x^2} = \sqrt{g(x)} = f[g(x)]$$

The function $h(x)$ is called the **composition** of the two functions f and g. We use the notation $h(x) = (f \circ g)(x)$.

The domain of h consists of those values in the domain of g for which $g(x)$ is in the domain of f. In the above-mentioned case the domain of g is $(-\infty, \infty)$, and the domain of f is the set of nonnegative numbers. Thus, the domain of h consists of those numbers x for which $g(x) = 25 - x^2$ is nonnegative. Thus, the domain of h consists of all numbers for which $x^2 \leq 25$ or the set $[-5, 5]$ (see Figure 1.43).

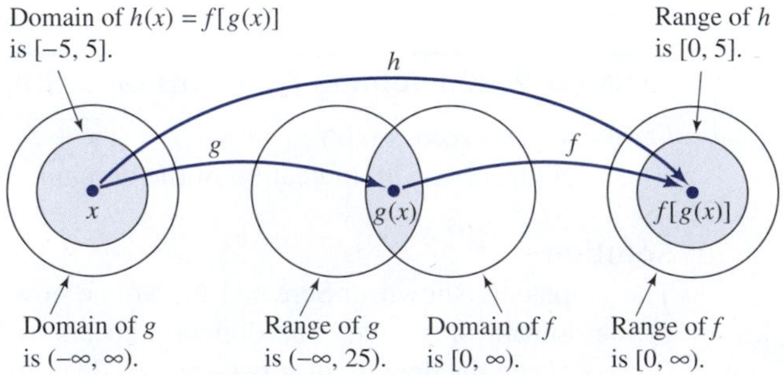

Figure 1.43
A pictorial of the composite function in which $f(x) = \sqrt{x}$ and $g(x) = 25 - x^2$.

We now define in general terms the composition of two functions (see Figure 1.44).

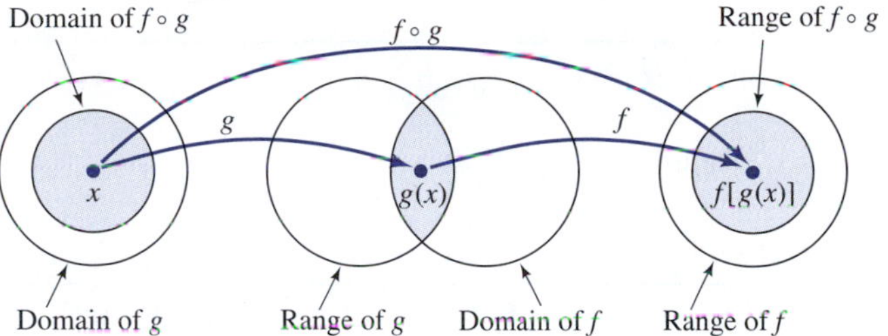

Figure 1.44
Notice that $g(x)$ must be in the domain of f for $f[g(x)]$ to be defined.

DEFINITION **Composite Function**

Let f and g be two functions. The composite function $f \circ g$ is defined by

$$(f \circ g)(x) = f[g(x)]$$

The domain of $f \circ g$ is the set of all x in the domain of g for which $g(x)$ is in the domain of f.

Sometimes all of the range of g is in the domain of f, as shown in Figure 1.45. In this special case the domain of $f \circ g$ is just the domain of g.

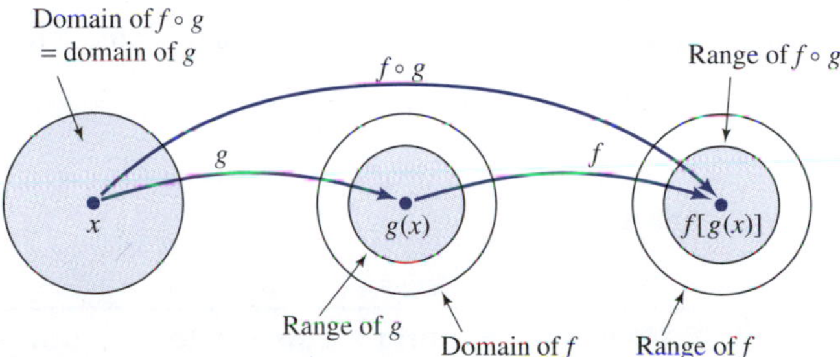

Figure 1.45
This is the case when the range of g is in the domain of f.

Another way of looking at the composition is shown in Figure 1.46. Here the conveyor belt inputs x into the function g, which outputs $g(x)$. The conveyor belt then inputs the value $g(x)$ into the function f, which then outputs $f[g(x)]$.

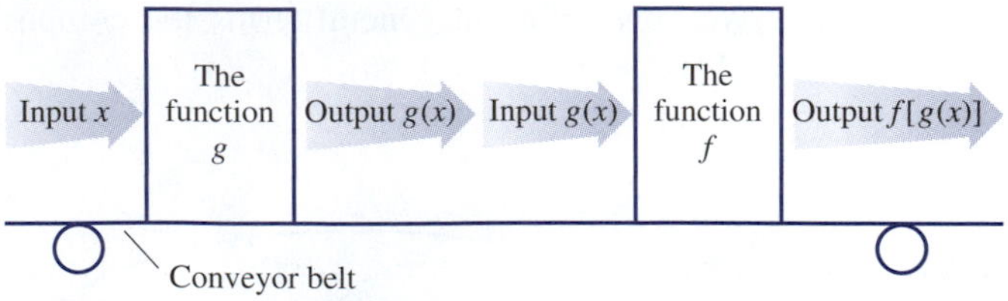

Figure 1.46

EXAMPLE 3 **Finding a Composite Function**

Let $f(x) = x + 1$ and $g(x) = x^3$. Find the domains of $f \circ g$ and $g \circ f$. Find $(f \circ g)(2)$ and $(g \circ f)(2)$. Find $(f \circ g)(x)$ and $(g \circ f)(x)$.

Solution

Since the domains of f and g are all the real numbers, the domains of $f \circ g$ and $g \circ f$ are also all the real numbers.

$$(f \circ g)(2) = f[g(2)] = f[2^3] = f[8] = 8 + 1 = 9$$
$$(g \circ f)(2) = g[f(2)] = g[2 + 1] = g[3] = 3^3 = 27$$
$$(f \circ g)(x) = f[g(x)] = f[x^3] = x^3 + 1$$
$$(g \circ f)(x) = g[f(x)] = g[x + 1] = (x + 1)^3$$

■

Remark Notice that $f \circ g$ and $g \circ f$ are two very different functions.

Exploration 2

Compositions on Your Grapher

(a) Let $y_1 = f(x) = \sqrt{x}$, $y_2 = g(x) = 25 - x^2$, $y_3 = \sqrt{y_2}$, and $y_4 = 25 - y_1^2$. Does y_3 correspond to $f \circ g$ or $g \circ f$? What about y_4?

(b) Confirm your conjectures by graphing y_3 and either $y = (f \circ g)(x)$ or $y = (g \circ f)(x)$ together on the same screen. Do the same with y_4.

In the next example we need to be more careful about the domain of the composite function.

EXAMPLE 4 **Finding a Composite Function and Its Domain**

Let $g(x) = \sqrt{x}$ and let $f(x) = x^2 + 5$. Find $(f \circ g)(x)$ and its domain.

Solution

$$(f \circ g)(x) = f[g(x)] = f[\sqrt{x}] = [\sqrt{x}]^2 + 5 = x + 5$$

One might naturally be inclined to think (incorrectly) that since one can take $x + 5$ for any x, the domain of $f \circ g$ is all real numbers. But notice that the domain of $f \circ g$ is all x in the *domain of* g, for which $g(x)$ is in the domain of f. But the domain of g is $[0, \infty)$, and the domain of f is all real numbers. Thus, the domain of $f \circ g$ is $[0, \infty)$.

■

Remark To avoid possible mistakes, keep in mind that the domain of $f \circ g$ is always a subset of the domain of g and that the range of $f \circ g$ is always a subset of the range of f.

Exploration 3

The Domain of the Composite Function

Confirm Example 4 on your grapher by graphing $y = (\sqrt{x})^2 + 5$.

EXAMPLE 5 **Writing a Function as the Composition Of Simpler Functions**

Let $h(x) = (3x + 1)^9$. Find two functions f and g such that $(f \circ g)(x) = h(x)$.

Solution

There are many possibilities. Perhaps the simplest is to notice that we are taking the ninth power of something. The something is $3x + 1$. Thus, we might set $g(x) = 3x + 1$ and $f(x) = x^9$. Then

$$(f \circ g)(x) = f[g(x)] = f[3x + 1] = (3x + 1)^9 = h(x)$$ ■

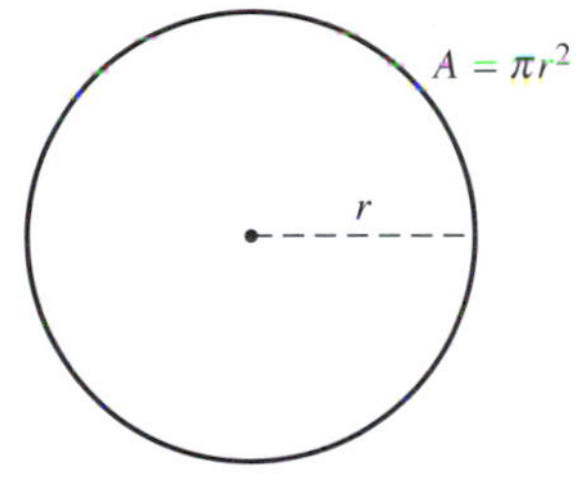

Figure 1.47
The area of a circle of radius is r is πr^2.

EXAMPLE 6 **Area of a Forest Fire**

Suppose a forest fire is spreading in a circular manner with the radius of the burnt area increasing at a steady rate of three feet per minute. Write the area of the circular burnt region as a function of the time?

Solution

The area of a circle of radius r is given by $A(r) = \pi r^2$ (see Figure 1.47). If r is the radius of the burnt region, then $r(t) = 3t$, where t is measured in minutes and r is measured in feet. Thus

$$A = A[r(t)] = A[3t] = \pi(3t)^2 = 9\pi t^2$$ ■

Warm Up Exercise Set 1.4

1. If $f(x) = x^2 + 1$ and $g(x) = x - 3$, find (a) $(f + g)(x)$, (b) $(f - g)(x)$, (c) $(f \cdot g)(x)$, (d) $(f/g)(x)$, and the domains of each.

2. If $f(x) = x^2 + 1$ and $g(x) = \sqrt{x - 3}$, find $(f \circ g)(x)$ and the domain.

Exercise Set 1.4

In Exercises 1 through 2, let $f(x) = 3x + 1$ and $g(x) = x^2$, and find the indicated quantity.

1. **a.** $(f + g)(2)$ **b.** $(f - g)(2)$ **c.** $(f \cdot g)(2)$ **d.** $(f/g)(2)$

2. **a.** $(g - f)(-3)$ **b.** $(g \cdot f)(-3)$ **c.** $(g/f)(-3)$ **d.** $(g + f)(-3)$

In Exercises 3 through 10 you are given a pair of functions, f and g. In each case, find $(f+g)(x)$, $(f-g)(x)$, $(f \cdot g)(x)$, and $(f/g)(x)$ and the domains of each.

3. $f(x) = 2x + 3$, $g(x) = x^2 + 1$

4. $f(x) = x^3$, $g(x) = 3$

5. $f(x) = 2x + 3$, $g(x) = x + 1$

6. $f(x) = x^3$, $g(x) = x - 1$

7. $f(x) = \sqrt{x+1}$, $g(x) = x + 2$

8. $f(x) = \sqrt{x+1}$, $g(x) = x - 3$

9. $f(x) = 2x + 1$, $g(x) = 1/x$

10. $f(x) = \sqrt{x+2}$, $g(x) = 1/x$

In Exercises 11 through 14 you are given a pair of functions, f and g. In each case, estimate the domain of $(f+g)(x)$ using your computer or graphing calculator. Confirm analytically.

11. $f(x) = \sqrt{x+3}$, $g(x) = \sqrt{x+1}$

12. $f(x) = \sqrt{1-x}$, $g(x) = \sqrt{3-x}$

13. $f(x) = \sqrt{2x+9}$, $g(x) = \sqrt{2x-5}$

14. $f(x) = \sqrt{2x-7}$, $g(x) = \sqrt{4x+15}$

In Exercises 15 and 16, let $f(x) = 2x + 3$ and $g(x) = x^3$. Find the indicated quantity.

15. **a.** $(g \circ f)(1)$ **b.** $(g \circ f)(-2)$

16. **a.** $(f \circ g)(1)$ **b.** $(f \circ g)(-2)$

In Exercises 17 through 20 you are given a pair of functions, f and g. In each case, find $(f \circ g)(x)$ and $(g \circ f)(x)$ and the domains of each.

17. $f(x) = 2x + 1$, $g(x) = 3x - 2$

18. $f(x) = 2x + 3$, $g(x) = x^3$

19. $f(x) = x^3$, $g(x) = \sqrt[3]{x}$

20. $f(x) = x^2 + x + 1$, $g(x) = x^2$

In Exercises 21 and 22, let $f(x) = 2x + 1$ and $g(x) = 2x^3$. Find the indicated quantity.

21. **a.** $(f \circ g)(1)$ **b.** $(g \circ f)(1)$
c. $(f \circ f)(1)$ **d.** $(g \circ g)(1)$

22. **a.** $(f \circ g)(x)$ **b.** $(g \circ f)(x)$
c. $(f \circ f)(x)$ **d.** $(g \circ g)(x)$

In Exercises 23 through 30, express each of the given functions as the composition of two functions. Find the two functions that seem the simplest.

23. $(x+5)^5$

24. $\sqrt{x^3+1}$

25. $\sqrt[3]{x+1}$

26. 2^{2x-3}

27. $|x^2 - 1|$

28. $1/(3x+2)$

29. $1/(x^2+1)$

30. e^{x^3+1}

31. Let the two functions f and g be given by the graphs in the figure. Find
a. $(f \circ g)(0)$ **b.** $(f \circ g)(1)$
c. $(g \circ f)(0)$ **d.** $(g \circ f)(1)$

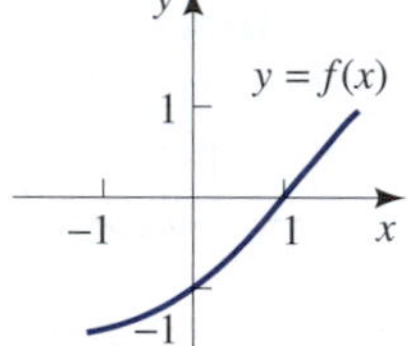

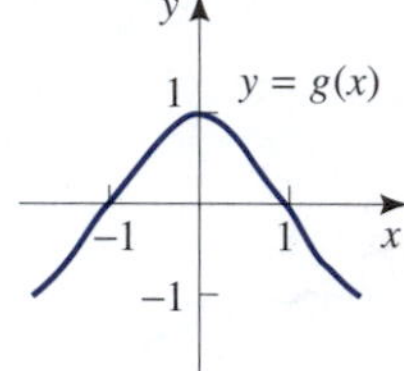

32. Let the two functions f and g be given by the graphs in the figure. Find the domains of $(f \circ g)(x)$ and $(g \circ f)(x)$.

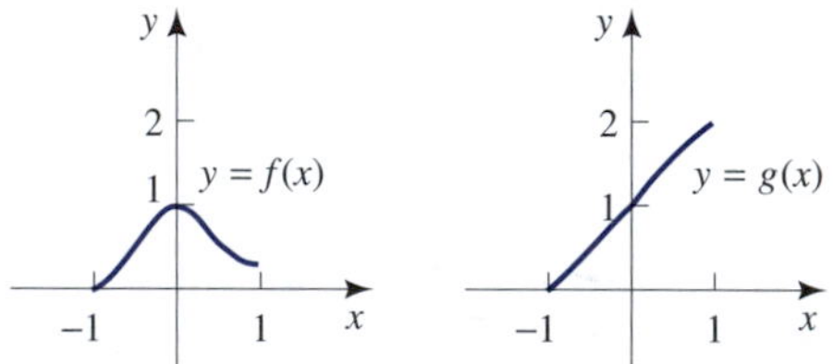

Applications and Mathematical Modeling

33. Cost A manufacturing firm has a daily cost function of $C(x) = 3x + 10$, where x is the number of thousands of an item produced and C is in thousands of dollars. Suppose the number of items that can be manufactured is given by $x = n(t) = 3t$, where t is measured in hours. Find $(C \circ n)(t)$, and state what this means.

34. Revenue A firm has a revenue function given by $R(p) = 10p - p^2$, where p is the price of a chocolate bar sold in dollars and R is measured in thousands of dollars per day. Suppose the firm is able to increase the price of each bar by 5 cents each year (without affecting demand). If t is time measured in years, write an equation for the revenue as a function of t if the price of a candy bar starts out at 25 cents.

35. Revenue and Cost If $R(x)$ is the revenue function and $C(x)$ is the cost function, what does the function $(R - C)(x)$ stand for?

36. Distance Two ships leave the same port at the same time. The first travels due north at 4 miles per hour, and the second travels due east at 5 miles per hour. Find the distance d between the ships as a function of time t measured in hours. (Hint: $d = \sqrt{x^2 + y^2}$.) Now find both x and y as functions of t.

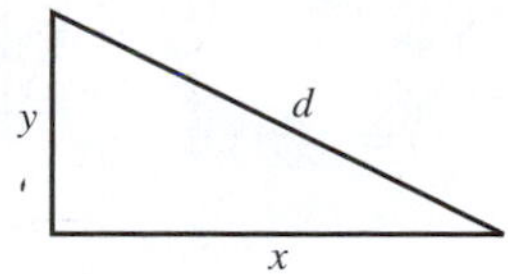

37. Medicine A cancerous spherical tumor that was originally 30 millimeters in radius is decreasing at the rate of 2 mm per month after treatment. Write an equation for the volume of the tumor as a function of time t measured in months. Note that the volume V of a sphere of radius r is given by $V(r) = \frac{4}{3}\pi r^3$.

38. Revenue Suppose that a firm's annual revenue function is given by $R(x) = 20x + 0.01x^2$, where x is the number of items sold and R is in dollars. The firm sells 1000 items now and anticipates that its sales will increase by 100 in each of the next several years. If t is the number of years from now, write the number of sales as a function of t and also write the revenue as a function of t.

39. Measurement Conversion Let x be the length of an object in furlongs, let r be the length in rods, and let y be the length in yards. Given that there are 40 rods to a furlong, find the function g such that $r = g(x)$. Given that there are 5.5 yards to a rod, find the function f such that $y = f(r)$. Now determine y as a function of x, and relate this to the composition of two functions. Explain your formula in words.

40. Measurement Conversion Let x be the size of a field in hectares, let a be the size in acres, and let y be the size in square yards. Given that there are 2.471 acres to a hectare, find the function g such that $a = g(x)$. Given that there are 4840 square yards to an acre, find the function f such that $y = f(a)$. Now determine y as a function of x and relate this to the composition of two functions. Explain your formula in words.

41. Growth Rates of Lake Trout Ruzycki and coworkers[74] studied the growth habits of lake trout in Bear Lake, Utah-Idaho. The mathematical model they created was given by the equation $L(t) = 960(1 - e^{-0.12[t+0.45]})$, where t is age in years and L is length in millimeters. Graph this equation. What seems to be happening to the length as the trout become older? Ruzycki and coworkers also created a mathematical model that connected length with weight and was given by the equation $W(L) = 1.30 \times 10^{-6} \cdot L^{3.31}$, where L is length in millimeters and W is weight in grams. Find the length of a 10-year old lake trout. Find the weight of a 10-year old lake trout. Find W as a function of t.

42. Growth Rates of Cutthroat Trout Ruzycki and coworkers[75] studied the growth habits of cutthroat trout in Bear Lake, Utah-Idaho. The mathematical model they created was given by the equation $L(t) = 650(1 - e^{-0.25[t+0.50]})$, where t is age in years and L is length in millimeters. Graph this equation. What seems to be happening to the length as the trout become older? Ruzycki and coworkers also created a mathematical model that connected length with weight and was given by the equation $W(L) = 1.17 \times 10^{-5} \cdot L^{2.93}$, where L is length in millimeters and W is weight in grams. Find the length of a 10-year old cutthroat trout. Find the weight of a 10-year old cutthroat trout. Find W as a function of t.

---

[74] James R. Ruzycki, Wayne A. Wurtsbaugh, and Chris Luecke. 2001. Salmonine consumption and competition for endemic prey fishes in Bear Lake, Utah-Idaho. *Trans. Amer. Fisheries Soc.* 130:1175–1189.

[75] Ibid.

More Challenging Exercises

In Exercises 43 through 48, you are given a pair of functions, f and g. In each case, use your grapher to estimate the domain of $(g \circ f)(x)$. Confirm analytically.

43. $f(x) = x^2$, $g(x) = \sqrt{x}$

44. $f(x) = |x|$, $g(x) = \sqrt{x}$

45. $f(x) = \sqrt{x+5}$, $g(x) = x^2$

46. $f(x) = \sqrt{x-5}$, $g(x) = x^2$

47. $f(x) = \sqrt{5-x}$, $g(x) = x^4$

48. $f(x) = \sqrt{5-x}$, $g(x) = x^3$

Solutions to WARM UP EXERCISE SET 1.4

1. If $f(x) = x^2 + 1$ and $g(x) = x - 3$, then

a. $(f+g)(x) = f(x) + g(x) = (x^2+1) + (x-3) = x^2 + x - 2$

b. $(f-g)(x) = f(x) - g(x) = (x^2+1) - (x-3) = x^2 - x + 4$

c. $(f \cdot g)(x) = f(x) \cdot g(x) = (x^2+1)(x-3) = x^3 - 3x^2 + x - 3$

d. $\left(\frac{f}{g}\right)(x) = \frac{f(x)}{g(x)} = \frac{x^2+1}{x-3}$

The domains of both f and g are the set of all real numbers. The domains of the first three functions are also the set of all real numbers. The domain of the quotient is the set of all real numbers except 3.

2. If $f(x) = x^2 + 1$ and $g(x) = \sqrt{x-3}$, then

$$\begin{aligned}(f \circ g)(x) &= f[g(x)] = (\sqrt{x-3})^2 + 1 \\ &= (x-3) + 1 = x - 2\end{aligned}$$

The domain of f is $(-\infty, \infty)$, and the domain of g is $[3, \infty)$. The domain of $(f \circ g)$ is then $[3, \infty)$.

1.5 Logarithms

Basic Properties of Logarithms
Solving Equations with Exponents and Logarithms
More on the Natural Logarithm
Enrichment: Half-Life

The Granger Collection

John Napier, 1550–1617

John Napier, Laird of Merchiston, was a Scottish nobleman who invented logarithms around 1594 to ease the burden of numerical calculations needed in astronomy. However, most of his time was spent engaged in the political and religious controversies of his day. He published an extremely widely read attack on the Roman Catholic Church, an act that he believed was his primary achievement. For relaxation he amused himself with the study of mathematics and science.

APPLICATION **Time Needed to Double an Account**

A person has \$1000 and can invest it in an account that earns an annual rate of 8% compounded quarterly. In five years \$1400 will be needed. Will this account have \$1400 within five years? See Example 3 on page 83 for the answer.

Basic Properties of Natural Logarithms

In Section 1.3 we faced a number of problems in which the unknown quantity in an equation was an exponent. For example, when we wanted the time required for \$1000 in an account growing at an annual rate of 9% compounding continuously to reach \$1500, we needed to solve for t in the equation

$$1000e^{0.09t} = 1500$$

Our only available method was to find an approximate solution using our grapher.

We now wish to develop an analytic method of solving such an equation. As a simplified example, suppose we want to solve for x in the equation $2 = 10^x$. To do this analytically, we will need to invent a function, which we temporarily call $l(x)$, that "undoes" the exponential. That is, we want $l(x)$ to have the property that for all x, $l(10^x) = x$. Thus, in Figure 1.48, the conveyor belt inputs x into the exponential function, which outputs 10^x. The term 10^x is then inputed into the l function, which outputs x. An equation such as $2 = 10^x$ can then be solved by applying l to each side, obtaining

$$l(2) = l(10^x) = x$$

The values of the function $l(x)$ can be tabulated or obtained on a computer or calculator. In this way we will solve for x.

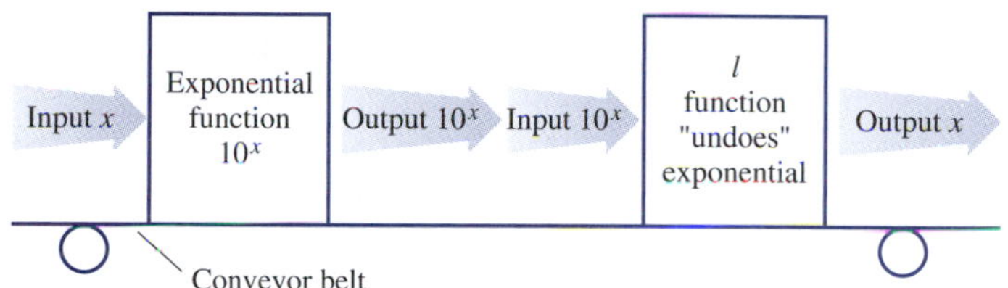

Figure 1.48

You are already familiar with a similar process. For example, to solve the equation $2 = x^3$, we apply the cube root function $g(x) = \sqrt[3]{x}$ to each side. The cube root function $g(x) = \sqrt[3]{x}$ "undoes" the cube function $f(x) = x^3$; that is, $g(x^3) = \sqrt[3]{x^3} = x$. Thus, we can solve the equation $2 = x^3$ for x by applying g to each side, obtaining

$$\sqrt[3]{2} = \sqrt[3]{x^3} = x$$

A table or calculator gives the value of $\sqrt[3]{2}$. In this way we solve for x.

When given any x, Figure 1.49 indicates geometrically how we can find $y = 10^x$. Since the function l is to "undo" the exponential function $y = 10^x$, then, given any $y > 0$, we see in Figure 1.50 how to obtain $l(y)$ by "reversing the arrows." That is, given any $y > 0$,

$$l(y) = x \qquad \text{if and only if} \qquad y = 10^x$$

Thus, $l(y)$ is that exponent to which 10 must be raised to obtain y. We normally write our functions with x as the independent variable. So interchanging x and y above gives, for any $x > 0$,

$$y = l(x) \qquad \text{if and only if} \qquad x = 10^y$$

Exploration 1

Evaluating $l(x)$

Use a computer or graphing calculator to solve for x to three decimal places in the equation $2 = 10^x$. (One way to do this graphically is to find the intersection of the two graphs of $y = 2$ and $y = 10^x$.) Check your answer.

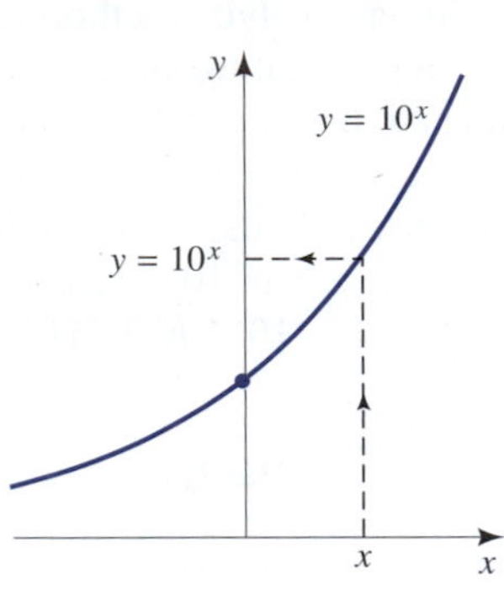

Figure 1.49
We see how to obtain 10^x graphically.

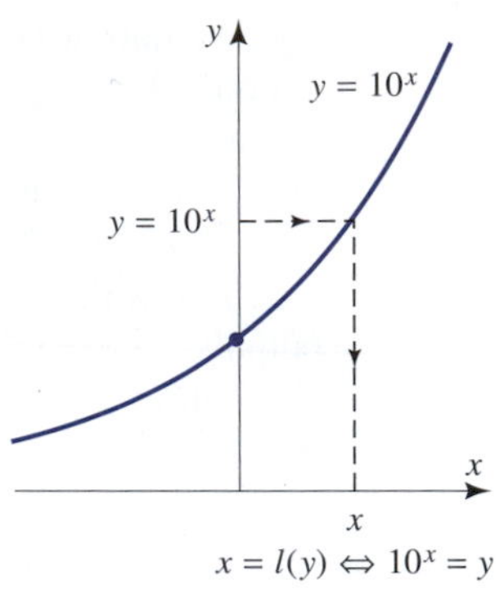

Figure 1.50
Given y, we see graphically how to find the x for which $10^x = y$.

The function $l(x)$ is usually written as $l(x) = \log_{10} x$ and is called the **logarithm to the base 10 of x**. In terms of this notation we have the following.

DEFINITION **Logarithm to the Base 10**

If $x > 0$, the **logarithm to the base 10 of x**, denoted by $\log_{10} x$, is defined as follows:

$$y = \log_{10} x \qquad \text{if and only if} \qquad x = 10^y$$

Instead of using the number 10, we can use any positive number $a \neq 1$.

DEFINITION **Logarithm to the Base a**

Let a be a positive number with $a \neq 1$. If $x > 0$, the **logarithm to the base a of x**, denoted by $\log_a x$, is defined as follows:

$$y = \log_a x \qquad \text{if and only if} \qquad x = a^y$$

EXAMPLE 1 **Using the Definition of Logarithm**

Solve for x in (a) $\log_{10} x = -\frac{3}{2}$, (b) $\log_3 3^2 = x$.

Solution

(a) $\log_{10} x = -\frac{3}{2}$ if and only if $x = 10^{-3/2}$.
(b) $\log_3 3^2 = x$ if and only if $3^x = 3^2$. This is true if and only if $x = 2$. ■

Remark Notice that we can have only $x > 0$ in the definition of the logarithm, since $x = a^y$ is always positive. Thus, the domain of the function $y = \log_a x$ is $(0, \infty)$.

Remark Notice that $y = \log_a x$ means that $x = a^y = a^{\log_a x}$. Thus, $\log_a x$ is an *exponent*. In fact $\log_a x$ *is that exponent to which a must be raised to obtain x.*

According to the definition of logarithm, for any $x > 0$, $y = \log_a a^x$ if and only if $a^x = a^y$. But this implies that $x = y = \log_a a^x$. This shows that the function $\log_a x$ "undoes" the exponential a^x.

Basic Properties of the Logarithmic Function

Property 1. $a^{\log_a x} = x$ if $x > 0$.

Property 2. $\log_a a^x = x$ for all x.

Remark It has already been mentioned that Property 2 means that the logarithmic function $\log_a x$ "undoes" the exponential function a^x. Property 1 says that the exponential function a^x "undoes" the logarithmic function $\log_a x$.

Remark To find any intercepts of the graph of $y = \log_a x$, set $0 = \log_a x$. Then we must have $x = a^0 = 1$. Thus, $\log_a 1 = 0$, and $(1, 0)$ is the only intercept of the graph of $y = \log_a x$.

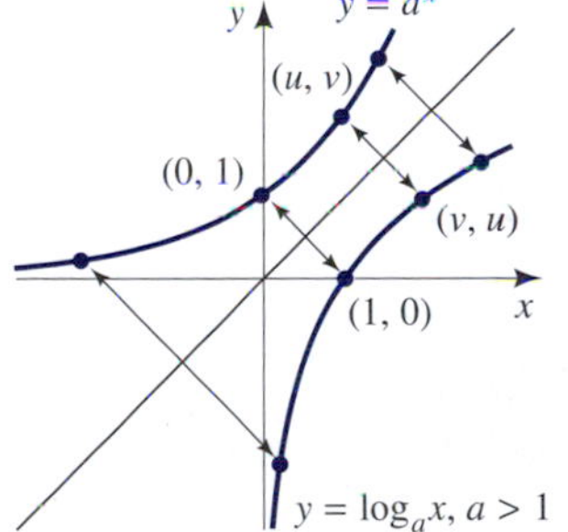

Figure 1.51
The graph of $y = \log_a x$ is the reflection about the line $y = x$ of the graph $y = a^x$.

EXAMPLE 2 Using the Basic Properties of Logarithms

Find (a) $\log_3 3^{-2}$, (b) $10^{\log_{10} \pi}$.

Solution

(a) Use Property 2 to obtain $\log_3 3^{-2} = -2$.
(b) Use Property 1 to obtain $10^{\log_{10} \pi} = \pi$. ■

We now consider the graphs of the functions $y = \log_a x$. The graphs are different depending on whether $a > 1$ or $0 < a < 1$. We are interested only in the case $a > 1$, since this covers the two important cases. The first of these is $a = 10$. This case is historically important. The logarithm $\log_{10} x$ is called the **common logarithm** and is also denoted by $\log x$. The second important case is $a = e = 2.71828$ (to five decimal places) mentioned in Section 1.3. This case is actually critically important and will be encountered in the next chapter and throughout the remainder of the text. We call the logarithm $\log_e x$ the **natural logarithm** and also denote it by $\ln x$.

First notice that whether $a > 1$ or $0 < a < 1$, a point (u, v) is on the graph of $y = a^x$ if and only if $v = a^u$. But according to the definition of the logarithm, $v = a^u$ if and only if $u = \log_a v$. This says that (v, u) is on the graph of $y = \log_a x$. Thus, (u, v) lies on the graph of $y = a^x$ if and only if (v, u) lies on the graph of $y = \log_a x$. This means that the graphs of $y = a^x$ and $y = \log_a x$ are symmetrical about the line $y = x$. Figure 1.51 shows the case $a > 1$. In this way we are able to construct the graph of $y = \log_a x$ shown in Figures 1.51 and 1.52, where the properties of $y = \log_a x$ are also given in the case $a > 1$.

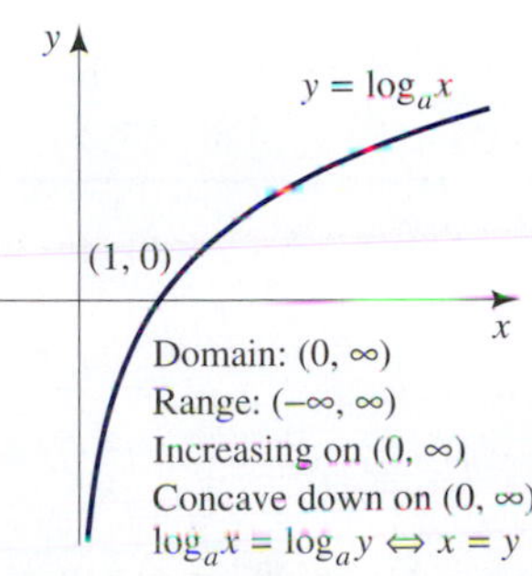

Figure 1.52

Interactive Illustration 1.8

Construction of the Graph of ln x

This interactive illustration allows you to explore the construction of the graph of the logarithm seen in Figure 1.51. Shown is the graph of $y = e^x$.

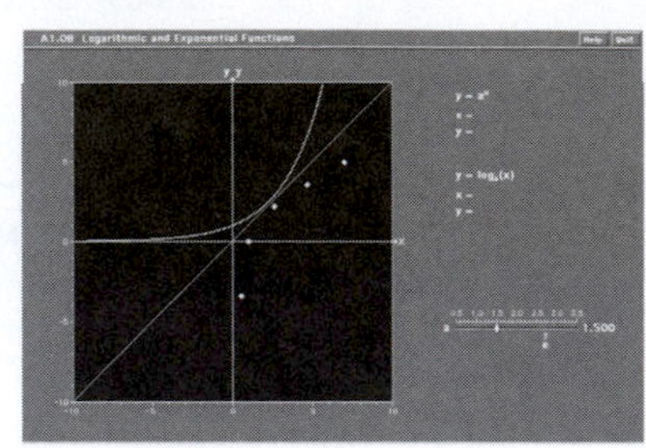

1. Click on points that you think are on the graph of $y = \ln x$. Pick at least eight points.
2. A "Show ln x" box now appears. Click on this box to give the graph of $\ln x$.

For any positive $a \neq 1$ the logarithm $\log_a x$ obeys the following rules.

Rules for the Logarithm

Rule 1. $\log_a xy = \log_a x + \log_a y$

Rule 2. $\log_a \frac{x}{y} = \log_a x - \log_a y$

Rule 3. $\log_a x^c = c \log_a x$

To establish these rules, we need to use $x = a^{\log_a x}$, $y = a^{\log_a y}$, $xy = a^{\log_a xy}$, and $x/y = a^{\log_a (x/y)}$.

To establish Rule 1, notice that

$$a^{\log_a xy} = xy = a^{\log_a x} a^{\log_a y} = a^{\log_a x + \log_a y}$$

Then equating exponents gives Rule 1.

We can establish Rule 2 in a similar manner. Notice that

$$a^{\log_a (x/y)} = \frac{x}{y} = \frac{a^{\log_a x}}{a^{\log_a y}} = a^{\log_a x - \log_a y}$$

Equating exponents gives Rule 2.

To establish Rule 3, notice that

$$a^{\log_a x^c} = x^c = \left(a^{\log_a x}\right)^c = a^{c \log_a x}$$

Equating exponents gives Rule 3.

Exploration 2

Determining Equalities by Graphing

Decide whether any of the following equations is true by finding the graph of each side of each equation in the same window.

(a) $\log_{10}(x + 3) = \log_{10} x + \log_{10} 3$

(b) $\log_{10} x^2 = (\log_{10} x)^2$

(c) $\log_{10} \frac{x}{2} = \frac{\log_{10} x}{\log_{10} 2}$

Solving Equations with Exponents And Logarithms

We now consider an important case in which we must solve for an unknown exponent.

EXAMPLE 3 **Solving an Equation with an Unknown Exponent**

A person has \$1000 and can invest it in an account that earns an annual rate of 8% compounded quarterly. In five years \$1400 will be needed. How long until this account reaches \$1400?

Solution

The value of this account in t quarters is $F = 1000(1 + 0.02)^t$. To find the required time t for this account to grow from \$1000 to \$1400, we need to solve for t in the equation $1400 = 1000(1 + 0.02)^t$:

$$
\begin{aligned}
1400 &= 1000(1.02)^t \\
1.4 &= (1.02)^t \\
\log 1.4 &= \log(1.02)^t = t \log 1.02 \qquad \text{Rule 3} \\
t &= \frac{\log 1.4}{\log 1.02} \\
&\approx \frac{0.1461}{0.0086} \approx 17
\end{aligned}
$$

Thus, it takes about 17 quarters, or 4.25 years, for this account to reach \$1400 (see Figure 1.53).

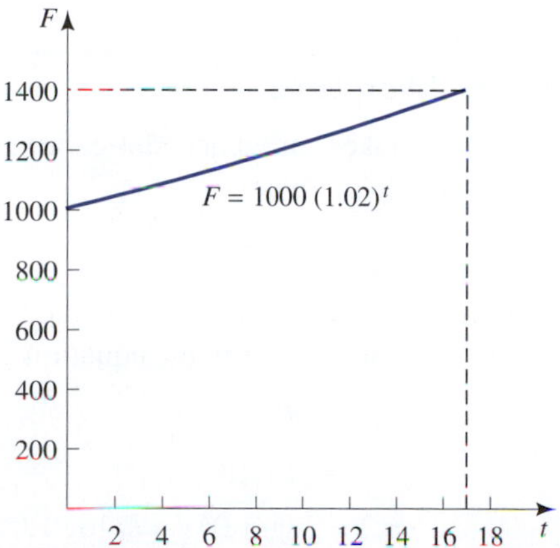

Figure 1.53
This account grows to \$1400 in about 17 quarters.

■

If we have an equation with a logarithm, then at an appropriate point we can use the definition of logarithm, as in the following example.

EXAMPLE 4 **Solving an Equation with a Logarithm**

Solve $2\log(2x+5)+6=0$.

Solution

$$\begin{aligned} 2\log(2x+5)+6 &= 0 \\ \log(2x+5) &= -3 \\ 2x+5 &= 10^{-3} = 0.001 \qquad \text{definition of logarithm} \\ x &= \tfrac{1}{2}(-4.999) = -2.4995 \end{aligned}$$

■

Remark In Example 4 we had $\log(2x+5)=-3$ at one point. We could have applied the exponential function to each side of this equation to "undo" the logarithm. Thus,

$$10^{-3} = 10^{\log(2x+5)} = 2x+5$$

by Property 1.

Exploration 3

No Analytic Solution

An equation with exponents cannot always be solved by taking logarithms. For example, the equation $1+x=5^x$ cannot be solved by taking logarithms. Try it! Use your computer or graphing calculator to approximate all solutions.

EXAMPLE 5 **Doubling Time**

Determine the time it takes for an account earning interest at an annual rate of 4% compounded annually to double.

Solution

Let t be the required time in years for the account to double, and let P be the initial amount. Then we must solve for t in the equation $2P = P(1+0.04)^t$:

$$\begin{aligned} 2P &= P(1.04)^t \\ 2 &= (1.04)^t \\ \log 2 &= \log(1.04)^t = t\log 1.04 \qquad \text{Rule 3} \\ t &= \frac{\log 2}{\log 1.04} \\ &\approx \frac{0.30103}{0.017} \approx 17.67 \end{aligned}$$

Thus, it takes nearly 18 years for this account to double (see Figure 1.54).

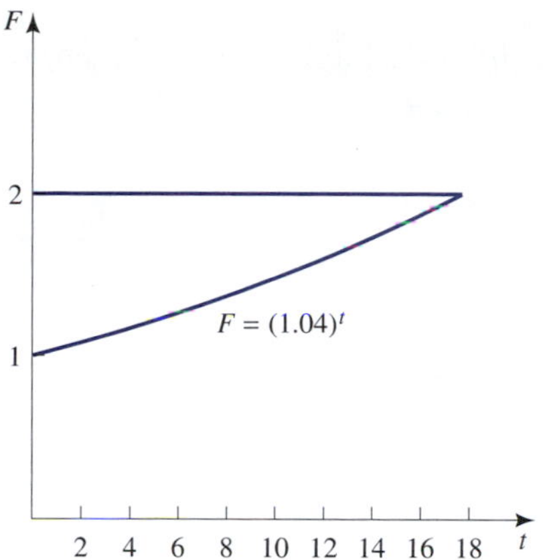

Figure 1.54
This account doubles in about 18 years.

Table 1.11

Interest Rate	Doubling Times (years)
4	17.67
5	14.21
6	11.90
7	10.24
8	9.01
9	8.04
10	7.27
15	4.96
20	3.80

Table 1.11 gives the doubling times for various annual interest rates for accounts that are compounding annually.

The interest rate i determines the rate at which the account is growing. The doubling times give an alternative way to assess the rate of growth.

More on the Natural Logarithm

We summarize here the definition of natural logarithm.

DEFINITION Natural Logarithm

If $x > 0$, the **natural logarithm of x**, denoted by $\ln x$, is defined as follows:

$$y = \ln x \qquad \text{if and only if} \qquad x = e^y$$

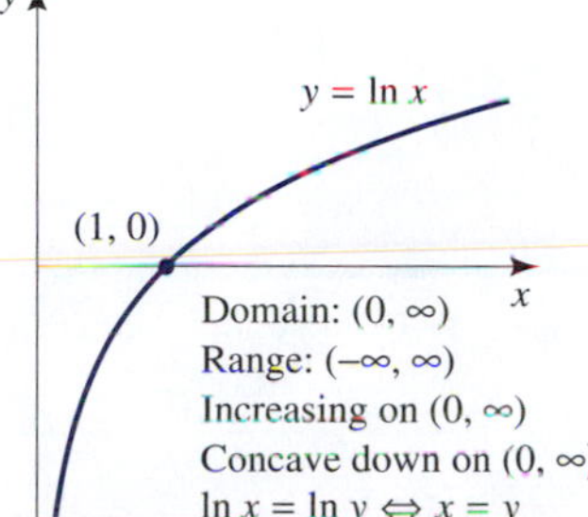

Figure 1.55

Since the base $e > 1$, we see that the graph of $y = \ln x$ is increasing as shown in Figure 1.55. Since the natural logarithm is a logarithm, it satisfies all rules of logarithms.

EXAMPLE 6 Doubling Time

An account with an initial amount of \$1000 is growing at 9% a year, compounding continuously. Find the time it takes for this account to double.

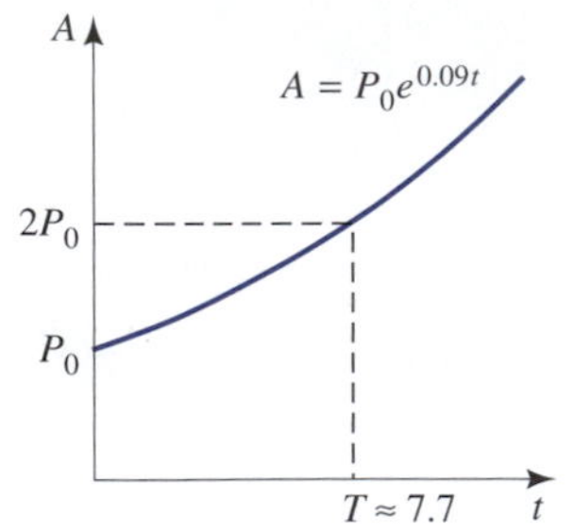

Figure 1.56
An account growing at 9% per year and compounding continuously doubles in value in about 7.7 years.

Solution

The amount at the end of t years is $A(t) = 1000e^{0.09t}$. Let T be the time in which the account doubles. Then

$$\begin{aligned} 2000 &= A(T) \\ &= 1000e^{0.09T} \\ 2 &= e^{0.09T} \\ \ln 2 &= \ln e^{0.09T} \\ &= 0.09T \\ T &= \frac{\ln 2}{0.09} \approx 7.7 \end{aligned}$$

Thus, it takes about 7.7 years for this account to double in value (see Figure 1.56). ■

Interactive Illustration 1.9

Compound Interest over Very Long Periods

This interactive illustration allows you to experiment with using different interest rates over long periods of time. We revisit Example 3 of Section 1.3. Recall that the Canarsie Indians sold Manhattan for \$24 in 1626 and have invested this money. Move the cursor to any time and see how much money was in the account at any time. A slider is provided so that you can change the interest rate.

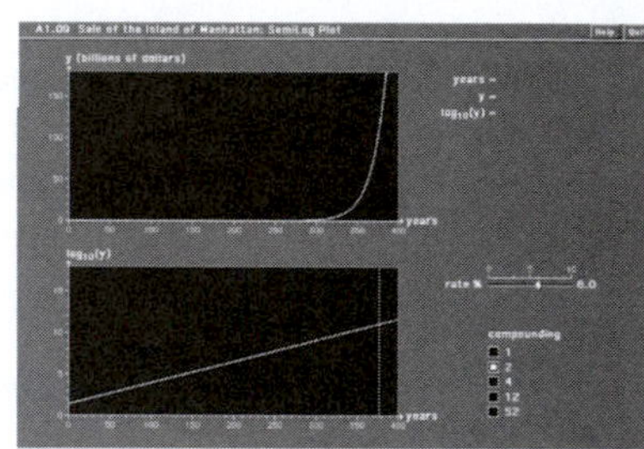

1. Change the rate a small amount, and see the effect over a long period. For example, take the rate to be 1%, 2%, and so on. What is the difference in the ending amount?
2. For each rate in (1), examine the bottom window. Notice that the vertical axis on the bottom window is measured in $\ln P$. How do the graphs differ?
3. Why do you think a graph such as shown in the bottom window is helpful?

We will now use $\ln x$ to write any exponential function $y = a^x$, with $a > 0$, as e^{kx} for some constant k. To see how to do this, notice that

$$a^x = (e^{\ln a})^x = e^{x \ln a}$$

We have shown the following.

Writing Exponential Functions in Terms of the Natural Exponent

Let $a > 0$. If $k = \ln a$, then

$$a^x = e^{kx}$$

EXAMPLE 7 **Writing an Exponential Function in Terms of the Natural Exponent**

In Section 1.3 we found that if \$1000 was deposited into an account with an annual yield of 8% and compounded quarterly, then the future value of this account after t quarters was $F = 1000(1.02)^t$. Write $(1.02)^t$ as e^{kt}, and rewrite the model using the exponent e.

Solution

Since $\ln 1.02 \approx 0.0198$, we have

$$\begin{aligned} F &= 1000(1.02)^t \\ &\approx 1000e^{0.02t} \end{aligned}$$

■

We can write the logarithm to one base in terms of the logarithm to another base.

Change of Base Theorem

$$\log_a x = \frac{\log_b x}{\log_b a}$$

We establish this by starting with $x = a^{\log_a x}$ and take the logarithm to the base b of each side as follows:

$$\begin{aligned} x &= a^{\log_a x} \\ \log_b x &= \log_b(a^{\log_a x}) \\ &= (\log_a x)\log_b a \\ \log_a x &= \frac{\log_b x}{\log_b a} \end{aligned}$$

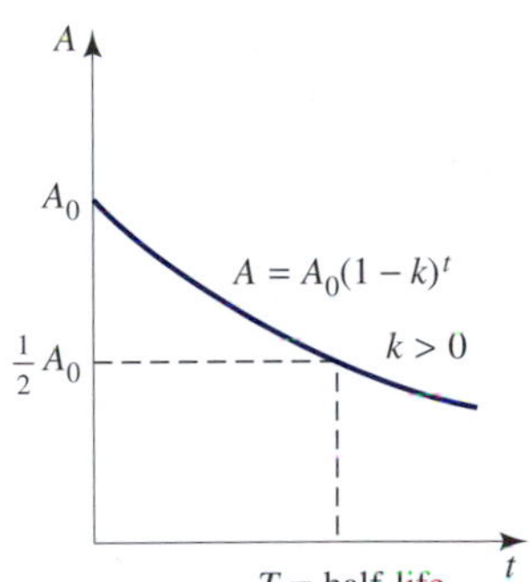

Figure 1.57
The half-life T is the time needed for the substance to decay to half its original size.

Enrichment: Half-Life

In Section 1.3 we considered the radioactive substance iodine-131, used in medical diagnosis. We found that if $A(t)$ is the amount after t days, then $A(t) = A_0(0.917)^t$, where A_0 is the initial amount. The decay constant is $k = 1 - 0.917 = 0.083$ and determines the rate at which the substance decays. An alternative way to assess the rate of decay is to determine the time until the substance decays to half of its original amount. This is called the **half-life** (see Figure 1.57).

EXAMPLE 8 **Half-Life**

A hospital has ordered 20 grams of iodine-131 and will need 10 grams within a week of ordering. Did they order enough?

Solution

We can answer this question by finding the half-life of iodine-131. Let T be the half-life of iodine-131. Then $0.5A_0 = A(T) = A_0(0.917)^T$. We must solve for T in this equation. We do this by taking common logarithms at an appropriate point. We have

$$\begin{aligned} 0.5A_0 &= A_0(0.917)^T \\ 0.5 &= (0.917)^T \\ \log 0.5 &= \log(0.917)^T \\ &= T \log 0.917 \\ T &= \frac{\log 0.5}{\log 0.917} \\ &= \frac{-0.30103}{-0.03763} \approx 8 \end{aligned}$$

Thus, the half-life of iodine-131 is about 8 days. As a consequence, 10 grams of the original 20 grams will be left in 8 days. Thus, more than 10 grams will still remain in 7 days or less. So the hospital has ordered enough (see Figure 1.58).

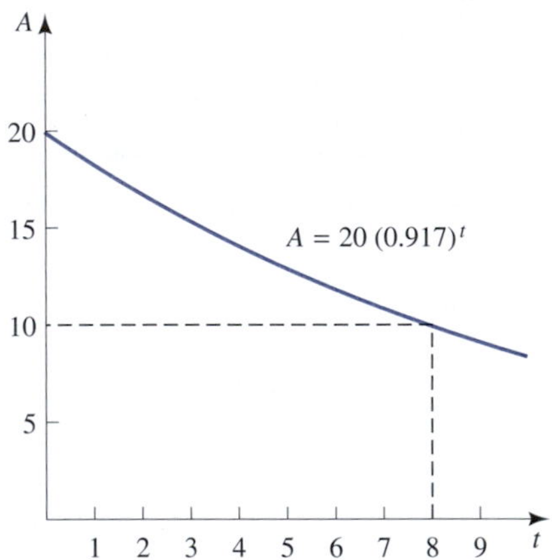

Figure 1.58
The half-life is about 8 days.

■

Warm Up Exercise Set 1.5

1. Solve for x in the equation $10^{x^3} = 5$.
2. Suppose inflation causes the value of a dollar to decrease by 3% a year. How long does it take for a dollar to be worth \$0.75?

Exercise Set 1.5

In Exercises 1 through 6, solve for x.

1. $\log x = 2$
2. $\log x = -3$
3. $\ln x = -\pi$
4. $\ln x = -\frac{3}{4}$
5. $3\log_3 x = 5$
6. $2\log_2 x = 9$

In Exercises 7 through 16, simplify.

7. $\log 10^4$
8. $\log \sqrt{10}$
9. $\log \frac{1}{10}$
10. $\log \frac{1}{\sqrt{10}}$
11. $10^{\log 2\pi}$
12. $10^{\log \sqrt{2}}$
13. $e^{3\ln 2}$
14. $e^{0.5\ln 9}$
15. $\left(5^{\log_5 3}\right)^4$
16. $\sqrt{3^{\log_3 2}}$

In Exercises 17 through 22, write the given quantity in terms of $\log x$, $\log y$, and $\log z$.

17. $\log x^2\sqrt{yz}$
18. $\log \sqrt{xyz}$
19. $\log \frac{\sqrt{xy}}{z}$
20. $\log \frac{xy^2}{z^3}$
21. $\log \sqrt{\frac{xy}{z}}$
22. $\log \frac{x^2y^3}{\sqrt{z}}$

In Exercises 23 through 28, write the given quantity as one logarithm.

23. $2\log x + \log y$
24. $2\log x - \log y$
25. $\frac{1}{2}\log x - \frac{1}{3}\log y$
26. $2\log x - \frac{1}{2}\log y + \log z$
27. $3\ln x + \ln y - \frac{1}{3}\ln z$
28. $\sqrt{2}\ln x - \ln y$

In Exercises 29 through 40, solve the equation for x.

29. $5 \cdot 10^{5x} = 3$
30. $2 \cdot 10^{3x} = 5$
31. $6 = 2 \cdot 10^{-2x}$
32. $3 = 4 \cdot 10^{-0.5x}$
33. $2 \cdot 10^{3x-1} = 1$
34. $3 \cdot 10^{2-5x} = 4$
35. $e^{x^2} = 4$
36. $e^{\sqrt{x}} = 4$
37. $2\log(x+7) + 3 = 0$
38. $\log x^2 = 9$
39. $\log(\log 4x) = 0$
40. $\log 3x = \log 6$

In Exercises 41 through 43, use basic graphing principles to sketch the graphs. (See Review Appendix A.8.) Then support your answers, using a grapher.

41. $y = \log x$, $y = \log(x+3)$, $y = \log(x-3)$
42. $y = \log x$, $y = 2\log x$, $y = 0.25\log x$
43. $y = \log x$, $y = -\log x$
44. Solve the equation $10^x = -\log x$ approximately on your computer or graphing calculator.
45. On your computer or graphing calculator, find all solutions to $\log x = x^2 - 2$.
46. Write 2^x as e^{kx} for some k.
47. Write 10^x as e^{kx} for some k.

Applications and Mathematical Modeling

48. **Doubling Time** If an account has an annual yield of 8% compounded monthly, how long before the account doubles?

49. **Decay of Advertising Effectiveness** Cotterill and Haller[76] showed in a recent study that the weekly decay rate of advertising in the breakfast cereal industry is 0.30. Thus, the equation $E = A(0.70)^t$, where t is time measured in weeks gives the effectiveness of the advertising. Find the amount of time it takes for the effectiveness to drop to 10% of its original.

50. **Inflation** Inflation, as measured by the U.S consumer price index[77] increased by 2.8% in the year 2001. If this rate were to continue for the next 10 years, determine how long before the value of a dollar would be reduced to 90 cents.

51. **Deflation** Inflation, as measured by Japan's consumer price index[78] *decreased* (thus the word *deflation*) by 0.7% in the year 2001. If this rate were to continue for the next 10 years, determine how long before the value of a typical item would be reduced to 95% of its value in 2001.

52. **Population of the World** The human population of the world was about 6 billion in the year 2000 and increasing at the rate of 1.3% a year.[79] Assume that this population will continue to grow at this rate and determine the year in which the population of the world will reach 7 billion.

53. **Population of India** The population of India is projected to grow from 0.998 billion in 1999 to 1.529 billion

[76] Ronald W. Cotterill and Lawrence E. Haller. 1997. An economic analysis of the demand for RTE cereal: product market definition and unilateral market power effects. Research Report No. 35. Food Marketing Policy Center, University of Connecticut.

[77] U.S. Bureau of Labor Statistics.

[78] Japan Information Network. http://www.jinjapan.org/stat/stats/05ECN32.html.

[79] U.N. report. The world at six billion. http:www.un.org/esa/population/publications/sixbillion.htm.

in the year 2050.[80] Assuming that the population is growing according to the model $P(t) = P_0e^{rt}$ over this time period, find r.

54. Population of the United States The population of the United States[81] is projected to grow from 276 million in 1999 to 349 million in the year 2050. Assuming that the population is growing according to the model $P(t) = P_0e^{rt}$ over this time period, find r.

55. Compounding An individual deposits \$2000 into an account with an annual yield of 6% compounded annually and \$1000 into an account with an annual yield of 9% compounded annually. How long will it take for the amount in the second account to equal the amount in the first account?

56. Population If a population grows according to $P(t) = P_0e^{kt}$ and if the population at time T is P_1, then show that

$$T = \frac{1}{k}\ln\frac{P_1}{P_0}$$

57. Exponential Decay Suppose that a quantity is decaying according to $y = P_0e^{-kt}$, where $k > 0$. Let T be the time it takes for the quantity to be reduced to half of its original amount. Show that $\ln 2 = kT$.

58. Exponential Growth Suppose a quantity is growing according to $y = P_0e^{kt}$, where $k > 0$. Let T be the time it takes for the quantity to double in size. Show that $\ln 2 = kT$.

59. Tripling Time A population grows according to $P(t) = P_0e^t$, where t is measured in years. How long before the population triples?

60. Demand for Water Timmins[82] created a mathematical model relating demand and supply of water in Delano, California. According to his model, $p = 1187 - 133\ln x$, where p is in dollars and x is in acre feet. Find the demand if p is \$800.

61. Gastric Evacuation Rate Hartman[83] created a mathematical model of the gastric evacuation rate of striped bass and gave the equation $P(t) = 100e^{-0.12t}$, where t is time in hours and P is the percentage of full stomach. How many hours until the stomach is half full?

62. Root Growth Johnson and Matchett[84] developed a mathematical model that related new root growth in tallgrass prairies in Kansas to the depth of the roots and gave the equation $y = 191.57e^{-0.3401x}$, where x is soil depth in centimeters and y is root growth in grams per square meter. Find the soil depth for which the root growth is one third of the amount at the surface.

63. Salmon Migration Coutant[85] created a mathematical model relating the percentage of juvenile salmon migrants passing through Wanapum (upper) and Priest Rapids (lower) dams on the Columbia River via spill in relation to the percentage of total flow spilled over spillways and gave the equation $y = 15.545\ln x$, where x is the percentage of river spilled and y is the percentage of fish passed through the spill. Determine the percentage of river spilled to have 50% fish pass through the spill.

64. Growth Rates of Graysby Groupers Potts and Manooch[86] studied the growth habits of graysby groupers. These groupers are important components of the commercial fishery in the Caribbean. The mathematical model they created was given by the equation $L(t) = 446(1 - e^{-0.13[t+1.51]})$, where t is age in years and L is length in millimeters. Graph this equation. Find the age at which graysby groupers reach 200 mm.

65. Recreational Demand Curve Shafer and colleagues[87] created a mathematical model of a demand function for recreational boating in the Three Rivers Area of Pennsylvania given by the equation $q = 65.64 - 12.11\ln p$, where q is the number of visitor trips, that is, the number of individuals who participated in any one recreational power boating trip, and p is the cost (or price) per person per trip. Using this demand equation, determine the price per visitor trip when 21 trips were taken. (This will give the actual average price per visitor trip, according to the authors.)

---

80 Ibid.

81 Ibid.

82 Christopher Timmins. 2002. Measuring the dynamic efficiency costs of regulators' preferences: municipal water utilities in the arid West. *Econometrica*. 70(2):603–629.

83 K. J. Hartman. 2000. Variability in daily ration estimates of age-0 striped bass in the Chesapeake Bay. *Trans. Amer. Fisheries Soc.* 129:1181–1186.

84 Loretta C Johnson and John R. Matchett. 2001. Fire and grazing regulate below ground processes in tallgrass prairie. *Ecology*. 82(12):3377–3389.

85 Charles C. Coutant. 2000. Fish behavior in relation to passage through hydropower turbines: a review. *Trans. Amer. Fisheries. Soc.* 129:351–380.

86 Jennifer C. Potts and Charles S. Manooch III. 1999. Observations on the age and growth of graysby and coney from the Southeastern United States. *Trans. Amer. Fisheries Soc.* 128:751–757.

87 Elwood L. Shafer, Arun Upneja, Wonseok Seo, and Jihwan Yoon. 2000. Economic values of recreational power boating resources in Pennsylvania. *Environ. Man.* 26(3):339–348.

More Challenging Exercises

66. Use the change of base theorem to find natural logarithms in terms of $\log_{10} x$.

67. Use the change of base theorem to find $\log_{10} x$ in terms of natural logarithms.

68. Determine the graph of $y = \log_a x$ in the case that $0 < a < 1$ in the same way as was done in the text for the case $a > 1$. That is, use the fact that the graphs of $y = a^x$ and $y = \log_a x$ are symmetric about the line $y = x$.

69. Can you define $\log_a x$ when $a = 1$? Explain why or why not.

70. Let $y_1 = \log x$ and $y_2 = x^{1/5}$. Determine which is larger for large values of x using your grapher.

71. Given any positive integer n, speculate on whether $y_1 = \ln x$ or $y_2 = x^{1/n}$ is larger for large x. Experiment on your grapher to decide. What does this say about how slow $\log x$ is increasing?

72. Biology The accompanying figure can be found in Deshmukh[88] and shows the dry weight of each shoot of 29 species of plants versus the density of shoots. The figure illustrates the **3/2 thinning law** in biology. The data points fall more or less in a linear pattern. The line drawn is the line found by using the least squares method given in Chapter 2. Explain where the "3/2" comes from.

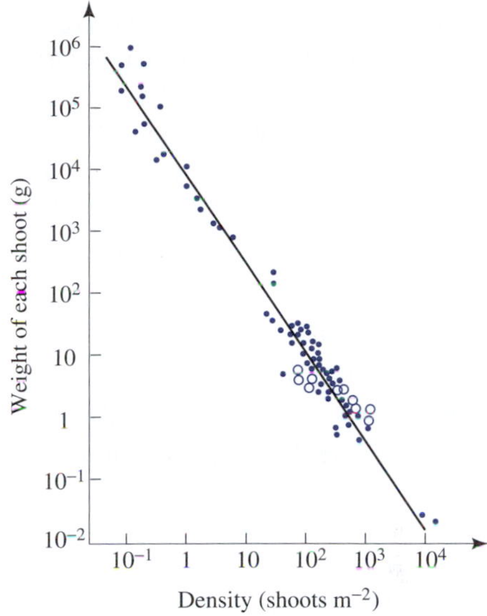

73. Growth Rates of Sturgeon Pine and Allen[89] studied the growth habits of sturgeon in the Suwannee River, Florida. The sturgeon fishery was once an important commercial fishery but, because of overfishing, was closed down in 1984, and the sturgeon is now both state and federally protected. The mathematical model that Pine and Allen created was given by the equation $L(t) = 222.273(1 - e^{-0.08042[t+2.18]})$, where t is age in years and L is length in centimeters. Graph this equation. Find the expected age of a 100-cm-long sturgeon algebraically.

74. Growth Rates of Atlantic Croaker Diamond and colleagues[90] studied the growth habits of the Atlantic croaker, one of the most abundant fishes of the southeastern United States. The mathematical model that they created for the ocean larva stage was given by the equation

$$L(t) = 0.26e^{2.876[1-e^{-0.0623t}]}$$

where t is age in days and L is length in millimeters. Graph this equation. Find the expected age of a 3-mm-long larva algebraically.

[88] Ian Deshmukh. 1986. *Ecology and Tropical Biology.* Boston: Blackwell Scientific Publ.

[89] William E. Pine III and Mike S. Allen. 2001. Population viability of the Gulf of Mexico sturgeon: inferences from capture-recapture and age-structured models. *Trans. Amer. Fisheries Soc.* 130:1164–1174.

[90] Sandra L. Diamond, Larry B. Crowder, and Lindsay G. Cowell. 1999. Catch and bycatch: the qualitative effects of fisheries on population vital rates of Atlantic croaker. *Trans. Amer. Fisheries Soc.* 128:1085–1105.

Solutions to WARM UP EXERCISE SET 1.5

1. Taking the logarithm of each side gives

$$\log 5 = \log 10^{x^3} = x^3$$

Thus, $x = \sqrt[3]{\log 5} = \sqrt[3]{0.69897} \approx 0.88747$.

2. The value $V(t)$ of the dollar is $V(t) = (0.97)^t$, where t is given in years. We must find t so that $0.75 = V(t) = (0.97)^t$. We have

$$\begin{aligned} 0.75 &= (0.97)^t \\ \log 0.75 &= \log(0.97)^t \\ &= t \log 0.97 \\ t &= \frac{\log 0.75}{\log 0.97} \approx 9.4 \end{aligned}$$

Thus, at this rate a dollar will be reduced in value to \$0.75 in about 9.4 years.

Summary Outline

- A **function** f from the set D to the set R is a rule that assigns to each element x in D one and only one element $y = f(x)$ in R. p. 7.
- The set D above is called the **domain**. p. 7.
- One thinks of the domain as the set of inputs and the values $y = f(x)$ as the outputs. p. 7.
- The set of all possible outputs is called the **range**. p. 7.
- The letter representing the elements in the domain is called the **independent variable**. p. 7.
- The letter representing the elements in the range is called the **dependent variable**. p. 7.
- The **graph** of the function f consists of all points (x, y) such that x is in the domain of f and $y = f(x)$. p. 9.
- A function is **increasing** (↗) on the interval (a, b) if the values of $f(x)$ increase as x increases. A function is **decreasing** (↙) on the interval (a, b) if the values of $f(x)$ decrease as x increases. p. 14.
- A function is **concave up** (⌣) on the interval (a, b) if the graph is bent upward there. A function is **concave down** (⌢) on the interval (a, b) if the graph is bent downward there. p. 14.
- **Vertical Line Test** A graph in the xy-plane represents a function, if and only if every vertical line intersects the graph in at most one place. p. 12.
- **Linear Cost, Revenue, and Profit Equations** p. 27. Let x be the number of items made and sold.

$$\begin{aligned} \textbf{variable cost} &= (\text{cost per item}) \times (\text{number of items produced}) \\ &= mx \\ C = \textbf{cost} &= (\text{variable cost}) + (\text{fixed cost}) \\ &= mx + b \\ R = \textbf{revenue} &= (\text{price per item}) \times (\text{number sold}) \\ &= px \\ P = \textbf{profit} &= (\text{revenue}) - (\text{cost}) \\ &= R - C \end{aligned}$$

- The quantity at which the profit is zero is called the **break-even quantity**. p. 30.
- Let x be the number of items made and sold and p the price of each item. A **linear demand equation**, which governs the behavior of the consumer, is of the form $p = mx + b$, where m must be negative. A **linear supply equation**, which governs the behavior of the producer, is of the form $p = mx + b$, where m must be positive. p. 32.
- The point at which supply equals demand is called the **equilibrium point**. The x-coordinate of the equilibrium point is called the **equilibrium quantity**, and the p-coordinate is called the **equilibrium price**. p. 32.
- The function $f(x) = ax^2 + bx + c$, $a \neq 0$ is called a **quadratic function**. p. 34.
- Every quadratic function $f(x) = ax^2 + bx + c$ with $a \neq 0$ can be put in **standard form** $f(x) = a(x - h)^2 + k$ by **completing the square**. The point (h, k) is called the **vertex**. p. 34.
- When $a < 0$, the graph of the quadratic $y = ax^2 + bx + c$ looks like ⌢ and assumes the maximum value k when $x = h$. p. 34.
- When $a > 0$, the graph of the quadratic $y = ax^2 + bx + c$ looks like ⌣ and assumes the minimum value k when $x = h$. p. 34.
- An **exponential** function is a function of the form $y = a^x$. p. 63.
- **Properties of a^x** p. 64.

$a > 1$	$0 < a < 1$
increasing on $(-\infty, \infty)$	decreasing on $(-\infty, \infty)$
domain $= (-\infty, \infty)$ range $= (0, \infty)$ $a^x = a^y$ if and only if $x = y$	

- **Compound Interest** Suppose a principal P earns interest at the annual rate of r, expressed as a decimal, and interest is compounded m times a year. Then the amount F after t years is $F = P(1 + i)^n = P\left(1 + \frac{r}{m}\right)^{mt}$, where $n = mt$ is the number of time periods and $i = \frac{r}{m}$ is the interest per period. p. 53.
- **Effective Yield** Suppose a sum of money is invested at an annual rate of r expressed as a decimal and is compounded m times a year. The effective yield r_e is $r_e = \left(1 + \frac{r}{m}\right)^m - 1$. p. 56.
- **Present Value** Suppose an account earns an annual rate of r expressed as a decimal and is compounded m times a year. Then the amount P, called the **present value**, needed currently in this account so that a future amount of F will be attained in t years is given by $P = F/(1 + r/m)^{mt}$. p. 57.
- **Logarithm to Base a** Let a be a positive number with $a \neq 1$. If $x > 0$, the logarithm to the base a of x, denoted by $\log_a x$, is defined as $y = \log_a x$ if and only if $x = a^y$. p. 80.

- **Properties of Logarithms** p. 81–82.

$$a^{\log_a x} = x \quad \text{if } x > 0 \qquad \log_a a^x = x \quad \text{for all } x$$

$$\log_a 1 = 0 \qquad \log_a xy = \log_a x + \log_a y$$

$$\log_a \frac{x}{y} = \log_a x - \log_a y \qquad \log_a x^c = c \log_a x$$

- The number e to five decimals is 2.71828. p. 57.
- An account with an initial amount of P that is growing at an annual rate of r (expressed as a decimal) **compounding continuously** becomes Pe^{rt} after t years. p. 59.
- If a sum of money is invested at an annual rate of r expressed as a decimal and is compounded continuously, the **effective yield** is $r_{eff} = e^r - 1$. p. 60.
- If an account earns an annual rate of r expressed as a decimal and is compounded continuously, then the amount P, called the **present value**, needed currently in this account now so that a future balance of A will be attained in t years is $P = Ae^{-rt}$. p. 60.
- **Natural Logarithm** If $x > 0$, the natural logarithm of x, denoted by $\ln x$, is defined as $y = \ln x$ if, and only if, $x = e^y$. p. 85.
- For any $a > 0$, $a^x = e^{kx}$ if $k = \ln a$. p. 86.
- If a quantity grows according to $Q(t) = Q_0(1 + k)^t$ or according to $Q(t) = Q_0 e^{kt}$, where $k > 0$, then k is called the **growth constant**. p. 66.
- If a quantity decays according to $Q(t) = Q_0(1 - k)^t$ or according to $Q(t) = Q_0 e^{kt}$, where $k > 0$, then k is called the **decay constant**. p. 66.
- The **half-life** of a radioactive substance is the time that must elapse for half of the material to decay. p. 87.
- The **sum** $f + g$ is defined by $(f + g)(x) = f(x) + g(x)$. p. 71.
- The **difference** $f - g$ is defined by $(f - g)(x) = f(x) - g(x)$. p. 71.
- The **product** $f \cdot g$ is defined by $(f \cdot g)(x) = f(x) \cdot g(x)$. p. 71.
- The **quotient** f/g is defined by $(f/g)(x) = f(x)/g(x)$, if $g(x) \neq 0$. p. 71.
- The **composition** $f \circ g$ is defined by $(f \circ g)(x) = f[g(x)]$ for all x in the domain of g for which $g(x)$ is in the domain of f. p. 73.

Review Exercises

1. Let $f(x) = 2x^2 + 3$. Find $f(1)$, $f(-2)$, $f(x + 1)$, and $f(x^2)$.

2. Let $f(x) = \dfrac{x - 1}{x^2 + 1}$. Find $f(1)$, $f(-2)$, $f(x + 2)$, and $f(x^3)$.

In Exercises 3 through 8, determine which graphs are graphs of functions.

3.
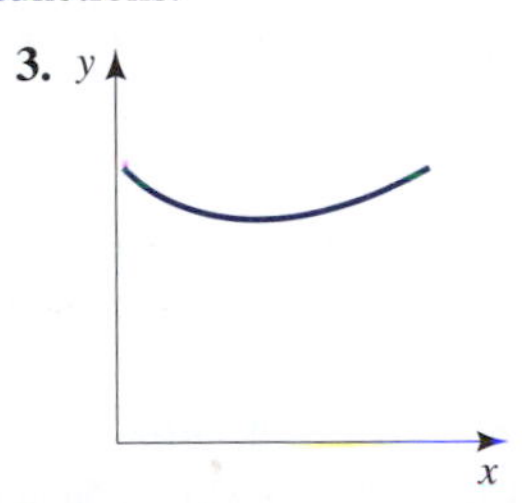

4.
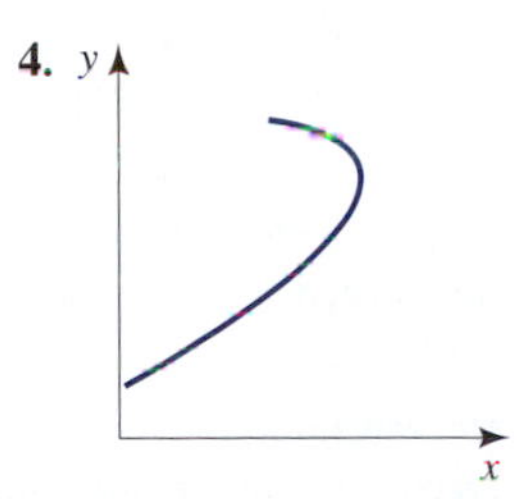

5.
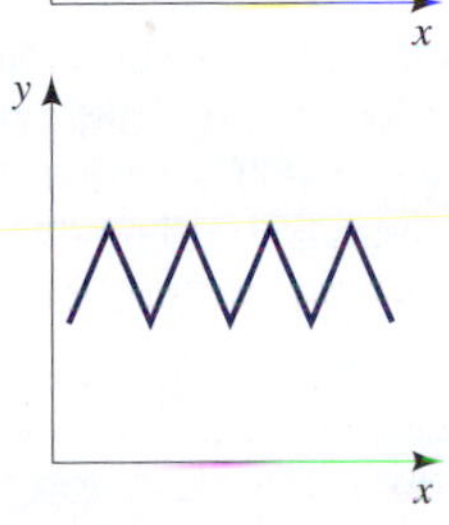

6.
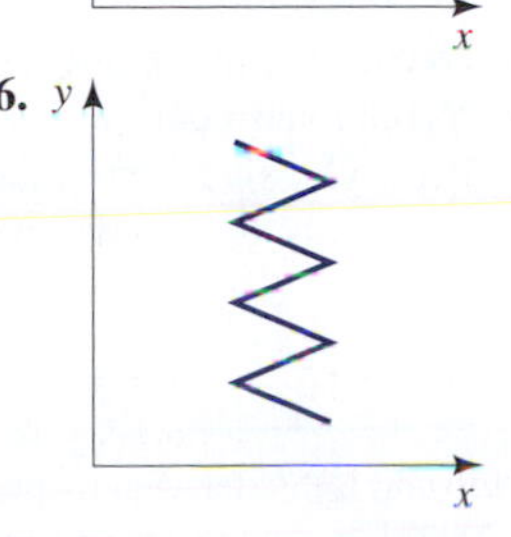

7.
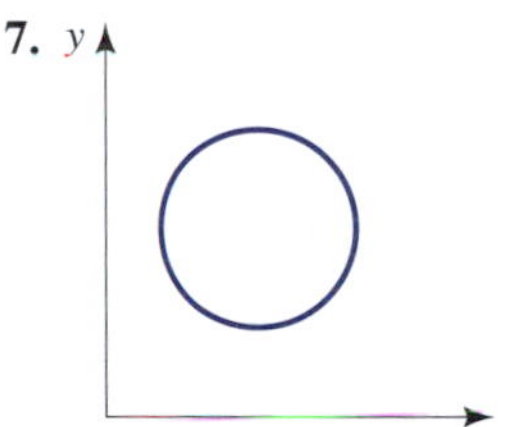

8.
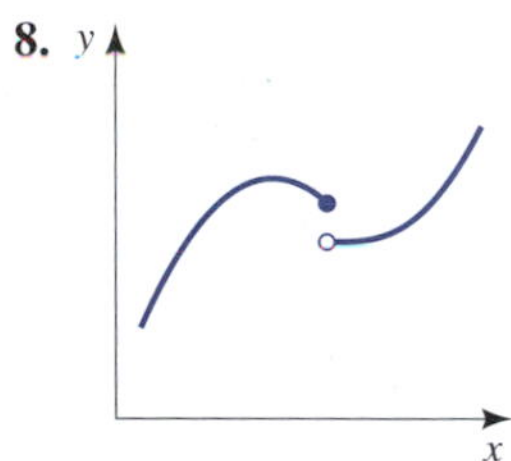

In Exercises 9 through 16, find the domain of each of the functions. Support your answer with a computer or graphing calculator.

9. $\dfrac{x + 1}{x - 3}$ 10. $\dfrac{x + 1}{\sqrt{x - 5}}$ 11. $\dfrac{x^2 - 2}{x^2 - 1}$

12. $\dfrac{x^3 + 1}{\sqrt{x^2 - 1}}$ 13. $x^{-1/7}$ 14. $x^{-1/8}$

15. $\sqrt{2 - x}$ 16. $\sqrt[3]{2 - x}$

17. Let $f(x) = \dfrac{1}{x^2 + 1}$. Evaluate $f(0)$, $f(\sqrt{2})$, $f(a - 1)$, and $f(x + 1)$.

18. Let $f(x) = \dfrac{1}{x + 1}$. Evaluate $f(0)$, $f(\sqrt{2})$, $f(a - 1)$, and $f\left(\dfrac{1}{x + 1}\right)$.

19. Graph

$$f(x) = \begin{cases} x^2 & \text{if } x \le 0 \\ x+1 & \text{if } 0 < x \le 1 \\ -x+2 & \text{if } x > 1 \end{cases}$$

20. Taxes A certain state has a tax on electricity of 1% of the monthly electricity bill, the first \$50 of the bill being exempt from tax. Let $P(x)$ be the percent one pays in taxes, and let x be the amount of the bill in dollars. Graph this function, and find where it is continuous. What is the significance of the points of discontinuity?

21. Interest A \$100,000 savings certificate has an annual interest of 8% compounded quarterly and credited to the account only at the end of each quarter. Determine the amount $A(t)$ in the account during the first year and find points at which the function is discontinuous. What is the significance of these points?

22. Costs A telephone company charges \$1.50 for the first minute for a certain long-distance call and \$0.25 for each additional minute. Graph this cost function and determine where the function is continuous. What is the significance of the points of discontinuity? Will some people be affected by these points?

23. Salary A certain computer salesman earns a base salary of \$20,000 plus 5% of the total sales if sales are above \$100,000 and 10% of total sales if sales are above \$1,000,000. Let $S(x)$ be his salary and let x be the amount in dollars of computers he sells. Find the discontinuities of $S(x)$. What is the significance of these discontinuities? Do you think that the salesman will be affected if his sales are almost at one of the discontinuities?

In Exercises 24 through 27, let $f(x) = x^3 + 4$ and $g(x) = 2x - 1$. Find the indicated quantity and the domains for each.

24. $(f+g)(x)$ **25.** $(f-g)(x)$

26. $(f \cdot g)(x)$ **27.** $(f/g)(x)$

In Exercises 28 and 29, find $(f \circ g)(x)$ and $(g \circ f)(x)$ for the given f and g and find the domains for each.

28. $f(x) = x^2 + 2$, $g(x) = 3x + 1$

29. $f(x) = \sqrt{x-1}$, $g(x) = x + 1$

In Exercises 30 through 35, graph.

30. $y = 3^x$ **31.** $y = 3^{-x}$ **32.** $y = 10^{2x}$

33. $y = 10^{-2x}$ **34.** $y = 10^{|x|}$ **35.** $\log |x|$

In Exercises 36 through 41, solve for x.

36. $9^x = 27$ **37.** $(1/3)^x = \frac{1}{27}$ **38.** $\ln x = 1$

39. $e^{4x} = e^{8x-2}$ **40.** $\log 10^x = 4$ **41.** $10^{\log x} = 5$

42. Yellow Perch Growth Power and Heuvel[91] studied yellow perch in Kimowin Lake in Alberta, Canada, and created a mathematical model given by the equation $L(D) = 72.45(1 - 18.54e^{-0.018D})$, where L is the length of perch in millimeters and D is the day of the year, January 1 being given by $D = 1$. Find the day of the year when the perch reach 50 mm in length.

[91] M. Power and M. R. van den Heuvel. 1999. Age-0 yellow perch growth and its relationship to temperature. *Trans. Amer. Fish. Soc.* 128:687–700.

43. Solve $e^{-x} = \log x$ to two decimal places using your computer or graphing calculator.

44. Cost Assuming a linear cost model, find the equation for the cost C, where x is the number produced, the cost per item is \$6, and the fixed costs are \$2000.

45. Revenue Assuming a linear revenue equation, find the revenue equation for R, where x is the number sold and the price per item is \$10.

46. Profit Assuming the cost and revenue equations in the previous two problems, find the profit equation. Also find the break-even quantity.

47. Profit Given that the cost equation is $C = 5x + 3000$ and the revenue equation is $R = 25x$, find the break-even quantity.

48. Supply and Demand Given the demand curve $p = -2x + 4000$ and the supply curve $p = x + 1000$, find the equilibrium point.

49. Transportation A plane travels a straight path from O to A and then another straight path from A to B. How far does the plane travel? Refer to the figure.

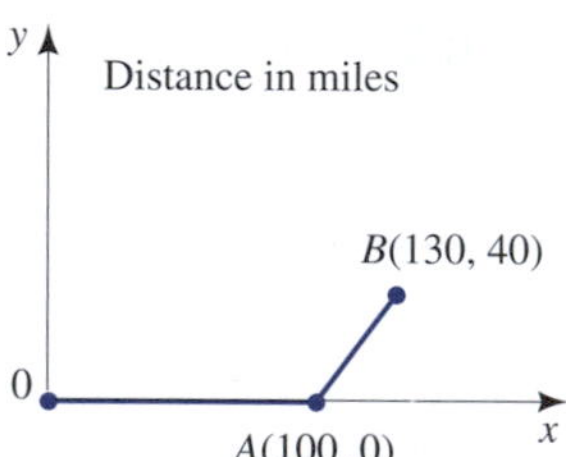

50. Transportation If the plane in the previous exercise travels 200 miles per hour from O to A and 250 miles per hour from A to B, how long does the trip take?

51. Demand A company notes from experience that it can sell 100,000 pens at \$1 each and 120,000 of the same pens at \$0.90 each. Find the demand equation, assuming that it is linear.

52. Recreational Demand Curve Shafer and colleagues[92] created a mathematical model of a demand function for recreational boating in Presque Isle Bay on Lake Erie in Pennsylvania given by the equation $q = 56.84 - 9.42 \ln p$, where q is the number of visitor trips, that is, the number of

[92] Elwood L. Shafer, Arun Upneja, Wonseok Seo, and Jihwan Yoon. 2000. Economic values of recreational power boating resources in Pennsylvania. *Environ. Man.* 26(3):339–348.

individuals who participated in any one recreational power boating trip, and p dollars is the cost (or price) per person per trip. Using this demand equation, determine the price per visitor trip when 25 trips were taken. (This will give the approximate actual average price per visitor trip, according to the authors.)

53. **Profit** It costs a publisher \$2 to produce each copy of a weekly magazine. The magazine sells for \$2.50 a copy, and the publisher obtains advertising revenue equal to 30% of the revenue from sales. How many copies must be sold to obtain a profit of \$15,000?

54. **Nutrition** A certain woman needs 15 mg of iron in her diet each day but cannot eat fish or liver. She plans to obtain all of her iron from kidney beans (4.5 mg per cup) and soybeans (5.5 mg per cup). If x is the number of cups of kidney beans and y is the number of cups of soybeans she consumes each day, what linear equation must x and y satisfy? Interpret the meaning of the slope of this equation.

55. **Cost, Revenue, and Profit Equations** In 1996 Rogers and Akridge[93] of Purdue University studied fertilizer plants in Indiana. For a typical large-sized plant they estimated fixed costs at \$447,917 and estimated that it cost \$209.03 to produce each ton of fertilizer. The plant sells its fertilizer output at \$266.67 per ton. Find the cost, revenue, and profit equations.

56. **Facility Location** A company is trying to decide whether to locate a new plant in Houston or Boston. Information on the two possible locations is given in the following table. The initial investment is in land, buildings, and equipment.

	Houston	Boston
Variable cost	\$0.25 per item	\$0.22 per item
Annual fixed costs	\$4,000,000	\$4,200,000
Initial investment	\$16,000,000	\$20,000,000

Suppose 10,000,000 items are produced each year.

a. Find which city has the lower annual total costs, not counting the initial investment.

b. Which city has the lower total cost over five years, counting the initial investment?

57. **Quadruped Size** Let l denote the length of a quadruped measured from hip to shoulder, and let h denote the average height of the trunk of the body. From physics it is known that for a uniform bar with similar dimensions the ratio $l{:}h^{2/3}$ is limited by some value. If the ratio exceeds a certain value, the rod will break. The following table lists quadrupeds whose ratios are among the highest. Find the ratio, and estimate the limiting value.

Quadruped	l	h	$l{:}h^{2/3}$
Ermine	12 cm	4 cm	
Dachshund	35 cm	12 cm	
Indian tiger	90 cm	45 cm	
Llama	122 cm	73 cm	
Indian elephant	153 cm	135 cm	

58. **Loudness** Loudness is measured by using the formula $L = 10\log(I/I_0)$, where I_0 is the lowest intensity that can be heard. How much would L increase if we multiplied the intensity I by 10? By 100? By 1000?

59. **Forestry** Klopzig and colleagues[94] noted that a common symptom of red pine decline is a large circular area of dead trees (pocket) ringed by trees showing reduced diameter and height growth. The equation $y = 0.305 - 0.023x$, where y is the proportion of dead roots and x is the distance from the pocket margin, approximates this effect. Interpret what the slope of this line means.

60. **Forestry** Harmer[95] indicated that the equation $y = 5.78 + 0.0275x$ approximated the relationship between shoot length x and bud number y for the plant *Quercus petraea*. Interpret what the slope of this line means.

61. **Cost of Irrigation Water** Using an argument that is too complex to give here, Tolley and Hastings[96] argued that if c is the cost in 1960 dollars per acre-foot of water in the area of Nebraska and x is the acre-feet of water available, then $c = 12$ when $x = 0$. They also noted that farms used about 2 acre-feet of water in the Ainsworth area when this water was free. If we assume (as they did) that the relationship between c and x is linear, then find the equation that c and x must satisfy.

62. **Agriculture** Tronstad and Gum[97] indicated that the weight w in pounds of a calf (at eight months) was related to the

[93] Duane S. Rogers and Jay T. Akridge. 1996. Economic impact of storage and handling regulations on retail fertilizer and pesticide plants. *Agribusiness* 12(4):327–337.

[94] K. D. Klopzig, K. F. Raffa, E. B. Smalley. 1991. Association of an insect-fungal complex with red pine decline in Wisconsin. *Forest Sci.* 37:1119–1139.

[95] R. Harmer. 1992. Relationship between shoot length, bud number and branch production in *Quercus petraea* (Matt.) Liebl. *Forestry* 65:1–13.

[96] G. S. Tolley and V. S. Hastings. 1960. Optimal water allocation: the North Platte River. *Quart. J. Econ.* 74:279–295.

[97] Russell Tronstad and Russell Gum. 1994. Cow culling decisions adapted for management with CART. *Amer. J. Agr. Econ.* 76:237–249.

age t in years of its mother according to $w = 412.34 + 28.05t - 2.05t^2$. Determine the age of the mother that results in a maximum calf weight, and determine this maximum. Support your answer with your grapher.

63. **Agriculture** The yield response of Texas coastal bend grain sorghum to nitrogen fertilizer has been given by Sri Ramaratnam[98] as $y = 2102 + 42.4x - 0.205x^2$, where y is in pounds per acre and x is pounds of nitrogen per acre. Using your graphing calculator, determine the amount of nitrogen that results in a maximum yield, and find this maximum. Confirm your answer algebraically.

64. **Forestry** To monitor the changes in forest growth, some estimate of leaf surface area y is needed. McIntyre and colleagues[99] give the estimate $y = 0.6249x^{1.8280}$, where y is leaf surface area in square meters and x is the (easy to measure) diameter at breast height in centimeters. Find y if $x = 20$, 30, and 45. Graph.

65. **Nursery Management** Klingeman and colleagues[100] created a mathematical model relating lace bug leaf damage on azaleas to buyer unwillingness to purchase given by the equation $P(x) = 99.2952(1 - 0.5e^{-0.0307(x-10.9502)})$, where x is the percentage of leaves with greater than 2% injury (the lower limit that can be normally seen) and P is the percentage of potential buyers unwilling to purchase such a plant. Find the percentage of leaves with greater than 2% injury for which 50% of potential buyers would be unwilling to purchase.

66. **Packaging** A box with top has length twice the width and height equal to the width. If x denotes the width and S the surface area, find S as a function of x.

67. **Postage** The postage on a first-class letter is \$0.37 for the first ounce, and \$0.23 for each additional ounce or any fraction thereof. If x denotes the number of ounces that a letter weighs and $C(x)$ the first-class postage, graph the function $C(x)$ for $0 < x \leq 3$. Plot $y = C(x)$ on your computer or graphing calculator.

68. **Spread of Infestation** A certain pine disease is spreading in a circular fashion through a pine forest, the radius of the infested region increasing at a rate of 2 miles per year. Write the area of the infested circular region as a function of time, t, measured in years.

69. **Population** According to the U.S. Bureau of Statistics, the developing world is growing at a rate of 2% a year. If we assume exponential growth $P_0(1.02)^t$, how long until this population increases by 30%?

70. **Population** According to the U.S. Bureau of Statistics, the population of India will double in 35 years. If we assume exponential growth $P_0(1 + k)^t$, what must be the growth constant k?

71. **Population Growth** According to the U.S. Bureau of the Census, the population of Indonesia was 181 million in 1990 and growing at about 1.9% a year, and the population of the United States was 249 million in 1990 and growing at about 0.9% a year. Find when the populations of these two countries will be the same, assuming that the two growth rates continue at the given values.

72. **Biology** Gowen[101] determined that the equation $y = y_0e^{-kr}$ approximated the number of surviving tobacco viruses on tobacco plants, where r is radiation measured by r that the virus was exposed to and k is a positive constant. What should be the radiation, that is, the amount of r, that will kill 90% of the virus? Your answer will be in terms of k.

73. **Biology** Power and Heuvel[102] developed a mathematical model given by the equation $L(D) = 72.45(1 - 18.85e^{-0.018D})$, where L is the length in millimeters of yellow perch hatched in the Spring in Kimowin Lake in Alberta and D is the day of the year, $D = 1$ corresponding to January 1. The researchers were interested in finding when the perch reached 80% of their final length of 72.45 millimeters. Find the time.

74. **Biology** Wiedenmann and coworkers[103] showed that the proportion y of a certain host larvae parasitized by *Cotesian flavipes* was approximated by $y = 1 - e^{-1.48x}$, where x is the parasite/host ratio. What ratio results in 50% of the host being parasitized?

75. **Soil Carbon Dynamics** Knopps and Tilman[104] studied the soil carbon dynamics of abandoned agricultural fields and created a mathematical model of soil carbon given by the equation $C(t) = 3.754/(1 + 7.981e^{-0.0217t})$, where t is the number of years after a agricultural field has been abandoned and $C(t)$ is the percentage of carbon in the soil. The researchers asked how many years it would take to bring the soil carbon percentage back to 95% of the original. The original carbon in the soil was estimated to be 3.754%. Answer this question.

---

[98] S. Sri Ramaratnam, David A. Bessler, M. Edward Rister, John E. Matocha, and James Novak. 1987. Fertilization under uncertainty: an analysis based on producer yield expectations. *Amer. J. Agr. Econ.* 69:349–357.

[99] Blodwyn M. McIntyre, Martha A. Scholl, and John T. Sigmon. 1990. A quantitative description of a deciduous forest canopy using a photographic technique. *Forest Sci.* 36:381–393.

[100] W. E. Klingman, S. K. Braman, and G. D. Buntin. 2000. Evaluating grower, landscape manager, and consumer perceptions of azalea lace bug feeding injury. *L. Econ. Entomol.* 93(1):141–148.

[101] J. W. Gowen. 1964. Effects of x-rays of different wave lengths on viruses. In: D. Kempthorne, T. A. Bancroft, J. W. Gowen, and J. L. Lush (Eds.), *Statistics and Mathematics in Biology*. New York: Hafner.

[102] M. Power and M. R. van den Heuvel. 1999. Age-0 yellow perch growth and its relationship to temperature. *Trans. Amer. Fish. Soc.* 128:687–700.

[103] Robert Wiedenmann et al. 1992. Laboratory rearing and biology of the parasite *Cotesian flavipes*. *Environ. Entomol.* 21:1160–1167.

[104] Dynamics of soil nitrogen and carbon accumulation for 61 years after agricultural abandonment. 2000. *Ecology* 81(1):88–98.

Chapter 1 Cases

CASE 1 **Portfolio Management**

A fundamental tenet in the modern theory of finance is that an investor must assume greater risks to obtain larger returns. In general, the greater the risk, the larger the return. In the theory of finance the "risk" of an asset is defined to be a measure of the variability of the asset. The basis of this definition rests on the observation that the greater the risk of an asset, the greater will be the daily fluctuations in the value of the asset.

If everyone has a great deal of confidence in what the future value of an investment is, then the value of the investment should vary little. For example, everyone has essentially complete confidence what the future performance of money in the bank will be. Thus, the value of this asset (the principal) does not vary in time. This reflects the fact that this investment is (essentially) risk free. Everyone has very high confidence in the future performance of a high-quality bond. However, there is some variability in price as investors change somewhat from week to week their assessment of the economy, inflation, and other factors that can affect the price of the bond. This reflects the fact that a bond does have some risk.

Moving on to stocks, the future earnings and dividend performance of a very large and regulated company, such as Bell South, is estimated by many analysts with reasonable confidence, and so the price of this security does not vary significantly. However, the future earnings of a small biotech firm are likely to be very unpredictable. Will the company be able to come up with new products? Will it be able to fend off competitors? Will some key personnel leave? Therefore, the weekly assessment of such a firm is likely to change substantially more than that of Bell South. As a consequence, the weekly price changes of this company will vary significantly more, reflecting the fact that an investment in the small company is more risky.

In the figure the letter M refers to the market, defined as, say, all stocks listed on the New York Stock Exchange. The number r_M is the risk of the market portfolio. The height R_T of the point T indicates the riskless rate of return of Treasury bills. Long-term studies indicate that this rate is about 4% per year on average and the return R_M on the Market portfolio is about 9% per year on average. Thus, the point T is placed lower than the point M.

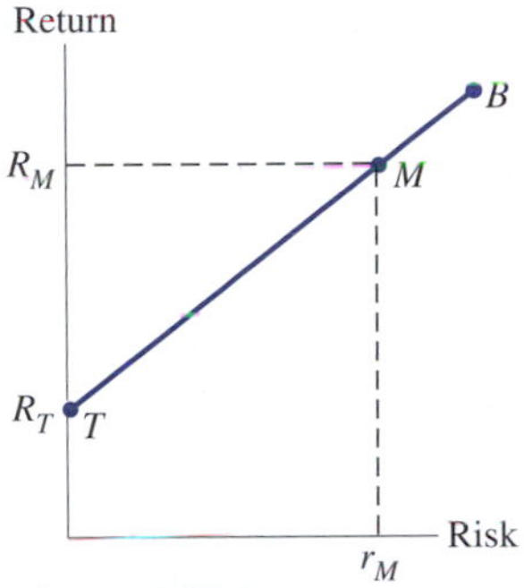

The line *TMB* is called the **capital market line** and plays a critical role in portfolio analysis. A person can construct a portfolio by investing a fraction of his or her funds in the Market portfolio and the remainder in Treasury bills. Such a portfolio is called a *lending portfolio*. Also a person can place all his or her funds in the market portfolio, then using this as collateral, borrow money, and invest the borrowed funds in the market portfolio. Such a portfolio is called a *leverage portfolio*. For convenience we assume that the rate charged on the borrowed money in the leveraged portfolio is R_T.

It turns out that the curve that relates the risk to the return of lending and leveraged portfolios is a straight line through the points T and M. To fully understand this requires a more careful definition of risk and additional mathematics beyond the scope of this text. Interested readers can refer to *The Stock Market: Theories and Evidence* by Lorie and Hamilton[105] an excellent text that presents the results of the academic research into the stock market.

(a) Label the risk-axis as r and the return-axis as R, and find the equation of the line *TMB*.
(b) Find R if $R_M = 0.09$, $R_T = 0.04$, and $r = 2r_M$.
(c) Find R if $R_M = 0.09$, $R_T = 0.04$, and $r = 0.5r_M$.
(d) Find R if $R_M = -0.10$, $R_T = 0.04$, and $r = 2r_M$.
(e) Find R if $R_M = -0.10$, $R_T = 0.04$, and $r = 0.5r_M$.

What conclusion can you make for lending and leveraged portfolios in rising markets? Falling markets?

[105] James H. Lorie and Mary T. Hamilton. *The Stock Market: Theories and Evidence*. 1977. Homewood, Ill.: Richard D. Irwin.

CASE 2 **Diversity**

Buoniorno and coworkers[106] gave the index of tree size diversity

$$H = -[p_1 \ln p_1 + p_2 \ln p_2 + \cdots + p_n \ln p_n]$$

where p_i is the proportion of trees in the ith size class.

(a) Is this index negative? Why or why not.

(b) Explain why this index does or does not equal

$$p_1 \ln \frac{1}{p_1} + p_2 \ln \frac{1}{p_2} + \cdots + p_n \ln \frac{1}{p_n}$$

(c) What is the value of this index if all the p_i's are equal?

(d) Suppose $n = 2$, $p_1 = p$, and $p_2 = 1 - p$. What happens to H if p is close to 0? To 1? For what value of p is the diversity a minimum? Use your graphing calculator to answer these questions. Explain why your answers makes sense in view of the fact that H measures diversity.

[106] Joseph Buongiorno, Sally Dahir, Hsien-Chih Lu, and Ching-Tong Lin. 1994. Tree size diversity and economic returns in uneven-aged forest stands. *Forest Sci.* 40:5–17.

CASE 3 **Island Biogeography**

A fundamental result in island biogeography is that the number of species N in island faunas as a whole is given by the equation $N = aA^b$, where A is the area of the island. One of the first pieces of empirical evidence was given by Preston in 1960. The following figure gives his data in graphical form and was taken from MacArthur and Wilson.[107] The vertical axis gives the number of species of herpetofauna (amphibians plus reptiles). Notice that each axis is marked off in *powers* of 10. The indicated line does not go exactly through each data point but is judged to be the line "closest" to the data points. Clearly, the dots very nearly fall on a line. (See Section 2.1 on least squares to see how this line is found.) From the graph, determine the equation of the indicated line. Use this to determine the parameters a and b in the equation $N = aA^b$. You will need to take logarithms of this last equation.

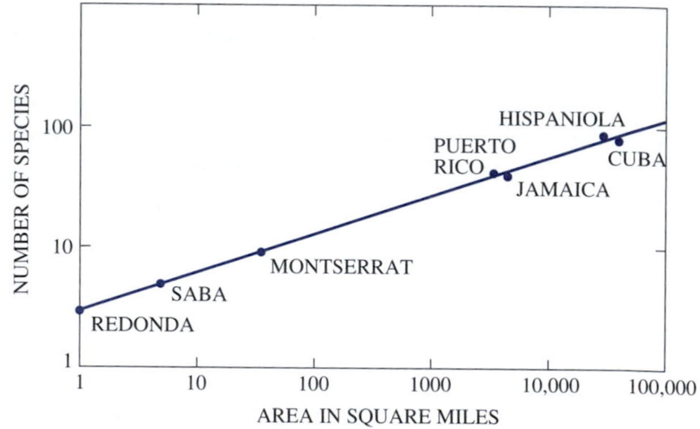

[107] Robert MacArthur and Edward O. Wilson. 1967. *The Theory of Island Biogeograhy.* Princeton, N.J.: Princeton University Press.

CASE 4 **Evaporation of a Pollutant into the Atmosphere**

Part of a pollutant in a body of water can, among other things, be absorbed by bottom sediment or be biodegraded through bacterial action. Other parts, however, may be released into the atmosphere by evaporation. This pollutant evaporation can represent a serious problem, since, as we will see, a very high percentage of certain pollutants pass into the atmosphere. Carried by the wind and air currents, these pollutants can be carried for long distances and dramatically pollute regions far from the source of the original contamination. The artic regions, for example, are suffering from this effect.

One model that is used is given by

$$\left(\frac{P}{SP_W} - 1\right) \ln \frac{M}{M_0} = \ln \frac{x}{x_0}$$

where M is the amount of water and pollutant in the body of water and M_0 the original amount, x the pollutant concentration in the aqueous phases with x_0 the original concentration, P is the vapor pressure of the pure pollutant, P_W is the vapor pressure of water, and S is the solubility of the pollutant in water.

Notice that $\frac{M}{M_0} \times 100$ is the percentage of water left, while $\frac{x}{x_0} \times 100$ is the percentage of pollutant left in the water.

Consider the example of mercury pollution. Mercury solubility at 25° C is 2.7×10^{-9} mole fraction. Mercury vapor pressure at the same temperature is 0.173 Pa. Finally, water vapor pressure at 25° C is 3170 Pa.

Assume that 0.01% of the body of water evaporates. (Note that this implies that $\frac{M}{M_0}$ is $\frac{1}{10^4}$.) Find the concentration change $\frac{x}{x_0}$. We are more interested in $\left(1 - \frac{x}{x_o}\right) \times 100$, the percentage of pollutant concentration in the vapor. Find this percentage. What can you say about this number? Does this surprise you? What does this say about the significance of pollution by evaporation?

CASE 5 **Ground Penetration from an Oil Spill**

Accidental oil spills on land (and water) are unfortunately becoming common. For an oil spill on land, the strategy for cleanup operations will depend on the depth of the oil penetration in a given time. The model presented here gives the time in seconds for the oil spill to penetrate a given number of cm. We have

$$t = \frac{\epsilon\mu}{k\rho g}\left[\frac{L}{1-\epsilon} - \frac{h}{(1-\epsilon)^2}\ln\left(1 + \frac{L(1-\epsilon)}{h}\right)\right]$$

where h is the initial height (thickness) of the oil level in cm, L is the depth of the porous soil in cm, ϵ is the porosity of the ground, k is the permeability of the ground in cm^2, μ is the viscosity of oil in kg/m^3 in Pa, ρ the density of oil in kg/m^3, and $g = 9.81 \times 10^{-4}$.

(a) Find the time if $h = 6$, $L = 0.04$, $\epsilon = 0.7$, $k = 10^{-6}$, $\mu = 0.02$, $\rho = 800$.

(b) Find the time if we have a denser soil and $h = 6$, $L = 0.04$, $\epsilon = 0.1$, $k = 10^{-8}$, $\mu = 0.02$, $\rho = 800$.

CASE 6 **Dynamics of Human Immunodeficiency Virus (HIV)**

We consider here an early model of HIV dynamics that was developed about ten years ago. We first give a brief description of the disease and its progress. A class of white blood cells known as T cells are attacked by the HIV infection. T cells are important for the proper function of the immune system. Following an infection, the immune system produces antibodies that can be detected in the blood. Individuals with such antibodies are considered to be HIV positive.

Over a period that may last for years, the viral concentration remains relatively constant while the T cell level slowly declines. At some point, however, the T cell decline may accelerate and the virus concentration climb again. When the T cell count drops below 200 cells/mm^3 (normal is 1000 cells/mm^3), the patient is classified as having acquired immune deficiency syndrome (AIDS).

According to this model the progress of the viral concentration V depends on the sign and magnitude of the roots of the following quadratic equation

$$x^2 + (e_T + e_V)x + e_T(e_V - NkU)$$

where U is the concentration of unaffected T cells, N is the number of viruses, k is a constant that determines the rate at which cells are infected, e_V is the rate of elimination or clearance of the virus, and e_T is the rate of elimination or clearance of infected T cells.

We assume conditions on all the constants that give rise to two different real roots of the quadratic. Let a and b be the real roots of the quadratic. Then the model indicates that V will be in the form $V = Ae^{at} + Be^{bt}$, where A and B are positive constants.

(a) Determine conditions on e_V, N, k, and U that result in both roots being negative and thus the concentration V of the virus steadily decreases. When you determine this condition, ask yourself if your condition makes sense by varying the constants and determining the consequences. For example, by making k smaller or e_V larger, you expect the virus to decrease, or at least spread more slowly. Is this true for your condition?

(b) Determine conditions on e_V, N, k, and U that result in one positive root and one negative root. What happens to V in this case? When you determine this condition, ask yourself if your condition makes sense by varying the constants and determining the consequences. For example, by making k larger or e_V smaller, you expect the concentration of the virus to eventually increase. Is this true for your condition? Comment: Notice from part (a) that AIDS is not an inevitable consequence of exposure to HIV.

2 Modeling with Least Squares

This entire chapter is exclusively about modeling using regression analysis based on least squares. All of Chapter 2 can be skipped without any loss of continuity in the subsequent development. The entire chapter is to be considered "enrichment."

CASE STUDY Economies of Scale

In 1996, Rogers and Akridge[1] carefully studied the fertilizer and pesticide industry in the state of Indiana and collected data for a medium-sized fertilizer plant relating the amount of fertilizer produced to the cost per ton. The data are found in the following table.

x	4.2	4.8	5.4	6.0	6.6	7.2	7.8
y	298	287	278	270	263	259	255

Quantity x is in thousands of tons of fertilizer produced; y is the (total) cost per ton in dollars.

What function do you think best models the data in the table? For the answer see Example 1 in Section 6 on page 131.

[1] Duane S. Rogers and Jay T. Akridge. 1996. Economic impact of storage and handling regulations on retail fertilizer and pesticide plants. Agribusiness 12(4):327–337.

2.1 Method of Least Squares

The Method of Least Squares
Correlation
Additional Examples

APPLICATION A Linear Model

In 1995 Cohen[2] published a study correlating corporate spending on communications and computers (as a percent of all spending on equipment) with annual

[2] Robert B. Cohen. 1995. The economic impact of information technology. *Bus. Econ.* Vol. XXX(4):21–25.

productivity growth. He collected data on 11 companies for the period from 1985 to 1989. This data is found in the following table.

x	0.06	0.11	0.16	0.20	0.22	0.25	0.33	0.33	0.47	0.62	0.78
y	−1.0	4.5	−0.6	4.2	0.4	0.1	0.4	1.4	1.1	3.4	5.5

Here x is the spending on communications and computers as a percent of all spending on equipment, and y is the annual productivity growth.

What is the equation of a line that best approximates this data? What can you conclude about the relationship between spending on communications and computers and annual productivity growth? See Example 3 on page 104 for the answer.

Let x be the number of items produced and sold, and let p be the price of each item. Suppose x_1 items were sold at a price of p_1 and x_2 items were sold at a price of p_2. If we then *assume* that the demand equation is linear, then of course there is only *one* straight line through these two points, and we can easily calculate the equation $y = ax + b$ of this line.

Table 2.1

x_i	1	2	3	5	9
p_i	10	9	8	7	5

But suppose that we have more than two data points. Suppose, as in Table 2.1, that we have five points available. Here, p_i are the prices in dollars for a certain product, and x_i is the corresponding demands for the product in the number of thousands of items sold per day.

David Eugene Smith Collection/Rare Book and Manuscript Library/Columbia University

Carl Gauss, the Missing Planet, And the Method of Least Squares

In 1781 Sir William Herschel discovered the planet Uranus, bringing the number of planets known at that time to seven. At that time astronomers were certain that an eighth planet existed between Mars and Jupiter, but the search for the missing planet had proved fruitless. On January 1, 1801, Giuseppe Piazzi, director of the Palermo Observatory in Italy, announced the discovery of a new planet Ceres in that location. (It was subsequently realized that Ceres was actually a large asteroid.) Unfortunately, he lost sight of it a few weeks later. Carl Gauss **(1777–1855)**, one of the greatest mathematicians of all times, took up the challenge of determining its orbit from the few recorded observations. Gauss had in his possession one remarkable mathematical tool—the method of least squares—which he formulated in 1794 (at the age of 16!) but had not bothered to publish. Using this method, he predicted the orbit. Astronomers around the world were astonished to find Ceres exactly where Gauss said it would be. This incident catapulted him to fame.

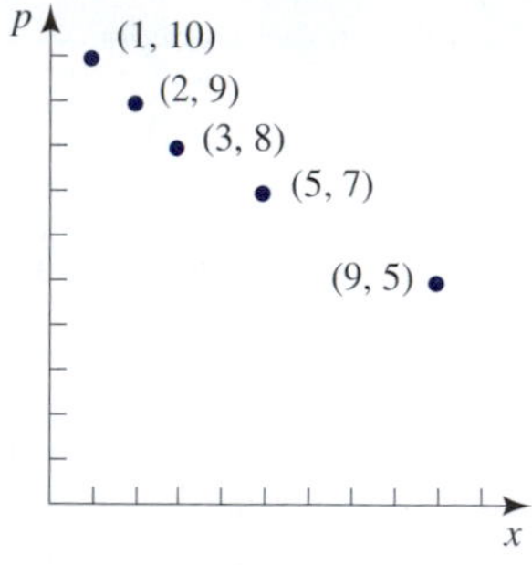

Figure 2.1
Discrete points on the demand curve.

These are plotted in Figure 2.1, which is called a **scatter diagram**.

If we examine the scatter diagram, we see clearly that the points do not lie on any straight line but seem to be scattered in a more or less linear fashion. Under such circumstances we might be justified in assuming that the demand equation was more or less a straight line. But what straight line? Any line that we draw will miss most of the points. We might then think to draw a line that is somehow closest to the data points. To actually follow such a procedure, we need to state exactly how we are to measure this closeness. In this section we will measure this closeness in a manner that will lead us to the **method of least squares**.

First notice that to be given a nonvertical straight line is the same as to be given two numbers a and b with the equation of the straight line given as $y = ax + b$. Suppose now that we are given n data points $(x_1, y_1), (x_2, y_2), \ldots, (x_n, y_n)$ and a line $y = ax + b$. We then define $d_1 = y_1 - (ax_1 + b)$, and note from the figure that $|d_1|$ is just the vertical distance from the first data point (x_1, y_1) to the line $y = ax + b$. Doing the same for all the data points, we then have

$$\begin{aligned} d_1 &= y_1 - (ax_1 + b) \\ d_2 &= y_2 - (ax_2 + b) \\ &\vdots \\ d_n &= y_n - (ax_n + b) \end{aligned}$$

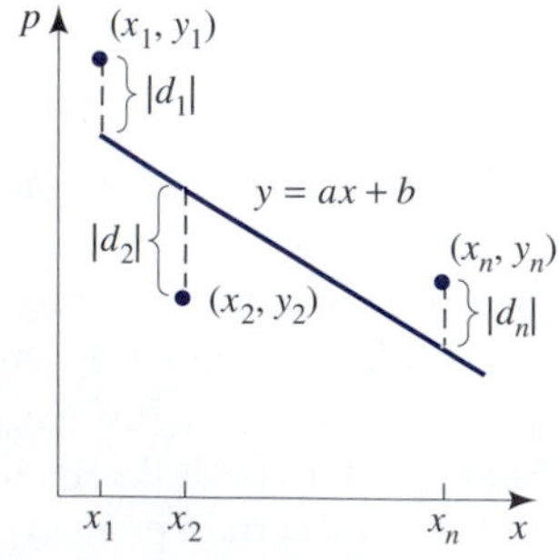

Figure 2.2
We measure the distance from each point to the line.

where $|d_2|$ is the vertical distance from the second data point (x_2, y_2) to the line $y = ax + b$, and so on. Refer to Figure 2.2.

Now if all the data points were on the line $y = ax + b$, then all the distances $|d_1|$, $|d_2|, \ldots, |d_n|$ would be zero. Unfortunately, this will rarely be the case. We then use the sum of the squares of these distances

$$d = d_1^2 + d_2^2 + \cdots + d_n^2$$

as a measure of how close the set of data points is to the line $y = ax + b$. Notice that this number d will be different for different straight lines: large if the straight line is far removed from the data points and small if the straight line passes close to all the data points. We then seek the line—or, what is the same thing, the two numbers a and b—that will make this sum of squares the least. Thus the name *least squares*.

Interactive Illustration 2.1

Finding the Best-Fitting Line

This interactive illustration allows you to change the line $y = ax + b$ and see the resulting sum of squares for the data in Table 2.1. There are two sliders: one for the variable a and one for b. Click or drag the mouse on the slider to change the values of a and b. You will see a vertical line connecting each data point with the line $y = ax + b$ and a meter that gives the sum of the squares of the lengths of all of these vertical lines as the height of the bar. Adjust the variables a and b to minimize the sum of the squares (the height of the meter).

You can check your result by clicking on the Best Fit box, which will graph the best-fitting line and give the minimum value of the sum of the squares.

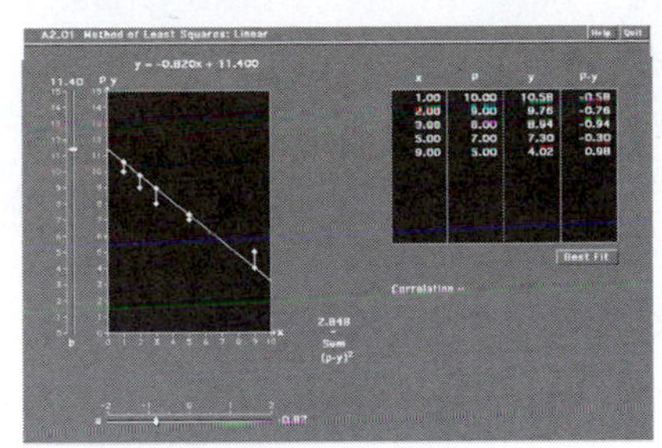

EXAMPLE 1 **Using the Method of Least Squares To Find a Demand Function**

(a) Find the best-fitting line through the data points in Table 2.1 and thus find a linear demand function.
(b) Estimate the price if the demand is 6000.

Solution

(a) Find the best-fitting line using the linear regression operation on your computer or graphing calculator. (See the Technology Resource Manual for complete details.) You will find that $a = -0.6$ and $b = 10.2$.
Thus, the equation of the best-fitting straight line that we are seeking is

$$y = p(x) = -0.6x + 10.2$$

The graph is shown Screen 2.1.

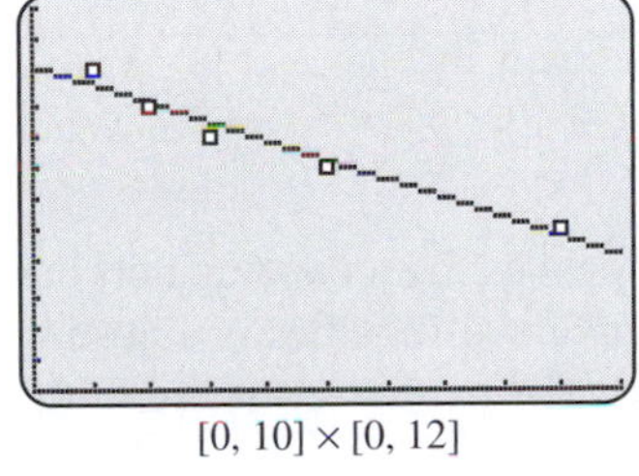

$[0, 10] \times [0, 12]$

Screen 2.1
The graph of the best-fitting straight line in Example 1.

(b) The answer to the second question is

$$p(6) = (-0.6)(6) + 10.2 = 6.6$$

That is, if 6000 items are to be sold, then the price should be \$6.60. ■

Correlation

We have just seen how to determine a functional relationship between two variables. This is called **regression analysis**. We now wish to determine the strength or degree of association between two variables. This is referred to as **correlation analysis**. The strength of association is measured by the **correlation coefficient**, which is defined as

$$r = \frac{n\sum x_i y_i - \sum x_i \sum y_i}{\sqrt{\left[n\sum x_i^2 - \left(\sum x_i\right)^2\right]\left[n\sum y_i^2 - \left(\sum y_i\right)^2\right]}}$$

where $\sum y_i$, for example, is $y_1 + y_2 + \cdots + y_n$. The value of this correlation coefficient ranges from $+1$ for two variables with perfect positive correlation to -1 for two variables with perfect negative correlation. See Figure 2.3 for examples.

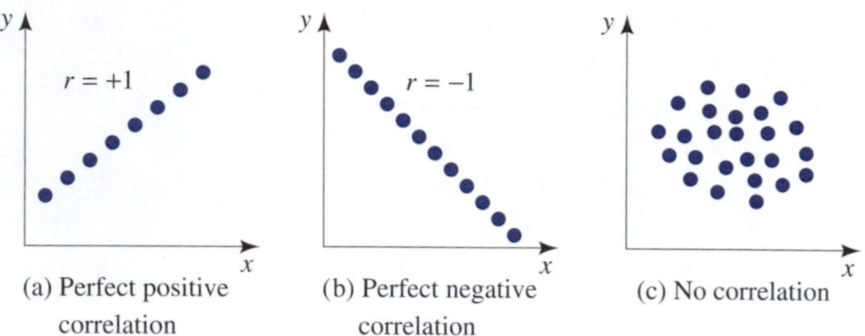

(a) Perfect positive correlation (b) Perfect negative correlation (c) No correlation

Figure 2.3

EXAMPLE 2 **Correlation**

Find the correlation coefficient for the data in Table 2.1.

Solution

We can use the above formula or let our computers or graphing calculators do the work. In doing Example 1, you will see that $r \approx -0.99$. This indicates a high negative correlation, which can easily be seen by observing the original data in the screen on the previous page. We conclude that there is a strong negative correlation between price and demand in this instance. Thus, we expect increases in prices to lead to decreases in demand. ■

Additional Examples

There has been a debate recently as to whether corporate investment in computers has a positive effect on productivity. Some suggest that workers' difficulties in adjusting to using computers might actually hinder productivity. The paper mentioned in the following example attacks this question.

EXAMPLE 3 **Least Squares and Correlation**

In 1995 Cohen[3] published a study correlating corporate spending on communications and computers (as a percent of all spending on equipment) with annual productivity growth. He collected data on 11 companies for the period from 1985 to 1989. This data is found in the following table.

x	0.06	0.11	0.16	0.20	0.22	0.25	0.33	0.33	0.47	0.62	0.78
y	−1.0	4.5	−0.6	4.2	0.4	0.1	0.4	1.4	1.1	3.4	5.5

Here x is the spending on communications and computers as a percent of all spending on equipment, and y is the annual productivity growth.

---

[3] Ibid.

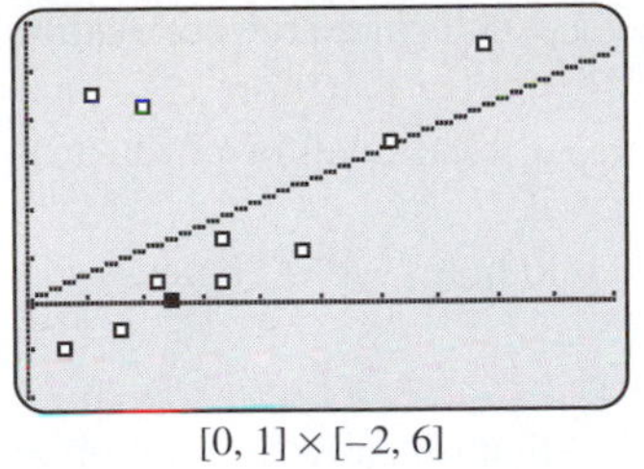
[0, 1] × [−2, 6]

Screen 2.2
The graph of the best-fitting straight line in Example 3.

Determine the best-fitting line using least squares. Also determine the correlation coefficient.

Solution

Using our computers or graphing calculators, we find that $a \approx 5.246$, $b \approx 0.080$, and $r \approx 0.518$. Thus, the best-fitting straight line is

$$y = ax + b = 5.246x + 0.080$$

The correlation coefficient is $r = 0.518$. This is somewhat significant, and we conclude, on the basis of this study, that investment in communications and computers increases the productivity of corporations. See Screen 2.2. ■

Interactive Illustration 2.2

Finding the Best-Fitting Line in Example 3

This interactive illustration allows you to change the line $y = ax + b$ and see the resulting sum of squares for the data in Example 3. There are two sliders: one for the variable a and one for b. Click or drag the mouse on the slider to change the values of a and b. You will see a vertical line connecting each data point with the line $y = ax + b$ and a meter that gives the sum of the squares of the lengths of all of these vertical lines as the height of the bar. Adjust the variables a and b to minimize the sum of the squares (the height of the meter).

You can check your result by clicking on the Best Fit box, which will graph the best-fitting line and give the minimum value of the sum of the squares.

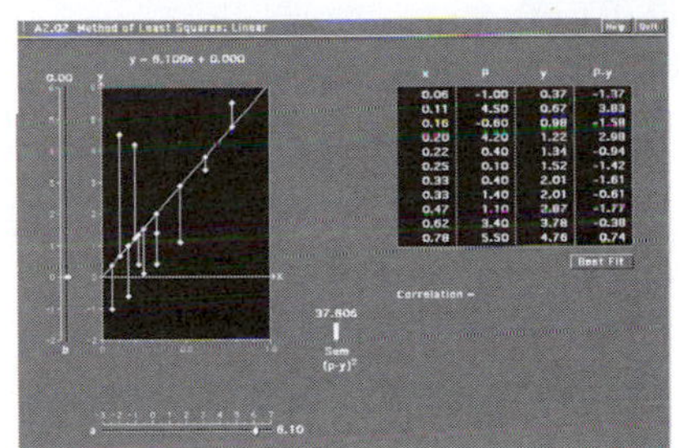

EXAMPLE 4 **Least Squares and Correlation**

In 1997 Fuller and coworkers[4] at Texas A & M University estimated the operating costs of cotton gin plants of various sizes. The operating costs of the smallest plant is shown in the following table.

x	2	4	6	8
y	109.40	187.24	273.36	367.60

Here x is annual number of thousands of bales produced, and y is the total cost in thousands of dollars.

[4] S. Fuller, M. Gillis, C. Parnell, A. Ramaiyer, and R. Childers. 1997. Effect of air pollution abatement on financial performance of Texas cotton gins. *Agribusiness* 13(5):521–532.

(a) Determine the best-fitting line using least squares. Also determine the correlation coefficient.
(b) The study noted that revenue was \$55.45 per bale. At what level of production will this plant break even?
(c) What are the profits or losses when production is 1000 bales? 2000 bales?

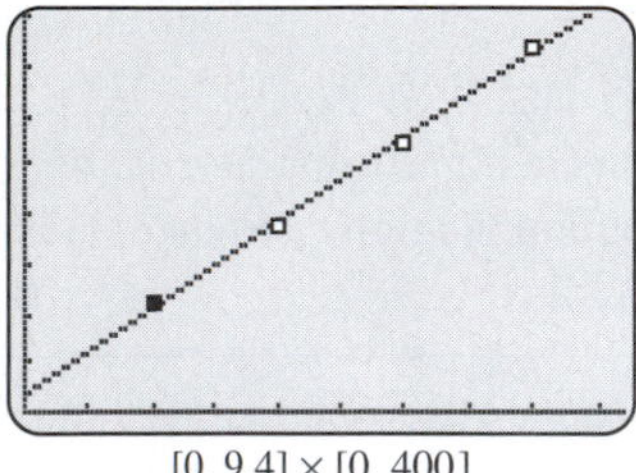

$[0, 9.4] \times [0, 400]$

Screen 2.3
The graph of the best-fitting straight line in Example 4.

Solution

(a) Using our computers or graphing calculators, we find that linear regression gives the cost equation as $y = C(x) = 43.036x + 19.22$ and $r = 0.9991$ (Screen 2.3). This is certainly significant.
(b) We set revenue equal to cost and obtain

$$\begin{aligned} R &= C \\ 55.45x &= 43.036x + 19.22 \\ 12.414x &= 19.22 \\ x &= 1.548 \end{aligned}$$

Thus, this plant will break even when production is set at approximately 1548 bales.
(c) Profit is revenue less cost. Doing this, we obtain

$$\begin{aligned} P &= R - C \\ &= 55.45x - (43.036x + 19.22) \\ &= 12.414x - 19.22 \end{aligned}$$

Then $P(1) = -6.806$ and $P(2) = 5.608$. Thus, there is a loss of \$6806 when production is at 1000 bales and a profit of \$5608 when production is set at 2000 bales. ■

Alternatively, parts (b) and (c) can be done on your calculator or computer. Check the Technology Resource Manual.

Interactive Illustration 2.3

Finding the Best-Fitting Line in Example 4

This interactive illustration allows you to change the line $y = ax + b$ and see the resulting sum of squares for the data in Example 4. There are two sliders: one for the variable a and one for b. Click or drag the mouse on the slider to change the values of a and b. You will see a vertical line connecting each data point with the line $y = ax + b$ and a meter that gives the sum of the squares of the lengths of all of these vertical lines as the height of the bar. Adjust the variables a and b to minimize the sum of the squares (the height of the meter).

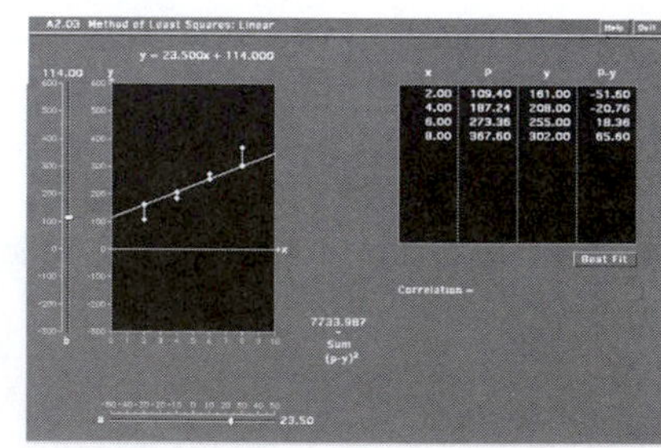

You can check your result by clicking on the Best Fit box, which will graph the best-fitting line and give the minimum value of the sum of the squares.

Warm Up Exercise Set 2.1

1. Using the method of least squares, find the best-fitting line through the three points (0, 0), (2, 2), (3, 2). Find the correlation coefficient.

Exercise Set 2.1

In Exercises 1 through 8, find the best-fitting straight line to the given set of data, using the method of least squares. Graph this straight line on a scatter diagram. Find the correlation coefficient.

1. (0, 0), (1, 2), (2, 1)
2. (0, 1), (1, 2), (2, 2)
3. (0, 0), (1, 1), (2, 3), (3, 3)
4. (0, 0), (1, 2), (2, 2), (3, 0)
5. (1, 4), (2, 2), (3, 2), (4, 1)
6. (0, 0), (1, 1), (2, 2), (3, 4)
7. (0, 4), (1, 2), (2, 2), (3, 1), (4, 1)
8. (0, 0), (1, 2), (2, 1), (3, 2), (4, 4)

Modeling Using Least Squares

9. **Cost, Revenue, and Profit** In 1997 Fuller and coworkers[5] at Texas A & M University estimated the operating costs of cotton gin plants of various sizes. The operating costs of the next to the smallest plant is shown in the following table.

x	2000	4000	6000
y	163,200	230,480	301,500
x	8000	10,000	12,000
y	376,160	454,400	536,400

Here x is annual number of bales produced, and y is the dollar total cost.

   a. Determine the best-fitting line using least squares. Also determine the correlation coefficient.
   b. The study noted that revenue was $63.25 per bale. At what level of production will this plant break even?
   c. What are the profits or losses when production is 3000 bales? 4000 bales?

10. **Economies of Scale in Advertising** *Strategic Management*[6] relates a study in economies of scale in the automobile tire industry. The data is found in the following table. The companies are Firestone, General, Goodrich, Goodyear, and Uniroyal.

x	14	1	4.9	17	4.3
y	4.2	5.8	4.7	3.7	4.8

Here x is the average annual replacement market volume for 1973–1977 in millions of units, and y is the cumulative dollars spent per tire.

   a. Determine the best-fitting line using least squares and the correlation coefficient.
   b. Is there an advantage to being a large company? Explain.
   c. What does this model predict the cumulative dollars spent per tire will be when the average replacement market is 10 million units?
   d. What does this model predict the average replacement market will be if the cumulative dollars spent per tire is 5.0?

11. **Economies of Scale in Advertising** *Strategic Management*[7] relates a study in economies of scale in the paper industry. The data is found in the following table.

x	50	60	73	82	90	215	230	260
y	1.8	1.6	1.5	1.25	0.95	0.60	0.50	0.54

Here x is the weighted average machine size in tons per hour, and y is the employee-hours per ton.

[5] Ibid.

[6] Alan J. Rowe, Richard O. Mason, Karl E. Dickel, Richard B. Mann, and Robert J. Mockler. 1994. *Strategic Management.* New York: Addison-Wesley.

[7] Ibid.

a. Determine the best-fitting line using least squares and the correlation coefficient.
b. Is there an advantage in using a large machine? Explain.
c. What does this model predict the employee-hours per ton will be when the weighted average machine size in tons per hour is 10 million units?
d. What does this model predict the weighted average machine size will be if the employee-hours per ton is 1.00?

12. **Economies of Scale in Plant Size** *Strategic Management*[8] relates a study in economies of scale in the machine tool industry. The data is found in the following table.

x	70	115	130	190	195	400	450
y	1.1	1.0	0.85	0.75	0.85	0.67	0.50

Here x is the plant capacity in thousands of units, and y is the employee-hours per unit.

a. Determine the best-fitting line using least squares and the correlation coefficient.
b. Is there an advantage in having a large plant? Explain.
c. What does this model predict the employee-hours per unit will be when the plant capacity is 300,000 units?
d. What does this model predict the plant capacity will be if the employee-hours per unit is 0.90?

13. **Demand Curve for Beef** *Managerial Economics*[9] gives a demand curve for beef. The data is given in the following table.

Year	1947	1948	1949	1950	1951
y	69.6	63.1	63.9	63.4	56.1
x	0.0358	0.0390	0.0335	0.0345	0.0385
Year	1952	1953	1954	1955	1956
y	62.2	77.6	80.1	82.0	85.4
x	0.0356	0.0261	0.0261	0.0243	0.0224
Year	1957	1958	1959	1960	
y	84.6	80.5	81.6	85.8	
x	0.0226	0.0253	0.0246	0.0228	

Here y is the price of beef divided by disposable income per capita, and x is beef consumption per capita in pounds.

Determine the best-fitting line using least squares and the correlation coefficient. Graph the line. Does it slope downward? What does this say about the price of beef and the demand for beef?

14. **Demand for Margarine** We would expect the demand for margarine to increase as the price of butter increases. In *Managerial Economics*[10] we find a study relating the price of butter to the demand for margarine. The data is given in the following table.

Year	1940	1941	1942	1943	1944	1945
x	29.5	34.3	40.1	44.8	42.3	42.8
y	2.4	2.8	2.8	3.9	3.9	4.1
Year	1946	1947	1948	1949	1950	1951
x	62.8	71.3	75.8	61.5	62.2	69.9
y	3.9	5.0	6.1	5.8	6.1	6.6
Year	1952	1953	1954	1955	1956	1957
x	73.0	66.6	60.5	58.2	59.9	60.7
y	7.9	8.1	8.5	8.2	8.2	8.6
Year	1958	1959	1960			
x	59.7	60.6	59.6			
y	9.0	9.2	9.4			

Here x is the price of butter in cents per pound, and y is margarine consumption in pounds per person.

Determine the best-fitting line using least squares and the correlation coefficient. Graph the line. Does it slope upward? Does this indicate that the demand for margarine increases as the price of butter increases?

15. **Cost Curve** Dean[11] made a statistical estimation of the cost-output relationship in a hosiery mill. The data is given in the following table.

x	16	31	48	57	63	103	110
y	30	60	100	130	135	230	230
x	114	116	117	118	123	126	
y	235	245	250	235	250	260	

Here x is the output in thousands of dozens, and y is the total cost in thousands of dollars.

---

[8] Ibid.

[9] Milton H. Spencer. 1968. *Managerial Economics.* Homewood, Ill.: Richard D. Irwin.

[10] Ibid., pages 143–144.

[11] Joel Dean. 1976. *Statistical Cost Estimation.* Bloomington, Ind.: Indiana University Press, page 215.

a. Determine the best-fitting line using least squares and the correlation coefficient. Graph the line.
b. What does this model predict the total cost will be when the output is 100,000 dozen?
c. What does this model predict the output will be if the total cost is $125,000?

16. Cost Curve Johnston[12] made a statistical estimation of the cost-output relationship for 40 firms. The data for the fifth firm is given in the following table.

x	180	210	215	230	260
y	130	180	205	190	215
x	290	340	400	405	430
y	220	250	300	285	305
x	430	450	470	490	510
y	325	330	330	340	375

Here x is the output in millions of units, and y is the total cost in thousands of pounds sterling.

a. Determine the best-fitting line using least squares and the correlation coefficient. Graph the line.
b. What does this model predict the total cost will be when the output is 300 million units?
c. What does this model predict the output will be if the total cost is 200,000 of pounds sterling?

17. Productivity Bernstein[13] studied the correlation between productivity growth and gross national product (GNP) growth of six countries: France (F), Germany (G), Italy (I), Japan (J), the United Kingdom (UK), and the United States (US). Productivity is given as output per employee-hour in manufacturing. The data they collected for the years 1950–1977 is given in the following table.

Country	US	UK	F	I	G	J
x	2.5	2.7	5.2	5.6	5.7	9.0
y	3.5	2.3	4.9	4.9	5.7	8.5

Here x is the productivity growth (%), and y is the GNP growth (%).

a. Determine the best-fitting line using least squares and the correlation coefficient.
b. What does this model predict the GNP growth will be when the productivity growth is 7%?
c. What does this model predict the productivity growth will be if the GNP growth is 7%?

18. Productivity Bernstein[14] also studied the correlation between investment as a percent of GNP and productivity growth of six countries: France (F), Germany (G), Italy (I), Japan (J), the United Kingdom (UK), and the United States (US). Productivity is given as output per employee-hour in manufacturing. The data they collected for the years 1960–1977 is given in the following table.

Country	US	UK	I	F	G	J
x	17	18	22	23	24	34
y	2.8	3.0	5.6	5.5	5.8	8.3

Here x is investment as percent of GNP, and y is the productivity growth (%).

a. Determine the best-fitting line using least squares and the correlation coefficient.
b. What does this model predict the productivity growth will be when investment is 20% of GNP?
c. What does this model predict investment as a percent of GNP will be if productivity growth is 7%?

19. Productivity Petralia[15] studied the correlation between investment as a percent of GNP and productivity growth of nine countries: Belgium (B), Canada (C), France (F), Germany (G), Italy (I), Japan (J), the Netherlands (N), the United Kingdom (UK), and the United States (US). Productivity is given as output per empoyee-hour in manufacturing. The data they collected for the years 1960–1976 is given in the following table.

Country	US	UK	B	I	C
x	14	17	20.5	21	21.5
y	2.2	3.3	6.4	5.8	3.8
Country	F	N	G	J	
x	22	24	25	32	
y	5.7	6.3	5.8	9.0	

Here x is investment as percent of GNP, and y is the productivity growth (%).

[12] J. Johnston. 1960. *Statistical Cost Analysis*. New York: McGraw-Hill, page 58.

[13] Peter L. Berstein. 1980. Productivity and growth: another approach. *Bus. Econ.* XV(1):68–71.

[14] Ibid.

[15] John W. Petralia. 1981. Effective use of econometrics in corporate planning. *Bus. Econ.* XVI(3):59–61.

a. Determine the best-fitting line using least squares and the correlation coefficient.
b. What does this model predict the productivity growth will be when investment is 30% of GNP?
c. What does this model predict investment as a percent of GNP will be if productivity growth is 7%?

20. **Productivity** Recall from Example 3 that Cohen[16] studied the correlation between corporate spending on communications and computers (as a percent of all spending on equipment) and annual productivity growth. In Example 3 we looked at his data on 11 companies for the period from 1985 to 1989. The data found in the following table is for the years 1977–1984.

x	0.03	0.07	0.10	0.13	0.14	0.17
y	−2.0	−1.5	1.7	−0.6	2.2	0.3
x	0.24	0.29	0.39	0.62	0.83	
y	1.3	4.2	3.4	4.0	−0.5	

Here x is the spending on communications and computers as a percent of all spending on equipment, and y is the annual productivity growth.

Determine the best-fitting line using least squares and the correlation coefficient.

21. **Environmental Entomology** The fall armyworm has historically been a severe agricultural pest. Rogers and Marti[17] studied how the age at first mating of the female of this pest affected its fecundity as measured by the number of viable larvae resulting from the eggs laid. They collected the data shown in following table.

x	1	1	3	4	4	6	6
y	1650	1450	550	1150	650	850	800
x	8	10	10	12	13	15	
y	450	900	500	100	100	200	

Here x is the age of first mating of the female in days, and y is the total number of viable larvae per female.

Determine the best-fitting line using least squares. Also determine the correlation coefficient.

22. **Environmental Entomology** The beet armyworm has become one of the most serious pests of the tomato in southern California. Eigenbrode and Trumble[18] collected the data shown in the table that relates the nine-day survival of larva of this insect to the percentage of fruit damaged.

x	39.2	29.6	95.4	113.7	179.3	42.4	60.0
y	0.5	0.1	4.9	4.0	10.0	5.3	1.8

Here x is the nine-day weight of the larvae in milligrams, and y is the percentage of fruit damaged.

a. Use linear regression to find the best-fitting line that relates the nine-day weight of the larvae to the percentage of fruit damaged.
b. Find the correlation coefficient.
c. Interpret what the slope of the line means.

23. **Horticultural Entomology** The brown citrus aphid and the melon aphid both spread the citrus tristeza virus to fruit and thus have become important pests. Yokomi and coworkers[19] collected the data found in the following table.

x	1	5	10	20
y	25	22	50	85
z	10	5	18	45

Here x is the number of aphids per plant, and y and z are the percentage of times the virus is transmitted to the fruit for the brown and melon aphid, respectively.

a. Use linear regression for each aphid to find the best-fitting line that relates the number of aphids per plant to the percentage of times the virus is transmitted to the fruit.
b. Find the correlation coefficients.
c. Interpret what the slope of the line means in each case.
d. Which aphid is more destructive? Why?

[16] Robert B. Cohen. 1995. The economic impact of information technology. *Bus. Econ.* XXX(4):21–25.

[17] C. E. Rogers and O. G. Marti, Jr. 1994. Effects of age at first mating on the reproductive potential of the fall armyworm. *Environ. Entomolog.* 23(2):322–325.

[18] Sanford D. Eigenbrode and John T. Trumble. 1994. Fruit-based tolerance to damage by beet armyworm in tomato. *Environ. Entomol.* 23(4):937–942.

[19] R. K. Yokomi, R. Lastra, M. B. Stoetzel, V. D. Damsteegt, R. F. Lee, S. M. Garnsey, T. R. Gottwald, M. A. Rocha-Pena, and C. L. Niblett. 1994. Establishment of the brown citrus aphid in Central America and the Caribbean basin and transmission of citrus tristeza virus. *J. Econ. Entomol.* 87(4):1078–1085.

24. Plant Resistance Talekar and Lin[20] collected the data shown in the table that relates the pod diameter (seed size) of soybeans to the percentage of pods damaged by the lima bean pod borer.

x	3.1	3.4	3.9	4.1	4.2	4.5
y	12	28	38	37	44	48

Here x is the pod diameter in millimeters, and y is the percentage of damaged pods.

a. Use linear regression to find the best-fitting line that relates the pod diameter to the percentage of pods damaged.
b. Find the correlation coefficient.
c. Interpret what the slope of the line means.

25. Plant Resistance Talekar and Lin[21] collected the data shown in the table that relates the number of trichomes (hairs) per 6.25 mm^2 on pods of soybeans to the percentage of pods damaged by the lima bean pod borer.

x	208	212	230	255	260	335
y	9	9	10	8	14	25

Here x is the number of trichomes per 6.25 mm^2 on the pods, and y is the percentage of damaged pods.

a. Use linear regression to find the best-fitting line that relates the number of trichomes on the pods to the percentage of pods damaged.
b. Find the correlation coefficient.
c. Interpret what the slope of the line means.

26. Biological Control Briano and colleagues[22] studied the host-pathogen relationship between the black imported fire ant and a microsporidium (*T. solenopsae*) that infects them. This ant represents a serious medical and agricultural pest. The study was to determine whether *T. solenopae* could be used as a biological control of the imported fire ants. The table includes data that they collected and relates the number of colonies per hectare of the ants with the percentage that are infected with *T. solenopsae*.

x	23	30	32	50	72	74	79
y	27	35	50	34	14	25	15
x	81	98	110	116	122	132	138
y	33	23	26	18	28	19	23
x	140	150	150	152	162		
y	22	18	24	21	22		

Here x is the colonies of ants per hectare, and y is the percentage of infected colonies.

a. Use linear regression to find the best-fitting line that relates the number of colonies of ants per hectare to the percentage of infected colonies.
b. Find the correlation coefficient.
c. Interpret what the slope of the line means.

[20] Narayan S. Talekar and Chih Pin Lin. 1994. Characterization of resistance to lima bean pod borer in soybean. *J. Econ. Entomol.* 87(3):821–825.

[21] Ibid.

[22] J. A. Briano, R. S. Patterson, and H. A. Cordo. 1995. Long-term studies of the black imported fire ant infected with a microsporidium. *Environ. Entomol.* 24(5):1328–1332.

Solutions to WARM UP EXERCISE SET 2.1

1. Input the data into your computer or graphing calculator. Using the linear regression operation, we find that $a \approx 0.714$, $b \approx 0.143$, and $r \approx 0.9449$. Thus, the best-fitting straight line is

$$y = ax + b = 0.714x + 0.143$$

The correlation coefficient is $r = 0.9449$. This is significant. See the screen.

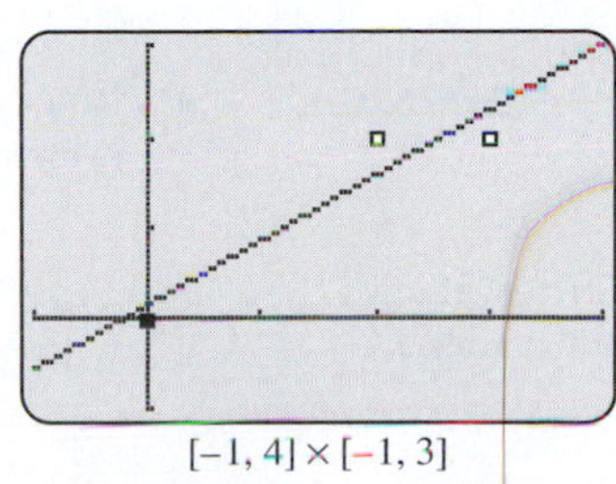

$[-1, 4] \times [-1, 3]$

2.2 Quadratic Regression

We have already found the line $y = ax + b$ closest to a set of discrete data points using least squares. However, instead of being more or less dispersed along a straight line, some data may be more or less dispersed along the graph of some quadratic $y = ax^2 + bx + c$. In such a case we wish to find the numbers a, b, and c, that is, the quadratic $y = ax^2 + bx + c$, that are closest to the data in the sense of least squares. We will find this quadratic on our computer or graphing calculators. The two interactive illustrations given in this section allow you to change the parameters in the quadratic to experimentally minimize the sum of the squares and find the approximate best-fitting quadratic. This is similar to what was done in the preceding section for finding the best-fitting line.

In Section 2.1 we defined the correlation coefficient r. The number r^2 is called the **coefficient of determination**. We will now only use r^2, since r^2 is given for *every* type of regression, whereas r is given only for some. Recall that r could be negative or positive for linear regression. This just indicated whether the best-fitting line had positive or negative slope. But every quadratic function $ax^2 + bx + c$, with $c \neq 0$, increases on one interval and decreases on another. So it makes no sense to have a *negative* r for quadratic regression. To avoid any confusion, we will now always use r^2.

EXAMPLE 1 **Quadratic Regression**

Dean[23] made a statistical estimation of the cost-output relationship for a shoe chain. The data for the firm is given in the following table.

x	4.5	9	15	20	20	35	50
y	5	7.2	8	10.2	12	18	30

Here x is the output in thousands of pairs of shoes, and y is the total cost in thousands of dollars.

(a) Determine the best-fitting line using least squares. Also determine the square of the correlation coefficient. Graph.
(b) Determine the best-fitting quadratic using least squares. Also determine the square of the correlation coefficient. Graph.
(c) Which curve is better? Why?

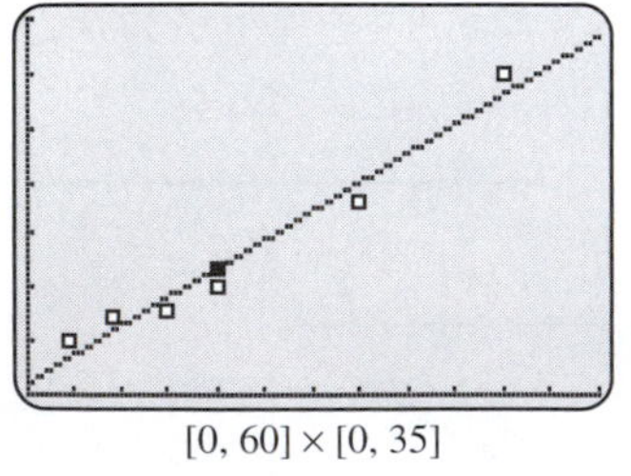

$[0, 60] \times [0, 35]$

Screen 2.4
The graph of the best-fitting straight line for Example 1.

Solution

(a) Enter the data into your computer or graphing calculator and first select linear regression. We obtain

$$y = ax + b = 0.5379x + 1.1180$$

with $r^2 = 0.9649$. There is a strong linear relationship. See Screen 2.4.

---

[23] Joel Dean. 1976. *Statistical Cost Estimation.* Bloomington, Ind.: Indiana University Press, page 337.

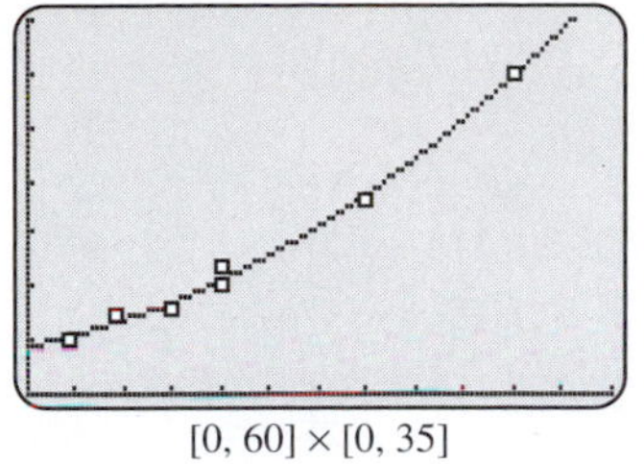

$[0, 60] \times [0, 35]$

Screen 2.5
The graph of the best-fitting quadratic for Example 1.

(b) If we look carefully at Screen 2.4, we notice that the cost curve bends slightly upward. This suggests that a quadratic model might be even better. Now select quadratic regression. We obtain

$$\begin{aligned} y &= ax^2 + bx + c \\ &= 0.00666x^2 + 0.1714x + 4.537 \end{aligned}$$

with $r^2 = 0.9921$. The graph in shown in Screen 2.5.

(c) The square of the correlation coefficient for the quadratic model is distinctly larger than that for the linear. It does appear from the graphs that the quadratic is a better fit. ■

Interactive Illustration 2.4

Finding the Best-Fitting Quadratic in Example 1

This interactive illustration allows you to change both the linear $y = mx + b$ and the quadratic $y = ax^2 + bx + c$ and see the resulting sum of squares. There are sliders for both functions. Click or drag the mouse on a slider to change the value of a parameter. You will see a box that contains the sum of the squares. Adjust the value of the parameters to minimize the sum of the squares. Clicking on the Best Fit box will graph the best-fitting line or quadratic and give the minimum value of the sum of the squares.

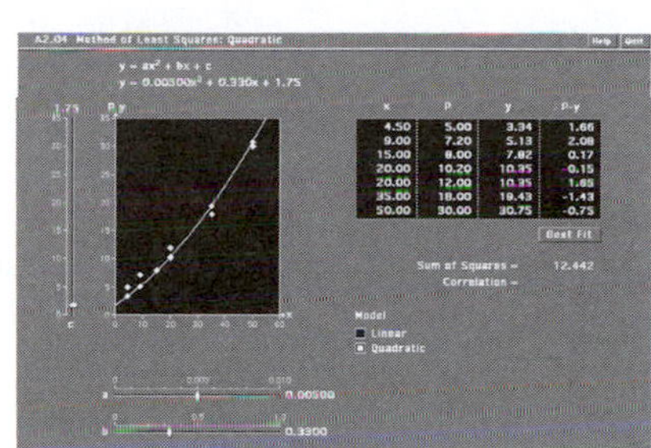

When we perform quadratic regression, we are looking for the best-fitting quadratic. But linear equations are among the quadratics, since they are given by $ax^2 + bx + c$, where $a = 0$. So the coefficient of determination for quadratic regression must be at least as large as the coefficient of determination for linear regression. We would pick a quadratic model over a linear model only if the coefficient of determination for quadratic regression is distinctly larger than for linear regression. Just how much larger will depend on the individual situation.

EXAMPLE 2 **Texas Cotton Gins**

In 1997 Fuller and coworkers[24] at Texas A & M University estimated the operating costs of cotton gin plants of various sizes. The operating costs of the smallest plant are shown in the following table.

[24] S. Fuller, M. Gillis, C. Parnell, A. Ramaiyer, R. Childers. 1997. Effect of air pollution abatement on financial performance of Texas cotton gins. *Agribusiness* 13(5):521–532.

x	2	4	6	8
y	109.4	187.24	273.36	367.6

Here x is the annual number of thousands of bales produced, and y is the total cost in thousands of dollars.

(a) Determine the best-fitting linear and quadratic functions and the square of the correlation coefficient of each. Which seems better, linear or quadratic regression?

(b) The study noted that revenue was $55.45 per bale. At what level of production will this plant break even?

(c) Determine the production level that will maximize the profits. Plant capacity is 8000 bales.

Solution

[0, 9.4] × [0, 400]

Screen 2.6
The graph of the best-fitting straight line for Example 2.

(a) We did linear regression in Example 4 of Section 2.1 and obtained $y = 43.036x + 19.22$ with $r^2 = 0.9991$. The graph is shown again in Screen 2.6. Quadratic regression gives $y_1 = 1.025x^2 + 32.786x + 39.72$ with $r^2 = 0.99999997$. The square of the correlation coefficient for the quadratic is larger than for linear and is very nearly perfect. The graph is shown in Screen 2.7.

(b) Now input $y_2 = 55.45x$, graph, and obtain Screen 2.8. Using available operations we find that the break-even quantity occurs at approximately 1919 bales.

(c) Now input $y_3 = y_2 - y_1$, graph, and obtain Screen 2.9. We see that the profit equation is increasing, and therefore profits are maximized at the maximum capacity of 8000 bales.

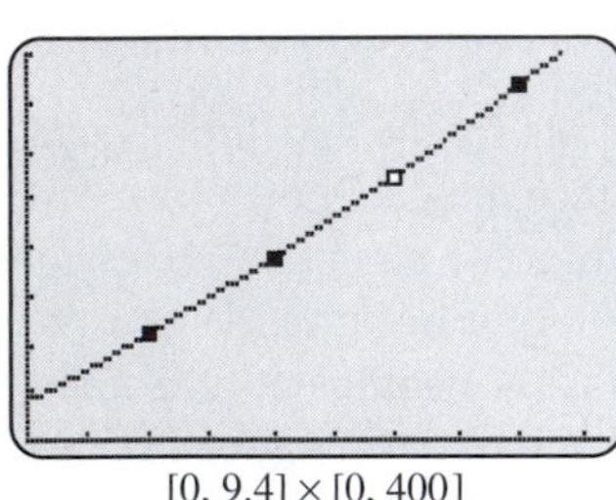

[0, 9.4] × [0, 400]

Screen 2.7
The graph of the best-fitting quadratic for Example 2.

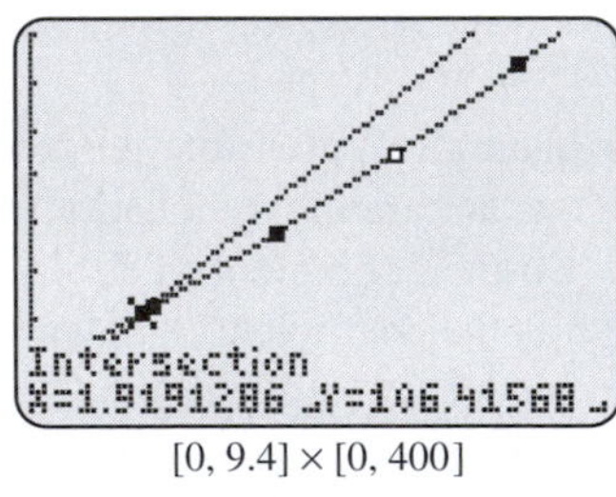

[0, 9.4] × [0, 400]

Screen 2.8
The intersection of the cost and revenue curves for Example 2.

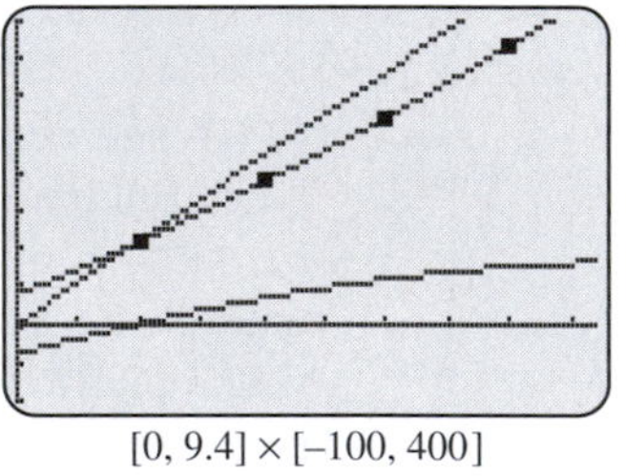

[0, 9.4] × [–100, 400]

Screen 2.9
The top two graphs are graphs of the cost and revenue equations. The bottom graph is the graph of the profit equation. We notice that profits increase throughout.

■

Remark Notice that the cost curve for the cotton gin plant is essentially linear. This means that the cost per bale is essentially constant for these plants no matter what the production level. This is not uncommon in industry.

Interactive Illustration 2.5

Finding the Best-Fitting Line in Example 2

This interactive illustration allows you to change both the linear $y = mx + b$ and the quadratic $y = ax^2 + bx + c$ and see the resulting sum of squares. There are sliders for both functions. Click or drag the mouse on a slider to change the value of a parameter. You will see a meter that gives the sum of the squares as the height of the bar. Adjust the values of the parameters to minimize the sum of the squares. Clicking on the Best Fit box will graph the best-fitting line or quadratic and give the minimum value of the sum of the squares.

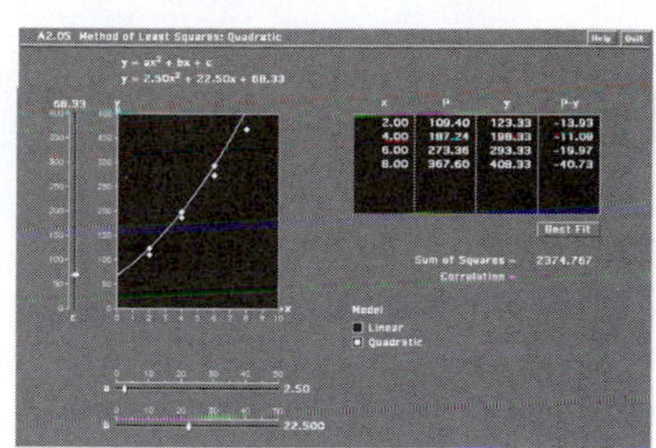

Exercise Set 2.2

1. **Cost Curve** Dean[25] made a statistical estimation of the cost-output relationship for a shoe chain. The data for the firm is given in the following table.

x	16	22	35	48	53	70	100	150
y	4.8	5.9	7	9	10	12	18	30

Here x is the output in thousands of dollars, and y is the total cost in thousands of dollars.

   a. Determine the best-fitting line (using least squares) and the square of the correlation coefficient. Graph.
   b. Determine the best-fitting quadratic (using least squares) and the square of the correlation coefficient. Graph.
   c. Which curve is better? Why?

2. **Cost Curve** Dean[26] made a statistical estimation of the cost-output relationship for a shoe store. The data for the firm is given in the following table.

x	4.5	7	9	10	15	20	33	50
y	3	3.3	3.4	3.5	4.5	5.5	7.5	12

Here x is the output in thousands of pairs of shoes, and y is the cost in thousands of dollars.

   a. Determine the best-fitting line (using least squares) and the square of the correlation coefficient. Graph.
   b. Determine the best-fitting quadratic (using least squares) and the square of the correlation coefficient. Graph.
   c. Which curve is better? Why?

3. **Cost Curve** Dean[27] made a statistical estimation of the cost-output relationship for a shoe chain for 1937. The data for the firm is given in the following table.

x	4.5	7	10	15	20	25	50
y	2.7	2.8	3.2	3.6	4.6	5.2	9.5

Here x is the output in thousands of pairs of shoes, and y is the cost in thousands of dollars.

   a. Determine both the best-fitting quadratic (using least squares) and the square of the correlation coefficient. Graph.
   b. What does this model predict the cost will be when the output is 30,000 pairs of shoes?
   c. What does this model predict the output will be if the cost is $4000?
   d. Determine the point on the curve where average cost is minimum.

4. **Shopping Behavior** Baker[28] collected the data found in the table below that correlates the number (N) of retail out-

[25] Joel Dean. 1976. *Statistical Cost Estimation.* Bloomington, Ind.: Indiana University Press, page 339.
[26] Ibid, page 342.
[27] Ibid, page 346.
[28] R. G. V. Baker. 1994. Multipurpose shopping behaviour at planned suburban shopping centers: a space-time analysis. *Environ. Planning A* 28:611–630.

lets in a suburban shopping center with the standardized number C of multipurpose shopping consumers.

N	70	80	90	90	90	125	205	205
C	20.5	23.5	18.5	13.0	12.5	11.0	17.5	13.5

a. Use quadratic regression to find C as a function of N.
b. Determine the number of retail outlets for which this model predicts the minimum standardized number of multipurpose shopping consumers.

5. Shopping Behavior Baker[29] collected the data found in the table below that correlates the number (N) of retail outlets in a suburban shopping center with the mean trip frequency f of consumers.

N	70	80	80	90	90
f	1.50	1.57	1.47	1.36	1.29
N	90	125	205	205	
f	1.13	1.16	1.30	1.47	

a. Use quadratic regression to find f as a function of N.
b. Determine the number of retail outlets for which this model predicts the minimum mean trip frequency.

6. Forest Birds Belisle and colleagues[30] studied the effects of forest cover on forest birds. They collected data found in the table relating the percent of forest cover with homing time (time taken by birds for returning to their territories).

Forest Cover (%)	20	25	30	40	72
Homing Time (hours)	35	105	67	60	56
Forest Cover (%)	75	80	82	89	93
Homing Time (hours)	22	80	10	15	20

a. Find the best-fitting quadratic (as the researches did) relating forest cover to homing time and the square of the correlation coefficient.
b. Find the forest cover percentage that minimizes homing time.

7. Moose Reproductive Effort Ericsson and colleagues[31] studied the effect of the age of a female moose on the mortality of her offspring. They collected data shown in the table relating the age of the female moose to offspring mortality during the hunting season.

Moose age	2	3	4	5
Mortality of Offspring	0.5	0.4	0.25	0.35
Moose age	6	7	8	9
Mortality of Offspring	0.35	0.5	0.37	0.35
Moose age	10	11	12	13
Mortality of Offspring	0.48	0.37	0.53	0.38
Moose age	14			
Mortality of Offspring	0.60			

a. Find the best-fitting quadratic (as the researches did) relating age to mortality of offspring and the square of the correlation coefficient.
b. Find the age at which mortality of offspring is minimized.

8. Texas Cotton Gins In 1997 Fuller and coworkers[32] at Texas A & M University estimated the operating costs of cotton gin plants of various sizes. The operating costs of the next to the smallest plant are shown in the following table.

x	2	4	6
y	163.20	240.48	301.50
x	8	10	12
y	376.16	454.40	536.40

Here x is the annual number of thousands of bales produced, and y is the total cost in thousands of dollars.

a. Determine the best-fitting quadratic and the square of the correlation coefficient. Compare with the square of the correlation coefficient found in Exercise 9 in Exercise Set 2.1. Which seems better, linear or quadratic regression?

[29] Ibid.

[30] Marc Belisle, Andre Desrochers, and Marie-Josee Fortin. 2001. Influence of forest cover on the movements of forest birds: a homing experiment. *Ecology* 82(7):1893–1904.

[31] Goran Ericsson, Kjell Wallin, John P. Ball, and Martin Broberg. 2001. Age-related reproductive effort and senescence in free-ranging moose. *Ecology* 82(6):1623–1620.

[32] S. Fuller, M. Gillis, C. Parnell, A. Ramaiyer, R. Childers. 1997. Effect of air pollution abatement on financial performance of Texas cotton gins. *Agribusiness* 13(5):521–532.

b. The study noted that revenue was \$63.25 per bale. Using the quadratic model, at what level of production will this plant break even?

c. Determine the production level that will maximize profit for such a plant using the quadratic model.

9. Texas Cotton Gins Fuller and coworkers[33] estimated the operating costs of a medium-sized cotton gin plant as shown in the following table.

x	2	4	6	8
y	217.6	271.12	325.5	380.8
x	10	12	15	20
y	437	494.16	581.7	732

Here x is the annual number of thousands of bales produced, and y is the total cost in thousands of dollars.

a. Determine the best-fitting linear and quadratic functions and the square of the correlation coefficient of each. Which seems better, linear or quadratic regression?

b. The study noted that revenue was \$63.25 per bale. At what level of production will this plant break even?

c. Determine the production level that will maximize the profits for such a plant.

10. Pest Management Hollingsworth and coworkers[34] collected the data shown in the table relating fungal prevalence y in aphids to time t measured in days.

x	0	6	6	7	10	12	13	15
y	0	3	4	8	18	20	25	32

a. Use quadratic regression to find y as a function of x.

b. Graph on the interval $[-1, 16]$. How would you describe this curve on this interval?

11. Population Ecology Lactin and colleagues[35] collected the following data relating the feeding rate y of the first-instar Colorado potato beetle and the temperature T in degrees Celsius.

T	14	17	20	23	29	32	38
y	0.15	0.35	1.05	1.15	1.55	1.55	1.45

a. Use quadratic regression to find y as a function of T.

b. Find the temperature for which the feeding rate is maximum.

12. Ecology Savopoulou-Soultani and coworkers[36] collected the data shown in the table relating the number y of days *Lobesia botrana* spent in the larvae stage to the percent x of brewer's yeast in its diet.

x	0	1.5	2.7	5.7	8.2	9.2	10.8
y	33.1	28.8	27.2	27.7	26.7	30.3	29.2

a. Use quadratic regression to find y as a function of x.

b. Find the percent of yeast that results in a minimum number of days spent in the larvae stage.

13. Ecological Entomology Elliott[37] studied the temperature effects on the alder fly. In 1967 he collected the data shown in the following table relating the temperature in degrees Celsius to the number of pupae successfully completing pupation.

t	8	10	12	16	20	22
y	18	27	43	44	37	5

Here t is the temperature in degrees Celsius, and y is the number of pupae successfully completing pupation.

a. Use quadratic regression to find y as a function of t.

b. Determine the temperature at which this model predicts the maximum number of successful pupations.

c. Determine the two temperatures at which this model predicts there will be no successful pupation.

---

33 Ibid.

34 R. G. Hollingsworth, D. C. Steinkraus, and R. W. McNew. 1995. Sampling to predict fungal epizotics in cotton aphids. *Environ. Entomol.* 24(6):1414–1421.

35 Derek J. Lactin, N. J. Holliday, and L. L. Lamari. 1993. Temperature dependence and constant-temperature duel aperiodicity of feeding by Colorado potato beetle larvae in short-duration laboratory trials. *Environ. Entomol.* 22(4):784–790.

36 M. Savopoulou-Soultani, D. G. Stavridid, A. Vassiliou, J. E. Stafilidis, and I. Iraklidis. 1994. Response of *Lobesia botrana* to levels of sugar and protein in artificial diets. *J. Econ. Entomol.* 87(1):84–90.

37 J. M. Elliott. 1996. Temperature-related fluctuations in the timing of emergence and pupation of Windermere alder-flies over 30 years. *Ecol. Entomol.* 21:241–247.

14. Milk Yield Rigout and colleagues[38] studied the impact of glucose on milk yields of dairy cows. The following table includes data they collected.

Glucose infused (%)	0	2	5
Milk Yield (kg/day)	30.0	31.7	31.5
Glucose infused (%)	8	12	14
Milk Yield (kg/day)	31.7	31.3	29.8

a. Find the best-fitting quadratic (as the researchers did) that relates glucose to milk yield and the square of the correlation coefficient. Graph.

b. Find the percent of glucose that maximizes milk yield.

15. Milk Yield Walters and colleagues[39] studied dairy cow milk yield with respect to days postpartum. They gathered the data in the table that relates days postpartum with daily milk yield in kilograms.

[38] S. Rigout, S. Lemosquet, J.E. van Eys, J.W. Blum, and H. Rulquin. 2002. Duodenal glucose increases glucose fluxes and lactose synthesis in grass silage-fed dairy cows. *J. Dairy Sci.* 85:595–606.

[39] A. H. Walters, T. L. Bailey, R. E. Pearson, and F. C. Gwazdauskas. 2002. Parity-related changes in bovine follicle and oocyte populations, oocyte quality, and hormones to 90 days postpartum. *J. Dairy Sci.* 85:824–832.

Days Postpartum	7	21	35	49	64	77
Milk Yield (kg)	25	27	33	33	32	32

a. Find the best-fitting quadratic (as the researchers did) that relates days postpartum to milk yield and the square of the correlation coefficient. Graph.

b. Find the number of days postpartum that maximizes milk yield.

16. Ecology Kelly and Harwell[40] studied the ecology of streams and collected data found in the table that related the biomass of stream bottom fauna to juvenile trout length.

Biomass of Bottom Fauna	.40	.55	1.05	1.17	1.60
Trout Length (cm)	.49	.53	.72	.61	.69

a. Find the best-fitting quadratic (as the researchers did) that relates biomass of bottom fauna to trout length and the square of the correlation coefficient. Graph.

b. Find the amount of bottom fauna that maximizes trout length.

[40] John Kelly and Mark A. Harwell. 1990. Indicators of ecosystem recovery. *Environ. Man.* 14(5):527–545.

2.3 Cubic, Quartic, and Power Regression

We have already found the best-fitting linear and quadratic functions for a set of discrete data using least squares. This can also be done for power functions and for a polynomial of any degree. We now consider the best-fitting power function ax^b and the best-fitting cubic $y = ax^3 + bx^2 + cx + d$ and will leave quartic regression to the exercise set.

Study the power functions in Figures 1.8, 1.9, 1.10, and 1.11. Notice that on the interval $(0, \infty)$, each is always increasing while on $(-\infty, 0)$ the first is always decreasing. Also, each is either always concave up or always concave down on the given intervals. Thus, if data is given for nonnegative x and the function is either always increasing or always decreasing and either always concave up or always concave down, consider using power regression.

A polynomial can increase on one interval and decrease on another and, unlike quadratics, can change concavity. See Figure 2.4 for some examples. Thus, if data changes concavity, consider using polynomials. In fact, cubics can change concavity at most once, and quartics at most twice. Thus, if data changes concavity once, consider a cubic model. If data changes concavity twice, consider a quartic model.

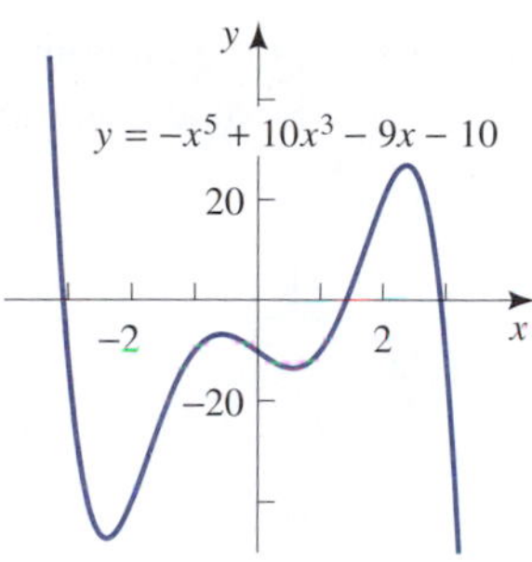

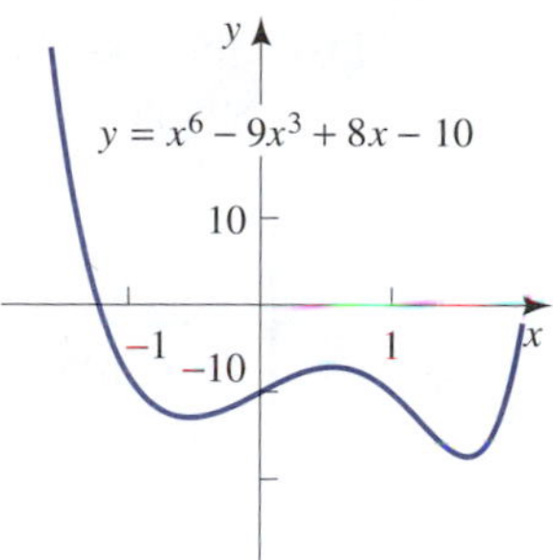

Figure 2.4
Graphs of typical polynomial functions.

EXAMPLE 1 **Power Regression**

Johnston[41] reports on a study relating output to total variable cost for 40 British firms. Given in the following table is a smaller (representative) set of his data.

x	20	25	25	50	75	100	190	230	270	470	650	790	900	920	1200
y	40	50	50	125	150	200	300	350	600	650	950	1150	1600	1300	1500

Here x is output in millions of units, and y is total variable cost in thousands of pounds sterling.

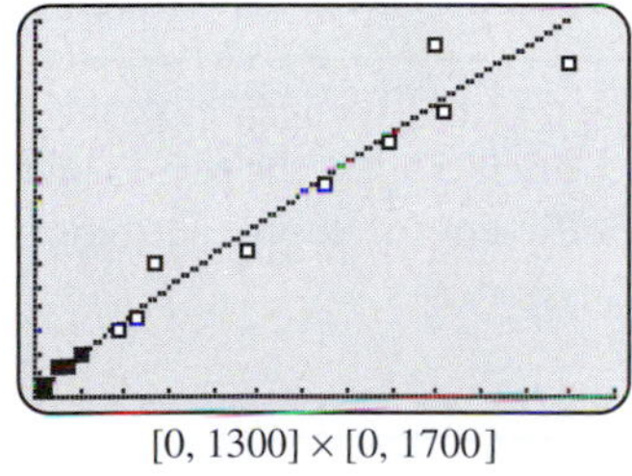
[0, 1300] × [0, 1700]

Screen 2.10
The graph of the best-fitting power function for Example 1.

Use power regression to find the best-fitting power function to the data and the square of the correlation coefficient. Graph.

Solution

Input the data into your computer or calculator, select power regression, and obtain

$$y = ax^b = 3.031x^{0.8937}$$

where the square of the correlation coefficient is $r^2 = 0.9892$. The graph is shown in Screen 2.10. ■

Remark You probably noticed that your computer or graphing calculator also gave $r = 0.9946$. Since power functions are either always increasing or always decreasing, a positive r means the best-fitting power function is increasing (positive correlation) and a negative r means the best-fitting power function is decreasing (negative correlation).

EXAMPLE 2 **Cubic Regression**

Grossman and Krueger[42] studied the relationship between per capita income and various environmental indicators in a variety of countries. The object was to determine whether environmental quality deteriorates steadily with growth. The following table

[41] J. Johnston. 1960. *Statistical Cost Analysis*. New York: McGraw-Hill, page 66.

[42] Gene M. Grossman and Alan B. Krueger. 1995. Economic growth and the environment. *Quart. J. Econ.* CX:352–377.

gives the data they collected relating the amount y of smoke in cities to the gross domestic product (GDP) per capita income x in thousands of dollars.

x	1	3	5	7	9	11
y	70	100	115	95	75	65

Use cubic regression to find the best-fitting cubic to the data and the square of the correlation coefficient. Graph. Find the square of the correlation coefficient for quadratic regression and compare to that of cubic. What conclusion do you arrive at?

Solution

Input the data into your computer or calculator, select cubic regression, and obtain

$$\begin{aligned} y &= ax^3 + bx^2 + cx + d \\ &= 0.27x^3 - 6.31x^2 + 39.87x + 34.78 \end{aligned}$$

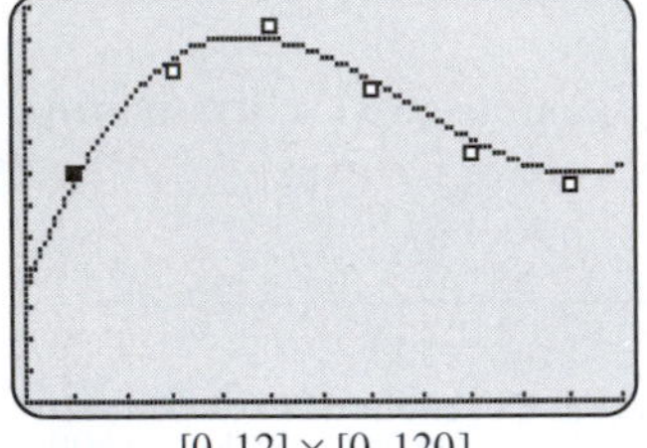

$[0, 12] \times [0, 120]$

Screen 2.11
The graph of the best-fitting cubic function for Example 2.

where the square of the correlation coefficient is $r^2 = 0.9702$. The graph is shown in Screen 2.11. Notice that the amount of smoke initially increases but at larger amounts of per capita GDP the amount of smoke decreases.

Quadratic regression yields the square of the correlation coefficient as $r^2 = 0.8182$. This is considerably less than the square of the correlation coefficient given for cubic regression. We conclude that the cubic polynomial is definitely a better fit to the data than a quadratic. ■

Remark When finding the best-fitting cubic, quadratics are included, the coefficient of x^3 is just zero. Thus, the square of the correlation coefficient for cubic regression will always be at least as large as that for quadratic regression.

Exercise Set 2.3

1. **Cost Curve** Wylie and Ezekiel[43] obtained actual data from the United States Steel Corporation relating the percent of capacity at which the plants operated to cost per ton of steel. The following table gives their data.

x	20	30	40	50	60
y	33.20	27.50	24.40	22.50	21.45
x	70	80	90	100	
y	20.80	20.35	20.10	19.95	

Here x is the percent of capacity at which the plants operated, and y is cost per ton.

   a. Use power regression to find the best-fitting power function to the data and the correlation coefficient. Graph.
   b. What does this model predict the cost per ton will be when plants are operated at 75%?
   c. What does this model predict the operating percent of capacity will be if the cost per ton is \$30.00?

2. **Cost Curve** Johnston[44] reports on a study of 40 firms relating the output to average fixed costs. Instead of using their 40 pieces of data, we use just their array means in the following table.

[43] Kathyrn H. Wylie and Mordecai Ezekiel. 1940. The cost curve for steel production. *J. Political Econ.* XLVIII(6):777–821.

[44] J. Johnston. 1960. *Statistical Cost Analysis.* McGraw-Hill. New York, pages 70–71.

x	50	160	250	400	650	875	1250	2000
y	4.6	4	3.1	3.2	3.3	2	2.7	2.5

Here x is output in millions of units, and y is average cost per unit of output (in millions).

Use power regression to find the best-fitting power function to the data and the correlation coefficient. Graph.

3. **Costs** In 1997, Fuller and coworkers[45] at Texas A & M University estimated the operating costs of cotton gin plants of various sizes. The operating costs per bale of the medium-sized plant is shown in the following table.

x	2	4	6	8
y	217.6	271.12	325.5	380.8
x	10	12	15	20
y	437	494.16	581.7	732

Here x is annual number of thousands of bales produced, and y is the total cost in thousands of dollars.

Revenue was estimated at $63.25 per bale.

a. In Exercise 9 of Section 2.2 you found the best-fitting line and quadratic and the square of the correlation coefficients. Now determine the best-fitting power function and the square of the correlation coefficient using least squares. Which seems better, linear, quadratic, or power regression?

b. The study noted that revenue was $63.25 per bale. Using the power model, determine the level of production at which this plant will break.

c. Determine the production level that will maximize the profits for such a plant using the power model. Production capacity is 20,000 bales.

4. **Economies of Scale** Cook and Duffield[46] collected data relating the size of assets of money market mutual funds with the average costs. Their data are presented in the following table. (Only a small number of representative data are used.)

x	10	20	40	60	90	110	500
y	1.5	1.1	0.90	0.75	0.70	0.70	0.60

Here x is assets in millions of dollars, and y is average cost in percent of total expenses to average assets.

Use power regression to find the best-fitting power function to the data and the correlation coefficient. Graph. Give an economic interpretation.

5. **Production Costs** Suarez-Villa and Karlsson[47] studied the relationship between the sales in Sweden's electronic industry and production costs (per unit value of product sales). Their data are presented in the following table.

x	7	13	14	15	17
y	0.91	0.72	0.91	0.81	0.72
x	20	25	27	35	45
y	0.90	0.77	0.65	0.73	0.70
x	45	63	65	82	
y	0.78	0.82	0.92	0.96	

Here x is product sales in millions of krona, and y is production costs per unit value of product sales.

a. Use cubic regression to find the best-fitting cubic to the data and the correlation coefficient. Graph.

b. Find the minimum production costs.

6. **Economic Growth and the Environment** Grossman and Krueger[48] studied the relationship between per capita income and various environmental indicators in a variety of countries. The object was to determine whether environmental quality deteriorates steadily with growth. The following table gives the data they collected relating the amount y of arsenic in rivers to the GDP per capita income x in thousands of dollars.

x	1	3	5	9
y	0.007	0.0075	0.011	0.008
x	11	13	15	
y	0.0025	0.001	0.0005	

[45] S. Fuller, M. Gillis, C. Parnell, A. Ramaiyer, and R. Childers. 1997. Effect of air pollution abatement on financial performance of Texas cotton gins. *Agribusiness* 13(5):521–532.

[46] Timothy Q. Cook and Jeremy G. Duffield. 1979. Average costs of money market mutual funds. *Econ. Rev. Federal Reserve Bank Richmond* 65(4):32–39.

[47] L. Suarez-Villa and C. Karlsson. 1996. The development of Sweden's R&D-intensive electronics industries: exports, outsourcing, and territorial distribution. *Environ. Planning A* 28:783–817.

[48] Gene M. Grossman and Alan B. Krueger. 1995. Economic growth and the environment. *Quart. J. Econ.* CX:352–377.

Use cubic regression to find the best-fitting cubic to the data and the correlation coefficient. Graph. Explain what is happening.

7. Cost Curve Nordin[49] obtained the following data relating output and total final cost for an electric utility in Iowa. (Instead of using the original data set for 541 eight-hour shifts, we have just given the array means.)

x	25	28	33	38	41	45	50	53
y	24	24	26	28	27	31	34	38
x	57	63	66	71	74	79	83	89
y	39	42	43	51	53	52	54	52

Here x is the output as a percent of capacity, and y is final total cost in dollars (multiplied by a constant not given to avoid showing the exact level of the costs.)

Use cubic regression to find the best-fitting cubic to the data and the correlation coefficient. Graph. Also find the correlation coefficients associated with both linear and quadratic regression, and compare them to that found for cubic regression.

8. Ecology Allen and coworkers[50] collected the data found in the following table relating the intrinsic rate y of increase of the citrus rust mite to the temperature T in degrees Celsius.

T	14	17	19	21	23
y	−0.05	−0.02	0.04	0.08	0.10
T	25	27	29	31	33
y	0.09	0.12	0.12	−0.08	−0.24

Use cubic regression to find the best-fitting cubic to the data and the correlation coefficient. Graph. Explain what is happening.

9. Population Ecology Wagner[51] collected the data found in the following table relating the percent mortality y of eggs of the sweet-potato whitefly and the temperature T in degrees Celsius.

T	20	22	24	26	28	30	32	34
y	7	6	3.5	5	7	4	6.8	18

Use quartic regression to find the best-fitting fourth-order polynomial to the data and the correlation coefficient. Graph.

10. Population Ecology Grace and colleagues[52] studied the population history of termite colonies. They noted an initial 10-year increase in population, followed by a 10-year static population, followed by a 15-year decline in population. In the paper cited they found a curious relationship between the average weight of individual foraging workers and the population of the colony during the 15-year decline in population. Their data are given in the following table.

x	2.75	2.79	2.84	2.98	2.98
y	4.1	7.0	5.2	4.5	3.0
x	3.08	3.72	4.21	6.26	
y	2.2	1.8	1.4	1.0	

Here x is the weight in milligrams of individual foraging workers, and y is the population of the colony in millions.

a. Use power regression to find the best-fitting power function to the data and the correlation coefficient. Graph.

b. What does this model predict the population will be when the average weight of individual foraging workers is 3.00 mg?

c. What does this model predict the average weight of individual foraging workers will be if the population is 6 million?

11. Population Ecology Grace and colleagues[53] found a correlation between the percent increase in the individual weight of foraging workers and the percent decrease in colony population during the latter stages of the life of the termite colony. Their data are found in the following table.

---

49 J. A. Nordin 1947. Note on a light plant's cost curve. *Econometrica* 15(1):231–235.

50 J. C. Allen, Y. Yang, and J. L. Knapp. 1995. Temperature effects on development and fecundity of the citrus rust mite. *Environ. Entomol.* 24(5):996–1004.

51 Terence L. Wagner. 1995. Temperature-dependent development, mortality, and adult size of sweet-potato whitefly biotype B on cotton. *Environ. Entomol.* 24(5):1179–1188.

52 J. Kenneth Grace, Robin T. Yamamoto, and Minoru Tamashiro. 1995. Relationship of individual worker mass and population decline in a Formosan subterranean termite colony. *Environ. Entomol.* 24(5):1258–1262.

53 Ibid.

Use power regression to find the best-fitting power function to the data and the correlation coefficient. Graph.

x	7	32	45	120
y	50	62	73	79

Here x is the percent increase in the weight of individual foraging workers in millimeters, and y is the percent decrease in the population of the colony.

2.4 Exponential and Logarithmic Regression

In this section we find the best-fitting exponential function ab^x or logarithmic function $a + b \log x$ to a set of discrete data using least squares. Recall from Figures 1.38 and 1.41 that the graph of $y = b^x$ is always increasing when $b > 1$ and always decreasing when $0 < b < 1$. Also notice that in either case the graph is concave up. Thus, data that is concave up and either always increasing or always decreasing is a possible candidate for exponential modeling.

EXAMPLE 1 **Exponential Regression**

The following table gives the population of the United States in thousands for some selected years.

Year	1790	1791	1792	1793	1794
Population	3926	4056	4194	4332	4469

(a) On the basis of the data given for the years 1790–1794, find the best-fitting exponential function using exponential regression. Graph.

(b) Use this model to estimate the population in 1805. Compare your answer to the actual population of 6,258,00 in 1805.

(c) According to the model found in part (a), when will the population reach 10,000,000?

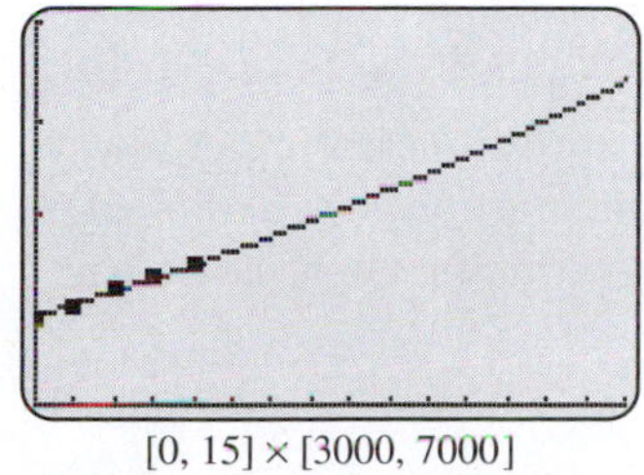

[0, 15] × [3000, 7000]

Screen 2.12
The graph of the best-fitting exponential function for Example 1.

Solution

(a) Input the data into your computer or graphing calculator with the year 1790 corresponding to $x = 0$, select exponential regression, and obtain

$$y = p(x) = ab^x = 3927.3(1.033)^x$$

where the square of the correlation coefficient is $r^2 = 0.9999$. The graph is shown in Screen 2.12.

Remark You probably noticed that your computer or graphing calculator also gave $r = 0.9999$. Since exponential functions are either always increasing or always decreasing, a positive r means that the best-fitting exponential function is increasing (positive correlation), and a negative r means that the best-fitting power function is decreasing (negative correlation).

(b) The year 1805 corresponds to $x = 15$. We note that $p(15) = 6391$. (We get 6394 if we do not round off the exponential function given by the computer or calculator.) So this model predicted the population to be 6,391,000 in 1805.

(c) Add $y_2 = 10,000$ to the graph found in part (a). Using available operations, we find the x-component of the intersection to be about 28.8. See Screen 2.13. Thus, according to this model, the population will reach 10,000,000 toward the latter part of 1818. ■

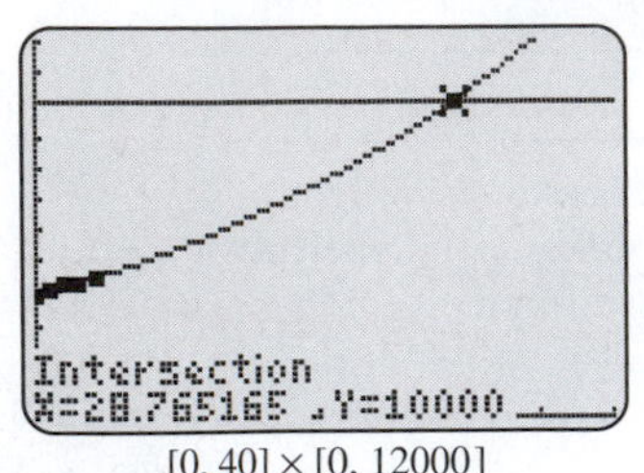

$[0, 40] \times [0, 12000]$

Screen 2.13
We see that the exponential reaches 10,000 when x is about 28.8.

EXAMPLE 2 **Logarithmic Regression**

Stein and Price[54] studied the relationship between the shoot age and shoot length of willows. They collected the data given in the following table.

Age (Years)	1	2	3	4	5	6	7	8	9	10	11
Shoot Length (mm)	160	125	115	130	85	75	81	80	70	60	63

(a) On the basis of the data given in the table, find the best-fitting logarithmic function using least squares. Graph.

(b) Use this model to estimate the average shoot lengths of a 12-year-old willow.

Solution

(a) Input the data into your computer or calculator, select logarithmic regression, and obtain

$$\begin{aligned} y = l(x) &= a + b \ln x \\ &= 160.75 - 41.38 \ln x \end{aligned}$$

where the square of the correlation coefficient is $r^2 = 0.9015$. The graph is shown in Screen 2.14.

(b) Then obtain $l(12) \approx 57.92$. ■

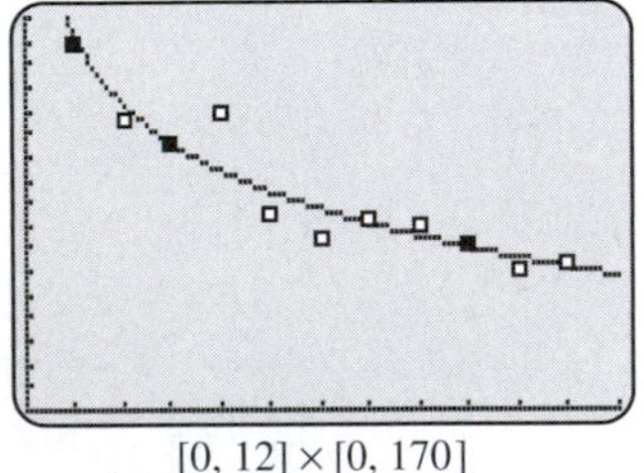

$[0, 12] \times [0, 170]$

Screen 2.14
The graph of the best-fitting logarithmic function for Example 2.

Remark You probably noticed that your computer or graphing calculator also gave $r = -0.9495$. Since logarithmic functions are either always increasing or always decreasing, a positive r means the best-fitting logarithmic function is increasing (positive correlation), and a negative r means the best-fitting power function is decreasing (negative correlation).

[54] Steven J. Stein and Peter W. Price. 1995. Relative effects of plant resistance and natural enemies by plant development and on sawfly preference and performance. *Environ. Entomol.* 24(4):909–916.

Exercise Set 2.4

1. **Medicare Benefit Payments** The following table gives the benefit payments for Medicare for selected years and can be found in Glassman.[55] Payments are in billions of dollars.

Year	1970	1975	1980	1985	1990	1993
Payments	7.0	15.5	35.6	70.5	108.7	150.4

On the basis of this data, find the best-fitting exponential function using exponential regression. Let $x = 0$ correspond to 1970. Graph. Use this model to estimate payments in 1997.

2. **Sales** The following table gives the sales of CD-ROM drives and compact disk players for selected years and can be found in Glassman.[56] Sales are in millions.

Year	1988	1990	1992	1993
Sales	0.075	0.24	1.5	4.8

On the basis of this data, find the best-fitting exponential function using exponential regression. Let $x = 0$ correspond to 1988. Graph. Use this model to estimate sales in 1997.

3. **The Internet** The following table gives the number of computers connected to the Internet for selected years and can be found in Glassman.[57] Numbers are in millions.

Year	1989	1990	1991	1992	1993	1995
Numbers	0.08	0.20	0.40	0.80	1.50	30

On the basis of this data, find the best-fitting exponential function using exponential regression. Let $x = 0$ correspond to 1989. Graph. Use this model to estimate the numbers in 1997.

4. **National Football League Salaries** The following table gives the average National Football League salary for selected years and can be found in Glassman.[58] Salary is in thousands of dollars.

Year	1986	1987	1988	1989
Salary	198	203	239	295
Year	1990	1991	1992	1993
Salary	352	415	488	737

On the basis of this data, find the best-fitting exponential function using exponential regression. Let $x = 0$ correspond to 1986. Graph. Use this model to estimate the average salary in 1997.

5. **Cellular Telephones** The following table gives the number of people with cellular telephone service for recent years and can be found in Glassman.[59] Number is in millions.

Year	1984	1985	1986	1987	1988
Number	0.2	0.5	0.8	1.4	2.0
Year	1989	1990	1991	1992	1993
Number	3.8	5.7	8	11	13.8

   a. On the basis of this data, find the best-fitting exponential function using exponential regression. Let $x = 0$ correspond to 1984. Graph. Use this model to estimate the numbers in 1997.

   b. Using the model in part (a), estimate when the number of people with cellular telephone service will reach 50 million.

6. **U.S. Jobs and Mexico** The following table gives the number of U.S. jobs supported by exports to Mexico for recent years and can be found in Glassman.[60] Number is in thousands.

Year	1986	1987	1988	1989
Number	274	300	400	500
Year	1990	1991	1992	
Number	540	610	716	

---

[55] Bruce S. Glassman. 1996. *The Macmillan Visual Almanac*. New York: Macmillan, page 61.

[56] Ibid., page 530.

[57] Ibid., page 535.

[58] Ibid., page 554.

[59] Ibid., page 161.

[60] Ibid., page 517.

a. On the basis of this data, find the best-fitting exponential function using exponential regression. Let $x = 0$ correspond to 1986. Graph. Use this model to estimate the number of jobs in 1997.
b. Using the model in part (a), estimate when the number of U.S. jobs supported by exports to Mexico will reach 1 million.

7. **Rollerblading** The following table gives the number of in-line skaters mounted for recent years and can be found in Glassman.[61] Number is in millions.

Year	1989	1990	1991	1992	1993
Number	3.0	4.3	6.2	9.4	12.4

a. On the basis of this data, find the best-fitting exponential function using exponential regression. Let $x = 0$ correspond to 1989. Graph. Use this model to estimate the number in 1997.
b. Using the model in part (a), estimate when the number of in-line skaters mounted will reach 20 million.

8. **U.S. Infant Mortality Rate** The following table gives the U.S. infant mortality rate per 1000 births for selected years and can be found in Glassman[62] and Elliot.[63] The rate is per 1000 births.

Year	1940	1950	1960	1970
Rate	47.0	29.2	26.0	20.0
Year	1980	1990	1994	
Rate	12.6	9.2	8.0	

a. On the basis of this data, find the best-fitting exponential function using exponential regression. Let $x = 0$ correspond to 1940. Graph. Use this model to estimate the rate in 1997.
b. Using the model in part (a), estimate when the U.S. infant mortality rate will reach 4 per thousand births.

9. **Number of Roman Catholic Priests** The following table gives the number of Roman Catholic ordinations per year in the United States for selected years and can be found in Glassman.[64]

Year	1967	1972	1977	1982	1987	1992
Number	932	705	613	453	365	289

a. On the basis of this data, find the best-fitting exponential function using exponential regression. Let $x = 0$ correspond to 1967. Graph. Use this model to estimate the number in 1997.
b. Using the model in part (a), estimate when the number of Roman Catholic ordinations per year will reach 150.

10. **Education** The following table gives the percentage for the selected years of the U.S. population with less than 12 years of school and can be found in Glassman.[65]

Year	1970	1980	1985
Percentage	47.7	33.5	26.1
Year	1989	1990	1991
Percentage	23.1	22.4	21.6

a. On the basis of this data, find the best-fitting exponential function using exponential regression. Let $x = 0$ correspond to 1970. Graph. Use this model to estimate the percentage in 1997.
b. Using the model in part (a), estimate when the percentage of the U.S. population with less than year 12 years of school will reach 15.

11. **Economic Entomology** Karr and Coats[66] studied the effects of several chemicals on the growth rate of the German cockroach. The following table gives their data for the percent of the chemical d-limonene in the diet of the cockroach.

Percent d-Limonene	1	10	25
Days to Adult Stage	123	113	108

[61] Ibid., page 569.
[62] Ibid., page 50.
[63] J. M. Elliott. 1996. Temperature-related fluctuations in the timing of emergence and pupation of Windermere alder-flies over 30 years. *Ecol. Entomol.* 21:241–247.

[64] Bruce S. Glassman. 1996. *The Macmillan Visual Almanac.* New York: Macmillan, p. 117.
[65] Ibid., page 584.
[66] L. L. Karr and J. R. Coats. 1992. Effects of four monoterpernoids on growth and reproduction of the German cockroach. *J. Econ. Entomol.* 85(2):424–429.

a. On the basis of the data given in the table, find the best-fitting logarithmic function using least squares. Graph.
b. Use this model to estimate the days to adult stage with a diet of 20% d-limonene.

12. **Economic Entomology** Karr and Coats[67] studied the effects of several chemicals on the growth rate of the German cockroach. The following table gives their data for the percent of the chemical α-terpineol in the diet of the cockroach.

Percent α-terpineol	1	10	25
Days to Adult Stage	129	113	115

a. On the basis of the data given in the table, find the best-fitting logarithmic function using least squares. Graph.
b. Use this model to estimate the days to adult stage with a diet of 20% α-terpineol.

13. **Forest Entomology** Rieske and Raffa[68] studied the relationship of population increase to previous year population size of *H. pales* trapped in pitfall traps baited with ethanol and turpentine. Their data are given in the following table.

1988 Population	220	320	360	410	620
Population Increase (%)	375	170	130	250	120

a. On the basis of the data given in the table, find the best-fitting logarithmic function using least squares. (Note that the authors used logarithms to the base 10.) Graph.
b. Use this model to estimate the population increase with a 1988 population of 500.

14. **Economic Entomology** Smitley and Davis[69] studied the changes in gypsy moth egg mass densities over one generation as a function of the initial egg mass density in a control plot and two treated plots. The data below are for the control plot.

Initial Egg Mass (per 0.04 ha)	50	75	100	160
Change in Egg Mass Density (%)	250	−100	−25	−25
Initial Egg Mass (per 0.04 ha)	175	180	200	
Change in Egg Mass Density (%)	−50	50	0	

a. On the basis of the data given in the table, find the best-fitting logarithmic function using least squares. (Note that the authors used logarithms to the base 10.) Graph.
b. Use this model to estimate the change in egg mass density with an initial egg mass of 150 per 0.04 ha.

[67] Ibid.

[68] L. K. Rieske and K. F. Raffa. 1993. Potential use of baited traps in monitoring pine root weevil populations and infestation levels. *J. Econ. Entomol.* 86(2):475–485.

[69] D. R. Smitley and T. W. Davis. 1993. Aerial application of *Bacillus thuringiensis* for suppression of gypsy moth in *Populus-Quercus* forests. *J. Econ. Entomol.* 86(4):1178–1184.

2.5 Logistic Regression

In this section we find the best-fitting logistic curve $c/(1+ae^{-bx})$ to a set of discrete data using least squares. Figure 2.5 shows an increasing logistic curve and a decreasing logistic curve. Notice that the first one is always increasing, initially concave up and then concave down. Also notice that the graph has both upper and lower limiting values. The second one is always decreasing, initially concave down and then concave up. Notice again that the graph has both upper and lower limiting values. If you have data that behaves in either way, then consider logistic modeling. The interactive illustration given in this section allow you to change the parameters in the logistic function to experimentally minimize the sum of the squares and find the approximate best-fitting logistic function. This is similar to what was done in Sections 2.1 through 2.4.

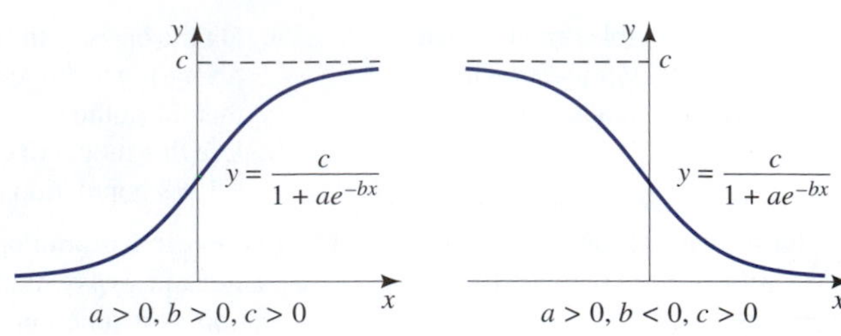

Figure 2.5
The first logistic curve rises quickly at first and then levels off. The second logistic curve falls quickly at first and then levels off.

EXAMPLE 1 **Modeling with Logistic Curves**

The following table gives the population of the United States in millions for some selected early years.

Year	1790	1810	1830	1850	1870	1890
Population	3.9	7.2	12.9	23.2	38.6	63.0

Y1=4.0963533600062*1.028
X=200 Y=1087.4506

[0, 200] × [−200, 1100]

Screen 2.15
The best-fitting exponential function $e(x)$. We see that $e(200) \approx 1087$.

(a) On the basis of the data given for the years 1790–1890, find the best-fitting exponential function using exponential regression. Determine the square of the correlation coefficient. Graph. Using this model, estimate the population in 1990.

(b) Now find the best-fitting logistic curve. Graph. Using this model, estimate the population in 1990.

Solution

(a) Input the data into your computer or calculator with the year 1790 corresponding to $x = 0$, select exponential regression, and obtain

$$y = e(x) = ab^x = 4.0964(1.0283)^x$$

where the square of the correlation coefficient is $r^2 = 0.9981$. The graph is shown in Screen 2.15. Notice the positive and very high correlation. We note that $e(200) \approx 1087$. So according to the exponential model, the population in 1990 will be 1087 million.

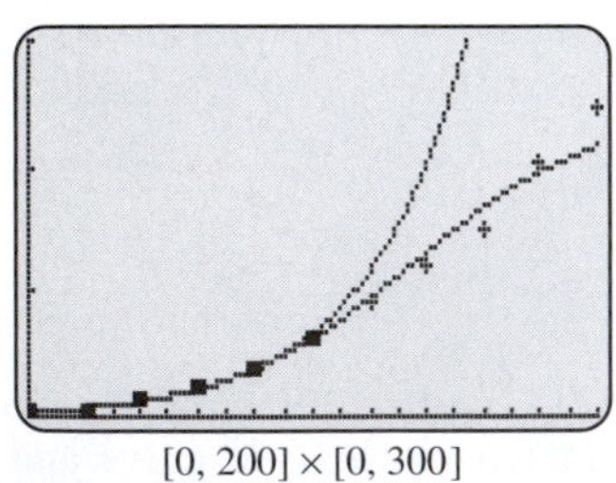

[0, 200] × [0, 300]

Screen 2.16
The top graph is the graph of the exponential function found in Screen 2.15. The bottom graph is the graph of the best-fitting logistic function. The crosses show the actual population. The logistic function was the better predictor.

(b) Logistic regression gives

$$y = l(x) = \frac{c}{1 + ae^{-bx}} = \frac{246.33}{1 + 60.36e^{-0.0303x}}$$

We note that $l(200) \approx 216$. The actual population of the United States in 1990 was approximately 249 million. Thus, the logistic curve was a much better predictor, despite the very high correlation for the exponential regression. Screen 2.16 shows the logistic curve and the exponential curve; the squares show the original data, and the crosses show the actual population subsequent to 1890. Notice how the exponential and logistic curves are nearly identical from 1790 to 1890 but then begin to diverge, the logistic curve staying much closer to the actual population.

Notice how the graph of the logistic curve increases exponentially initially; then the increase slows in later years to the point of very little growth. We will have more to say about logistic curves in Chapter 5. ■

Interactive Illustration 2.6

Finding the Best-Fitting Logistic Function in Example 1

This interactive illustration allows you to change the logistic function $y = c/(1 + ae^{-bx})$ and see the resulting sum of squares. There are three sliders: one for each of the variable a, b, and c. Click or drag the mouse on any of the sliders to change the values of the parameters. You will see a meter that gives the sum of the squares as the height of the bar. Adjust the values of the parameters to minimize the sum of the squares (the height of the meter).

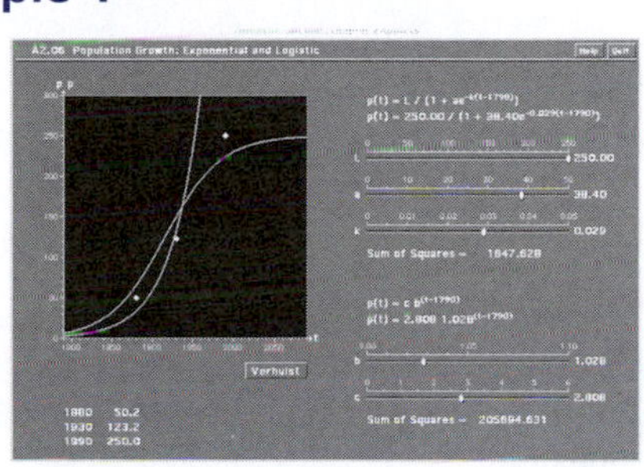

Exercise Set 2.5

1. **Population of Northeast** The table gives the population in millions of the northeastern part of the United States for some selected early years.[70]

Year	1790	1810	1830
Population	2.0	3.5	5.5
Year	1850	1870	1890
Population	8.6	12.3	17.4

   a. On the basis of the data given for the years 1790–1890, find the best-fitting exponential function using exponential regression. Determine the correlation coefficient. Graph. Using this model, estimate the population in 1990.
   b. Now find the best-fitting logistic curve. Graph. Using this model, estimate the population in 1990. Note that the actual population of the Northeast in 1990 was 50.8 million.

2. **Automobile Sales** The following table gives the factory sales in the United States in thousands of passenger cars for some selected early years.[71]

Year	1900	1910	1920	1930	1940	1950
Sales	4	181	1906	2787	3717	6666

   a. On the basis of the data given for the years 1900–1950, find the best-fitting exponential function using exponential regression. Determine the correlation coefficient. Graph. Using this model, estimate factory sales in 1990.
   b. Now find the best-fitting logistic curve. Graph. Using this model, estimate factory sales in 1990. Note that the actual factory sales in 1990 were 6.050 million.

3. **VCR Sales** The following table gives the sales of VCRs in the United States in millions for some selected years.[72]

Year	1978	1980	1982	1984	1986	1988
Sales	0.20	0.84	2.53	8.88	30.92	51.39

   a. On the basis of the data given for the years 1978 to 1988, find the best-fitting exponential function using exponential regression. Determine the correlation coefficient. Graph. Using this model, estimate sales in 1992.
   b. Now find the best-fitting logistic curve. Graph. Using this model, estimate sales in 1992. Note that the actual factory sales in 1992 was 66.78 million.

[70] John W. Wright. 1993. *The Universal Almanac*. New York: Andrews and McMeel.

[71] Ibid.

[72] Ibid.

4. TV Ownership The following table gives the households with televisions in the United States in millions for some selected years.[73]

Year	1950	1955	1960	1965	1970	1975
Number	3.8	32.0	45.2	53.8	60.1	71.5

a. On the basis of the data given for the years 1950–1975, find the best-fitting exponential function using exponential regression. Determine the correlation coefficient. Graph. Using this model, estimate households with TVs in 1992.

b. Now find the best-fitting logistic curve. Graph. Using this model, estimate sales in 1992. Note that the actual number of households with TVs in 1992 was 92.1 million.

5. U.S. Energy Consumption The following table gives the consumption of energy in the United States in quadrillions of Btu for some selected years.[74]

Year	1950	1955	1960	1965	1970	1975
Energy	33.1	38.9	44.8	52.7	66.4	70.6

a. On the basis of the data given for the years 1950–1975, find the best-fitting exponential function using exponential regression. Determine the correlation coefficient. Graph. Using this model, estimate energy consumption in 1992.

b. Now find the best-fitting logistic curve. Graph. Using this model, estimate sales in 1992. Note that the actual consumption of energy in 1992 was 82.4 quadrillions of Btu.

6. Answering Machine Sales The following table gives the sales of answering machines in the United States in millions for some selected years.[75]

Year	1983	1984	1985	1986	1987	1988
Sales	2.2	3.0	4.22	6.45	8.8	11.1

a. On the basis of the data given for the years 1983 to 1988, find the best-fitting exponential function using exponential regression. Determine the correlation coefficient. Graph. Using this model, estimate sales in 1991.

b. Now find the best-fitting logistic curve. Graph. Using this model, estimate sales in 1991. Note that the actual factory sales in 1991 were 14.5 million.

7. U.S. Limiting Population The following table gives the population of the United States in millions for some selected years.[76]

Year	1790	1830	1870
Population	3.9	12.9	38.6
Year	1910	1950	1990
Population	92.2	151.3	248.7

Find the best-fitting logistic curve. Graph on [0, 400]. Using this model, estimate the limiting population.

8. Limiting Population of the Northeast The following table gives the population of the Northeast in millions for some selected years.[77]

Year	1790	1830	1870
Population	2.0	5.5	12.3
Year	1910	1950	1990
Population	25.9	39.5	50.8

Find the best-fitting logistic curve. Graph on [0, 400]. Using this model, estimate the limiting population.

[73] Ibid.
[74] Ibid.
[75] Ibid.
[76] Ibid.
[77] Ibid.

2.6 Selecting the Best Model

Some Guidelines for Selecting a Mathematical Model
Examples

Some Guidelines for Selecting A Mathematical Model

The best mathematical model to use on the x-interval between the smallest and largest values of the x-coordinates of the data points is the one that yields the largest coefficient of determination r^2. You can check all of the possibilities on your computer or graphing calculator. Nonetheless, the following are some useful guidelines on selecting the appropriate model.

1. If the data points on the graph appear to lie more or less on a straight line, select a linear function as a model. Nonetheless, you can still try quadratic, cubic, exponential, and power regression to see whether you obtain a better fit.
2. If the data seems to increase and then decrease or to decrease and then increase, consider quadratics, cubics, or quartics. Linear, power, exponential, logarithmic, and logistic functions do not have this property.
3. If the data is in the first quadrant and is concave down, then you need not try exponential regression, as exponential functions do not have this property.
4. If the data is in the first quadrant and is concave up on part of an interval and concave down on another part, try cubic, quartic, or logistic regression. Linear, quadratic, logarithmic, and exponential functions do not have this property.
5. If the concavity changes twice, select quartic regression.

Examples

EXAMPLE 1 **Economies of Scale**

In 1996 Rogers and Akridge[78] carefully studied the fertilizer and pesticide industry in the state of Indiana and collected data for a medium-sized fertilizer plant relating the amount of fertilizer produced to the cost per ton. The data is found in the following table.

x	4.2	4.8	5.4	6.0	6.6	7.2	7.8
y	298	287	278	270	263	259	255

Here x is in thousands of tons of fertilizer produced, and y is the (total) cost per ton in dollars.

Find the best mathematical model.

[78] Duane S. Rogers and Jay T. Akridge. 1996. Economic impact of storage and handling regulations on retail fertilizer and pesticide plants. *Agribusiness* 12(4):327–337.

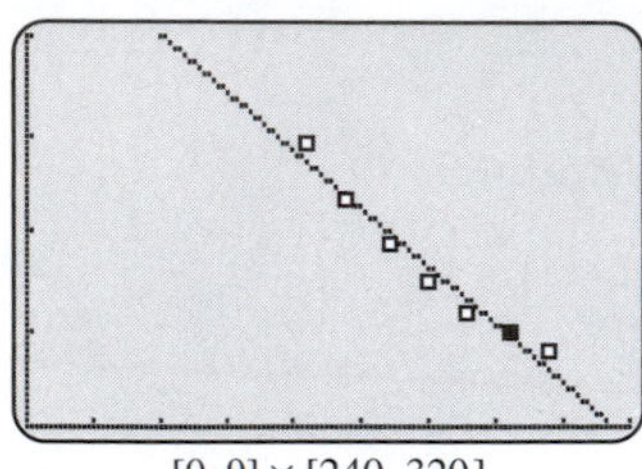

$[0, 9] \times [240, 320]$

Screen 2.17
The graph of the best-fitting straight line in Example 1.

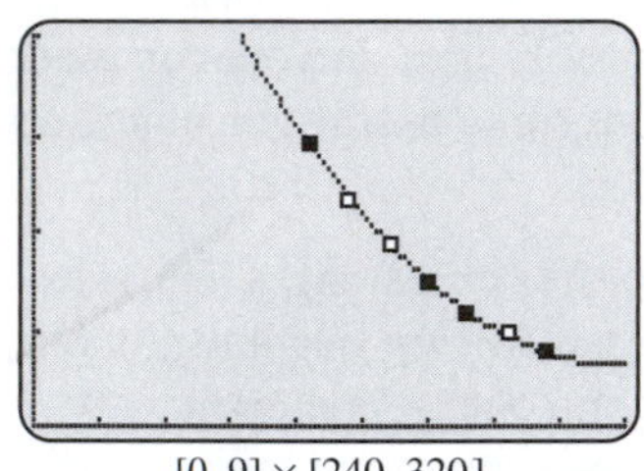

$[0, 9] \times [240, 320]$

Screen 2.18
The graph of the best-fitting quadratic function in Example 1. We clearly see a better fit than that of the linear function shown in Screen 2.17.

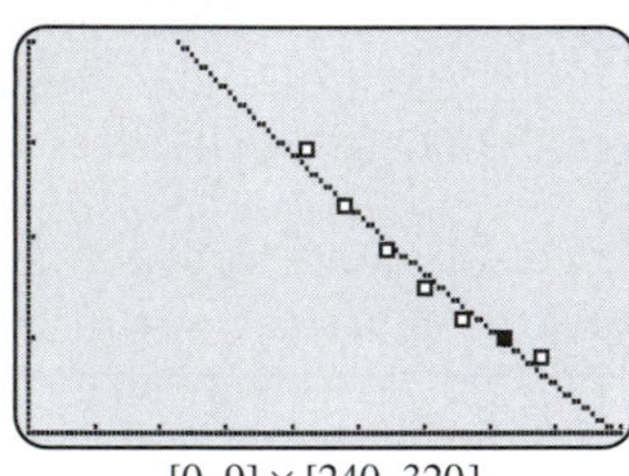

$[0, 9] \times [240, 320]$

Screen 2.19
The graph of the best-fitting exponential function in Example 1.

Solution

Linear regression gives $y = -11.905x + 344.29$ and $r^2 = 0.9686$. Screen 2.17 shows that linear looks good, but the data is slightly concave upward and always decreasing. This suggests also trying quadratic, exponential, power, and logarithmic regression.

Quadratic regression gives $y = 2.050x^2 - 36.508x + 415.14$, $r^2 = 0.9996$. This is larger than the $r^2 = 0.9686$ found for linear regression and therefore is a distinct improvement over linear regression. We readily see from Screen 2.18 that the quadratic appears to be a better fit.

Exponential regression gives $y = 353.42(0.9576)^x$, $r^2 = 0.9753$. This is larger than the r^2 for linear regression but not as large as the r^2 for quadratic regression. Thus, exponential regression gives a better fit than linear regression but not as good a fit as quadratic regression. See Screen 2.19.

Power regression gives $y = 428.29x^{-0.2554}$, $r^2 = 0.9948$. This r^2 is distinctly larger than the r^2 for exponential regression and very slightly smaller than the r^2 for quadratic regression. Thus, power regression gives a distinctly better fit than exponential regression and a fit very nearly as good as that for quadratic regression. See Screen 2.20.

Logarithmic regression gives $y = 397.297 - 70.265 \ln x$, $r^2 = 0.9917$. This is not as large as the r^2 for quadratic regression. Thus, logarithmic regression also gives a fit very nearly as good as for quadratic regression. See Screen 2.21.

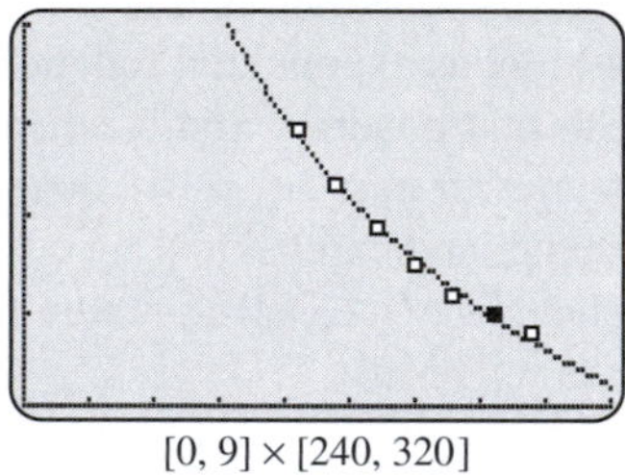

$[0, 9] \times [240, 320]$

Screen 2.20
The graph of the best-fitting power function in Example 1.

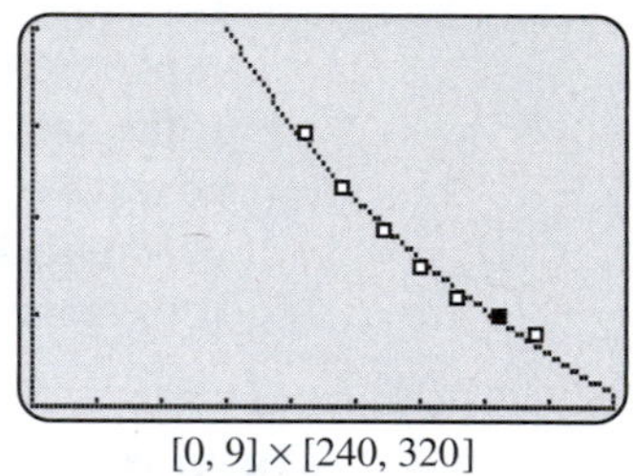

$[0, 9] \times [240, 320]$

Screen 2.21
The graph of the best-fitting logarithmic function in Example 1.

Naturally, cubic regression will give a cubic model that is at least as good a fit as a quadratic model. (It gives $r^2 = 0.9996$.) This is true, since all quadratics are included among all cubics. For the same reason a quartic model will give at least as good a fit as a cubic. (Quartic regression gives $r^2 = 0.9997$.) Notice, however, that r^2 for the quadratic regression is already 0.9996. We should not use a much more complicated model such as a cubic or quartic to gain such a small amount of additional fitness. In general, we should keep our model as simple as possible.

So what is the best mathematical model? The answer is not entirely obvious. First, suppose we want a simple mathematical model that *interpolates*, that is, one that evaluates y for values of x on the interval [4.2, 7.8] between the x-coordinates of the first and last data point. Then we would use the quadratic model, since the quadratic model gives a very large r^2 while still being simple.

But suppose we wish to use our mathematic model to *extrapolate*, or to evaluate y for values of x outside the interval [4.2, 7.8]. (The endpoints of this interval are the x-coordinates of the first and last data points.) Then we need to know what general

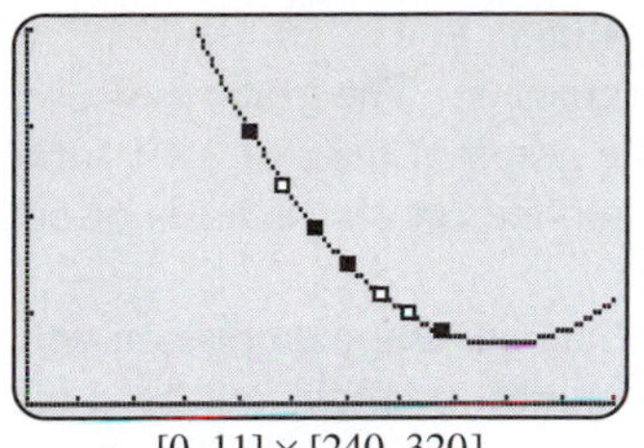

[0, 11] × [240, 320]

Screen 2.22
The graph of the best-fitting quadratic function has been extended. Notice that the graph turns up at the right.

characteristics the graph should have outside this interval and use this to select an appropriate model.

Screen 2.22 shows the graph of the quadratic on the interval [0, 11]. Notice that the graph of the quadratic turns up at the end of this interval. But we expect economies of scale; therefore, we expect the graph to continue to decrease. This is indeed the case for the other three models. Thus, we would probably prefer the power function model, since the graph of the power function continues to decrease and also had an excellent fit on the original interval. ■

EXAMPLE 2 **Average Cost**

Dean[79] made a statistical estimation of the average cost–output relationship for a shoe company. The data for the firm is given in the following table.

x	4	7	13	17	19	27	33	49
y	110	75	65	67	55	45	55	60

Here x is the output in thousands of pairs of shoes, and y is the average cost in dollars.

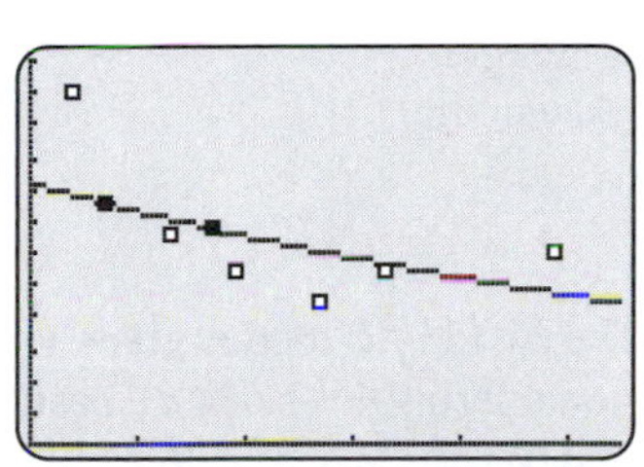

[0, 55] × [0, 120]

Screen 2.23
The graph of the best-fitting exponential function shows a poor fit.

Determine the best mathematical model. Justify your answer.

Solution

In selecting a mathematical model, we want not only a curve that fits the data, but also one that makes sense from a business perspective. For example, average costs should be excessive if only a very small number of items is made. If we have a huge shoe factory but produce only one pair of shoes, the average cost should be astronomical! As we produce more shoes, we expect the obvious economies of scale and therefore expect average cost to decrease. For example, when we produce in bulk, we can purchase raw materials in bulk and naturally expect suppliers to give discounts. This reduces our average costs. On the other hand, if we push our factory to its limits of production, we might expect average costs to go up. For example, production workers would need to work overtime. Since overtime hourly wages may be 50% higher than regular hourly wages, average cost for wages escalates. To increase production to the very limits, older, less efficient machinery may be used. Such inefficiencies also increase average cost.

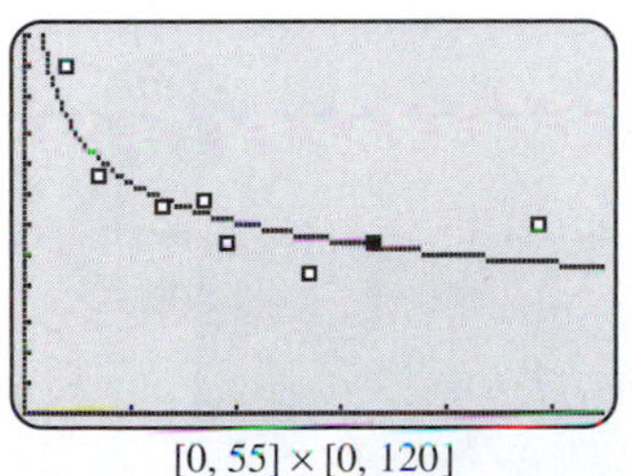

[0, 55] × [0, 120]

Screen 2.24
The best-fitting power function is a better fit than the exponential function in Screen 2.23.

We would not expect an exponential model $y = ab^x$ to be a good fit. First, ab^x does not become huge for small x. Second, ab^x for $0 < b < 1$ will always decrease. Thus, this model will not increase for large x. Indeed, exponential regression gives $y = 81.52(0.9889)^x$ with $r^2 = 0.3890$. This is not a good fit. See Screen 2.23.

We would also not expect to be entirely happy with a power model $y = ax^b$. If $b < 0$, the power function ax^b does become huge when x becomes small. However, such a power function always decreases; therefore, the graph will not turn up for large x. Power regression gives $y = 136.74x^{-0.2698}$ with $r^2 = 0.6956$. This is an improvement over exponential regression. See Screen 2.24.

---

[79] Joel Dean. 1976. *Statistical Cost Estimation.* Bloomington, Ind.: Indiana University Press, page 340.

Quadratic regression gives $y = 0.06204x^2 - 4.063x + 112.76$ with $r^2 = 0.8280$. This is a distinct improvement over power regression. The graph is shown in Screen 2.25. Notice the good fit, and notice that the graph of the quadratic turns up at the right. The graph looks very much as we would expect on the basis of our business insights.

Naturally, cubic regression will lead to a better fit than quadratic regression will. Cubic regression gives $r^2 = 0.8770$. However, this might or might not be a big enough improvement to justify using the more complicated cubic model. The graph of the cubic polynomial obtained is shown in Screen 2.26. Notice how the graph of the cubic polynomial turns downward at the right! This does not make business sense. We certainly would not use this cubic to extrapolate to the right.

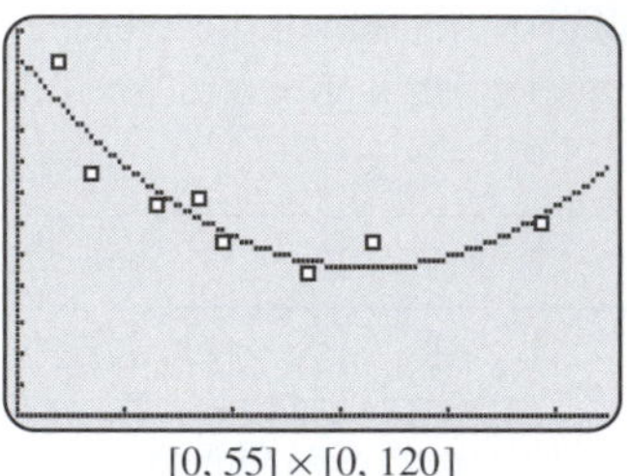

$[0, 55] \times [0, 120]$

Screen 2.25
The best-fitting quadratic function is a better fit than the power function in Screen 2.24.

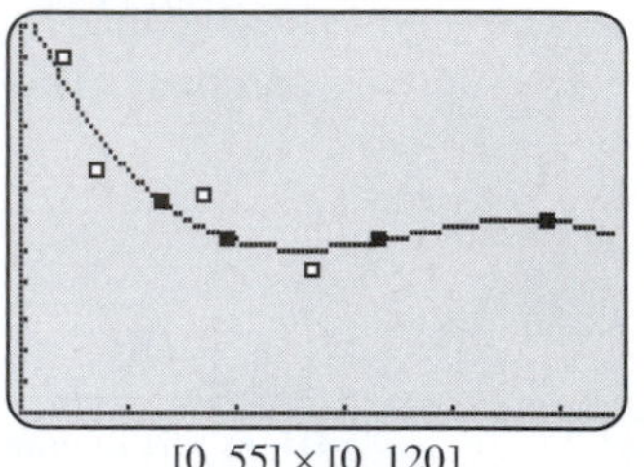

$[0, 55] \times [0, 120]$

Screen 2.26
The graph of the best-fitting cubic function turns downward at the right.

We conclude that the quadratic model is best. The quadratic model gives an excellent fit, is simple, and gives a graph that is reasonable on the basis of basic business considerations. In particular, the quadratic model gives reasonable results in extrapolating outside the original interval [4, 49]. ■

Interactive Illustration 2.7

Finding the Best-Fitting Function in Example 1

This interactive illustration allows you to experimentally select the best-fitting function. You can click on to the particular function you want and then adjust the parameters for that particular function.

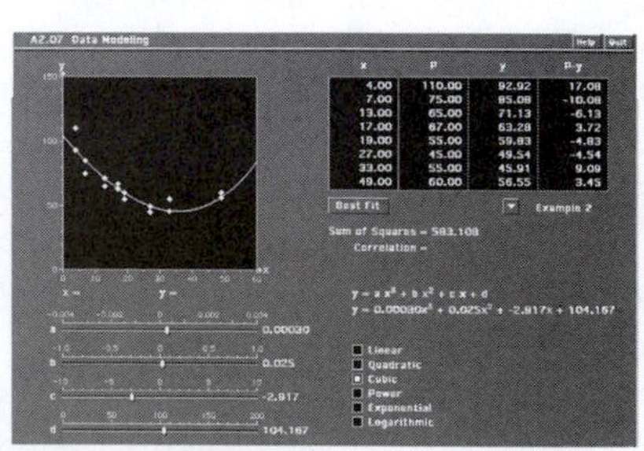

Exercise Set 2.6

1. **Average Cost** Dean[80] made a statistical estimation of the average cost–output relationship for the same shoe company as in Example 2 but for a different year. The data for the firm is given in the following table.

x	4	7	9	12	14	17
y	110	90	75	65	60	65
x	22	27	33	40	50	
y	60	45	52	40	61	

Here x is the output in thousands of pairs of shoes, and y is the average cost in dollars.

Determine the best mathematical model. Justify your answer.

2. **Cost Curve** Dean[81] made a statistical estimation of the labor cost–output relationship for a furniture factory warehouse. The data for the firm is given in the following table.

x	5	6	11	16	22
y	680	720	760	950	1450
x	26	29	32	36	42
y	1650	1450	1650	1750	1950

Here x is the output in thousands of dollars of warehouse value, and y is the total cost of labor in dollars.

Determine the best mathematical model. Justify your answer.

3. **Economic Growth and the Environment** Grossman and Krueger[82] studied the relationship between per capita income and various environmental indicators in a variety of countries. The object was to determine whether environmental quality deteriorates steadily with growth. The table gives the data they collected relating the units y of coliform in waters to the GDP per capita income x in thousands of dollars.

x	1	3	5	7	9	11	13
y	1.8	2.8	2.5	3.7	1	3.5	6

Determine the best mathematical model. Justify your answer.

4. **Environmental Entomology** Allen and coworkers[83] found the relationship between the temperature and the number of eggs laid by the female citrus rust mite. The data is shown in the following table.

x	14	17	19	21	23
y	2.22	6.10	7.93	14.69	11.52
x	25	27	29	31	33
y	10.55	15.44	11.58	11.01	2.64

Here x is the temperature in degrees Celsius, and y is the total eggs per female.

Determine the best mathematical model. Justify your answer.

5. **Commerce on the Internet** The following table gives the estimated commerce on the Internet worldwide in millions of dollars from May 1996 to May 1997.[84]

Month	M	J	J	A	S
Commerce	200	240	260	300	350
Month	O	N	D	J	F
Commerce	380	440	510	550	625
Month	M	A	M		
Commerce	730	790	875		

Let May 1996 correspond to $x = 1$. Determine the best mathematical model that gives sensible extrapolation in both the past and the future. Justify your answer.

[80] Ibid., page 340.

[81] Ibid., page 121.

[82] Gene M. Grossman and Alan B. Krueger. 1995. Economic growth and the environment. *Quart. J. Econ.* CX:352–377.

[83] J. C. Allen, Y. Yang, and J. L. Knapp. 1995. Temperature effects on development and fecundity of the citrus rust mite. *Environ. Entomol.* 24(5):996–1004.

[84] *The Wall Street Journal*, May 22, 1997.

6. **Cost Curve** Dean[85] made a statistical estimation of the salary cost per account in a finance chain with the total number of open accounts. The data is given in the following table.

x	1200	1450	1900	2400	3200
y	0.72	0.61	0.46	0.45	0.43

Here x is the total number of open accounts, and y is the salary cost per account.
Determine the best mathematical model with the requirement that the graph of the function be always decreasing. Justify your answer.

7. **Population Ecology** Fantinou and associates[86] studied the development of the larvae of the corn stalk borer under two regimens. (See the paper for details.) We consider the second regimen in this exercise. The first regimen is considered in Exercise 8. The table gives the instar head capsule width in millimeters. (Only every other piece of data is given.)

Instar	1	3	5	7
Head Width	0.33	0.81	1.76	2.53
Instar	9	11	13	
Head Width	2.69	2.74	2.68	

Determine the best mathematical model that yields an always increasing head width on the x-interval [0, 15]. (There are actually 15 instars.) Justify your answer.

8. **Population Ecology** Refer to Exercise 7. In the second regimen, Fatinou[87] and associates gave the data in the following table.

Instar	1	2	3	4	5	6
Head Width	0.33	0.52	0.81	1.21	1.74	2.32

Determine the best mathematical model that yields the 15th instar with head width less than 5 mm. Justify your answer.

9. **Biological Control** Briano and colleagues[88] studied the host-pathogen relationship between the black imported fire ant and a microsporidium (*T. solenopsae*) that infects them. This ant represents a serious medical and agricultural pest. The following table includes data collected and relates the amount of rainfall in millimeters to the percentage of ants that are infected with *T. solenopsae*.

x	50	65	75	80	85	90	92
y	18	22	15	13	19	23	22
x	94	97	103	115	127	155	
y	25	26	18	26	26	33	

Here x is the rainfall in, and y is the percentage of infected colonies.

Determine the best mathematical model that gives the percentage of infested colonies in terms of the rainfall.

[85] Joel Dean. 1976. *Statistical Cost Estimation.* Bloomington, Ind.: Indiana University Press, page 317.

[86] Argyro A. Fantinou, John A Tsitsipis, and Michael G. Karandinos, 1996. Effects of short- and long-day photoperiods on growth and development of *Sesamia nonagrioides*. *Environ. Entomol.* 25(6):1337–1343.

[87] Ibid.

[88] J. A. Briano, R. S. Patterson, and H. A. Cordo. 1995. Long-term studies of the black imported fire ant infected with a microsporidium. *Environ. Entomol.* 24(5):1328–1332.

3 Limits and the Derivative

This chapter begins the study of calculus. The notion of limit is discussed first. Then the derivative is introduced with applications.

CASE STUDY A Cost Function in a Belt Shop

Joel Dean[1] made an extensive study of cost functions in various industries. We consider now the study he made of cost functions in a belt shop. Rigorous empirical investigations of cost functions have not been numerous owing to the difficulties of obtaining confidential cost data of firms. A firm would divulge important secrets to competitors by giving out approximate cost curve estimates. The following is quoted from Dean:

> In this investigation "cost" is taken to mean the "expenses of production" an entrepreneur incurs in operating an enterprise. The cost theory developed from this point of view is concerned with the magnitude of the cost associated with different levels of operation of a given enterprise. The simplicity of this relation should not be overemphasized, nor its importance underestimated. This cost behavior has a crucial role in determining the most profitable adjustment of the individual enterprise to its economic environment. Consequently, the business executive is also vitally concerned with the response of cost to changes in output.

For a belt shop Dean found that if x is the output in thousands of square feet and y is the total cost in thousands of dollars, then the equation $y = C(x) = -12.995 + 1.330x - 0.0062x^2 + 0.000022x^3$ approximates the relationship between x and y for $50 \leq x \leq 117$. Figure 3.1 shows his data and the graph of the cubic cost function that approximates the data. What important *economic* information can we learn from the slopes of the tangent lines to this cost curve? What economic implications can we infer if as x increases, the slopes of the tangent line to the cost curve decrease? What if they increase? How can we determine unit costs? What is happening to unit costs as x increases? All of these questions are answered in Example 4 of Section 3.4.

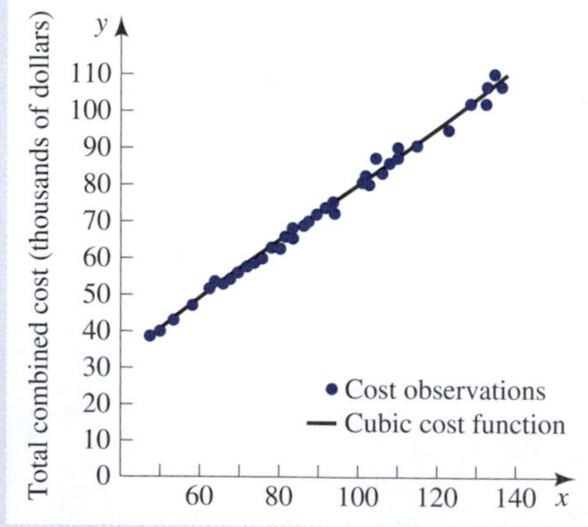

Figure 3.1
Combined cost function for a belt shop.

[1] Joel Dean. 1976. *Statistical Cost Estimation.* Bloomington, Ind.: Indiana University Press, page 215.

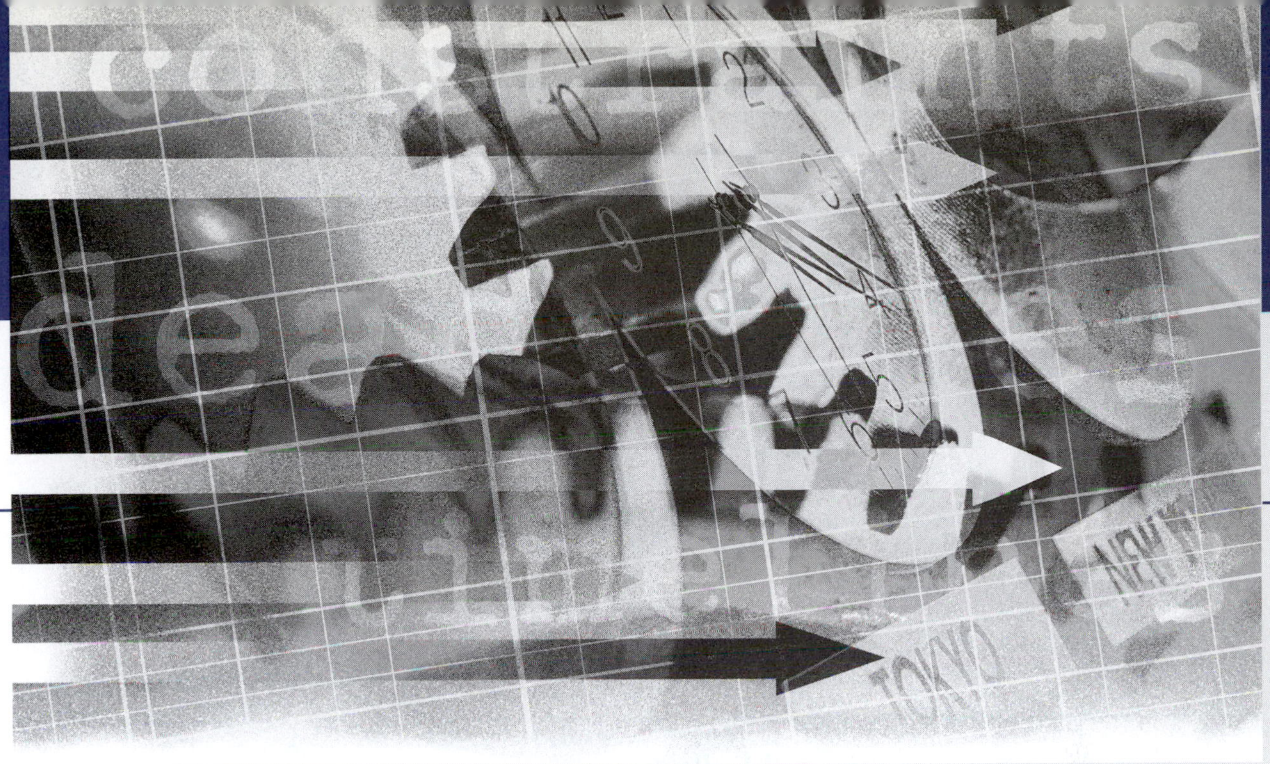

3.0 Introduction to Calculus

Except possibly for the creation of Euclidean geometry, the calculus represents the greatest creation in all of mathematics. Armed with the calculus, we were able to solve major scientific problems of the 17th century. Even today, the use of calculus continues to spread into new areas such as business, economics, and biology.

Four major 17th century problems were solved by the calculus. The first was finding a means of calculating the (instantaneous) velocity of a moving body. To everyone who walks, runs, rides a horse, or flies in a plane it is physically apparent that a moving body has an instantaneous velocity. If the function $s = s(t)$ gives the distance a body has moved in the time t, how can one calculate the (instantaneous) velocity if the velocity is different at different times? One can calculate the average velocity as distance divided by time, but since no distance or time elapses in an instant, this formula results in the meaningless expression $\frac{0}{0}$ if used to calculate the instantaneous velocity.

The second problem was to find the slope of the line that is tangent to a curve at a given point (Figure 3.2). This problem was not just a problem in pure geometry but was also of considerable scientific interest in the 17th century. At that time men such as Pierre de Fermat and Christiaan Huygens were interested in the construction of lenses, which required a knowledge of the angle between the ray of light and the normal line to the lens shown in Figure 3.3. Since the normal line is perpendicular to the tangent line, knowledge of the tangent line will immediately give the normal line.

However, a serious problem arose as to what *tangent* actually meant. For example, if we look at Figure 3.4, it is easy to agree that a tangent to a figure, such as a circle, is that line that touches the figure at only one point and lies entirely on one side of the figure. This was known by the Greeks. But more complex curves such as the one shown in Figure 3.5 were being considered in the 17th century. Is the line shown in this figure a tangent? (We will see later that it is.)

Isaac Newton, 1642–1727

Newton is considered the inventor of the calculus. He also made very significant discoveries in the sciences and is often considered the most accomplished scientific thinker of recent centuries. He attended public schools in England, where his performance was undistinguished. Indeed, he was reported to be idle and inattentive. He blossomed quickly, however, and while confining himself to his room for two years from 1665 to 1667 to avoid the plague that was raging at that time, he laid the foundations for his future accomplishments. During these extremely productive years he set down the basis for the calculus, discovered that white light can be split into different colors, and discovered the law of universal gravitation. Any one of these discoveries would have made Newton an outstanding figure in the history of science.

Gottfried Leibniz, 1646–1716

Leibniz is considered the coinventor of the calculus. He was a precocious child and mentioned once that at age 14 he would go for walks in the woods, "comparing and contrasting the principles of Aristotle with those of Democritus." He invented a calculating machine that could multiply and divide, commenting that "It is unworthy of excellent men to lose hours like slaves in the labor of calculation." He received a degree in law and served in the diplomatic service, only later taking up mathematics. He developed much of the currently used notation in the calculus and a number of the elementary formulas.

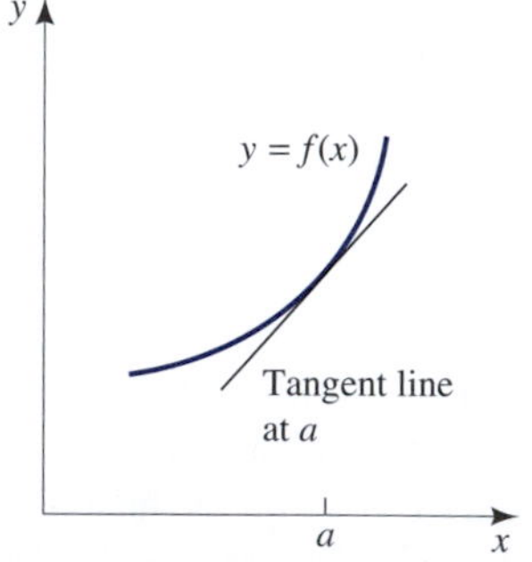

Figure 3.2
Finding the slope of the tangent line was a problem of major importance.

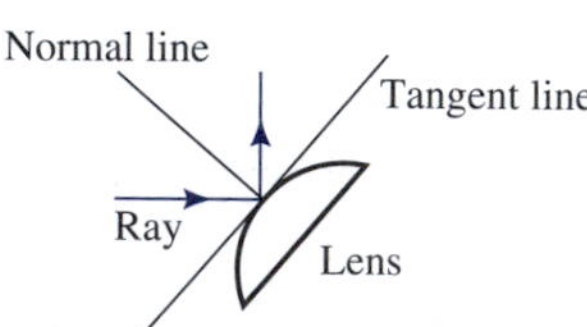

Figure 3.3
Making lenses requires knowledge of the tangent line.

The third problem was finding the maximum or minimum value of a function on some given interval. A contemporary example of such a problem is finding the point at which maximum profits occur (Figure 3.6).

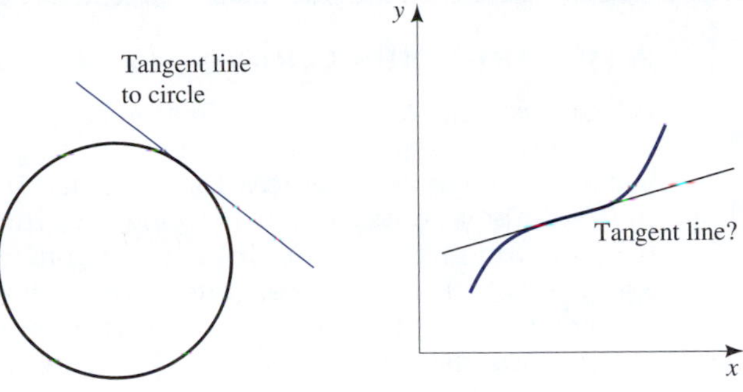

Figure 3.4
It is clear what a tangent line to a circle is.

Figure 3.5
Do we want to call this line a tangent line? We do.

The fourth type was finding the area under a curve (Figure 3.7). We will see in a later chapter why this is so important in the applications.

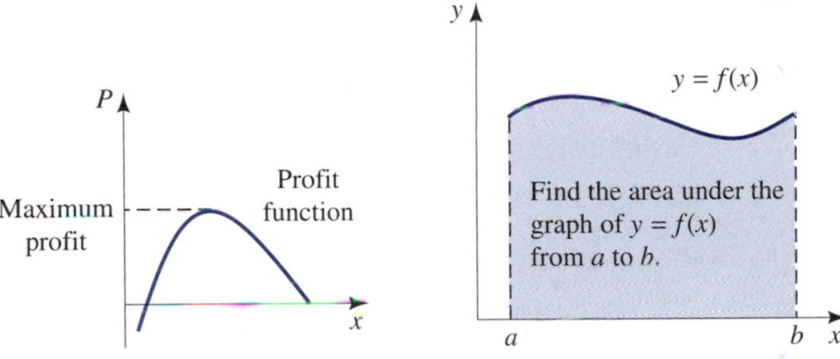

Figure 3.6
We want to find maxima and minima of functions.

Figure 3.7
We want to find the area between the graph of $y = f(x)$ and the x-axis.

In the first problem considered above, velocity measures the rate at which distance changes with respect to a change in time. In exact analogy, the calculus will then enable us to determine, among other things, the rate at which profits change with respect to the change in the number of items produced and sold, the rate at which a population grows with respect to change in the time, and the rate at which demand for an item drops with respect to increases in its price. Thus, we will see that the calculus can be used not only to solve the great scientific problems of the past centuries, but also to solve many important problems in modern business, economics, and biology.

Augustin Louis Cauchy, 1789–1857

It is interesting to note that Newton and Leibniz, who are given credit for the discovery of the calculus, did not develop the theory on the basis of a clear and logical foundation. Indeed, they and the many individuals who expanded the calculus in the subsequent 150 years used arguments that were intuitively plausible rather than mathematically exact. From the beginning the calculus, as presented, evoked considerable controversy, many critics pointing out (correctly) that many fundamental concepts were vague. The success of the calculus in solving many physical problems is what gave individuals the confidence that the mathematics must somehow be correct. It was Cauchy who finally placed the calculus on a sound footing. He developed an acceptable theory of limits and used this theory as the logical underpinnings of the calculus. Thus, Cauchy has become known as the creator of the modern calculus.

3.1 Limits

APPLICATION The Unemployment Rate and the Inflation Rate

A. W. Phillips[2] collected extensive data for the United Kingdom that indicated a clear relationship between the unemployment rate and the rate of inflation: The lower the unemployment rate, the higher the rate of inflation. Phillips originally published data for the years 1861–1913. If x is the unemployment rate in percent and $y = f(x)$ is the percentage change in inflation, then Phillips found that $f(x)$ was approximated by

$$y = f(x) = -1 + \frac{10}{x^{1.4}}$$

He then showed that observations for the years 1948–1957 lay close to his original curve. According to this model, what happens to the inflation rate as the unemployment rate gets closer and closer to zero? For the answer see Example 8 on page 150.

[2] A. W. Phillips. 1958. The relation between unemployment and the rate of change of money wage rates in the United Kingdom, 1861–1957. *Economica* 25:283–299.

Introduction to Limits

Before beginning to read this section, you will find it very helpful to do the following exploration.

Exploration 1

Finding a Limit

Let $f(x) = \dfrac{x^3 + x^2 - x - 1}{x - 1}$, where $x \neq 1$. What happens to $f(x)$ as x gets closer and closer to 1? Filling out the following table will help you answer this question.

x	$f(x)$	x	$f(x)$
1.1		0.9	
1.01		0.99	
1.001		0.999	
1.0001		0.9999	
1.00001		0.99999	

The foundation of calculus rests on the notion of limit. To understand the fundamental ideas in calculus, we need an intuitive understanding of limits.

Toward this end, consider the graph of some function $y = f(x)$ shown in Figure 3.8. Notice that the function is not defined for the value $x = 1$. We seek to understand what is happening to $f(x)$ as x approaches (gets closer and closer to) 1 without x actually being equal to 1.

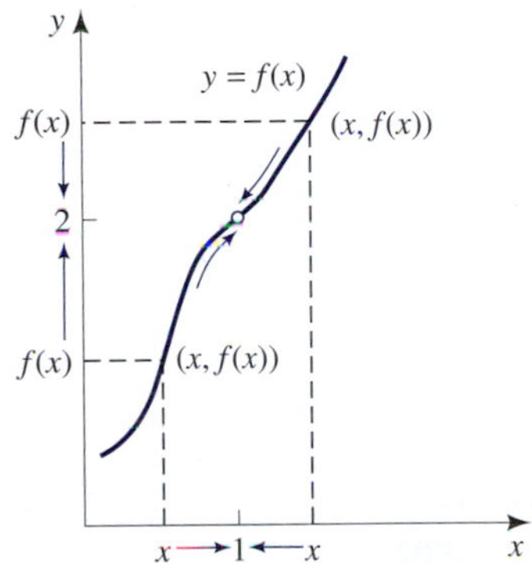

Figure 3.8
No matter how $x \to 1$, $f(x) \to 2$. Thus, $\lim_{x \to 1} f(x) = 2$.

First notice that it is meaningless to have x actually equal the value 1, since $f(1)$ is not even defined. In Figure 3.8 we can see that as x approaches 1 from the right, the point $(x, f(x))$ slides along the graph and approaches the point $(1, 2)$ and $f(x)$ approaches the number 2. Also, we can see that as x approaches 1 from the left, the point $(x, f(x))$ slides along the graph and approaches the point $(1, 2)$, and $f(x)$ again approaches the number 2. We then see that no matter how x approaches 1, $f(x)$ approaches the same number $L = 2$.

In such a case we say that $L = 2$ is the **limit** of $f(x)$ as x approaches 1, and we write

$$\lim_{x \to 1} f(x) = 2$$

We now give the following intuitive definition of limit of a function.

DEFINITION **Limit of a Function**

Suppose that the function $f(x)$ is defined for all values of x near a but not necessarily at a. If as x approaches a (without actually attaining the value a), $f(x)$ approaches the number L, then we say that L is the limit of $f(x)$ as x approaches a and write

$$\lim_{x\to a} f(x) = L$$

We say that this definition is intuitive, since the definition does not define the term "approaches" in a strictly mathematical sense.

Remark Notice that in this definition, $f(a)$ need not be defined. Furthermore, even if $f(a)$ is defined, the limit process never requires knowledge of $f(a)$, since this process requires only knowledge of $f(x)$ for x near a but not *equal* to a. Thus, even if $f(a)$ is defined, this value has nothing to do with the limit as x approaches a.

In determining the limit of $f(x)$ as x approaches 1, we found it convenient to consider the limit of $f(x)$ as x approaches 1 from the left and also the limit of $f(x)$ as x approaches 1 from the right. These two limits are written as

$$\lim_{x\to 1^-} f(x) \qquad \text{and} \qquad \lim_{x\to 1^+} f(x)$$

respectively. In the preceding example we discovered that both of these limits exist and equal the very same number, $L = 2$. We then concluded that the limit of $f(x)$ as x approaches 1 existed and equaled $L = 2$. This reasoning is true in general. We then have the following theorem.

Existence of a Limit

If the function $f(x)$ is defined near $x = a$ but not necessarily at $x = a$, then

$$\lim_{x\to a} f(x) = L$$

if and only if both the limits

$$\lim_{x\to a^-} f(x) \qquad \text{and} \qquad \lim_{x\to a^+} f(x)$$

exist and are equal to the same number L.

The following example should further clarify the idea of the limit.

EXAMPLE 1 **Finding the Limit Graphically**

Define the function $g(x)$ to be the very same function as $f(x)$, except that $g(1) = 1.5$ (Figure 3.9). Find $\lim_{x\to 1} g(x)$ if it exists.

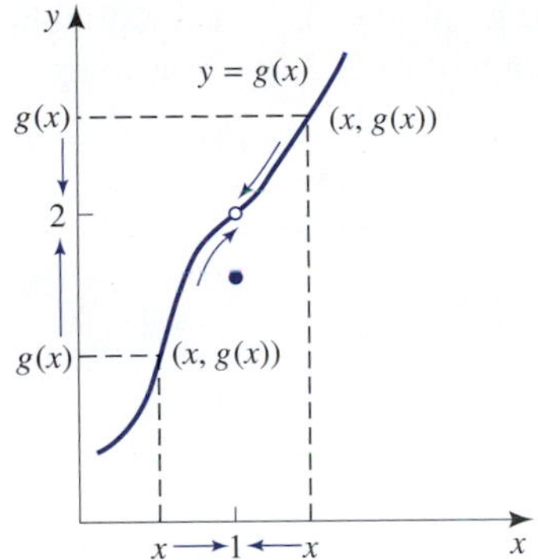

Figure 3.9
$\lim_{x \to 1} g(x) = 2$ even though $g(1) = 1.5$. The value of $g(x)$ at $x = 1$ is irrelevant to the limit as $x \to 1$.

Solution
Notice that from Figure 3.9 that we still have

$$\lim_{x \to 1^-} g(x) = 2 \qquad \text{and} \qquad \lim_{x \to 1^+} g(x) = 2$$

Thus,

$$\lim_{x \to 1} g(x) = 2.$$

■

Remark Notice that this last statement is true despite the fact that $g(1) = 1.5$, since the limit process does not involve the value of the function at $x = 1$.

The two functions $f(x)$ and $g(x)$ given above are not actually the same function, since they differ for the single value $x = 1$. They are equal for all other values, however, and therefore the limits as x approaches 1 of these two functions must be equal.

We now look at two limits that we can find using our intuitive knowledge of limits. In each case we will support our answer both numerically and graphically.

EXAMPLE 2 **Finding the Limit Intuitively and Supporting It Numerically and Graphically**

Find $\lim_{x \to 2} (3x + 1)$ if it exists.

Solution
It seems intuitive that as x approaches 2, $3x$ approaches 6, and $3x + 1$ approaches $6 + 1 = 7$. This is supported by the calculations shown in Table 3.1.

Table 3.1

x	1.9	1.99	1.999	→	2	←	2.001	2.01	2.1
$3x + 1$	6.7	6.97	6.997	→	7	←	7.003	7.03	7.3

Furthermore, the graph $y = 3x + 1$ is a straight line and is shown in Figure 3.10. We can see that as x approaches 2, y seems to approach 7. Thus,

$$\lim_{x \to 2} (3x + 1) = 7$$

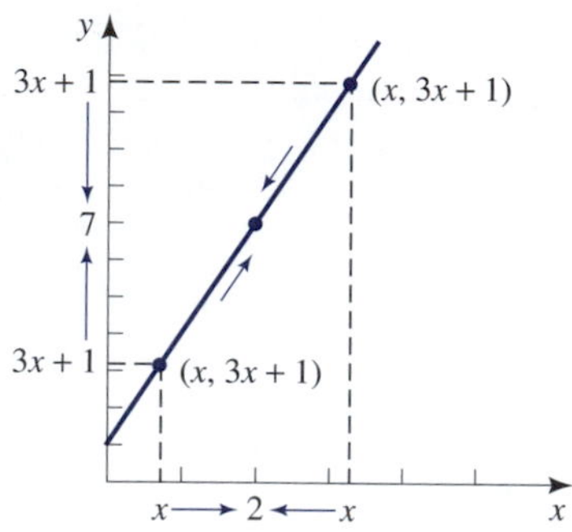

Figure 3.10
As $x \to 2$, $3x + 1 \to 3(2) + 1 = 7$. Thus, $\lim_{x \to 2} (3x + 1) = 7$.

■

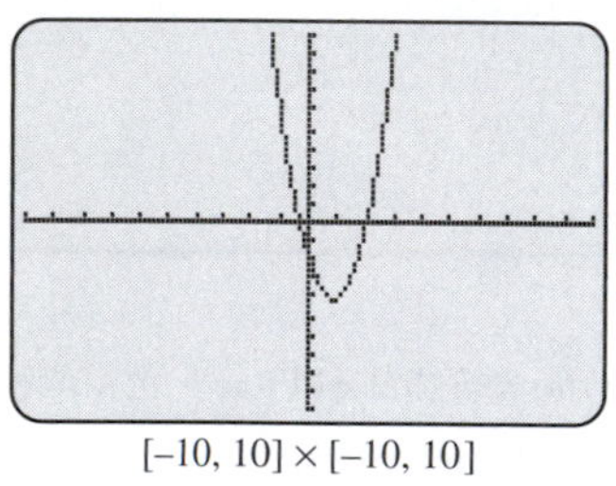

[−10, 10] × [−10, 10]

Screen 3.1
From the graph of $y = 3x^2 - 5x - 2$ we see that $y \to 0$ as $x \to 2$.

EXAMPLE 3 **Finding the Limit Intuitively and Supporting It Graphically and Numerically**

Find $\lim_{x \to 2} (3x^2 - 5x - 2)$ based on your intuitive understanding of limits. Support your answer graphically and numerically.

Solution

It seems intuitive that as x approaches 2, $x^2 = x \cdot x$ approaches $2 \cdot 2 = 4$, $3x^2$ approaches $3 \cdot 4 = 12$, and $3x^2 - 5x - 2$ approaches $12 - 5(2) - 2 = 0$.

The graph $y = 3x^2 - 5x - 2$ is shown in Screen 3.1. Viewing the graph, we can see that as x approaches 2, y seems to approach 0.

This is also supported by the numerical calculations shown in Table 3.2.

Table 3.2

x	1.9	1.99	1.999	→	2	←	2.001	2.01	2.1
$3x^2 - 5x - 2$	−0.670	−0.697	−0.007	→	0	←	0.007	0.0703	0.730

Thus,

$$\lim_{x \to 2} (3x^2 - 5x - 2) = 0$$

■

The limits in Examples 2 and 3 were rather transparent. The following limit is not so transparent.

EXAMPLE 4 **Finding the Limit Using Cancellation**

Find $\lim_{x \to 2} \left(\dfrac{3x^2 - 5x - 2}{x - 2} \right)$ if it exists.

Solution

First notice that the function

$$f(x) = \frac{3x^2 - 5x - 2}{x - 2}$$

is not defined at $x = 2$, since setting $x = 2$ in the fraction yields $\frac{0}{0}$. We saw in Example 3 that the numerator approaches 0 as x approaches 2. On the other hand, the denominator, $x - 2$, also approaches 0 as x approaches 2. So the fraction approaches $\frac{0}{0}$. At this point it is not clear whether the limit exists.

To get some idea of what is happening as x approaches 2, we should evaluate the function for various values of x. The results are shown in Table 3.3.

Table 3.3

x	1.9	1.99	1.999	$\rightarrow$	2	$\leftarrow$	2.001	2.01	2.1
$f(x)$	6.7	6.97	6.997	$\rightarrow$	7	$\leftarrow$	7.003	7.03	7.3

As Table 3.3 indicates, $f(x)$ seems to approach the number 7 as x approaches $x = 2$ from the right and also seems to approach the number 7 as x approaches $x = 2$ from the left.

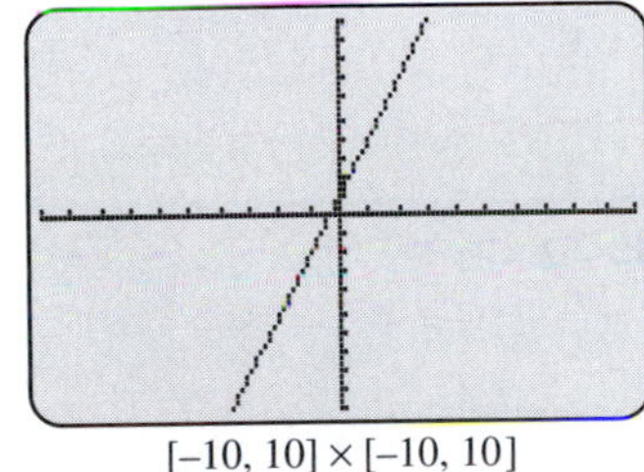

[−10, 10] × [−10, 10]

Screen 3.2
From the graph of $y = \frac{3x^2 - 5x - 2}{x - 2}$ we see that $y \rightarrow 7$ as $x \rightarrow 2$.

Furthermore, Screen 3.2 shows a graph of

$$y = \frac{3x^2 - 5x - 2}{x - 2}$$

From this graph it appears also that as x approaches 2, y approaches 7. Thus, we have every reason to suspect that the limit is equal to 7.

We can confirm this suspicion algebraically. First notice that the numerator can be factored as

$$3x^2 - 5x - 2 = (3x + 1)(x - 2)$$

Therefore,

$$f(x) = \frac{(3x + 1)(x - 2)}{(x - 2)} = 3x + 1 \qquad \text{for} \quad x \neq 2$$

This is just a line, except that the function is not defined at $x = 2$. A graph is shown in Figure 3.11, where the circle at the point (2, 7) indicates that this point is not on

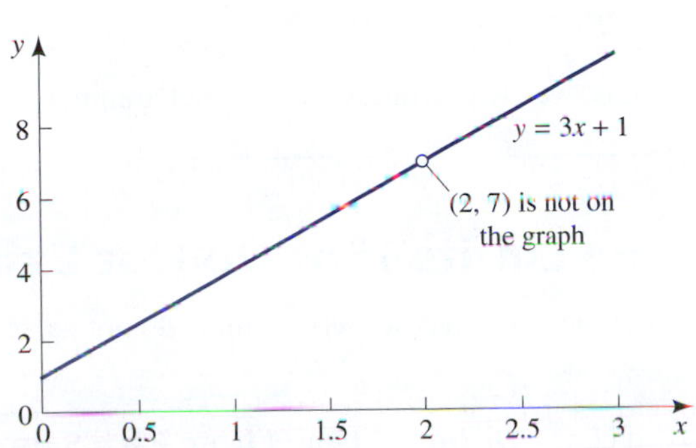

Figure 3.11

the graph of the function. Then

$$\begin{aligned} \lim_{x\to 2} \frac{3x^2 - 5x - 2}{x - 2} &= \lim_{x\to 2} \frac{(3x+1)(x-2)}{(x-2)} && \text{Factor} \\ &= \lim_{x\to 2} (3x+1) && \text{Cancel} \\ &= 7 && \text{Example 2} \end{aligned}$$

where we can cancel the factor $(x - 2)$, since taking the limit does not involve the value $x = 2$. ■

Exploration 2

Find Graphically, Confirm Algebraically

Find $\lim_{x\to 1} \left(\dfrac{2x^2 - x - 1}{x - 1}\right)$ graphically if it exists.
Confirm algebraically.

In Example 5 we present another limit that is not at all transparent. Although we will be able to give strong graphical and numerical support for the limit in that example, we do not have the necessary mathematical tools at this time to confirm our answer algebraically.

[0, 2] × [0, 4]

Screen 3.3
From the graph of $y = x^x$ we see that $y \to 1$ as $x \to 0^+$.

EXAMPLE 5 Finding a Limit Graphically and Supporting It Numerically

Find $\lim_{x\to 0^+} x^x$.

Solution

You probably do not have the vaguest idea of what the limit might be or even whether the limit exists. Using a screen with dimensions [0, 2] by [0, 4] we obtain Screen 3.3. Screen 3.3 seems to indicate that 1 is the limit of $y = x^x$ as x goes to 0^+.

The following table lends numerical support.

x	0	←	0.0001	0.001	0.01	0.1	1
y	1	←	0.99908	0.99312	0.95499	0.79433	1

Thus, we very strongly suspect, but are unable to prove, that $\lim_{x\to 0^+} x^x = 1$. ■

Some Limits That Do Not Exist

We now look at two cases where limits do not exist.

EXAMPLE 6 Different Limits from Each Side

A furniture store charges \$20 to deliver a couch to your house if you live 10 miles or less from the store and charges \$2.50 per mile from the store if you live more than 10

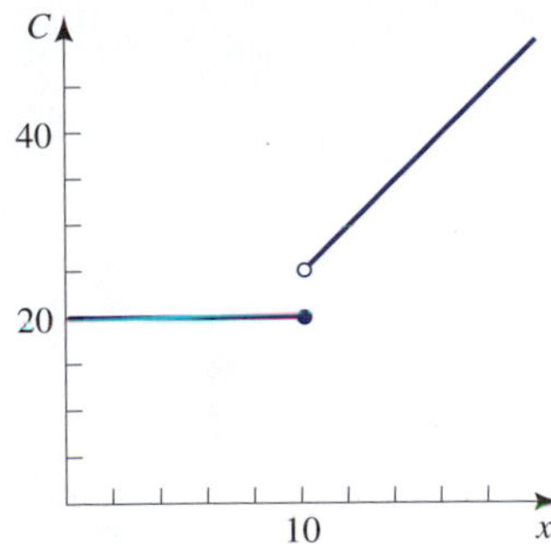

Figure 3.12
$\lim_{x \to 10^-} C(x) = 20$ while $\lim_{x \to 10^+} C(x) = 25$. Because these two limits are not equal, $\lim_{x \to 10} C(x)$ does not exist.

miles from the store. Find the function $C(x)$ that gives the delivery cost, where x is measured in miles. Find $\lim_{x \to 10} C(x)$, if the limit exists.

Solution

The function $C(x)$ is given by

$$C(x) = \begin{cases} 20 & \text{if } x \leq 10 \\ 2.5x & \text{if } x > 10 \end{cases}$$

This function is graphed in Figure 3.12. Notice that

$$\begin{aligned} \lim_{x \to 10^-} C(x) &= \lim_{x \to 10^-} 20 \\ &= 20 \\ \lim_{x \to 10^+} C(x) &= \lim_{x \to 10^+} 2.5x \\ &= 25 \end{aligned}$$

Because the limits from each side are not the same, $\lim_{x \to 10} C(x)$ does not exist. ■

EXAMPLE 7 **Limit Does Not Exist If the Function Is Unbounded**

Determine what is happening to

$$f(x) = \frac{1}{1 - x}$$

as x approaches 1 from the left. In particular, does $\lim_{x \to 1^-} f(x)$ exist?

Solution

To help us answer the questions, we evaluate $f(x)$ for values of x close to but less than $x = 1$. This is summarized in Table 3.4.

Table 3.4

x	0.9	0.99	0.999	0.9999	0.99999	→	1
$f(x)$	10	100	1000	10,000	100,000	→	?

We see that as x gets closer and closer to 1, $f(x)$ becomes large without bound. See Figure 3.13. Therefore, since $f(x)$ is not approaching any particular number L, $\lim_{x \to 1^-} f(x)$ does not exist. ■

One practical place where vertical asymptotes arise is in pollution abatement. Hernandez and associates[3] studied the total costs of removing a pollutant. Their graph is shown in Figure 3.14. Notice how the total costs begin to explode once 99% of the pollutant—benzene, in this case—is removed. Although it is difficult to see in

[3] O. Romero-Hernandez, E. N. Pistikopoulos, and A. G. Livingston. 1998. Waste treatment and optimal degree of pollution abatement. *Environ. Progress* 17(4):270–277.

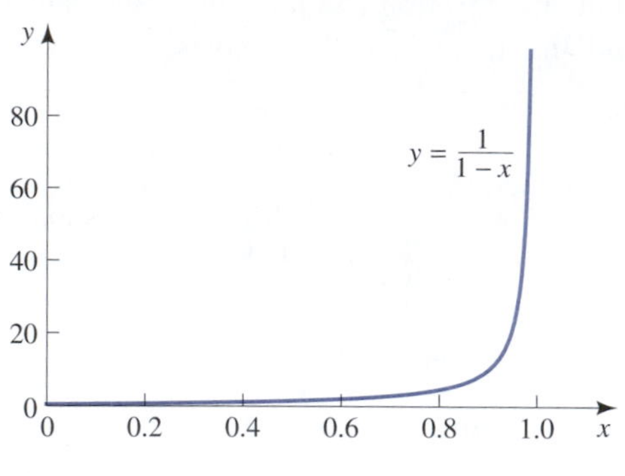

Figure 3.13

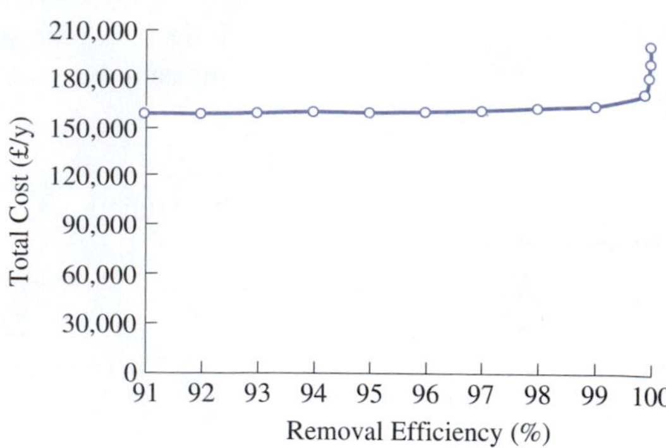

Figure 3.14
Treatment costs as a function of benzene removed.

the graph, the researchers report that increasing the removal from 99.9% to 99.9999% increases the cost by 30% and provides little or no reduction in environmental impact. Viewed another way, the money might be better spent reducing environmental impact elsewhere. We can summarize mathematically by saying that as the removal approaches 100%, the cost becomes large without bound. If $C(x)$ is the total cost, then $\lim_{x \to 100^-} C(x)$ does not exist.

Whenever the graph of a curve, such as that shown in Figure 3.13, approaches a vertical line $x = x_0$, we say that $x = x_0$ is a **vertical asymptote**.

DEFINITION **Vertical Asymptote**

Suppose as $x \to c^-$ or $x \to c^+$ the function $y = f(x)$ becomes large (either positively or negatively) without bound, that is, the graph of $y = f(x)$ approaches the vertical line $x = c$, then the line $x = c$ is called a **vertical asymptote**.

EXAMPLE 8 **Limit Does Not Exist If the Function Is Unbounded**

A. W. Phillips[4] collected extensive data for the United Kingdom that indicated a clear relationship between the unemployment rate and the rate of inflation: The lower the unemployment rate, the higher the rate of inflation. He originally published data for the years 1861–1913. If x is the unemployment rate in percent and $y = f(x)$ is the percentage change in inflation, then Phillips found that $f(x)$ was approximated by

$$y = f(x) = -1 + \frac{10}{x^{1.4}}$$

He then showed that observations for the years 1948–1957 lay close to his original curve. According to this model, what happens to the inflation rate as the unemployment rate gets closer and closer to zero?

Solution

In Table 3.5, we see the result of evaluating $f(x)$ at values of x that approach 0 from the right. The values of the function $f(x)$ increase without bound and therefore do

[4] A. W. Phillips. 1958. The relation between unemployment and the rate of change of money wage rates in the United Kingdom, 1861–1957. *Economica* 25:283–1299.

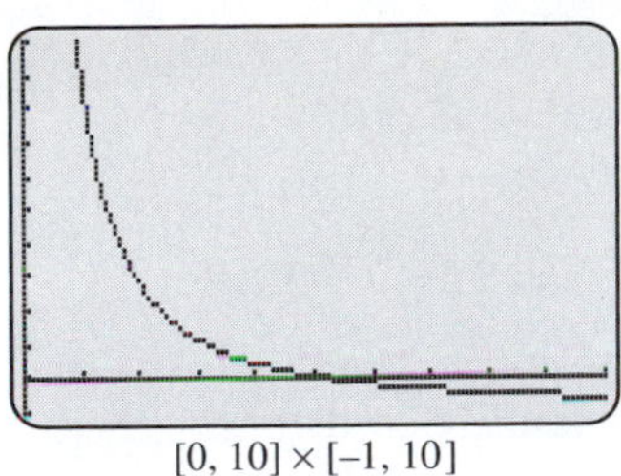
$[0, 10] \times [-1, 10]$

Screen 3.4
According to the Phillips curve, as unemployment (x-axis) decreases inflation (y-axis) increases.

CONNECTION
Phillips Curve

Up until the late 1960s the Phillips curve had received widespread and largely uncriticized acceptance by the monetary authorities in the United States: When the unemployment rate is low, the inflation rate will be high, and vice versa. Many policy makers assumed that their task was to choose a particular inflation-unemployment target and adopt policies to achieve it. For an interesting early history of the Phillips curve, see Humphrey.[5]

The Phillips curve has recently become controversial. Two excellent recent articles on the status of the Phillips curve controversy can be found in the first- and second-quarter issues of the *Economic Review of the Federal Reserve Bank of Atlanta*.[6,7]

[5] Thomas M. Humphrey. 1985. The early history of the Phillips' curve. *Econ. Rev. Fed. Res. Bank Richmond* 71(5):17–24.

[6] Roberto Chang. 1997. Is low unemployment inflationary? *Econ. Rev. Fed. Res. Bank Atlanta* 82(1):4–25.

[7] Marco A. Espinosa-Vega and Steven Russell. 1997 History and theory of the NAIRU: a capital review. *Econ. Rev. Federal Reserve Bank Atlanta* 82(2):4–25.

Table 3.5

x	0	$\leftarrow$	0.0001	0.001	0.01	0.1	1
$f(x)$	?	$\leftarrow$	3,981,071	158,488	6309	250	9

not approach any particular number L. This conclusion is supported in Screen 3.4. Thus, $\lim_{x\to 0^+} f(x)$ does not exist. ■

Remark We have now seen two basic ways in which a limit at $x = a$ might not exist.

1. If a function "blows up" near $x = a$, such as in Figure 3.13 for $a = 1$
2. If the limits from each side of $x = a$ exist but are unequal, as in Figure 3.12 at $a = 10$

Rules for Limits

We have just seen that graphs and tables can be used to find limits or to determine that limits do not exist. Rules concerning limits can be used to evaluate limits efficiently, however. We now give a number of such rules. For convenience, let us use the notation $x \to a$ to represent the phrase "x approaches a" and $f(x) \to L$ to represent the phrase "$f(x)$ approaches L."

First consider $\lim_{x\to a} p(x)$, where $p(x)$ is a polynomial. Take, for example, the polynomial in Example 3, $p(x) = 3x^2 - 5x - 2$, $a = 2$. It seems intuitive that as $x \to 2$, $p(x) \to 3(2)^2 - 5(2) - 2$, which is $p(2)$. Thus, $\lim_{x\to 2} p(x) = p(2)$. This means that we can evaluate $\lim_{x\to 2} p(x)$ by *substituting* 2 for x in $p(x)$. This same reasoning applies to any polynomial function.

The Limit of a Polynomial Function

If $p(x)$ is any polynomial and a is any number, then $\lim_{x\to a} p(x) = p(a)$.

Let us now consider some other rules. Let $f(x)$ and $g(x)$ be two functions, and suppose that as $x \to a$, $f(x) \to L$ and $g(x) \to M$. Then it is reasonable that $cf(x) \to cL$, $f(x) \pm g(x) \to L \pm M$, $f(x)g(x) \to LM$, $\dfrac{f(x)}{g(x)} \to \dfrac{L}{M}$ (if $M \neq 0$), and $[f(x)]^n \to L^n$ when L^n makes sense. This then gives the following five rules.

Rules for Limits

Assume that

$$\lim_{x\to a} f(x) = L \quad \text{and} \quad \lim_{x\to a} g(x) = M. \text{ Then}$$

Rule 1. $\lim_{x\to a} cf(x) = c \lim_{x\to a} f(x) = cL$

Rule 2. $\lim_{x\to a} (f(x) \pm g(x)) = \lim_{x\to a} f(x) \pm \lim_{x\to a} g(x) = L \pm M$

Rule 3. $\lim_{x \to a} (f(x) \cdot g(x)) = \left(\lim_{x \to a} f(x)\right) \cdot \left(\lim_{x \to a} g(x)\right) = L \cdot M$

Rule 4. $\lim_{x \to a} \frac{f(x)}{g(x)} = \frac{\lim_{x \to a} f(x)}{\lim_{x \to a} g(x)} = \frac{L}{M}$ if $\lim_{x \to a} g(x) = M \neq 0$

Rule 5. $\lim_{x \to a} (f(x))^n = L^n$, n any real number, L^n defined, $L \neq 0$

Remark The following is a verbal interpretation of the above rules.

1. The limit of a constant times a function is the constant times the limit of the function.
2. The limit of a sum or difference equals the sum or difference of the limits.
3. The limit of a product is the product of the limits.
4. The limit of a quotient is the quotient of the limits if the limit of the denominator is not zero.
5. The limit of a function raised to a power is the limit of the function raised to the power provided that the terms make sense.

Remark In Rule 5, L^n will not be defined if, for example, $L < 0$ and $n = 1/2$, $1/4$, and so forth or if $L = 0$ and $n \leq 0$. Refer to Exercise 52 on page 158 for further comments on Rule 5.

Consider also the following: If $f(x) = c$, then since $f(x)$ is always the constant c, the limit as $x \to a$ must also be c. Thus, $\lim_{x \to a} c = c$. Furthermore, it is intuitively clear that $\lim_{x \to a} x = a$.

EXAMPLE 9 **Using the Rules of Limits**

Evaluate

(a) $\lim_{x \to 3} \frac{x^2 + 11}{x^2 - 4}$ (b) $\lim_{x \to 1} (x^2 - 9)^{1/3}$ (c) $\lim_{x \to 4} \frac{x^2 - 16}{x^2 - 2x - 8}$

Solution

(a) $\lim_{x \to 3} \frac{x^2 + 11}{x^2 - 4} = \frac{\lim_{x \to 3} (x^2 + 11)}{\lim_{x \to 3} (x^2 - 4)}$ Rule 4

$= \frac{(3)^2 + 11}{(3)^2 - 4} = \frac{20}{5} = 4$ Limit of polynomial functions

(b) $\lim_{x \to 1} (x^2 - 9)^{1/3} = \left(\lim_{x \to 1} (x^2 - 9)\right)^{1/3}$ Rule 5

$= (-8)^{1/3} = -2$ Limit of polynomial functions

(c) Using Rule 4 and substitution gives $\frac{0}{0}$. Therefore, we must try to factor the numerator and denominator to find a common factor:

$$\begin{aligned}
\lim_{x\to 4}\frac{x^2-16}{x^2-2x-8} &= \lim_{x\to 4}\frac{(x+4)(x-4)}{(x+2)(x-4)} && \text{Factor}\\
&= \lim_{x\to 4}\frac{(x+4)}{(x+2)} && \text{Cancel}\\
&= \frac{\lim_{x\to 4}(x+4)}{\lim_{x\to 4}(x+2)} && \text{Rule 4}\\
&= \frac{4+4}{4+2}=\frac{4}{3} && \text{Limit of polynomial functions}
\end{aligned}$$

The graph for part(c) is shown in Figure 3.15.

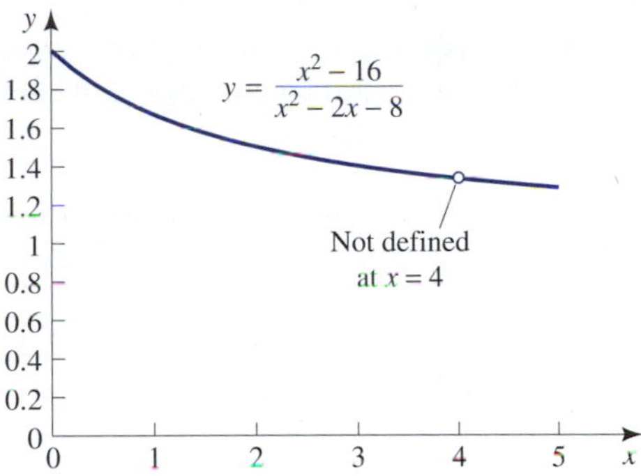

Figure 3.15

■

EXAMPLE 10 **Finding a Limit Graphically**

Graphically find $\lim_{x\to 1}\dfrac{\sqrt{x}-1}{x-1}$.

Solution

First notice that both the numerator and denominator are heading for 0 as $x \to 1$. Therefore, we cannot use Rule 4. Using a window of dimension $[0, 2]$ by $[0, 1]$, we obtain Screen 3.5. From this it appears that the limit is 0.5. ■

$[0, 2] \times [0, 1]$

Screen 3.5

From the graph of $y = \frac{\sqrt{x}-1}{x-1}$ we see that $y \to 0.5$ as $x \to 1$.

You are asked in Exercise 55, where a hint is given, to confirm algebraically the limit found in Example 10.

Enrichment: Continuity As Defined By a Limit

In Section 1.1 we said that a function $y = f(x)$ was continuous on an interval if the graph had no breaks, jumps, or holes in that interval. We now give a more precise definition in terms of limits.

We already noticed that we can calculate limits of polynomials by using substitution. Thus, if $p(x)$ is a polynomial, then $\lim_{x\to a} p(x) = p(a)$. Any function with this property is said to be **continuous** at $x = a$.

DEFINITION **Limit Definition of a Function Continuous at a Point**

The function $f(x)$ is **continuous** at $x = a$ if $\lim_{x\to a} f(x) = f(a)$.

The rules for continuity are very similar to those for limits. For example, the quotient of two continuous functions is continuous at points at which the denominator is not zero. This follows immediately from the rule for the limit of a quotient. We can use this rule to determine at which points a rational function is continuous. Because a rational function is a quotient of polynomials and because a polynomial is continuous at all points, a rational function is continuous at every point at which the denominator is not zero. We then have the following.

Continuity of Polynomial and Rational Functions

A polynomial function is continuous everywhere.
A rational function is continuous at every point at which the denominator is not zero.

It is helpful to notice that the definition of continuity implies that a function is continuous at $x = a$ only if

1. $\lim_{x\to a} f(x)$ exists.
2. $f(a)$ is defined.
3. $\lim_{x\to a} f(x) = f(a)$.

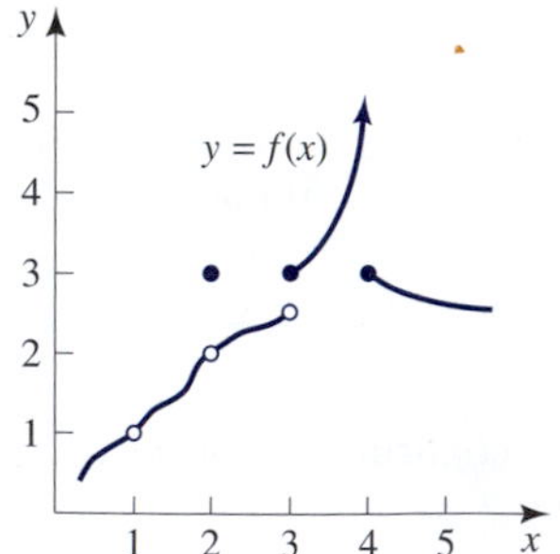

Figure 3.16
$y = f(x)$ is not continuous at $x = 1, 2, 3,$ and 4.

EXAMPLE 11 **Continuity**

Determine the points at which the function whose graph is shown in Figure 3.16 is not continuous.

Solution

Since $f(1)$ is not defined, the function is not continuous at $x = 1$. We notice that $f(2)$ is defined and $f(2) = 3$. Also, $\lim_{x\to 2} f(x)$ exists and equals 2. However, $f(x)$ is not continuous at $x = 2$, since $\lim_{x\to 2} f(x) = 2 \neq 3 = f(2)$. Finally, $f(x)$ is not continuous at $x = 3$ and $x = 4$, since neither $\lim_{x\to 3} f(x)$ nor $\lim_{x\to 4} f(x)$ exist. ■

Notice from Figure 3.16 that the function is not continuous at precisely the points where the graph has "holes" or "gaps" or "breaks." This is true in general and is consistent with how we defined continuity in Section 1.1. Thus, *a function will be continuous at a point if the graph of the function has no hole or gap there.*

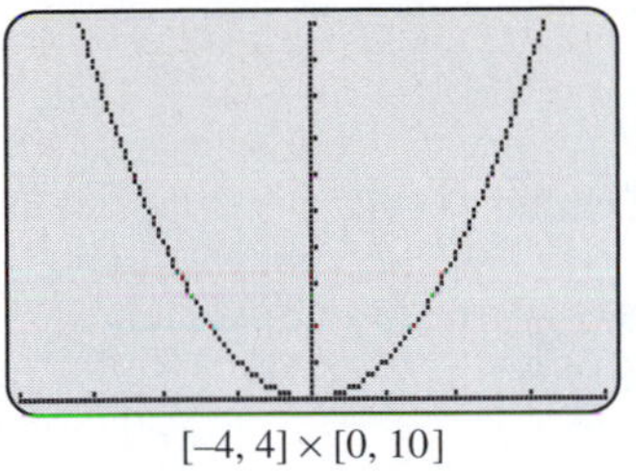

Screen 3.6
Careful examination of the graph of $y = x^2$ on a grapher shows many breaks. However, the actual graph has no breaks and is continuous.

It is not possible to use your computer or graphing calculator to determine points of continuity. We know, for example, that $f(x) = x^2$ is continuous everywhere. However, the graph of this function shown by the graphing calculator (Screen 3.6) has many breaks! (The same can occur on a computer.) This of course is not accurate, as the actual graph of this continuous function has no breaks. What went wrong? The computer or graphing calculator graphs by using *discrete* pixels of light with a small distance separating each pixel. This results in graphs that are "jerky" and discontinuous looking.

EXAMPLE 12 **Continuity**

Determine all points at which the function

$$f(x) = \begin{cases} x^5 - x^4 + x^3 + 2x - 1 & \text{if } x \leq 1 \\ \dfrac{x^2 + x + 2}{x + 1} & \text{if } x > 1 \end{cases}$$

is continuous.

Solution

First notice that $f(x)$ is the polynomial $x^5 - x^4 + 2x - 1$ when $x < 1$. Since polynomials are continuous everywhere, $f(x)$ is continuous for $x < 1$. Now notice that $f(x)$ is the rational function $\dfrac{x^2 + x + 2}{x + 1}$ when $x > 1$. Since rational functions are continuous at every point where the denominator is not zero, $f(x)$ is certainly continuous for $x > 1$. We need only check the point $x = 1$. We need to show that $\lim_{x \to 1} f(x) = f(1)$. First, notice that $f(1)$ is defined and

$$f(1) = (1)^5 - (1)^4 + (1)^3 + 2(1) - 1 = 2$$

To show that $\lim_{x \to 1} f(x)$ exists, we need to consider the limits from the left and right. Since $f(x) = x^5 - x^4 + x^3 + 2x - 1$ when $x < 1$, we have

$$\begin{aligned} \lim_{x \to 1^-} f(x) &= \lim_{x \to 1^-} (x^5 - x^4 + x^3 + 2x - 1) \\ &= (1)^5 - (1)^4 + (1)^3 + 2(1) - 1 \\ &= 2 \end{aligned}$$

Since $f(x) = \dfrac{x^2 + x + 2}{x + 1}$ when $x > 1$, we have

$$\begin{aligned} \lim_{x \to 1^+} f(x) &= \lim_{x \to 1^+} \frac{x^2 + x + 2}{x + 1} \\ &= \frac{(1)^2 + (1) + 2}{1 + 1} \\ &= 2 \end{aligned}$$

Thus, $\lim_{x \to 1} f(x) = 2$. Since this also equals $f(1)$, f is continuous at $x = 1$. Therefore, f is continuous everywhere. ■

Warm Up Exercise Set 3.1

1. For the function in the figure, find $\lim_{x\to a^-} f(x)$, $\lim_{x\to a^+} f(x)$, $\lim_{x\to a} f(x)$ for $a = 1, 2, 3, 4$.

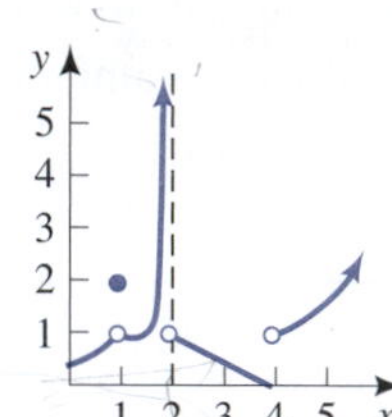

2. Find $\lim_{x\to 1^-} f(x)$, $\lim_{x\to 1^+} f(x)$, $\lim_{x\to 1} f(x)$ if

$$f(x) = \begin{cases} x^4 - x + 1 & \text{if } x < 1 \\ x^2 & \text{if } x > 1 \end{cases}$$

3. Find $\lim_{x\to 3} \dfrac{x^2 + x - 12}{x - 3}$.

4. Find $\lim_{x\to 2} \sqrt{\dfrac{x^2 - 8}{x^3 - 9}}$.

Exercise Set 3.1

In Exercises 1 through 6, refer to the figure. Find $\lim_{x\to a^-} f(x)$, $\lim_{x\to a^+} f(x)$, and $\lim_{x\to a} f(x)$ at the indicated value for the function given in the figure.

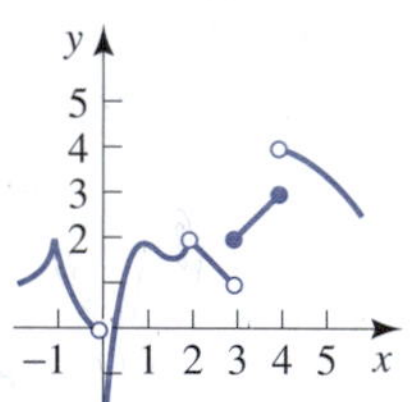

1. $a = -1$ **2.** $a = 0$ **3.** $a = 1$

4. $a = 2$ **5.** $a = 3$ **6.** $a = 4$

In Exercises 7 through 13, fill in the table and then estimate the limit of the given function at the given value or decide that the limit does not exist. Support your answer using your grapher.

7. $a = 1$ and $f(x) = 3x + 2$.

x	0.9	0.99	0.999	→	1	←	1.001	1.01	1.1
$f(x)$				→	?	←			

8. $a = 2$ and $f(x) = 2x + 3$.

x	1.9	1.99	1.999	→	2	←	2.001	2.01	2.1
$f(x)$				→	?	←			

9. $a = 2$ and $f(x) = x^3 - 4$.

x	1.9	1.99	1.999	→	2	←	2.001	2.01	2.1
$f(x)$				→	?	←			

10. $a = -1$ and $f(x) = \dfrac{|x + 1|}{x + 1}$.

x	−1.1	−1.01	−1.001	→	−1	←	−0.999	−0.99	−0.9
$f(x)$				→	?	←			

11. $a = 2$ and $f(x) = \dfrac{1}{x - 2}$.

x	1.9	1.99	1.999	→	2	←	2.001	2.01	2.1
$f(x)$				→	?	←			

12. $a = 4$ and $f(x) = \dfrac{\sqrt{x} - 2}{x - 2}$.

x	3.9	3.99	3.999	→	4	←	4.001	4.01	4.1
$f(x)$				→	?	←			

13. $a = 4$ and $f(x) = \dfrac{x - 4}{\sqrt{x} - 2}$.

x	3.99	3.999	→	4	←	4.001	4.01	4.1
$f(x)$			→	?	←			

In Exercises 14 through 25, find $\lim_{x\to a^-} f(x)$, $\lim_{x\to a^+} f(x)$, and $\lim_{x\to a} f(x)$ at the indicated value for the indicated function. Do not use a computer or graphing calculator.

14. $a = 1,\ f(x) = \begin{cases} x & \text{if } x < 1 \\ 2 & \text{if } x = 1 \\ x - 1 & \text{if } x > 1 \end{cases}$

15. $a = 0,\ f(x) = \begin{cases} x^4 - x + 1 & \text{if } x < 0 \\ x^2 - x & \text{if } x \geq 0 \end{cases}$

16. $a = 1,\ f(x) = \begin{cases} x^3 - x + 1 & \text{if } x < 1 \\ x^4 + x^2 - 1 & \text{if } x > 1 \end{cases}$

17. $a = 1,\ f(x) = \begin{cases} x^2 & \text{if } x < 1 \\ 2 & \text{if } x = 1 \\ x & \text{if } x > 1 \end{cases}$

18. $a = 1,\ f(x) = \begin{cases} -x + 2 & \text{if } x < 1 \\ 0 & \text{if } x = 1 \\ x^2 & \text{if } x > 1 \end{cases}$

19. $a = 0,\ f(x) = \begin{cases} x & \text{if } x < 0 \\ \dfrac{1}{x} & \text{if } x > 0 \end{cases}$

20. $a = 2,\ f(x) = \dfrac{x - 2}{|x - 2|}$

21. $a = -1,\ f(x) = \dfrac{x + 1}{|x + 1|}$

22. $a = 0,\ f(x) = \dfrac{1}{x^2}$

23. $a = 1,\ f(x) = \dfrac{1}{x - 1}$

24. $a = 1,\ f(x) = \begin{cases} -x + 1 & \text{if } x < 1 \\ \dfrac{1}{1 - x} & \text{if } x > 1 \end{cases}$

25. $a = 1,\ f(x) = \begin{cases} x^5 + x^4 + x^2 + 1 & \text{if } x < 1 \\ \dfrac{1}{x - 1} & \text{if } x > 1 \end{cases}$

In Exercises 26 through 36, use the rules of limits to find the indicated limits if they exist. Support your answer using a computer or graphing calculator.

26. $\lim_{x \to 1} \pi$

27. $\lim_{x \to 0} 13$

28. $\lim_{x \to -1} (3x - 2)$

29. $\lim_{x \to 2} (3x^3 - 2x^2 - 2x + 1)$

30. $\lim_{x \to 0} \dfrac{x}{x^2 + x + 1}$

31. $\lim_{x \to 1} \dfrac{x^2 + 2}{x^2 - 2}$

32. $\lim_{x \to 2} \dfrac{x + 3}{x - 3}(2x^2 - 1)$

33. $\lim_{x \to 1} \dfrac{\sqrt{x}}{x - 3} \dfrac{x + 2}{x^2 + 1}$

34. $\lim_{x \to 2} \dfrac{x^2 + 3}{x - 1}$

35. $\lim_{x \to 2} \sqrt{x^2 - 3}$

36. $\lim_{x \to 0} \sqrt[3]{x - 8}$

In Exercises 37 through 46, find the limits graphically. Then confirm algebraically.

37. $\lim_{x \to 2} \dfrac{x^2 - 4}{x - 2}$

38. $\lim_{x \to 2} \dfrac{x^2 - x - 2}{x - 2}$

39. $\lim_{x \to 1} \dfrac{x^2 - 1}{x - 1}$

40. $\lim_{x \to -1} \dfrac{x^2 + 3x + 2}{x + 1}$

41. $\lim_{x \to -1} \dfrac{x^2 - 1}{x + 1}$

42. $\lim_{x \to -2} \dfrac{x^2 + 4}{x + 2}$

43. $\lim_{x \to 0} \dfrac{x + 1}{x^2 + x}$

44. $\lim_{x \to 0} \dfrac{x^3 + x}{x}$

45. $\lim_{x \to 4} \dfrac{x - 4}{\sqrt{x} - 2}$

46. $\lim_{h \to 0} \dfrac{(2 + h)^2 - 4}{h}$

Applications and Mathematical Modeling

47. **Postal Rates** First-class postage in 2003 was \$0.37 for the first ounce (or any fraction thereof) and \$0.23 for each additional ounce (or fraction thereof). If $C(x)$ is the cost of postage for a letter weighing x ounces, then for $x \leq 3$,

$$C(x) = \begin{cases} 0.37 & \text{if } 0 < x \leq 1 \\ 0.60 & \text{if } 1 < x \leq 2 \\ 0.83 & \text{if } 2 < x \leq 3 \end{cases}$$

Graph and find

a. $\lim_{x \to 1^-} C(x)$, $\lim_{x \to 1^+} C(x)$, and $\lim_{x \to 1} C(x)$.

b. $\lim_{x \to 1.5^-} C(x)$, $\lim_{x \to 1.5^+} C(x)$, and $\lim_{x \to 1.5} C(x)$.

48. **Average Cost** An electric utility company estimates that the cost, $C(x)$, of producing x kilowatts of electricity is given by $C(x) = a + bx$, where a represents the fixed costs and b represents the unit costs. The *average* cost, $\bar{C}(x)$, is defined as the total cost of producing x kilowatts divided by x, that is, $\bar{C}(x) = \dfrac{C(x)}{x}$. What happens to the average cost if the amount of kilowatts produced, x, heads toward zero? That is, find $\lim_{x \to 0^+} \bar{C}(x)$. How do you interpret your answer in economic terms?

49. **Cost** It is estimated that the cost, $C(x)$, in billions of dollars, of maintaining the atmosphere in the United States at an average x percent free of a chemical toxin is given by $C(x) = \dfrac{10}{100 - x}$. Find the cost of maintaining the atmosphere at levels 90%, 99%, 99.9%, and 99.99% free of this toxin. What is happening as $x \to 100^-$? Does your answer make sense?

More Challenging Exercises

50. Determine graphically and numerically whether or not any of the following limits exist.

a. $\lim_{x\to 0^-}(x \cdot 3^{1/x})$ **b.** $\lim_{x\to 0^+}(x \cdot 3^{1/x})$

c. $\lim_{x\to 0}(x \cdot 3^{1/x})$

51. Determine graphically and numerically whether or not any of the following limits exist.

a. $\lim_{x\to 0}(2^{-|x|})$ **b.** $\lim_{x\to 0^+}(x^{2x})$

c. $\lim_{x\to 0^+}(x \ln x)$ **d.** $\lim_{x\to 0^+}\left(\ln \frac{1}{x}\right)^x$

e. $\lim_{x\to 1}\left(\frac{x-1}{\ln x}\right)$

52. Rule 5 on limits states that $\lim_{x\to a}(f(x))^n = L^n$, n any real number, L^n defined, $L \neq 0$. This exercise explores the possibility that this rule is true if $L = 0$.

a. Is it true that $\lim_{x\to 0}\sqrt{x^2} = \sqrt{0} = 0$?

b. Explain why $\lim_{x\to 1^-}\sqrt{x-1}$ does not exist.

c. Explain why $\lim_{x\to 0}\sqrt{-x^2}$ does not exist.

53. Let $f(x) = \frac{|x|}{x}$ if $x \neq 0$. Is it possible to define $f(0)$ so that $\lim_{x\to 0} f(x)$ exists? Why or why not?

54. Suppose $y = f(x)$ is defined on $(-1, 1)$. Of what importance is knowledge of $f(0)$ to finding $\lim_{x\to 0} f(x)$? Explain.

55. Confirm algebraically the limit found in Example 10. (Hint: Write

$$\frac{\sqrt{x}-1}{x-1} = \frac{\sqrt{x}-1}{x-1} \cdot \frac{\sqrt{x}+1}{\sqrt{x}+1}$$

simplify the right-hand side, and use the rules of limits.)

56. Is it true that if $\lim_{x\to 0} f(x) = 0$ and $\lim_{x\to 0} g(x) = 0$, then $\lim_{x\to 0}\left(\frac{f(x)}{g(x)}\right)$ does *not* exist? Explain why this is true or give an example that shows it is not true.

57. Is it true that if both $\lim_{x\to a} f(x)$ and $\lim_{x\to a} g(x)$ do *not* exist, then $\lim_{x\to a}(f(x) + g(x))$ does *not* exist? Explain why this is true or give an example that shows it is not true.

58. Finance The following figure appeared in the *Business Review for the Federal Reserve Bank of Philadelphia.*[8] Explain why you think the figure is drawn so that the curve has a vertical asymptote at z, which is the maximum amount of available space.

59. Average Production In a study on the output of small farms in Kenya, Carter and Wiebe[9] gave the following graph based on their empirical study. Explain why it is reasonable that the graph is drawn asymptotic to $y = 0$.

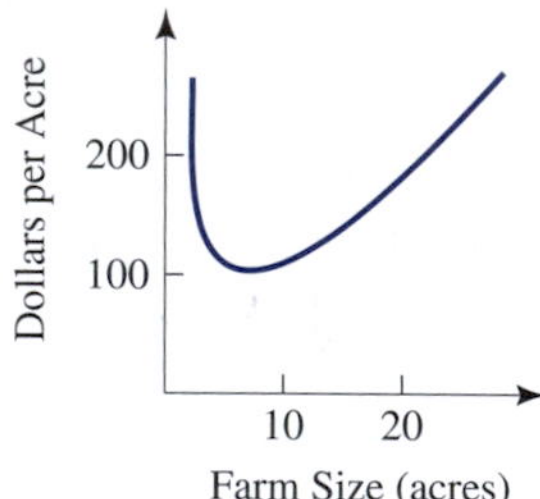

60. Medicine Two rules have been suggested by Cowling and also by Young to adjust adult drug dosage levels for young children. Let a denote the adult dosage, and let t be the age (in years) of the child. The two rules are given by

$$C = \frac{t+1}{24}a \quad \text{and} \quad Y = \frac{t}{t+12}a$$

respectively. Find the limit as $t \to 0^+$ for both of these. Which seems more realistic for a newborn baby? Why?

[8] Theodore M. Crone. 1989. Office vacancy rates: how should we interpret them? *Bus. Rev. Fed. Res. Bank Philadelphia* May/June: 3–12.

[9] Michael R. Carter and Keith D. Wiebe. 1990. Access to capital and its impact on agrarian structure and productivity in Kenya. *Amer. J. Agr. Econ.* 72:1146–1150.

Enrichment Exercises on Limit Definition of Continuity

In Exercises 61 through 77, find, without graphing, where each of the given functions is continuous.

61. $f(x) = |x - 2|$

62. $f(x) = \begin{cases} x & \text{if } x < 1 \\ 2 & \text{if } x = 1 \\ x - 1 & \text{if } x > 1 \end{cases}$

63. $f(x) = \begin{cases} -x + 1 & \text{if } x < 0 \\ x^2 & \text{if } x \geq 0 \end{cases}$

64. $f(x) = \begin{cases} -x + 1 & \text{if } x < 1 \\ x^2 - 1 & \text{if } x > 1 \end{cases}$

65. $f(x) = \begin{cases} x^2 & \text{if } x < 1 \\ 2 & \text{if } x = 1 \\ x & \text{if } x > 1 \end{cases}$

66. $f(x) = \begin{cases} -x + 2 & \text{if } x < 1 \\ 0 & \text{if } x = 1 \\ x^2 & \text{if } x > 1 \end{cases}$

67. $f(x) = \begin{cases} x & \text{if } x < 0 \\ \dfrac{1}{x} & \text{if } x > 0 \end{cases}$

68. $f(x) = \dfrac{x - 2}{|x - 2|}$

69. $5x^7 - 3x^2 + 4$

70. $\dfrac{x}{x^2 + x}$

71. $\dfrac{x}{x^2 - 1}$

72. $\dfrac{x - 1}{x^2 - 1}$

73. $\dfrac{x - 1}{x^2 + 1}$

74. $f(x) = \begin{cases} \dfrac{x^2 + x + 2}{x - 1} & \text{if } x < 0 \\ x^5 + x^3 + 2x - 2 & \text{if } x \geq 0 \end{cases}$

75. $f(x) = \begin{cases} x^4 + 5x^3 + 8x - 1 & \text{if } x \leq 0 \\ \dfrac{x^3 + 4x - 3}{x + 3} & \text{if } x > 0 \end{cases}$

76. $f(x) = \begin{cases} x^4 + 2x^3 + 4x - 1 & \text{if } x \leq 1 \\ \sqrt{x} + 5 & \text{if } x > 1 \end{cases}$

77. $f(x) = \begin{cases} \sqrt{x + 8} & \text{if } -8 \leq x \leq 1 \\ \dfrac{x^3 - x + 6}{x + 1} & \text{if } x > 1 \end{cases}$

78. Let $f(x) = \begin{cases} 2x + 1 & \text{if } x \leq 1 \\ kx^2 - 1 & \text{if } x > 1 \end{cases}$
Find k so that $f(x)$ is continuous everywhere.

79. Let $f(x) = \begin{cases} 2x + 1 & \text{if } x \leq 0 \\ x - k & \text{if } x > 0 \end{cases}$
Find k so that $f(x)$ is continuous everywhere.

80. Explain the difference between requiring a function $f(x)$ to be continuous at $x = a$ and requiring $\lim_{x \to a} f(x)$ to exist.

81. Explain why no polynomial could have a graph like the one shown in the figure.

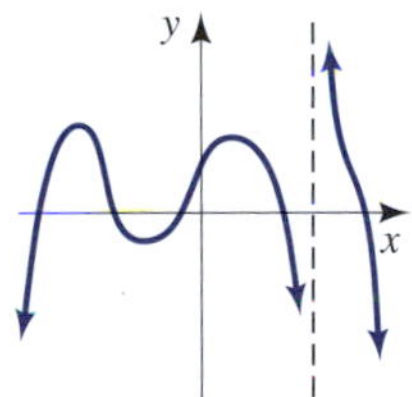

Solutions to WARM UP EXERCISE SET 3.1

1. From the graph

$$\lim_{x \to 1^-} f(x) = 1, \quad \lim_{x \to 1^+} f(x) = 1, \quad \lim_{x \to 1} f(x) = 1$$

$$\lim_{x \to 2^-} f(x) \text{ does not exist}, \quad \lim_{x \to 2^+} f(x) = 1,$$

$$\lim_{x \to 2} f(x) \text{ does not exist} \quad \lim_{x \to 3^-} f(x) = \tfrac{1}{2},$$

$$\lim_{x \to 3^+} f(x) = \tfrac{1}{2}, \quad \lim_{x \to 3} f(x) = \tfrac{1}{2}$$

$$\lim_{x \to 4^-} f(x) = 0, \quad \lim_{x \to 4^+} f(x) = 1,$$

$$\lim_{x \to 4} f(x) \text{ does not exist}$$

2. We have

$$\lim_{x \to 1^-} f(x) = \lim_{x \to 1^-} (x^4 - x + 1) = 1 - 1 + 1 = 1$$

$$\lim_{x \to 1^+} f(x) = \lim_{x \to 1^+} x^2 = 1$$

Since the limit from the left equals the limit from the right, we then have $\lim_{x \to 1^+} f(x) = 1$.

3. We have

$$\lim_{x\to 3}\frac{x^2+x-12}{x-3}=\lim_{x\to 3}\frac{(x+4)(x-3)}{x-3}$$
$$=\lim_{x\to 3}(x+4)$$
$$=7$$

4. $$\lim_{x\to 2}\sqrt{\frac{x^2-8}{x^3-9}}=\sqrt{\lim_{x\to 2}\frac{x^2-8}{x^3-9}} \quad \text{Rule 5}$$
$$=\sqrt{\frac{\lim_{x\to 2}(x^2-8)}{\lim_{x\to 2}(x^3-9)}} \quad \text{Rule 4}$$
$$=\sqrt{\frac{-4}{-1}}=2 \quad \text{Limit of polynomial functions}$$

3.2 Rates of Change

APPLICATION Instantaneous Rate of Change

A wholesale Christmas tree farm sells evergreen seedlings to growers and has revenue $R(x)$ in tens of thousands of dollars given by $R(x)=-x^2+20x$, where x is the number of millions of seedlings sold. How fast are revenues changing with respect to changes in the number of seedlings sold when 5 million seedlings are sold? See Example 5 on page 170 for the answer.

Average Rate of Change

Figure 3.17 shows the graph of a linear function $y=f(x)$ through the two points $(a, f(a))$ and $(b, f(b))$. We denote the change in y on the interval $[a, b]$ by Δy (pronounced "delta y"). We denote the change in x by Δx. The slope is defined as

$$\text{slope}=\frac{\text{rise}}{\text{run}}=\frac{\Delta y}{\Delta x}=\frac{f(b)-f(a)}{b-a}$$

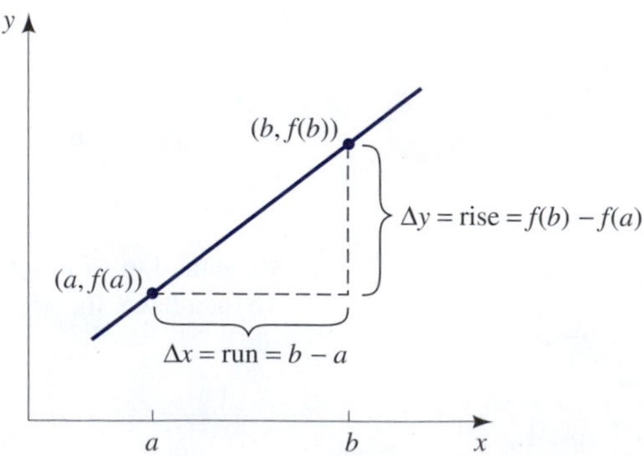

Figure 3.17

The term $\frac{f(b)-f(a)}{b-a}$ is also called the *average rate of change on the interval* $[a, b]$. Notice for linear functions that the average rate of change is a constant, since it is always equal to the slope of the line. The units of average rate of change are units of y divided by units of x or units of output divided by units of input.

$$\text{units of average rate of change} = \frac{\text{units of output}}{\text{units of input}}$$

EXAMPLE 1 **Average Rate of Change for a Linear Function**

Rogers and Akridge[10] of Purdue University studied fertilizer plants in Indiana. For a typical medium-sized plant, they estimated the cost $C(x)$ in dollars to be $C(x) = 366,821 + 208.52x$ where x is tons of fertilizer produced. Find the average rate of change of costs with respect to the amount of fertilizer produced on any interval.

Solution

Figure 3.18 shows the graph of the cost $C(x) = 366,821 + 208.52x$. The average rate of change of costs with respect to the amount of fertilizer produced on any interval (a, b) is

$$\frac{C(b) - C(a)}{b - a}$$

This is just the slope of the line, or 208.52. Units are

$$\frac{\text{units of cost}}{\text{units of fertilizer}}$$

or dollars per ton. So, on average, it costs \$208.52 to produce each ton of fertilizer.

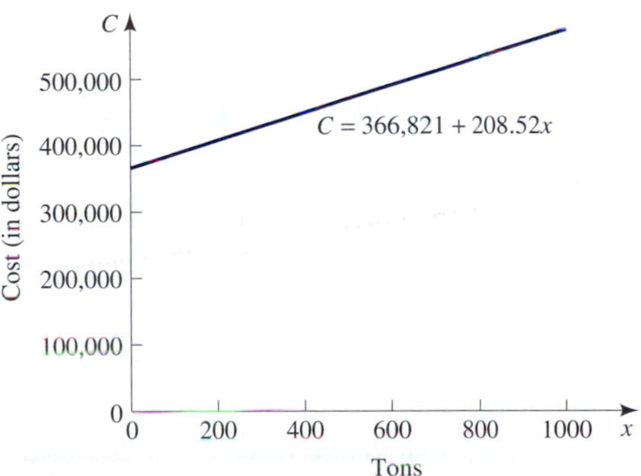

Figure 3.18

■

---

[10] Duane S. Rogers and Jay T. Akridge. 1996. Economic impact of storage and handling regulations on retail fertilizer and pesticide plants. *Agribusiness* 12(4):327–337.

Interactive Illustration 3.1

Average Rates of Change of a Linear Funtion

This interactive illustration allows you to find both graphically and numerically the average rates of change of the functions $y = mx$ at a variety of points. Select a value for m by using the slider provided. As you move the slider on x, you will see a triangle with base h and height $f(x+h) - f(x)$. There are several possible values for h. Thus, the slope of the hypotenuse of the triangle is the average rate of change. A numerical value is also shown.

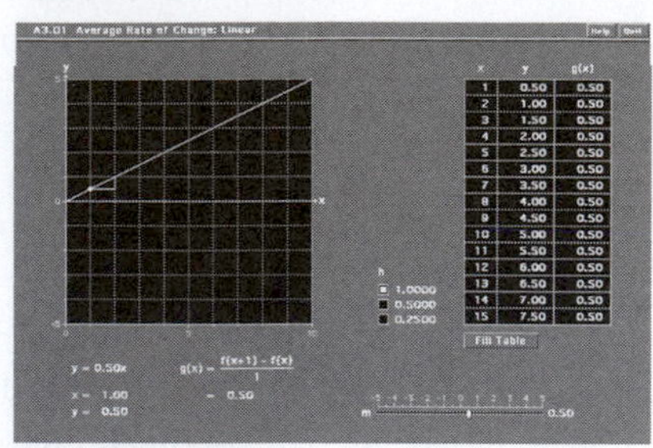

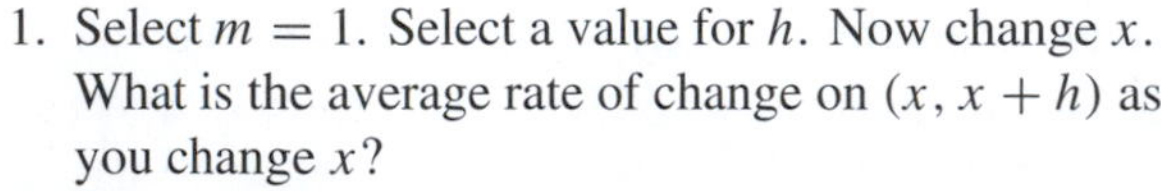

1. Select $m = 1$. Select a value for h. Now change x. What is the average rate of change on $(x, x+h)$ as you change x?
2. Repeat (1) with $m = 2$.
3. Repeat (1) with $m = -1$.
4. Repeat (1) with your own selections of m.
5. What do you conclude about the rate of change of a linear function?

Figure 3.19 shows the graph of a nonlinear function. The average rate of change is defined by the same quotient as before, but as we shall see, now the average rate of change need not be constant.

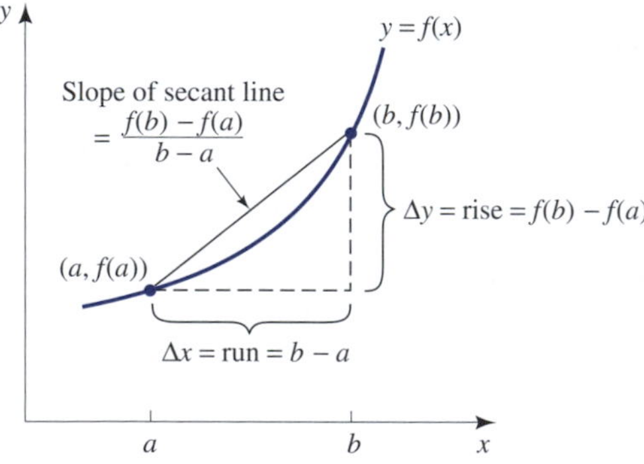

Figure 3.19

DEFINITION **Average Rate of Change**

The average rate of change of $y = f(x)$ with respect to x from a to b is the quotient

$$\frac{\text{change in } y}{\text{change in } x} = \frac{\Delta y}{\Delta x} = \frac{f(b) - f(a)}{b - a}$$

From Figure 3.19 we see that the average rate of change of $y = f(x)$ from a to b is the slope of the secant line from $P(a, f(a))$ to $P(b, f(b))$.

Average Rate of Change Is Slope of Secant Line

The average rate of change of $y = f(x)$ from a to b is the slope of the secant line from $P(a, f(a))$ to the point $Q(b, f(b))$.

EXAMPLE 2 Average Rate of Change

The graph in Figure 3.20 shows the revenue $R(x)$ in millions of dollars of a lawn tractor manufacturer, where x is the number of thousands of tractors sold. Find the average rate of change of revenue with respect to number of tractors sold on the x-intervals (a) $(1, 4)$, (b) $(4, 7)$, and (c) $(5, 7)$.

Solution

(a) On the interval from 1000 to 4000 tractors the average rate of change of revenue with respect to the number of tractors sold is

$$\frac{R(4) - R(1)}{4 - 1} = \frac{9 - 3}{3} = \frac{6}{3} = 2$$

The units are

$$\frac{\text{units of } R}{\text{units of } x} = \frac{\text{millions of dollars}}{\text{thousands of tractors}}$$

or thousands of dollars per tractor. On average, the company had an increase in revenue of \$2000 per tractor when between 1000 and 4000 tractors were sold. See Figure 3.21.

(b) On the interval from 4000 to 7000 tractors the average rate of change of revenue with respect to the number of tractors sold is

$$\frac{R(7) - R(4)}{7 - 4} = \frac{9 - 9}{3} = \frac{0}{3} = 0$$

As in part (a), the units are in thousands of dollars per tractor. On average, the company had no increase or decrease in revenue when between 4000 and 7000 tractors were sold. See Figure 3.22.

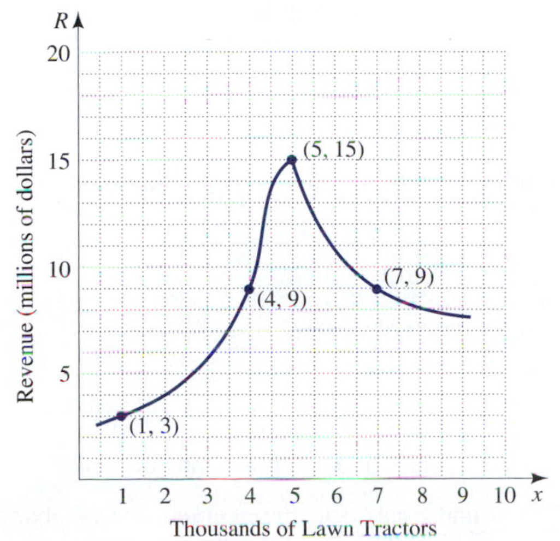

Figure 3.20

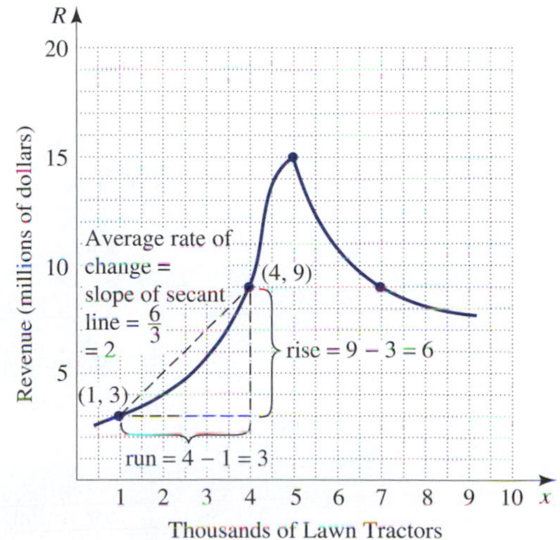

Figure 3.21

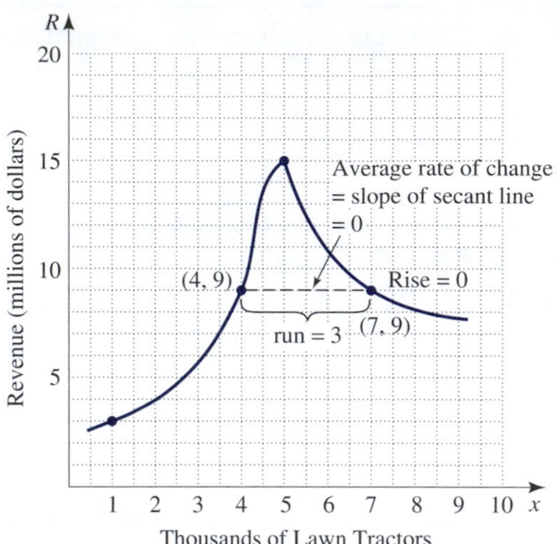

Figure 3.22

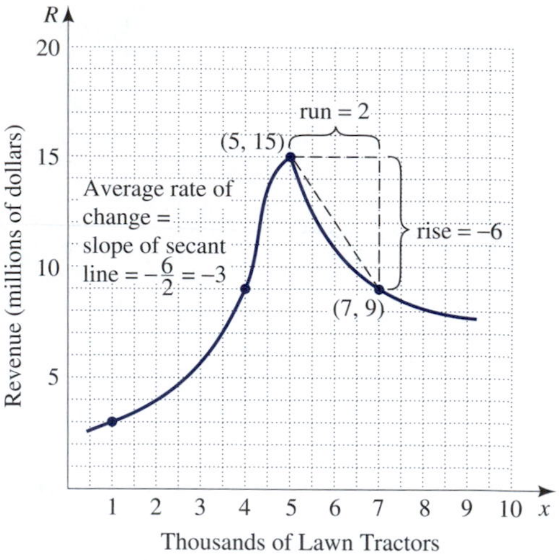

Figure 3.23

(c) On the interval from 5000 to 7000 tractors the average rate of change of revenue with respect to the number of tractors sold is

$$\frac{R(7) - R(5)}{7 - 5} = \frac{9 - 15}{2} = \frac{-6}{2} = -3$$

The units are, as before, in thousands of dollars per tractor. On average, the company had an decrease in revenue of \$3000 per tractor when between 5000 and 7000 tractors were sold. See Figure 3.23. ■

EXAMPLE 3 **Average Rate of Change**

The demand function for a commodity is given by $p(x) = \frac{18}{x}$, where x is the number of thousands of items made and p is the price in dollars of an item. Find the average rate of change of price with respect to demand (a) on the x interval $[1, 3]$ and (b) on $[3, 6]$.

Solution

(a) The average rate of change of price with respect to demand on the x-interval $[1, 3]$ is

$$\frac{p(3) - p(1)}{3 - 1} = \frac{6 - 18}{2} = -6$$

See Figure 3.24. The units are

$$\frac{\text{units of } p}{\text{units of } x} = \frac{\text{dollars}}{\text{thousands of items}}$$

or dollars per thousand items. So the average rate of change of price on this interval is $-\$6$ per thousand items. So on the interval $[1, 3]$ the price is dropping on average by \$6 for each additional thousand items sold.

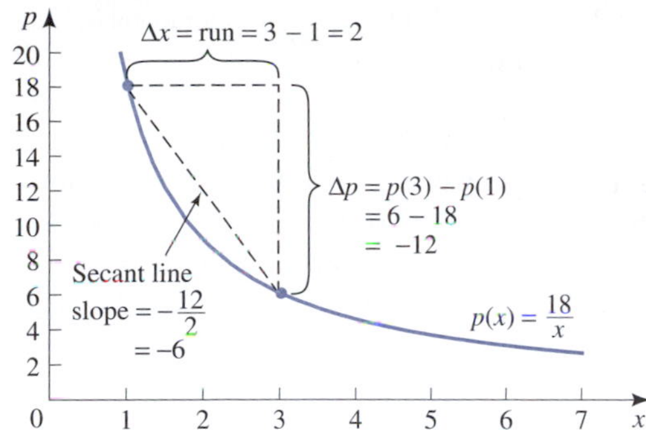

Figure 3.24

Figure 3.25

(b) The average rate of change of price with respect to demand on the x-interval [3, 6] is

$$\frac{p(6) - p(3)}{6 - 3} = \frac{3 - 6}{3} = -1$$

or −$1 per thousand items. See Figure 3.25. So on the interval [3, 6] the price is dropping on average by $1 for each additional thousand items sold. ■

Interactive Illustration 3.2

Average Rates of Change of Nonlinear Functions

This interactive illustration allows you to find both graphically and numerically the average rates of change for a number of functions at a variety of points. You can explore the average rate of change for several functions on various intervals. Select a function and a value for h. As you move the slider on x, you will see a triangle with base h and height $f(x + h) - f(x)$. Thus, the slope of the hypotenuse of the triangle is the average rate of change.

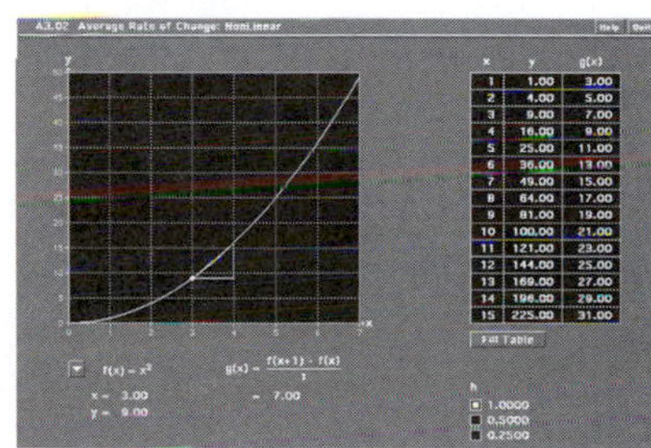

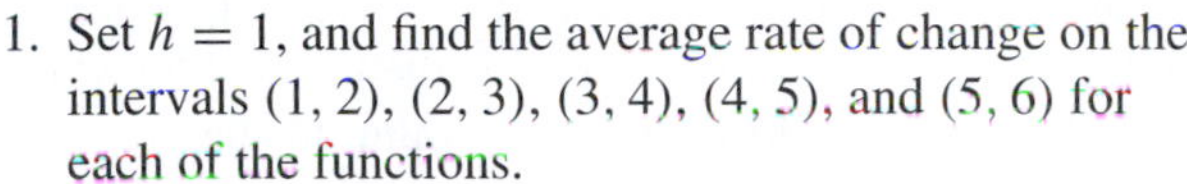

1. Set $h = 1$, and find the average rate of change on the intervals (1, 2), (2, 3), (3, 4), (4, 5), and (5, 6) for each of the functions.
2. Which functions always have a positive rate of change, no matter what the values of x and h?
3. Which function always has a negative rate of change, no matter what the values of x and h?
4. Which function has both positive and negative rates of change?
5. Which functions always have a rate of change that increases as x increases?
6. Which function always has a rate of change that decreases as x increases?

Instantaneous Rate of Change

If a baseball is dropped from the side of a building as in Figure 3.26, it can be shown, by using basic principles of physics, that the distance s in feet that the ball falls is given approximately by $s = s(t) = 16t^2$, where t is measured in seconds. How can we find the velocity $v(t)$ at some instant, say, $t = 1$? The answer is actually fundamental to

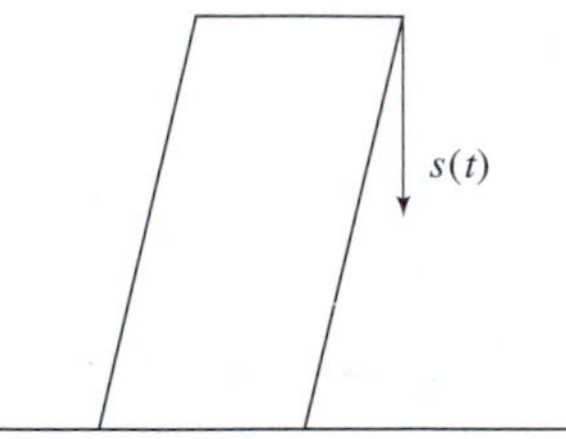

Figure 3.26
Path of a baseball dropped from the top of a building.

the foundations of calculus and yields important applications in business, economics, biology, physics, and a host of other fields.

First, it is intuitively clear that the ball does have an (instantaneous) velocity at any instant. Since we know that the ball falls faster and faster as it drops, this instantaneous velocity is changing every instant. But the question remains: How do we find this instantaneous velocity? Also, although it might be clear that the ball has an instantaneous velocity, just how is it in fact defined?

We can first try something well known. The formula

$$\text{velocity} = \frac{\text{distance}}{\text{time}}$$

is familiar. But no time elapses in the "instant" $t = 1$, and no distance is traveled in this instant of time. The formula for velocity then becomes $\frac{0}{0}$ and is meaningless.

We need an alternative approach. How do the police determine your (instantaneous) velocity while you are driving down the highway? They use radar instruments that send out two beams of radio signals in quick succession. These beams bounce off your vehicle and return to the radar instrument, which carefully records the time of arrival of each signal. Since the speed of the radar signals is known, the radar instrument actually determines the distance your vehicle has traveled in a very short fraction of a second. The instrument then measures your *average velocity* over that short time interval. The assumption is then made that this is very close to your *instantaneous velocity* and, for legal purposes, is considered to be your instantaneous velocity.

We can use this as a good clue on how to define and find the instantaneous velocity. We simply take the *average* velocity over smaller and smaller time intervals and see what number these average velocities are heading for. That number is then defined to be the instantaneous velocity.

Before we continue, let us carefully distinguish between speed and velocity. If an object moves along a line, pick a direction and label this direction "positive." If the object moves in the positive direction, we say that the velocity is positive. If the object moves in the negative direction, we say that the velocity is negative. The speed of the object is defined to be the absolute value of the velocity and so is always positive or zero. In the example of the baseball being dropped, let us consider the downward direction to be positive.

For the baseball dropping from the building we first evaluate the average velocity on any interval of time $[1, 1 + h]$. See Figure 3.27. We will do this and then take h smaller and smaller. The smaller the value of h, the smaller is the interval on which we are finding the average velocity and the closer we are coming to the "instant" $t = 1$. We have

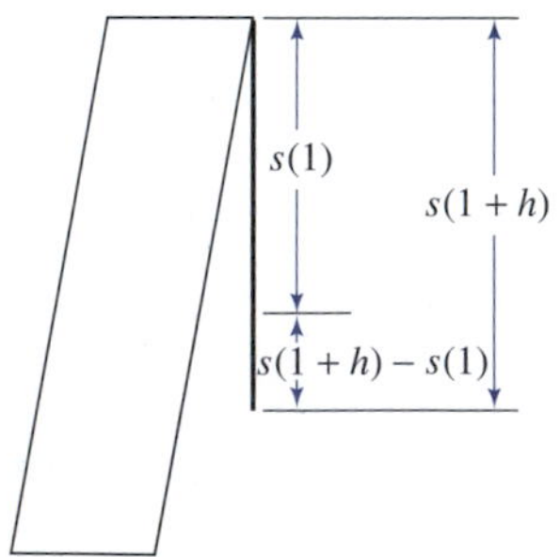

Figure 3.27
The expression $s(1 + h) - s(1)$ is the distance the ball travels during the time interval $[1, 1 + h]$.

$$\begin{aligned}\text{average velocity on } [1, 1 + h] &= \frac{\text{change in distance}}{\text{change in time}} \\ &= \frac{\Delta s}{\Delta t} \\ &= \frac{s(1 + h) - s(1)}{h} \\ &= \frac{16(1 + h)^2 - 16}{h}\end{aligned}$$

Refer to Figure 3.28, where $h > 0$. Figure 3.28 shows the important observation, noted earlier in this section, that the average velocity on the interval $[1, 1 + h]$ is also the slope of the secant line from $(1, 16)$ to $(1 + h, s(1 + h))$. Table 3.6 shows the

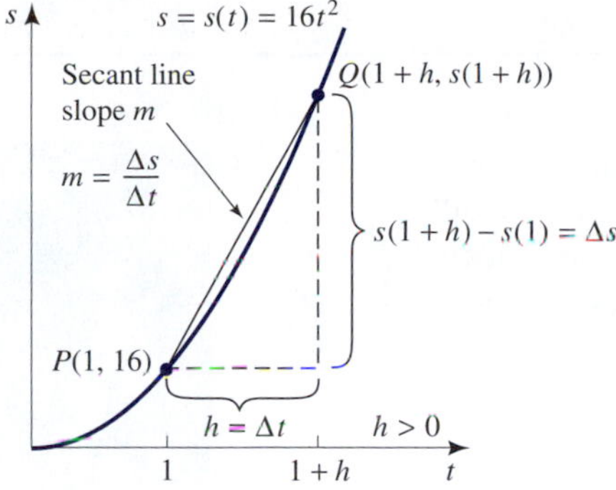

Figure 3.28
The change in s, $s(1+h)-s(1)$, is denoted by Δs. The change in t is h and is denoted by Δt.

average velocity using selected values of h. (Taking h to be -0.1, -0.01, -0.001, and -0.0001, the respective average velocities are 30.4, 31.84, 31.984, and 31.9984. Verify this as an exercise.) As we see, the average velocity seems to be heading for the limiting value of 32.

Table 3.6

h	Interval	Average Velocity $= \dfrac{\Delta s}{\Delta t}$	
0.1	[1,1.1]	$\dfrac{16(1.1)^2-16}{0.1} =$	33.6000
0.01	[1,1.01]	$\dfrac{16(1.01)^2-16}{0.01} =$	32.1600
0.001	[1,1.001]	$\dfrac{16(1.001)^2-16}{0.001} =$	32.0160
0.0001	[1,1.0001]	$\dfrac{16(1.0001)^2-16}{0.0001} =$	32.0016

We then *define* the limiting value of 32 to be the instantaneous velocity at $t = 1$ and denote it by $v(1)$. That is, we define

$$v(1) = \lim_{h \to 0} \frac{s(1+h)-s(1)}{h}$$

In general we have the following definition.

DEFINITION **Average and Instantaneous Velocity**

Suppose $s = s(t)$ describes the position of an object at time t. The **average velocity** from c to $c+h$ is

$$\text{average velocity} = \frac{s(c+h)-s(c)}{h}$$

The **instantaneous velocity** (or simply velocity) $v(c)$ at time c is

$$v(c) = \lim_{h \to 0}(\text{average velocity}) = \lim_{h \to 0} \frac{s(c+h)-s(c)}{h}$$

if this limit exists.

Interactive Illustration 3.3

Instantaneous Velocity

In this interactive illustration you can experimentally find the instantaneous velocity of two functions at a number of different times. Select a function and click on the graph to select a time t. As you click on each smaller value of h, one at a time, the average rate of change on $(t, t+h)$ is shown. You can also click in the Fill Table box and obtain a table similar to Table 3.6.

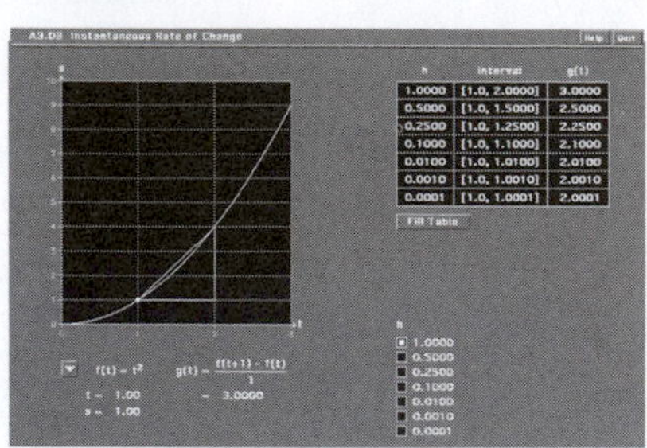

1. What is the instantaneous velocity at $t = 1$ when $s(t) = t^2$? At $t = 2$?
2. What is the instantaneous velocity at $t = 1$ when $s(t) = t^3$? At $t = 2$?
3. Pick other points for each of the functions. What is the instantaneous velocity at these points?

EXAMPLE 4 **Calculating the Instantaneous Velocity**

For the baseball dropped from the building shown in Figure 3.26, find analytically the instantaneous velocity $v(t)$ at $t = 1$.

Solution

First find the average velocity on any interval $[1, 1+h]$ with $h > 0$. We obtain

$$\begin{aligned}
\text{average velocity} &= \frac{\Delta s}{\Delta t} \\
&= \frac{s(1+h) - s(1)}{h} && \text{Definition} \\
&= \frac{16(1+h)^2 - 16}{h} && (1+h)^2 = 1 + 2h + h^2 \\
&= \frac{16 + 32h + 16h^2 - 16}{h} && \text{Simplify} \\
&= \frac{(32 + 16h)(h)}{h} && \text{Cancel } h \\
&= 32 + 16h
\end{aligned}$$

where the h can be canceled, since h will never be zero. (We get the same formula when $h < 0$.) Then

$$v(1) = \lim_{h \to 0} (32 + 16h) = 32$$

This confirms our previous work. Thus, the instantaneous velocity when $t = 1$ is 32 feet per second. ■

Exploration 1

Limit

Verify graphically that $\lim_{h \to 0} \frac{16(1+h)^2 - 16}{h} = 32$.

Now we will see how to connect the instantaneous velocity with the slope of a certain tangent line. Figure 3.29 shows the tangent line to the curve $s = 16t^2$ at $(1, 16)$. Careful examination of Figure 3.29 indicates that the slopes of the secant lines are approaching the slope of the tangent line to the curve at $(1, 16)$, where $m_{\tan}(1)$ is the slope of the tangent line at the point $(1, 16)$, $m_{\sec}(1)$ is the slope of the secant line from the point $(1, 16)$ to Q_1, and so on. Thus, the instantaneous velocity at $t = 1$ is the slope of the line tangent to the curve $s = 16t^2$ at $t = 1$.

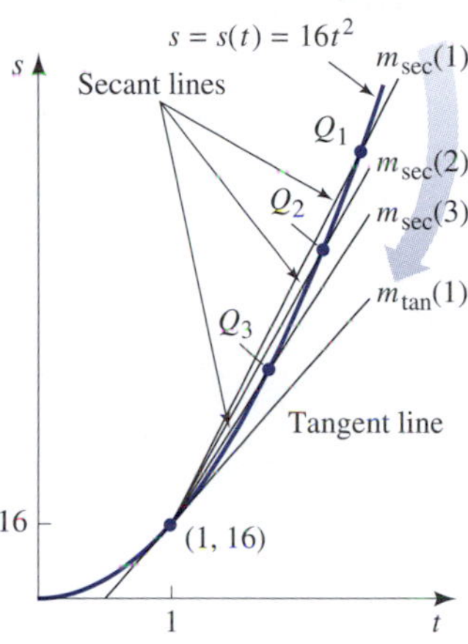

Figure 3.29
As the point slides along the graph and heads for $(1, 16)$, the slopes of the corresponding secant lines head for the slope of the tangent line at $t = 1$.

Interactive Illustration 3.4

Illustrating Figure 3.29

This interactive illustration allows you to illustrate how the slopes of the secant lines in Figure 3.29 will approach the slope of the tangent line. Sliders are provided for x and h. Slide x to where you want. Slide h to zero, and notice how the slopes of the secant lines approach the slope of the tangent line. First try $t = 1$.

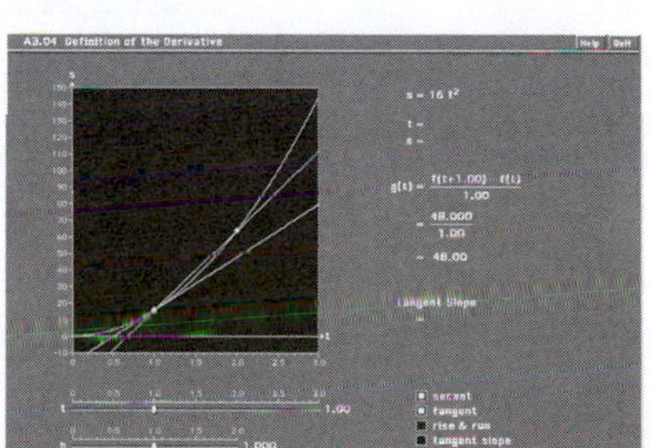

1. Set $t = 2$, and let h go to zero. Toward what value do the slopes of the secant lines seem to be heading?
2. Set $t = 1.5$, and let h go to zero. Toward what value do the slopes of the secant lines seem to be heading?
3. Select other values of t, and let h go to zero. Toward what value do the slopes of the secant lines seem to be heading in each case?

Exploration 2

Tangent Line

Graph $y = 16x^2$ on a screen with dimensions $[0, 1.88]$ by $[0, 64]$. Draw tangent lines to the graph of the curve when $x = 1$, and note the slope. Does your answer agree with that in Example 4?

We can take the instantaneous rate of change for any function $y = f(x)$. We now do this and take care to note any geometric interpretations.

Consider the graph of the function $y = f(x)$ shown in Figure 3.30. The term $h = \Delta x$ represents the change in x on the interval $[c, c+h]$, and $f(c+h) - f(c) = \Delta y$ represents the corresponding change in y on this interval. Recall that the average rate of change of $y = f(x)$ with respect to x from c to $c+h$ is the ratio

$$\frac{\text{change in } y}{\text{change in } x} = \frac{\Delta y}{\Delta x} = \frac{f(c+h) - f(c)}{h}$$

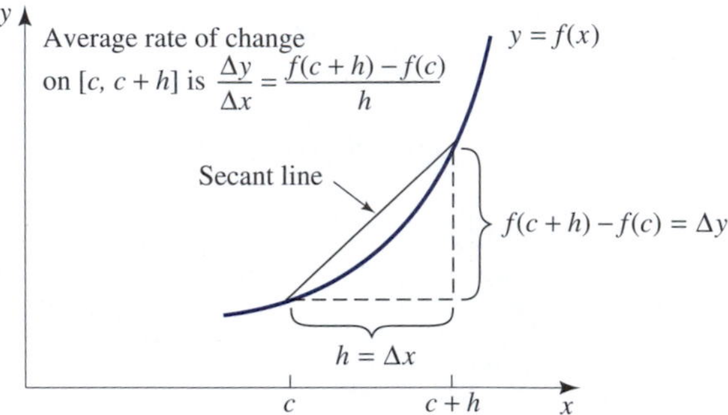

Figure 3.30
The average rate of change is the slope of the secant line.

In analogy with instantaneous velocity, taking the limit of this last expression as $h \to 0$ will yield the **instantaneous rate of change** at $x = c$. We have the following.

DEFINITION **Instantaneous Rate of Change**

Given a function $y = f(x)$, the instantaneous rate of change of y with respect to x at $x = c$ is given by

$$\lim_{h \to 0} \frac{f(c+h) - f(c)}{h}$$

if this limit exists.

EXAMPLE 5 **Finding the Instantaneous Rate of Change**

A wholesale Christmas tree farm sells evergreen seedlings to growers and has revenue $R(x)$ in tens of thousands of dollars given by $R(x) = -x^2 + 20x$, where x is the number of millions of seedlings sold. A graph is shown in Figure 3.31. Find the instantaneous rate of change of revenue with respect to number of seedlings sold when (a) 5 million seedlings are sold, (b) 15 million are sold. Interpret your answer.

Solution

(a) First, we have

$$R(5) = -(5)^2 + 20(5) = -25 + 100 = 75$$

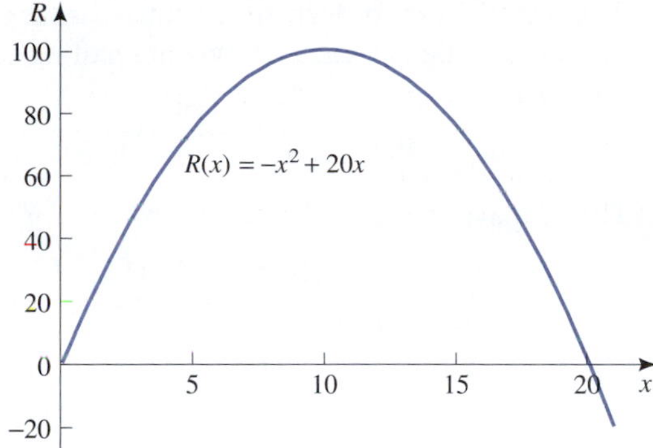

Figure 3.31

Then the average rate of change of revenue R as x changes from 5 to $5 + h$ is

$$\begin{aligned}\frac{R(5+h)-R(5)}{h} &= \frac{[-(5+h)^2+20(5+h)]-[75]}{h} \\ &= \frac{-25-10h-h^2+100+20h-75}{h} \\ &= \frac{10h-h^2}{h} \\ &= 10-h\end{aligned}$$

See Figure 3.32. Taking the limit as $h \to 0$ of the expression for the average rate of change gives the instantaneous rate of change:

$$\begin{aligned}\lim_{h\to 0}\frac{R(5+h)-R(5)}{h} &= \lim_{h\to 0}(10-h) \\ &= 10\end{aligned}$$

Units are

$$\frac{\text{units of revenue}}{\text{units of numbers sold}} = \frac{10{,}000 \text{ dollars}}{\text{million seedlings}}$$

or 1¢ per seedling sold. Thus, at $x = 5$, that is, when 5 million seedlings are sold, revenue is increasing at the instantaneous rate of 10¢ per seedling. This

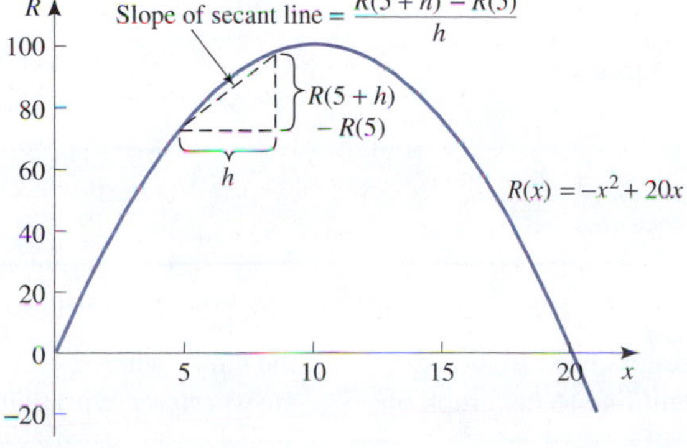

Figure 3.32

means that if revenue were to continue to increase at this rate, then revenue would increase by 10¢ per each additional seedling sold.

(b) First, we have

$$R(15) = -(15)^2 + 20(15) = -225 + 300 = 75$$

Then the average rate of change of revenue R as x changes from 15 to $15 + h$ is

$$\begin{aligned}\frac{R(15+h) - R(15)}{h} &= \frac{[-(15+h)^2 + 20(15+h)] - [75]}{h} \\ &= \frac{-225 - 30h - h^2 + 300 + 20h - 75}{h} \\ &= \frac{-10h - h^2}{h} \\ &= -10 - h\end{aligned}$$

See Figure 3.33. Taking the limit as $h \to 0$ of the expression for the average rate of change gives the instantaneous rate of change:

$$\begin{aligned}\lim_{h\to 0} \frac{R(15+h) - R(15)}{h} &= \lim_{h\to 0} (-10 - h) \\ &= -10\end{aligned}$$

We saw in part (a) that units were in cents per seedling sold. Thus, at $x = 15$, that is, when 15 million seedlings are sold, revenue is decreasing at the instantaneous rate of 10¢ per seedling. This means that if revenue were to continue to decrease at this rate, then revenue would decrease by 10¢ per each additional seedling sold.

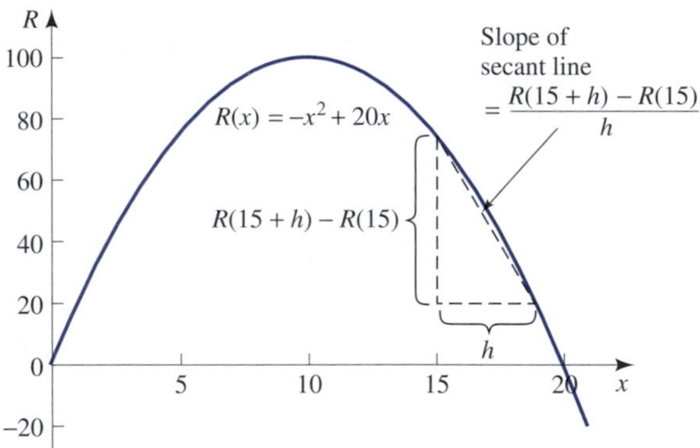

Figure 3.33

■

Exploration 3

Tangent Line

Graph $y = -x^2 + 20x$ on a screen with dimensions [0, 20] by [0, 100]. Draw a tangent line to the graph of the curve when $x = 8$, and note the slope. Does your answer agree with that in Example 5?

Slope of the Tangent Line

Recall from Figure 3.30 that the average rate of change from c to $c+h$ is also the slope m_{sec} of the secant line from the point $P(c, f(c))$ to the point $Q(c+h, f(c+h))$.

Exploration 4

Tangent Line

Graph $y_1 = x^2$ and $y_2 = 2x - 1$ on the same screen of dimensions $[0, 2]$ by $[-1, 5]$. Explain what you see.

Figure 3.34 shows what happens to the slopes of the secant lines as h approaches 0. Close examination of this figure indicates that m_{sec} approaches $m_{\tan}(c)$, the slope of the tangent line to the curve at the point c. That is,

$$\lim_{h \to 0} \frac{f(c+h) - f(c)}{h} = \lim_{h \to 0} m_{\text{sec}}$$
$$= m_{\tan}(c)$$

provided that the limit exists.

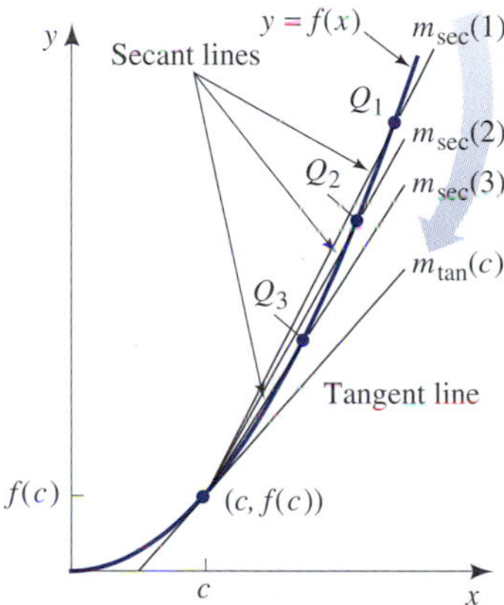

Figure 3.34
As the point slides along the graph and heads for $(c, f(c))$, the slopes of the corresponding secant lines head for the slope of the tangent line at $x = c$.

Interactive Illustration 3.5

Illustrating Figure 3.34

This interactive illustration allows you to illustrate how the slopes of the secant lines in Figure 3.34 will approach the slope of the tangent line. Select one of the functions. Sliders are provided for x and h. Use the slider on x to select a desired value of x. Slide h to zero, and notice how the slopes of the secant lines approach the slope of the tangent line. Numerical values for the slopes of the secant lines are also shown.

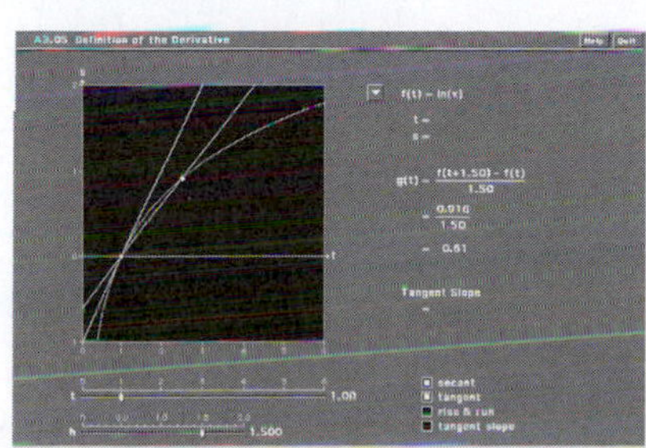

1. For each function, take $x = 1$. As h goes to zero, what are the slopes of the secant lines heading for in each case?
2. Repeat (1) for $x = -1$ for the first two functions and for $x = 2$ for the next three functions.
3. Repeat (1) for other values of x.

DEFINITION **Tangent Line**

The **tangent line** to the graph of $y = f(x)$ at $x = c$ is the line through the point $(c, f(c))$ with slope

$$m_{\tan}(c) = \lim_{h \to 0} \frac{f(c+h) - f(c)}{h}$$

provided that this limit exists.

Notice now that the above limit (if it exists) that defines the slope $m_{\tan}(c)$ of the tangent line is the very same limit that defines the instantaneous rate of change at c. We then have the following.

Instantaneous Rate of Change and the Slope of the Tangent Line

If the instantaneous rate of change of $f(x)$ with respect to x exists at a point c, then it is the slope of the tangent line at that point.

EXAMPLE 6 **Slope of Tangent Line**

Find the slope of the line tangent to the curve in Example 5 at $x = 15$.

Solution

We saw above that for the revenue function $R(x) = -x^2 + 20x$ in Example 5, the instantaneous rates of change at $x = 15$ was -10. Thus, $m_{\tan}(15) = -10$. This is indicated in Figure 3.35.

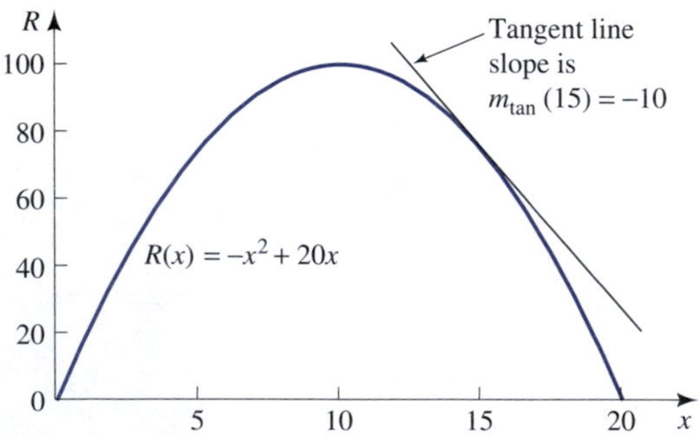

Figure 3.35

■

Interactive Illustration 3.6

Slope of Tangent Line

This interactive illustration allows you to draw the tangent line to the curve in Example 5 at any point and to find its value. A slider is provided for x. As you change x, you see the tangent line change.

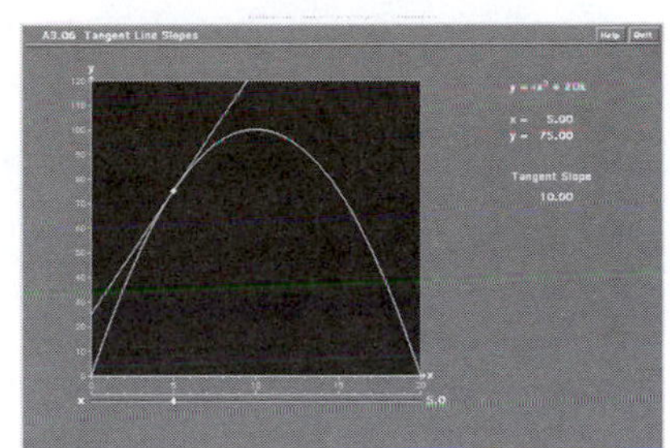

1. For what values of x is the instantaneous rate of change positive?
2. For what values of x is the instantaneous rate of change negative?
3. For what value of x is the instantaneous rate of change zero?

EXAMPLE 7 **Equation of Tangent Line**

Find the equation of the line tangent to the curve in Example 5 at $x = 15$. Support the answer using a computer or graphing calculator.

Solution

From Example 6 we know that the slope is $m_{\tan}(15) = -10$. At $x = 15$, $y = R(15) = 75$. Thus, the equation of the line tangent to the curve $y = R(x)$ at $(15, 75)$ is

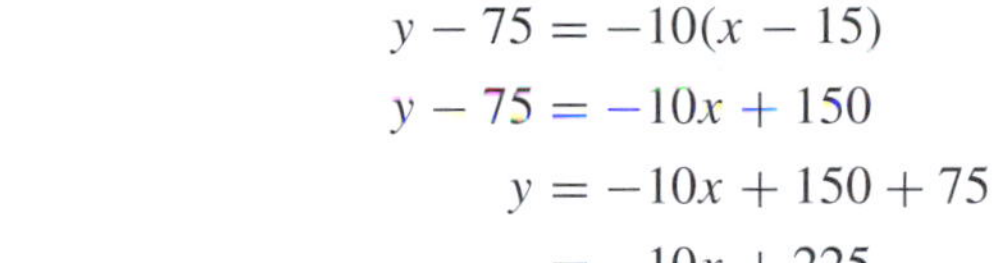

$$\begin{aligned} y - 75 &= -10(x - 15) \\ y - 75 &= -10x + 150 \\ y &= -10x + 150 + 75 \\ &= -10x + 225 \end{aligned}$$

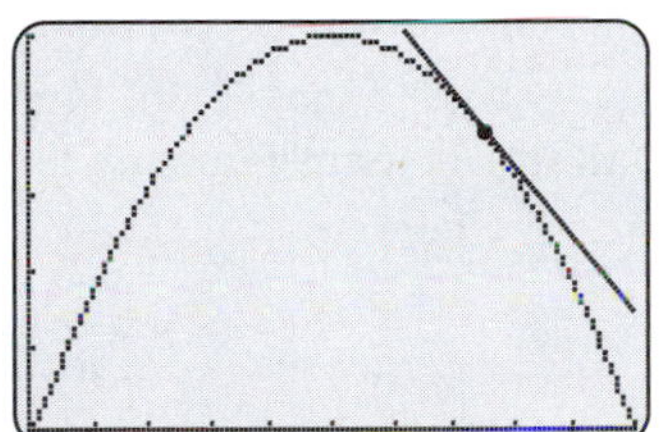

Screen 3.7

Screen 3.7 shows a graph of $y_1 = -x^2 + 20x$ and $y_2 = -10x + 225$ on a window of dimension $[0, 20]$ by $[0, 100]$. The line is tangent to the curve. ■

Further Interpretations

We have seen that the slope of the tangent line at one point is the same as the instantaneous rate of change at that point. In Example 4 we had a function $s = s(t)$, where s was measured in feet and t in seconds. We found that the instantaneous rate of change when $t = 1$ was 32 feet per second. In terms of slope, this gives $m_{\tan}(1) = 32$. This means that the baseball was dropping at a speed of 32 feet per second at the instant when $t = 1$. This means that if the baseball continued to move at this speed for an entire second, the baseball would move 32 feet. The speed is changing from instant to instant, however, so it does not actually remain constant for any length of time.

EXAMPLE 8 **Interpretation of Instantaneous Rate of Change**

Suppose the cost and revenue functions for a plant that produces flooring nails are given by $C(x)$ and $R(x)$, respectively, where x is measured in pounds of nails and $C(x)$ and $R(x)$ are given in cents. Suppose that when $x = 1000$, the instantaneous rate of change of C with respect to x is 50 and the instantaneous rate of change of R

with respect to x is 52. What are the units of the instantaneous rate of change, and what do they mean?

Solution

The instantaneous rate of change in both cases is given in units of cents per pound. When 1000 pounds are being produced, the cost is increasing at the rate of 50 cents per pound, and the revenue is increasing at the rate of 52 cents per pound. This means that if the cost and revenue were to continue at the given respective constant rates, then the cost of producing the next pound of nails would be 50 cents, and the revenue resulting from selling the next pound of nails would be 52 cents. ■

Enrichment: A Biology Example

EXAMPLE 9 Average Rate of Change of a Mathematical Model

Wu and coworkers[11] determined a mathematical model that captured the relationship between stream temperature and juvenile steelhead trout density in Oregon. They found that if t is the stream temperature measured in degrees Celsius and $n = f(t)$ is the number of trout in a typical 100 m^3 volume, the equation

$$f(t) = -440 + 89t - 3t^2$$

approximated the relationship in Camp Creek. A graph is shown in Figure 3.36. Find the average rate of change of trout density with respect to stream temperature on the interval $[10, 10 + h]$ for any $h > 0$.

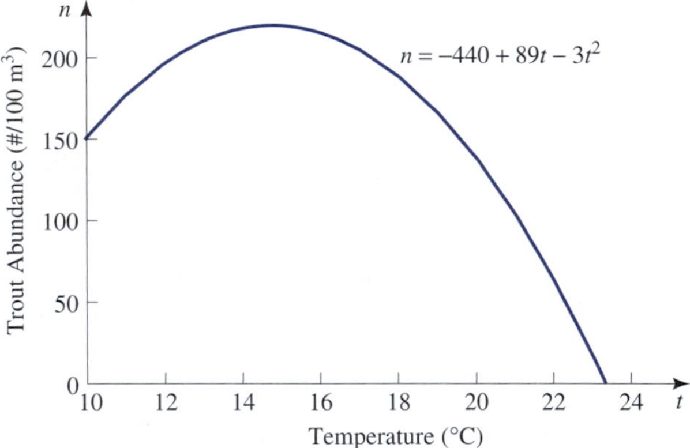

Figure 3.36

Solution

The average rate of change on the interval $[10, 10 + h]$ is

$$\begin{aligned}\frac{\Delta n}{\Delta t} &= \frac{f(10+h) - f(10)}{(10+h) - (10)} \\ &= \frac{[-440 + 89(10+h) - 3(10+h)^2] - [-440 + 89(10) - 3(10)^2]}{h}\end{aligned}$$

[11] Junjie Wu, Richard M. Adams, and William G. Boggess. 2000. Cumulative effects and optimal targeting of conservation efforts: steelhead trout habitat enhancement in Oregon. *Amer. J. Agr. Econ.* 82:400–413.

$$= \frac{[-440 + 890 + 89h - 300 - 60h - 3h^2] - [150]}{h}$$
$$= \frac{29h - 3h^2}{h}$$
$$= 29 - 3h$$

See Figure 3.37. For $h = 5$ this becomes $29 - 3(5) = 14$. So on the temperature interval [10, 15] the number of trout in a typical 100 m^3 volume is increasing on average by 14 per degree Celsius.

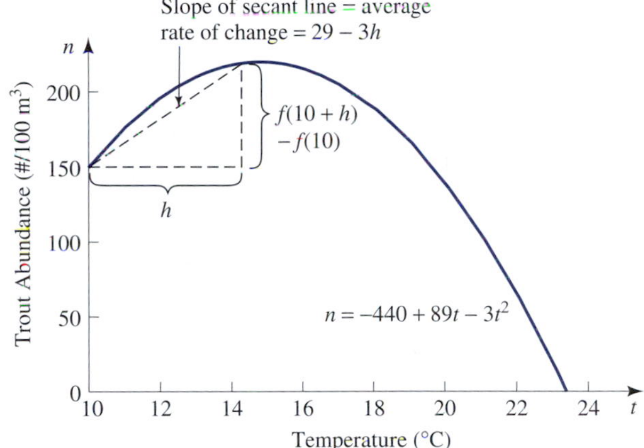

Figure 3.37

■

CONNECTION
Wise Use of Limited Resources

Why is a graph such as shown in Figure 3.36 important for maximizing our fishery stocks with limited financial resources? Figure 3.38 shows the trout density versus water temperature in two streams: Camp Creek, already shown in Figure 3.36, and Middle Fork. Suppose we find ourselves in an environment in which the two stream temperatures are 25°C. Then we would not have any trout in either stream. We might wish to increase the numbers of trout in these two streams but have only limited resources. Suppose we have resources that, if equally split between the two streams, would allow us to improve conditions so that the temperature of each stream could be lowered by 3°C. Then from the figure we see that Camp Creek would sustain some trout but Middle Fork still would not. In effect we have wasted half of our resources on Middle Fork. With such limited resources available, we would obtain a larger trout fishery by putting all of these limited resources into Camp Creek.

Remark We see that the trout are most plentiful at a maximum temperature of about 15°C. As the temperature moves farther away from this value in either direction, the numbers of trout decline. Clearly, the trout cannot survive in water that is too cold or water that is too warm.

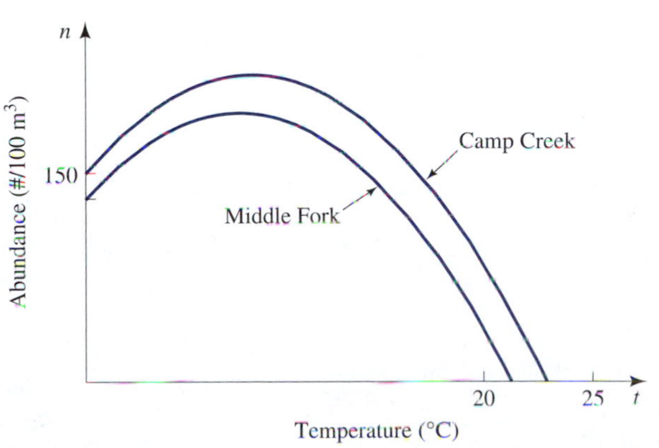

Figure 3.38

EXAMPLE 10 **Finding the Instantaneous Rate of Change**

Referring to Example 9, find the instantaneous rate of change of trout density with respect to stream temperature when $t = 10$.

Solution

Recall from Example 9 that we found that the average rate of change on the interval $[10, 10 + h]$ for any $h \neq 0$ was $29 - 3h$. To find the instantaneous rate of change when $t = 10$, we need only take the limit of this last expression as $h \to 0$. Doing this, we obtain

$$\begin{aligned}\lim_{h\to 0} \text{average rate of change} &= \lim_{h\to 0} \frac{f(10+h) - f(10)}{h} \\ &= \lim_{h\to 0} (29 - 3h) \\ &= 29 - 3(0) \\ &= 29\end{aligned}$$

So in a typical 100 m^3 volume the number of trout were increasing at the rate of 29 per degree increase in temperature when $t = 10$. This means that if the rate were to remain at this value, then the number of trout would increase by 29 in a typical 100 m^3 volume when the temperature increased by 1°C. ■

Warm Up Exercise Set 3.2

1. In 2002 Timmins[12] determined a mathematical model for a demand curve for water in Delano, California. Located in the southern end of the San Joaquin Valley, this city owned and operated a water utility that served its municipal population. Timmins estimated that if demand x was given in acre-feet of water (the amount of water needed to cover 1 acre of ground at a depth of 1 foot) and p in dollars, then

$$p = 1187 - 133 \ln x$$

See the figure. Find the average rate of change of price with respect to demand on the intervals [2, 5] and [5, 10]. Explain what is happening on these two intervals.

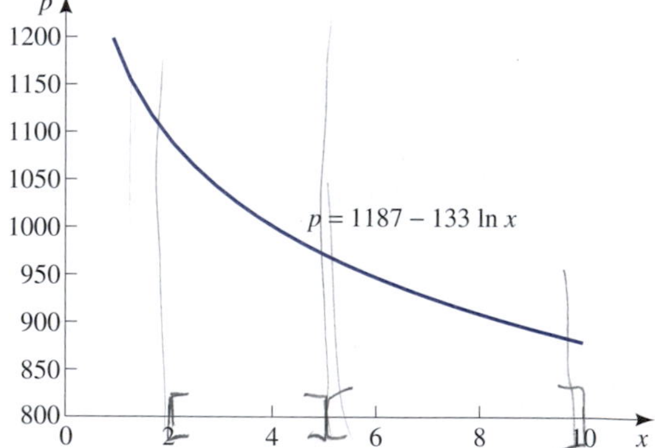

2. For $s(t) = 16t^2$, find the instantaneous velocity $v(2)$ at $t = 2$.

3. Suppose a ball is thrown straight upward with an initial velocity (that is, velocity at the time of release) of 96 ft/sec. Suppose, furthermore, that the point at which the ball is released is considered to be at zero height. Then from physics it is known that the height $s(t)$ in feet of the ball at time t in seconds is given by $s(t) = -16t^2 + 96t$. Find the instantaneous velocity $v(1)$ at $t = 1$.

4. Find the equation of the line tangent to the curve $y = f(x) = x^2$ at the point $(-1, 1)$.

[12] Christopher Timmins. 2002. Measuring the dynamic efficiency costs of regulators' preferences: municipal water utilities in the arid West. *Econometrica* 70(2):603–629.

Exercise Set 3.2

In Exercises 1 through 8, find the average rate of change of the given function on the given interval(s).

1. $f(x) = x^2$; $(-2, 0)$, $(0, 2)$
2. $f(x) = -x^2$; $(-3, 0)$, $(0, 3)$
3. $f(x) = 1/x$; $(1, 4)$, $(2, 10)$
4. $f(x) = \ln x$; $(1, 3)$, $(0.5, 1)$
5. $f(x) = x^2 - 8x + 16$; $(0, 2)$, $(2, 4)$
6. $f(x) = -x^2 + 10x$; $(0, 5)$, $(5, 10)$
7. $f(x) = -x^3$; $(0, 2)$, $(0, 3)$
8. $f(x) = -x^4$; $(-2, 0)$, $(-2, 2)$
9. Find the average rate of change of y with respect to x (a) on the interval $(1, 3)$, (b) on the interval $(3, 5)$.

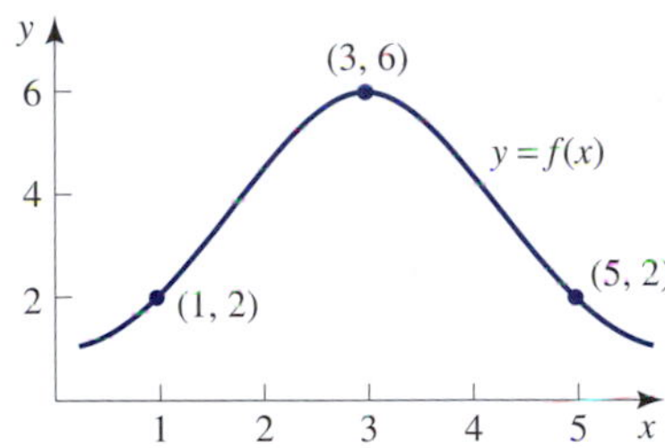

10. Refer to the figure in Exercise 9. Find the average rate of change of y with respect to x on the interval $(1, 5)$.
11. If a baseball is dropped from the top of a building, the distance s in feet that the ball has fallen is given by $s(t) = 16t^2$, where t is in seconds. Find the instantaneous velocity of the ball at $t = 3$.
12. For the ball in Exercise 11, find the instantaneous velocity when $t = 4$.

In Exercises 13 through 18, refer to the following. Suppose a ball is thrown straight upward with an initial velocity (that is, velocity at the time of release) of 128 ft/sec and that the point at which the ball is released is considered to be at zero height. Then the height $s(t)$ in feet of the ball at time t in seconds is given by $s(t) = -16t^2 + 128t$. Let $v(t)$ be the instantaneous velocity at time t. Find $v(t)$ for the indicated values of t.

13. $v(1)$
14. $v(3)$
15. $v(4)$
16. $v(5)$
17. $v(6)$
18. $v(7)$
19. Find the equation of the line tangent to the curve $s = -16t^2 + 64t$ at the point $(1, 48)$. Support your answer by graphing the equation and the tangent line on the same screen.
20. Find the equation of the line tangent to the curve $s = -16t^2 + 64t$ at the point $(3, 48)$. Support your answer by graphing the equation and the tangent line on the same screen.

In Exercises 21 through 32, find the instantaneous rates of change of the given functions at the indicated points.

21. $f(x) = 2x + 3$, $c = 2$
22. $f(x) = -3x + 4$, $c = 3$
23. $f(x) = x^2 - 1$, $c = 1$
24. $f(x) = -x^2 + 3$, $c = 2$
25. $f(x) = -2x^2 + 3$, $c = 2$
26. $f(x) = -3x^2 - 1$, $c = 1$
27. $f(x) = x^2 + 2x + 3$, $c = -1$
28. $f(x) = x^2 - 3x + 4$, $c = -2$
29. $f(t) = -2t^2 - 4t - 2$, $c = 3$
30. $f(r) = 3r^2 - 4r + 2$, $c = 3$
31. $f(u) = u^3$, $c = 1$
32. $f(v) = -v^3 + 1$, $c = 1$
33. Between which pair of consecutive points on the curve is the average rate of change positive? Negative? Zero?

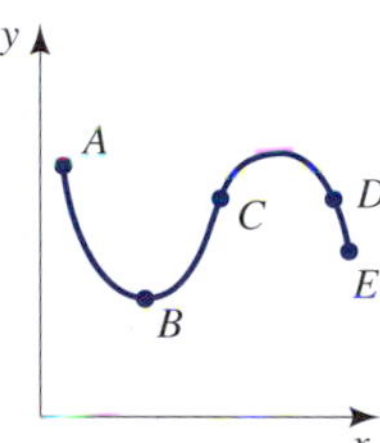

34. Between which pair of consecutive points on the curve is the average rate of change largest? Smallest?

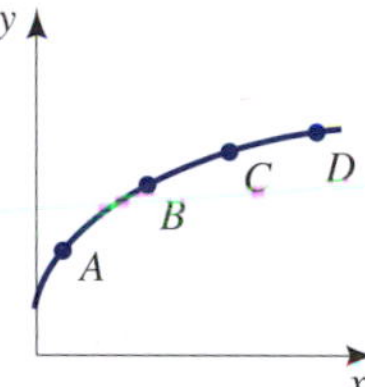

35. Identify the points on the curve for which the slope of the tangent line is positive. Negative. Zero.

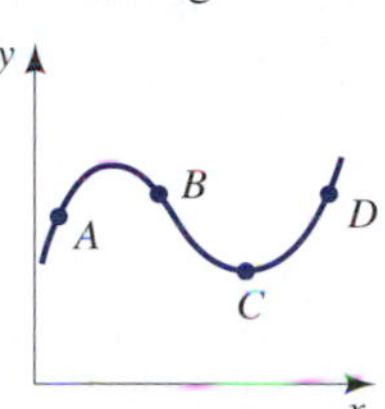

36. Which point on the curve has the tangent line with largest slope? Smallest slope?

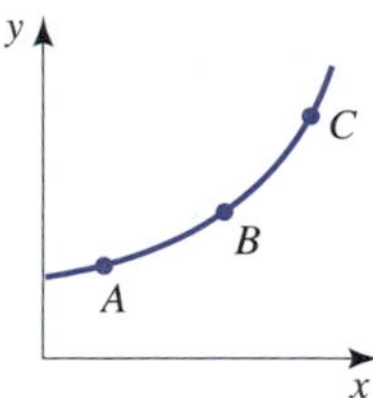

37. The following table gives some values (rounded to five decimal places) of $f(x) = \sqrt{x}$ near $x = 1$. From this, estimate the slope of the tangent line to $y = \sqrt{x}$ at $x = 1$.

x	0.98	0.99	1	1.01	1.02
$\sqrt{x}$	0.98995	0.99499	1	1.00499	1.00995

38. The following table gives some values of $f(x) = \ln x$ near $x = 1$. From this, estimate the slope of the tangent line to $y = \ln x$ at $x = 1$.

x	0.980	0.990	1	1.010	1.020
$\ln x$	−0.020	−0.010	0	0.010	0.020

39. Find the equation of the line tangent to the curve $s = 2x^2 - x$ at the point $(2, 6)$. Support your answer by graphing the equation and the tangent line on the same screen.

40. Find the equation of the line tangent to the curve $s = -3x^2 + 2x$ at the point $(2, -8)$. Support your answer by graphing the equation and the tangent line on the same screen.

Applications and Mathematical Modeling

41. Population The table gives the populations of Peoria, Illinois, and Springfield, Illinois, in 1980 and 1997.[13] What was the average rate of change of the population of each of these cities from 1980 to 1997? Give units and interpret your answer.

City/Year	1980	1997
Peoria	124,160	113,504
Springfield	100,054	105,417

42. Mortgage Rates The average annual interest rates on 30-year fixed-rate mortgages[14] went from 7.5% in 1972 to 16.5% in 1981 to 6.5% in 2002. What was the average rate of change in these mortgage rates (a) from 1972 to 1981? (b) From 1981 to 2002? Give units and interpret your answer.

43. Phillips Curve There has long been a debate among economists as to whether or not there is a relationship between the rate of unemployment and the rate of inflation. In 2001 Hamilton[15] showed that in the United States the equation $i(u) = -1.2u + 12$ approximated the relationship between the unemployment rate u and the inflation rate i during the period from 1984 to 1997, both expressed as a percent. According to this equation, what was the average rate of change of inflation with respect to unemployment from 1984 to 1997? Give units and interpret your answer.

44. Price of Rice According to Dawe[16] the price of rice in real adjusted 1997 dollars can be roughly approximated by the equation $p = 1012 - 690t$, where t is the number of years since 1950 and p is in dollars per metric ton. According to this equation, what was the average rate of change of the price of rice with respect to the number of years since 1950? Give units and interpret your answer.

45. Anchoveta Length Castro and Hernandez[17] formulated a mathematical model given by the equation $L = 2.9 + 0.79a$, where a $(0 < a < 20)$ is the age in days of larval anchoveta along the coastal zone of central Chile and L is the length in millimeters. Find the average rate of change of length with respect to age on any interval on $(0, 20)$. Give units and interpret your answer.

46. Growing Season Start Jobbagy and colleagues[18] developed a mathematical model of the start of the growing

[13] Robert Famighetti. *The World Almanac and Book of Facts 1997*. Mahwah, N.J.: K-III Reference Corp.

[14] Ronald J. Alsop. 1998. *The Wall Street Journal Almanac*. New York: Dow Jones and Company.

[15] James D. Hamilton. 2002. A parametric approach to flexible nonlinear inference. *Econometrica* 69(3):537–573.

[16] David Dawe. 1998. Reenergizing the green revolution in rice. *Amer. J. Agr. Econ.* 80(5):948–953.

[17] Leonardo R. Castro and Eduardo H. Hernandez. 2000. Early life survival of the anchoveta *Engraulis ringens* off central Chile during 1995 and 1996 winter spawning seasons. *Trans. Amer. Fisheries Soc.* 129:1107–1117.

[18] Esteban G. Jobbagy, Osvaldo E. Sala, and Jose M. Paruelo. 2002. Patterns and control of primary production in the Patagonian steppe: a remote sensing approach. *Ecology.* 83(2):307–319.

season on the Patagonian steppe given by the equation $S(t) = 304 - 4.6t$, where S is the starting date of the growing season in Julian days (day 1 = January 1) and t is the mean annual temperature in degrees Centigrade. Find the average rate of change of the start of the growing season with respect to mean annual temperature on $(-2, 10)$. Give units and interpret your answer.

47. Abundance of Warblers Martin[19] determined two mathematical models that related the numbers of Virginia's warblers and orange-crowned warblers to the May plus June precipitation. The two graphs that he obtained are shown in the accompanying figures. Find the average rate of change of the number of each warbler to the amount of precipitation in centimeters. Give units and interpret your answer. Comment on what you have found.

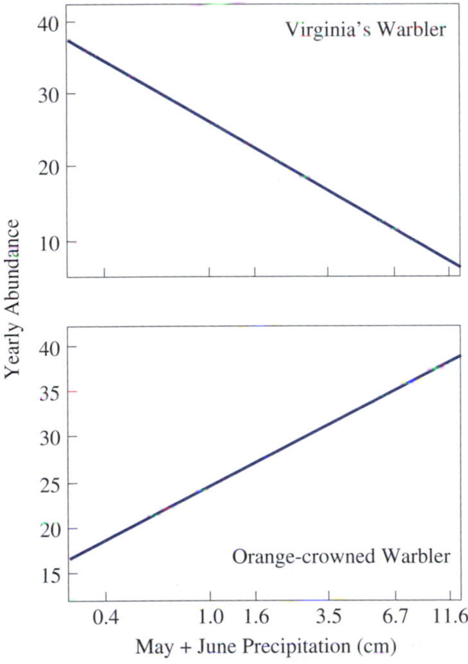

48. Conservation Lant[20] studied the amount of streamside land farmers were willing to set aside as a function of the amount of money they would be paid by government agencies to do so. Their graph is shown. From this graph, estimate the average rate of change of acreage enrolled in the government programs with respect to the annual amount per acre paid to the farmer. Give units and interpret your answer.

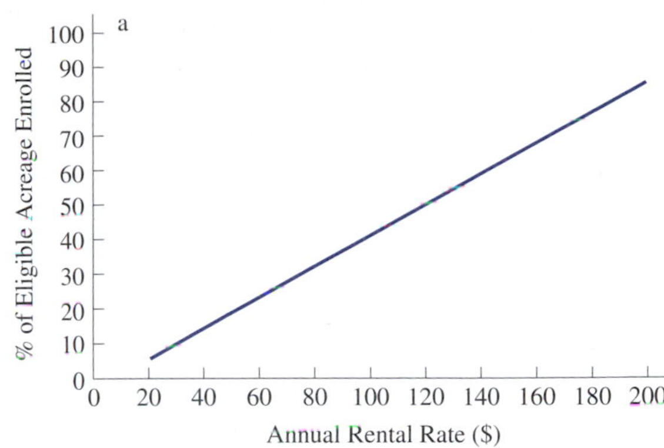

49. Moose Reproduction Ericsson and coworkers[21] studied a Swedish population of moose. They created a mathematical model given by the graph in the accompanying figure. Using this graph, estimate the rate of change of offspring birth mass in kilograms with respect to female age in years. Give units and interpret your answer.

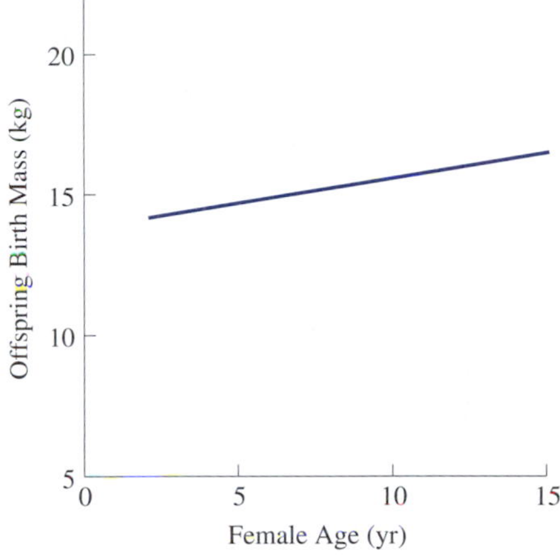

Birth mass for moose offspring in relation to female age and litter size.

50. Breeding Success in Red-Winged Blackbirds In 2001 Weatherhead and Sommerer[22] constructed a mathematical model that was based on a linear relationship between the age of a female red-winged blackbird and the number of fledglings in her nest. They found, for example, that one-year-old females had on average two fledglings in their nest, while eight-year-old females had on average one fledgling in their nest. Find the linear function that describes this relationship, the average rate of change on any interval $[a, b]$, and what this rate of change means.

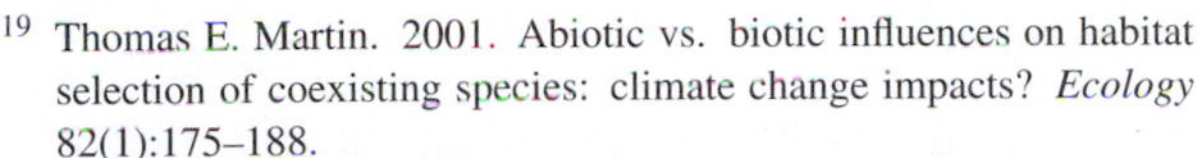

[19] Thomas E. Martin. 2001. Abiotic vs. biotic influences on habitat selection of coexisting species: climate change impacts? *Ecology* 82(1):175–188.

[20] Christopher L. Lant. 1991. Potential of the conservation reserve program to control agriculture surface water pollution. *Environ. Manag.* 15(4):507–518.

[21] Goran Ericsson, Kjell Wallin, John P. Ball, and Martin Broberg. 2001. Age-related reproductive effort and senescence in free-ranging moose. Ecology 82(6):1620–1623.

[22] Patrick J. Weatherhead and Sophie Sommerer. 2001. Breeding synchrony and nest predation in red-winged Blackbirds. *Ecology* 82(6):1632–1641.

51. Hourly Wages In 1997 Murphy and Topel[23] used microeconomic data on over 800,000 men to establish a relationship between hourly wages and weeks worked. The figure shows the graph they presented in their study. Using this graph, roughly estimate the average rate of change of hourly wages with respect to weeks worked for the two intervals [37, 44] and [48, 50]. Explain what is happening.

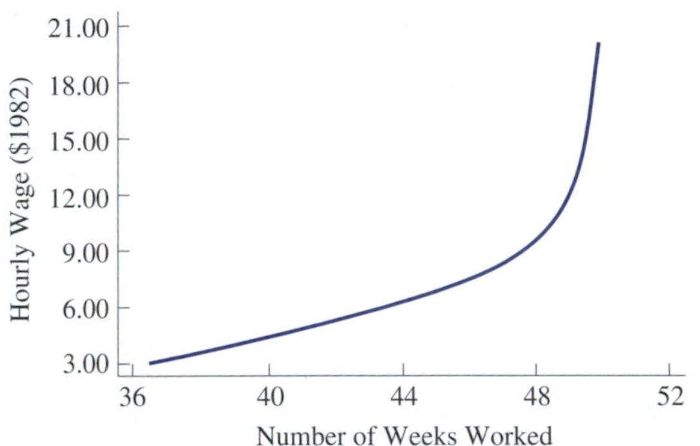

52. Investor Trading Odean[24] was given private access to tens of thousands of brokerage accounts. He noted that the average rate of return of securities bought was 5.69% over the next 252 trading days, whereas the average rate of return of the securities sold was 9.00% over the same period. Find the daily average rate of gain (loss) that these investors incurred by trading the new security for the old one.

53. Private-Equity Firms The figure,[25] entitled "After the Binge," shows the number of U.S. private-equity firms that provided funds by quarter. From the figure, roughly estimate the average rate of change with respect to quarters for (a) the first quarter of 1999 to the first quarter of 2000 and (b) the fourth quarter of 2000 to the fourth quarter of 2001. Explain what is happening.

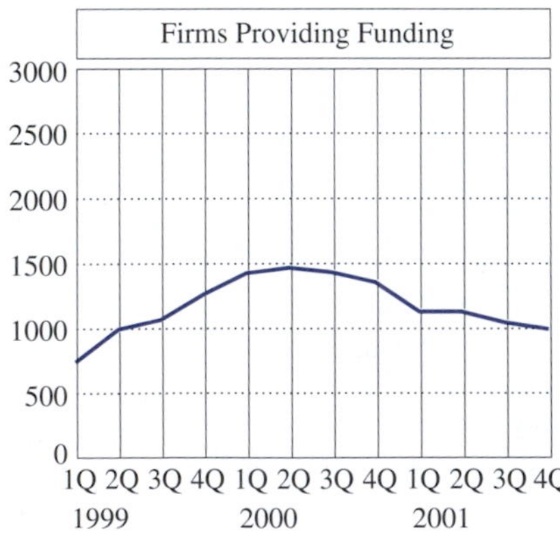

54. Growth-Temperature Relationship Edsall[26] collected through experiments the data given in the following table of the length and weight gains over 55 days of juvenile lake whitefish for various temperatures.

Temp. (°C)	Length (at start, mm)	Length (at end, mm)	Weight (at start, g)	Weight (at end, g)
5.0	89.6	101.1	4.78	7.52
10.1	88.5	115.7	4.54	12.07
15.0	91.7	132.8	5.19	21.14
18.1	92.0	136.9	5.20	24.16
21.0	91.8	132.4	5.12	21.03
24.1	93.9	105.0	5.40	9.95

a. Find the average rate of change of length with respect to days for each of the temperatures given. Give units and interpret your answer. What is happening? Explain.

b. Find the average rate of change of weight with respect to days for each of the temperatures given. Give units and interpret your answer. What is happening? Explain.

55. Price of Housing Near a City The accompanying figure from Voith[27] summarizes his work on the relationship between the distance from a large city, King of Prussia, Pennsylvania, and the percent difference in the value of a similar house. Using this figure, estimate the average rate of change in the percent difference in the value of a house with respect to distance in miles from the city for (a) 1 to 10 miles from the city and (b) 10 to 29 miles from the city. Give units and interpret your answer.

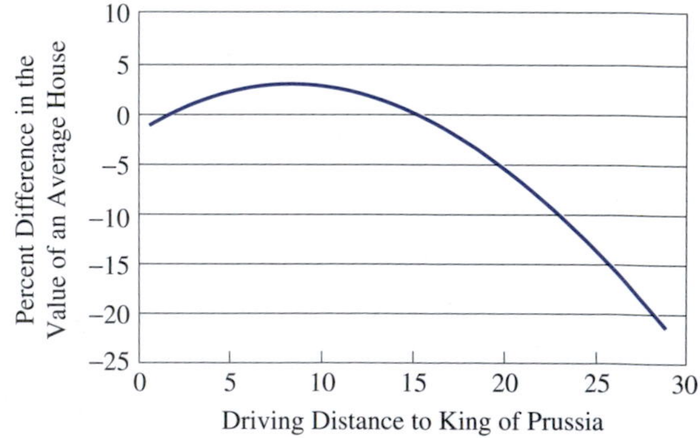

[23] Kevin M. Murphy and Robert Topel. 1997. Unemployment and nonemployment. *Amer. Econ. Rev.* 87(2):295–300.

[24] Terrance Odean. 1999. Do investors trade too much? *Amer. Econ. Rev.* 89(5):1279–1298.

[25] *The Wall Street Journal*, March 27, 2002.

[26] Thomas A. Edsall. 1999. The growth-temperature relation of juvenile lake whitefish. *Trans. Amer. Fish. Soc.* 128:962–964.

[27] Richard Voith. 2000. Has suburbanization diminished the importance of access to center city? *Bus. Rev. Fed. Res. Bank Philadelphia* March/April 200:17–29.

56. Time Worked and Age The accompanying figure can be found in Baker and Benjamin.[28] Using this figure, estimate the average rate of change in the number of hours per year worked by husbands with respect to years of age for (a) for an age between 20 and 40 and (b) for an age between 40 and 65. Give units and interpret your answer.

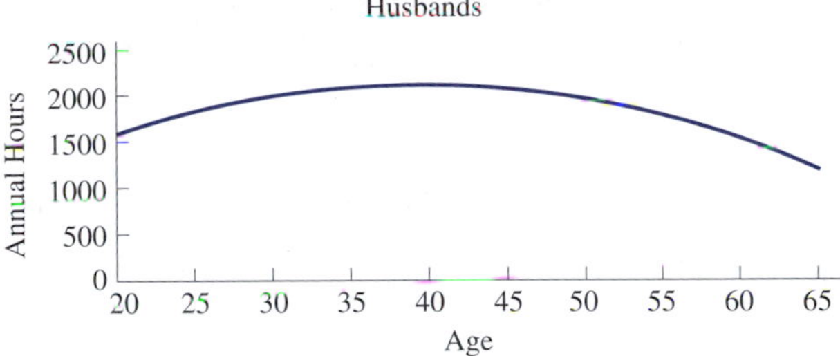

57. Income and Age The accompanying figure can be found in Gourinchas and Parker.[29] Using this figure, estimate the average rate of change in salary in dollars per year with respect to years of age for (a) for an age between 26 and 50 and (b) for an age between 50 and 65. Give units and interpret your answer.

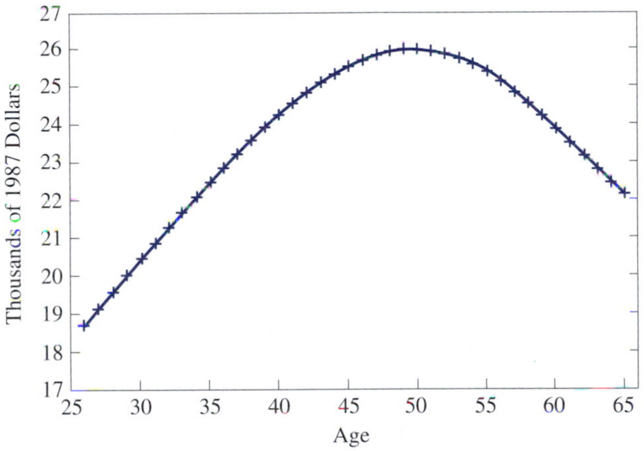

Household consumption and income over the life cycle.

58. Spread of Technological Innovation The accompanying figure can be found in Hornstein.[30] Using this figure, estimate the average rate of change in percentage of diesel locomotives with respect to years for (a) for the years between 1930 and 1950, (b) for the years between 1950 and 1955, and (c) for the years between 1955 and 1960. Give units and interpret your answer.

Diesel Locomotion in the U.S. Railroad Industry, 1925–66: Diffusion

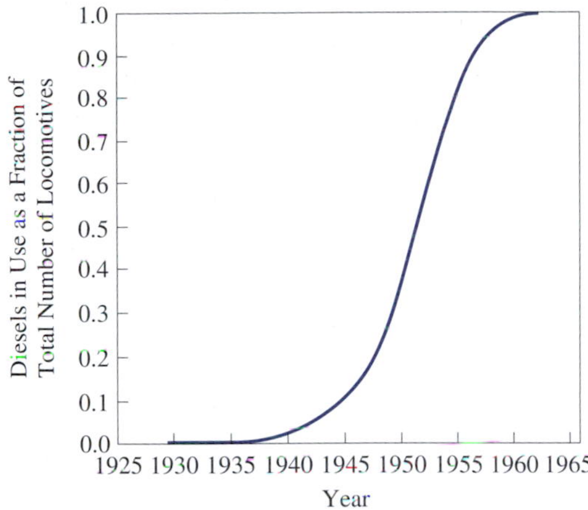

59. Body Temperature Merola-Zwartjes and Ligon[31] studied the Puerto Rican Tody. Even though these birds live in the relatively warm climate of the tropics, they face thermoregulatory challenges due to their extremely small size. The researchers created a mathematical model given approximately by the equation $B(T) = 36 - 0.28T + 0.01T^2$, where T is the ambient temperature in degrees Celsius and B is body temperature in degrees Celsius for $10 \leq T \leq 42$. Graph. Find the instantaneous rate of change of body temperature with respect to ambient temperature when (a) $T = 20$, (b) $T = 30$. Give units and interpret your answer.

60. Lamprey Growth Griffiths and colleagues[32] studied the larval sea lamprey in the Great Lakes, where these creatures represent a serious threat to the region's fisheries. They created a mathematical model given approximately by the equation $G(T) = -1 + 0.3T - 0.02T^2$, where T is the mean annual water temperature with $6.2 \leq T \leq 9.8$ and G is the growth in millimeters per day. Graph. Find the instantaneous rate of change of growth with respect to water temperature when (a) $T = 7$, (b) $T = 9$. Give units and interpret your answer.

61. Corn Yield per Acre Atwood and Helmers[33] studied the effect of nitrogen fertilizer on corn yields in Nebraska. Nitrate contamination of groundwater in Nebraska has become such a serious problem that actions such as

---

[28] Michael Baker and Dwayne Benjamin. 1997. The role of the family in immigrant's labor-market activity: an evaluation of alternative explanations. *Amer. Econ. Rev.* 87(4):705–727.

[29] Pierre-Olivier Gourinchas and Jonathan A. Parker. 2002. Consumption over the life cycle. *Econometrica* 70(1):47–89.

[30] Andreas Hornstein. 1999. Growth accounting with technological revolutions. *Economic Quart. Fed. Res. Bank Richmond* 85(3): 1–10.

[31] Michele Merola-Zwartjes and J. David Ligon. 2000. Ecological energetics of the Puerto Rican Tody: heterothermy, torpor, and intra-island variation. *Ecology* 81(4):990–1003.

[32] Ronald W. Griffiths, F. W. Beamish, B. J. Morrison, and L. A. Barker. 2001. Factors affecting larval sea lamprey growth and length at metamorphosis in lampricide-treated streams. *Trans. Amer. Fish. Soc.* 130:289–306.

[33] Joseph A. Atwood and Glenn A. Helmers. 1998. Examining quantity and quality effects of restricting nitrogen applications to feedgrains. *Amer. J. Agr. Econ.* 80:369–381.

nitrogen rationing and taxation are under consideration. Nitrogen fertilizer is needed to increase yields, but an excess of nitrogen has been found not to help increase yields. Atwood and Helmers created a mathematical model given approximately by the equation $Y(N) = 59 + 0.8N - 0.003N^2$, where N is the amount of fertilizer in pounds per acre and Y is the yield in bushels per acre. Graph this equation on the interval $[50, 200]$. Find the instantaneous rate of change of yield with respect to the amount of fertilizer when (a) $N = 100$ (common rate), (b) $N = 200$. Give units and interpret your answer.

62. Lactation Curves Cardellino and Benson[34] studied lactation curves for ewes rearing lambs. They created the mathematical model given approximately by the equation $P(D) = 273 + 5.4D - 0.13D^2$, where D is day of lactation and P is milk production in grams. Graph this equation on the interval $[6, 60]$. Find the instantaneous rate of change of milk production with respect to the day of lactation when (a) $D = 10$, (b) $D = 50$. Give units and interpret your answer.

63. Growing Season Start Jobbagy and colleagues[35] determined a mathematical model given by the equation $G(T) = 278 - 7.1T - 1.1T^2$, where G is the start of the growing season in the Patagonian steppe in Julian days (day $1 =$ January 1) and T is the mean July temperature in degrees Celsius. Graph this equation on the interval $[-3, 5]$. Find the instantaneous rate of change of the day of the start of the growing season with respect to the mean July temperature when (a) $T = -3$, (b) $T = 0$. Give units and interpret your answer.

64. Carbon Dynamics Kaye and colleagues[36] noted that for some experimental tree plantations in Hawaii, *Albizia* trees are used as an intercrop to increase nitrogen. They created the mathematical model given approximately by the equation $C(P) = 9490 + 386P - 3.8P^2$, where P is the percentage of *Albizia* planted and C is the above-ground tree carbon in grams per square meter. Graph the equation on the interval $[0, 100]$. Find the instantaneous rate of change of the above-ground tree carbon with respect to the percentage of *Albizia* when (a) $P = 40$, (b) $P = 80$. Give units and interpret your answer.

65. Protein in Milk Crocker and coworkers[37] studied the northern elephant seal in Ano Nuevo State Reserve, California. They created a mathematical model given approximately by the equation $L(D) = 18 + 2.9D - 0.06D^2$, where D is days postpartum and L is the percentage of lipid in the milk. Graph the equation on the interval $[0, 28]$. Find the instantaneous rate of change of the percentage of lipid in the milk with respect to days postpartum when (a) $D = 10$, (b) $D = 25$. Give units and interpret your answer.

66. Finance A study of Dutch manufacturers[38] found that the total cost C in thousands of guilders incurred by a company for hiring (or firing) x workers was approximated by $C = 0.0071x^2$. Find the rate of change of costs with respect to workers hired when 100 workers are hired. Give units and interpret your answer.

67. Cost Function A steel plant has a cost function $C(x)$, where x is measured in tons of steel and C is measured in dollars. Suppose that when $x = 150$, the instantaneous rate of change of cost with respect to tons is 300. Explain what this means.

68. Revenue Function A steel plant has a revenue function $R(x)$, where x is measured in tons of steel and R is measured in dollars. Suppose that when $x = 150$, the instantaneous rate of change of revenue with respect to tons is 310. Explain what this means.

69. Tax Revenue Suppose that $R(x)$ gives the revenue in billions of dollars for a certain state when the income tax is set at $x\%$ of taxable income. Suppose that when $x = 3$, the instantaneous rate of change of R with respect to x is 2. Explain what this means.

70. Number of Species in a Tropical Forest Let $N = f(S)$ be the number of species that will exist in a tropical forest of area S square miles. Suppose that when $S = 100$, the instantaneous rate of change of f with respect to S was 3. Explain what this means both for increasing the size of the forest and for decreasing the size of the forest.

---

[34] R. A. Cardellino and M. E. Benson. 2002. Lactation curves of commercial ewes rearing lambs. *J. Anim. Sci.* 80:23–27.

[35] Esteban G. Jobbagy, Osvaldo E. Sala, and Jose M. Paruelo. 2002. Patterns and control of primary production in the Patagonian steppe: a remote sensing approach. *Ecology* 83(2):307–319.

[36] Jason P. Kaye, Sigrid C. Resh, Margot W. Kaye, and Rodney A. Chimner. 2000. Nutrient and carbon dynamics in a replacement series of *Eucalyptus* and *Albizia* trees. *Ecology* 81(12):3267–3273.

[37] Daniel E. Crocker, Jeannine D. Williams, Daniel P. Costa, and Burney J. Le Boeup. 2001. Maternal traits and reproductive effort in northern elephant seals. *Ecology* 82(12):3541–3555.

[38] Gerard A. Pfann and Bart Verspagen. 1989. The structure of adjustment costs for labour in the Dutch manufacturing sector. *Econ. Lett.* 29:365–371.

More Challenging Exercises

71. Cost Curve Dean[39] made a statistical estimation of the average cost–output relationship for a shoe chain for 1938. He found that if x is the output in thousands of pairs of shoes and y is the average cost in dollars, y was approximately given by $y = 0.06204x^2 - 4.063x + 112.76$. Graph on your computer or graphing calculator using a screen with dimensions $[0, 75.2]$ by $[0, 120]$. Have your computer or graphing calculator obtain the tangent line when x is 12, 20, 32, 40, and 52. In each case, relate the slope of the tangent to the rate of change. Interpret what each of these numbers means. On the basis of this model, what happens to average costs as output increases?

72. Environmental Entomology Allen and coworkers[40] found the relationship between the temperature and the number of eggs laid by the female citrus rust mite. They found that if x is the temperature in degrees Celsius and y is the total eggs per female, y was given approximately by the equation $-0.00574x^3 + 0.292x^2 - 3.632x + 11.661$. Graph using a window with dimensions $[10, 38.2]$ by $[0, 20]$. Have your computer or graphing calculator draw tangent lines to the curve when x is 19, 23.5, 25, 28, and 31. Note the slope and relate this to the rate of change. Interpret what each of these numbers means. On the basis of this model, describe what happens to the rate of change of number of eggs as temperature increases.

73. Economic Growth and the Environment Grossman and Krueger[41] studied the relationship in a variety of countries between per capita income and various environmental indicators. The object was to determine whether environmental quality deteriorates steadily with growth. They found that the equation

$$y = f(x) = 0.27x^3 - 6.31x^2 + 39.87x + 34.78$$

approximated the relationship between x given as GDP per capita income in thousands of dollars and y given as units of smoke in cities. Graph on your computer or graphing calculator using a screen with dimensions $[0, 9.4]$ by $[0, 150]$. Have your computer or graphing calculator draw tangent lines to the curve when x is 2, 3, 5, and 6. Note the slope, and relate this to the rate of change. Interpret what each of these numbers means. On the basis of this model, determine whether environmental quality deteriorates with economic growth.

74. Population Ecology Lactin and colleagues[42] collected data relating the feeding rate in y units of the second-instar Colorado potato beetle and the temperature T in degree Celsius. They found that the equation $y = -0.0239T^2 + 1.3582T - 14.12$ was approximately true. Graph using a window with dimensions $[10, 57]$ by $[0, 7]$. Have your computer or graphing calculator draw tangent lines to the curve when x is 20, 25, 29, 32, and 38. Note the slope and relate this to the rate of change. Interpret what each of these numbers means. On the basis of this model, describe what happens to the rate of change of feeding rate as temperature increases.

75. On a large sheet of paper, draw a smooth curve. Pick a point on the curve, and with a ruler draw a tangent line to the curve at this point. Now draw a sequence of secant lines as was done in Figure 3.34. Demonstrate that the slopes of the secant lines are heading for the slope of the tangent line.

76. On your computer or graphing calculator, draw graphs of the three exponential functions $y_1 = 2^x$, $y_2 = e^x$, $y_3 = 3^x$ on the same screen. Decide which graph has a tangent line with the largest slope at $x = 0$ and which has the smallest.

39 Joel Dean. 1976. *Statistical Cost Estimation.* Bloomington, Ind.: Indiana University Press, page 340.

40 J. C. Allen, Y. Yang, and J. L. Knapp. 1995. Temperature effects on development and fecundity of the citrus rust mite. *Environ. Entomol.* 24(5):996–1004.

41 Gene M. Grossman and Alan B. Krueger. 1995. Economic growth and the environment. *Quart. J. Econ.* CX:352–377.

42 Derek J. Lactin, N. J. Holliday, and L. L. Lamari. 1993. Temperature dependence and constant-temperature duel aperiodicity of feeding by Colorado potato beetle larvae in short-duration laboratory trials. *Environ. Entomol.* 22(4):784–790.

Modeling Using Least Squares

77. Cost Curve Dean[43] made a statistical estimation of the average cost–output relationship for a shoe chain for 1937. The data for the firm is given in the following table.

x	4	7	9	12	14	17	22	27	33	40
y	110	90	75	65	60	65	60	45	52	35

Here x is the output in thousands of pairs of shoes, and y is the average cost in dollars.

43 Joel Dean. 1976. *Statistical Cost Estimation.* Bloomington, Ind.: Indiana University Press, page 340.

a. Use quadratic regression to find the best-fitting quadratic polynomial using least squares.
b. Graph on your graphing calculator or computer using a screen with dimensions [0, 56.4] by [0, 120]. Have your grapher draw tangent lines to the curve when x is 18, 27, 36, 42, and 48. Note the slope, and relate this to the rate of change. On the basis of this model, describe what happens to the rate of change of average cost as output increases.

78. Number of Related People Living Together The percentage of families with two or more related people living together is given in the table for selected years.[44]

x	1970	1980	1990	1992	1993
p	81.2	73.7	70.8	70.2	74.8

a. Use quadratic regression to find p as a function of x. Let 1970 correspond to $x = 0$. Graph using a window with dimensions [0, 28.2] by [60, 90].
b. Have your graphing calculator or computer draw tangent lines to the curve when x is 6, 12, 18, 21, and 24. Note the slope and relate this to the rate of change. Interpret what each of these numbers means. On the basis of this model, describe what happens to the rate of change of percentage as time increases.

79. Economic Growth and the Environment Grossman and Krueger[45] studied the relationship in a variety of countries between per capita income and various environmental indicators. The object was to see if environmental quality deteriorates steadily with growth. The table gives the data they collected relating the units y of coliform in waters to the GDP per capita income x in thousands of dollars.

x	1	3	5	7	9	11	13
y	1.8	2.8	2.5	3.7	1	3.5	6

a. Use cubic regression to find the best-fitting cubic to the data and the correlation coefficient.
b. Graph on your graphing calculator or computer using a screen with dimensions [0, 14.1] by [0, 5]. Have your grapher draw tangent lines to the curve when x is 1.5, 2.7, 3.9, 6, 8.4, 9.6, and 12. Note the slope, and relate this to the rate of change. Interpret what each of these numbers means. On the basis of this model, describe what happens to the rate of change of coliform as income increases.

80. Cost Curve Dean[46] made a statistical estimation of the cost-output relationship for a shoe chain for 1937. The data for the firm is given in the following table.

x	18	24	35	45	61	78	120	170
y	5	5.5	6.9	8.2	12	14	18	30

Here x is the sales in thousands of dollars, and y is the cost in thousands of dollars.

a. Use quadratic regression to find the best-fitting quadratic polynomial using least squares.
b. Graph on your graphing calculator or computer using a screen with dimensions [0, 188] by [0, 35]. Have your grapher draw tangent lines to the curve when x is 30, 60, 90, and 120. Note the slope and relate this to the rate of change. Interpret what each of these numbers means. On the basis of this model, describe what happens to the rate of change of cost as output increases.

[44] Bruce S. Glassman. 1996. *The Macmillan Visual Almanac*. New York: Macmillan.

[45] Gene M. Grossman and Alan B. Krueger. 1995. Economic growth and the environment. *Quart. J. Econ.* CX:352–377.

[46] Joel Dean. 1976. *Statistical Cost Estimation*. Bloomington, Ind.: Indiana University Press, page 339.

Solutions to WARM UP EXERCISE SET 3.2

1. The average rate of change on the interval [2, 5] is the slope of the secant line from (2, 1095) to (5, 973), or

$$\frac{p(5) - p(2)}{5 - 2} = \frac{973 - 1095}{3} = -40.67$$

The units are

$$\frac{\text{units of price}}{\text{units of demand}} = \frac{\text{dollars}}{\text{acre-foot}}$$

or dollars per acre-foot. Thus, on this interval, for each additional acre of water demanded the price dropped by \$40.67. The average rate of change on the interval [5, 10] is

$$\frac{p(10) - p(5)}{10 - 5} = \frac{881 - 973}{5} = -18.40$$

The units are in dollars per acre-foot. Thus on this interval, for each additional acre of water demanded the price dropped by \$18.40. Notice that on the second interval the average rate of change of price was less than on the first interval. This is indicated in the figure by the fact that the secant line associated with the second interval has a slope less negative than that of the first interval.

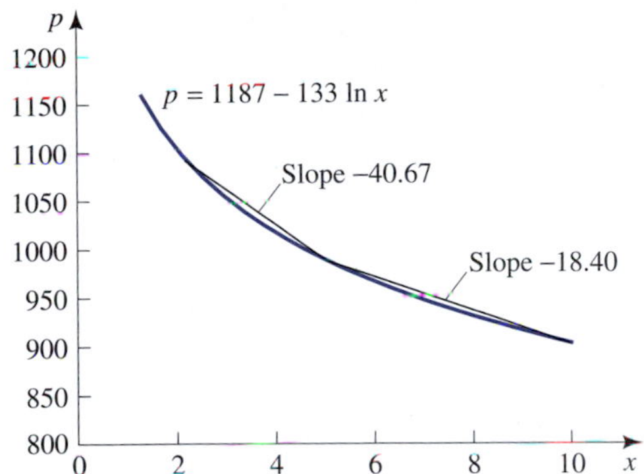

2. First find the average velocity

$$\begin{aligned}\text{average velocity} &= \frac{s(2+h)-s(2)}{h}\\ &= \frac{16(2+h)^2-64}{h}\\ &= \frac{64+64h+16h^2-64}{h}\\ &= 64+16h\end{aligned}$$

Now take the limit of this expression and obtain

$$v(2) = \lim_{h\to 0}(\text{average velocity}) = \lim_{h\to 0}(64+16h) = 64$$

3. Since $s(t) = 16t(6-t)$, the function is easily graphed.

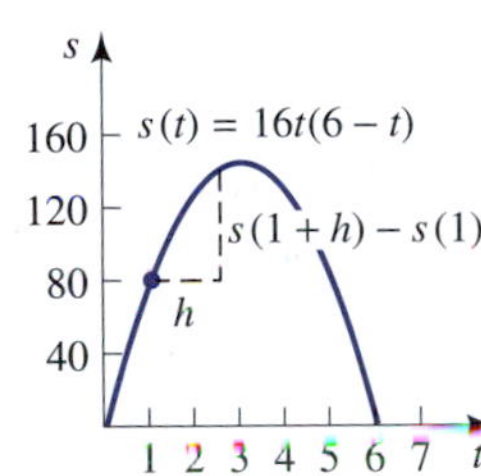

a. The average velocity on the interval $[1, 1+h]$ is

$$\begin{aligned}\frac{s(1+h)-s(1)}{h} &= \frac{[-16(1+h)^2+96(1+h)]-[80]}{h}\\ &= \frac{[-16(1+h)^2+96(1+h)]-[80]}{h}\\ &= \frac{-16-32h-16h^2+96+96h-80}{h}\\ &= \frac{h(64-16h)}{h}\\ &= 64-16h\end{aligned}$$

b. Taking the limit as $h \to 0$ of the expression for the average velocity found above gives

$$\lim_{h\to 0}\frac{s(1+h)-s(1)}{h} = \lim_{h\to 0}(64-16h) = 64$$

4. First find the slope of the secant line

$$\begin{aligned}m_{\text{sec}} &= \frac{f(-1+h)-f(-1)}{h}\\ &= \frac{(-1+h)^2-(-1)^2}{h}\\ &= \frac{1-2h+h^2-1}{h}\\ &= -2+h\end{aligned}$$

Now take the limit of this expression and obtain

$$m_{\tan}(-1) = \lim_{h\to 0} m_{\text{sec}} = \lim_{h\to 0}(-2+h) = -2$$

The equation of the tangent line at $(-1, 1)$ is then $y - 1 = -2(x+1)$.

3.3 The Derivative

The Derivative
Nonexistence of the Derivative
Graphing the Derivative Function
Interpreting the Derivative
Enrichment: The Derivative of $\sqrt{x}$

APPLICATION Instantaneous Rate of Change

Suppose that $p(x) = \frac{1}{x}$ is the price in dollars of a pen, where x is the number of these pens sold per year (in millions). Find the instantaneous rate of change of the price with respect to the number of pens sold per year (in millions) when 2 million pens are sold. See Example 3 on page 190 for the answer.

The Derivative

Given a function $y = f(x)$ for which

$$\lim_{h \to 0} \frac{f(c+h) - f(c)}{h}$$

exists, we have seen that this limit represents the instantaneous rate of change of y with respect to x at $x = c$ and also the slope of the line tangent to the graph of the function at the point $(c, f(c))$. This limit (when it exists) is called the **derivative** and is denoted by $f'(c)$.

As we noted above,

$$f'(c) = \lim_{h \to 0} \frac{f(c+h) - f(c)}{h}$$

when the limit exists. Replacing the letter c with x, we then have the following definition.

DEFINITION Derivative

If $y = f(x)$, the derivative of $f(x)$, denoted by $f'(x)$, is defined to be

$$f'(x) = \lim_{h \to 0} \frac{f(x+h) - f(x)}{h}$$

if this limit exists.

Remark Other notations for this derivative are y', $\dfrac{dy}{dx}$, and $\dfrac{d}{dx}f(x)$.

The general procedure for finding the derivative is summarized as the following three-step process.

To Find the Derivative of $f(x)$

1. Find $\dfrac{f(x+h)-f(x)}{h}$.
2. Simplify.
3. Take the limit as $h \to 0$ of the simplified expression. That is,

$$f'(x) = \lim_{h\to 0} \frac{f(x+h)-f(x)}{h}$$

EXAMPLE 1 **Finding the Derivative of a Function**

Let $y = f(x) = x^2$. Find $f'(x)$.

Solution

First find an expression for the average rate of change of y with respect to x from x to $x+h$ and simplify. This gives

$$\begin{aligned}
\frac{f(x+h)-f(x)}{h} &= \frac{\Delta y}{\Delta x} && \frac{\text{change in } y}{\text{change in } x} \\
&= \frac{(x+h)^2 - x^2}{h} \\
&= \frac{x^2 + 2xh + h^2 - x^2}{h} \\
&= \frac{2xh + h^2}{h} \\
&= \frac{h(2x+h)}{h} \\
&= 2x + h
\end{aligned}$$

Now to find the derivative, take the limit of this expression as $h \to 0$, obtaining

$$\begin{aligned}
f'(x) &= \lim_{h\to 0} \frac{f(x+h)-f(x)}{h} \\
&= \lim_{h\to 0}(2x+h) \\
&= 2x
\end{aligned}$$

■

EXAMPLE 2 **Finding the Derivative of a Function**

Let $y = f(x) = 1/x$, $x \neq 0$. Find $f'(x)$.

Solution

First find an expression for the average rate of change of y with respect to x from x to $x+h$ and simplify. This gives

$$\begin{aligned}
\frac{f(x+h)-f(x)}{h} &= \frac{\Delta y}{\Delta x} && \frac{\text{change in } y}{\text{change in } x} \\
&= \frac{\dfrac{1}{x+h} - \dfrac{1}{x}}{h}
\end{aligned}$$

$$
\begin{aligned}
&= \frac{1}{h}\left[\frac{1}{x+h} - \frac{1}{x}\right] \\
&= \frac{1}{h}\left[\frac{x-(x+h)}{x(x+h)}\right] \\
&= \frac{1}{h}\left[\frac{-h}{x(x+h)}\right] \qquad \text{Cancel the } h \\
&= -\frac{1}{x(x+h)}
\end{aligned}
$$

Now to find the derivative, take the limit of this expression as $h \to 0$, obtaining

$$
\begin{aligned}
f'(x) &= \lim_{h\to 0} \frac{f(x+h)-f(x)}{h} \\
&= \lim_{h\to 0} \frac{-1}{x(x+h)} \\
&= -\frac{1}{x^2}
\end{aligned}
$$

provided that $x \neq 0$. ■

Notice that the derivative $f'(x)$ is itself a *function*. In Example 2, the original function was $f(x) = \frac{1}{x}$, whereas the derivative of this function is the function $f'(x) = -\frac{1}{x^2}$, which for any x gives the slope of the tangent line at x to the graph of $y = f(x)$ (or the instantaneous rate of change at x).

EXAMPLE 3 **Finding an Instantaneous Rate of Change**

Suppose that $p(x) = 1/x$ is the price in dollars of a pen, where x is the number of these pens sold per year (in millions). Find the instantaneous rate of change of the price with respect to the number of pens sold per year (in millions) when 2 million pens are sold. Check your answer by finding the numerical derivative on your grapher.

Solution

Since the instantaneous rate of change is the same as the derivative, the instantaneous rate of change of price with respect to the number of millions of pens sold per year when 2 million pens are sold is $p'(2)$, where $p(x) = 1/x$. From Example 2, we have $p'(2) = -1/(2)^2 = -0.25$. Thus, when 2 million pens are being sold, the price is dropping at a rate of \$0.25 per million sold per year. ■

A project at the end of this chapter gives details on how your computer or graphing calculator finds the numerical derivative. The following exploration gives an alternative way of finding the numerical derivative.

Exploration 1

Finding *dy/dx* on a Computer or Graphing Calculator

Graph $y = p(x) = 1/x$ on a screen with dimensions $[0, 4.7]$ by $[0, 4]$. Using your computer or graphing calculator, find an approximation to $p'(2)$. Consult the Technology Resource Manual for details.

At this time we do not have the mathematical tools to determine analytically the derivative of $f(x) = \ln x$. But realizing that the derivative is a *function* and using our computer or graphing calculator, we can see what the derivative must be. By the limit definition of derivative we know that

$$f'(x) = \lim_{h \to 0} \frac{f(x+h) - f(x)}{h}$$
$$= \lim_{h \to 0} \frac{\ln(x+h) - \ln(x)}{h}$$

If h is small, say, $h = 0.001$, then from our knowledge of limits,

$$f'(x) \approx \frac{\ln(x + 0.001) - \ln x}{0.001}$$

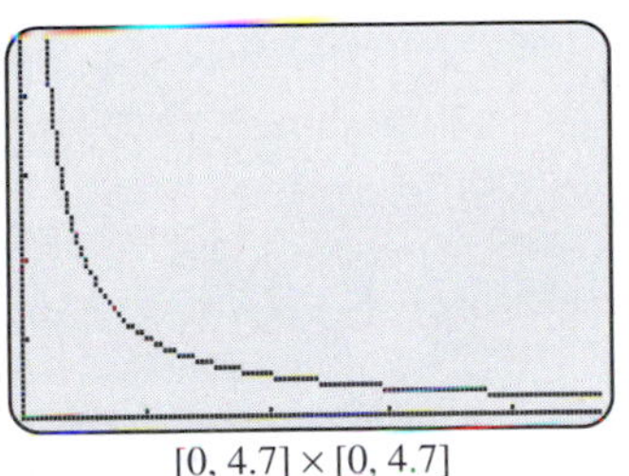

$[0, 4.7] \times [0, 4.7]$

Screen 3.8
Since the derivative of $\ln x$ is approximately $y_1 = \dfrac{\ln(x+0.001) - \ln x}{0.001}$ and the graph of this function appears to be the same as the graph of $y_2 = \dfrac{1}{x}$, we strongly suspect that $\dfrac{d}{dx}(\ln x) = \dfrac{1}{x}$.

Screen 3.8 shows a graph of

$$y_1 = g(x) = \frac{\ln(x + 0.001) - \ln x}{0.001}$$

using a window of dimension [0, 4.7] by [0, 4.7]. (This can also be done by using Interactive Illustration 3.7.) We must have $f'(x) \approx g(x)$. We evaluate this function to create the following table.

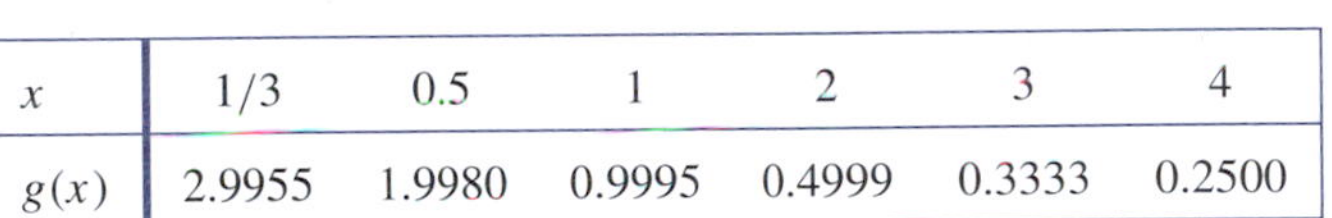

x	1/3	0.5	1	2	3	4
$g(x)$	2.9955	1.9980	0.9995	0.4999	0.3333	0.2500

It is rather apparent that $g(x) = 1/x$. If we graph $y_2 = 1/x$ on the same screen, we obtain nothing new. This verifies graphically that $g(x) \approx 1/x$. Thus, we have ample graphical and numerical evidence to believe that

$$\boxed{\frac{d}{dx} \ln x = \frac{1}{x}}$$

Interactive Illustration 3.7

Finding the Derivative of ln x Experimentally

This interactive illustration allows you to experimentally create a graph of the derivative of $y = \ln x$ by following the ideas in the previous two paragraphs. Click on a value of x in the top window. What do you conclude from the table? For this value of x the value of

$$y = g(x) = \frac{\ln(x + 0.001) - \ln x}{0.001} \approx f'(x)$$

appears in the table and is also used to approximate the slope of the tangent line at that point. The value of this slope is then graphed in the bottom window. By clicking on a variety of points x, you can construct the graph of $y = g(x)$. From the table and the graph, determine what this function is. This function should be $f'(x)$.

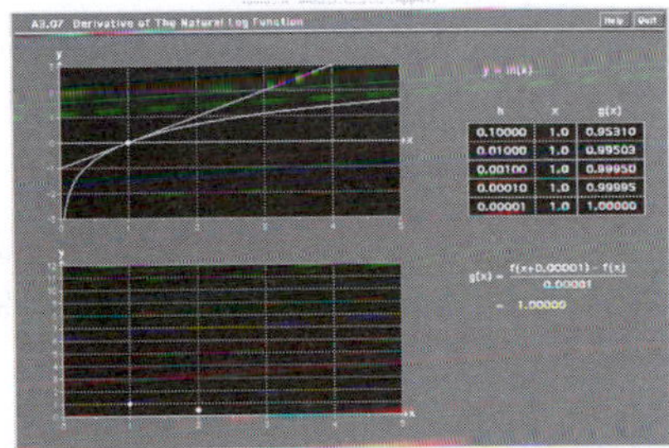

You can also fix x and create a table that shows the average rate of change on $[x, x + h]$ as h gets smaller.

1. Using the table, what do you guess is $f'(0.5)$?
2. $f'(1)$?
3. $f'(2)$?
4. $f'(4)$?

In Section 3.4 we show analytically that this is indeed the case.

EXAMPLE 4 **Finding the Equation of the Tangent Line**

Find the equation of the tangent line to $y = f(x) = \ln x$ at the point where $x = 2$. See Figure 3.39. Support your answer by graphing $y = \ln x$ and the tangent line on the same screen.

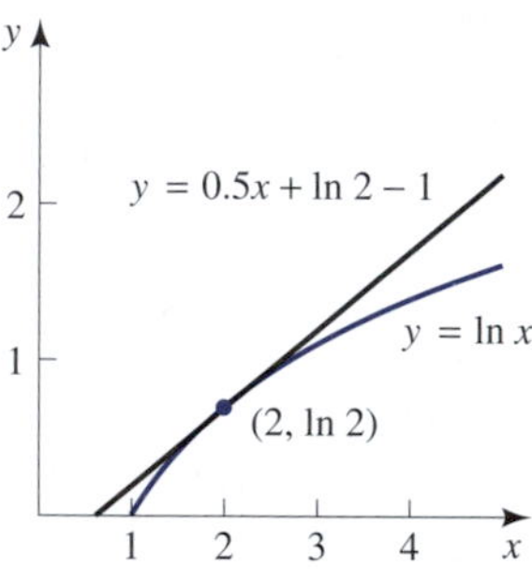

Figure 3.39
Shown is the graph of the tangent line $y = 0.5x + \ln(2) - 1$ at $x = 2$ to the graph of $y = \ln x$.

$[0, 4] \times [-2, 2]$

Screen 3.9
$y_2 = 0.5x - 1 + \ln 2$ is tangent to $y_1 = \ln x$ at $x = 2$.

Solution

Here $c = 2$ and $f(c) = f(2) = \ln 2$. Also from above, $f'(x) = 1/x$ and $f'(2) = 1/2$. Thus, the slope of the tangent line is $m = f'(2) = 1/2$. The equation of the tangent line is then

$$y - \ln 2 = \tfrac{1}{2}(x - 2)$$

Screen 3.9 shows the graphs of $y_1 = \ln x$ and the tangent line $y_2 = 0.5x - 1 + \ln 2$ using a window of dimension $[0, 4]$ by $[-2, 2]$. ■

Exploration 2

Finding an Approximation to $f'(c)$ on Your Computer or Graphing Calculator

Let $y = f(x) = x^5 + x^3 - x + 1$. Find an approximation to $f'(-1)$ on your computer or graphing calculator.

Nonexistence of the Derivative

EXAMPLE 5 **A Function Whose Derivative Does Not Exist at One Point**

Let $f(x) = |x|$. Does $f'(0)$ exist?

Solution

Refer to Figure 3.40. We first note that

$$\lim_{h\to 0^+}\frac{f(0+h)-f(0)}{h} = \lim_{h\to 0^+}\frac{h-0}{h} = \lim_{h\to 0^+} 1 = 1$$

On the other hand,

$$\lim_{h\to 0^-}\frac{f(0+h)-f(0)}{h} = \lim_{h\to 0^-}\frac{-h-0}{h} = \lim_{h\to 0^-} -1 = -1$$

Since the limit from the left does not equal the limit from the right,

$$\lim_{h\to 0}\frac{f(0+h)-f(0)}{h}$$

does not exist; therefore, $f'(0)$ does not exist. ■

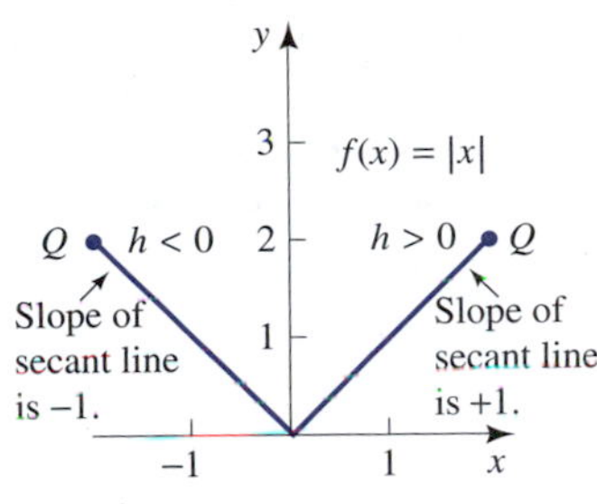

Figure 3.40 $\lim_{h\to 0^-}\frac{f(0+h)-f(0)}{h} = -1$, while $\lim_{h\to 0^+}\frac{f(0+h)-f(0)}{h} = 1$. Since these two limits are not equal, $\lim_{h\to 0}\frac{f(0+h)-f(0)}{h}$ and $f'(0)$ do not exist.

Remark The very same situation occurs whenever a function has a corner. Thus, when the graph of a function has a corner, such as in Figure 3.41, $\lim_{h\to 0^-}\frac{f(x+h)-f(x)}{h}$ does not equal $\lim_{h\to 0^+}\frac{f(x+h)-f(x)}{h}$. Thus, $\lim_{h\to 0}\frac{f(x+h)-f(x)}{h}$ does not exist; therefore, $f'(c)$ does not exist.

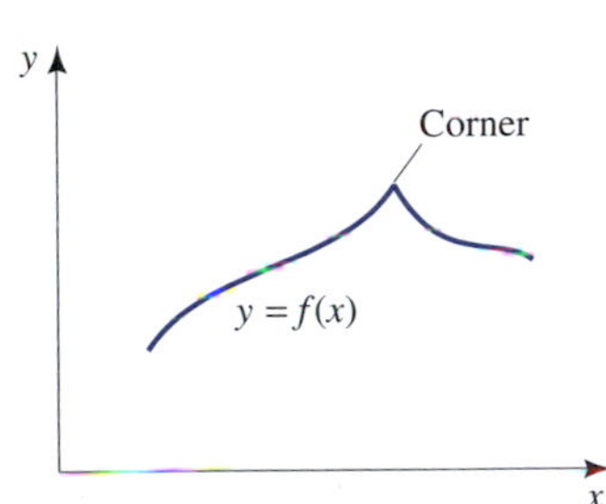

Figure 3.41

Remark If you let $f(x) = |x|$ and find the numerical derivative on your computer or graphing calculator, you will incorrectly obtain that $f'(0) = 0$. This is another way in which your graphing calculator can give a wrong answer. See Case 1 at the end of this chapter for further details.

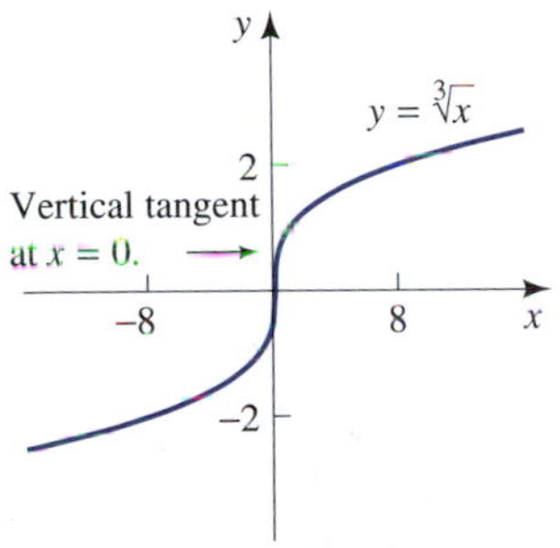

EXAMPLE 6 **A Function with a Vertical Tangent**

If $f(x) = \sqrt[3]{x}$, determine whether $f'(0)$ exists.

Solution

A very carefully drawn graph of this function, such as the one shown in Figure 3.42, indicates that the curve *does* have a tangent line at $x = 0$. However, since the tangent line is *vertical*, the slope does not exist. Therefore, $f'(0)$ cannot exist. ■

Figure 3.42 The derivative does not exist if the tangent is vertical.

Exploration 3

Vertical Tangent

Verify Figure 3.42 on your grapher.

There is an important connection between the existence of the derivative and continuity, as detailed in the following theorem.

A Differentiable Function Is Continuous

If $y = f(x)$ has a derivative at $x = c$, then $f(x)$ is continuous at $x = c$.

The converse of this theorem is not true. That is, a function may be continuous at points where it is not differentiable. For example, the functions graphed in Figures 3.40, 3.41, and 3.42 are continuous at all points in their domains, but each has one value for which the function is not differentiable.

This theorem does say that if a function is not continuous at a point $x = c$, then $f'(c)$ cannot exist.

We now summarize the circumstances for which the derivative does not exist.

When the Derivative Fails to Exist

The derivative fails to exist in the following three circumstances.

1. The graph of the function has a corner.
2. The graph of the function has a vertical tangent.
3. The graph of the function has a break (discontinuity).

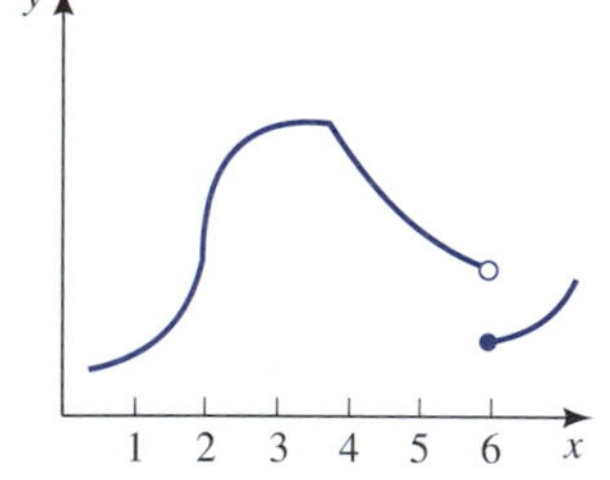

Figure 3.43
The derivative does not exist at the vertical tangent at $x = 2$, at the corner at $x = 4$, and at the point of discontinuity at $x = 6$.

EXAMPLE 7 **Determining Graphically When the Derivative Exists**

Find the points where the derivative of the function whose graph is shown in Figure 3.43 fails to exist.

Solution

The derivative does not exist at $x = 2, 4$, and 6, since at $x = 2$, the tangent is vertical, at $x = 4$ the graph has a corner, and at $x = 6$ the function is not continuous. ■

Exploration 4

Additional Graphical Support for Example 6

Use your computer or graphing calculator to find the numerical derivative of the function in Example 6, $f(x) = \sqrt[3]{x}$, and then graph. What is happening near $x = 0$? Does this support the conclusion in Example 6 that the tangent is vertical at $x = 0$? Explain.

Graphing the Derivative Function

The following example emphasizes again that the derivative $f'(x)$ is itself a function and therefore can be graphed.

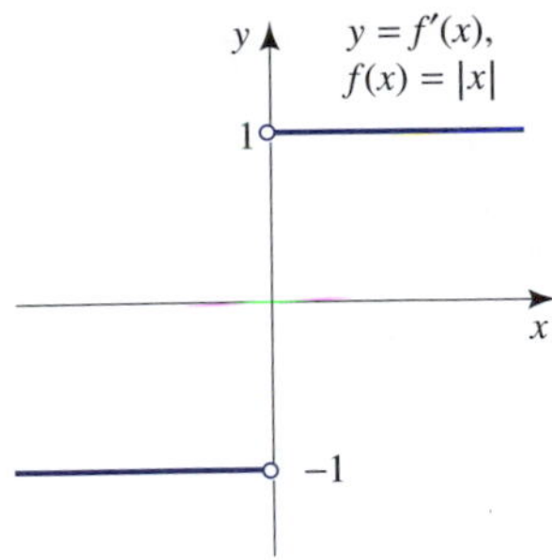

Figure 3.44
We can draw the graph of $y = f'(x)$ by realizing that $f'(x)$ is the slope of the tangent line to the graph of $y = f(x)$.

EXAMPLE 8 **Drawing a Graph of $y = f'(x)$**

Find $f'(x)$ and graph $y = f'(x)$ if $f(x) = |x|$.

Solution

From Figure 3.40 we notice that for $x > 0$, $f(x) = |x| = x$ is a line with slope equal to 1. Thus, for $x > 0$, $f'(x) = m_{\tan}(x) = 1$. In a similar fashion we have for $x < 0$, $f'(x) = m_{\tan}(x) = -1$. The graph is shown in Figure 3.44, in which we notice that no value is given for $x = 0$, since we saw in Example 5 that $f'(0)$ does not exist. ■

We now consider another function.

EXAMPLE 9 **Graphing the Derivative Function**

Figure 3.45 shows the graph of a function $y = f(x)$. From this graph, estimate $f'(x)$ and sketch a graph of $y = f'(x)$.

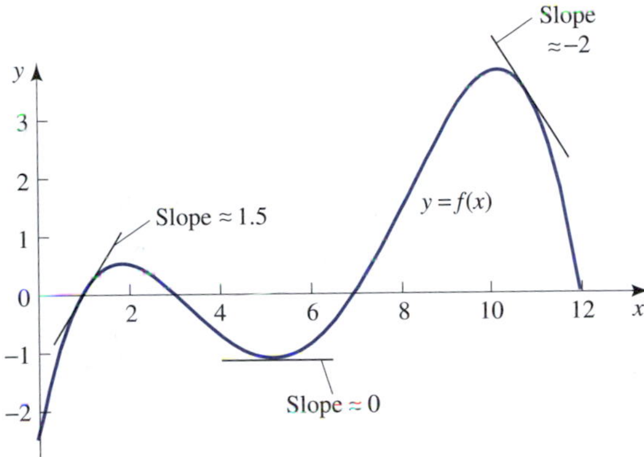

Figure 3.45

Solution

Recall that the derivative at any point is the slope of the tangent line at that point. So we place a straightedge on the curve in Figure 3.45 so that it forms a tangent at a point. The slope of the straightedge is then the slope of the tangent line at that point and hence the derivative at that point. At $x = 1$ we note that the slope is about 1.5, hence $f'(1) \approx 1.5$. At about $x = 5.2$ the slope is zero, so $f'(5.2) \approx 0$. At $x = 11$ the slope is negative and about -2. Thus, $f'(11) \approx -2$. Continuing in this way we can make the following table.

x	0	1	2	3	5.2	7	9	10.2	11	12
$f'(x)$	4	1.5	0	−1	0	1	1	0	−2	−4

Using this information, we can sketch the graph of $y = f'(x)$ shown in Figure 3.46.

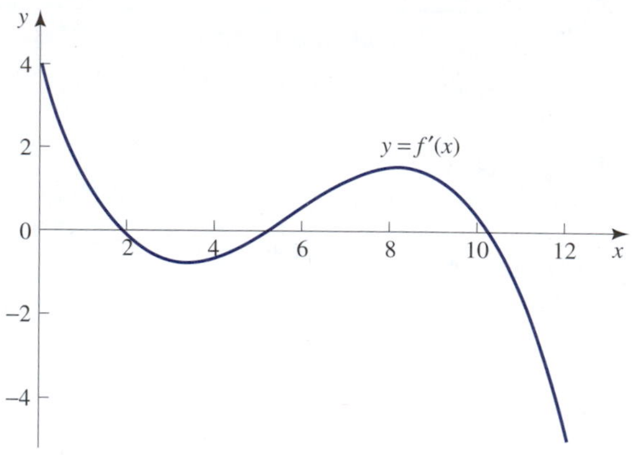

Figure 3.46

■

Interactive Illustration 3.8

Determining the Graph of $y = f'(x)$ from the Graph of $y = f(x)$

This interactive illustration allows you to determine the graph of the derivative of a function using the graph of the function. Select a function. As you move the cursor in the bottom window, notice that a line appears in the top window with slope determined by the y-value of the cursor. Adjust the cursor so that the line is tangent, and then click. Do this at a number of points to determine the graph of $y = f'(x)$. Click on the Graph f' box to check your work.

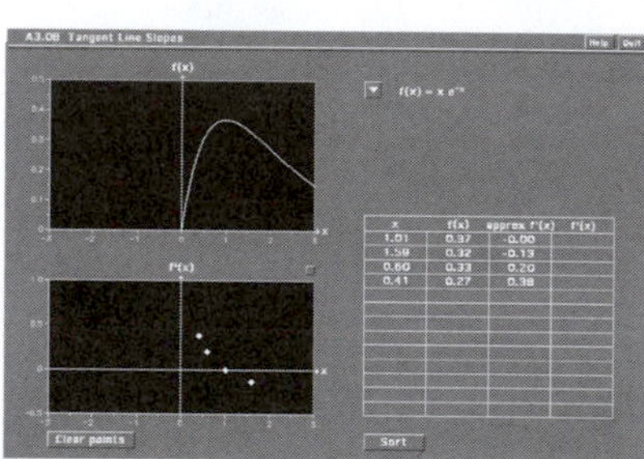

Interpreting the Derivative

We know that given a differentiable function $y = f(x)$, the derivative $f'(c)$ is also the instantaneous rate of change of y with respect to x when $x = c$, and we know that this is also the slope of the line tangent to the curve $y = f(x)$ at $x = c$. We gave some interpretations of the instantaneous rate of change in Section 3.2, which must be the same interpretations for the derivative. For example, we considered $s(t) = 16t^2$, where t is measured in seconds and s in feet. We labeled the instantaneous velocity at time t as $v(t)$. Using the derivative notation, we must have $v(t) = s'(t)$. We found that $s'(1) = v(1) = 32$. The units are in feet per second. Thus, if the ball continued to drop at this rate for an entire second, the ball would drop 32 feet during that second.

EXAMPLE 10 **Interpretation of the Derivative**

Suppose the cost and revenue functions for a plant that produces flooring nails is given by $C(x)$ and $R(x)$, respectively, where x is measured in pounds of nails and $C(x)$ and $R(x)$ are given in cents. Suppose that $C'(1000) = 50$ and $R'(1000) = 52$. What are the units of the two derivatives, and what do they mean?

Solution

The derivative is the instantaneous rate of change in both cases and is given in units of cents per pound. When 1000 pounds are being produced, the cost is increasing at the rate of 50 cents per pound, and the revenue is increasing at the rate of 52 cents per pound. This means that if the cost and revenue were to continue at the given respective constant rates, then the cost of producing the next pound of nails would be 50 cents and the revenue resulting from selling the next pound of nails would be 52 cents. ■

Enrichment: The Derivative of $\sqrt{x}$

To find the derivative of $\sqrt{x}$ requires some algebraic manipulations.

EXAMPLE 11 **Finding the Derivative of a Function**

Find the derivative of $y = f(x) = \sqrt{x}$.

Solution

Doing the first two steps gives

$$\begin{aligned}
\frac{f(x+h)-f(x)}{h} &= \frac{\sqrt{x+h}-\sqrt{x}}{h} \\
&= \frac{\sqrt{x+h}-\sqrt{x}}{h} \cdot \frac{\sqrt{x+h}+\sqrt{x}}{\sqrt{x+h}+\sqrt{x}} && \text{Rationalize the numerator} \\
&= \frac{x+h-x}{h(\sqrt{x+h}+\sqrt{x})} \\
&= \frac{h}{h(\sqrt{x+h}+\sqrt{x})} && \text{Cancel the } h \\
&= \frac{1}{\sqrt{x+h}+\sqrt{x}}.
\end{aligned}$$

Now take the limit as $h \to 0$ of the last expression, using the rules of limits. This gives

$$\begin{aligned}
\frac{dy}{dx} &= \lim_{h\to 0} \frac{f(x+h)-f(x)}{h} \\
&= \lim_{h\to 0} \frac{1}{\sqrt{x+h}+\sqrt{x}} \\
&= \frac{1}{2\sqrt{x}} \quad \text{if } x > 0.
\end{aligned}$$

■

Warm Up Exercise Set 3.3

1. If $y = f(x) = \frac{1}{x+1}$, find $f'(x)$ when $x \neq -1$.

2. Find the equation of the line tangent to the curve $y = \sqrt{x}$ at the point $(4, 2)$. Use the derivative found in Example 11.

Exercise Set 3.3

In Exercises 1 through 6, find $f'(x)$ using the limit definition of $f'(x)$.

1. $f(x) = 5x - 3$

2. $f(x) = -2x + 3$

3. $f(x) = x^2 + 4$

4. $f(x) = 2x^2 - x$

5. $f(x) = 3x^2 + 3x - 1$

6. $f(x) = -3x^2 - x + 1$

In Exercises 7 through 12, find dy/dx using the limit definition.

7. $y = -2x^3 + x - 3$

8. $y = 3x^3 - 4x + 1$

9. $y = \dfrac{1}{x+2}, x \neq -2$

10. $y = \dfrac{2}{3x+4}, x \neq -\dfrac{4}{3}$

11. $y = \dfrac{1}{2x-1}, x \neq \dfrac{1}{2}$

12. $y = \dfrac{3}{2-3x}, x \neq \dfrac{2}{3}$

13. In each of Exercises 1, 3, and 5, find the equation of the tangent line to the curve at the point $(1, f(1))$. Support your answer by using a computer or graphing calculator to graph on the same screen the function and the tangent line that was requested.

14. In each of Exercises 2, 4, and 6, find the equation of the tangent line to the curve at the point $(-1, f(-1))$. Support your answer by using a computer or graphing calculator to graph on the same screen the function and the tangent line that was requested.

15. Find the equation of the tangent line to the curve $y = \ln x$ at the point $(1, 0)$. Recall that $\frac{d}{dx} \ln x = \frac{1}{x}$. Support your answer by using a computer or graphing calculator to graph on the same screen the function and the tangent line that was requested.

16. Find the equation of the tangent line to the curve $y = \sqrt{x}$ at the point $(4, 2)$. Recall that $\frac{d}{dx}\sqrt{x} = \frac{1}{2\sqrt{x}}$. Support your answer by using a computer or graphing calculator to graph on the same screen the function and the tangent line that was requested.

In Exercises 17 through 22, sketch a graph of $y = f(x)$ to find where $m_{\tan}(x)$ and thus $f'(x)$ exists.

17. $f(x) = \begin{cases} 0 & \text{if } x \leq 0 \\ x & \text{if } x > 0 \end{cases}$

18. $f(x) = -|x|$

19. $f(x) = -|x - 1|$

20. $f(x) = \begin{cases} x & \text{if } x \leq 1 \\ -x + 2 & \text{if } x > 1 \end{cases}$

21. $f(x) = \begin{cases} x & \text{if } x \leq 1 \\ 1 & \text{if } x > 1 \end{cases}$

22. $f(x) = \begin{cases} -x & \text{if } x \leq 0 \\ 0 & \text{if } x > 0 \end{cases}$

23. A graph of the function $y = f(x)$ is shown. For what values of x does $f'(x)$ not exist?

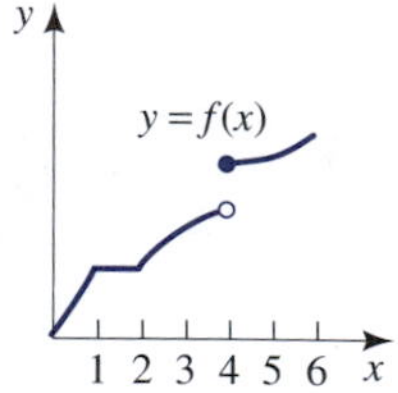

24. A graph of the function $y = g(x)$ is shown. For what values of x does $g'(x)$ not exist?

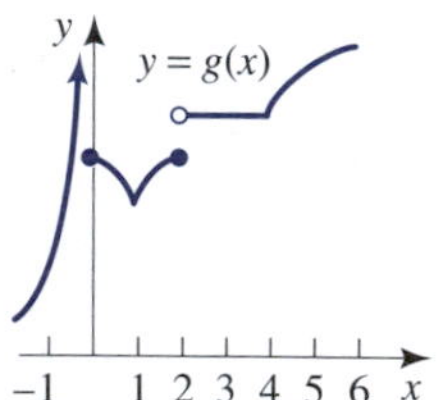

25. Consider the graph of $y = f(x)$.

a. Where is $f'(x) > 0$?

b. Where is $f'(x) < 0$?

c. What is happening to $f'(x)$ as x becomes large? As $x \to 0^+$? As $x \to 0^-$?

d. On what intervals is $f'(x)$ an increasing function? A decreasing function?

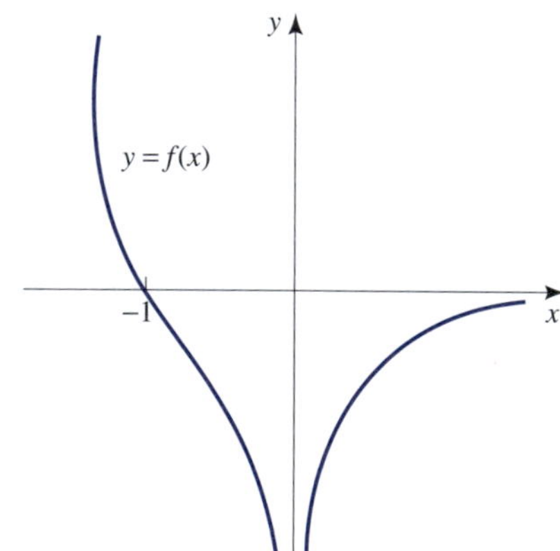

26. Consider the graph of $y = f(x)$.

a. Where is $f'(x) > 0$?

b. Where is $f'(x) < 0$?

c. On what intervals is $f'(x)$ an increasing function? A decreasing function?

For Exercises 27 through 32, consider the graphs of the following six functions.

27.

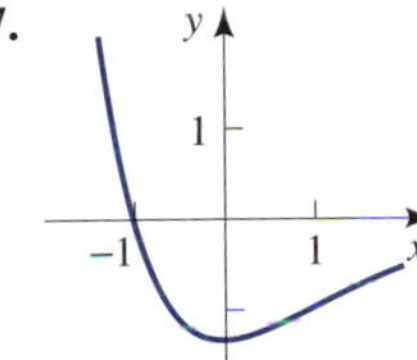

28.

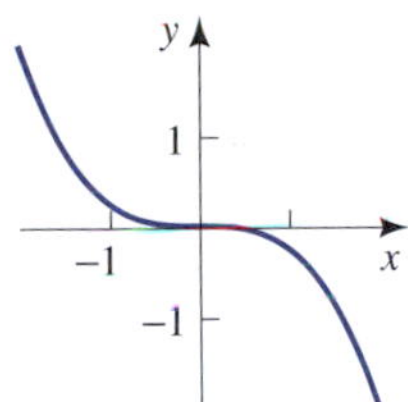

29.

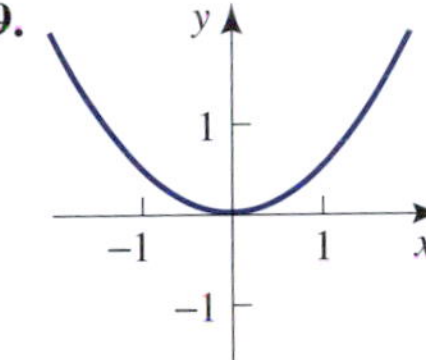

30.

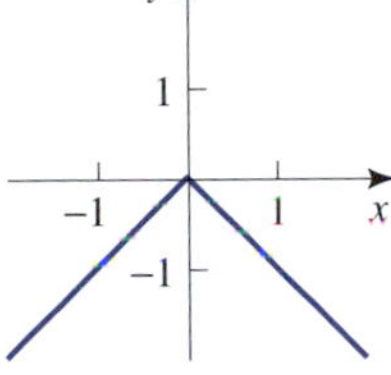

31.

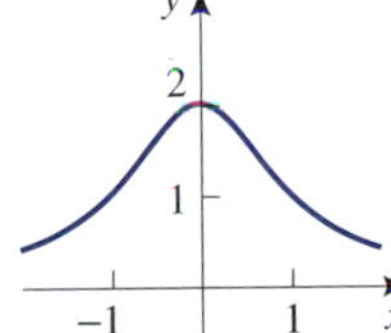

32.

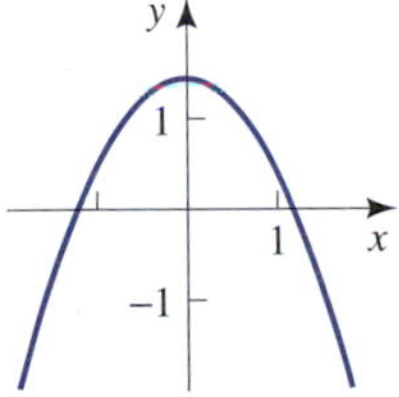

Each of the following figures represent the graph of the derivative of one of the above functions but in a different order. Match the graph of the function to the graph of its derivative.

(a)

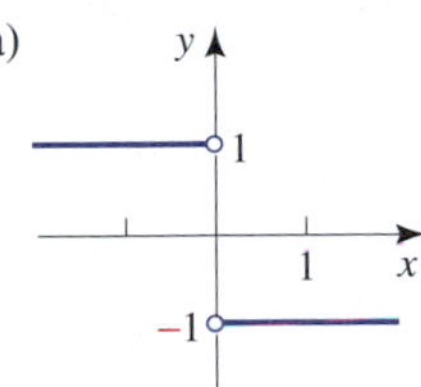

(b)

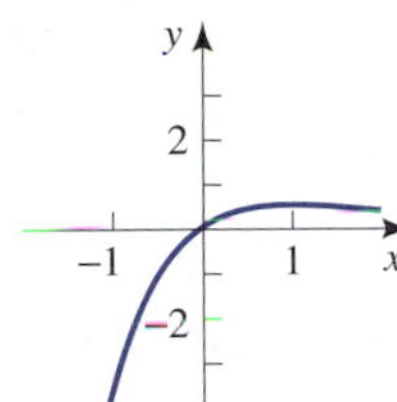

(c)

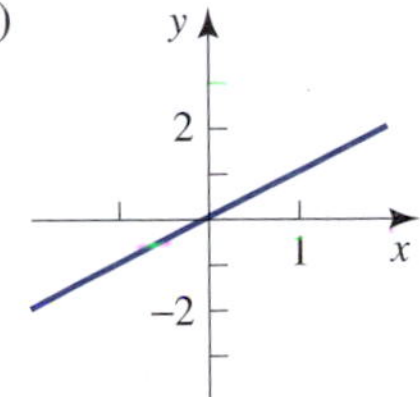

(d)

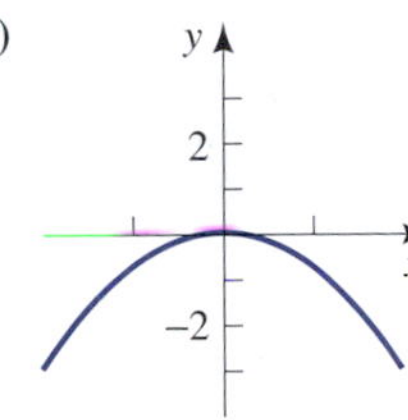

(e)

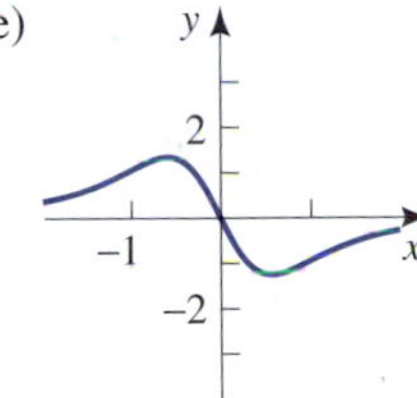

(f)

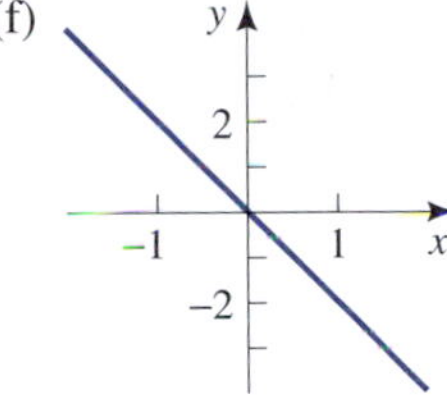

In Exercises 33 and 34, a graph of $y = f(x)$ is given together with several tangent lines. Estimate $f'(x)$ at the points where the tangent lines are shown and use this as an aid in graphing $y = f'(x)$.

33.

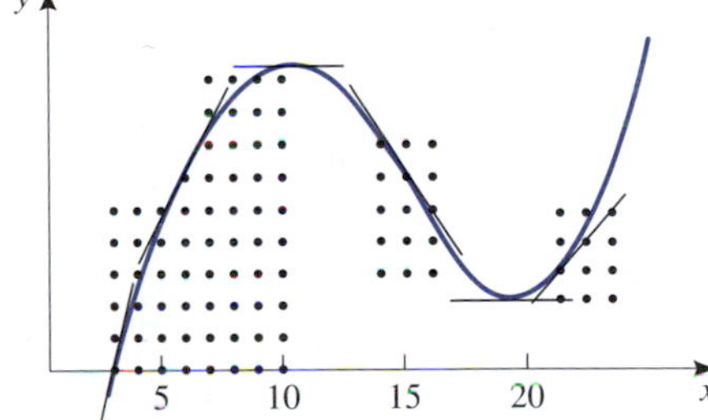

34.

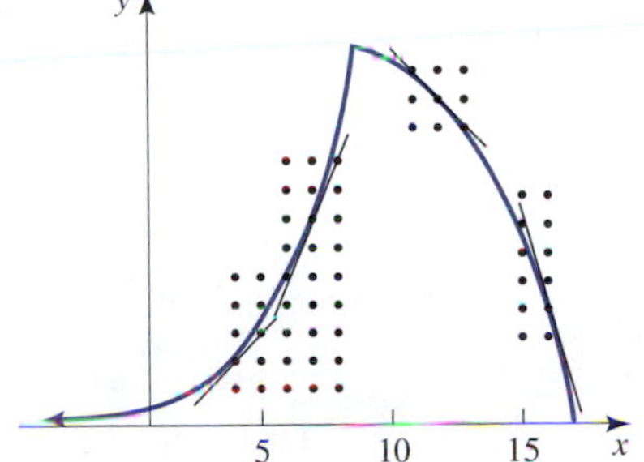

35. Let $f(x) = \sqrt[5]{x}$. Use your computer or graphing calculator to determine whether $f'(0)$ exists by graphing $y = f(x)$ and trying to estimate the tangent at $x = 0$.

36. The graph of $y = f(x)$ is shown.

a. On what interval(s) is $f'(x) > 0$?

b. On what interval(s) is $f'(x) < 0$?

c. For what values of x is $f'(x) = 0$?

d. $f'(3)$ is closest to which of the following values: 2.5, -1.5, 0, 1.5?

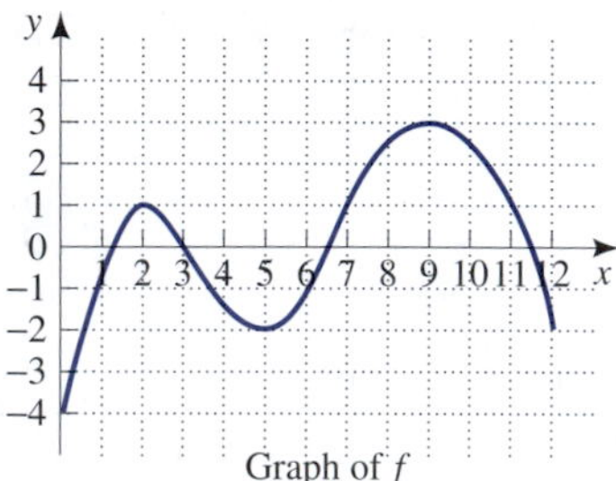

Graph of f

In Exercises 37 through 40 are given the graphs of functions. Sketch a graph of the derivative function.

37.

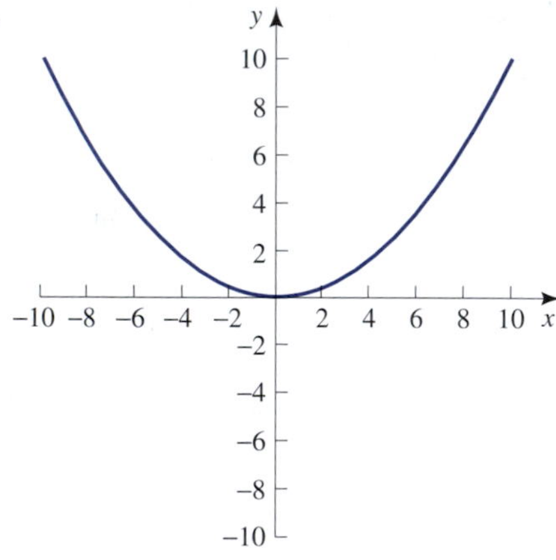

38.

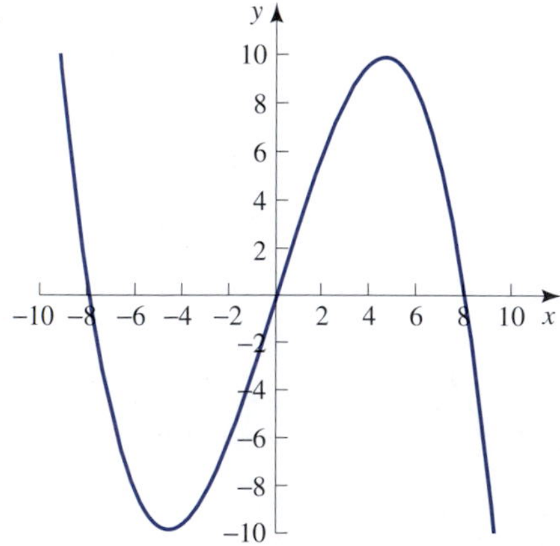

39.

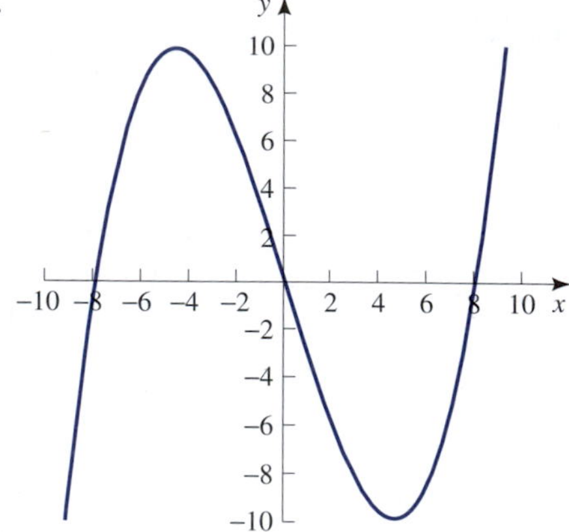

40.

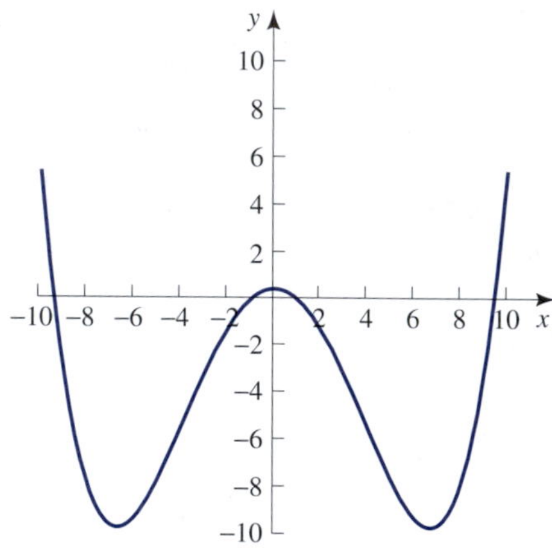

Applications and Mathematical Modeling

41. Body Temperature Merola-Zwartjes and Ligon[47] studied the Puerto Rican Tody. Even though they live in the relatively warm climate of the tropics, these organisms face thermoregulatory challenges due to their extremely small size. The researchers created a mathematical model given approximately by the equation $B(T) = 36 - 0.28T + 0.01T^2$, where T is the ambient temperature in degrees Celsius and B is body temperature in degrees Celsius for $10 \leq T \leq 42$. Graph. Find the instantaneous rate of change of body temperature with respect to ambient temperature for any T on the given interval. Give units and interpret your answer.

42. Lamprey Growth Griffiths and colleagues[48] studied the larval sea lamprey in the Great Lakes, where these creatures represent a serious threat to the region's fisheries. They created a mathematical model given approximately by the equation $G(T) = -1 + 0.3T - 0.02T^2$, where T is the mean annual water temperature with $6.2 \leq T \leq 9.8$ and G is the growth in millimeters per day. Graph. Find the instantaneous rate of change of growth with respect to water temperature for any T on the given interval. Give units and interpret your answer.

43. Corn Yield per Acre Atwood and Helmers[49] studied the effect of nitrogen fertilizer on corn yields in Nebraska. Nitrate contamination of groundwater in Nebraska has become such a serious problem that actions such as nitrogen rationing and taxation are under consideration. Nitrogen fertilizer is needed to increase yields, but an excess of nitrogen has been found not to help increase yields. Atwood and Helmers created a mathematical model given approximately by the equation $Y(N) = 59 + 0.8N - 0.003N^2$, where N is the amount of fertilizer in pounds per acre and Y is the yield in bushels per acre. Graph this equation on the interval $[50, 200]$. Find the instantaneous rate of change of yield with respect to the amount of fertilizer for any N on the given interval. Give units and interpret your answer.

44. Lactation Curves Cardellino and Benson[50] studied lactation curves for ewes rearing lambs. They created the mathematical model given approximately by the equation $P(D) = 273 + 5.4D - 0.13D^2$, where D is day of lactation and P is milk production in grams. Graph this equation on the interval $[6, 60]$. Find the instantaneous rate of change of milk production with respect to the day of lactation for any D. Give units and interpret your answer.

45. Growing Season Start Jobbagy and colleagues[51] determined a mathematical model given by the equation $G(T) = 278 - 7.1T - 1.1T^2$, where G is the start of the growing season in the Patagonian steppe in Julian days (day 1 = January 1) and T is the mean July temperature in degrees Celsius. Graph this equation on the interval $[-3, 5]$. Find the instantaneous rate of change of the day of the start of the growing season with respect to the mean July temperature for any T. Give units and interpret your answer.

46. Carbon Dynamics Kaye and colleagues[52] noted that for some experimental tree plantations in Hawaii, *Albizia* trees are used as an intercrop to increase nitrogen. They created the mathematical model given approximately by the equation $C(P) = 9490 + 386P - 3.8P^2$, where P is the percentage of *Albizia* planted and C is the above-ground tree carbon in grams per square meter. Graph the equation on the interval $[0, 100]$. Find the instantaneous rate of change of the above-ground tree carbon with respect to the percentage of *Albizia* for any P. Give units and interpret your answer.

47. Protein in Milk Crocker and coworkers[53] studied the northern elephant seal in Ano Nuevo State Reserve, California. They created a mathematical model given approximately by the equation $L(D) = 18 + 2.9D - 0.06D^2$, where D is days postpartum and L is the percentage of lipid in the milk. Graph the equation on the interval $[0, 28]$. Find the instantaneous rate of change of the percentage of lipid in the milk with respect to days postpartum for any D. Give units and interpret your answer.

---

[47] Michele Merola-Zwartjes and J. David Ligon. 2000. Ecological energetics of the Puerto Rican Tody: heterothermy, torpor, and intra-island variation. *Ecology* 81(4):990–1003.

[48] Ronald W. Griffiths, F. W. Beamish, B. J. Morrison, and L. A. Barker. 2001. Factors affecting larval sea lamprey growth and length at metamorphosis in lampricide-treated streams. *Trans. Amer. Fish. Soc.* 130:289–306.

[49] Joseph A. Atwood and Glenn A. Helmers. 1998. Examining quantity and quality effects of restricting nitrogen applications to feedgrains. *Amer. J. Agr. Econ.* 80:369–381.

[50] R. A. Cardellino and M. E. Benson. 2002. Lactation curves of commercial ewes rearing lambs. *J. Anim. Sci.* 80:23–27.

[51] Esteban G. Jobbagy, Osvaldo E. Sala, and Jose M. Paruelo. 2002. Patterns and control of primary production in the Patagonian steppe: a remote sensing approach. *Ecology* 83(2):307–319.

[52] Jason P. Kaye, Sigrid C. Resh, Margot W. Kaye, and Rodney A. Chimner. 2000. Nutrient and carbon dynamics in a replacement series of *Eucalyptus* and *Albizia* trees. *Ecology* 81(12):3267–3273.

[53] Daniel E. Crocker, Jeannine D. Williams, Daniel P. Costa, and Burney J. Le Boeup. 2001. Maternal traits and reproductive effort in northern elephant seals. *Ecology* 82(12):3541–3555.

48. Finance Granger[54] in Great Britain found that the relationship between yield y on 20-year government bonds was approximated by $y = -3.42 + \dfrac{15.23}{x}$, where x is demand deposits divided by gross national product. Find the rate of change of y with respect to x. Give units and interpret your answer.

49. Cost Curve Dean[55] made a statistical estimation of the labor cost-output relationship for a furniture factory warehouse. He found that if x is the output in thousands of dollars of warehouse value and y is total labor cost in dollars, y was given approximately by $y = -0.02464x^3 + 1.4807x^2 + 13.404x + 555.65$. Graph on your computer or graphing calculator using a screen with dimensions $[0, 47]$ by $[400, 2100]$. Obtain the approximate derivative when x is 5, 10, 15, 20, 25, 30, 35, and 40 by using available operations. Relate the derivative to the slope of the curve and the rate of change of costs. Interpret what these numbers mean. On the basis of this model, what happens to the rate of increase of labor costs as output increases?

50. Cost Curve Dean[56] made a statistical estimation of the average cost–output relationship for a shoe chain for 1938. He found that if x is the output in thousands of pairs of shoes and y is the average cost in dollars, y was given approximately by $y = 0.0648x^2 - 4.3855x + 116.67$. Graph on your computer or graphing calculator using a screen with dimensions $[0, 56.4]$ by $[0, 120]$. Have your grapher find the approximate derivative when x is 18, 27, 36, 42, and 48 using the dy/dx operation. Note the slope and relate this to the rate of change. On the basis of this model, describe what happens to the rate of change of average cost as output increases.

51. Margins Versus Size Barton and colleagues[57] studied how the size of agricultural cooperatives were related to their margin ratio (gross margins over sales as a percent). They found that if x is the assets in millions of dollars and y is the margin ratio, y was given approximately by $0.146x^2 - 1.387x + 11.269$. Graph on your computer or graphing calculator using a screen with dimensions $[0, 7.05]$ by $[5, 10]$. Find the approximate derivative when x is 1.5, 3, 4.5, 6, and 6.6, using available operations. Note the slope, and relate this to the rate of change. On the basis of this model, describe what happens to the rate of change of the margin ratio as assets increase.

52. Population Ecology Lactin and colleagues[58] studied the relationship between the feeding rate in y units of the third-instar Colorado potato beetle and the temperature T in degrees Celsius. They found that y was given approximately by $-0.07286x^2 + 4.1911x - 42.214$. Have your computer or graphing calculator find the approximate derivative when x is 20, 25, 29, 32, and 38 using available operations. Relate this number to the slope of the tangent line to the curve and to the rate of change. Interpret what each of these numbers means. On the basis of this model, describe what happens to the rate of change of the feeding rate as temperature increases.

---

[54] Clive W. J. Granger. 1993. What are we learning about the long-run? *The Econ. J.* 103:307–317.

[55] Joel Dean. 1976. Statistical cost estimation. Bloomington, Ind.: Indiana University Press, page 121.

[56] Ibid., page 340.

[57] David G. Barton, Ted C. Schroeder, and Allen M. Featherstone. 1993. Evaluating the feasibility of local cooperative consolidation: a case study. *Agribusiness* 9(3):281–294.

[58] Derek J. Lactin, N. J. Holliday, and L. L. Lamari. 1993. Temperature dependence and constant-temperature diel aperiodicity of feeding by Colorado potato beetle larvae in short-duration laboratory trials. *Environ. Entomol.* 22(4):784–790.

Enrichment Exercises

In Exercises 53 through 56, find dy/dt using the limit definition.

53. $y = \sqrt{t+1}$, $t > -1$ **54.** $y = \sqrt{t-3}$, $t > 3$

55. $y = \sqrt{2t+5}$, $t > -\frac{5}{2}$ **56.** $y = \sqrt{1-t}$, $t < 1$

More Challenging Exercises

57. In Example 6 and Exploration 3, we indicated graphically why the derivative of $f(x) = \sqrt[3]{x^2}$ does not exist at $x = 0$. Confirm this analytically, using the limit definition of the derivative.

58. Use the limit definition of derivative to show that $f'(0)$ does not exist if $f(x) = \sqrt[5]{x^2}$.

59. This example indicates another way in which your grapher can mislead you. Let $f(x) = \sqrt{x^2 + 0.000001}$.

a. Graph on a window with dimensions $[-10, 10]$ by $[0, 10]$. What does the graph appear to tell about the existence of $f'(0)$?

b. Support your answer in part (a) graphically and numerically as follows. Graph

$$y = \frac{f(0+h) - f(0)}{h} = \frac{\sqrt{h^2 + 0.000001} - 0.001}{h}$$

on a window with dimensions $[-0.01, 0.01]$ by $[-1, 1]$. What do you conclude about the limit of this expression as h goes to zero? What does this say about $f'(0)$? Support your answer numerically by evaluating the expression for values of h such as 0.0001, 0.00001, and 0.000001.

c. Graph $y = f(x)$ again on your graphing calculator using a window of dimensions $[-10, 10]$ by $[-1, 10]$. Take windows with smaller and smaller dimensions about the point $(0, 0.001)$. Explain what you have observed. From this, what do you conclude $f'(0)$ is?

60. Use the limit definition of derivative to show that $f'(0)$ does exist if $f(x)$ is the function given in the previous exercise.

61. Let $f(x) = 0$ if $x < 0$ and x^2 if $x \geq 0$. Explore graphically and numerically whether $f'(0)$ exists. Confirm analytically, that is, using the limit definition of derivative.

62. Explain in complete sentences the circumstances under which the derivative of a function does not exist.

63. On your computer or graphing calculator, graph $y = f(x) = \sin x$ in radian mode using a window with dimensions $[-6.14, 6.14]$ by $[-1, 1]$. As you see, this function moves back and forth between -1 and 1. We wish to estimate $f'(0)$. For this purpose, graph using a window with dimensions $[-0.5, 0.5]$ by $[-0.5, 0.5]$. From the graph, estimate $f'(0)$.

64. Let $y = f(x) = \sin x$. Graph with your grapher in radian mode. As you can readily see, $\sin(0) = 0$. By definition

$$f'(0) = \lim_{h \to 0} \frac{\sin(0+h) - \sin(0)}{h} = \lim_{h \to 0} \frac{\sin h}{h}$$

Graph $\frac{\sin x}{x}$ on your grapher to estimate the limit as $x \to 0$. Support numerically. Compare your answer to what you found in the previous exercise.

65. On your computer or graphing calculator, graph $y = f(x) = \cos x$ in radian mode, using a window with dimensions $[-6.14, 6.14]$ by $[-1, 1]$ to familiarize yourself with this function. As you see, this function moves back and forth between -1 and 1. We wish to estimate $f'(\pi/2)$, where $\pi/2 \approx 1.57$. For this purpose, graph using a window with dimensions $[1.07, 2.07]$ by $[-0.5, 0.5]$. From the graph, estimate $f'(1.57)$.

66. Let $y = f(x) = \cos x$. Graph with your computer or graphing calculator in radian mode. As you readily can see, $\cos(\pi/2) = 0$. By definition,

$$f'(\pi/2) = \lim_{h \to 0} \frac{\cos(\pi/2 + h) - \cos(\pi/2)}{h} = \lim_{h \to 0} \frac{\cos(\pi/2 + h)}{h}$$

Graph $\frac{\cos(\pi/2 + x)}{x}$ on your grapher to estimate the limit as $x \to 0$. Support numerically. Compare your answer to what you found in Exercise 65.

67. With your computer or graphing calculator in radian mode, graph $y_1 = \sin x$ and $y_2 = \cos x$, and familiarize yourself with these functions. Now replace $y_1 = \sin x$ with $y_1 = \frac{\sin(x + 0.001) - \sin x}{0.001}$ and graph. This latter function is approximately the derivative of $\sin x$. How does the graph of this latter function compare with the graph of $\cos x$? Does this show that $\frac{d}{dx}(\sin x) = \cos x$?

68. With your computer or graphing calculator in radian mode, graph $y_1 = \cos x$ and $y_2 = -\sin x$, and familiarize yourself with these functions. Now replace $y_1 = \cos x$ with $y_1 = \frac{\cos(x + 0.001) - \cos x}{0.001}$ and graph. This latter function is approximately the derivative of $\cos x$. How does the graph of this latter function compare with the graph of $-\sin x$? Does this show that $\frac{d}{dx}(\cos x) = -\sin x$?

Modeling Using Least Squares

69. Cost Curve Dean[59] made a statistical estimation of the salary cost per account in a finance chain with the total number of open accounts. The data is given in the following table.

x	1200	1450	1900	2400	3200
y	0.72	0.61	0.46	0.45	0.43

Here x is the total number of open accounts, and y is the salary cost per account.

[59] Joel Dean. 1976. *Statistical Cost Estimation.* Bloomington, Ind.: Indiana University Press, page 317.

a. Use power regression to find the best-fitting power function using least squares.

b. Graph on your computer or graphing calculator using a screen with dimensions [1000, 4760] by [0, 0.9]. Have your grapher find the derivative when x is 1200, 1600, 2000, 2400, 2800, and 3200. Relate the derivative to the slope of the curve and to the rate of change. On the basis of this model, describe what happens to the rate of change of salary cost as the total number of accounts increases.

70. Cost Curve Dean[60] made a statistical estimation of the cost-output relationship for a shoe chain. The data for the firm is given in the following table.

x	4.7	6.5	7.8	10	15	20	30	38	50
y	4.7	5.5	6.4	6.8	8.5	12	20	16	30

Here x is the output in thousands of pairs of shoes, and y is the total cost in thousands of dollars.

a. Use quadratic regression to find the best-fitting quadratic polynomial using least squares.

b. Graph on your computer or graphing calculator using a screen with dimensions [0, 56.4] by [0, 40]. Have your grapher find the derivative when x is 6, 18, 24, 30, and 36. Note the slope, and relate this to the rate of change. On the basis of this model, describe what happens to the rate of change of average cost as output increases.

71. Variable Cost Barton and colleagues[61] studied how the size of agricultural cooperatives were related to their variable cost ratio (variable cost over assets as a percent). The data is given in the following table.

x	2.48	3.51	2.50	3.37	3.77	6.33	4.74
y	11.7	16.8	25.4	16.9	17.9	22.8	16.7

Here x is the assets in millions of dollars, and y is the variable cost ratio.

a. Use quadratic regression to find the best-fitting quadratic function using least squares.

b. Graph on your computer or graphing calculator using a screen with dimensions [0, 9.4] by [0, 30]. Have your grapher find the derivative when x is 1.5, 3, 4.5, 6, and 6.6. Note the slope, and relate this to the rate of change. On the basis of this model, describe what happens to the rate of change of the variable cost ratio as assets increase.

72. Ecological Entomology Elliott[62] studied the temperature affects on the alder fly. He collected in his laboratory in 1969 the data shown in the following table relating the temperature in degrees Celsius to the number of pupae successfully completing pupation.

t	8	10	12	16	20	22
y	15	29	41	40	31	6

Here t is the temperature in degrees Celsius, and y is the number of pupae successfully completing pupation.

a. Use quadratic regression to find y as a function of T. Graph using a window with dimensions [6, 24.8] by [0, 60].

b. Have your computer or graphing calculator find the numerical derivative when x is 10, 12, 15, 18, and 20. Relate this number to the slope of the tangent line to the curve and to the rate of change. Interpret what each of these numbers means. On the basis of this model, describe what happens to the rate of change of number of pupae completing pupation as temperature increases.

---

[60] Ibid., page 337.

[61] David G. Barton, Ted C. Schroeder, and Allen M. Featherstone. 1993. Evaluating the feasibility of local cooperative consolidation: a case study. *Agribusiness* 9(3):281–294.

[62] J. M. Elliott. 1996. Temperature-related fluctuations in the timing of emergence and pupation of Windermere alder-flies over 30 years. *Ecol. Entomol.* 21:241–247.

Solutions to WARM UP EXERCISE SET 3.3

1. First find an expression for the average rate of change of y with respect to x from x to $x+h$ and simplify. This gives

$$\begin{aligned}\frac{f(x+h)-f(x)}{h} &= \frac{\dfrac{1}{x+1+h}-\dfrac{1}{x+1}}{h} \\ &= \frac{1}{h}\left[\frac{1}{x+1+h}-\frac{1}{x+1}\right] \\ &= \frac{1}{h}\left[\frac{x+1-(x+1+h)}{(x+1)(x+1+h)}\right] \\ &= \frac{1}{h}\left[\frac{-h}{(x+1)(x+1+h)}\right] \\ &= -\frac{1}{(x+1)(x+1+h)}\end{aligned}$$

Now to find the derivative, take the limit of this expression as $h \to 0$, obtaining

$$\begin{aligned}f'(x) &= \lim_{h\to 0}\frac{f(x+h)-f(x)}{h} \\ &= \lim_{h\to 0}\frac{-1}{(x+1)(x+1+h)} = -\frac{1}{(x+1)^2}\end{aligned}$$

provided that $x \neq -1$.

2. From Example 11 of the text we have that $\frac{d}{dx}\left(\sqrt{x}\right) = \frac{1}{2\sqrt{x}}$. Thus,

$$m_{\tan}(4) = f'(4) = \frac{1}{2\sqrt{4}} = \frac{1}{4}$$

Thus, the equation of the tangent line at $x = 4$ is

$$y - 2 = \tfrac{1}{4}(x-4)$$

3.4 Local Linearity

APPLICATION Marginal Cost

Dean[63] made an extensive study of cost functions in various industries. For a belt shop he found that if x is the output in thousands of square feet and y is the total cost in thousands of dollars, then the equation $y = C(x) = -12.995 + 1.330x - 0.0062x^2 + 0.000022x^3$ approximated the relationship between x and y for $50 \leq x \leq 117$. Find $C'(70)$, $C'(90)$, and $C'(110)$. Interpret what is happening. See Example 4 on page 209 for the answer.

[63] Joel Dean. 1976. *Statistical Cost Estimation.* Bloomington, Ind.: Indiana University Press, page 215.

Local Linearity

Given a function $y = f(x)$ differentiable at $x = c$, we know that

$$f'(c) = \lim_{h\to 0}\frac{f(c+h)-f(c)}{h}$$

From our knowledge of limits we know that if h is small, then

$$f'(c) \approx \frac{f(c+h)-f(c)}{h}$$

Multiplying both sides of this expression by h then yields

$$f(c+h) - f(c) \approx f'(c)h$$

The term $f(c+h) - f(c)$ is the actual change in y and is shown in Figure 3.47. The term $f'(c)h$ gives the change in y that occurs when moving along the tangent line. According to Figure 3.47, we expect the change in y when moving along the tangent line to be an increasingly better approximation to the actual change in y as h becomes smaller. The term $E = [f(c+h) - f(c)] - f'(c)h$ represents the error in using the term $f'(c)h$ to approximate the actual change in y. We see that as h approaches zero, this error E will also approach zero.

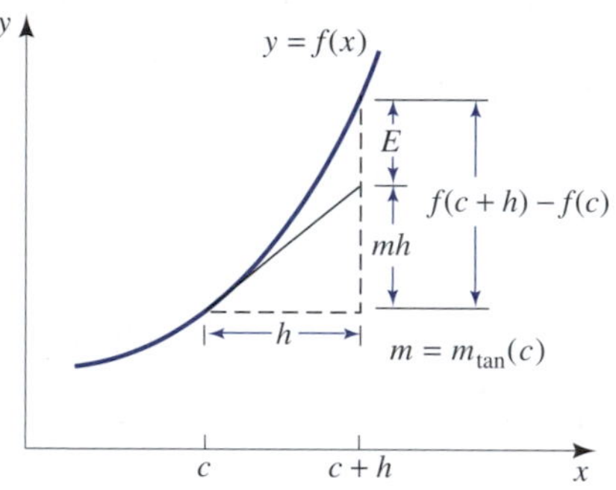

Figure 3.47
If $m_{\tan}(c) = m$ exists, $f(c+h) - f(c) \approx mh$ for small h.

[0.75, 1.25] × [0.532, 1.532]

Screen 3.10
The graph of $y_1 = x^2$ is nearly the same as the graph of the tangent line $y_2 = 2x - 1$ near $x = 1$.

Approximating the Change in *y*

If $y = f(x)$ is differentiable at $x = c$, then the change in y, given by $f(c+h) - f(c)$, can be approximated by the linear function $f'(c)h$. That is,

$$f(c+h) - f(c) \approx f'(c)h \qquad (3.1)$$

if h is small.

EXAMPLE 1 The Tangent Line Approximation

Graph $y_1 = x^2$ and the tangent line to the graph of this curve at the point $(1, 1)$. Check to see how close the tangent line is to the curve.

Solution

Screen 3.10 shows graphs of $y_1 = x^2$ and the tangent line $y_2 = 2x - 1$ using a window with dimensions [0.75, 1.25] by [0.532, 1.532]. We can already see that the graph of the tangent line nearly merges with the graph of the curve on the interval [0.8, 1.2]. Using a window with dimensions [0.9375, 1.0625] by [0.875, 1.125], we obtain Screen 3.11. Notice that in this last window the graph of $y = x^2$ appears to be a straight line! What we see is that under magnification the graph of the curve near $x = 1$ is about the same as the graph of the tangent line. ■

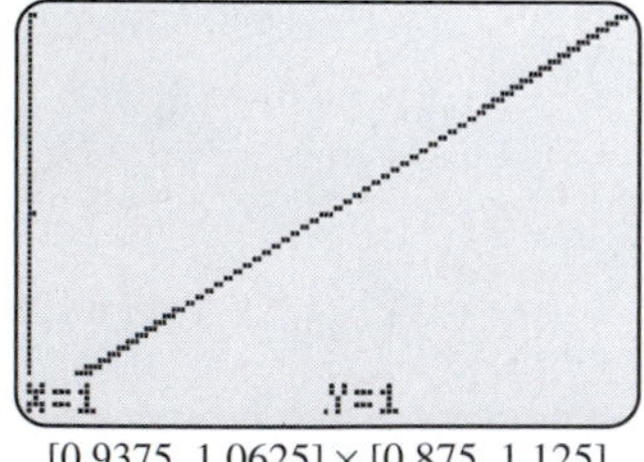

[0.9375, 1.0625] × [0.875, 1.125]

Screen 3.11
As we ZOOM in even further, we see that the graph of $y_1 = x^2$ is about the same as the graph of the tangent line $y_2 = 2x - 1$ near $x = 1$.

Interactive Illustration 3.9

The Tangent Line Approximation

This interactive illustration allows you to zoom in on the graph of a function to see that at any point, the function looks like a straight line (the tangent line) when blown up sufficiently. Select a function. The top window has a graph of this function, click on a point of the graph. The lower window shows a blow-up of a segment of the graph. Click on the Zoom In box and zoom until the graph of the function appears to be a straight line. Use this method to determine $f'(x)$ at the various points x.

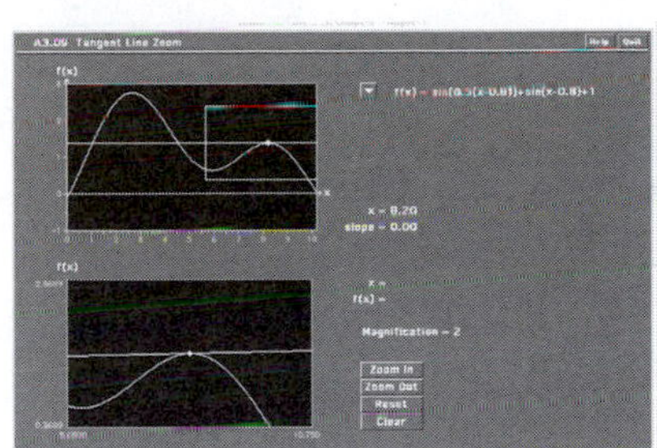

The following example illustrates a way in which this approximation can be used.

Figure 3.48

EXAMPLE 2 **Approximating the Change in Area**

A square field with sides of 100 feet is to be increased by 1 foot around the entire square. See Figure 3.48. Use Equation (3.1) to approximate the increase in area.

Solution

The area of a square x feet by x feet is $A(x) = x^2$. According to Figure 3.48, the length of the sides increases from 100 feet to 102 feet. Referring to Equation (3.1), we will take $c = 100$ and $h = 2$. Recall from Section 3.3 that the derivative of x^2 is $2x$, so $A'(x) = 2x$ and $A'(100) = 200$. Thus, the approximate change in the area using Equation (1) yields

$$A(102) - A(100) \approx A'(100) \cdot h = 200 \cdot 2 = 400$$

Compare this approximation with the exact answer $(102)^2 - (100)^2 = 404$.

The Tangent Line Approximation

We now turn our attention to establishing a fundamentally important geometrical fact discovered in Example 1 concerning the behavior of a function near a point where the function is differentiable. Briefly stated, if $f'(x)$ exists at $x = c$, then for values of x near to c, the graph of the function $y = f(x)$ is approximately the same as the graph of the tangent line through $(c, f(c))$.

We now see analytically why this is true. In Equation (3.1), move the $f(c)$ term to the right-hand side and obtain

$$f(c + h) \approx f(c) + f'(c)h$$

Think of c as fixed and h as the variable, and let $x = c + h$. Then the last equation becomes

$$f(x) \approx f(c) + f'(c)(x - c)$$

Recall that the right-hand side of this last equation is the equation of the line tangent

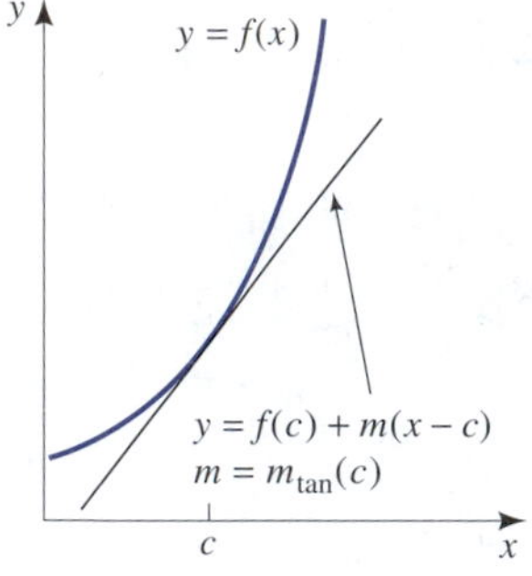

Figure 3.49
If $m_{\tan}(c)$ exists, the graph of $y = f(x)$ is approximately the same as the graph of the tangent line for x near c.

to the curve at c. Thus, we are saying that the graph of $y = f(x)$ is approximately the same as the graph of the line tangent to the curve at $x = c$ (Figure 3.49).

Tangent Line Approximation

If $y = f(x)$ is differentiable at $x = c$, then for values of x near c,

$$f(x) \approx f(c) + f'(x)(x - c) \tag{3.2}$$

Thus, for values of x near c, the graph of the curve $y = f(x)$ is approximately the same as the graph of the tangent line through the point $(c, f(c))$.

Another way of viewing this approximation is to realize that using the tangent line to approximate the function $y = f(x)$ in Figure 3.49 near c is the same as assuming that the function changes at a constant rate of change near c given by the instantaneous rate of change at c.

In the enrichment subsection in Section 3.3 we showed that $\frac{d}{dx}\sqrt{x} = \frac{1}{2\sqrt{x}}$. If you have not done this, we note this formula here so that it can be used in the next example.

EXAMPLE 3 **The Tangent Line Approximation**

Use the tangent line approximation (3.2) to find an approximation to $\sqrt{4.1}$ without resorting to a calculator.

Solution

Let $f(x) = \sqrt{x}$, $x = 4.1$, and $c = 4$. Then $f'(x) = \frac{1}{2\sqrt{x}}$. Using the approximation (3.2), we have

$$\begin{aligned}\sqrt{4.1} &= f(4.1)\\ &\approx f(4) + f'(4)(4.1 - 4)\\ &= \sqrt{4} + \frac{1}{2\sqrt{4}}(0.1)\\ &= 2 + 0.025\\ &= 2.025\end{aligned}$$

■

One should compare this to the actual $\sqrt{4.1}$, which to four decimal places is 2.0248. This approximation is off by only 0.0002. In Figure 3.47 this means that $E \approx -0.0002$.

Exploration 1

Tangent Line Approximation

On your grapher, graph $y_1 = \sqrt{x}$ and the tangent line to this graph at (4, 2) on the same screen of dimensions [3, 5] by [1, 3]. Notice already how close the graph of the tangent line is to the graph of the curve. Now take a screen with smaller dimensions about the point (4, 2) and observe how close the graph of the curve is to the graph of the tangent line.

Marginal Analysis

Suppose now we look again at Figure 3.47 and replace $f(x)$ with a revenue function $R(x)$, where x now represents the number of units sold. If c is now a "large" number, say, 1000, then $h = 1$ unit will be "small." Under these circumstances the approximation shown in Figure 3.47 and also given by Equation (3.1) should apply. The new figure for $R(x)$ is shown in Figure 3.50. Equation (3.1) then becomes

$$R(c + 1) - R(c) \approx R'(c)$$

But $R(c + 1) - R(c)$ is the revenue generated by the next unit, that is, of the $(c + 1)$st unit. We now have an important *economic* interpretation of the derivative $R'(x)$: It is approximately the revenue generated by the next unit. In a similar fashion $C'(x)$ is approximately the cost generated by the next unit, and $P'(x)$ is approximately the profit generated by the next unit. Also, $C'(x)$, $R'(x)$, and $P'(x)$ are referred to as the **marginal cost**, **marginal revenue**, and **marginal profit**, respectively.

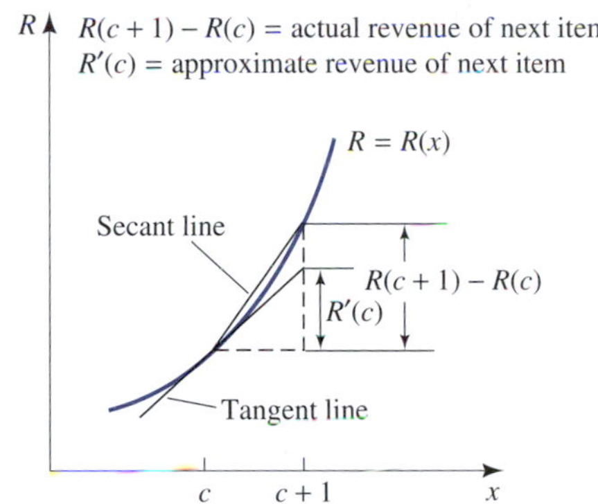

Figure 3.50
The revenue of the next item, $R(c + 1) - R(c)$ is approximately $R'(c)$.

DEFINITION Economic Interpretation of Derivative

If $C(x)$, $R(x)$, and $P(x)$ give the cost, revenue, and profit functions, respectively, then under reasonable circumstances $C'(x)$, $R'(x)$, and $P'(x)$ are the approximate cost, revenue, and profit, respectively, associated with the $(x + 1)$st item. The terms $C'(x)$, $R'(x)$, and $P'(x)$ are referred to as the **marginal cost**, **marginal revenue**, and **marginal profit**, respectively.

EXAMPLE 4 Marginal Cost

Dean[64] made an extensive study of cost functions in various industries. For a belt shop he found that if x is the output in thousands of square feet and y is the total cost in thousands of dollars, then the equation $y = C(x) = -12.995 + 1.330x -$

[64] Ibid.

$0.0062x^2 + 0.000022x^3$ approximated the relationship between x and y for $50 \le x \le 117$. Find $C'(70)$, $C'(90)$, and $C'(110)$. Interpret what is happening. Find where $C'(x)$ reaches a minimum.

Solution

Figure 3.51 shows the data used by Dean and the cubic cost function. We can use our computer or graphing calculator to graph an approximation to the derivative. This is shown in Screen 3.12. (Using formulas found in Chapter 4, we can also give the marginal cost to be exactly $C'(x) = 1.330 - 0.0124x + 0.000066x^2$.) Then we find that

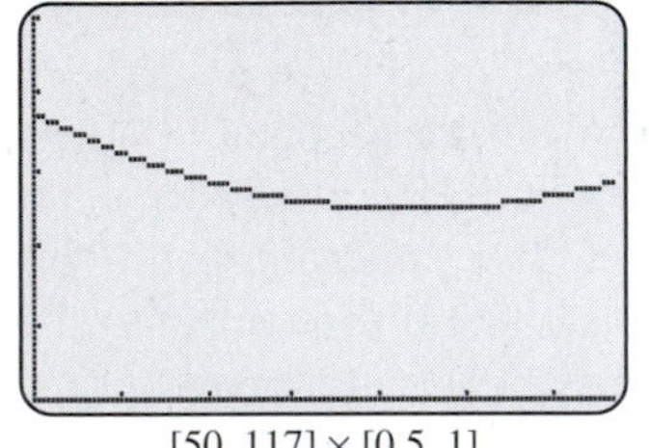

[50, 117] × [0.5, 1]

Screen 3.12
We see the graph of marginal cost. We note that the graph decreases at first but then increases at the right.

$$C'(70) = 0.7854$$
$$C'(90) = 0.7486$$
$$C'(110) = 0.7646$$

Thus, when 70,000 square feet of belt is produced, cost increases at the rate of \$785.40 per 1000 ft^2, and we infer from the discussion above that the cost of the next 1000 ft^2 is approximately \$785.40. When 90,000 ft^2 of belt is produced, cost increases at the rate of \$748.60 per 1000 ft^2, and we infer that the cost of the next 1000 ft^2 is approximately \$748.60. When 100,000 ft^2 of belt is produced, cost increases at the rate of \$764.60 per 1000 ft^2, and we infer that the cost of the next 1000 ft^2 is approximately \$764.60.

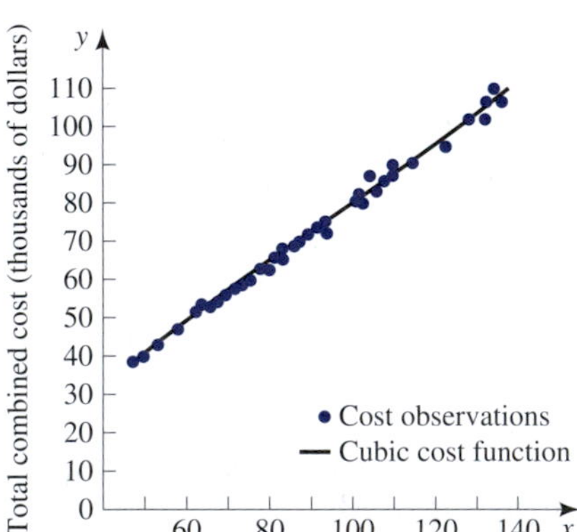

Figure 3.51
Combined cost function for a belt shop.

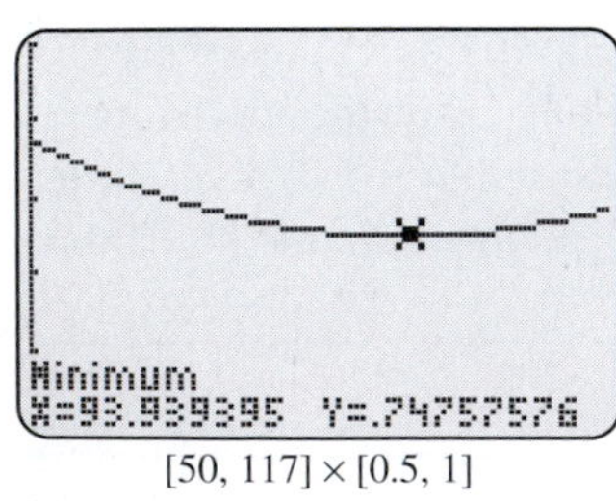

[50, 117] × [0.5, 1]

Screen 3.13
The marginal cost reaches a minimum of approximately 0.7476.

Notice that marginal cost has decreased going from 70,000 to 90,000 units but increases going from 90,000 to 110,000 units. This can also be seen in Screen 3.13, where we see that marginal cost decreases initially and then, at some value, the marginal cost begins to increase. This is typical of cost functions. Because of economies of scale, unit costs normally will decrease initially. But if production is increased too much, unit costs will increase owing to the use of less efficient equipment, payment of overtime, and so on.

We can use the available operations on our computers or graphing calculators to find the point at which $C'(x)$ is at a minimum. We find that the minimum occurs at approximately $x_m = 93.94$ with $C'(x_m) \approx 0.7476$. See Screen 3.13. ■

Interactive Illustration 3.10

Marginal Cost

This interactive illustration allows you to explore the marginal cost by observing the tangent line to a cost function and the value of the slope. By clicking on a value of x on the slider at the bottom, you will see a tangent line in the top window and a graph of the point $(x, f(x))$ in the bottom window. Do this for a number of values of x. Explain what happens to the slope. Is this similar to Example 4?

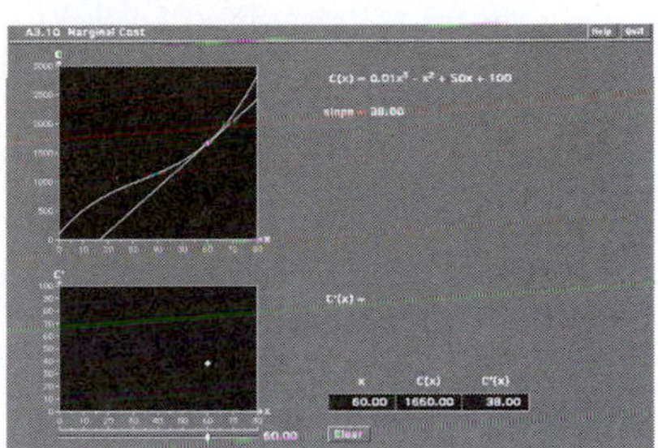

CONNECTION

Production Disruption Costs

A dramatic example of large costs associated with stretching plants to unusually high production occurred in 1997. In that year Boeing Corporation, the giant aircraft manufacturer, incurred large "production disruption costs" while attempting to increase its production of planes from 18 a month in 1996 to 43 a month in 1997.[65] Problems became so severe that Boeing found it necessary to shut down the line for its 747-model jumbo jets and 737-model planes for about a month to unsnarl production flow and would write off production disruption costs of over $1 billion.

[65] *The Wall Street Journal*, October 24, 1997.

Warm Up Exercise Set 3.4

1. The outside of a cube with edges 10 inches long is painted with a paint that is 0.10 inch thick. Use the tangent line approximation to estimate the volume of paint used. Use a computer or graphing calculator to estimate any slope of any tangent line that you need.

Exercise Set 3.4

In Exercises 1 through 12, use the linear function $f(c) + f'(c)(x - c)$ to approximate the given quantity. Recall from the previous section that $\frac{d}{dx}(\sqrt{x}) = \frac{1}{2\sqrt{x}}$, $\frac{d}{dx}(\ln x) = \frac{1}{x}$, and $\frac{d}{dx}(x^{-1}) = -x^{-2}$. Use your computer or calculator to find the actual answer, and compare your approximation to this. (Of course, the answer given by your computer or calculator is also an approximation.)

1. $\sqrt{16.24}$ **2.** $\sqrt{25.5}$ **3.** $\sqrt{24.7}$

4. $\sqrt{15.84}$ **5.** $\ln 0.95$ **6.** $\ln 1.05$

7. $\ln 2.02$ **8.** $\ln 1.98$ **9.** $\frac{1}{1.05}$

10. $\frac{1}{0.96}$ **11.** $\frac{1}{1.96}$ **12.** $\frac{1}{2.08}$

In Exercises 13 through 18, graph the function on your grapher using a screen with smaller and smaller dimensions about the point $(c, f(c))$ until the graph looks like a straight line. Find the approximate slope of this line. What is $f'(c)$?

13. $y = \sqrt[3]{x}$, $c = 1$ **14.** $y = x^3$, $c = 2$

15. $y = e^x \ln x$, $c = 1$ **16.** $y = e^x$, $c = 1$

17. $f(x) = 0.5x^4 - 2\sqrt{x} + 1$, $c = 1$

18. $f(x) = xe^x$, $c = 0$

In Exercises 19 through 24, use the local linearity approximation to approximate the given quantity. Use the results of Exercises 13 through 18. Use a computer or calculator to find the actual answer, and compare your approximation to this. (Of course, the answer on your calculator is also an approximation.)

19. $\sqrt[3]{1.1}$ **20.** $(1.8)^3$

21. $e^{0.9}\ln 0.9$ **22.** $e^{1.1}$

23. $0.5(1.1)^4 - 2\sqrt{1.1} + 1$ **24.** $0.1e^{0.1}$

25. If the side of a square decreases from 4 inches to 3.8 inches, use the linear approximation formula to estimate the change in area.

26. If the side of a cube decreases from 10 feet to 9.6 feet, use the linear approximation formula to estimate the change in volume. Use a computer or graphing calculator to estimate any slope that you need.

27. Let $C(x)$ be a cost function for a firm, with x the number of units produced and $C(x)$ the total cost in dollars of producing x units. If $C'(1000) = 4$, what is the approximate cost of the 1001st unit?

28. Let $R(x)$ be a revenue function for a firm, with x the number of units sold and $R(x)$ the total revenue in dollars from selling x units. If $R'(1000) = 6$, what is the approximate revenue from the sale of the 1001st unit?

29. Let $P(x)$ be a profit function for a firm, with x the number of units produced and sold and $P(x)$ the total profit in dollars from selling x units. If $P'(1000) = 2$, what is the approximate profit from selling the 1001st unit?

30. Let $P(x)$ be a profit function for a firm, with x the number of units produced and sold and $P(x)$ the total profit in dollars from selling x units. If $P'(1000) = -2$, what is the approximate profit from selling the 1001st unit?

Applications and Mathematical Modeling

31. Demand The demand equation for a certain product is given by $p(x) = \dfrac{10}{\sqrt{x+1}}$. Use the linear approximation formula to estimate the change in price p as x changes from 99 to 102. Find the slope of the tangent line by using your computer or graphing calculator to plot the tangent line at the appropriate point or use a screen with smaller and smaller dimensions until the graph looks like a straight line. (See the Technology Resource Manual.)

32. Revenue The revenue equation for a certain product is given by $R(x) = 10\sqrt{x}(100 - x)$. Use the linear approximation formula to estimate the change in revenue as x changes from 400 to 402. Find the slope of the tangent line by using your computer or graphing calculator to plot the tangent line at the appropriate point or use a screen with smaller and smaller dimensions until the graph looks like a straight line. (See the Technology Resource Manual.)

33. Cost The cost equation for a certain product is given by $C(x) = \sqrt{x^3}(x + 100) + 1000$. Use the linear approximation formula to estimate the change in costs as x changes from 900 to 904. Find the slope of the tangent line by using your computer or graphing calculator to plot the tangent line at the appropriate point or use a screen with smaller and smaller dimensions until the graph looks like a straight line. (See the Technology Resource Manual.)

34. Profit The profit equation for a certain product is given by $P(x) = \sqrt{x}(1000 - x) - 5000$. Use the linear approximation formula to estimate the change in profits as x changes from 25 to 27. Find the slope of the tangent line by using your computer or graphing calculator to plot the tangent line at the appropriate point or use a screen with smaller and smaller dimensions until the graph looks like a straight line. (See the Technology Resource Manual.)

35. Biology Use the linear approximation formula to estimate the change in volume of a spherical bacterium as its radius increases from 2 microns to 2.1 microns. Find the slope of the tangent line by using your computer or graphing calculator to plot the tangent line at the appropriate point, or use a screen with smaller and smaller dimensions until the graph looks like a straight line. (See the Technology Resource Manual.)

36. Learning The time T it takes to memorize a list of n items is given by $T(n) = 3n\sqrt{n-3}$. Find the approximate change in time required from memorizing a list of 12 items to memorizing a list of 14 items. Find the slope of the tangent line by using your computer or graphing calculator to plot the tangent line at the appropriate point, or use a screen with smaller and smaller dimensions until the graph looks like a straight line. (See the Technology Resource Manual.)

37. Cost Function Dean[66] found that the cost function for indirect labor in a furniture factory was approximated by $y = C(x) = 0.4490x - 0.01563x^2 + 0.000185x^3$ for $0 \le x \le 50$. Find the marginal cost for any x. Find $C'(10)$, $C'(30)$, and $C'(40)$. Interpret what is happening. Graph marginal cost on a screen with dimensions $[0, 47]$ by $[0, 0.3]$.

38. Cost Function Dean[67] found that the cost function for direct materials in a furniture factory was approximated by $y = C(x) = 0.667x - 0.00467x^2 + 0.000151x^3$ for $0 \le x \le 50$. Find the marginal cost for any x. Find $C'(5)$, $C'(10)$, and $C'(20)$. Interpret what is happening. Graph marginal cost on a screen with dimensions $[0, 47]$ by $[0.6, 1.4]$.

---

[66] Joel Dean. 1976. *Statistical Cost Estimation*. Bloomington, Ind.: Indiana University Press.

[67] Ibid.

More Challenging Exercises

39. In Exercise 67 of Section 3.3 you found that $d/dx\,(\sin x) = \cos x$. Notice from your computer or calculator that $\sin 0 = 0$ and $\cos 0 = 1$. Use all of this and the tangent line approximation to approximate sin 0.02. Compare your answer to the value your computer or calculator gives.

40. You observe an object moving in a straight line and note that at precisely noon the instantaneous velocity is 45 ft/sec. Approximately how far has the object moved in the next 5 seconds? Explain carefully what facts learned in this section you are using to obtain your answer.

41. Biology An anatomist wishes to measure the surface area of a bone that is assumed to be a cylinder. The length l can be measured with great accuracy and can be assumed to be known exactly. However, the radius of the bone varies slightly throughout its length, making it difficult to decide what the radius should be. Discuss how slight changes in the radius r will approximately affect the surface area.

42. A cone has a height of 6 inches. The radius has been measured as 1 inch with a possible error of 0.01 inch. Estimate the maximum error in the volume ($\pi r^2 h/3$) of the cone. Use your computer or calculator to estimate any slope that you need.

Modeling Using Least Squares

43. Cost Curve Johnston[68] made a statistical estimation of the cost-output relationship for 40 firms. The data for one of the firms is given in the following table.

x	2.5	3	4	5	6	7
y	20	26	20	27	30	29
x	11	22.5	29	37	43	
y	42	53	75	73	78	

Here x is the output in millions of units, and y is the total cost in thousands of pounds sterling.

a. Determine both the best-fitting line using least squares and the square of the correlation coefficient. Graph.
b. Determine both the best-fitting quadratic using least squares and the square of the correlation coefficient. Graph.
c. Which curve is better? Why?
d. Using the quadratic cost function found in part (b) and the approximate derivative found on your computer or graphing calculator, graph the marginal cost. What is happening to marginal cost as output increases? Explain what this means.

44. Cost Curve Dean[69] made a statistical estimation of the cost-output relationship for a hosiery mill. The data for the firm is given in the following table.

x	45	56	62	70	74	78
y	14	17	19	20.5	21.5	22.5

Here x is production in hundreds of dozens, and y is the total cost is thousands of dollars.

a. Determine both the best-fitting quadratic using least squares and the square of the correlation coefficient. Graph.
b. Using the quadratic cost function found in part (a) and the approximate derivative found on your computer or graphing calculator, graph the marginal cost. What is happening to marginal cost as output increases? Explain what this means.

---

[68] J. Johnston. 1960. *Statistical cost analysis.* New York: McGraw-Hill, page 61.

[69] Joel Dean. 1976. *Statistical Cost Estimation.* Bloomington, Ind.: Indiana University Press, page 218.

Solutions to WARM UP EXERCISE SET 3.4

1. The volume of a cube with edges of length x is given by $V(x) = x^3$. The paint increases the length of each edge by twice 0.10, or by 0.20. Thus, $c = 10$ and $h = 0.20$. Then $V(10.2) - V(10) \approx V'(10)(0.2)$. Graphing $y = x^3$ and ZOOMing about the point (10, 1000) a few times gives a straight line with slope about 300. Thus,

$$V(10.2) - V(10) \approx V'(10)(0.2) = 300(0.2) = 60$$

Compare this to $V(10.2) - V(10) \approx 61$.

Summary Outline

- **Limit of a function** If as x approaches a, $f(x)$ approaches the number L, then we say that L is the limit of $f(x)$ as x approaches a, and write $\lim_{x\to a} f(x) = L$. p. 144.
- **Limits from left and right** If as x approaches a from the left, $f(x)$ approaches the number M, then we write $\lim_{x\to a^-} f(x) = M$. If as x approaches a from the right, $f(x)$ approaches the number N, then we write $\lim_{x\to a^+} f(x) = N$. p. 144.
- **Existence of a limit** $\lim_{x\to a} f(x) = L$ if, and only if, both of the limits $\lim_{x\to a^-} f(x)$ and $\lim_{x\to a^+} f(x)$ exist and equal L. p. 144.
- **Replacement theorem** Suppose $f(x) = g(x)$ for all $x \neq a$. If $\lim_{x\to a} g(x)$ exists, then $\lim_{x\to a} f(x)$ exists and $\lim_{x\to a} f(x) = \lim_{x\to a} g(x)$. p. 145.
- **Rules for limits** p. 151. Assume that $\lim_{x\to a} f(x) = L$ and $\lim_{x\to a} g(x) = M$. Then

$$\lim_{x\to a} cf(x) = cL$$

$$\lim_{x\to a} (f(x) \cdot g(x)) = L \cdot M$$

$$\lim_{x\to a} (f(x))^n = L^n, \text{ if } L^n \text{ is defined}$$

$$\lim_{x\to a} (f(x) \pm g(x)) = L \pm M$$

$$\lim_{x\to a} \frac{f(x)}{g(x)} = \frac{L}{M} \text{ if } M \neq 0$$

- **Limit of polynomial function** If $p(x)$ is a polynomial function, then $\lim_{x\to a} p(x) = p(a)$. p. 151.
- **Continuity** The function $f(x)$ is continuous at $x = a$ if $\lim_{x\to a} f(x) = f(a)$. p. 154.
- A polynomial function is continuous everywhere. p. 154.
- A rational function is continuous at every point at which the denominator is not zero. p. 154.
- **Average and instantaneous velocity** These are $\frac{s(t+h) - s(t)}{h}$ and $\lim_{h\to 0} \frac{s(t+h) - s(t)}{h}$, respectively, where $s(t)$ represents position at time t. p. 167.
- **The slope of the tangent line** to the graph of $y = f(x)$ at $x = x_0$ is $\lim_{h\to 0} \frac{f(x_0 + h) - f(x_0)}{h}$ if this limit exists. p. 174.
- The instantaneous rate of change of $f(x)$ at x_0 equals the slope of the tangent line to the graph of $f(x)$ at x_0. p. 174.
- **Definition of derivative** The derivative of a function $y = f(x)$, denoted by $f'(x)$ or dy/dx, is defined to be $f'(x) = \lim_{h\to 0} \frac{f(x+h) - f(x)}{h}$ if this limit exists. p. 188.
- A differentiable function is continuous. p. 194.
- **Tangent Line Approximation** If $y = f(x)$ is a differentiable function at c, then $f(c+h) - f(c) \approx f'(c)h$ if h is small. p. 206.
- If $C(x)$, $R(x)$, and $P(x)$ are cost, revenue, and profit functions, respectively, then $C'(x)$, $R'(x)$, and $P'(x)$ are the **marginal cost**, **marginal revenue**, and **marginal profit**, respectively. p. 209.
- The terms $C'(x)$, $R'(x)$, $P'(x)$ approximate the cost, revenue, and profit, respectively, associated with the next item. p. 209.

Review Exercises

1. For the function in the figure, find

$$\lim_{x\to a^-} f(x), \qquad \lim_{x\to a^+} f(x), \qquad \lim_{x\to a} f(x)$$

for $a = -1, 0, 1, 2$.

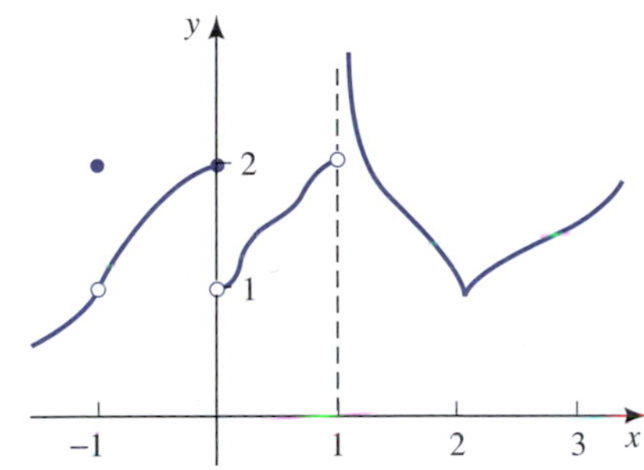

2. For the function in the figure, find all points where $f(x)$ is not continuous.

3. For the function in the figure, find all points where $f(x)$ does not have a derivative.

In Exercises 4 through 23, find the limits if they exist.

4. $\lim_{x\to 1} 12.7$

5. $\lim_{x\to 1} \pi^3$

6. $\lim_{x\to 1} (x^2 + 2x - 3)$

7. $\lim_{x\to -1} (x^3 - 3x - 5)$

8. $\lim_{x\to 0} (3x^5 + 2x^2 - 4)$

9. $\lim_{x\to 1} \dfrac{2x^2 + 1}{x + 2}$

10. $\lim_{x\to 2} \dfrac{x + 5}{x - 3}$

11. $\lim_{x\to 1} \dfrac{x^2 + x - 2}{x - 1}$

12. $\lim_{x\to 3} \dfrac{x^2 + x - 12}{x - 3}$

13. $\lim_{x\to 2} \dfrac{x^2 + 1}{x - 2}$

14. $\lim_{x\to -2} \dfrac{x + 3}{x + 2}$

15. $\lim_{x\to 2} \sqrt{x^3 - 4}$

16. $\lim_{x\to -1} \sqrt{10 - x^2}$

17. $\lim_{x\to 4} \left(\sqrt{x} - \dfrac{1}{\sqrt{x}}\right)$

18. $\lim_{x\to -2} \sqrt{1 - x^2}$

19. $\lim_{x\to 0} \dfrac{2x^3 + x - 3}{x^3 + 3}$

20. $\lim_{x\to -1} \dfrac{x^3 + x + 2}{x^4 + 1}$

21. $\lim_{x\to 0} \dfrac{1}{1 + e^x}$

22. $\lim_{x\to 0} (1 + e^{-2x})$

23. $\lim_{x\to 0^+} \dfrac{1}{1 + \ln x}$

In Exercises 24 through 31, use the limit definition of derivative to find the derivative.

24. $f(x) = 2x^2 + 1$

25. $f(x) = 2 - x^2$

26. $f(x) = 2x^2 + x - 3$

27. $f(x) = x^2 + 5x + 1$

28. $f(x) = \dfrac{1}{x^2}$

29. $f(x) = \dfrac{1}{2x + 5}$

30. $f(x) = \sqrt{x + 7}$

31. $f(x) = \sqrt{2x - 1}$

In Exercises 32 through 35, find the equation of the tangent line to the graph of each of the functions at the indicated point.

32. $y = x^3 + 1$, $x_0 = 2$

33. $y = x^{3/2} + 2$, $x_0 = 4$

34. $y = x^5 + 3x^2 + 2$, $x_0 = 1$

35. $y = \dfrac{1}{\sqrt{x}}$, $x_0 = 9$

In Exercises 36 through 39, take screens of smaller and smaller dimensions about the point $(c, f(c))$ until the graph looks like a straight line. Estimate $f'(c)$ by estimating the slope of the straight line.

36. $f(x) = \sqrt[5]{x}$, $c = 1$

37. $f(x) = \sqrt[3]{x^2}$, $c = 8$

38. $f(x) = \dfrac{1}{x^3}$, $c = 2$

39. $f(x) = \sqrt{x} + \dfrac{1}{\sqrt{x}}$, $c = 4$

In Exercises 40 through 43, use the derivatives found in the previous four exercises and the linear approximation to find approximations of the following quantities.

40. $\sqrt[5]{1.05}$

41. $\sqrt[3]{(7.7)^2}$

42. $\dfrac{1}{(2.08)^3}$

43. $\sqrt{4.016} + \dfrac{1}{\sqrt{4.016}}$

44. Match the graphs of the functions shown in figures a–d with the graphs of their derivatives in figures i–iv.

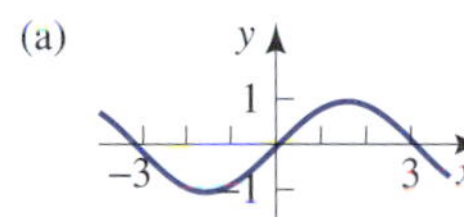

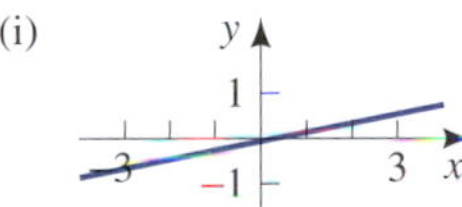

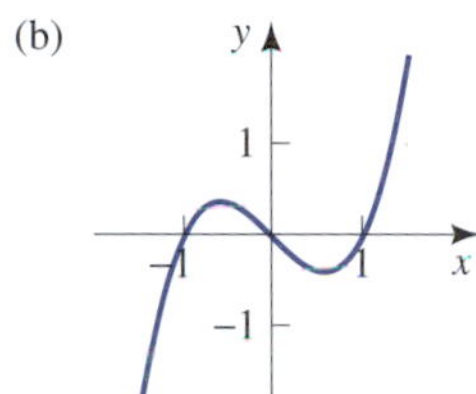

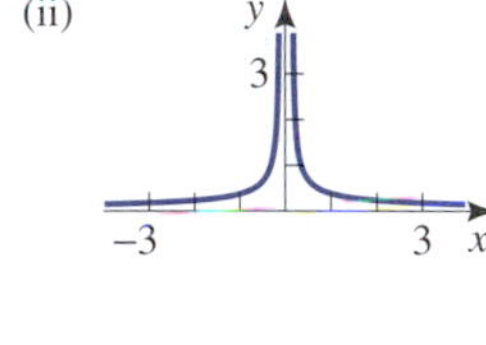

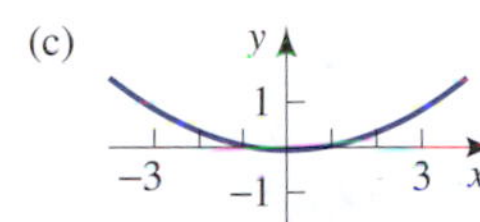

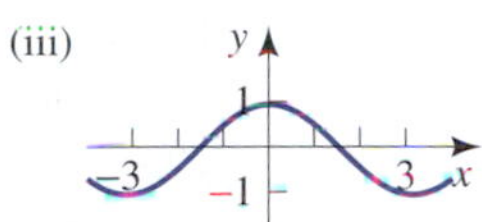

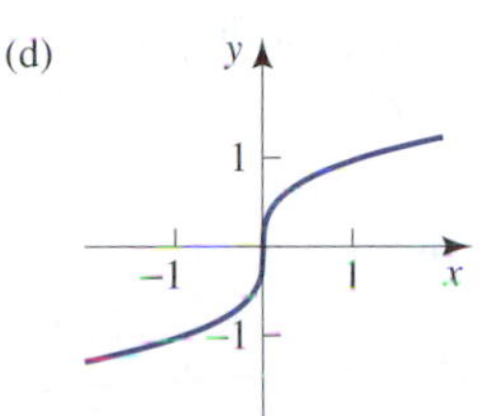

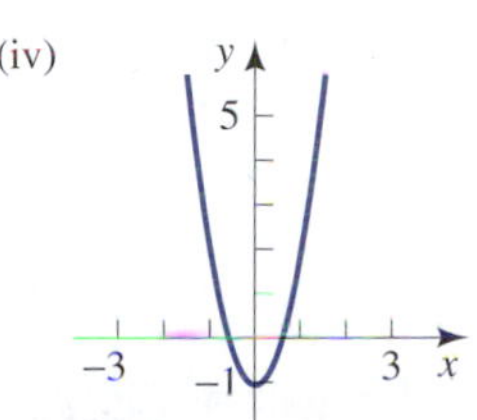

45. Below are the graphs a–f of six functions, $y = f_1(x), \ldots, y = f_6(x)$. Below them are the graphs i–vi of the derivatives of these six functions but in a different order. Match the graph of the function with its derivative.

(a)
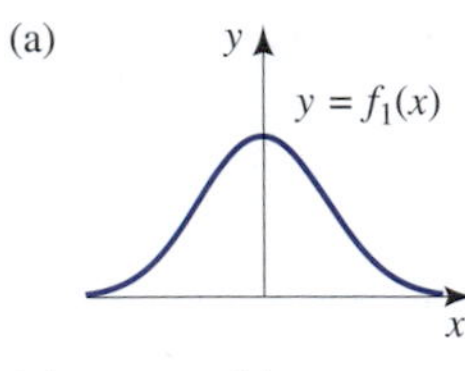

(b)
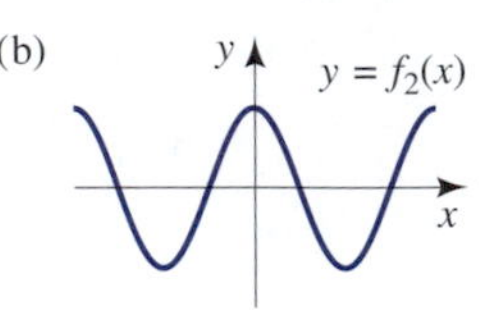

(c)
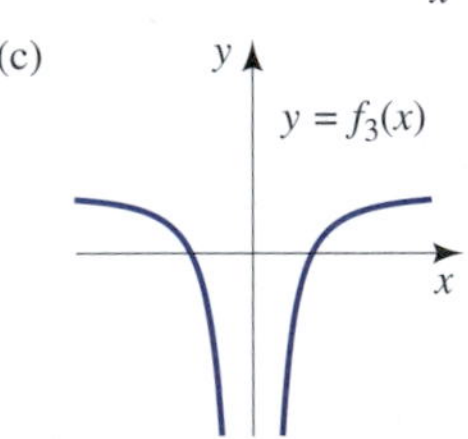

(d)
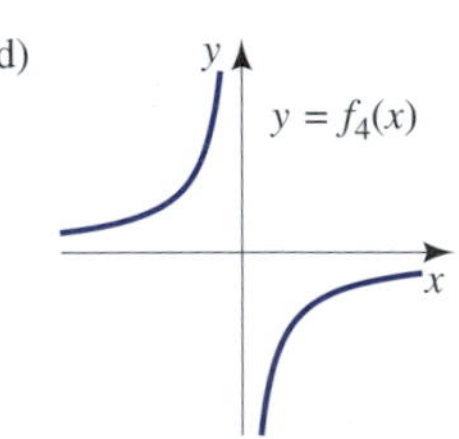

(e)
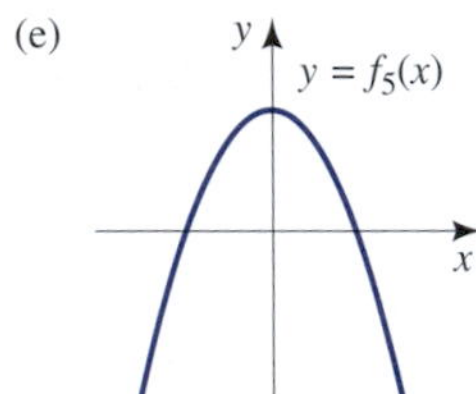

(f)
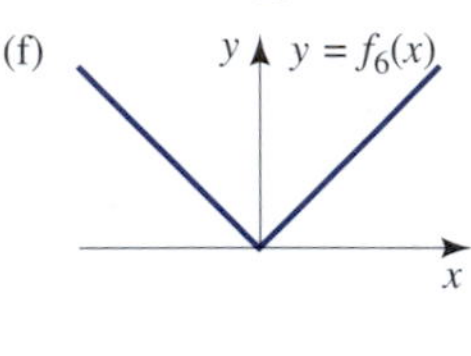

(i)
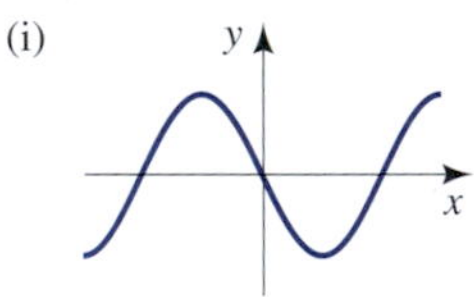

(ii)
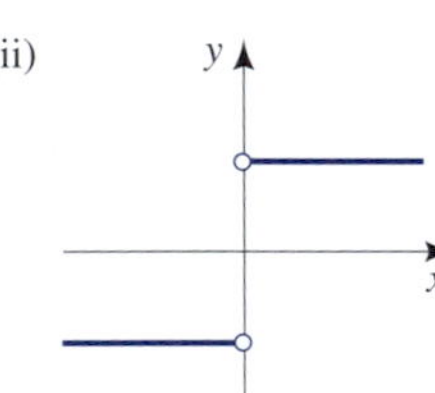

(iii)
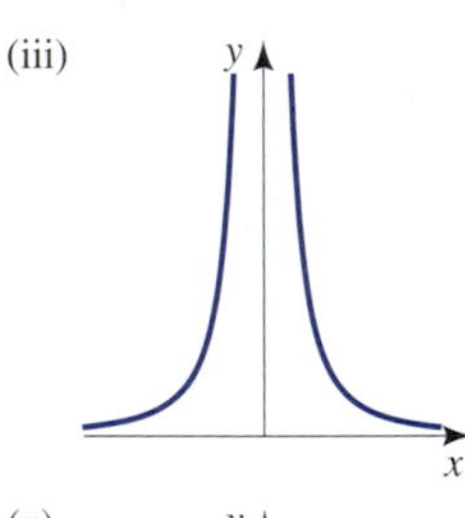

(iv)
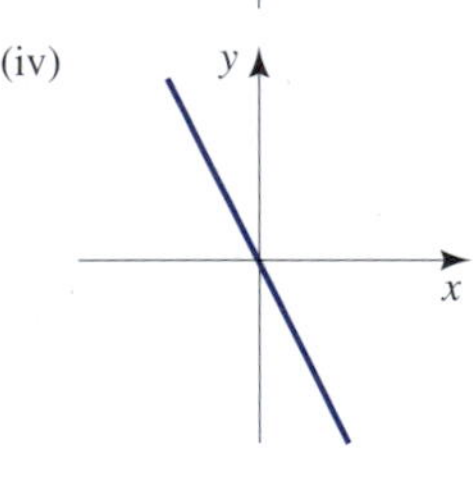

(v)
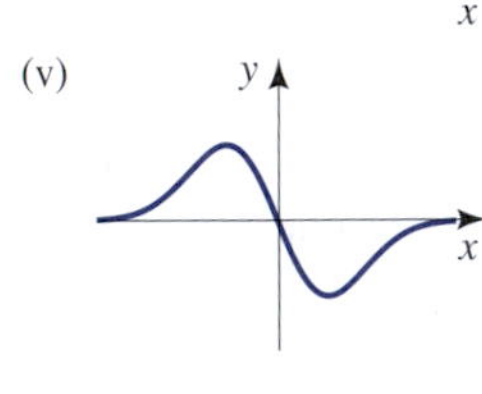

(vi)
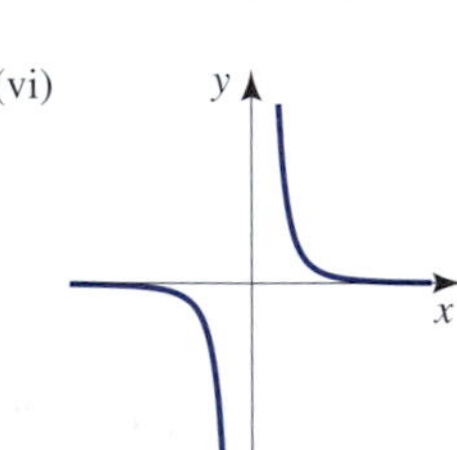

46. Determine by numerical and graphical means whether $f'(0)$ exists with $f(x) = \sqrt[5]{x^2}$.

47. Show analytically, using the limit definition of derivative, that $f'(0)$ does not exist if $f(x) = \sqrt[5]{x^2}$.

48. If $f(2) = 3$ and $f(x) = \frac{x^2-x-2}{x-2}$ if $x \neq 2$, is $f(x)$ continuous at $x = 2$? Why or why not?

49. Show analytically, using the limit definition of derivative, that $f'(0)$ exists if $f(x) = 0$ when $x < 0$ and $f(x) = x^3$ when $x \geq 0$.

50. Costs The weekly total revenue function for a company in dollars is $R(x)$. If $R'(100) = 50$, what is the approximate revenue from the sale of the 101st item?

51. Biology According to Wiess's law, the intensity of an electric current required to excite a living tissue (threshold strength) is given by $i = \frac{A}{t} + B$, where t is the duration of the current and A and B are positive constants. What happens as $t \to 0^+$? Does your answer seem reasonable?

52. Medicine Suppose that the concentration c of a drug in the bloodstream t hours after it is taken orally is given by $c = \frac{5t}{t^2+2}$. What happens to the drug concentration as $t \to 0^+$? Does this make sense?

53. Costs For a certain long-distance call, a telephone company charges \$1.00 for the first minute or fraction thereof and \$0.50 for each additional minute or fraction thereof. Graph this cost function, and determine where the function is continuous.

54. Profits The daily profits in dollars of a firm is given by $P = 100(-x^2 + 8x - 12)$, where x is the number of items sold. Find the instantaneous rate of change when x is (a) 3, (b) 4, and (c) 5. Interpret your answer.

55. Velocity A particle moves according to the law $s(t) = t^2 - 8t + 15$, where s is in feet and t in seconds. Find the instantaneous velocity at $t = 3, 4$, and 5. Interpret your answers.

56. Biology Use the numerical derivative feature on your computer or graphing calculator and the linear approximation formula to estimate the change in volume of a cylindrical bacterium as the radius increases from 2 microns to 2.1 microns while its length stays at 10 microns.

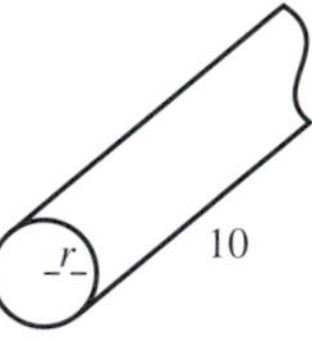

57. Hispanic Businesses The number of Hispanic-owned businesses[70] was about 100,000 in 1969 and was estimated to be about 800,000 in 2000. What was the average rate of change in the number of Hispanic businesses from 1969 to 2000? Give units and interpret your answer.

[70] Bruce S. Glassman. 1996. *The Macmillan Visual Almanac*. New York: Macmillan.

58. Waste The total waste[71] generated in the United States in pounds per person per day went from 2.66 pounds in 1960 to 4.00 pounds in 1988. What was the average rate of change in waste during this period? Give units and interpret your answer.

59. Number of Farms According to Drabenstott,[72] the number of farms primarily growing corn in the United States in 1947 was about 6000 and the number in 1997 was about 2000. Using this data, find the average rate of change in the number of farms with respect to years from 1947. Give units and interpret your answer.

60. Labor Supply Biddle and Zarkin[73] determined that male workers who worked 1000 hours a year earned \$5.75 per hour, while those who worked 3400 hours per year earned \$13.50 per hour. Find the average rate of change of hourly wage with respect to number of hours worked annually. Give units and interpret your answer.

61. Light Exposure on Milk Whited and associates[74] determined the effects of light exposure on vitamin A degradation in whole milk. They gave the equation $y = 0.7387x + 1.2652$, where x is the hours exposed to light and y is the percent of vitamin A lost. Find the average rate of change of vitamin A loss with respect to hours of light exposure on any interval on $(0, 16)$. Give units and interpret your answer.

62. Protein in Milk Crocker and coworkers[75] studied the northern elephant seal in Ano Nuevo State Reserve, California. They created a mathematical model given by the equation $P(t) = 12.7 - 0.09t$, where t is days postpartum and P the percentage of protein in the milk. Find the rate of change of the percentage of protein in the milk with respect to days postpartum. Give units and interpret your answer.

63. Light Exposure on Milk Whited and associates[76] determined the effects of light exposure on vitamin A degradation in nonfat milk. They gave the equation $y = 2.4722x + 10.809$, where x is the hours exposed to light and y is the percent of vitamin A lost. Find the average rate of change of vitamin A loss with respect to hours of light exposure on any interval on $(0, 16)$. Give units and interpret your answer.

64. White Bass Fecundity Knight and coworkers[77] created a mathematical model relating the length of female white bass and fecundity. They used the formula $Y = -1,717,324 + 6558X$, where X is the length of female white bass in mm and Y is the number of eggs per female. Find the average rate of change of fecundity with respect to length on any interval. Give units and interpret your answer.

65. Turtle Clutch Size Hellgren and colleagues[78] studied turtles in the Tamaulipan Biotic Province in southern Texas. They created a mathematical model given by the equation $S(L) = -4.21 + 0.042L$, where S is clutch size and L is carapace length in millimeters. Find the average rate of change of clutch size with respect to length of carapace on any interval. Give units and interpret your answer.

66. Length of Largemouth Bass Prey Yako and Mather[79] created a mathematical model that related the length of largemouth bass and their prey length in Santuit and Coonamessett ponds. The equation they used was $y = 7.44 + 0.19x$, where x is the length of largemouth bass in millimeters and y is the length of prey in millimeters. Find the average rate of change of length of bass with respect to length of their prey on any interval. Give units and interpret your answer.

67. Diversity of Fish Species Schultz and colleagues[80] collected data on 60 Florida lakes ranging in size from 2 hectares to more than 12,000 hectares. They gave the accompanying figure, which relates the size of the lake in hectares to the number of fish species. Estimate the rate of change in the number of fish species with respect to the size of the lake for size ranging from 10 hectares to 100,000 hectares. Give units and interpret your answer.

---

[71] Allen Hammond. 1992. *Environmental Almanac*. Boston: Houghton and Mifflin.

[72] Mark Drabenstott. 1999. Consolidation in U.S. agriculture: the new rural landscape and public policy. *Econ. Rev. Fed. Res. Bank Kansas City* 84(1):63–71.

[73] Jeff E. Biddle and Gary A. Zarkin. 1989. Choice among wage-hours packages: an empirical investigation of male labor supply. *J. Labor Econ.* 7(4):415–428.

[74] L. J. Whited, B. H. Hammond, K. W. Chapman, and K. J. Boor. 2002. Vitamin A degradation and light-oxidized flavor defects in milk. *J. Dairy Sci.* 85:351–354.

[75] Daniel E. Crocker, Jeannine D. Williams, Daniel P. Costa, and Burney J. Le Boeup. 2001. Maternal traits and reproductive effort in northern elephant seals. *Ecology.* 82(12):3541–3555.

[76] L. J. Whited, B. H. Hammond, K. W. Chapman, and K. J. Boor. 2002. Vitamin A degradation and light-oxidized flavor defects in milk. *J. Dairy Sci.* 85:351–354.

[77] Charles L. Knight, Michael T. Bur, and John L. Forney. 2000. Reduction in recruitment of white bass in Lake Erie after invasion of white perch. *Trans. Amer. Fisheries Soc.* 129:1340–1353.

[78] Eric C. Hellgren, Richard T. Kazmaier, Donald C. Ruthven III, and David R. Synatzske. 2000. Variation in tortoise life history: demography of *Gopherus berlandieri*. *Ecology* 81(5):1297–1310.

[79] Lisa A. Yako and Martha E. Mather. 2000. Assessing the contribution of anadromous herring to largemouth bass growth. *Trans. Amer. Fish. Soc.* 129:77–88.

[80] Eric J. Schultz, Mark V. Hoyer, and Daniel E. Canfield, Jr. 1999. An index of biotic integrity: a test with limnological and fish data from sixty Florida lakes. *Trans. Amer. Fish. Soc.* 128:564–577.

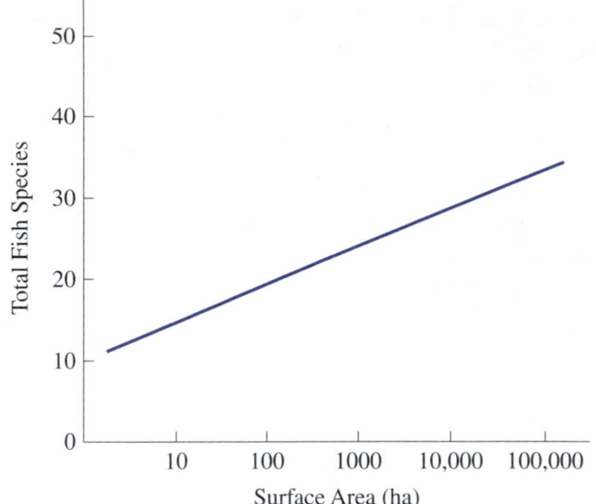

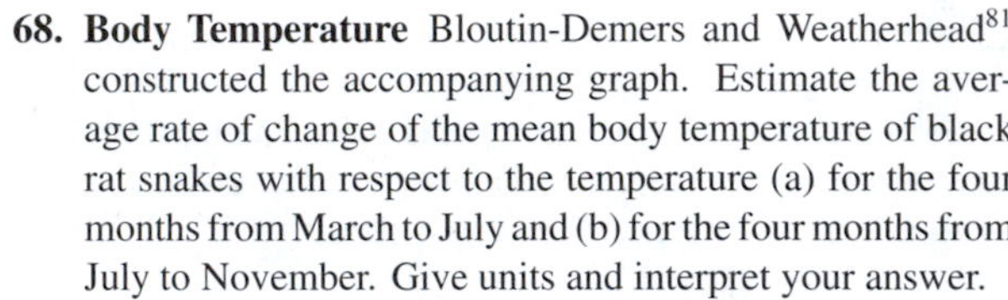

68. Body Temperature Bloutin-Demers and Weatherhead[81] constructed the accompanying graph. Estimate the average rate of change of the mean body temperature of black rat snakes with respect to the temperature (a) for the four months from March to July and (b) for the four months from July to November. Give units and interpret your answer.

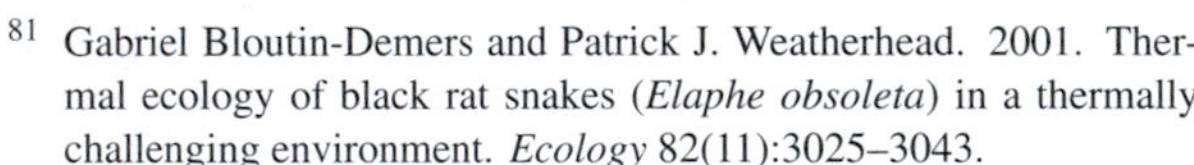

[81] Gabriel Bloutin-Demers and Patrick J. Weatherhead. 2001. Thermal ecology of black rat snakes (*Elaphe obsoleta*) in a thermally challenging environment. *Ecology* 82(11):3025–3043.

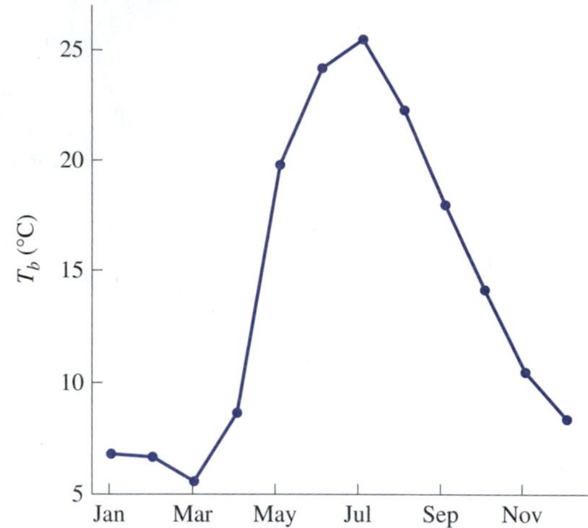

69. The graph of a function is shown. Sketch a graph of the derivative function.

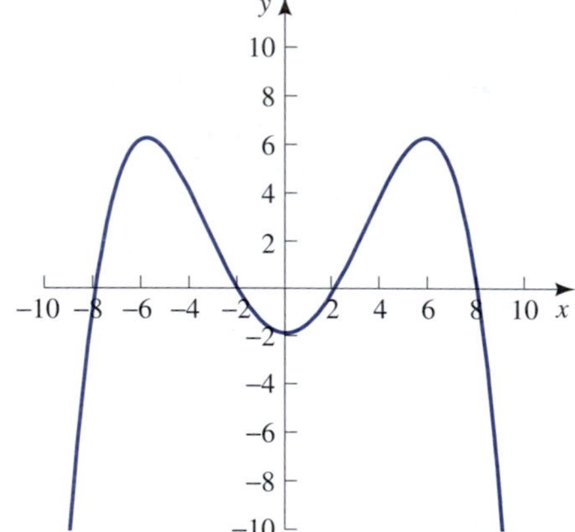

Chapter 3 Cases

CASE 1 A Close Look at the Numerical Derivative

We defined the derivative of $f(x)$ at $x = a$ to be

$$f'(a) = \lim_{h \to 0} \frac{f(a+h) - f(a)}{h}$$

The fraction

$$\frac{f(a+h) - f(a)}{h} \tag{1}$$

is the slope of the secant line through $(a, f(a))$ and $(a + h, f(a+h))$. The slope of this secant line approaches the slope of the line tangent to the curve $y = f(x)$ at the point $(a, f(a))$ as h approaches zero.

Now consider the expression

$$\frac{f(a+h) - f(a-h)}{2h} \tag{2}$$

It is this expression that your grapher uses to find the numerical derivative.

(a) This also is the slope of a secant line. On a graph, draw this line.

(b) Geometrically, explain why

$$f'(a) = \lim_{h \to 0} \frac{f(a+h) - f(a-h)}{2h}$$

(c) Your grapher uses the expression $\dfrac{f(a+h)-f(a-h)}{2h}$ for small h to find the numerical derivative. Call this nDeriv. Use nDeriv to find the numerical derivative of $f(x)=|x|$ at $x=0$. Explain why you obtain the answer 0 when we know from Example 5 in Section 3.3 that $f'(0)$ does not exist.

(d) Consider $f(x)=|x^{1/3}|$. From reviewing Example 6 in Section 3.3 we can see that the slope of the tangent line to the curve $y=f(x)$ at $x=0$ is vertical, and therefore $f'(0)$ does not exist. Now use nDeriv to find $f'(0)$. Why is nDeriv insisting that $f'(0)=0$?

(e) Find the limit in part (b) if $f(x)=x^2$.

CASE 2 **Drug Kinetics**

The figure shows a model, referred to as the *two-compartment model*, that is frequently used in drug kinetics. The figure indicates that a drug enters the bloodstream, either by injection or orally. The drug flows between the blood and tissue at different rates and is cleared at another rate, all rates depending on the particular drug. From certain mathematical theory, not presented here, the concentrations of the drug in the bloodstream and in the tissues is approximated by an equation of the form

$$c(t)=Ae^{-\alpha t}+Be^{-\beta t} \qquad (1)$$

where α and β are positive constants and $\alpha<\beta$. It is quite remarkable that in physiological and pharmaceutical applications such a model gives excellent approximations in describing the kinetics.

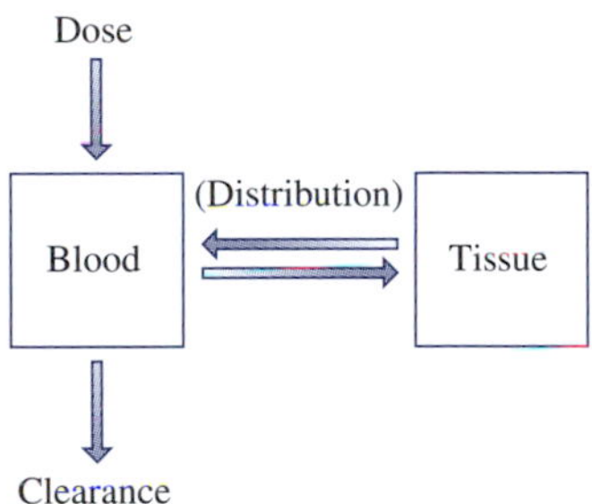

The practical problem for researchers is to determine the constants A, B, α, and β for a given drug. In this project we will consider the *method of exponential peeling* to estimate these constants.

To understand how and why this method works, we need to understand what is happening to $c(t)$ in Equation (1) for large t. We will then use these insights to tackle a specific problem.

(a) Using a window with dimensions [0, 20] by [0, 5], graph $y_1=2e^{-0.1x}+3e^{-0.2x}$, $y_2=2e^{-0.1x}$, and $y_3=3e^{-0.2x}$.

(b) Describe what you see. In particular, which of the two functions y_2 or y_3 is approximately equal to y_1 for large t?

(c) Now give a convincing mathematical explanation for your answer in part (b) that uses the exponents -0.1 and -0.2.

(d) Now explain what you think $c(t)$ in Equation (1) will approximate for large values of t.

(e) In part (d) you should have concluded that $c(t)\approx Ae^{-\alpha t}$ for large t. The method of peeling is based on this observation. Show for large t that we must have $\ln c(t)\approx \ln A-\alpha t$. Notice that the graph of $y=\ln A-\alpha t$ is a *straight line* with slope $-\alpha$ and y-intercept $\ln A$. Thus, we can determine α and A by determining the slope and t-intercept of this line.

(f) Creatinine is a component of urine that is originally formed from creatinine phosphate, a substance that supplies energy to the muscles. In an experiment carried out by Sepirstein and colleagues[82] 2 g of creatinine were injected into the bloodstream of dogs, and subsequent concentrations in the blood are shown in the table. Using this data and assuming that the concentration satisfies an equation of the form of equation (1), use the method of peeling to determine approximately the values of A and α.

t (min)	10	20	30	40
$c(t)$(g)	0.14390	0.04273	0.01316	0.00416
t (min)	50	60	70	
$c(t)$(g)	0.00134	0.00044	0.00014	

(g) Now estimate the constants B and β.

[82] L. A. Sepirstein, D. G. Vidt, M. J. Mandel, and G. Hanusek. 1955. Volumes of distributions and clearances of intravenously injected creatinine in the dog. *Amer. J. Phys.* 181:330–336.

CASE 3 Limits

(a) The text mentions two fundamentally different ways in which a limit might not exist. There is a third way. This project explores this third way. Familiarize yourself with the function $f(x) = \sin x$ by graphing this function in a window with dimension $[-10, 10]$ by $[-1, 1]$. As you see, this function moves back and forth between 1 and -1. We wish to determine whether or not $\lim_{x\to 0} \sin(1/x)$ exists. Graph this function in a window with dimensions $[-0.314, 0.314]$ by $[-1, 1]$. What is happening? Graph again in a window with dimensions $[-0.0314, 0.0314]$ by $[-1, 1]$. What is happening? Graph again in a window with dimensions $[-0.00314, 0.00314]$ by $[-1, 1]$. What is happening? What do you conclude about the limit?

(b) Determine graphically and numerically whether or not $\lim_{x\to\infty} x \sin \frac{1}{x}$ exists.

4 Rules for the Derivative

In the previous chapter we learned how to find the derivatives of functions by taking the limit, that is, by using the definition of derivative. In this chapter we develop some basic rules for derivatives that can then be used to find the derivatives of a wide variety of functions without having to take limits.

CASE STUDY How Do Changes In Price Effect Changes In Demand

Braley and Nelson[1] studied the effect of price increases on school lunch participation in the city of Pittsburgh. They estimated the demand equation for school lunches to be $x = 16{,}415.21 - 262.743p$, where x is the number of lunches purchased and p is the price in cents. If the price of a lunch was 25 cents, what would happen to demand if the price was increased by a given percentage? If the price of a lunch was 42 cents, what would happen to demand if the price was increased by a given percentage?

You can see how it is important in business to know what effects increasing the price of an item will have on demand. We shall see in Section 4.5 that when the price of a lunch in the above example is at 25 cents, then increasing the price by 1% will decrease demand by only 0.67%. But when the price of a lunch in the above example is at 42 cents, then increasing the price by 1% will decrease demand by 2.05%. The numbers 0.67 and 2.05 are called *elasticities of demand*.

Economists have estimated elasticities in a large number of cases. For example, Wold[2] estimated the elasticity of demand for milk to be 0.31, and Schultz[3] estimated the elasticity of demand for wheat to be 0.03. These are less than 1 and small, indicating that these items are needed and therefore the demand will not change much if the price is changed.

On the other hand, Stone and Rowe[4] estimated the elasticity of demand for furniture to be 3.04. This is bigger than 1 and rather large. This indicates that raising the price of furniture will considerably decrease the demand, since new furniture is something people can normally get along without.

In Section 4.5 we will see how to precisely define and determine elasticities of demand when we have the demand equation. We will also see the importance of whether the elasticity is greater or less than 1.

[1] George A. Braley and P. E. Nelson. 1973. Effect of a controlled price increase on a school lunch participation: Pittsburgh. *Amer. J. Agr. Econ.* February 1975:90–96.

[2] Herman Wold. 1953. *Demand Analysis*. New York: Wiley.

[3] Henry Schultz. 1938. *The Theory and Measurement of Demand*. Chicago: University of Chicago Press.

[4] Richard D. Stone and D. A. Rowe. 1960. The durability of consumers' durable goods. *Econometrica* 28:407–416.

4.1 Derivatives of Powers, Exponents, and Sums

Derivatives of Constants
Derivatives of Powers
The Derivative of the Exponential
Derivative of a Constant Times a Function
Derivatives of Sums and Differences
Rates of Change in Business and Economics
Enrichment: Proofs

APPLICATION A Cost Function for a Manufacturing Plant

Johnston[5] made an extensive study of cost functions for 40 firms in Great Britain. For one firm he found that if x is the output in millions of units and y is the total cost in thousands of pounds sterling, then the equation $y = C(x) = 16.311 + 2.3401x - 0.02009x^2$ approximated the relationship between x and y for $0 \le x \le 50$. Make a table of values for marginal cost when x is 5, 15, ..., 45. What is the meaning of these numbers? What is happening to the rate of change of costs as output increases? Does this make sense? See Example 8 on page 234 for the answer.

[5] J. Johnston. 1960. *Statistical Cost Analysis*. New York: McGraw-Hill.

Derivatives of Constants

In Chapter 3, we learned how to find the derivatives of functions by taking the limit. In this section we discover some basic rules for derivatives that can then be used to find the derivatives of a wide variety of functions without having to take limits. We begin by finding the derivative of a constant function.

Suppose $f(x)$ is a constant function, that is, $f(x) = c$, where c is a constant. Since the value of the function never changes, the instantaneous rate of change must be zero. Also, the graph of this function is given in Figure 4.1 and is just a horizontal straight line. The slope of any such horizontal line is zero. Thus, we have two reasons to suspect that the derivative should be zero.

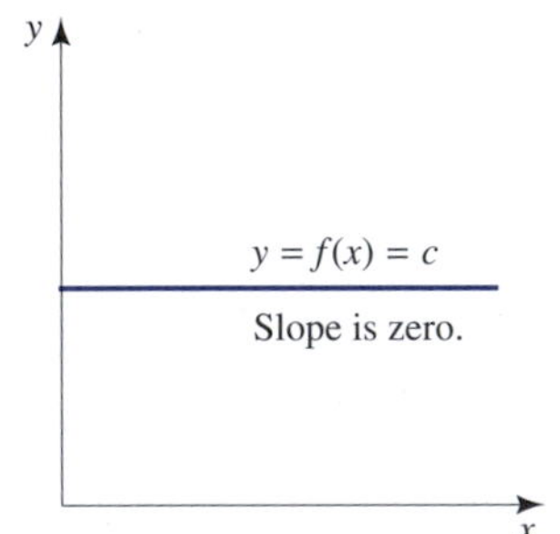

Figure 4.1
The slope of a horizontal line is zero. Thus $\frac{d}{dx}(c) = 0$.

We have the following, a proof of which can be found in the Enrichment subsection at the end of this subsection.

Derivative of a Constant

For any constant c,

$$\frac{d}{dx}(c) = 0$$

EXAMPLE 1 Finding the Derivative of Constants

Find

(a) $\dfrac{d}{dx}(3.4)$, (b) $f'(x)$ if $f(x) = -\pi^2$, (c) $\dfrac{dy}{dx}$ if $y = \dfrac{1}{3}$.

Solution

(a) $\dfrac{d}{dx}(3.4) = 0$

(b) $f'(x) = \dfrac{d}{dx}(-\pi^2) = 0$

(c) $\dfrac{dy}{dx} = \dfrac{d}{dx}\left(\dfrac{1}{3}\right) = 0$ ■

Derivatives of Powers

In Chapter 3 we found the derivatives of some powers. For example, we found that $\frac{d}{dx}(x^2) = 2x$ and $\frac{d}{dx}(x^{-1}) = -x^{-2}$.

Furthermore, since the function $f(x) = x$ represents a straight line with constant slope equal to 1 (Figure 4.2), we have $\frac{d}{dx}(x) = 1$.

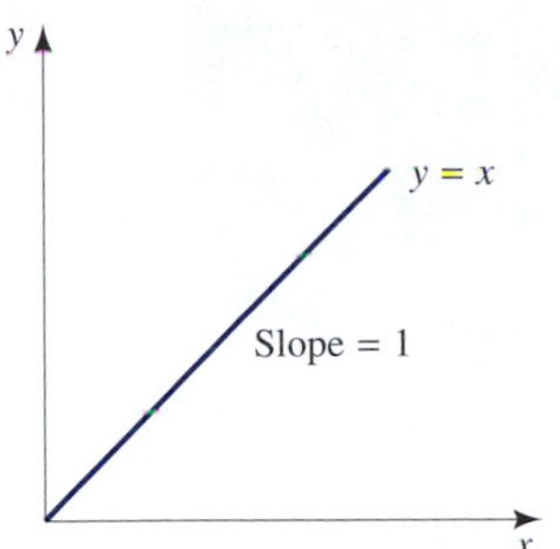

Figure 4.2
The slope of the line $y = x$ is 1. Thus $\frac{d}{dx}(x) = 1$.

Let us now see whether we can determine the derivative of $f(x) = x^3$. (This also can be done by using Interactive Illustration 4.1.) We know that by definition,

$$f'(x) = \lim_{h\to 0} \frac{(x+h)^3 - x^3}{h}$$

If h is small, say, $h = 0.001$, then we should have

$$f'(x) \approx \frac{(x+0.001)^3 - x^3}{0.001}$$

Therefore, the graph of $y = f'(x)$ should be approximated by the graph of

$$y_1 = \frac{(x+0.001)^3 - x^3}{0.001}$$

The graph of this latter function on a screen of dimensions $[-4.7, 4.7]$ by $[-10, 10]$ is shown in Screen 4.1. We see that the graph appears to be a parabola with vertex at the origin. So we suspect that the function is of the form kx^2. We can readily see that the graph goes through the point $(1, 3)$, so we must have $k = 3$. Taking $y_2 = 3x^2$ and graphing this on the same screen, we see that this graph is the same as that for y_2. Since the two graphs are the same, we strongly suspect that $f'(x) = 3x^2$.

We now have $\frac{d}{dx}(x) = 1$, $\frac{d}{dx}(x^2) = 2x$, and $\frac{d}{dx}(x^3) = 3x^2$. At this point you might guess that if $f(x) = x^4$, $f'(x) = 4x^3$. Graphing the two functions $y_1 = \frac{(x+0.001)^4 - x^4}{0.001}$ and $y_2 = 4x^3$ gives Screen 4.2. Since both graphs are approximately the same, we have graphical support for our guess.

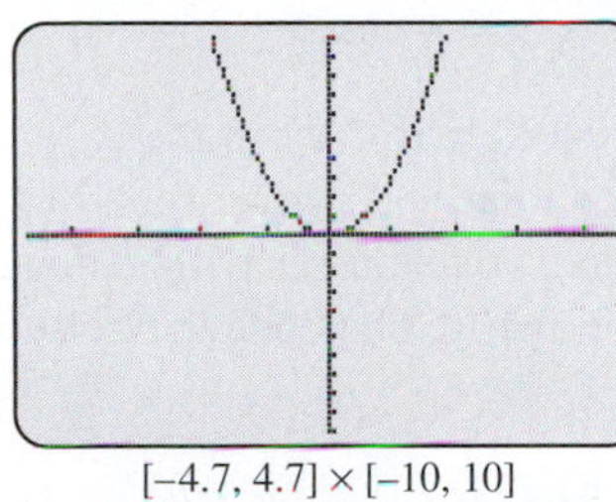

$[-4.7, 4.7] \times [-10, 10]$

Screen 4.1
Since the derivative of x^3 is approximately $y_1 = \frac{(x+0.001)^3 - x^3}{0.001}$ and the graph of this functions appears to be the same as the graph of $y_2 = 3x^2$, we strongly suspect that $\frac{d}{dx}(x^3) = 3x^2$.

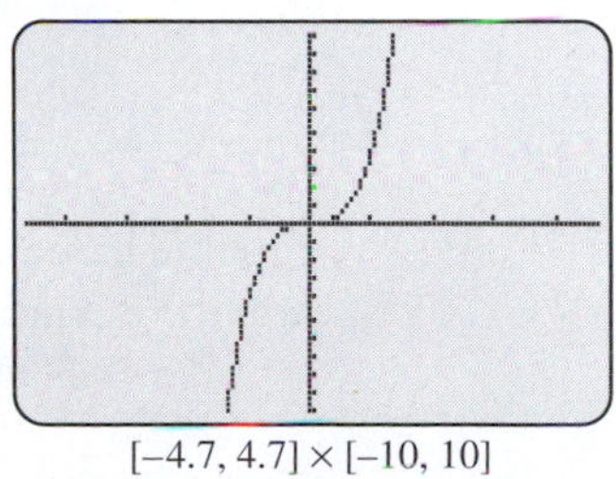

$[-4.7, 4.7] \times [-10, 10]$

Screen 4.2
Since the derivative of x^4 is approximately $y_1 = \frac{(x+0.001)^4 - x^4}{0.001}$ and the graph of this function appears to be the same as the graph of $y_2 = 4x^3$, we strongly suspect that $\frac{d}{dx}(x^4) = 4x^3$.

Interactive Illustration 4.1

Finding Derivatives Experimentally

This interactive illustration allows you to find the derivatives of a number of functions experimentally. Select a function. Click on the graph in the top window. An approximation to the tangent line for the given value of x and h will appear using

$$g(x) = \frac{f(x+h) - f(x)}{h} \approx f'(x)$$

as the approximate slope. This number will also appear in the table and will be plotted on the graph in the second window. Click on various values of x until you have a sketch of the graph of $y = g(x)$. From the table and the graph, guess what $g(x)$ is and thus $f'(x)$. In particular answer the following questions.

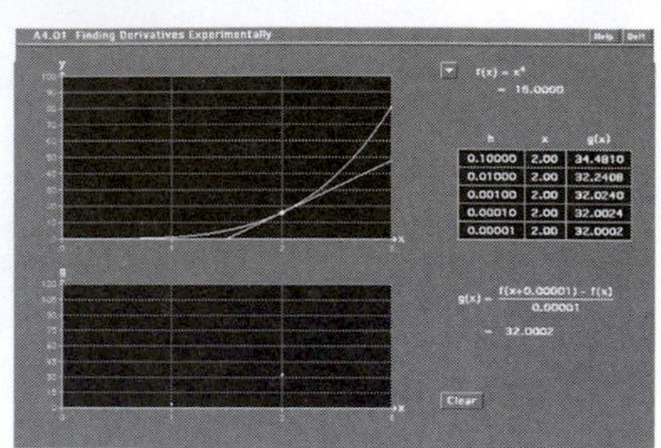

1. What is $\frac{d}{dx}(x(6-x))$?
2. What is $\frac{d}{dx}(x^3 - 2x)$?
3. What is $\frac{d}{dx}\left(\frac{1}{x}\right)$?

Thus, it appears that if $f(x) = x^n$, then $f'(x) = nx^{n-1}$. This is indeed true.

Derivative of a Power

If n is any real number (n may or may not be an integer),

$$\frac{d}{dx}(x^n) = nx^{n-1}$$

Remark You might think of the derivative of a power rule as: Power in front, reduce power by 1.

We shall give the proof here only in the case that $n = 3$, the case for a general positive integer being similar and can be found in the Enrichment subsection. The proof in the case n is a negative integer will be done in the next section and the proof when n is a rational number will be done in Section 4.3.

If $n = 3$, then

$$\begin{aligned}
\frac{f(x+h) - f(x)}{h} &= \frac{(x+h)^3 - x^3}{h} \\
&= \frac{(x^3 + 3x^2h + 3xh^2 + h^3) - x^3}{h} \\
&= \frac{3x^2h + 3xh^2 + h^3}{h} \\
&= \frac{h(3x^2 + 3xh + h^2)}{h} \\
&= 3x^2 + 3xh + h^2
\end{aligned}$$

Now to find $f'(x)$, we let $h \to 0$ in this last expression to obtain

$$\begin{aligned} f'(x) &= \lim_{h\to 0} \frac{f(x+h)-f(x)}{h} \\ &= \lim_{h\to 0} \frac{(x+h)^3 - x^3}{h} \\ &= \lim_{h\to 0} (3x^2 + 3xh + h^2) \\ &= 3x^2 + 0 + 0 \\ &= 3x^2 \end{aligned}$$

This completes the proof for the derivative of x^3.

EXAMPLE 2 **Finding the Derivatives of Some Powers**

Find (a) $\frac{d}{dx}(x^7)$, (b) $f'(t)$ if $f(t) = t^{-4}$, (c) $\frac{dy}{dv}$ if $y = v^{4/3}$.

Solution

(a) $\dfrac{d}{dx}(x^7) = 7x^{7-1} = 7x^6$

(b) $f'(t) = \dfrac{d}{dt}(t^{-4}) = -4t^{-4-1} = -4t^{-5}$

(c) $\dfrac{dy}{dv} = \dfrac{d}{dv}(v^{4/3}) = \dfrac{4}{3}v^{(4/3)-1} = \dfrac{4}{3}v^{1/3}$ ■

Sometimes a function $f(x)$ is actually x to some power, without this being apparent at first.

EXAMPLE 3 **Finding the Derivatives of Some Powers**

Find (a) $\dfrac{dy}{dx}$ if $y = \sqrt[3]{x^2}$, (b) $\dfrac{d}{dx}f(x)$ if $f(x) = 1/x^2$.

Solution

(a) $\dfrac{dy}{dx} = \dfrac{d}{dx}(\sqrt[3]{x^2}) = \dfrac{d}{dx}(x^{2/3}) = \dfrac{2}{3}x^{-1/3}$

(b) $\dfrac{d}{dx}f(x) = \dfrac{d}{dx}(1/x^2) = \dfrac{d}{dx}(x^{-2}) = -2x^{-3}$ ■

The Derivative of the Exponential

We now consider the derivative of e^x. To determine the derivative of $f(x) = e^x$, we first use the fact that

$$f'(x) \approx \frac{f(x+h)-f(x)}{h}$$

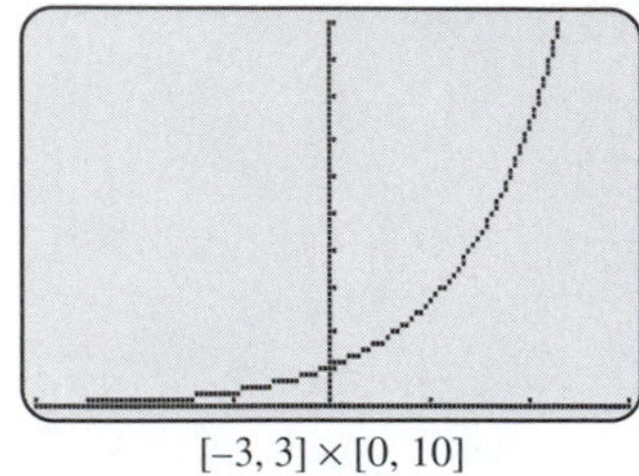

$[-3, 3] \times [0, 10]$

Screen 4.3

Since the derivative of e^x is approximately $y_1 = \dfrac{e^{x+0.001} - e^x}{0.001}$ and the graph of this function appears to be the same as the graph of $y_2 = e^x$, we strongly suspect that $\dfrac{d}{dx}(e^x) = e^x$.

if h is small. If we take $h = 0.001$, the graph of $y = f'(x)$ is approximated by the graph of

$$
\begin{aligned}
y_1 &= \frac{f(x+h) - f(x)}{h} \\
&= \frac{e^{x+h} - e^x}{h} \\
&= \frac{e^{x+0.001} - e^x}{0.001}
\end{aligned}
$$

The graph of this last function is shown in Screen 4.3 using a window of dimensions $[-3, 3]$ by $[0, 10]$. The graph looks like that of an exponential function, $y = e^{kx}$, $k > 0$. Trying $k = 1$ and graphing $y_2 = e^x$ on the same screen does not give anything different, indicating that the two graphs are the same. We thus strongly suspect that

$$
\frac{d}{dx}\left(e^x\right) = e^x
$$

Interactive Illustration 4.2

Finding the Derivative of e^x Experimentally

This interactive illustration allows you to find the derivatives of e^x experimentally. Click on the graph in the top window. An approximation to the tangent line for the given value of x and h will appear when

$$
g(x) = \frac{f(x+h) - f(x)}{h} \approx \frac{d}{dx}(e^x)
$$

is used as the approximate slope. This number will also appear in the table and will be plotted on the graph in the second window. Click on various values of x until you have a sketch of the graph of $y = g(x)$. From the table and the graph, guess what $g(x)$ is and thus $(e^x)'$.

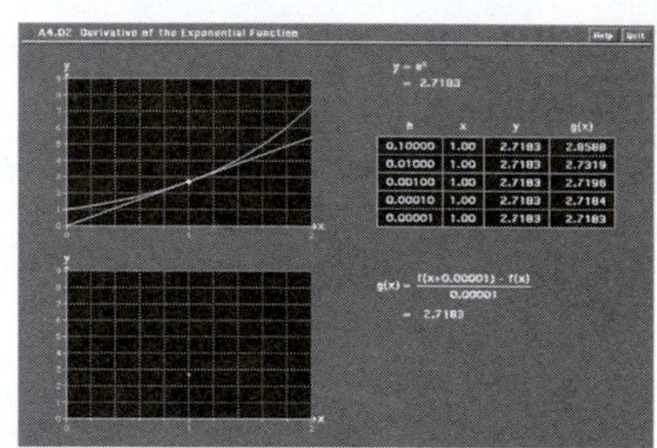

Also, if $f(x) = e^x$, find the following:

1. $f'(0)$ 2. $f'(1)$ 3. $f'(2)$

The Derivative of e^x

The derivative of e^x is again e^x, that is,

$$
\frac{d}{dx}\left(e^x\right) = e^x
$$

Further support for this exponential rule can be found in the Enrichment subsection at the end of this section.

Derivative of a Constant Times a Function

We now consider a rule for finding the derivative of a constant times a function. Consider the two functions whose graphs are shown in Figure 4.3. The function

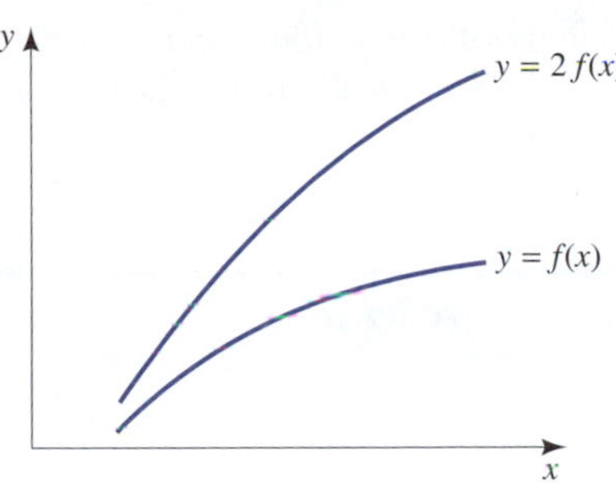

Figure 4.3

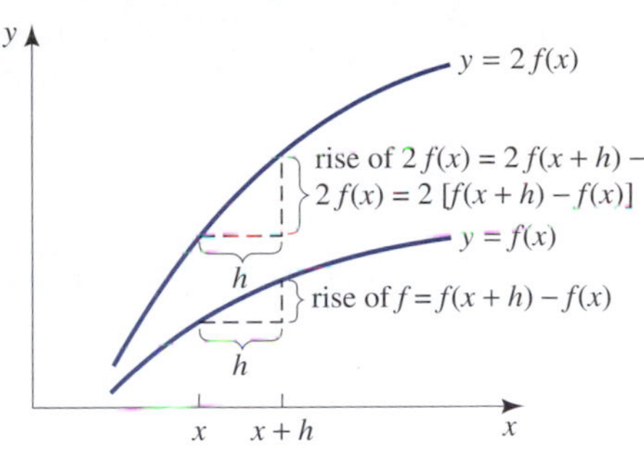

Figure 4.4

$y = 2f(x)$ stretches the graph of $y = f(x)$ by a factor of 2 for any x. We might then think that the rate of change of $2f$ is then twice that of f. Indeed, Figure 4.4 shows that for the same "run" of h the change in f is $f(x+h) - f(x)$, whereas

$$\begin{aligned} \text{change in } 2f &= 2f(x+h) - 2f(x) \\ &= 2[f(x+h) - f(x)] \\ &= 2[\text{change in } f] \end{aligned}$$

That is, the change in $2f$ is twice the change in f for the same run h. Thus, the average rate of change for $2f$ is twice the average rate of change of f. Since this is true for any run h, this implies that the instantaneous rate of change of $2f$ is twice that of f. Finally, this is the same as saying $\frac{d}{dx}[2f(x)] = 2\frac{d}{dx}[f(x)]$. We can do the same analysis for any constant c and conclude that $\frac{d}{dx}[cf(x)] = c\frac{d}{dx}[f(x)]$.

Interactive Illustration 4.3

Experimentally Determining $\frac{d}{dx}[cf(x)] = c\frac{d}{dx}[f(x)]$

This interactive illustration demonstrates experimentally that $\frac{d}{dx}[cf(x)] = c\frac{d}{dx}[f(x)]$. Select one of the functions. Now select 4 different values of x. Compare the values of $\frac{d}{dx}[2f(x)] = 2\frac{d}{dx}[f(x)]$. What appears to be the case? Do this again for another function.

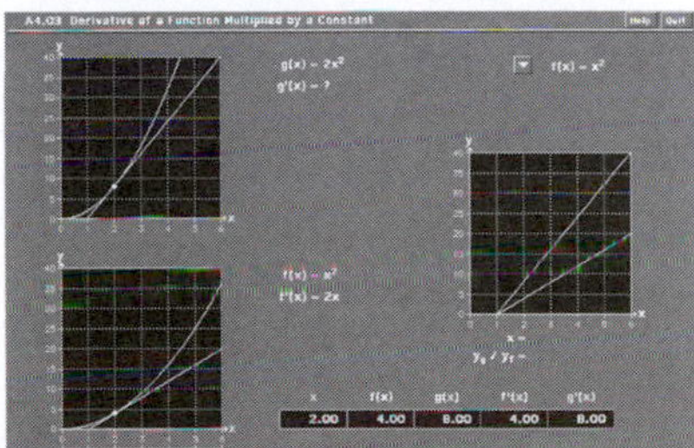

We then have a rule for finding the derivative of a constant times a function.

Derivative of a Constant Times a Function

If $f'(x)$ exists, then

$$\frac{d}{dx}[cf(x)] = c\frac{d}{dx}[f(x)]$$

The complete proof can be found in the Enrichment subsection.

In Section 3.3 we graphically found the derivative of $\ln x$. Since we will now be using this formula, we recall it here.

The Derivative of ln x

If $x > 0$, then

$$\frac{d}{dx}(\ln x) = \frac{1}{x}$$

EXAMPLE 4 **Finding the Derivative of a Constant Times a Function**

Find the derivatives of (a) $-5x^3$, (b) $\sqrt{2}e^x$, (c) $\frac{3}{x^8}$, (d) $\ln x^4$.

Solution

(a)

$$\begin{aligned}\frac{d}{dx}(-5x^3) &= -5\frac{d}{dx}(x^3)\\ &= -5(3x^{3-1})\\ &= -15x^2\end{aligned}$$

(b)

$$\begin{aligned}\frac{d}{dx}(\sqrt{2}e^x) &= \sqrt{2}\frac{d}{dx}(e^x)\\ &= \sqrt{2}e^x\end{aligned}$$

(c)

$$\begin{aligned}\frac{d}{dx}\left(\frac{3}{x^8}\right) &= \frac{d}{dx}(3x^{-8})\\ &= 3\frac{d}{dx}(x^{-8})\\ &= 3(-8x^{-9})\\ &= -24x^{-9}\end{aligned}$$

(d)

$$\begin{aligned}\frac{d}{dx}(\ln x^4) &= \frac{d}{dx}(4\ln x)\\ &= 4\frac{d}{dx}(\ln x)\\ &= 4\frac{1}{x}\end{aligned}$$

■

Derivatives of Sums and Differences

The last rule in this section is for derivatives of sums and differences. Consider the two functions $f(x) = 2 + 2x$ and $g(x) = x^2$ and their sum $[f + g](x) = 2 + 2x + x^2$. The table indicates the changes on the interval [1, 2].

	$f(x) = 2 + 2x$	$g(x) = x^2$	$[f + g](x) = 2 + 2x + x^2$
$x = 1$	4	1	5
$x = 2$	6	4	10
Change	2	3	5

Notice that the change on the interval [1, 2] is 2 for f, 3 for g, and 5 for $[f + g]$. Since $5 = 2 + 3$, the change in $[f + g]$ is the change in f plus the change in g. Indeed, for any run h and any functions f and g,

$$\begin{aligned}\text{change in } [f + g] &= [f + g](x + h) - [f + g](x) \\ &= [f(x + h) + g(x + h)] - [f(x) + g(x)] \\ &= [f(x + h) - f(x)] + [g(x + h) - g(x)] \\ &= [\text{change in } f] + [\text{change in } g]\end{aligned}$$

In other words, the change in the sum is the sum of the changes. This then implies that the instantaneous rate of change of $[f + g]$ equals the instantaneous rate of change of f plus the instantaneous rate of change of g. A similar analysis can be done for the difference of two functions. We then have the following.

Derivative of Sum or Difference

If $f'(x)$ and $g'(x)$ exist, then $\frac{d}{dx}[f(x) \pm g(x)]$ exists, and

$$\frac{d}{dx}[f(x) \pm g(x)] = \frac{d}{dx}[f(x)] \pm \frac{d}{dx}[g(x)]$$

A complete proof is given in the Enrichment subsection.

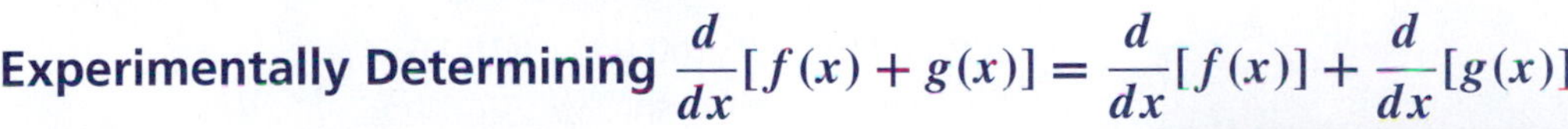

Interactive Illustration 4.4

Experimentally Determining $\frac{d}{dx}[f(x) + g(x)] = \frac{d}{dx}[f(x)] + \frac{d}{dx}[g(x)]$

This interactive illustration demonstrates experimentally that $\frac{d}{dx}[f(x) + g(x)] = \frac{d}{dx}[f(x)] + \frac{d}{dx}[g(x)]$. Select a function. Select 4 different values of x. Compare the values of $\frac{d}{dx}[f(x) + g(x)]$ and $\frac{d}{dx}[f(x)] + \frac{d}{dx}[g(x)]$. What can you conclude?

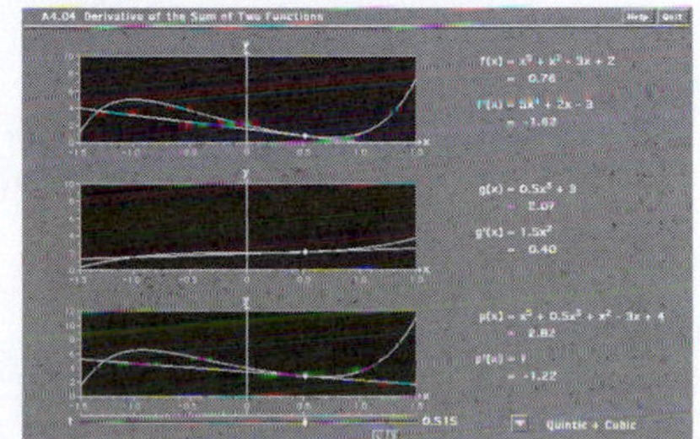

EXAMPLE 5 **Derivatives of Sums and Differences**

Find the derivatives of (a) $x^4 + x^3$, (b) $-\dfrac{2}{x^3} - 3\sqrt{x^3} + \pi$.

Solution

(a)

$$\begin{aligned}\frac{d}{dx}(x^4 + x^3) &= \frac{d}{dx}(x^4) + \frac{d}{dx}(x^3)\\ &= 4x^3 + 3x^2\end{aligned}$$

(b)

$$\begin{aligned}\frac{d}{dx}\left(-\frac{2}{x^3} - 3\sqrt{x^3} + \pi\right) &= -2\frac{d}{dx}(x^{-3}) - 3\frac{d}{dx}(x^{3/2}) + \frac{d}{dx}(\pi)\\ &= -2(-3x^{-4}) - 3\left(\frac{3}{2}x^{1/2}\right) + 0\\ &= 6x^{-4} - \frac{9}{2}x^{1/2}\end{aligned}$$

■

We know that if $R(x)$, $C(x)$, and $P(x)$ are the revenue, cost, and profit functions, respectively, then $P(x) = R(x) - C(x)$. Now that we know that the derivative of the difference is the difference of the derivatives, we have the important formula

$$P'(x) = R'(x) - C'(x)$$

EXAMPLE 6 **Finding Marginal Profit**

Suppose the cost and revenue functions for a plant that produces flooring nails is given by $C(x)$ and $R(x)$, respectively, where x is measured in pounds of nails and $C(x)$ and $R(x)$ are given in cents. Suppose $C'(1000) = 50$ and $R'(1000) = 52$. Find $P'(1000)$, the units, and what it means.

Solution

In view of the comments above, since $P(x) = R(x) - C(x)$, we have

$$\begin{aligned}P'(x) &= [R(x) - C(x)]'\\ &= R'(x) - C'(x)\\ P'(1000) &= R'(1000) - C'(1000)\\ &= 52 - 50\\ &= 2\end{aligned}$$

When 1000 pounds nails are sold, profits are increasing at the rate of 2 cents per pound. This also implies that the profits on the next pound of nails will be approximately 2 cents.

■

EXAMPLE 7 **Finding Horizontal Tangents**

Graph $y = x^4 + x^3$ in the standard viewing window.

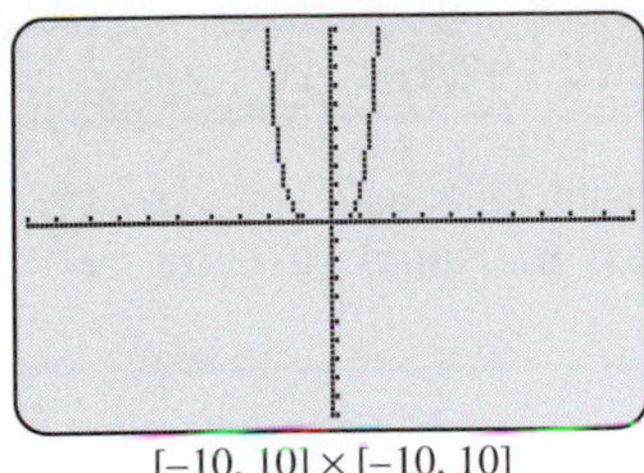

$[-10, 10] \times [-10, 10]$

Screen 4.4
The graph of $y = x^4 + x^3$ seems to have a horizontal tangent at $x = 0$.

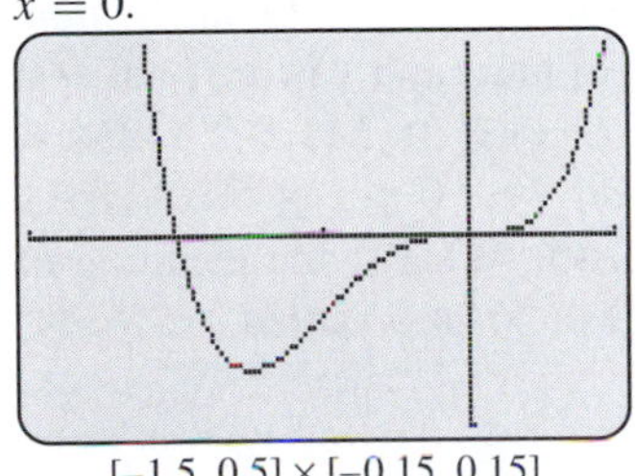

$[-1.5, 0.5] \times [-0.15, 0.15]$

Screen 4.5
A closer look at the graph of $y = x^4 + x^3$ indicates hidden behavior not seen on Screen 4.4.

(a) Roughly estimate the value(s) of x at which the graph of $y = f(x)$ has a horizontal tangent.
(b) Confirm your answer(s) in part (a) using calculus.

Solution

(a) The graph is shown in Screen 4.4. We see that there appears to be a horizontal tangent at approximately $x = 0$.
(b) The tangent will be horizontal when the slope is zero, that is, at points where the derivative is zero. From Example 5a we have

$$\begin{aligned} 0 &= f'(x) \\ &= 4x^3 + 3x^2 \\ &= x^2(4x + 3) \end{aligned}$$

Thus, $x = 0$, but also $x = -0.75$. We missed the second value in part (a) because our graph missed some hidden behavior! Screen 4.5 shows the graph of $y = f(x)$ on a screen of dimension $[-1.5, .5]$ by $[-0.15, 0.15]$. On this screen we clearly see the behavior that was hidden before. ■

Example 7 demonstrates that we cannot rely completely on our computer or graphing calculator. Only by using calculus can we be certain that we have found *all* the hidden behavior.

Exploration 1

Finding Horizontal Tangents

On your computer or graphing calculator graph $y = -\dfrac{2}{x^3} - 3\sqrt{x^3} + \pi$.

(a) From the graph roughly estimate the value(s) of x at which the graph of $y = f(x)$ has a horizontal tangent.
(b) Find the value(s) in (a) to two decimal places using available operations.
(c) Confirm your answer(s) in (a) and (b) using calculus. (Refer to Example 5b.)

Rates of Change in Business and Economics

To make intelligent business and economic decisions, one needs to know the effect of changes in production and sales on costs, revenues and profits. That is, one needs to know the rate of change of cost, revenue, and profits. For example, knowing whether the rate of change of profits with respect to an increase in production is positive or negative would be very useful information. Recall from the last section that in economics, the word *marginal* refers to an instantaneous rate of change, that is, to the derivative. Thus, if $C(x)$ is the cost function, the marginal cost is $C'(x)$; the same is the case for revenue and profit.

In addition, we found in Section 3.4 that $C'(x)$, $R'(x)$, and $P'(x)$ is approximately the cost, revenue, and profit, respectively, of the next item.

DEFINITION Marginal Cost, Revenue, and Profit

If $C(x)$, $R(x)$, and $P(x)$ are the cost, revenue, and profit functions, respectively, then $C'(x)$, $R'(x)$, and $P'(x)$ are the **marginal cost**, **marginal revenue**, and **marginal profit**.

EXAMPLE 8 Marginal Cost

Johnston[6] made an extensive study of cost functions for 40 firms in Great Britain. For one firm he found that if x is the output in millions of units and y is the total cost in thousands of pounds sterling, then the equation $y = C(x) = 16.311 + 2.3401x - 0.02009x^2$ approximated the relationship between x and y for $0 \le x \le 50$. Make a table of values for marginal cost when x is 5, 15, . . . , 45. What is the meaning of these numbers? What is happening to the rate of change of costs as output increases? Does this make sense?

$[0, 47] \times [0, 100]$

Screen 4.6
We see the graph of the cost function $C(x) = 16.311 + 2.3401x - 0.02009x^2$.

Solution

Screen 4.6 shows a graph of $y = C(x)$ using a window of dimensions [0, 47] by [0, 100]. We notice that naturally, total costs increase as more units are made. The marginal cost is $C'(x) = 2.3401 - 0.04018x$. Screen 4.7 shows a graph of $y = C'(x)$ using a window of dimensions [0, 47] by [0, 3]. We readily find that

$$C'(5) = 2.3401 - 0.04018(5) \approx 2.139$$

Thus, cost is increasing at about the rate of 2139 pounds sterling per million units when 5 million units are being produced.

The table gives the other values for $C'(x)$.

x	5	10	15	20	25	30	35	40	45
$C'(x)$	2.139	1.938	1.737	1.537	1.336	1.135	0.934	0.733	0.532

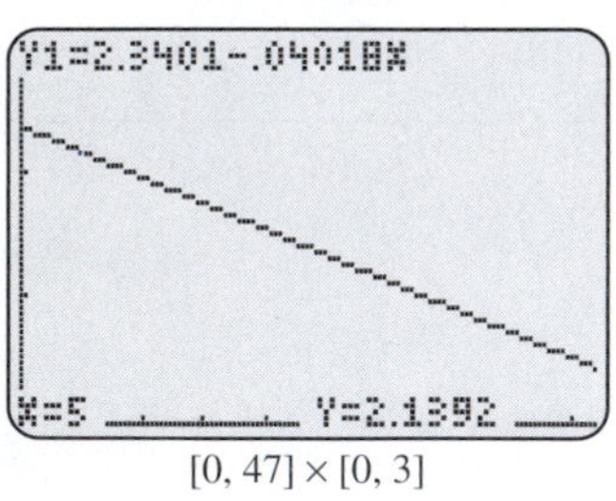

$[0, 47] \times [0, 3]$

Screen 4.7
We see the graph of the marginal cost using the cost function in Screen 4.6. Notice that marginal costs are decreasing.

For example, cost is increasing at the rate of £2139 per million units when 5 million units are produced, at the smaller rate of £1938 per million units when 10 million units are produced, at the even smaller rate of £1737 per million units when 15 million units are produced, and so on. The table shows the rate of change of costs steadily decreasing over the range under consideration. This is typical of cost functions. Of course, if more units are made, the total cost is greater. However, because of economies of scale, the rate of increase of costs will itself normally decrease, at least initially. That is, the manufacturing process will become more cost efficient as more units are made. (If production is increased too much, the rate of increase of costs might actually begin to *increase* owing to the use of less efficient equipment, payment of overtime, and so on.) ■

[6] Ibid.

Exploration 2

Tangent Line

On your grapher, graph on the same screen the cost function given in the last example and the tangent line to this curve at $x = 10$. How is the slope of this tangent line related to the marginal cost $C'(10)$ found in Example 8?

Enrichment: Proofs

We outline a proof for the following.

Derivative of a Constant

For any constant c,

$$\frac{d}{dx}(c) = 0$$

Using the definition of derivative, we have

$$\begin{aligned} f'(x) &= \lim_{h\to 0} \frac{f(x+h) - f(x)}{h} \\ &= \lim_{h\to 0} \frac{c-c}{h} \\ &= \lim_{h\to 0} \frac{0}{h} \\ &= \lim_{h\to 0} 0 \\ &= 0 \end{aligned}$$

Derivative of a Power

If n is any real number (n may or may not be an integer),

$$\frac{d}{dx}(x^n) = nx^{n-1}$$

The proof was given earlier in this section in the case that $n = 3$ and in the last section when $n = 2$. We now give the proof for n any positive integer.

To establish this result, we will need to use the binomial theorem. (See Appendix A.2.) The binomial theorem gives an expansion of the power $(a+b)^n$. In our case we will need to expand $(x+h)^n$. We have already done this for $n = 2$ and $n = 3$. For convenience we list these two cases and also the case for $n = 4$.

$$\begin{aligned} (x+h)^2 &= x^2 + 2xh + h^2 \\ (x+h)^3 &= x^3 + 3x^2h + 3xh^2 + h^3 \\ (x+h)^4 &= x^4 + 4x^3h + 6x^2h^2 + 4xh^3 + h^4 \end{aligned}$$

In general we have

$$(x+h)^n = x^n + nx^{n-1}h + \tfrac{1}{2}n(n-1)x^{n-2}h^2 + \cdots + h^n$$

We can now take the average rate of change from x to $x+h$ and simplify:

$$\begin{aligned}
\frac{f(x+h)-f(x)}{h} &= \frac{(x+h)^n - x^n}{h} \\
&= \frac{[x^n + nx^{n-1}h + \tfrac{1}{2}n(n-1)x^{n-2}h^2 + \cdots + h^n] - x^n}{h} \\
&= \frac{nx^{n-1}h + \tfrac{1}{2}n(n-1)x^{n-2}h^2 + \cdots + h^n}{h} \\
&= nx^{n-1} + \tfrac{1}{2}n(n-1)x^{n-2}h + \cdots + h^{n-1}
\end{aligned}$$

Now to find $f'(x)$, we let $h \to 0$ in this last expression to obtain

$$\begin{aligned}
f'(x) &= \lim_{h\to 0} \frac{f(x+h)-f(x)}{h} \\
&= \lim_{h\to 0} \frac{(x+h)^n - x^n}{h} \\
&= \lim_{h\to 0} (nx^{n-1} + \frac{1}{2}n(n-1)x^{n-2}h + \cdots + h^{n-1}) \\
&= nx^{n-1} + 0 + \cdots + 0 \\
&= nx^{n-1}
\end{aligned}$$

This completes the proof for the derivative of x^n, where n is a positive integer. We gave some evidence earlier that

$$\frac{d}{dx}(e^x) = e^x.$$

We now present further support for this conclusion. With $f(x) = e^x$, use the limit definition of the derivative and obtain

$$\begin{aligned}
f'(x) &= \lim_{h\to 0} \frac{f(x+h)-f(x)}{h} \\
&= \lim_{h\to 0} \frac{e^{x+h} - e^x}{h} \\
&= \lim_{h\to 0} \frac{e^x e^h - e^x \cdot 1}{h} \\
&= \lim_{h\to 0} e^x \frac{e^h - 1}{h} \\
&= e^x \lim_{h\to 0} \frac{e^h - 1}{h}
\end{aligned}$$

To continue, we must evaluate

$$\lim_{h\to 0} \frac{e^h - 1}{h}$$

To obtain some idea what this might be, we use our computer or graphing calculator to graph

$$y = \frac{e^x - 1}{x}$$

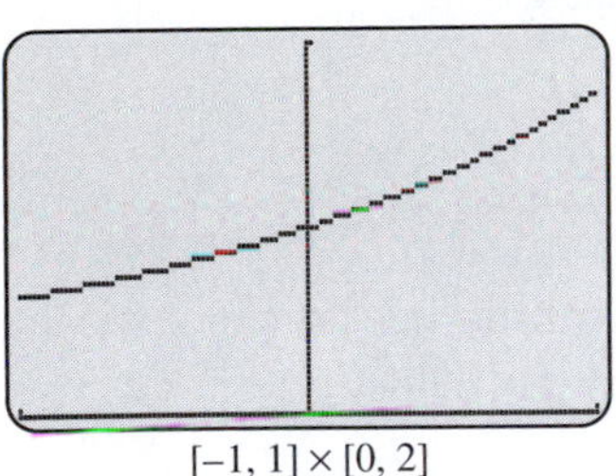

$[-1, 1] \times [0, 2]$

Screen 4.8
As $x \to 0$, it appears that $y = \dfrac{e^x - 1}{x} \to 1$.

using a window of dimensions $[-1, 1]$ by $[0, 2]$ and obtain Screen 4.8. The limit appears to be 1.

We then support this numerically by evaluating $\dfrac{e^h - 1}{h}$ for ever smaller values of h. This is shown in the following table.

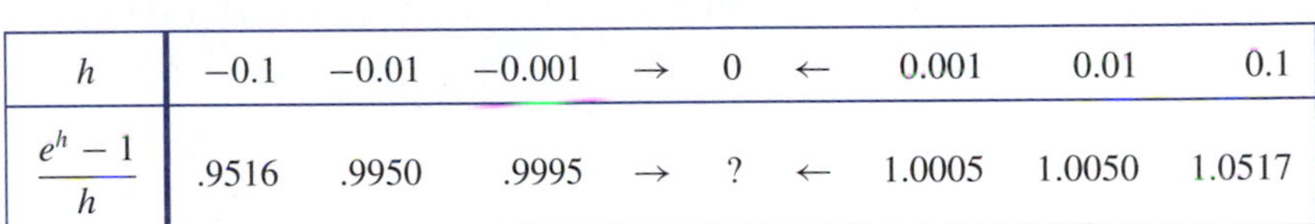
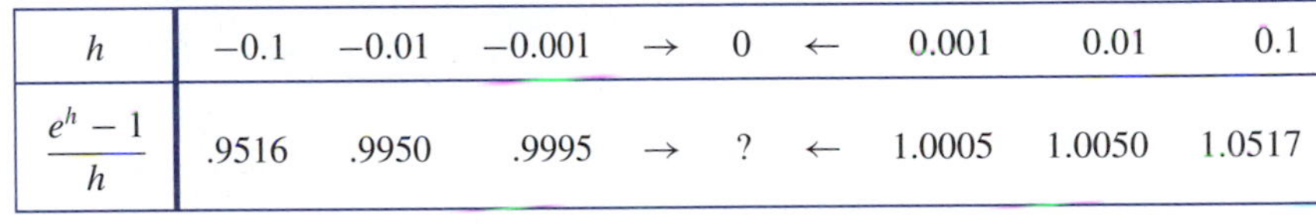

h	-0.1	-0.01	-0.001	$\to$	0	$\leftarrow$	0.001	0.01	0.1
$\dfrac{e^h - 1}{h}$	.9516	.9950	.9995	$\to$	?	$\leftarrow$	1.0005	1.0050	1.0517

The table then supports our conjecture that

$$\lim_{h \to 0} \frac{e^h - 1}{h} = 1$$

Of course, none of this actually constitutes a proof. Although too difficult to do here, it can be proven that this limit is indeed 1.

Using this result, we then have

$$\begin{aligned} f'(x) &= e^x \cdot \lim_{h \to 0} \frac{e^h - 1}{h} \\ &= e^x \cdot 1 \\ &= e^x \end{aligned}$$

We now consider a proof of $\dfrac{d}{dx}[cf(x)] = c\dfrac{d}{dx}[f(x)]$.

Derivative of a Constant Times a Function

If $f'(x)$ exists, then

$$\frac{d}{dx}[cf(x)) = c\frac{d}{dx}[f(x)]$$

We use the definition of derivative to give a proof of this rule.

$$\begin{aligned} \frac{d}{dx}[cf(x)] &= \lim_{h \to 0} \frac{cf(x+h) - cf(x)}{h} \\ &= \lim_{h \to 0} c\frac{f(x+h) - f(x)}{h} \\ &= c \lim_{h \to 0} \frac{f(x+h) - f(x)}{h} \\ &= cf'(x) \end{aligned}$$

We now give a proof that $\dfrac{d}{dx}[f(x) \pm g(x)] = \dfrac{d}{dx}[f(x)] \pm \dfrac{d}{dx}[g(x)]$.

Derivative of Sum or Difference

If $f'(x)$ and $g'(x)$ exist, then $\frac{d}{dx}[f(x) \pm g(x)]$ exists, and

$$\frac{d}{dx}[f(x) \pm g(x)] = \frac{d}{dx}f(x) \pm \frac{d}{dx}g(x)$$

We show only the proof for the derivative of the sum, but the proof for the derivative of the difference is similar. If $s(x) = f(x) + g(x)$, then

$$\begin{aligned} s'(x) &= \lim_{h\to 0} \frac{s(x+h) - s(x)}{h} \\ &= \lim_{h\to 0} \frac{[f(x+h) + g(x+h)] - [f(x) + g(x)]}{h} \\ &= \lim_{h\to 0} \frac{[f(x+h) - f(x)] + [g(x+h) - g(x)]}{h} \\ &= \lim_{h\to 0} \frac{f(x+h) - f(x)}{h} + \lim_{h\to 0} \frac{g(x+h) - g(x)}{h} \\ &= f'(x) + g'(x) \end{aligned}$$

Warm Up Exercise Set 4.1

1. Find the derivative of each of the following.

a. $x^7 + 4x^{2.5} - 3e^x$ **b.** $\sqrt{t} + \frac{1}{\sqrt{t}} + 5\ln x$

2. Let $f(x) = 0.125x^2 + 0.50$.

a. Find $f'(x)$.

b. Find the equation of the line tangent to the curve $y = f(x)$ at the point $x = 2$.

3. Suppose the cost function is given by $C(x) = 10 + 0.02x^2$, where x is the number of items produced and $C(x)$ is the cost of producing x items. Find the marginal cost for any x. Find $C'(30)$. Interpret what this means.

4. Lightning strikes a tree in a forest and ignites a fire that spreads in a circular manner. What is the instantaneous rate of change of the burnt area with respect to the radius r of the burnt area when $r = 50$ feet?

Exercise Set 4.1

In Exercises 1 through 27, find $\frac{d}{dx}f(x)$.

1. $f(x) = 13$

2. $f(x) = \sqrt{2}$

3. $f(x) = e + \pi$

4. $f(x) = 4\pi^2$

5. $f(x) = x^{23}$

6. $f(x) = x^7$

7. $f(x) = x^{1.4}$

8. $f(x) = x^{3.1}$

9. $f(x) = x^e$

10. $f(x) = x^{0.03}$

11. $f(x) = x^{4/3}$

12. $f(x) = x^{2/5}$

13. $f(x) = 3x^{-1.7}$

14. $f(x) = \frac{1}{3}e^x$

15. $f(x) = 0.24\sqrt{x}$

16. $f(x) = 0.24\sqrt[3]{x}$

17. $f(x) = 2e^x - 1$

18. $f(x) = 4 - 3\ln x$

19. $f(x) = \pi x + 2\ln x$

20. $f(x) = 3e^x - \pi^2 x$

21. $f(x) = 4 - x - x^2$

22. $f(x) = 3x^2 + x - 1$

23. $f(x) = \frac{1}{\pi}x^2 + \pi x - 1$

24. $f(x) = \pi^2 x^2 + \frac{1}{\sqrt{2}}x + \sqrt{3} + 2\ln x$

25. $f(x) = 1 - x - x^2 - x^3 - 10\ln x$

26. $f(x) = 0.02e^x + 2x^3 - 2x^2 + 3x - 10$

27. $f(x) = x^5 - 5e^x + 1$

In Exercises 28 through 33, find $f'(x)$.

28. $f(x) = 1 - x^4 - x^8$

29. $f(x) = 0.35x^2 - 0.3e^x + 0.45$

30. $f(x) = \ln 3 + e^x - 0.001x^2$

31. $f(x) = e^4 - \ln x - 0.01x^2 - 0.03x^3$

32. $f(x) = 3e^x - 0.003x^4 + 0.01x^3$

33. $f(x) = \dfrac{e^x + x^2 - 2x + 4}{6}$

In Exercises 34 through 39, find $f'(x)$.

34. $f(x) = \dfrac{x^2 - 3x - 6}{3x}$

35. $f(x) = 3x^3 + \dfrac{3}{x^3}$

36. $f(x) = \dfrac{1}{x} + x$

37. $f(x) = x^2 - \dfrac{1}{x}$

38. $f(x) = x^{-2} + x^{-3} + 3\ln\left(\dfrac{x}{4}\right)$

39. $f(x) = 4\ln(5x) - 3x^{-3} + 4x^2$

In Exercises 40 through 45, find $\dfrac{d}{dt} f(t)$.

40. $f(t) = (2t)^5 + e^{x+2}$

41. $f(t) = (-3t)^3 - e^{t-3}$

42. $f(t) = \dfrac{1}{t^2} + \dfrac{2}{t^3}$

43. $f(t) = \dfrac{1}{t} + \dfrac{1}{t^2} + \dfrac{1}{t^3}$

44. $f(t) = \sqrt{t} - \dfrac{1}{\sqrt{t}}$

45. $f(t) = \sqrt[3]{t} - \dfrac{1}{\sqrt[3]{t}}$

In Exercises 46 through 51, find $\dfrac{d}{du} f(u)$.

46. $f(u) = 2u^{3/2} + 4u^{3/4}$

47. $f(u) = 3u^{5/3} - 3u^{2/3}$

48. $f(u) = (u-1)^3$

49. $f(u) = (u+1)(2u+1)$

50. $f(u) = \dfrac{(u+1)^2}{u}$

51. $f(u) = \dfrac{u + 3/u}{\sqrt{u}}$

In Exercises 52 through 64, find $\dfrac{dy}{dx}$.

52. $y = \dfrac{x^2 + x - 2}{x}$

53. $y = \sqrt{4x} - \dfrac{1}{\sqrt{9x}}$

54. $y = x^{1.5} - x^{-1.5} + \ln x^{-2}$

55. $y = \pi x^{\pi} + \ln x^2$

56. $y = \dfrac{1}{\pi} x^{1/\pi}$

57. $y = x^{\sqrt{2}} - \dfrac{1}{x^{\sqrt{2}}}$

58. $y = 3x^{-3/5} - 6x^{-5/2}$

59. $y = x^{-3/2} + x^{-4/5} + \ln\sqrt{x}$

60. $y = x^{-2} + 3x^{-4}$

61. $y = \sqrt[3]{x^2}$

62. $y = \sqrt{x^3}$

63. $y = \dfrac{1}{\sqrt[5]{x^2}}$

64. $y = \dfrac{1}{\sqrt[3]{x^4}}$

In Exercises 65 and 66, find the equation of the tangent line of the given function at the indicated point. Support your answer using a computer or graphing calculator.

65. $y = f(x) = x^2 + e^x + 1$, $x_0 = 0$

66. $y = f(x) = x^{-1} + x^{-2} + \ln x$, $x_0 = 1$

In Exercises 67 through 68, graph the functions on your computer or graphing calculator and roughly estimate the values where the tangent to the graph of $y = f(x)$ is horizontal. Confirm your answer using calculus.

67. $y = f(x) = x^3 - 9x + 3$

68. $y = f(x) = 7 + 12x - x^3$

In Exercises 69 through 72, estimate the cost of the next item or the revenue or profits from the sale of the next item for the indicated functions and the indicated x.

69. $C(x) = \sqrt{x}(x + 12) + 50$, $x = 36$

70. $R(x) = \sqrt{x}(x - 24)$, $x = 16$

71. $P(x) = x^2 - 17x - 60$, $x = 50$

72. $P(x) = \sqrt[3]{x}(x - 54) - 50$, $x = 27$

Applications and Mathematical Modeling

73. Body Temperature Merola-Zwartjes and Ligon[7] studied the Puerto Rican Tody. Even though they live in the relatively warm climate of the tropics, these organisms face thermoregulatory challenges owing to their extremely small size. The researchers created a mathematical model given by the equation $B(T) = 36.0 - 0.28T + 0.01T^2$, where T is the ambient temperature in degrees Celsius and B is body temperature in degrees Celsius for $10 \le T \le 42$. Find $B'(T)$. Find $B'(25)$. Give units.

74. Lamprey Growth Griffiths and colleagues[8] studied the larval sea lamprey in the Great Lakes, where they represent a serious threat to the region's fisheries. They created a mathematical model given by the equation $G(T) = -0.991 + 0.273T - 0.0179T^2$, where T is the mean annual water temperature with $6.2 \le T \le 9.8$ and G is the growth in millimeters per day. Find $G'(T)$. Find $G'(8)$. Give units.

[7] Michele Merola-Zwartjes and J. David Ligon. 2000. Ecological energetics of the Puerto Rican Tody: heterothermy, torpor, and intra-island variation. *Ecology* 81(4):990–1003.

[8] Ronald W. Griffiths, F. W. Beamish, B. J. Morrison, and L. A. Barker. 2001. Factors affecting larval sea lamprey growth and length at metamorphosis in lampricide-treated streams. *Trans. Amer. Fish. Soc.* 130:289–306.

75. Corn Yield per Acre Atwood and Helmers[9] studied the effect of nitrogen fertilizer on corn yields in Nebraska. Nitrate contamination of groundwater in Nebraska has become a serious problem. So serious is this problem that actions such as nitrogen rationing and taxation are under consideration. Nitrogen fertilizer is needed to increase yields, but an excess of nitrogen has been found not to help increase yields. Atwood and Helmers created a mathematical model given by the equation $Y(N) = 58.51 + 0.824N - 0.00319N^2$, where N is the amount of fertilizer in pounds per acre and Y is the yield in bushels per acre. Find $Y'(N)$. Find $Y'(150)$. Give units.

76. Lactation Curves Cardellino and Benson[10] studied lactation curves for ewes rearing lambs. They created the mathematical model given by the equation $P(D) = 273.23 + 5.39D - 0.1277D^2$, where D is day of lactation and P is milk production in grams. Find $P'(D)$. Find $P'(30)$. Give units.

77. Growing Season Start Jobbagy and colleagues[11] determined a mathematical model given by the equation $G(T) = 278 - 7.1T - 1.1T^2$, where G is the start of the growing season in the Patagonian steppe in Julian days (day $1 =$ January 1) and T is the mean July temperature in degrees Centigrade. Find $G'(T)$. Find $G'(-1)$. Give units.

78. Protein in Milk Crocker and coworkers[12] studied the northern elephant seal in Ano Nuevo State Reserve, California. They created a mathematical model given by the equation $W(D) = 62.3 - 2.68D + 0.06D^2$, where D is days postpartum and W is the percentage of water in the milk. Find $W'(D)$. Find $W'(10)$. Give units.

79. Marginal Cost, Revenue, Profit In 1997, Fuller and coworkers[13] at Texas A & M University estimated the operating costs of cotton gin plants of various sizes. A quadratic model of cost in thousands of dollars for the largest plant is given by $C(x) = 0.026689x^2 + 21.7397x + 377.3865$, where x is the annual quantity of bales in thousands produced if annual production is at most 50,000 bales. Revenue was estimated at \$63.25 per bail.

a. Find the marginal cost.
b. Find the marginal revenue.
c. Find the marginal profit.

80. Marginal Cost, Revenue, Profit In 1997, Fuller and coworkers[14] at Texas A & M University estimated the operating costs of cotton gin plants of various sizes. A power model of cost in thousands of dollars for the next-to-the-largest plant is given by $C(x) = 165.46x^{-0.4789}$, where x is the annual quantity of bales in thousands produced if annual production is at most 30,000 bales. Revenue was estimated at \$63.25 per bale.

a. Find the marginal cost.
b. Find the marginal revenue.
c. Find the marginal profit.

81. Finance Fry[15] studied 85 developing countries and found that the average percentage growth rate y in each country was approximated by the equation $y = g(r) = -0.033r^2 + 0.008r^3$, where r is the real interest rate in the country. Find $g'(r)$, and explain what this term means. Find $g'(-2)$ and $g'(2)$. Explain in each case what these two numbers mean. What is the significance of the sign of each number?

82. Weight-Length of Paddlefish Timmons and Hughbanks[16] studied the paddlefish in the lower Tennessee and Cumberland rivers. Paddlefish roe is processed as a high-quality substitute for sturgeon caviar. These researches created a mathematical model given by the equation $\log W = -5.68 + 3.29 \log L$, where L is the length in millimeters and W is the weight in grams. Find $W'(L)$. Find $W'(1000)$. Give units and interpret your answer.

83. Biology Zonneveld and Kooijman[17] determined that the equation $y = 2.81L^2$ approximated the lettuce intake versus the shell length of a pond snail. Find the rate of change of lettuce intake with respect to shell length.

84. Biology Kaakeh and coworkers[18] artificially exposed one-year-old apple trees to a controlled aphid infestation. They showed that the average shoot length y in centimeters was approximated by $y = f(x) = 62.8 + 0.486x - 0.003x^2$, where x is the number of days since the aphid infestation.

---

[9] Joseph A. Atwood and Glenn A. Helmers. 1998. Examining quantity and quality effects of restricting nitrogen applications to feedgrains. *Amer. J. Agr. Econ.* 80:369–381.

[10] R. A. Cardellino and M. E. Benson. 2002. Lactation curves of commercial ewes rearing lambs. *J. Anim. Sci.* 80:23–27.

[11] Esteban G. Jobbagy, Osvaldo E. Sala, and Jose M. Paruelo. 2002. Patterns and control of primary production in the Patagonian steppe: a remote sensing approach. *Ecology* 83(2):307–319.

[12] Daniel E. Crocker, Jeannine D. Williams, Daniel P. Costa, and Burney J. Le Boeup. 2001. Maternal traits and reproductive effort in northern elephant seals. *Ecology* 82(12):3541–3555.

[13] S. Fuller, M. Gillis, C. Parnell, A. Ramaiyer, and R. Childers. 1997. Effect of air pollution abatement on financial performance of Texas cotton gins. *Agribusiness* 13(5):521–532.

[14] Ibid.

[15] Maxwell J. Fry. 1997. In favour of financial liberalization. *Econ. J.* 107:754–770.

[16] Tom J. Timmons and Tyrone A. Hughbanks. 2000. Exploitation and mortality of paddlefish in the lower Tennessee and Cumberland rivers. *Trans. Amer. Fish. Soc.* 129:1171–1180.

[17] C. Zonneveld and S. Kooijman. 1989. The application of a dynamic energy budget model to Lymnaea stagnalis. *Funct. Ecol.* 3:269–278.

[18] Walid Kaakeh, Douglas Pfeiffer, Richard Marini. 1992. Combined effects of spirea aphid and nitrogen fertilization on shoot growth in young apple trees. *J. Econ. Entomol.* 85:496–505.

Find $f'(x)$. Find $f'(60)$ and $f'(110)$, and explain what these numbers mean.

85. **Biology** Miller[19] determined that the equation $y = 0.12(L^2 + 0.0026L^3 - 16.8)$ approximated the reproduction rate, measured by the number of eggs per year, of the rock goby versus its length in centimeters. Find the rate of change of reproduction with respect to length.

86. **Medicine** Poiseuille's law (see Clow and Urquhart[20]) states that the total resistance R to blood flow in a blood vessel of constant length l and radius r is given by $R = \frac{al}{r^4}$, where a is a positive constant. Find the rate of change of resistance with respect to the radius when $r = 3$.

87. **Biology** The free water vapor diffusion coefficient D in soft woods[21] is given by

$$D(T) = \frac{167.2}{P}\left(\frac{T}{273}\right)^{1.75}$$

where T is the temperature and P is the constant atmospheric pressure measured in appropriate units. Find the rate of change of D with respect to T.

88. **Biology** The chemical d-limonene is extremely toxic to some insects. Karr and Coats[22] showed that the number of days y for a German cockroach nymph (*Blatta germanica*) to reach the adult stage when given a diet containing this chemical was approximated by $y = f(x) = 125 - 3.43 \ln x$, where x is the percent of this chemical in the diet. Find $f'(x)$ for any $x > 0$. Interpret what this means. What does this say about the effectiveness of this chemical as a control of the German cockroach nymph? Graph $y = f(x)$.

19 P. J. Miller. 1961. Age, growth, and reproduction of the rock goby. *J. Mar. Biol. Ass. U.K.* 41:737–769.

20 Duane J. Clow and N. Scott Urquhart. 1974. *Mathematics in Biology*, New York: Ardsley House, page 243.

21 Ibid.

22 L. L. Karr and J. R. Coats. 1992. Effects of four monoterpernoids on growth and reproduction of the German cockroach. *J. Econ. Entomol.* 85(2):424–429.

More Challenging Exercises

89. Graph the function $y = f(x) = x^5 + 3x^2 - 9x + 4$ on your graphing calculator.
    a. Roughly estimate the values where the tangent to the graph of $y = f(x)$ is horizontal.
    b. Now graph $y = f'(x)$ and solve $f'(x) = 0$ using your computer or graphing calculator.
    c. How do you relate the solutions of $f'(x) = 0$ found in part (b) to the values of x found in part (a)?
    d. Can you confirm your answers found in part (b) using calculus? Why or why not?

90. Let L_1 be the line tangent to the graph of $y = 1/x^2$ at $x = 1$, and let L_2 be the line tangent to the graph of $y = -x^2 - 1.5x + 1$ at $x = -1$. Show that the two tangent lines are perpendicular.

91. Let L_1 be the tangent line to the graph of $y = x^4$ at $x = -1$, and let L_2 be the tangent line to the graph of $y = x^4$ at $x = 1$. Show that these two tangent lines intersect on the y-axis.

92. Let $f(x) = g(x) + 4x^3$, and suppose that $g'(2)$ does *not* exist. Can $f'(2)$ exist? Explain why or why not.

93. Suppose both $f(x)$ and $g(x)$ are *not* differentiable at $x = a$. Does this imply that $h(x) = f(x) + g(x)$ is *not* differentiable at $x = a$? Explain why this is true, or give an example that shows it is not true.

94. **Biology** Potter and colleagues[23] showed that the percent mortality y of a New Zealand thrip was approximated by $y = f(T) = 81.12 + 0.465T - 0.828T^2 + 0.04T^3$, where T is the temperature measured in degrees Celsius. Graph on your grapher using a window of dimensions [0, 20] by [0, 100].
    a. Estimate the value of the temperature where the tangent line to the curve $y = f(T)$ is horizontal.
    b. Check your answer using calculus. (You will need to use the quadratic formula.)
    c. Did you miss any points in part (a)?
    d. Is this another example of how your computer or graphing calculator can mislead you? Explain.

95. Let $f(x) = x^3 + 4\cos x - 3\sin x$. Find $f'(x)$. (Recall from Exercises 67 and 68 in Section 3.3 that $\frac{d}{dx}(\sin x) = \cos x$ and $\frac{d}{dx}(\cos x) = -\sin x$.)

96. **Economies of Scale in Food Retailing** Using data from the files of the National Commission on Food Retailing concerning the operating costs of thousands of stores,

23 Murray Potter, Alan Carpenter, Adrienne Stocker, and Sandy Wright. 1994. Controlled atmospheres for the postharvest disinfestation of *Thrips obscuratus*. *J. Econ. Entomol.* 87:1251–1255.

Smith[24] showed that the sales expense as a percent of sales S was approximated by the equation $S(x) = 0.4781x^2 - 5.4311x + 16.5795$, where x is in sales per square foot and $x \leq 7$.

a. Find $S'(x)$ for any x.

b. Find $S'(4)$, $S'(5)$, $S'(6)$, and $S'(7)$. Interpret what is happening.

c. Graph the cost function on a screen with dimensions $[0, 9.4]$ by $[0, 12]$. Also graph the tangent lines at the points where x is 4, 5, 6, and 7. Observe how the slope of the tangent line is changing, and relate this to the observations made above concerning the rates of change.

d. Use the available operations on your computer or graphing calculator to the find where the function attains a minimum.

[24] National Commission on Food Marketing. 1966. *Organization and Competition in Food Retailing*. Technical Study 7.

97. Marginal Cost Suppose the cost function is given by $C(x) = .01x^3 - x^2 + 50x + 100$, where x is the number of items produced and $C(x)$ is the cost in dollars to produce x items.

a. Find the marginal cost for any x.

b. Find $C'(20)$, $C'(40)$, and $C'(60)$. Interpret what is happening.

c. Graph the cost function on a screen with dimensions $[0, 90]$ by $[0, 5000]$. Also graph the tangent lines at the points where x is 20, 40, and 60. Observe how the slope of the tangent line is changing, and relate this to the observations made above concerning the rates of change.

Modeling Using Least Squares

98. Cost Curve Dean[25] made a statistical estimation of the cost-output relationship for a hosiery mill. The data for the firm is given in the following table.

x	46	65	62	70	74	78
y	11	14.5	16	17.5	18.5	19

Here x is production in hundreds of dozens, and y is total labor cost in thousands of dollars.

a. Determine both the best-fitting quadratic using least squares and the correlation coefficient. Graph.

b. Using the quadratic cost function found in (a), graph the marginal labor cost. What is happening to marginal labor cost as output increases?

99. Cost Curve Dean[26] made a statistical estimation of the cost-output relationship for a department store. The data for the firm is given in the following table. (To keep absolute cost magnitudes confidential, both cost and output measures were transformed into index numbers.)

x	35	50	80	100	140	170	200
y	55	70	95	100	130	145	145

Here x is the transaction index, and y is the cost index.

a. Determine both the best-fitting quadratic using least squares and the correlation coefficient. Graph.

b. Using the quadratic cost function found in part (a), graph the marginal cost. What is happening to marginal cost as output increases?

100. Economies of Scale Riew[27] studied the economies of scale in operating a public high school in Wisconsin. See the table for his data.

[25] Joel Dean. 1976. *Statistical Cost Estimation*. Bloomington, Ind.: Indiana University Press, page 219.

[26] Ibid., page 285.

[27] John Riew. 1966. Economies of scale in high school operation. *Rev. Econ. Stat.* 48(3):280–287.

Average Daily Attendance	Midpoint of Values in First Column	Expenditure per Pupil
143–200	171	\$531.9
201–300	250	480.8
301–400	350	446.3
401–500	450	426.9
501–600	550	442.6
601–700	650	413.1
701–900	800	374.3
901–1100	1000	433.2
1101–1600	1350	407.3
1601–2400	2000	405.6

a. Find both the best-fitting linear model and the correlation coefficient using least squares. Are economies of scale indicated? Why?

b. Find both the best-fitting quadratic model and the square of the correlation coefficient using least squares. Is quadratic better than linear? Does the graph turn up at the end? What does this mean in terms of economies of scale?

Solutions to WARM UP EXERCISE SET 4.1

1. a.
$$\frac{d}{dx}(x^7 + 4x^{2.5} - 3e^x) = \frac{d}{dx}(x^7) + 4\frac{d}{dx}(x^{2.5}) - 3\frac{d}{dx}(e^x) = 7x^6 + 10x^{1.5} - 3e^x$$

b.
$$\frac{d}{dx}\left(\sqrt{t} + \frac{1}{\sqrt{t}} + 5\ln x\right) = \frac{d}{dt}(t^{1/2} + t^{-1/2} + 5\ln x) = \frac{1}{2}t^{-1/2} - \frac{1}{2}t^{-3/2} + 5\frac{1}{x}$$

2. a. $f'(x) = \frac{d}{dx}(0.125x^2 + 0.50) = 0.125\frac{d}{dx}(x^2) + \frac{d}{dx}(0.50) = 0.25x$

b. Since $x_0 = 2$, $y_0 = f(x_0) = f(2) = 1$. The slope of the tangent line at $x = 2$ is $f'(2) = 0.25(2) = 0.50$. Thus, the equation of the line tangent to the curve at $x = 2$ is

$$y - 1 = 0.50(x - 2)$$

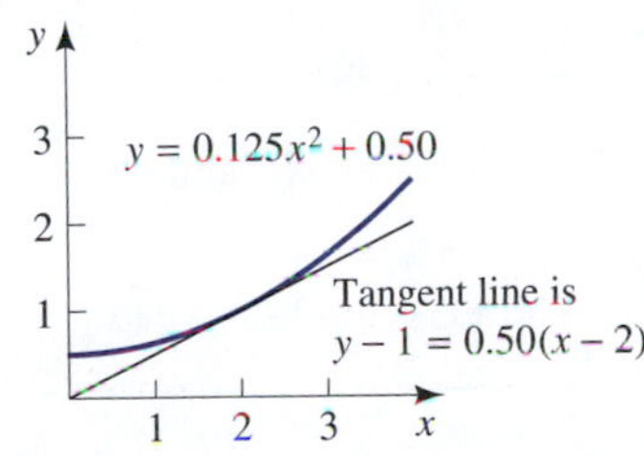

3. The marginal cost for any x is

$$C'(x) = \frac{d}{dx}(10 + 0.02x^2) = \frac{d}{dx}(10) + 0.02\frac{d}{dx}(x^2) = 0.04x$$

Then $C'(30) = 0.04(30) = 1.20$. Thus, the cost is increasing at a rate of \$1.20 per item when 30 items are being produced.

4. The area of a circle of radius r is given by $A = A(r) = \pi r^2$. Then

$$\frac{dA}{dr} = A'(r) = 2\pi r$$

and $A'(50) = 2\pi(50) \approx 314$. Thus, the burnt area is increasing at the rate of approximately 314 square feet per 1 foot increase in r when $r = 50$.

4.2 Derivatives of Products And Quotients

APPLICATION Marginal Revenue

Suppose the demand equation for a certain product is given by $p = e^{-x}$, where x is the number of items sold (in thousands) and p is the price in dollars. What is the marginal revenue? For what values of x is the marginal revenue positive? Negative? Zero? Interpret what this means. See Example 4 on page 247 for the answer.

Product Rule

We continue to develop formulas that permit us to take derivatives without having to resort to using the definition of the derivative. In this section we will develop the rules for differentiating the product of two functions and the quotient of two functions.

Since the derivative of the sum is the sum of the derivatives, one might suspect that the derivative of the product is the product of the derivatives. This is *not* true in general. For example, on the one hand,

$$\frac{d}{dx}\left(x^3 \cdot x^2\right) = \frac{d}{dx}(x^5) = 5x^4$$

but on the other hand,

$$\frac{d}{dx}(x^3) \cdot \frac{d}{dx}(x^2) = 3x^2 \cdot 2x = 6x^3$$

The correct rule for taking the derivative of the product of two functions follows.

Product Rule

If both $f'(x)$ and $g'(x)$ exist, then $\frac{d}{dx}(f(x) \cdot g(x))$ exists, and

$$\frac{d}{dx}[f(x) \cdot g(x)] = f(x) \cdot \frac{d}{dx}g(x) + g(x) \cdot \frac{d}{dx}f(x)$$

We give some support for the product rule at the end of this section.

Remark You might think of the product rule as: first times derivative of second plus second times derivative of first.

We now consider two examples that use the product rule.

EXAMPLE 1 **Using the Product Rule**

Find $h'(x)$ if $h(x) = e^x \ln x$.

Solution

To differentiate $h(x)$, think of $h(x)$ as a product of the two functions: $f(x) = e^x$ and $g(x) = \ln x$. Then

$$\begin{aligned} h(x) &= f(x) \cdot g(x) \\ h'(x) &= f(x) \cdot \frac{d}{dx} g(x) + g(x) \cdot \frac{d}{dx} f(x) \\ &= e^x \cdot \frac{d}{dx}(\ln x) + (\ln x) \cdot \frac{d}{dx} e^x \\ &= e^x \frac{1}{x} + (\ln x) e^x \\ &= e^x \left(\frac{1}{x} + \ln x \right) \end{aligned}$$

■

EXAMPLE 2 **Using the Product Rule**

Find $h'(x)$ if $h(x) = (x^4 + 3x^2 + 2x) \ln x$.

Solution

We think of $h(x)$ as $f(x) \cdot g(x)$, where

$$f(x) = (x^4 + 3x^2 + 2x), \qquad g(x) = \ln x$$

Then

$$\begin{aligned} h(x) &= f(x) \cdot g(x) \\ h'(x) &= f(x) \cdot \frac{d}{dx} g(x) + g(x) \cdot \frac{d}{dx} f(x) \\ &= (x^4 + 3x^2 + 2x) \cdot \frac{d}{dx}(\ln x) + (\ln x) \cdot \frac{d}{dx}(x^4 + 3x^2 + 2x) \\ &= (x^4 + 3x^2 + 2x) \cdot \left(\frac{1}{x} \right) + (\ln x) \cdot (4x^3 + 6x + 2) \\ &= x^3 + 3x + 2 + (4x^3 + 6x + 2) \ln x \end{aligned}$$

■

Quotient Rule

Now that we know that the derivative of the product is not the product of the derivatives, we suspect that the derivative of the quotient is not the quotient of the derivatives. This is indeed the case. We can use the product rule to find the derivative of the quotient. Let $Q(x) = \dfrac{f(x)}{g(x)}$. Assuming that $Q(x)$ is differentiable, we can apply the product rule to $f(x) = Q(x)g(x)$ and obtain

$$f'(x) = Q'(x)g(x) + Q(x)g'(x)$$

$$= Q'(x)g(x) + \frac{f(x)}{g(x)}g'(x)$$

$$Q'(x) = \frac{f'(x) - \frac{f(x)}{g(x)}g'(x)}{g(x)}$$

To simplify, multiply the numerator and denominator by $g(x)$ and obtain

$$\frac{d}{dx}\left[\frac{f(x)}{g(x)}\right] = \frac{f'(x)g(x) - f(x)g'(x)}{[g(x)]^2}$$

Quotient Rule

If $f'(x)$ and $g'(x)$ exist, then $\frac{d}{dx}\left[\frac{f(x)}{g(x)}\right]$ exists, and

$$\frac{d}{dx}\left[\frac{f(x)}{g(x)}\right] = \frac{g(x)\cdot\frac{d}{dx}f(x) - f(x)\cdot\frac{d}{dx}g(x)}{[g(x)]^2}$$

at points where $g(x) \neq 0$.

EXAMPLE 3 **Using the Quotient Rule**

Find $h'(x)$ if $h(x) = \frac{e^x}{x^2+1}$.

Solution

Here we think of $h(x)$ as the quotient of two functions $\frac{f(x)}{g(x)}$, where

$$f(x) = e^x, \qquad g(x) = x^2 + 1$$

Then

$$h(x) = \frac{f(x)}{g(x)}$$

$$h'(x) = \frac{g(x)\cdot\frac{d}{dx}f(x) - f(x)\cdot\frac{d}{dx}g(x)}{[g(x)]^2}$$

$$= \frac{(x^2+1)\cdot\frac{d}{dx}(e^x) - e^x\cdot\frac{d}{dx}(x^2+1)}{(x^2+1)^2}$$

$$= \frac{(x^2+1)e^x - e^x(2x)}{(x^2+1)^2}$$

$$= \frac{(x^2-2x+1)e^x}{(x^2+1)^2}$$

■

EXAMPLE 4 Instantaneous Rate of Change of Revenue

Suppose the demand equation for a certain product is given by $p = e^{-x}$, where x is the number of items sold (in thousands) and p the price in dollars. What is the marginal revenue? For what values of x is the marginal revenue positive? Negative? Zero? Interpret what this means.

Solution

We have $R(x) = xp = xe^{-x} = \frac{x}{e^x}$. We think of $R(x)$ as the quotient of two functions, $\frac{f(x)}{g(x)}$, where $f(x) = x$ and $g(x) = e^x$. Then

$$\begin{aligned}
r(x) &= \frac{f(x)}{g(x)} \\
R'(x) &= \frac{g(x)\cdot\frac{d}{dx}f(x) - f(x)\cdot\frac{d}{dx}g(x)}{[g(x)]^2} \\
&= \frac{e^x\cdot\frac{d}{dx}(x) - (x)\cdot\frac{d}{dx}e^x}{(e^x)^2} \\
&= \frac{e^x(1) - xe^x}{e^{2x}} \\
&= \frac{(1-x)e^x}{e^{2x}} \\
&= \frac{1-x}{e^x}
\end{aligned}$$

Since e^x is always positive, the sign of $R'(x)$ is the same as the sign of the factor $(1-x)$. Thus, $R'(x) > 0$ when $0 < x < 1$, $R'(1) = 0$, and $R'(x) < 0$ when $x > 1$. Screen 4.9 shows a graph of $y = R(x)$ using a window of dimensions $[0, 7]$ by $[0, 0.5]$. Notice that the revenue curve is rising on $(0, 1)$ and falling on $(1, \infty)$. ■

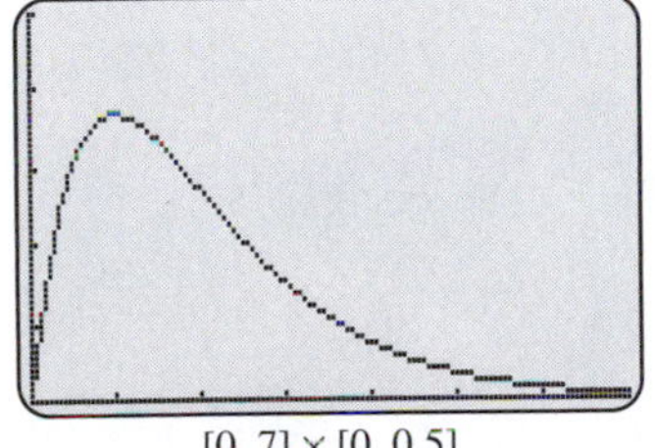

$[0, 7] \times [0, 0.5]$

Screen 4.9
The revenue $y = xe^{-x}$ initially increases, then decreases as the number of items sold increases.

We now consider average cost, use the quotient rule to develop a formula used in economics relating marginal average cost with marginal cost, and give a graphical demonstration of this formula.

DEFINITION Average Cost

If $C(x)$ is the cost of producing x items, then the **average cost**, denoted by $\bar{C}(x)$, is the cost of all the items divided by the number of items, that is,

$$\bar{C}(x) = \frac{C(x)}{x}$$

The marginal average cost is given by $\frac{d}{dx}\bar{C}(x)$.

EXAMPLE 5 Application of the Quotient Rule

Use the quotient rule to find a formula for the marginal average cost in terms of the cost and the marginal cost.

Solution

Using the quotient rule, we have

$$\begin{aligned}\frac{d}{dx}\bar{C}(x) &= \frac{d}{dx}\left(\frac{C(x)}{x}\right)\\ &= \frac{x\cdot\frac{d}{dx}C(x) - C(x)\cdot\frac{d}{dx}(x)}{x^2}\\ &= \frac{xC'(x) - C(x)}{x^2}\\ &= \frac{1}{x}\left(C'(x) - \frac{C(x)}{x}\right)\\ &= \frac{1}{x}\left(C'(x) - \bar{C}(x)\right)\end{aligned}$$

■

In words, this formula says that marginal average cost is found by taking marginal cost less average cost and dividing the result by the number of items sold.

EXAMPLE 6 **Application of the Quotient Rule**

If the cost function is given by $C(x) = 2x^3 - x^2 + 1$, find (a) marginal cost, (b) average cost, (c) marginal average cost.

Solution

Marginal cost is given by

$$C'(x) = 6x^2 - 2x$$

Average cost is given by

$$\begin{aligned}\bar{C}(x) &= \frac{1}{x}C(x)\\ &= \frac{1}{x}\left(2x^3 - x^2 + 1\right)\\ &= \frac{2x^3}{x} - \frac{x^2}{x} + \frac{1}{x}\\ &= 2x^2 - x + \frac{1}{x}\end{aligned}$$

Marginal average cost is given by

$$\frac{d}{dx}\bar{C} = \frac{d}{dx}\left(2x^2 - x + \frac{1}{x}\right) = 4x - 1 - \frac{1}{x^2}$$

To obtain marginal average cost, we could also use the formula given in Example 5. This gives

$$\begin{aligned}\frac{d}{dx}\bar{C} &= \frac{1}{x}\left((6x^2 - 2x) - \left(2x^2 - x + \frac{1}{x}\right)\right)\\ &= \frac{1}{x}\left(4x^2 - x - \frac{1}{x}\right)\\ &= 4x - 1 - \frac{1}{x^2}\end{aligned}$$

■

Exploration 1

An Important Point in Economics

Notice that the result of Example 5 indicates that marginal average cost is zero when marginal cost equals the average cost. For the cost function given in Example 6, graph the average and marginal cost functions on a screen of dimension [0, 3] by [0, 5].

Notice carefully that the graph of $y = C'(x)$ cuts the graph of $y = \bar{C}(x)$ at the point where the graph of $y = \bar{C}(x)$ has a tangent line with slope zero, which is the point where marginal average cost is zero.

Enrichment: Support for The Product Rule

Let $p(x) = f(x)g(x)$. We would like to establish the product rule of differentiation. From the definition of derivative we have

$$\begin{aligned} p'(x) &= \lim_{h\to 0} \frac{p(x+h) - p(x)}{h} \\ &= \lim_{h\to 0} \frac{f(x+h)g(x+h) - f(x)g(x)}{h} \end{aligned}$$

We now consider the term $f(x+h)g(x+h) - f(x)g(x)$. In Figure 4.5 the smaller unshaded rectangle has sides $f(x)$ and $g(x)$ and so has area $f(x)g(x)$. The large rectangle has sides $f(x+h)$ and $g(x+h)$ and area $f(x+h)g(x+h)$. (We are assuming that both f and g are increasing functions and that h is positive for the purposes of this figure.) For convenience we use the notation $\Delta f = f(x+h) - f(x)$ and $\Delta g = g(x+h) - g(x)$. Then from Figure 4.5 we see that

$$\begin{aligned} f(x+h)g(x+h) - f(x)g(x) &= (\text{area of large rectangle}) \\ &\quad - (\text{area of unshaded rectangle}) \\ &= \text{area of three shaded rectangles} \\ &= \Delta f \cdot g(x) + f(x) \cdot \Delta g + \Delta f \cdot \Delta g \end{aligned}$$

Dividing by h then gives

$$\frac{f(x+h)g(x+h) - f(x)g(x)}{h} = \frac{\Delta f}{h} \cdot g(x) + f(x) \cdot \frac{\Delta g}{h} + \frac{\Delta f \cdot \Delta g}{h}$$

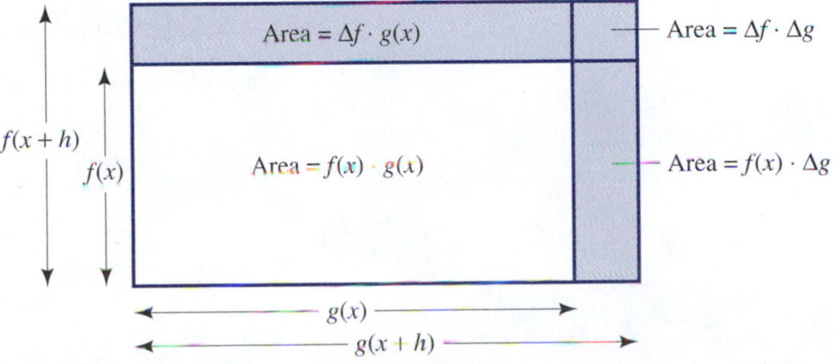

Figure 4.5

We now consider each of the three terms on the right of the last equation and notice that

$$\begin{aligned}
\lim_{h\to 0} \frac{\Delta f}{h} \cdot g(x) &= \lim_{h\to 0} \frac{f(x+h)-f(x)}{h} \cdot g(x) \\
&= f'(x)g(x) \\
\lim_{h\to 0} f(x) \cdot \frac{\Delta g}{h} &= \lim_{h\to 0} f(x) \cdot \frac{g(x+h)-g(x)}{h} \\
&= f(x)g'(x)
\end{aligned}$$

Also

$$\begin{aligned}
\lim_{h\to 0} \frac{\Delta f \cdot \Delta g}{h} &= \lim_{h\to 0} \frac{\Delta f}{h} \cdot \frac{\Delta g}{h} \cdot h \\
&= \lim_{h\to 0} \frac{f(x+h)-f(x)}{h} \cdot \lim_{h\to 0} \frac{g(x+h)-g(x)}{h} \cdot \lim_{h\to 0} h \\
&= f'(x) \cdot g'(x) \cdot 0 \\
&= 0
\end{aligned}$$

Therefore, we have

$$\begin{aligned}
p'(x) &= \lim_{h\to 0} \frac{f(x+h)g(x+h) - f(x)g(x)}{h} \\
&= \lim_{h\to 0} \left(\frac{\Delta f}{h} \cdot g(x) + f(x) \cdot \frac{\Delta g}{h} + \frac{\Delta f \cdot \Delta g}{h} \right) \\
&= \lim_{h\to 0} \frac{\Delta f}{h} \cdot g(x) + \lim_{h\to 0} f(x) \cdot \frac{\Delta g}{h} + \lim_{h\to 0} \frac{\Delta f \cdot \Delta g}{h} \\
&= f'(x)g(x) + f(x)g'(x)
\end{aligned}$$

Enrichment: Derivative of Power When Exponent Is a Negative Integer

In the Enrichment subsection of Section 4.1, a proof was given for the derivative of x^n being nx^{n-1} in the case that n is a positive integer. We will now give a proof that this rule holds when n is a negative integer. Suppose, then, that n is a negative integer, and let $m = -n$. Then m is a positive integer. Let $f(x) = x^n = x^{-m}$. Then using the product rule for differentiation and the power rule when the power is a positive integer, we have

$$\begin{aligned}
x^m f(x) &= 1 \\
x^m f'(x) + mx^{m-1} f(x) &= 0 \\
xf'(x) + mf(x) &= 0 \\
f'(x) &= -\frac{mf(x)}{x} \\
&= -\frac{mx^{-m}}{x} \\
&= -mx^{-m-1} \\
&= nx^{n-1}
\end{aligned}$$

Warm Up Exercise Set 4.2

Find the derivative of each of the following. Do not simplify.

1. $(x^2 - e^x)(x^3 + x + \ln x)$

2. $\dfrac{x^{3/2} + 1}{x^4 + 1}$

Exercise Set 4.2

In Exercises 1 through 8, use the product rule to find the derivative.

1. $x^3 e^x$
2. $\sqrt{x}e^x$
3. $\sqrt{x}\ln x$
4. $x^4 \ln x$
5. $e^x(x^2 - 3)$
6. $(2x^3 + 3)\ln x$
7. $(4e^x - x^5)\ln x$
8. $(x^3 - 3\ln x)(2e^x + 3x)$

In Exercises 9 through 12, use the product rule to find $f'(x)$.

9. $f(x) = (e^x + 1)(\sqrt{x} + 1)$
10. $f(x) = (x^2 - 3)(\sqrt[3]{x} - \ln x)$
11. $f(x) = (2x - 3\ln x)\left(x + \dfrac{1}{x}\right)$
12. $f(x) = \left(e^x + \dfrac{1}{x}\right)\left(1 + \dfrac{1}{x^2}\right)$

In Exercises 13 through 24, use the quotient rule to find the derivative.

13. $f(x) = e^{-x}\ln x$
14. $f(x) = \dfrac{\ln x}{x^2 + 3}$
15. $f(x) = \dfrac{1}{\ln x}$
16. $f(x) = (x + 3)e^{-x}$
17. $f(x) = \dfrac{x}{x + 2}$
18. $f(x) = \dfrac{x + 1}{x - 2}$
19. $f(x) = \dfrac{x - 3}{x + 5}$
20. $f(x) = \dfrac{3}{x + 3}$
21. $f(x) = \dfrac{e^x}{x - 2}$
22. $f(x) = \dfrac{2x - 3}{4x - 1}$
23. $f(x) = \dfrac{2 - 3x}{2x - 1}$
24. $f(x) = \dfrac{3 - 2e^x}{1 - 2x}$

In Exercises 25 through 32, use the quotient rule to find $\dfrac{dy}{du}$.

25. $y = \dfrac{3e^u}{u^2 + 1}$
26. $y = \dfrac{-4\ln u}{u^4 + 3}$
27. $y = \dfrac{u + 1}{u^2 + 2}$
28. $y = \dfrac{2u - 1}{u^3 - 1}$
29. $y = \dfrac{u^3 - 1}{u^2 + 3}$
30. $y = \dfrac{\sqrt{u}}{u^2 + e^u + 1}$
31. $y = \dfrac{\sqrt[3]{u}}{u^3 - e^u - 1}$
32. $y = \dfrac{\sqrt[4]{u}}{u^2 + 1}$

Applications and Mathematical Modeling

33. **Revenue** Suppose the demand equation for a certain product is given by $p = \dfrac{1}{1 + x^2}$, where x is the number of items sold and p the price in dollars. Find the marginal revenue.

34. **Cost** It is estimated that the cost, $C(x)$, in millions of dollars, of maintaining the toxic emissions of a certain chemical plant x% free of the toxins is given by

$$C(x) = \frac{4}{100 - x}$$

Find the marginal cost.

35. **Cost** A large corporation already has cornered 20% of the disposable diaper market and now wishes to undertake a very extensive advertising campaign to increase its share of this market. It is estimated that the advertising cost, $C(x)$, in billions of dollars, of attaining an x% share of this market is given by

$$C(x) = \frac{1}{10 - 0.1x}$$

What is the marginal cost?

36. **Aphid Injury in Citrus** Mendoza and colleagues[28] studied aphid injury in citrus. They created a mathematical model given by the equation $y = 57.09\dfrac{x - 195}{x + 2183.33}$, where x is the maximum number of aphids per square meter and y is the percentage yield loss of clementines. Find $\dfrac{dy}{dx}$. What is the sign of the derivative? Does this make sense? Explain. What is happening to the derivative as x becomes larger?

[28] A. Hermoso de Mendoza, B. Belliure, E. A. Carbonell, and V. Real. 2001. *J. Econ. Entomol.* 94(2):439–444.

37. Demand for Northern Cod Grafton[29] created a mathematical model of demand for northern cod and formulated the demand equation

$$p(x) = \frac{173213 + 0.2x}{138570 + x}$$

where p is the price in dollars and x is in kilograms. Find $p'(x)$. What is the sign of the derivative? Is the sign consistent with how a demand curve should behave? Explain.

38. Fishery Revenue Grafton[30] created a mathematical model of demand for northern cod and formulated the demand equation

$$p(x) = \frac{173213 + 0.2x}{138570 + x}$$

where p is the price in dollars and x is in kilograms. Find marginal revenue. What is the sign of the derivative for $x \leq 400{,}000$? Is the sign consistent with how a revenue curve should behave? Explain.

39. Medicine Suppose that the concentration c of a drug in the blood t hours after it is taken orally is given by

$$c = \frac{3t}{2t^2 + 5}$$

What is the rate of change of c with respect to t?

40. Chemistry Salt water with a concentration 0.1 pounds of salt per gallon flows into a large tank that initially holds 100 gallons of pure water. If 5 gallons of salt water per minute flows into the tank, show that the concentration of salt in the tank is given by

$$c(t) = \frac{t}{200 + 10t}$$

where t is measured in minutes. What is the rate of change of c with respect to t?

41. Propensity to Save The amount of money a family saves is a function $S(I)$ of its income I. The *marginal propensity to save* is $\frac{dS}{dI}$. If

$$S(I) = \frac{2I^2}{5(I + 50{,}000)}$$

find the marginal propensity to save.

42. Biology Theoretical studies of photosynthesis[31] assume that the gross rate R of photosynthesis per unit of leaf area is given by

$$R(E) = \frac{aE}{1 + bE}$$

where E is the incident radiation per unit leaf area, and a and b are positive constants. Find the rate of change of R with respect to E.

43. Fishery The cost function for wild crawfish was estimated by Bell[32] to be

$$C(x) = \frac{7.12}{13.74x - x^2}$$

where x is the number of millions of pounds of crawfish caught and C is the cost in millions of dollars. Find the marginal cost.

44. Biology Pilarska[33] used the equation

$$y = 15.97\frac{x}{1.47 + x}$$

to model the ingestion rate y of an individual female rotifer as a function of the density x of green algae it feeds on. Determine the rate of change of the ingestion rate with respect to the density. What is the sign of this rate of change. What is the significance of this sign?

---

29 R. Quentin Grafton, Leif K. Sandal, and Stein Ivar Steinhamn. 2000. How to improve the management of renewable resources: the case of Canada's northern cod fishery. *Amer. J. Agr. Econ.* 82:570–580.

30 Ibid.

31 Duane J. Clow and N. Scott Urquhart. 1974. *Mathematics in Biology*. New York: Ardsley Houser, page 400.

32 Frederick W. Bell. 1986. Competition from fish farming in influencing rent dissipation: the crawfish fishery. *Amer. J. Agr. Econ.* 68:95–101.

33 J. Pilarska. 1977. Eco-physiological studies on Brachionus rubens. *Pol. Arch. Hydrobiol.* 24:319–328.

More Challenging Exercises

45. **a.** Differentiate e^{2x} by writing $e^{2x} = e^x e^x$.
b. Now differentiate e^{3x} by writing $e^{3x} = e^{2x} e^x$ and the result in part (a).
c. Do you see a pattern? What do you think $\frac{d}{dx} e^{nx}$ is when n is an integer?

46. **a.** Differentiate e^{-x} by writing $e^{-x} = \frac{1}{e^x}$.
b. Now differentiate e^{-2x} by writing $e^{-2x} = e^{-x} e^{-x}$ and using part (a).
c. Now differentiate e^{-3x} by writing $e^{-3x} = e^{-2x} e^{-x}$ and using part (b).
d. Do you see a pattern. What do you think $\frac{d}{dx} e^{-nx}$ is when n is an integer?

47. Suppose $f(x) = (x - a)^2 g(x)$ and assume $g(x)$ is differentiable at $x = a$. Show that the x-axis is tangent to the graph of $y = f(x)$ at $x = a$.

48. Show that $\frac{d}{dx}(f \cdot g \cdot h) = f' \cdot g \cdot h + f \cdot g' \cdot h + f \cdot g \cdot h'$, if f, g, and h are differentiable.

49. Let $h(x) = |x| \cdot x$. Can you use the product rule of differentiation to find $h'(0)$? Explain why or why not.

50. **Biological Control** Lysyk[34] studied the effect of temperature on various life history parameters of a parasitic wasp for the purposes of pest management. He constructed a mathematical model given approximately by

$$L(t) = \frac{1}{-1 + 0.03t + \frac{11}{t}}$$

where t is temperature in degrees Celsius and L is median longevity in days of the female. Find $L'(t)$.

In Exercises 51 through 55, find $f'(x)$. (Recall from Exercises 67 and 68 in Section 3.3 that $\frac{d}{dx}(\sin x) = \cos x$ and $\frac{d}{dx}(\cos x) = -\sin x$.)

51. $f(x) = x \sin x$

52. $f(x) = x^2 \cos x$

53. $f(x) = e^x \cos x$

54. $f(x) = (\ln x)(\sin x)$

55. $f(x) = \frac{\sin x}{x}$

56. $f(x) = \frac{\cos x}{e^x}$

57. The trigonometric function $\tan x$ is defined as $\tan x = \frac{\sin x}{\cos x}$. Find $\frac{d}{dx} \tan x$.

58. The trigonometric function $\cot x$ is defined as $\cot x = \frac{\cos x}{\sin x}$. Find $\frac{d}{dx} \cot x$.

[34] T. J. Lysyk. 2001. Relationship between temperature and life history parameters of *Muscidifurax zaraptor*. *Environ. Entomol.* 30(1):147–156.

Modeling Using Least Squares

59. **Costs** Rogers and Akridge[35] of Purdue University studied fertilizer plants in Indiana. For a typical medium-sized plant, they gave data (see their Figure 1) that related average total cost, $\bar{C}(x)$, with cost in dollars per ton. Their data is presented in the following table.

x	4.2	4.8	5.4	6.0	6.6	7.2	7.8
y	298	287	278	270	263	259	255

Here x is production in thousands of tons, and y is average total cost in dollars per ton.

a. Determine the best-fitting quadratic, $q(x)$, using least squares and the square of the correlation coefficient. Graph.
b. Using the quadratic average cost function $q(x)$ found in part (a), find $q'(6)$. Interpret what this means.
c. Let $C(x)$ be total cost. Given that x is given in thousands of tons and that $q(x)$ and $\bar{C}(x)$ are given in dollars per ton, show that $C(x) = 1000\bar{C}(x) \cdot x$. Now replace $\bar{C}(x)$ with $q(x)$, and use the product rule of differentiation to find $C'(6)$. Interpret what this means.
d. Multiply the y values in the table by $1000x$ to obtain values for total cost. Now using this modified data, find the best-fitting quadratic, $Q(x)$, and the square of the correlation coefficient. Then $Q'(6)$ gives $C'(6)$. Compare this value with the one found in part (c).

60. **Costs** Repeat all four parts of Exercise 59, replacing the best-fitting quadratic with the best-fitting logarithmic function, $a + b \ln x$.

[35] Duane S. Rogers and Jay T. Akridge. 1996. Economic impact of storage and handling regulations on retail fertilizer and pesticide plants. *Agribusiness* 12(4):327–337.

61. Costs Repeat all four parts of Exercise 59, replacing the best-fitting quadratic with the best-fitting power function, ax^b. In addition, determine whether quadratic or power regression is best. Which way gives the closest agreement between $C'(6)$ and $Q'(6)$?

Solutions to WARM UP EXERCISE SET 4.2

1. $\frac{d}{dx}[(x^2 - e^x)(x^3 + x + \ln x)]$

$$= (x^2 - e^x)\frac{d}{dx}(x^3 + x + \ln x)$$

$$+ (x^3 + x + \ln x)\frac{d}{dx}(x^2 - e^x)$$

$$= (x^2 - e^x)\left(3x^2 + 1 + \frac{1}{x}\right) + (x^3 + x + \ln x)(2x - e^x)$$

2. $\frac{d}{dx}\left(\frac{x^{3/2} + 1}{x^4 + 1}\right)$

$$= \frac{(x^4 + 1)\frac{d}{dx}(x^{3/2} + 1) - (x^{3/2} + 1)\frac{d}{dx}(x^4 + 1)}{(x^4 + 1)^2}$$

$$= \frac{(x^4 + 1)\left(\frac{3}{2}x^{1/2}\right) - (x^{3/2} + 1)(4x^3)}{(x^4 + 1)^2}$$

4.3 The Chain Rule

The Chain Rule
The General Power Rule
Derivatives of Complex Expressions
Enrichment: Derivative of a Power Function with Rational Exponent

APPLICATION Spread of the Black Death

On the basis of some theoretical analysis, Murray[36] determined that the velocity in miles per year of the spread of the Black Death in Europe from 1347 to 1350 was approximated by the equation $V = 200\sqrt{0.008x - 15}$ if $x \geq 1875$ and $V(x) = 0$ if $0 \leq x \leq 1875$, where x is the population density measured in people per square mile. What is $\frac{dV}{dx}$ when $x > 1875$, the sign of this quantity when $x > 1875$, and the significance of this sign? For the answer, see Example 4 on page 259.

[36] J. D. Murray. 1989. *Mathematical Biology*. New York: Springer-Verlag.

The Chain Rule

Consider the situation in Figure 4.6, where $y = (g \circ f)(x) = g[f(x)]$ is the composition of the two functions $f(x)$ and $g(x)$. We wish to find a formula for the derivative of $(g \circ f)(x)$ in terms of the derivatives of $f(x)$ and $g(x)$. This formula is perhaps the most important rule of differentiation.

To gain some insight into how this formula can be obtained, let us set $u = f(x)$. Then we have $y = g(u)$. Since we are looking for the derivative of y with respect

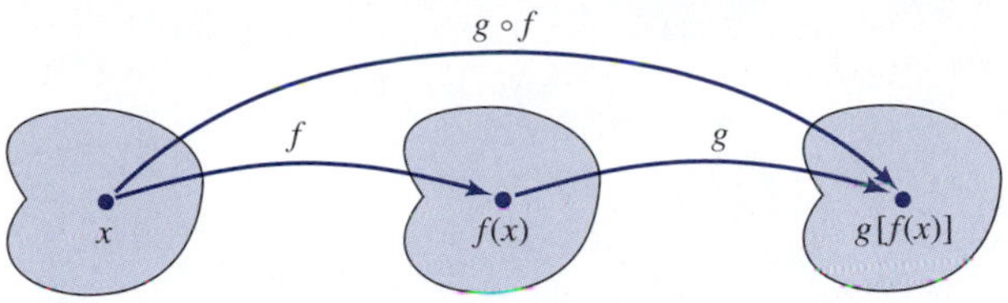

Figure 4.6
The composition of two functions.

to x, we need to determine the relationship between the change in y and the change in x. We can write

$$\begin{aligned}\frac{\Delta y}{\Delta x} &= \frac{\text{change in } y}{\text{change in } x}\\ &= \frac{\text{change in } y}{\text{change in } u} \cdot \frac{\text{change in } u}{\text{change in } x}\\ &= \frac{\Delta y}{\Delta u} \cdot \frac{\Delta u}{\Delta x}\end{aligned}$$

This formula suggests that

$$\frac{dy}{dx} = \frac{dy}{du} \cdot \frac{du}{dx}$$

This can be shown to be true, but a correct proof is too complicated to include here. We now state two forms of the chain rule.

Chain Rule

If $y = g(u)$ and $u = f(x)$ and both of these functions are differentiable, then the composite function $y = g[f(x)]$ is differentiable, and

$$\frac{dy}{dx} = \frac{dy}{du} \cdot \frac{du}{dx}$$

or

$$\frac{d}{dx} g[f(x)] = g'[f(x)] \cdot f'(x)$$

To obtain some geometric insight to the chain rule, let us specify x in the last formula to be $x = c$. Let $h(x) = g[f(x)]$. Then we have

$$h'(c) = (g \circ f)'(c) = g'[f(c)] \cdot f'(c)$$

Figure 4.7a shows a graph of $u = f(x)$ and the tangent line at $(c, f(c))$. The slope of this tangent line is $f'(c)$. Figure 4.7b shows a graph of $y = g(u)$ and the tangent line at $u = f(c)$. The slope of this tangent line is $g'(f(c))$. Now $h'(c) = (g \circ f)'(c)$ is just the product of these two slopes, that is,

$$\begin{aligned}h(c) &= g(f(c))\\ h'(c) &= g'[f(c)] \cdot f'(c)\end{aligned}$$

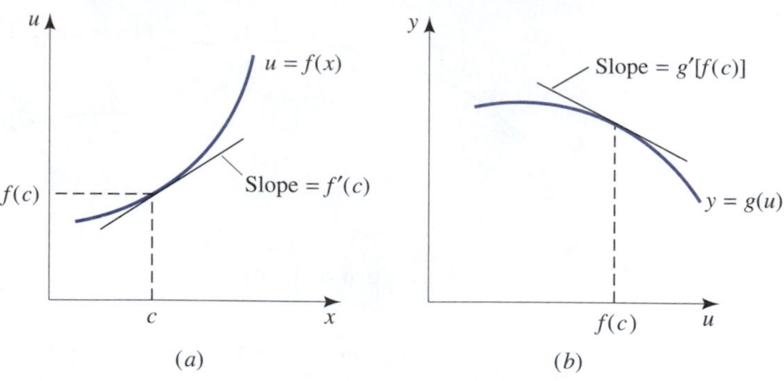

Figure 4.7
If $h(c) = g(f(c))$, then $h'(c) = g'(f(c)) \cdot f'(c)$.

Interactive Illustration 4.5

Chain Rule

This interactive illustration allows you to better visualize the chain rule formula. You are given $y = g(u) = u^2$, $u = f(x) = 1/x$, and $y = h(x) = g(f(x)) = (1/x)^2$. You will be able to follow on the graphics how the composition function operates. Click on the Chain Rule box. Take specific values of x such as 0.8, 1.0, 1.2, and 1.5. Follow the trail of the composition and compare the two sides of the chain rule formula.

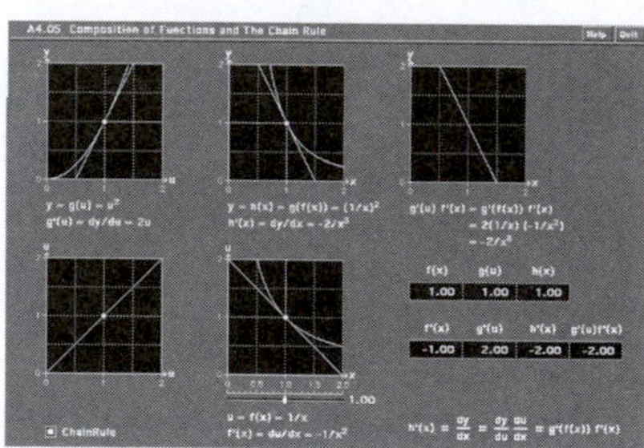

EXAMPLE 1 **Verifying the Chain Rule**

The function $y = 8x - 28 = 4(2x - 7)$ can be written as $y = g[f(x)]$, where $y = g(u) = 4u$ and $u = f(x) = 2x - 7$. Verify the chain rule in this case.

Solution

We have

$$\frac{dy}{dx} = 8, \qquad \frac{dy}{du} = 4, \qquad \frac{du}{dx} = 2$$

Since $8 = 4 \cdot 2$,

$$\frac{dy}{dx} = \frac{dy}{du} \cdot \frac{du}{dx}$$

■

EXAMPLE 2 **Chain Rule and Units**

Ruzycki and coworkers[37] studied the growth habits of lake trout in Bear Lake, Utah-Idaho. They created a mathematical model that gave the weight $W(L)$ of the trout

[37] James R. Ruzycki, Wayne A. Wurtsbaugh, and Chris Luecke. 2001. Salmonine consumption and competition for endemic prey fishes in Bear Lake, Utah-Idaho. *Trans Amer. Fish. Soc.* 130: 1175–1189.

in grams as a function of the length in millimeters. They also gave an equation that gave the length $L(t)$ of the trout in millimeters as a function of age t in years. Using their equations, we have that

$$\begin{aligned} L(10) &= 686 \text{ mm} \\ W(686) &= 3178 \text{ g} \\ L'(10) &= 33 \text{ mm/year} \\ W'(686) &= 15 \text{ g/mm} \end{aligned}$$

(a) Find the weight of a 10-year-old trout.
(b) Find the instantaneous rate of change of weight with respect to age for a 10-year-old trout.

Solution

(a) We must have $W(t) = W(L(t))$, so

$$\begin{aligned} W(10) &= W(L(10)) \\ &= W(686) \\ &= 3178 \end{aligned}$$

grams.

(b) From the chain rule we have

$$W'(t) = \frac{dW}{dt} = \frac{dW}{dL} \cdot \frac{dL}{dt}$$

Since $L(10) = 686$, we then have

$$\begin{aligned} W'(10) &= \frac{dW}{dL} \cdot \frac{dL}{dt} \\ &= \left(15 \, \frac{\text{grams}}{\text{mm}}\right) \cdot \left(33 \, \frac{\text{mm}}{\text{year}}\right) \\ &= 495 \, \frac{\text{grams}}{\text{year}} \end{aligned}$$

That is a 10-year-old trout is gaining weight at the rate of 495 grams per year. ■

EXAMPLE 3 **Using the Chain Rule**

Find the derivative of $y = (x^2 + 3)^{78}$.

Solution

We could expand this function using the binomial theorem, obtain a polynomial of degree 156, and then differentiate, a very time-consuming process. An alternative way of finding the derivative is to use the chain rule. First recognize that this function is *something* raised to the 78th power, where the something is $(x^2 + 3)$. This suggests that y can be viewed as the composition of two functions; the first function takes x into $x^2 + 3$, and the second function raises this quantity to the 78th power. In terms of the chain rule, set $y = g(u) = u^{78}$ and $u = f(x) = x^2 + 3$. Then

$$y = g(u) = g[f(x)] = [f(x)]^{78} = (x^2 + 3)^{78}$$

Since

$$\frac{dy}{du} = \frac{d}{du}(u^{78}) = 78u^{77} \quad \text{and} \quad \frac{du}{dx} = \frac{d}{dx}(x^2 + 3) = 2x$$

the chain rule yields

$$\begin{aligned} \frac{dy}{dx} &= \frac{dy}{du} \cdot \frac{du}{dx} \\ &= 78u^{77} \cdot (2x) \\ &= 78(x^2 + 3)^{77} 2x \\ &= 156x(x^2 + 3)^{77} \end{aligned}$$

■

The General Power Rule

Example 3 suggests how to take the derivative of any *function* raised to any power. Now we see how to find the derivative of $[f(x)]^n$, where n is any real number. Here we recognize $y = [f(x)]^n$ as the composition of two functions, the first function takes x to $f(x)$ and the second function raises $f(x)$ to the nth power. If we set $y = g(u) = u^n$ and $u = f(x)$, then

$$y = g[f(x)] = [f(x)]^n$$

Then the chain rule yields

$$\begin{aligned} \frac{dy}{dx} &= \frac{dy}{du} \cdot \frac{du}{dx} \\ &= nu^{n-1} \cdot \frac{d}{dx} f(x) \\ &= n[f(x)]^{n-1} \cdot \frac{d}{dx} f(x) \end{aligned}$$

We state this result as the general power rule.

General Power Rule

If n is any real number and $f'(x)$ exists, then

$$\frac{d}{dx}[f(x)]^n = n[f(x)]^{n-1} \cdot \frac{d}{dx} f(x)$$

Remark You might think of the general power rule as: Power in front of function, reduce power of function by one, times derivative of function.

Warning A common error is to ignore the $\frac{d}{dx} f(x)$ term in the general power formula.

EXAMPLE 4 Using the General Power Rule

On the basis of some theoretical analysis, Murray[38] determined that the velocity in miles per year of the spread of the Black Death in Europe from 1347 to 1350 was approximated by the equation $V = 200\sqrt{0.008x - 15}$ if $x \geq 1875$ and $V(x) = 0$ if $0 \leq x \leq 1875$, where x is the population density measured in people per square mile. Find $\frac{dV}{dx}$ when $x > 1875$. Determine both the sign when $x > 1875$ and the significance of this sign. Graph.

Solution

We first note that $0.008x - 15 \geq 0$ if $x \geq 1875$. Rewrite V as $V = 200(0.008x - 15)^{1/2}$. Using the general power rule when $f(x) = 0.008x - 15$ and $n = 1/2$ yields for $x > 1875$

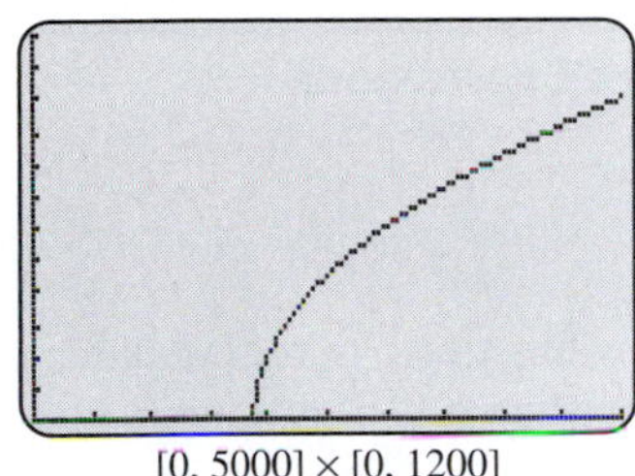

$[0, 5000] \times [0, 1200]$

Screen 4.10
As the density $y = 200\sqrt{0.008x - 15}$ of population increases, the plague begins to spread and with increasing velocity.

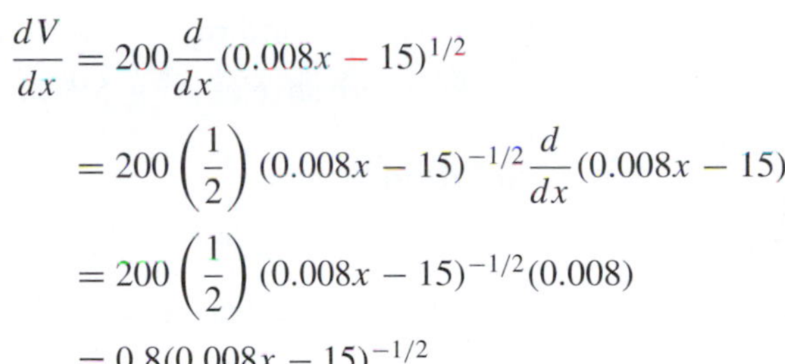

$$\begin{aligned}\frac{dV}{dx} &= 200\frac{d}{dx}(0.008x - 15)^{1/2}\\ &= 200\left(\frac{1}{2}\right)(0.008x - 15)^{-1/2}\frac{d}{dx}(0.008x - 15)\\ &= 200\left(\frac{1}{2}\right)(0.008x - 15)^{-1/2}(0.008)\\ &= 0.8(0.008x - 15)^{-1/2}\end{aligned}$$

When $x > 1875$, the derivative is positive, indicating that the velocity increases with an increase in density. This is reasonable for an infectious disease. (To transmit this disease, fleas must jump from rats to humans and humans must be close enough to infect other humans.) Screen 4.10 shows a graph using a window of dimensions $[0, 5000]$ by $[0, 1200]$. ■

EXAMPLE 5 Using the General Power Rule

Find the derivative of $\frac{1}{(x^3 + x - 5)^4}$ using the general power rule.

Solution

We must recognize this as some function raised to a power. The power is *negative* four. Then we can use the general power formula to obtain

$$\begin{aligned}\frac{d}{dx}\left[\frac{1}{(x^3 + x - 5)^4}\right] &= \frac{d}{dx}\left[\left(x^3 + x - 5\right)^{-4}\right]\\ &= -4\left(x^3 + x - 5\right)^{-5}\frac{d}{dx}\left(x^3 + x - 5\right)\\ &= -4(x^3 + x - 5)^{-5}(3x^2 + 1)\end{aligned}$$ ■

Notice that the general power rule reduces to the ordinary power rule discussed in a previous section. Thus, using the general power rule yields

[38] J. D. Murray. 1989. *Mathematical Biology*. New York: Springer-Verlag.

$$\begin{aligned}\frac{d}{dx}(x^n) &= nx^{n-1}\frac{d}{dx}(x)\\ &= nx^{n-1}(1)\\ &= nx^{n-1}\end{aligned}$$

Derivatives of Complex Expressions

We now consider some examples that use the general power rule together with other differentiation rules learned earlier.

EXAMPLE 6 **Finding the Derivatives of Complex Expressions**

Find the derivative of $(x^3+5)^{10}\ln x$.

Solution

We first think of this expression as the product of two functions $f(x)g(x)$, where $f(x)=(x^3+5)^{10}$ and $g(x)=\ln x$, and then use the product rule of differentiating. We then use the general power rule where necessary. Thus, we obtain

$$\begin{aligned}f'(x) &= \frac{d}{dx}\left((x^3+5)^{10}\cdot\ln x\right)\\ &= (x^3+5)^{10}\frac{d}{dx}(\ln x)+(\ln x)\frac{d}{dx}(x^3+5)^{10} && \text{Product rule}\\ &= (x^3+5)^{10}\frac{1}{x}+(\ln x)10(x^3+5)^9\frac{d}{dx}(x^3+5) && \text{General power rule}\\ &= (x^3+5)^{10}\frac{1}{x}+(\ln x)10(x^3+5)^9(3x^2)\\ &= (x^3+5)^{10}\frac{1}{x}+30x^2(x^3+5)^9\ln x\end{aligned}$$

■

EXAMPLE 7 **Finding the Derivative of Complex Expressions**

Find the marginal cost if $C(x)=\dfrac{e^x}{\sqrt{x^2+1}}$.

Solution

$$\begin{aligned}C'(x) &= \frac{d}{dx}\left(\frac{e^x}{(x^2+1)^{1/2}}\right)\\ &= \frac{(x^2+1)^{1/2}\frac{d}{dx}(e^x)-e^x\frac{d}{dx}\left[(x^2+1)^{1/2}\right]}{\left[(x^2+1)^{1/2}\right]^2} && \text{Quotient rule}\\ &= \frac{(x^2+1)^{1/2}e^x-e^x\frac{1}{2}(x^2+1)^{-1/2}\frac{d}{dx}(x^2+1)}{(x^2+1)} && \text{General power rule}\end{aligned}$$

$$= \frac{(x^2+1)^{1/2}e^x - e^x\frac{1}{2}(x^2+1)^{-1/2}(2x)}{(x^2+1)}$$

$$= \frac{(x^2+1)^{1/2}e^x - xe^x(x^2+1)^{-1/2}}{x^2+1}$$

$$= e^x(x^2+1)^{-1/2}\frac{(x^2+1)-x}{x^2+1}$$

$$= e^x\frac{x^2-x+1}{(x^2+1)^{3/2}}$$ ■

To find the derivative of the cost function $\frac{e^x}{\sqrt{x^2+1}}$ in Example 7, we viewed this expression as a quotient and then used the quotient rule for differentiation. Actually, any quotient $\frac{f(x)}{g(x)}$ can be viewed as the *product* $f(x)\cdot[g(x)]^{-1}$ of the two function $f(x)$ and $[g(x)]^{-1}$. To find the derivative, you can then use the *product* rule of differentiation. You are invited to do this in the next Exploration.

Exploration 1

Alternative Way of Doing Example 7

Rewrite the cost function $\frac{e^x}{\sqrt{x^2+1}}$ in Example 7 as the product $e^x(x^2+1)^{-1/2}$. Now find marginal cost using the product rule of differentiation.

Enrichment: Derivative of Power with Rational Exponent

We can now give a proof of the power rule when the power is a rational number. That is, we now give a proof of $\frac{d}{dx}(x^r) = rx^{r-1}$, when r is a rational number. Let $r = n/m$, where n and m are integers, and let $f(x) = x^r$. Then $f(x) = x^{n/m}$, and

$$[f(x)]^m = x^n$$

$$m[f(x)]^{m-1}\cdot f'(x) = nx^{n-1}$$

$$m[x^{n/m}]^{m-1}\cdot f'(x) = nx^{n-1}$$

$$f'(x) = \frac{n}{m}x^{n-1}(x^{n/m})^{1-m}$$

$$= \frac{n}{m}x^{n-1}x^{n/m}x^{-n}$$

$$= \frac{n}{m}x^{n/m-1}$$

$$= rx^{r-1}$$

Warm Up Exercise Set 4.3

1. Find the derivative of $\sqrt{3x^2 + 2e^x}$.
2. Use the general power rule and the product rule to derive the quotient rule. Hint: Write

$$\frac{f(x)}{g(x)} = f(x) \cdot [g(x)]^{-1}$$

Exercise Set 4.3

In Exercises 1 through 22, find the derivative.

1. $(2x + 1)^7$
2. $(3x - 2)^{11}$
3. $(x^2 + 1)^4$
4. $(1 - x^2)^5$
5. $(2e^x - 3)^{3/2}$
6. $(3 - x)^{2/3}$
7. $\sqrt{e^x}$
8. e^{4x}
9. $\sqrt{2x + 1}$
10. $\sqrt{1 - 3x}$
11. $6\sqrt[3]{x^2 + 1}$
12. $-15\sqrt[5]{2x^3 - 3}$
13. $\dfrac{1}{e^x + 1}$
14. $\dfrac{4}{x^2 + 1}$
15. $-\dfrac{2}{x^3 + 1}$
16. $\dfrac{1}{\sqrt{x + 1}}$
17. $\dfrac{\ln x}{\sqrt{x^2 + 1}}$
18. $\dfrac{3}{\sqrt{e^x + 1}}$
19. $(2x + 1)(3x - 2)^5$
20. $(2x - 3)^5(4x + 7)$
21. $(\ln x)^4$
22. $1/\sqrt{\ln x}$

In Exercises 23 through 34, find $\dfrac{dy}{dx}$.

23. $y = e^x(x^2 + 1)^8$
24. $y = (1 - x^3)^4 \ln x$
25. $y = (1 - x)^4(2x - 1)^5$
26. $y = (7x + 3)^3(x^2 - 4)^6$
27. $y = \sqrt{x}(4x + 3)^3$
28. $y = -3\sqrt{x}(1 - x^3)^3$
29. $y = e^x(\ln x)^2$
30. $y = e^{2x}\sqrt{\ln x}$
31. $y = \dfrac{3}{(x + 1)^4}$
32. $y = \dfrac{-3}{(3e^x + 1)^3}$
33. $y = \dfrac{e^x + 1}{(2x + 3)^3}$
34. $y = \dfrac{x^3 + x^2 - 1}{(x + 2)^5}$
35. $f(x) = \sqrt{\sqrt{x} + 1}$
36. $f(x) = \sqrt[3]{\sqrt{x} + 1}$

Applications and Mathematical Modeling

37. **Demand** The demand equation for a certain product is given by

$$p = \frac{1}{\sqrt{1 + x^2}}$$

where x is the number of items sold and p is the price in dollars. Find the instantaneous rate of change of p with respect to x.

38. **Revenue** Find the instantaneous rate of change of revenue with respect to the number sold if the demand equation is given in the previous problem.

39. **Growth of the Northern Cod Fishery** Grafton[39] created a mathematical model of the growth rate for northern cod and formulated the equation

$$g(x) = 0.30355x\left(1 - \frac{x}{3.2}\right)^{0.35865}$$

where x is the biomass of northern cod in millions of tons and $g(x)$ is the growth of the biomass over the next year. Find where $g(x)$ is positive, negative, and zero. What does this say about the growth of the fishery? Explain. Find $g'(x)$. Determine where $g'(x)$ is positive and where it is negative. What does this have to say about the growth of the fishery. Explain.

40. **Forestry** The volume, V, in board feet of certain trees was found to be given by $V = 10 + .007(d - 5)^3$ for $d > 10$ where d is the diameter in inches.[40] Find the rate of change of V with respect to d when $d = 12$.

41. **Fishery** Let n denote the number of prey in a school of fish, let r denote the detection range of the school by a predator, and let c denote a constant that defines the average spacing of fish within the school. Biologists[41] have given the volume of the region within which a predator can detect the school, called the *visual volume*, as

$$V = \frac{4\pi r^3}{3}\left(1 + \frac{c}{r}n^{1/3}\right)^3$$

[39] R. Quentin Grafton, Leif K. Sandal, and Stein Ivar Steinhamn. 2000. How to improve the management of renewable resources: the case of Canada's northern cod fishery. *Amer. J. Agr. Econ.* 82:570–580.

[40] Duane J. Clow and N. Scott Urquhart. 1974. *Mathematics in Biology.* New York: Ardsley House.

[41] Colin W. Clark. 1976. *Mathematical Bioeconomics.* New York: John Wiley & Sons.

With c and r considered constants, one then has $V = V(n)$. Find the rate of change of V with respect to n.

42. Fishery In Exercise 41, suppose c and n are constants. Thus, $V = V(r)$. Find the rate of change of V with respect to r.

43. Psychology The psychologist L. L. Thurstone suggested that the relationship between the learning time T of memorizing a list of length n is a function of the form $T = an\sqrt{n-b}$, where a and b are positive constants. Find the rate of change of T with respect to n.

44. Biology Chan et al.[42] showed that the pupal weight y in milligrams of the Mediterranean fruit fly was approximated by $y = 5.5\sqrt{\ln x}$, where x is the percent protein level in the diet. Find the rate of change of weight with respect to change in percent protein level.

[42] Harvey Chan, James Hansen, and Stephen Tam. 1990. Larval diets from different protein sources for Mediterranean fruit flies. *J. Econ. Entomol.* 83:1954–1958.

More Challenging Exercises

45. Find $\dfrac{dy}{dx}$ if $y = \sqrt{\sqrt{\sqrt{x}+1}+1}$.

46. Suppose $y = f(x)$ is a differentiable function and $f'(4) = 7$. Let $h(x) = f(x^2)$. Find $h'(2)$.

47. Study the solution to Warm Up Exercise 2, and then do Exercise 33 *without* using the quotient rule.

48. Study the solution to Warm Up Exercise 2. Now do Exercise 34 without using the quotient rule.

For Exercises 49 through 51, recall the derivatives of $\sin x$ and $\cos x$ you found in Exercises 67 and 68 in Section 3.3 and find $f'(x)$.

49. $f(x) = (\sin x)^3$

50. $f(x) = (\cos x)^4$

51. $f(x) = \sqrt{\cos x}$

52. Biological Control Lysyk[43] studied the effect of temperature on various life history parameters of a parasitic wasp for the purposes of pest management. He constructed a mathematical model given approximately by

$$D(t) = 0.00007t(t-13)(36-t)^{0.4}$$

where t is temperature in degrees Celsius and D is the reciprocal of development time in days. Find $D'(t)$. (Hint: You might wish to refer to the formula $\dfrac{d}{dx}(f \cdot g \cdot h) = f' \cdot g \cdot h + f \cdot g' \cdot h + f \cdot g \cdot h'$ found in Exercise 48 of Section 4.2.)

53. Biological Control Lysyk[44] (refer to previous exercise) also constructed a mathematical model given approximately by the equation $r(t) = 0.0002t(t-14)\sqrt{36-t}$, where t is temperature in degrees Celsius and r is the intrinsic rate of increase (females per females per day). Find $r'(t)$.

[43] T. J. Lysyk. 2001. Relationship between temperature and life history parameters of *Muscidifurax zaraptor*. *Environ. Entomol.* 30(1):147–156.

[44] Ibid.

Modeling Using Least Squares

54. Darwin's Finch Productivity Grant and colleagues[45] studied the effect of rainfall on Darwin's finch productivity. They collected data found in the table relating the amount of rainfall with hatchling success.

x	24.5	38.5	44.7	54.6	60.3
p	0	0.27	0.25	0.75	0.23
x	81.5	90.0	134.3	141.2	181.3
p	0.45	0.83	0.65	0.74	0.65
x	165.1	897.8	1408.1		
p	0.63	0.59	0.40		

where x is rainfall in millimeters and p is the fraction of successful hatchlings. (The last four data points were years of El Niño conditions.)

[45] Peter R. Grant, B. Rosemary Grant, Lukas F. Keller, and Kenneth Petren. 2000. Effects of El Nino events on Darwin's finch productivity. *Ecology* 81(9):2442–2457.

a. Take the natural logarithm of all the x values, giving $u = \ln x$, as the researchers did. Then find the best-fitting quadratic of hatchling success versus the natural logarithm of rainfall, as the researchers did, obtaining $q(u) = au^2 + bu + c$.

b. From part (a) you have a best-fitting quadratic $q(x) = a(\ln x)^2 + b\ln x + c$ as a function of rainfall x. Find $q'(x)$.

Solutions to WARM UP EXERCISE SET 4.3

1. We must recognize this expression as some function raised to a power. In this case the power is $n = \frac{1}{2}$. Using the general power rule, we obtain

$$\begin{aligned}\frac{d}{dx}\left(\sqrt{3x^2+2e^x}\right) &= \frac{d}{dx}\left[\left(3x^2+2e^x\right)^{1/2}\right] \\ &= \frac{1}{2}\left(3x^2+2e^x\right)^{-1/2}\frac{d}{dx}\left(3x^2+2e^x\right) \\ &= \frac{1}{2}\left(3x^2+2e^x\right)^{-1/2}(6x+2e^x) \\ &= (3x+e^x)\left(3x^2+2e^x\right)^{-1/2}\end{aligned}$$

2. First apply the product rule, and then use the general power rule when appropriate to obtain

$$\begin{aligned}&\frac{d}{dx}\left(\frac{f(x)}{g(x)}\right) \\ &= \frac{d}{dx}\left(f(x)\cdot[g(x)]^{-1}\right) \\ &= [g(x)]^{-1}\cdot\frac{d}{dx}f(x) + f(x)\cdot\frac{d}{dx}\left([g(x)]^{-1}\right) \qquad \text{Product rule} \\ &= [g(x)]^{-1}\cdot\frac{d}{dx}f(x) + f(x)\left(-[g(x)]^{-2}\frac{d}{dx}g(x)\right) \qquad \text{General power rule} \\ &= \frac{\frac{d}{dx}f(x)}{g(x)}\cdot\frac{g(x)}{g(x)} - \frac{f(x)\frac{d}{dx}g(x)}{g^2(x)} \\ &= \frac{g(x)\cdot\frac{d}{dx}f(x) - f(x)\cdot\frac{d}{dx}g(x)}{[g(x)]^2}\end{aligned}$$

4.4 Derivatives of Exponential and Logarithmic Functions

Derivatives of Exponential Functions

Derivatives of Logarithmic Functions

Enrichment: Differentiating Using Properties of Logarithms

Enrichment: Proofs

APPLICATION Survival Curves

A survival curve for white males in the United States in the period 1969–1971 was determined by Elandt-Johnson and Johnson[46] to be approximately $y = 0.99e^{-0.001t-0.0001t^2}$, where t is measured in years. What is the rate of change with respect to t and what is the sign of this rate of change? What is the significance of this sign? See Example 2 on page 265 for the answer.

[46] R. C. Elandt-Johnson and N. L. Johnson. 1980. *Survival Models and Data Analysis*. New York: John Wiley & Sons.

Derivatives of Exponential Functions

We already know that the derivative of e^x is again e^x, that is,

$$\frac{d}{dx}\left(e^x\right) = e^x$$

We need to be able to differentiate functions of the form $y = e^{f(x)}$. To do this, set $y = e^u$ where $u = f(x)$; then

$$\frac{dy}{du} = \frac{d}{du}e^u = e^u$$

Now use the chain rule and obtain

$$\begin{aligned}\frac{d}{dx}\left(e^{f(x)}\right) &= \frac{dy}{dx} \\ &= \frac{dy}{du} \cdot \frac{du}{dx} \\ &= e^u \frac{du}{dx} \\ &= e^{f(x)} f'(x)\end{aligned}$$

The Derivative of $e^{f(x)}$

$$\frac{d}{dx}\left(e^{f(x)}\right) = e^{f(x)} f'(x)$$

EXAMPLE 1 **The Derivatives of e Raised to a Function**

Find the derivative of $f(x) = e^{x^3+4x^2+x-5}$.

Solution

We have

$$\begin{aligned}f'(x) &= \frac{d}{dx}\left(e^{x^3+4x^2+x-5}\right) \\ &= e^{x^3+4x^2+x-5} \frac{d}{dx}(x^3 + 4x^2 + x - 5) \\ &= e^{x^3+4x^2+x-5}(3x^2 + 8x + 1)\end{aligned}$$

■

EXAMPLE 2 **The Derivatives of e Raised to a Function**

A survival curve for white males in the United States in the period 1969–1971 was determined by Elandt-Johnson and Johnson[47] to be $y = 0.99e^{-0.001t-0.0001t^2}$, where t is measured in years. Determine the rate of change with respect to t and the sign of this rate of change. What is the significance of this sign? Use your grapher to graph.

[47] Ibid.

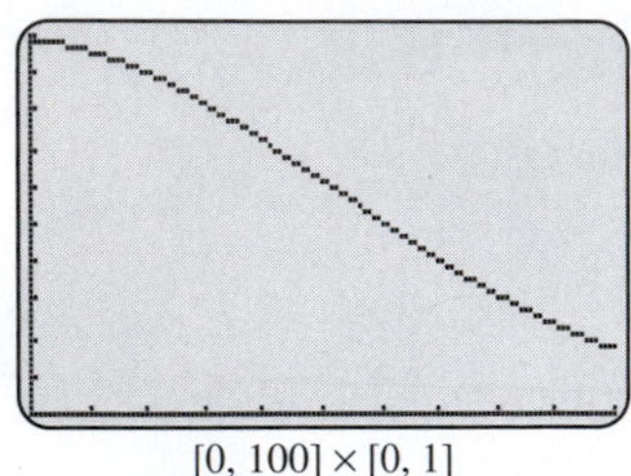

[0, 100] × [0, 1]

Screen 4.11
Notice how the curve $y = 0.99e^{-0.001t-0.0001t^2}$ drops increasingly faster and then levels out.

Solution

Using the above formula with $f(t) = -0.001t - 0.0001t^2$ yields

$$\begin{aligned}\frac{d}{dt}\left(0.99e^{-0.001t-0.0001t^2}\right) &= 0.99e^{-0.001t-0.0001t^2}\frac{d}{dt}(-0.001t - 0.0001t^2)\\ &= 0.99e^{-0.001t-0.0001t^2}(-0.001 - 0.0002t)\\ &= -0.99(0.001 + 0.0002t)e^{-0.001t-0.0001t^2}\end{aligned}$$

The rate of change is always negative. This simply means that the number of survivors is decreasing at all times. Using a window with dimensions [0, 100] by [0, 1] we obtain Screen 4.11. Curves such as these are used by insurance companies. ■

EXAMPLE 3 **Derivative of Function Containing Exponentials**

The equation

$$p(t) = \frac{200}{1 + 50e^{-0.03t}}$$

was used about 150 years ago by the mathematical-biologist P. Verhulst to model the population $p(t)$ of the United States in millions, where t is the calendar year and $t = 0$ corresponds to 1790. Find the rate of change of this population with respect to time and the sign of this rate of change. What is the significance of this sign? Use your computer or graphing calculator to graph.

Solution

Using the power rule for derivatives, we have

$$\begin{aligned}\frac{d}{dt}\left(\frac{200}{1 + 50e^{-0.03t}}\right) &= \frac{d}{dt}\left(200(1 + 50e^{-0.03t})^{-1}\right)\\ &= 200(-1)(1 + 50e^{-0.03t})^{-2}\frac{d}{dt}(1 + 50e^{-0.03t})\\ &= -200(1 + 50e^{-0.03t})^{-2}(0 + 50e^{-0.03t}\frac{d}{dt}(-0.03t))\\ &= -200(1 + 50e^{-0.03t})^{-2}50e^{-0.03t}(-0.03)\\ &= \frac{(200)(50)(.03)e^{-0.03t}}{(1 + 50e^{-0.03t})^2}\\ &= \frac{300e^{-0.03t}}{(1 + 50e^{-0.03t})^2}\end{aligned}$$

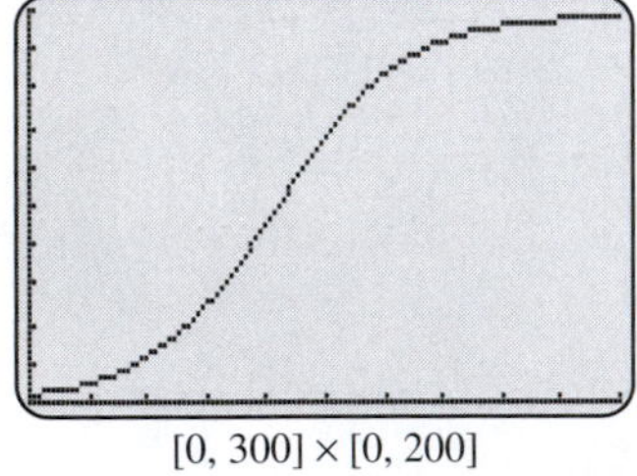

[0, 300] × [0, 200]

Screen 4.12
The population $y = \dfrac{200}{1 + 50e^{-0.03t}}$ initially increases fast, but then levels out.

Note that this quantity is positive for all time. This indicates that the population is increasing at all times. A graph is shown in Screen 4.12 using a window with dimensions [0, 300] by [0, 200]. ■

EXAMPLE 4 **Marginal Revenue**

Let the demand equation be given $p(x) = e^{-0.02x}$. Find the marginal revenue. Find where the marginal revenue is zero.

Solution

We have $R(x) = xp(x) = xe^{-0.02x}$. Thus,

$$\begin{aligned} R'(x) &= \frac{d}{dx}\left(x \cdot e^{-0.02x}\right) \\ &= e^{-0.02x}\frac{d}{dx}(x) + x\frac{d}{dx}\left(e^{-0.02x}\right) \\ &= e^{-0.02x}(1) + x(-0.02e^{-0.02x}) \\ &= e^{-0.02x}(1 - 0.02x) \end{aligned}$$

This is the marginal revenue. Now recall that e^z is never zero, no matter what the value of z is. Thus, marginal revenue $R'(x) = e^{-0.02x}(1 - 0.02x)$ will be zero only when $(1 - 0.02x) = 0$ or $x = 50$. ■

We have already seen how to differentiate the expression $e^{f(x)}$. We now see how to differentiate the expression $a^{f(x)}$ for any positive value of a. First recall that for any $a > 0$, $a^x = e^{x \ln a}$. Differentiating this yields

$$\begin{aligned} \frac{d}{dx}(a^x) &= \frac{d}{dx}\left(e^{x \ln a}\right) \\ &= e^{x \ln a}\frac{d}{dx}(x \ln a) \\ &= e^{x \ln a}(\ln a) \\ &= a^x \ln a \end{aligned}$$

Using the chain rule, we can write the more general form of this as follows.

The Derivative of $a^{f(x)}$

$$\frac{d}{dx}\left(a^{f(x)}\right) = a^{f(x)} f'(x) \ln a$$

EXAMPLE 5 **Finding the Derivative of $a^{f(x)}$**

Find $\frac{d}{dx}\left(2^{x^2+x+1}\right)$.

Solution

We have from the previous formula

$$\begin{aligned} \frac{d}{dx}\left(2^{x^2+x+1}\right) &= 2^{x^2+x+1}\frac{d}{dx}(x^2 + x + 1) \cdot \ln 2 \\ &= 2^{x^2+x+1}(2x + 1) \ln 2 \end{aligned}$$

■

Derivatives of Logarithmic Functions

We have already determined graphically that the derivative of $\ln x$ is $1/x$. We now see how to determine the derivative analytically using the fact that the derivative of e^x is again e^x.

To show this, suppose that $x > 0$; then $x = e^{\ln x}$. Now we differentiate this last expression and obtain

$$\frac{d}{dx}(x) = \frac{d}{dx}e^{\ln x}$$

which is just

$$1 = e^{\ln x}\frac{d}{dx}(\ln x)$$

Then

$$\begin{aligned}\frac{d}{dx}(\ln x) &= \frac{1}{e^{\ln x}} \\ &= \frac{1}{x}\end{aligned}$$

We can use this formula to find the derivative of an expression such as $y = \ln f(x)$. The details are given in an Enrichment subsection. We find the following.

Derivative of Logarithmic Function

$$\frac{d}{dx}(\ln|f(x)|) = \frac{f'(x)}{f(x)}$$

Remark Notice that the derivative of an expression such as $\ln|f(x)|$ has *no* logarithm in the answer.

Exploration 1

Derivatives of Logarithmic Functions

Take the derivative of $f(x) = \ln(kx^n)$, where $k > 0$, $x > 0$, and n is any real number. Where does the k appear? Surprised?

EXAMPLE 6 **Derivative of a Logarithmic Function**

Find the derivative of $\ln|x^2 + x + 1|$.

Solution

Use the above formula with $f(x) = x^2 + x + 1$. Then

$$\begin{aligned}\frac{d}{dx}\left(\ln|x^2 + x + 1|\right) &= \frac{\frac{d}{dx}(x^2 + x + 1)}{x^2 + x + 1} \\ &= \frac{2x + 1}{x^2 + x + 1}\end{aligned}$$

■

The last differentiation formula we give in this section is the following.

$$\frac{d}{dx}\log_a |f(x)| = \frac{f'(x)}{f(x)\ln a}$$

EXAMPLE 7 **Derivative of Logarithm to Base *a***

Find $\dfrac{d}{dx}\log x^2$.

Solution

We use the last formula and obtain

$$\begin{aligned}\frac{d}{dx}\log x^2 &= \frac{d}{dx}\log_{10} x^2\\ &= \frac{2x}{x^2\ln 10}\\ &= \frac{2}{x\ln 10}\end{aligned}$$

Enrichment: Differentiating Using Properties of Logarithm

We can make the task of differentiating expressions involving the logarithm easier by exploiting the properties of logarithms. The following example illustrates this.

EXAMPLE 8 **Differentiating Using Properties of Logarithm**

Find the derivative of $\ln\dfrac{x+1}{x^2+1}$ for $x > -1$.

Solution

Use the fact that

$$\ln\frac{x+1}{x^2+1} = \ln(x+1) - \ln(x^2+1)$$

Then

$$\begin{aligned}\frac{d}{dx}\left(\ln\frac{x+1}{x^2+1}\right) &= \frac{d}{dx}(\ln(x+1)) - \frac{d}{dx}\left(\ln(x^2+1)\right)\\ &= \frac{\frac{d}{dx}(x+1)}{x+1} - \frac{\frac{d}{dx}(x^2+1)}{x^2+1}\\ &= \frac{1}{x+1} - \frac{2x}{x^2+1}\end{aligned}$$

Enrichment: Proofs

We now establish the following formula:

$$\frac{d}{dx}(\ln|f(x)|) = \frac{f'(x)}{f(x)}$$

We first let $u = f(x)$ and notice that

$$\frac{dy}{du} = \frac{d}{du}\ln u = \frac{1}{u}$$

Now the chain rule shows that

$$\begin{aligned}\frac{d}{dx}(\ln f(x)) &= \frac{dy}{dx}\\ &= \frac{dy}{du}\cdot\frac{du}{dx}\\ &= \frac{1}{u}\frac{du}{dx}\\ &= \frac{1}{f(x)}f'(x)\end{aligned}$$

Thus,

$$\frac{d}{dx}(\ln f(x)) = \frac{f'(x)}{f(x)}$$

This formula can be used to show that

$$\frac{d}{dx}\ln|x| = \frac{1}{x}$$

To see this, consider $y = \ln(-x)$ where $x < 0$ and set $f(x) = -x$. Then

$$\begin{aligned}\frac{d}{dx}(\ln(-x)) &= \frac{f'(x)}{f(x)}\\ &= \frac{-1}{-x}\\ &= \frac{1}{x}\end{aligned}$$

This in turn can be used to establish the more general formula.

We now establish the following.

$$\boxed{\frac{d}{dx}\log_a|f(x)| = \frac{f'(x)}{f(x)\ln a}}$$

To establish this formula, first recall the change of base theorem

$$\log_a x = \frac{\log_b x}{\log_b a}$$

given on page 87, Section 1.5. Using this with $b = e$, we have

$$\log_a |x| = \frac{\ln |x|}{\ln a}$$

$$\begin{aligned}\frac{d}{dx}\log_a |x| &= \frac{d}{dx}\left(\frac{\ln |x|}{\ln a}\right)\\ &= \frac{1}{\ln a}\cdot\frac{d}{dx}\ln |x|\\ &= \frac{1}{\ln a}\cdot\frac{1}{x}\\ &= \frac{1}{x\ln a}\end{aligned}$$

From the chain rule we then have the formula.

Warm Up Exercise Set 4.4

1. Find $f'(x)$ if $f(x) = e^{-0.50x^2}$. Find where $f'(x)$ is zero.

2. Find $g'(x)$ if $g(x) = x\ln x$ for $x > 0$. Find where $g'(x)$ is zero.

Exercise Set 4.4

In Exercises 1 through 42, find the derivative.

1. e^{4x} **2.** $e^{-0.1x}$ **3.** e^{-3x}

4. $4e^{\pi x}$ **5.** e^{2x^2+1} **6.** e^{4x^3-x-1}

7. $e^{\sqrt{x}}$ **8.** $e^{\sqrt[3]{x}}$ **9.** x^2e^x

10. $\sqrt[3]{x}e^x$ **11.** x^5e^x **12.** $(x^3+1)e^x$

13. $\dfrac{x}{e^x}$ **14.** $(x^3-2)e^x$ **15.** $x^2e^{x^2}$

16. $\dfrac{e^x}{x^2+1}$ **17.** $\sqrt{e^x+1}$ **18.** $(e^x+2)^8$

19. $\sqrt{x}e^{\sqrt{x}}$ **20.** $\sqrt{x^3}e^{3x}$ **21.** $\dfrac{x}{1+e^x}$

22. $\dfrac{x}{1+e^{-x}}$ **23.** $\dfrac{e^{2x}-e^{-2x}}{e^{2x}+e^{-2x}}$

24. $(e^{3x}+1)^4$ **25.** $\sqrt{e^{3x}+2}$ **26.** $\sqrt{2+e^{-3x}}$

27. 5^x **28.** 3^{3x} **29.** $3^{\sqrt{x}}$

30. 7^{x^7} **31.** $x3^x$ **32.** $3^x\cdot 5^x$

33. $\ln|x^3+x^2+1|$ **34.** $\ln(x^4+x^2+3)$

35. $x^2\ln(x^2)$ **36.** $\dfrac{\ln(x^2)}{x+1}$

37. $\sqrt{\ln|x^2+x+1|}$ **38.** $\ln(x+1)\sqrt{x^2+1}$

39. $\ln\left|\dfrac{x}{x+1}\right|$ **40.** $\ln\sqrt{\left|\dfrac{x}{x+1}\right|}$

41. $e^{-x^2}\ln(x^2)$ **42.** $\dfrac{\ln(x^2)}{e^x+e^{-x}}$

In Exercises 43 through 50, take $x > 0$ and find the derivatives.

43. $x^2\ln x$ **44.** $(x^3+2)\ln x$

45. $\dfrac{\ln x}{x^2+1}$ **46.** $\sqrt{\ln x}$ **47.** $(\ln x)^{11}$

48. $e^{x^2}\ln x$ **49.** $\ln(x^3)$ **50.** $\ln(x^7)$

In Exercises 51 through 62, find the derivative, and find where the derivative is zero. Assume that $x > 0$ in 59 through 62.

51. $y = xe^{-3x}$ **52.** $y = xe^x$ **53.** $y = e^{-x^2}$

54. $y = e^{-x^4}$ **55.** $y = x^2e^x$ **56.** $y = x^4e^x$

57. $y = x^2e^{-x}$ **58.** $y = x^2e^{-0.5x^2}$

59. $y = x\ln x^2$ **60.** $y = x - \ln x$

61. $y = \dfrac{\ln x}{x^2}$ **62.** $y = (\ln x)^3$

Applications and Mathematical Modeling

63. Revenue Suppose the price and demand of a commodity is related by $p(x) = e^{-2x}$. Find the marginal revenue $R(x)$, and find where the marginal revenue is zero.

64. Profits Suppose the profit equation is given by $P(x) = xe^{-0.5x^2}$. Find the marginal profit, and find where the marginal profit is zero.

65. Cost If the cost function is given by $C(x) = \ln(x^2 + 5) + 100$, find the marginal cost. Find where the marginal cost is positive.

66. Bumble Bee Activity Morandin and colleagues[48] studied the pollination of tomato plants by bumble bees. They created the mathematical model given by the equation $y = 454.3(1 - e^{-0.000492x})$, where x is the activity given by the mean number of bee trips per hectare per day and y is the mean number of pollen grains per tomato flower stigma. Find $\frac{dy}{dx}$ and the sign. Does the sign make sense? Explain.

67. Growth Rates of Graysby Groupers Potts and Manooch[49] studied the growth habits of graysby groupers. These groupers are important components of the commercial fishery in the Caribbean. The mathematical model they created was given by the equation $L(t) = 446(1 - e^{-0.13[t+1.51]})$, where t is age in years and L is length in millimeters. Graph this equation. Find $L'(t)$. What is the sign of $L'(t)$? Is this reasonable? Explain.

68. Biology Pearl and colleagues[50] determined that the equation $y = 34.53e^{-0.018x}x^{-0.658}$ approximated the rate of reproduction of fruit flies, where x is the density of the fruit flies in flies per bottle. Find the rate of change of the reproduction rate, and determine the sign of this rate of change. What is the significance of this sign. Use your grapher to graph.

69. Soil Nitrogen Dynamics Knops and Tilman[51] studied the accumulation of nitrogen in abandoned agricultural fields. They developed a mathematical model of nitrogen dynamics given by the equation

$$N(t) = \frac{0.172}{1 + 2.979e^{-0.0223t}}$$

where t is time in years and N is the percent of nitrogen in the soil. Find the rate of change of the percentage of nitrogen in the soil with respect to time and determine the sign of this rate of change. What is the significance of this sign. What is happening to this rate of change in time? Use your grapher to graph.

70. Clutch Size of Darwin's Finches Grant and colleagues[52] studied Darwin's finches on several of the Galapagos islands and created a mathematical model given by the equation $y = f(x) = -3.401 + 2.302 \ln x - 0.168(\ln x)^2$, where x is the rainfall in millimeters and y is clutch size. Find $f'(x)$, and find where this is positive and where it is negative. What does this say about the impact of rain on clutch size?

48 L. A. Morandin, T. M. Laverty, and P. G. Kevan. 2001. Bumble bee activity and pollination levels in commercial tomato greenhouses. *J. Econ. Entomol.* 94(2):462–467.

49 Jennifer C. Potts and Charles S. Manooch III. 1999. Observations on the age and growth of graysby and coney from the Southeastern United States. *Trans Amer. Fisheries Soc.* 128:751–757.

50 R. Pearl, J. R. Miner, and S. L. Parker. 1927. Experimental studies on the duration of life. *Am. Nat.* 61:289–318.

51 Johannes M. H. Knops and David Tilman. 2000. Dynamics of soil nitrogen and carbon accumulation for 61 years after agricultural abandonment. *Ecology* 81(1):88–98.

52 Peter R. Grant, B. Rosemary Grant, Likas F. Keller, and Kenneth Petren. 2000. Effects of El Nino events on Darwin's finch productivity. *Ecology* 81(9):2442–2457.

More Challenging Exercises

71. Find $f'(x)$ if $f(x) = \log_2(e^{3x} + x^2)$

72. Biology The Gompertz growth curve is given by $P(t) = ae^{-be^{-kt}}$, where a, b, and k are positive constants.[53] Show that $P'(t)$ is always positive.

In Exercises 73 and 74, find $f'(x)$.

73. $f(x) = e^{e^{e^x}}$ **74.** $f(x) = \ln[\ln(\ln x)]$

75. In Example 3, the text considered a logistic model of the population of the United States formulated by P. Verhulst about 150 years ago. The following model is an updated version. If $p(t)$ is the population of the United States in millions and t is the time in years with $t = 0$ corresponding

53 Duane J. Clow and N. Scott Urquhart. 1974. *Mathematics in Biology.* New York: Ardsley House.

to the year 1790, then

$$p(t) = \frac{500}{1 + 124e^{-0.024t}}$$

a. What happens to $p(t)$ when t becomes very large?
b. Find $p'(t)$.
c. Determine the sign of $p'(t)$ found in part (b). Interpret the significance of this sign.
d. Graph $y = p(t)$ on your computer or graphing calculator using a window with dimensions $[0, 400]$ by $[0, 500]$. Does your graph verify your answers in parts (a) and (c)? Explain.
e. The population of the United States in 1790 was about four million, and it grew during the earlier years at about 2.4% per year. Graph the exponential function $y = 4e^{0.025t}$ on the same screen as that in part (d). How does the graph of this exponential function compare with the graph of the logistic function in part (d)? What do you conclude about the logistic model $y = p(t)$ during the first 50 years as compared to the exponential model given here?

76. We showed in the text that

$$\frac{d}{dx}\log_a |f(x)| = \frac{f'(x)}{f(x) \cdot \ln a}$$

What value of a makes this formula as simple as possible? Explain. What is the formula in this simplest case?

77. You found in Exercises 67 and 68 in Section 3.3 that

$$\frac{d}{dx}(\sin x) = \cos x \qquad \text{and} \qquad \frac{d}{dx}(\cos x) = -\sin x$$

Use the chain rule to show that

$$\frac{d}{dx}(\sin f(x)) = (\cos f(x)) \cdot f'(x)$$

$$\frac{d}{dx}(\cos f(x)) = -(\sin f(x)) \cdot f'(x)$$

In Exercises 78 through 84, use the formulas found in Exercise 77 to find the indicated derivatives.

78. $\frac{d}{dx}\sin e^x$

79. $\frac{d}{dx}\cos x^3$

80. $\frac{d}{dx}\cos e^{x^2}$

81. $\frac{d}{dx}\sin(\ln x)$

82. $\frac{d}{dx}\sin(\cos x)$

83. $\frac{d}{dx}\cos(\sin x)$

84. $\frac{d}{dx}\ln(\sin x)$

85. The Ricker Growth Model The Ricker growth model is used extensively[54] in population dynamics, particularly for insect growth. It is given by the general equation $g(x) = rxe^{-bx}$, where r and b are positive constants, x is the current population, and g is the population of the next generation. Find $g'(x)$, and find where g' is positive and where it is negative. Relate this to the biology, that is, to what happens to the population in the next generation when the current population is small and when it is large.

86. Modeling Seasonal Temperature Power and Heuvel[55] developed a mathematical model that profiled seasonal temperatures around several Canadian lakes. They used the general equation $T(D) = aD^b e^{-cD}$, where a, b, and c are all positive constants, D is the day of the year, and T is the temperature in degrees Celsius. Find $T'(D)$. When is this zero?

87. Growth Rates of Lake Trout Ruzycki and coworkers[56] studied the growth habits of lake trout in Bear Lake, Utah-Idaho. The mathematical model they created was given by the equation $L(t) = 960(1 - e^{-0.12[t+0.45]})$, where t is age in years and L is length in millimeters. They also created a mathematical model that connected length with weight and was given by the equation $W(L) = 1.30 \times 10^{-6} \cdot L^{3.31}$, where L is in millimeters and W is in grams. Find the rate of change of W with respect to t *without* first finding W as a function of t. Instead, use the chain rule.

88. Growth Rates of Cutthroat Trout Ruzycki and coworkers[57] studied the growth habits of cutthroat trout in Bear Lake, Utah-Idaho. The mathematical model they created was given by the equation $L(t) = 650(1 - e^{-0.25[t+0.50]})$, where t is age in years and L is length in millimeters. Graph this equation. What seems to be happening to the length as the trout become older? They also created a mathematical model that connected length with weight and was given by the equation $W(L) = 1.17 \times 10^{-5} \cdot L^{2.93}$, where L is in millimeters and W is in grams. Find the rate of change of W with respect to t *without* first finding W as a function of t. Instead, use the chain rule.

89. Herring Abundance Kosa and Mather[58] studied various factors that affected the abundance of herring, a species on the decline. They formulated a mathematical model given approximately by the equation $P(D) = 10^{0.03D^2 - 0.4D + 1}$, where D is the depth of water in meters, and P is the numbers (per unit volume). Find $P'(D)$. Where is $P'(D) > 0$?

---

[54] Nora Underwood and Mark D. Rausher. 2000. The effects of host-plant genotype on herbivore population dynamics. *Ecology* 81(6):1565–1576.

[55] M. Power and M. R. van den Heuvel. 1999. Age-0 yellow perch growth and its relationship to temperature. *Trans. Amer. Fish. Soc.* 128:687–700.

[56] James R. Ruzycki, Wayne A. Wurtsbaugh, and Chris Luecke. 2001. Salmonine consumption and competition for endemic prey fishes in Bear Lake, Utah-Idaho. *Trans Amer. Fish. Soc.* 130:1175–1189.

[57] Ibid.

[58] Jarrad T. Kosa and Martha E. Mather. 2001. Processes contributing to variability in regional patterns of juvenile river herring abundance across small coastal systems. *Trans. Amer. Fish. Soc.* 130:600–619.

Explain what this means. Where is $P'(D) < 0$? Explain what this means.

90. Herring Abundance Kosa and Mather[59] studied various factors that affected the abundance of herring, a species that is on the decline. They formulated a mathematical model given approximately by the equation $P(x) = 10^{-2x^2+32x-127}$, where x is the pH level and P is the numbers (per unit volume). Find $P'(x)$.

91. Biological Control Lysyk[60] studied the effect of temperature on various life history parameters of a parasitic wasp for the purposes of pest management. He constructed a mathematical model given approximately by the equation

$$p(t) = \left(1 - \frac{3-t}{2.95}\right)^{[(3-t)/2.95]^{1.3}}$$

where t is normalized time (time/time to 50% oviposition) and p is the proportion of progeny produced. Finding $p'(t)$ is important but is difficult, since we have a *function* raised to a power, which is also a function. For this exercise, consider $f(t) = (1-t)^t$ and find $f'(t)$. (Hint: Take natural logarithms of both sides, and then take the derivative.)

[59] Ibid.

[60] T. J. Lysyk. 2001. Relationship between temperature and life history parameters of *Muscidifurax zaraptor*. *Environ. Entomol.* 30(1):147–156.

Modeling Using Least Squares

92. Costs Rogers and Akridge[61] of Purdue University studied fertilizer plants in Indiana. For a typical large-sized plant, they gave data (see their Figure 1) that related average total cost, $\bar{C}(x)$, with cost in dollars per ton. Their data is presented in the following table.

x	6.3	7.2	8.1	9.0	9.9	10.8	11.7
y	282	273	265	259	255	251	248

Here x is production in thousands of tons, and y is average total cost in dollars per ton.

a. Determine the best-fitting exponential function, $e(x) = a \cdot x^b$, using least squares and the square of the correlation coefficient. Graph.

b. Using the exponential average cost function $e(x)$ found in part (a), find $e'(9)$. Interpret what this means.

c. Let $C(x)$ be total cost. Given that x is given in thousands of tons and that $e(x)$ and $\bar{C}(x)$ are given in dollars per ton, show that $C(x) = 1000\bar{C}(x) \cdot x$. Now replace $\bar{C}(x)$ with $e(x)$, and use the product rule of differentiation to find $C'(9)$. Interpret what this means.

d. Multiply the y values in the table by $1000x$ to obtain values for total cost. Now using this modified data, find the best-fitting exponential, $E(x)$, and the square of the correlation coefficient. Then $E'(9)$ gives $C'(9)$. Compare this value with the one found in part (c).

[61] Duane S. Rogers and Jay T. Akridge. 1996. Economic impact of storage and handling regulations on retail fertilizer and pesticide plants. *Agribusiness* 12(4):327–337.

93. Population Growth The table gives the population of the United States in thousands for some selected years[62]. On the basis of the data given for the years 1790–1794, find the best-fitting exponential function $e(t) = a \cdot b^t$ and the square of the correlation coefficient. (Take $t = 0$ to correspond to the year 1790.) Graph. Find $e'(3.5)$. Interpret what this means.

Year	1790	1791	1792	1793	1794
Population	3926	4056	4194	4332	4469

94. Herring Abundance Kosa and Mather[63] studied various factors that affected the abundance of herring, a species on the decline. They collected the data given in the table relating the common logarithm of abundance with discharge in cubic meters per sample.

[62] John W. Wright. 1993. *The Universal Almanac*. New York: Andrews and McMeel.

[63] Jarrad T. Kosa and Martha E. Mather. 2001. Processes contributing to variability in regional patterns of juvenile river herring abundance across small coastal systems. *Trans. Amer. Fish. Soc.* 130: 600–619.

Discharge	0	0	.02	.03	.04	.05
log (Abundance)	1.0	1.4	3.4	2.7	2.4	4.0
Discharge	.07	.12	.15	.22	.31	
log (Abundance)	3.0	3.5	5.2	4.0	2.3	

a. Use quadratic regression (as the researches did) to find a model of log (abundance) as a function of discharge.

b. Use (a) to find abundance as a function of discharge, $A(D)$, and find $A'(D)$.

Solutions to WARM UP EXERCISE SET 4.4

1. We have

$$f'(x) = \frac{d}{dx}\left(e^{-0.50x^2}\right)$$
$$= e^{-0.50x^2}\frac{d}{dx}(-0.50x^2) = -xe^{-0.50x^2}$$

Since $e^{-0.50x^2}$ is positive for all x, $f'(x)$ is zero only when $x = 0$.

2. Since $g(x) = x \ln x$, we have, using the product rule,

$$g'(x) = x \cdot \frac{d}{dx}(\ln x) + (\ln x) \cdot \frac{d}{dx}(x)$$
$$= x \cdot \frac{1}{x} + (\ln x) \cdot (1)$$
$$= 1 + \ln x$$

This will be zero when $\ln x = -1$. This will happen, according to the definition of the logarithm, if and only if $x = e^{-1}$.

4.5 Elasticity of Demand

Elasticity of Demand
Applications
Enrichment

APPLICATION The Effect of Price Increases

In 1973 Braley and Nelson[64] studied the effect of price increases on school lunch participation in the city of Pittsburgh. They estimated the demand equation for school lunches to be $x = 16{,}415.21 - 262.743p$, where x is the number of lunches purchased and p is the price in cents. When the cost of a lunch is 25 cents, how does an increase in the price of the lunch effect the demand? When the cost of a lunch is 42 cents, how does an increase in the price of the lunch effect the demand? See Example 4 on page 279 for the answer.

[64] George A. Braley and P. E. Nelson, Jr. 1975. Effect of a controlled price increase on school lunch participation: Pittsburgh. *Amer. J. Agr. Econ.* February, 1975:90–96.

Elasticity of Demand

Let us now consider gauging the sensitivity of the demand for a product to changes in the price of the product. In the past we have written the price p as a function of

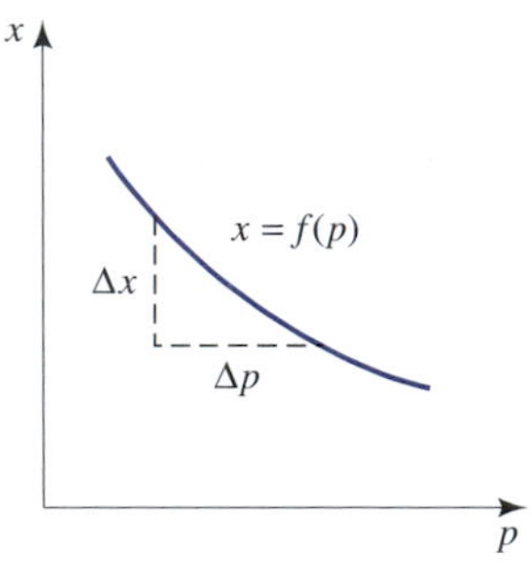

Figure 4.8
A typical demand curve slopes downward.

demand x. It will be convenient in this section to write x as a function of p. Any reasonable demand curve given by $x = f(p)$ will be a *decreasing* function. See Figure 4.8. That is, an increase in price will lead to a decrease in demand. We wish to measure by just how much an increase in prices will decrease demand.

We expect the amount of change will vary from product to product. For example, if a product such as furniture has a substantial increase in price, we would expect demand to fall sharply, since most people can normally postpone or delay purchasing this product. If, on the other hand, the price of milk went up, most people would still buy the same amount of milk and probably forgo purchasing something else of less importance.

We probably would not choose to measure these changes in dollar amounts, since increasing the cost of a sofa by a dollar is very different from increasing the price of a gallon of milk by one dollar. Instead, we will use percentages. For example, a merchant might ask, "If the price of sofas sold are increased by 10%, what would be the percentage decrease in the number of sofas sold?" Say the decrease was 30%. Then the ratio of percentage decrease in demand to percentage increase in price would be $30/10 = 3$. This is the ratio that economists work with. The fact that this number is greater than 1 is important; it indicates that the percentage drop in demand is greater than the percentage increase in price. On the other hand increasing the price of milk by 10% might reduce the demand for milk by only 1%. Then the ratio of percentage decrease in demand to percentage increase in price would be $1/10 = 0.1$. The fact that this number is less than 1 is important; it indicates that the percentage drop in demand is less than the percentage increase in price.

In general, we will need a demand equation, given in the form of $x = f(p)$. As usual, we let Δp represent the change in price and let Δx represent the change in demand. This is shown in Figure 4.8. Notice that if Δp is positive, then Δx must be *negative* since the function is decreasing. Also, if Δp is negative, then Δx must be positive. (That is, a decrease in price will lead to greater demand.) We conclude that Δp and Δx must be of opposite signs.

The percentage change in demand is given by the expression $\frac{\Delta x}{x}100$, and the percentage change in price is given by the expression $\frac{\Delta p}{p}100$. The ratio is then

$$\frac{\frac{\Delta x}{x}100}{\frac{\Delta p}{p}100} = \frac{p}{x}\frac{\Delta x}{\Delta p}$$

As was noted above, Δp and Δx must be of opposite signs, and therefore, this ratio is *negative*. Since economists prefer to work with positive numbers, they actually work with

$$-\frac{p}{x}\frac{\Delta x}{\Delta p}$$

A large ratio means that a small change in price leads to a relatively large change in demand. A small ratio (say, 0.1) means that a large change in price is needed to effect a relatively small change in demand.

Now if $x = f(p)$ is differentiable, then

$$\lim_{\Delta p \to 0} \frac{\Delta x}{\Delta p} = \frac{dx}{dp}$$

Then we define the term E by

$$E = \lim_{\Delta p \to 0} -\frac{p}{x}\frac{\Delta x}{\Delta p} = -\frac{p}{x}\frac{dx}{dp}$$

This term is called the *elasticity of demand* and measures the instantaneous sensitivity of demand to price.

We say that demand is *inelastic* if $E < 1$, *elastic* if $E > 1$ and unit elasticity if $E = 1$.

Thus, if the percentage change in demand is less than the percentage change in price, demand is inelastic, and $E < 1$. If the percentage change in demand is greater than the percentage change in price, demand is elastic, and $E > 1$. Finally, if the percentage change in demand is equal to the percentage change in price, demand is of unit elasticity and $E = 1$.

DEFINITION **Elasticity of Demand**

If the demand equation is given by $x = f(p)$, then

$$E = -\frac{p}{x}\frac{dx}{dp}$$

is called the **elasticity of demand**.

1. If $E > 1$, then demand is **elastic**.
2. If $E < 1$, then demand is **inelastic**.
3. If $E = 1$, then demand has **unit elasticity**.

EXAMPLE 1 **Finding Elasticity of Demand**

A restaurant owner sells 100 dinner specials for \$10 each. After raising the price to \$11, she noticed that only 80 specials were sold. What is the elasticity of demand?

Solution

When the price was increased by 10%, the demand decreased by 20%. So the elasticity is the ratio $20/10 = 2$. ■

Notice in the previous example that the original revenue was \$10 times 100 specials, or \$1000. After the price increase, the revenue was \$11 times 80 specials, or only \$880. We will see later that revenue always decreases when prices are increased if the elasticity of demand is greater than 1.

EXAMPLE 2 **The Effect of Elasticity of Demand**

Alice has a restaurant across the street from the one in Example 1 and sells 200 dinner specials for \$5 each. An economist informs her that at the current price of her specials, her elasticity of demand is 0.50. Will increasing the price of each special to \$6 increase her revenue?

Solution

Increasing the price of her specials to \$6 from \$5 represents a 20% increase in price. But she knows that the elasticity of demand is 0.50, so demand will drop by half of the percentage increase in price. Thus, she will sell 10% fewer specials, or 180 in total. Her revenue before her price increase was \$5 times 200 specials or \$1000. After the price increase, her revenue is \$6 times 180 specials, or \$1080. Her revenue has increased! ■

We will see later that revenue always increases when prices are increased if the elasticity of demand is less than 1.

We normally expect the elasticity of demand to change with changes in price. We will see this in the next example.

EXAMPLE 3 **Finding Elasticity of Demand**

A small grocery store makes their own candy bars. Their demand equation is given by $p = 2 - x$, where p is the price of each candy bar in dollars and x is the number of hundreds of candy bars they sell in one week. See Figure 4.9. Find the elasticity of demand when (a) $p = 0.75$, (b) $p = 1.50$, (c) $p = 1.00$.

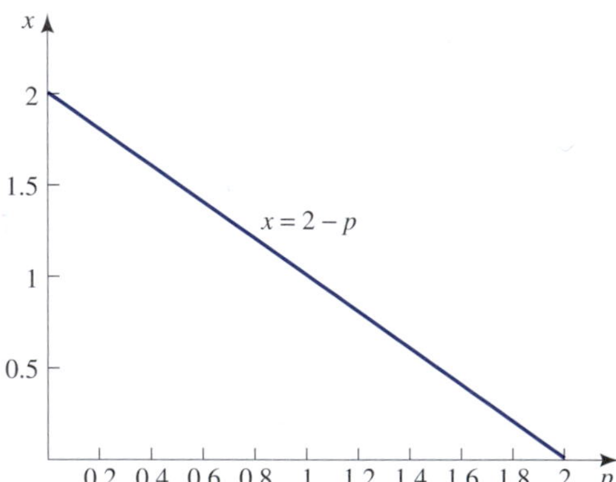

Figure 4.9

Solution

We first need to write x as a function of p. Since $p = 2 - x$, $x = 2 - p$. Then

$$\begin{aligned} E &= -\frac{p}{x}\frac{dx}{dp} \\ &= -\frac{p}{2-p}(-1) \\ &= \frac{p}{2-p} \end{aligned}$$

(a) When $p = 0.75$,

$$E = \frac{p}{2-p} = \frac{0.75}{2-0.75} = \frac{0.75}{1.25} = \frac{75}{125} = 0.60$$

Demand is inelastic. If the price is increased by 1%, demand will decrease by 0.60%.

(b) When $p = 1.50$,

$$E = \frac{p}{2-p} = \frac{1.50}{2-1.50} = \frac{1.50}{0.50} = 3$$

Demand is elastic. If the price is increased by 1%, demand will decrease by 3%.

(c) When $p = 1.00$,

$$E = \frac{p}{2-p} = \frac{1.00}{2-1.00} = \frac{1.00}{1.00} = 1$$

Demand has unit elasticity. If the price is increased by 1%, demand will decrease by 1.0%. ■

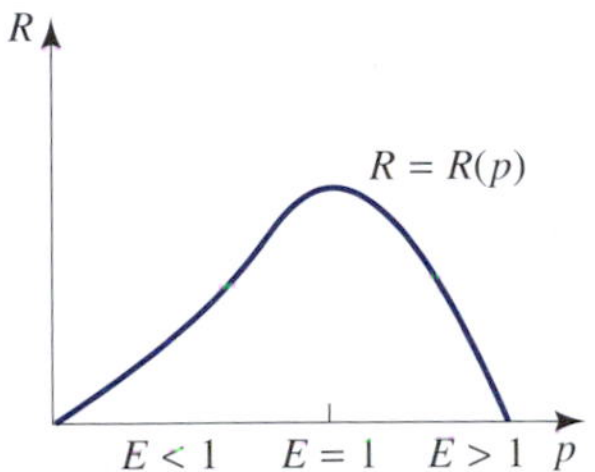

Figure 4.10
The graph shows that if $E < 1$, increasing price increases revenue. But, if $E > 1$, increasing price decreases revenue.

We mentioned earlier that increasing prices might or might not increase revenue. What actually happens depends on the elasticity. We now state an important fact. The proof is given in the Enrichment subsection.

Theorem

If $E < 1$, increasing price increases revenue. If $E > 1$, increasing price decreases revenue. Revenue is optimized when $E = 1$. See Figure 4.10.

The next Interactive Illustration allows you to discover this fact using a variety of demand equations.

Interactive Illustration 4.6

Elasticity of Demand

This interactive illustration allows you to explore elasticity of demand for a range of linear demand equations. There are sliders on the two parameters in the demand equation. Move the sliders to give different demand and supply equations. Change the demand (q), and note both the revenue and price elasticity.

1. For a given set of parameters, find where increasing demand leads to increasing revenues. What is the elasticity?
2. For a given set of parameters, find where increasing demand leads to decreasing revenues. What is the elasticity?

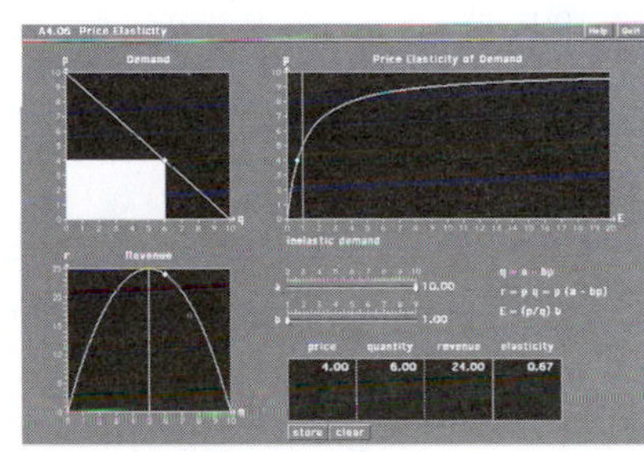

3. What is the elasticity when revenue is maximized?

Applications

EXAMPLE 4 **Finding Elasticity of Demand**

In 1973 Braley and Nelson[65] studied the effect of price increases on school lunch participation in the city of Pittsburgh. They estimated the demand equation for school

[65] Ibid.

lunches to be $x = 16{,}415.21 - 262.743p$, where x is the number of lunches purchased and p is the price in cents. Find the elasticity of demand when p equals (a) 25 cents, (b) 42 cents. (c) Find the approximate price of a lunch that gives a unit elasticity.

Solution

Since $\dfrac{dx}{dp} = -262.743$,

$$E = -\frac{p}{x}\frac{dx}{dp} = -\frac{p}{16{,}415.21 - 262.743p}(-262.743) = \frac{262.743p}{16{,}415.21 - 262.743p}$$

(a) For $p = 25$,

$$E = \frac{262.743(25)}{16{,}415.21 - 262.743(25)} \approx 0.67$$

Demand is inelastic at this price. A percentage increase in price results in about a two thirds percentage decrease in demand.

(b) For $p = 42$,

$$E = \frac{262.743(42)}{16{,}415.21 - 262.743(42)} \approx 2.05$$

Demand is elastic at this price. A percentage increase in price results in about a 2.05 percentage decrease in demand.

(c) We graph the equation for E as

$$y_1 = \frac{262.743p}{16{,}415.21 - 262.743p}$$

and also graph $y_2 = 1$ on a screen with dimensions [0, 47] by [0, 3], use the available operations, and we obtain Screen 4.13. We see that $E = 1$ when the price is approximately 31 cents. (You can also do this algebraically.) ■

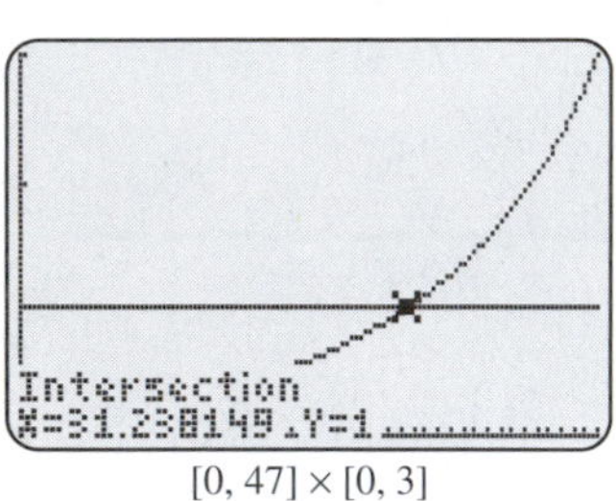

$[0, 47] \times [0, 3]$

Screen 4.13
We see the graphs of the elasticity of demand $y_1 = \dfrac{262.743p}{16{,}415.21 - 262.743p}$, the graph of $y_2 = 1$, and the intersection when p is approximately 31 cents.

Interactive Illustration 4.7

Elasticity of Demand for Example 4

This interactive illustration allows you to explore the graphs mentioned in Example 4. Click on the first function. Use the slider to move x, and note what is happening to the rectangle, the revenue, and the elasticity.

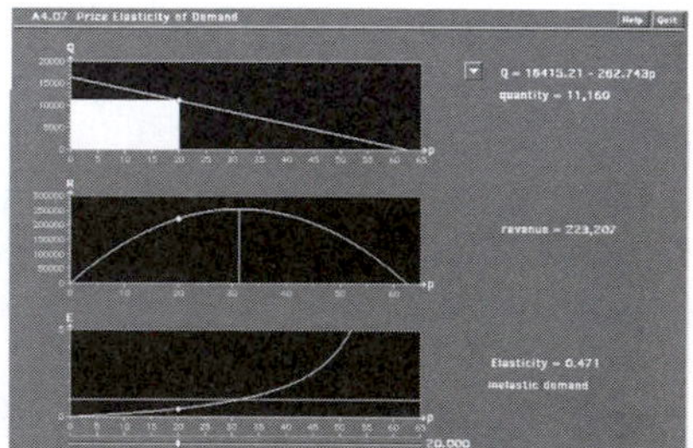

1. When is demand elastic? Inelastic? Unit elastic?
2. How are your answers in (1) related to the revenue graph shown in the second window?

3. Is there a connection between the demand that yields maximum revenue and that which yields unit elasticity?

EXAMPLE 5 **Finding Elasticity of Demand**

In a major recent study of the breakfast cereal industry, Cotterill and Haller[66] found that the price p of Shredded Wheat was related to the quantity x sold by the equation $x = Ap^{-1.7326}$, where A is a constant. Find the elasticity of demand and explain what it means.

Solution

We have $x = f(p) = Ap^{-1.7326}$. So

$$\begin{aligned} E &= -\frac{p}{x}\frac{dx}{dp} \\ &= -\frac{p}{Ap^{-1.7326}}A(-1.7326)p^{-2.7326} \\ &= \frac{1.7326p^{-1.7326}}{p^{-1.7325}} \\ &= 1.7326 \end{aligned}$$

So an increase of 1% in price should result in a 1.7326% decrease in demand. Furthermore, the elasticity of demand does not depend on price but stays at the constant value of 1.7326. ■

You can explore Example 5 by looking again at Interactive Illustration 4.7. Click on the second function, and explore the revenue and price elasticity.

Notice that when the relationship between demand and price is linear, such as is found in Example 3, the elasticity changes as the price chances. (Screen 4.13 shows this clearly. See also Example 6 below.) But when the demand is a power function, as in Example 4, the elasticity is constant. The two most important demand functions that are used are linear and power functions.

Enrichment

EXAMPLE 6 **Elasticity for Linear Demand**

Let the demand equation be the linear function $x = a - bp$, where a and b are positive constants. Discuss the elasticity for various values of price.

Solution

A graph of the demand equation is shown in Figure 4.11. We have

$$\begin{aligned} E &= -\frac{p}{x}\frac{dx}{dp} \\ &= -\frac{p}{a - bp}(-b) \\ &= \frac{pb}{a - bp} \end{aligned}$$

[66] Ronald W. Cotterill and Lawrence E. Haller. 1997. An economic analysis of the demand for RTE cereal: product market definition and unilateral market power effects. Research Report No. 35. Food Marketing Policy Center. University of Connecticut.

We note that $\lim_{p\to 0^+} E = 0$ and $\lim_{p\to (a/b)^-} E = \infty$. We also note that $E = 1$ when $bp = a - bp$, or $p = \frac{1}{2}\frac{a}{b}$. In fact, it appears that the elasticity E may take all values on $(0, \infty)$. To see this clearly, note that by the quotient rule

$$\begin{aligned}\frac{dE}{dp} &= \frac{(a-bp)b - bp(-b)}{(a-bp)^2} \\ &= \frac{ab}{(a-bp)^2}\end{aligned}$$

and this is always positive. As we will see in Section 5.1 this implies that E as a function of p steadily increases from 0 to ∞ as p goes from 0 to $\frac{a}{b}$. See Figure 4.12. ■

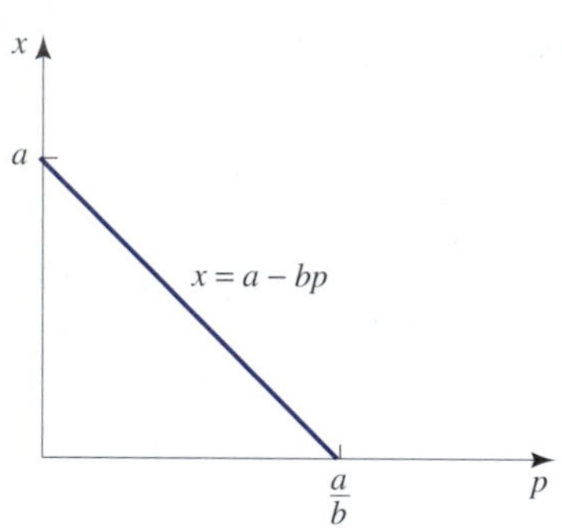

Figure 4.11

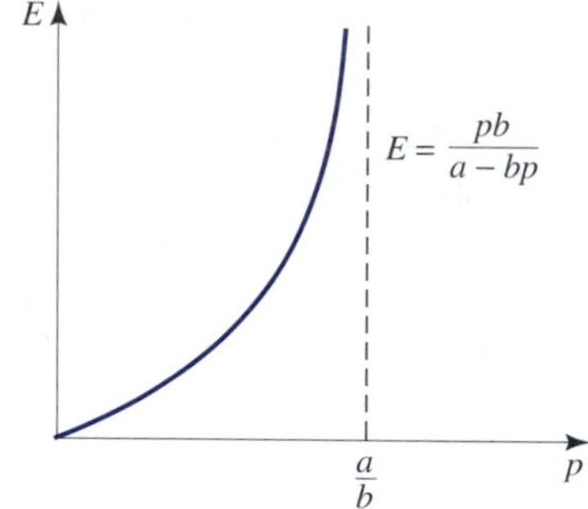

Figure 4.12

We now derive the interesting connection between the revenue R and elasticity E that we mention earlier. Suppose we have $x = f(p)$. Then since $R = px$,

$$\begin{aligned}\frac{dR}{dp} &= x + p\frac{dx}{dp} \\ &= x + x\frac{p}{x}\frac{dx}{dp} \\ &= x - xE \\ &= x(1-E)\end{aligned}$$

If demand is inelastic, $E < 1$, and then $R'(p) > 0$; therefore, $R(p)$ is increasing. If demand is elastic, $E > 1$, and then $R'(p) < 0$; therefore, $R(p)$ is decreasing.

Therefore, if $E < 1$, increasing price increases revenue. If $E > 1$, increasing price decreases revenue. Revenue is optimized when $E = 1$. See Figure 4.10.

Demand for a product not only depends on price but also depends on advertising, couponing, and so on. Economists define an elasticity of demand for these variables in a manner similar to ours for price. This is explored in Exercises 28 and 29.

Warm Up Exercise Set 4.5

1. Let $x = 82 - p^2$. Find the elasticity of demand if
   a. $p = 1$
   b. $p = 9$.

Exercise Set 4.5

1. According to Schultz,[67] the price elasticity of corn is 0.77. If the price of corn is increased by 10%, what will happen to the demand for corn?

2. According to Schultz,[68] the price elasticity of wheat is 0.03. If the price of wheat is increased by 10%, what will happen to the demand for wheat?

3. According to Wold,[69] the price elasticity of beef is 0.50. If the price of beef is increased by 10%, what will happen to the demand for beef?

4. According to Wold,[70] the price elasticity of butter is 0.70. If the price of butter is increased by 10%, what will happen to the demand for butter?

5. According to Wold,[71] the price elasticity of milk is 0.31. If the price of milk is increased by 10%, what will happen to the demand for milk?

6. According to Frisch,[72] the price elasticity of air transportation is 1.10. If the price of air transportation is increased by 10%, what will happen to the demand for air transportation?

7. According to Ferguson and Polasek,[73] the price elasticity of wool is 1.32. If the price of wool is increased by 10%, what will happen to the demand for wool?

8. According to Stone and Rowe,[74] the price elasticity of furniture is 3.04. If the price of furniture is increased by 10%, what will happen to the demand for furniture?

9. According to Gum and Martin,[75] the price elasticity of outdoor recreation is 0.56. If the price of outdoor recreation is increased by 10%, what will happen to the demand for outdoor recreation?

10. A restaurant owner sells 100 dinner specials for \$10 each. After raising the price to \$11, she noticed that only 70 specials were sold. What is the elasticity of demand?

11. A restaurant owner sells 100 dinner specials for \$10 each. After raising the price to \$11, she noticed that only 95 specials were sold. What is the elasticity of demand?

12. A restaurant owner sells 100 dinner specials for \$10 each. After raising the price to \$11, she noticed that only 90 specials were sold. What is the elasticity of demand?

In Exercises 13 through 22, find the elasticity E at the given points and determine whether demand is inelastic, elastic, or unit elastic.

13. $x = 10 - 2p$,
 a. $p = 1$ b. $p = 2.5$ c. $p = 4.5$

14. $x = 20 - 4p$,
 a. $p = 1$ b. $p = 2.5$ c. $p = 4$

15. $x = 24 - 3p$,
 a. $p = 1$ b. $p = 4$ c. $p = 6$

16. $x = 20 - 5p$,
 a. $p = 2$ b. $p = 3$ c. $p = 1$

17. $x = \dfrac{1}{10 + p}$
 a. $p = 5$ b. $p = 10$ c. $p = 100$

18. $x = 10 + \dfrac{1}{p}$
 a. $p = 0.1$ b. $p = 1$

19. $x = 10 + \dfrac{1}{p^2}$
 a. $p = 0.01$ b. $p = 1$

20. $x = \dfrac{10}{p}$
 a. $p = 2$ b. $p = 3$

21. $x = \dfrac{10}{p^2}$
 a. $p = 1$ b. $p = 3$

22. $x = \dfrac{10}{\sqrt{p}}$
 a. $p = 1$ b. $p = 3$

In Exercises 23 and 24, find (a) the value, if any, of x for which $E = 1$ (b) the value(s), if any, of x for which the revenue is maximized.

23. $x = 30 - 10p$

24. $x = \dfrac{1}{p^3}$

---

[67] Henry Schultz. 1938. *The Theory and Measurement of Demand*. Chicago: University of Chicago Press. Chicago.

[68] Ibid.

[69] Herman Wold. 1953. *Demand Analysis*. New York: Wiley.

[70] Ibid.

[71] Ibid.

[72] Ragnar Frisch. 1959. A complete scheme for computing all direct cross demand elasticities in a model with many sectors. *Econometrica* 28:177–96.

[73] C. E. Ferguson and M. Polasek. 1962. The elasticity of import demand for raw apparel wool in the United States. *Econometrica* 30:670–699.

[74] Richard D. Stone and D. A. Rowe. 1960. The durability of consumers' durable goods. *Econometrica* 28:407–416.

[75] Russell L. Gum and W. E. Martin. 1975. Problems and solutions in estimating the demand for and value of rural outdoor recreation. *Amer. J. Agr. Econ.* November 1975:558–566.

25. Elasticity of Beef Spencer[76] noted that at one time if x were beef consumption per capita in pounds and p is the price of beef divided by disposable income per capita, then $x = 126.5 - 1800x$.

a. Find E as a function of p, and graph it.

b. Determine E when $p = 0.03. 0.05$.

26. Using the demand function in the previous exercise, find the approximate price at which elasticity is unity.

[76] Milton H. Spencer. 1968. *Managerial Economics*. Homewood, Ill.: Richard D. Irwin.

27. Elasticity in the Breakfast Cereal Industry Recently, Cotterill and Haller[77] found that the price p of the breakfast cereal Grape Nuts was related to the quantity x sold by the equation $x = Ap^{-2.0711}$, where A is a constant. Find the elasticity of demand and explain what it means.

[77] Ronald W. Cotterill and Lawrence E. Haller. 1997. An economic analysis of the demand for RTE cereal: product market definition and unilateral market power effects. Research Report No. 35. Food Marketing Policy Center. University of Connecticut.

More Challenging Exercises

28. Elasticity of Demand Related to Advertising In this section we considered demand x as a function $x = f(p)$ of price and then defined the elasticity as $E = -\frac{p}{x}\frac{dx}{dp}$. But demand is also a function of other variables. For example, Cotterill and Haller recently found that the demand x for the breakfast cereal Shredded Wheat was approximately related to advertising dollars spent a by $x = Ba^{0.0295}$, where B is a constant. Define an elasticity with respect to advertising in a way analogous to what was done for demand with respect to price. Find the elasticity with respect to advertising in this case, and explain in words what it means.

[78] Ibid.

29. Elasticity of Demand Related to Couponing In this section we considered demand x as a function $x = f(p)$ of price and then defined the elasticity as $E = -\frac{p}{x}\frac{dx}{dp}$. But demand is also a function of other variables. For example, Cotterill and Haller[79] recently found that the demand x for the breakfast cereal Shredded Wheat was approximately related to the amount a of coupons issued by $x = Ba^{0.0229}$, where B is a constant. Define an elasticity with respect to coupons in a way analogous to what was done for demand with respect to price. Find the elasticity with respect to couponing in this case, and explain in words what it means.

[79] Ibid

Solutions to WARM UP EXERCISE SET 4.5

1. We have

$$E = -\frac{p}{x}\frac{dx}{dp} - \frac{p}{82 - p^2}(-2p) = \frac{2p^2}{82 - p^2}.$$

a. When $p = 1$, $E = \frac{2}{81} < 1$. Thus demand is inelastic.

b. When $p = 9$, $E = 162 > 1$. Thus demand is elastic.

4.6 Management of Renewable Natural Resources

Enrichment: Management of Renewable Natural Resources

Enrichment: Harvesting

Enrichment: Management of Renewable Natural Resources

As another important application, we now consider a mathematical model of a fishery. We can also apply the ideas presented here to other renewable natural resources. To avoid confusion, we emphasize that we are interested in the maximization of productive resources rather than the preservation of natural environments.

A fishery can consist of all or part of an ocean, a lake, a river, a natural or artificial pond, or even an area of water that has been fenced in. To fix a particular situation in our minds, we assume that a corporation that is engaged in aquaculture (fish farming) has the sole rights to the fish in a fairly small lake. Of course, the corporation wishes to harvest the maximum number of fish with the minimum effort and cost. We assume that the corporation does no active management of the fish other than simply harvesting.

Let $y = f(x)$, where x represents the total pounds (**biomass**) of fish at the beginning of a year and y represents the total increase in the biomass of fish over the next year. Thus, the term $x + y$ represents the total biomass 1 year later.

For convenience we always harvest at the end of the year. We are interested in the *increase* in population, since it is this increase that we wish to harvest. We want to predict how much of the resource we can harvest and still allow the resource to replenish itself. The largest possible increase in biomass is called the **maximum sustainable yield**. We want to know what the biomass of fish should be to obtain the maximum sustainable yield.

We know that if the biomass is zero, then the increase in the biomass will also be zero. Mathematically, this says that $f(0) = 0$. The presence of some fish in the lake results in an increase in the population and biomass, and the presence of even more fish results in even more of an increase in the biomass. Thus, f should be increasing, at least initially.

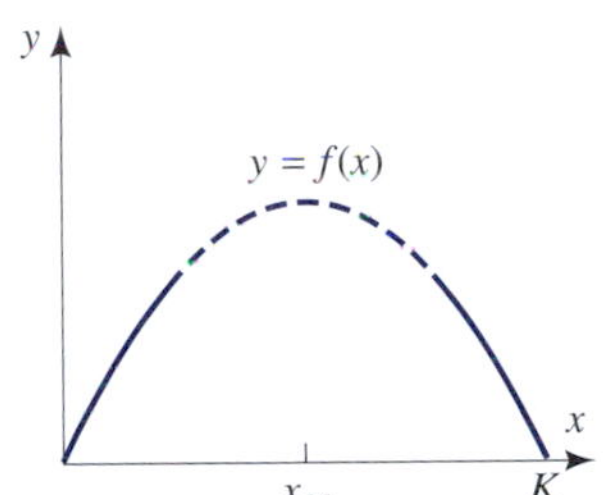

Figure 4.13
As the number of fish increases, the yearly change in the fish population itself increases, but then at some point begins to decrease.

Clearly, our small lake has only a finite amount of resources to sustain the fish and can hold only so much biomass. This number is referred to as the **carrying capacity**, which we label K. Since K is the saturation level of the species, there can be *no* increase in biomass beyond this level. Thus, $f(K) = 0$. If the biomass, x, is a little less than K, then we do expect some increase in biomass. Thus, moving from right to left starting at $x = K$, y should increase. Beginning just to the left of K and moving left to right, the function f should be decreasing.

At this point in the discussion the function looks like the solid curve in Figure 4.13. Biologists now normally make the assumption that the actual curve at the intermediate values is described by the dotted curve in Figure 4.13.

It is to be expected that the increase in the biomass, $y = f(x)$, becomes larger and larger (increases) until some value is reached, say, $x = x_M$, after which the increase in the biomass begins to get smaller and smaller (decreases). The number $M = f(x_M)$ is then the maximum sustainable yield. For the corporation that is interested in the long-term benefits, the harvesting should be managed so that $x = x_M$. Thus, given these assumptions, if the biomass is at $x = x_M$ at the beginning of the year, the corporation can harvest $f(x_M)$ pounds of fish at the end of the year and see the population renew itself over the subsequent year so that the biomass returns once again a year later to x_M. Furthermore, this can go on indefinitely.

In the following problem, for convenience we take x to be the *population* of whales.

EXAMPLE 1 **Finding the Maximum Sustainable Yield**

Dasgupta and Heal[80] studied the blue whale fishery and estimated the function $f(x)$ given in the previous discussion to be $f(x) = x^{0.8204}(8.356 - x^{0.1996})$, where x is the number of blue whales and $f(x)$ is the increase in the number of blue whales over

[80] P. S. Dasgupta and G. M. Heal. 1979. *Economic Theory and Exhaustible Resources*. Cambridge, England: Cambridge University Press.

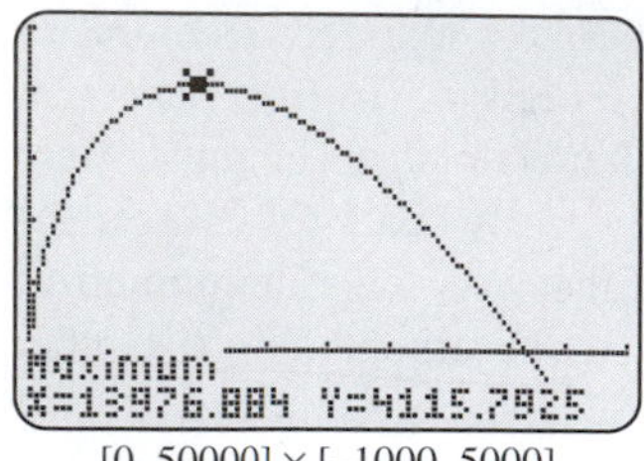

[0, 50000] × [–1000, 5000]

Screen 4.14
The function $y = x^{0.8204}(8.356 - x^{0.1996})$ assumes a maximum of about 13,977 when x is approximately 4116.

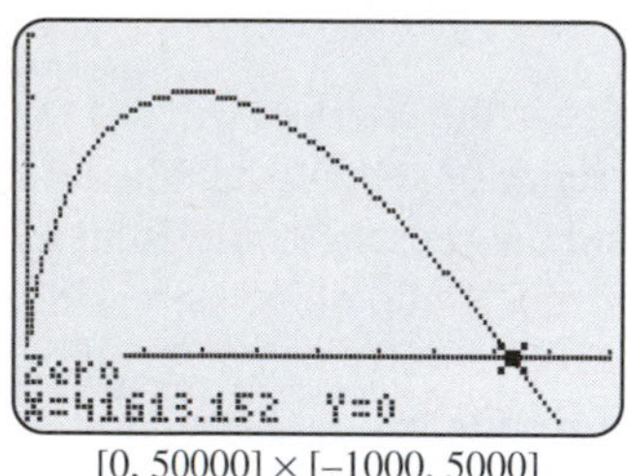

[0, 50000] × [–1000, 5000]

Screen 4.15
The function in Screen 4.14 is zero when x is approximately 41,613.

the next year. Find the carrying capacity and the maximum sustainable yield. What should regulators do if there are 4000 whales? 12,000?

Solution

Graph on your computer or graphing calculator using a screen with dimensions [0, 50000] by [−1000, 5000]. Use the available operations and obtain Screen 4.14. As we see, the maximum occurs at $x_M \approx 13{,}977$ with $f(13{,}977) \approx 4116$. The maximum sustainable yield is $f(13{,}977) = 4116$, or 4116 whales.

Using available operations, we find that $y = 0$ when $x \approx 41{,}613$. Thus, the carrying capacity is $K = 41{,}613$, or 41,613 whales. See Screen 4.15.

If the blue whale population is 4000, then we find that $f(4000) \approx 2814$. Thus, at the end of the year there will be $x + y = 6814$ whales. Since this is well below $x_M = 13{,}917$, no harvesting should be done.

If the blue whale population is 12,000, then we find that $f(12{,}000) \approx 4079$. Thus, at the end of the year there will be $x + y = 16{,}079$ whales. Since this is well above $x_M = 13{,}917$, $16{,}079 - 13{,}917 = 2162$ whales could be harvested, resulting in a population of 13,917. ■

Sometimes, an argument is made that the amount of fish (or some other animal species) should be kept at the carrying capacity. One might speculate on the condition of the fish in the fishery if the biomass were at the carrying capacity. Since the resources are such that no additional pound of fish can survive an additional year, the fish must be obtaining the absolute minimal amount of food for survival. This clearly does not represent pleasant living conditions for the fish or sufficient resources for the fish to grow well and reproduce.

We decided that in a relatively small lake a small population of fish would have enough resources to breed and produce more fish. If the lake is relatively large, however, with only a small number of fish, the fish might not be able to find each other to breed. (For some species of whales the feeding grounds amount to as much as 10 million square miles.) In this circumstance, instead of the population and biomass *increasing*, they would actually *decrease* owing to other natural causes.

Now assume as before that the biomass in pounds of fish is x and the increase in the biomass over the next year is given by $y = f(x)$. The biomass 1 year later is therefore $x + y$, and we wish to harvest y.

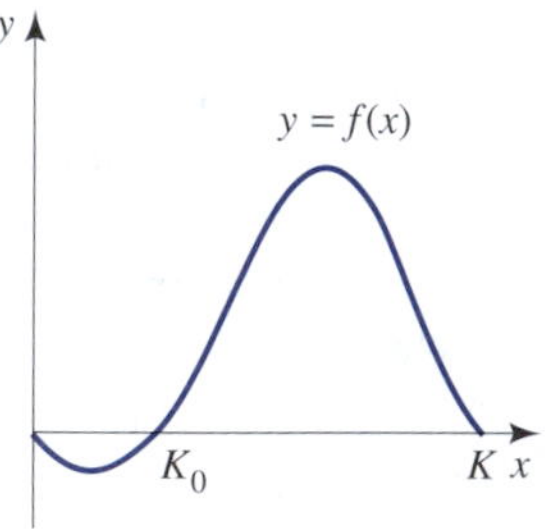

Figure 4.14
If the population becomes less than K_0, then population decreases.

Figure 4.14 indicates that in a relatively large lake the increase in the biomass is *negative*, that is, the biomass decreases for small numbers of fish for the reasons indicated earlier. For some larger amount of fish, indicated by K_0, the increase in biomass moves to zero, after which the biomass then increases.

The number K_0, called the **minimum viable biomass level**, is very important. If the biomass of fish in the lake ever goes *below* K_0, then the population will head eventually to zero. Thus, overfishing to the point where the biomass goes below the minimum viable biomass level results in an irrevocable catastrophe (assuming no intervention).

In the following problem we take for convenience x to be the *population* of whales.

EXAMPLE 2 **Management of a Fishery**

Suppose the fishery consisting of the Antarctic blue whale has the function $f(x)$ in the above discussion given by

$$f(x) = 0.000002(-2x^3 + 303x^2 - 600x)$$

where x is in thousands of whales. Find the maximal sustainable yield. Also estimate the minimum viable population level and the carrying capacity.

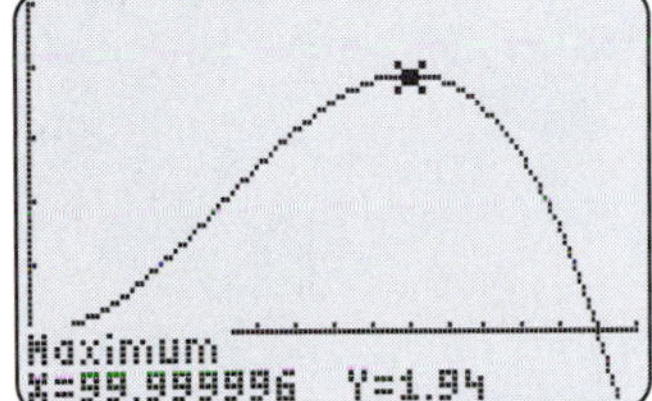

[0, 160] × [−0.5, 2.5]

Screen 4.16
The function $y = 0.000002(-2x^3 + 303x^2 - 600x)$ assumes a maximum of about 1.94 when x is approximately 100.

Solution

Graph on your computer or graphing calculator using a screen with dimensions [0, 160] by [−0.5, 2.5]. Use the available operations and obtain Screen 4.16. As can be seen, the maximum occurs at $x \approx 100$ with $f(100) \approx 1.94$. Thus, the maximum sustainable yield is $f(100) = 1.940$, or 1940 whales.

We also can obtain Screen 4.17, where we see that $y = 0$ when $x \approx 149.5$. Thus, the carry capacity is about 150,000 whales.

But what is the minimal viable population level? The graphs given by our computer or graphing calculator are not complete, since they do not show the fact that $f(x)$ is negative for small values of x. Thus, if we just relied on our computer or graphing calculator, we could be misled into thinking that K_0 shown in Figure 4.14 did not exist.

To algebraically find the zeros of f, set $f(x) = 0$, and obtain

$$2x^2 - 303x + 600 = 0$$

Using the quadratic formula yields $x \approx 2$ and $x \approx 150$. Thus, the minimum viable population level is about 2000 whales. We can verify this on our graphing calculators by graphing using a window with dimensions [0, 3] by [−0.001, 0.001]. See Screen 4.18.

Whaling scientists have estimated the above numbers for the carrying capacity and the maximum sustainable yield of the Antarctic blue whale. The minimum viable population level is unknown.

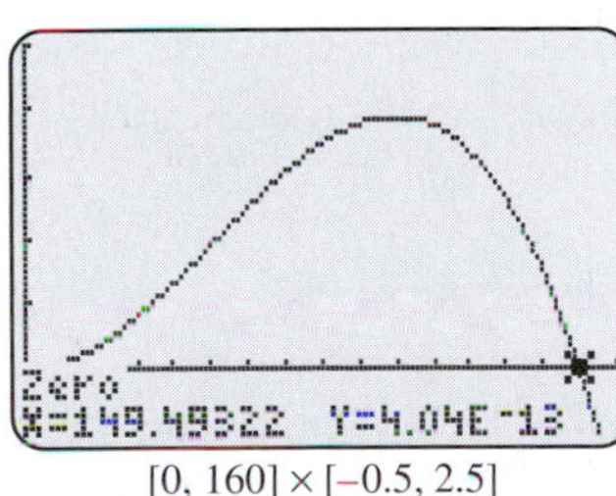

[0, 160] × [−0.5, 2.5]

Screen 4.17
The function in Screen 4.16 is zero when x is approximately 149.5.

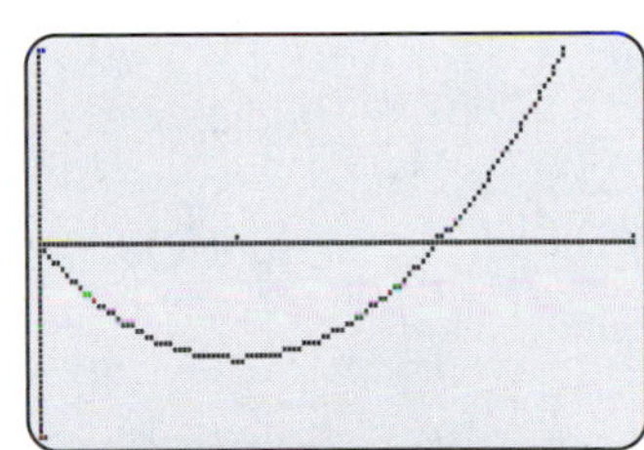

[0, 3] × [−0.001, 0.001]

Screen 4.18
We see a graph of the function in Screen 4.16 for small values of x. We note that the function is zero when x is approximately 2.

■

CONNECTION
The Antarctic Blue Whale Fishery

In the 1960s the population of the Antarctic blue whale was reduced to levels that were thought to lead to extinction. Under the protection of the International Whaling Commission the population increased to almost 10,000 in about 10 years and to about 50,000 in 1994.

CONNECTION
Fish Catches by Region

Fish Catches by Region, 1988 (in metric tons, live weight catches)

Region	Total Catch
World Total	**97,985,400**
Africa	5,310,300
Asia	43,601,200
Europe	12,877,400
North America	9,567,900
Oceania	887,300
South America	14,412,200
(Former) USSR	11,332,100

CONNECTION
The Pacific Halibut Fishery

Halibut is a highly regarded table fish, and the halibut fishery is second only to the Pacific salmon fishery. Halibut was heavily fished until the population dropped dramatically below its maximum sustainable yield. Under the supervision of the Pacific Halibut Commission, founded in 1924, the overfished halibut stocks were steadily built to near maximal sustainable yields by the 1950s. However, recent incursions by Japanese and Russians have seriously depleted the halibut stocks again.

For addition material on mathematical bioeconomics and the optimal management of renewable resources, see Clark[81].

Enrichment Harvesting

EXAMPLE 3 **Harvesting a Renewable Resource**

Grafton and colleagues[82] studied Canada's northern cod fishery and created the mathematical model given by the equation

$$f(x) = 0.30355x\left(1 - \frac{x}{3.2}\right)^{0.35865}$$

where x is the biomass of the population in millions of tons of cod and $f(x)$ is the growth of the biomass in millions of tons per year when no cod are harvested. Before the 1950s the industry harvested the cod at the rate of $h(x) = 0.20$, that is, 200,000 tons of cod per year were harvested no matter how large or small the biomass. A graph of these two functions is shown in Figure 4.15. Let p be the change in the population of the cod under this harvesting policy. What will happen to the growth of the population when the biomass is at each of the two levels a and b, shown in the graph. Indicate whether the growth rate will increase, decrease, or neither.

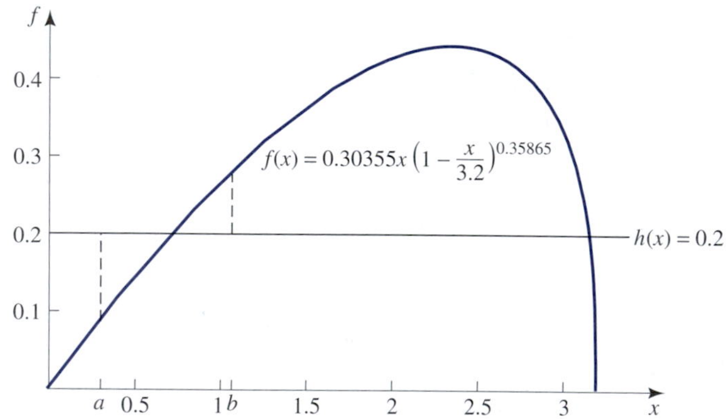

Figure 4.15

Solution

The population under the given harvesting regime is just $p = f - h$. We notice from the graph that at $x = a$, h is larger than f, so p is negative. This means that the total population of cod is decreasing. To continue to deplete the cod resource at this rate with the given biomass of a would lead to a serious depletion in time.

We notice from the graph that at $x = b$, h is smaller than f, so p is positive. This means that the total population of cod is increasing. Depleting the cod resource at this rate with the given biomass of b would be sustainable. ■

[81] Colin Clark. 1976. *Mathematical Bioeconomics*. New York: John Wiley and Sons.

[82] R. Quentin Grafton, Leif K. Sandal, and Stein Ivar Steinshamn. 2000. How to improve the management of renewable resources: the case of Canada's northern cod fishery. *Amer. J. Agr. Econ.* 82:570–580.

Remark Since the cod fishery remained healthy before the 1950s, we can only assume that the total cod population was roughly between 1 and 3 million tons.

CONNECTION
The Harp Seal

Scientists with Canada's Department of Fisheries and Oceans (DFO) have been studying harp seals to produce a model of harp seal population dynamics. The model assists the government in setting the total allowable catch (TAC). At this time the DFO considers harp seals to be abundant with an estimated population of 5.2 million. As a consequence, the government has set the TAC at a relatively generouos 975,000 for the three years from 2003 to 2005. However, the government is committed to keeping the population above 3.85 million. If the population dips below this mark, the TAC will be lowered. If the population hits the critical point of 1.65 million, the hunt will be shut down. National Geograhic, March 2004.

CONNECTION
Northern Cod Moratorium

The northern cod was so heavily fished that the Canadian Department of Fisheries and Oceans instituted a fishing moratorium in July 1992 that still remains in force. The estimated exploitable biomass was estimated at only 21,000 tons, or just 1% of its level 10 years earlier.

Exercise Set 4.6

Applications and Mathematical Modeling

1. **Fishery** Clark[83] reports that for the Eastern Pacific yellowfin tuna the function

$$f(x) = 2.61x\left(1 - \frac{x}{1.34 \times 10^8}\right)$$

has been used. Here x is in kilograms. Graph $y = f(x)$. Find the carrying capacity and the maximum sustainable yield.

2. **Fishery** Clark[84] reports that for the Antarctic fin whale the function $f(x)$ is estimated to be

$$f(x) = 0.08x\left(1 - \frac{x}{400{,}000}\right)$$

where x is the number of Antarctic fin whales. Graph $y = f(x)$. Find the carrying capacity and the maximum sustainable yield.

3. **Fishery** Clark[85] reports that a typical function $f(x)$ that is used is

$$f(x) = rx\left(1 - \frac{x}{K}\right)$$

where x is the biomass. Graph $y = f(x)$. Show that the carrying capacity is K and the maximum sustainable yield is $K/2$.

4. **Harvesting** Referring to Example 3 of the text, Grafton reports that the total catch of northern cod peaked in 1968 at 800,000 tons. Graph this harvesting function together with the $f(x)$ given in Example 3, and determine what should happen according to this model.

5. Let $y = f(x) = 16x^3 - x^4$ and let x represent the thousands of pounds of fish at the beginning of a year and y the thousands of pounds of increase in the biomass over the next year. Find the carrying capacity and the maximum sustainable yield.

6. Let $y = f(x) = 20\sqrt[3]{x} - \sqrt[3]{x^4}$ and let x represent the thousands of pounds of fish at the beginning of a year and y the thousands of pounds of increase in the biomass over the next year. Find the carrying capacity and the maximum sustainable yield.

7. Let $y = f(x) = -x^3 + 15x^2 - 27x$ and let x represent the millions of pounds of fish at the beginning of a year and y the millions of pounds of increase in the biomass over the next year. Find the maximum sustainable yield, the carrying capacity, and the minimal viable population level.

8. **Fishery** Redo Exercise 7 if $f(x) = -x^5 + 5x^4 - 4x^3$.

[83] Colin W. Clark. 1976. *Mathematical Bioeconomics*. New York: John Wiley and Sons.
[84] Ibid.
[85] Ibid.

Summary Outline

- **Rules For Derivatives** If $f(x)$ and $g(x)$ are continuous, c a constant, and n a real number, then

$$\frac{d}{dx}(c) = 0.$$

$$\frac{d}{dx}(cf(x)) = c\frac{d}{dx}f(x)$$

$$\frac{d}{dx}(f(x)\cdot g(x)) = f(x)\cdot\frac{d}{dx}g(x) + g(x)\cdot\frac{d}{dx}f(x)$$

$$\frac{d}{dx}\left(\frac{f(x)}{g(x)}\right) = \frac{g(x)\cdot\frac{d}{dx}f(x) - f(x)\cdot\frac{d}{dx}g(x)}{[g(x)]^2} \text{ if } g(x) \neq 0.$$

$$\frac{d}{dx}(f[g(x)]) = f'[g(x)]\cdot\frac{d}{dx}g(x)$$

$$\frac{d}{dx}(e^x) = e^x$$

$$\frac{d}{dx}\left(e^{f(x)}\right) = e^{f(x)}f'(x)$$

$$\frac{d}{dx}(\ln x) = \frac{1}{x}$$

$$\frac{d}{dx}(\ln f(x)) = \frac{f'(x)}{f(x)}$$

$$\frac{d}{dx}(x^n) = nx^{n-1}$$

$$\frac{d}{dx}(f(x) \pm g(x)) = \frac{d}{dx}f(x) \pm \frac{d}{dx}g(x)$$

$$\frac{d}{dx}([g(x)]^n) = n[g(x)]^{n-1}\cdot\frac{d}{dx}g(x)$$

$$\frac{d}{dx}\left(a^{f(x)}\right) = a^{f(x)}f'(x)\ln a$$

- **Elasticity of Demand** If the demand equation is given by $x = f(p)$, then

$$E = -\frac{p}{x}\frac{dx}{dp}$$

is called the **elasticity of demand**. If $E > 1$, then demand is **elastic**. If $E < 1$, then demand is **inelastic**. If $E = 1$, then demand has **unit elasticity**. p. 277.

Review Exercises

In Exercises 1 through 34, find the derivative.

1. $f(x) = x^6 - 3x^2 + 1$

2. $f(x) = 4x^{6/5} + 5x^{2/5} + 3$

3. $f(x) = 2x^{-3} + 3x^{-2/3}$

4. $f(x) = 2x^{1.3} + 3x^{-2.4}$

5. $f(x) = \pi^2x^2 + 2x - \pi$

6. $f(x) = \pi^3x^3 + e^2x^2 - \pi^2$

7. $f(x) = \sqrt[3]{x^2}$ **8.** $f(x) = \sqrt[3]{x^{-5}}$

9. $f(x) = (x+3)(x^3+x+1)$

10. $f(x) = (x^2+5)(x^5-3x+2)$

11. $f(x) = (\sqrt{x})(x^4 - 2x^2 + 3)$

12. $f(x) = (\sqrt[3]{x})(x^3 - 3x + 5)$

13. $f(x) = (x^{3/2})(x^2 + 4x + 7)$

14. $f(x) = (x^{-7/2})(x^3 + 2x + 1)$

15. $f(x) = \dfrac{x-1}{x+1}$ **16.** $f(x) = \dfrac{\sqrt{x}}{2x+1}$

17. $f(x) = \dfrac{x^3}{x^2+1}$ **18.** $f(x) = \dfrac{x^2+x+1}{3x-2}$

19. $f(x) = \dfrac{x^2-1}{x^4+1}$ **20.** $f(x) = (3x+1)^{10}$

21. $f(x) = (x^2+4)^{20}$ **22.** $f(x) = (\sqrt{x} - 5)^{5/2}$

23. $f(x) = \sqrt{x^2+x+1}$

24. $f(x) = x^3(x^2+x+2)^5$

25. $f(x) = (x^2+3)(x^4+x+1)^5$

26. $f(x) = \dfrac{x^2+1}{(x^4+x+1)^4}$

27. $f(x) = \sqrt{x^{3/2}+1}$ **28.** $f(x) = \sqrt[3]{\sqrt{x}+1}$

29. e^{-7x} **30.** e^{2x^3+1}

31. $(x^2+x+2)e^x$ **32.** $\ln(x^2+1)$

33. $e^{x^2}\ln x$ **34.** $\ln(x^2+e^x)$

In Exercises 35 through 38, find dy/dx.

35. $y = \sqrt[5]{x}$ **36.** $y = \sqrt[3]{x^2}$

37. $y = \dfrac{1}{x^3}$ **38.** $y = \sqrt{x} + \dfrac{1}{\sqrt{x}}$

39. Learning Curve Chen and McGarrah[86] report on a study of the KD 780 camera assembly operation at the Foster

[86] Gordon K. Chen and Robert E. McGarrah. 1982. *Productivity Management Text and Cases*. New York: The Dryden Press.

Optical Equipment company. They found that the direct labor hours y required to produce the xth camera was approximated by the equation $y = h(x) = 716x^{-0.474}$. Graph. Find values for $h'(x)$ at $x = 100, 200, 300$, and 400. Interpret what these numbers mean. What is happening? Explain in terms of a learning process.

40. Economics of Scale In 1955 Surdis[87] obtained records from a utility company regarding its trench digging operations. The records show that the unit cost $C(n)$ per foot of earth removed by the mechanical trencher is given approximately by

$$C(n) = \frac{15.04 + 0.74n}{25n}$$

where n is the number of hours worked per day.

a. Graph. Find values for $C'(n)$ at $x = 2, 4, 6$, and 8. Interpret what these numbers mean. What is happening? Units costs for hand digging was found to be \$0.60.

b. Approximate the number of hours worked at which using the trench digging machinery is more cost effective than hand digging.

41. Economies of Scale In 1996 Kebede and Schreiner[88] reported on their study of the economies of scale in dairy marketing cooperatives in the Rift Valley of Kenya. They estimated that the average cost C in Kenyan shillings per kilogram of a plant was given by $C(Q) = 1.2530Q^{-0.07}$, where Q is in units of 1000 kg. (The equation given in the paper is in error.) Graph. Find values for $C'(Q)$ at $Q = 100, 200, 400$, and 800. Interpret what these numbers mean. What is happening? What does this mean in terms of economies of scale?

42. Transportation Costs Koo and coworkers[89] used production cost data from the U.S. Department of Agriculture to estimate transportation costs for durum wheat. They found that the rail transportation cost R in dollars per ton from producing regions to durum mills was approximated by the equation $R(D) = 0.0143D^{0.67743}$, where D is the distance in miles. Graph. Find values for $R'(D)$ at $D = 100, 200, 400$, and 800. Interpret what these numbers mean. What is happening? What does this mean in terms of economies of scale?

43. Profits The daily profits in dollars of a firm are given by $P = -10x^3 + 800x - 10e^x$, where x is the number of items sold. Find the instantaneous rate of change when x is (a) 2, (b) 3, (c) 4, (d) 5. Interpret your answers.

44. Velocity A particle moves according to the law $s(t) = -2t^4 + 64t + 15$, where s is in feet and t in seconds. Find the instantaneous velocity at $t = 1, 2, 3$. Interpret your answers.

45. Biology Schotzko and Smith[90] showed that the number y of wheat aphids in an experiment was given approximately by $y = f(t) = 5.528 + 1.360t^{2.395}$, where t is the time measured in days. Find $f'(t)$ and explain what this means.

46. Biology Talekar and coworkers[91] showed that the number y of the eggs of *O. furnacalis* per mung bean plant was approximated by $y = f(x) = -57.40 + 63.77x - 20.83x^2 + 2.36x^3$, where x is the age of the mung bean plant and $2 \le x \le 7$. Find $f'(x)$, and explain what this means. Use the quadratic formula to show that $f'(x) > 0$ for $2 \le x \le 7$. What does this say about the preference of *O. furnacalis* to laying eggs on older plants?

47. Biology Buntin and colleagues[92] showed that the net photosynthetic rate y of tomato leaves was approximated by the equation $y = f(x) = 8.094(1.053)^{-x}$, where x is the number of immature sweet potato whiteflies per square centimeter on tomato leaflets. Find $f'(x)$, and explain the significance of the sign of the derivative.

48. Costs The weekly total cost function for a company in dollars is $C(x)$. If $C'(100) = 25$, what is the approximate cost of the 101st item?

49. Elasticity Given the demand curve $x = 10 - 2p$, determine whether the demand is elastic, inelastic, or unit elastic if (a) $p = 2$, (b) $p = 2.5$, (c) $p = 3$.

50. Demand If the demand for a product is given by $\frac{300}{3x + 1}$ find the marginal demand, where p and x have the usual meanings.

51. Economies of Scale Christensen and Greene[93] studied economies of scale in U.S. electric power generation. A graph of their results is shown. Explain in words what this graph says about economies of scale.

---

[87] John Surdis. 1957. Trench digging. *J. Industrial Econ.* 5:233–238.

[88] Ellene Kebede and Dean F. Schreiner. 1996. Economies of scale in dairy marketing cooperatives in Kenya. *Agribusiness* 12(4):395–402.

[89] Won W. Koo, Joel T. Golz, and Seung R. Yang. 1993. Competitiveness of the world durum wheat and durum milling industries under alternative trade policies. *Agribusiness* 9(1):1–14.

[90] D. J. Schotzko and C. M. Smith. 1991. Effects of host plant on the between-plant spatial distribution of the Russian wheat aphid. *J. Econ. Entomol.* 84:1725–1734.

[91] Narayan Talekar, Chih Pin Lin, Yii Fei Yin, Ming Yu Ling, Yi De Wang, and David Chang. 1991. Characteristics of infestation by *Ostrinia furnacalis* in mungbean. *J. Econ. Entomol.* 84:1499–1502.

[92] G. Davis Buntin, David Gilbertz, Ronald Oetting. 1993. Chlorophyll loss and gas exchange in tomato leaves after feeding injury by *Bemisia tabace*, *J. Econ. Entomol.* 86:517–522.

[93] L. H. Christensen and W. H. Greene. 1976. Economies of scale in U.S. electric power generation. *J. Political Econ.* 84(4).

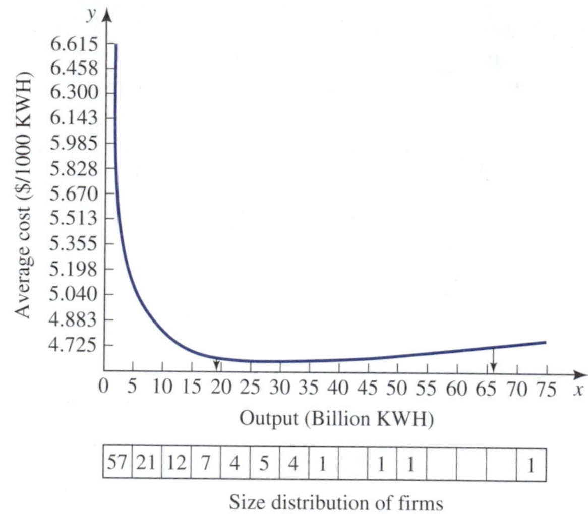

52. **Cost Function** McGee[94] studied the average costs of steel ingot production. A graph of his result is shown. Explain what this graph has to say about efficiencies in this industry.

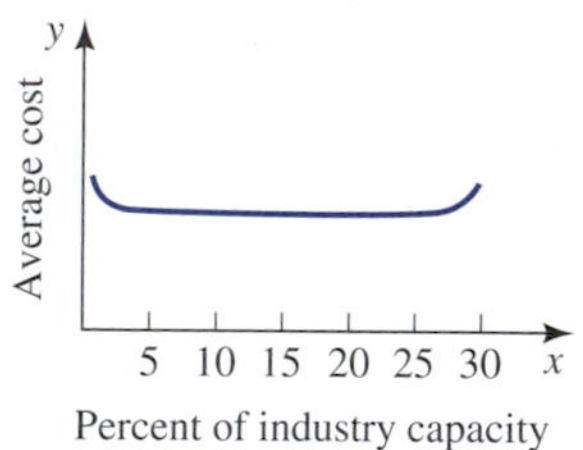

[94] J. S. McGee. 1974. *Efficiency and Economies of Size. Industrial Concentration: The New Learning*. Boston: Little, Brown.

53. Taplin[95] showed that if Y is dollars spent overseas, then the equation $Y = AX^{0.7407}$ approximately held, where X is household disposable income in dollars per week and A is a constant. Explain in words what the exponent 0.7407 means.

54. Bell[96] estimated that if Q was the quantity of haddock sold in New England, then $Q = Ap^{-2.174}$, where p is the price per pound of haddock. What is the elasticity of demand?

55. Shown are the graphs of two demand curves. For the same value of p_0 which has the larger elasticity of demand? Explain.

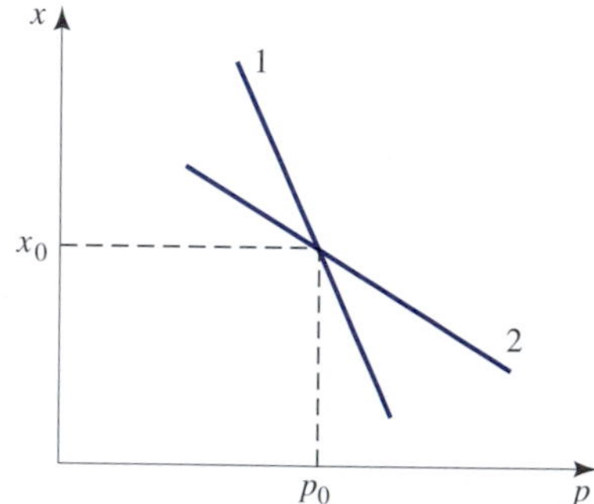

56. Let demand be given by $x = ap^{-n}$ where a and n are positive constants. Find the elasticity of demand, E.

[95] John Taplin. 1980. A coherence approach to estimates of price elasticities in the vacation travel market. *J. Transport. Econ. Policy* 14.

[96] F. W. Bell. 1958. The Pope and the price of fish. *Amer. Econ. Rev.* December.

Chapter 4 Cases

CASE 1 Loss of Species Due to Deforestation

In one part of Edwin Wilson's wonderful book *The Diversity of Life*[97] he attempts to approximate the number of species lost due to the deforestation of the tropics. He starts with the mathematical model $N = aS^b$, which approximates the number N of different species that exist in an area S of tropical forest. The positive constants a and b will depend on the specific region. The constant a is not important to this discussion. Wilson notes that empirical studies indicate that b ranges from about 0.15 to 0.35 and that 0.30 is typical. Thus, we will take $N = f(S) = aS^{0.30}$.

Notice that since N is an increasing function of S, this formula states that the number of species in an area will increase with the size of the area. The formula for N attempts to quantify this relationship.

Estimates from both the ground and the air indicate that the deforestation of the tropics is taking place at about 1.8% per year.

(a) From the formula for N, estimate the percent decrease per year in the number of species that will occur.

(b) Now estimate the actual number of species lost per year using the conservative estimate of $b = 0.15$ and make the conservative estimate that there are 10,000,000 species.

[97] Edward O. Wilson. 1992. *The Diversity of Life*. New York: W. W. Norton.

CASE 2 **Polynomial Extrapolation**

Often in applications an experiment or survey yields the values of some function f at a finite number of discrete points. Suppose, then, we know the values $f(x_1), f(x_2), \ldots, f(x_n)$, at the n distinct (nodal) points $x_1, x_2, \ldots, x_n$. We always assume $x_1 < x_2 < \cdots < x_n$. A theorem in mathematics states that there exists a unique polynomial p of degree $\leq n-1$ such that $p(x_i) = f(x_i)$, for $i = 1, 2, \ldots, n$. Such a polynomial is called an **interpolating polynomial**.

We can use our graphing calculators to find this interpolating polynomial. We then can use the interpolating polynomial to approximate the values of f at nonnodal points on the interval (x_1, x_n). This is called **interpolation**. We also can use the interpolating polynomial to approximate the function f outside the interval (x_1, x_n). This is called **extrapolation**.

Although the interpolating polynomial can sometimes be useful in both interpolating and extrapolating, its use in both interpolating and extrapolating is fraught with dangers. Sometimes, grossly incorrect answers can be obtained!

Consider again the annual population of the United States from 1790 to 1794. In Section 1.3, we noted that the population increased at almost exactly 3.3% a year. We then created a mathematical model based on the exponential function to model the population. We then used this model to estimate the population in 1804. Biologists have observed many populations, both human and nonhuman, following the exponential model, at least when the population is small enough not to be inhibited by a lack of resources. Thus, had we formulated this exponential model in 1794, we would have had some confidence in using it to project the population over the next 10 years. Recall that the model gave a reasonable approximation of the population in 1804.

Now let us use the interpolating polynomial to extrapolate the population to 1804. Let t be the year and $f(t)$ the population in thousands in the year t. Recall we have $f(1790) = 3926$, $f(1791) = 4056$, $f(1792) = 4194$, $f(1793) = 4332$, $f(1794) = 4469$. Let the year 1790 correspond to $t = 0$. Enter the data (0,3926), (1,4056), (2,4194), (3,4332), (4,4469) into the statistical section. Select quadratic regression, and obtain the interpolating polynomial.

(a) Graph on [1790, 1804] on your graphing calculator.

(b) Find $p(1804)$. Compare with the actual population in 1804 of 6,065,000.

(c) Find $p'(t)$ approximately at $t = 1791$, 1792, 1793, and 1794. What do your answers mean in terms of the growth of the population? Do your answers seem reasonable? Why or why not?

(d) Graph on [1780, 1804], and find the population this model predicts for the year 1780. Find p' for this year. Are your answers believable? Do you think this model gives an accurate model of the population from 1780 to 1790? Why or why not?

(e) What do you think went wrong? Was there any *biological* basis to have confidence in this interpolating polynomial to model the population? Was there any biological basis to have confidence in the exponential model used in Section 1.5? What are they?

5 Curve Sketching and Optimization

In this chapter we learn how to sketch the graph of a function and how to find the maximum and the minimum of a function on a given interval. To mention just a few applications, we see how to find the quantity of sales that will produce the most profit, the quantity of sales that will produce the most revenue, and the average velocity a fleet of vans should maintain to minimize costs. In the last section we see how to find the derivative when given an equation in two variables, only one of which can be solved for.

CASE STUDY Federal Tax Revenue

In the early 1980s, the Reagan administration was struggling with the consequences of lowering the tax rates. Some had the view that if the tax rate is lowered, the government's total income is lowered. Others argued, however, that lowering the tax rate *increases* total tax revenue.

One of the latter was economist Arthur Laffer of the University of Southern California. He argued as follows. Let $R = R(x)$, where x is the percentage of income taken by the federal government in the form of taxes and $R(x)$ the total revenue from taxes from a rate of x percent. Obviously, if the rate is set at zero, the government has no revenue. That is, $R(0) = 0$. If the rate x is increased somewhat, surely the revenue $R(x)$ increases somewhat. So $R(x)$ must be an increasing function at least for small values of x. On the other hand, it is reasonable to expect that if $x = 100$, that is, the government takes everything, then probably nobody will wish to earn any money and federal revenue will be zero. Thus, $R(100) = 0$. If x is a little less than 100, then some people will probably earn some money, resulting in some income for the federal government. So one can assume that for x near 100, $R(x)$ will be decreasing. At this point the graph of the curve $R = R(x)$ looks like the solid curve in Figure 5.1.

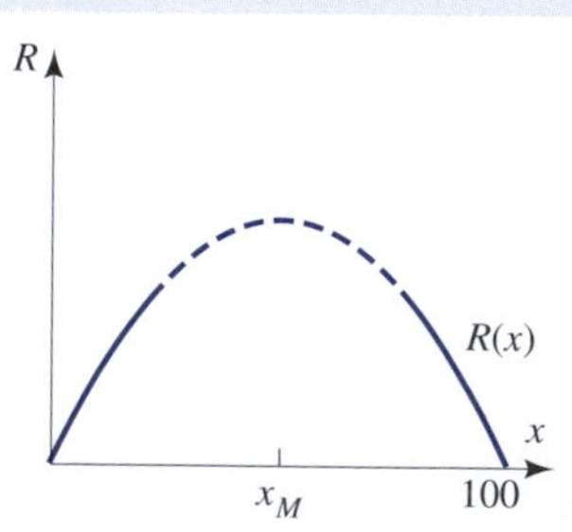

Figure 5.1
As the tax rate increases, tax revenue initially increases but then begins to decrease.

Now Laffer made the assumption that the rest of the curve probably looks like the dotted curve in Figure 5.1 at the intermediate values. (This curve is called a Laffer curve.) Label as x_M the value at which this curve changes direction from increasing to decreasing.

Suppose currently $x = c$. The argument was then made that x_M was *less* than c, as in Figure 5.2. Since $R(x)$ is *decreasing* near the value c, increasing x results in a *decrease* of revenue. Decreasing x therefore results in an increase of revenue.

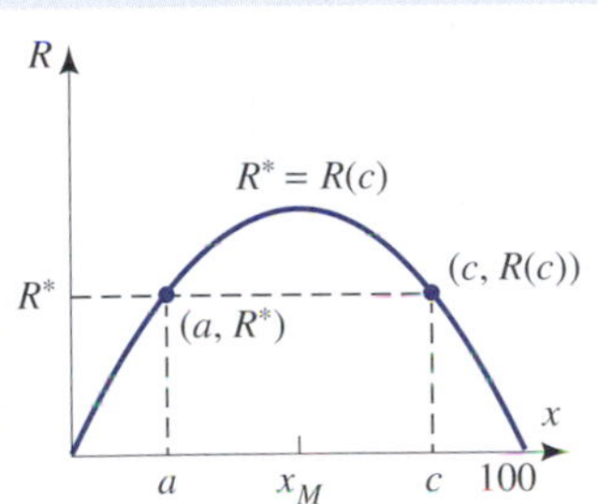

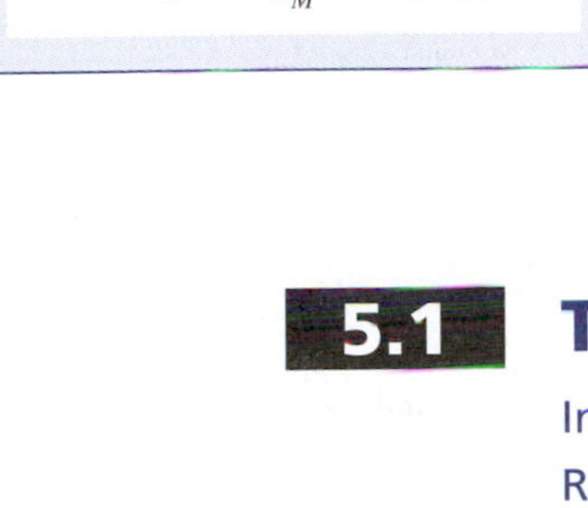

Figure 5.2
According to the Laffer curve, the same revenue can be obtained for two tax rates.

The Reagan administration wished only to lower the tax rate, not to increase federal tax revenue. Drawing a horizontal line at height $R^* = R(c)$ in Figure 5.2 results in this line intersecting the graph of $y = R(x)$ at *two* points: (c, R^*), of course, but also (a, R^*), as shown in the figure. Thus, according to this model, the income tax rate could be decreased all the way down to $x = a$, and the government would still have the same revenue.

Suppose in the above discussion that $R(x) = \sqrt[3]{x}(100 - x)$, where $R(x)$ is measured in billions of dollars. For what value of x does R increase, and for what values of x does R decrease? For the answers, see Example 8 in Section 5.1.

5.1 THE FIRST DERIVATIVE

Increasing and Decreasing Functions
Relative Extrema

APPLICATION **Finding Maximum Revenue**

Let the demand equation be given by $p(x) = e^{-0.02x}$. For what values of x does revenue increase? Decrease? For the answer, see Example 6 on page 304.

Increasing and Decreasing Functions

In this section we examine how to use the sign of the first derivative to determine whether a function is increasing or decreasing.

We already know that for straight lines a positive slope indicates the line is rising and therefore the function is increasing. But if the slope is negative, the line is falling, and therefore, the function is decreasing. A similar situation exists for curves that are not lines.

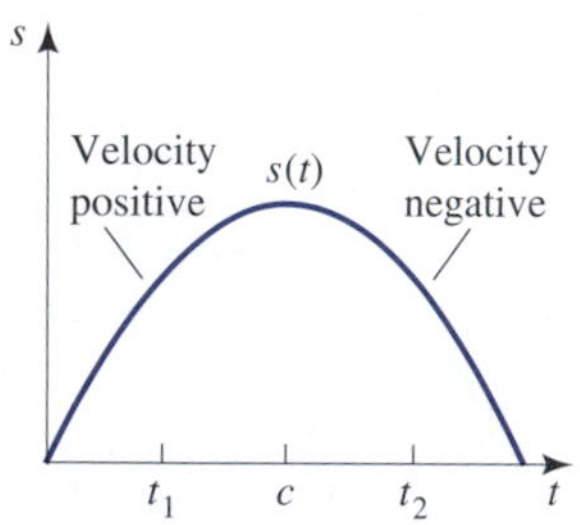

Figure 5.3
Positive velocity means that the ball is rising; negative velocity means that the ball is falling.

Suppose that a ball is thrown upward, and the function $s(t)$ gives the height of the ball in feet, where t is measured in seconds (Figure 5.3). We already know that the derivative $s'(t)$ gives the velocity $v(t)$ at time t and also the slope of the tangent line to the graph of the curve at t. If the slope or velocity $s'(t_1)$ is positive, then we know that the ball is rising at t_1. If the slope or velocity $s'(t_2)$ is negative, then we know that the ball is falling at t_2. We can see in Figure 5.3 that for values of t less than c the slope, or derivative, is positive and the function is increasing, whereas for $t > c$ the slope or derivative is negative and the function is decreasing.

Interactive Illustration 5.1

Increasing and Decreasing

This interactive illustration allows you to explore the relationship between the slope of the curve and whether the function is increasing or decreasing. A graph is shown of a function encountered in the last chapter, $R(x) = -x^2 + 20x$. Move the cursor, and you will see tangent lines on the graph and values of the slope.

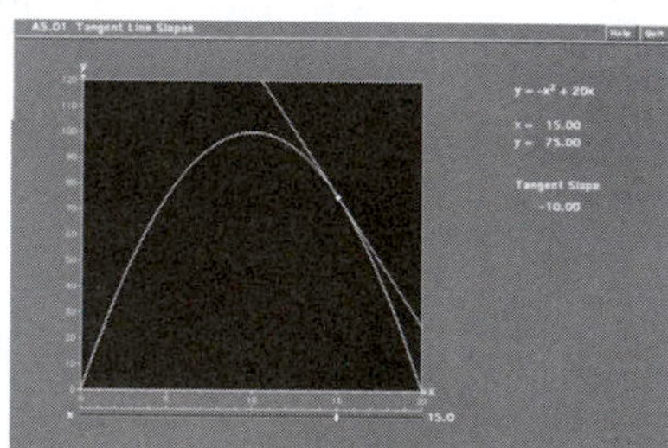

1. Find all points x where the slope of the tangent line is positive (and thus $f'(x)$ positive). Is the function increasing?
2. Find all points x where the slope of the tangent line is negative (and thus $f'(x)$ negative). Is the function decreasing?
3. What is happening when $f'(x) = 0$?

Exploration 1

The Sign of the Derivative

Graph the function $y_1 = f(x) = x^3 - 6x^2 + 9x + 2$ and its derivative $y_2 = 3x^2 - 12x + 9$ on the same screen with dimensions $[-1, 5]$ by $[-10, 10]$. Locate the two intervals on which $y_1 = f(x)$ is increasing. What is true about the sign of $f'(x)$ on these intervals? Now locate the interval on which $y_1 = f(x)$ is decreasing. What is true about the sign of $f'(x)$ on this interval? Does this make sense to you? Explain.

It is hoped that you just discovered the following theorem, a proof of which we do not give here.

Test for Increasing or Decreasing Functions

If for all $x \in (a, b)$, $f'(x) > 0$, then $f(x)$ is increasing (↗) on (a, b).
If for all $x \in (a, b)$, $f'(x) < 0$, then $f(x)$ is decreasing (↘) on (a, b).

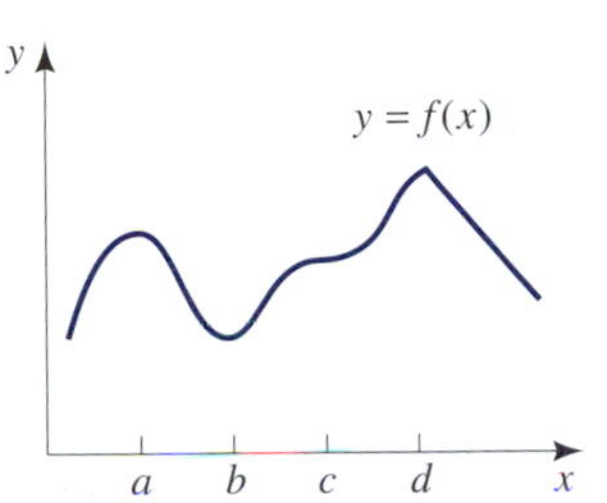

Figure 5.4
The critical values are at $x = a, b, c$, and d.

To find the intervals on which $f(x)$ is either always increasing or always decreasing, for a function like the one graphed in Figure 5.4, we need to find the intervals on which $f'(x)$ is either always positive or always negative. The simplest way to accomplish this is first to find all values where the derivative is neither positive nor negative, since normally only a small number of such values exist. Obviously, all values for which the derivative is *zero* fall into this category, but also in this category are all values for which the derivative does *not exist*. These two types of values are so important that they are given a special name: **critical values**.

DEFINITION **Critical Value**

A value $x = c$ is a **critical value** for a function $f(x)$ if

1. c is in the domain of the function $f(x)$ and
2. $f'(c) = 0$ or $f'(c)$ does not exist.

EXAMPLE 1 **Finding Critical Values Graphically**

Find the critical values for the function shown in Figure 5.4.

Solution

Since the derivative of the function is zero at the values a, b, and c, these three values are critical values. Because the derivative does not exist at the value d, this is also a critical value. ■

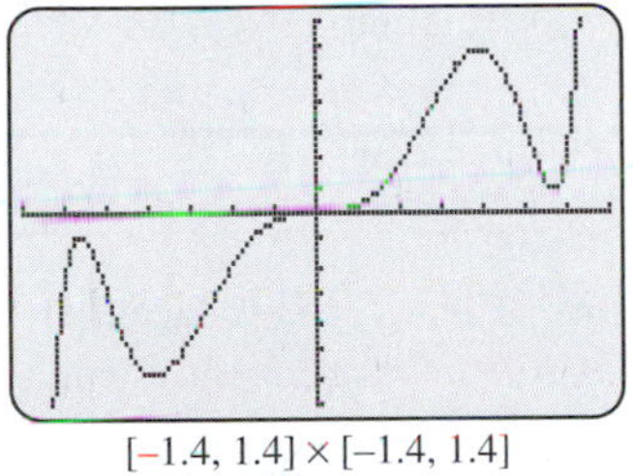
$[-1.4, 1.4] \times [-1.4, 1.4]$

Screen 5.1
Critical values for the function $y = 1.6x^9 - 8x^5 + 7x^3$ appear to be at about $x = \pm 1.15$, $x = \pm 0.8$, and $x = 0$.

EXAMPLE 2 **Finding Critical Values Graphically**

Graph $y = f(x) = 1.6x^9 - 8x^5 + 7x^3$ on your computer or graphing calculator using a window of dimension $[-1.4, 1.4]$ by $[-1.4, 1.4]$, and roughly estimate the critical values. Can you confirm your answer analytically?

Solution

The graph is shown in Screen 5.1. (The tick marks on the x-axis in the screen are spaced at intervals of 0.2). Very roughly, the critical values appear to be at $x = \pm 1.15$, $x = \pm 0.8$, and $x = 0$, since, at these points, the derivative appears to be zero. We can only confirm our $x = 0$ answer analytically, since we are unable to solve $f'(x) = 14.4x^8 - 40x^4 + 21x^2 = 0$ for the other solutions! ■

Exploration 2

Finding Critical Values Graphically

For the function in Example 2, use your computer or graphing calculator to graph $y = f'(x)$ on a screen with dimensions $[-1.4, 1.4]$ by $[-6, 6]$. Use available operations to find the approximate critical values.

Before we can use the first derivative to help graph a function, we need an important fact about continuous functions. Suppose $f(x)$ is continuous at every point on the interval (a, b) and never zero on this interval. Then we will now show that $f(x)$ is either always positive on this interval or always negative on this interval. To see why this must be true, suppose this were not the case and there were two points, c and d, on the interval with $f(c)$ and $f(d)$ of opposite signs. For the sake of definiteness, suppose $f(c) > 0$ and $f(d) < 0$. Refer to Figure 5.5. Since we are assuming that $f(x) \neq 0$, then the graph of $y = f(x)$ can never touch the x-axis. But if we draw a graph connecting the two points $(c, f(c))$ and $(d, f(d))$ with*out* touching the x-axis, our graph will need to "jump" over the x-axis. This will create a hole or gap in the graph, which is contrary to the definition of a continuous function. From this contradiction we then see that the function $f(x)$ can never change sign on the interval (a, b). Therefore, $f(x)$ is either always positive on the interval or always negative on the interval. We call this the **constant sign theorem**.

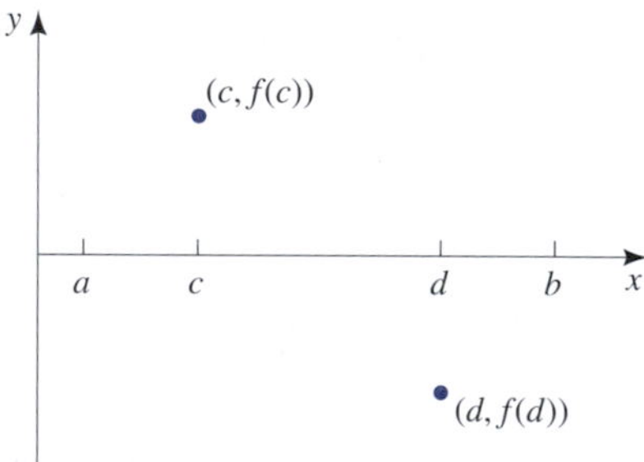

Figure 5.5

Constant Sign Theorem

If $f(x)$ is continuous on (a, b) and $f(x) \neq 0$ for any x in (a, b), then either $f(x) > 0$ for all x in (a, b) or $f(x) < 0$ for all x in (a, b).

We now use the first derivative to help us graph functions.

EXAMPLE 3 **Finding Where *f* Is Increasing or Decreasing**

Find the critical values and the intervals where $f(x) = x^3 - 3x^2 + 1$ is increasing or decreasing. Use this information to sketch a graph.

Solution

Step 1. Find the critical values.

(a) Notice that $f'(x) = 3x^2 - 6x$ exists everywhere. Thus, the only critical values are values at which $f'(x) = 0$.

(b) Setting $f'(x) = 3x^2 - 6x = 3x(x - 2) = 0$ gives $x = 0$ and $x = 2$. Thus, the critical values are $x = 0$ and $x = 2$.

Step 2. Find the subintervals on which $f'(x)$ has constant sign.

The two critical values divide the real line into the three subintervals: $(-\infty, 0)$, $(0, 2)$, and $(2, \infty)$. The derivative is continuous and never zero on each of these subintervals. By the constant sign theorem the sign of $f'(x)$ must therefore be constant on each of these subintervals.

Step 3. Determine the sign of f' on each subinterval.

Since the sign of $f'(x)$ is constant on the subinterval $(-\infty, 0)$, we need only evaluate f' at one (test) point of this subinterval to determine the sign on the entire subinterval. The same principal applies to the other two subintervals. So take the three test values to be $x_1 = -1$, $x_2 = 1$ and $x_3 = 3$. These three values fall respectively in the three subintervals given in the preceding step (Figure 5.6). Evaluating f' at these three values gives $f'(-1) = +9$, $f'(1) = -3$, $f'(3) = +9$. Thus, $f'(x) > 0$ on the first interval, $f'(x) < 0$ on the second interval, and $f'(x) > 0$ on the third interval.

Step 4. Use the test for increasing or decreasing functions.

We conclude that $f(x)$ is increasing ($\nearrow$) on the first interval, decreasing ($\searrow$) on the second, and increasing ($\nearrow$) on the third.

Now, using this information and the fact that $f(0) = 1$ and $f(2) = -3$, we obtain the graph in Figure 5.7.

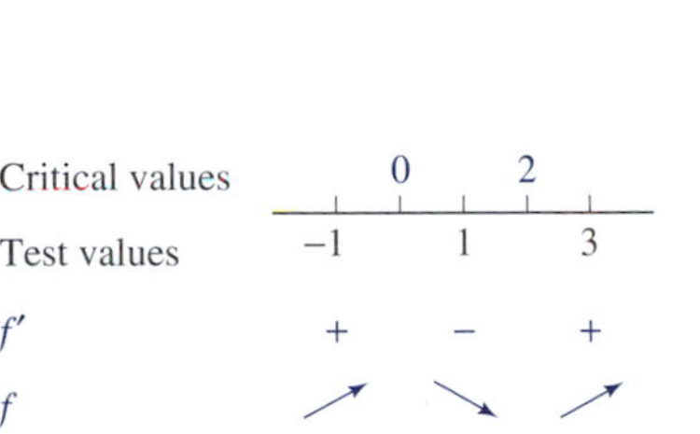

Figure 5.6

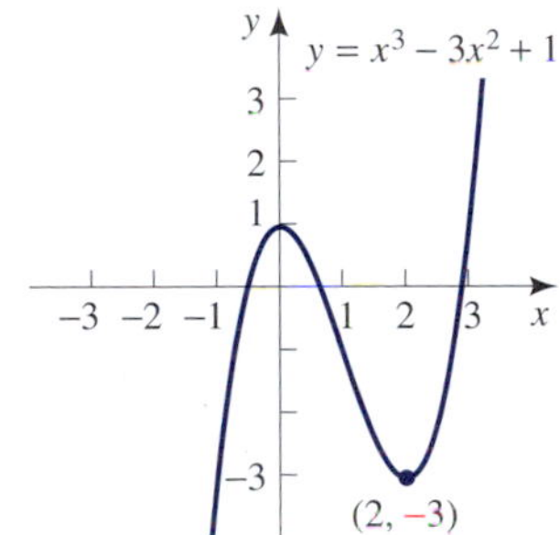

Figure 5.7
The function increases on $(-\infty, 0)$ and $(2, \infty)$ and decreases on $(0, 2)$.

■

Warning In Step 1 it is easy to overlook the solution $x = 0$. A common mistake is to set $f'(x) = 3x^2 - 6x = 0$, cancel x, and obtain $3x - 6 = 0$ or $x = 2$, as the *only* solution.

Procedure for Finding Where a Function Is Increasing or Decreasing

Step 1. Find the critical values. To do this:
(a) Locate all values c for which $f'(c) = 0$.
(b) Locate all values c for which $f'(c)$ does not exist but $f(c)$ does.
Step 2. Find the subintervals on which $f'(x)$ has constant sign.
Step 3. Select a convenient test value on each subinterval found in the previous step and evaluate f' at this value. The sign of f' at this test value will be the sign of f' on the whole subinterval.
Step 4. Use the test for increasing or decreasing functions to determine whether f is increasing or decreasing on each subinterval.

Relative Extrema

We now are interested in studying the peaks and troughs on the graph of a continuous function. For example, the graph in Figure 5.7 has a peak at (0, 1) and a trough at (2, −3). We call the peaks **relative maxima** and the troughs **relative minima**.

DEFINITION **Relative Maximum And Relative Minimum**

We say that the quantity $f(c)$ is a relative maximum if $f(x) \leq f(c)$ for all x in some open interval (a, b) that contains c.

We say that the quantity $f(c)$ is a relative minimum if $f(x) \geq f(c)$ for all x in some open interval (a, b) that contains c.

For convenience we also make the following definition.

DEFINITION **Relative Extremum**

We say that $f(c)$ is a **relative extremum** if $f(c)$ is a relative maximum or a relative minimum.

Figure 5.6 indicates what happens at relative extrema. Notice that $f'(x)$ is positive to the left of the critical value $x = 0$ and negative to the right. This indicates that f is increasing (↗) to the left of $x = 0$ and decreasing (↘) to the right. This gives ↗↘ and indicates a relative maximum. Notice also that $f'(x)$ is negative to the left of the critical value $x = 2$ and positive to the right. This indicates that f is decreasing (↘) to the left of $x = 2$ and increasing (↗) to the right. This gives ↘↗ and indicates a relative minimum.

This is not the entire story. As the graph of $y = f(x)$ shows in Figure 5.8, $f'(x)$ is negative on both sides of the critical value x_2 and so does not have a relative extrema at $x = x_2$. Also, $f'(x)$ is positive on both sides of the critical value x_4 and so does not have a relative extrema at $x = x_4$. Thus, only when the derivative *changes sign* at a critical value does a relative extremum occur at this value.

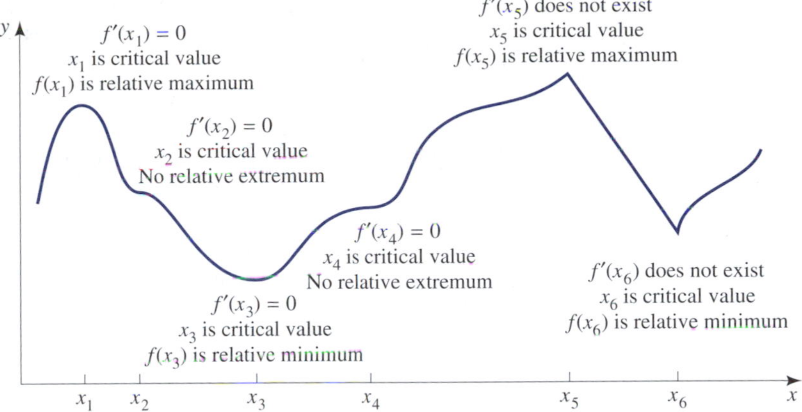

Figure 5.8
Only when the derivative changes sign at a critical point does a relative extremum occur.

We summarize this discussion as the first derivative test for finding relative extrema.

First Derivative Test

Suppose f is defined on (a, b) and c is a critical value in the interval (a, b).

1. If $f'(x) > 0$ for x near and to the left of c and $f'(x) < 0$ for x near and to the right of c, then we have $\nearrow\searrow$ and $f(c)$ is a relative maximum.
2. If $f'(x) < 0$ for x near and to the left of c, and $f'(x) > 0$ for x near and to the right of c, then we have $\searrow\nearrow$ and $f(c)$ is a relative minimum.
3. If the sign of $f'(x)$ is the same on both sides of c, then $f(c)$ is not a relative extremum.

Figure 5.9 shows all the possible cases.

Interactive Illustration 5.2

First Derivative Test

This interactive illustration allows you to explore the possibilities shown in Figure 5.8. Click on one of the functions. Move the cursor, and explore what happens to the slope as x moves through a critical point c. Notice how the slope changes when x moves through a relative minimum and a relative maximum.

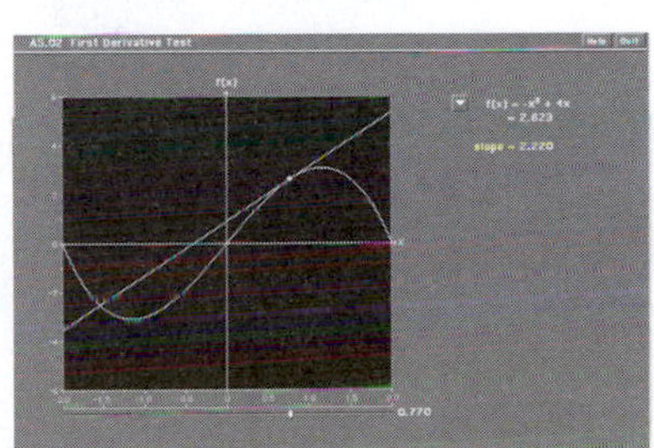

1. For each function locate values c at which the function has a relative maximum. What is the slope for x a little less than c and a little greater than c?
2. For each function locate values c at which the function has a relative minimum. What is the slope for x a little less than c and a little greater than c?
3. Locate a function and a point c where the slope goes from positive to zero to positive. Is there a relative extrema there?

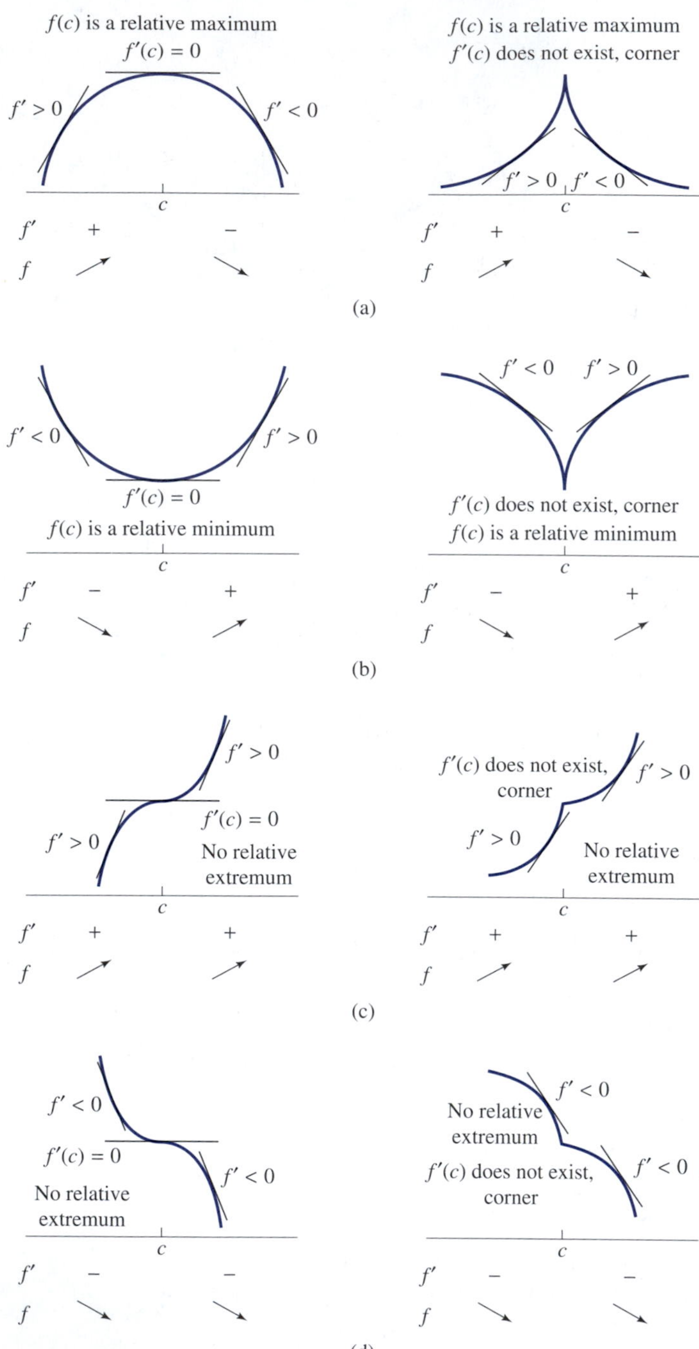

Figure 5.9
All possible cases for the first derivative test.

EXAMPLE 4 **Finding Relative Extrema Graphically**

On your computer or graphing calculator, graph $y = f(x) = 3x^5 - 20x^3 + 10$ and estimate the relative extrema.

Solution

Using a window with dimensions $[-3, 3]$ by $[-100, 100]$, we obtain a graph that looks very much like Figure 5.10. This could be the correct graph; however, there might be some hidden behavior we are not seeing. No matter how many places we might ZOOM in on the graph, we can never be certain that some relative extrema do not still remain hidden. Also, no matter how large we make the dimensions of the x-axis, we can never be certain that relative extrema do not lie outside our screen. But on the basis of what we see in the window, we believe that the critical values occur at about $x = \pm 2$ and $x = 0$.

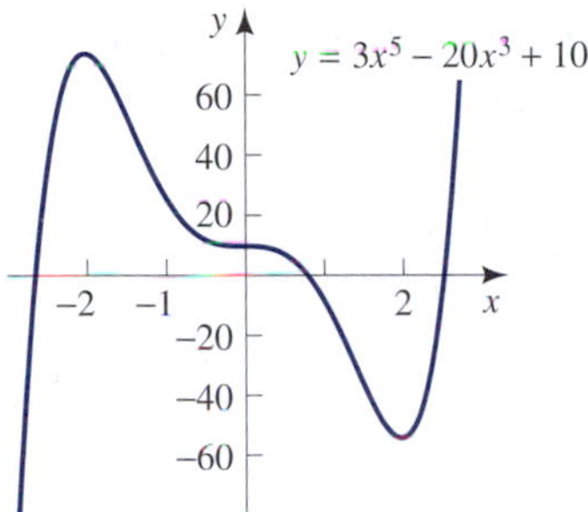

Figure 5.10
We see a relative maximum at $x = -2$, a relative minimum at $x = 2$, but no relative extrema at the critical point $x = 0$.

■

Only by using calculus can we hope to find all the relative extrema, and then only if we can find the solutions to $f'(x) = 0$. As we saw in Example 2, we might need to resort to a combination of graphical and numerical procedures to find them. Thus, if we want to solve a variety of problems, we will need to know graphical, numerical, and analytical methods.

EXAMPLE 5 **Finding Relative Extrema Using Calculus**

Use calculus to confirm the work in Example 4.

Solution

Step 1. Find the critical values.

(a) The derivative $f'(x) = 15x^4 - 60x^2$ exists and is continuous everywhere. Thus, the only critical values are values at which $f'(x) = 0$.

(b) Setting $f'(x) = 15x^4 - 60x^2 = 15x^2(x^2 - 4) = 0$ yields $x = 0$ and $x = \pm 2$. Thus, the critical values are at -2, 0, and $+2$.

Step 2. These three critical values divide the real line into the four subintervals $(-\infty, -2)$, $(-2, 0)$, $(0, 2)$, and $(2, \infty)$.

Step 3. Use the test values $-3, -1, 1$, and 3 (Figure 5.11). Then notice that $f'(-3) > 0$, $f'(-1) < 0$, $f'(1) < 0$, $f'(3) > 0$. This leads to the sign chart in Figure 5.11.

Step 4. (a) Since $f'(x) > 0$ to the left of $c = -2$ and $f'(x) < 0$ to the right of -2, $f(x)$ is increasing ($\nearrow$) on $(-\infty, -2)$ and decreasing ($\searrow$) on $(-2, 0)$, and thus, we have $\nearrow\searrow$ and f has a relative maximum at $x = -2$.

Critical values		−2		0		2	
Test values	−3		−1		1		3
f'	+		−		−		+
f	↗		↘		↘		↗

Figure 5.11

(b) Since $f'(x) < 0$ on both sides of $c = 0$, $f(x)$ is decreasing (↘) on $(-2, 0)$ and $(0, 2)$; thus, f is actually decreasing on the entire interval $(-2, 2)$ and has no relative extremum at $c = 0$.

(c) Since $f'(x) < 0$ to the left of $c = 2$ and $f'(x) > 0$ to the right of 2, f is decreasing (↘) on $(0, 2)$ and increasing (↗) on $(2, \infty)$; thus, we have ↘↗ and f has a relative minimum at $x = 2$.

Notice $f(-2) = 74$, $f(0) = 10$, and $f(2) = -54$. The graph is then given in Figure 5.10. ■

Remark Sometimes calculating the *numerical* value of the derivative at the test values can be tedious. Observe, however, that we are interested only in the *sign* of the derivative. Calculating the *sign* is usually less tedious if we have f' factored. Thus, in the previous example, one factor of f' was $15x^2$. This factor is clearly positive at all of the test values, however; thus, the sign of f' at the test values is determined by the sign of the remaining factors $(x + 2)(x - 2)$. Therefore, we need only evaluate $(x + 2)(x - 2)$ at the test values. In more complicated situations this alternative procedure can be very useful.

EXAMPLE 6 **Finding Relative Extrema**

Let the demand equation be given by $p(x) = e^{-0.02x}$. Find where revenue is increasing and where it is decreasing.

	0		50	
$(1 - 0.02x)$		+		−
$R'(x)$		+		−
R		↗		↘

Figure 5.12

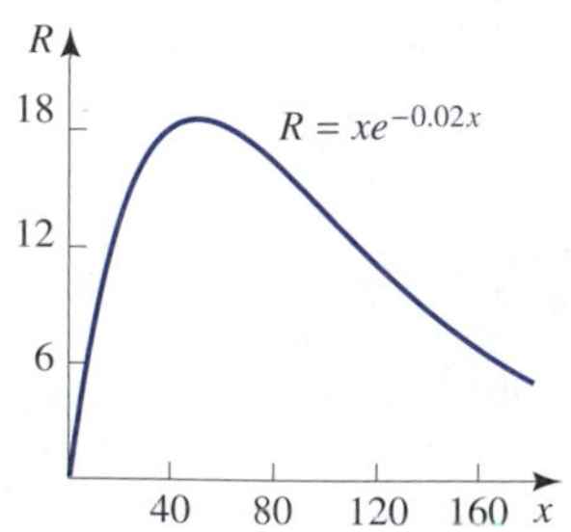

Figure 5.13
A relative maximum at $x = 50$.

Solution

Revenue is $R(x) = xp(x) = xe^{-0.02x}$. We found $R'(x)$ in Example 4 of Section 4.4 to be $(1 - 0.02x)e^{-0.02x}$. For convenience, we repeat the calculation here.

$$\begin{aligned}\frac{d}{dx}\left(xe^{-0.02x}\right) &= e^{-0.02x}\frac{d}{dx}(x) + x\frac{d}{dx}\left(e^{-0.02x}\right)\\ &= e^{-0.02x}(1) + x(-0.02e^{-0.02x})\\ &= e^{-0.02x}(1 - 0.02x)\end{aligned}$$

From this factorization we see that since $e^{-0.02x}$ is never zero, the only critical point is $x = 50$. Also from this factorization we see that the sign of $R'(x)$ is the same as the sign of the factor $(1 - 0.02x)$. Thus, $R'(x) > 0$ when $0 < x < 50$ and $R'(x) < 0$ when $x > 50$. Therefore, $R(x)$ is increasing on the interval $(0, 50)$ and decreasing on $(50, \infty)$. See Figure 5.12. The graph is shown in Figure 5.13. ■

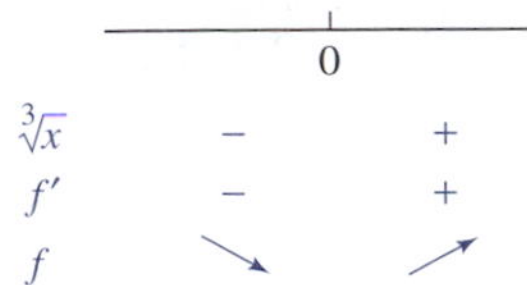

Figure 5.14

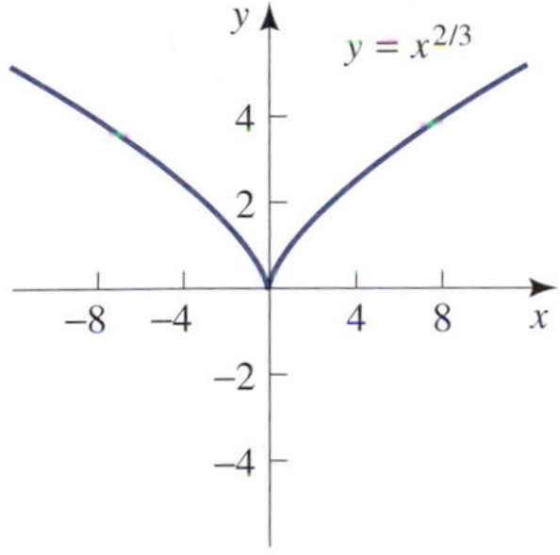

Figure 5.15
A relative minimum at $x = 0$.

EXAMPLE 7 **Finding Relative Extrema**

Find the critical values, the intervals on which $f(x) = \sqrt[3]{x^2}$ is increasing or decreasing, and the relative extrema.

Solution

Step 1. Find the critical values.

(a) Since $f(x) = x^{2/3}$,

$$f'(x) = \frac{2}{3}x^{-1/3} = \frac{2}{3\sqrt[3]{x}}$$

Note that $f'(x)$ exists and is continuous everywhere except at $x = 0$. Since $f'(0)$ does not exist, and $f(0)$ does exist, $x = 0$ is a critical value.

(b) There are no values for which $f'(x) = 0$; thus, $x = 0$ is the only critical value.

Steps 2, 3, and 4. From the simple form of $f'(x)$ we see that the sign of f' is the same as the sign of $\sqrt[3]{x}$. Thus, $f'(x) < 0$ when $x < 0$ and $f'(x) > 0$ when $x > 0$. Thus, $f(x)$ is decreasing (↘) on $(-\infty, 0)$ and increasing (↗) on $(0, \infty)$. See Figure 5.14. Then f has a relative minimum at $x = 0$ (Figure 5.15). ■

Exploration 3

Graphical Support for Example 7

Graph $y = f(x) = x^{2/3}$. Observe what happens near $x = 0$. (To obtain the correct answer on your grapher, you might need to write $x^{2/3}$ as $(x^{1/3})^2$ or as $(x^2)^{1/3}$.)

Before reading the next example, review the material on federal tax revenue and the Laffer curve found at the very beginning of this chapter.

EXAMPLE 8 **Finding Where Tax Revenue Increases or Decreases**

Suppose in the discussion on federal tax revenue found earlier in this chapter that $R(x) = \sqrt[3]{x}(100 - x)$, where $R(x)$ is measured in billions of dollars and where x is the income tax rate in percent. Find where revenue is increasing and where it is decreasing.

Solution

We need to find the derivative. We have

$$\begin{aligned} R(x) &= \sqrt[3]{x}(100 - x) \\ &= 100x^{1/3} - x^{4/3} \\ R'(x) &= \frac{100}{3}x^{-2/3} - \frac{4}{3}x^{1/3} \\ &= \frac{4}{3}\left(25x^{-2/3} - x^{1/3}\right) \end{aligned}$$

$$= \frac{4}{3}x^{-2/3}(25 - x)$$
$$= \frac{4}{3}\frac{1}{\sqrt[3]{x^2}}(25 - x)$$

The term $\frac{4}{3}\frac{1}{\sqrt[3]{x^2}}$ is positive when $x > 0$, so the sign of R' is the same as the sign of the factor $(25 - x)$. From this we see that $R'(x) > 0$ on $(0, 25)$ and $R'(x) < 0$ on $(25, 100)$. Therefore, R is increasing on $(0, 25)$ and decreasing on $(25, 100)$. See Figure 5.16. The graph is then shown in Figure 5.17.

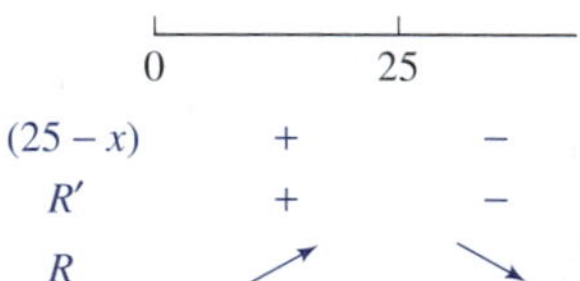

Figure 5.16

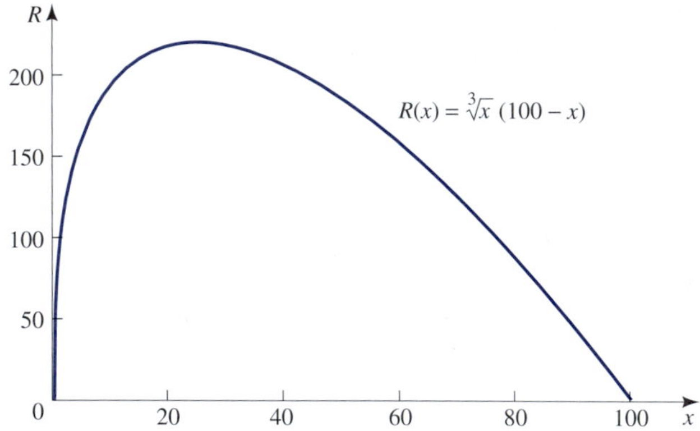

Figure 5.17

■

CONNECTION
Decreasing Federal Tax Rates

When the Kennedy administration lowered the tax rate, federal tax revenue *increased.* The Reagan administration, however, had inconclusive results, perhaps owing to other factors operating in the economy at that time.

CONNECTION
Philadelphia Fiscal Crisis

Robert Inman,[1] writing in the *Business Review of the Federal Reserve Bank of Philadelphia,* drew possible Laffer curves applied to the city of Philadelphia. The issue was whether raising tax rates would give Philadelphia the extra revenue needed to rescue it from a fiscal crisis, or whether the increase in tax rates would result in a *decrease* in economic activity, leading to a *decrease* in tax revenue.

[1] Robert Inman. 1992. Can Philadelphia escape its fiscal crisis with another tax increase? *Bus. Rev. Fed. Res. Bank Philadelphia,* Sept/Oct:5–20.

CONNECTION
Capital Gains Tax Rates

After a sweeping victory for the Republican Party in the 1994 congressional elections, the new congressional leaders proposed a *decrease* in the capital gains tax rates. They argued that one consequence of lower rates would be increased trading activity and an *increase* in taxes collected from capital gains. Therefore, Congress passed some decreases in capital gains rates that went into effect starting in 1998.

Warm Up Exercise Set 5.1

1. Suppose a firm has a monopoly on VCRs and x is the number of millions of VCRs sold in a year, with $P(x) = -0.1x^2 + 1.1x - 1.0$ the profit in tens of millions of dollars for selling x million VCRs.
   a. Find where $P(x)$ is increasing and find where it is decreasing. Of what significance is this to the firm?
   b. Find the break-even values.
   c. Draw a graph of $P = P(x)$.
   d. Is this model realistic?
2. Find the critical values and the intervals on which $f(x) = 3x^4 + 8x^3 + 5$ is increasing or decreasing. Draw a graph. Must f' change sign at a critical value?
3. Suppose that $f'(x) > 0$ everywhere. Show that $g(x) = e^{f(x)}$ is increasing everywhere.

Exercise Set 5.1

In Exercises 1 through 6 various graphs are shown. In each case, find all values where the function attains a relative maximum. A relative minimum.

1.

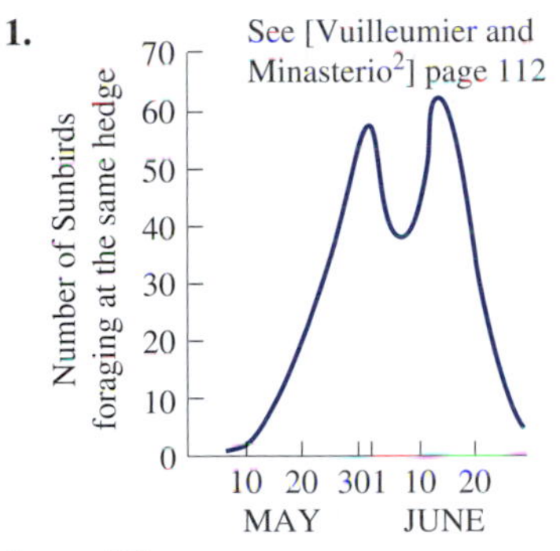

2.

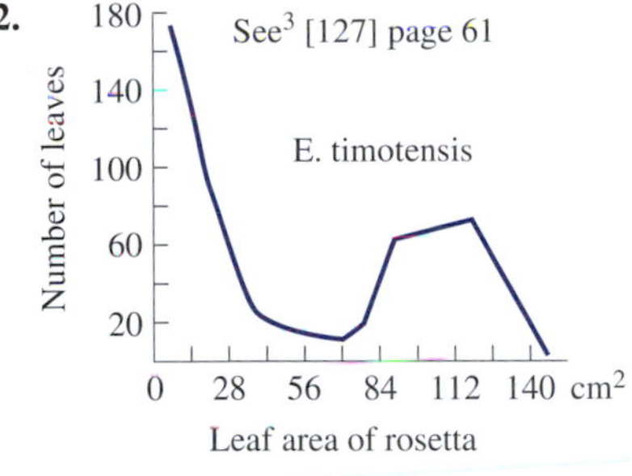

3.

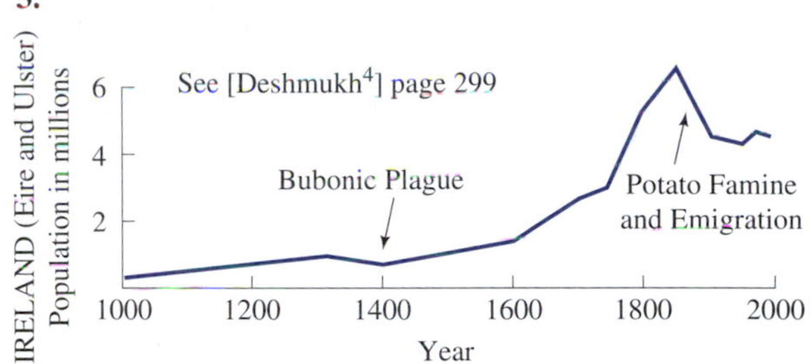

4.

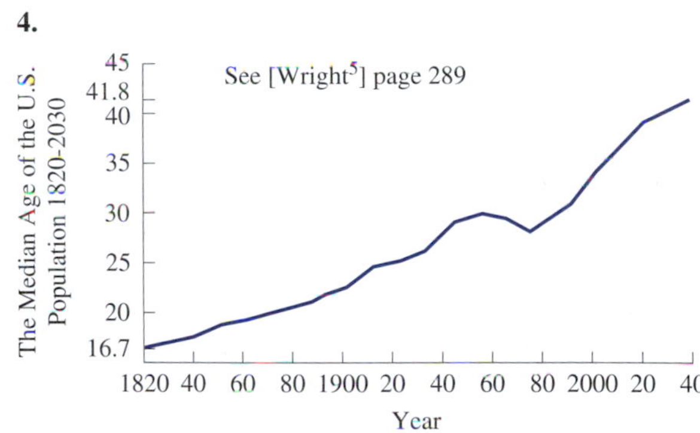

[2] F. Vuilleumier and M. Monasterio. 1986. *High Altitude Tropical Biogeography*. New York: Oxford University Press.

[3] Ibid.

[4] Ian Deshmukh. 1986. *Ecology and Tropical Biology*. Boston: Blackwell Scientific Publications.

[5] John W. Wright. 1992. *The Universal Almanac 1993*. New York: Universal Press.

5. 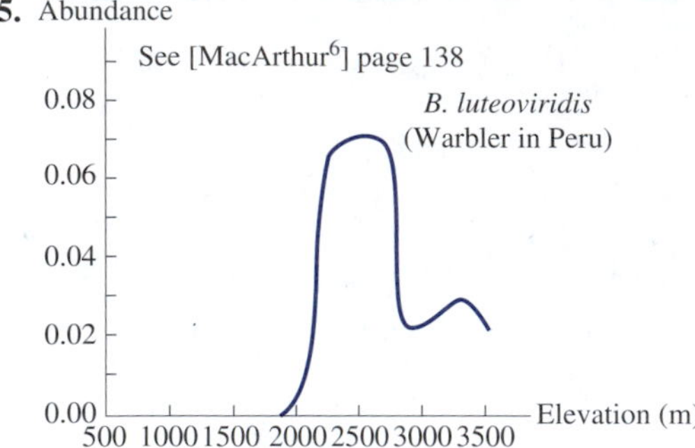

6. 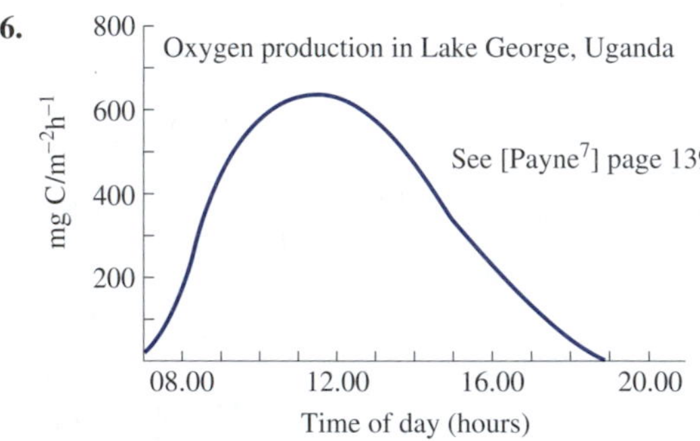

In Exercises 7 through 12, $f'(x)$ is given in factored form, and $p(x)$ is a function that is always positive. Find all critical values of f, the largest open intervals on which f is increasing, the largest open intervals on which f is decreasing, and all relative maxima and all relative minima. Using the additional information given for each function, sketch a rough graph of f.

7. $f'(x) = p(x)(x-2)(x-4)$, $f(2) = 2$, $f(4) = -2$

8. $f'(x) = p(x)(x-1)(x-3)(x-5)$, $f(1) = 1$, $f(3) = 4$, $f(5) = 0$

9. $f'(x) = p(x)(x-1)(x-3)(x-5)^2$, $f(1) = 1$, $f(3) = -2$, $f(5) = 5$

10. $f'(x) = p(x)(x+1)^2(x-1)^3(x-2)$, $f(-1) = -2$, $f(1) = 3$, $f(2) = 0$

11. $f'(x) = p(x)(x+1)(x-1)^3(x-2)^5$, $f(-1) = 1$, $f(1) = 3$, $f(2) = 0$

12. $f'(x) = p(x)(x+1)^2(x-1)^4(x-2)$, $f(-1) = 5$, $f(1) = 3$, $f(2) = 0$

In Exercises 13 through 42, find all critical values, the largest open intervals on which f is increasing, the largest open intervals on which f is decreasing, and all relative maxima and minima. Sketch a rough graph of f. In Exercises 37 through 42, assume that the constants a and b are positive.

13. $f(x) = x^2 - 4x + 1$

14. $f(x) = -2x^2 + 8x - 1$

15. $f(x) = x^3 + 3x + 1$

16. $f(x) = 1 - 5x - x^5$

17. $f(x) = x^4 + 2x^2 + 1$

18. $f(x) = -4x^6 - 6x^4 + 5$

19. $f(x) = 1 - (1-x)^{2/3}$

20. $f(x) = (x+3)^{2/3}$

21. $f(x) = x^3 - 3x + 1$

22. $f(x) = x^3 + 3x^2 - 9x + 3$

23. $f(x) = 8x^3 - 24x^2 + 18x + 6$

24. $f(x) = -8x^3 + 6x - 1$

25. $f(x) = x^4 - 8x^2 + 3$

26. $f(x) = 2x^6 - 3x^4 + 1$

27. $f(x) = -x^5 + 5x - 1$

28. $f(x) = 5 + 20x - x^5$

29. $f(x) = 3x^5 - 20x^3 + 5$

30. $f(x) = 3x^5 - 5x^3 + 1$

31. $f(x) = x - \ln x$

32. $f(x) = 1 - \ln x$

33. $f(x) = 1/(1 + e^{x^2})$

34. $f(x) = e^x + e^{-x}$

35. $f(x) = x^2 e^{-x}$

36. $f(x) = e^{-x^2}$

37. $f(x) = x^2 - 4bx + a$

38. $f(x) = x^4 - 8bx^2 + a$

39. $f(x) = x^4 + 2bx^2 + a$

40. $f(x) = -x^5 + 5b^2x + a$

41. $f(x) = x^3 - 3bx + a$

42. $f(x) = 3x^5 - 20bx^3 + a$

43. For the special cubic equation $f(x) = ax^3 + bx + c$, show that $f(x)$ is always increasing if both a and b are positive and that $f(x)$ is always decreasing if both a and b are negative.

44. Consider the function $\dfrac{x+a}{x+b}$. If $b > a$, show that this function is increasing on $(-\infty, -b)$ and $(-b, \infty)$. If $b < a$, show that this function is decreasing on $(-\infty, -b)$ and $(-b, \infty)$.

In Exercises 45 through 48, use your computer or graphing calculator to graph the function and its derivative on the same screen. Verify that the function increases on intervals where the derivative is positive and decreases on intervals where the derivative is negative.

45. $y = x^3 + 2x^2 - x - 2$

46. $y = x^3 - 4x$

47. $y = x^4 - 5x^2 + 4$

48. $y = x^5 - 5x^3 + 4x$

[6] Robert MacArthur. 1972. *Geographical Ecology*. New York: Harper & Row.

[7] A. I. Payne. 1986. *The Ecology of Tropical Lakes and Rivers*. New York: John Wiley & Sons.

Applications and Mathematical Modeling

49. Revenue If the revenue function for a firm is given by $R(x) = 10x - 0.01x^2$, find the value of x for which revenue is maximum.

50. Revenue If the demand equation for a firm is given by $p = -0.2x + 16$, find the value of x at which the revenue is maximum.

51. Profit Let the cost function for a firm be given by $C(x) = 5 + 8x$, and let the demand equation be the same as that in Exercise 50. Find the value at which the profit is maximum.

52. Revenue If the revenue function for a firm is given by $R(x) = -x^3 + 36x$, find the value of x at which the revenue is maximized.

53. Profit If the cost function for a firm is given by $C(x) = 9x + 30$ and the revenue function is the same as in Exercise 52, find the value of x at which the profit is maximized.

54. Average Revenue The total revenue function for producing x items is given by $R(x) = 100x^2 - 20x^3$. The average revenue per item is $\bar{R}(x) = R(x)/x$. How many items should be produced to maximize the average revenue per item?

55. Cost Function Noreen and Soderstrom[8] suggested that a reasonable cost function for hospitals is $C(x) = px^{0.70}$, where x is in appropriate units of service, p is a positive constant, and $C(x)$ is cost. For what values of x is $C(x)$ increasing? Sketch a possible graph.

56. Firm Size Philips[9] created a mathematical model given by the equation $y = 2.43 - 2.55x + 2.86x^2$, where y is the number of research personnel in a firm in one sector of the food, beverage, and tobacco industry and x is the number of thousands of total personnel, with x restricted to the interval $[0.5, 2.5]$. Sketch a graph on the given interval. Where is the function increasing on this interval?

57. Cost Function Dean[10] found that the cost function for direct materials in a furniture factory was approximated by $y = C(x) = 0.667x - 0.00467x^2 + 0.000151x^3$ for $0 \le x \le 50$, where x is output in thousands of dollars. Use calculus to find where marginal cost is maximized.

58. Cost Function Dean[11] found that the cost function for indirect labor in a furniture factory was approximated by $y = C(x) = 0.4490x - 0.01563x^2 + 0.000185x^3$, for $0 \le x \le 50$, where x is output in thousands of dollars. Use calculus to find where marginal cost is minimized.

59. Development Rate In 1997 Got and coworkers[12] reported on a mathematical model they constructed relating the temperature to the developmental rate of the corn borer *Ostrinia nubilalis*, a major pest of corn. For the Logan mathematical model they found that the developmental rate $v(\theta)$ was approximated by the equation $v(\theta) = 0.00498\{e^{0.1189\theta} - e^{[(0.1189)(40.92)-(40.92-\theta)/5.8331]}\}$, where θ is measured in degrees Celsius. Use your grapher to find the approximate temperature at which the developmental rate is maximized.

60. Development Rate In their 1997 report on corn borers, Got and coworkers[13] stated that for the "normal" mathematical model, the developmental rate $v(\theta)$ was approximated by the equation $v(\theta) = 0.08465e^{-0.5(\theta-30.27)/8.14}$, where θ is measured in degrees Celsius. Use your grapher to find the approximate temperature at which the development rate is maximized.

61. Agriculture Paris[14] created a mathematical model showing plant responses to phosphorus fertilizer given by the equation $y = -0.057 - 0.417x + 0.852\sqrt{x}$, where y is the yield and x the units of nitrogen. Use calculus to find the value of x that maximizes yield.

62. Agriculture Feinerman and colleagues[15] created a mathematical model that related the yield response of corn to nitrogen fertilizer was given by the equation $y = -5.119 + 0.099x - 0.004x^{1.5}$, where y is the yield measured in tons per hectare and x is in kilograms per hectare. Use calculus to find the value of x that maximizes yield.

63. Agriculture Dai and colleagues[16] created a mathematical model that showed the yield response of corn in Indiana to soil moisture was approximated by the equation $y = -2600 + 2400x - 70x^2 + \frac{2}{3}x^3$, where y is measured in bushels per acre and x is a soil moisture index. Find the value of the soil moisture index on the interval $[0, 40]$ that maximizes yield.

---

[8] Eric Noreen and Naomi Soderstrom. 1994. Are overhead costs strictly proportional to activity? *J. Account. Econ.* 17:255–278.

[9] Louis Philips. 1971. *Effects of Industrial Concentration.* Amsterdam-London: North Holland.

[10] Joel Dean. 1976. *Statistical Cost Estimation.* Bloomington, Ind.: Indiana University Press.

[11] Ibid.

[12] B. Got, S. Piry, A. Miggeon, and J. M. Labatte. 1997. Comparisons of different models for predicting development time of the European corn borer. *Environ. Entomol.* 26(1):46–60.

[13] Ibid.

[14] Quirino Paris. 1992. The von Liebig hypothesis. *Amer. Agr. Econ.* 74:1019–1028.

[15] E. Feinerman, E. Choi, and S. Johnson. 1990. Uncertainty and split nitrogen application in corn production. *Amer. J. Agr. Econ.* 72:975–984.

[16] Q. Dai, J. Fletcher, and J. Lee. 1993. Incorporating stochastic variables in crop response models. *Amer. J. Agr. Econ.* 75:377–386.

64. Medicine It is well known[17] that when a person coughs, the diameter of the trachea and bronchi shrinks. It has been shown that a mathematical model of the flow of air F through the windpipe is given by $F(r) = k(R - r)r^4$, where r is the radius of the windpipe under the pressure of the air being released from the lungs, R is the radius with no pressure, and k is a positive constant. Find the radius at which the flow is maximum.

65. Autocatalytic Reaction The rate V of a certain autocatalytic reaction is proportional to the amount x of the product and also proportional to the remaining substance being catalyzed. Thus, if I is the initial amount of the substance, then $V = kx(I - x)$, where k is a positive constant.[18] At what concentration x will the rate be maximum?

66. Biology Lactin and coworkers[19] created a mathematical model that showed the feeding rate y of the Colorado potato beetle larvae in square millimeters per hour per larva was approximated by $y = -3.97 + 0.374T - 0.00633T^2$, where T is the temperature in degrees Centigrade. Find the temperature at which the feeding rate is maximum.

67. Biology Arthur and Zettler[20] created a mathematical model that showed that the percentage mortality y of the red flour beetle was approximated by $y = f(x) = 9.1 + 86.3e^{-0.64x}$, where x is the number of weeks after a malathion treatment. Use calculus to show that $f(x)$ is a decreasing function. What happens in the long-term? Sketch a graph. Check using your grapher.

68. Biology Cherry and associates[21] created a mathematical model that showed that the number y of eggs laid per female of the sugarcane grub was approximated by $y = 14.6e^{-(x-116.6)/60.9}$, where x is a measure of soil moisture. Sketch a graph. Find where the number of eggs is increasing and where they are decreasing. Check using your grapher.

69. Biology Eller and coauthors[22] created a mathematical model that showed that the percentage y of the females of a parasite ovipositing was approximated by $y = -298.1 + 241.1x - 51.9x^2 + 3.4x^3$, where $x = \ln V$ and V is the volume of an insecticide in appropriate units. Graph y as a function of V on your grapher. Find the value of V that maximizes y on the interval $[0, 400]$.

70. Biology The formula $w = w_0 N^{-a}$ is a mathematical model used by plant biologists (see Shainsky and Radosevich[23]), where w is the plant size, w_0 is the maximum plant size in absence of competitors, N is plant density, and a is a positive constant. According to this formula, for what values of N is plant size decreasing? Sketch a possible graph.

71. Biology Webb and colleagues[24] created a mathematical model that showed that the population y of gypsy moths was approximated by $y = f(x) = 10^{3.8811-0.1798\sqrt{x}}$, where x was the dose in grams per hectare of the insecticide racemic disparlure. Find $f'(x)$, and determine whether this function is decreasing. Sketch a graph. What happens if the dose x becomes very large?

---

[17] Duane J. Clow and N. Scott Urquhart. 1974. *Mathematics in Biology*. New York: Ardsley House.

[18] Ibid.

[19] Derek Lactin, N. J. Holliday, and L. L. Laman. 1993. Temperature dependence of feeding by Colorado potato beetle. larvae. *Environ. Entomol.* 22:784–790.

[20] Frank Arthur and J. Larry Zettler. 1991. Malathion resistance in *Tribolium castancum*. *J. Econ. Entomol.* 84:721–726.

[21] R. H. Cherry, F. J. Coale, and P. S. Porter. 1990. Oviposition and survivorship of sugarcane grubs at different soil moisture. *J. Econ. Entomol.* 83:1355–1359.

[22] F. J. Eller, R. R. Heath, and S. M. Ferkovich. 1990. Factors affecting oviposition by the parasitoid *Microplitis croceipes*. *J. Econ. Entomol.* 83:398–404.

[23] Lauri J. Shainsky and Stevin R. Radosevich. 1991. Analysis of yield-density relationships in experimental stands of Douglass fir and red alder seedlings. *Forest Sci.* 37(2):574–592.

[24] R. E. Webb, B. A. Leonhardt, J. R. Plimmer, K. M. Tatman, V. K. Boyd, D. L. Cohen, C. P. Schwalbe, and L. W. Douglass. 1990. Effect of racemic disparlure released from grids of plastic ropes on mating success of gypsy moth as influenced by dose and by population density. *J. Econ. Entomol.* 83(3):910–916.

More Challenging Exercises

In Exercises 72 through 75, assume that $f'(x)$ is continuous everywhere and that $f(x)$ has one and only one critical value at $x = 0$. Use the additional given information to determine whether $y = f(x)$ attains a relative minimum, a relative maximum, or neither at $x = 0$. Explain your reasoning. Sketch a possible graph in each case.

72. $f(-1) = 1$, $f(0) = 3$, $f(2) = 4$

73. $f(-5) = -50$, $f(10) = 100$ and $f(0) = 0$

74. $f(0) = 1$, $f(5) = 5$, and $f(-5) = 0$

75. $f(-1) = 3$, $f(0) = 10$, $f(2) = 4$

76. Suppose that $f'(x)$ is continuous everywhere and $f(x)$ has two and only two critical values. Explain why it is not possible for f to have the following values: $f(0) = 10$, $f(1) = 5$, $f(2) = 8$, $f(3) = 6$, $f(4) = 9$.

77. Show that $x^{13} - x^{12} = 10$ has one and only one solution.

78. Show that $x^{12} - x^{11} = 10$ has two and only two solutions by using an analytic argument.

79. Using the properties of $f(x) = e^x$, show that $e^x \geq 1 + x$ for all x.

80. Suppose that $f(x)$ is positive everywhere. Let $g = 1/f$. If f has a relative maximum at $x = a$, what can you say about g?

81. Suppose that $f(x)$ and $g(x)$ are differentiable and negative everywhere. If $f'(x) > 0$ and $g'(x) > 0$ everywhere, show that $h(x) = f(x)g(x)$ is a decreasing function.

82. Suppose that $f(x)$ and $g(x)$ are differentiable and positive everywhere. If $f'(x) < 0$ and $g'(x) > 0$ everywhere, show that $h(x) = f(x)/g(x)$ is a decreasing function.

83. Give conditions on the constants b, c, and d so that the polynomial function $f(x) = x^3 + bx^2 + cx + d$ will be increasing on the interval $(-\infty, \infty)$.

84. Use calculus and properties of cubic polynomials to explain why any polynomial function of the form $f(x) = x^4 + ax^3 + bx^2 + cx + d$ cannot be increasing on all of $(-\infty, \infty)$ or decreasing on all of $(-\infty, \infty)$.

85. The figure shows a graph of the derivatives of two functions. What can you say about a possible relative extremum of each of them at $x = 1$? Justify your answer.

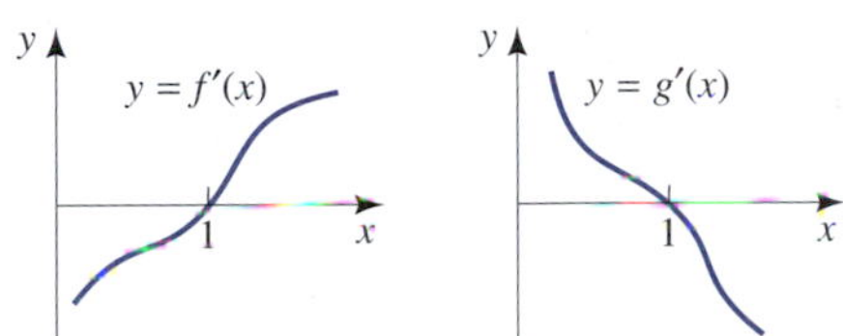

86. a. Suppose that the function $y = f'(x)$ is decreasing on the interval $(0, 2)$ and $f'(1) = 0$. What can you say about a possible relative extremum of $f(x)$ at $x = 1$? Justify your answer.

b. Suppose that the function $y = g'(x)$ is increasing on the interval $(0, 2)$ and $g'(1) = 0$. What can you say about a possible relative extremum of $g(x)$ at $x = 1$? Justify your answer.

87. Suppose that $f(x)$ is continuous on $(0, 2)$ and never zero there and $f(1) > 0$. What can you say about the sign of f on $(0, 2)$?

88. Find a function $f(x)$ defined on $[0, 1]$ that changes its sign on $[0, 1]$ but is never zero on $[0, 1]$. Can this function be continuous on $[0, 1]$?

89. Suppose $f(x)$ is continuous on $(0, 2)$ and never zero there, and $f(1) < 0$. What can you say about the sign of f on $(0, 2)$?

90. Use the constant sign theorem to show that the polynomial $y = x^5 - 4x^2 - x + 3$ has a root in the interval $(1.25, 2)$.

91. Price Discrimination The following equation arose in a problem of price discrimination in economics found in an article by Sailors and colleagues.[25]

$$1 - \left(\frac{m}{b}\right)^{x-1} = \left(1 - \frac{k}{m}\right)(x - 1)$$

where $x > 1$ and $b > k$. Show that there is a unique solution for m on the interval (k, b).

92. Accounting Thornton and Moore[26] used a total audit fee function as $F(w, y) = p(y)q(w)$, where p is the hourly fee charged by the auditor, q is the number of hours required to perform the audit, y is the auditor quality, and w is a measure of internal controls. Then the authors state that $p'(y) > 0$ and $q'(w) < 0$. Explain why this is reasonable.

93. Biology Laudelout[27] used the equation $Q(t) = at + be^{-kt}$ to mathematically model the amount Q of carbon dioxide at time t from microbial biomass in soil. Take $a = b = 1$ and $k = 5$, graph on your grapher, and locate the approximate value of x for which the amount of carbon dioxide changes from decreasing to increasing. Verify using calculus.

94. The Labor Market Rosenthal[28] indicated that the equation $y = a + b(1 - e^{-ct})$, mathematically models the relationship between the proportion y of individuals who have regained employment after being laid off and the time t of unemployment measured in weeks. Sketch a graph. Explain in words what the value a signifies. What happens as t becomes very large? Explain in words what it would mean if $a + b = 1$.

95. Sales of New Products Dodds[29] created a mathematical model that related the sales S of color television sets to the number t of years after introduction and given by

$$S(t) = 418{,}000 \frac{e^{-0.842t}}{\left(1 + 154e^{-0.842t}\right)^2}$$

Determine the time for which $S(t)$ reaches its maximum.

---

[25] Joel Sailors, M. L. Greenhut, and H. Ohta. 1985. Reverse dumping: a form of spacial price discrimination. *J. Indust. Econ.* 34(2):167–181.

[26] Daniel B. Thornton and Giora Moore. 1993. Auditor choice and audit fee determinants. *J. Bus. Finance Account.* 20(3):333–349.

[27] H. Laudelout. 1993. Chemical and microbiological effects of soil liming in a broad-leaved forest ecosystem. *Forest Ecol. Manage.* 61:247–261.

[28] Leslie Rosenthal. 1990. Time to re-establishment of equilibrium for a group of redundant workers. *Applied Econ.* 22:83–95.

[29] Wellesley Dodds. 1973. An application of the bass model in long-term new product forecasting. *J. of Marketing Res.* 10:308–315.

96. Biology Ring and Benedict[30] created a mathematical model of the yield response of cotton to injury by the bollworm. Normalized yield y (yield with injury divided by yield without injury) was shown to be approximated by $\ln y = 0.1494 \ln x - 5.5931x + 7.12x^2$, where x is the number of injured reproductive organs (flower buds plus capsules) per 100. Graph on your grapher, and determine the approximate value of x for which y attains a maximum on the interval $[0, 0.05]$. Confirm, using calculus.

[30] Dennis Ring and John Bennedict. 1993. Comparison of insect injury-cotton yield response functions and economic injury levels for *Helicoverpa zea* and *Heliothis virescens* in the lower Gulf coast of Texas. *J. Econ. Entomol.* 86(4):1228–1235.

Modeling Using Least Squares

97. Production Costs Suarez-Villa and Karlsson[31] studied the relationship between the sales in Sweden's electronic industry and production costs (per unit value of product sales). Their data is presented in the following table.

x	7	13	14	15	17
y	0.91	0.72	0.91	0.81	0.72
x	20	25	27	35	45
y	0.90	0.77	0.65	0.73	0.70
x	45	63	65	82	
y	0.78	0.82	0.92	0.96	

Here x is product sales in millions of krona, and y is production costs per unit value of product sales.

a. Use cubic regression to find the best-fitting cubic to the data. Graph.
b. Find where the function is increasing and where it is decreasing. What does this say about economies of scale?

98. Temperature Effects on Trout Selong and colleagues[32] studied the temperature effects on growth rates of bull trout. The table gives the data they collected.

Temperature (°C)	8	10	12	14
Growth rate (g/day)	0.087	0.119	0.143	0.140
Temperature (°C)	16	18	20	
Growth rate (g/day)	0.140	0.086	0.041	

a. Find the best-fitting quadratic (as the researchers did) relating temperature to the growth rate of bull trout.
b. Find where the function attains a maximum.

99. Rare Plants Pavlik and Enberg[33] studied the effect of temperature on a rare species of geyser panic grass. They collected data on temperature effects on the seed emergence of these grasses given in the table.

Soil Surface Temperature (°C)	11	13	14
Number of seeds emerging	0	35	45
Soil Surface Temperature (°C)	15	20	23
Number of seeds emerging	45	105	72

a. Find the best-fitting quadratic (as the researchers did) relating temperature to the emergence of these grasses.
b. Find where the function attains a maximum.

100. Rare Plants Pavlik and Enberg[34] studied the effect of temperature on a rare species of geyser panic grass. They collected data on temperature effects on the mortality rate of these grasses given in the table.

[31] L. Suarez-Villa and C. Karlsson. 1996. The development of Sweden's R&D-intensive electronics industries: exports, outsourcing, and territorial distribution. *Environment and Planning A* 28:783–817.

[32] Jason H. Selong, Thomas E. McMahon, Alexander V. Zale, and Frederic T. Barrows. 2001. Effect of temperature on growth and survival of bull trout, with application of an improved method of determining thermal tolerance in fish. *Trans. Amer. Fish. Soc.* 130:1026–1037.

[33] Bruce M. Pavlik and Andrew Enberg. 2001. Developing an ecosystem perspective from experimental monitoring programs: demographic responses of a rare geothermal grass to soil temperature. *Environ. Manage.* 28(2):225–242.

[34] Ibid.

Soil Surface Temperature (°C)	18	20	20	23
Mortality rate (% per week)	2.4	0.2	3.9	0.1
Soil Surface Temperature (°C)	24	27	28	29
Mortality rate (% per week)	0.8	0.2	0.1	0.7
Soil Surface Temperature (°C)	30	38	39	44
Mortality rate (% per week)	0.1	0.6	1.7	6.2

a. Find the best-fitting quadratic (as the researches did) relating temperature to the mortality of these grasses.
b. Find where the function attains a maximum.

Solutions to WARM UP EXERCISE SET 5.1

1. a. Taking the derivative gives

$$P'(x) = -0.2x + 1.1$$

Notice that $P'(x)$ is continuous everywhere. Thus, the only critical values are values at which $P'(x) = 0$. Setting $P'(x) = -0.2x + 1.1 = 0$ gives $x = 5.5$ as the only critical value. This divides the interval $(0, \infty)$ into the 2 intervals $(0, 5.5)$ and $(5.5, \infty)$. Take the test values to be, say, $x_1 = 1$ and $x_2 = 10$. Then

$$P'(1) = -0.2(1) + 1.1 = 0.9 > 0,$$
$$P'(10) = -0.2(10) + 1.1 = -0.9 < 0$$

Thus, $P(x)$ is increasing on $(0, 5.5)$ and decreasing on $(5.5, \infty)$.

Thus, if $0 < x < 5.5$, $P(x)$ is increasing. The firm would be wise to produce and sell more items since profits would increase. If $5.5 < x$, $P'(x) < 0$ and $P(x)$ is decreasing. Producing and selling more items means less profits.

b. Since

$$P(x) = -0.1(x^2 - 11x + 10) = -0.1(x - 10)(x - 1)$$

$P(1) = 0$ and $P(10) = 0$. This means that profits will be zero if the firm sells 1 million or 10 million VCRs in a year. (Apparently, a very substantial price reduction will be necessary to sell such a large number of VCRs.)

c. The graph is given in the figure.

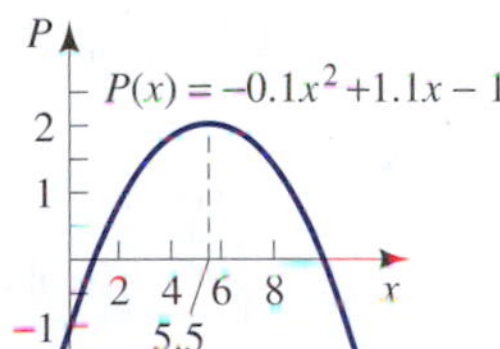

d. Is this model realistic? It is clearly very unrealistic to think that any firm would know the explicit form of the function $P(x)$. However, experience has shown that the general shape of the graph of the profit function is realistic. Firms are well aware of the fact that flooding the market with their products will lead to *decreased* profits and even losses.

2. Taking the derivative gives

$$f'(x) = 12x^3 + 24x^2 = 12x^2(x + 2)$$

The critical values are $x = -2, 0$ where f' is zero. Since the derivative exists everywhere, there are no other critical values.

We divide the real line into three intervals: $(-\infty, -2)$, $(-2, 0)$, $(0, \infty)$.

Pick the three test values -3, -1, and 1 that lie in the intervals given in the preceding step. See the figure. Notice $f'(-3) = -108$, $f'(-1) = +12$, $f'(1) = +36$. Thus, $f'(x)$ is negative on the first interval and positive on the second and third intervals; therefore, f is decreasing on the first interval and increasing on the second and third intervals. The function has a relative minimum at $x = -2$.

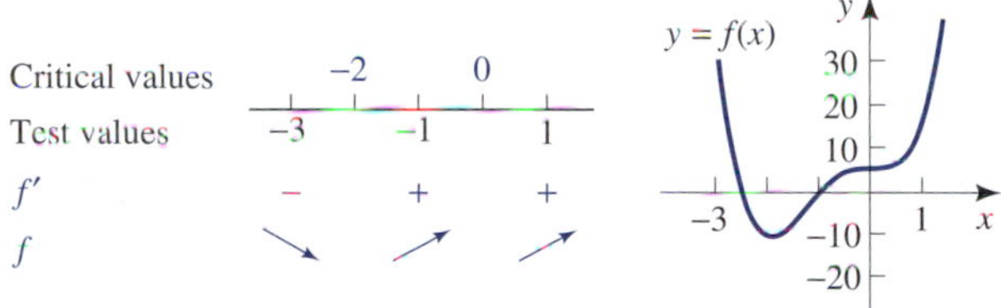

Notice that $f(-2) = -11$ and $f(0) = 5$. The graph is given in the figure.

Note that $f'(0) = 0$ but f' does not change sign at this value.

3. We have $g'(x) = f'(x)e^{f(x)}$. But $e^{f(x)}$ is positive no matter what $f(x)$ is. Then $f'(x) > 0$ implies that $g'(x) = f'(x)e^{f(x)} > 0$. Thus, $y = g(x)$ is increasing.

5.2 The Second Derivative

Concavity

The Second Derivative Test

APPLICATION Economies of Scale

To determine whether any economies of scale exist, Haldi and Whitcomb[35] studied over 200 industrial plants and found that a typical cost function was $C(x) = ax^{0.7}$, where x is output in appropriate units, a is a positive constant, and $C(x)$ is total cost. How can you use calculus to decide whether there are economies of scale? See Example 2 on page 317 for the answer.

[35] J. Haldi and D. Whitcomb. 1967. Economies of scale in industrial plants. *J. Political Econ.* 75:373–385.

Concavity

In this section we see that the second derivative measures the rate at which the first derivative is increasing or decreasing. This can then be used in graphing and also to give another test for relative extrema.

Given a function $y = f(x)$, the second derivative, denoted by $f''(x)$, is defined to be the derivative of the first derivative.

DEFINITION Second Derivative

Given a function $y = f(x)$, the **second derivative**, denoted by $f''(x)$, is defined to be the derivative of the first derivative. Thus,

$$f''(x) = \frac{d}{dx}(f'(x))$$

Remark The second derivative is also denoted by $\dfrac{d^2}{dx^2}f(x)$ or $\dfrac{d^2y}{dx^2}$.

For example, given $y = f(x) = x^4$,

$$\frac{d^2y}{dx^2} = f''(x) = \frac{d}{dx}[f'(x)] = \frac{d}{dx}(4x^3) = 12x^2$$

Remark Notice that the second derivative is a function, just like the first derivative is a function.

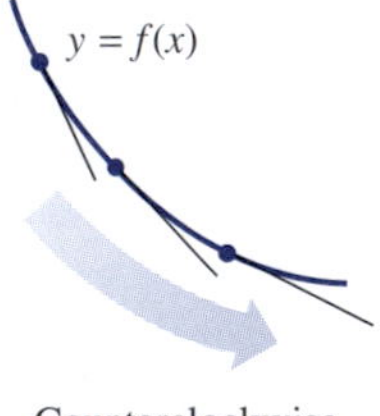

Counterclockwise

(a)

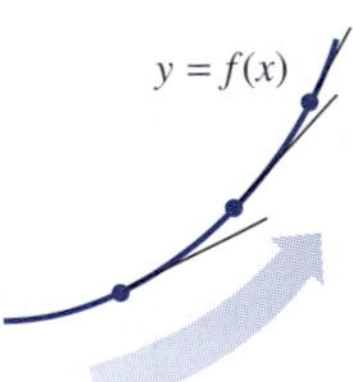

Counterclockwise

(b)

Figure 5.18
When the slope increases, the curve is concave up.

In Figures 5.18a and 5.18b, we see two functions that have one characteristic in common. They both have derivatives that are *increasing*. Notice that this means that the slopes are rotating *counterclockwise* as x increases. When this happens we say that the curve is **concave up** (denoted by $\smile$). Also notice that in such a case the curve always lies *above* its tangent.

Exploration 1

Concavity and the Second Derivative

Graph $y_1 = f(x) = 1.5x^5 - 5x^3 + 5x$ and the *sign* of the second derivative, $y_2 = sign(30x^3 - 30x)$, on a screen with dimensions $[-2, 2]$ by $[-4, 4]$. (If your computer or calculator does not have a sign function, then graph $y_2 = f''(x)/|f''(x)|$.) Determine the intervals on which $y = f(x)$ is concave up and check the sign of the second derivative on these intervals. Now determine the intervals on which $y = f(x)$ is concave down and check the sign of the second derivative on these intervals. What conclusions do you draw?

We now see that a curve will be concave up if $f''(x) > 0$. (We hope this is one of the conclusions that you came to earlier.) Recall that if the derivative of a function is positive, then the function must be increasing. The second derivative is the derivative of the first derivative, however, so if

$$f''(x) = \frac{d}{dx}\left(f'(x)\right) > 0$$

then the function $f'(x)$ is increasing. Therefore the curve is concave up ($\smile$).

Similarly in Figures 5.19a and 5.19b, we see two functions that both have derivatives that are *decreasing*, that is the slopes are rotating *clockwise* as x increases. When this happens, we say that the curve is **concave down** (denoted by $\frown$). In such a case the curve always lies *below* its tangent.

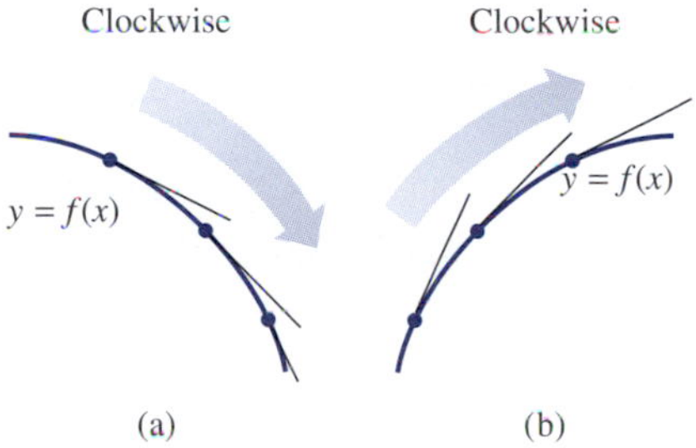

Figure 5.19
When the slope decreases, the curve is concave down.

Remark The reader should carefully note that concavity has nothing to do with the function itself increasing or decreasing. In Figures 5.18a and 5.18b the graphs of both functions are concave up, but the first function is decreasing while the second function is increasing. In Figures 5.19a and 5.19b the graphs of both functions are concave down, but the first function is decreasing while the second function is increasing.

We summarize all of this in the following definition and theorem, where we assume that $f'(x)$ exists on (a, b).

DEFINITION **Concave Up and Down**

1. We say that the graph of f is concave up ($\smile$) on (a, b) if $f'(x)$ is increasing on (a, b).
2. We say that the graph of f is concave down ($\frown$) on (a, b) if $f'(x)$ is decreasing on (a, b).

Test for Concavity

1. If $f''(x) > 0$ on (a, b), then the graph of f is concave up ($\smile$) on (a, b).
2. If $f''(x) < 0$ on (a, b), then the graph of f is concave down ($\frown$) on (a, b).

A point $(c, f(c))$ on the graph of f where the concavity changes is called an **inflection point**, and c is called an **inflection value**. More precisely we have the following definition.

DEFINITION **Inflection Point**

A point $(c, f(c))$ on the graph of f is an **inflection point** and c is an **inflection value** if $f(c)$ is defined and the concavity of the graph of f changes at $(c, f(c))$.

Figure 5.20 shows four inflection points for four functions. Notice that in each case the graph of the function must cross the tangent line.

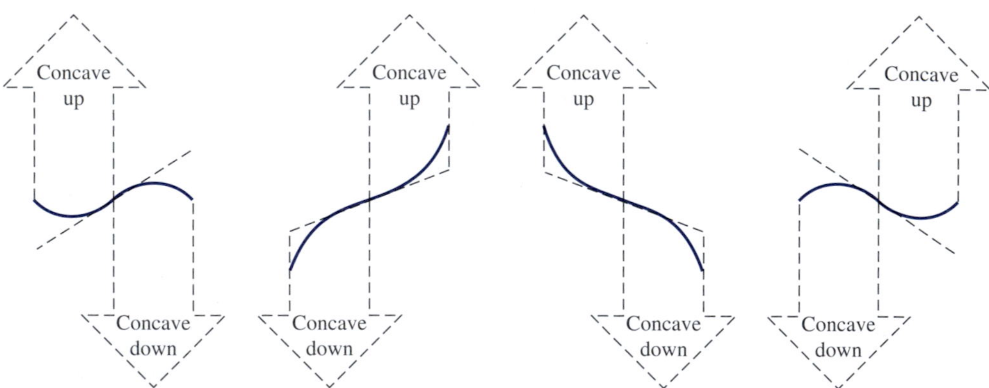

Figure 5.20
At the point where the concavity changes, the curve crosses the tangent line.

EXAMPLE 1 **Inflection Values**

On your computer or graphing calculator find a rough approximation for the inflection values for $f(x) = 1.3x^9 - 8x^5 + 7x^3 + 0.6x$.

Solution

Graph $y = f(x)$ on your grapher using a window of dimension $[-1.4, 1.4]$ by $[-1.8, 1.8]$, and obtain Screen 5.2. (The tick marks are set at 0.2). Very roughly, the inflection values seem to be $x = \pm 0.5$, $x = \pm 1.07$, and $x = 0$. A graph of $y = f''(x) = 93.6x^7 - 160x^3 + 42x$ shown in Screen 5.3 with a window of dimensions $[-1.4, 1.4]$ by $[-40, 40]$ shows the second derivative changing sign at $x = 0$ and roughly $x = \pm 0.5$ and $x = \pm 1.07$.

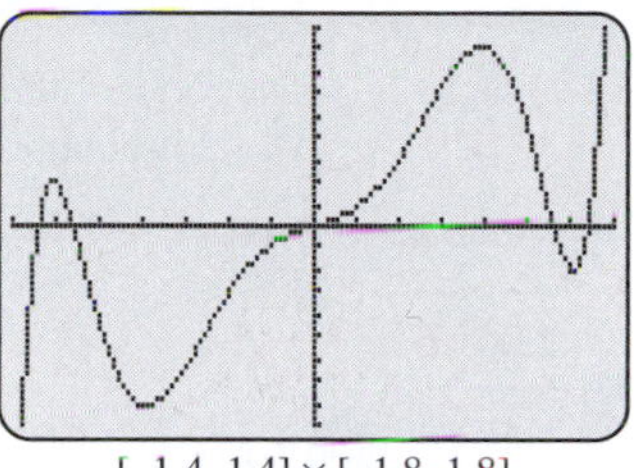

$[-1.4, 1.4] \times [-1.8, 1.8]$

Screen 5.2
The inflection values for the function $y = 1.3x^9 - 8x^5 + 7x^3 + 0.6x$ appear to be at about $x = \pm 0.5$, $x = \pm 1.07$, and $x = 0$.

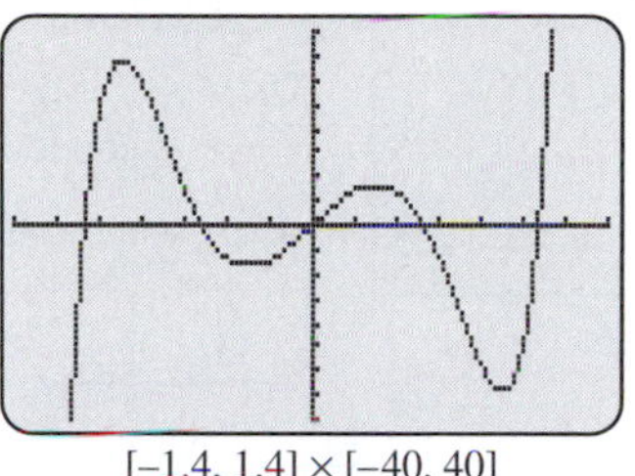

$[-1.4, 1.4] \times [-40, 40]$

Screen 5.3
This is a graph of $y = 93.6x^7 - 160x^3 + 42x$ of the second derivative of the function in Screen 5.2. The second derivative changes sign at the inflection points noted in Screen 5.2.

■

To find where the graph of f is concave up and where the graph of f is concave down, we need to find the values where $f''(x)$ changes sign. This will occur at values c where $f''(c) = 0$ or where $f''(c)$ does not exist.

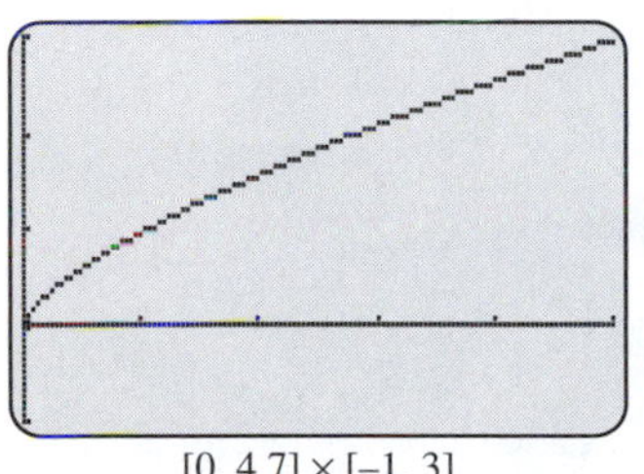

$[0, 4.7] \times [-1, 3]$

Screen 5.4
Shown is the graph of $C(x) = x^{0.7x}$.

EXAMPLE 2 **Cost Function**

To determine whether there were any economies of scale, Haldi and Whitcomb[36] studied over 200 industrial plants and showed that a typical cost function was approximated by $C(x) = x^{0.7}$, where x is output in appropriate units and $C(x)$ is the total cost. The exponent 0.7 is a measure of scale economies. Graph this function, and show that the marginal cost is decreasing, $C''(x) < 0$, and the cost curve is concave down. What does the concavity tell you about the economy of scale?

Solution

A graph of the cost function is shown in Screen 5.4. The graph seems to bend downward and thus seems to be concave down. Tangents are drawn to the cost curve at $x = 0.5$, 1, and 2 in Screens 5.5, 5.6, and 5.7, respectively. As can be seen, the slopes are smaller for larger values of x, indicating that f' appears to be decreasing. Table 5.1 shows the numerical values; a decreasing derivative means concave down.

The marginal cost is $C'(x) = 0.7x^{-0.3}$. This is definitely a decreasing function and confirms our previous work. A graph is shown in Screen 5.8.

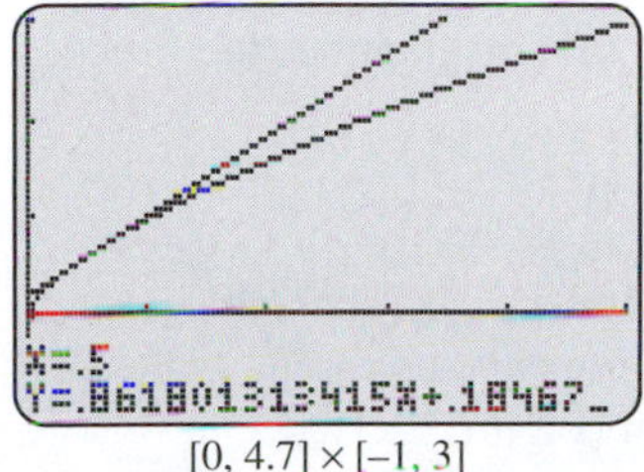

$[0, 4.7] \times [-1, 3]$

Screen 5.5
The slope of the tangent line at $x = 0.5$ is about 0.86.

[36] Ibid.

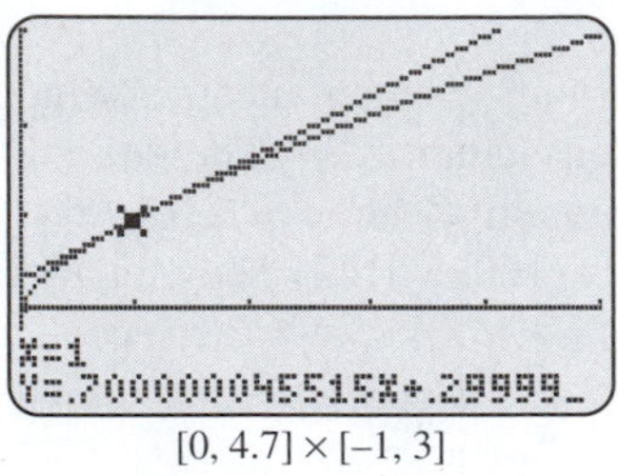
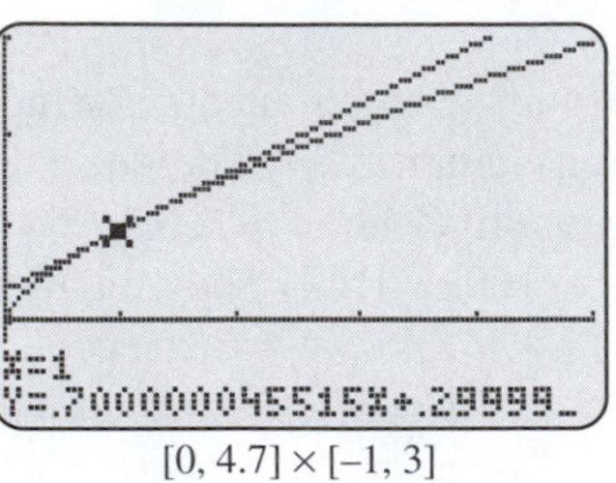

[0, 4.7] × [−1, 3]

Screen 5.6
The slope of the tangent line at $x = 1$ is about 0.70.

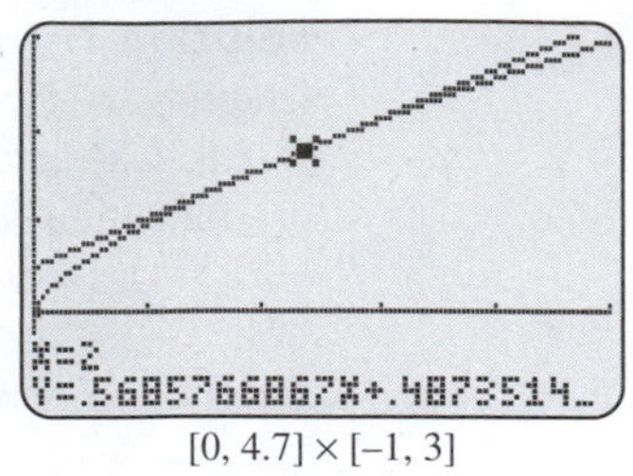

[0, 4.7] × [−1, 3]

Screen 5.7
The slope of the tangent line at $x = 2$ is about 0.57.

Table 5.1

x	0.50	1.00	2.00
$C'(x)$	0.86	0.70	0.57

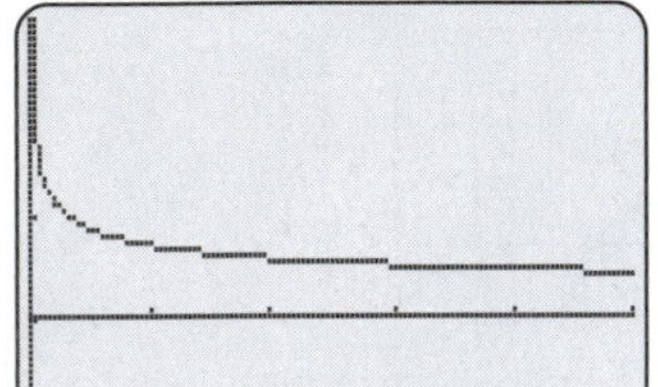

[0, 4.7] × [−1, 3]

Screen 5.8
The graph shows that the marginal cost of the function in Screen 5.4 is decreasing.

Finally, we note that $C''(x) = -0.21x^{-1.3} < 0$. Since marginal cost is decreasing for these plants, greater output leads to reduced marginal costs, indicating economies of scale. ■

Interactive Illustration 5.3

Marginal Cost

This interactive illustration allows you to see geometrically that marginal cost is decreasing for the cost function in Example 2. We know that $C'(x)$ is both marginal cost and also the slope of the tangent line at the point x. As you move the cursor observe what happens to the tangent line. What do you conclude is happening to the slope of the tangent line, and thus to marginal cost, as x increases? Specifically, find the slopes at $x = 1, 2, 3$ and 4.

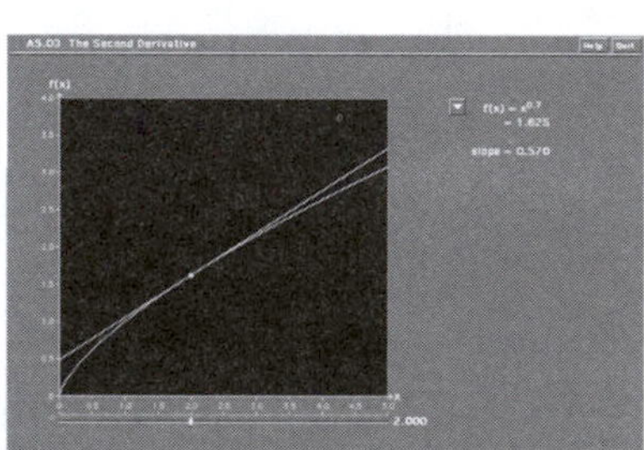

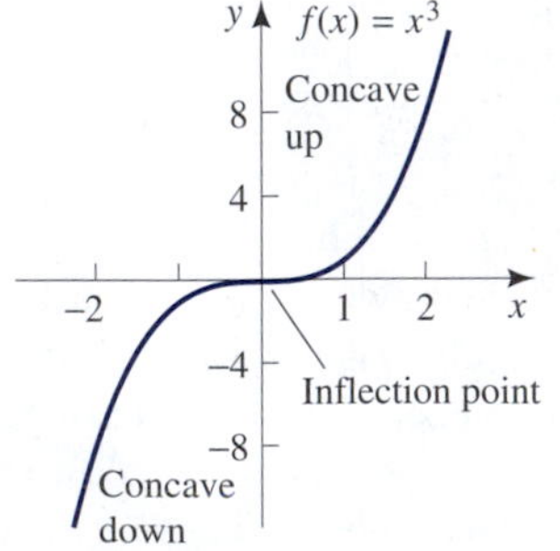

Figure 5.21
This function is concave down on $(-\infty, 0)$ and concave up on $(0, \infty)$, with $(0, 0)$ as an inflection point.

EXAMPLE 3 Determining Concavity

Discuss the concavity of f if $f(x) = x^3$. What does this mean about the function $y = f'(x)$?

Solution

Since $f''(x) = 6x$, $f''(x) < 0$ if $x < 0$, and $f''(x) > 0$ if $x > 0$. Thus, the graph of f is concave down for all values of x on $(-\infty, 0)$ and concave up for all values of x on $(0, \infty)$, and $(0, f(0)) = (0, 0)$ is an inflection point (Figure 5.21).

Since the function $y = f(x)$ is concave down on $(-\infty, 0)$ and concave up on $(0, \infty)$, $y = f'(x)$ must be a decreasing function on $(-\infty, 0)$ and increasing on $(0, \infty)$. Figure 5.22 shows two tangent lines at $x = -2$ and $x = -1$. Note how the slope decreases as x increases on the interval $(-\infty, 0)$. Figure 5.22 also shows two tangent lines at $x = 1$ and $x = 2$. Note how the slope increases as x increases on the interval $(0, \infty)$.

A graph of $y = f'(x) = 3x^2$ is shown in Figure 5.23. This function is indeed decreasing on $(-\infty, 0)$ and increasing on $(0, \infty)$.

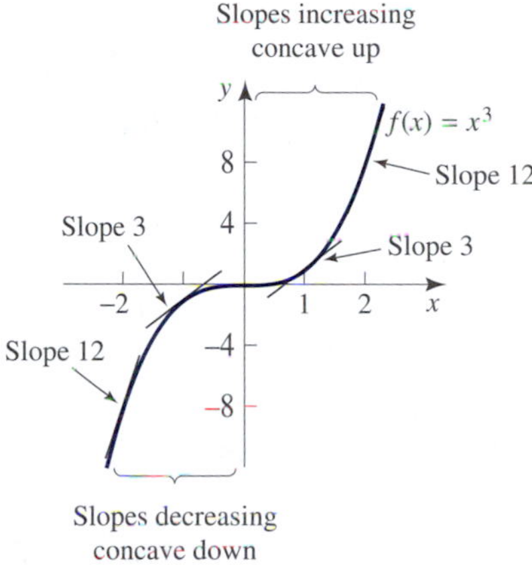

Figure 5.22
The derivative is decreasing when the function is concave down; the derivative is increasing when the function is concave up.

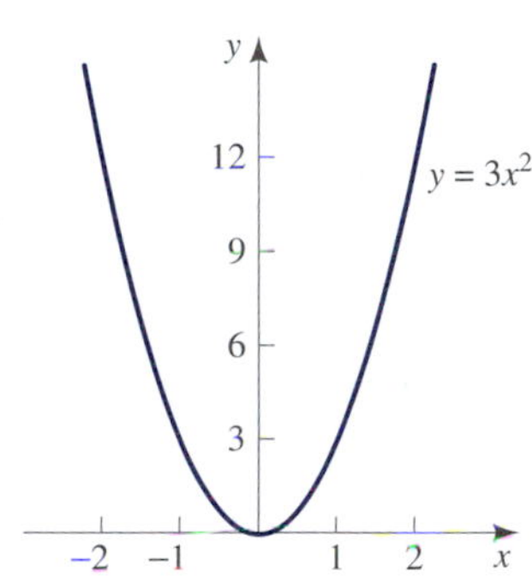

Figure 5.23
The graph of the derivative of $f(x) = x^3$.

Interactive Illustration 5.4

Concavity

This interactive illustration allows you to observe how the slope of the tangent line is changing as you vary x and how this relates to concavity. Experiment by moving the cursor and observing the change in the slope of the tangent line. Then answer the following.

1. Select a function.
2. In the first window locate the intervals on which the slope of the tangent line is increasing (decreasing). Select at least 5 points on this interval and make a table that shows each point and the corresponding slope.
3. Check the second window to see if $f'(x)$ is indeed increasing (decreasing) on the intervals found in part (2).

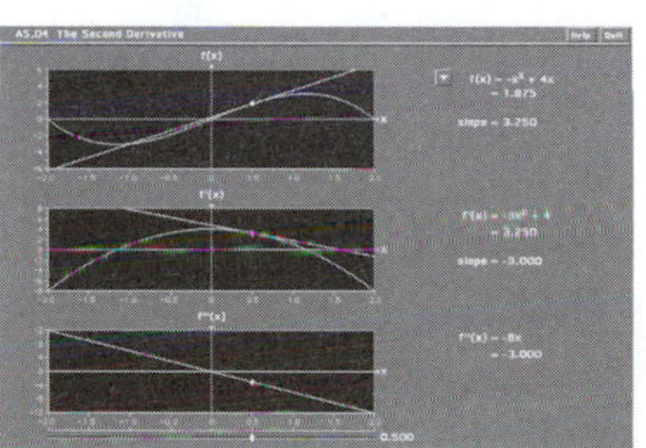

4. On the intervals found in part (2) is the sign of $f''(x)$ shown in the third window positive (negative)?

Exploration 2

Concavity

Discuss the concavity of $f(x) = 9x^{1/3}$ and what this means for $y = f'(x)$. Follow an analysis similar to that done in Example 3.

EXAMPLE 4 **Determining Concavity**

Suppose a firm believes that the percentage $P(t)$ of the market that it can attain for its product by spending a certain amount on advertising each year for the next 3 years is given by $P(t) = -t^3 + 6t^2 + 10$, where t is given in years. Find where the graph of the function is concave up and where it is concave down. Find the inflection point. How might you characterize the inflection point in this case?

Solution

We have

$$P'(t) = -3t^2 + 12t = 3t(4 - t)$$

Thus, P' is continuous everywhere and positive on the interval $(0, 3)$. So P is increasing throughout this interval. Also

$$P''(t) = -6t + 12 = 6(2 - t)$$

For the interval $(0, 3)$ we see that $P''(t) > 0$ on $(0, 2)$ and $P''(t) < 0$ on $(2, 3)$. Thus, the graph of P on the interval $(0, 3)$ is concave up for all values of t on $(0, 2)$ and concave down for all values of t on $(2, 3)$. See Figure 5.24. Since $P(2)$ is defined and the second derivative changes sign at $t = 2$, $(2, P(2)) = (2, 26)$ is an inflection point. See Figure 5.25. We see then that the rate at which profits are increasing on the interval $(0, 2)$ is itself increasing. (The slopes of tangent lines become larger as you move from $t = 0$ to $t = 2$.) So $P(x)$ is increasing ever faster on the interval $(0, 2)$. We also see that the rate at which profits are increasing on the interval $(2, 3)$ is itself decreasing. (The slopes of tangent lines become smaller as you move from $t = 2$ to $t = 3$.) So $P(t)$ is increasing ever slower on the interval $(2, 3)$. We would characterize the inflection point as **the point of diminishing returns**.

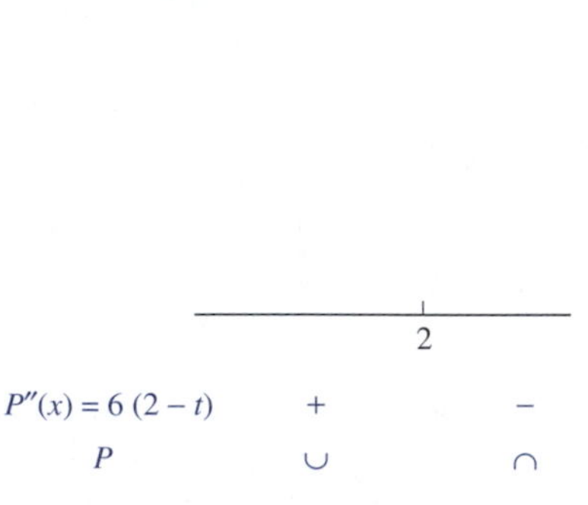

Figure 5.24

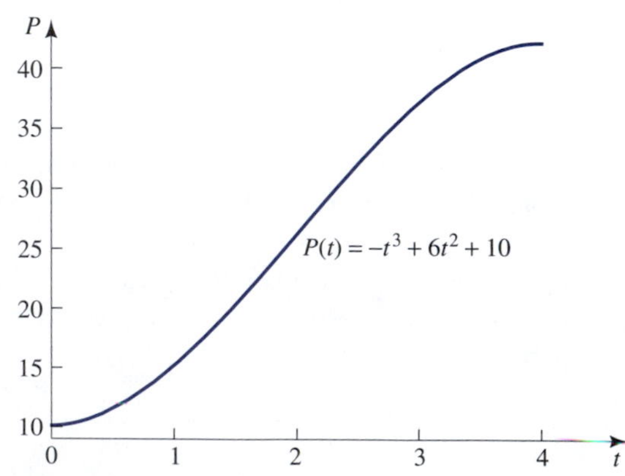

Figure 5.25

Another way to look at this is to recall that $P'(t)$ is approximately the profit of the next item. The profit per item then is increasing on the interval $(0, 2)$ but decreasing on $(2, 3)$. ■

Interactive Illustration 5.5

Concavity and the Point of Diminishing Return

This interactive illustration allows you to observe how the slope of the tangent line is changing as you vary x, indicating how this is related to concavity. Experiment moving the cursor and observing the change in the slope of the tangent line and then answer the following.

1. Locate the interval on which the slope of the tangent lines is decreasing and therefore where $P'(t)$ is a decreasing function. What must this imply about concavity on this interval?
2. Locate the interval on which the slope of the tangent lines is increasing and therefore where $P'(t)$ is an increasing function. What must this imply about concavity on this interval?

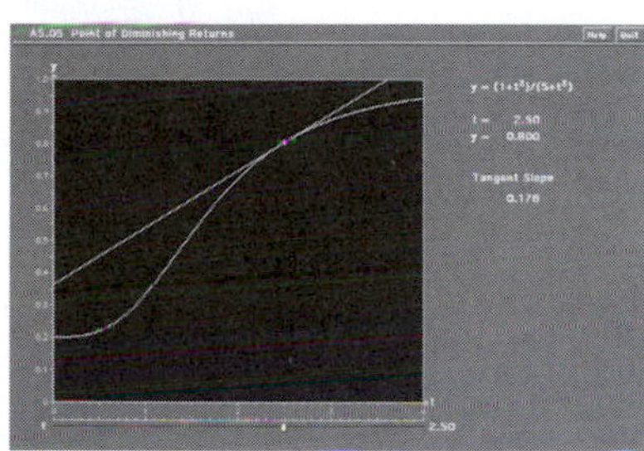

3. Approximately locate the point of diminishing returns.

The Second Derivative Test

We shall now see that the second derivative can sometimes be used to determine whether a point is a relative extremum. Suppose (a, b) is the domain of definition of f and $c \in (a, b)$. If the second derivative exists and is not zero at a point c where $f'(c) = 0$, then we shall see that the sign of $f''(c)$ will determine whether the value c is a relative minimum or a relative maximum.

Consider, for example, the situation shown in Figure 5.26, in which $f'(c) = 0$ and $f''(c) > 0$. We will now show that in this case $f(c)$ is a relative minimum. To see this, notice that since $f''(c) > 0$, $f'(x)$ is increasing for x near c, and so $f'(x)$ must be smaller than $f'(c) = 0$ for x just to the left of c and must be larger than $f'(c) = 0$ for x just to the right of c. Thus, $f'(x) < 0$ for x to the left of c and $f'(x) > 0$ for x to the right of c. The sign chart of f' is then $-0+$, and we therefore conclude from the first derivative test that $f(c)$ is a relative minimum.

For a second example, consider the situation shown in Figure 5.27 in which we also have $f'(c) = 0$, but this time $f''(c) < 0$. Then, using a similar argument, we see that the sign chart of f' must be $+0-$, and $f(c)$ must be a relative maximum.

We summarize this as *the second derivative test.*

Second Derivative Test

Suppose that f is defined on (a, b), $f'(c) = 0$, and $c \in (a, b)$.

1. If $f''(c) > 0$, then $f(c)$ is a relative minimum.
2. If $f''(c) < 0$, then $f(c)$ is a relative maximum.

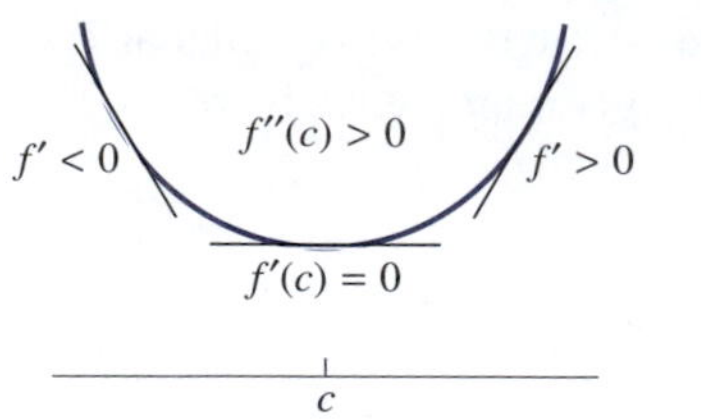

Figure 5.26
$f'(c) = 0$ and $f''(c) > 0$ implies a relative minimum.

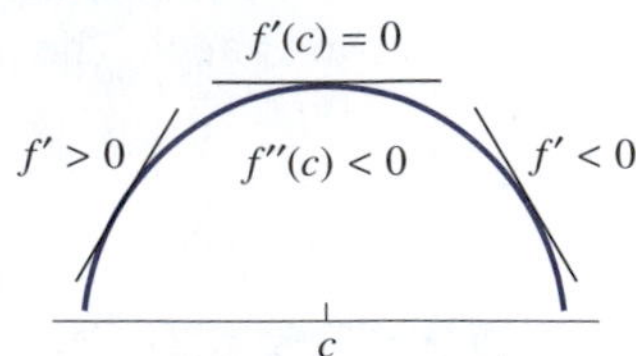

Figure 5.27
$f'(c) = 0$ and $f''(c) < 0$ implies a relative maximum.

EXAMPLE 5 **Using the Second Derivative Test**

Consider the function $f(x) = 1 + 9x + 3x^2 - x^3$. Use the second derivative test to find the relative extrema. Graph.

Solution

We first find the derivative

$$\begin{aligned} f'(x) &= (1 + 9x + 3x^2 - x^3)' \\ &= 9 + 6x - 3x^2 \\ &= -3(x^2 - 2x - 3) \\ &= -3(x - 3)(x + 1) \end{aligned}$$

Setting $f'(x) = 0$ then gives the two critical values $x = -1$ and $x = 3$. Taking the second derivative, we have

$$f''(x) = (9 + 6x - 3x^2)' = 6 - 6x$$

Since $f''(-1) = 6 - 6(-1) = 12 > 0$, the second derivative test indicates that $f(-1)$ is a relative minimum. Since $f''(3) = -12 < 0$, the second derivative test indicates that $f(3)$ is a relative maximum. The graph is shown in Figure 5.28.

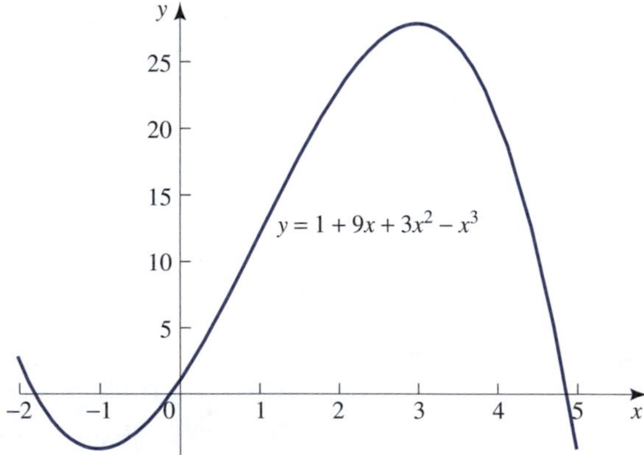

Figure 5.28

■

Remark The second derivative test can be useful if the second derivative can be easily calculated. A serious problem with the second derivative test is that it can fail to work. For example, if the second derivative is zero or does not exist at a point where the first derivative is zero, then the second derivative fails to supply any information. Exercise 63 on page 327 addresses these issues.

Warm Up Exercise Set 5.2

1. Suppose a firm believes that the percent, $p(t)$, of a product's market that can be attained by spending a certain amount on advertising each year for the next six years is given by $p(t) = -\frac{1}{3}t^3 + 3t^2 + 14$, where t is given in years. Find where the graph of the function is concave up and where it is concave down. Discuss the graph of the curve and interpret.

2. Suppose the second derivative of some function $y = f(x)$ is given as

$$f''(x) = (x-1)(x-2)^2(x-3)$$

Find the intervals where the graph of the function is concave up and where it is concave down. Find all inflection values.

Exercise Set 5.2

In Exercises 1 through 6, find $f''(x)$.

1. $f(x) = 4x^5 + x^2 + x + 3$

2. $f(x) = x^3 + x$

3. $f(x) = 1/(x+1)$

4. $f(x) = 1/(x^2+1)$

5. $f(x) = \sqrt{2x+1}$

6. $f(x) = \sqrt{x^2+1}$

Define $f'''(x) = \dfrac{d}{dx}(f''(x))$ and $f^{iv}(x) = \dfrac{d}{dx}(f'''(x))$. In Exercises 7 through 12, find $f'''(x)$ and $f^{iv}(x)$ for the given functions.

7. $f(x) = 4x^5 + x^2 + x + 3$

8. $f(x) = x^3 + x$

9. $f(x) = 1/x$

10. $f(x) = \sqrt{x}$

11. $f(x) = \sqrt[3]{x^2}$

12. $f(x) = 1/\sqrt{x} + \sqrt{x}$

In Exercises 13 through 20, find all inflection values, find the largest open intervals on which the graph of f is concave up, and find the largest open intervals on which the graph of f is concave down for the given functions.

13.

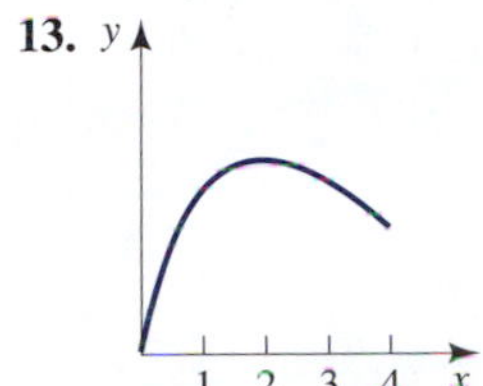

14.

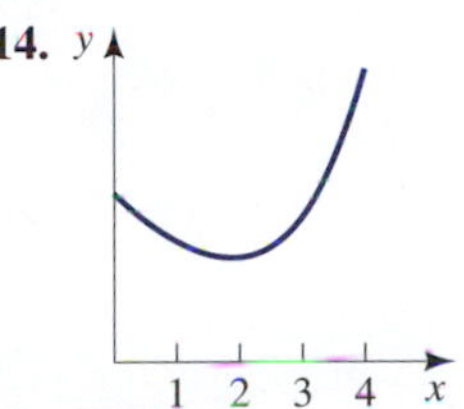

15.

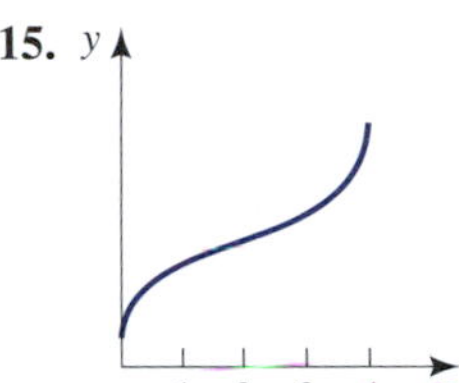

16.

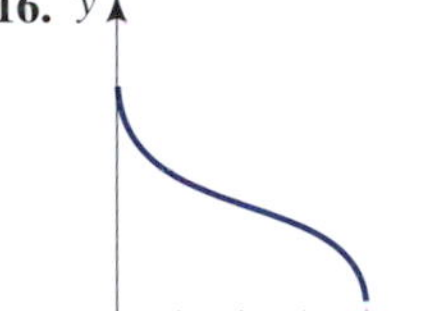

17.

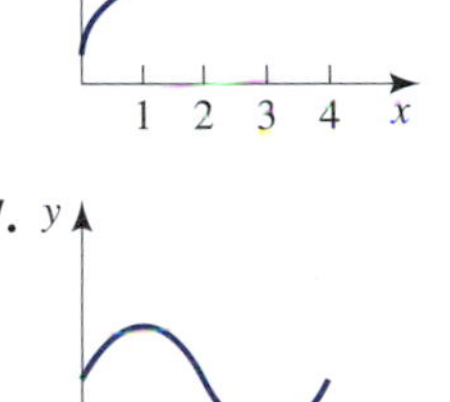

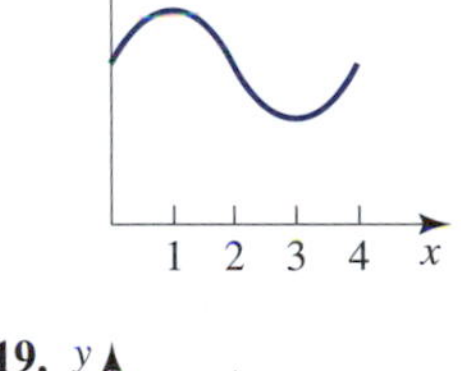

18.

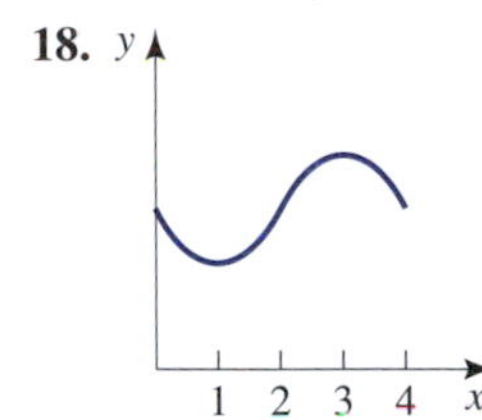

19.

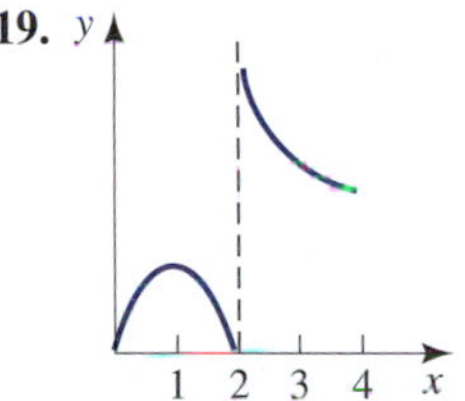

20.

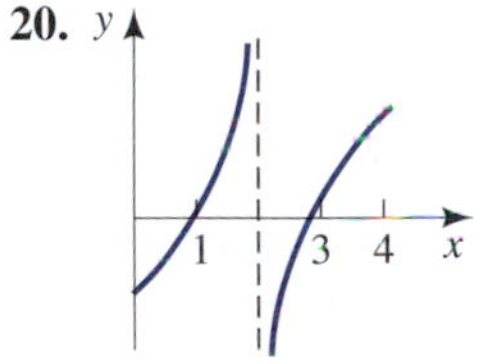

In Exercises 21 through 24, $f''(x)$ is given in factored form. Find all inflection values, find the largest open intervals on which the graph of f is concave up, and find the largest open intervals on which the graph of f is concave down.

21. $f''(x) = (x-1)(x-3)(x-5)$

22. $f''(x) = x(x-2)^4(x-4)$

23. $f''(x) = (x+2)x^2(x-2)^2$

24. $f''(x) = (x+2)x^3(x-2)^5$

In Exercises 25 through 36, find where the function is increasing, decreasing, concave up, and concave down. Find critical points, inflection points, and where the function attains a relative minimum or relative maximum. Then use this information to sketch a graph.

25. $f(x) = x^3 - 3x^2 + 2$

26. $f(x) = 1 - 9x + 6x^2 - x^3$

27. $f(x) = 4x^5 - 5x^4 + 1$

28. $f(x) = 6x^7 - 7x^6 + 1$

29. $f(x) = x^4 + 6x^2 - 2$

30. $f(x) = xe^x$

31. $f(x) = 5 + 8x^3 - 3x^4$

32. $f(x) = 3x^4 - 16x^3 + 3$

33. $f(x) = x^2e^x$

34. $f(x) = 3 - x^2 - x^4$

35. $f(x) = e^{-x^4}$

36. $f(x) = e^x - e^{-x}$

37. Show that the graph of the quadratic function $f(x) = ax^2 + bx + c$, $a \neq 0$, is always concave up if $a > 0$ and always concave down if $a < 0$.

38. Show that the graph of the quadratic function $f(x) = ax^2 + bx + c, a \neq 0$, has a minimum at $x = -\dfrac{b}{2a}$ if $a > 0$ and has a maximum at $x = -\dfrac{b}{2a}$ if $a < 0$.

39. Show that the graph of the cubic function $f(x) = ax^3 + bx^2 + cx + d, a \neq 0$, has one and only one inflection point. Find this point.

40. Show that the special cubic function $f(x) = x^3 - 3bx + c$, $b > 0$, has exactly one relative minimum and exactly one relative maximum. Locate and identify them.

41. Consider the graphs of the following four functions.

i.

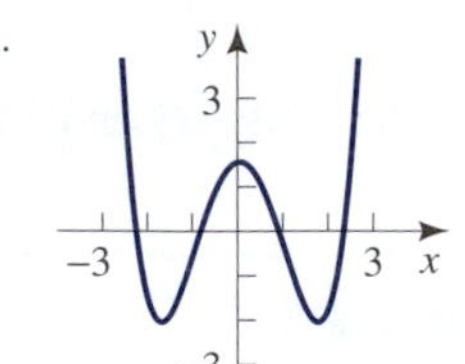

ii.

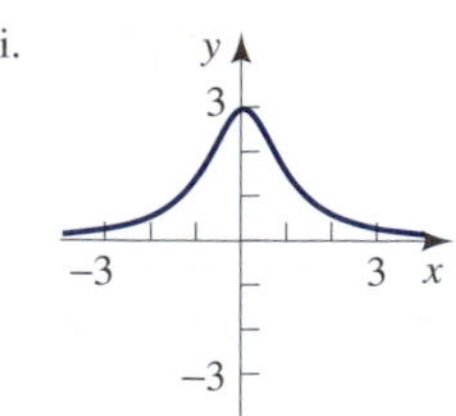

iii.

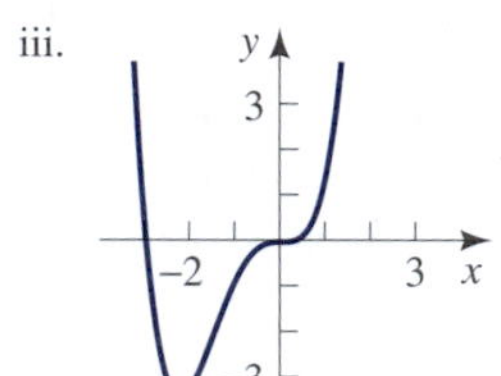

iv.

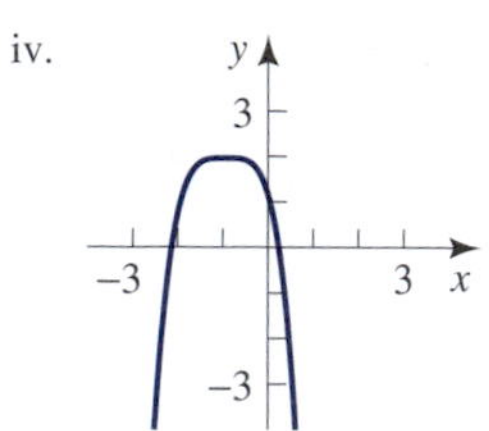

Each of the following figures represents the graph of the second derivative of one of the above functions but in a different order. Match the graph of the function to the graph of its second derivative.

a.

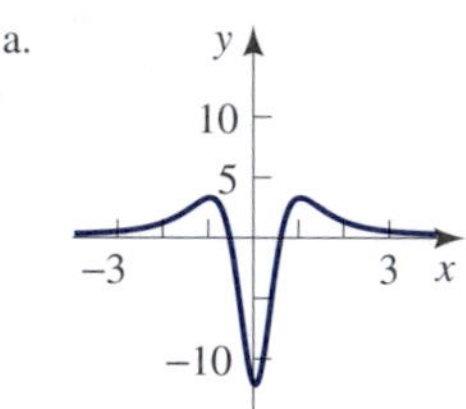

b.

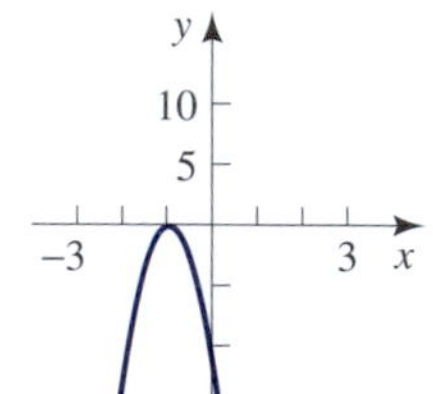

c.

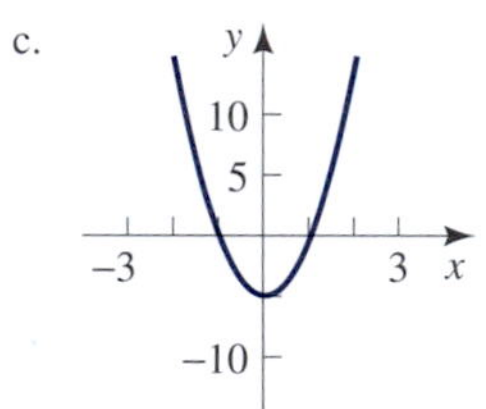

d.

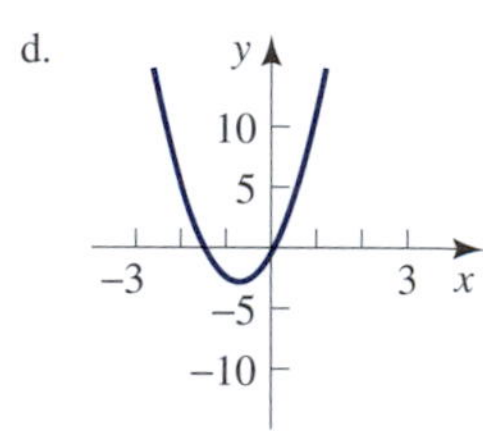

In Exercises 42 through 44 you will see that you cannot solve the equation $f'(x) = 0$ exactly to find the critical points. Use the available operations on your computer or graphing calculator to find the approximate solution to the equation $f'(x) = 0$, and then use the second derivative test to determine the relative extrema. Now sketch a graph of $y = f(x)$. Check your graph by using your grapher.

42. $f(x) = 0.20x^5 + 0.50x^2 - x$

43. $f(x) = x^4 + x^3 + x + 1$

44. $f(x) = x^4 - 4x^3 + 2x + 1$

Applications and Mathematical Modeling

Diminishing Returns In Examples 45 through 48 you are given a profit equation for a firm, where t is time in years. In each case graph the function on the given interval, and find the point of diminishing returns. Explain what is happening to profits on the given time interval.

45. $P(t) = -t^3 + 9t^2 + t + 5$, $[0, 6]$

46. $P(t) = -2t^3 + 24t^2 + t + 5$, $[0, 8]$

47. $P(t) = -2t^3 + 45t^2 + 2t + 5$, $[0, 10]$

48. $P(t) = -2t^3 + 60t^2 + 3t + 5$, $[0, 19]$

49. Economies of Scale The figure shows the graph of the cost function for a firm where x is the number of units produced. Is this firm experiencing economies of scale? Explain.

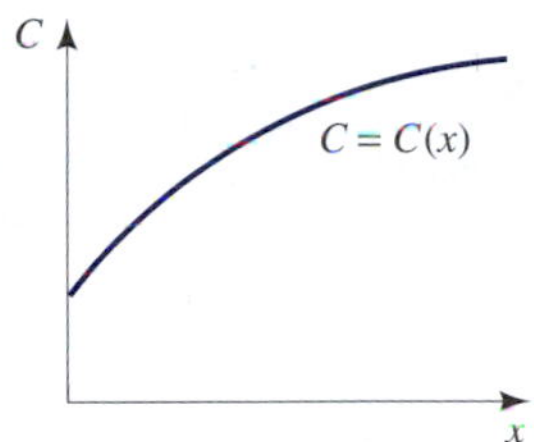

50. Economies of Scale The cost function for a firm is given by $C(x) = -0.1x^2 + 2x + 10$, where x is the number of units produced. Is this firm experiencing economies of scale? Explain.

51. Economies of Scale The cost function for a firm is given by $C(x) = 0.1x^2 + 2x + 5$, where x is the number of units produced. Is this firm experiencing economies of scale? Explain.

52. Economies of Scale The figure shows the graph of the cost function for a firm, where x is the number of units produced. Is this firm experiencing economies of scale? Explain.

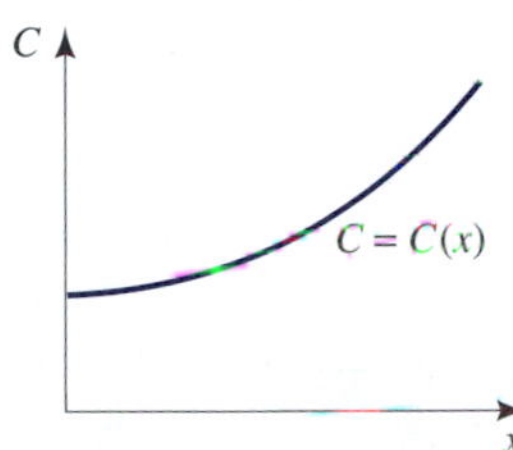

53. Sales A firm has a successful product it has been manufacturing when a consumer group issues a safety warning against the product. In the next several months the company makes changes to the product to make it safer and conducts an advertising campaign to combat the bad publicity. The sales equation in the 2 years subsequent to the consumer group announcement is given by $S(t) = 2t^3 - 60t^2 + 10,000$, where t is measured in months and S is measured in units. Graph and explain what happened.

54. Sheep Ranching Freetly and colleagues[37] studied Finnsheep ewes. In their paper they gave the graph shown here that relates the age in weeks of Finnsheep ewes to their weight in kilograms. What is happening to the weight of the ewes as they age? What is the concavity of this curve? Use your answer to explain what is happening to the rate of increase of weight of these ewes as they age.

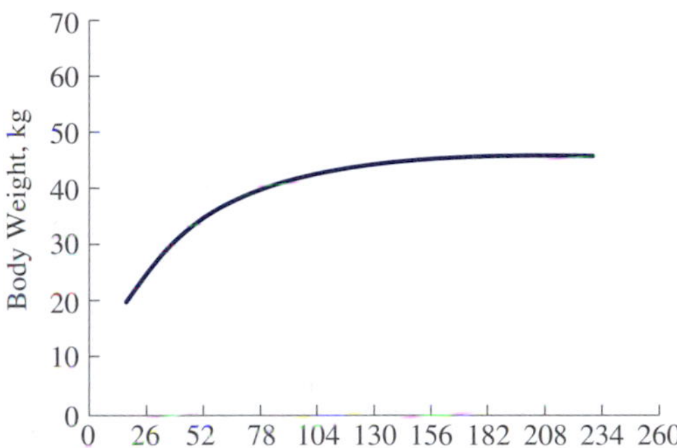

55. Economies of Size Jacques and colleagues[38] studied economies of size in school district consolidation in Oklahoma. In their paper they presented the graph shown in the figure, where the total cost included the cost of instruction, administration, and transportation. What is happening to the per student costs as the size of the school district increases? What is the concavity of this curve? Use your answer to explain what is happening to the rate of decrease in per student costs as the size of the school district increases.

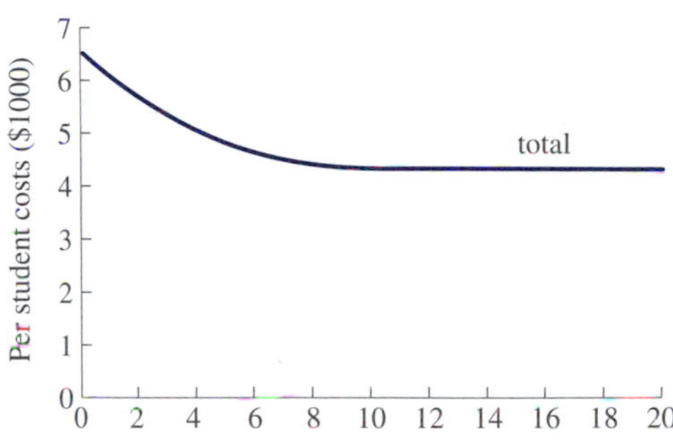

[37] H. C. Freetly, J. A. Nienaber, and T. Brown-Brandl. 2002. Relationships among heat production, body weight, and age in Finnsheep and Rambouillet ewes. *J. Anim. Sci.* 80:825–832.

[38] Charles Jacques, B. Wade Brorsen, and Francisca G. C. Richter. 2000. Consolidating rural school districts: potential savings and effects on student achievement. *J. Agr. Appl. Econ.* 32(3):573–583.

56. Growth Rates Scharf[39] studied juvenile red drum in nine estuaries along the Texas Gulf Coast. The paper gave the graph shown in the figure for Galveston. (The capture date extends beyond 365 days, since the growing season for these fish extends from September of one calendar year through July of the next.) What is happening to the length of these juvenile fish as they age? What is the concavity of the curve? Use your answer to explain what is happening to the rate of growth in the length as they age during this time period.

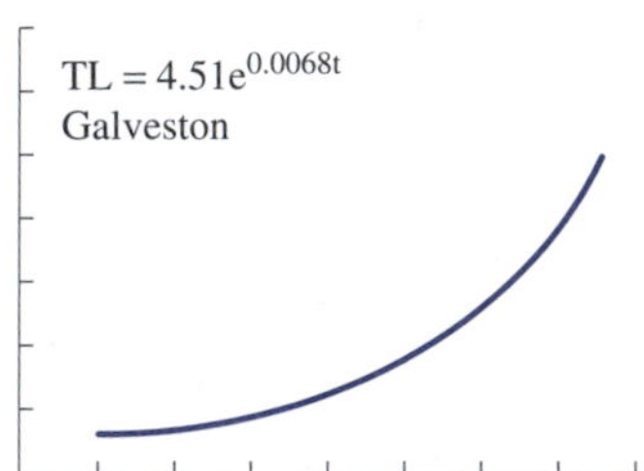

57. Aphid Injury in Citrus Mendoza and colleagues[40] studied aphid injury in citrus. The figure reproduces the graph they showed in their paper. What is the concavity of this curve? Use your answer to explain what is happening to the rate of increase in the yield loss as the maximum number of aphids per meter squared increases.

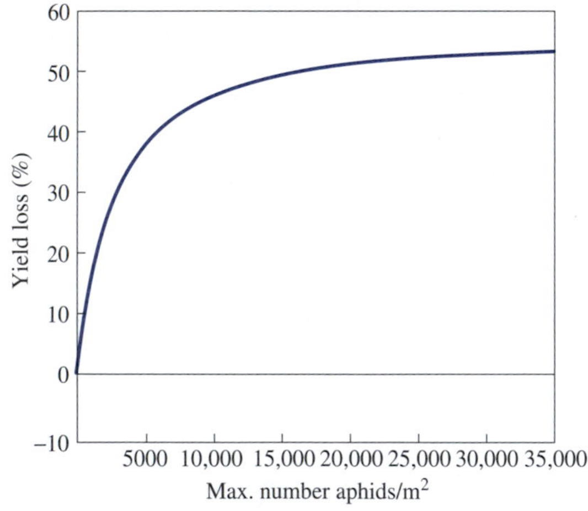

58. Demand for Recreational Fishing Boats The graph shows a demand curve estimated by Li.[41] The horizontal axis is labeled by q, the number of boat trips. The owners of these boats take out parties of recreational anglers on day trips for snapper and other game fish. On the vertical axis is p, the cost the owner charges the party for using the boat on the day trip. What is the concavity of this curve? Use your answer to explain what is happening to the rate of decrease in the cost as the number of boat trips increases.

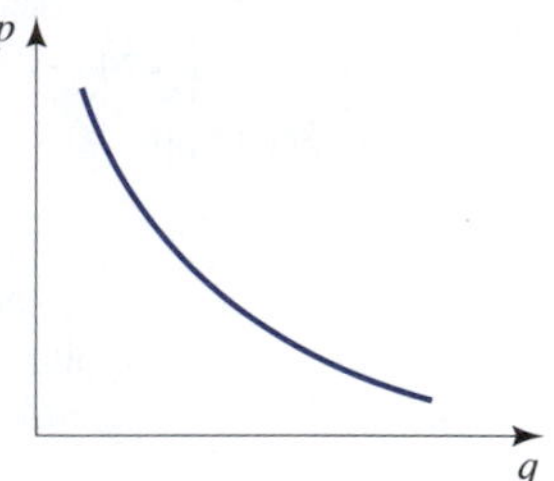

59. Biology The following figure from Deshmukh[42] shows the result of the classical experiment of Gause using the protozoan *Paramecium caudatum* in the laboratory. On approximately what day did the growth rate of the population change from increasing to decreasing?

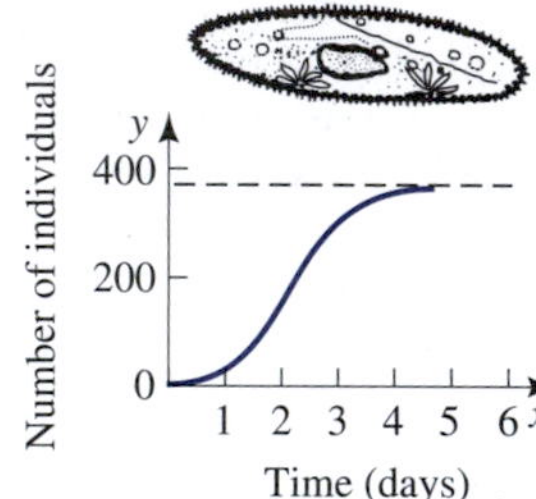

60. Acceleration If $s(t)$ is the distance traveled in time t, then the velocity, $v(t)$, is given by $v(t) = s'(t)$. The **acceleration**, denoted by $a(t)$, is defined as the rate of change of velocity. Thus, $a(t) = v'(t) = s''(t)$. A body moves according to the law $s(t) = t^4 - t^2 + 1$. Find the acceleration, and find where it is positive and where it is negative.

61. Learning Curve Cost engineers have noted that the efficiency of workers in a plant or on a machine increases over time as the work force gains efficiency and skill in performing the repeated manual tasks. Hirsch[43] reported that the

[39] Frederick S. Scharf. 2000. Patterns in abundance, growth, and mortality of juvenile red drum across estuaries on the Texas Coast with implications for recruitment and stock enhancement. *Amer. Fish. Soc. Trans.* 129:1207–1222.

[40] A. Hermoso de Mendoza, B. Belliure, E. A. Carbonell, and V. Real. 2001. *J. Econ. Entomol.* 94(2):439–444.

[41] Eric A. L. Li. 1999. A travel cost demand model for recreational snapper angling in Port Phillip Bay, Australia. *Trans. Amer. Fish. Soc.* 128:639–647.

[42] Ian Deshmukh. 1986. *Ecology and Tropical Biology*. Boston: Blackwell Scientific Publications, page 93.

[43] Werner Hirsch. 1956. Firm progress ratios. *Econometrica* 24:137–138.

relationship between the accumulative average amount of direct labor input y on a textile machine and the cumulative number of units produced x is given by $y = ax^{-0.175}$, where a is a constant.

a. Find the first and second derivatives and sketch a graph.

b. If productions doubles, what happens to the average amount of needed labor? What does this mean in terms of learning?

62. Portfolio Management Christensen and Sorensen[44] showed that the yield i was approximated by $i(t) = 0.09 + 0.0025(30 - t) - 0.00005(30 - t)^2$, where t is the time in years on a 30-year zero-coupon bond. Use calculus to sketch a graph on $[0, 30]$.

[44] Peter Oue Christensen and Gjarne G. Sorensen. 1994. Duration, convexity, and time value. *J. Portfolio Manage.* 20(2):51–60.

More Challenging Exercises

63. Failure of the Second Derivative Test Try to use the second derivative test on the following functions.

a. x^4 **b.** $-x^4$ **c.** x^3 **d.** $9x^{4/3}$

Explain what went wrong. Use the first derivative test to find the relative extrema.

64. Spread of Technology When a new, useful technology is introduced into a particular industry, it naturally spreads through this industry (see Mansfield[45]). A certain model (very similar to that of Mansfield) of this effect indicates that if t is the number of years after the introduction of the new technology, the percentage of the firms in this industry using the new technology is given by

$$p(t) = \frac{100t^2}{t^2 + 300}$$

a. Find where $p(t)$ is increasing and where it is decreasing.

b. Find where the graph of $p(t)$ is concave up and where it is concave down. Find all inflection points.

c. Sketch a graph.

In Exercises 65 through 68, find where the function is increasing, decreasing, concave up, and concave down. Find critical points, inflection points, and where the function attains a relative minimum or relative maximum. Then use this information to sketch a graph. Assume that the constants a and b are positive.

65. $f(x) = 4x^5 - 5bx^4 + a$

66. $f(x) = x^3 - 3bx^2 + a$

67. $f(x) = -3x^4 + 8bx^3 + a$

68. $f(x) = x^4 + 6bx^2 + a$

69. If $f''(a)$ exists, explain why $y = f(x)$ must be continuous at $x = a$.

70. Use the concavity of $f(x) = \ln x$ to show that $\ln x \leq x - 1$.

71. Assume that a polynomial has exactly one relative maximum and two relative minima.

a. Sketch a possible graph of $y = f(x)$.

b. What is the largest number of zeros that f could have?

c. What is the least number of inflection values that f could have?

72. Assume that a polynomial has exactly one relative maximum and one relative minima. Sketch a possible graph of $y = f(x)$.

73. Suppose that $y = f(x)$ is a polynomial with two and only two critical points at $x = -2$ and $x = 3$. Also, $f''(-2) > 0$ and $f''(3) < 0$.

a. Draw a possible graph.

b. Must there be an inflection point? Explain.

74. Suppose that $y = f(x)$ is a polynomial with three and only three critical points at $x = -2$, $x = 3$, and $x = 6$. Also, $f''(-2) > 0$, $f''(3) < 0$, and $f''(6) > 0$.

a. Draw a possible graph.

b. Must there be an inflection point? Explain.

75. Suppose that $f''(x)$ is continuous and positive everywhere and $f(x)$ is negative everywhere. Let $g = 1/f$. Show that $g(x)$ is concave down everywhere.

76. Let $f(x) = x^2 p(x)$, where $p(x)$ is a polynomial with $p(0) > 0$. Show that $f(x)$ attains a relative minimum at $x = 0$. What happens if $p(0) < 0$?

77. Let $f(x) = 0$ if $x < 0$ and $f(x) = x^3$ if $x \geq 0$. Use the limit definition to show that $f'(0)$ exists. Use the limit definition to show that $f''(0)$ exists.

78. Sketch a graph of $y = P'(x)$ below the given graph of $y = P(x)$. Explain carefully what happens to the graph of $y = P'(x)$ when it crosses the two vertical dotted lines.

[45] E. Mansfield. 1961. Technical change and the rate of imitation. *Econometrica* 29(4):741–766.

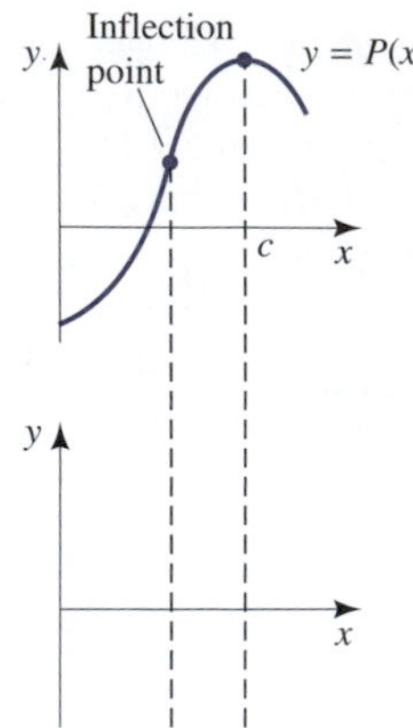

79. Recall that $\frac{d}{dx}\sin x = \cos x$ and $\frac{d}{dx}\cos x = -\sin x$

   a. Find $\frac{d^2}{dx^2}(\sin x)$ and $\frac{d^2}{dx^2}(\cos x)$.

   b. On your computer or graphing calculator, graph $y_1 = \sin x$ and $y_2 = \cos x$ on a screen with dimensions $[0, 2\pi]$ by $[-1, 1]$. Determine where $\sin x$ is positive and where it is negative. Do the same for $\cos x$. Use this information together with the second derivatives found in part (a) to determine where the functions $y_1 = \sin x$ and $y_2 = \cos x$ are concave up and concave down. Verify by closely examining the graphs of these functions.

80. Shown is a graph of the derivative $y = f'(x)$. Find the critical values. Find where the function is increasing, decreasing, concave up, and concave down. Find any inflection values.

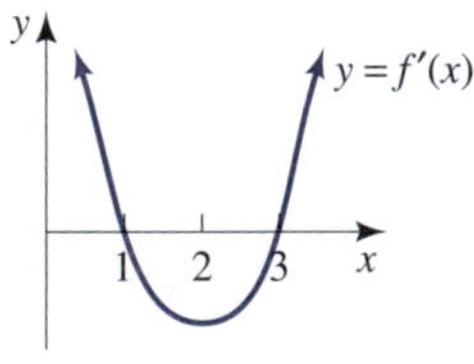

81. Shown is a graph of the derivative $y = f'(x)$. Find the critical values. Find where the function is increasing, decreasing, concave up, and concave down. Find any inflection values.

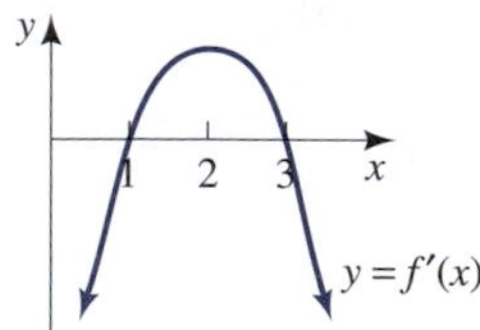

82. **Biological Control** Lysyk[46] studied the effect of temperature on various life history parameters of a parasitic wasp for the purposes of pest management. He constructed a mathematical model given approximately by the equation

$$p(t) = \left(1 - \frac{3-t}{2.95}\right)^{[(3-t)/2.95]^{1.3}}$$

where t is normalized time (time/time to 50% oviposition) and p is the proportion of progeny produced. Graph on your grapher, and find the approximate inflection point. Explain what is happening.

[46] T. J. Lysyk. 2001. Relationship between temperature and life history parameters of *Muscidifurax zaraptor*. *Environ. Entomol.* 30(1):147–156.

Modeling Using Least Squares

83. **Cost Curve** Johnston[47] made a statistical estimation of the cost-output relationship for 40 firms. The data for one of the firms is given in the following table.

x	2.5	3	4	5	6	7
y	20	26	20	27	30	29
x	11	22.5	29	37	43	
y	42	53	75	73	78	

Here x is the output in millions of units, and y is the total cost in thousands of pounds sterling.

[47] J. Johnston. 1960. *Statistical Cost Analysis*. New York: McGraw-Hill.

a. Determine the best-fitting quadratic using least squares. Graph.
b. Determine the concavity, and use this to decide whether the firm is experiencing economies of scale? Explain.

84. Cost Curve Dean[48] made a statistical estimation of the cost-output relationship for a hosiery mill. The data for the firm is given in the following table.

x	45	56	62	70	74	78
y	14	17	19	20.5	21.5	22.5

Here x is production in hundreds of dozens, and y is total cost in thousands of dollars.

a. Determine the best-fitting quadratic using least squares. Graph.
b. Determine the concavity, and use this to decide whether the firm is experiencing economies of scale. Explain.

85. Forestry Pearce[49] gave the data found in the table for the net stumpage values for a typical stand of British Columbia Douglas fir trees. Find a model that best fits the data. Then explain what the model implies. In particular, make note of concavity and what this implies.

Age (years)	Net Stumpage Value
40	$43
50	$143
60	$303
70	$497
80	$650
90	$805
100	$913
110	$1000
120	$1075

[48] Joel Dean. 1976. *Statistical Cost Estimation.* Bloomington, Ind.: Indiana University Press.

[49] P. Pearce. 1967. The optimal forest rotation. *Forestry Chron.* 43:178–195.

86. Cost Curve Nordin[50] obtained the following data relating output and total final cost for an electric utility in Iowa. (Instead of using the original data set for 541 eight-hour shifts, we have given just the array means.)

x	25	28	33	38	41	45	50	53
y	24	24	26	28	27	31	34	38
x	57	63	66	71	74	79	83	89
y	39	42	43	51	53	52	54	52

Here x is the output as a percent of capacity, and y is final total cost in dollars (multiplied by a constant not given to avoid showing the exact level of the costs.)

a. Use cubic regression to find the best-fitting cubic to the data. Graph.
b. Determine concavity and estimate the point of inflection. Discuss what all of this implies for costs per unit and economies of scale.

[50] J. A. Nordin 1947. Note on a light plant's cost curve. *Econometrica* 15(1):231–235.

Solutions to WARM UP EXERCISE SET 5.2

1. Since

$$p'(t) = -t^2 + 6t = t(6 - t)$$

p is always increasing on the interval $(0, 6)$. Since

$$p''(t) = -2t + 6 = 2(3 - t)$$

$p''(t) > 0$, and thus the graph of p is concave up on $(0, 3)$, and $p''(t) < 0$, and thus the graph of p is concave down on $(3, 6)$. On the interval $(0, 3)$, p is increasing ever faster, while on $(3, 6)$, p is increasing ever slower. We might characterize the interval $(3, 6)$ as the period of diminishing returns.

2. According to the given factorization of $f''(x)$, $f''(x) > 0$ on the intervals $(-\infty, 1)$ and $(3, \infty)$, while $f''(x) < 0$ on $(1, 2)$ and $(2, 3)$. Thus, the graph of f is concave up on the intervals $(-\infty, 1)$ and $(3, \infty)$, and concave down on $(1, 3)$.

Since the second derivative changes sign at $x = 1$ and $x = 3$, these values are inflection values.

5.3 Limits at Infinity

Limits at Infinity

Enrichment: Polynomial Behavior for Large $|x|$

Enrichment: Difficulties with a Grapher

Enrichment: The Base e and Continuous Compounding

APPLICATION **Insect Infestation**

Asante[51] showed that the proportion y of apple trees infested with the wooly apple aphid was approximated by

$$y = f(x) = \frac{0.33x}{x + 41}$$

where x is the aphid density measured by the average number of aphids per tree. What proportion of trees will be infested as the aphid density increases without bound? See Example 1 on page 331 for the answer.

[51] Stephen Asante. 1993. Spatial and temporal distribution patterns of *Eriosoma lanigerum* on apple. *Environ. Entomol.* 22:1060–1065.

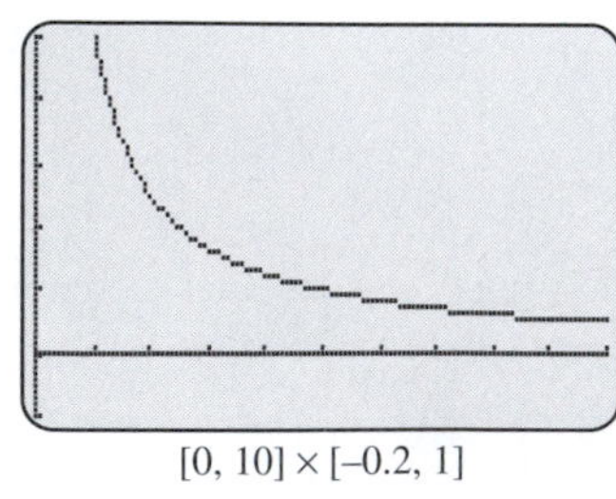

$[0, 10] \times [-0.2, 1]$

Screen 5.9
As x becomes large, the graph of $y = \dfrac{1}{x}$ approaches the x-axis.

Limits at Infinity

Suppose we are given a demand equation $f(p) = \dfrac{1}{p}$, where p is the price of an item and $f(p)$ is the number of items sold at a price equal to p. Let us see what happens to f as p becomes very large.

From Table 5.2, as p becomes large, $f(p)$ approaches zero. In such a case we say that *the limit of $f(p)$ as p goes to ∞ is zero*, and we write $\lim\limits_{p\to\infty} \dfrac{1}{p} = 0$. This is confirmed in Screen 5.9.

Table 5.2

p	1	10	100	1000	→	∞
$1/p$	1	0.1	.01	.001	→	0

This makes sense. If we set the price very high, then the number of items sold should be very small.

We have the following definitions.

DEFINITION **Limits at Infinity**

1. If $f(x)$ approaches the number L as x becomes large without bound, then we say that L is the limit of $f(x)$ as x approaches ∞, and we write

$$\lim_{x \to \infty} f(x) = L$$

2. If $f(x)$ approaches the number K as x becomes a large negative number without bound, then we say that K is the limit of $f(x)$ as x approaches $-\infty$, and we write

$$\lim_{x \to -\infty} f(x) = K$$

We refer to the lines $y = L$ and $y = K$ as **horizontal asymptotes**.

Thus, for $y = f(p) = 1/p$, $y = 0$ is a horizontal asymptote.

EXAMPLE 1 **Insect Infestation**

Asante[52] showed that the proportion y of apple trees infested with the wooly apple aphid was approximated by

$$y = f(x) = \frac{0.33x}{x + 41}$$

where x is the aphid density measured by the average number of aphids per tree. What proportion of trees will be infested as the aphid density increases without bound?

Solution

We can obtain a good idea of the answer by evaluating $f(x)$ using an extremely large value of x, say, x equals one billion. In this case we have

$$f(\text{ billion}) = \frac{0.33(\text{ billion})}{(\text{ billion}) + 41}$$

However, when we check the denominator, since the number 41 is insignificant when compared to the one billion, the denominator is approximately equal to one billion. Then

$$\begin{aligned} f(\text{ billion}) &= \frac{0.33(\text{ billion})}{(\text{ billion}) + 41} \\ &\approx \frac{0.33(\text{ billion})}{(\text{ billion})} \\ &= 0.33 \end{aligned}$$

Thus, $f(x)$ is approximately 0.33, and we suspect that the limit is 0.33.

[52] Ibid.

The following is a formal proof. First we note that to determine the limit as $x \to \infty$ in a rational expression, we need to *divide the numerator and denominator by x to the highest power occurring in the denominator*. In this case, this means divide numerator and denominator by x. Since limits as $x \to \infty$ satisfy the usual rules for limits, we obtain

$$\begin{aligned}
\lim_{x\to\infty}\left(0.33\frac{x}{x+41.30}\right) &= \lim_{x\to\infty}\left(0.33\frac{x}{x+41.30}\cdot\frac{1/x}{1/x}\right) \\
&= 0.33\lim_{x\to\infty}\left(\frac{1}{1+\dfrac{41.30}{x}}\right) \\
&= 0.33\frac{1}{1+\lim\limits_{x\to\infty}\left(\dfrac{41.30}{x}\right)} \\
&= 0.33\frac{1}{1+0} \\
&= 0.33
\end{aligned}$$

Thus, about one third of the trees will become infested as the aphid density increases without bound. ■

In Example 1 the highest power of the numerator of the rational function was the same as the highest power of the denominator. We now consider an example for which the highest power of the numerator is less than the highest power of the denominator.

EXAMPLE 2 **Limit at Infinity**

Find $\lim\limits_{x\to\infty}\dfrac{3x}{x^2+6}$.

Solution

Again first let's try to make an intelligent guess at the answer. To do this, let us evaluate the function

$$f(x) = \frac{3x}{x^2+6}$$

for a very large value of x, say, x equals one billion. Then

$$f(\text{billion}) = \frac{3(\text{billion})}{(\text{billion})^2+6}$$

In the denominator the number 6 is insignificant when compared to the square of one billion. So we can say that the denominator is approximately equal to $(\text{billion})^2$. We then have

$$\begin{aligned} f(\text{ billion}) &= \frac{3(\text{ billion})}{(\text{ billion})^2 + 6} \\ &\approx \frac{3(\text{ billion})}{(\text{ billion})^2} \\ &= \frac{3}{(\text{ billion})} \\ &\approx 0 \end{aligned}$$

Thus, we have that when x is one billion f is approximately equal to 0. So, we suspect that the limit as $x \to \infty$ is also 0.

We now present a formal proof. First recall that to determine the limit as $x \to \infty$ in a rational expression, we should *divide the numerator and denominator by x to the highest power occurring in the denominator*. In this case, this means dividing the numerator and denominator by x^2. Since limits as $x \to \infty$ satisfy the usual rules for limits, we obtain

$$\begin{aligned} \lim_{x\to\infty} \frac{3x}{x^2+6} &= \lim_{x\to\infty} \frac{3x}{x^2+6} \cdot \frac{1/x^2}{1/x^2} \\ &= \lim_{x\to\infty} \frac{\frac{3x}{x^2}}{\frac{x^2}{x^2} + \frac{6}{x^2}} \\ &= \lim_{x\to\infty} \frac{\frac{3}{x}}{1 + \frac{6}{x^2}} \\ &= \frac{\lim_{x\to\infty} \frac{3}{x}}{1 + \lim_{x\to\infty} \frac{6}{x^2}} \\ &= \frac{0}{1+0} \\ &= 0 \end{aligned}$$

■

We can also say that $y = 0$ is a horizontal asymptote.

The values of the function $f(x) = x^2$ become large without bound as x becomes large without bound. We then use the notation $\lim_{x\to\infty} x^2 = \infty$ to describe this situation. (This notation is not intended to suggest that the limit exists, since it does not.) We can also make corresponding definitions for limits when $x \to -\infty$. For example, with this notation we have $\lim_{x\to-\infty} x^3 = -\infty$.

We now consider an example for which the highest power of the numerator of a rational function is greater than the highest power of the denominator.

EXAMPLE 3 **Limit at Infinity**

Find $\lim_{x\to-\infty} \dfrac{3x^2}{x+6}$

Solution

Again first let's try to make an intelligent guess at the answer. To do this let us evaluate the function

$$f(x) = \frac{3x^2}{x+6}$$

for a very large negative value of x, say, x equals negative one billion. Then

$$f(-\text{ billion}) = \frac{3(-\text{ billion})^2}{(-\text{ billion}) + 6}$$

In the denominator the number 6 is insignificant when compared to negative one billion. So we can say that the denominator is approximately equal to (– billion). We then have

$$\begin{aligned} f(-\text{ billion}) &= \frac{3(-\text{ billion})^2}{(-\text{ billion}) + 6} \\ &\approx \frac{3(-\text{ billion})^2}{(-\text{ billion})} \\ &= 3(-\text{ billion}) \end{aligned}$$

Thus, we have that when x is negative one billion f is approximately equal to negative three billion, which is a very large negative number. So we suspect that the limit as $x \to -\infty$ is $-\infty$.

We now present a formal proof. First recall that to determine the limit as $x \to -\infty$ in a rational expression, we should *divide the numerator and denominator by x to the highest power occurring in the denominator*. In this case this means divide the numerator and denominator by x. Since limits as $x \to -\infty$ satisfy the usual rules for limits, we obtain

$$\begin{aligned} \lim_{x\to-\infty}\left(\frac{3x^2}{x+6}\right) &= \lim_{x\to-\infty}\left(\frac{3x^2}{x+6}\cdot\frac{1/x}{1/x}\right) \\ &= \lim_{x\to-\infty}\left(\frac{\frac{3x^2}{x}}{\frac{x}{x}+\frac{6}{x}}\right) \\ &= \lim_{x\to-\infty}\left(\frac{3x}{1+\frac{6}{x}}\right) \\ &= \frac{\lim_{x\to-\infty} 3x}{1+\lim_{x\to-\infty}\left(\frac{6}{x}\right)} \end{aligned}$$

Since the numerator is $-\infty$ and the denominator is 1, the quotient must be $-\infty$. ■

We also use the notation $\lim_{x\to 0^+} \frac{1}{x} = \infty$ to describe the fact that $\frac{1}{x}$ becomes large without bound when x approaches 0 from the right and $\lim_{x\to 0^-} \frac{1}{x} = -\infty$ to describe the fact that $\frac{1}{x}$ becomes negatively large without bound when x approaches 0 from the left.

EXAMPLE 4 **Limits at Infinity**

Power and Heuvel[53] developed a mathematical model given approximately by the equation $L(D) = 72(1 - 19e^{-0.02D})$, where L is the length in millimeters of yellow perch hatched in the spring in Kimowin Lake in Alberta and D is the day of the year with $D = 1$ corresponding to January 1. Sketch a graph. According to this model, what will be the ultimate length of these perch?

Solution

We have

$$L'(D) = -(72)(19)e^{-0.02D}(-0.02) = 27.36e^{-0.02D}$$
$$L''(D) = 27.36e^{-0.02D}(-0.02) = -0.5472e^{-0.02D}$$

Notice that $L'(D) > 0$ and $L''(D) < 0$. So the graph is always increasing and always concave down. The fact that the derivative is positive means that the perch are growing day by day. The fact that the second derivative is negative indicates that the growth rate of the perch slows up as they get older. A graph is shown in Figure 5.29. To see what ultimately happens to the length of this perch, we need to take the limit as D goes to infinity.

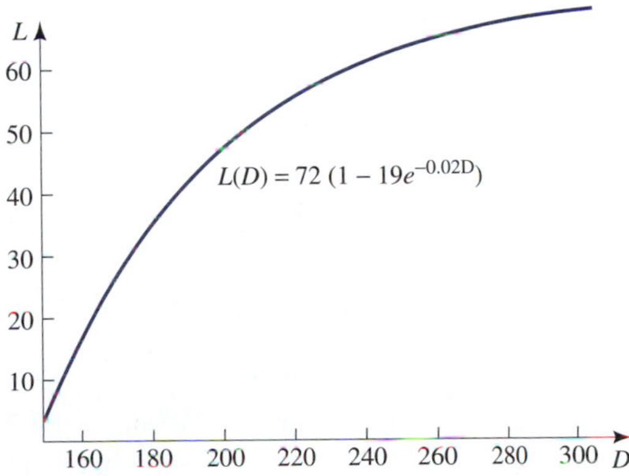

Figure 5.29

53 M. Power and M. R. van den Heuvel. 1999. Age-0 yellow perch growth and its relationship to temperature. *Trans. Amer. Fish. Soc.* 128:687–700.

First notice that $e^{-0.02D} = \frac{1}{e^{0.02D}}$. When D is large, $e^{0.02D}$ is large and the reciprocal is small. We then have

$$\lim_{D\to\infty} e^{-0.2D} = 0$$

Thus,

$$\lim_{D\to\infty} 72(1 - 19e^{-0.02D}) = 72(1 - 0) = 72$$

That is, their ultimate length is 72 mm. ■

Enrichment: Polynomial Behavior for Large $|x|$

We will see that polynomial behavior for large $|x|$ falls into only four cases. Knowing this can help greatly in graphing.

Consider the polynomial $p(x) = x^5 - 10x^4 + 9x - 3$. Let us consider the behavior of this polynomial for large x. Suppose that x is huge, say, a billion. Then the first term x^5 can be written as $x \cdot x^4$, or one billion times x^4. The second term $-10x^4$ can be thought of as -10 times x^4. Since -10 is insignificant compared to one billion, the term $-10x^4$ is insignificant compared to x^5. The other terms $9x$ and -3 are even less significant. So the polynomial for huge x should be approximately equal to the leading term x^5.

Looking at this algebraically, we can factor out the x^5 term and obtain

$$p(x) = x^5\left(1 + \frac{14}{x} + \frac{9}{x^4} - \frac{3}{x^5}\right)$$

Since the term in parenthesis is approximately 1 for large $|x|$, we then have that $p(x) \approx x^5$ for large $|x|$.

Formally, we have

$$\begin{aligned}
\frac{p(x)}{x^5} &= \left(1 + \frac{14}{x} + \frac{9}{x^4} - \frac{3}{x^5}\right) \\
\lim_{x\to\pm\infty} \frac{p(x)}{x^5} &= \lim_{x\to\pm\infty}\left(1 + \frac{14}{x} + \frac{9}{x^4} - \frac{3}{x^5}\right) \\
&= 1 + 0 + 0 - 0 \\
&= 1
\end{aligned}$$

A similar analysis leads us to the conclusion that for large $|x|$ the polynomial $a_nx^n + a_{n-1}x^{n-1} + \cdots + a_1x + a_0$ is approximately equal to the leading term a_nx^n. (We assume that $a_n \neq 0$.) Thus, there are only four cases according to whether n is odd and a_n is positive or negative or n is even and a_n is positive or negative. Graphs of the four cases are shown in Figure 5.30.

n odd, $a_n > 0$

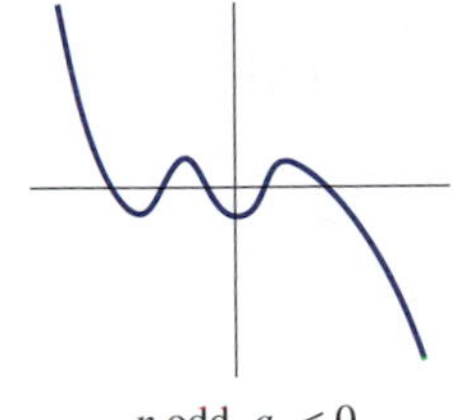

n odd, $a_n < 0$

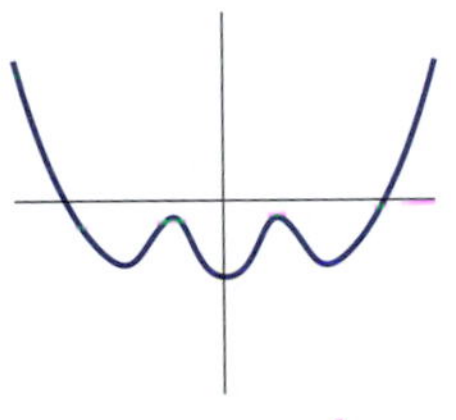

n even, $a_n > 0$

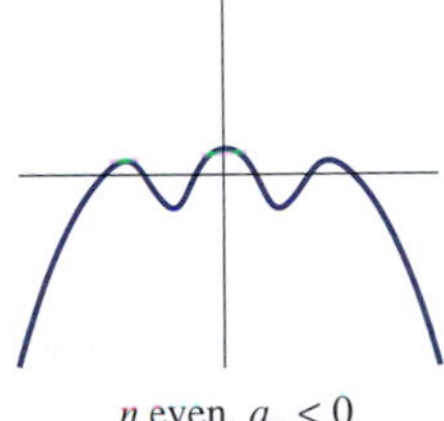

n even, $a_n < 0$

Figure 5.30

Enrichment: Difficulties with a Grapher

The following example illustrates how your graphing calculator can mislead you in attempting to evaluate a limit.

EXAMPLE 5 **Difficulties with a Grapher**

Determine $\lim_{x \to \infty} \dfrac{x^{2.646} + 1}{x^{\sqrt{7}}}$.

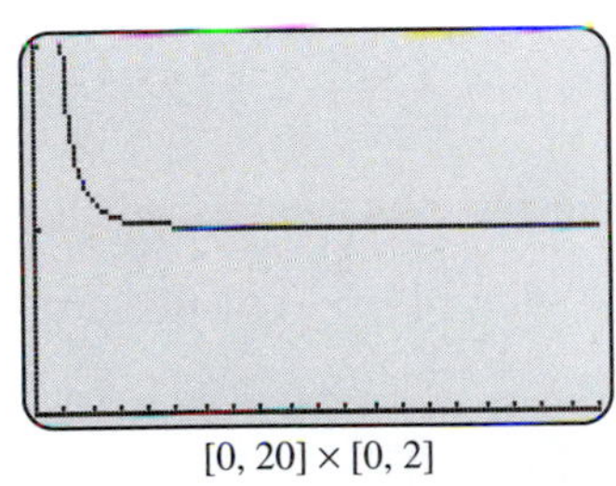

$[0, 20] \times [0, 2]$

Screen 5.10
The grapher indicates that $\lim_{x \to \infty} \dfrac{x^{2.646} + 1}{x^{\sqrt{7}}} = 1$. However, a mathematical analysis shows otherwise!

Solution

A graph is shown in Screen 5.10. It certainly appears that the limit is 1. However, a mathematical analysis gives

$$
\begin{aligned}
\lim_{x \to \infty} \frac{x^{2.646} + 1}{x^{\sqrt{7}}} &= \lim_{x \to \infty} \left(\frac{x^{2.646}}{x^{\sqrt{7}}} + \frac{1}{x^{\sqrt{7}}} \right) \\
&= \lim_{x \to \infty} \left(x^{2.646 - \sqrt{7}} + \frac{1}{x^{\sqrt{7}}} \right) \\
&= \infty
\end{aligned}
$$

since $2.646 > \sqrt{7} \approx 2.64575$. ■

Enrichment: The Base e and Continuous Compounding

We noted in Section 1.3 that for an account earning compound interest, the more we compound each year, the larger the amount at the end of each year. For example, \$1000 deposited into an account that yields 6% annually becomes \$1061.36 in one year when compounded quarterly and the larger amount of \$1061.68 when compounded monthly. Suppose we compound monthly, then daily, then by the hour, then by the minute, then by the second, and so on indefinitely. What would happen to the amount in the account at the end of a year? We refer to this process as **continuous compounding**.

Recall from Section 1.3 that if an account with an initial principal P and an annual rate of r, where r is expressed as a decimal, is compounded m times a year, then the future amount in the account after t years is

$$F(t) = P\left(1 + \frac{r}{m}\right)^{mt}$$

We now wish to see what happens if the number of compoundings increases without bound, that is, as $m \to \infty$. We wish to show that

$$\lim_{m\to\infty} F(t) = \lim_{m\to\infty} P\left(1 + \frac{r}{m}\right)^{mt} = Pe^{rt}$$

First, recall again from Section 1.3 that if an account with an initial principal P and an annual rate of r, where r is expressed as a decimal, is compounded m times a year, then the future amount in the account after t years is

$$F(t) = P\left(1 + \frac{r}{m}\right)^{mt}$$

It is helpful to set $n = m/r$ in the above formula for F. This gives $m = rn$ and $r/m = 1/n$. Then

$$\begin{aligned} F(t) &= P\left(1 + \frac{r}{m}\right)^{mt} \\ &= P\left(1 + \frac{1}{n}\right)^{rnt} \\ &= P\left[\left(1 + \frac{1}{n}\right)^{n}\right]^{rt} \end{aligned}$$

Since $n = m/r$, as the number m of compoundings increases without bound, that is, $m \to \infty$, then also $n \to \infty$. Thus, we wish to find $\lim_{n\to\infty}\left(1 + \frac{1}{n}\right)^n$. Table 5.3 indicates that this limit is 2.71828 (to five decimal places).

Table 5.3

n	10	100	1000	10,000	100,000	1,000,000
$\left(1 + \frac{1}{n}\right)^n$	2.593743	2.704814	2.716924	2.718146	2.718268	2.718281

This is the number we defined to six decimal places in Section 1.3.

DEFINITION The Number e

The number e is defined to be $e = \lim_{n\to\infty}\left(1 + \frac{1}{n}\right)^n$, and $e = 2.71828\cdots$

Using this in the above formula for $F(t)$ then yields

$$\begin{aligned}\lim_{n\to\infty} F(t) &= \lim_{n\to\infty} P\left[\left(1+\frac{1}{n}\right)^n\right]^{rt}\\ &= P\left[\lim_{n\to\infty}\left(1+\frac{1}{n}\right)^n\right]^{rt}\\ &= P[e]^{rt}\\ &= Pe^{rt}\end{aligned}$$

Warm Up Exercise Set 5.3

1. Find $\lim_{x\to\infty} \dfrac{x^2}{x^3+1}$.
2. The weekly sales of a new product satisfy $S(t) = 1000 - 800e^{-0.1t}$, where t is time measured in months from the time of introduction. Indicate what happens in the long term.

Exercise Set 5.3

In Exercises 1 through 18, find the limits.

1. $\lim_{x\to\infty} \dfrac{2x^2+1}{3x^2-1}$
2. $\lim_{x\to-\infty} \dfrac{4x^2-x-1}{3x^2+1}$
3. $\lim_{x\to\infty} \dfrac{3x^2+x+1}{2x^4-1}$
4. $\lim_{x\to\infty} \dfrac{x^3-1}{2x^4+1}$
5. $\lim_{x\to-\infty} (x^2+1)$
6. $\lim_{x\to\infty} \sqrt{x}$
7. $\lim_{x\to-\infty} \sqrt[3]{x}$
8. $\lim_{x\to-\infty} \dfrac{x^4-x^2+x-1}{x^3+1}$
9. $\lim_{x\to\infty} (1+2e^{-x})$
10. $\lim_{x\to-\infty} (3-2e^{x})$
11. $\lim_{x\to\infty} \dfrac{10}{2+e^{-x}}$
12. $\lim_{x\to\infty} \dfrac{2}{1+e^{x}}$
13. $\lim_{x\to\infty} \dfrac{e^x-1}{e^x+1}$
14. $\lim_{x\to\infty} \dfrac{e^x-e^{-x}}{e^x+e^{-x}}$
15. $\lim_{x\to\infty} \dfrac{\sqrt{x}+1}{x+2}$
16. $\lim_{x\to\infty} \dfrac{x^2-1}{\sqrt{x}-3}$
17. $\lim_{x\to\infty} \sqrt{4-\dfrac{1}{x}}$
18. $\lim_{x\to-\infty} \sqrt{1-\dfrac{1}{x^2}}$

Applications and Mathematical Modeling

19. **Nutrition** The Morgan-Mercer-Flodid model[54] $r = \dfrac{ab+cx^k}{b+x^k}$, characterizes the nutritional responses of higher organisms, where r is the weight gain and x is the nutrient intake. What happens to the weight gain as the nutrient intake becomes large without bound?

20. **Biology** According to Wiess's law, the intensity of an electric current required to excite a living tissue (threshold strength) is given by $i = \dfrac{A}{t} + B$, where t is the duration of the current and A and B are positive constants. What happens as $t \to \infty$? Does your answer seem reasonable?

21. **Spread of Technology** When a new useful technology is introduced into a particular industry, it naturally spreads through this industry. A certain model (see Mansfield[55]) of this effect indicates that if t is the number of years after the introduction of the new technology, the percentage of the firms in this industry using the new technology is given by

$$p(t) = \frac{100}{1+2e^{-0.15t}}$$

To what extent does this new technology spread throughout this industry in the long-term?

[54] Paul Morgan, L. Preston Mercer, Nestor Flodin. 1975. General model for nutritional responses of higher organisms. *Proc. Nat. Acad. Sci.* 72:4327–4331.

[55] E. Mansfield. 1961. Technical change and the rate of imitation. *Econometrica* 29(4):741–766.

22. Biology Wiedenmann and coworkers[56] showed that the proportion y of a certain host larvae parasitized by *Cotesian flavipes* was approximated by $y = 1 - e^{-1.48x}$, where x is the parasite/host ratio. What proportion of the larvae are parasitized as the parasite/host ratio becomes large without bound?

23. Biology Reed and Semtner[57] showed that the proportional yield loss y of flue-cured tobacco was approximated by $y = f(x) = 0.259(1 - e^{-0.000232x})$, where x is cumulative aphid-days in thousands. Use calculus to draw a graph. What happens to the proportional yield as the cumulative aphid-days becomes large without bound?

24. Growth Rates of Sturgeon Pine and Allen[58] studied the growth habits of sturgeon in the Suwannee River, Florida. The sturgeon fishery was once an important commercial fishery but, owing to overfishing, was closed down in 1984, and the sturgeon is now both state and federally protected. The mathematical model they created was given by the equation $L(t) = 222.273(1 - e^{-0.08042[t+2.18]})$, where t is age in years and L is length in centimeters. Graph this equation. Find the ultimate length of these sturgeon.

25. Growth Rates of Atlantic Croaker Diamond and colleagues[59] studied the growth habits of the Atlantic croaker, one of the most abundant fishes of the southeastern United States. The mathematical model they created for the ocean larva stage was given by the equation

$$L(t) = 0.26e^{2.876[1-e^{-0.0623t}]}$$

where t is age in days and L is length in millimeters. Find an upper bound on the length of these larva.

26. Average Cost Suppose the cost function is given by $C(x) = a + bx^2$, where a and b are positive constants. Sketch a graph of the average cost. What is happening?

27. Limited Growth A jug of milk at a temperature of 40°F is placed in a room held at 70°F. The temperature of the milk in degrees Fahrenheit satisfies $T(t) = 70 - 30e^{-0.40t}$, where t is in hours.

a. Use the model to determine what happens to the temperature of the milk in the long-term. Does your experience confirm your answer?

b. Find the time it takes for the milk to reach 50°F.

28. Consider $f(t) = L - ae^{-kt}$, where L, a, and k are positive constants. Establish the following.

a. $f'(t) > 0$ for $t \geq 0$

b. $\lim_{t\to\infty} f(t) = L$

c. $f''(t) < 0, t \geq 0$

Now sketch a graph.

29. Biology Flamm and colleagues[60] encountered the function $y = 1.27(1 - e^{-6.77x^{0.44}})$, when studying the Southern pine bark beetle. What happens as $x \to \infty$?

30. Sheep Ranching Freetly and colleagues[61] studied Rambouillet ewes. They created a mathematical model relating the body weight of Rambouillet ewes with age and given by the equation $W(t) = 53.9 - 57.5e^{-0.043t}$, where W is weight in kilograms and t is age in weeks. What happens to the weight as the ewes age?

31. Diminishing Returns The profit equation for a firm after the introduction of a new product is given by

$$P(t) = \frac{4t^3 + 0.1}{t^3 + 1}$$

where P is profit in millions of dollars and t is years. What happens for large time?

32. Diminishing Returns The profit equation for a firm after the introduction of a new product is given by

$$P(t) = \frac{5t^4 + 0.1}{t^4 + 1}$$

where P is profit in millions of dollars and t is years. What happens for large time?

[56] Robert Wiedenmann et al. 1992. Laboratory rearing and biology of the parasite *Cotesian flavipes*. *Environ. Entomol.* 21:1160–1167.

[57] David Reed and Paul Semtner. 1992. Effects of tobacco aphid populations on flue-cured tobacco production. *J. Econ. Entomol.* 85(5):1963–1971.

[58] William E. Pine III and Mike S. Allen. 2001. Population viability of the Gulf of Mexico sturgeon: inferences from capture-recapture and age-structured models. *Trans. Amer. Fish. Soc.* 130:1164–1174.

[59] Andra L. Diamond, Larry B. Crowder, and Lindsay G. Cowell. 1999. Catch and bycatch: the qualitative effects of fisheries on population vital rates of Atlantic croaker. *Trans. Amer. Fish. Soc.* 128:1085–1105.

[60] Richard Flamm, Paul Pulley, and Robert Coulson. 1993. Colonization of disturbed trees by the southern pine bark beetle guild. *Env. Entomol.* 22(1):62–70.

[61] H. C. Freetly, J. A. Nienaber, and T. Brown-Brandl. 2002. Relationships among heat production, body weight, and age in Finnsheep and Rambouillet ewes. *J. Anim. Sci.* 80:825–832.

More Challenging Exercises

33. Find $\lim_{x\to\infty} \sqrt{x^2+1} - x$.

$\left(\text{Hint: Multiply by } \frac{\sqrt{x^2+1}+x}{\sqrt{x^2+1}+x}.\right)$

34. Find $\lim_{x\to\infty} \sqrt{x^2+1} - \sqrt{x^2-1}$.

$\left(\text{Hint: Multiply by } \frac{\sqrt{x^2+1}+\sqrt{x^2-1}}{\sqrt{x^2+1}+\sqrt{x^2-1}}.\right)$

35. Match the given polynomials with their graphs. There can be only one possible match.
 a. $p_1(x) = x^{14} + g_1(x)$, where $g_1(x)$ is a polynomial of order 13.
 b. $p_2(x) = -x^{18} + g_2(x)$, where $g_2(x)$ is a polynomial of order 17.
 c. $p_3(x) = x^{15} + g_3(x)$, where $g_3(x)$ is a polynomial of order 14.
 d. $p_4(x) = -x^{17} + g_4(x)$, where $g_4(x)$ is a polynomial of order 16.

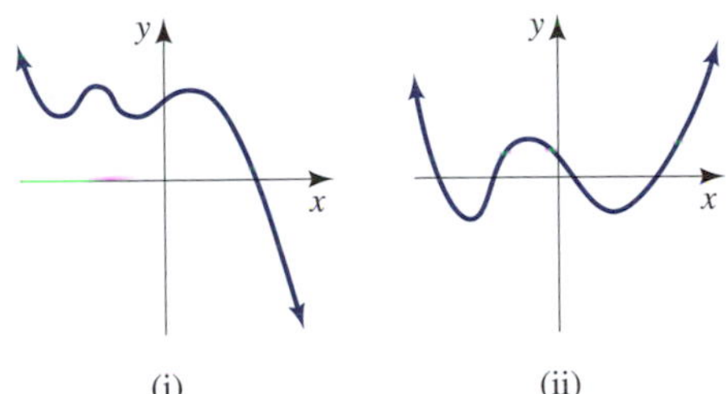

(i) (ii)

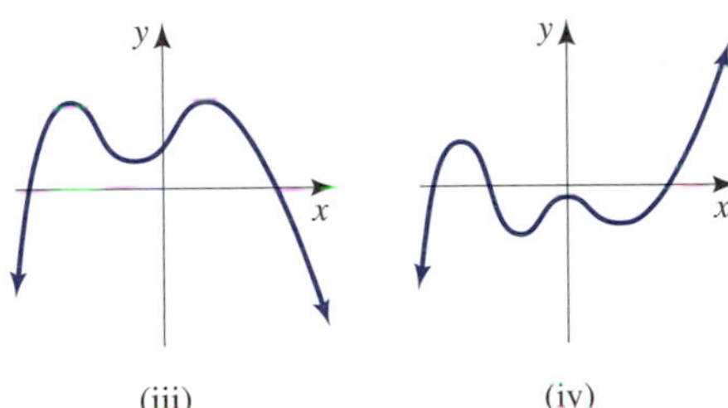

(iii) (iv)

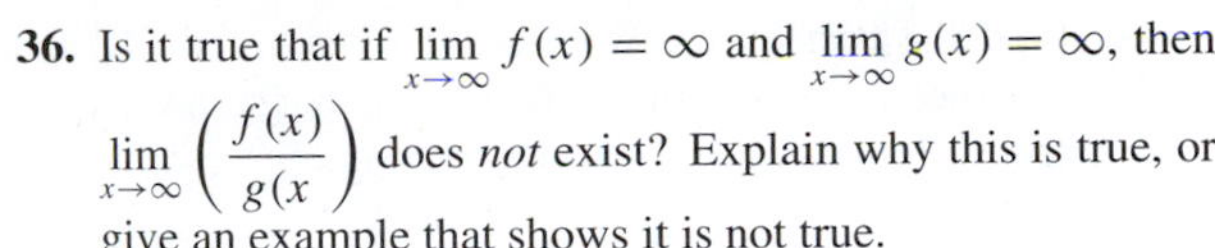

36. Is it true that if $\lim_{x\to\infty} f(x) = \infty$ and $\lim_{x\to\infty} g(x) = \infty$, then $\lim_{x\to\infty} \left(\frac{f(x)}{g(x}\right)$ does *not* exist? Explain why this is true, or give an example that shows it is not true.

37. Explain why no polynomial could have a graph like the one shown in the figure.

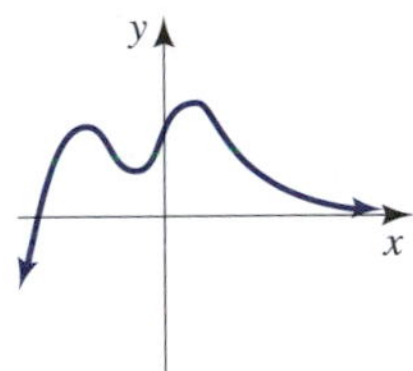

Modeling Using Least Squares

38. **Density Dependent Population Growth** Mueller and colleagues[62] studied the relationship between fecundity y of *Drosophila* and the density of adults x. Fecundity was defined as the number of eggs per female. The following table gives their data on the reciprocal of fecundity versus the number of adults (per unit of area).

x	2	4	8
$1/y$	0.0161	0.0172	0.0200
x	16	32	64
$1/y$	0.0227	0.0263	0.0357

 a. Find the best-fitting line relating the reciprocal of fecundity to density.
 b. Graph fecundity versus density. What do you observe is the concavity? Use your answer to explain what is happening to the rate of decrease of fecundity as the density of adults increases.
 c. What seems to happening as $x \to \infty$?

39. **Bass Feeding** Hartman[63] collected data on the rate of stomach content evacuation for age-0 striped bass found in the table.

Time post feeding (hours)	0	1	3
Stomach content weight (g)	0.009	0.010	0.007
Time post feeding (hours)	5	8	11
Stomach content weight (g)	0.009	0.005	0.003
Time post feeding (hours)	14	18	
Stomach content weight (g)	0.003	0.002	

62 Laurence D. Mueller, Amitabh Joshi, and Daniel J. Borash. 2000. Does population stability evolve? *Ecology* 81(5):1273–1285.

63 K. J. Hartman 2000. Variability in daily ration estimates of age-0 bass in the Chesapeake Bay. *Trans. Amer. Fish. Soc.* 129:1181–1186

a. Find the best-fitting exponential function (as the researcher did) that relates time to weight of stomach contents. Graph.

b. What happens as $t \to \infty$?

Solutions to WARM UP EXERCISE SET 5.3

1. Let's first see if we can make an intelligent guess at the answer. To do this let us evaluate the function $f(x) = \dfrac{x^2}{x^3+1}$ for a very large value of x, say, x equals one billion. Then

$$f(\text{billion}) = \frac{(\text{billion})^2}{(\text{billion})^3 + 1}$$

In the denominator the number 1 is insignificant when compared to the cube of one billion. So we can say that the denominator is approximately equal to $(\text{billion})^3$. We then have

$$\begin{aligned} f(\text{billion}) &= \frac{(\text{billion})^2}{(\text{billion})^3 + 1} \\ &\approx \frac{(\text{billion})^2}{(\text{billion})^3} \\ &= \frac{1}{(\text{billion})} \\ &\approx 0 \end{aligned}$$

Thus, we have that when x is one billion f is approximately equal to 0. So, we suspect that the limit as $x \to \infty$ is also 0.

We now present a formal proof. First recall that to determine the limit as $x \to \infty$ in a rational expression, we should *divide the numerator and denominator by x to the highest power occurring in the denominator*. In this case this means divide numerator and denominator by x^3. Since limits as $x \to \infty$ satisfy the usual rules for limits, we obtain

$$\begin{aligned} \lim_{x\to\infty}\left(\frac{x^2}{x^3+1}\right) &= \lim_{x\to\infty}\left(\frac{x^2}{x^3+1}\cdot\frac{1/x^3}{1/x^3}\right) \\ &= \lim_{x\to\infty}\left(\frac{\frac{x^2}{x^3}}{\frac{x^3}{x^3}+\frac{1}{x^3}}\right) \\ &= \lim_{x\to\infty}\left(\frac{\frac{1}{x}}{1+\frac{1}{x^3}}\right) \\ &= \frac{\lim_{x\to\infty}\frac{1}{x}}{1+\lim_{x\to-\infty}\frac{1}{x^3}} \\ &= \frac{0}{1+0} \\ &= 0 \end{aligned}$$

We can also say that $y = 0$ is a horizontal asymptote.

2. We have

$$\lim_{t\to\infty}(1000 - 800e^{-0.1t}) = 1000 - 800(0) = 1000$$

5.4 Additional Curve Sketching

APPLICATION Limited Growth

The annual sales $S(t)$ of a new product satisfy

$$S(t) = \frac{3t^2 + 0.2}{t^2 + 1}$$

where t is measured in years from the time of introduction. What is happening to sales? For the answer see Example 3 on page 346.

In the first two sections of this chapter we saw how to use both the first and the second derivative to sketch the graph of a function. In the third section we learned some additional techniques on graphing. We now consider a more systematic approach to sketching the graph of a function.

The following checklist is a useful procedure for graphing a function.

Checklist for Graphing a Function

A. Use $f(x)$ to

1. Determine the domain of the function and the intervals on which the function is continuous.
2. Determine whether the function is symmetric about the y-axis or the origin.
3. Find all vertical asymptotes.
4. Find all horizontal asymptotes.
5. Find where the function crosses the axes.

B. Use $f'(x)$ to

6. Find the critical values.
7. Find intervals where the function is increasing and decreasing.
8. Find all relative extrema.

C. Use $f''(x)$ to

9. Find intervals where the graph of the function is concave up and concave down.
10. Find all inflection values.

D. (Final step.) Use Steps A, B, C, and the values of f at the critical values and inflection values to graph.

We will see how to apply this checklist by working through the following examples. Not all the items might apply in graphing any one particular function.

EXAMPLE 1 **Sketching a Graph**

Sketch a graph of the function $f(x) = x^4 - 6x^2 + 4$ using calculus. Support using your grapher.

Solution

1. The domain of this polynomial is the entire real line. The function is continuous everywhere.
2. Since $f(-x) = f(x)$, the function is symmetric about the y-axis. Thus, we need only consider the set of points $x \geq 0$. (However, for practice and purposes of illustration, negative values of x will be considered.)
3. There are no vertical asymptotes.
4. We notice that for very large values of $|x|$,

$$f(x) = x^4\left(1 - \frac{1}{x^2} + \frac{4}{x^4}\right) \approx x^4$$

Thus, as $x \to \infty$ or as $x \to -\infty$, $f(x)$ becomes a large positive number without bound. Thus,

$$\lim_{x\to\infty} f(x) = \infty, \qquad \lim_{x\to-\infty} f(x) = \infty$$

5. The curve crosses the y-axis when $x = 0$ at the point $(0, 4)$. Although normally one cannot hope to find the zeros of a fourth-degree polynomial, in this case it is possible. Setting $z = x^2$ and then setting $f = 0$ gives $z^2 - 6z + 4 = 0$. Thus, by the quadratic formula, $z = 3 \pm \sqrt{5}$ and $x = \pm\sqrt{(3 \pm \sqrt{5})}$. These latter numbers are approximately ± 2.3, ± 0.9.
6. Since

$$f'(x) = 4x^3 - 12x = 4x(x^2 - 3) = 4x(x + \sqrt{3})(x - \sqrt{3})$$

there are three critical values: $-\sqrt{3}, 0, +\sqrt{3}$.
7. The three critical values divide the real line into the four intervals $(-\infty, -\sqrt{3})$, $(-\sqrt{3}, 0)$, $(0, \sqrt{3})$, and $(\sqrt{3}, \infty)$. Figure 5.31 gives the critical values and the test values. At the test values $f'(-2) = -8$, $f'(-1) = +8$, $f'(1) = -8$, $f'(2) = +8$. This then implies the rest of Figure 5.31. This figure indicates that on the successive intervals $(-\infty, -\sqrt{3})$, $(-\sqrt{3}, 0)$, $(0, \sqrt{3})$, and $(\sqrt{3}, \infty)$, $f'(x)$ is negative, positive, negative, and positive. Thus, on the four intervals, $f(x)$ is successively decreasing (↘), increasing (↗), decreasing (↘), and increasing (↗).
8. By the *first* derivative test, $f(-\sqrt{3})$ and $f(\sqrt{3})$ are relative minima and $f(0)$ is a relative maximum.
9. Now notice that

$$f''(x) = 12x^2 - 12 = 12(x + 1)(x - 1)$$

The second derivative exists and is continuous everywhere and is zero at the two values -1 and 1. These two values divide the real line into the three subintervals, $(-\infty, -1)$, $(-1, 1)$, and $(1, \infty)$. Figure 5.32 lists the values where the second derivative is zero and the test values for the three intervals. At the test values, $f''(-2) = +36$, $f''(0) = -12$, and $f''(2) = +36$. From this is constructed a sign chart for f''. From the sign chart of f'' we see that $f''(x)$ is positive on the intervals $(-\infty, -1)$ and $(1, \infty)$; thus, the graph of f is concave up ($\smile$) on these two intervals. From the sign chart of f'' we see that $f''(x)$ is negative on the interval $(-1, 1)$; thus, the graph of f is concave down ($\frown$) on this interval.

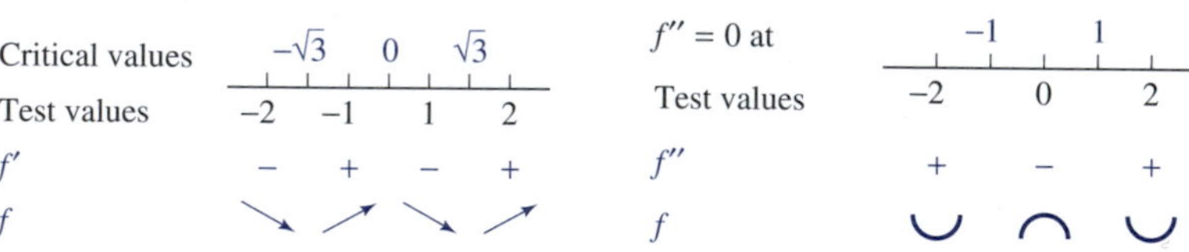

Critical values		$-\sqrt{3}$		0		$\sqrt{3}$	
Test values	-2		-1		1		2
f'	$-$		$+$		$-$		$+$
f	↘		↗		↘		↗

Figure 5.31

$f'' = 0$ at		-1		1	
Test values	-2		0		2
f''	$+$		$-$		$+$
f	∪		∩		∪

Figure 5.32

10. Since $f''(x)$ changes sign at $x = -1$ and $x = +1$, the inflection values are ± 1. Finally, notice that the values of f at the critical values $-\sqrt{3}$, 0, and $\sqrt{3}$ are -5, 4, and -5. The values of f at the inflection values -1 and $+1$ are -1 and -1, respectively.
11. This information is then put together to give the graph shown in Figure 5.33.

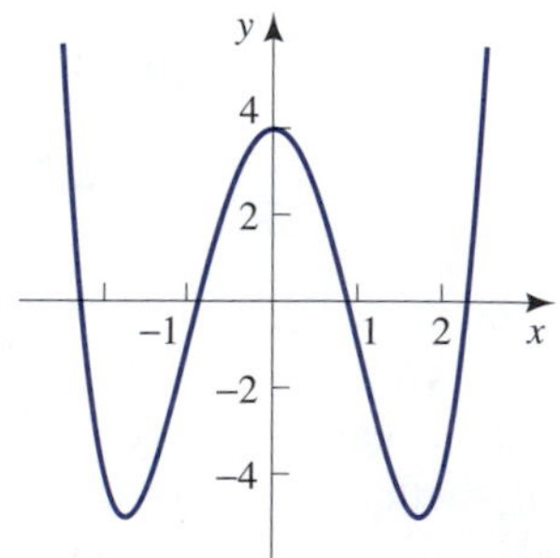

Figure 5.33
Relative minima at $x = \pm\sqrt{3}$ and a relative maximum at $x = 0$.

We can support this on our grapher using a window with dimensions $[-3, 3]$ by $[-10, 10]$ and obtain the graph shown in Figure 5.33. ∎

In Example 1, the *second* derivative test could have been used to find the relative extrema. Since $f''(-\sqrt{3}) > 0$ and $f''(\sqrt{3}) > 0$, the second derivative test indicates that $f(-\sqrt{3})$ and $f(\sqrt{3})$ are relative minima. Since $f''(0) < 0$, the second derivative test indicates that $f(0)$ is a relative maximum.

EXAMPLE 2

Sketch a graph of $f(x) = \dfrac{x^2}{x^2 - 1}$.

Solution

Use the checklist for graphing a function.

1. The domain of f is all real values except for $x = \pm 1$. The function is continuous at all values except for $x = \pm 1$.
2. Since $f(-x) = f(x)$, the curve is symmetric about the y-axis.
3. We have $\lim_{x \to -1^-} f(x) = \infty$ and $\lim_{x \to -1^+} f(x) = -\infty$ while $\lim_{x \to 1^-} f(x) = -\infty$ and $\lim_{x \to 1^+} f(x) = \infty$. Certainly, $x = \pm 1$ are two vertical asymptotes.
4. Notice that
$$\lim_{x \to \pm\infty} \frac{x^2}{x^2 - 1} = \lim_{x \to \pm\infty} \frac{1}{1 - \frac{1}{x^2}} = \frac{1}{1} = 1$$
Thus, $y = 1$ is a horizontal asymptote.
5. Setting $x = 0$ gives $y = 0$. Setting $y = 0$ gives $x = 0$. Thus, the curve crosses the axes at the point $(0, 0)$.
6. Since
$$f'(x) = \frac{2x(x^2 - 1) - x^2(2x)}{(x^2 - 1)^2} = \frac{-2x}{(x^2 - 1)^2}$$
the only critical value is $x = 0$.
7. Since the denominator in the above expression for $f'(x)$ is always positive (except for $x = \pm 1$ where f' is not defined), the sign of f', where it exists, is the same as the sign of $-x$. Thus on the two intervals $(-\infty, -1)$ and $(-1, 0)$, $f'(x) > 0$ and f is increasing. On the two intervals $(0, 1)$ and $(1, \infty)$, $f'(x) < 0$ and f is decreasing. See Figure 5.34.

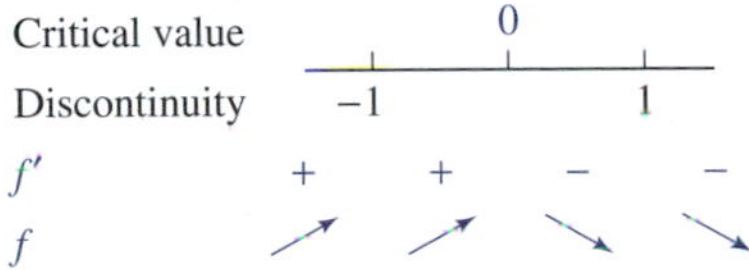

Figure 5.34

8. Since f' is positive to the left of 0 and negative to the right, the first derivative test indicates that $f(0) = 0$ is a relative maximum.
9. We have
$$f''(x) = \frac{-2(x^2 - 1)^2 + (2x)2(x^2 - 1)2x}{(x - 1)^4} = \frac{6x^2 + 2}{(x^2 - 1)^3}$$
The numerator of this expression is always positive. Thus the sign of f'' is the same as the sign of $(x^2 - 1)$. Refer to Figure 5.35. Thus, $f''(x)$ is positive on the two intervals $(-\infty, -1)$ and $(1, \infty)$ and negative and on the interval $(-1, 1)$.

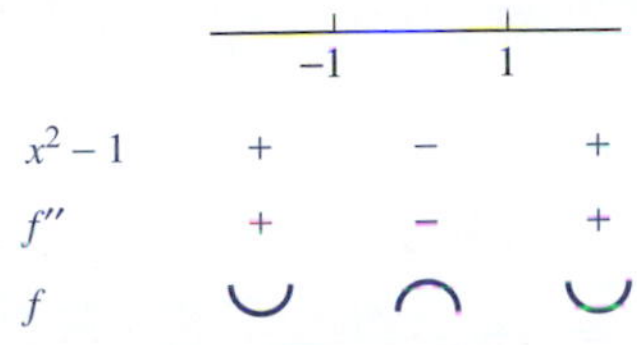

Figure 5.35

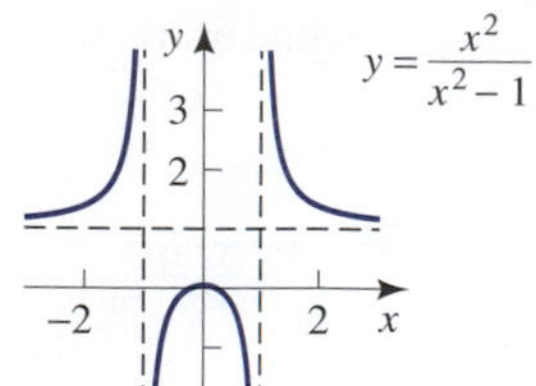

Figure 5.36

Thus, the graph of f is concave up on the first two of these intervals and concave down on the last one.

10. There are no points of inflection, since $x = +1$ and -1 are not in the domain of definition of f.
11. $f(0) = 0$. Putting all this information together gives Figure 5.36. ■

EXAMPLE 3 **Limited Growth**

The annual sales $S(t)$ in thousands of a new product satisfy

$$S(t) = \frac{3t^2 + 0.2}{t^2 + 1}$$

where t is measured in years from the time of introduction. What is happening to sales?

Solution

We first calculate $S'(t)$ and obtain

$$\begin{aligned} S'(t) &= \frac{(t^2+1)(3t^2+0.2)' - (3t^2+0.2)(t^2+1)'}{(t^2+1)^2} \\ &= \frac{(t^2+1)(6t) - (3t^2+0.2)(2t)}{(t^2+1)^2} \\ &= \frac{6t^3 + 6t - 6t^3 - 0.4t}{(t^2+1)^2} \\ &= \frac{5.6t}{(t^2+1)^2} \end{aligned}$$

This is always positive for $t > 0$, so sales are always increasing. We now find $S''(t)$.

$$\begin{aligned} S''(t) &= \frac{(t^2+1)^2(5.6) - 5.6t(2)(t^2+1)(2t)}{(t^2+1)^4} \\ &= 5.6\frac{(t^2+1) - 4t^2}{(t^2+1)^3} \\ &= 5.6\frac{1 - 3t^2}{(t^2+1)^3} \end{aligned}$$

The sign of S'' is then the same as the sign of $(1 - 3t^2)$. So $S''(t) > 0$ on $(0, 1/\sqrt{3})$ and $S''(t) < 0$ and on $(1/\sqrt{3}, \infty)$. Therefore, S is concave up on $(0, 1/\sqrt{3})$ and concave down on $(1/\sqrt{3}, \infty)$. Thus, sales are increasing ever faster on $(0, 1/\sqrt{3})$ and increasing ever slower on $(1/\sqrt{3}, \infty)$. Finally, we have

$$\begin{aligned} \lim_{t\to\infty} S(t) &= \lim_{t\to\infty} \frac{3t^2 + 0.2}{t^2 + 1}\frac{1/t^2}{1/t^2} \\ &= \lim_{t\to\infty} \frac{3 + \frac{0.2}{t^2}}{1 + \frac{1}{t^2}} \end{aligned}$$

$$= \frac{3+0}{1+0}$$
$$= 3$$

Thus, sales head for 3000 in the long term. A graph is shown in Figure 5.37.

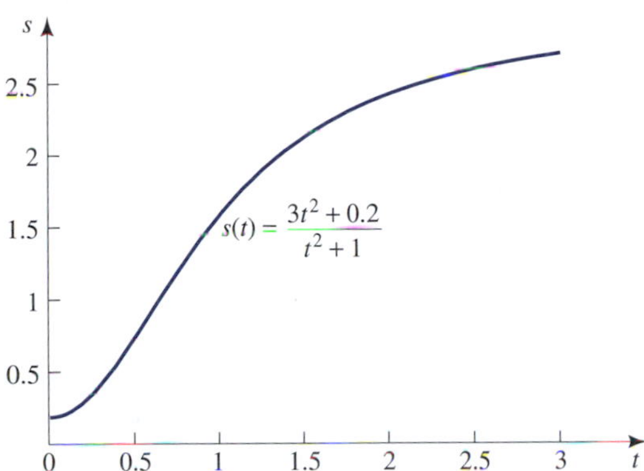

Figure 5.37

■

Warm Up Exercise Set 5.4

1. Graph $y = f(x) = 6\sqrt[3]{x^5} + 15\sqrt[3]{x^2}$.

Exercise Set 5.4

In Exercises 1 through 36, go through the checklist on page 343 and obtain as much information as possible; then sketch a graph of the given function. Support your answer with a computer or graphing calculator.

1. $f(x) = x^4 + 4x^3 + 1$
2. $f(x) = 3x^4 - 16x^3 + 24x^2 - 1$
3. $f(x) = x^4 - 32x^2 + 10$
4. $f(x) = x^4 - 2x^3 + x^2$
5. $f(x) = -3x^5 + 10x^3 - 5$
6. $f(x) = 3x^5 - 40x^3 + 20$
7. $f(x) = \dfrac{1}{x-1}$
8. $f(x) = \dfrac{x}{x-1}$
9. $f(x) = \dfrac{x+1}{x-1}$
10. $f(x) = \dfrac{x+1}{x+2}$
11. $f(x) = \dfrac{1}{x^2+1}$
12. $f(x) = \dfrac{1}{x^4+1}$
13. $f(x) = x + \dfrac{9}{x}$
14. $f(x) = x + \dfrac{25}{x}$
15. $f(x) = x + \dfrac{1}{3x^3}$
16. $f(x) = x + \dfrac{16}{3x^3}$
17. $f(x) = \dfrac{x^2}{x-1}$
18. $f(x) = -\dfrac{x^2}{x+2}$
19. $f(x) = \dfrac{x}{x^2+1}$
20. $f(x) = 2 - \dfrac{1}{x^2+4}$
21. $f(x) = \dfrac{x-1}{x^2}$
22. $f(x) = \dfrac{x^2}{1+x^2}$
23. $f(x) = \dfrac{x^2+2x-4}{x^2}$
24. $f(x) = \dfrac{1}{x^2-x-6}$
25. $f(x) = 2\sqrt[3]{x^5} - 5\sqrt[3]{x^2}$
26. $f(x) = 6\sqrt[3]{x^2} - 4x$
27. $f(x) = 7\sqrt[7]{x^6} - 6x$
28. $f(x) = 5\sqrt[5]{x^4} - 4x$
29. $f(x) = x^2e^{-0.5x^2}$
30. $f(x) = xe^{-x^2}$
31. $f(x) = \dfrac{2}{e^x + e^{-x}}$
32. $f(x) = x\ln x$
33. $f(x) = (\ln x)/x^2$
34. $f(x) = (\ln x)^3$
35. $f(x) = \dfrac{2}{e^x - e^{-x}}$
36. $f(x) = \dfrac{e^x + e^{-x}}{e^x - e^{-x}}$

Applications and Mathematical Modeling

37. Productivity A plant manager has worked hard over the past years using various innovative means to increase the number of items produced per employee in his plant and has noticed that the number of items $n(t)$ produced per employee per day over the past 10 years is given by

$$n(t) = 9 \cdot \frac{t}{t+1}$$

where t is the number of years from 10 years ago. Sketch the graph of the function on $(0, \infty)$. What is happening?

38. Photosynthesis The rate of production of photosynthesis is related to the light intensity I by the formula

$$P(I) = \frac{aI}{b + I^2}$$

where a and b are positive constants. Sketch a graph of this function. What is happening?

39. Biology Mayer[64] showed that the number y of screwworm flies a distance x km from animals was approximated by

$$y = a\left[\pi b\left(1 + \left(\frac{x}{b}\right)^2\right)\right]^{-1}$$

where a and b are positive constants. Use calculus to sketch a graph.

40. Biology Overholt and coworkers[65] showed that the proportion y of infested corn plants was approximated by $y = 1 - e^{-x \ln[2.77x^{0.206}/(2.77x^{0.206}-1)]}$, where x is the number of corn borers per corn plant. Use your grapher to graph. Using your grapher estimate where the function is increasing. Concave down. What happens as $x \to 0^+$ and $x \to +\infty$?

41. Diminishing Returns The profit equation for a firm after the introduction of a new product is given by

$$P(t) = \frac{4t^3 + 0.1}{t^3 + 1}$$

where P is profit in millions of dollars and t is years. Find the point of diminishing returns. Graph.

42. Diminishing Returns The profit equation for a firm after the introduction of a new product is given by

$$P(t) = \frac{5t^4 + 0.1}{t^4 + 1}$$

where P is profit in millions of dollars and t is years. Find the point of diminishing returns. Graph.

---

[64] David G. Mayer. 1993. Estimation of dispersal distances. *Env. Entomol.* 22(2):368–380.

[65] W. A. Overholt et al. 1990. Distribution and sampling of southwestern corn borer in preharvest corn. *J. Econ. Entomol.* 83(4):1370–1375.

More Challenging Exercises

In Exercises 43 through 46, go through the checklist on page 343 and obtain as much information as possible; then sketch a graph of the given function. Support your answer with a computer or graphing calculator.

43. $f(x) = x^4 - 32bx^2 + a$

44. $f(x) = 3x^5 - 40bx^3 + a$

45. $f(x) = x + \dfrac{b^2}{3x^3}$

46. $f(x) = x^2 + \dfrac{b^2}{x^2}$

In Exercises 47 through 50, you are given some conditions that a function must satisfy. In each case graph a function that satisfies the given conditions.

47. $f'(x) > 0$ and $f''(x) > 0$ on $(-\infty, 1)$, $f'(x) < 0$ and $f''(x) > 0$ on $(1, \infty)$, $f(0) = 0$, $\lim_{x\to-\infty} f(x) = -\infty$, $\lim_{x\to 1^-} f(x) = \infty$, $\lim_{x\to 1^+} f(x) = \infty$, $\lim_{x\to\infty} f(x) = 0$

48. $f'(x) > 0$ and $f''(x) > 0$ on $(-\infty, 0)$, $f'(x) < 0$ and $f''(x) < 0$ on $(0, \infty)$, $f(0) = 0$, $\lim_{x\to-\infty} f(x) = -1$, $\lim_{x\to\infty} f(x) = 1$

49. $f'(x) > 0$ on $(-2, -1)$ and $(-1, 0)$, $f'(x) < 0$ on $(-\infty, -2)$ and $(0, 1)$ and $(1, \infty)$, $f''(x) > 0$ on $(-\infty, -1)$ and $(1, \infty)$, $f''(x) < 0$ on $(-1, 1)$, $f(-2) = 1$, $f(0) = 0$, $\lim_{x\to-\infty} f(x) = \infty$, $\lim_{x\to-1^-} f(x) = \infty$, $\lim_{x\to-1^+} f(x) = -\infty$, $\lim_{x\to 1^-} f(x) = -\infty$, $\lim_{x\to 1^+} f(x) = \infty$, $\lim_{x\to\infty} f(x) = 0$

50. $f'(x) > 0$ on $(-1, 0)$ and $(0, 1)$, $f'(x) < 0$ on $(-\infty, -1)$ and $(1, \infty)$, $f''(x) > 0$ on $(-\infty, 0)$ and $(2, \infty)$, $f''(x) < 0$ on $(0, 2)$, $f(-1) = 1$, $f(1) = 1$, $\lim_{x\to-\infty} f(x) = \infty$, $\lim_{x\to 0^-} f(x) = \infty$, $\lim_{x\to 0^+} f(x) = -\infty$, $\lim_{x\to\infty} f(x) = -1$

51. Medicine The function $C = k(e^{-at} - e^{-bt})$ represents the relationship between the time t and the concentration C of a drug or dye in the blood after injection[66] where k, a, and b are positive constants and $b > a$. Determine what $C(t)$ looks like for large t.

[66] Duane J. Clow and N. Scott Urquhart. 1974. *Mathematics in Biology.* New York: Ardsley House.

52. Profits The profit equation for a firm's main product after the introduction of a competing product from another firm is given by

$$P(t) = \frac{4t^3 + 10}{t^3 + 1}$$

where P is profit in millions of dollars and t is years. Determine what is happening.

Solutions to WARM UP EXERCISE SET 5.4

1. Use the checklist for graphing a function.

1. The domain of f is $(-\infty, \infty)$ where it is continuous.

2–4. There are no symmetries, vertical or horizontal asymptotes.

5. Note that $f(x) = x^{2/3}(6x + 15)$ implies crossing the x-axes at the points $x = 0$ and $x = -2.5$. Since $f(0) = 0$, the graph crosses the y-axis at the origin.

6. **a.** The derivative $f'(x) = 10x^{2/3} + 10x^{-1/3}$ exists and is continuous everywhere except at $x = 0$, where f' does not exist, but f does. Thus, $x = 0$ is a critical value.

b. Now setting $f'(x) = 10x^{2/3} + 10x^{-1/3} = 0$ yields

$$10x^{2/3} + 10x^{-1/3} = 0$$

$$x^{2/3} = -\frac{1}{x^{1/3)}}$$

$$x^{2/3}(x^{1/3}) = -\frac{1}{x^{1/3}}(x^{1/3})$$

$$x = -1$$

The critical values are then $x = -1$ and $x = 0$.

5. These two critical values divide the line into the three subintervals $(-\infty, -1)$, $(-1, 0)$, and $(0, \infty)$, on each of which f' is continuous and not zero.

Use test values -8, $-\frac{1}{8}$, and 1. See the figure. Then

$$f'(-8) = +35, \quad f'\left(-\frac{1}{8}\right) = -\frac{35}{2}, \quad f'(1) = +20$$

This leads to the rest of the sign chart given in the figure. We see that f is increasing on $(-\infty, -1)$ and $(0, \infty)$ and that f is decreasing on $(-1, 0)$.

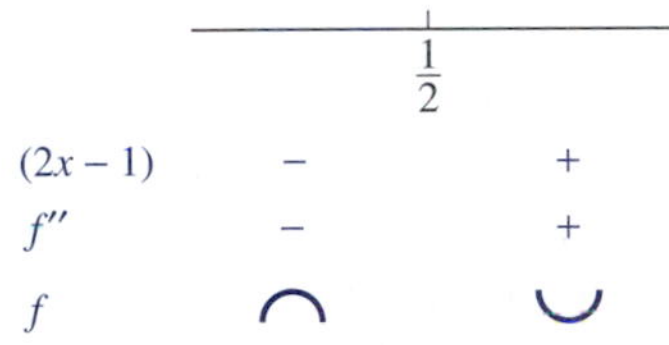

6. The first derivative test indicates that $f(-1)$ is a relative maximum and $f(0)$ is a relative minimum.

7. We have

$$f''(x) = \frac{20}{3}x^{-1/3} - \frac{10}{3}x^{4/3} = \frac{10}{3x^{4/3}}(2x - 1)$$

Since $x^{4/3}$ is never negative, f is concave down on $(-\infty, 0)$ and $(0, \frac{1}{2})$ and concave up on $(\frac{1}{2}, \infty)$. See the figure.

		$\frac{1}{2}$	
$(2x - 1)$	−		+
f''	−		+
f	∩		∪

8. Thus, $x = \frac{1}{2}$ is an inflection value.

9. Noting further that $f(-1) = 9$ and $f(0) = 0$, all this information together gives the following figure.

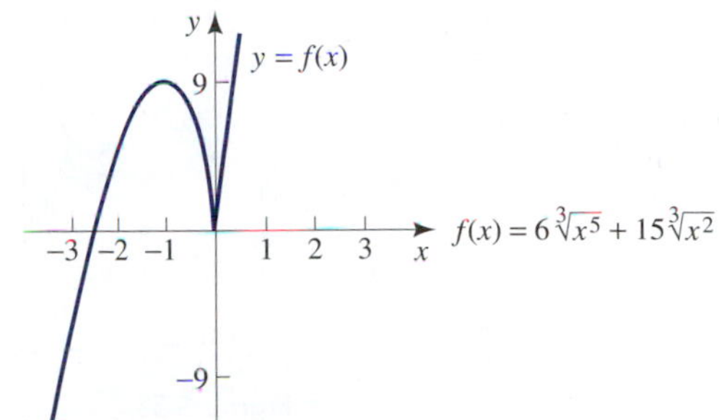

5.5 Absolute Extrema

APPLICATION **Minimizing Vehicle Operating costs**

In England, Frost and Spence[67] studied the operating costs of automobiles based on the average vehicle speed. They found that operating costs, measured in pence per kilometer, were approximated by the equation

$$C(v) = 6 + \frac{118}{v} + 0.0002v^2$$

Find the speed at which operating costs are minimized and find the minimum cost. See Example 5 on page 355 for the answer.

[67] M. E. Frost and N. A. Spence. 1995. The rediscovery of accessibility and economic potential: the critical issue of self-potential. *Environ. Plan.* 27:1833–1848.

Introduction to Absolute Extrema

In previous sections we studied relative extrema and gave the first and second derivative tests to determine them. In this section we study absolute extrema and find conditions that ensure that a continuous function attains its absolute extrema on a given interval.

Notice that the function defined on $[0, b]$ in Figure 5.38 attains its smallest value at the left endpoint and its largest value at x_M.

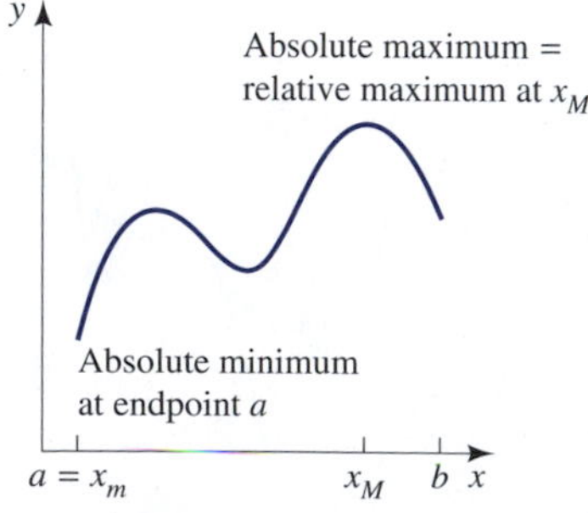

Figure 5.38
The absolute maximum occurs at the critical value x_m, but the absolute maximum occurs at an endpoint.

We refer to the smallest value as the **absolute minimum** and to the largest value as the **absolute maximum**. An **absolute extremum** is either an absolute minimum or an absolute maximum. We have the following definitions.

DEFINITION **Absolute Maximum and Minimum**

Suppose $f(x)$ is defined on some interval I.

1. We say that $f(x_M)$ is an absolute maximum on the interval I if $x_M \in I$ and
$$f(x) \leq f(x_M) \quad \text{for all } x \in I$$
2. We say that $f(x_m)$ is an absolute minimum on the interval I if $x_m \in I$ and
$$f(x_m) \leq f(x) \quad \text{for all } x \in I$$

Absolute Extrema on Closed And Bounded Intervals

We first consider closed and bounded intervals since continuous functions on a closed and bounded interval have the following very important property.

Extreme Value Theorem

If $f(x)$ is continuous on the closed and bounded interval $[a, b]$, then f attains both its absolute maximum and its absolute minimum on $[a, b]$.

From Figure 5.38 we can see that an absolute extremum for a continuous function can be a relative extremum or *one of the endpoints*. There can be no other possibilities. If $f'(x_0) > 0$ or $f'(x_0) < 0$ and $x_0 \in (a, b)$, then we know from the previous sections that $f(x)$ will be increasing in the first case and decreasing in the second. Thus, in neither case can $f(x)$ attain an absolute extremum. The only values that are left are the critical values and the endpoints.

The procedure for finding the absolute extrema of a continuous function on a closed and bounded interval can then be given as follows.

To Find the Absolute Extrema of a Continuous Function *f*(*x*) on [*a*, *b*]

1. Locate all critical values.
2. Evaluate $f(x)$ at all the critical values and also at the two values a and b.
3. The absolute maximum of $f(x)$ on $[a, b]$ will be the largest number found in Step 2, while the absolute minimum of $f(x)$ on $[a, b]$ will be the smallest number found in Step 2.

EXAMPLE 1 **Finding the Absolute Extrema**

Find the absolute extrema of $f(x) = 3x^4 + 4x^3 - 12x^2 + 1$ on $[-1, 1]$ if they exist. Repeat for the interval $[-1, 2]$.

Solution

Since the function is continuous on each of the intervals, f will attain both an absolute minimum and an absolute maximum on each of the intervals.

Since

$$f'(x) = 12x^3 + 12x^2 - 24x = 12x(x^2 + x - 2) = 12x(x - 1)(x + 2)$$

the function is differentiable everywhere, and the critical values are -2, 0, and 1, with only the latter two on the given intervals.

(a) The absolute maximum and the absolute minimum must be among the list of numbers, $f(-1)$, $f(0)$, and $f(1)$.

x	f	Comment
-1	-12	absolute minimum
0	1	absolute maximum
1	-4	

Refer to Figure 5.39.

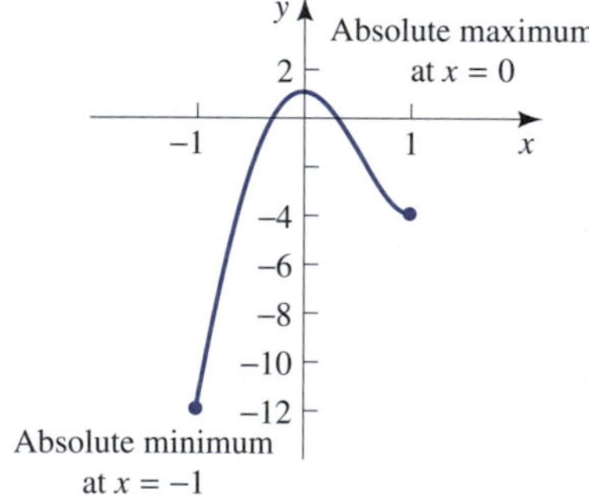

Figure 5.39
The absolute maximum occurs at the critical value $x = 0$, but the absolute minimum occurs at an endpoint $x = -1$.

(b) The absolute maximum and the absolute minimum must be among the list of numbers, $f(-1)$, $f(0)$, $f(1)$, and $f(2)$.

x	f	Comment
-1	-12	absolute minimum
0	1	
1	-4	
2	33	absolute maximum

Refer to Figure 5.40.

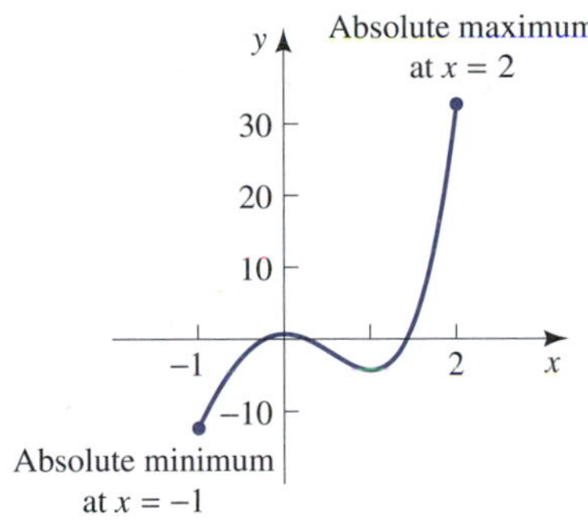

Figure 5.40
The absolute maximum and the absolute minimum occur at endpoints.

■

EXAMPLE 2 **Finding the Absolute Extrema**

Use your computer or graphing calculator to determine the absolute extrema for the function $y = f(x) = 1 + (x + 3)e^x$ on the interval $[-5, 1]$. Verify your work using calculus.

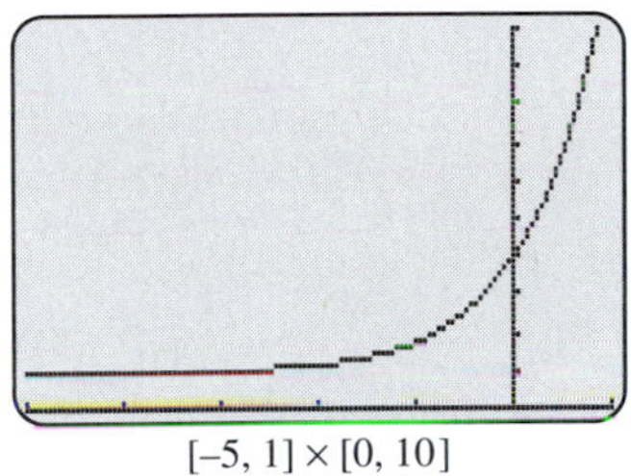

$[-5, 1] \times [0, 10]$

Screen 5.11
It appears from the grapher that the function is increasing on the entire interval and that the function assumes its minimum at the left endpoint. This is not the case, as a mathematical analysis reveals!

Solution

Using a window with dimensions $[-5, 1]$ by $[0, 10]$, we obtain Screen 5.11. It seems apparent that f is increasing on $[-5, 1]$. Thus, f attains an absolute minimum at $x = -5$ and attains an absolute maximum at $x = 1$.

Using calculus, we have

$$f'(x) = (x + 3)e^x + e^x = (x + 4)e^x$$

But this is negative for $x < -4$ and positive for $x > -4$. Thus, we see that f is actually decreasing on the interval $(-5, -4)$ (contrary to what we thought we saw on the grapher) and increasing on the interval $(-4, 1)$. This shows that f attains an absolute minimum at $x = -4$. A calculation readily shows that $f(1) > f(-5)$. Thus, f attains an absolute maximum at $x = 1$. We see that using a computer or graphing calculator did not give us a correct answer. ■

Intervals That Are Not Closed and Bounded

When we have a continuous function on a closed interval we are guaranteed that the function has a relative maximum and also a relative minimum on the interval. If the interval is not closed and bounded or if the function is not continuous, there may not be an absolute maximum or minimum. Consider now the next two examples.

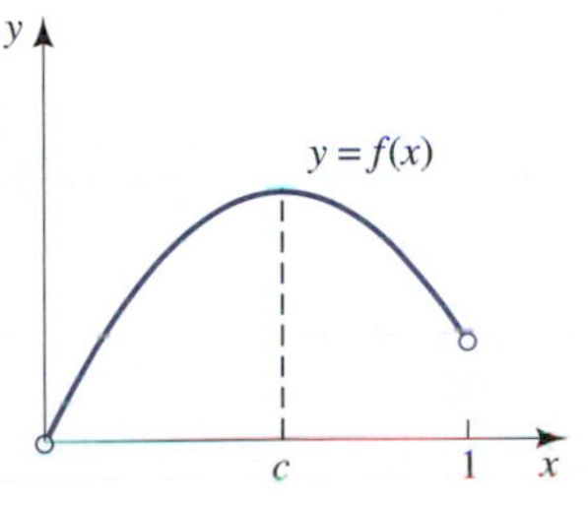

Figure 5.41

EXAMPLE 3 **No Absolute Extrema**

Let f be the continuous function shown in Figure 5.41. Does $f(x)$ have an absolute maximum on the interval $(0, 1)$? Does $f(x)$ have an absolute minimum on this interval?

Solution

The function attains an absolute maximum at $x = c$ but has no absolute minimum since $x = 0$ is not included in the interval. ■

EXAMPLE 4 **No Absolute Extrema**

Let $f(x) = 1/x$ on $(0, \infty)$. Is $f(x)$ continuous on this interval? Does $f(x)$ have an absolute maximum on this interval? Does $f(x)$ have an absolute minimum on this interval?

Solution

We are familiar with this function. A graph is shown in Figure 5.42. The function f is continuous on $(0, \infty)$. Since $\lim_{x\to 0^+} f(x) = \infty$, f has no absolute maximum on $(0, \infty)$.

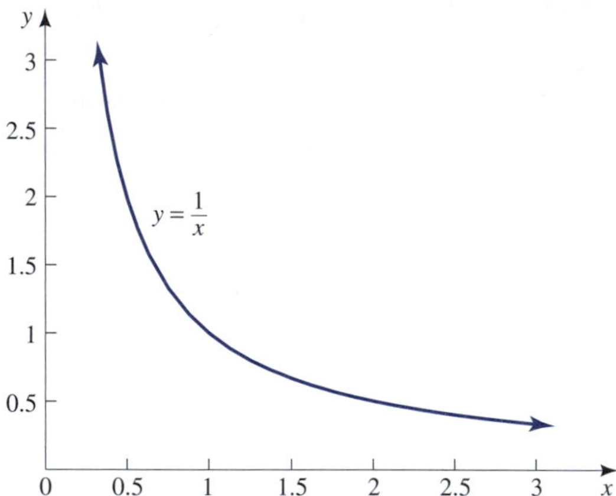

Figure 5.42

Since the function f decreases throughout the interval $(0, \infty)$ (since $f'(x) = -\frac{1}{x^2} < 0$ there), the function f has no absolute minimum. This is true despite the fact that $\lim_{x\to\infty} f(x) = 0$. ■

When the interval is not closed and bounded or the function not continuous, the safest and surest way to find absolute extrema (if they exist) is to graph the function using calculus.

Exploration 1

Absolute Extrema for a Function Not Continuous

Sketch a function that is not continuous throughout a closed and bounded interval that has neither an absolute maximum nor an absolute minimum.

An Application Using a Mathematical Model

EXAMPLE 5 **Vehicle Operating Costs**

In England, Frost and Spence[68] studied the operating costs of automobiles based on the average vehicle speed. They found that operating costs, measured in pence per kilometer, were approximated by the equation

$$C(v) = 6 + \frac{118}{v} + 0.0002v^2$$

Find the speed at which operating costs are minimized, and find the minimum cost.

Solution

We need to look at the v-interval $(0, \infty)$. Notice that $v = 0$ is a vertical asymptote. We next need to find the derivative. We have

$$\begin{aligned} C'(v) &= -\frac{118}{v^2} + 0.0004v \\ &= \frac{0.0004}{v^2}\left(v^3 - \frac{118}{0.0004}\right) \\ &= \frac{0.0004}{v^2}\left(v^3 - 295,000\right) \end{aligned}$$

Since $\sqrt[3]{295,000} \approx 66.6$, we see that $C'(x) < 0$ on $(0, 66.6)$ and $C'(x) > 0$ on $(66.6, \infty)$. This implies that C is decreasing on $(0, 66.6)$ and increasing on $(66.6, \infty)$. Thus, the absolute minimum occurs at $v = 66.6$. The graph is shown in Figure 5.43. Since $\lim_{v \to 0^+} C(v) = \infty$ there is no absolute maximum. (We also have $\lim_{v \to \infty} C(v) = \infty$.)

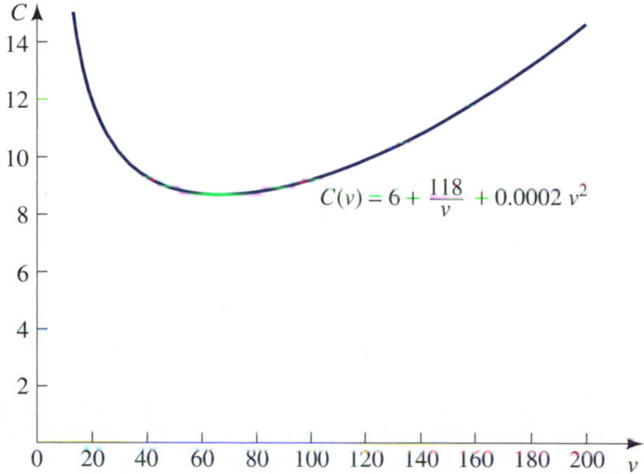

Figure 5.43

■

[68] Ibid.

Remark The situation in Example 5 is representative of many applications. The function decreases until some point c is reached and then increases afterward. Then $f(c)$ will be the absolute minimum. Also, if the function increases until some point c is reached and then decreases afterward, then $f(c)$ will then be the absolute maximum.

Warm Up Exercise Set 5.5

1. Find the absolute extrema of $f(x) = x^3 - 12x + 1$ on $[-3, 5]$.

Exercise Set 5.5

In Exercises 1 through 6 you will find various graphs. In each case, find all values where the function attains an absolute maximum. An absolute minimum.

1.

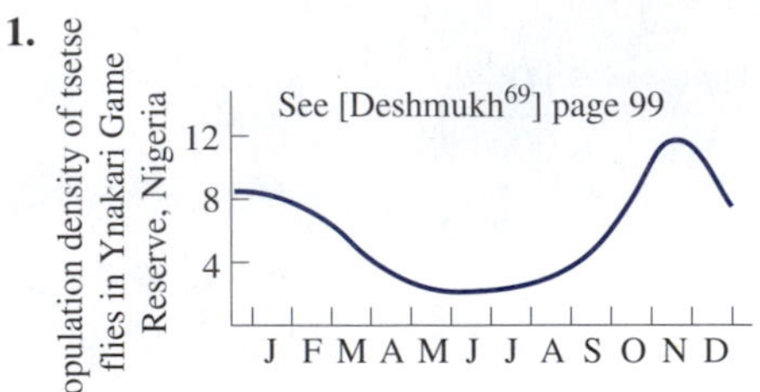

2.

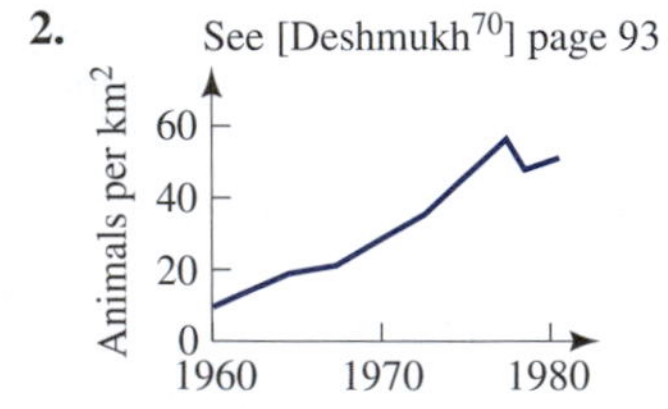

3.

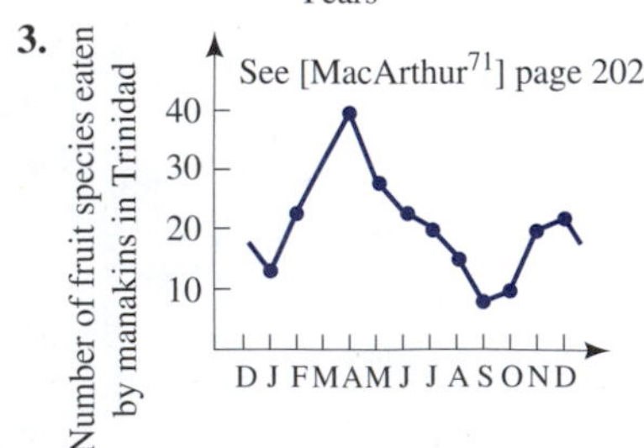

4.

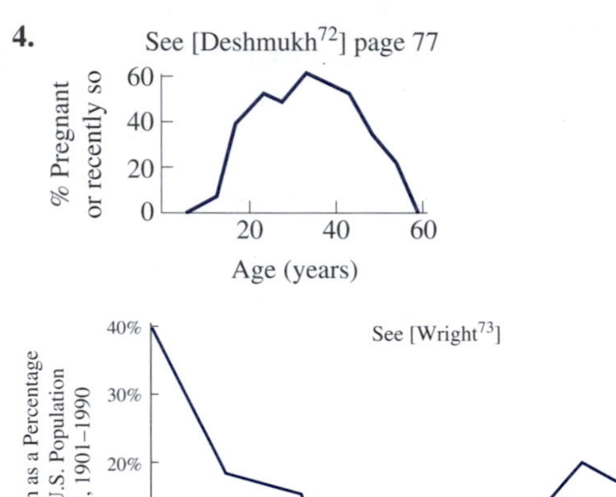

5.

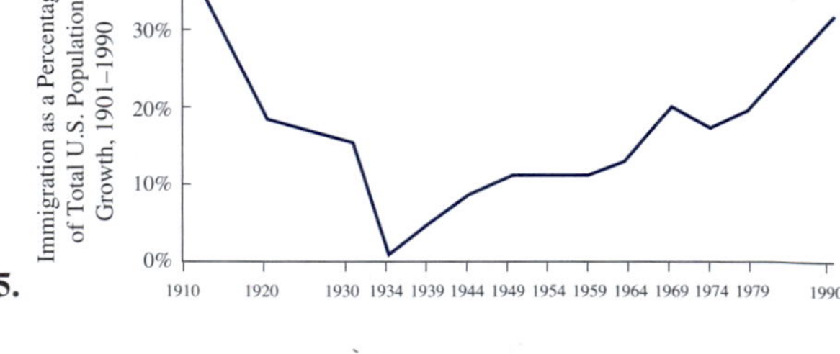

6.

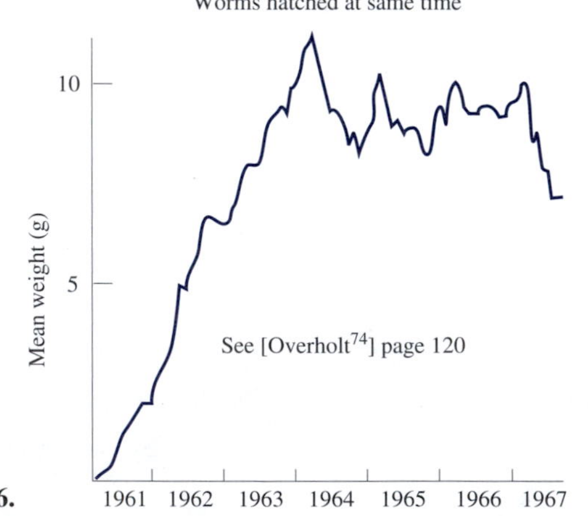

[69] Ian Deshmukh. 1986. *Ecology and Tropical Biology*. Boston: Blackwell Scientific Publications.

[70] Ibid.

[71] Robert MacArthur. 1972. *Geographical Ecology*. New York: Harper & Row.

[72] Ian Deshmukh. 1986. *Ecology and Tropical Biology*. Boston: Blackwell Scientific Publications.

[73] John W. Wright. 1993. *The Universal Almanac*. New York: Andrews and McMeel.

[74] W. A. Overholt et al. 1990. Distribution and sampling of southwestern corn borer in preharvest corn. *J. Econ. Entomol.* 83(4):1370–1375.

In Exercises 7 and 8, let $f(x) = x^3 - 3x + 1$. Locate the value(s) where f attains an absolute maximum and the value(s) where f attains an absolute minimum on each interval.

7. a. $[2, 4]$ **b.** $[0, 2]$ **c.** $[-2, 2]$

8. a. $[-1, 1]$ **b.** $[-2, 0]$ **c.** $[-3, 2]$

In Exercises 9 and 10, let $f(x) = -2x^3 + 3x^2$. Locate the value(s) where f attains an absolute maximum and the value(s) where f attains an absolute minimum on each interval.

9. a. $[-2, 0]$ **b.** $[0, 2]$ **c.** $[-2, 2]$

10. a. $[0, 1]$ **b.** $[-1, 1]$ **c.** $[-2, 1]$

In Exercises 11 through 28, locate the value(s) where each function attains an absolute maximum and the value(s) where the function attains an absolute minimum, if they exist, of the given function on the given interval.

11. $f(x) = x^2 - 4x + 1$ on $[-1, 1]$

12. $f(x) = x^2 - 2x + 1$ on $[2, 4]$

13. $f(x) = -x^2 + 4x - 2$ on $[0, 3]$

14. $f(x) = x^2 + 2x + 1$ on $[-3, 0]$

15. $f(x) = x^3 - 3x + 1$ on $[-1, 3]$

16. $f(x) = x^3 - 3x^2 + 2$ on $[0, 4]$

17. $f(x) = x^3 + 3x^2 - 2$ on $[-3, 2]$

18. $f(x) = -8x^3 + 6x - 1$ on $[-1, 1]$

19. $f(x) = x^4 - 8x^2 + 3$ on $[-3, 3]$

20. $f(x) = -x^4 + 8x^2 + 1$ on $[-1, 3]$

21. $f(x) = x^2 + 16x^{-2}$ on $[1, 8]$

22. $f(x) = x^2 - 16x^{-2}$ on $[1, 8]$

23. $f(x) = 3 - x + x^2$ on $(-\infty, \infty)$

24. $f(x) = 4 - 2x - x^2$ on $(-\infty, \infty)$

25. $f(x) = x^4 + 6x^2 - 2$ on $(-\infty, \infty)$

26. $f(x) = 3 - x^2 - x^4$ on $(-\infty, \infty)$

27. $f(x) = \sqrt{x^2 + 1}$ on $(-\infty, \infty)$

28. $f(x) = (x - 3)^4$ on $(-\infty, \infty)$

In Exercises 29 and 30, let $f(x) = 3x + x^{-3}$. On each interval, locate the value(s) (if they exist) where f attains an absolute extremum.

29. a. $[0.5, 2]$ **b.** $[-1, 3]$ **c.** $(0, 3]$ **d.** $(-\infty, 0)$

30. a. $[-2, -0.5]$ **b.** $[-2, 3]$ **c.** $[-3, 0)$ **d.** $(0, \infty)$

31. Find two numbers whose sum is 16 and whose product is maximum.

32. Find two numbers whose sum is 24 and whose product is maximum.

33. Find two numbers whose sum is 20 and for which the sum of the squares is a minimum.

34. Find two numbers whose sum is 32 and for which the sum of the squares is a minimum.

35. Find two nonnegative numbers x and y with $2x + y = 30$ for which the term xy^2 is maximized.

36. Find two nonnegative numbers x and y with $x + y = 60$ for which the term x^2y is maximized.

37. Find two numbers x and y with $x - y = 20$ for which the term xy is minimized.

38. Find two numbers x and y with $x - y = 36$ for which the term xy is minimized.

39. What is the area of the largest rectangle that can be enclosed by a circle of radius a?

40. What is the area of the largest rectangle that can be enclosed by a semicircle of radius a?

41. Find the coordinates of the point P that maximizes the area of the rectangle shown in the figure.

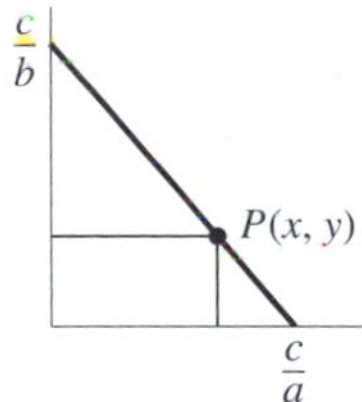

42. A triangle has base on the x-axis, one side on the y-axis, and the third side goes through the point $(1, 2)$ as shown in the figure. Find the slope of the line through $(1, 2)$ if the area of the triangle is to be a minimum.

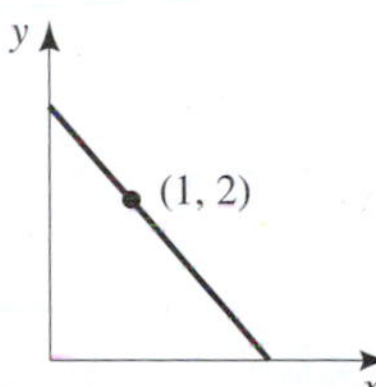

Applications and Mathematical Modeling

43. Revenue If the revenue function for a firm is given by $R(x) = -x^3 + 75x$, find where the revenue is maximized.

44. Profit If the cost function for a firm is given by $C(x) = 48x + 5$ and the revenue function is the same as in the previous exercise, find where the profit is maximized.

45. Spread of Rumors It has been estimated that a rumor spreads at a rate R that is proportional both to the ratio r of individuals who have heard the rumor and to the ratio $(1 - r)$ who have not. Thus, $R = kr(1 - r)$, where k is a positive constant. For what value of r does the rumor spread fastest?

46. Biology When predators are not stimulated to search very hard for prey, the rate of change R of the number N of prey attacked per unit of area with respect to the number N of prey is given by $R(x) = kx(ax - N)$, where x is the number of predators, k is the per capita effectiveness of the predators, and a is the maximum per capita effectiveness of the predators over an extended period.[75] Find the number of predators that minimizes R.

47. Maximizing Corn Yields Roberts[76] formulated a mathematical model of corn yield response to nitrogen fertilizer in high-yield response land given by the equation $Y(N) = 25 + 1.2N - 0.003N^2$, where Y is bushels of corn per acre and N is pounds of nitrogen per acre. Determine the amount of nitrogen per acre that gives a maximum corn yield, and find this maximum corn yield.

48. Maximizing Corn Yields Roberts[77] formulated a mathematical model of corn yield response to nitrogen fertilizer in low-yield response land given by the equation $Y(N) = 15 + 0.5N - 0.0016N^2$, where Y is bushels of corn per acre and N is pounds of nitrogen per acre. Determine the amount of nitrogen per acre that gives a maximum corn yield, and find this maximum corn yield.

49. Modeling Seasonal Temperature Power and Heuvel[78] developed a mathematical model that profiled seasonal temperatures around several Canadian lakes. They used the general equation $T(D) = aD^b e^{-cD}$, where a, b, and c are all positive constants, D is the day of the year, and T is the temperature in degrees Celsius. Find D for which temperature is maximum. In one of their models, $\dfrac{b}{c} = 200$. With $D = 1$ corresponding to January 1, does this make sense? Explain.

50. Biological Control Lysyk[79] studied the effect of temperature on various life history parameters of a parasitic wasp for the purposes of pest management. He constructed a mathematical model given approximately by

$$L(t) = \frac{1}{-1 + 0.03t + \dfrac{11}{t}}$$

where t is temperature in degrees Celsius and L is median longevity in days of the female. Graph, and find where L attains a maximum.

51. Biological Control Lysyk[80] studied the effect of temperature on various life history parameters of a parasitic wasp for the purposes of pest management. He constructed a mathematical model given approximately by

$$D(t) = 0.00007t(t - 13)(36 - t)^{0.4}$$

where t is temperature in degrees Celsius and D is the reciprocal of development time in days. Find $D'(t)$. (Hint: You might wish to refer to the formula $(f \cdot g \cdot h)' = f' \cdot g \cdot h + f \cdot g' \cdot h + f \cdot g \cdot h'$ found in Exercise 48 of Section 4.2.) Graph and find where D attains a maximum.

52. Biological Control Lysyk[81] also constructed a mathematical model given approximately by the equation $r(t) = 0.0002t(t - 14)\sqrt{36 - t}$, where t is temperature in degrees Celsius and r is the intrinsic rate of increase (females per females per day). Graph, and using your grapher find where r attains a maximum.

[75] Duane J. Clow and N. Scott Urquhart. 1974. *Mathematics in Biology*. New York: Ardsley House.

[76] Roland K. Roberts, Burton C. English, and S. B. Mahajashetti. 2000. Evaluating the returns to variable rate nitrogen application. *J. Agr. Appl. Econ.* 32(1):133–143.

[77] Ibid.

[78] M. Power and M. R. van den Heuvel. 1999. Age-0 yellow perch growth and its relationship to temperature. *Trans. Amer. Fish. Soc.* 128:687–700.

[79] T. J. Lysyk. 2001. Relationship between temperature and life history parameters of *Muscidifurax zaraptor*. *Environ. Entomol.* 30(1):147–156.

[80] Ibid.

[81] Ibid.

Modeling Using Least Squares

53. Shopping Behavior Baker[82] collected the data found in the table below that correlates the number (N) of retail outlets in a suburban shopping center with the number of high disposable income I consumers.

[82] R. G. V. Baker. 1994. Multipurpose shopping behaviour at planned suburban shopping centers: a space-time analysis. *Environment and Planning A* 28:611–630.

N	70	80	80	90	90
I	10.4	13.0	6.0	14.2	10.4
N	90	125	205	205	
I	9.2	7.5	15.5	14.7	

Here N is the number of retail outlets, and I is the number of high disposable income consumers.

a. Use quadratic regression to find I as a function of N.
b. Determine the number of retail outlets for which this model predicts the minimum number of high-disposable-income consumers.

54. Shopping Behavior Baker[83] studied multipurpose shopping behavior. He created a *gravity coefficient* β that measures the loyalty or bonding of consumers to a particular shopping center. He collected the data found in the following table, which correlates the number (N) of retail outlets in a suburban shopping center with the loyalty index β.

N	70	80	80	90	90
β	0.84	0.93	0.96	0.46	0.53
N	90	125	205	205	
β	0.68	0.38	0.55	0.47	

a. Use quadratic regression to find β as a function of N.
b. Determine the number of retail outlets for which this model predicts the minimum loyalty coefficient.

55. Herring Abundance Kosa and Mather[84] studied various factors that affected the abundance of herring, a species on the decline. They collected the data given in the following table relating the log (to base 10) of abundance with depth in meters.

Depth	1.5	3	9	9
log (abundance)	0.25	0.45	0.05	0.14
Depth	10	11	27.4	
log (abundance)	0.0	0.9	15	

a. Use quadratic regression (as the researchers did) to find a model of log (abundance) as a function of depth.
b. Use part (a) to find the depth at which abundance is minimum.

56. Herring Abundance Kosa and Mather[85] studied various factors that affected the abundance of herring, a species on the decline. They collected the data given in the following table relating the log (to base 10) of abundance with pH.

pH	7.2	7.6	8.0	8.0	8.0	8.1
log (abundance)	2.2	2.3	5.0	3.6	2.7	3.5
pH	8.3	8.3	8.9	9.1	9.1	
log (abundance)	4.0	3.6	1.5	1.0	3.0	

a. Use quadratic regression (as the researchers did) to find a model of log (abundance) as a function of pH.
b. Use part (a) to find the pH at which abundance is maximum.

57. Biological Control Lysyk[86] studied the effect of temperature on various life history parameters of a parasitic wasp for the purposes of pest management. He collected data given in the following table that related temperature to the proportion that survived.

Temperature (°C)	15	20	25	30	33
Proportion Survival	0.31	0.87	0.96	0.94	0.74

a. Find the best-fitting quadratic (as the researchers did).
b. Find the temperature at which survival was maximum.

58. Biological Control Lysyk[87] studied the effect of temperature on various life history parameters of a parasitic wasp for the purposes of pest management. He collected data given in the following table that related temperature to the lifetime progeny per female.

Temperature (°C)	15	20	25	31	33
Proportion Survival	40	70	95	65	50

a. Find the best-fitting quadratic (as the researchers did).
b. Find the temperature at which lifetime progeny was maximum.

83 Ibid.

84 Jarrad T. Kosa and Martha E. Mather. 2001. Processes contributing to variability in regional patterns of juvenile river herring abundance across small coastal systems. *Trans. Amer. Fish. Soc.* 130:600–619.

85 Ibid.

86 T. J. Lysyk. 2001. Relationship between temperature and life history parameters of *Muscidifurax zaraptor*. *Environ. Entomol.* 30(1):147–156.

87 Ibid.

Solutions to WARM UP EXERCISE SET 5.5

1. We have

$$f'(x) = 3x^2 - 12 = 3(x^2 - 4) = 3(x + 2)(x - 2)$$

Thus f' is continuous everywhere on the interval $[-3, 5]$, and the extreme value theorem ensures that the function attains both an absolute minimum and an absolute maximum on this interval. Thus, the extrema must be attained at the endpoints, -3 or 5, or the critical values -2 or $+2$. Evaluating f at these 4 values yields

$$f(-3) = 10$$
$$f(-2) = 17$$
$$f(2) = -15 \quad \leftarrow \text{absolute minimum}$$
$$f(5) = 66 \quad \leftarrow \text{absolute maximum}$$

One can do some further work and construct the following graph.

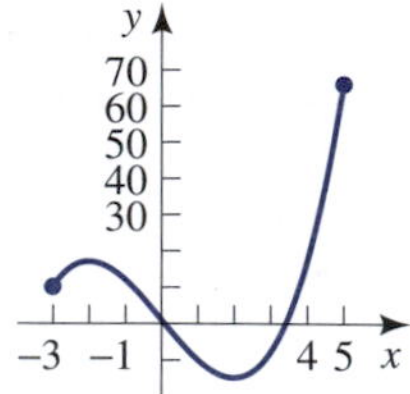

5.6 Optimization and Modeling

A First Example of Optimization and Mathematical Modeling
Mathematical Modeling
More Optimization and Mathematical Modeling
Enrichment: Inventory Cost Model

APPLICATION Minimizing Total Cost

Suppose the start-up costs of each production run for a manufactured item is \$50, and suppose it costs \$5 to manufacture each item and \$0.50 to store each item for 1 year. Determine the number of items in each run and the number of runs required to minimize the total cost if the total number of items to be produced and sold is 5000. See Example 4 on page 365 for the answer.

A First Example of Optimization And Mathematical Modeling

EXAMPLE 1 Construction of Fencing

A farmer has 500 yards of fencing with which to fence in three sides of a rectangular pasture. A straight river will form the fourth side. Find the dimensions of the pasture of greatest area that the farmer can fence.

Solution

The situation is shown in Figure 5.44. Since we do not know either of the dimensions of the rectangle, we label them x and y, as shown in the figure. It is these two dimensions that we must determine, where we take the dimensions in yards.

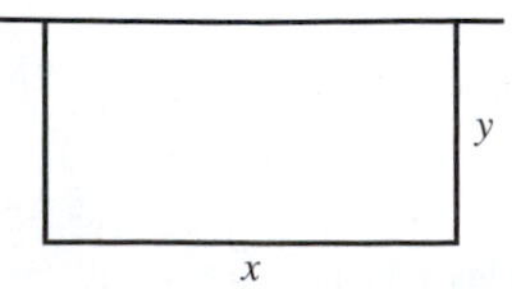

Figure 5.44

The area A of the pasture is then just $A = xy$. It is A that we wish to maximize. We cannot maximize A in this form, however; we need to have A as a function of *one* variable, not two.

But the two variables x and y are *not* both independent. Rather, these two variables are connected by the fact that on the one hand, we are given that the total amount of fencing is 500 yards, and on the other hand, Figure 5.44 indicates that the total amount of fencing is $x + 2y$. Thus,

$$500 = x + 2y$$

We can now solve this equation for one of the variables, say, y, obtaining

$$y = 250 - \frac{1}{2}x$$

Now that we have y in terms of x, we can substitute this value of y in the formula for area, $A = xy$, and obtain

$$\begin{aligned} A &= xy \\ &= x(250 - \frac{1}{2}x) \\ &= 250x - \frac{1}{2}x^2 \end{aligned}$$

Finally, this gives us A as a function of one variable x.

Before proceeding as in the previous sections, we should first determine the domain of the function $A(x)$. Naturally, we must have $x \geq 0$. On the other hand, we must have $x \leq 500$, since $y \geq 0$ and $500 = x + 2y$.

Let us now summarize what we have. We wish to find the maximum value of the *continuously differentiable* function $A = A(x)$ given above on the interval $[0, 500]$. This is now just like problems we have worked in the previous sections.

We then differentiate, obtaining

$$A'(x) = 250 - x$$

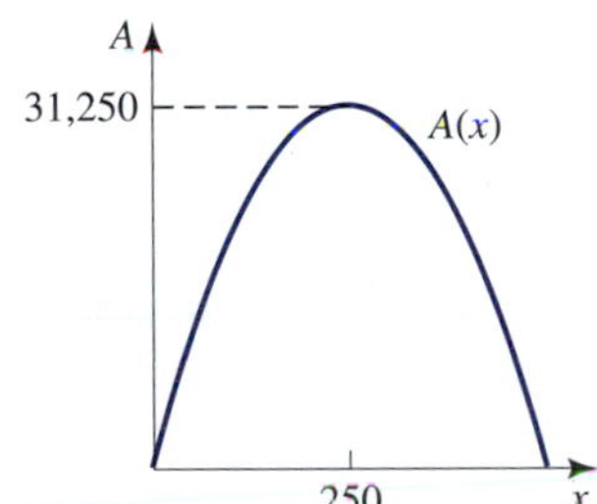

Figure 5.45
The absolute maximum of 31,250 occurs at $x = 250$.

It follows that $A'(x) > 0$ and $A(x)$ is increasing on $(0, 250)$, while $A'(x) < 0$ and $A(x)$ is decreasing on $(250, 500)$. Thus, $A(x)$ will attain its absolute maximum at $x = 250$ (Figure 5.45).

Since $y = 250 - \frac{1}{2}x$, setting $x = 250$ will yield

$$y = 250 - \frac{1}{2}(250) = 125$$

Thus, $x = 250$ and $y = 125$.

The maximum area is

$$A = xy = 250 \times 125 = 31,250$$ ■

Mathematical Modeling

It is not possible to give an explicit procedure for creating mathematical models and for solving the problems we encounter in this section so that the solution will be automatic. However, the following six-step procedure for modeling should be helpful.

Six-Step Procedure for Mathematical Modeling

Step 1. Identify the unknowns and denote each with a symbol.
Step 2. Identify the variable that is to be maximized or minimized and express it as a function of the other variables.
Step 3. Find any relationships between the variables.
Step 4. Express the quantity to be maximized or minimized as a function of only one variable. To do this, you might need to solve for one of the variables in one of the equations in Step 3 and substitute this into the formula found in Step 2.
Step 5. Determine the interval over which the function is to be minimized or maximized.
Step 6. Steps 1 through 5 leave you with a function of a single variable on a certain interval to be maximize or minimized. Proceed now as in previous sections.

The six-step procedure applies to Exercise 1 as follows.

1. We identified the two sides of the rectangle and the area as unknowns.
2. We identified A as the variable to be maximized and wrote $A = xy$.
3. We noted that $500 = x + 2y$.
4. We solved for y in the formula $500 = x + 2y$ and substituted the result into the formula for $A = xy$ to obtain A as a function of x only.
5. We noted that x must lie between 0 and 500.
6. We noted that $A(x) = 250x - \frac{1}{2}x^2$ was to be maximized on the interval $[0, 500]$.

CONNECTION
Profit Maximization

If a producer raises prices and the demand drops only modestly, then an increase in profits is possible. In a recent lawsuit the court forced a major breakfast cereal manufacturer to produce documents and data that gives us a rare insight into actual profit maximization. Using data provided by the company, Cotterill and Haller[88] calculated that an increase of 5% in the price of cereal was expected to yield an almost 1% increase in profits. This is a key reason why the price of breakfast cereal rose sharply at one time.

[88] Ronald W. Cotterill and Lawrence E. Haller. 1997. An economic analysis of the demand for RTE cereal: product market definition and unilateral market power effects. Research Report No. 35. Food Marketing Policy Center. University of Connecticut.

More Optimization and Mathematical Modeling

In the next example we see what happens if the price of a ticket is increased.

EXAMPLE 2 **Maximizing Revenue**

A theater owner charges \$5 per ticket and sells 250 tickets. By checking other theaters, the owner decides that for every one dollar she raises the ticket price, she will lose 10 customers. What should she charge to maximize revenue?

Solution
If p is the price of a ticket, n is the number of tickets sold, and R is the revenue, the formula for revenue is given by

$$\text{revenue} = (\text{price per ticket}) \times (\text{number of tickets sold})$$

or

$$R = pn$$

Thus, her current revenue is

$$(5) \times (250) = 1250$$

If she raises the price by x dollars, the price of a ticket will be $p = 5 + x$, the number of tickets sold will be $n = 250 - 10x$, and the revenue will be

$$\begin{aligned} R(x) &= pn \\ &= (5 + x)(250 - 10x) \\ &= 1250 + 200x - 10x^2 \end{aligned}$$

This is an everywhere continuously differentiable function. Since the owner clearly wishes to have the price of each ticket *positive*, we must have $5 + x = p > 0$ or $x > -5$. Also, the owner clearly wants to sell at least one ticket, that is, $250 - 10x = n > 0$ or $x < 25$.

We thus wish to maximize the function R given above on the interval $(-5, 25)$. Since

$$\begin{aligned} R'(x) &= 200 - 20x \\ &= 20(10 - x) \end{aligned}$$

there is one critical value at $x = 10$. Also, $R'(x) > 0$ on $(-5, 10)$ and $R'(x) < 0$ on $(10, 25)$. Thus, R attains an absolute maximum at $x = 10$ (Figure 5.46). It is thus clear that the revenue, R, will be maximum when $x = 10$. This means that the price of each ticket should be $5 + 10 = 15$ dollars.

The maximum revenue will be

$$R(10) = (5 + 10)(250 - 10 \times 10) = 2250$$

■

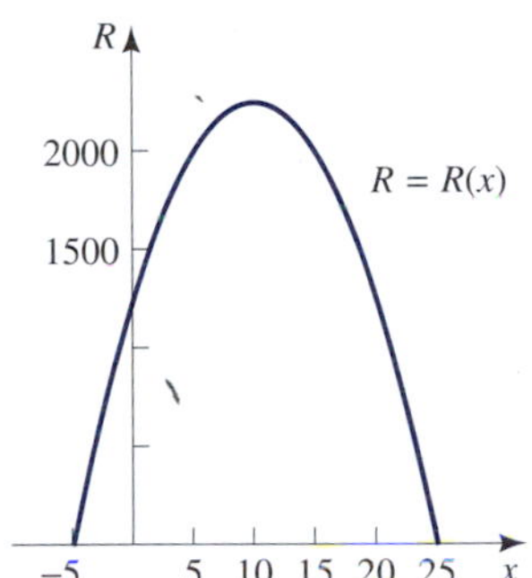

Figure 5.46
The absolute maximum occurs at $x = 10$.

EXAMPLE 3 **Minimizing the Cost of Materials**

A manufacturer must produce a sturdy rectangular container with a square base and a volume of 128 cubic feet. The cost of materials making up the top and four sides is \$2 per square foot, while the cost of the materials making up the bottom (which must be reinforced) is \$6 per square foot. Find the dimensions of the box that minimizes the cost of the materials. Use your grapher to estimate the minimum. Confirm analytically.

Solution

In Figure 5.47, each side of the square base is labeled x, and the height is labeled h. We wish to minimize the total cost. The total cost is the cost of the bottom plus the cost of the top plus the cost of the four sides.

The area of the bottom is x^2, so the cost of the bottom is $6(x^2) = 6x^2$.
The area of the top is x^2, so the cost of the top is $2(x^2) = 2x^2$.
The area of the four sides is $4xh$, so the cost of the four sides is $2(4xh) = 8xh$.
Thus, the total cost, $C(x)$, is given by

$$C(x) = 6x^2 + 2x^2 + 8xh = 8x^2 + 8xh$$

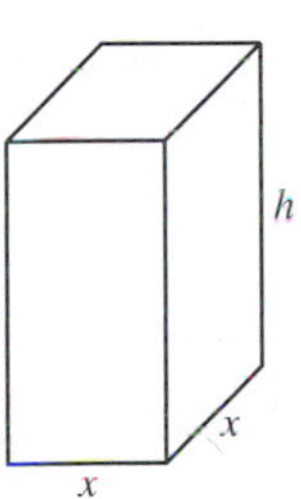

Figure 5.47

The volume must be 128, and by Figure 5.47 the volume is also x^2h. Thus,

$$128 = V = x^2h$$

Solving this for h gives

$$h = \frac{128}{x^2}$$

We substitute this into the formula for $C(x)$ and obtain

$$C(x) = 8x^2 + 8x\left(\frac{128}{x^2}\right) = 8x^2 + 1024x^{-1}$$

Now we have C as a function of one variable x, and we wish to find the absolute minimum of C on the interval $(0, \infty)$. We have

$$\begin{aligned} C(x) &= 8x^2 + 1024x^{-1} \\ C'(x) &= 16x - 1024x^{-2} \\ &= 16x^{-2}\left(x^3 - 64\right) \\ &= \frac{16}{x^2}\left(x^3 - 64\right) \end{aligned}$$

We see that $C'(x) < 0$ on $(0, 4)$ and that $C'(x) > 0$ on $(4, \infty)$. Therefore, a minimum occurs at $x = 4$. The graph is shown in Figure 5.48.

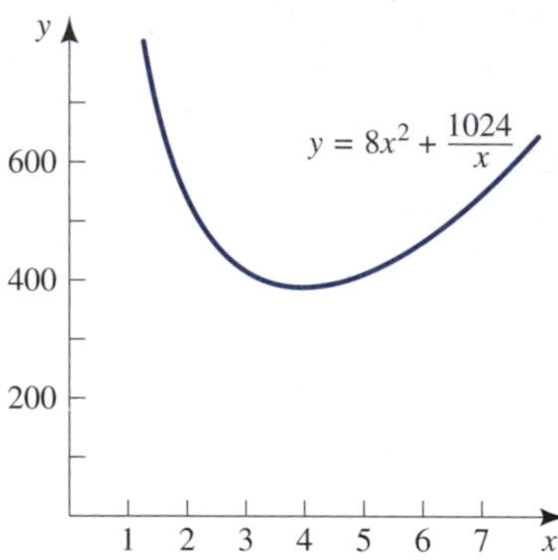

Figure 5.48
The minimum cost occurs at $x = 4$.

The minimum cost is

$$C(4) = 8(4)^2 + \frac{1024}{4} = 384$$

The dimensions of the box that minimizes cost are 4' × 4' × 8'. ■

Exploration 1

Graphing a Sum

The function $y = C(x)$ in Example 3 is the sum of two simpler functions: $f(x) = 8x^2$ and $g(x) = \dfrac{1024}{x}$. Sketch a rough graph (without using your grapher) of $y = f(x)$ and $y = g(x)$ on the same graph. The graphs of these two functions are probably familiar. Now sketch a graph of the sum $y = f(x) + g(x) = C(x)$. Notice that since $g(x)$ is extremely large for small x, the sum must be extremely large. Since $f(x)$ is extremely large for large x, the sum must be extremely large. The minimum must occur at some intermediate value. Confirm all of this on your grapher.

Enrichment: Inventory Cost Model

Suppose a manufacturer intended to produce and sell a large number of items during the next year. All of the items could be manufactured at the start of the year. Because of the economies of mass production, this would certainly hold down the manufacturing costs. However, this would result in a large cost of storing the large number of items until they could be sold. If, on the other hand, a large number of production runs were made and spread out over the year, there would be substantially fewer unsold items that would require storage. Thus, storage costs would be much smaller. However, because of the lack of economies inherent in small production runs, the manufacturing costs would be large. Presumably, there will be some middle ground that would minimize the total costs.

We make the simplifying assumption that sales are made at a steady rate throughout the year and the production run occurs when the number of unsold items has been reduced to zero. We assume that each run produces the same number of items, and we designate this number by x. Figure 5.49 shows the number of items in storage. From this it is apparent that on average there are $\frac{1}{2}x$ items in storage at any time. If s is the cost of storing an item for 1 year, then the cost of storage will be given by the cost of storing one item for one year times the average number in storage or

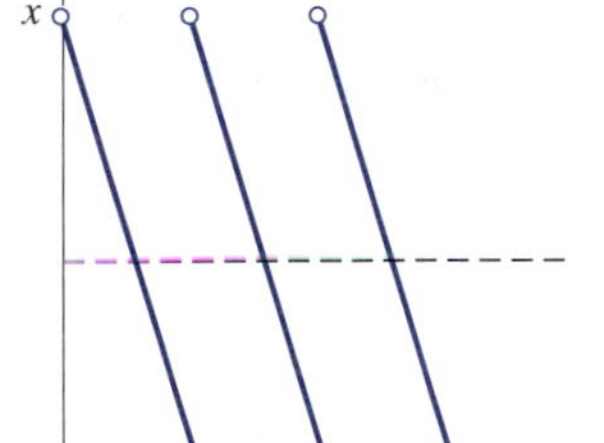

Figure 5.49
Number of items in storage.

$$\text{cost of storage} = s\left(\frac{1}{2}x\right)$$

EXAMPLE 4 **Minimizing Total Cost**

Suppose the start-up cost of each production run is \$50, that it costs \$5 to manufacture each item, and that it costs \$0.50 to store each item for one year. Determine the number of items in each run and the number of runs to minimize the total cost if the total number of items to be produced and sold is 5000.

Solution

As we saw above,

$$\text{cost of storage} = s\left(\frac{1}{2}x\right) = 0.50\left(\frac{1}{2}x\right) = \frac{1}{4}x$$

where x is the number of items produced in each run.

Let the number of runs be given by n. Then the total number of items produced must be given by the number of runs times the number of items in each run or

$$5000 = nx$$

or

$$n = \frac{5000}{x}$$

The total cost of preparing the plant for n runs is the cost of preparing the plant for one run times the number of runs or

$$\text{total preparation cost} = 50n = 50\frac{5000}{x} = 250000\frac{1}{x}$$

The total cost of manufacturing the 5000 items is given by 5000 times the cost of each item or

$$\text{total manufacturing cost} = 5000(5) = 25000$$

The total cost, $C(x)$, is the sum of the previous three costs, or

$$C(x) = 25000 + \frac{1}{4}x + 250000\frac{1}{x}$$

We then wish to find the absolute minimum of this function on the interval $[1, 5000]$.

Taking the derivative, we have

$$C'(x) = \frac{1}{4} - \frac{250,000}{x^2} = \frac{1}{4x^2}\left(x^2 - 1,000,000\right)$$

Since $C'(x) < 0$ on $(0, 1000)$ and $C'(x) > 0$ on $(1000, \infty)$, the minimum occurs at $x = 1000$. The graph is shown in Figure 5.50.

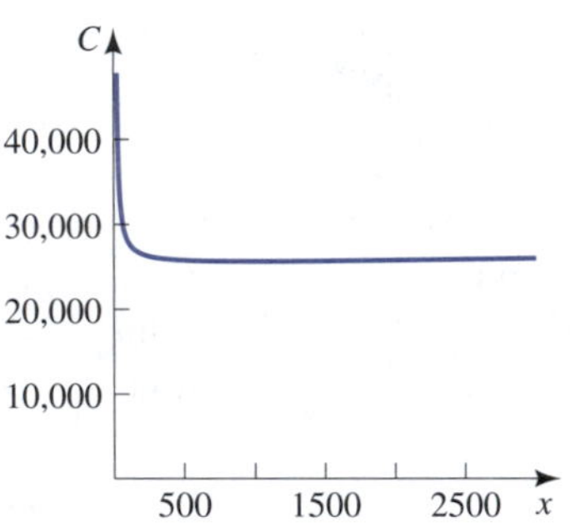

Figure 5.50
An absolute minimum occurs at $x = 1000$. The function increases slowly after $x = 1000$.

The number of runs is

$$n = \frac{5000}{x} = \frac{5000}{1000} = 5$$

The minimum cost is

$$C(1000) = 25,000 + \frac{1}{4}(1000) + \frac{250,000}{1000}$$
$$= 25,500$$

■

Warm Up Exercise Set 5.6

1. A package mailed in the United States must have length plus girth of no more than 108 inches. Find the dimensions of a rectangular package with square base of greatest volume that can be mailed. (The girth is the perimeter of a rectangular cross section.)

2. The figure shows a lake. A medical team is located at point A and must get to an injured person at point D as quickly as possible. The team can row a boat directly to point D, or can row the boat directly across the lake to point B and then run along the shore to D, or can row to some intermediate point C and then run down the shore to point D. If the team can row at 6 mph and run at 10 mph, where on the opposite shore should they land the boat?

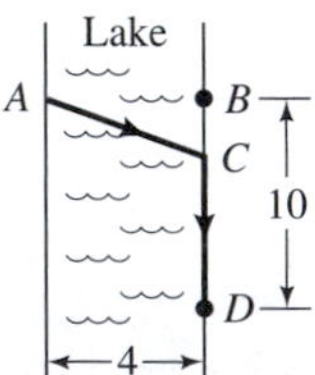

3. A farmer wishes to enclose a rectangular field of area 200 square feet using an existing wall as one of the sides. The cost of the fence for the other three sides is $1 per foot. Find the dimensions of the rectangular field that minimizes the cost of the fence.

Exercise Set 5.6

Applications and Mathematical Modeling

1. **Fencing** Where a river makes a right-angle turn, a farmer with 400 feet of fencing wishes to construct a rectangular fenced-in pasture that uses the river for two sides. What should be the dimensions of the rectangle to maximize the area enclosed?

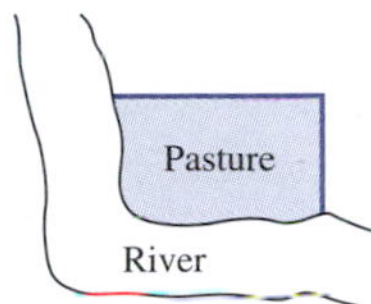

2. **Fencing** A farmer with 400 feet of fencing wishes to construct a rectangular pasture with maximum area. What should the dimensions be?

3. **Packaging** From a 9-inch by 9-inch piece of cardboard, square corners are cut out so that the sides can be folded up to form a box with no top. What should x be to maximize the volume?

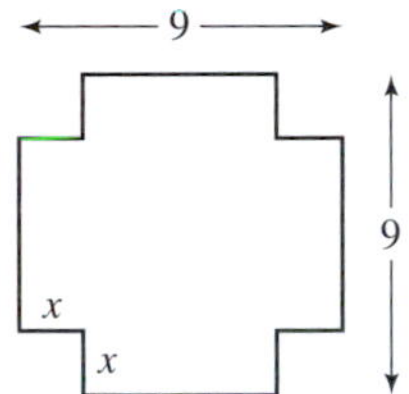

4. **Packaging** Find the dimensions of a cylindrical package with circular base of largest volume that can be mailed in this country.

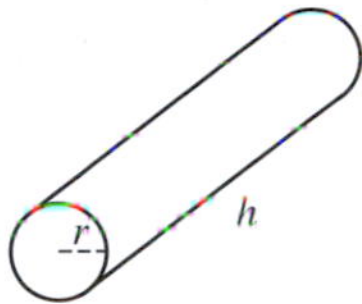

5. **Cost** A fence is to be built around a 200-square-foot rectangular field. Three sides are to be made of wood costing \$10 per foot, while the other side is made of stone costing \$30 per foot. Find the dimensions of the enclosure that minimizes total cost.

6. **Cost** A farmer wishes to enclose a rectangular field of area 450 square feet using an existing wall as one of the sides. The cost of the fence for the other three sides is \$3 per foot. Find the dimensions of the rectangular field that minimizes the cost of the fence.

7. **Cost** What are the dimensions of the rectangular field of 20,000 square feet that will minimize the cost of fencing if one side costs three times as much per unit length as the other three?

8. **Fencing** A rectangular corral of 32 square yards is to be fenced off and then divided by a stone fence into two sections. The outer fence costs \$10 per yard, while the special dividing stone fence costs \$20 per yard. Find the dimensions of the corral that minimizes the cost.

9. **Revenue** A bus company charges \$10 per person for a sightseeing trip if 30 people travel in a group. If for each person above 30 the company reduces the charge per person by \$0.20, how many people will maximize the total revenue for the bus company?

10. **Revenue** A hotel has 10 luxury units that it will rent out during the peak season at \$300 per day. From experience management knows that one unit will become vacant for each \$50 increase in charge per day. What rent should be charged to maximize revenue?

11. **Profit** Suppose that in Exercise 10 the daily cost of maintaining and cleaning a rented luxury room was \$100 per day. What rent should be charged to maximize profit?

12. **Profit** An apple orchard produces annual revenue of \$50 per tree when planted with 1000 trees. Because of overcrowding, the annual revenue per tree is reduced by 2 cents for each additional tree planted. If the cost of maintaining each tree is \$10 per year, how many trees should be planted to maximize total profit from the orchard?

13. **Packaging** A company wishes to design a rectangular box with square base and no top that will have a volume of 32 cubic inches. What should the dimensions be to yield a minimum surface area? What is the minimum surface area?

14. **Packaging** A company wishes to design a rectangular box with square base and no top that will have a volume of 27 cubic inches. The cost per square inch of the bottom is twice the cost of the sides. What should the dimensions be to minimize the cost? What is the minimum cost?

15. **Cost** A closed rectangular box of volume 324 cubic inches is to be made with a square base. If the material for the bottom costs twice per square inch as much as the material for the sides and top, find the dimensions of the box that minimize the cost of materials.

16. Construction Find the most economical proportions for a closed cylindrical can that will hold 16 cubic inches.

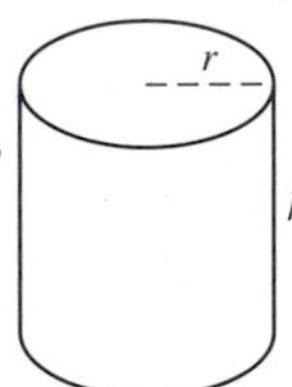

17. Maximizing Profit Roberts[89] formulated a mathematical model of corn yield response to nitrogen fertilizer in high-yield response land given by the equation $Y(N) = 25 + 1.2N - 0.003N^2$, where Y is bushels of corn per acre and N is pounds of nitrogen per acre. They estimated that the farmer obtains \$2.42 for a bushel of corn and pays \$0.22 a pound for nitrogen fertilizer. Assuming that the only cost to the farmer is the cost of nitrogen fertilizer, find the amount of nitrogen fertilizer that maximizes profits.

18. Maximizing Profit Roberts[90] formulated a mathematical model of corn yield response to nitrogen fertilizer in low-yield response land given by the equation $Y(N) = 15 + 0.5N - 0.0016N^2$, where Y is bushels of corn per acre and N is pounds of nitrogen per acre. They estimated that the farmer obtains \$2.42 for a bushel of corn and pays \$0.22 a pound for nitrogen fertilizer. Assuming that the only cost to the farmer is the cost of nitrogen fertilizer, find the amount of nitrogen fertilizer that maximizes profits.

89 Roland K. Roberts, Burton C. English, and S. B. Mahajashetti. 2000. Evaluating the returns to variable rate nitrogen application. *J. Agr. Appl. Econ.* 32(1):133–143.

90 Ibid.

More Challenging Exercises

19. Inventory Control Suppose the start-up cost of each production run is \$1000 and that it costs \$20 to manufacture each item and \$2 to store each item for 1 year. Determine the number of items in each run and the number of runs to minimize total cost if the total number of items to be produced and sold is 16,000.

20. Inventory Control Suppose the start-up cost of each production run is \$2500 and that it costs \$20 to manufacture each item and \$2 to store each item for one year. Determine the number of items in each run and the number of runs to minimize total cost if the total number of items to be produced and sold is 10,000.

21. Construction Pipe is to be laid connecting A with B in the figure. The cost along the level stretch from A to C is \$10 per foot, whereas the cost along the difficult (away from the road) stretch from C to B is \$20 per foot. The distance from A to D is 40 feet, and the distance from B to D is 30 feet. At what point C will the cost of laying pipe from A to B to C be minimum? Use your grapher to find an approximate solution. Confirm analytically.

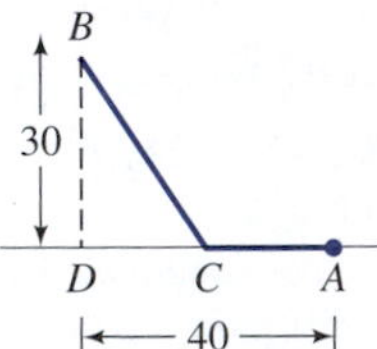

22. Construction A wire 12 inches long is cut into two pieces with one piece used to construct a square and the other piece used to construct a circle. Where should the wire be cut to minimize the sum of the areas enclosed by the two figures? Use your grapher to find an approximate solution. Confirm analytically.

23. Construction A wire 12 inches long is cut into two pieces with one piece used to construct a square and the other piece used to construct an equilateral triangle. Where should the wire be cut to minimize the sum of the areas enclosed by the two figures? (The area of an equilateral triangle is $\frac{\sqrt{3}}{4}r^2$, where r is the length of a side.) Use your grapher to find an approximate solution. Confirm analytically.

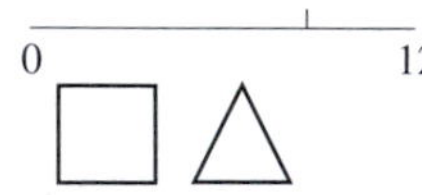

24. Construction An individual is planning to construct two fenced enclosures, one square and one circular. The cost per yard of constructing the circular enclosure is \$8, and the cost of constructing the square enclosure is \$4. If the total cost is fixed at \$1600, what should be the dimensions of the two figures to minimize the area enclosed by the two figures? Use your grapher to find an approximate solution. Confirm analytically.

25. Construction An individual is planning to construct two fenced enclosures, one square and one an equilateral triangle. The cost per yard of constructing the triangular enclosure is \$6, and the cost of constructing the square enclosure is \$3. If the total cost is fixed at \$1000, what should be the length of each side of the triangular enclosure to minimize the area enclosed by the two figures? (The area of an equilateral triangle is $\frac{\sqrt{3}}{4}r^2$, where r is the length of a side.) Use your grapher to find an approximate solution. Confirm analytically.

26. Inventory Control In the discussion on inventory control in the text, let T be the total number of items to be produced and sold, k is the amount of money needed to prepare each production run, and s is the cost of storing one item for an entire year. Show that the number x of items in each run needed to minimize the total cost is $x = \sqrt{2kT/s}$. Notice that the answer does not depend on the cost p of producing each item.

In Exercises 27 and 28, graph using your grapher to estimate the value where the function attains its absolute minimum and the value where the function attains its absolute maximum. Verify using calculus.

27. $f(x) = 9 + (x - 3)e^{-x}$ on $[-1, 5]$

28. $f(x) = 50x^4 - x^2 + 10$ on $[0, 1]$

Solutions to WARM UP EXERCISE SET 5.6

1. See the figure where the length of the package is given as l and each side of the square base is given by s. We wish to maximize the volume, V, which is given by $V = s^2l$. Here V is a function of *two* variables. We need to make V a function of only *one* variable.

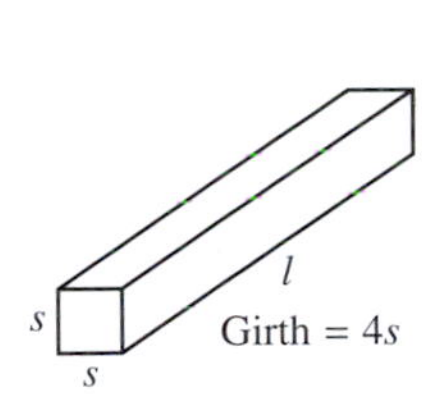

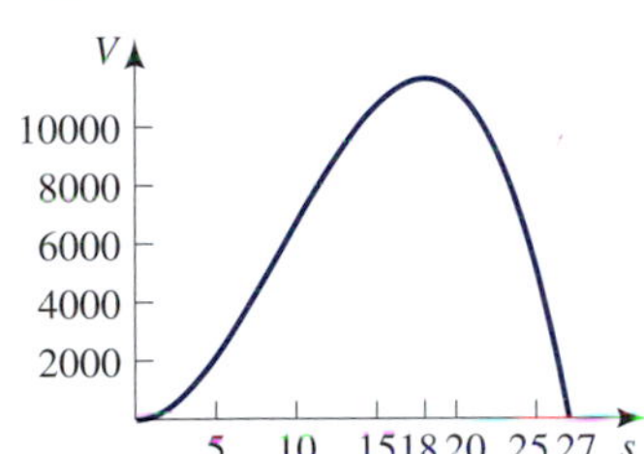

But the two variables s and l are connected by the fact that the length plus the girth can be at most 108 inches. It is clear that if we are to maximize the volume, then we should take the length plus the girth to be as large as possible, which is 108. Thus, take

$$108 = \text{length} + \text{girth} = l + 4s$$

From this equation we can solve for l and obtain

$$l = 108 - 4s$$

Now substitute this into the formula for V, and obtain

$$V = s^2l = s^2(108 - 4s) = 108s^2 - 4s^3$$

We now have V as a function of one variable.
This function is continuous and differentiable everywhere. Naturally, we must have $s \geq 0$ and also $s \leq 108/4 = 27$ inches.

In summary, we wish to find the absolute maximum of the continuously differentiable function $V = V(s) = 108s^2 - 4s^3$ on the interval $[0, 27]$.

To do this, we first differentiate $V(s)$ obtaining

$$V'(s) = 216s - 12s^2 = 12s(18 - s)$$

Since $V'(s) = 0$ implies that $s = 0$ or $s = 18$ inches, the only critical value of $V(s)$ on $(0, 27)$ is at $s = 18$ inches. Furthermore, from the factorization of $V'(s)$ we see that $V'(s) > 0$ on $(0, 18)$ and $V'(s) < 0$ on $(18, 27)$. Thus, V is increasing on $(0, 18)$ and decreasing on $(18, 27)$. From this it is clear that $V(18)$ is an absolute maximum on the interval $[0, 27]$. See the figure. Then

$$l = 108 - 4s = 108 - 4(18) = 36$$

The maximum volume is

$$V = s^2l = 18^2 \times 36 = 11,664 \text{ in.}^3$$

The dimensions of the box are $18 \times 18 \times 36$ inches.

2. Let x be the distance from B to C. Since

$$\text{time} = \frac{\text{distance}}{\text{rate}}$$

the time spent in the boat is $\dfrac{\sqrt{x^2+16}}{6}$, and the time spent running along the shore is $\dfrac{10-x}{10}$. Thus, the total time of the trip is given by

$$T(x) = \frac{\sqrt{x^2+16}}{6} + \frac{10-x}{10}$$

We wish to find the minimum of T on the interval $[0, 10]$. We have

$$T'(x) = \frac{x}{6\sqrt{x^2+16}} - \frac{1}{10}$$

Then $T'(x) = 0$ if and only if $5x = 3\sqrt{x^2+16}$, or $25x^2 = 9x^2 + 144$, or $16x^2 = 144$ or $x = 3$. Evaluating $T(x)$ at the critical value $x = 3$ and the endpoints $x = 0$ and $x = 10$ gives $T(0) = \frac{5}{3} \approx 1.67$, $T(3) = \frac{5}{6} + \frac{7}{10} \approx 1.53$, $T(10) = \frac{\sqrt{116}}{6} \approx 1.80$. Thus, the minimum occurs at $x = 3$. This means that the medical team should row to a point C located 3 miles below point B and run the remaining 7 miles to the injured person at point D.

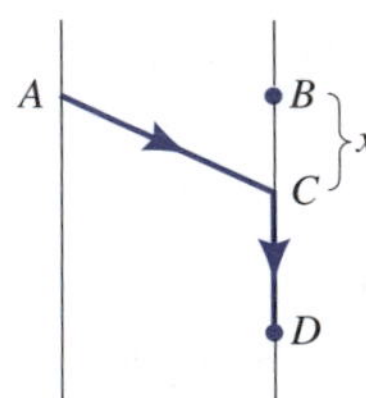

3. Let x and y denote the sides of the rectangular enclosure shown in the figure. Then the area is given by $A = xy = 200$. The cost in dollars is the same as the number of feet of fence. Thus,

$$
\begin{aligned}
C(x) &= x + 2y \\
&= x + 2\frac{200}{x} \\
&= x + \frac{400}{x} \\
C'(x) &= 1 - \frac{400}{x^2} \\
&= \frac{1}{x^2}\left(x^2 - 400\right)
\end{aligned}
$$

Since $\sqrt{400} = 20$, $C'(x) < 0$ on $(0, 20)$ and $C'(x) > 0$ on $(20, \infty)$. Therefore, $C(x)$ has an absolute minimum at $x = 20$. Since $xy = 200$, we have $20y = 200$ or $y = 10$. Thus, the area should be 20 by 10.

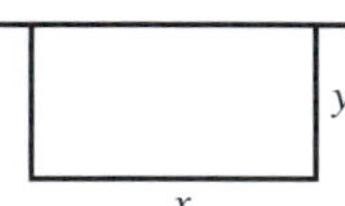

5.7 The Logistic Model

Enrichment: Modeling

Enrichment: The Logistic Curve

Enrichment: Logistic Curve Continued

APPLICATION Soil Carbon Dynamics

Knopps and Tilman[91] studied the soil carbon dynamics of abandoned agricultural fields and created a mathematical model of soil carbon given by the equation

$$C(t) = \frac{3.754}{1 + 7.981e^{-0.0217t}}$$

where t is the number of years after an agricultural field has been abandoned and $C(t)$ is the percent of carbon in the soil. Graph and discuss what this model is saying. See Example 2 on page 375 for an answer.

[91] Dynamics of soil nitrogen and carbon accumulation for 61 years after agricultural abandonment. 2000. *Ecology* 81(1):88–98.

Enrichment: Modeling

Let us consider some examples from diverse areas that have a particular growth pattern.

Consider the classical experiment of the growth of a population of protozoa in a petri dish containing some nutrients. Figure 5.51 is taken from Deshmukh.[92] The population explodes at first—in fact, it grows exponentially—but then levels off. What happened? The protozoa ran out of nutrients and space. As the nutrients were

Figure 5.51

[92] Ian Deshmukh. 1986. *Ecology and Tropical Biology.* Boston: Blackwell Scientific Publications.

used up the population growth slowed up. The amount of nutrients was not infinite, but rather finite.

Figure 5.52 shows a graph of the human population of the world found in a United Nations report.[93] Notice how the population grows rapidly until recently. Notice how the population increase is expected to slow up in the future. Figure 5.53 is from the same report and shows the annual percentage increases in the population of the world and projections out to 2050. Notice how the rate of growth is itself decreasing in recent years. Actually, the report estimates that the "world population will nearly stabilize at just above 10 billion after 2200."

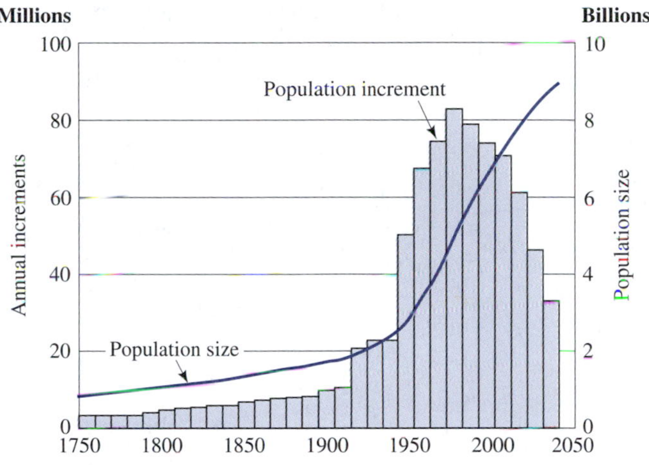

Source: United Nations Population Division.

Figure 5.52
Long-term world population growth, 1750–2050.

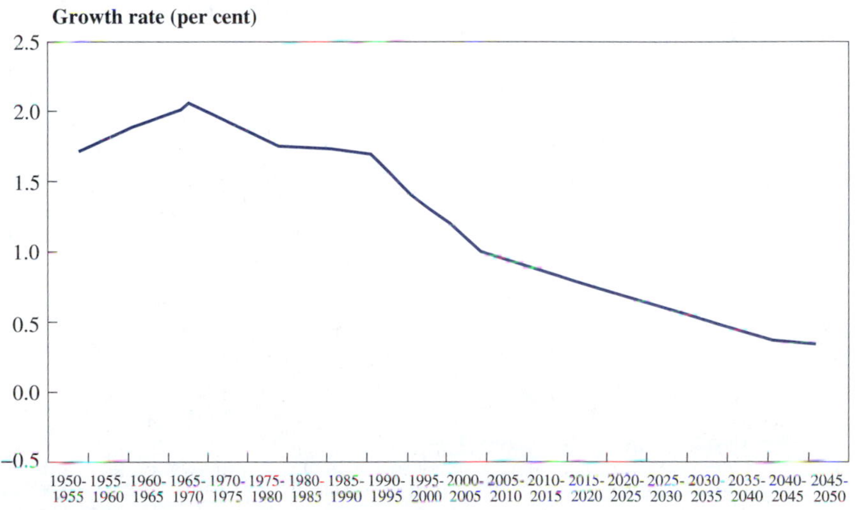

Source: United Nations Population Division.

Figure 5.53
World population growth rates.

93 United Nations. *The world at six billion.* http://www.un.org/esa/population/publications/sixbillion.htm

Consider next, examples from plant science. Many tree species in tropical rain forests grow at a spectacular rate. For example, balsa can grow more than 15 feet a year.[94] Yet we all know that trees, even balsa trees, can only grow to some *limiting* height. Whitmore[95] draws a graph of a typical freely growing tree and is shown in Figure 5.54. Figure 5.55 reproduces a graph given by Pienaar and Turnbull[96] that shows the growth of a spruce tree. We see in these cases a rapidly increasing exponential growth rate at first, followed by a decreasing rate of growth, with the height leveling off to some limiting value.

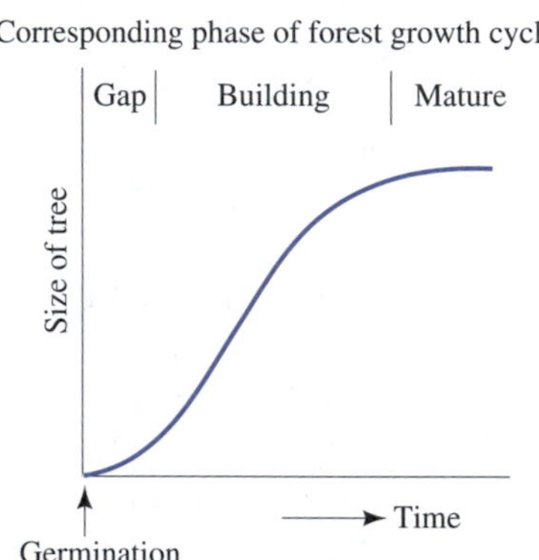

Figure 5.54

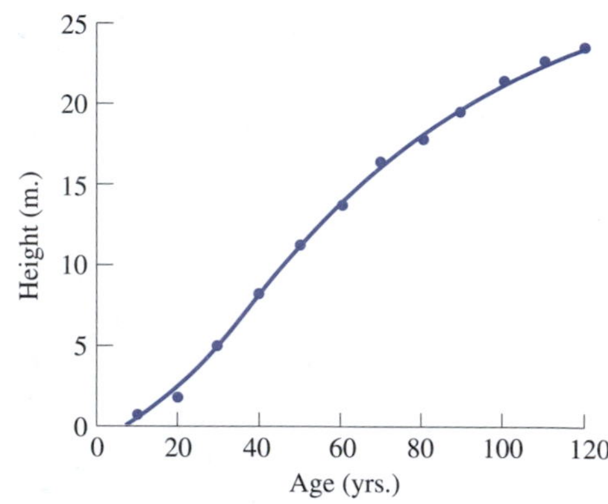

Figure 5.55

In business, sales of a new product often rise quickly immediately after introduction and then level off later. See Figure 5.56 for an example. When a new technology is introduced in an industry, this technology often spreads rapidly through the industry after introduction and then levels off later. Hornstein[97] gives the graph shown in Figure 5.57 of the example of the diesel locomotive in the U.S. railroad industry.

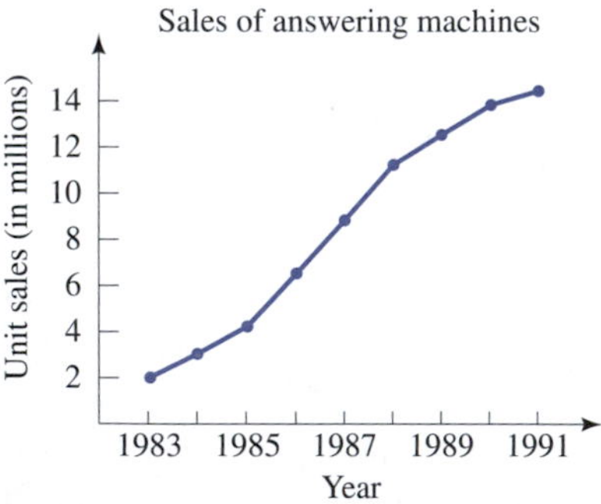

Figure 5.56
The graph of the sales of a new innovation looks like a logistic curve.

[94] T. C. Whitmore. 1990. *An Introduction to Tropical Rain Forests*. New York: Oxford University Press.
[95] Ibid.
[96] L. V. Pienaar and K. J. Turnbull. 1973. The Chapman-Richards generalization of von Bertalanffy's growth model for basal area growth and yield in even-ages stands. *Forest Sci.* 19(1):2–22.
[97] Andreas Hornstein. Growth accounting with technological revolutions. *Econ. Quart. Fed. Res. Bank Richmond* 85(3):6–10.

An equation that models these situations is the logistic equation given by

$$P(t) = \frac{L}{1 + ae^{-kt}}$$

Figure 5.58 gives a graph in which L, a, and k are all positive. Exercises 15 through 19 indicate that $P(t)$ is always increasing, $\lim_{t\to\infty} P(t) = L$, and $P(t)$ is concave up on $(0, c)$ and concave down on (c, ∞) with $P(c) = \frac{L}{2}$ and $c = \frac{1}{k}\ln a$.

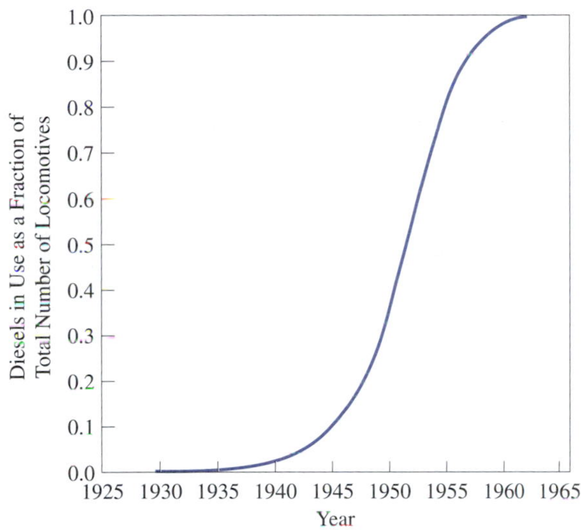

Figure 5.57
Diesel Locomotion in the U.S. Railroad Industry, 1925–1966: Diffusion.

P
L
$P(t) = \frac{L}{1 + ae^{-kt}}$, k > 0
$P(t)$ always increasing
$P(t) \to L$ as $t \to \infty$
$P(t)$ concave up on $(0, c)$
$P(t)$ concave down on (c, ∞)
$c = \frac{1}{k}\ln a$
t

Figure 5.58

Interactive Illustration 5.6

Logistic Curve

This interactive illustration allows you to observe the tangent line and its slope on a logistic curve. Shown is an example of a graph of a logistic equation. Move the cursor and observe the tangent line.

1. When is the slope increasing? Decreasing?
2. What do your answers in (1) have to do with concavity?

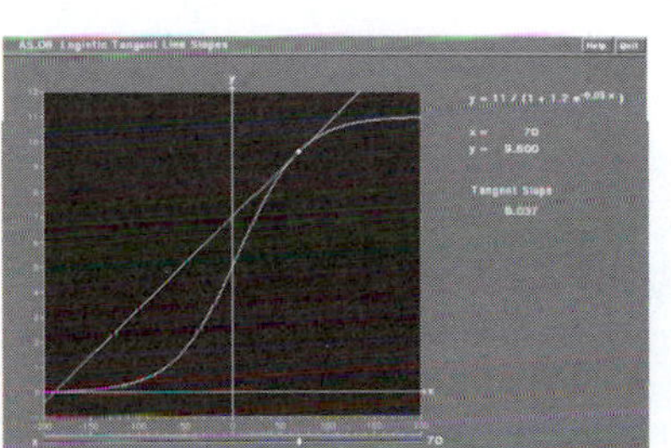

Recall that in Section 1.3 that we gave a mathematical model of the population of the United States from the year 1790. This was the exponential model $p(t) = 3.929(1.033)^t$, where p is in millions and t is years from 1790. We developed this model based on population data from 1790 to 1794. We then used this model to predict the population in 1804, and the model came close. This model predicts the population in 1830 to be about 14.4 million. The actual population was 12.9 million. The model predicts a population of 101 million for 1890, whereas the actual population was only 62.9 million. Finally, the model predicts a population of 3.6 *trillion* in the year 2000, whereas the actual population was only 281 million. Something has gone terribly wrong with our model. With the initial exponential growth and then a slowing in growth, perhaps the population of the United States follows a logistic growth pattern.

Actually, in about 1850 the mathematical-biologist P. Verhulst used a logistic model to predict the future population of the United States. We now look at his model and see just how successful it was.

Enrichment: The Logistic Curve

EXAMPLE 1 **Population of the United States in the Long Term**

About 150 years ago the mathematical-biologist P. Verhulst used the equation

$$p(t) = \frac{197.27}{1 + 49.58e^{-0.03134(t-1790)}}$$

to model the population of the United States $p(t)$ in millions, where t is the calendar year. Find the population this model predicts for 1880, 1930, and 1990. Compare with the actual population of 50.2, 123.2, and 250. What does this model predict will happen to the population in the long term?

Solution

We have

$$\begin{aligned} 1880{:}\quad & p(90) = 49.9 \\ 1930{:}\quad & p(140) = 122.0 \\ 1990{:}\quad & p(200) = 180.3 \end{aligned}$$

Thus, this model gave extraordinarily accurate projects of the population of the United States for over 100 years after being formulated. Only in recent years has the model gone a bit astray. To determine the long-term prediction of the model, we need to find $\lim_{t\to\infty} p(t)$. Using the rules of limits, we have

$$\begin{aligned} \lim_{t\to\infty} p(t) &= \lim_{t\to\infty} \frac{197.27}{1 + 49.58e^{-0.03134(t-1790)}} \\ &= \frac{197.27}{1 + 49.58 \lim_{t\to\infty} e^{-0.03134(t-1790)}} \\ &= \frac{197.27}{1 + 49.58(0)} \\ &= 197.27 \end{aligned}$$

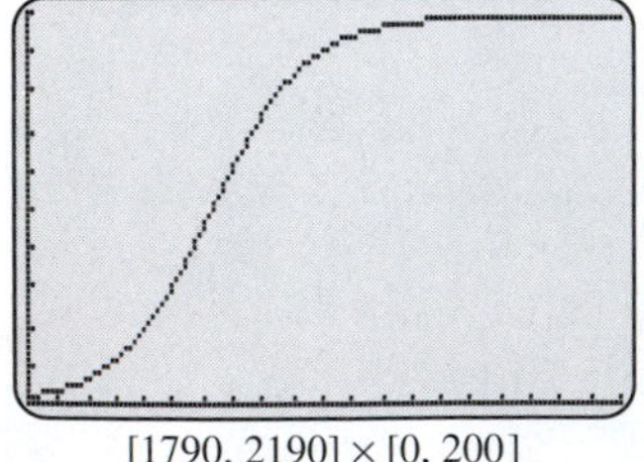

[1790, 2190] × [0, 200]

Screen 5.12
The population $y = \dfrac{197.27}{1 + 49.58e^{-0.03134(x-1790)}}$ grows fast at first and then levels out and heads for a limiting value.

Thus, this model, formulated in the year 1837, predicted a limiting population of the United States of very nearly 200 million. Screen 5.12 shows a graph of $y = p(t)$ using a window with dimensions [1790, 2190] by [0, 200].

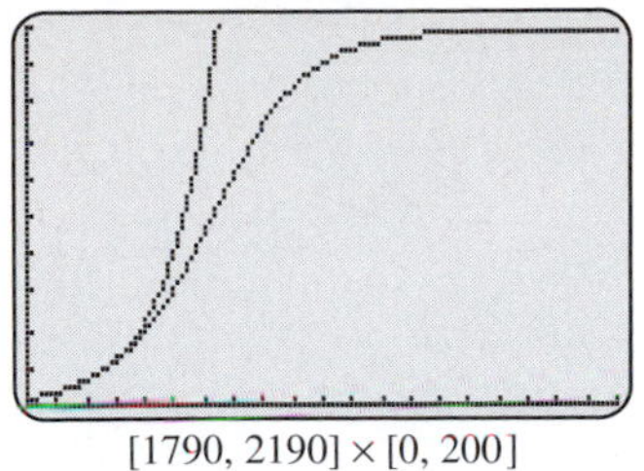
$[1790, 2190] \times [0, 200]$

Screen 5.13
During the earlier years, the bottom graph of the logistic curve
$$y_1 = \frac{197.27}{1 + 49.58e^{-0.03134(x-1790)}}$$
is about the same as the top graph of the exponential function $y_2 = 3.926(1.033)^{x-1790}$.

We showed in Section 1.3 that the population of the United States in the years from 1790 to about 1805 was growing exponentially according to $p(t) = 3.926(1.033)^t$, where $t = 0$ corresponds to the year 1790. Does this contradict the equation given in Example 1? To check this, we graph this last exponential function on the same screen as the logistic equation. This is shown in Screen 5.13. Notice the close agreement for the first 50 years after 1790. Thus, *the logistic curve is approximately an exponential curve during the early years.* ■

CONNECTION
Carbon Sequestration

The Kyoto Protocol to the United Nations Framework Convention on Climate Change (1997) establishes emission reduction targets for the United States and other industrialized nations. The Kyoto Protocol states that carbon sequestration can be used by participating nations to achieve their targets.

We turn now to an important area of concern for the environment: carbon dioxide in the atmosphere and the consequent threat of global climate change. Many laws have been passed to curb carbon emissions to the atmosphere from industrial plants and vehicles. But another important way of reducing carbon in the atmosphere is to create "sinks," that is something that will absorb carbon from the atmosphere and store it. This is, of course, plants.

Increased use of land for agricultural purposes (and the often destruction of forests to obtain land for agriculture) over the past 100 years has led to a decrease in carbon stored in soils and a new release of carbon into the atmosphere.[98] We look at one study of soil carbon dynamics following abandonment of agriculture fields on a Minnesota sand plain.

EXAMPLE 2 **Soil Carbon Dynamics**

Knopps and Tilman[99] studied the soil carbon dynamics of abandoned agricultural fields and created a mathematical model of soil carbon given by the equation

$$C(t) = \frac{3.754}{1 + 7.981e^{-0.0217t}}$$

where t is the number of years after an agricultural field has been abandoned and $C(t)$ is the percent of carbon in the soil. Graph and discuss what this model is saying.

Solution

A graph is shown in Figure 5.59. Notice that the graph of this logistic curve is concave up for about the first 100 years. This indicates that the rate at which carbon is accumulating in the soil is increasing with time. During the next 100 years the graph is concave down. This indicates that the rate at which carbon is accumulating in the soil is decreasing with time. We see that the graph eventually levels out.

To determine the long-term prediction of the model, we need to find $\lim_{t\to\infty} C(t)$.

[98] W. H. Schlesinger. 1986. Soil organic matter: a source of atmospheric CO_2. In: G. M. Woodwell (Ed.), *The Role of Terrestrial Vegetation in the Global Carbon Cycle*. New York: Wiley. pages 111–127.

[99] Knopps and Tilman. 2000. Dynamics of soil nitrogen and carbon accumulation for 61 years after agricultural abandonment. *Ecology* 81(1):88–98.

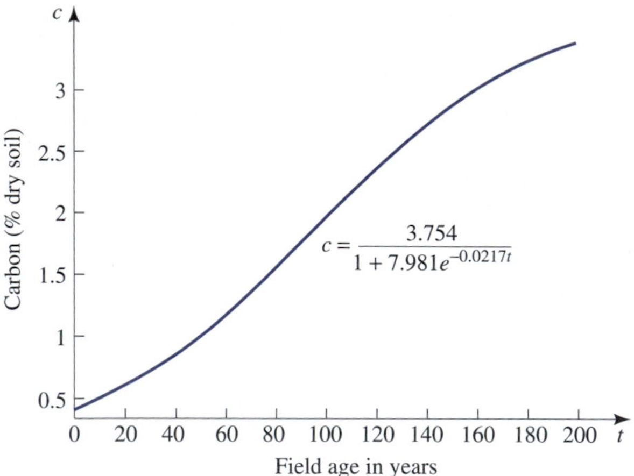

Figure 5.59

Using the rules of limits, we have

$$\begin{aligned}\lim_{t\to\infty} C(t) &= \lim_{t\to\infty} \frac{3.754}{1 + 7.981e^{-0.0217t}} \\ &= \frac{3.754}{1 + 7.981 \lim_{t\to\infty} e^{-0.03217t}} \\ &= \frac{3.754}{1 + 7.981(0)} \\ &= 3.754\end{aligned}$$

So the model predicts that the percentage of carbon in the soil will tend toward 3.754 in time. ■

CONNECTION
Importance of Forests in Storing CO_2

Forests also can play important roles in storing CO_2 and thus removing it from the atmosphere. By one estimate,[100] the entire CO_2 content of the atmosphere passes through the terrestrial vegetation every 7 years, about 70% of the entire exchange occurring through forest ecosystems.

[100] Paul Schroeder. 1991. Can intensive management increase carbon storage in forests? *Environ. Manage.* 15(4):475–481.

EXAMPLE 3 Spread of Technological Innovation

Let $P(t)$ denote the fraction of firms in an industry that are currently using a technological innovation. Assume that $P(t)$ satisfies a logistic curve model,

$$P(t) = \frac{L}{1 + ae^{-kt}}$$

and that this innovation will eventually spread throughout the entire industry. Initially, 10% of the firms used the innovation; two years later, 20% did. Find the logistic equation that $P(t)$ satisfies.

Solution

Since eventually the entire industry uses the innovation, we must have $L = 1$. Also we have $P(0) = P_0 = 0.10$. Thus,

$$\begin{aligned}0.1 &= P(0) \\ &= \frac{L}{1 + ae^0} \\ &= \frac{L}{1 + a}\end{aligned}$$

Solving for a gives

$$a = \frac{L}{P_0} - 1 = \frac{L}{0.1L} - 1 = 9$$

Thus,

$$P(t) = \frac{L}{1 + ae^{-kt}} = \frac{1}{1 + 9e^{-kt}}$$

Also $P(2) = 0.20$, and this gives

$$\begin{aligned} 0.20 = P(2) &= \frac{1}{1 + 9e^{-2k}} \\ 1 + 9e^{-2k} &= 5 \\ \frac{4}{9} &= e^{-2k} \\ \ln \frac{4}{9} &= \ln e^{-2k} = -2k \\ k &= \frac{1}{2} \ln \frac{9}{4} \approx 0.4055 \end{aligned}$$

Thus,

$$P(t) = \frac{1}{1 + 9e^{-0.4055t}}$$

■

We summarized some important facts about logistic curves in Figure 5.58. First comes the initial exponential growth explosion with both f and f' increasing and the curve concave up. Then comes the inflection point which indicates the point at which the concavity changes. Finally comes the leveling off phase with f' now decreasing and the curve concave down.

Remark Sometimes you will see the equation

$$P(t) = \frac{L}{1 + e^{b-kt}}$$

Since $e^{b-kt} = e^b e^{-kt}$, we can see that this is just the logistic equation

$$P(t) = \frac{L}{1 + ae^{-kt}}$$

with $a = e^b$.

Exercise Set 5.7

Applications and Mathematical Modeling

1. **Spread of Technology** When a new useful technology is introduced into a particular industry, it naturally spreads through this industry. A certain model (see Mansfield[101]) of this effect indicates that if t is the number of years after the introduction of the new technology, the percentage of the firms in this industry using the new technology is given by

$$p(t) = \frac{100}{1 + 2e^{-0.15t}}$$

Graph. To what extent does this new technology spread throughout this industry in the long term?

2. **Banking** Keeley and Zimmerman[102] used the model

$$P(t) = \frac{.21}{1 + e^{2.63-1.80t}}$$

to describe the growth in the percentage of deposits in money market accounts, $P(t)$, in time t in years. Graph using a window of dimensions [0, 5] by [0, .25]. Does this look like a logistic curve? Graphically find the inflection point.

3. **Biology** Lescano and coworkers[103] showed that the average mortality y at pupation of the Tephritid fruit fly was approximated by

$$y = \frac{e^{3.908-0.064t}}{1 + e^{3.908-0.064t}}$$

where t is the age at irradiation. Graph on your grapher. Graphically find the value of t at which the concavity changes.

4. **Biology** Roltsch and Mayse[104] showed that the cumulative percent emergence y of the moth *H. brillians* was approximated by

$$y = \frac{100}{1 + e^{4.9376-0.02351x}}$$

where x is the cumulative degree days. Graph on your grapher using a window with dimensions [0, 500] by [0, 100]. Graphically find the value of x at which the concavity changes.

5. **Soil Nitrogen Dynamics** Knopps and Tilman[105] studied the soil nitrogen dynamics of abandoned agricultural fields and created a mathematical model of soil nitrogen given by the equation

$$N(t) = \frac{0.172}{1 + 2.979e^{-0.0223t}}$$

where t is the number of years after an agricultural field has been abandoned and $N(t)$ is the percent of nitrogen in the soil. Graph and discuss what this model is saying.

6. **Carbon Nitrogen Dynamics** In the study given in Exercise 5 the researchers asked how many years would it take to bring the soil nitrogen percentage back to 95% of the original, estimated to be 0.172%. Answer this question.

7. **Citrus Growth** Yang and colleagues[106] created a mathematical model of citrus growth given by the equation

$$y = \frac{146.3346}{1 + e^{4.389115-0.023039t}}$$

where y is the fruit surface area in square centimeters and t is time in days, where $t = 1$ corresponds to January 1. Graph and discuss what is happening.

8. **Citrus Damage** Yang and colleagues[107] created a mathematical model of citrus damage by mites given by the equation

$$y = \frac{100}{1 + e^{7.230067-0.010659t}}$$

where y is the cumulative percentage fruit drop and t is time in days where $t = 1$ corresponds to January 1. Graph and discuss what is happening.

---

[101] E. Mansfield. 1961. Technical change and the rate of imitation. *Econometrica* 29(4):741–766.

[102] Michael C. Keeley and Gary C. Zimmerman. 1985. Competition for money market deposit accounts. *Econ. Rev. Fed. Res. Bank San Francisco* Spring(2):5-27.

[103] H. G. Lescano, B. C. Congdon, and N. W. Heather. 1994. *J. Econ. Entomol.* 87(5):1256–1261.

[104] William Roltsch and Mark Mayse. 1993. Simulation phenology model for the western grapeleaf skeletonizer. *Environ. Entomol.* 22(3):577–586.

[105] Knopps and Tilman. 2000. Dynamics of soil nitrogen and carbon accumulation for 61 years after agricultural abandonment. *Ecology* 81(1):88–98.

[106] Y. Yang, J. C. Allen, J. L. Knapp, and P. A. Stanley. 1995. Relationship between population density of citrus rust mite and damage to 'Hamlin' orange fruit. *Environ. Entomol.* 24(5):1024–1031.

[107] Y. Yang, J. C. Allen, J. L. Knapp, and P. A. Stanley. 1994. Citrus rust mite and damage effects on 'Hamlin' orange fruit growth and drop. *Environ. Entomol.* 23(2):244–247.

9. Population Ecology Fantinou and associates[108] studied the development of the larvae of the corn stalk borer under two regimens. (See the paper for details.) We consider the second regimen in this exercise. The first regimen is considered in the next exercise. They created the mathematical model given by the equation

$$y = \frac{2.75}{1 + e^{3.17 - 0.77x}}$$

where x is the successive instars (stage of life of the insect) and y is the head capsule width of the successive instars. Graph and discuss what is happening.

10. Population Ecology Fantinou and associates[109] studied the development of the larvae of the corn stalk borer under two regimens. (See the paper for details.) We consider the first regimen in this exercise. The second regimen was considered in the previous exercise. They created the mathematical model given by the equation

$$y = \frac{4.68}{1 + e^{3.11 - 0.51x}}$$

where x is the successive instars (stage of life of the insect) and y is the head capsule width of the successive instars. Graph and discuss what is happening.

11. Profits The profits from the introduction of a new product often explode at first and then level off, and thus can sometimes be modeled by a logistic equation. Suppose the annual profits $P(t)$ in millions of dollars t years after the introduction of a certain new product satisfies the logistic equation

$$P(t) = \frac{10}{1 + 20e^{-0.2t}}$$

Graph $P(t)$ on your grapher. Find the point where the concavity changes. (This is called the **point of diminishing returns.**) What is the limiting value of the annual profits?

12. Spread of Rumors In a corporation, rumor spreads of large impending layoffs. We assume that eventually everyone will hear the rumor. The percentage of the workforce that has heard the rumor satisfies a logistic equation. If initially 0.1% of the workforce hears the rumor and 12.5% of the workforce hears the rumor after the first day, what percentage will have heard the rumor after 6 days from the start of the rumor?

13. Population Studies indicate that a small state forest with no deer could hold a maximum of 100 deer. Five deer are transplanted into this forest; 1 year later the population has increased to 10. Assuming that the deer population satisfies a logistic equation, how long will it take for the deer population to reach 75?

14. Spread of Disease Jim Smith is a member of a small, completely isolated community of 100 residents. Jim leaves and returns with an infectious disease to which everyone in the community is susceptible. One day after returning, Jim gives the disease to one other resident. If the number of persons who get this disease satisfies a logistic equation, find how many of these residents will have contracted the disease 10 days after Jim returns.

[108] Argyro A. Fantinou, John A. Tsitsipis, and Michael G. Karandinos. 1996. Effects of short- and long-day photoperiods on growth and development on *Sesamia nonagrioides*. *Environ. Entomol.* 25(6):1337–1343.

[109] Ibid.

More Challenging Exercises

In Exercises 15 through 20, let

$$P(t) = \frac{L}{1 + ae^{-kt}}$$

where a, k, and L are all positive constants. Establish each statement analytically using calculus.

15. $P'(t)$ is positive for $t \geq 0$.

16. $\lim_{t \to \infty} P(t) = L$

17. $c = \frac{1}{k} \ln a$ is an inflection point.

18. $P(t)$ is concave up on $(0, c)$ and concave down on (c, ∞).

19. $P(c) = \frac{L}{2}$, where c is given in Exercise 17.

20. $a = \frac{L}{P(0)} - 1$

21. Technological Innovation Suppose all firms in an industry will eventually institute a technological innovation and the percentage of firms at time t that use the innovation is given by the logistic equation

$$P(t) = \frac{100}{1 + ae^{-kt}}$$

If initially, at time $t = 0$, the percentage of firms that use the innovation is $P_0 = P(0)$ and T years later is Q, show that

$$k = \frac{1}{T} \ln \frac{Q(1 - P_0)}{P_0(1 - Q)}$$

Modeling Using Least Squares

22. Forestry Pearce[110] gave the data found in the table for the net stumpage values for a typical stand of British Columbia Douglas fir trees. Find a model that best fits the data. Then explain what the model implies.

Age (years)	Net Stumpage Value
40	\$43
50	\$143
60	\$303
70	\$497
80	\$650
90	\$805
100	\$913
110	\$1000
120	\$1075

[110] P. Pearce. 1967. The optimal forest rotation. *Forestry Chron.* 43:178–195.

5.8 Implicit Differentiation and Related Rates

Enrichment: Implicit Differentiation
Enrichment: Applications
Enrichment: Related Rates

APPLICATION Spread of an Oil Slick

An oil slick is spreading in a circular fashion. The radius of the circular slick is observed to be increasing at 0.1 mile per day. What is the rate of increase of the area of the slick when $r = 2$ miles? See Example 4 on page 384 for the answer.

Enrichment: Implicit Differentiation

We often have an expression such as $y^6 + y^5 + xy = 0$ in which we are unable to solve for y as an explicit function of x. Nevertheless, we may still need to find $\frac{dy}{dx}$ at some specific point $x = x_0$. In this section we shall see how to find $\frac{dy}{dx}$ at a point without knowing y as an explicit function of x.

We usually have encountered functions of the form $y = f(x)$, which express y explicitly as a function of x. But we have had to deal with equations such as

$$x^2 + y^2 = 1$$

that do not give y as an explicit function of x. Still, this last equation can define y as a function of x. If we set x equal to some specific value, then the resulting equation can be solved for one or more values of y. If we specify which value of y to take, we have a function. We then say that y is an **implicit function** of x.

For example, if $x^2 + y^2 = 1$, and we set $x = 0$, then there results the equation $y^2 = 1$, which has solutions $y = -1$ and $y = +1$.

If we were interested in finding the slope to the curve $x^2 + y^2 = 1$ at the point $(1/\sqrt{2}, 1/\sqrt{2})$, we would have to solve for y as an *explicit* function of x and then find the derivative. In this case we would have

$$y = f(x) = \sqrt{1 - x^2}$$

and

$$f'(x) = -\frac{x}{\sqrt{1 - x^2}}$$

Then

$$f'(1/\sqrt{2}) = -1$$

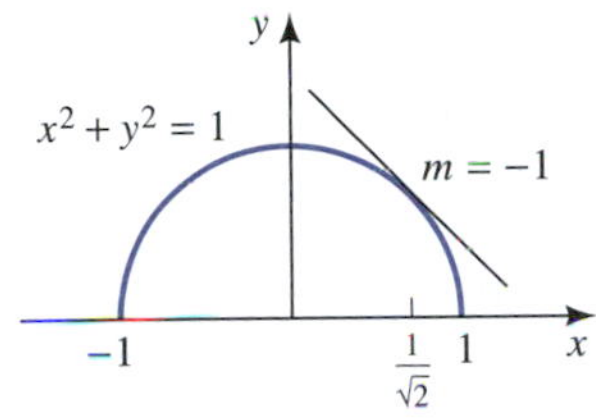

Figure 5.60
The slope of the tangent line at $x = 1/\sqrt{2}$ is -1.

See Figure 5.60.

However, there are equations such as

$$x^4 - 2xy^3 + y^5 = 32$$

for which it is not possible to solve for y explicitly in terms of x. In such cases we still need to know the slope of the curve at a given point. It might seem impossible to ever be able to find $\frac{dy}{dx}$ at a point if we do not know y as a function of x. But the technique of *implicit differentiation* will enable us to do precisely this.

Let us now see how we can find the slope of the tangent line to the curve $x^2 + y^2 = 1$ given above at the point $(1/\sqrt{2}, 1/\sqrt{2})$ without finding y as an explicit function of x. We first *assume* that it is possible at least theoretically to solve for y explicitly as, say, $y = f(x)$. Then

$$x^2 + [f(x)]^2 = x^2 + y^2 = 1$$

Thus,

$$x^2 + [f(x)]^2 = 1$$

We now differentiate this last equation using the rules we have already learned and obtain

$$\frac{d}{dx}x^2 + \frac{d}{dx}[f(x)]^2 = \frac{d}{dx}(1)$$

This gives

$$2x + 2f(x)f'(x) = 0$$

We then solve for $f'(x)$ and obtain

$$\frac{dy}{dx} = f'(x) = -\frac{x}{f(x)} = -\frac{x}{y}$$

At the point under consideration, $x = 1/\sqrt{2}$ and also $y = 1/\sqrt{2}$. Using this in the previous equation yields

$$f'(1/\sqrt{2}) = -\frac{x}{y} = -\frac{1/\sqrt{2}}{1/\sqrt{2}} = -1$$

Notice that we were able to find $f'(1/\sqrt{2})$ without ever knowing what $f(x)$ was explicitly!

EXAMPLE 1 **Finding the Derivative of an Implicitly Defined Function**

Let $x^4 - 2xy^3 + y^5 = 32$.

(a) Find $\dfrac{dy}{dx}$.

(b) Find the slope of the tangent line to the curve at $x_0 = 0$.

Solution

(a) It would be extremely difficult, if not impossible, to solve for y *explicitly* as a function of x. We then need to differentiate *implicitly*. We now assume that y is a function of x. We have

$$x^4 - 2xy^3 + y^5 = 32$$

Differentiating, using the product rule and the rule for a function raised to a power, we obtain

$$4x^3 - 2y^3 - (2x)3y^2y' + 5y^4y' = 0$$

We now solve this equation for y' and obtain

$$\left(-6xy^2 + 5y^4\right)y' = 2y^3 - 4x^3$$

and finally

$$y' = \frac{2y^3 - 4x^3}{-6xy^2 + 5y^4}$$

or

$$\frac{dy}{dx} = \frac{2y^3 - 4x^3}{-6xy^2 + 5y^4}$$

(b) We wish to find y' when $x = 0$. When $x = 0$, the original equation becomes

$$0^4 - 2(0)y^3 + y^5 = 32$$

or $y = 2$. Thus, we substitute $x = 0$ and $y = 2$ in the answer to part (a) and obtain

$$\left.\frac{dy}{dx}\right|_{x=0} = \frac{2(2)^3 - 0}{0 + 5(2)^4} = \frac{1}{5}$$ ■

Remark In the above analysis we *assumed* that we could at least theoretically solve for $y = f(x)$ for some $f(x)$. There is a theorem called the *implicit function theorem* that gives specific conditions under which this assumption is guaranteed. We do not go into these conditions in this text.

Remark Notice in the previous work that the equation involving $f'(x)$ was *linear* in f' and therefore it was easy to solve for f'. *This is always the case.*

Enrichment: Applications

EXAMPLE 2 **Finding the Derivative of an Implicitly Defined Function**

The demand function for a certain commodity is given by

$$p = \frac{13}{1 + x^2 + x^3}$$

Find the instantaneous rate of change of the number sold with respect to the price when $x = 2$.

Solution

It would be extremely difficult, if not impossible, to solve for x explicitly as a function of p to find $\frac{dx}{dp}$. Therefore, we try implicit differentiation.

We assume that x is a function p. Then

$$p = \frac{13}{1 + x^2 + x^3}$$

Clearing the fraction, we obtain

$$p + px^2 + px^3 = 13$$

Now differentiate this with respect to p, let $x' = \frac{dx}{dp}$, and obtain

$$1 + x^2 + 2pxx' + x^3 + 3px^2x' = 0$$

This is a *linear* equation in x'. Solve for x', and obtain

$$x' = -\frac{1 + x^2 + x^3}{2px + 3px^2}$$

Now when $x = 2$,

$$p = \frac{13}{1 + 2^2 + 2^3} = 1$$

Then when $x = 2$ and $p = 1$,

$$x' = -\frac{1 + 2^2 + 2^3}{2 \cdot 1 \cdot 2 + 3 \cdot 1 \cdot 2^2} = -\frac{13}{16}$$ ■

Enrichment: Related Rates

In this subsection the rate of change of one quantity is related to the rate of change of another quantity by the chain rule.

It is often the case when given a function, such as a cost function, $C = C(x)$, that the number of items sold is a function of time. For example, a company might estimate that their sales may increase on a yearly basis at the same rate as the increase in population. Thus, the number sold, x, is actually a function of the time t or $x = f(t)$.

If we now wish to find the rate of change of costs with respect to time, then we are interested in the quantity $\frac{dC}{dt}$. But since $C = C(f(t))$, the chain rule then gives

$$\frac{dC}{dt} = \frac{dC}{dx}\frac{dx}{dt}$$

EXAMPLE 3 **Finding a Related Rate**

Suppose the cost function in thousands of dollars is given by $C(x) = 100 + x^2$, where x is in thousands of units. The company estimates that its rate of change of sales with respect to time will be 0.04. What will be the rate of increase of costs with respect to time if the company produces and sells 10,000 items?

Solution

Since the rate of change of sales with respect to time is 0.40,

$$\frac{dx}{dt} = 0.04$$

Then

$$\frac{dC}{dt} = \frac{dC}{dx}\frac{dx}{dt} = 2x(0.04)$$

When $x = 10$, this yields

$$\frac{dC}{dt} = 2(10)(0.04) = 0.8$$

That is, costs are increasing at an annual rate of \$800 when 10,000 items are being produced. ■

EXAMPLE 4 **Finding a Related Rate of the Spread of an Oil Slick**

An oil slick is spreading in a circular fashion. The radius of the circular slick is observed to be increasing at 0.1 mile per day. What is the rate of increase of the area of the slick when $r = 2$ miles?

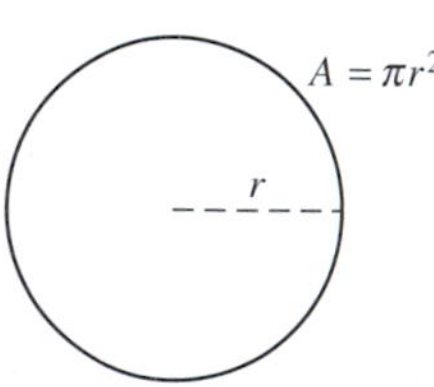

Figure 5.61
Area of a circle with radius r.

Solution

If r is the radius of the circular slick in miles, then the area, as shown in Figure 5.61, is given by $A = \pi r^2$. Also $\frac{dr}{dt} = 0.1$. Thus,

$$\frac{dA}{dt} = \frac{dA}{dr}\frac{dr}{dt} = 2\pi r(0.1)$$

When $r = 2$, this becomes

$$\frac{dA}{dt} = 2\pi 2(0.1) \approx 1.26$$

Thus, when the radius of the slick is 2 miles, the area of the slick is increasing at a rate of 1.26 square miles per day. ■

EXAMPLE 5 Change of Distance between Two Ships

Two ships leave the same port at 1:00 P.M. The first ship heads due north, and at 2:00 P.M. (one hour later) is observed to be 3 miles due north of port and going at 6 miles per hour at that instant. The second ship leaves port at the same time as the first ship, and heads due east, and 1 hour later is observed to be 4 miles due east of port and going at 7 miles per hour. At what rate is the distance between the two ships changing at 2:00 P.M.?

Solution

Let $a(t)$ and $b(t)$ be the respective distances that the first and second ships are from port at time t after leaving. See Figure 5.62.

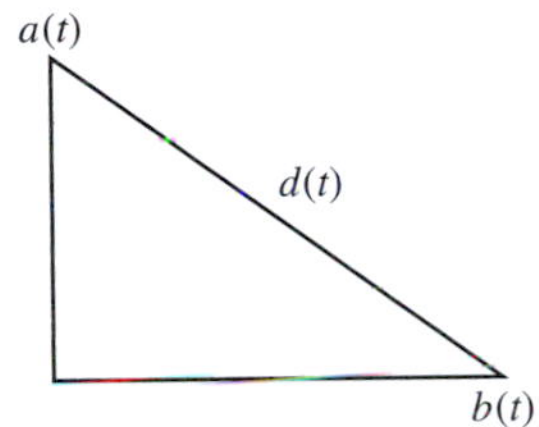

Figure 5.62
The distance between the ships is $d(t)$.

If $d(t)$ is the distance between the ships, then

$$d^2(t) = a^2(t) + b^2(t)$$

Differentiating this expression gives

$$2d(t)d'(t) = 2a(t)a'(t) + 2b(t)b'(t)$$

Thus,

$$d'(t) = \frac{a(t)a'(t) + b(t)b'(t)}{d(t)}$$

Now we are given that $a(1) = 3$, $b(1) = 4$, $a'(1) = 6$, and $b'(1) = 7$. It follows that

$$d(1) = \sqrt{3^2 + 4^2} = 5$$

Therefore,

$$d'(1) = \frac{3 \cdot 6 + 4 \cdot 7}{5} = 9.2$$

The ships are moving apart at 9.2 miles per hour at 2:00 P.M. ■

Notice how we were able to do this problem without knowing explicitly the two functions $a(t)$ and $b(t)$.

Warm Up Exercise Set 5.8

1. If $2x^2\sqrt{y} + y^3 + x^4 = 27$, find $\frac{dy}{dx}$.
2. Find the slope of the tangent line to the curve in the previous example at the point $x_0 = 0$.
3. After using an experimental drug, the radius of a spherical tumor is observed to be decreasing at the rate of 0.1 cm per week when the radius is 2 cm. At what rate is the volume of the tumor decreasing?

Exercise Set 5.8

In Exercises 1 through 14, differentiate implicitly and find the slope of the curve at the indicated point.

1. $xy + 2x + y = 6$, $(1, 2)$
2. $xy + x + y = -1$, $(-2, -1)$
3. $xy + x + y^2 = 7$, $(1, 2)$
4. $x^2y + x - y^3 = -1$, $(2, -1)$
5. $3x^2 - y^2 = 23$, $(3, 2)$
6. $x^3 - 2y^3 = -15$, $(1, 2)$

7. $x^2y - 2x + y^3 = 1$, $(2, 1)$

8. $\sqrt{x}y + x^2 - y^4 = 17$, $(4, 1)$

9. $x^2 + y^2 + xy^3 = 4$, $(0, 2)$

10. $\sqrt{x} - \sqrt{y} = 1$, $(9, 4)$

11. $x^2y^2 + xy^4 = 2$, $(1, 1)$

12. $xy^2 + xy + y^3 = 5$, $(2, 1)$

13. $x\sqrt{y} - y^2 + x^2 = -13$, $(1, 4)$

14. $\sqrt{xy} + x^3y^2 = 84$, $(1, 9)$

Applications and Mathematical Modeling with Implicit Differentiation

In Exercises 15 through 22, the price of a commodity is given as a function of the demand x. Use implicit differentiation to find $\frac{dx}{dp}$ for the indicated x.

15. $p = -2x + 15$, $x = 3$

16. $p = -3x + 20$, $x = 5$

17. $p = \frac{3}{1 + x}$, $x = 2$

18. $p = \frac{5}{1 + x^2}$, $x = 2$

19. $p = \frac{15}{3 + x + x^2}$, $x = 3$

20. $p = \frac{5}{3 + x + x^3}$, $x = 1$

21. $p = \sqrt{25 - x^2}$, $x = 4$

22. $p = \sqrt{25 - x^4}$, $x = 2$

23. Demand Suppose the demand x and the price p are related by the equation $16x^2 + 100p^2 = 500$. Find dx/dp at the point where $p = 1$.

24. Demand Suppose the demand x and the price p are related by the equation $x^2 + 100p^2 = 20{,}000$. Find dp/dx at the point where $x = 100$.

25. Medicine Poiseuille's law states that the total resistance R to blood flow in a blood vessel of constant length l and radius r is given by $R = \frac{al}{r^4}$, where a is a positive constant. Use implicit differentiation to find $\frac{dr}{dR}$, where $R = 1$.

26. Medicine It is well known that when a person coughs, the diameter of the trachea and bronchi shrinks. It has been shown that the flow of air F through the windpipe is given by $F = k(R - r)r^4$, where r is the radius of the windpipe under the pressure of the air being released from the lungs, R is the radius with no pressure, and k is a positive constant. Suppose $k = 1$ and $R = 2$. Use implicit differentiation to find $\frac{dr}{dF}$, where $F = 1$.

27. Biology Biologists have proposed that the rate of production P of photosynthesis is related to the light intensity I by the formula $P = \frac{aI}{b + I^2}$, where a and b are positive constants. Suppose that $a = 6$ and $b = 8$. Use implicit differentiation to find $\frac{dI}{dP}$, where $P = 1$ and $I = 2$.

28. Biology The free water vapor diffusion coefficient D in soft woods is given by

$$D = \frac{167.2}{P}\left(\frac{T}{273}\right)^{1.75}$$

where T is the temperature and P is the constant atmospheric pressure measured in appropriate units. Suppose $P = 1$. Use implicit differentiation to find $\frac{dT}{dD}$ when $D = 1$.

Exercises 29 through 40 involve related rates. In these exercises find $\frac{dy}{dt}$ given the indicated information.

29. $y = 3 + 2x$, $\frac{dx}{dt} = 3$.

30. $y = 3x - 4$, $\frac{dx}{dt} = 2$.

31. $y = 2 + x^2$, $\frac{dx}{dt} = 4$, $x = 2$.

32. $y = 3 - 2x^2$, $\frac{dx}{dt} = 2$, $x = 3$.

33. $y = 3 - 2x^3$, $\frac{dx}{dt} = -2$, $x = 1$.

34. $y = 1 - x^4$, $\frac{dx}{dt} = -1$, $x = 2$.

35. $y = \frac{1 - x}{1 + x}$, $\frac{dx}{dt} = 2$, $x = 1$.

36. $y = \frac{1 - x}{1 + x^2}$, $\frac{dx}{dt} = -2$, $x = -1$.

37. $x^2 + y^2 = 5$, $\frac{dx}{dt} = 2$, $x = 1$, $y = 2$.

38. $x^2 - y^2 = 3$, $\frac{dx}{dt} = -2$, $x = 2$, $y = -1$.

39. $x^3 + xy + y^2 = 7$, $\frac{dx}{dt} = 2$, $x = 1$, $y = 2$.

40. $x + x^2y - x^2 = -1$, $\frac{dx}{dt} = 3$, $x = -1$, $y = 1$.

Applications and Mathematical Modeling with Related Rates

41. **Demand** If the demand equation is given by $x = 10 - 0.1p$ and the number of items manufactured (and sold) is increasing at the rate of 20 per week, find the rate of change of p with respect to time.

42. **Cost** If the cost equation is given by $C(x) = 8 + 0.2x^2$ and the number of items manufactured is increasing at the rate of 20 per week, find the rate of change of C with respect to time when $x = 4$.

43. **Revenue** If the demand equation is given by $x = 10 - 0.1p$ and the number of items manufactured (and sold) is increasing at the rate of 20 per week, find the rate of change of revenue with respect to time when $x = 4$.

44. **Profits** If the demand and cost equations are the same as in the two previous problems, find the rate of change of profits with respect to time.

45. **Distance** A ship is observed to be 4 miles due north of port and traveling due north at five miles per hour. At the same time another ship is observed to be 3 miles due west of port and traveling due *east* on its way back to port at 4 miles per hour. What is the rate at which the distance between the ships is changing?

46. **Distance** Answer the same question as in the previous problem if the first ship is traveling due *south* at 5 miles per hour, everything else remaining the same.

47. **Physics** A 10-foot ladder leans against a wall and slides down, with the foot of the ladder observed to be moving away from the wall at 4 ft/sec when it is 6 feet from the wall. At what speed is the top of the ladder moving downward?

48. **Medicine** After using an experimental drug, the volume of a spherical tumor is observed to be decreasing at the rate of 2 cubic cm per month when the radius is 5 cm. At what rate is the radius of the tumor decreasing?

49. **Physics** A water tank in the shape of a circular cone (see figure) has a radius of 4 yards and a height of 10 yards. If water is being poured into the tank at the rate of 5 cubic yards per minute, find the rate at which the water level is rising when the water level is at 2 yards. (Hint: $V = \frac{1}{3}\pi r^2 h$.) Use similar triangles to find r as a function of h and substitute this into the last equation. Then differentiate the latter expression.

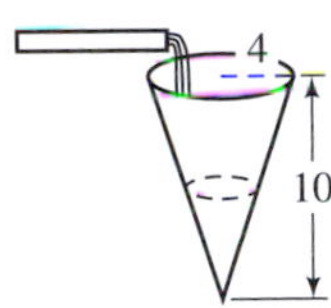

50. **Geometry** If the volume of a cube is increasing at 6 ft^3/sec, what is the rate at which the sides are increasing when the sides are 10 ft long?

51. **Geometry** If the sides of a square are increasing at 2 ft/sec, what is the rate at which the area is changing when the sides are 10 ft long?

More Challenging Exercises

52. In Section 4.1 we indicated how to use the limit definition of derivative to find the derivative of x^n, where n is a positive integer, and said that a proof for general n was somewhat difficult. You can now construct a proof in the case that r is a rational number. Let $y = f(x) = x^{p/q}$ where p and q are positive integers. Then $y^q = x^p$. Differentiate and show that

$$\frac{dy}{dx} = \frac{p}{q}x^{(p/q)-1}$$

53. Suppose two functions $f(x)$ and $g(x)$ satisfy $f(g(x)) = x$ and are both differentiable. An example of two such functions is e^x and $\ln x$. If $y_0 = g(c)$ and $g'(c) \neq 0$, show that

$$f'(y_0) = \frac{1}{g'(c)}$$

54. Verify the formula in Exercise 53 if $f(x) = x^2$ and $g(x) = \sqrt{x}$. Take $c = 2$.

55. Verify the formula in Exercise 53 if $g(x) = x^3$ and $f(x) = \sqrt[3]{x}$. Take $c = 2$.

Solutions to WARM UP EXERCISE SET 5.8

1. Assume that we have $y = f(x)$; then

$$2x^2\sqrt{f(x)} + [f(x)]^3 + x^4 = 27$$

Now differentiate the above expression and obtain

$$4x[f(x)]^{1/2} + x^2[f(x)]^{-1/2}f'(x) + 3[f(x)]^2 f'(x) + 4x^3 = 0$$

Solving for $f'(x)$ gives

$$f'(x) = -\frac{4x[f(x)]^{1/2} + 4x^3}{x^2[f(x)]^{-1/2} + 3[f(x)]^2} = -4\frac{x\sqrt{y} + x^3}{\frac{x^2}{\sqrt{y}} + 3y^2} = -4\frac{xy + x^3\sqrt{y}}{x^2 + 3y^{5/2}}$$

2. Setting $x_0 = 0$ in the original equation gives

$$0 + y^3 + 0 = 27$$

or $y = 3$. Substituting $x = 0$ and $y = 3$ in the answer to the previous exercise yields

$$f'(0) = -4\frac{(0)(3) + (0)(\sqrt{3})}{0 + (3)3^{5/2}} = 0$$

3. The relationship between the volume V in cubic centimeters and the radius r in centimeters of the tumor is given by $V = \frac{4}{3}\pi r^3$. Thus,

$$\frac{dV}{dt} = 4\pi r^2 \frac{dr}{dt}$$

where t is measured in weeks. We are given that $\frac{dr}{dt} = -0.1$ when $r = 2$. Putting this information into the last displayed line gives

$$\frac{dV}{dt} = 4\pi(2)^2(-0.1) \approx -5.03$$

cubic centimeters per week.

Summary Outline

- The number c is a **critical value** for the function $f(x)$ if $f'(c) = 0$ or if $f'(c)$ does not exist but $f(c)$ is defined. p. 297.
- $f(x_M)$ is a **relative maximum** for the function $f(x)$ if $f(x) \le f(x_M)$ for all x on some interval (a, b) that contains x_M and $f(x_m)$ is a **relative minimum** if $f(x) \ge f(x_m)$ for all x on some interval (a, b) that contains x_m. p. 300.
- **First Derivative Test** The sign charts for f' on either side of a critical value c have the following consequences: $+-$ implies a relative maximum, $-+$ implies a relative minimum, and $++$ and $--$ both indicate no extremum. p. 301.
- **Constant sign theorem** If $f(x)$ is continuous on (a, b) and $f(x) \ne 0$ on (a, b), then $f(x)$ is either always positive or always negative on (a, b). p. 298.
- The **second derivative** $f''(x)$ of $y = f(x)$ is the derivative of the first derivative, that is, $f''(x) = \frac{d}{dx}[f'(x)]$. We also use the notation $f''(x) = \frac{d^2y}{dx^2}$. p. 314.
- The graph of a function $f(x)$ is **concave up** $\smile$ if the derivative $f'(x)$ is an increasing function. This will happen if $f''(x) > 0$. The graph of a function $f(x)$ is **concave down** $\frown$ if the derivative $f'(x)$ is a decreasing function. This will happen if $f''(x) < 0$. p. 316.
- An **inflection point** is a point on the graph of f where the concavity changes sign. p. 316.
- **Second Derivative Test** Let $f'(c) = 0$. If $f''(c) > 0$, then f assumes a relative minimum at c. If $f''(c) < 0$, then f assumes a relative maximum at c. p. 321.
- If the graph of a function $y = f(x)$ approaches the horizontal line $y = a$, then the line $y = a$ is a **horizontal asymptote**. If the graph approaches the vertical line $x = b$, then the line $x = b$ is a **vertical asymptote**. p. 331.
- $f(x_M)$ is the **absolute maximum** of the function $f(x)$ on the interval I if $f(x) \le f(x_M)$ for all $x \in I$, and an **absolute minimum** on the interval I if $f(x) \ge f(x_m)$ for all $x \in I$. p. 351.
- **Extreme Value Theorem** A continuous function on a closed and bounded interval attains it maximum and its minimum. p. 351.
- A **logistic equation** is any equation of the form $P(t) = \frac{L}{1 + ae^{-kt}}$. p. 373.

Review Exercises

1. For the function whose graph is shown in the accompanying figure, find all critical values and the largest open intervals on which the function is increasing and the largest open intervals on which the function is decreasing.

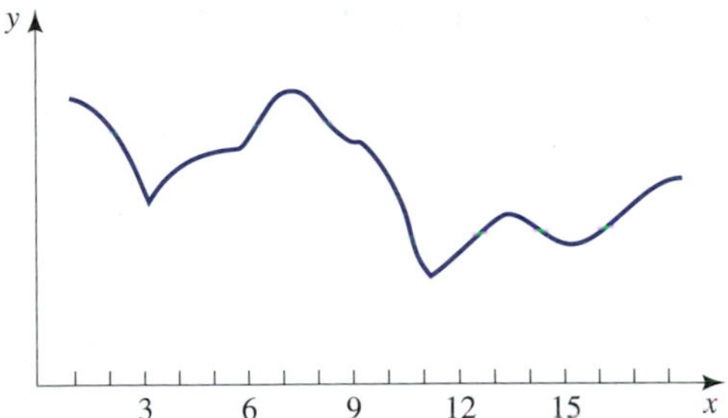

2. Suppose the derivative of a function $y = f(x)$ is given by $f'(x) = (x + 2)(x - 1)$. Find all the largest open intervals on which the function is increasing, the largest open intervals on which the function is decreasing, all critical values, and all relative extrema.

3. Repeat Exercise 2 if $f'(x) = (x + 4)(x - 2)^2(x - 5)$.

In Exercises 4 through 21, find the intervals on which the function is increasing, the intervals on which the function is decreasing, the critical values, the intervals on which the graph of the function is concave up, the intervals on which the graph of the function is concave down, the inflection points, the vertical asymptotes, the horizontal asymptotes; then sketch a graph.

4. $y = f(x) = 2x^2 + 3x - 4$
5. $y = f(x) = -x^2 + 2x + 5$
6. $y = f(x) = -3x^2 - 6x + 1$
7. $y = f(x) = 3x^2 + 12x - 10$
8. $y = f(x) = \frac{1}{3}x^3 - x + 1$
9. $y = f(x) = -x^3 + 3x + 2$
10. $y = f(x) = x^4 - 4x^3 + 4$
11. $y = f(x) = x^5 - 15x^4 + 5$
12. $y = f(x) = x^4 - 18x^2 + 15$
13. $y = f(x) = -x^5 + 15x^3 - 10$
14. $y = f(x) = x + \frac{4}{x^2}$
15. $y = f(x) = x^2 + \frac{1}{x^2}$
16. $y = f(x) = x - \frac{1}{x^2}$
17. $y = f(x) = x^2 - \frac{27}{x^2}$
18. $y = f(x) = \frac{x}{x - 5}$
19. $y = f(x) = \frac{x - 1}{x - 5}$
20. $y = f(x) = \sqrt{x^6 + 1}$
21. $y = f(x) = (9 - x)^{1/5}$
22. $y = e^{-x^2-1}$
23. $y = 1 - e^{-x^2}$
24. $\ln(x^2 + 1)$
25. $\ln(1 + e^x)$

In Exercises 26 through 35, find the location of all absolute maximums and absolute minima on the given intervals.

26. $f(x) = \frac{x}{3 + x^4}$, $(-\infty, \infty)$
27. $f(x) = 1 - \frac{x}{48 + x^4}$, $(-\infty, \infty)$
28. $f(x) = x - \frac{1}{x}$, $[1, \infty)$
29. $f(x) = \frac{1}{x^3} - x$, $[1, \infty)$
30. $f(x) = 2x^3 - 3x^2 + 1$, $[-1, 2]$
31. $f(x) = 2x^3 - 9x^2 + 12x + 1$, $[0, 3]$
32. $f(x) = (x - 2)^2$, $[0, 3]$
33. $f(x) = x + \frac{4}{x^2}$, $[1, 3]$
34. $f(x) = x^2 + \frac{16}{x^2}$, $[1, 3]$
35. $f(x) = x - \frac{1}{x^2}$, $[1, 4]$

In Exercises 36 through 39, find $\frac{dy}{dx}$ at any point (x, y), and then find the slope of the tangent line at the given point.

36. $x^3y^4 - xy = 6$, $(2, 1)$
37. $\sqrt{x} + x\sqrt{y} + y^2 = 7$, $(4, 1)$
38. $\frac{x + y}{x - y} = 3y^2$, $(2, 1)$
39. $\sqrt[3]{y^2 + 7} - x^2y^3 = 2$, $(0, 1)$
40. Show that the graph of the special fourth-order polynomial function $f(x) = ax^4 + bx^2 + cx + d$ is always concave up if both a and b are positive and always concave down if both a and b are negative.
41. If $b > a$ and c is any constant, show that the polynomial function $f(x) = -2x^3 + 3(a + b)x^2 - 6abx + c$ has exactly one relative minimum and exactly one relative maximum. Locate and identify them.

42. Suppose $f''(x) > 0$ everywhere. Show that $y = g(x) = e^{f(x)}$ is concave up everywhere.

43. Suppose that $f(x) > 0$ and $f''(x) < 0$ everywhere. Show that $y = g(x) = \ln f(x)$ is concave down.

44. Average Cost An electric utility company estimates that the cost, $C(x)$, of producing x kilowatts of electricity is given by $C(x) = a + bx^2$, where a represents the fixed costs and b is the unit cost. The *average* cost, $\bar{C}(x)$, is defined to be the total cost of producing x kilowatts divided by x, that is, $\bar{C}(x) = \dfrac{C(x)}{x}$. Find where the average cost increases and where it decreases.

45. Average Cost The total cost function for producing x items is given by $C(x) = 1000 + 10x^2$. How many items should be produced to minimize $\bar{C}(x) = C(x)/x$, the average cost per item?

46. Average Cost In the previous problem, show that average cost is minimized at the value at which $\bar{C} = C'$.

47. Average Cost Suppose the total cost function for producing x items is given by $C(x)$ and the average cost per item is given by $\bar{C}(x) = C(x)/x$. If average cost is at a minimum at some value on $(0, \infty)$, then show that at this value $\bar{C} = C'$.

48. Biology In the attack equation the rate of change R of the number attacked with respect to the number N vulnerable to attack is given by $R(x) = x^{1-a}(kx - N)$, where x is the number of attackers, a and k are positive constants with $a < 1$. Find the number of attackers that minimizes R.

49. Average Cost An electric utility company estimates that the cost, $C(x)$, of producing x kilowatts of electricity is given by $C(x) = a + bx^3$, where a represents the fixed costs and b is a positive constant. The *average* cost, $\bar{C}(x)$, is defined to be the total cost of producing x kilowatts divided by x, that is, $\bar{C}(x) = \dfrac{C(x)}{x}$. Find where the average cost attains a minimum.

50. Cost An individual is planning to construct two fenced enclosures, one square and one an equilateral triangle. The cost per yard of constructing the triangle is twice that of the square enclosure. If the total cost is fixed at T dollars, what should be the dimensions of the two figures in order to maximize the area enclosed by the two figures? (The area of an equilateral triangle is $\sqrt{3}r^2/4$, where r is the length of a side.)

51. Revenue A bus company charged \$60 per person for a sightseeing trip and obtained 40 people for the trip. The company has data that indicates that for the same trip, each \$2 increase in the price above \$60 results in the loss of one customer. What should the company charge to maximize revenue?

52. Profits The bus company in the previous exercise has fixed costs of \$3000 for each trip and in addition there are costs of \$4 per customer. What should the company charge to maximize profits?

53. Packaging A manufacturer wishes to make a cylindrical can of 108 cubic inches. The top and bottom will cost twice as much per square inch as the sides. What should the dimensions of the can be to minimize the cost?

54. Inventory Control A company produces 1000 cases of perfume a year. The start-up cost of each production run is \$1000, it costs \$50 to manufacture each case, and \$1 to store each case for a year. Determine the number of cases to manufacture in each run and the number of runs to minimize the total cost.

55. Cost Assume that a firm has a cost function $C = b + ax$, where x is the number of items made and sold. Suppose $p = p(x)$, where p is the price of each item and assume $p'(x) < 0$. Then profit is given by $P(x) = xp(x) - b - ax$. By taking $P'(x)$ and setting this equal to zero, show mathematically that at equilibrium the adage "relatively high-cost firms produce relatively little" is true. (See Kimmel.[111])

56. Population An isolated lake was stocked with 50 lake trout. Two years later it was estimated that there were 100 of these trout in this lake. It is estimated that this lake can support 500 of these trout. Assuming that the trout population satisfies a logistic equation, find how many lake trout will be in this lake in 4 years.

57. Physics A 30-foot-ladder leans against a wall and slides down with the top of the ladder observed to be moving down the wall at 4 ft/sec when 6 feet above the ground. At what speed is the base of the ladder moving away from the wall?

58. Forest Fires McAlpine and Wakimoto[112] showed that $R = f(t) = 70.13e^{-2.1645/t}$, where R is head fire rate of spread at time t and t is time from ignition. Find $\lim_{t\to 0^+} f(t)$ and $\lim_{t\to\infty} f(t)$. Also find where $f(t)$ is increasing. Graph. Does the graph make sense?

59. Biology Boyce and coworkers[113] showed that the needle temperature $T(u)$ in degree Centigrade of red spruce was approximated by $T(u) = T_{air} + 0.0061e^{-2.0087u}$, where T_{air} is air temperature and u is wind velocity in meters per second. Find where $T(u)$ is decreasing. Does this agree with your own experience, at least in cold weather?

---

111 Sheldon Kimmel. 1992. Effects of cost changes on oligopolists' profits. *J. Indust. Econ.* 40(4):442–449.

112 R. S. McAlpine and R. H. Wakimoto. 1991. The acceleration of fire from point source to equilibrium spread. *Forest Sci.* 37(5):1314–1337.

113 Richard L. Boyce, Andrew J. Friedland, Elizabeth T. Webb, and Graham T. Herrick. 1991. Modeling the effect of winter climate on high-elevation red spruce shoot water contents. *Forest Sci.* 37(6):1567–1580.

60. Biology Biologists have used the mathematical model $Q(t) = at + be^{-kt}$, where $Q(t)$ is the CO_2 evolved at time t from microbial biomass in the soil. (See Laudelout.[114]) What condition must the constants a, b, and k satisfy so that $Q(t)$ is always an increasing function?

61. Environmental Pollution McCune[115] estimated that the foliar chlorine content in soybeans as a percent of dry matter was approximated by the cubic polynomial $y = f(x) = 0.059 + 0.073x + 0.004x^2 - 0.0004x^3$, where x is the dose of salts in kilograms per hectare per week. Determine the approximate absolute maximum of this function for $0 \le x \le 20$.

62. Sales versus Advertising Chang and Kinnucan[116] studied the economics of butter advertising in Canada. They found that the sales S per pound per person per quarter was approximated by the equation $S(a) = 2.6a^{0.0228}$, where a is the advertising expenditure in cents per person per quarter. Graph this function, determine the concavity, and interpret in economic terms what the concavity implies.

63. Money Market Mutual Fund Expenses Cook and Duffield[117] studied the average cost of money market mutual funds as a function of their total assets. Shown is the graph of average total costs. What is the concavity of this curve and what is the economic interpretation?

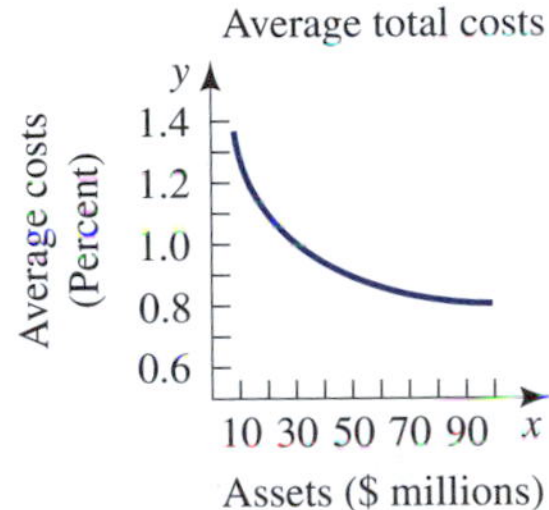

64. Household Distribution of Earnings Parker[118] determined that for two-earner households the earnings E of the greatest earner in the household was related to the earnings e (both in pounds Sterling) of the secondary worker's earnings by the cubic polynomial $E(e) = -9.551 + 0.813e - 0.001e^2 + 0.0000006e^3$ on the interval [0, 1243]. (The paper has an error in the coefficient of e^3.) Determine where this curve is concave up and concave down. Give an economic interpretation of the concavity.

In Exercise 65, match the equation with the given graphs in (i) through (vi).

65. a. $y = 2 - e^{-t}$ **b.** $y = e^{-t}$
c. $y = 1/(1 + e^{-t})$ **d.** $y = t^{0.75}$
e. $y = 1/t^2$ **f.** $y = e^t$

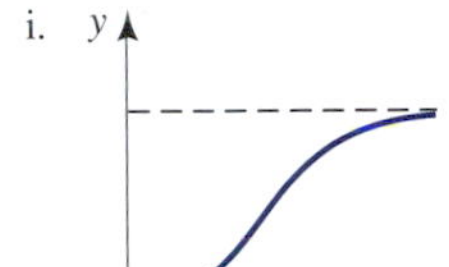

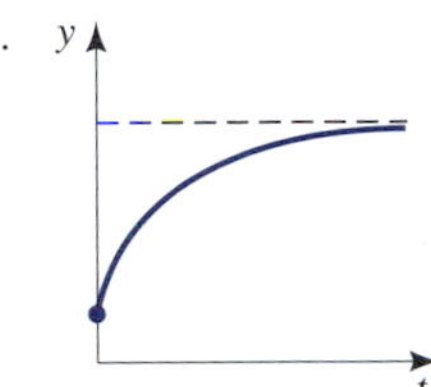

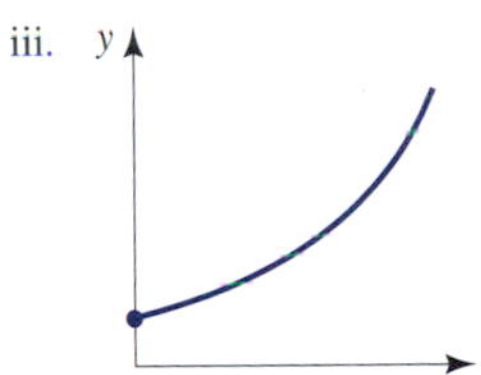

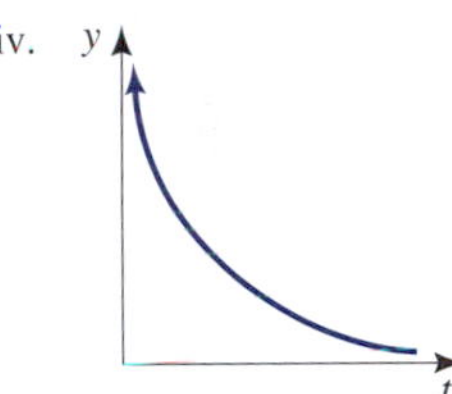

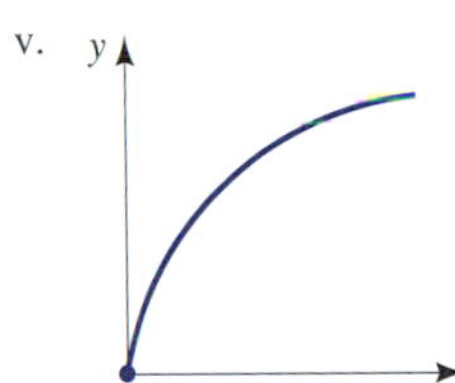

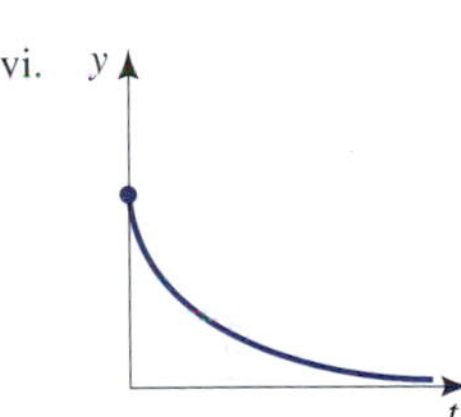

In Exercises 66 through 71, find the limits if they exist.

66. $\lim_{x\to\infty} \dfrac{2x^3 + x - 3}{x^3 + 3}$

67. $\lim_{x\to-\infty} \dfrac{x^3 + x + 2}{x^4 + 1}$

68. $\lim_{x\to\infty} \dfrac{x^4 + x^2 + 1}{x^3 + 1}$

69. $\lim_{x\to\infty} \dfrac{1}{1 + e^x}$

70. $\lim_{x\to\infty} (1 + e^{-2x})$

71. $\lim_{x\to\infty} \dfrac{1}{1 + \ln x}$

72. Diminishing Returns Suppose a firm believes that the percent, $p(t)$, of the market that it can attain of its product by spending a certain amount on advertising each year for the next ten years is given by

$$p(t) = 0.1\frac{t^4 + 125}{t^4 + 135}$$

where t is given in years. Find where the graph of the function is concave up and where it is concave down. Find the inflection point. Where is the point of diminishing returns?

---

[114] H. Laudelout. 1993. Chemical and microbiological effects of soil liming in a broad-leaved forest ecosystem. *Forest Ecol. Manage.* 61(3):247–261.

[115] D. C. McCune. 1991. Effects of airborne saline particles on vegetation in relation to variables of exposure and other factors. *Environ. Pollution* 74:176–203.

[116] Hui-Shung Chang and Henry Kinnucan. 1990. Generic advertising of butter in Canada: optimal advertising levels and returns to producers. *Agribusiness* 6(4):345–354.

[117] Timothy Q. Cook and Jeremy G. Duffield. 1979. Average costs of money market mutual funds. *Econ. Rev. Fed. Res. Bank Richmond* 65:32–39.

[118] Simon Charles Parker. 1995. The structure of the earnings distribution: how are households and individuals related? *Bull. Econ. Res.* 47(2):127–139.

Chapter 5 Cases

CASE 1 A Problem in Price Discrimination

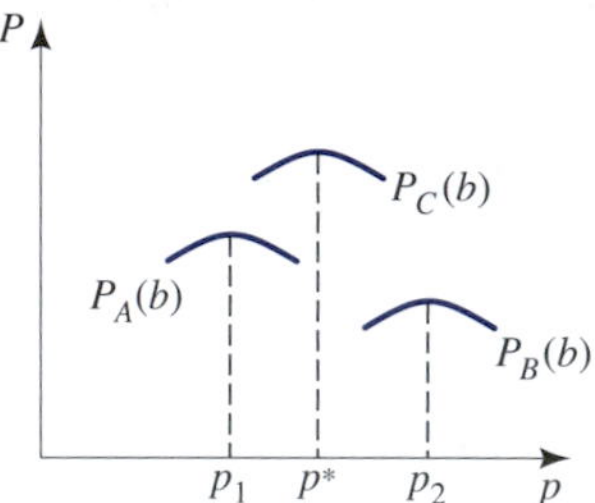

In this project we consider a problem in price discrimination that economists have consider over the years.

In the type of price discrimination we consider here we assume that the seller can divide customers into two groups, each of which has its own demand function.

Suppose that in Market A the demand equation is given by $x = f(p)$ and in Market B the demand equation is given by $x = g(p)$, where x is a measure of the units of the commodity and p is the price in dollars of each unit. The seller wants to ensure that the last unit of output sold in Market A adds the same amount to total revenue as the last unit sold in Market B. To accomplish this, we must sum the two demand curves to obtain the demand curve for the combined Market C. Thus the combined demand equation is $x = h(p) = f(p) + g(p)$.

The cost function in all markets is assumed to be $C(x) = cx$. Then, for example, the profit function in Market A is

$$P_A(p) = px - cx = pf(p) - cf(p).$$

Then the profit functions in Market B and C are

$$P_B(p) = pg(p) - cg(p), \qquad P_C(p) = ph(p) - ch(p)$$

respectively.

(a) A basic result that remained unquestioned for many years in the economic literature is that when given the two groups of buyers mentioned above, discrimination raises the price for one group and lowers it for the other. Establish this result mathematically when the profit curves are all concave down. See the figure. Notice that if the seller discriminates, then the seller will simply set the price p_1 in Market A to maximize $P_A(p)$ and will set the price p_2 in Market B to maximize $P_B(p)$. However, if the seller does not discriminate and sets the same price in both markets, then establish the result by showing that the price p^* that maximizes $P_C(p)$ must be between p_1 and p_2 when all the profit curves are concave down.

(b) This part of the project requires a graphing calculator or computer.

It has been recently reported by Nahata[119] that the above situation need not be true if the profit functions are no longer assumed to be concave down.

The following counterexample was given:

$$f(p) = -0.25p^3 + 2.0001p^2 - 5.5p + 10$$
$$g(p) = -0.2561p^3 + 2.7p^2 - 9.5p + 12$$

Show in this case that $p^* > p_1 > p_2$ by using your graphing calculator or a computer to find the (single) zero of each of the marginal profit functions.

[119] Babu Nahuta, Krzysztof Ostaszewski, and P. K. Sahoo. 1990. Direction of price changes in third-degree price discrimination. *Amer. Econ. Rev.* 80:1254–1258.

CASE 2 Drug Treatment of Irregular Heartbeat

Ventricular arrhythmia (irregular heartbeat) are frequently treated by the drug *lidocaine*. (See Cullen.[120]) To be effective in the treatment of arrhythmias, the drug must have a concentration above 1.5 mg/liter in the bloodstream. However, a concentration above 6 mg/liter in the bloodstream can produce serious side effects and even death.

[120] M. R. Cullen. 1985. *Linear Models in Biology*. New York: John Wiley & Sons.

Let t be the time in minutes from injection of the drug lidocaine, and let $c(t)$ be the concentration at time t in the bloodstream.

1. If a steady infusion rate of 2.52 mg/liter of the drug into the bloodstream is maintained, then the approximate concentration is given by the formula

$$c(t) = 3.5 - 0.5095e^{-0.12043t} - 2.9905e^{-0.0075t}$$

(a) Use calculus to show that the concentration is always increasing.

(b) Use calculus to find $\lim_{t\to\infty} c(t)$.

(c) Verify parts (a) and (b) by graphing on your graphic calculator.

(d) Determine the approximate time for which the concentration reaches the effective concentration of 1.5 mg/liter.

2. If in addition to the infusion rate of 2.5 mg/liter there is a load dose of 100 mg, then the concentration is given by

$$c(t) = 3.5 + 1.92513e^{-0.12043t} - 2.0918e^{-0.00757t}$$

(a) Graph this function on your graphic calculator. Does the concentration stay above the minimum therapeutic level and at a safe level?

(b) Graph $y = c'(t)$ on your calculator, and find where $y = c(t) = 0$. Use this information to find the minimum level of the concentration.

CASE 3 Tracking Radioactive Fallout in a Tropical Rain Forest

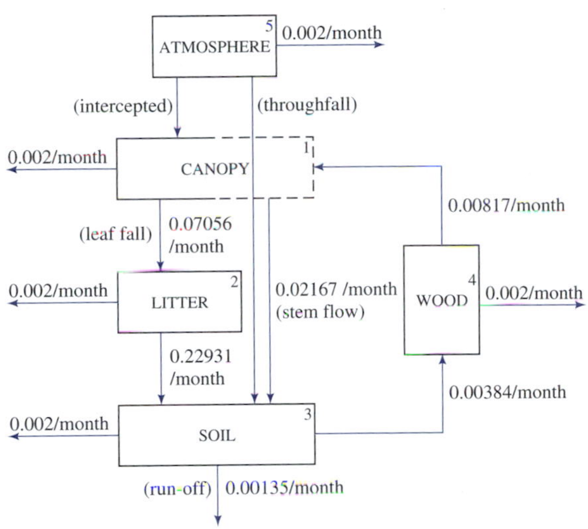

Strontium cycling in a tropical forest.

The figure shows a model for the cycling of radioactive strontium 90 (from large-scale nuclear testing) in a rain forest. The arrows pointing to the left and right indicate the losses due to natural radioactive decay. The transfers indicated by arrows pointing up or down must be estimated by researchers. For details of how this was done see Jordan and colleagues.[121] The litter consists of leaf, twig, and fruit-fall. The rainwater runoff carries some of the strontium 90 away. The canopy intercepts 16% of the radioactive strontium 90 with the remaining 84% reaching the soil.

The following four functions approximate the percentage of strontium 90 in the canopy, litter, soil, and wood, respectively, at time t measured in months:

$$\begin{aligned} p_1(t) &= -2.35(0.0011)e^{-0.2314t} + 32.65(0.0885)e^{-0.0029t} \\ &\quad + 8.69(-0.5243)e^{-0.0150t} - 20.22(1.0179)e^{-0.0936t} \\ &\quad + 100(0.2226)e^{-0.0508t} \end{aligned}$$

$$\begin{aligned} p_2(t) &= -2.35(-0.9978)e^{-0.2314t} + 32.65(0.0273)e^{-0.0029t} \\ &\quad + 8.69(-0.1711)e^{-0.0150t} - 20.22(0.5217)e^{-0.0936t} \\ &\quad + 100(0.088)e^{-0.0508t} \end{aligned}$$

$$\begin{aligned} p_3(t) &= -2.35(1.0205)e^{-0.2314t} + 32.65(1.8858)e^{-0.0029t} \\ &\quad + 8.69(6.4448)e^{-0.0150t} - 20.22(-1.6392)e^{-0.0936t} \\ &\quad + 100(-1.4836)e^{-0.0508t} \end{aligned}$$

$$\begin{aligned} p_4(t) &= -2.35(-0.0177)e^{-0.2314t} + 32.65(0.9894)e^{-0.0029t} \\ &\quad + 8.69(-5.0822)e^{-0.0150t} - 20.22(0.0754)e^{-0.0936t} \\ &\quad + 100(0.1336)e^{-0.0508t} \end{aligned}$$

(a) Graph all of these functions on your graphing calculator, and determine approximately where each have relative maxima and minima; also determine approximately what the maxima and minima are. (You will have some difficulty in determining the dimensions of an appropriate window, since you will initially have little idea of the values of t at which the maxima and minima occur, nor the values of the maxima and minima.)

(b) What can you say about all of these functions for the first couple of years? What is true about all of these functions after about 15 years?

(c) Make an attempt to use only calculus to find the maxima and minima. What difficulties do you encounter?

121 Carl F. Jordan, Jerry R. Kline, and Donald S. Sasser. 1973. A simple model of strontium and maganese dynamics in a tropical rain forest. *Health Phys.* 24:477–489.

CASE 4 The Relationship between Wing Length and Altitude

(This project requires a solver routine on a calculator or computer.) By considering the per gram hovering costs of a hummingbird and the air density dependence on elevation, Wolf and Hill[122] showed that the relationship between the length l in centimeters of the wing of a hummingbird and the altitude z in meters that the hummingbird lives is given by

$$l + 0.404l^{0.6} = 1.404(1 - 0.0065z/288)^{-2.128}$$

This equation gives l as an implicit function of z. Algebraically, we cannot solve for l. We must use some graphical or numerical method. Input this equation into your graphing calculator or computer. Take z equal to 500, 1000, 1500, and so on, and use the solver routine to determine l. Graph l as a function of z. Approximately what kind of curve do you obtain? Is $l(z)$ an increasing function? Is your answer reasonable? Why?

[122] Larry Wolf and Frank Hill. 1986. Physiological and ecological adaptations of high montane sunbirds and hummingbirds. In: F. Vuilleumier and M. Monasterio (Eds.), *High Altitude Tropical Biogeography*. New York: Oxford University Press.

6 Integration

In this second part of the study of calculus we introduce the integral and see in what sense the integral is the opposite of the derivative. The integral enables us to the find the areas between curves, which, we will see, has many concrete applications. We also discuss how to recover the cost, revenue, and profit functions when given the rates of change of these quantities; how to find the population when given a measurement of the density; how to find important economic quantities such as consumers' and producers' surplus; and even how to measure to what extent an economic system is distributing income equitably.

CASE STUDY Income Inequality

In his book *The Great Wave*, David Hackett Fischer[1] of Stanford University identified four great worldwide inflationary waves during the last 800 years, interspersed with three great deflationary waves. The four great inflationary waves were the medieval price revolution, in 1180–1350; the 16th century price revolution, in 1475–1660; the 18th century price revolution, in 1729–1820; and the 20th century price revolution, from 1896 until the present. Between these four great inflationary waves were three great deflationary waves: the equilibrium of the Renaissance, from about 1400 to 1470; the equilibrium of the Enlightenment, in 1660–1730; and the Victorian equilibrium, in 1820–1896.

[1] David Hackett Fischer. 1996. *The Great Wave (Price Revolutions and the Rhythm of History)*. New York: Oxford University Press.

Each of these great waves saw a similar pattern develop. For example, as inflation continued in the mid-13th century, money wages began to lag behind. As a consequence real wages (that is wages adjusted for inflation) fell, slowly at first and then with increasing momentum. At the same time that real wages fell, rents and interest rose sharply. Returns to landowners generally kept pace with inflation and even exceeded it. This growing gap between returns to labor and capital was typical of price revolutions in modern history. So also was its social result: a rapid growth of inequality that appeared in the later stages of every long inflationary trend.

Not surprisingly, during the later stages of these price revolutions, social upheavals went to extremes, with acts of collective violence commonplace and great misery inflicted on common people. For example,

there were often no resources available to feed convicts in jails. Starving inmates in the 14th century would ferociously attack new prisoners and devour them half alive. Condemned criminals were cut down from the gallows, butchered, and eaten.

Fortunately, in the later price revolutions, governments developed sensitivities that resulted in some social welfare programs that at least kept people from starvation.

It is important to note that Fischer contends that our very philosophy and outlook on life itself might be overwhelmingly determined by which stage of which wave we find ourselves in. Notice, for example, that the Renaissance and the Enlightenment occurred during respites from inflationary surges.

One way of objectively measuring the inequalities in a society is to measure the income distribution. For example, the table shows the income distribution in the town of Santa Maria Impruneta, Tuscany, for some selected years. How can we use this table to develop a measure of the inequality of income distribution? For the answer, see Example 6 on page 467 and the discussion preceding this example.

Year	Percent by Decile									
1307	1.7	2.7	4.2	5.8	6.2	7.0	8.8	12.3	17.6	33.7
1330	1.2	1.3	2.4	3.2	5.5	7.8	10.3	12.7	18.3	37.3
1427	0.0	0.0	0.5	1.9	3.6	7.0	8.7	11.0	18.6	48.7
	Taxable Wealth by Decile									

6.1 Antiderivatives

APPLICATION Finding Cost Given Marginal Cost

What is the cost function $C(x)$ if the marginal cost is $10e^x + 40x - 100$ and fixed costs are 1000? See Example 8 on page 405 for the answer.

Antiderivatives

Given a function $f(x) = x^2$, can we find some function $F(x)$ whose derivative is x^2? We know that in differentiating x to a power, the power is reduced by one. Thus we might tentatively try the function x^3. But $\frac{d}{dx}(x^3) = 3x^2$. Thus, we have *three* times our function x^2. Dividing by this number we obtain $\frac{d}{dx}\left(\frac{1}{3}x^3\right) = x^2$. This is one answer. But are there others? Recalling that the derivative of a constant is zero, we readily see that

$$\frac{d}{dx}\left(\frac{1}{3}x^3 + C\right) = x^2$$

for any constant C. This gives us the family of answers

$$F(x) = \frac{1}{3}x^3 + C$$

See Figure 6.1.

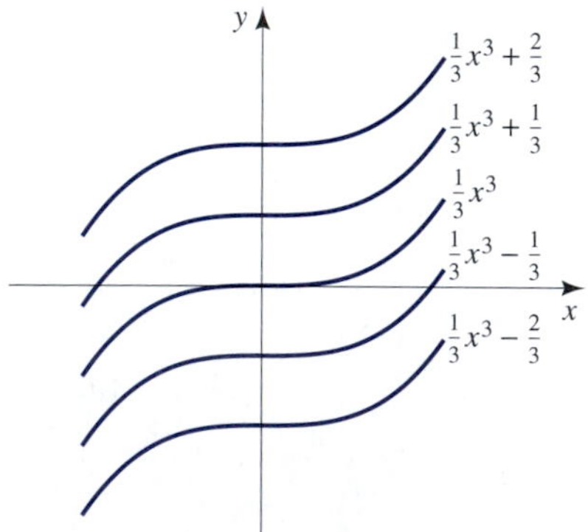

Figure 6.1
Shown are some members of the family $\frac{1}{3}x^3 + C$, all of which are antiderivatives of x^2.

We must now address the following general problem. Given a function $f(x)$, we would like to find a function $F(x)$ such that $F'(x) = f(x)$. We call $F(x)$ an

antiderivative of $f(x)$. We use the term *anti*derivative to refer to the fact that finding $F(x)$ is the *opposite* of differentiating.

DEFINITION **Antiderivative**

If $F'(x) = f(x)$, then $F(x)$ is called an antiderivative of $f(x)$. (The proof is omitted.)

We saw that $\frac{1}{3}x^3 + C$, where C is any constant, represents antiderivatives of x^2. Are there still others? The following theorem indicates that the answer is no. (The proof is omitted.)

Theorem 1

If $F'(x) = G'(x)$ on (a, b), then on (a, b),

$$F(x) = G(x) + C$$

for some constant C.

That is, if two functions have the same derivative on (a, b), then they differ by a constant on (a, b).

EXAMPLE 1 **Finding All the Antiderivatives**

Find all antiderivatives of x^2.

Solution

From the above theorem and work, we know that all antiderivatives of x^2 must be of the form

$$\frac{1}{3}x^3 + C$$

where C is any constant. ■

When we include the arbitrary constant C, we refer to the function $\frac{1}{3}x^3 + C$ as the *general form* of the antiderivative. By this we mean that the collection of all antiderivatives of x^2 can be obtained by adding all possible constants to $\frac{1}{3}x^3$.

The collection of all antiderivatives of a function $f(x)$ is called the **indefinite integral** and is denoted by $\int f(x)\,dx$. This is read as *the indefinite integral of* $f(x)$ or, more carefully, as *the indefinite integral of* $f(x)$ *with respect to* x. We refer to the symbol $\int$ as the **integral sign** and $f(x)$ as the **integrand**. The symbol $\int$ looks like an S and in fact stands for *sum*. We shall later see the important connection between the integral and a certain sum.

If we know of one function $F(x)$ for which $F'(x) = f(x)$, then Theorem 1 indicates that the collection of all antiderivatives of $f(x)$ must be of the form $F(x) + C$, where C is a constant. Thus, we have the following.

DEFINITION The Indefinite Integral

The collection of all antiderivatives of a function $f(x)$ is called the **indefinite integral** and is denoted by $\int f(x)\,dx$.

If we know one function $F(x)$ for which $F'(x) = f(x)$, then

$$\int f(x)\,dx = F(x) + C$$

where C is an arbitrary constant and called the **constant of integration**.

From Example 1 above we see that

$$\int x^2\,dx = \frac{1}{3}x^3 + C$$

The following is also true:

$$\int x^3\,dx = \frac{1}{4}x^4 + C$$

This can be readily verified by differentiating the right hand side to obtain the integrand. Thus

$$\frac{d}{dx}\left(\frac{1}{4}x^4 + C\right) = x^3$$

Rules of Integration

It would be convenient if we had some rules to find indefinite integrals. The following is the first such rule.

Power Rule

If $n \neq -1$, then

$$\int x^n\,dx = \frac{1}{n+1}x^{n+1} + C$$

In words, increase the power by one and divide by the resulting number, then add an arbitrary constant.

To prove the power rule formula we need only differentiate the right-hand side and note that we obtain the integrand. Thus,

$$\frac{d}{dx}\left(\frac{1}{n+1}x^{n+1} + C\right) = x^n$$

EXAMPLE 2 **Using the Power Rule**

Find (a) $\int x^5\,dx$ and (b) $\int \frac{1}{\sqrt{x}}dx$. Check the answers.

Solution

(a) Here we set $n = 5$ in the power rule formula and obtain

$$\int x^5\,dx = \frac{1}{5+1}x^{5+1} + C = \frac{1}{6}x^6 + C$$

Our answer checks, since

$$\frac{d}{dx}\left(\frac{1}{6}x^6 + C\right) = x^5$$

(b) Here we first note that $\frac{1}{\sqrt{x}} = x^{-1/2}$ and set $n = -1/2$ in the power rule formula; then we obtain

$$\int \frac{1}{\sqrt{x}}dx = \int x^{-1/2}\,dx = \frac{1}{-1/2+1}x^{-1/2+1} + C = 2x^{1/2} + C$$

Our answer checks, since

$$\frac{d}{dx}\left(2x^{1/2} + C\right) = x^{-1/2} = \frac{1}{\sqrt{x}}$$ ■

Remark As in the previous example, *always remove radical signs before integrating.*

Since

$$\frac{d}{dx}(kF(x)) = k\frac{d}{dx}F(x)$$

we have the following rule.

Constant Times Function Rule

For any constant k,

$$\int kf(x)\,dx = k\int f(x)\,dx$$

In words, the integral of a constant times a function is the constant times the integral of the function.

EXAMPLE 3 **Using the Constant Times Function Rule**

Find (a) $\int 3t^9\,dt$ and $\int \frac{\sqrt{x}}{\pi}\,dx$. Check the answers.

Solution

(a) Since 3 is a constant,

$$\int 3t^9\,dt = 3\int t^9\,dt = \frac{3}{10}t^{10} + C$$

The answer checks, since

$$\frac{d}{dt}\left(\frac{3}{10}t^{10} + C\right) = 3t^9$$

(b) Since $\frac{1}{\pi}$ is a constant,

$$\int \frac{\sqrt{x}}{\pi}\,dx = \int \frac{1}{\pi}\sqrt{x}\,dx = \frac{1}{\pi}\int \sqrt{x}\,dx = \frac{1}{\pi}\int x^{1/2}\,dx$$

$$= \frac{1}{\pi}\,\frac{1}{\frac{1}{2}+1}x^{1/2+1} + C$$

$$= \frac{2}{3\pi}x^{3/2} + C$$

This checks, since

$$\frac{d}{dx}\left(\frac{2}{3\pi}x^{3/2} + C\right) = \frac{1}{\pi}x^{1/2} = \frac{\sqrt{x}}{\pi}$$ ■

Since

$$\frac{d}{dx}(F(x) \pm G(x)) = \frac{d}{dx}F(x) \pm \frac{d}{dx}G(x)$$

we have the following rule.

Sum or Difference Rule

$$\int [f(x) \pm g(x)]\,dx = \int f(x)\,dx \pm \int g(x)\,dx$$

In words, the integral of the sum (or difference) is the sum (or difference) of the integrals.

EXAMPLE 4 **Using the Sum and Difference Rules**

Find (a) $\int (u^2 + u^3)\,du$ and (b) $\int \left(2x^4 - \frac{1}{2}x^2\right)\,dx$. Check the answers.

Solution

(a)

$$\int (u^2 + u^3)\,du = \int u^2\,du + \int u^3\,du$$

$$= \frac{1}{3}u^3 + \frac{1}{4}u^4 + C$$

Our answer checks since

$$\frac{d}{du}\left(\frac{1}{3}u^3 + \frac{1}{4}u^4 + C\right) = u^2 + u^3$$

(b)

$$\begin{aligned}\int\left(2x^4-\frac{1}{2}x^2\right)\,dx &= \int 2x^4\,dx-\int \frac{1}{2}x^2\,dx\\ &= 2\int x^4\,dx-\frac{1}{2}\int x^2\,dx\\ &= \frac{2}{5}x^5-\frac{1}{6}x^3+C\end{aligned}$$

Our answer checks, since

$$\frac{d}{dx}\left(\frac{2}{5}x^5-\frac{1}{6}x^3+C\right)=2x^4-\frac{1}{2}x^2$$ ■

As we have seen so far, for every rule on derivatives there is a corresponding rule on integrals. For example,

$$\frac{d}{dx}\left(e^x\right)=e^x \qquad \text{implies} \qquad \int e^x\,dx=e^x+C$$

We then have the following rule for integration of the exponential function.

Indefinite Integral of Exponential Function

$$\int e^x\,dx=e^x+C$$

EXAMPLE 5 **Integrating Exponential Functions**

Find $\int(3e^x-4x)\,dx$. Check the answer.

Solution

$$\begin{aligned}\int(3e^x-4x)\,dx &= 3\int e^x\,dx-4\int x\,dx\\ &= 3e^x+4\frac{1}{2}x^2+C\\ &= 3e^x-2x^2+C\end{aligned}$$

The answer checks, since

$$\frac{d}{dx}\left(3e^x-2x^2+C\right)=3e^x-4x$$ ■

Recall from Section 4.4 that

$$\frac{d}{dx}\left(\ln|x|\right)=\frac{1}{x}$$

Therefore,

$$\int \frac{1}{x}\,dx = \ln|x| + C$$

We then have the following rule.

Indefinite Integral of $x^{-1} = \frac{1}{x}$

$$\int \frac{1}{x}\,dx = \ln|x| + C$$

Recall that in the power rule for integrating x^n we had to restrict $n \neq -1$. (If we try to use this formula with $n = -1$, we obtain a zero in the denominator.) With this last rule we can now find the indefinite integral of x^n for *any* number n.

EXAMPLE 6 **Integrating x^{-1}**

Find $\int \frac{2}{x}\,dx$. Check the answer.

Solution

$$\int \frac{2}{x}\,dx = 2\int \frac{1}{x}\,dx$$
$$= 2\ln|x| + C$$

The answer checks, since

$$\frac{d}{dx}(2\ln|x| + C) = \frac{2}{x}$$

■

We summarize the integration rules found in this section.

Integration Rules

$$\int kf(x)\,dx = k\int f(x)\,dx$$

$$\int [f(x) \pm g(x)]\,dx = \int f(x)\,dx \pm \int g(x)\,dx$$

$$\int x^n\,dx = \frac{1}{n+1}x^{n+1} + C, n \neq -1$$

$$\int e^x\,dx = e^x + C$$

$$\int \frac{1}{x}\,dx = \ln|x| + C$$

Applications

EXAMPLE 7 **Finding Revenue Given Marginal Revenue**

Find the revenue function $R(x)$ if the marginal revenue is $2x - 10$.

Solution

We are given that $R'(x) = 2x - 10$. Thus,

$$R(x) = \int (2x - 10)\, dx$$
$$= x^2 - 10x + C$$

where C is some constant not yet determined.

But we know that zero sales result in zero revenue, that is, $R(0) = 0$. Using this gives

$$0 = R(0) = (0)^2 - 10(0) + C = C$$

Thus,

$$R(x) = x^2 - 10x$$

■

EXAMPLE 8 **Finding Cost Given Marginal Cost**

Find the cost function $C(x)$ if the marginal cost is $10e^x + 40x - 100$ and fixed costs are 1000.

Solution

We are given that $C'(x) = 10e^x + 40x - 100$. Thus,

$$C(x) = \int (10e^x + 40x - 100)\, dx$$
$$= 10e^x + 20x^2 - 100x + K$$

where K is some constant not yet determined.

If fixed costs are 1000, this translates into $C(0) = 1000$. Then

$$1000 = C(0) = 10e^0 + 20(0) - 100(0) + K = 10 + K$$

Thus, $K = 990$, and

$$C(x) = 10e^x + 20x^2 - 100x + 990$$

■

Warm Up Exercise Set 6.1

1. Find $\int \left(1 + \sqrt[3]{x} + \frac{3}{x} - 2e^x\right) dx$.
2. If a ball is thrown upward with an initial velocity of 20 feet per second, then from physics it can be shown that the velocity (neglecting air resistance) is given by $v(t) = -32t + 20$. Find $s(t)$ if the ball is thrown from 6 feet above ground level and $s(t)$ measures the height of the ball in feet.

Exercise Set 6.1

In Exercises 1 through 38, find the antiderivatives.

1. $\int x^{99}\,dx$
2. $\int x^{0.01}\,dx$
3. $\int x^{-99}\,dx$
4. $\int x^{-0.01}\,dx$
5. $\int 5\,dx$
6. $\int dx$
7. $\int \frac{5}{y^3}\,dy$
8. $\int \frac{1}{2y^5}\,dy$
9. $\int \frac{y^{3/2}}{\sqrt{2}}\,dy$
10. $\int \pi y^{2/3}\,dy$
11. $\int \sqrt[3]{u^2}\,du$
12. $\int \sqrt[5]{u^3}\,du$
13. $\int (6x^2 + 4x)\,dx$
14. $\int (8x^3 + 1)\,dx$
15. $\int (x^2 + x + 1)\,dx$
16. $\int (3u^2 + 4u^3)\,du$
17. $\int (\sqrt{2}u^{0.1} - 0.1u^{1.1})\,du$
18. $\int (\frac{3}{t^2} - 6t^2)\,dt$
19. $\int (t^2 - 1)\,dt$
20. $\int (t^{-2} + 3)\,dt$
21. $\int (\sqrt{t} - \sqrt[3]{t^5})\,dt$
22. $\int (t^3 + 4t^2 + 5)\,dt$
23. $\int (6t^5 - 4t^3 + 1)\,dt$
24. $\int (x + \frac{1}{x})\,dx$
25. $\int (1 + \frac{3}{x})\,dx$
26. $\int \frac{2}{\pi x}\,dx$
27. $\int (\pi + \frac{1}{x})\,dx$
28. $\int \frac{t^4 + 3}{t^2}\,dt$
29. $\int \frac{t+1}{\sqrt{t}}\,dt$
30. $\int \sqrt{t}(t+1)\,dt$
31. $\int (e^x - 3x)\,dx$
32. $\int (e^x - \frac{1}{x})\,dx$
33. $5\int e^x\,dx$
34. $\int (6 - e^x)\,dx$
35. $\int (5e^x - 4)\,dx$
36. $\int (6x - e^x)\,dx$
37. $\int \frac{x+1}{x}\,dx$
38. $\int \frac{e^{-x} + 1}{e^{-x}}\,dx$

Applications and Mathematical Modeling

39. **Revenue** Find the revenue function for a knife manufacturer if the marginal revenue, in dollars, is given by $30 - 0.5x$, where x is the number of knives sold.

40. **Revenue** Find the revenue function for a paper plate manufacturer if the marginal revenue, in dollars, is given by $200 - 0.26x$, where x is the number of plates sold.

41. **Demand** Find the demand function for a salt shaker manufacturer if marginal demand, in dollars, is given by $p'(x) = -x^{-3/2}$, where x is the number of salt shakers sold and $p(1) = 3$.

42. **Demand** Find the demand function for a table cloth manufacturer if marginal demand, in dollars, is given by $p'(x) = -x^{-3/2}$, where x is the number of thousands of table cloths sold. Assume that $p(1) = 100$.

43. **Cost** Find the cost function for an envelope manufacturer if the marginal cost, in dollars, is given by $100 - 0.1e^x$, where x is the number of thousands of envelopes produced and fixed costs are \$1000.

44. **Cost** Find the cost function for an adhesive tape manufacturer if the marginal cost, in dollars, is given by $150 - 0.01e^x$, where x is the number of cases of tape produced. Assume that $C(0) = 100$.

45. **Cost** Find the cost function for a computer disk manufacturer if the marginal cost, in dollars, is given by $30\sqrt{x} - 6x^2$, where x is the number of thousands of disks sold. Assume that there are no fixed costs.

46. **Cost** Find the cost function for a spark plug manufacturer if the marginal cost, in dollars, is given by $30x - 4e^x$, where x is the number of thousands of plugs sold. Assume no fixed costs.

47. **Velocity** If a ball is thrown upward with an initial velocity of 30 ft/sec, then from physics it can be shown that the velocity (neglecting air resistance) is given by $v(t) = -32t + 30$. Find $s(t)$ if the ball is thrown from 15 feet above ground level and $s(t)$ measures the height of the ball in feet.

48. **Acceleration** If a ball is thrown upward, then from physics it can be shown that the acceleration is given by $a(t) = -32$ ft/sec. If a ball is thrown upward with an initial velocity of 10 ft/sec, find $v(t)$.

49. **Velocity** If in the previous problem, the ball is at an initial height of 6 feet, find $s(t)$, the height of the ball after t seconds.

50. **Population** A certain insect population is increasing at the rate given by $P'(t) = 2000e^t$, where t is time in years measured from the beginning of 1994 when the population was 200,000. Find the population at the end of 1996.

51. Appreciation A painting by one of the masters is purchased by a museum for \$1,000,000 and increases in value at a rate given by $V'(t) = 100e^t$, where t is measured in years from the time of purchase. What will the painting be worth in 10 years?

52. Marginal Propensity to Consume Let $C(x)$ represent national consumption in trillions of dollars, where x is disposable national income in trillions of dollars. Suppose that the marginal propensity to consume, $C'(x)$, is given by

$$C'(x) = 1 + \frac{2}{\sqrt{x}}$$

and that consumption is \$0.1 trillion when disposable income is zero. Find the national consumption function.

53. Marginal Propensity to Save Let S represent national savings in trillions of dollars. We can assume that disposable national income equals national consumption plus national savings, that is, $x = C(x) + S(x)$. Thus, $C'(x) = 1 - S'(x)$, where $S'(x)$ is the marginal propensity to save. If the marginal propensity to save is given by $0.5x$ and consumption is \$0.2 trillion dollars when disposable income is zero, find the consumption function.

54. Sum-of-Years-Digit Depreciation The sum-of-years-digit depreciation method depreciates a piece of equipment more in the early years of its life and less in the later years. In this method the value decreases at a rate given by $\frac{d}{dt}V(t) = -k(T - t)$, where T is the useful life in years of the equipment, t is the time in years since purchase, and k is a positive constant. (Notice that as t gets closer to T, $V'(t)$ becomes smaller and thus the value decreases less.) If a piece of machinery is purchased for I dollars and has no salvage value, show that

$$k = \frac{2I}{T^2}$$

and write an expression for $V(t)$ in terms of t, I, and T.

55. Agriculture Carter and Wiebe[2] studied small farms in the African country of Kenya. Their data indicated that the relationship $L'(x) = 0.28 + 0.05x - 0.0001x^2$, held approximately, where x was the size in acres of the farm and L the amount of labor measured in appropriate units. Find $L(x)$.

[2] Michael R. Carter and Keith D. Wiebe. 1990. Access to capital and its impact on agrarian structure and productivity in Kenya. *Amer. J. Agr. Econ.* 72:1146–1150.

More Challenging Exercises

56. Explain in complete sentences the difference between an antiderivative of $f(x)$ and the indefinite integral $\int f(x)\, dx$.

57. Find the antiderivatives of e^{2x}. (Hint: Start by finding the derivative of e^{2x}.)

58. In Exercises 67 and 68 of Section 3.3 we noted that $\frac{d}{dx}(\sin x) = \cos x$ and $\frac{d}{dx}(\cos x) = -\sin x$. Using this, find

a. $\int \sin x\, dx$ **b.** $\int \cos x\, dx$

59. Find the antiderivatives of $\cos 3x$. (Hint: Start by finding the derivative of $\sin 3x$ by recalling from Exercise 77 of Section 4.4 that $\frac{d}{dx}(\sin f(x)) = (\cos f(x)) \cdot f'(x)$.)

60. Find the antiderivatives of $\sin 3x$. (Hint: Start by finding the derivative of $\cos 3x$ by recalling from Exercise 77 of Section 4.4 that $\frac{d}{dx}(\cos f(x)) = -(\sin f(x)) \cdot f'(x)$.)

Solutions to WARM UP EXERCISE SET 6.1

1. First remove the radical and obtain

$$\begin{aligned}\int \left(1 + x^{1/3} + \frac{3}{x} - 2e^x\right) dx &= \int 1\, dx + \int x^{1/3}\, dx + 3\int \frac{1}{x}\, dx - 2\int e^x\, dx \\ &= x + \frac{1}{4/3}x^{4/3} + 3\ln|x| - 2e^x + C \\ &= x + \frac{3}{4}x^{4/3} + 3\ln|x| - 2e^x + C\end{aligned}$$

2. Since $\frac{ds}{dt}(t) = v(t)$ and $s(0) = 6$, we have

$$\begin{aligned}s(t) = \int (-32t + 20)\, dt &= -32\int t\, dt + 20\int 1\, dt \\ &= -16t^2 + 20t + C\end{aligned}$$

Then since

$$6 = s(0) = -16(0)^2 + 20(0) + C$$

$C = 6$, and

$$s(t) = -16t^2 + 20t + 6$$

6.2 Substitution

APPLICATION Production of Oil

The rate at which a natural resource is extracted tends to increase initially and then to fall off later after the easily accessible material has been extracted. Suppose the rate, in hundreds of thousands of barrels per year, at which oil is being extracted from a field during the early and intermediate stages of the life of the field is given by

$$P'(t) = \frac{t}{(t^2+1)^2} + 1$$

where t is in years. Find the amount of oil extracted during the first 3 years. See Example 6 on page 413 for the answer.

The Method of Substitution

The methods discussed in the previous section are inadequate to find all the indefinite integrals that we will need. In this section we will develop the method of *substitution*.

The method of substitution is based on reversing the chain rule. Recall that if $h = g(u)$, $u = f(x)$, and so $h(x) = g(f(x))$, then by the chain rule we have

$$h'(x) = g'(f(x)) \cdot f'(x)$$

For example, suppose $h(x) = \frac{1}{100}\left(x^2+1\right)^{100}$. We set $u = f(x) = x^2 + 1$ and $g(u) = \frac{1}{100}u^{100}$; then

$$\frac{d}{dx}\frac{1}{100}\left(x^2+1\right)^{100} = (x^2+1)^{99} \cdot 2x$$

It is important to notice that the term

$$g'(f(x)) \cdot f'(x) = (x^2+1)^{99} \cdot 2x$$

consists of the product of two factors. The first factor is $g'(f(x))$ or $(x^2+1)^{99}$ in the above example. Think of this as $f(x) = x^2 + 1$ raised to a power. The second factor is $f'(x)$, or $2x$ in the above example. So when we look at the term $(x^2+1)^{99} \cdot 2x$, we should notice that the first factor is (x^2+1) raised to a power and the second factor, $2x$, is the derivative of (x^2+1). Once we recognize this, we set $u = f(x) = x^2 + 1$.

We will see that in this section $g(u)$ is either u^n, e^u, or $\ln u$. So, if we have a function to a power, e raised to a function, or the logarithm of a function, we set u equal to the function.

EXAMPLE 1 Using the Method of Substitution

Find $\int 2x(x^2+1)^{99}\, dx$.

Solution

We could actually solve this problem using the techniques of the previous section by expanding the term $(x^2+1)^{99}$ by the binomial theorem and then multiplying the result by $2x$. We would have an expression with 100 terms to integrate.

Instead we try the method of substitution. One should clearly understand that this method is in practice a trial-and-error method.

Following our previous discussion, we note that the integrand is the product of two factors. The first is (x^2+1) raised to a power, and the second, $2x$, is the derivative of this function. So we try the substitution $u = x^2+1$. Then $\dfrac{du}{dx} = 2x$. Then $du = 2x\,dx$. (We treat du and dx as if they were Δu and Δx.) Substitute all of this into the integral, and obtain

$$\begin{aligned}\int 2x(x^2+1)^{99}\,dx &= \int (x^2+1)^{99} 2x\,dx \\ &= \int u^{99} du \\ &= \frac{1}{100}u^{100} + C \\ &= \frac{1}{100}\left(x^2+1\right)^{100} + C\end{aligned}$$

■

How can we be certain we have the correct answer? Simply differentiate the alleged answer to see whether we obtain the integrand. Since

$$\frac{d}{dx}\left(\frac{1}{100}\left(x^2+1\right)^{100} + C\right) = \frac{1}{100}(100)(x^2+1)^{99}(2x) + 0 = 2x(x^2+1)^{99}$$

our answer is correct.

How can we possibly know what to substitute? The preceding process indicates that if you encounter an integrand that is of the form of a function raised to a power, then set u equal to the function as a first try. It just might work! (Fortunately, the above integrand has the factor $2x$ in it; otherwise, we would be stuck and would have to resort to some other as yet unlearned technique.)

Notice that after making the substitution in the last example, we were left with

$$\int u^n\,du = \frac{1}{n+1}\,u^{n+1} + C$$

We refer to this as the **general power rule**.

EXAMPLE 2 Using the General Power Rule

Find $\displaystyle\int \frac{3x^2}{\sqrt{x^3+3}}\,dx$.

Solution

Rewrite the integral as $\displaystyle\int (x^3+3)^{-1/2} \cdot 3x^2\,dx$. We notice that the integrand is the product of two functions. The first is (x^3+3) raised to the power $-1/2$ and the second is the derivative of this function $3x^2$. Thus, we try $u = x^3+3$ and hope for the best. With u defined in this way, we then have $du = 3x^2\,dx$. Substitute all of this into the integral, and obtain

$$\begin{aligned}\int 3x^2(x^3+3)^{-1/2}\,dx &= \int (x^3+3)^{-1/2}3x^2\,dx \\ &= \int u^{-1/2}\,du \\ &= 2u^{1/2}+C \\ &= 2\left(x^3+3\right)^{1/2}+C\end{aligned}$$

■

To check this answer, we simply differentiate and notice that it does indeed equal the integrand. That is,

$$\frac{d}{dx}\left(2\left(x^3+3\right)^{1/2}+C\right) = 3x^2(x^3+3)^{-1/2}$$

EXAMPLE 3 **Using the General Power Rule**

Find $\displaystyle\int \frac{4\ln x}{x}\,dx$.

Solution

This is by no means immediately obvious! By some trial and error, we rewrite the integral as

$$4\int (\ln x)\cdot\frac{1}{x}\,dx$$

Now we have the integrand as a product of two factors. The first factor is a function, $\ln x$, raised to a power, the first power. The second factor, $\frac{1}{x}$, is the derivative of this function. So we set $u = \ln x$, then $du = \frac{1}{x}\,dx$. Then

$$\begin{aligned}\int \frac{4\ln x}{x}\,dx &= 4\int \ln x\cdot\frac{1}{x}\,dx \\ &= 4\int u\,du \\ &= 2u^2+C \\ &= 2(\ln x)^2+C\end{aligned}$$

■

This is the correct answer, since

$$\frac{d}{dx}\left(2(\ln x)^2+C\right) = 2\cdot 2\,(\ln x)\,\frac{1}{x} = \frac{4\ln x}{x}$$

Remark Notice that in each of the previous three examples the integrand was in the form $(f(x))^n\cdot f'(x)$. We then set $u = f(x)$.

EXAMPLE 4 **Using the Method of Substitution**

Find $\displaystyle\int \frac{x+1}{x^2+2x}\,dx$.

Solution

Rewrite the integral as

$$\int \frac{1}{x^2+2x}\cdot(x+1)\,dx$$

We recognize the integrand as a product of two factors. The first factor is a function, x^2+2x, raised to the -1 power. The second factor is $x+1$. Although this is not the derivative of the function x^2+2x, it is a *constant* times the derivative. So we go ahead and let $u = x^2+2x$. We will see that this will work. Then $du = (2x+2)\,dx = 2(x+1)\,dx$ and $\frac{1}{2}\,du = (x+1)\,dx$. Now formally substitute all of this into the integral, and obtain

$$\begin{aligned}
\int \frac{x+1}{x^2+2x}\,dx &= \int \frac{1}{x^2+2x}\cdot(x+1)\,dx \\
&= \int \frac{1}{u}\,\frac{1}{2}\,du \qquad (x+1)dx = \frac{1}{2}du \\
&= \frac{1}{2}\int \frac{du}{u} \\
&= \frac{1}{2}\ln|u| + C \\
&= \frac{1}{2}\ln|x^2+2x| + C
\end{aligned}$$

■

This is the correct answer, since

$$\frac{d}{dx}\left(\frac{1}{2}\ln|x^2+2x| + C\right) = \frac{1}{2}\,\frac{2x+2}{x^2+2x} = \frac{x+1}{x^2+2x}$$

Remark Notice in Example 4 that the integrand was in the form $\frac{1}{f(x)}\cdot f'(x)$. We then set $u = f(x)$. After this substitution we then had

$$\int \frac{du}{u} = \ln|u| + C$$

We refer to this as the **general logarithmic rule**.

EXAMPLE 5 **The General Exponential Rule**

Find $\displaystyle\int (x^2+1)e^{x^3+3x+1}\,dx$.

Solution

We rewrite the integral as

$$\int e^{x^3+3x+1}\cdot(x^2+1)\,dx$$

We then see the integrand as the product of two factors. The first factor is e raised to a function. The second factor is a constant times the derivative of this function. So we let $u = x^3 + 3x + 1$. Then $du = (3x^2 + 3)\,dx = 3(x^2 + 1)\,dx$, and $\frac{1}{3}\,du = (x^2 + 1)\,dx$. Then

$$\begin{aligned}
\int (x^2 + 1)e^{x^3+3x+1}\,dx &= \int e^{x^3+3x+1} \cdot (x^2 + 1)\,dx \\
&= \int e^u \frac{1}{3}\,du \\
&= \frac{1}{3}\int e^u\,du \\
&= \frac{1}{3}e^u + C \\
&= \frac{1}{3}e^{x^3+3x+1} + C
\end{aligned}$$

■

This is the correct answer, since

$$\frac{d}{dx}\left(\frac{1}{3}e^{x^3+3x+1} + C\right) = (x^2 + 1)e^{x^3+3x+1}$$

which is the integrand.

Remark Notice that the integrand in Example 5 was in the form $e^{f(x)} \cdot f'(x)$. We then set $u = f(x)$. After this substitution we then had

$$\int e^u\,du = e^u + C$$

We refer to this as the **general exponential rule**.

We now summarize the formulas we have learned in this section.

Integration Formulas

If $u = f(x)$, then $du = f'(x)\,dx$ and

$$\int u^n\,du = \frac{1}{n+1}\,u^{n+1} + C$$

$$\int \frac{du}{u} = \ln|u| + C$$

$$\int e^u\,du = e^u + C$$

CONNECTION
Historic Oil Well

The first development oil well in Alaska's Prudhoe Bay, Number DS 1-1, produced over 19 million barrels of oil during its 19-year life. It went into production in December 1970.

Applications

EXAMPLE 6 **Production of Oil**

The rate at which a natural resource is extracted tends to increase initially and then fall off later after the easily accessible material has been extracted. (See Dorner and El-Shafie.[3]) Suppose the rate, in hundreds of thousands of barrels per year, at which oil is being extracted during the early and intermediate stages of the life of the field is given by

$$P'(t) = \frac{t}{(t^2+1)^2} + 1$$

where t is in years. Refer to Figure 6.2. Find the amount of oil extracted during the first 3 years.

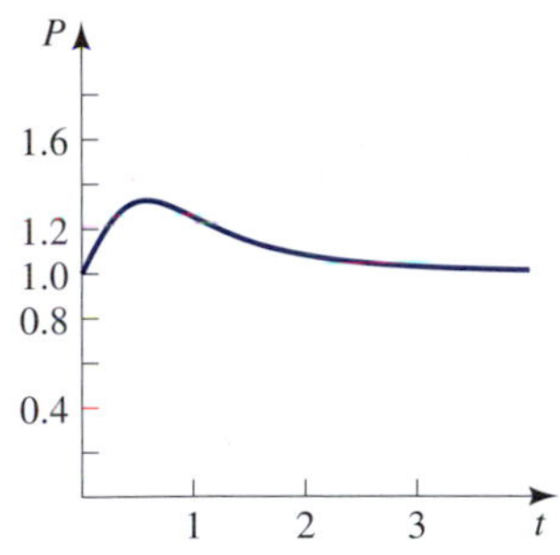

Figure 6.2
Graph of the rate of extraction of oil from a field.

Solution

We must have

$$P(t) = \int \left(\frac{t}{(t^2+1)^2} + 1 \right) dt = \int (t^2+1)^{-2} \cdot t\, dt + \int dt$$

The second integral is $t + C$. The first integral requires the substitution $u = t^2 + 1$. Then $du = 2t\, dt$, and

$$\begin{aligned} \int (t^2+1)^{-2} t\, dt + \int dt &= \int u^{-2} \frac{1}{2}\, du + t + C \\ &= -\frac{1}{2} u^{-1} + t + C \\ &= -\frac{1}{2} \frac{1}{t^2+1} + t + C \end{aligned}$$

There can be no production of oil in no time; thus,

$$0 = P(0) = -\frac{1}{2} + C$$

This implies that $C = \frac{1}{2}$. Thus,

$$P(3) = -\frac{1}{2} \frac{1}{3^2+1} + 3 + \frac{1}{2} = 3.45$$

Thus, 345,000 barrels of oil are extracted in the first 3 years. ■

[3] Peter Dorner and Mahmoud El-Shafie. 1980. *Resources and Development*. Madison, Wisc.: The University of Wisconsin Press.

Warm Up Exercise Set 6.2

1. Find $\int \frac{2x}{\sqrt[3]{x^2+4}}\,dx$

2. Find $\int \frac{e^{2x}}{e^{2x}+1}\,dx$

3. Find $\int \frac{e^{1/x}}{x^2}\,dx$

Exercise Set 6.2

In Exercises 1 through 34, find the indefinite integral.

1. $\int 6(3x+1)^{10}\,dx$

2. $\int 8(3-2x)^5\,dx$

3. $\int x(3-x^2)^7\,dx$

4. $\int 10x(2x^2+1)^3\,dx$

5. $\int 8(x+1)(2x^2+4x-1)^{3/2}\,dx$

6. $\int (x^3+2)(x^4+8x+3)^{1/3}\,dx$

7. $\int (x^3+x^2+x+1)(3x^4+4x^3+6x^2+12x+1)^3\,dx$

8. $\int (x^4+x^2+2)(3x^5+5x^3+30x+1)^5\,dx$

9. $\int 2\sqrt{x+1}\,dx$

10. $\int \sqrt[3]{x+1}\,dx$

11. $\int x\sqrt{x^2+1}\,dx$

12. $\int x\sqrt{3x^2+1}\,dx$

13. $\int \frac{x}{\sqrt[3]{x^2+1}}\,dx$

14. $\int \frac{x}{\sqrt[5]{2x^2+5}}\,dx$

15. $\int \frac{\sqrt{x^{1/3}+1}}{x^{2/3}}\,dx$

16. $\int \frac{\ln 2x}{x}\,dx$

17. $\int \frac{\sqrt{\ln x}}{x}\,dx$

18. $\int e^{2x+1}\,dx$

19. $\int e^{1-x}\,dx$

20. $\int 4xe^{x^2-1}\,dx$

21. $\int xe^{1-x^2}\,dx$

22. $\int \frac{e^{1/x^2}}{x^3}\,dx$

23. $\int \frac{e^{\sqrt{x}}}{\sqrt{x}}\,dx$

24. $\int \frac{1}{2x+1}\,dx$

25. $\int \frac{1}{3x+5}\,dx$

26. $\int \frac{x^3}{(x^4+4)^2}\,dx$

27. $\int \frac{x}{(x^2+3)^4}\,dx$

28. $\int \frac{e^x}{e^x+1}\,dx$

29. $\int \frac{e^{-x}}{e^{-x}+1}\,dx$

30. $\int \frac{e^x-e^{-x}}{e^x+e^{-x}}\,dx$

31. $\int \frac{e^{2x}-e^{-2x}}{e^{2x}+e^{-2x}}\,dx$

32. $\int \frac{1}{x\ln x}\,dx$

33. $\int \frac{1}{x\ln x^2}\,dx$

34. $\int \frac{1}{x\ln\sqrt{x}}\,dx$

Applications and Mathematical Modeling

35. Demand Find the demand function for a clock manufacturer if marginal demand, in dollars, is given by $p'(x) = -24x/(3x^2+1)^2$ and $p(1) = 10$, where x is the number of thousands of clocks sold.

36. Demand Find the demand function for a cardboard box manufacturer if marginal demand, in dollars, is given by $p'(x) = -4xe^{-x^2}$ and $p(1) = 10$, where x is the number of thousands of boxes sold.

37. Revenue Find the revenue function for a shoulder bag manufacturer if the marginal revenue, in dollars, is given by $4x(10-x^2)$, where x is the number of hundreds of bags sold.

38. Revenue Find the revenue function for a hand purse manufacturer if the marginal revenue, in dollars, is given by $(1-x)e^{2x-x^2}$, where x is the number of thousands of purses sold.

39. Cost Find the cost function for a lipstick manufacturer if the marginal cost, in dollars, is given by $4x/(x^2+1)$, where x is the number of cases of lipstick produced and fixed costs are \$1000.

40. Cost Find the cost function for a hairbrush manufacturer if the marginal cost, in dollars, is given by $x^3(100+x^4)$, where x is the number of cases of brushes produced and fixed costs are \$1000.

41. Production Suppose the rate in tons per year at which copper is being extracted from a mine during the early and intermediate stages of the life of the mine is given by

$$P'(t) = \frac{-54t^2}{(t^3+1)^2} + 11$$

Find the amount of copper extracted during the first 2 years.

42. Depreciation The value of a piece of machinery purchased for \$10,000 is decreasing at the rate given by $V'(t) = -1000te^{-0.25t^2}$, where t is measured in years from purchase. How much will the machine be worth in 2 years?

43. Population The population of a certain city for the next several years is increasing at the rate given by $P'(t) = 2000(t+10)e^{0.05t^2+t}$, where t is time in years measured from the beginning of 1994 when the population was 100,000. Find the population at the end of 1995.

44. Sales After the introduction of a new camera, the rate of sales in thousands per month is given by $N'(t) = 10(1+2t)^{-5/2}$, where t is in months. Find the number sold in the first 4 months.

45. Productivity Suppose the number of items produced on a certain piece of machinery by an average employee is increasing at a rate given by

$$N'(t) = \frac{1}{2t+1}$$

where t is measured in hours since being placed on the machine for the first time. How many items are produced by the average employee in the first 3 hours?

46. Memorization Suppose the rate at which an average person can memorize a list of items is given by

$$N'(t) = \frac{15}{\sqrt{3t+1}}$$

where t is the number of hours spent memorizing. How many items does the average person memorize in 1 hour?

47. Depreciation The value of a piece of machinery purchased for \$20,000 is decreasing at the rate given by $V'(t) = -1000e^{-0.1t}$, where t is measured in years from purchase. How much will the machine be worth in 4 years?

48. Radioactive Decay A certain radioactive substance initially weighing 100 grams decays at a rate given by $A'(t) = -20e^{-0.5t}$, where t is measured in years. How much of the substance is there after 10 years?

More Challenging Exercises

49. Try to find $\int (x^5 + x^4 + 1)^2(5x^3 + 4x^2)\,dx$ by using the method of substitution with $u = x^5 + x^4 + 1$. Show that the method fails since there is no way of solving the equation $u = x^5 + x^4 + 1$ for x.

50. Evaluate the integral in the previous exercise. Do *not* try any form of substitution.

In Exercises 51 through 54, use the fact that $\frac{d}{dx}(\sin u) = (\cos u)\frac{du}{dx}$ and $\frac{d}{dx}(\cos u) = (-\sin u)\frac{du}{dx}$.

51. $\int 2x \sin x^2\,dx$

52. $\int e^x \cos e^x\,dx$

53. $\int x^2 \cos x^3\,dx$

54. $\int (8x+2)\sin(4x^2+2x)\,dx$

Solutions to WARM UP EXERCISE SET 6.2

1. Write the integral as $\int (x^2+4)^{-1/3} \cdot 2x\,dx$ and use the substitution $u = x^2 + 4$. Then $du = 2x\,dx$, and

$$\begin{aligned}\int (x^2+4)^{-1/3} 2x\,dx &= \int u^{-1/3}\,du \\ &= \frac{3}{2}u^{2/3} + C \\ &= \frac{3}{2}(x^2+4)^{2/3} + C\end{aligned}$$

2. Write the integral as $\int (e^{2x}+1)^{-1} \cdot e^{2x}\,dx$. This is a function raised to the power $n = -1$. Let $u = e^{2x} + 1$, then $du = 2e^{2x}\,dx$. Then

$$\begin{aligned}\int \frac{e^{2x}}{e^{2x}+1}\,dx &= \int \frac{\frac{1}{2}\,du}{u} = \frac{1}{2}\int \frac{du}{u} \\ &= \frac{1}{2}\ln|u| + C \\ &= \frac{1}{2}\ln(e^{2x}+1) + C\end{aligned}$$

3. First write the integral as $\int e^{1/x} \cdot x^{-2}\,dx$. Let $u = 1/x = x^{-1}$. Then $du = -x^{-2}\,dx$, and

$$\begin{aligned}\int \frac{e^{1/x}}{x^2}\,dx &= \int e^{1/x}x^{-2}\,dx \\ &= \int e^u(-\,du) = -\int e^u\,du \\ &= -e^u + C = -e^{1/x} + C\end{aligned}$$

6.3 Estimating Distance Traveled

Estimating Distance Traveled

Enrichment: Accuracy of Approximation

CONNECTION
Method of Exhaustion

Around 370 B.C., the Greek geometer Eudoxus of Cnidus made the method of exhaustion a rigorous procedure. To find the area of a circle, he first inscribed an equilateral triangle P_0, shown in Figure 6.3a, and labeled E the area outside the triangle but inside the circle. He then placed a regular hexagon P_1, as shown in Figure 6.3b, and noted that, from simple geometrical principals, the area outside P_1 and inside the circle was less than $\frac{1}{2}E$. Continuing in this manner, as shown in Figure 6.3c, the area outside the regular 12-sided polygon P_2 and inside the circle is less than $\left(\frac{1}{2}\right)^2 E$. In

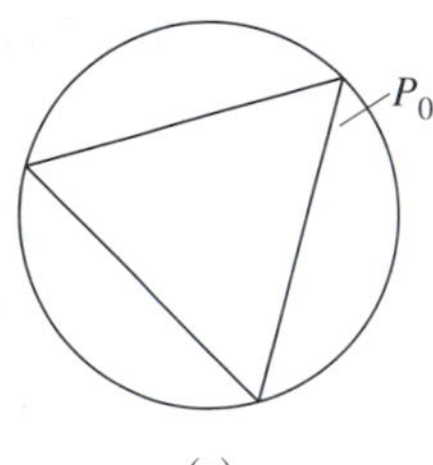

(a)

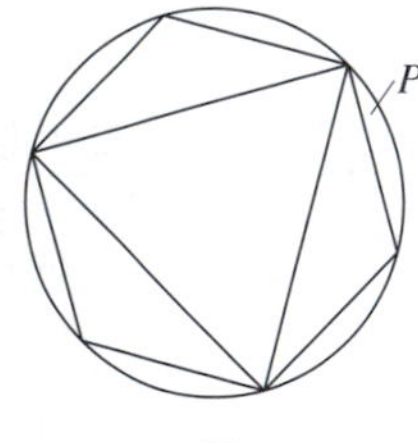

(b)

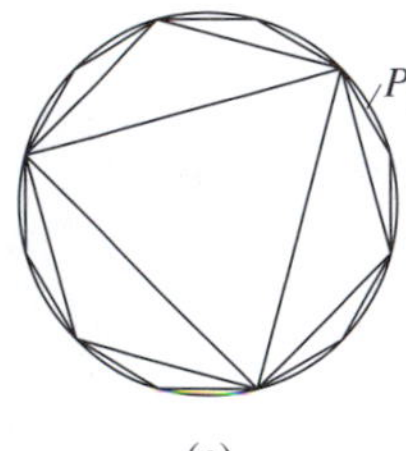

(c)

Figure 6.3

APPLICATION Distance Traveled

An object travels with velocity $v = t^2$, where v is in feet per second and t is in seconds. Find estimates, to any desired accuracy, for the distance the object travels during the time interval [0, 1]. For the answer, see Example 1 on page 420.

Estimating Distance Traveled

As our first goal, we wish to estimate the distance an object has traveled in one second if we know that the velocity function is given by $v = f(t) = t^2$ in feet per second on the t-interval [0, 1]. We are all familiar with monitoring the velocity of an object, since when driving an automobile we keep our eyes on the speedometer from time to time. As we shall soon see, this is a problem that models basic problems in business, economics, biology, and other areas. Thus, a clear understanding of this velocity problem will lead to a deeper understanding of basic problems in other areas.

If our car travels at a constant velocity of 55 miles per hour for 2 hours, then, since distance is the product of velocity with time ($d = v \cdot t$), we have traveled $55 \cdot 2 = 110$ miles. It is useful to represent the distance traveled as a certain area. In Figure 6.4 the velocity is constant. Notice, in this case, that the distance traveled in 2 hours, $55 \cdot 2$, is also the area of the rectangle with height equal to the constant velocity of 55 and base equal to the length, 2, of the interval [0, 2]. Thus, in this case, distance traveled is the area of the region between the graph of $v = 55$ and the x-axis on the interval [0, 2]. But what happens if velocity varies from instant to instant? How can we determine the distance traveled? This is a fundamental problem that we must address.

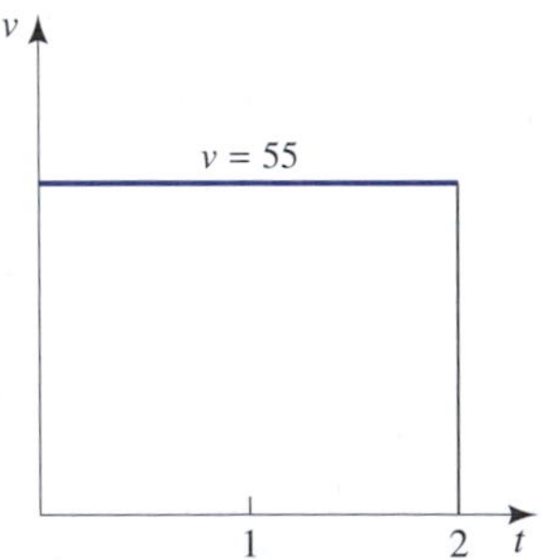

Figure 6.4
The area of the rectangle is base $\times$ height $= 2 \cdot 55 = 110$. This area is also the distance the car travels in 2 hours.

Bernhard Riemann, 1826–1866

Bernhard Riemann clarified the concept of the integral by developing the Riemann sums, which led later to the more important "Lebesgue" integral. He also developed a non-Euclidean geometry called Riemanian geometry that was used by Albert Einstein as a basis for Einstein's relativity theory. He also is known for the Riemann zeta function and the associated Riemann hypothesis, a celebrated unproven conjecture. His achievements were all the more remarkable, since he died from tuberculosis at the age of 39.

such a manner, regular $3 \cdot 2^n$−sided polygons P_n could be inscribed in a circle with the area outside the polygon P_n and inside the circle less than $\left(\frac{1}{2}\right)^n E$. Since $\left(\frac{1}{2}\right)^n E \to 0$ as $n \to \infty$, the areas of P_n must approach the area of the circle. Since he knew the areas of P_n, Eudoxus could find the area of the circle to any desired accuracy.

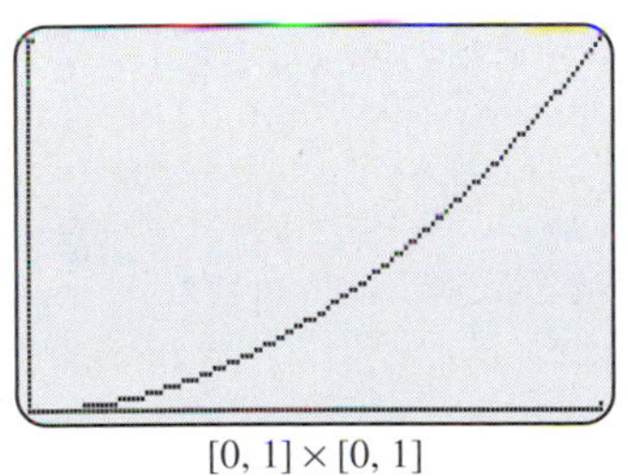
$[0, 1] \times [0, 1]$

Screen 6.1
A graph of $y = x^2$.

The object mentioned earlier has a variable velocity $v = f(t) = t^2$ on the t interval [0, 1] (Screen 6.1). We first note that the velocity is positive on the interval (0, 1], and therefore, the object always moves in the positive direction. We also note that the velocity is *increasing* on this interval. For starters, let's divide the interval [0, 1] into $n = 5$ subintervals of equal length and create the following table of velocities.

Time (seconds)	0	0.2	0.4	0.6	0.8	1
Velocity (feet per second) Velocity =	0 0	$(0.2)^2$ 0.04	$(0.4)^2$ 0.16	$(0.6)^2$ 0.36	$(0.8)^2$ 0.64	$(1)^2$ 1

Let us first determine an upper estimate of the distance traveled. Since the velocity of our object is always *increasing*, the velocity at any time during the subinterval [0, 0.2] is less than the velocity at the *right-hand* endpoint $t = 0.2$. The velocity at $t = 0.2$ is $(0.2)^2 = 0.04$. Now imagine another object, which we call the fast object, that travels with a constant velocity of 0.04 during the same fifth of a second. A graph is shown in Figure 6.5. Since the velocity of our object is always less or equal to that of the fast object, the distance our object travels during the first fifth of a second is *less or equal* to the distance the fast object travels. This latter distance is $v \cdot t = (0.04)(0.2) = 0.008$ feet, and is also the area of the rectangle shown in Figure 6.5.

Likewise, the velocity during the second fifth of a second is less than at the time $t = 0.4$, which is located at the right-hand endpoint of the second interval. The velocity at $t = 0.4$ is $(0.4)^2 = 0.16$. Therefore, during the second fifth of a second the distance our object travels is less or equal to that of a fast object traveling with a constant velocity of 0.16 indicated in Figure 6.6. This latter distance is $v \cdot t = (0.16)(0.2) = 0.032$ feet and is also the area of the second rectangle indicated in Figure 6.6.

Continuing in this way we obtain Figure 6.7. An upper estimate of the distance traveled on the interval [0, 1] is then

$$\begin{aligned} \text{upper estimate} &= (0.04)(0.2) + (0.16)(0.2) + (0.36)(0.2) \\ &\quad + (0.64)(0.2) + (1)(0.2) \\ &= 0.44 \end{aligned}$$

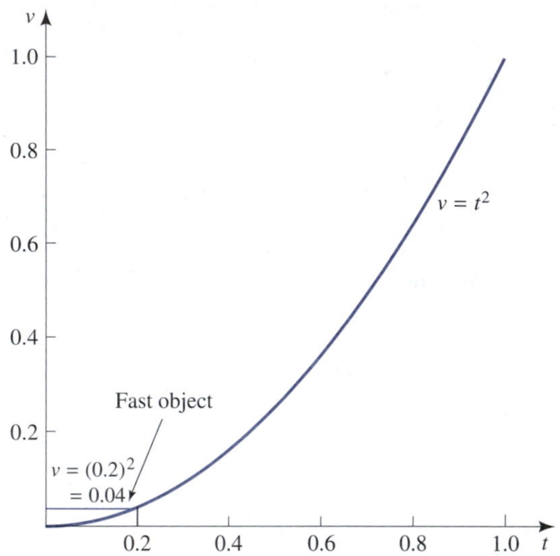

Figure 6.5
The area of the rectangle is base $\times$ height $= (0.2) \cdot (0.2)^2 = 0.008$. This area is also the distance the fast object travels during the first fifth of a second.

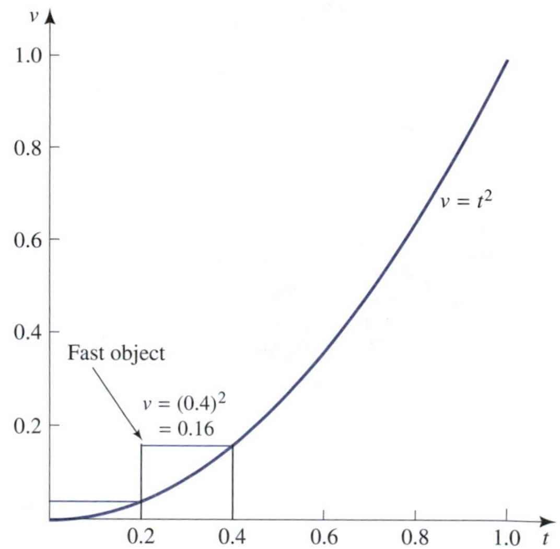

Figure 6.6
The area of the rectangle is $(0.2) \cdot (0.4)^2 = 0.032$. This area is also the distance the fast object travels during the second fifth of a second.

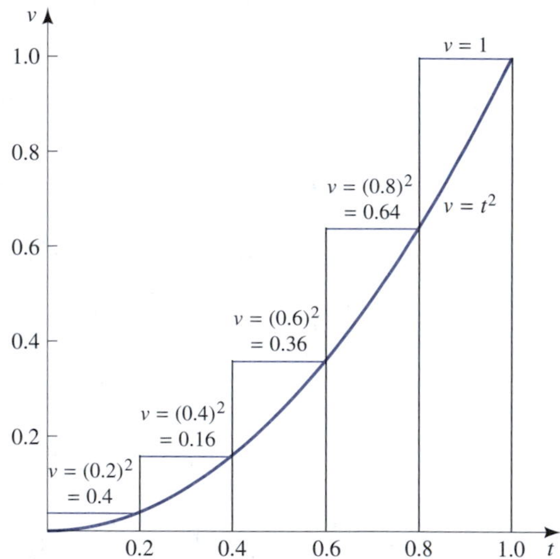

Figure 6.7
The distance the fast object travels is given by the sum of the areas of the five rectangles. This must be greater than the distance our object travels on [0, 1].

You can obtain Figure 6.7 on your computer or graphing calculator. (See the Technology Resource Manual.) You can also obtain this figure in Interactive Illustration 6.1, which follows Example 1.

Let us now determine a lower estimate of the distance traveled. Since the velocity of our object is always greater or equal to zero at any time during the time interval [0, 0.2], the distance our object travels during the first fifth of a second is greater or equal to 0. For future reference we notice that our object has 0 velocity at the *left-hand* endpoint of the first interval, and we can visualize the distance 0 as the area of a rectangle with height 0 and width 0.2.

Since the velocity of our object is always *increasing*, the velocity during the second fifth of a second is greater than at time $t = 0.2$, which is located at the left-hand endpoint of the second interval. The velocity at $t = 0.2$ is $(0.2)^2 = 0.04$. Therefore, during the second fifth of a second the distance our object travels is greater or equal to that of a slow object traveling with a constant velocity of 0.04 indicated in Figure 6.8. This latter distance is $v \cdot t = (0.04)(0.2) = 0.008$ foot and is also the area of the rectangle indicated in Figure 6.8. (The first rectangle on the first subinterval has height equal to zero, and therefore does not show in the figure.)

Continuing in this way, we obtain Figure 6.9. A lower estimate of the distance traveled on the interval [0, 1] is then:

$$\begin{aligned}\text{lower estimate} &= (0)(0.2) + (0.04)(0.2) + (0.16)(0.2) + (0.36)(0.2) + (0.64)(0.2) \\ &= 0.24\end{aligned}$$

You can obtain Figure 6.9 on your computer or graphing calculator. (See the Technology Resource Manual.)

In summary we have

$$0.24 \leq \text{distance traveled} \leq 0.44$$
$$\text{upper estimate} - \text{lower estimate} = 0.44 - 0.24 = 0.2$$

Remark We have an upper estimate and a lower estimate and know that the distance traveled is between these two estimates. It is natural to take as a final estimate (for $n = 5$) to be the average of these two. We obtain

$$\frac{0.22 + 0.44}{2} = 0.33$$

This would be our best guess at the distance traveled at this point.

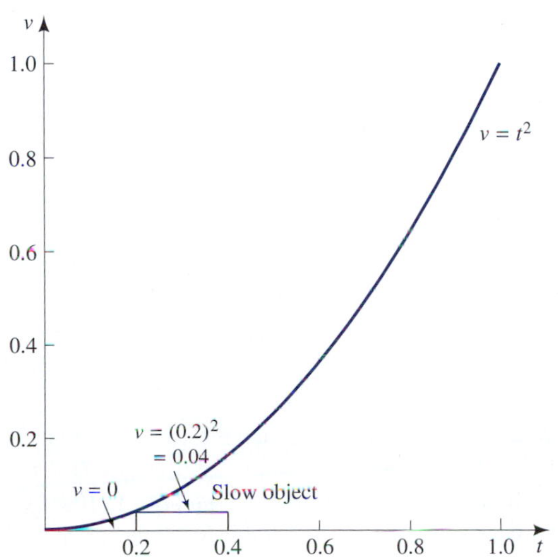

Figure 6.8
The area of the rectangle is $(0.2) \cdot (0.2)^2 = 0.008$. This area is also the distance the slow object travels during the second fifth of a second.

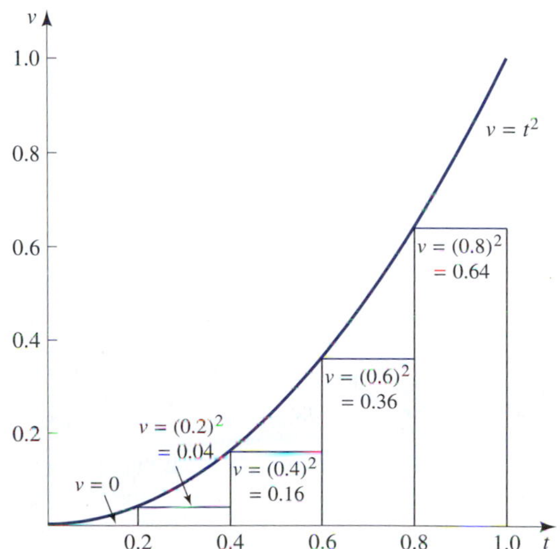

Figure 6.9
The distance the slow object travels is given by the sum of the areas of the five rectangles. (The first rectangle has height 0.) This must be less than the distance our object travels on [0, 1].

Figure 6.10 overlays Figure 6.9 with Figure 6.7. The total area of the five shaded rectangles is precisely the lower estimate of the distance traveled by our object, while the total area of the five tall rectangles is precisely the upper estimate of the distance traveled by the object. The total area of the five unshaded small rectangles is the difference between the upper and lower estimates. If we slide these unshaded rectangles off to the right as indicated in Figure 6.10, we see that the total area of these rectangles must be

$$[f(1) - f(0)](0.2) = [1 - 0](0.2) = 0.2$$

This is the value we obtained earlier.

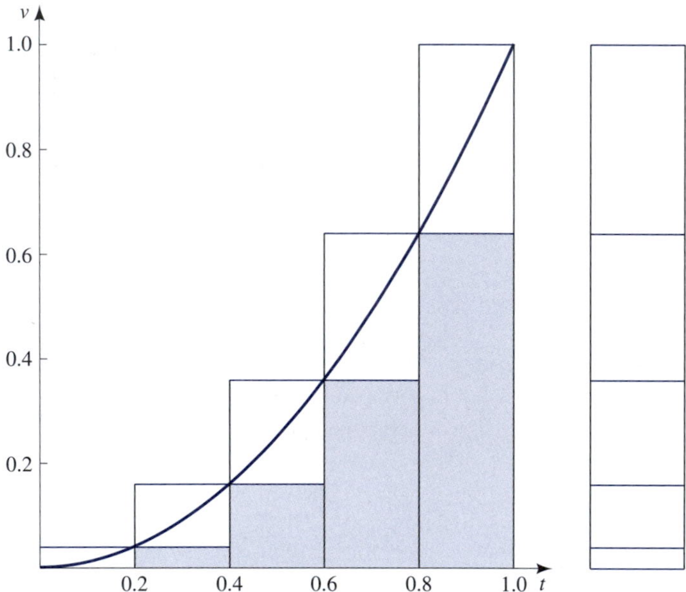

Figure 6.10
The total area of the shaded rectangles equals the left-hand sum, and the total area of the bigger rectangles equals the right-hand sum. The area of the rectangle on the right is the difference between the right- and left-hand sums.

EXAMPLE 1 **Estimating Distance Traveled**

Now divide the t interval [0, 1] into 10 subintervals and repeat the above process.

Solution

Let $t_0 = 0$, $t_1 = 0.1$, $t_2 = 0.2, \ldots, t_{10} = 1$. Since the velocity at any time during the first tenth of a second must be less than the velocity at time $t = 0.1$, the distance traveled during the first tenth of a second must be less than $f(t_1) \cdot (0.1) = (0.1)^2(0.1) = 0.001$, and so on. An upper estimate of the distance traveled is then

$$\text{upper estimate} = f(t_1) \cdot (0.1) + f(t_2) \cdot (0.1) + \cdots + f(t_{10}) \cdot (0.1)$$

In a similar fashion,

$$\text{lower estimate} = f(t_0) \cdot (0.1) + f(t_1) \cdot (0.1) + \cdots + f(t_9) \cdot (0.1)$$

Instead of calculating these by hand, we use our graphing calculators or computers. The two screens obtained are overlaid and shown in Figure 6.11.

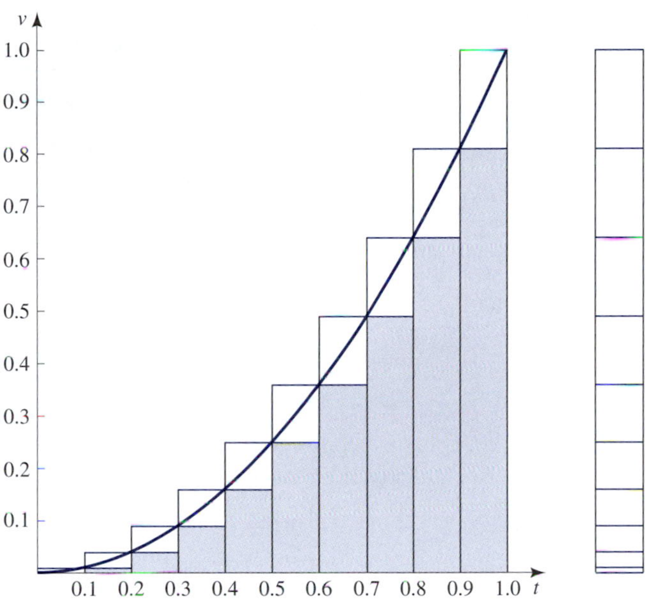

Figure 6.11
Notice that doubling the number of rectangles from that in Figure 5.10 halves the difference between right- and left-hand sums.

The numerical values obtained are as follows:

$$\text{upper estimate} = 0.285$$
$$\text{lower estimate} = 0.385$$

Therefore, we have

$$0.285 \leq \text{distance traveled} \leq 0.385$$
$$\text{upper estimate} - \text{lower estimate} = 0.385 - 0.285 = 0.1$$ ■

Remark We have an upper estimate and a lower estimate using $n = 10$ and know that the distance traveled is between these two estimates. It is natural to take as a final estimate (for $n = 10$) to be the average of these two. We obtain

$$\frac{0.285 + 0.385}{2} = 0.335$$

This would be our best guess at the distance traveled at this point.

Notice that by doubling the number of subintervals, we have halved the difference between the upper estimate and the lower estimate. This difference is the total area of the 10 lightly shaded rectangles in Figure 6.11. This is seen to be

$$[f(1) - f(0)](0.1) = [1 - 0](0.1) = 0.1$$

confirming the value given above.

Interactive Illustration 6.1

Left- and Right-Hand Sums

This interactive illustration produces the left- and right-hand sums in the previous example and allows you to take other values of n.

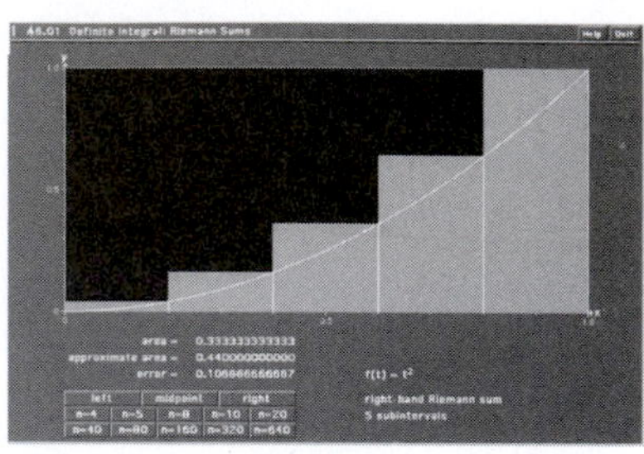

1. Take $n = 5$, and reproduce the associated rectangles for the left- and right-hand sums and their numerical values.
2. Take $n = 10$, and reproduce the associated rectangles for the left- and right-hand sums and their numerical values.
3. Take larger values of n. Comment on what you see.

In general, suppose we have an object traveling at a velocity given by a nonnegative continuous function $v = f(t)$ on the interval $[a, b]$. We divide the interval into n subintervals of equal length $\Delta t = \dfrac{b-a}{n}$. The endpoints of the subintervals are $a = t_0, t_1, t_2, \ldots, t_n = b$. We then define

$$\begin{aligned}\text{right-hand sum} &= f(t_1)\Delta t + f(t_2)\Delta t + \cdots + f(t_n)\Delta t \\ \text{left-hand sum} &= f(t_0)\Delta t + f(t_1)\Delta t + \cdots + f(t_{n-1})\Delta t\end{aligned}$$

We use the term **right-hand sum** because we use the times at the right-hand ends of the subintervals and use the term **left-hand sum** because we use the times at the left-hand ends of the subintervals.

EXAMPLE 2 **Estimating Distance Traveled**

Suppose an object travels with a velocity in feet per second given by $v = f(t) = \dfrac{1}{t}$ on the t interval $[1, 2]$ where t is in seconds. Estimate the distance traveled taking left- and right-hand sums for (a) $n = 4$ and (b) $n = 8$. Sketch the rectangles associated with these sums.

Solution

(a) We divide the interval $[1, 2]$ into $n = 4$ subintervals of equal length

$$\Delta t = \frac{b-a}{n} = \frac{2-1}{4} = 0.25$$

The endpoints of the subintervals are then $t_0 = 1$, $t_1 = 1.25$, $t_2 = 1.5$, $t_3 = 1.75$, and $t_4 = 2$. The right-hand sum is

$$\begin{aligned}\text{right-hand sum} &= f(t_1)\Delta t + f(t_2)\Delta t + f(t_3)\Delta t + f(t_4)\Delta t \\ &= f(1.25)(0.25) + f(1.5)(0.25) + f(1.75)(0.25) + f(2)(0.25) \\ &= \frac{1}{1.25}(0.25) + \frac{1}{1.5}(0.25) + \frac{1}{1.75}(0.25) + \frac{1}{2}(0.25) \\ &= 0.6345\end{aligned}$$

to four decimal places. The corresponding rectangles are shown in Figure 6.12. This is also shown in interactive illustration 6.2, which follows this example.

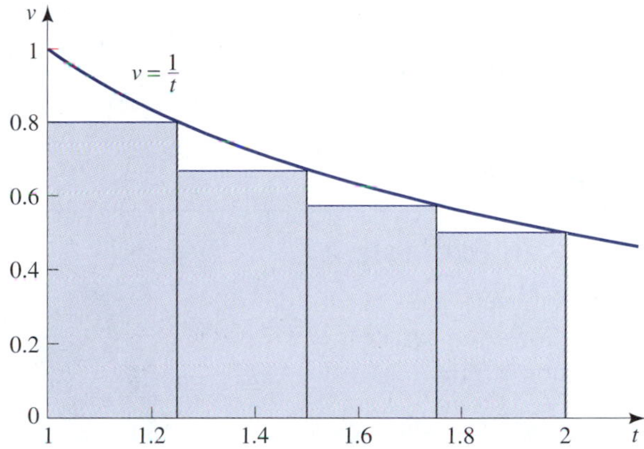

Figure 6.12

The left-hand sum is

$$\begin{aligned} \text{left-hand sum} &= f(t_0)\Delta t + f(t_1)\Delta t + f(t_2)\Delta t + f(t_3)\Delta t \\ &= f(1)(0.25) + f(1.25)(0.25) + f(1.5)(0.25) + f(1.75)(0.25) \\ &= \frac{1}{1}(0.25) + \frac{1}{1.25}(0.25) + \frac{1}{1.5}(0.25) + \frac{1}{1.75}(0.25) \\ &= 0.7595 \end{aligned}$$

to four decimal places. The corresponding rectangles are shown in Figure 6.13. Notice that since the velocity is always *decreasing*, the right-hand sum is a lower estimate, while the left-hand sum is an upper estimate. Summarizing, we have

$$\text{right-hand sum} = 0.6345 \leq \text{distance traveled} \leq 0.7595 = \text{left-hand sum}$$

We also have

$$\text{upper estimate} - \text{lower estimate} = 0.7595 - 0.6345 = 0.125$$

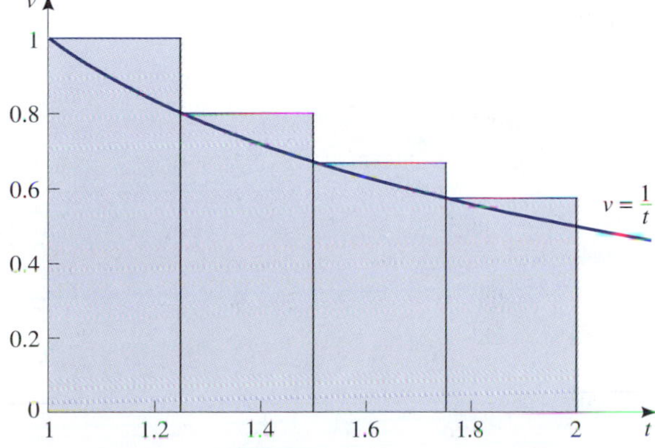

Figure 6.13

Remark Again we have an upper estimate and a lower estimate and know that the distance traveled is between these two estimates. Taking as a final estimate (for $n = 4$) to be the average of these two, we obtain

$$\frac{0.6345 + 0.7595}{2} = 0.697$$

This would be our best guess at the distance traveled at this point.

Interactive Illustration 6.2

Left- and Right-Hand Sums in Example 2

This interactive illustration produces the left- and right-hand sums in Example 2 and allows you to take other values of n. Click on the function $f(t) = 1/t$.

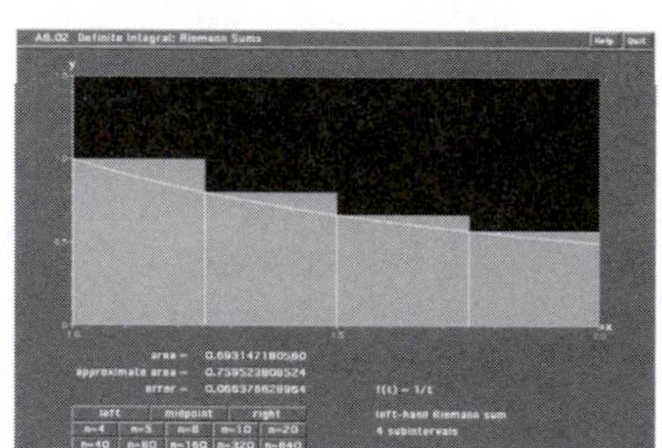

1. Take $n = 4$, and reproduce the associated rectangles for the left- and right-hand sums and their numerical values.
2. Take $n = 8$, and reproduce the associated rectangles for the left- and right-hand sums and their numerical values.
3. Take larger values of n. Comment on what you see.

Figure 6.14 overlays Figure 6.12 with Figure 6.13. The total area of the four shaded rectangles is precisely the lower estimate of the distance traveled by our object, while the total area of the four tall rectangles is precisely the upper estimate of the distance traveled by the object. The total area of the four unshaded small rectangles is the difference between the upper and lower estimates. If we slide these unshaded rectangles off to the right as indicated in Figure 6.14, we see that the total area of these rectangles must be

$$|f(2) - f(1)|(0.25) = |0.5 - 1|(0.25) = 0.125$$

This is the value we obtained earlier.

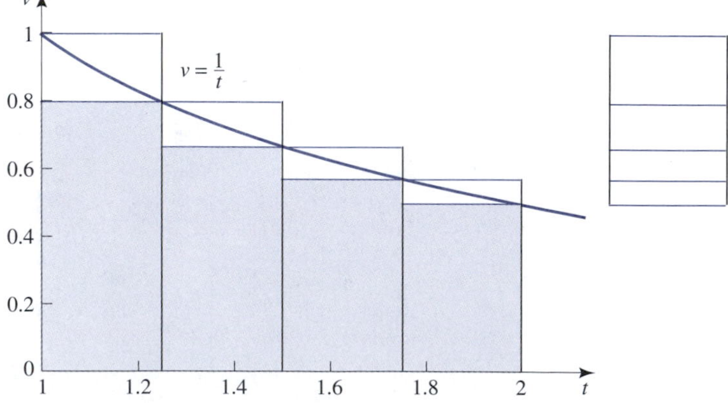

Figure 6.14

(b) We divide the interval $[1, 2]$ into $n = 8$ subintervals of equal length

$$\Delta t = \frac{b-a}{n} = \frac{2-1}{8} = 0.125$$

The endpoints of the subintervals are then $t_0 = 1, t_1 = 1.125, t_2 = 1.25, \ldots, t_7 = 1.875$, and $t_8 = 2$. The right-hand sum is

$$\begin{aligned}
&\text{right-hand sum} \\
&\quad = f(t_1)\Delta t + f(t_2)\Delta t + \cdots + f(t_7)\Delta t + f(t_8)\Delta t \\
&\quad = f(1.125)(0.125) + f(1.25)(0.125) + \cdots + f(1.875)(0.125) + f(2)(0.125) \\
&\quad = \frac{1}{1.125}(0.125) + \frac{1}{1.25}(0.125) + \cdots + \frac{1}{1.875}(0.125) + \frac{1}{2}(0.125) \\
&\quad = 0.6629
\end{aligned}$$

to four decimal places. The corresponding rectangles are shown in Figure 6.15.

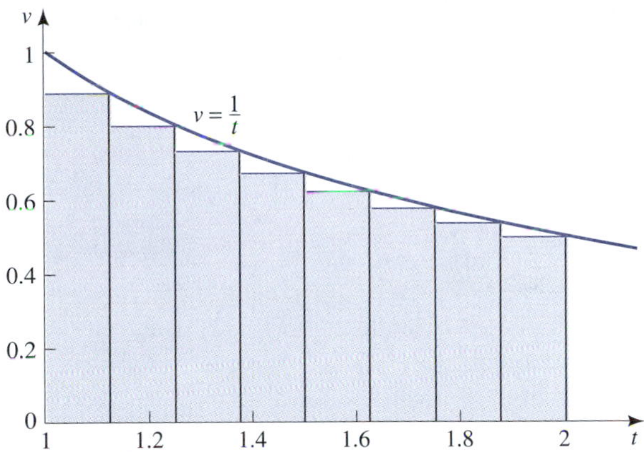

Figure 6.15

The left-hand sum is

$$\begin{aligned}
&\text{left-hand sum} \\
&\quad = f(t_0)\Delta t + f(t_1)\Delta t + \cdots + f(t_6)\Delta t + f(t_7)\Delta t \\
&\quad = f(1)(0.125) + f(1.125)(0.125) + \cdots + f(1.75)(0.125) + f(1.875)(0.125) \\
&\quad = \frac{1}{1}(0.125) + \frac{1}{1.125}(0.125) + \cdots + \frac{1}{1.75}(0.125) + \frac{1}{1.875}(0.125) \\
&\quad = 0.7254
\end{aligned}$$

to four decimal places. The corresponding rectangles are shown in Figure 6.16. Notice again that since the velocity is always *decreasing*, the right-hand sum is a lower estimate, while the left-hand sum is an upper estimate. Summarizing, we have

$$\text{right-hand sum} = 0.6629 \le \text{distance traveled} \le 0.7254 = \text{left-hand sum}$$

We also have

$$\text{upper estimate} - \text{lower estimate} = 0.7254 - 0.6629 = 0.0625$$

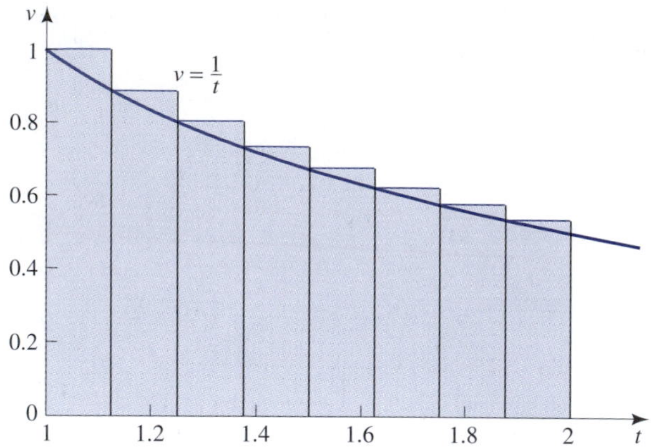

Figure 6.16

Remark Again we have an upper estimate and a lower estimate and know that the distance traveled is between these two estimates. Taking as a final estimate (for $n = 8$) to be the average of these two, we obtain

$$\frac{0.6629 + 0.7254}{2} = 0.69415$$

This would be our best guess at the distance traveled at this point.

Figure 6.17 overlays Figure 6.15 with Figure 6.16. The total area of the eight shaded rectangles is precisely the lower estimate of the distance traveled by our object, while the total area of the eight tall rectangles is precisely the upper estimate of the distance traveled by the object. The total area of the eight unshaded small rectangles is the difference between the upper and lower estimates. If we slide these unshaded rectangles off to the right as indicated in Figure 6.17, we see that the total area of these rectangles must be

$$|f(2) - f(1)|(0.125) = |0.5 - 1|(0.125) = 0.0625$$

This is the value we obtained earlier.

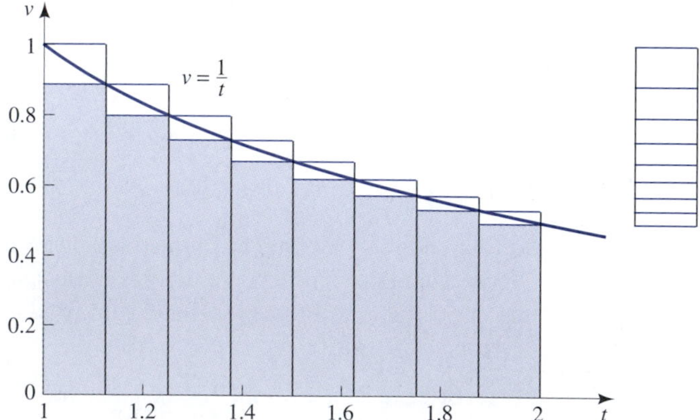

Figure 6.17

■

Exploration 1

Bounds on the Distance

By drawing graphs, show that if $v = f(t)$ is always increasing, then we must have

left-hand sum $\leq$ distance traveled $\leq$ right-hand sum

Show also that if $v = f(t)$ is always decreasing, then we must have

right-hand sum $\leq$ distance traveled $\leq$ left-hand sum

In the two examples we have considered so far, the functions were either always increasing or always decreasing on the interval of interest. Thus the left- and right-hand sums always bounded the area, giving a very convenient estimate of the error of our estimates. This is not necessarily the case for all functions.

EXAMPLE 3 **Estimating Distance Traveled**

Suppose an object travels with a velocity in feet per second given by $v = f(t) = t^2 - 2t + 2$ on the t-interval $[0, 2]$ where t is in seconds. Estimate the distance traveled, taking the right-hand sum for $n = 5$. Sketch the rectangles associated with this sum.

Solution

We divide the interval $[0, 2]$ into $n = 5$ subintervals of equal length

$$\Delta t = \frac{b-a}{n} = \frac{2-0}{5} = 0.4$$

The endpoints of the subintervals are then $t_0 = 0$, $t_1 = 0.4$, $t_2 = 0.8$, $t_3 = 1.2$, $t_4 = 1.6$, and $t_5 = 2$. The right-hand sum is

$$\begin{aligned}
\text{right-hand sum} \\
&= f(t_1)\Delta t + f(t_2)\Delta t + f(t_3)\Delta t + f(t_4)\Delta t + f(t_5)\Delta t \\
&= f(0.4)(0.4) + f(0.8)(0.4) + f(1.2)(0.4) + f(1.6)(0.4) + f(2)(0.4) \\
&= (1.36)(0.4) + (1.04)(0.4) + (1.04)(0.4) + (1.36)(0.4) + (2)(0.4) \\
&= 2.72
\end{aligned}$$

■

The corresponding rectangles are shown in Figure 6.18. This also can be seen by using Interactive Illustration 6.3. Notice that some of the rectangles are above the velocity curve and some are below. Thus, we do not know whether the right-hand sum is an upper estimate or a lower estimate.

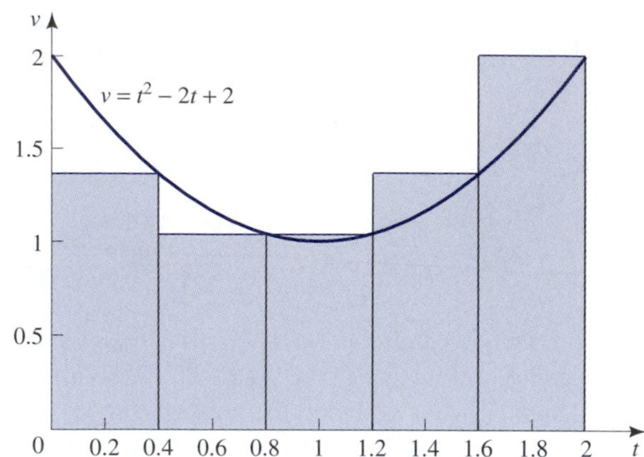

Figure 6.18

Interactive Illustration 6.3

Left- and Right-Hand Sums in Example 3

This interactive illustration produces the left- and right-hand sums in Example 3 and allows you to take other values of n. Click on the function $f(t) = t^2 - 2t + 2$.

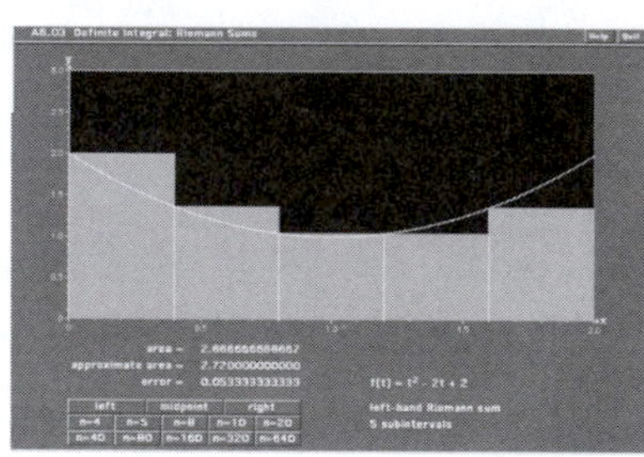

1. Take $n = 5$ and reproduce the associated rectangles for the left- and right-hand sums and their numerical values.
2. Take $n = 10$ and reproduce the associated rectangles for the left- and right-hand sums and their numerical values.
3. Take larger values of n. Comment on what you see.

Enrichment: Accuracy of Approximation

We obtain an interesting result if we calculate the difference between the right-hand sum and the left-hand sum.

We have

$$\begin{aligned}\text{right-hand sum} - \text{left-hand sum} &= [f(t_1) + f(t_2) + \cdots + f(t_n)]\Delta t \\ &\quad - [f(t_0) + f(t_1) + \cdots + f(t_{n-1})]\Delta t)] \\ &= f(t_n)\Delta t - f(t_0)\Delta t \\ &= [f(b) - f(a)]\Delta t\end{aligned}$$

Thus, we have the following.

Error for Any Function

For any function $v = f(t)$,

$$|\text{right-hand sum} - \text{left-hand sum}| = |f(b) - f(a)|\Delta t$$

EXAMPLE 4 **Accuracy of Approximation**

Consider the object in Example 1. What would be the difference between the upper and lower estimates for the object in Example 1 if we divided the interval into 100 subintervals? Into 1000 subintervals?

Solution

Since the function $v = f(t)$ is increasing, we must have that the right-hand sum minus the left-hand sum equals

$$[f(1) - f(0)]\Delta t = [1 - 0]\frac{1}{n} = \frac{1}{n}$$

Thus, if $n = 100$, the difference is 0.01, while if $n = 1000$, the difference is 0.001. ■

It is interesting to notice that for the function in Example 1 the difference between the right-hand sum and the left-hand sum was $\frac{1}{n}$. If $n \to \infty$, then this difference goes to zero. Thus, as $n \to \infty$, both the left- and right-hand sums must converge to the exact distance. We will explore this idea in the next section.

Warm Up Exercise Set 6.3

1. An object decelerates with a velocity $v = f(t) = \frac{1}{t}$ on the interval $[1, 2]$. Give an upper and a lower estimate for the distance traveled by dividing the interval $[1, 2]$ into five equal subintervals and calculate by hand the left- and right-hand sums. Sketch a graph similar to Figure 6.14 illustrating the left- and right-hand sums.
2. Identify the region whose area is exactly the distance traveled by the object in Exercise 1 on the interval $[1, 2]$.

Exercise Set 6.3

In Exercises 1 through 10 we will consider approximations to the distance traveled by an object with velocity $v = f(t)$ on the given interval $[a, b]$. For each of these exercises, do the following:

(a) For $n = 5$, make a sketch that illustrates the left- and right-hand sums, showing clearly the five rectangles and x_0, x_1, x_2, x_3, x_4, and x_5.

(b) For $n = 5$, find the left- and right-hand sums. Also calculate the difference between the upper and lower estimates. Calculate the average of the two sums.

(c) Repeat part (b) for $n = 10$.

1. $v = f(t) = 5 - 2t$, $[0, 2]$
2. $v = f(t) = 2t + 1$, $[0, 2]$
3. $v = f(t) = t^2$, $[0, 2]$
4. $v = f(t) = 1/t$, $[1, 3]$
5. $v = f(t) = t^2$, $[1, 3]$
6. $v = f(t) = 1/t$, $[1, 1.5]$
7. $v = f(t) = t^2 + 1$, $[-2, 3]$
8. $v = f(t) = 10t^2$, $[-3, 2]$
9. $v = f(t) = \sqrt{t}$, $[0, 5]$
10. $v = f(t) = t^3$, $[-2, 3]$

In Exercises 11 through 16 we will consider approximations to the distance traveled by an object with velocity $v = f(t)$ on the given interval $[a, b]$. The answers are most easily obtained by using the Integration Kit on the disk provided. For each of these exercises, do the following:

(a) For $n = 4$, make a sketch that illustrates the left- and right-hand sums, showing clearly the four rectangles and x_0, x_1, x_2, x_3, and x_4.

(b) For $n = 4$, find the left- and right-hand sums. Also calculate the difference between the upper and lower estimates. Calculate the average of the two sums.

(c) Repeat parts (a) and (b) with $n = 8$ and 40. Calculate the average of the two sums. Compare your answers with those in part (b).

11. $v = f(t) = \frac{1}{\sqrt{2\pi}} e^{-0.5t^2}$, $[0, 3]$

12. $v = f(t) = te^{-t}$, $[0, 8]$

13. $v = f(t) = \frac{2}{1+e^{-t}}$, $[0, 5]$

14. $v = f(t) = 0.0001t^3 - 0.03t^2 + 0.3t + 1$, $[0, 25]$

15. $v = f(t) = t^3 - 6t^2 + 9t - 1$, $[0, 4]$

16. $v = f(t) = t \sin t$, $[0, 2\pi]$

Applications and Mathematical Modeling

17. Refer to the figure. Estimate the distance traveled on the interval $[0, 10]$. Explain how you obtained your estimate.

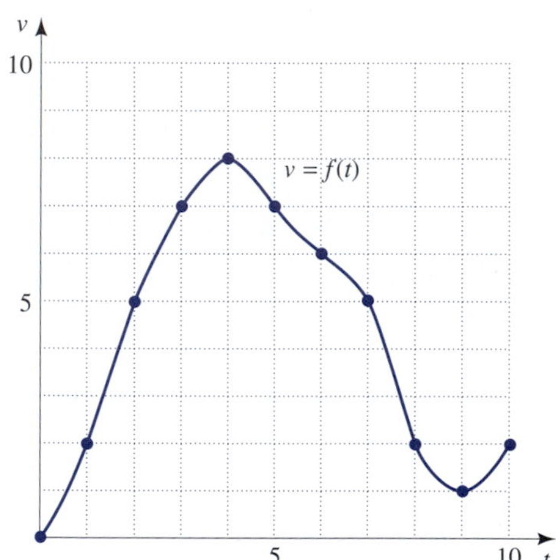

18. Refer to the figure. Estimate the distance traveled on the interval $[0, 10]$. Explain how you obtained your estimate.

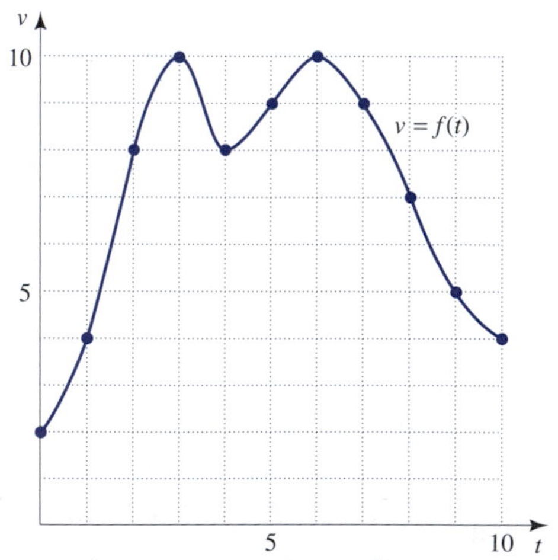

19. An object travels with a velocity function given by the figure. Use the grid to obtain an upper and a lower estimate of the distance traveled by the object. Explain what you are doing.

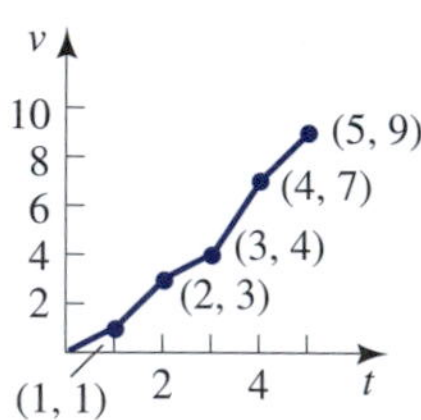

20. An object travels with a velocity function given by the figure. Use the grid to obtain an upper and a lower estimate of the distance traveled by the object. Explain what you are doing.

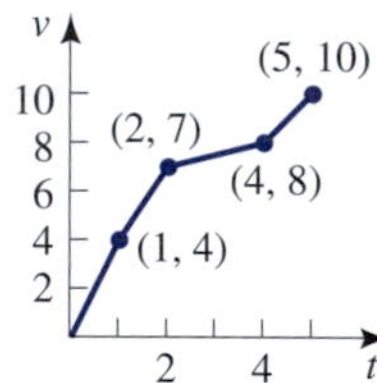

21. The following table gives the velocity at 1-second intervals of an accelerating automobile.

Time (seconds)	0	1	2	3	4
Velocity (feet per second)	10	20	25	40	50

Give an upper and a lower estimate for the distance the car has traveled during the 4-second interval. Explain what you are doing.

22. A person slams on the brakes of an automobile. The following table gives the velocity at 1-second intervals.

Time (seconds)	0	1	2	3	4
Velocity (feet per second)	50	40	25	10	0

Give an upper and a lower estimate for the distance the car has traveled during the 4-second interval. Explain what you are doing.

23. An object travels with a velocity function given by $v = \sqrt{t}$, where t is measured in hours and v is measured in miles per second. What would be the difference between the upper and lower estimates for the object if we divided the interval $[0, 1]$ into 100 subinterval? Into 1000 subintervals?

24. An object travels with a velocity function given by $v = 4 - \sqrt{t}$, where t is measured in hours and v is measured in miles per second. What would be the difference between the upper and lower estimates for the object if we divided the interval $[0, 16]$ into 100 subintervals? Into 1000 subintervals?

25. If $f(a) = f(b)$, then show that the left-hand sum of f on $[a, b]$ equals the right-hand sum.

More Challenging Exercises

26. Refer to Exercise 24. How large must you take n to ensure that the difference between the upper and lower estimates is less than 0.1? 0.01?

27. Refer to Exercise 23. How large must you take n to ensure that the difference between the upper and lower estimates is less than 0.01?

28. Figure 6.3 outlines the method of exhaustion formulated by Eudoxus. Let E be the area of the region inside the circle but outside the triangle P_0 shown in Figure 6.3a. Explain why the area inside the circle but outside the hexagon P_1 shown in Figure 6.3b is less than one half of E.

Solutions to WARM UP EXERCISE SET 6.3

1. We divide the interval $[1, 2]$ into five subintervals of equal length. Then $\Delta t = 0.20$ and $t_0 = 1$, $t_1 = 1.2$, $t_2 = 1.4$, $t_3 = 1.6$, $t_4 = 1.8$, and $t_5 = 2$. From the figure we see that the function $v = 1/t$ is *decreasing*, and therefore, the right-hand sum will be a *lower* estimate and the left-hand sum will be an *upper* estimate of the area we are seeking. Then

right-hand sum

$$= f(t_1)\Delta t + f(t_2)\Delta t + f(t_3)\Delta t + f(t_4)\Delta t + f(t_5)\Delta t$$

$$= f(1.2)(0.20) + f(1.4)(0.20) + f(1.6)(0.20) + f(1.8)(0.20) + f(2)(0.20)$$

$$= 0.20\left(\frac{1}{1.2} + \frac{1}{1.4} + \frac{1}{1.6} + \frac{1}{1.8} + \frac{1}{2}\right)$$

$$\approx 0.64563$$

This is the area of the shaded rectangles in the figure.

left-hand sum

$$= f(t_0)\Delta t + f(t_1)\Delta t + f(t_2)\Delta t + f(t_3)\Delta t + f(t_4)\Delta t$$

$$= f(1)(0.20) + f(1.2)(0.20) + f(1.4)(0.20) + f(1.6)(0.20) + f(1.8)(0.20)$$

$$= 0.20\left(\frac{1}{1} + \frac{1}{1.2} + \frac{1}{1.4} + \frac{1}{1.6} + \frac{1}{1.8}\right)$$

$$\approx 0.74563$$

This is the area of the tall rectangles in the figure. We therefore have

$$0.64563 \le \text{distance traveled} \le 0.74563$$

upper estimate − lower estimate $= 0.74563 - 0.64563 = 0.10$

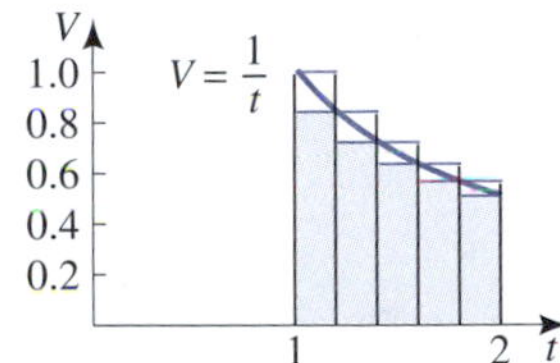

2. The exact distance traveled by the object is the area of the region between the graph of $v = 1/t$ and the t-axis on the interval $[1, 2]$.

6.4 The Definite Integral

APPLICATION The Bombay Plague of 1905–1906

Kermack and McKendrick[4] found that the number $D(t)$ of deaths per week during the Bombay plague of 1905–1906 was approximated by $D(t) = 3960(e^{0.2t-3.4} + e^{-0.2t+3.4})^{-2}$, where t is in weeks from the outbreak. What was the approximate number of deaths during the first 30 weeks of this plague? For the answer, see Example 3 on page 438.

[4] W. O. Kermack and A. G. McKendrick. 1933. Contributions to the mathematical theory of epidemics. *Proc. Roy. Soc. A* 141:94–122.

Finding Distance Exactly

If the nonnegative function $v = f(t)$ gives the velocity of an object on the interval $[a, b]$, we saw in the last section how the left- and right-hand sums approximate the distance traveled by the object on the time interval $[a, b]$. We also saw that the absolute value of the difference of the left- and right-hand sums for monotonic functions is bounded by $|f(b) - f(a)|\Delta t$, making for a convenient error estimate for this important class of functions.

We also noticed that for the velocity function $v = t^2$ on the interval $[0, 1]$ the difference between the right- and left-hand sums was $\dfrac{1}{n}$. Since this goes to zero as $n \to \infty$, we see that the left- and right-hand sums must be heading for the exact distance traveled as $t \to \infty$. Let us now explore this geometrically.

Figure 6.19 overlays the rectangles associated with the right-hand sums for $n = 10$ with those for $n = 5$. Notice that the rectangles associated with the right-hand sums for $n = 10$ (solid line) are always lower or at the same height as those for $n = 5$ (dashed line). First, this clearly implies that the total area under the rectangles for $n = 10$ must be less than that for $n = 5$. We found this to be the case in the last section (0.385 compared to 0.44). Second, we see that the rectangles for $n = 10$ are "snuggling" closer to the graph of the curve than those for $n = 5$.

Figure 6.20 overlays the rectangles associated with the right-hand sums for $n = 20$ with those for $n = 10$. Notice that the rectangles associated with the right-hand sums for $n = 20$ (solid line) are always lower or at the same height as those for $n = 10$ (dashed line). This clearly implies that the total area under the rectangles for $n = 20$ must be less than that for $n = 10$. (We can use our calculators or computers to obtain 0.35875 compared to 0.385). Second, we see that the rectangles for $n = 20$ are "snuggling" closer to the graph of the curve than those for $n = 10$.

We then can see that if we continued to double n, the right-hand sums will continue to become smaller (while always larger than the area under the curve), and

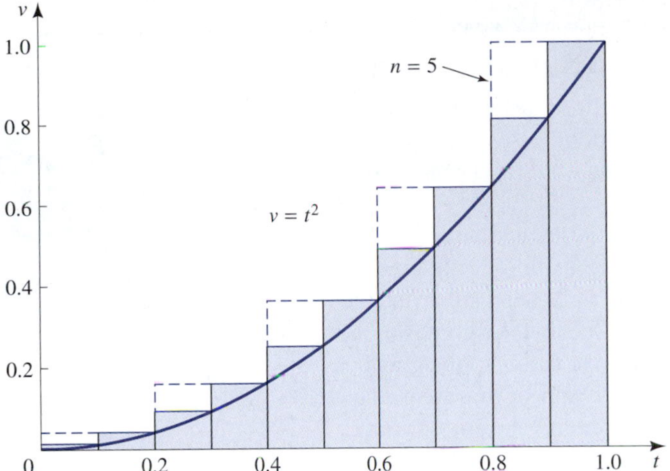

Figure 6.19

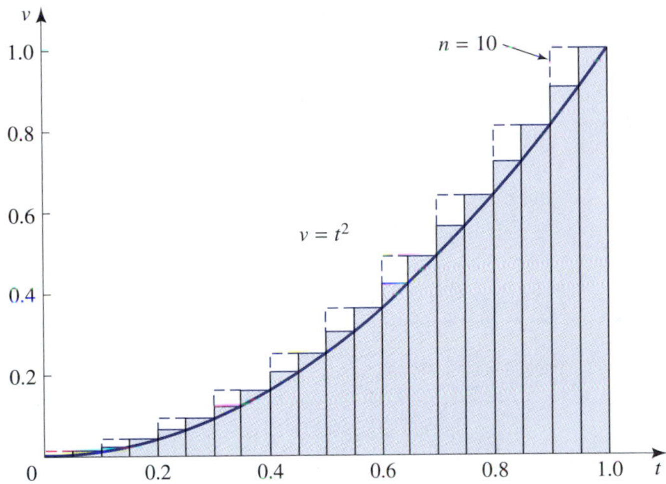

Figure 6.20

the rectangles will "snuggle" ever closer to the curve itself. As $n \to \infty$ the area under the right-hand sums seems to be approaching the area under the curve while at the same time (as we noted above) also approaching the exact distance traveled. This is nicely illustrated in interactive illustration 6.4.

Interactive Illustration 6.4

Left- and Right-Hand Sums for Large n

This interactive illustration illustrates what happens to the rectangles associated with the left- and right-hand sums for $f(t) = t^2$ on the interval [0, 1] as n becomes larger.

Take n to be increasingly larger and notice the relationship between the graphs of the rectangles associated with the left- and right-hand sums and the graph of $y = t^2$. Do you notice that the graphs of the rectangles are "snuggling" ever closer to the graph of the velocity function? Does it appear that the area under the rectangles is getting closer and closer to the area under the curve?

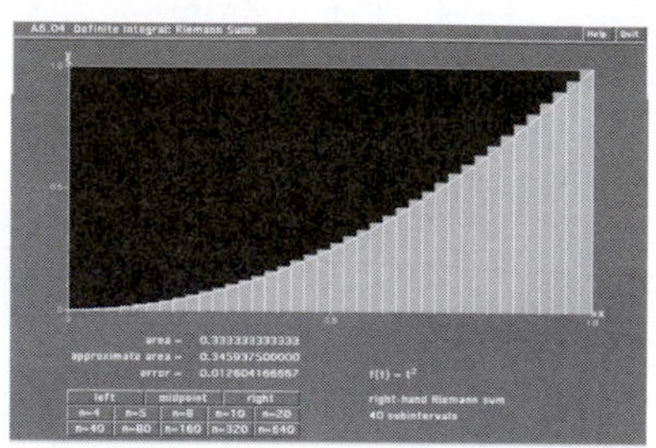

We have thus reached an important conclusion. As $n \to \infty$, the right-hand sums approach both the area under the curve and the distance traveled. Thus, the area under the curve is precisely the distance traveled.

A similar analysis applies to the left-hand sums for the increasing velocity function considered here. As n increases, they also increase, while always less than the area under the curve. As $n \to \infty$, the left-hand sums also approach the area under the curve and the total distance traveled.

Using a graphing calculator or a computer we then create Table 6.1.

Table 6.1

n	Left-Hand Sum	Right-Hand Sum	Difference of Sums	Average of Sums
5	0.24000	0.44000	0.20000	0.34000
10	0.28500	0.38500	0.10000	0.33500
20	0.30875	0.35875	0.05000	0.33375
100	0.32835	0.33835	0.01000	0.33335
1000	0.33283	0.33383	0.00100	0.33333
10,000	0.33328	0.33338	0.00010	0.33333

Table 6.1 confirms our geometric insight. Both the left- and right-hand sums seem to be approaching a limit. The limit appears to be $1/3$. Thus, the area under the curve $v = t^2$ on the interval [0, 1] appears to be $1/3$; therefore, the distance traveled by the object appears to be $1/3$.

Furthermore, notice in Table 6.1 that the left-hand sums are increasing, while the right-hand sums are decreasing, and the limit of $1/3$ is between the two sums.

Also notice from the fourth column in Table 6.1 that the difference between the sums is always $\frac{1}{n}$. We saw in the previous section that this must be the case.

We can do a similar analysis for a velocity function that is always decreasing. The only difference is that in this case the right-hand sum will be a lower estimate, and the left-hand sum will be an upper estimate.

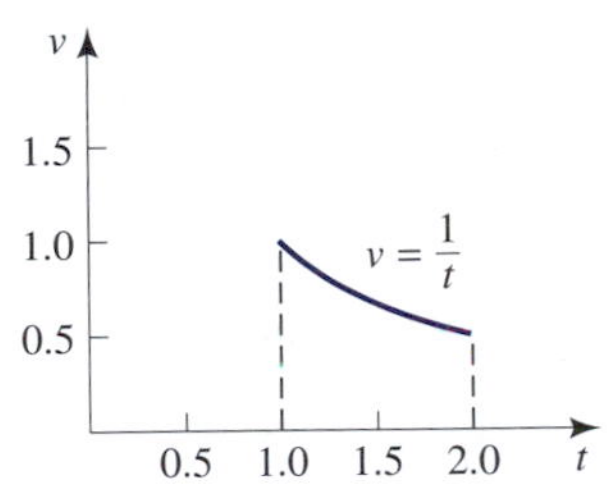

Figure 6.21
The distance traveled is the area under the curve. The only way to find this area is to find the limiting value of the left- and right-hand sums.

EXAMPLE 1 **Finding Distance Exactly**

Suppose an object travels with a velocity in feet per second given by $v = f(t) = \frac{1}{t}$ on the t interval $[1, 2]$ where t is in seconds. Determine the exact distance traveled.

Solution

See Figure 6.21 for a graph of $v = \frac{1}{t}$. Using our computer or graphing calculator, we obtain Table 6.2.

Table 6.2

n	Left-Hand Sum	Right-Hand Sum	Difference of Sums
5	0.74563	0.64563	0.10000
10	0.71877	0.66877	0.05000
100	0.69565	0.69065	0.00500
1000	0.69340	0.69290	0.00050
10,000	0.69317	0.69312	0.00005

Both the left- and right-hand sums are heading for 0.6931 to four decimal places. You just might recognize this number! When we looked in Section 1.5 at the time it takes for an account that is compounding continuously to double, we encountered the number $\ln 2 \approx 0.6931$. Apparently, the area under $y = 1/x$ on the interval $[1, 2]$ is converging to $\ln 2$. We will be able to confirm this in the next section. ■

Notice in Table 6.2 that since $v = f(t) = 1/t$ is a decreasing function, the left-hand sums decrease, while the right-hand sums increase with increasing n. Also, recall from the previous section that the difference between the sums should be bounded by

$$[f(1) - f(2)]\Delta x = \left(1 - \frac{1}{2}\right)\left(\frac{1}{n}\right) = \frac{1}{2n}$$

This is also confirmed in Table 6.2.

Interactive Illustration 6.5

Left- and Right-Hand Sums for Large *n*

This interactive illustration illustrates what happens to the rectangles associated with the left- and right-hand sums for $f(t) = \frac{1}{t}$ on the interval $[1, 2]$ as n becomes larger.

Take n to be increasingly larger, and notice the relationship between the graphs of the rectangles associated with the left- and right-hand sums and the graph of $y = \frac{1}{t}$. Do you notice that the graphs of the rectangles are "snuggling" ever closer to the graph of the velocity function? Does it appear that the area under the rectangles is getting closer and closer to the area under the curve?

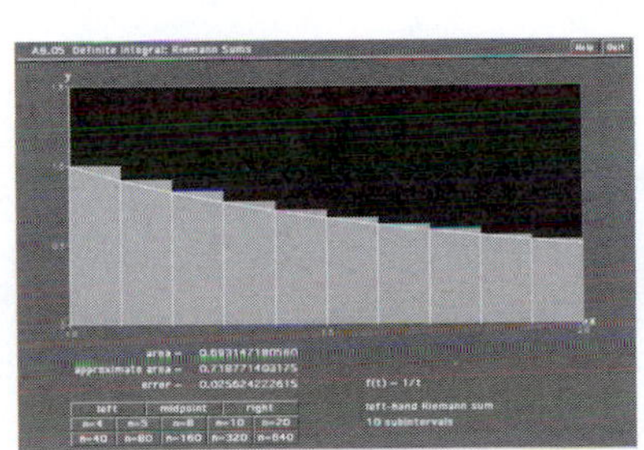

For functions that are not monotonic, we still will have the left- and right-hand sums approaching the area under the curve, and hence the total distance traveled. The left- and right-hand sums might not, however, be bounds on the area or distance traveled. We have the following.

Total Distance Traveled

Suppose an object travels at a velocity given by the nonnegative continuous function $v = f(t)$ on the interval $[a, b]$. Divide the interval into n subintervals of equal length $\Delta t = \dfrac{b-a}{n}$. The endpoints of the subintervals are $a = t_0, t_1, t_2, \ldots, t_n = b$. We then define

$$\text{right-hand sum} = f(t_1)\Delta t + f(t_2)\Delta t + \cdots + f(t_n)\Delta t$$
$$\text{left-hand sum} = f(t_0)\Delta t + f(t_1)\Delta t + \cdots + f(t_{n-1})\Delta t$$

Furthermore, the limit as $n \to \infty$ of the left-hand sum equals the limit as $t \to \infty$ of the right-hand sum. This common limit is the total distance traveled by the object on the interval $[a, b]$. This distance is also the area of the region between the graph of $v = f(t)$ and the t-axis on the interval $[a, b]$.

We now know that the distance traveled by an object on the t-interval $[a, b]$ with velocity $v = f(t)$ equals the area of the region between the graph of $v = f(t)$ and the t-axis on the interval $[a, b]$. In certain instances we can find this area by using known geometric formulas. The following example is such a case.

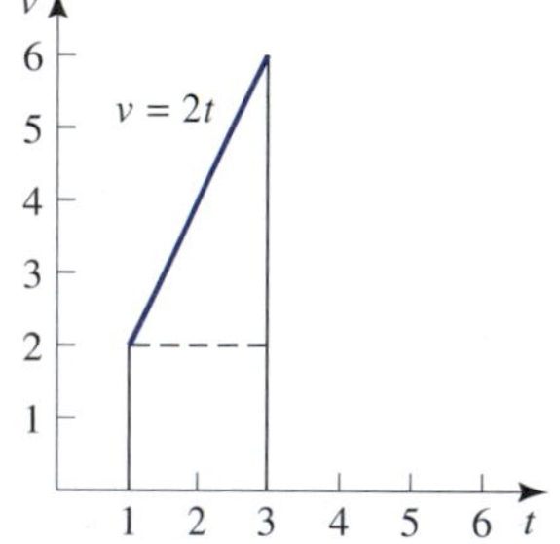

Figure 6.22
The distance traveled on the interval [1, 3] is the area under the graph.

EXAMPLE 2 **Finding Distance Exactly**

Find the distance traveled by the object with a velocity of $v = 2t$ on the interval $[1, 3]$ by finding the area of an appropriate geometric region.

Solution

A graph of $v = 2t$ is shown in Figure 6.22. The distance traveled by the object on the interval $[1, 3]$ is the area of the region between the graph of the line $v = 2t$ and the t-axis on the interval $[1, 3]$. The region is a triangle sitting on a rectangle of dimensions 2×2. The area of the rectangle is 4, and the area of the triangle is 4. Thus, the total area and the distance traveled by the object is 8. ■

Exploration 1

Estimating π

The equation $x^2 + y^2 = 4$ is the equation of a circle of radius 2 centered at $(0, 0)$. The area of this circle is $(2)^2\pi = 4\pi$. The upper half of this circle is the graph of the function $y = f(x) = \sqrt{4 - x^2}$. The area under the graph of this function in the first quadrant is then just $\frac{1}{4}(4\pi) = \pi$ (Figure 6.23). Find an estimate of π by estimating the area of the region between the graph of $y = \sqrt{4 - x^2}$ and the x-axis on the interval $[0, 2]$. Do this by reproducing Table 6.2 for this function. Note that $\Delta x = \dfrac{2-0}{n} = \dfrac{2}{n}$. Alternatively, you can use Interactive Illustration 6.6 for the values of n provided there.

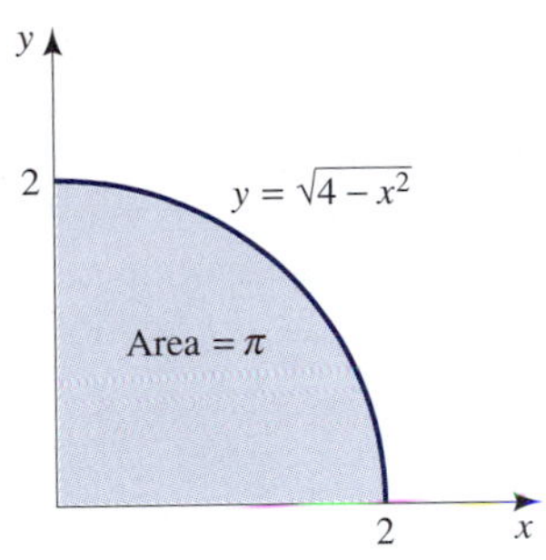

Figure 6.23
The area under the curve $y = \sqrt{4 - x^2}$ on $[0, 2]$ is π.

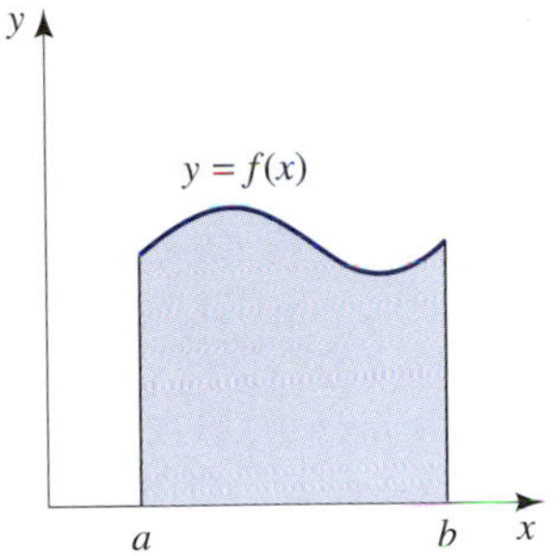

Figure 6.24
We want to find the area between the graph of $y = f(x)$ and the x-axis on $[a, b]$.

The Definite Integral

We have seen that we can assign an important *physical* interpretation to the area of the region between the graph of a nonnegative function and the horizontal axis on some interval. But we can also assign important economic, business, and biological interpretations to such areas. Therefore, we now consider finding the area between the graph of a nonnegative function $y = f(x)$ and the x-axis on the interval $[a, b]$ (Figure 6.24).

Before continuing, let us introduce the suggestive $\sum$ (sigma or sum) notation for the left- and right-hand sums. We divide the x interval $[a, b]$ into n subintervals of equal length

$$\Delta x = \frac{b-a}{n}$$

Let the endpoints of these subintervals be $a = x_0, x_1, x_2, \dots, x_n = b$. We have the following.

$$\begin{aligned}\text{right-hand sum} &= f(x_1)\Delta x + f(x_2)\Delta x + \cdots + f(x_n)\Delta x \\ &= \sum_{k=1}^{n} f(x_k)\Delta x\end{aligned}$$

$$\begin{aligned}\text{left-hand sum} &= f(x_0)\Delta x + f(x_1)\Delta x + \cdots + f(x_{n-1})\Delta x \\ &= \sum_{k=0}^{n-1} f(x_k)\Delta x\end{aligned}$$

DEFINITION **Definite Integral**

Suppose $f(x)$ is a continuous function on the finite interval $[a, b]$. Let $\Delta x = \frac{b-a}{n}$. Then the right-hand sum $\sum_{k=1}^{n} f(x_k)\Delta x$ and the left-hand sum $\sum_{k=0}^{n-1} f(x_k)\Delta x$ satisfy

$$\lim_{n\to\infty} \sum_{k=1}^{n} f(x_k)\Delta x = \lim_{n\to\infty} \sum_{k=0}^{n-1} f(x_k)\Delta x$$

We refer to this common limit as the **definite integral** of f from a to b and write it as

$$\int_a^b f(x)\, dx$$

We refer to a as the **lower limit** of integration and to b as the **upper limit** of integration and the interval $[a, b]$ as the **interval of integration.** If $f(x)$ is nonnegative on the interval $[a, b]$, we interpret the definite integral $\int_a^b f(x)\, dx$ as the area of the region between the graph of $y = f(x)$ and the x-axis on the interval $[a, b]$. The sums $\sum_{k=1}^{n} f(x_k)\Delta x$ and $\sum_{k=0}^{n-1} f(x_k)\Delta x$ are called **Riemann sums.**

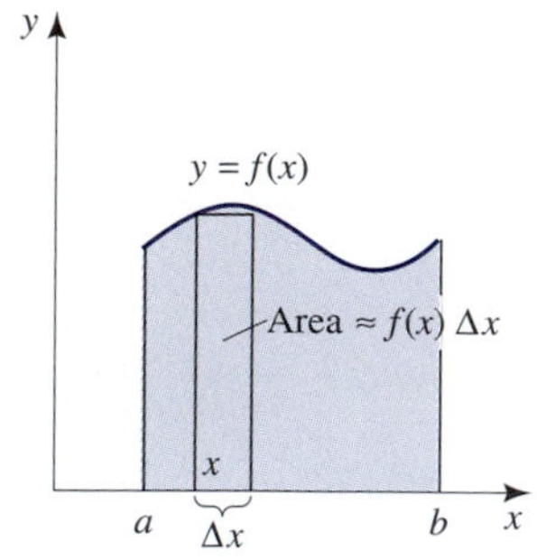

Figure 6.25

The symbol $\int$ suggests summation and the term $f(x)\,dx$ suggests $f(x)\,\Delta x$, which is the area of a thin rectangle. Thus, $\int_a^b f(x)\,dx$ suggests summing up the areas of the thin rectangles from a to b.

The notation for the definite integral, $\int_a^b f(x)\,dx$, is very suggestive of a Riemann sum. One can think of the integral sign $\int$ as an elongated S (or $\sum$) that stands for "sum." Also, the term $f(x)\,dx$ suggests $f(x)\,\Delta x$, which is an area of a rectangle (Figure 6.25). The notation for the definite integral thus suggests summing the area of such rectangles from a to b. This interpretation is widely used in the applications and can yield significant insights.

EXAMPLE 3 **The Bombay Plague of 1905–1906**

Kermack and McKendrick[5] found that the number $D(t)$ of deaths per week during the Bombay plague of 1905–1906 was approximated by $D(t) = 3960(e^{0.2t-3.4} + e^{-0.2t+3.4})^{-2}$, where t is in weeks from the outbreak. Find the approximate number of deaths during the first 30 weeks of this plague.

Solution

If $N(t)$ is the number of deaths t weeks after the outbreak, then $D(t) = \frac{d}{dt}N(t)$. The term $D(t)$ is the rate of change of the number of deaths with respect to time. Thus, $D(t)$ is analogous to the velocity $v(t)$ of an object, since $v(t)$ is the rate of change of distance with respect to time.

We can then find the number of deaths in a way very similar to how we have found the distance traveled when given the velocity. If we divide the t interval $[0, 30]$ into n equal subintervals with $0 = t_0, t_1, \ldots, t_n = 30$ denoting the endpoints, then $D(t_1)\,\Delta t$ approximates the number of deaths during the first interval of time $[t_0, t_1]$, $D(t_2)\,\Delta t$ approximates the number of deaths during the second interval of time $[t_1, t_2]$, and so on. (For example, if $\Delta t = 1$, then $D(t_1)\,\Delta t$ approximates the number of deaths in the first week.) Thus, the right-hand sum

$$\sum_{k=1}^{n} D(t_k)\,\Delta t = D(t_1)\,\Delta t + D(t_2)\,\Delta t + \cdots + D(t_n)\,\Delta t$$

approximates the number of deaths between the time $t = 0$ and $t = 30$. Then, $\int_0^{30} D(t)\,dt$ will give, according to the model, the exact number of deaths from the time $t = 0$ to $t = 30$. We assume that no deaths occurred at the very instant of the outbreak. Therefore, the number of deaths during the first 30 weeks of the outbreak is given by the definite integral

$$\int_0^{30} 3960(e^{0.2t-3.4} + e^{-0.2t+3.4})^{-2}\,dx$$

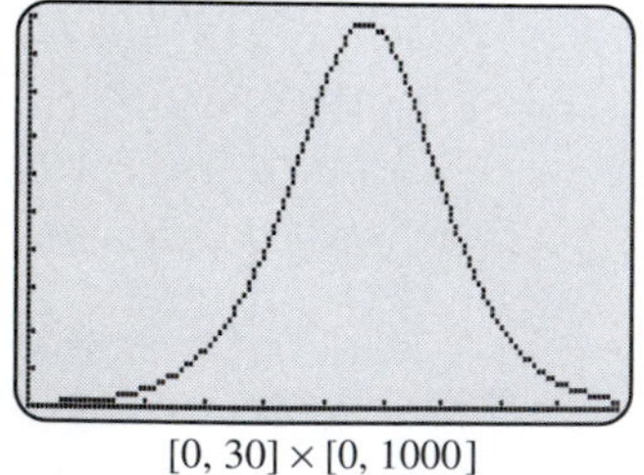

[0, 30] × [0, 1000]

Screen 6.2

A graph of $y = 3960(e^{0.2t-3.4} + e^{-0.2t+3.4})^{-2}$ giving the deaths per week during the Bombay plague of 1905–1906.

To determine this integral, we first use a computer or graphing calculator to construct Table 6.3.

The graph is shown in Screen 6.2 using a window with dimensions $[0, 30]$ by $[0, 1000]$. The function is not monotonic on the interval $[0, 30]$. We see from Table 6.3 that both the left- and right-hand sums are converging to 9835, rounded to the nearest integer. Thus, $\int_0^{30} 3960(e^{0.2t-3.4} + e^{-0.2t+3.4})^{-2}\,dx \approx 9835$. According to this model there were about 9835 deaths due to this plague during the first 30 weeks. Since $D(t)$ is nonnegative on $[0, 30]$, this is also the area between the graph of this

---

[5] Ibid.

Table 6.3

n	Left-Hand Sum	Right-Hand Sum
2	12,772.2	13,030.3
4	9,688.8	9,817.9
8	9,789.8	9,854.3
30	9,825.2	9.842.4
100	9,832.0	9,837.2
200	9,833.4	9,835.9
2000	9,834.5	9,834.8

function and the t-axis on $[0, 30]$. Notice that neither the left-hand sums nor the right-hand sums always increase or always decrease. Notice that the area 9835 is not bracketed by each of the left- and right-hand sums. ■

Interactive Illustration 6.7

Left- and Right-Hand Sums for Large n

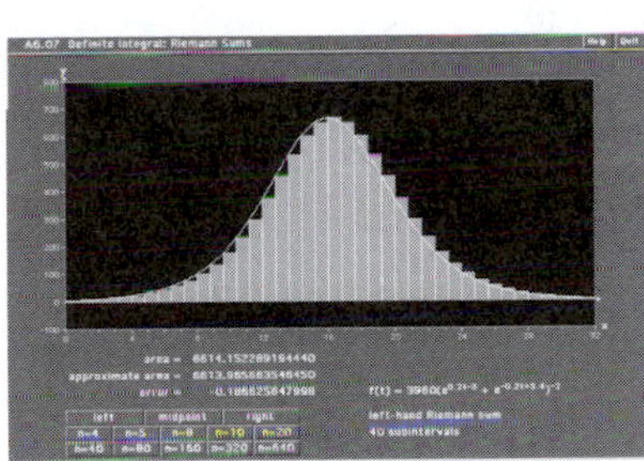

This interactive illustration illustrates what happens to the rectangles associated with the left- and right-hand sums for the function in Example 3 as n becomes larger.

Take n to be increasingly larger, and notice the relationship between the graphs of the rectangles associated with the left- and right-hand sums and the graph of the function. Do you notice that the graphs of the rectangles are "snuggling" ever closer to the graph of the velocity function? Does it appear that the area under the rectangles is getting closer and closer to the area under the curve?

EXAMPLE 4 **The Indefinite Integral and Area**

Find $\int_1^3 (6 - 2x)\, dx$.

Solution

Figure 6.26 shows the graph of $y = 6 - 2x$. The definite integral $\int_1^3 (6 - 2x)\, dx$ is the area between the line $y = 6 - 2x$ and the x-axis on $[1, 3]$. But this is just a triangle. Since the area of a triangle is one half the base times the height,

$$\int_1^3 (6 - 2x)\, dx = \frac{1}{2}(2)(4) = 4$$ ■

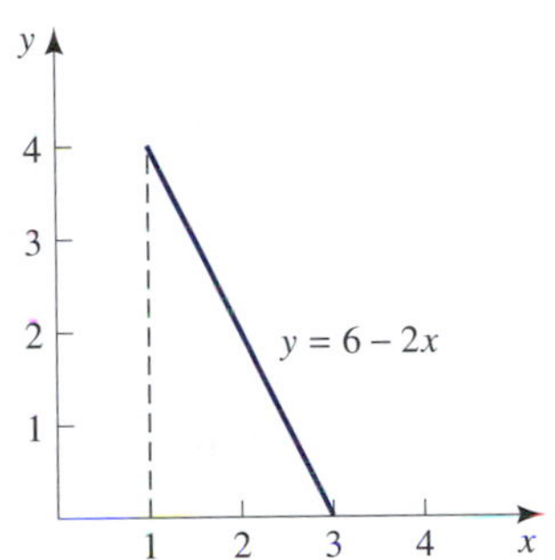

Figure 6.26
$\int_1^3 (6 - 2x)\, dx$ is the area between the line and the x-axis on the interval $[1, 3]$.

Exploration 2

Verifying Example 4 Numerically

Verify the result in Example 4 by reproducing Table 6.3 using the function $f(x) = 6 - 2x$ on the interval $[1, 3]$.

Order Properties of the Integral

We now state three fundamental properties of definite integrals.

Order Properties of Definite Integrals

Assume that all the integrals given below exist and that $a \leq b$. Then

Property 1. If $f(x) \geq 0$, then $\int_a^b f(x)\,dx \geq 0$.

Property 2. If $f(x) \leq g(x)$ for $a \leq x \leq b$, then $\int_a^b f(x)\,dx \leq \int_a^b g(x)\,dx$.

Property 3. If $m \leq f(x) \leq M$ for $a \leq x \leq b$, then

$$m(b-a) \leq \int_a^b f(x)\,dx \leq M(b-a)$$

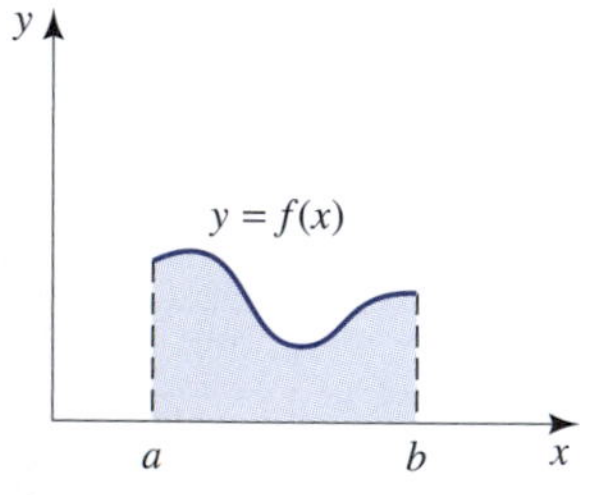

Figure 6.27

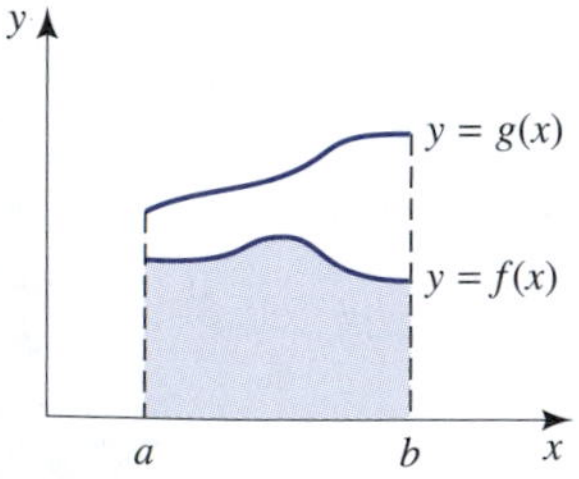

Figure 6.28

To see why Property 1 must be true, examine Figure 6.27. Notice that if the function $f(x)$ is nonnegative on $[a, b]$, then $\int_a^b f(x)\,dx$ is the area under the curve $y = f(x)$ from a to b and therefore is nonnegative. But the area under the curve is $\int_a^b f(x)\,dx$. This establishes Property 1.

Figure 6.28 shows the graphs of the functions $y = f(x)$ and $y = g(x)$, where $f(x) \leq g(x)$ for $a \leq x \leq b$. Notice that since the graph of $y = f(x)$ is below that of $y = g(x)$, the area under the graph of $y = f(x)$, which is $\int_a^b f(x)\,dx$, must be less than or equal to the area under the graph of $y = g(x)$, which is $\int_a^b g(x)\,dx$. This establishes Property 2.

To see why Property 3 is true, refer to Figure 6.29, where we see the graph of $y = f(x)$ bounded below by the line $y = m$ and bounded above by the line $y = M$.

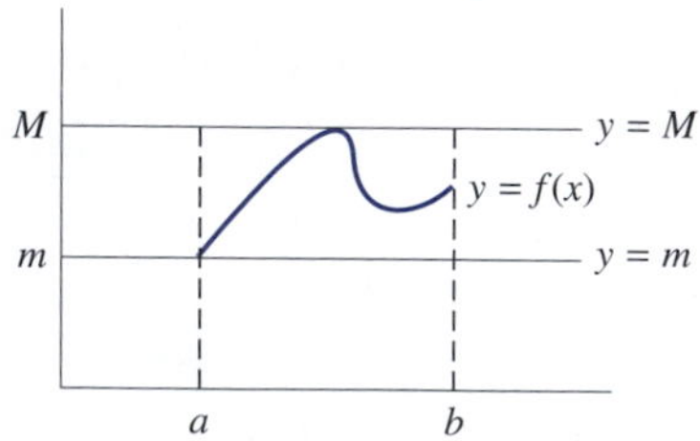

Figure 6.29

Using Property 2, we have

$$\int_a^b m\,dx \leq \int_a^b f(x)\,dx \leq \int_a^b M\,dx$$

Now $\int_a^b m\,dx$ is just the area under the line $y = m$ from a to b. This is a rectangle with base of length $b - a$ and height m. The area of this rectangle is $m(b - a)$. In the same way $\int_a^b M\,dx = M(b - a)$. We then have

$$m(b - a) = \int_a^b m\,dx \leq \int_a^b f(x)\,dx \leq \int_a^b M\,dx = M(b - a)$$

This establishes Property 3.

EXAMPLE 5 **Using the Order Properties**

Use Property 3 to show that $7 \leq \int_1^8 \sqrt[3]{x}\,dx \leq 14$.

Solution

Since $\sqrt[3]{x}$ is an increasing function, it assumes its absolute minimum at the left endpoint 1 and its absolute maximum at the right endpoint 8. See Figure 6.30. Thus, the absolute minimum is $m = \sqrt[3]{1} = 1$, and the absolute maximum is $M = \sqrt[3]{8} = 2$. Then Property 3 gives

$$1 \cdot (8 - 1) \leq \int_1^8 \sqrt[3]{x}\,dx \leq 2(8 - 1)$$

or

$$7 \leq \int_1^8 \sqrt[3]{x}\,dx \leq 14$$

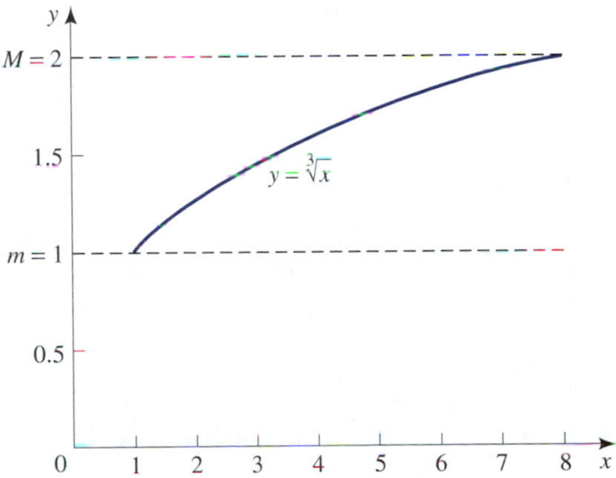

Figure 6.30

■

EXAMPLE 6 **Using the Order Properties**

Use Property 2 to show that $\int_0^1 x^2\,dx \le 0.5$.

Solution

On the interval [0, 1], $x^2 \le x$. See Figure 6.31. Then from Property 2,

$$\int_0^1 x^2\,dx \le \int_0^1 x\,dx$$

But $\int_0^1 x\,dx$ is just the area under the graph of $y = x$ from 0 to 1. From Figure 6.32 we see that this is a triangle with base 1 and height 1. This has area 0.5. Thus,

$$\int_0^1 x^2\,dx \le \int_0^1 x\,dx = 0.5$$

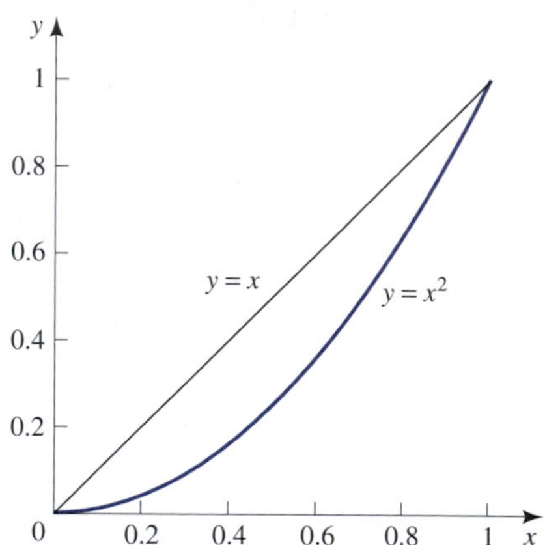

Figure 6.31

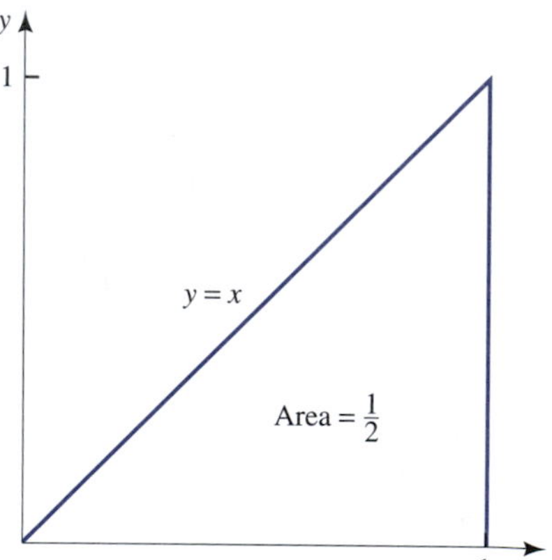

Figure 6.32

■

Enrichment: More General Riemann Sums

The left- and right-hand Riemann sums that we have been studying are actually specific examples of more general Riemann sums that we now consider.

Suppose we have a function $f(x)$ that is continuous on an interval $[a, b]$. As usual divide the x interval $[a, b]$ into n subintervals of equal length

$$\Delta x = \frac{b-a}{n}$$

Let the endpoints of these subintervals be $a = x_0, x_1, x_2, \ldots, x_n = b$. Denote I_1 as the first subinterval $[x_0, x_1]$, denote I_2 as the second subinterval $[x_1, x_2]$, and in general denote I_k as the kth subinterval $[x_{k-1}, x_k]$. Now for each interval I_k, pick a point $x_k^* \in I_k$. The sum

$$\sum_{k=1}^{n} f(x_k^*)\,\Delta x = f(x_1^*)\,\Delta x + f(x_2^*)\,\Delta x + \cdots + f(x_n^*)\,\Delta x$$

is called a **Riemann sum**. The term $f(x_k^*)\ \Delta x$ is the area of a rectangle with base Δx and height $f(x_k^*)$ (Figure 6.33).

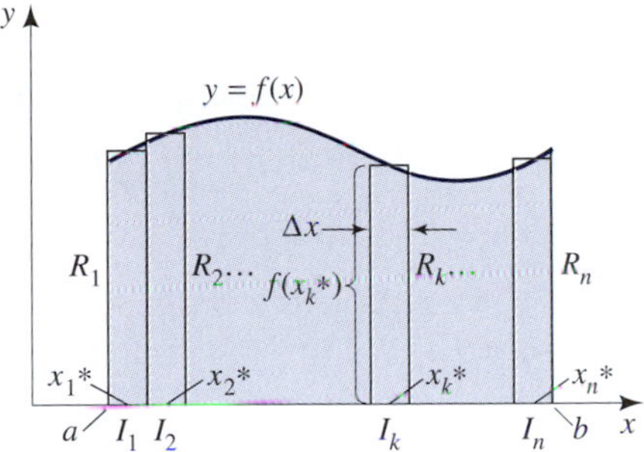

Figure 6.33
A general Riemann sum sums up the areas of the rectangles where x_k^* is any point on the kth subinterval.

If for each k we pick x_k^* as the right-hand point of the subinterval $I_k = [x_{k-1}, x_k]$, then $x_k^* = x_k$, and the Riemann sum $\sum_{k=1}^{n} f(x_k^*)\ \Delta x$ becomes equal to

$$\sum_{k=1}^{n} f(x_k)\ \Delta x = f(x_1)\ \Delta x + f(x_2)\ \Delta x + \cdots + f(x_n)\ \Delta x$$

This is the right-hand sum we have already considered.

If for each k we pick x_k^* as the left-hand point of the subinterval $I_k = [x_{k-1}, x_k]$, then $x_k^* = x_{k-1}$, and the Riemann sum $\sum_{k=1}^{n} f(x_k^*)\ \Delta x$ becomes equal to

$$\sum_{k=1}^{n} f(x_{k-1})\ \Delta x = f(x_0)\ \Delta x + f(x_1)\ \Delta x + \cdots + f(x_{n-1})\ \Delta x$$

This is the left-hand sum that we have already considered.

An important theorem on Riemann sums states that if $f(x)$ is continuous on the interval $[a, b]$ and x_k^* is any point in the subinterval I_k, then $\lim_{n\to\infty} \sum_{k=1}^{n} f(x_k^*)\ \Delta x$ exists and is the same limit independent of how the points x_k^* are chosen in the kth subinterval. Thus for continuous functions

$$\lim_{n\to\infty} \sum_{k=1}^{n} f(x_k^*)\ \Delta x = \int_a^b f(x)\, dx$$

no matter how the points x_k^* are picked in the subinterval I_k.

We have already seen that picking x_k^* as the left endpoint of the subinterval I_k gives the left-hand limit and picking x_k^* as the right endpoint of the subinterval I_k gives the right-hand limit. A third important way to obtain a Riemann sum is to pick x_k^* as the midpoint of the interval I_k.

For positive functions that are either always increasing or always decreasing on the interval $[a, b]$, we saw that the left- and right-hand sums bracketed the area $\int_a^b f(x)\, dx$. In such circumstances we naturally expect a better approximation of the area by using rectangles with heights given by the values of the function at the midpoints of the subintervals. We refer to the Riemann sum using midpoints of the subintervals as the **midpoint rule**.

EXAMPLE 7 A Riemann Sum Using the Midpoint Rule

Use rectangles with heights given by the values of the function at the midpoints of the subintervals for $n = 1000$ to approximate $\int_1^2 \frac{1}{x}\,dx$. Compare your answer to the correct answer ln 2 and to the left- and right-hand sums found in Table 6.2.

Solution

Using a computer or graphing calculator and $n = 1000$, we obtain $\int_1^2 \frac{1}{x}\,dx \approx 0.69315$. Referring to Table 6.2 we see that the left-hand sum is 0.69340 and the right-hand sum is 0.69290. The integral is exactly ln 2, which to five decimal places is 0.69315. Thus, the midpoint sum for $n = 1000$ gives the correct answer to five decimal places. ■

Interactive Illustration 6.8

Midpoint Sums for Large n

This interactive illustration illustrates what happens to the midpoint sums for $f(t) = \frac{1}{t}$ on the interval [1, 2] as n becomes larger.

Take n to be increasingly larger and notice how much more accurate the midpoint sums are when compared to either the left-hand sums or to the right-hand sums.

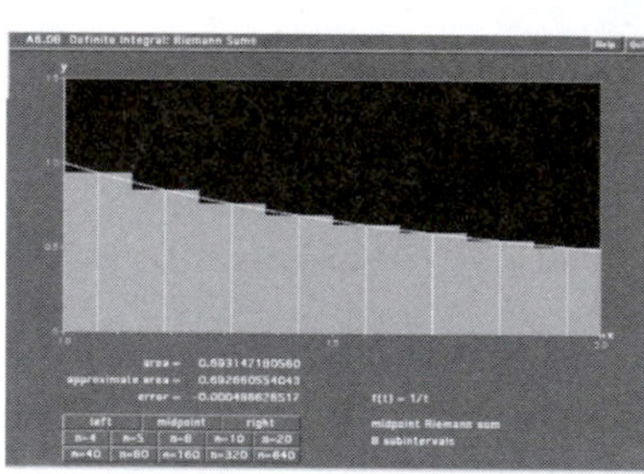

Warm Up Exercise Set 6.4

1. Find the definite integral $\int_0^1 2x\,dx$ by finding the area of an appropriate geometric region.

2. Confirm your answer in Exercise 1 by finding the left- and right-hand sums for $n = 10$, 100, and 1000 and determining the limits.

Exercise Set 6.4

In Exercises 1 through 6, estimate the given definite integrals by finding the left- and right-hand sums for $n = 10$, 100, 1000. (Notice that all of the integrands are monotonic on the interval of integration.)

1. $\int_0^1 \sqrt{x}\,dx$

2. $\int_0^1 (x^2 + x + 1)\,dx$

3. $\int_0^1 x^3\,dx$

4. $\int_0^1 x^4\,dx$

5. $\int_{-2}^0 (4 - x^2)\,dx$

6. $\int_{-1}^0 (1 - x^2)\,dx$

In Exercises 7 through 12, estimate the given definite integrals by finding the left- and right-hand sums for $n = 10$, 100, and 1000. (Notice that all of the integrands are monotonic on the interval of integration.)

7. $\displaystyle\int_1^3 \frac{1}{x}\,dx$ **8.** $\displaystyle\int_1^4 \frac{1}{x}\,dx$

9. $\displaystyle\int_0^1 e^x\,dx$ **10.** $\displaystyle\int_0^1 e^{-x}\,dx$

11. $\displaystyle 0.5\int_1^2 (\ln x + 1)\,dx$ **12.** $\displaystyle\int_1^2 x\ln x\,dx$

In Exercises 13 through 18, estimate the given definite integrals by finding the left- and right-hand sums for $n = 80$, 160, 320, and 640. These exercises are most easily done by using the interactive illustration found in the Integration Kit.

13. $v = f(t) = \dfrac{1}{\sqrt{2\pi}e^{0.5t^2}}$, $[0, 3]$

14. $v = f(t) = te^{-t}$, $[0, 8]$

15. $v = f(t) = \dfrac{2}{1+e^{-t}}$, $[0, 5]$

16. $v = f(t) = 0.001t^3 - 0.03t^2 + 0.3t + 1$, $[0, 25]$

17. $v = f(t) = t^3 - 6t^2 + 9t - 1$, $[0, 4]$

18. $v = f(t) = t\sin t$, $[0, 2\pi]$

In Exercises 19 through 22, estimate the given definite integrals by finding the left- and right-hand sums for $n = 10$, 100, and 1000. (Notice that none of the integrands are monotonic on the interval of integration.)

19. $\displaystyle\int_{-2}^2 (4 - x^2)\,dx$ **20.** $\displaystyle\int_{-1}^1 (1 - x^2)\,dx$

21. $\displaystyle\int_{-1}^2 \sqrt[3]{|x|}\,dx$ **22.** $\displaystyle\int_{-1}^2 |x^3|\,dx$

In Exercises 23 through 32, find the given definite integrals by finding the areas of the appropriate geometric region.

23. $\displaystyle\int_{-1}^3 5\,dx$ **24.** $\displaystyle\int_{-10}^{-2} 4\,dx$

25. $\displaystyle\int_0^2 3x\,dx$ **26.** $\displaystyle\int_1^3 2x\,dx$

27. $\displaystyle\int_1^4 (4 - x)\,dx$ **28.** $\displaystyle\int_{-1}^4 (4 - x)\,dx$

29. $\displaystyle\int_{-1}^1 \sqrt{1 - x^2}\,dx$ **30.** $\displaystyle\int_0^3 \sqrt{9 - x^2}\,dx$

31. $\displaystyle\int_0^2 \sqrt{1 - (x-1)^2}\,dx$ **32.** $\displaystyle\int_1^2 \sqrt{1 - (x-1)^2}\,dx$

In Exercises 33 through 42, find the distance traveled by the object on the given interval by finding the areas of the appropriate geometric region.

33. $v = f(t) = 4$, $[1, 5]$ **34.** $v = f(t) = 3$, $[2, 7]$

35. $v = f(t) = t$, $[0, 1]$ **36.** $v = f(t) = t$, $[0, 2]$

37. $v = f(t) = 2t + 1$, $[0, 2]$

38. $v = f(t) = 3t - 1$, $[1, 2]$

39. $v = f(t) = 6 - 2t$, $[1, 3]$

40. $v = f(t) = 10 - 3t$, $[1, 3]$

41. $v = f(t) = |1 - t|$, $[0, 2]$

42. $v = f(t) = |2 - t|$, $[1, 4]$

For Exercises 43 through 50, use the order properties of the definite integral to establish the inequalities.

43. $\displaystyle 3 \le \int_1^4 \sqrt{x}\,dx \le 6$

44. $\displaystyle 15 \le \int_1^{16} \sqrt[4]{x}\,dx \le 30$

45. $\displaystyle 0.5 \le \int_1^2 \frac{1}{x}\,dx \le 1$

46. $\displaystyle 0.25 \le \int_1^2 \frac{1}{x^2}\,dx \le 1$

47. $\displaystyle\int_0^1 x^3\,dx \le 0.5$

48. $\displaystyle\int_1^5 \sqrt{1 + x^2}\,dx \ge 12$

49. $\displaystyle 2\sqrt{3} \le \int_{-1}^1 \sqrt{3 + x^2}\,dx \le 4$

50. $\displaystyle 24 \le \int_{-4}^4 \sqrt{9 + x^2}\,dx \le 40$

The *only* way of evaluating the definite integrals of functions such as e^{-x^2} or $\dfrac{1}{\ln x}$ is by some approximation technique such as Riemann sums. The definite integral of e^{-x^2} is of critical importance in the applications of probability.

51. Use the left- and right-hand sums for $n = 1000$ to estimate $\displaystyle\int_0^1 e^{-x^2}\,dx$. Using a graph of e^{-x^2}, show which approximation must be less than the integral and which greater.

52. Use the left- and right-hand sums for $n = 1000$ to estimate $\displaystyle\int_2^3 \frac{1}{\ln x}\,dx$. Using a graph of $1/\ln x$, show which approximation must be less than the integral and which greater.

Applications and Mathematical Modeling

53. An object travels with a velocity function given by $v = 2t$, where t is measured in seconds and v is measured in feet per second. Find a formula that gives the exact distance this object travels during the first t seconds. (Hint: Consider the area of an appropriate geometric region.)

54. An object travels with a velocity function given by $v = 3t + 1$, where t is measured in seconds and v is measured in feet per second. Find a formula that gives the exact distance this object travels during the first t seconds. (Hint: Consider the area of an appropriate geometric region.)

More Challenging Exercises

In Exercises 55 through 60, determine an upper and lower estimate of the given definite integral so that the difference of the estimates is at most 0.1.

55. $\int_0^1 \sqrt[3]{x}\, dx$

56. $\int_0^1 (x^2 + 2x + 3)\, dx$

57. $\int_0^2 x^3\, dx$

58. $\int_0^2 x^4\, dx$

59. $\int_{-2}^0 (4 - x^2)\, dx$

60. $\int_{-1}^0 (1 - x^2)\, dx$

61. Explain carefully what you think is the difference (if any) between the definite integrals $\int_1^4 10x^2(4 - x)\, dx$ and $\int_1^4 10t^2(4 - t)\, dt$.

62. In Example 1 we found the area of the region between the graph of the curve $y = x^2$ and the x-axis on the interval $[0, 1]$ by first finding the left- and right-hand sums. It appeared to us that the sums had the common limit $1/3$. In this exercise you are asked to verify analytically that the right-hand sums converge to $1/3$. To do this, divide the interval $[0, 1]$ into n subintervals with right endpoints at $1/n, 2/n, 3/n, \ldots, n/n = 1$.

a. Then show that the right-hand sum S_n^r is

$$S_n^r = \frac{1}{n}\left(\frac{(1)^2}{n^2} + \frac{(2)^2}{n^2} + \frac{(3)^2}{n^2} + \cdots + \frac{(n)^2}{n^2}\right)$$

$$= \frac{1}{n^3}\left(1 + 2^2 + 3^2 + \cdots + n^2\right)$$

b. Now use the formula

$$1 + 2^2 + 3^2 + \cdots + n^2 = \frac{n(n+1)(2n+1)}{6}$$

and show that

$$S_n^r = \frac{1}{3} + \frac{1}{2n} + \frac{1}{6n^2}$$

c. Now take the limit of this expression as $n \to \infty$.

Enrichment Exercises

63. Use the midpoint rule for $n = 1000$ to approximate $\int_1^3 \frac{1}{x}\, dx$. Compare your answer to the correct answer ln 3 and to the left- and right-hand sums.

64. In this section we noted that $\int_0^2 \sqrt{4 - x^2}\, dx = \pi$. Use the midpoint rule for the values of n found in Table 6.2. Compare your answers to the actual value of π and to the approximations used in Exploration 1.

65. In Example 3 we found that, according to the given model, about 9835 deaths occurred during the first 30 weeks of the Bombay plague of 1905–1906. From Table 6.3 we needed to take $n = 2000$ before the left- and right-hand sums rounded to the same integer. Naturally, we expect greater accuracy using rectangles with heights given by the values of the function at the midpoints of the subintervals. Use the midpoint rule on the integral found in Example 3 and find the smallest value of n used in Table 6.3 for which these sums round to 9835. Compare this value of n to the value $n = 2000$.

66. In Exercise 63 you were asked to compare your answer to the correct answer ln 3. What does your calculator give as ln 3? Is this the exact answer? Explain.

67. The figure shows the velocity of two runners racing against each other. (They start together at the same time and run the same track.)

a. Which runner is ahead after the first minute? Explain your answer.
b. Which runner is ahead after the first 5 minutes? Explain your answer.

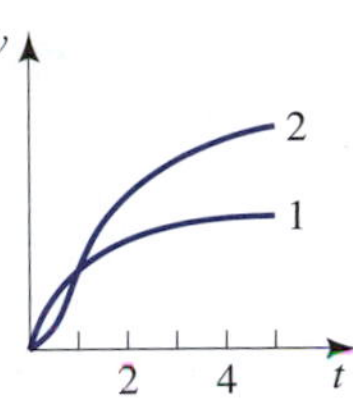

Solutions to WARM UP EXERCISE SET 6.4

1. The definite integral $\int_0^1 2x\,dx$ is the area between the line $y = 2x$ and the x-axis on $[0, 1]$. See the figure. This is a triangle with area one half the height times the base or $\frac{1}{2}(2)(1) = 1$.

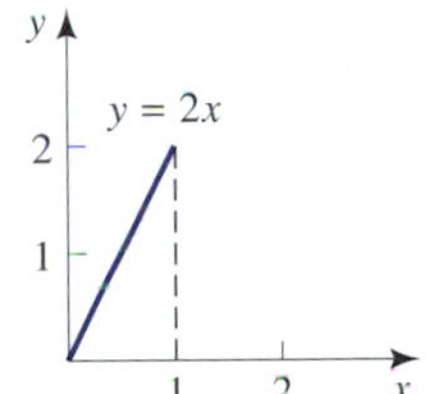

2. The table shows the results for $f(x) = 2x$ on $[0, 1]$

n	Left-Hand Sum	Right-Hand Sum
10	0.9000	1.1000
100	0.9900	1.0100
1000	0.9990	1.0010

As we see both the left- and right-hand sums appear to approach 1.

6.5 The Fundamental Theorem of Calculus

Properties of the Definite Integral
The Fundamental Theorem of Calculus, Part 1
The Fundamental Theorem of Calculus, Part 2
Average Value
Geometric Interpretation of Average Value
Enrichment: Proofs

APPLICATION Extracting Natural Gas From a Field

The rate at which a natural resource is extracted often increases at first until the easily accessible part of the resource is exhausted. Then the rate of extraction tends to decline. Suppose natural gas is being extracted from a new field at a rate of $\frac{4t}{t^2+1}$ billions of cubic feet a year where t is in years. If this field has an estimated 10 billion cubic feet, how long will it take to exhaust this field at the given rate of extraction? See Example 5 on page 453 for the answer.

Properties of the Definite Integral

We gave three (order) properties for the definite integral in Section 6.4. We now state five more properties for the definite integral. We assume that $f(x)$ and $g(x)$ are continuous on $[a, b]$.

Properties of Definite Integrals

4. $\int_a^a f(x)\,dx = 0.$
5. $\int_a^b kf(x)\,dx = k\int_a^b f(x)\,dx, \qquad k \text{ constant.}$
6. $\int_a^b [f(x) \pm g(x)]\,dx = \int_a^b f(x)\,dx \pm \int_a^b g(x)\,dx.$
7. $\int_a^b f(x)\,dx = \int_a^c f(x)\,dx + \int_c^b f(x)\,dx, \qquad a < c < b.$
8. $\int_b^a f(x)\,dx = -\int_a^b f(x)\,dx, a < b.$

When both $f(x)$ and $g(x)$ are positive, the definite integrals are areas under curves. Thus, Property 4 says that the area under a single point is zero. Property 5 says that the area under kf is k times the area under f. One part of Property 6 says that the area under $f + g$ is the area under f plus the area under g. Property 7 says that, according to Figure 6.34, the area under the curve from a to b is the area under the curve from a to c plus the area under the curve from c to b. The actual proofs require the use of Riemann sums and can be shown to be valid without the restriction that f and g be nonnegative. The proofs are not given here. Property 8 is actually a *definition*.

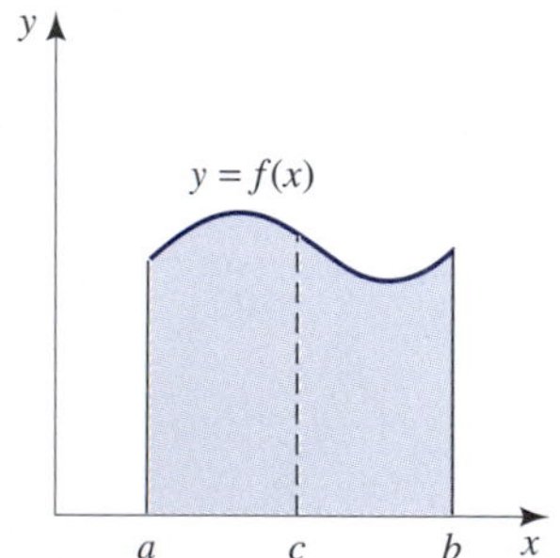

Figure 6.34
The area under the curve from a to b is the area under the curve from a to c plus the area under the curve from c to b.

The Fundamental Theorem Of Calculus, Part 1

The Fundamental Theorem of Calculus connects differential calculus with integral calculus. As we shall see, the derivative and the integral are inverse operations of each other. In a way that will be made clear, the derivative undoes the integral, and the integral undoes the derivative.

To obtain some insight into these ideas, consider the integral

$$g(x) = \int_a^x f(t)\,dt$$

where $f(t) = 1$ and x is considered a variable. Then from Figure 6.35, the integral is the area of the rectangle with height 1 and base $x - a$. This area is just $x - a$. Therefore, $g'(x) = 1$. But this is $f(x)$. So in this case differentiating the integral $\int_a^x f(t)\,dt$ gives $f(x)$.

Consider another example. Let

$$g(x) = \int_0^x f(t)\,dt$$

where $f(t) = t$. From Figure 6.36 we see that the integral is just the area of the triangle with height x and base x. Since the area of the triangle is one half the base

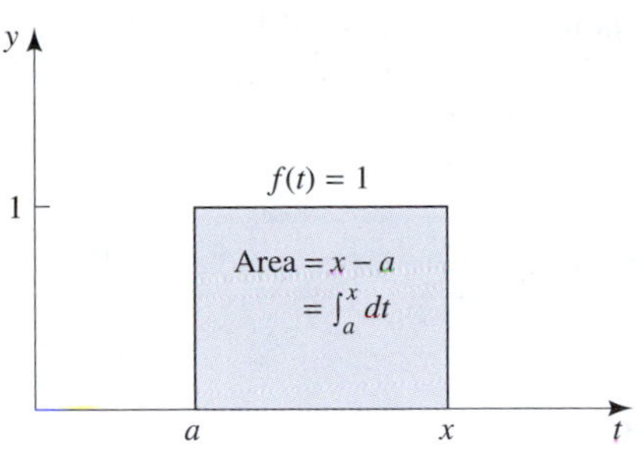

Figure 6.35

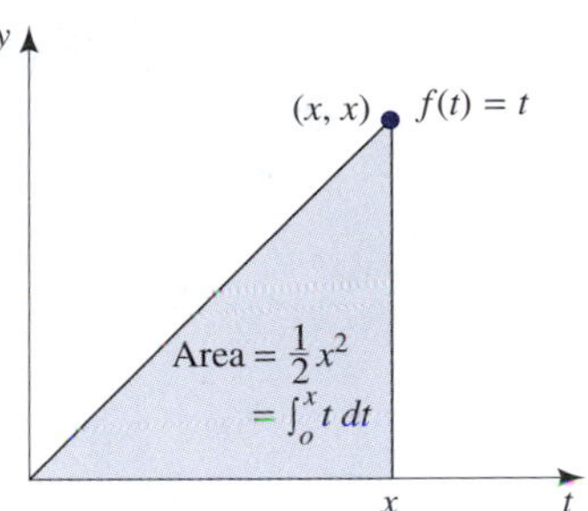

Figure 6.36

times the height, the area is $\frac{1}{2}x^2$. So $g(x) = \frac{1}{2}x^2$. Then $g'(x) = x$ and $g'(x) = f(x)$. So once again, differentiating the integral $\int_a^x f(t)\,dt$ gives $f(x)$. So in both cases g defined by $g(x) = \int_a^x f(t)\,dt$ turns out to be an antiderivative of f.

This turns out to be the case in general. To see why this might be true, consider a function f that is continuous on $[a, b]$ with $f(x) \geq 0$ there. From Figure 6.37 we see that $g(x) = \int_a^x f(t)\,dt$ is the area under the graph of f from a to x. We wish to find $g'(x)$. Consider any $h > 0$; then from Figure 6.38 and Property 4 we have

$$\int_a^x f(t)\,dt + \int_x^{x+h} f(t)\,dt = \int_a^{x+h} f(t)\,dt$$

or

$$\int_a^{x+h} f(t)\,dt - \int_a^x f(t)\,dt = \int_x^{x+h} f(t)\,dt$$

But the left-hand side of this last equation is $g(x + h) - g(x)$, so

$$g(x + h) - g(x) = \int_x^{x+h} f(t)\,dt$$

But if h is small, Figure 6.38 indicates that $\int_x^{x+h} f(t)\,dt$ is approximately the area of the rectangle with height $f(x)$ and base h. This is $f(x)h$. So if h is small, we have

$$g(x + h) - g(x) \approx f(x)h$$

and then

$$\frac{g(x + h) - g(x)}{h} \approx f(x)$$

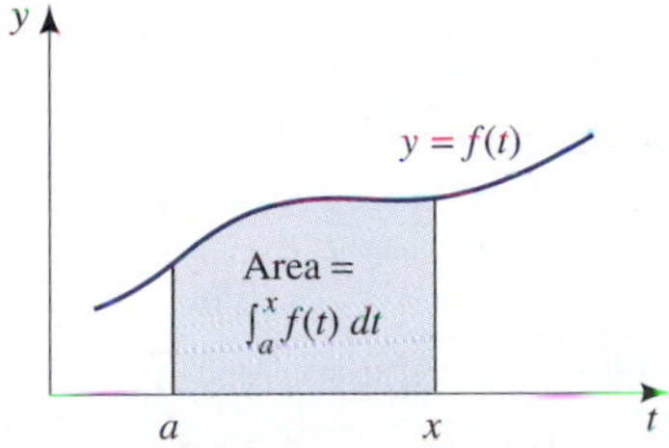

Figure 6.37

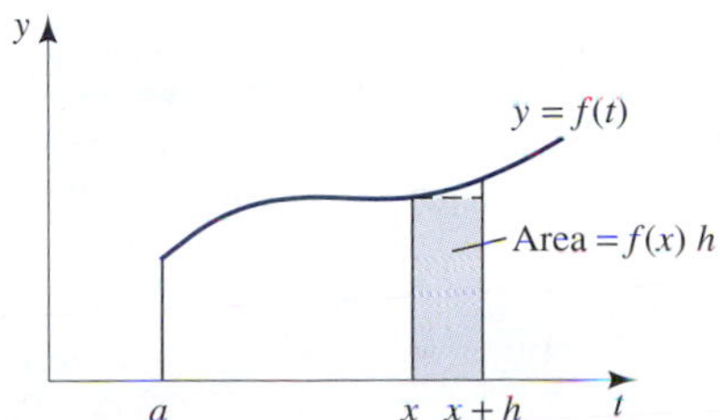

Figure 6.38

We would then expect $g'(x) = f(x)$. This is indeed true, and we will give the details of the proof in the Enrichment subsection. We now state the first part of the Fundamental Theorem of Calculus.

Fundamental Theorem of Calculus, Part 1

Suppose f is continuous on $[a, b]$, and let

$$g(x) = \int_a^x f(t)\,dt \qquad a \le x \le b$$

Then $g'(x) = f(x)$.

Notice that this says that

$$\frac{d}{dx}\int_a^x f(t)\,dt = f(x)$$

and that $\int_a^x f(t)\,dt$ is an antiderivative of f.

EXAMPLE 1 **Using the Fundamental Theorem of Calculus**

Find the derivative of $g(x) = \int_0^x \sqrt{1+t^3}\,dt$

Solution

Since $f(t) = \sqrt{1+t^3}$ is continuous, the first part of the Fundamental Theorem of Calculus gives $g'(x) = \sqrt{1+x^3}$ ■

The Fundamental Theorem of Calculus, Part 2

Earlier in this chapter we computed definite integrals by taking the limits of Riemann sums. In the second part of the Fundamental Theorem of Calculus, we will see how to evaluate definite integrals in a much simpler way.

Fundamental Theorem of Calculus, Part 2

Suppose f is continuous on $[a, b]$, then

$$\int_a^b f(x)\,dx = F(b) - F(a)$$

where F is any antiderivative of f.

To establish this theorem, let $g(x) = \int_a^x f(t)\,dt$. Then we know from the first part of the Fundamental Theorem of Calculus that $g'(x) = f(x)$, that is, g is an antiderivative of f. But recall from Section 6.1 that any two antiderivatives of the same function must differ by a constant. Thus, if F is any other antiderivative, then we must have

$$F(x) = g(x) + C$$

for some constant C. Notice from Property 4 that

$$g(a) = \int_a^a f(t)\, dt = 0$$

So

$$\begin{aligned} F(b) - F(a) &= [g(b) + C] - [g(a) + C] \\ &= g(b) - g(a) \\ &= g(b) \\ &= \int_a^b f(t)\, dt \end{aligned}$$

For convenience we write

$$F(x)|_a^b = F(b) - F(a)$$

Using the limit of Riemann sums, we showed numerically in the last section that

$$\int_0^1 x^2\, dx = \frac{1}{3}$$

The fundamental theorem of calculus gives us a much easier way of obtaining this answer.

EXAMPLE 2 **Using the Fundamental Theorem**

Use the fundamental theorem of calculus to find $\int_0^1 x^2\, dx$.

Solution

Since $F(x) = \frac{1}{3}x^3$ is an antiderivative of x^2, according to the fundamental theorem of calculus

$$\int_0^1 x^2\, dx = \frac{1}{3}\, x^3 \bigg|_1^3 = \frac{1}{3}(1)^3 - \frac{1}{3}(0)^3 = \frac{1}{3}$$ ■

EXAMPLE 3 **Finding a Definite Integral**

Find $\int_1^2 (1 + 6x^2)\, dx$.

Solution

Using Properties 5 and 6,

$$\begin{aligned} \int_1^2 (1 + 6x^2)\, dx &= \int_1^2 dx + 6\int_1^2 x^2\, dx \\ &= x\,|_1^2 + 2x^3\,|_1^2 \\ &= [2 - 1] + 2[(2)^3 - (1)^3] = 15 \end{aligned}$$ ■

EXAMPLE 4 **Finding Total Cost**

Suppose a company has a marginal cost function $y = C'(x) = x\sqrt{9 + x^2}$, where x is the number of thousands of items sold and the cost C is in thousands of dollars. Graph the function $y = C'(x)$ on your graphing calculator. If fixed costs are \$10,000, find the total cost of manufacturing the first 4000 items.

Solution

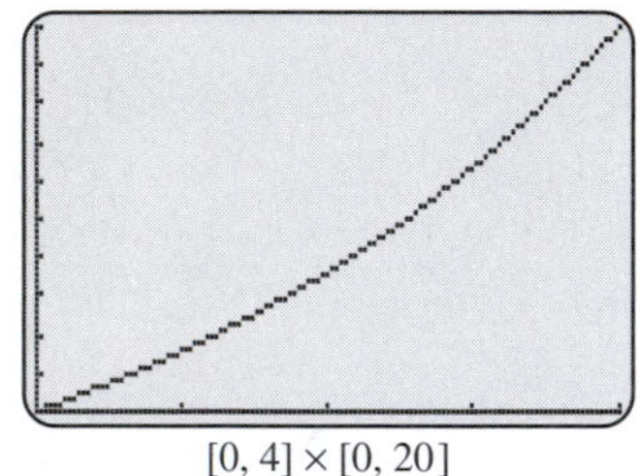

$[0, 4] \times [0, 20]$

Screen 6.3
A graph of a marginal cost function $y = x\sqrt{9 + x^2}$.

The graph of $y = C'(x)$ is shown in Screen 6.3 using a window with dimensions [0, 4] by [0, 20]. Since cost is measured in thousands of dollars, we have $C(0) = 10$. We are seeking $C(4)$. By the fundamental theorem of calculus,

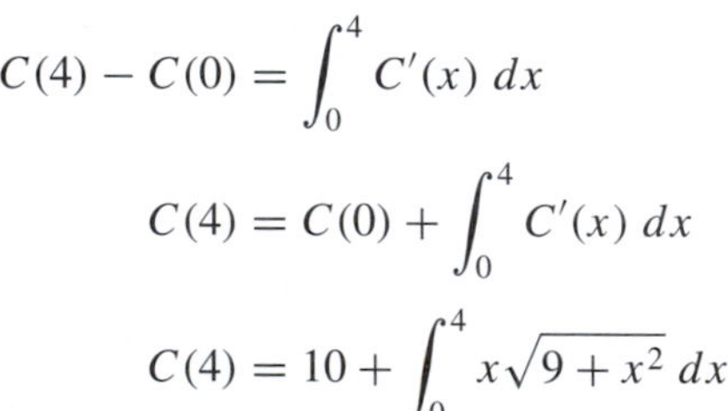

$$C(4) - C(0) = \int_0^4 C'(x)\,dx$$

$$C(4) = C(0) + \int_0^4 C'(x)\,dx$$

$$C(4) = 10 + \int_0^4 x\sqrt{9 + x^2}\,dx$$

Use the method of substitution. Let $u = 9 + x^2$. Then $du = 2x\,dx$. When the integration is with respect to x, then x varies from 0 to 4. But if we are to integrate with respect to u, we need to use limits of integration that correspond to the change in u. We have $u(0) = 9$ and $u(4) = 25$. So when $x = 0$, $u = 1$; and when $x = 4$, $u = 25$. Thus,

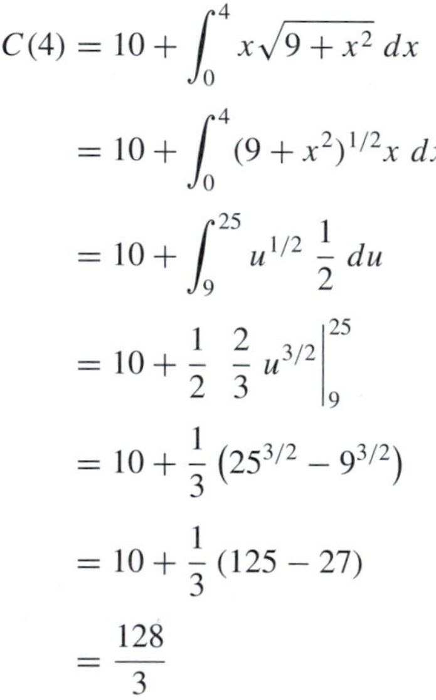

$$\begin{aligned} C(4) &= 10 + \int_0^4 x\sqrt{9 + x^2}\,dx \\ &= 10 + \int_0^4 (9 + x^2)^{1/2} x\,dx \\ &= 10 + \int_9^{25} u^{1/2}\,\frac{1}{2}\,du \\ &= 10 + \frac{1}{2}\,\frac{2}{3}\,u^{3/2}\Big|_9^{25} \\ &= 10 + \frac{1}{3}\left(25^{3/2} - 9^{3/2}\right) \\ &= 10 + \frac{1}{3}(125 - 27) \\ &= \frac{128}{3} \end{aligned}$$

or \$42,666.67. ■

Notice carefully from Example 4 what happens to the limits of integration when the method of substitution is used. If we are evaluating the integral $\int_a^b f(x)\,dx$ by using the method of substitution with the substitution $u = g(x)$, then the limits of integration on the new integral go from $g(a)$ to $g(b)$. In Example 4 we used the substitution $u = g(x) = 9 + x^2$. Since $a = 0$ and $b = 4$, the new lower limit of integration became $g(0) = 9 + (0)^2 = 9$, while the upper limit of integration became $g(4) = 9 + (4)^2 = 25$.

EXAMPLE 5 **Extracting Natural Gas from a Field**

The rate at which a natural resource is extracted often increases at first until the easily accessible part of the resource is exhausted. Then the rate of extraction tends to decline. Suppose natural gas is being extracted from a new field at a rate of $A'(t) = \dfrac{4t}{t^2+1}$ billions of cubic feet a year, where t is in years since the field was opened. Graph this function on your computer or graphing calculator. If this field has an estimated 10 billion cubic feet of gas, how long will it take to exhaust this field at the given rate of extraction?

Solution

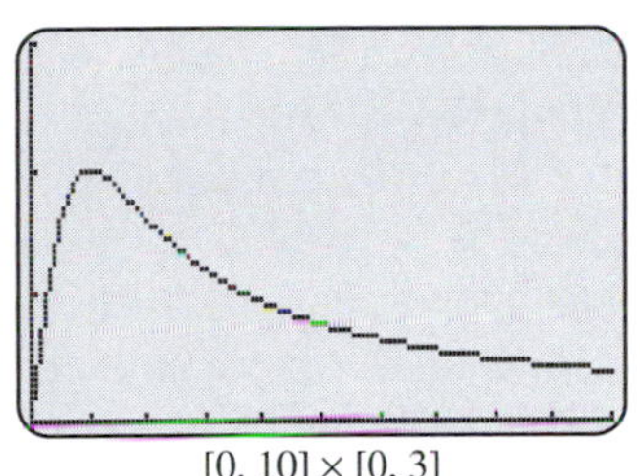

$[0, 10] \times [0, 3]$

Screen 6.4

A graph of $y = \dfrac{4t}{t^2+1}$ giving the rate of extraction of natural gas from a new field.

The graph of $y = A'(t)$ is shown in Screen 6.4 using a window of dimensions [0, 10] by [0, 3]. Since the field opens at $t = 0$, $A(0) = 0$. Using the fundamental theorem of calculus, we find that the amount extracted during the first T years is

$$A(T) = A(0) + \int_0^T \frac{4t}{t^2+1}\,dt = \int_0^T \frac{4t}{t^2+1}\,dt$$

We are seeking the time T for which this definite integral equals 10. Let $u = t^2 + 1$. Then $du = 2t\,dt$. As t goes from 0 to T, u goes from $u(0) = 1$ to $u(T) = T^2 + 1$. Then

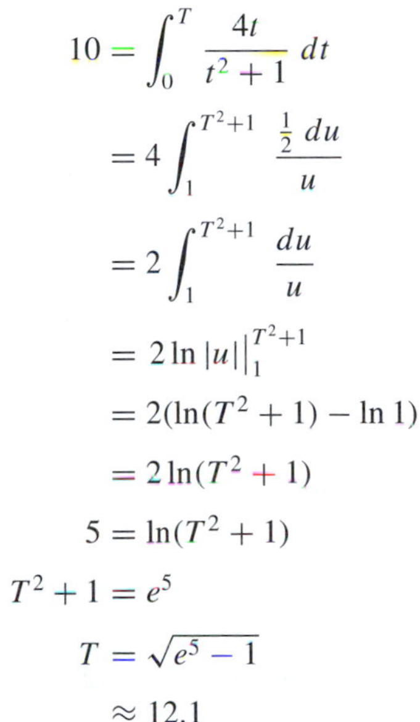

$$\begin{aligned}
10 &= \int_0^T \frac{4t}{t^2+1}\,dt \\
&= 4\int_1^{T^2+1} \frac{\frac{1}{2}\,du}{u} \\
&= 2\int_1^{T^2+1} \frac{du}{u} \\
&= 2\ln|u|\Big|_1^{T^2+1} \\
&= 2(\ln(T^2+1) - \ln 1) \\
&= 2\ln(T^2+1) \\
5 &= \ln(T^2+1) \\
T^2 + 1 &= e^5 \\
T &= \sqrt{e^5 - 1} \\
&\approx 12.1
\end{aligned}$$

or about 12.1 years. ■

Average Value

As we shall see, the indefinite integral can represent many different quantities in business, economics, and science. To recognize these quantities as a definite integral, we must be able to write them as a limit of Riemann sums. We first consider the average of a function.

The average of a set of n numbers $y_1, y_2, \ldots, y_n$, is

$$\frac{y_1 + y_2 + \cdots + y_n}{n}$$

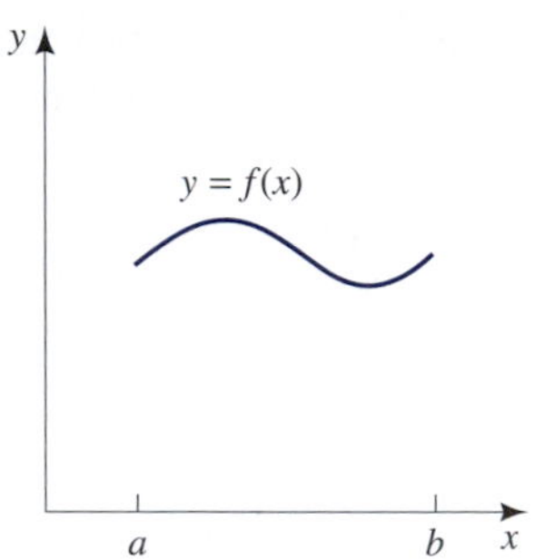

Figure 6.39
We want to define the average value of f on $[a, b]$.

But what is the average value of a continuous function such as seen in Figure 6.39 on the interval $[a, b]$? Here there are an *infinite* number of values of x.

To answer this question, let us start with an average such as

$$\frac{f(x_1) + f(x_2) + \cdots + f(x_n)}{n}$$

where n is very large and the numbers $x_1, x_2, \ldots, x_n$, are equally spaced throughout $[a, b]$. Then, if n is very large, the last expression is an average of a very large number of values of $f(x)$ at points evenly distributed throughout $[a, b]$. Our strategy will be to let $n \to \infty$ and see what happens to this average.

To simplify, let $[a, b] = [1, 3]$. Also let $\Delta x = (b - a)/n = 2/n$. Then

$$\begin{aligned}
\text{average value} &\approx \frac{f(x_1) + f(x_2) + \cdots + f(x_n)}{n} \\
&= \frac{1}{2}\left[2\left(\frac{f(x_1) + f(x_2) + \cdots + f(x_n)}{n}\right)\right] \\
&= \frac{1}{2}\left[f(x_1)\frac{2}{n} + f(x_2)\frac{2}{n} + \cdots + f(x_n)\frac{2}{n}\right] \\
&= \frac{1}{2}[f(x_1)\Delta x + f(x_2)\Delta x + \cdots + f(x_n)\Delta x] \\
&= \frac{1}{2}\sum_{i=1}^{n} f(x_i)\Delta x
\end{aligned}$$

We recognize this as $1/2$ times a Riemann sum of $f(x)$ over $[1, 3]$. If we let $n \to \infty$, the Riemann sum goes to an integral, and we obtain the average value:

$$\begin{aligned}
\text{average value} &= \lim_{n\to\infty} \frac{1}{2}\sum_{i=1}^{n} f(x_i)\Delta x \\
&= \frac{1}{2}\int_1^3 f(x)\,dx
\end{aligned}$$

We can do the same for any interval $[a, b]$. We have the following.

DEFINITION Average Value of $f(x)$ over $[a,b]$

If $f(x)$ is continuous on $[a, b]$, we define the **average value of $f(x)$ on $[a, b]$** to be

$$\frac{1}{b-a}\int_a^b f(x)\,dx$$

Notice that we do not require that $f(x)$ be a nonnegative function.

EXAMPLE 6 **Average Amount in an Account**

An account with an initial amount of $1000 is compounding continuously at an annual rate of 10% per year. What is the average amount of money in the account over the first 10 years?

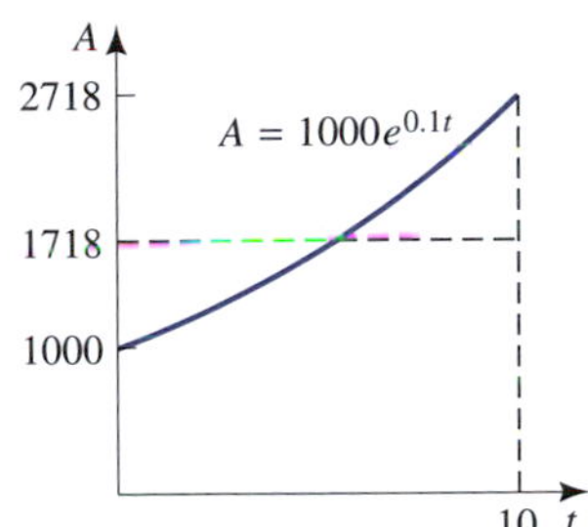

Figure 6.40
The average value of $1000e^{0.1t}$ on [0, 10] is $\frac{1}{10}\int_0^{10} 1000e^{0.1t}\,dt \approx 1718$.

Solution

From Section 1.3 the amount in the account after t years is $f(t) = 1000e^{0.1t}$. The average amount on [0, 10] is given by

$$\begin{aligned}\frac{1}{b-a}\int_a^b f(t)\,dt &= \frac{1}{10-0}\int_0^{10} 1000e^{0.1t}\,dt\\ &= 100\int_0^{10} e^{0.1t}\,dt\\ &= 1000e^{0.1t}\Big|_0^{10}\\ &= 1000[e-1]\\ &\approx 1718\end{aligned}$$

or $1718. Refer to Figure 6.40. ■

Geometric Interpretation of Average Value

When the function $f(x)$ is nonnegative on the interval $[a, b]$, one can give a simple geometric interpretation of the average value. Let us denote the average value of $f(x)$ on $[a, b]$ as $\bar{y}$. Then

$$\bar{y} = \frac{1}{b-a}\int_a^b f(x)\,dx$$

and thus

$$\bar{y}(b-a) = \int_a^b f(x)\,dx$$

Since $f(x)$ is nonnegative, the definite integral $\int_a^b f(x)\,dx$ is the area under the curve $y = f(x)$ on $[a, b]$. Thus, if we construct a rectangle with base on the interval $[a, b]$ and height $\bar{y}$, then this rectangle will have area equal to $\bar{y}(b-a)$, which by the above formula is also the area under the curve $y = f(x)$ on $[a, b]$. See Figure 6.41.

Enrichment: Proofs

We now provide the details for the proof of the first part of the fundamental theorem of calculus. We have a function f continuous on the interval $[a, b]$, and we wish to show that $g(x) = \int_a^x f(t)\,dt$ is an antiderivative of f, that is, $g'(x) = f(x)$.

We have already noted that

$$g(x+h) - g(x) = \int_x^{x+h} f(t)\,dt$$

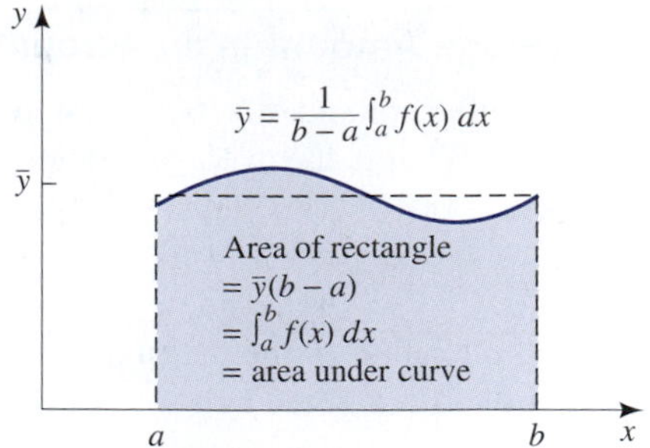

Figure 6.41
The area between the graph of $y = f(x)$ and the x-axis from a to b is $\bar{y}(b-a)$, where $\bar{y}$ is the average value of f on $[a, b]$.

So for any $h \neq 0$,

$$\frac{g(x+h) - g(x)}{h} = \frac{1}{h}\int_x^{x+h} f(t)\,dt$$

For convenience, assume that $h > 0$. Since f is continuous on $[x, x+h]$ the extreme value theorem stated in Section 5.5 says that there are numbers x_m and x_M in $[x, x+h]$ for which f assumes its absolute minimum, m, and its absolute maximum, M. So $f(x_m) = m$ and $f(x_M) = M$. Refer to Figure 6.42. Since $m \leq f(t) \leq M$, by Property 3 of integrals, stated in Section 6.4,

$$mh \leq \int_x^{x+h} f(t)\,dt \leq Mh$$

or

$$f(x_m)h \leq \int_x^{x+h} f(t)\,dt \leq f(x_M)h$$

Since $h > 0$, we have, after dividing by h,

$$f(x_m) \leq \frac{1}{h}\int_x^{x+h} f(t)\,dt \leq f(x_M)$$

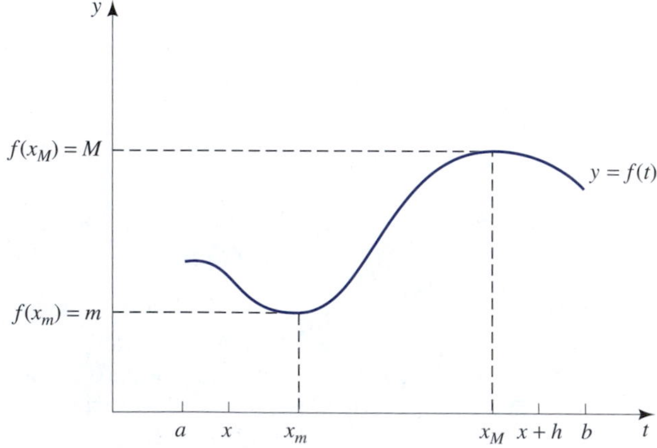

Figure 6.42

(This last inequality also holds if $h < 0$.) Now as we let $h \to 0$, $x_m \to x$ and $x_M \to x$, since both x_m and x_M are between x and $x + h$. Since f is continuous, we then have $\lim_{h\to 0} f(x_m) = f(x)$ and $\lim_{h\to 0} f(x_M) = f(x)$. Since we have trapped $\dfrac{g(x+h) - g(x)}{h}$ between two bounds ($f(x_m)$ from below and $f(x_M)$ from above), both of which are heading for $f(x)$ as $h \to 0$, we must have

$$g'(x) = \lim_{h\to 0} \frac{g(x+h) - g(x)}{h} = f(x)$$

Warm Up Exercise Set 6.5

1. Evaluate $\displaystyle\int_{-1}^{0} 4xe^{x^2+2}\,dx$.
2. Suppose the rate of sales of a certain brand of bicycle by a retailer in thousands of dollars per week is given by

$$\frac{d}{dt}S(t) = 10t - 0.30t^2, \qquad 0 \le t \le 5,$$

where t is the number of weeks after an advertising campaign has begun. Find the amount of sales during the third and fourth weeks.

Exercise Set 6.5

In Exercises 1 through 24, evaluate the definite integrals.

1. $\displaystyle\int_{1}^{2} 4x^3\,dx$
2. $\displaystyle\int_{-2}^{-1} 6x^2\,dx$
3. $\displaystyle\int_{-2}^{-2} 3x^4\,dx$
4. $\displaystyle\int_{-2}^{2} 2x^5\,dx$
5. $\displaystyle\int_{-1}^{0} (9x^2 - 1)\,dx$
6. $\displaystyle\int_{-1}^{1} (1 - x^4)\,dx$
7. $\displaystyle\int_{0}^{2} (4x^3 - 2x + 1)\,dx$
8. $\displaystyle\int_{1}^{4} \left(\frac{3}{\sqrt{x}} - \frac{6}{\sqrt{x}}\right)dx$
9. $\displaystyle\int_{1}^{2} (x^{-2} + 3x^{-4})\,dx$
10. $\displaystyle\int_{-2}^{-1} (x^{-5} + 1)\,dx$
11. $\displaystyle\int_{-2}^{-1} e^{2x}\,dx$
12. $\displaystyle\int_{-1}^{0} e^{-x}\,dx$
13. $\displaystyle\int_{1}^{3} \frac{1}{2x}\,dx$
14. $\displaystyle\int_{2}^{4} \frac{3}{x}\,dx$
15. $\displaystyle\int_{0}^{1} (2x - 1)^9\,dx$
16. $\displaystyle\int_{-1}^{0} (1 + 2x)^5\,dx$
17. $\displaystyle\int_{0}^{1} x(x^2 - 1)^7\,dx$
18. $\displaystyle\int_{0}^{1} x\sqrt[3]{8 + x^2}\,dx$
19. $\displaystyle\int_{0}^{4} \sqrt{2x + 1}\,dx$
20. $\displaystyle\int_{0}^{4} \frac{1}{\sqrt{2x + 1}}\,dx$
21. $\displaystyle\int_{-1}^{1} xe^{x^2+1}\,dx$
22. $\displaystyle\int_{1}^{4} \frac{e^{\sqrt{x}}}{\sqrt{x}}\,dx$
23. $\displaystyle\int_{1}^{2} \frac{1}{2x + 1}\,dx$
24. $\displaystyle\int_{-2}^{0} \frac{x}{x^2 + 1}\,dx$

In Exercises 25 through 34, find the average value of each of the given functions on the given interval.

25. $f(x) = 6$ on $[0, 10]$
26. $f(x) = -3$ on $[-10, -2]$
27. $f(x) = x$ on $[0, 10]$
28. $f(x) = x$ on $[-4, 4]$
29. $f(x) = 2x$ on $[-2, 2]$
30. $f(x) = x^2$ on $[0, 3]$
31. $f(x) = x^3$ on $[-1, 1]$
32. $f(x) = e^x$ on $[0, \ln 2]$
33. $f(x) = x(x - 1)$ on $[0, 2]$
34. $f(x) = x(x - 1)$ on $[-1, 2]$

In Exercises 35 through 38, write as a single definite integral.

35. $\displaystyle\int_{1}^{3} f(x)\,dx + \int_{3}^{5} f(x)\,dx$
36. $\displaystyle\int_{4}^{8} f(x)\,dx + \int_{0}^{4} f(x)\,dx$
37. $\displaystyle\int_{5}^{10} f(x)\,dx - \int_{5}^{8} f(x)\,dx$
38. $\displaystyle\int_{0}^{8} f(x)\,dx + \int_{4}^{0} f(x)\,dx$
39. Refer to the figure. If $F(0) = 1$, what is $F(2)$? $F(4)$?

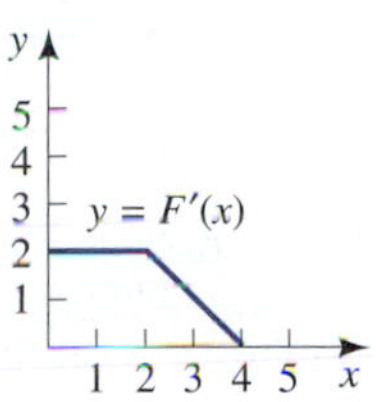

40. Refer to the figure. Suppose $F(1) = 3$ and the area of the shaded region is 10. Find $F(4)$.

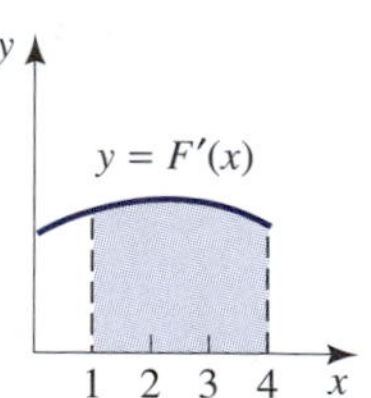

Applications and Mathematical Modeling

41. Revenue In the following figure the rate $R'(t)$ of revenue in millions of dollars per week is given, where t is in weeks. Find an upper and a lower estimate for the revenue received during this 4-week period. Explain what you are doing.

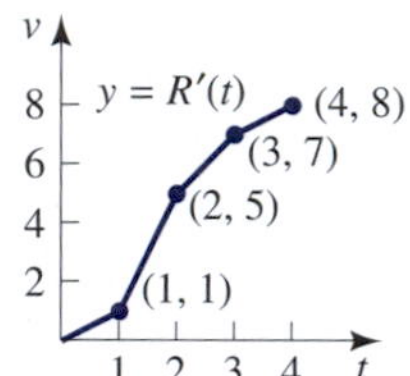

42. Sales In the following figure the rate $R'(t)$ of sales for an entire industry in billions of dollars is given, where t is in weeks. Find an upper and a lower estimate for the sales received during this 5-week period. Explain what you are doing.

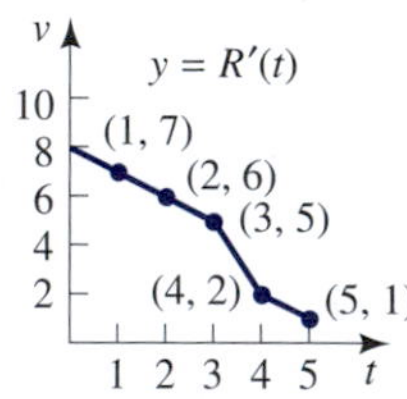

43. Population In the following table the rate of growth $M'(t)$ of a bacteria has been estimated by measuring the rate of increase in mass, where t is in hours. Find two estimates for the change in population during this 20-hour period. Explain what you are doing.

t_i	0	2	4	6	8	10
$M'(t_i)$	10	14	20	30	50	100
t_i	12	14	16	18	20	
$M'(t_i)$	100	80	50	20	0	

44. Velocity In the following table the velocity $v(t) = s'(t)$ is given for discrete equally spaced points $0 = t_0, t_1, \ldots, t_5 = 1$, where v is in feet per second and t is in seconds. Find two estimates for the change in position during this period. Explain what you are doing.

t_i	0	0.2	0.4	0.6	0.8	1.0
$v(t_i)$	1.0	1.6	2.0	1.8	1.5	1.0

45. Production Suppose oil is being extracted from a field at a rate given by $\frac{d}{dt}P(t) = 0.2e^{0.1t}$, where $P(t)$ is measured in millions of barrels and t is measured in years. At this rate how much oil will be extracted during the second 10-year period?

46. Production Suppose copper is being extracted from a certain mine at a rate given by $\frac{d}{dt}P(t) = 100e^{-0.2t}$, where $P(t)$ is measured in tons of copper and t is measured in years. At this rate how much copper will be extracted during the third year?

47. Sales Suppose the rate of sales of an item is given by $\frac{d}{dt}S(t) = -3t^2 + 36t$, where t is the number of weeks after an advertising campaign has begun. How many items were sold during the third week?

48. Sales A firm estimates that its sales will increase at a rate given by $\frac{d}{dt}S(t) = 10\sqrt{t}$, where t is measured in years. Find the total sales during the second and third years.

49. Natural Resource Depletion Suppose a new oil reserve has been discovered with an estimated 10 billion barrel capacity. Suppose production is given in billions of barrels and proceeds at a rate given by $\frac{d}{dt}P(t) = 1.5e^{-0.1t}$, where t is measured in years. At this rate, how long will it take to exhaust this resource?

50. Natural Resource Depletion Suppose a resource such as oil, gas, or some mineral, has total reserves equal to R. Suppose the resource is being extracted at a rate given by $\frac{d}{dt}P(t) = ae^{-kt}$, where a and k are positive constants and t is time. Show that at this rate the resource will be exhausted in time $T = -\frac{\ln(1 - kR/a)}{k}$.

51. Chemical Spill A tank holding 10,000 gallons of a polluting chemical breaks at the bottom and spills out at the rate given by $f'(t) = 400e^{-0.01t}$, where t is measured in hours. How much spills during the first day?

52. Chemical Spill In Exercise 51, how long before the tank is emptied?

53. Sum-of-Years-Digit Depreciation The sum-of-years-digit depreciation method depreciates a piece of equipment more in the early years of its life and less in the later years. In this method the value changes at a rate given by $\frac{d}{dt}V(t) = -k(T - t)$, where T is the useful life in years of the equipment, t is the time in years since purchase, and k is a positive constant.

a. If the initial value of the equipment is I, show that the scrap value of the equipment is given by $I - kT^2/2$.

b. Show that the number of years for the equipment to be worth half of its value is given by $T - \sqrt{T^2 - I/k}$.

54. Sum-of-Years-Digit Depreciation Suppose in the previous problem a certain computer with an initial value of \$51,000 has $T = 10$ and $k = 1000$.

a. Find the scrap value of this computer.

b. Find the number of years it takes for the computer to be worth half its initial value using this method of depreciation.

55. Sales From the time a firm introduced a new product, sales in thousands of this product have been increasing at a rate given by $S'(t) = 2 + t$, where t is the time since the introduction of the new product. At the beginning of the third year, an advertising campaign was introduced, and sales then increased at a rate given by $S_2'(t) = t^2$, where $t \geq 2$. Find the total sales during the first four years.

56. Production Suppose copper is being extracted from a certain mine at a rate given by $\frac{d}{dt}P(t) = 100e^{-0.2t}$, where $P(t)$ is measured in tons of copper and t is measured in years from opening the mine. At the beginning of the sixth year, a mining innovation is introduced to boost production to a rate given by $Q'(t) = 500/t$. Find the total production during the first 10 years.

57. Average Amount of Money An account with an initial amount of \$1000 is compounding continuously at an annual rate of 10% per year. What is the average amount of money in the account over the first 20 years?

58. Average Price Given the demand equation $p = 100e^{-0.2x}$ find the average price over the demand interval $[10, 20]$.

59. Average Profit The profits in millions of dollars of a certain firm is given by $P(t) = 6(t - 1)(t - 2)$, where t is measured in years. Find the average profits per year over the period of time $[0, 4]$.

60. Average Revenue The revenue in thousands of dollars is given by $R(x) = x^2 - 2x$, where x is the number of thousands of items sold. Find the average revenue for the first 10,000 items sold.

61. Average Air Pollution A study has indicated that the level $L(x)$ of air pollution a distance of x miles from a certain factory is given by $L(x) = e^{-0.1x} + 10$. Find the average level of pollution between 10 and 20 miles from the factory.

62. Average Blood Pressure The blood pressure in an artery of a healthy individual can change substantially over only a few seconds. The average pressure is used in some studies of blood pressure. Suppose the function $p(t) = 40t^4 - 160t^3 + 160t^2 + 80$ gives the blood pressure in mm Hg in an artery over an interval $[0, 2]$ of time measured in seconds. Find the average blood pressure in this artery over the 2-second interval $[0, 2]$.

63. Average Population The population in a certain city is projected to be given by $P(t) = 100,000e^{0.05t}$, where t is given in years from now. Find the average population over the next 10 years.

More Challenging Exercises

64. Revenue Two competing retail stores open in the same mall at time $t = 0$ and have the marginal revenue functions given in the figure. The marginal revenue is measured in thousands of dollars per day. Estimate the point in time, other than $t = 0$, for which the total sales of the two stores are equal. Justify your answer. (Hint: Use the fundamental theorem of calculus, and roughly compare the areas of two regions.)

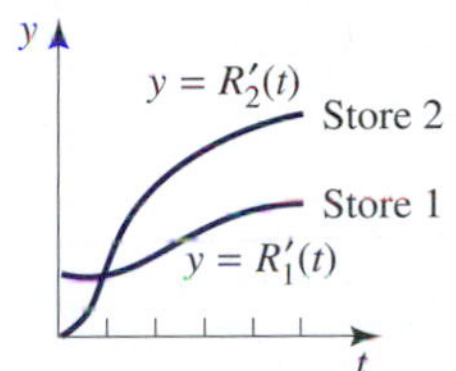

65. Revenue Consider the two competing retail stores in Exercise 64.

a. Which store had more revenue during the first year? Explain carefully why you think so.

b. Which store had more revenue during the first three years? Explain carefully why you think so. (Hint: Use the fundamental theorem of calculus, and roughly compare the areas of two regions.)

66. Suppose for any t that $v(t)$ is the velocity in feet per second of an object at time t measured in seconds. Suppose also that on the t interval $[0, 1]$, $v(t)$ is a continuous function that takes both negative and positive values. Explain, using Riemann sums, why $\int |v(t)|\, dt$ is the total distance traveled by the object on the t interval $[0, 1]$.

67. Explain in complete sentences the difference between the definite integral $\int_a^b f(x)\,dx$ and the indefinite integral $\int f(x)\,dx$. Take a specific function $f(x)$ and specific limits of integration, and illustrate.

68. Transportation Costs A plant has a constant market share s, serving a circular market area of radius R with uniform demand density D per square mile, and freight rate of T per mile. The transport cost per unit is then Tr. Determine the demand arising in the area at a radius between r and $r + \Delta r$, and sum all these demands up in a radius R. (See Hay and Morris.[6]) Use this Riemann sum to find a definite integral that gives the transportation cost over the circular market of radius R.

69. Demand The price p of an item located a distance D in miles from a mill is $p = m + tD$, where m is the mill price and t is the transportation cost per mile. (See Sailors.[7]) The demand curve is $x = a/p^2 = a/(m + tD)^2$, where a is a positive constant. The aggregated demand X over an assumed lined (one-dimensional) market, where $q(p)$ is uniformly distributed, is then $X = \int_0^b a/(m + tD)^2\,dD$, where b measures the extent of the market. Find this integral.

70. Optimum Mill Price The price p of an item located a distance D in miles from a mill is $p = m + tD$, where m is the mill price and t is the transportation cost per mile. (See Hsu.[8]) The demand curve is $x = e^{-ap} = e^{-a(m+tD)}\,dD$, where a is a positive constant. Then the profit is $P(m) = \int_0^b (m - c)e^{-a(m+tD)}\,dD - F$, where c is (constant) marginal cost, F is fixed cost, and b measures the extent of the market. Find the value of m that maximizes P.

71. Using Riemann sums, outline a proof that if $f(x)$ is continuous on $[a, b]$, $\int_a^b cf(x)\,dx = c\int_a^b f(x)\,dx$. Hint: Note that $\sum_{k=1}^{n} cf(x_k)\Delta x = c\sum_{k=1}^{n} f(x_k)\Delta x$.

72. Using Riemann sums, outline a proof that if $f(x)$ and $g(x)$ are continuous on $[a, b]$, $\int_a^b [f(x) + g(x)]\,dx = \int_a^b f(x)\,dx + \int_a^b g(x)\,dx$. (Hint: Note that $\sum_{k=1}^{n}[f(x_k) + g(x_k)]\Delta x = \sum_{k=1}^{n} f(x_k)\Delta x + \sum_{k=1}^{n} g(x_k)\Delta x$.)

6 Donald A. Hay and Derek J. Morris. 1979. *Industrial Economics*. New York: Oxford University Press.

7 Joel Sailors, M. L. Greenhut, and H. Ohta. 1985. Reverse dumping: a form of spacial price discrimination. *J. Indust. Econ.* 34(2):167–181.

8 Soon-Ken Hsu. 1979. Monopoly output under alternative spatial pricing techniques: comment. *Amer. Econ. Rev.* 69:678–679.

Solutions to WARM UP EXERCISE SET 6.5

1. Let $u = x^2 + 2$. Then $du = 2x\,dx$. As x goes from -1 to 0, u goes from $u(-1) = 3$ to $u(0) = 2$. Then

$$\begin{aligned}\int_{-1}^{0} 4xe^{x^2+2}\,dx &= 2\int_{-1}^{0} e^{x^2+2}2x\,dx\\ &= 2\int_{3}^{2} e^{u}\,du\\ &= 2(e^2 - e^3)\end{aligned}$$

2. We are seeking the definite integral $\int_2^4 (10t - 0.30t^2)\,dt$. Then

$$\begin{aligned}\int_2^4 (10t - 0.30t^2)\,dt &= (5t^2 - 0.10t^3)\Big|_2^4\\ &= [80 - 6.4] - [20 - 0.80] = 54.4\end{aligned}$$

or \$54,400.

6.6 Area Between Two Curves

Area between Two Curves

Lorentz's Curves

APPLICATION Measuring Income Inequality

At the beginning of this chapter we noted that David Hackett Fischer[9] has identified four great worldwide inflationary waves during the last 800 years, interspersed with three great deflationary waves. One social result that was observed was a rapid growth of inequality that appeared in the later stages of every long inflationary wave. How can we measure the inequality of income distribution? See Example 6 on page 467 for the answer.

[9] David Hackett Fischer. 1996. *The Great Wave (Price Revolutions and the Rhythm of History)*. New York: Oxford University Press.

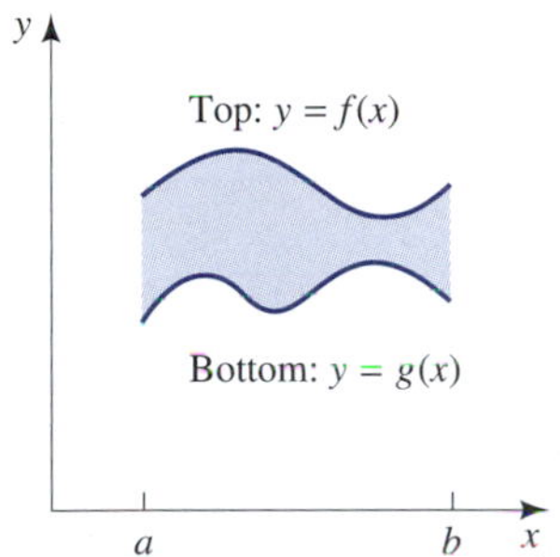

Figure 6.43
The area between the graphs of the two curves from a to b is $\int_a^b [f(x) - g(x)]\, dx$.

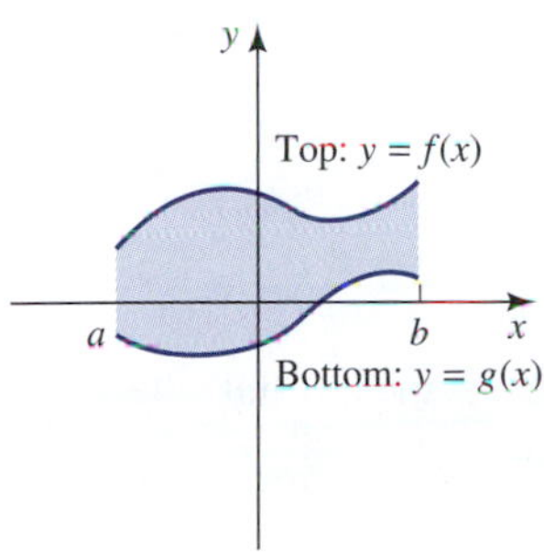

Figure 6.44
The area between the graphs of the two curves from a to b is $\int_a^b [f(x) - g(x)]\, dx$.

Area Between Two Curves

Suppose we have two continuous functions $f(x)$ and $g(x)$ on the interval $[a, b]$ with $f(x) \geq g(x) \geq 0$. See Figure 6.43. We wish to find the area between the two curves.

Clearly, the area between the two curves in Figure 6.43 is the area under f less the area under g. That is, the area between the two curves is given by

$$\int_a^b f(x)\, dx - \int_a^b g(x)\, dx = \int_a^b [f(x) - g(x)]\, dx$$

Consider now the more general situation shown in Figure 6.44. We still have $f(x) \geq g(x)$, but both functions can change sign.

It is apparent (see Figure 6.44) that there exists a positive constant large enough that the graph of $y = g(x) + k$ remains *above* the x-axis. We also add k to f and let $F(x) = f(x) + k$ and $G(x) = g(x) + k$. The graphs of these functions are shown in Figure 6.45. But adding the very same constant to each function does not change the area between the resulting functions. Thus, it is apparent that the area between F and G is the same as the area between f and g.

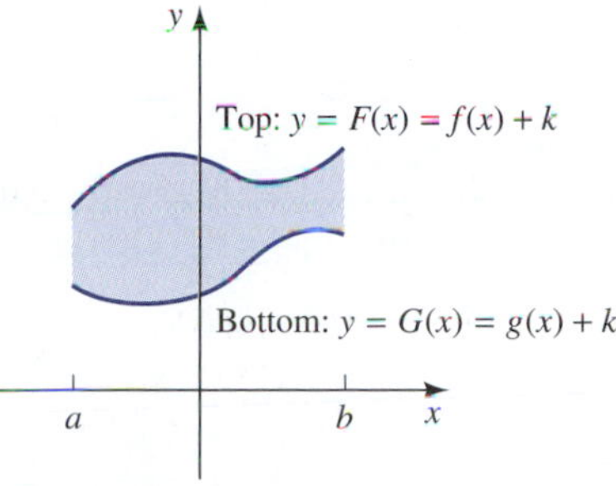

Figure 6.45

But now $F(x) \geq G(x) > 0$. Thus, the area between F and G is just

$$\begin{aligned}\int_a^b [F(x) - G(x)]\, dx &= \int_a^b [(f(x) + k) - (g(x) + k)]\, dx \\ &= \int_a^b [f(x) - g(x)]\, dx\end{aligned}$$

which is exactly the same formula as found before.

Area Between Two Curves

Let $y = f(x)$ and $y = g(x)$ be two continuous functions with $f(x) \geq g(x)$ on $[a, b]$. Then the area between the graphs of the two curves on $[a, b]$ is given by the definite integral

$$\int_a^b [f(x) - g(x)]\, dx$$

We can think of this as the area between the *top* curve and the *bottom* curve or as

$$\int_a^b [\text{top} - \text{bottom}]\, dx$$

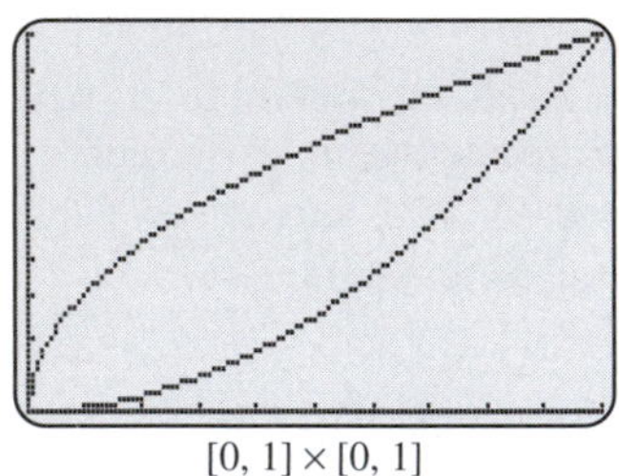

[0, 1] × [0, 1]

Screen 6.5
The graph has a top $y_1 = \sqrt{x}$ and a bottom $y_2 = x^2$.

EXAMPLE 1 Finding the Area between Two Curves

Find the area under the curve $y = \sqrt{x}$ and above the curve $y = x^2$ between $x = 0$ and $x = 1$.

Solution

Graphs of the two curves are shown in Screen 6.5 using a window with dimensions [0, 1] by [0, 1]. The graph indicates that the top of the region is given by the function $y = \sqrt{x}$ and the bottom is given by $y = x^2$ (Figure 6.46). According to the previous discussion, we are seeking

$$\begin{aligned}\int_a^b [\text{top} - \text{bottom}]\, dx &= \int_0^1 \left[x^{1/2} - x^2\right] dx \\ &= \left(\frac{2}{3}x^{3/2} - \frac{1}{3}x^3\right)\Bigg|_0^1 \\ &= \frac{2}{3} - \frac{1}{3} = \frac{1}{3}\end{aligned}$$

■

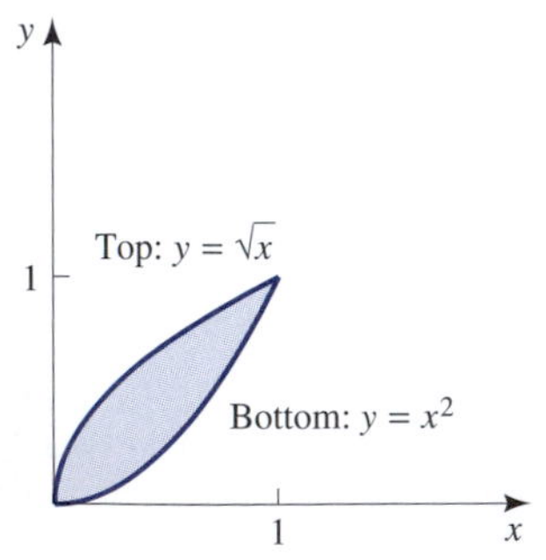

Figure 6.46
The area between the top $y = \sqrt{x}$ and the bottom $y = x^2$ between 0 and 1 is $\int_0^1 [\sqrt{x} - x^2]\, dx$.

Notice that a graph is critical to determine exactly what is the top and what is the bottom curve and to discern what is the interval of integration.

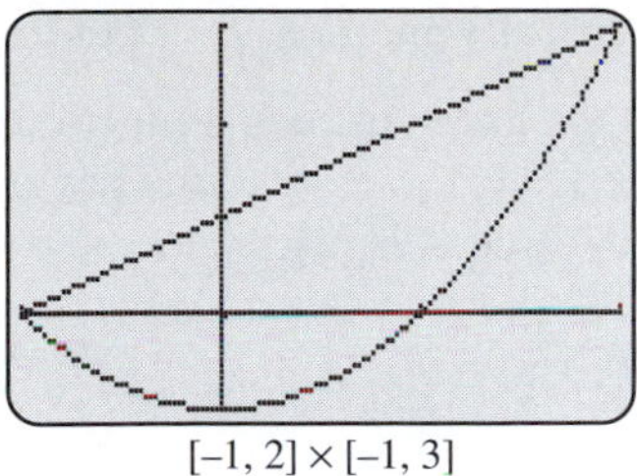
[−1, 2] × [−1, 3]

Screen 6.6
The region has a top $y_1 = x + 1$ and a bottom $y_2 = x^2 - 1$.

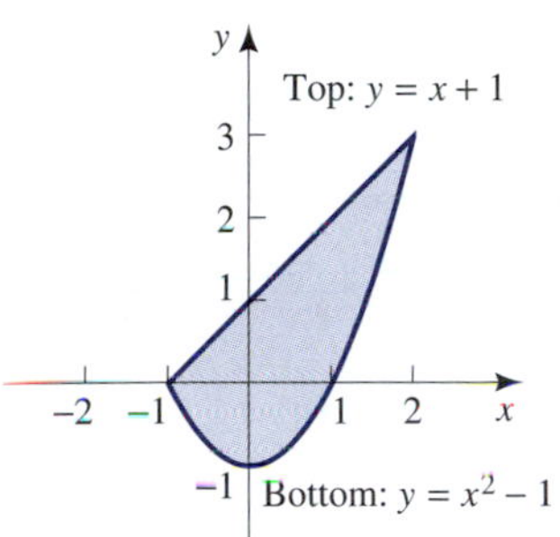

Figure 6.47
The area between the top $y = x + 1$ and the bottom $y = x^2 - 1$ between −1 and 2 is $\int_{-1}^{2} [(x + 1) - (x^2 - 1)]\, dx$.

EXAMPLE 2 **Finding the Area between Two Curves**

Find the area between the two curves $y = x + 1$ and $y = x^2 - 1$.

Solution

It is critical that we draw a graph. This is indicated in Screen 6.6, which uses a window with dimensions $[-1, 2]$ by $[-1, 3]$. The curve $y = x + 1$ is the top and the curve $y = x^2 - 1$ is the bottom (Figure 6.47). To determine where the two curves intersect, set $x^2 - 1 = x + 1$. This gives $0 = x^2 - x - 2 = (x + 1)(x - 2)$. Thus, $x = -1$ and $x = 2$ are the x-coordinates of the points of intersection. Then

$$\begin{aligned}
\int_a^b [\text{top} - \text{bottom}]\, dx &= \int_{-1}^{2} \left[(x + 1) - (x^2 - 1)\right] dx \\
&= \int_{-1}^{2} \left[2 + x - x^2\right] dx \\
&= \left(2x + \frac{1}{2}x^2 - \frac{1}{3}x^3\right)\Bigg|_{-1}^{2} \\
&= \left[4 + 2 - \frac{8}{3}\right] - \left[-2 + \frac{1}{2} + \frac{1}{3}\right] = \frac{9}{2}
\end{aligned}$$

■

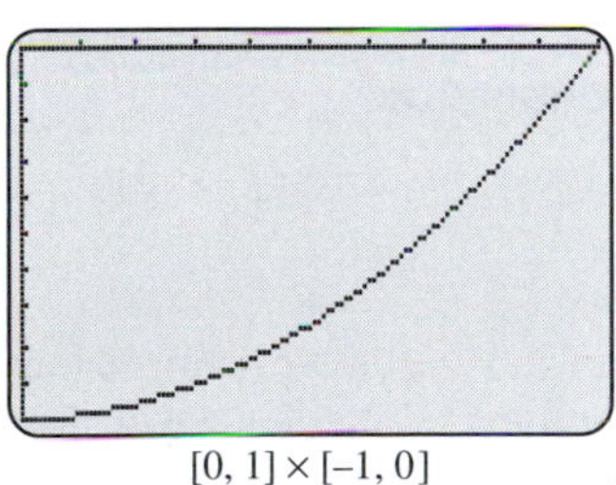
[0, 1] × [−1, 0]

Screen 6.7
The region has a top $y_1 = 0$ and a bottom $y_2 = x^2 - 1$.

EXAMPLE 3 **Finding the Area of a Curve below the *x*-Axis**

Find the area of the region in the fourth quadrant bounded by the x-axis, y-axis, and $y = x^2 - 1$.

Solution

We graph $y = x^2 - 1$ using a window with dimensions $[0, 1]$ by $[-1, 0]$. Refer to Screen 6.7. The curve $y = x^2 - 1$ is the bottom. The top is given by $y = 0$ (Figure 6.48). The region of integration is easily seen to be $[0, 1]$. Then

$$\begin{aligned}
\int_a^b [\text{top} - \text{bottom}]\, dx &= \int_0^1 \left[(0) - (x^2 - 1)\right] dx \\
&= \int_0^1 \left[1 - x^2)\right] dx \\
&= \left(x - \frac{1}{3}x^3\right)\Bigg|_0^1 \\
&= \left(\frac{2}{3}\right) - (0) = \frac{2}{3}
\end{aligned}$$

■

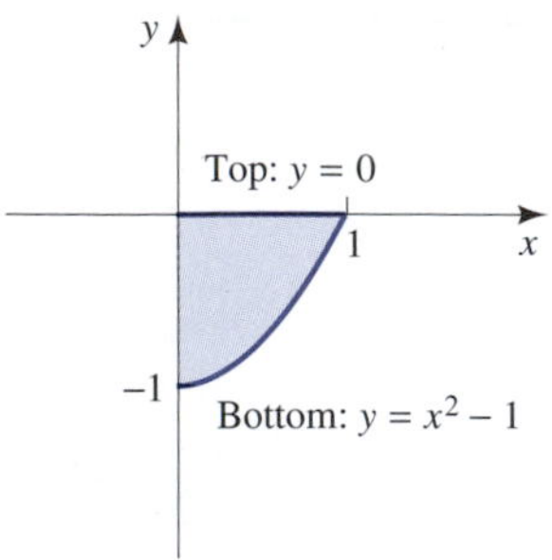

Figure 6.48
The area between the top $y = 0$ and the bottom $y = x^2 - 1$ between 0 and 1 is $\int_0^1 [(0) - (x^2 - 1)]\, dx$.

The previous example illustrates how we can geometrically interpret $\int_a^b f(x)\, dx$ when $f(x) \leq 0$ on $[a, b]$. In Figure 6.49 we are given the graph of such a function. The area of the region between the x-axis and the curve $y = f(x)$ from a to b has as the top $y = 0$ and has as the bottom $y = f(x)$. Thus, the area A is just

$$A = \int_a^b [\text{top} - \text{bottom}]\, dx$$

$$= \int_a^b [0 - f(x)]\, dx$$

$$= -\int_a^b f(x)\, dx$$

Therefore, we have the following.

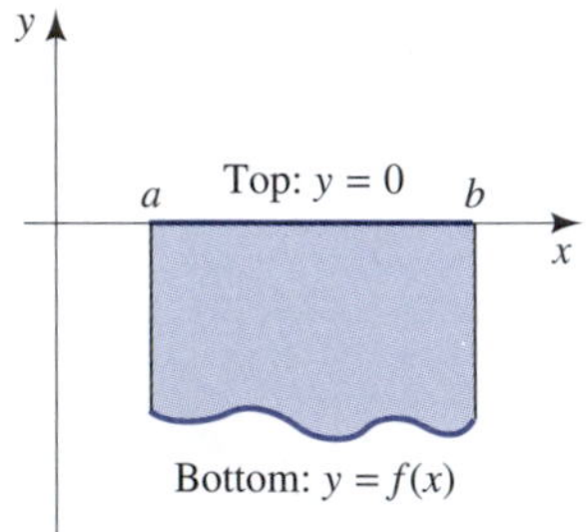

Figure 6.49
If $f(x) \leq 0$ on $[a, b]$ then the area between the x-axis and the graph of the curve $y = f(x)$ between a and b is $-\int_a^b f(x)\, dx$.

Area of a Curve under the x-Axis

If the graph of $y = f(x)$ is below the x-axis on $[a, b]$, then the area below the x-axis and above the graph of $y = f(x)$ on $[a, b]$ is

$$\text{area} = -\int_a^b f(x)\, dx$$

From another point of view, when $f(x) \leq 0$ on $[a, b]$, $\int_a^b f(x)\, dx$ represents the *negative* of the area between the x-axis and the curve $y = f(x)$ from a to b.

Sometimes the region of integration must be divided into subregions.

EXAMPLE 4 **Graphs with Multiple Points of Intersection**

Find the area enclosed between the curves $y = -x$ and $y = 2 - x^2$ on $[-1, 3]$.

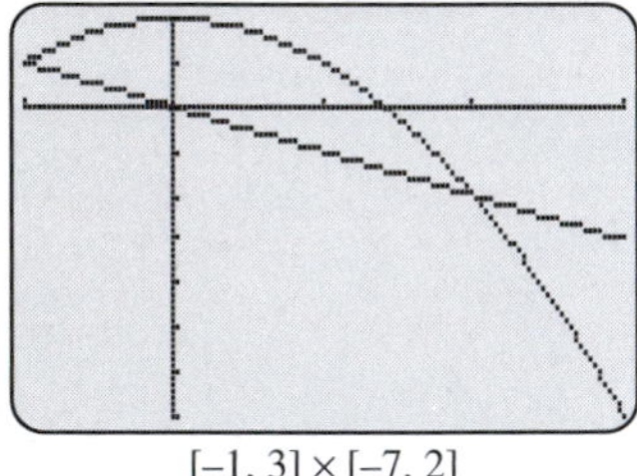

[−1, 3] × [−7, 2]

Screen 6.8
The equation $y_1 = -x$ is the bottom of the left region but the top of the right region. The equation $y_2 = 2 - x^2$ is the top of the left region but the bottom of the right region.

Solution

Graphs of the two functions are shown in Screen 6.8, using a window with dimensions $[-1, 3]$ by $[-7, 2]$. Notice that the top of the region is *not* given by one single equation. (In fact neither is the bottom.) Refer to Figure 6.50. For example, the top of the region R_1 is given by $y = 2 - x^2$ while the top of the region R_2 is given by $y = -x$. In such a "bow-tie" region we must divide the region into two subregions to use our formulas.

To do this, we must find the point where the two functions cross. We then set $2 - x^2 = y = -x$ or $0 = x^2 - x - 2 = (x + 1)(x - 2)$. Thus, the two curves cross when $x = -1$ and $x = 2$. So the two intervals of integration will be $[-1, 2]$ and $[2, 3]$. Let A be the area we are seeking, let A_1 be the area of R_1, and let A_2 be the area of R_2; then

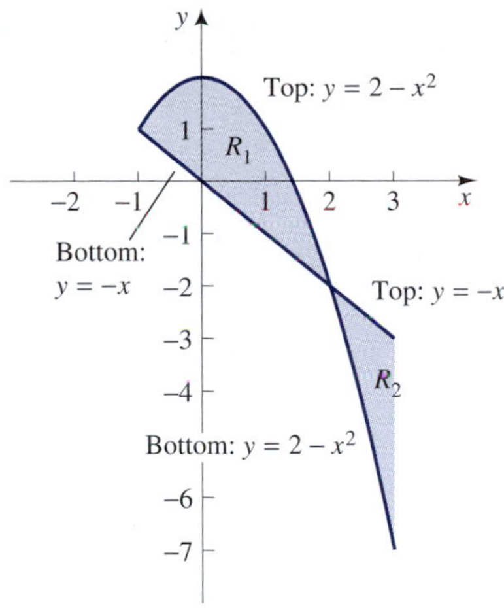

Figure 6.50
The area between the graphs of the two curves from -1 to 3 is the area of R_1 plus the area of R_2.

$$
\begin{aligned}
A &= A_1 + A_2 \\
&= \int_{-1}^{2} [\text{top of } R_1 - \text{bottom of } R_1]\, dx + \int_{2}^{3} [\text{top of } R_2 - \text{bottom of } R_2]\, dx \\
&= \int_{-1}^{2} [(2 - x^2) - (-x)]\, dx + \int_{2}^{3} [(-x) - (2 - x^2)]\, dx \\
&= \int_{-1}^{2} [2 + x - x^2]\, dx + \int_{2}^{3} [x^2 - x - 2]\, dx \\
&= \left(2x + \frac{1}{2}x^2 - \frac{1}{3}x^3\right)\Bigg|_{-1}^{2} + \left(\frac{1}{3}x^3 - \frac{1}{2}x^2 - 2x\right)\Bigg|_{2}^{3} \\
&= \left[\left(4 + 2 - \frac{8}{3}\right) - \left(-2 + \frac{1}{2} + \frac{1}{3}\right)\right] + \left[\left(9 - \frac{9}{2} - 6\right) - \left(\frac{8}{3} - 2 - 4\right)\right] \\
&= \left[\left(\frac{10}{3}\right) - \left(\frac{-7}{6}\right)\right] + \left[\left(\frac{-3}{2}\right) - \left(\frac{-10}{3}\right)\right] \\
&= \frac{9}{2} + \frac{11}{6} \\
&= \frac{19}{3}
\end{aligned}
$$

■

EXAMPLE 5 **Graphs with Multiple Points of Intersection**

Find an approximation to the area enclosed between the curves $y = f(x) = x^7 + x^2 - 3x$ and $y = 0.5x^3$.

Solution

Screen 6.9 shows a graph of the two functions $y_1 = f(x)$ and $y_2 = 0.5x^3$ using a window of dimensions $[-2, 2]$ by $[-3, 3]$. We see that our area is a bow-tie region.

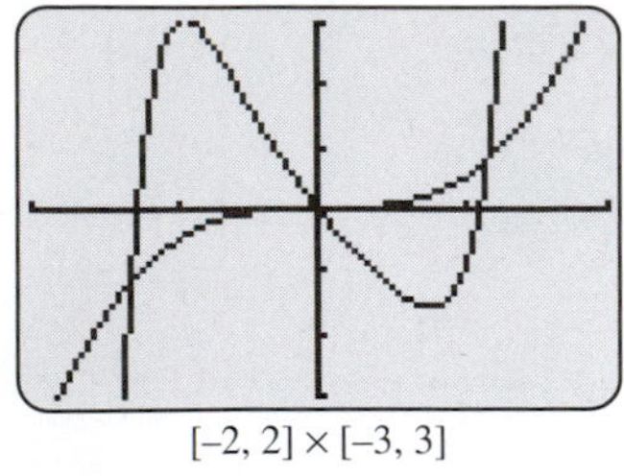
[−2, 2] × [−3, 3]

Screen 6.9
We need to use the "intersect" operation on our graphing calculator to find the x-coordinates of two of the points of intersection of the graphs of $y_1 = x^7 + x^2 - 3x$ and $y_2 = 0.5x^3$. We obtain $x_1 \approx -1.315$ and $x_2 \approx 1.166$.

Using available operations, we can locate the x coordinates of the two points of intersection as $x = -1.315$ and $x = 1.166$ to three decimal places. Using the average of the left- and right-hand sums for $n = 64$, we find that

$$\begin{aligned} A &= A_1 + A_2 \\ &= \int_{-1.315}^{0} [\text{top of } R_1 - \text{bottom of } R_1]\, dx + \int_{0}^{1.166} [\text{top of } R_2 - \text{bottom of } R_2]\, dx \\ &= \int_{-1.315}^{0} [(x^7 + x^2 - 3x) - (0.5x^3)]\, dx + \int_{0}^{1.166} [(0.5x^3) - (x^7 + x^2 - 3x)]\, dx \\ &\approx 2.607 + 1.314 = 3.921 \end{aligned}$$

■

Remark To find the points of intersection of the two curves $y = x^7 + x^2 - 3x$ and $y = 0.5x^3$ requires finding all the solutions to the equation $y = x^7 + x^2 - 3x = 0.5x^3$. However, there is no known analytical way of finding all of these solutions. Therefore, the best we can hope for is *approximate* solutions found by graphical or numerical methods.

Lorentz Curves

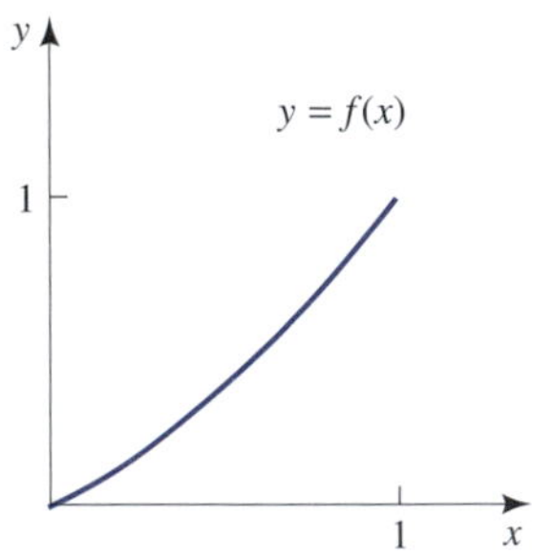

Figure 6.51
A typical Lorentz curve.

For a given population, let $f(x)$ be the proportion of total income that is received by the lowest-paid $100x\%$ of income recipients. Thus, $f(0.3) = 0.2$ means that the lowest-paid 30% of income recipients receive 20% of the total income, or if $f(0.6) = 0.25$, this means that the lowest-paid 60% of income recipients receive 25% of the total income. We must have $0 \le x \le 1$ and $0 \le f(x) \le 1$.

We include only income recipients; thus, $f(0) = 0$. Since all income is received by 100% of the recipients, $f(1) = 1$. Also $f(x) \le x$, since the lowest $100x\%$ of the income recipients cannot receive more than $100x\%$ of the total income. The graph of such an income distribution is called a **Lorentz curve**. See Figure 6.51 for a typical such curve.

Suppose a Lorentz curve is given by $f(x) = x^2$. Since $f(0.5) = 0.25$, the lowest-paid 50% receive 25% of the total income. Since $f(0.1) = 0.01$, the lowest-paid 10% receive 1% of the total income.

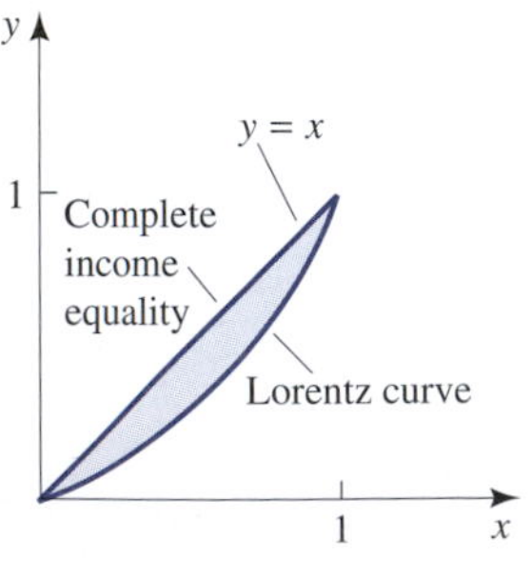

Figure 6.52
The coefficient of inequality is the ratio of the area under the graph of $y = f(x)$ on [0, 1] with the area under $y = x$ on [0, 1]. This equals $2\int_0^1 [x - f(x)]\, dx$.

A perfect equality of income distribution is represented by the curve $f(x) = x$. This is true, since in this case the lowest-paid 1% would receive 1% of total income, the lowest-paid 10% would receive 10% of total income, and so on. Any deviation from this perfect equality represents an inequality of income distribution. See Figure 6.52.

A convenient and intuitive measure of this deviation is the *area* between the curve $y = x$ and $y = f(x)$. Economists then define the **coefficient of inequality** to be the ratio of this area to the area under the curve $y = x$ that represents perfect equality of income distribution. Since of course the area under $y = x$ on [0, 1] is $\frac{1}{2}$, the ratio will be two times the area between $y = x$ and $y = f(x)$. We have the following.

DEFINITION **Coefficient of Inequality**

The coefficient of inequality L of a Lorentz curve $y = f(x)$ is given by

$$L = 2\int_0^1 [x - f(x)]\, dx$$

EXAMPLE 6 **Calculating the Coefficient of Inequality**

At the beginning of this chapter we noted that David Hackett Fischer[10] has identified four great worldwide inflationary waves during the last 800 years, interspersed with three great deflationary waves. One social result that was observed was a rapid growth of inequality that appeared in the later stages of every long inflationary wave. The following table gives income data for the town of Santa Maria Impruneta in Tuscany during one period of inflation.

Year	Percent by Decile									
1307	1.7	2.7	4.2	5.8	6.2	7.0	8.8	12.3	17.6	33.7
1330	1.2	1.3	2.4	3.2	5.5	7.8	10.3	12.7	18.3	37.3
1427	0.0	0.0	0.5	1.9	3.6	7.0	8.7	11.0	18.6	48.7
Taxable Wealth by Decile										

Determine the approximate coefficient of inequality for each year, and determine from these three numbers what was happening to the distributions of income during this inflationary period.

Solution

To obtain a Lorentz curve, we must convert the numbers in the table to cumulative and to decimals. For example, $f(0.40)$ is the proportion of total income received by the lowest 40% of income recipients. Thus, for the year 1427, $f(0.40) = 0.005 + 0.019 = 0.024$. Doing this, we have the following table.

Year	Cumulative Percent by Decile									
1307	0.017	0.044	0.086	0.144	0.206	0.276	0.364	0.487	0.663	1.000
1330	0.012	0.025	0.049	0.081	0.136	0.214	0.317	0.444	0.627	1.000
1427	0.000	0.000	0.005	0.024	0.060	0.130	0.217	0.327	0.513	1.000
Taxable Wealth by Decile										

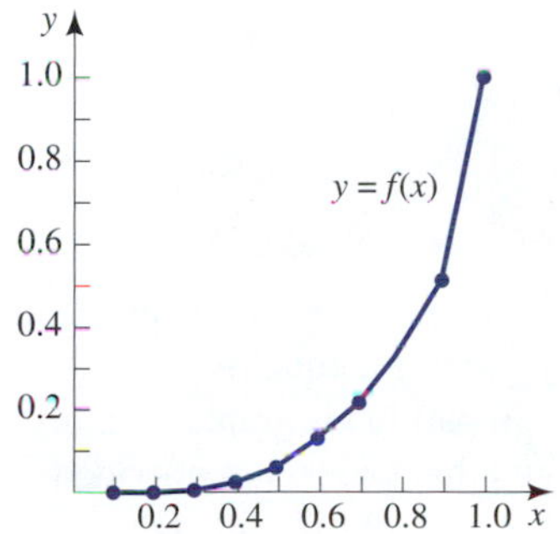

Figure 6.53
The graph of a Lorentz curve.

Figure 6.53 gives a graph of the data points for the year 1427. The points are connected with straight lines. The connected line segments then form the graph of a function that we call $f(x)$ and is a Lorentz curve. We must evaluate

$$2\int_0^1 [x - f(x)]\,dx = 2\int_0^1 x\,dx - 2\int_0^1 f(x)\,dx = 1 - 2\int_0^1 f(x)\,dx$$

Since we do not actually know what happened between the data points, we will approximate the integral $\int_0^1 f(x)\,dx$ by taking an average of the left- and right-hand

[10] Ibid.

sums. We have $\Delta x = 0.1$.

left-hand sum

$$= \Delta x[f(0) + f(0.1) + \cdots + f(0.9)]$$
$$= 0.1[0.000 + 0.000 + 0.005 + 0.024 + 0.060 + 0.130 + 0.217 + 0.327 + 0.513]$$
$$= 0.1276$$

right-hand sum

$$= \Delta x[f(0.1) + f(0.2) + \cdots + f(1.0)]$$
$$= 0.1[0.000 + 0.005 + 0.024 + 0.060 + 0.130 + 0.217 + 0.327 + 0.513 + 1.000]$$
$$= 0.2276$$

$$\int_0^1 f(x)\,dx \approx [0.1276 + 0.2276]/2 = 0.1776$$

$$L = 1 - 2\int_0^1 f(x)\,dx \approx 1 - 2(0.1776) = 0.6448$$

Doing the same for 1307 gives $L = 0.4443$, and doing the same for 1330 gives $L = 0.5202$. Thus, the income inequality increases. ■

In Case 2 at the end of this chapter we will calculate the coefficients of inequality for some recent years in the United States. For example, the coefficient of inequality for 1995 in the United States was approximately 0.4316. Surprisingly, this is only slightly lower than that for Santa Maria Impruneta in 1307. Case 2 will show the inequality of income distribution is worsening in the United States during recent years.

Interactive Illustration 6.9

Lorentz Curves

This interactive illustration allows you to explore some Lorentz curves.

1. Look at the coefficients of inequality for the various countries during different years. What was the trend in the USA from 1950 to 1997? Is this what you expected? Why or why not? Which country had the smallest coefficient of inequality? Is this surprising?
2. Consider the Lorentz curve $f(x) = ax + (1 - a)x^b$ where a and b are parameters. Use integral calculus to find an expression for the coefficient of inequality in terms of a and b.
3. According to your expression found in part (2) what should happen to the coefficient of inequality as a becomes larger and heads for 1? As b becomes larger?

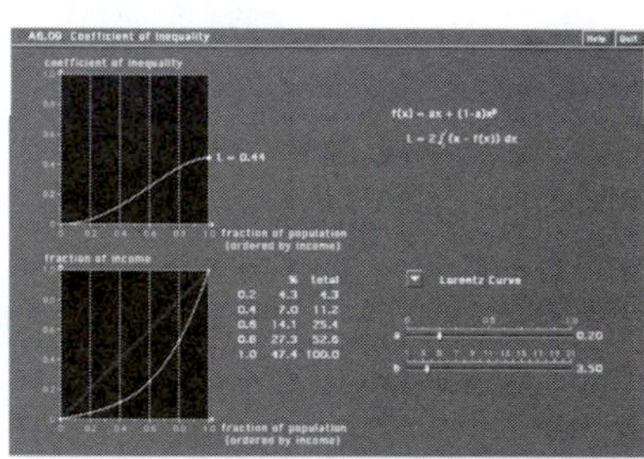

4. Take $a = 0.5$ and $b = 4$ and $b = 5$ in your expression and determine the coefficient of inequality.
5. Notice that the function in part (2) is graphed in this interactive illustration and that sliders are provided for a and b. Now check your work in parts (2) through (4) by moving the sliders for a and b appropriately.

Warm Up Exercise Set 6.6

1. Find the area between the two curves $y = x + 2$ and $y = x^2$.
2. Find the area enclosed by the graphs of the two curves $y = x$ and $y = f(x) = \sqrt[3]{x}$.

Exercise Set 6.6

In Exercises 1 through 36, find the area enclosed by the given curves.

1. $y = x^3$, $y = 0$, $x = 1$, $x = 2$
2. $y = x^2$, $y = 0$, $x = -1$, $x = 2$
3. $y = e^{0.1x}$, $y = 1$, $x = 0$, $x = 10$
4. $y = e^{-x}$, $y = 3$, $x = -1$, $x = 0$
5. $y = x^2$, $y = x$, $x = 0$, $x = 1$
6. $y = \sqrt{x}$, $y = x$, $x = 0$, $x = 1$
7. $y = x$, $y = 1/x$, $x = e$, $x = e^2$
8. $y = 4$, $y = x^2$
9. $y = x^2 - 2x + 1$, $y = x + 1$
10. $y = x^2$, $y = 8 - x^2$
11. $y = x^3$, $y = -x^3$, $x = 0$, $x = 1$
12. $y = x$, $y = -1/x$, $x = 1/e$, $x = e$
13. $y = -x^2$, $y = -3x - 1$, $x = 0$, $x = 2$
14. $y = x^2 - 2x$, $y = x - 4$, $x = 0$, $x = 1$
15. $y = x^2 + 1$, $y = x$, $x = -1$, $x = 1$
16. $y = x + 2$, $y = \sqrt[3]{x}$, $x = -1$, $x = 1$
17. $y = x^2 + 4x + 4$, $y = 4 - x$
18. $y = x^2 - 2x + 1$, $y = 1 + 2x - x^2$
19. $y = x(x - 4)$, $y = x$
20. $y = x(x - 2)$, $y = x$
21. $y = e^{|x|}$, $y = 0$, $x = -1$, $x = 2$
22. $y = e^{-|x|}$, $y = 0$, $x = -1$, $x = 2$
23. $y = x^2$, $y = 4$, $x = 0$, $x = 4$
24. $y = x^3$, $y = 2$, $x = 0$, $x = 2$
25. $y = x^3$, $y = -x$, $x = -1$, $x = 1$
26. $y = e^x$, $y = 2$, $x = 0$, $x = 1$
27. $y = -\sqrt[3]{x}$, $y = -x$, $x = -1$, $x = 1$
28. $y = \sqrt[3]{x}$, $y = -x$, $x = -1$, $x = 1$
29. $y = e^x$, $y = e^{-x}$, $x = -\ln 2$, $x = \ln 2$
30. $y = e^{2x}$, $y = e^{-2x}$, $x = -1$, $x = 2$
31. $y = x^3 - 3x$, $y = x$
32. $y = x^3 - 3x$, $y = 2x^2$
33. $y = |x|$, $y = 3$, $x = -1$, $x = 6$ (triple region)
34. $y = |x|$, $y = \frac{1}{2}x + 2$, $x = -2$, $x = 4$ (triple region)
35. $y = \sqrt[3]{x}$, $y = x$, $x = -8$, $x = 1$ (triple region)
36. $y = -\sqrt[3]{x}$, $y = -x$, $x = -1$, $x = 8$ (triple region)

In Exercises 37 through 40, graph the given pair of curves in the same viewing window of your grapher. Find the points of intersection to two decimal places. Then estimate the area enclosed by the given pairs of curves by taking the average of the left- and right-hand sums for $n = 100$.

37. $y = x^5 + x^2 - 3x$, $y = x$
38. $y = x^5 + x^4 - 3x$, $y = x$
39. $y = x^5 + x^4 - 3x$, $y = 0.50x^3$
40. $y = x^5 + x^4 - 3x$, $y = 3x - x^2 - x^5$
41. In the figure, the area of the region R_1 is 10, and that of R_2 is 6. Determine the following:

   a. $\int_1^2 f(x)\,dx$ b. $\int_2^3 f(x)\,dx$

   c. $\int_1^3 f(x)\,dx$ d. $\int_3^2 f(x)\,dx$

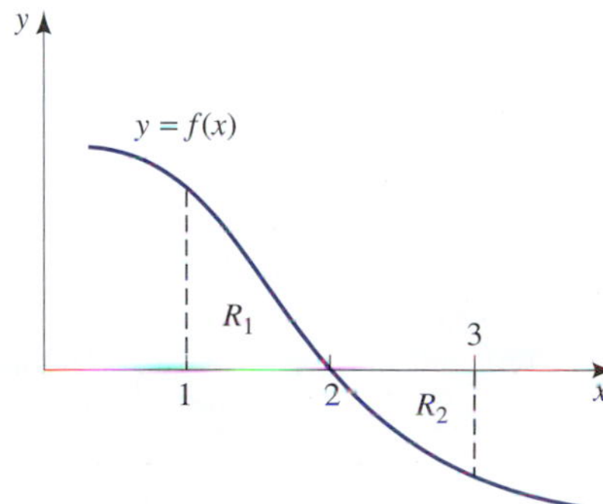

42. Refer to the figure. Estimate $\int_0^{10} f(x)\,dx$. Explain how you obtained your estimate.

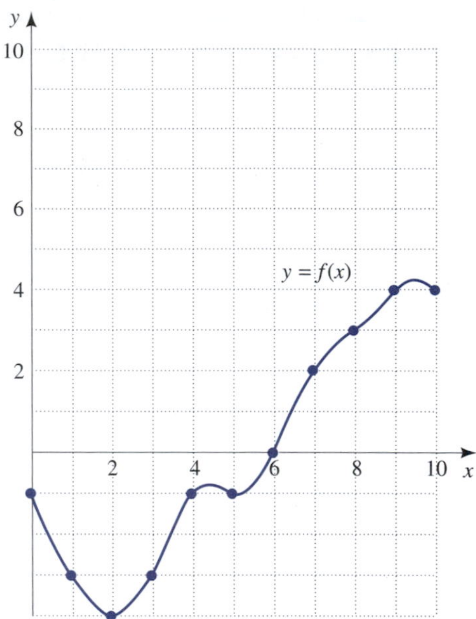

Applications and Mathematical Modeling

In Exercises 43 through 50, show that the curves are Lorentz curves. (You will need to find $f''(x)$.) Then find the coefficient of inequality.

43. $f(x) = x^2$
$f(x) = x^5$

44. $f(x) = \frac{1}{4}x^2 + \frac{3}{4}x$

45. $f(x) = \frac{3}{4}x^2 + \frac{1}{4}x$

46. $f(x) = 0.3x^2 + 0.7x$

47. $f(x) = 0.7x^2 + 0.3x$

48. $f(x) = 0.01x^2 + 0.99x$

49. $f(x) = 0.99x^2 + 0.01x$

50. Costs A firm has found that its costs in billions of dollars have been increasing (owing primarily to inflation) at a rate given by $C_1'(t) = 1.2t$, where t is the time measured in years since production of the item began. At the beginning of the second year an innovation in the production process resulted in costs increasing at a rate given by $C_2'(t) = 0.9\sqrt{t}$. Find the total costs during the first 4 years. Assume $C_1(0) = 0$.

51. Revenue A firm has found that its revenue in billions of dollars has been increasing at a rate given by $R'(t) = 3t^2$, where t is the time measured in years since the item was introduced. At the beginning of the third year, new competition has resulted in revenue increasing only at a rate given by $R_1'(t) = 12t$. Find the total revenue during the first 4 years.

52. Sales From the time a firm introduced a new product, sales in thousands of this product have been increasing at a rate given by $S_1'(t) = 2 + t$, where t is the time since the introduction of the new product. At the beginning of the third year an advertising campaign was introduced, and sales then increased at a rate given by $S_2'(t) = t^2$. Find the increase in sales due to the advertising campaign during the first 2 years of this campaign over what the sales would have been without the campaign.

53. Production Suppose copper is being extracted from a certain mine at a rate given by $\frac{d}{dt}P(t) = 100e^{-0.2t}$, where $P(t)$ is measured in tons of copper and t is measured in years from opening the mine. At the beginning of the sixth year a new mining innovation is introduced to boost production to a rate given by $Q'(t) = 500/t$. Find the increase in production of copper due to this innovation during the second 5 years of its use over what copper production would have been without its use.

More Challenging Exercises

55. Given the graph of $y = f'(x)$, sketch a graph of $y = f(x)$ if $f(0) = 10$ and if the areas of R_1, R_2, R_3, and R_4 are 10, 12, 3, and 2, respectively. Give the x- and y-coordinates of critical points and inflection points.

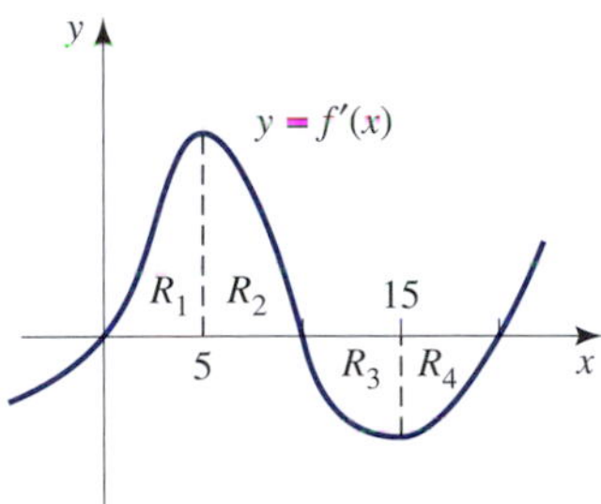

56. Find the area between the graph of the curve $y = \cos x$ and the x-axis on the interval $[-\pi/2, \pi/2]$.

57. Wage Distribution The following graph appeared recently in a paper in the *Economic Review Federal Reserve Bank of Dallas* by Phillips[11] and gave a Lorentz's curve for the wage distribution for year-round full-time workers in the Texas goods sector for the years 1978 and 1989. What conclusions can you come to about wages in these two years?

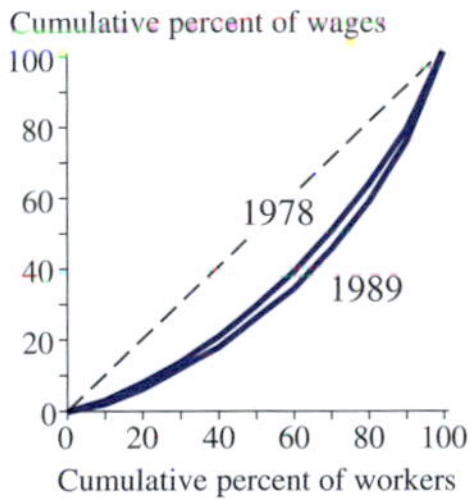

58. We learned in this section that the area of the shaded region in the accompanying figure is

$$\int_a^c [f(x) - g(x)]\,dx + \int_c^b [g(x) - f(x)]\,dx$$

Explain carefully why the area of the shaded region is also given by $\int_a^b |f(x) - g(x)|\,dx$.

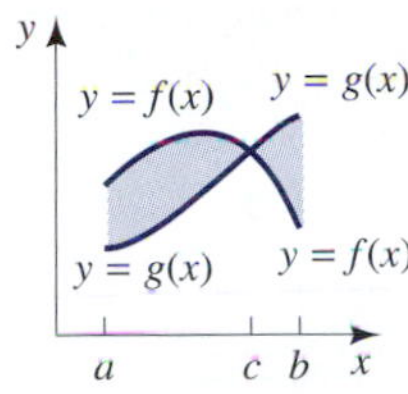

[11] Keith R. Phillips. 1991. The effects of the growing service sector on wages in Texas. *Econ. Rev. Fed. Res. Bank Dallas* Nov:15–28.

59. Refer to the figure. Suppose $F(2) = 5$ and the area of the shaded region is 10. Find the maximum value attained by F.

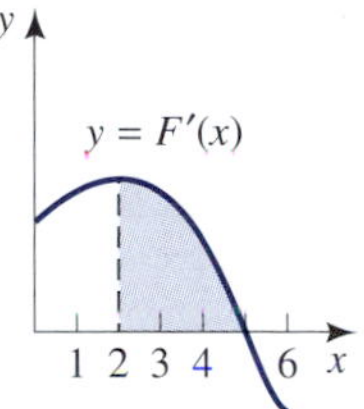

60. On your computer or graphing calculator, draw graphs of $y_1 = f(x) = \sin x$ and $y_2 = 0.5$ using a window with dimensions $[0, 3.14]$ by $[0, 1]$. By looking at the graphs and doing no calculations, explain why the average value of f on the interval $[0, 3.14]$ must be between 0.5 and 1.

61. The figure shows a graph of $y = f'(x)$. The area of region R_1 is 10, and that of R_2 is 7. If $f(1) = 12$, find $f(3)$ and $f(5)$.

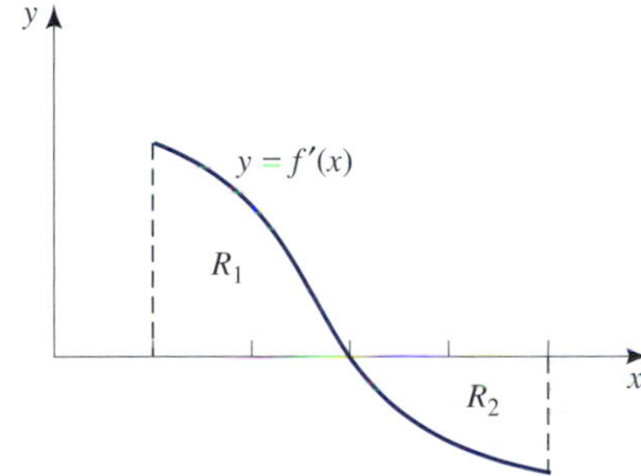

62. Coefficient of Inequality The following table gives the earnings in dollars of U.S. households in earnings quartiles. (See De Nardi and colleagues.[12])

1st	2nd	3rd	4th
6009	24,494	43,425	89,184

Determine the coefficient of inequality by taking the average of the right- and left-hand sums.

63. Coefficient of Inequality The following table gives the earnings in dollars of Canadian households in earnings quartiles. (See De Nardi and colleagues.[13])

[12] Mariacristina De Nardi, Liqian Ren, and Chao Wei. 2000. Income inequality and redistribution in five countries. *Econ. Perspect. Fed. Res. Bank Chicago* XXV(2):2–20.

[13] Ibid.

1st	2nd	3rd	4th
5008	22,753	37,856	68,148

Determine the coefficient of inequality by taking the average of the right- and left-hand sums.

64. Coefficient of Inequality The following table gives the earnings in dollars of German households in earnings quartiles. (See De Nardi and colleagues.[14])

1st	2nd	3rd	4th
9174	30,275	45,496	80,412

Determine the coefficient of inequality by taking the average of the right- and left-hand sums.

[14] Ibid.

65. Coefficient of Inequality The following table gives the earnings in dollars of Swedish households in earnings quartiles. (See De Nardi and colleagues.[15])

1st	2nd	3rd	4th
7010	28,120	44,315	76,646

Determine the coefficient of inequality by taking the average of the right- and left-hand sums.

[15] Ibid.

Solutions to WARM UP EXERCISE SET 6.6

1. A graph is given in the figure. To determine the interval of integration, we need to determine the values of x for which the two curves intersect. This will happen when $x + 2 = y = x^2$ or $0 = x^2 - x - 2 = (x + 1)(x - 2)$. Thus, $x = -1$ and $x = 2$. Then

$$\begin{aligned}\int_a^b [\text{top} - \text{bottom}]\, dx &= \int_{-1}^{2} [(x+2) - (x^2)]\, dx \\ &= \left(\frac{1}{2}x^2 + 2x - \frac{1}{3}x^3\right)\Bigg|_{-1}^{2} \\ &= \left[2 + 4 - \frac{8}{3}\right] - \left[\frac{1}{2} - 2 + \frac{1}{3}\right] = \frac{9}{2}\end{aligned}$$

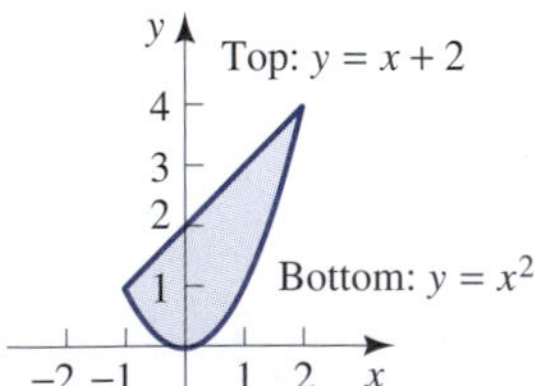

2. The region between the graphs of the two curves forms a bow-tie region, so we must divide the region into two subregions to use our formulas. If we let A be the area we are seeking, let A_1 be the area of R_1, and let A_2 be the area of R_2, then

$$\begin{aligned}A &= A_1 + A_2 \\ &= \int_{-1}^{0} [\text{top of } R_1 - \text{bottom of } R_1]\, dx \\ &\quad + \int_0^1 [\text{top of } R_2 - \text{bottom of } R_2]\, dx \\ &= \int_{-1}^{0} [(x) - (x^{1/3})]\, dx + \int_0^1 [(x^{1/3}) - (x)]\, dx \\ &= \left(\frac{1}{2}x^2 - \frac{3}{4}x^{4/3}\right)\Bigg|_{-1}^{0} + \left(\frac{3}{4}x^{4/3} - \frac{1}{2}x^2\right)\Bigg|_0^1 \\ &= -\left(\frac{1}{2} - \frac{3}{4}\right) + \left(\frac{3}{4} - \frac{1}{2}\right) = \frac{1}{2}\end{aligned}$$

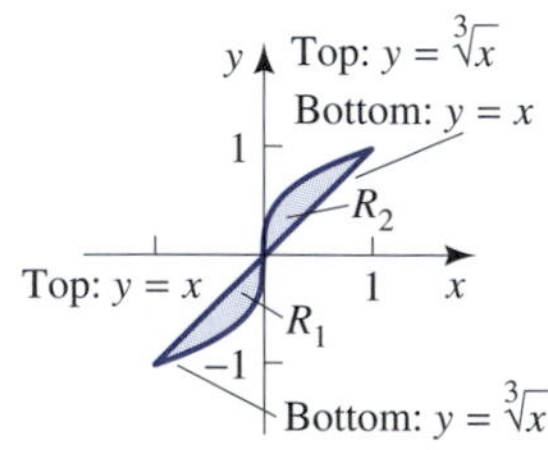

6.7 Additional Applications of the Integral

APPLICATION **Price of an Oil Well**

Suppose the rate of change in thousands of dollars of total income from an oil well is projected to be $f(t) = 100e^{0.1t}$, where t is measured in years. If the oil well is being offered for sale for \$500,000 and has a useful life of 10 years, and current annual interest rates are at 10%, should you buy it? See Example 4 on page 476 for the answer.

Continuous Money Flow

We are already familiar with the idea of *continuous* compounding. When money is being compounded continuously, we are not suggesting that the bank is actually continuously placing money into our account. Rather, we calculate what is in our account as if this were in fact happening. We then can think of our account as a **continuous flow of money**. This is not unlike the electric utility company that has a meter on our house continuously totaling the amount of electricity that they are selling to us. Since a price has been set for the electricity, this can be thought of as a continuous flow of income for the utility company.

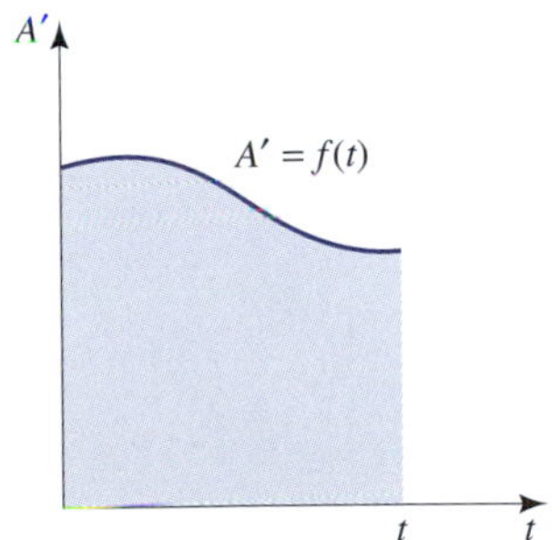

Figure 6.54
The graph of the rate of change of a typical income flow.

We then assume that we have a continuous flow of money and that the positive function $f(t)$ represents the *rate of change* of this flow. (We could think of $f(t)$ as the velocity of the flow.) If we let $A(t)$ be the total amount of income obtained from this flow, then we are saying that $A'(t) = f(t)$. We will always assume that the amount of money from this flow at time zero is zero; that is, we will always assume that $A(0) = 0$. Then the fundamental theorem of calculus tells us that

$$A(t) = \int_0^t f(x)\,dx$$

This is the area under the curve $f(x)$ on the interval $[0, t]$. See Figure 6.54.

EXAMPLE 1 **Finding Income from an Income Flow**

Suppose an investment brings in an income of \$1000 per year. How much income is obtained in 3 years?

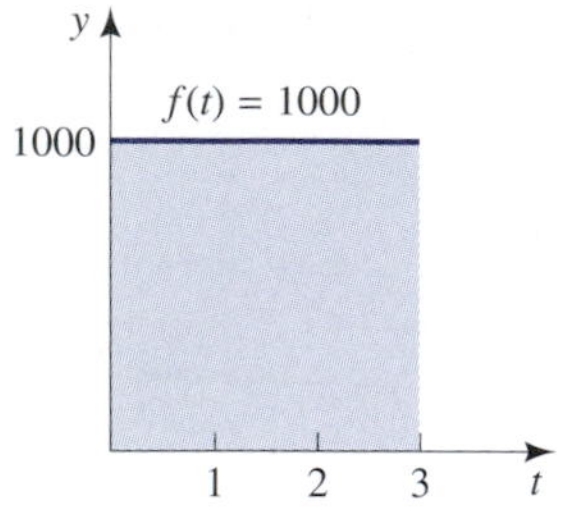

Figure 6.55
The area between the graph of $y = 1000$ and the t-axis from 0 to 3, $\int_0^3 1000\,dt = 3000$, is the total income obtained.

Solution

Obviously, the answer is \$3000. But this can also be viewed as an income flow given by $f(t) = 1000$ dollars per year on the interval $[0, 3]$. The area under this curve is the definite integral

$$\int_0^3 1000\,dx = 3000$$

See Figure 6.55. ■

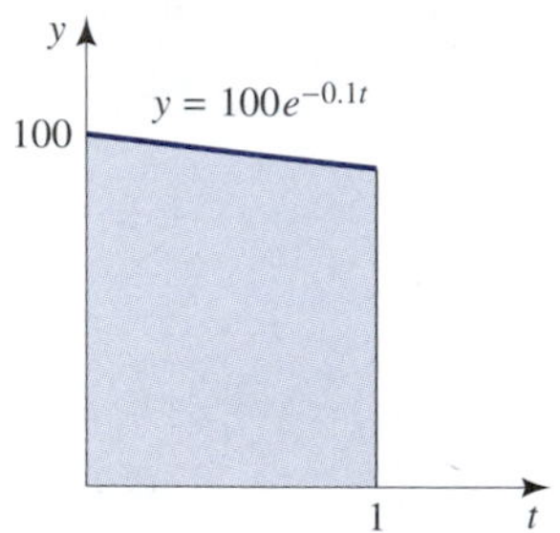

Figure 6.56
The area between the graph of $y = 100e^{-0.1t}\,dt$ and the t-axis from 0 to 1 is the total income obtained.

EXAMPLE 2 **Finding Income From an Income Flow**

Suppose the rate of change of income in thousands of dollars per year from an oil well is projected to be $f(t) = 100e^{-0.1t}$, where t is measured in years. Find the total amount of money from this well during the first year. See Figure 6.56.

Solution

We have

$$\begin{aligned}A(1) &= \int_0^1 f(x)\,dx \\ &= \int_0^1 100e^{-0.1t}\,dt \\ &= 100(-10)e^{-0.1t}\Big|_0^1 \\ &= 1000\left[1 - e^{-0.1}\right] \approx 95\end{aligned}$$

or approximately \$95,000. ■

Present Value of a Continuous Income Flow

We will now take up the subject of **present value** of an income flow and find a formula that yields the present value. This will permit us to compare different income flows.

We again assume that we have some continuous flow of income, perhaps from oil or gas wells, accounts with continuous compounding of interest, or other accounts that are assumed to be compounding continuously. We assume that $f(t)$ gives the rate of change of the total flow of funds with t measured in some appropriate units of time that, for convenience, we will take to be years.

For any $T > 0$ we wish to define a number that we call the present value of this income flow on the interval $[0, T]$. *Intuitively, we want this number to be the present amount of money needed to be able to generate the same income flow over the time interval* $[0, T]$ *as is generated by the given flow.* To do this, we will have to assume that the present money can be invested at a certain given rate (the current available rate) over that time interval, say, a rate equal to r compounded continuously.

Recall that if an amount P is invested at an annual rate of r (given as a decimal) and compounded continuously, then the amount becomes equal to $A = Pe^{rt}$ after t years. Solving for P gives the present value $P = Ae^{-rt}$. This is the present amount that is needed to attain the amount A after t years of continuous compounding at the annual rate of r.

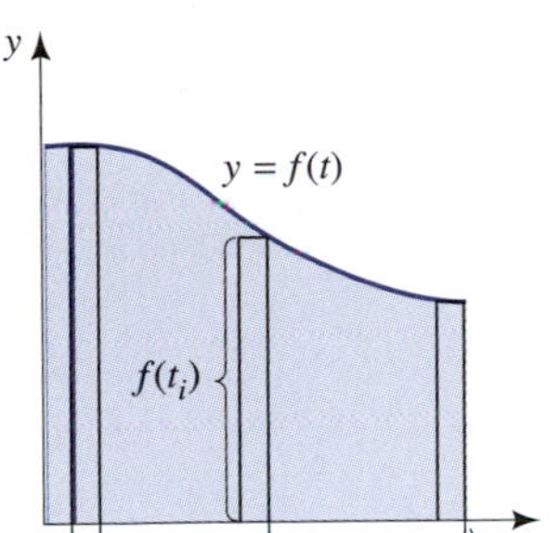

Figure 6.57
Divide the t-interval into n equal subintervals, and in each of these subintervals pick t_i to be the right-hand endpoint.

To proceed, divide the interval $[0, T]$ into n equal subintervals of length Δt as in Figure 6.57. In a typical interval I_i we pick some point t_i, say, the right-hand endpoint. The area under $f(t)$ on the interval I_i is precisely the income generated by the given flow over the time interval I_i. This area A_i can be approximated by the area of the rectangle given by $f(t_i)\Delta t$. The present value of this income A_i is

$$P_i = A_i e^{-rt_i} \approx f(t_i)e^{-rt_i}\Delta t$$

Summing all of these present values gives the total present value, P. This is also approximately the sum of all the areas of all n rectangles or

$$P = \sum_{i=1}^{n} P_i \approx \sum_{i=1}^{n} f(t_i)e^{-rt_i}\Delta t$$

which is a Riemann sum of $f(t)e^{-rt}$ on the interval $[0, T]$. If we take n larger and larger, the approximations will be better and better. Then letting $n \to +\infty$, we should obtain the exact value. But this is

$$\lim_{n\to+\infty} \sum_{i=1}^{n} f(t_i)e^{-rt_i}\Delta t$$

which is precisely the definition of the definite integral

$$\int_0^T f(t)e^{-rt}\,dt$$

We thus have the following.

Present Value of a Continuous Income Flow

Let $f(t)$ be the continuous rate of change of total income on the interval $[0, T]$, and suppose that we can invest current money at an interest rate of r, where r is a decimal, compounded continuously over the time interval $[0, T]$. Then the present value $P_V(T)$ of this continuous income flow over $[0, T]$ is given by

$$P_V(T) = \int_0^T f(t)e^{-rt}\,dt$$

EXAMPLE 3 **Finding the Present Value of a Continuous Income Flow**

Suppose that the rate of change of an income flow is given by the constant (and continuous) function $f(t) = 1000$ dollars per year. If the rate r is given by $r = 0.10$, find $P_V(3)$.

Solution

$$
\begin{aligned}
P_V(T) &= \int_0^T f(t)e^{-rt}\,dt \\
P_V(3) &= \int_0^3 1000e^{-0.1t}\,dt \\
&= 1000(-10)e^{-0.1t}\Big|_0^3 \\
&= 10,000\left[1 - e^{-0.3}\right] \approx 2592
\end{aligned}
$$

■

EXAMPLE 4 **Determining the Price of an Oil Well**

If the oil well in Example 2 is being offered for sale for \$500,000 and has a useful life of 10 years, would you buy it if (a) $r = 0.1$, (b) $r = 0.05$?

Solution

(a) In this case

$$
\begin{aligned}
P_V(T) &= \int_0^T f(t)e^{-rt}\,dt \\
P_V(10) &= \int_0^{10} 100e^{-0.1t}e^{-0.1t}\,dt \\
&= 100\int_0^{10} e^{-0.2t}\,dt \\
&= 100(-5)e^{-0.2t}\Big|_0^{10} \\
&= 500\left[1 - e^{-2}\right] \approx 432
\end{aligned}
$$

or \$432,000. This is not a good deal, since you can generate more income from \$500,000 from an investment at current interest rates.

(b) In this case

$$
\begin{aligned}
P_V(T) &= \int_0^T f(t)e^{-rt}\,dt \\
P_V(10) &= \int_0^{10} 100e^{-0.10t}e^{-0.05t}\,dt \\
&= 100\int_0^{10} e^{-0.15t}\,dt \\
&= -\frac{100}{0.15}e^{-0.15t}\Big|_0^{10} \\
&= \frac{100}{0.15}\left[1 - e^{-1.5}\right] \approx 518
\end{aligned}
$$

or \$518,000. This is a good deal, since you can generate more income from this oil well than you can from \$500,000 from an investment at current interest rates.

■

Remark Naturally, an important consideration in any investment is the riskiness of the investment. This has not been taken into account in the decision to buy in the previous example.

Continuous Reinvestment of Income

As we noted before, the formula $A(t) = Pe^{rt}$ relates the amount $A(t)$ in an account after t years with initial amount P and earning an annual rate r compounded continuously. We now suppose that the continuous income stream from the flow is continuously invested; furthermore, we assume that throughout the interval under consideration this income is always invested at the very same annual rate of r compounded continuously. If we let $A^*(T)$ be the amount of money after T years if all the income has been continuously invested at an annual rate of r compounded continuously, then

$$A^*(T) = e^{rT} P_V(T)$$

To see why this is true, recall from the setup in Figure 6.57 that the amount of income obtained during the time interval I_i was $A_i \approx f(t_i)\Delta t$. If this money is then invested at an annual rate of r compounded continuously for the next $T - t_i$ years, this money will then become

$$A_i^*(T) = A_i e^{r(T-t_i)} \approx f(t_i)e^{r(T-t_i)}\Delta t$$

or

$$A_i^*(T) \approx e^{rT} f(t_i)e^{-rt_i}\Delta t.$$

Summing all of these up gives the amount at time T, designated as $A^*(T)$, which, according to this last formula, also is approximately equal to

$$\sum_{i=1}^{n} e^{rT} f(t_i)e^{-rt_i}\Delta t$$

which can be written as

$$e^{rT}\sum_{i=1}^{n} f(t_i)e^{-rt_i}\Delta t$$

Letting $n \to +\infty$ the latter becomes

$$e^{rT}\int_0^T f(t)e^{-rt}\,dt = e^{rT}P_V(T)$$

Continuous Investment of Money

Let $A^*(T)$ be the amount of money after T years if all the income has been continuously invested at an annual rate of r compounded continuously; then

$$A^*(T) = e^{rT}P_V(T) = e^{rT}\int_0^T f(t)e^{-rt}\,dt$$

EXAMPLE 5 **Finding the Total Amount when Continuously Reinvested**

Suppose in Example 3 that the income is continuously invested at the rate r given there. How much money is there after 3 years?

Solution

Using the formula for $A^*(T)$ and the result found in Example 3, we obtain

$$A^*(3) = e^{(.1)(3)} P_V(3) \approx e^{0.3}(2592) \approx 3499$$

that is, approximately \$3499. ■

Naturally, the amount in Example 5 is greater than the amount in Example 3, since in the latter case the money is being continuously reinvested as it comes in.

Consumers' and Producers' Surplus

Suppose you have just landed your first real job and need some dress pants. You really need one pair of pants and are willing to pay \$60 for it. You would prefer to have another pair of pants and are willing to pay \$50 for this second pair. Finally, you are not sure just how badly you need a third pair, so you will pay only \$40 for the third pair. You go shopping and find that the dress pants you need are selling for \$50 each. So you buy two pairs and spend a total of \$100. You were willing to pay \$60 for the first one and \$50 for the second one, for a total of \$110 for two of them. You then "saved" \$110 – \$100 = \$10. Adding up all such savings of all consumers is called the *consumer surplus*.

To see how this works in general, let $p = D(x)$ be the demand equation, where p is the unit price of a commodity and x is the quantity demanded by the consumers at that price. As we know, this function must be decreasing (Figure 6.58).

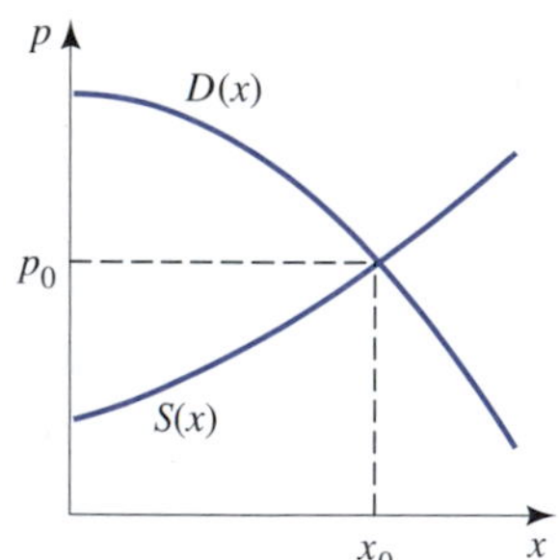

Figure 6.58
The demand and supply curves intersect at (x_0, p_0).

Let $p = S(x)$ be the supply equation, where p is the unit price of a commodity and x is the quantity made available by the producers at that price. As we know, this function must be increasing (see Figure 6.58). We assume that the demand curve and the supply curve intersect at $x = x_0$. Thus, $D(x_0) = p_0 = S(x_0)$. With (x_0, p_0) the point of intersection, p_0, is the equilibrium price. At the equilibrium price, consumers will purchase the same number of the commodity as the producers will supply.

As can be seen in Figure 6.58, $D(x) > p_0$ when $x < x_0$. This means that there are consumers who are willing to pay a *higher* price for the commodity than p_0. These consumers are then actually experiencing a *savings*. The total amount of such "savings" is called the **consumers' surplus**.

We shall now show that the consumer surplus is given by the area between the curve $p = D(x)$ and the straight line $p = p_0$ on the interval $[0, x_0]$, which is the definite integral

$$\int_0^{x_0} [D(x) - p_0]\, dx$$

To see this, we divide the interval $[0, x_0]$ into n subintervals of equal length Δx. Thus, Δx is the number of consumers in each subinterval. In a typical interval I_i we pick a point x_i, say, the right-hand endpoint. Thus, x_i represents a "typical" consumer in the subinterval I_i. The number $p_i - p_0 = D(x_i) - p_0$ then represents the "typical" savings experienced by consumers on this interval. The total savings of consumers on this interval is then approximately the "typical" savings times the number in the interval, or

$$[D(x_i) - p_0]\,\Delta x$$

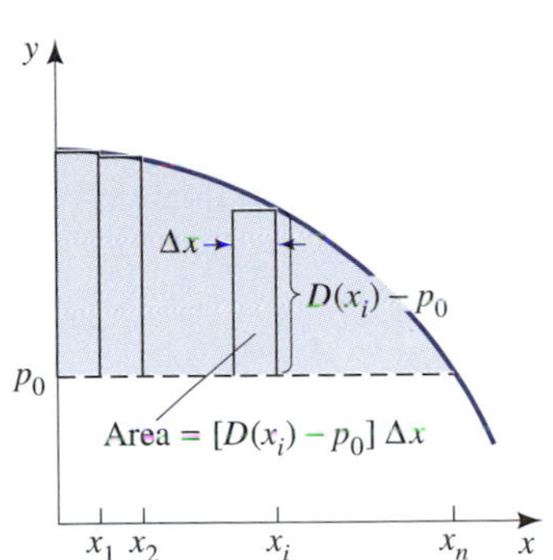

Figure 6.59
The total consumer savings on the ith interval is approximately the area of the indicated rectangle.

which is also the area of the rectangle in Figure 6.59.

To find an approximation of the total consumers' surplus, we sum n such terms and obtain

$$\sum_{i=1}^{n} [D(x_i) - p_0]\,\Delta x$$

This is also a right-handed Riemann sum of $[D(x) - p_0]$ over $[a, b]$. For n larger and larger, this approximation will become better and better. Taking the limit as $n \to +\infty$, that is,

$$\lim_{n\to+\infty} \sum_{i=1}^{n} [D(x_i) - p_0]\,\Delta x$$

we obtain the exact consumers' surplus. But by the definition of the definite integral this limit is

$$\int_0^{x_0} [D(x) - p_0]\; dx$$

We thus have the following.

DEFINITION Consumers' Surplus

If $p = D(x)$ is the demand equation, p_0 is the equilibrium price of the commodity, and x_0 is the equilibrium demand, then the **consumers' surplus** is given by

$$\int_0^{x_0} [D(x) - p_0]\; dx$$

We also see from Figure 6.58 that $p = S(x) < p_0$ if $x < x_0$. This means that there are producers who are willing to sell the commodity for *less* than the going price. For these producers the current price of p_0 represents a *savings*. The total of all such savings is called the **producers' surplus**.

In exact analogy with consumers' surplus, the producers' surplus is given by the area between the straight line $p = p_0$ and the curve $p = S(x)$ on the interval $[0, x_0]$, which is the definite integral $\int_0^{x_0} [p_0 - S(x)]\; dx$. We then have the following.

DEFINITION Producers' Surplus

If $p = S(x)$ is the supply equation, p_0 is the equilibrium price of the commodity, and x_0 is the equilibrium demand, then the **producers' surplus** is given by

$$\int_0^{x_0} [p_0 - S(x)]\; dx$$

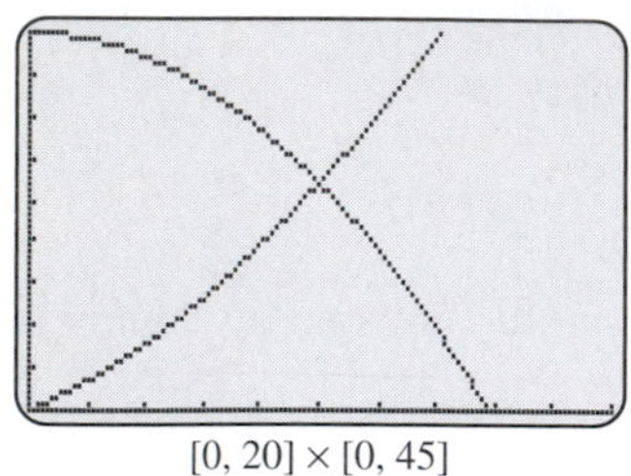

[0, 20] × [0, 45]

Screen 6.10
The supply and demand curves $y_1 = 0.12x^2 + 1.5x$ and $y_2 = 45 - 0.18x^2$ intersect at one point.

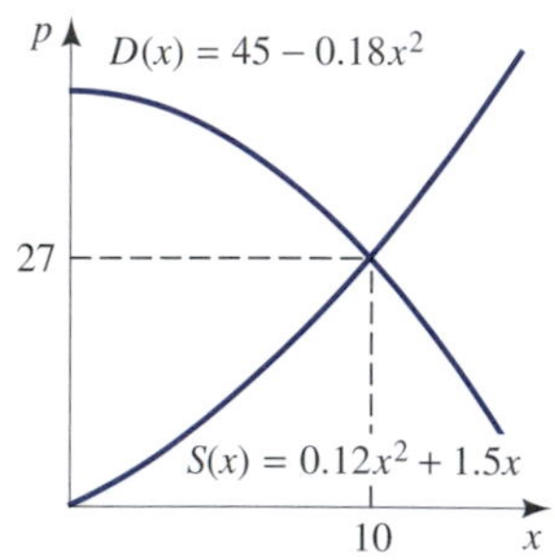

Figure 6.60
The demand and the supply curves intersect at (10, 27).

EXAMPLE 6 **Finding the Consumers' Surplus**

If the demand equation is given by $p = D(x) = 45 - 0.18x^2$ and the supply equation is given by $p = S(x) = 0.12x^2 + 1.5x$, find the consumers' surplus.

Solution

Graphs of the demand and supply curves are shown in Screen 6.10 using a window of dimensions [0, 20] by [0, 45]. We need to first find the equilibrium quantity. To do this, set $S(x) = D(x)$, and obtain

$$\begin{aligned} 0.12x^2 + 1.5x &= 45 - 0.18x^2 \\ 0 &= 0.30x^2 + 1.5x - 45 \\ &= 0.30\left[x^2 + 5x - 150\right] \\ &= 0.30(x + 15)(x - 10) \end{aligned}$$

Since the equilibrium quantity cannot be negative, we ignore the solution $x_0 = -15$. Thus, the equilibrium quantity is $x_0 = 10$. For this value we have $p_0 = D(x_0) = D(10) = 27$ (Figure 6.60).

The consumers' surplus is then given by

$$\begin{aligned} \int_0^{x_0} [D(x) - p_0]\, dx &= \int_0^{10} \left[45 - 0.18x^2 - 27\right] dx \\ &= \left[18x - 0.06x^3\right]\Big|_0^{10} \\ &= 120 \end{aligned}$$

Interactive Illustration 6.10

Consumers' and Producers' Surplus

This interactive illustration allows you to explore what happens to the consumers' and producers' surplus when a tax is added.

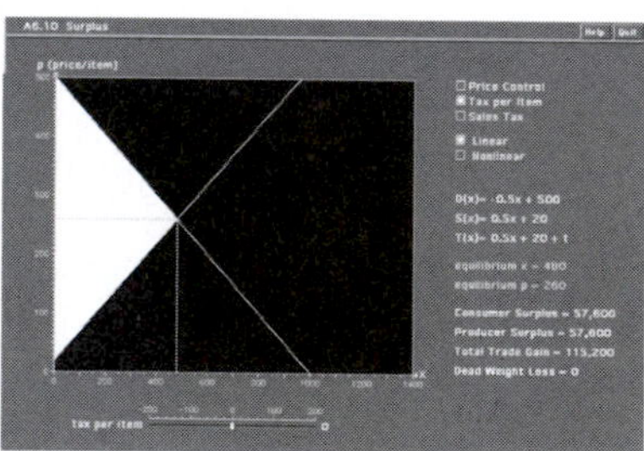

1. For both the linear and nonlinear examples shown, write down both the consumers' and producers' surplus when the product has no tax. Also locate on the graph the area each of these numbers corresponds to.
2. (a) Now ask yourself this question. What would be the difference in the consumers' surplus if a product sold for \$11 or if the product sold for \$10 and had a \$1 tax?
   (b) Now use the slider provided in the interactive illustration to add a tax. What happened to the consumers' surplus? Did it increase, decrease, or stay the same? Does this support your answer in part 2(a)?
3. (a) Now ask yourself this question. What would be the difference in the producers' surplus if a product sold for \$11 or if the product sold for \$10 and had a \$1 tax?
   (b) Now use the slider provided in the interactive illustration to add a tax. What happened to the producers' surplus? Did it increase, decrease, or stay the same? Does this support your answer in part 3(a)?

Warm Up Exercise Set 6.7

1. For the oil well in Example 2, find the total amount $A(t)$ of income over 10 years.
2. If the oil well in Example 2 of the text is offered for sale for \$500,000 and has a useful life of 8 years and $r = 0.05$, would you buy it?
3. If the demand equation is given by $p = D(x) = 45 - 0.18x^2$ and the supply equation is given by $p = S(x) = 0.12x^2 + 1.5x$, find the producers' surplus. (See Example 6.)

Applications and Mathematical Modeling

In Exercises 1 through 10, $f(t)$ is the rate of change of total income per unit time. Find (a) the total amount of income at the given time T, (b) the present value $P_V(T)$ at the given time T for the given interest rate r that is compounding continuously, and (c) the amount at the given time T if all income is reinvested continuously at the given annual rate of r compounded continuously. In Exercises 7 through 10, use your computer or calculator to estimate the integrals.

1. $f(t) = 20$, $T = 5$, $r = 10\%$
2. $f(t) = 30$, $T = 10$, $r = 10\%$
3. $f(t) = 100e^{0.05t}$, $T = 20$, $r = 5\%$
4. $f(t) = 100e^{-0.05t}$, $T = 40$, $r = 5\%$
5. $f(t) = 20e^{-0.05t}$, $T = 10$, $r = 10\%$
6. $f(t) = 20e^{0.05t}$, $T = 10$, $r = 10\%$
7. $f(t) = t$, $T = 10$, $r = 10\%$
8. $f(t) = t + 1$, $T = 10$, $r = 5\%$
9. $f(t) = t^2$, $T = 10$, $r = 10\%$
10. $f(t) = t^2 + 1$, $T = 10$, $r = 8\%$
11. **Investment Decision** An investment that has a continuous return of \$150,000 per year for 10 years is being offered for sale for \$850,000. Would you buy if the current interest rate r that can be compounded continuously over the next 10 years is (a) 0.10? (b) 0.15?
12. **Investment Decision** An investment that has a continuous return of \$240,000 per year for 10 years is being offered for sale for \$1,500,000. Would you buy if the current interest rate r that can be compounded continuously over the next 10 years is (a) 0.08? (b) 0.12?
13. **Oil Well** An oil well is being offered for sale for \$310,000 and has a rate of change of income per year given by $f(t) = 100e^{-0.1t}$, in thousands of dollars, and a useful life of 5 years. Would you buy if the current interest rate r that can be compounded continuously over the next 5 years is (a) 0.10? (b) 0.12?
14. **Oil Well** An oil well is being offered for sale for \$420,000 and has a rate of change of income per year given by $f(t) = 100e^{-0.1t}$, in thousands of dollars, and a useful life of 10 years. Would you buy if the current interest rate r that can be compounded continuously over the next 10 years is (a) 0.10? (b) 0.12?

In Exercises 15 through 18, find the consumers' surplus, using the given demand equations and the equilibrium price p_0.

15. $D(x) = 20 - x^2$, $p_0 = 4$
16. $D(x) = 30 - x^2$, $p_0 = 5$
17. $D(x) = e^{-x}$, $p_0 = 0.1$
18. $D(x) = 20 - x$, $p_0 = 10$

In Exercises 19 through 22, find the producers' surplus, using the given supply equations and the equilibrium price p_0.

19. $S(x) = 100x$, $p_0 = 10$
20. $S(x) = x^3$, $p_0 = 8$
21. $S(x) = 9x^2$, $p_0 = 1$
22. $S(x) = e^x$, $p_0 = e^2$

In Exercises 23 through 26 you are given the demand and supply equation. Find the equilibrium point, and then calculate both the consumers' surplus and the producers' surplus.

23. $D(x) = 12 - x$ and $S(x) = 2x$
24. $D(x) = 50 - x^2$ and $S(x) = 5x$
25. $D(x) = 20 - x^2$ and $S(x) = x$
26. $D(x) = 64 - x^2$ and $S(x) = 3x^2$
27. **Consumer Surplus** Shafer and colleagues[16] created a mathematical model of a demand function for recreational boating in the Three Rivers Area of Pennsylvania that is given by the equation $p(q) = 226e^{-0.0826}$ where q is the number of visitor trips, that is, the number of individuals who participated in any one recreational power boating trip, and p is the cost (or price) per person per trip. The average number of visitor trips was 21. Find, as the researchers did, the consumer surplus.

[16] Elwood L. Shafer, Arun Upneja, Wonseok Seo, and Jihwan Yoon. 2000. Economic values of recreational power boating resources in Pennsylvania. *Environ. Manage.* 26(3):339–348.

28. Consumer Surplus Shafer and colleagues[17] created a mathematical model of a demand function for recreational boating in the Lake Erie/Presque Isle Bay Area of Pennsylvania that is given by the equation $p(q) = 417e^{-0.1062}$, where q is the number of visitor trips, that is, the number of individuals who participated in any one recreational power boating trip, and p is the cost (or price) per person per trip. The average number of visitor trips was 25. Find, as the researchers did, the consumer surplus.

[17] Ibid.

Solutions to WARM UP EXERCISE SET 6.7

1. $A(T) = \int_0^T f(t)\, dt$

$$A(10) = \int_0^{10} 100e^{-0.1t}\, dt$$

$$= 100(-10)e^{-0.1t}\Big|_0^{10}$$

$$= 1000(1 - e^{-1}) \approx 632$$

or \$630,000.

2. In this case

$$P_V(T) = \int_0^T f(t)e^{-rt}\, dt$$

$$P_V(8) = \int_0^8 100e^{-0.05t}e^{-0.1t}\, dt$$

$$= 100\int_0^8 e^{-0.15t}\, dt$$

$$= -\frac{100}{0.15}e^{-0.15t}\Big|_0^8$$

$$= \frac{100}{0.15}\left[1 - e^{-1.2}\right] \approx 466$$

or \$466,000. This is not a good deal, since you can generate less income from this oil well than you can from \$500,000 from an investment at current interest rates.

3. The equilibrium demand was found to be $x_0 = 10$ with an equilibrium price of $p_0 = 27$ in Example 6. The producers' surplus is then given by

$$\int_0^{x_0} [p_0 - S(x)]\, dx = \int_0^{10} \left[27 - 0.12x^2 - 1.5x\right] dx$$

$$= \left[27x - 0.04x^3 - 0.75x^2\right]\Big|_0^{10}$$

$$= 155$$

Summary Outline

- If $F'(x) = f(x)$, then $F(x)$ is called the **antiderivative** of $f(x)$. p. 399.
- If two functions have the same derivative on (a, b), then they differ by a constant on (a, b). p. 399.
- If $F'(x) = f(x)$, then the **indefinite integral** is $\int f(x)\, dx = F(x) + C$, where C is an arbitrary constant called the **constant of integration**. p. 400.
- **Rules of Integration**

$$\int x^n\, dx = \frac{1}{n+1}x^{n+1} + C,\ n \neq -1$$

$$\int [f(x) \pm g(x)]\, dx = \int f(x)\, dx \pm \int g(x)\, dx$$

$$\int \frac{1}{x}\, dx = \ln|x| + C$$

$$\int kf(x)\, dx = k\int f(x)\, dx$$

$$\int e^{kx}\, dx = \frac{1}{k}e^{kx} + C$$

- **Left- and Right-Hand Sums** Divide the x-interval $[a, b]$ into n subintervals of equal length $\Delta x = \frac{b-a}{n}$. Let the endpoints of these subintervals be $a = x_0, x_1, x_2, \ldots, x_n = b$. Then

$$\text{right-hand sum} = f(x_1)\Delta x + f(x_2)\Delta x + \cdots + f(x_n)\Delta x = \sum_{k=1}^{n} f(x_k)\Delta x$$

$$\text{left-hand sum} = f(x_0)\Delta x + f(x_1)\Delta x + \cdots + f(x_{n-1})\Delta x = \sum_{k=0}^{n-1} f(x_k)\Delta x \qquad \text{p. 437}$$

- **The Definite Integral** Suppose $f(x)$ is a continuous function on the finite interval $[a, b]$. Then the right-hand sum $\sum_{k=1}^{n} f(x_k)\Delta x$ and the left-hand sum $\sum_{k=0}^{n-1} f(x_k)\Delta x$ satisfy

$$\lim_{n\to\infty} \sum_{k=1}^{n} f(x_k)\Delta x = \lim_{n\to\infty} \sum_{k=0}^{n-1} f(x_k)\Delta x$$

We refer to this common limit as the **definite integral** of f from a to b and write it as $\int_a^b f(x)\,dx$. We refer to a as the **lower limit** of integration and to b as the **upper limit** of integration and the interval $[a, b]$ as the **interval of integration**. p. 437.

- **Riemann Sum** Divide the interval $[a, b]$ into n subintervals of equal length Δx and let x_i be any point in the ith subinterval. Then $S_n = \sum_{i=1}^{n} f(x_i)\Delta x$ is called a general **Riemann sum**. p. 442.

- **Common Limit of Riemann Sums** Suppose that $f(x)$ is continuous on the interval $[a, b]$. If this interval is divided into n subintervals of equal length Δx and x_i is any point in the ith subinterval, then $\lim_{n\to+\infty} \sum_{i=1}^{n} f(x_i)\Delta x$ exists and is the same limit independent of how the points x_i are chosen in the ith interval. This limit is the definite integral $\int_a^b f(x)\,dx$. p. 443.

- **Area under The Curve** If $f(x)$ is nonnegative and continuous on $[a, b]$, we say the **area under the curve $y = f(x)$ from a to b** is given by the indefinite integral $\int_a^b f(x)\,dx$. p. 437.

- **Properties of Definite Integrals**

$$\int_a^a f(x)\,dx = 0$$

$$\int_a^b kf(x)\,dx = k\int_a^b f(x)\,dx, \qquad k \text{ constant}$$

$$\int_a^b [f(x) \pm g(x)]\,dx = \int_a^b f(x)\,dx \pm \int_a^b g(x)\,dx$$

$$\int_a^b f(x)\,dx = \int_a^c f(x)\,dx + \int_c^b f(x)\,dx, \quad a < c < b$$

$$\int_b^a f(x)\,dx = -\int_a^b f(x)\,dx, \quad a < b. \qquad \text{p. 448.}$$

- **Fundamental Theorem of Calculus** Suppose $f(x)$ is continuous on $[a, b]$. If $F(x)$ is an antiderivative of $f(x)$, then $\int_a^b f(x)\,dx = F(b) - F(a)$. p. 450.

- **Average Value** If $f(x)$ is continuous on $[a, b]$ we define the **average value of $f(x)$ on $[a, b]$** to be $\frac{1}{b-a}\int_a^b f(x)\,dx$. p. 454.

- **Area between the Graphs of Two Curves** Let $y = f(x)$ and $y = g(x)$ be two continuous functions with $f(x) \geq g(x)$ on $[a, b]$. Then the **area between the graphs of the two curves** on $[a, b]$ is given by the definite integral $\int_a^b [f(x) - g(x)]\,dx$. p. 462.

- **Consumers' Surplus** If $p = D(x)$ is the demand equation, p_0 is the equilibrium price of the commodity, and x_0 the current demand, then the **consumers' surplus** is given by $\int_0^{x_0} [D(x) - p_0]\,dx$. p. 479.

- **Producers' Surplus** If $p = S(x)$ is the supply equation, p_0 is the equilibrium price of the commodity, and x_0 the current demand, then the **producers' surplus** is given by $\int_0^{x_0} [p_0 - S(x)]\,dx$. p. 479.

- A **Lorentz curve** is the graph of any function $y = f(x)$ defined on $[0, 1]$ for which $f(x) \leq x$, $f(0) = 0$, and $f(1) = 1$. p. 466.

- The **coefficient of inequality** L of a Lorentz curve $y = f(x)$ is given by $L = 2\int_0^1 [x - f(x)]\,dx$. p. 466.

- **Present Value** Let $f(t)$ be the continuous rate of change of income on the interval $[0, T]$, and suppose that one can invest current money at an interest rate of r, where r is a decimal, compounded continuously over the time interval $[0, T]$. Then the **present value** $P_V(T)$ of this continuous income flow over $[0, T]$ is given by

$$P_V(T) = \int_0^T f(t)e^{-rt}\,dt. \qquad \text{p. 475.}$$

- Let $A^*(T)$ be the amount of money after T years if all the income has been continuously invested at an annual rate of r compounded continuously; then $$A^*(T) = e^{rT} P_V(T) = e^{rT} \int_0^T f(t)e^{-rt}\, dt.$$ p. 477.

Review Exercises

In Exercises 1 through 9, find the antiderivatives.

1. $\int x^9\, dx$
2. $\int x^{0.50}\, dx$
3. $\int \frac{2}{y^2}\, dy$
4. $\int \sqrt{3}y^2\, dy$
5. $\int (6x^2 + 8x)\, dx$
6. $\int \left(2x - \frac{3}{x}\right) dx$
7. $\int \frac{t^2+1}{\sqrt{t}}\, dt$
8. $\int e^{3x}\, dx$
9. $\int \frac{e^{2x} + e^{-2x}}{e^{3x}}\, dx$

In Exercises 10 through 18, use the method of substitution to evaluate the integrals.

10. $\int 4(2x+1)^{20}\, dx$
11. $\int 60x^2(2x^3+1)^9\, dx$
12. $\int \frac{x^2}{\sqrt{x^3+1}}\, dx$
13. $\int \frac{x+1}{x^2+2x}\, dx$
14. $\int xe^{x^2+5}\, dx$
15. $\int \frac{10x}{(x^2+1)^8}\, dx$
16. $\int \frac{1}{\sqrt{x}e^{\sqrt{x}}}\, dx$
17. $\int \frac{e^x}{e^x+1}\, dx$
18. $\int \frac{(\ln x)^3}{x}\, dx$

19. Find the left- and right-hand sums for $f(x) = 2x^2 + 1$ on the interval $[0, 1]$ for $n = 10, 100, 1000$. Determine $\lim_{n\to\infty} \sum_{k=1}^{n} f(x_k)\Delta x$ and relate this to some definite integral.

20. Find the left- and right-hand sums for $f(x) = x^3 + 1$ on the interval $[0, 1]$ for $n = 10, 100, 1000$. Determine $\lim_{n\to\infty} \sum_{k=1}^{n} f(x_k)\Delta x$ and relate this to some definite integral.

21. Find $\int_0^3 \sqrt{9 - x^2}\, dx$ by finding the area of an appropriate geometric figure.

22. Since $\int_1^3 \frac{1}{x}\, dx = \ln 3$ use the left-hand Riemann sum with $n = 4$ to approximate ln 3. Using a graph of $y = \frac{1}{x}$, show why your answer must be *greater* than ln 3.

23. Use the definite integral in Exercise 22 to approximate ln 3 using the right-hand Riemann sum with $n = 4$. Using a graph of $y = \frac{1}{x}$, show why your answer must be *less* than ln 3.

In Exercises 24 through 31, evaluate the definite integrals.

24. $\int_1^3 6x^2\, dx$
25. $\int_{-3}^{-1} 2x\, dx$
26. $\int_{-1}^{1} (9x^2 - 2x + 1)\, dx$
27. $\int_1^2 \left(e^{2x} + \frac{1}{x}\right) dx$
28. $\int_0^1 4x(x^2+1)^9\, dx$
29. $\int_0^2 x^2 e^{x^3+1}\, dx$
30. $\int_0^2 \frac{1}{2x+1}\, dx$
31. $\int_0^3 \frac{1}{\sqrt{2x+3}}\, dx$

32. Write an expression for the average value of $y = e^{t^4}$ on the interval $[0, 2]$. Do not attempt to evaluate.

33. Find the average value of $f(x) = x^3$ on $[-2, 2]$.

34. Find the average value of $f(x) = \frac{1}{x}$ on $[1, 3]$.

In Exercises 35 through 46, find the area enclosed by the given curves.

35. $y = x^2,\ y = 0,\ x = 0,\ x = 3$
36. $y = x^2 + 1,\ y = x,\ x = 0,\ x = 2$
37. $y = 2\sqrt{x},\ y = x,\ x = 0,\ x = 4$
38. $y = \sqrt{x},\ y = -x^2,\ x = 0,\ x = 1$
39. $y = x^2 + 1,\ y = 1 - 2x,\ x = 0,\ x = 2$
40. $y = x^2,\ y = 18 - x^2$
41. $y = x^2 - 4x + 3,\ y = 2x - 2$
42. $y = x^2 - 5x + 4,\ y = -x^2 + 5x - 4.$
43. $y = \sqrt[5]{x},\ y = x$
44. $y = x^2 - 4x + 3,\ y = 2x - 2,\ x = 0,\ x = 5$
45. $y = \frac{2}{x},\ y = 1,\ x = 1,\ x = 3$
46. $y = x^2 - 5x + 4,\ y = -x^2 + 5x - 4,\ x = 0,\ x = 4$

47. **Revenue** If marginal revenue is given by $2x - 10$, find the revenue function.

48. Cost If marginal cost is given by $3x^2 + 1$ and fixed costs are 1000, find the cost function.

49. Depreciation A piece of machinery purchased for \$100,000 is decreasing at a rate given by $V'(t) = -500e^{-0.20t}$, where t is measured in years from date of purchase. How much will the machine be worth in 5 years?

50. Sales After an advertising campaign ends, the rate of change of sales (in thousands) of a product is given by $S'(t) = 4te^{-t^2}$, where t is in years. How many will be sold in the next 2 years?

51. Sales In the following table the rate $R'(t)$ of sales for a firm in millions of dollars per month is increasing and is given in monthly intervals. Assuming that $R(0) = 0$, give upper and lower estimates for the amount of sales during the 5-month period.

t_i	0	1	2	3	4	5
$R'(t_i)$	2	3	5	6	10	8

52. Population The population of a certain city is increasing at the rate given by $P'(t) = 2000e^{0.04t}$, where t is time in years measured from the beginning of 1992. Find the change in the population from the beginning of 1994 to the beginning of 1997.

53. Medicine A tumor is increasing in volume at a rate given by $V'(t) = 4e^{0.1t}$, where t is measured in months from the time of diagnosis. How much will the tumor increase in volume during the third and fourth months after diagnosis?

54. Average Population The population of a certain city is given by $10,000e^{0.1t}$, where t is given in years. Find the average population during the next 10 years.

In Exercises 55 through 57, suppose the demand equation is given by $p = D(x) = 10 - x^2$ and the supply equation is given by $p = S(x) = 3x$.

55. Supply and Demand Find the equilibrium point.

56. Consumers' Surplus Find the consumers' surplus.

57. Producers' Surplus Find the producers' surplus.

58. Lorentz Curve Show that $f(x) = x^4$ is a Lorentz curve and find the coefficient of inequality.

59. Lorentz Curve The figure, found in the work of Bishop and colleagues,[18] shows two Lorentz curves. What does this say about the income distribution in United Kingdom compared to Norway?

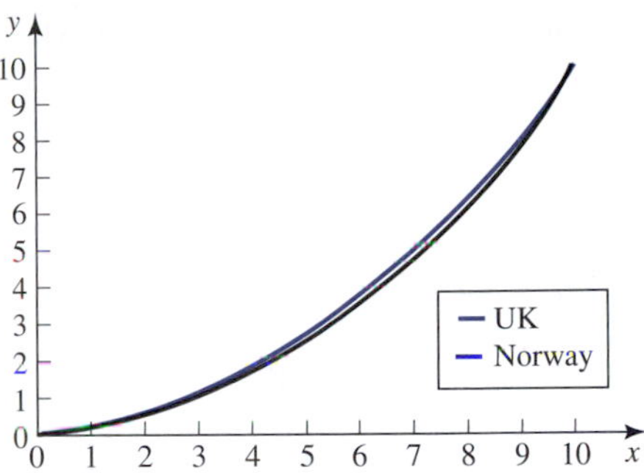

60. Lorentz Curve The figure, found in the work of Doiron and Barrett,[19] shows two Lorentz curves. What does this say about the income distribution for females compared to males?

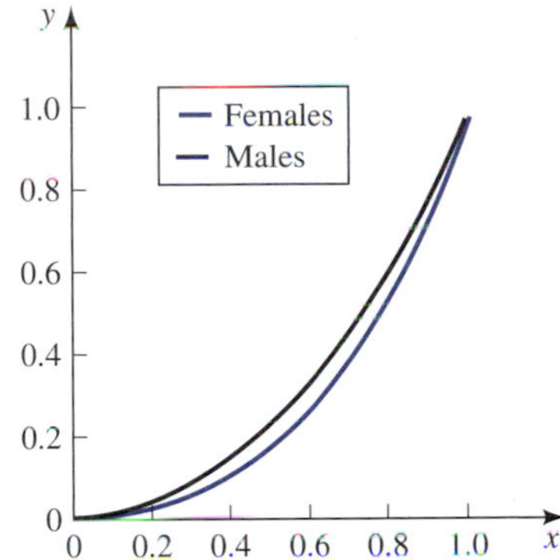

In Exercises 61 through 63, assume that $f(t) = 10e^{-0.04t}$ is the rate of change of total income per year and the interest rate is 8% per year compounded continuously.

61. Find the total amount of income over the first 10 years.

62. Find the present value $P_V(10)$.

63. Find the amount in the account after 10 years if all income is continuously reinvested at the given interest rate.

64. A mine produces a rate of change of total income in thousands of dollars per year given by $f(t) = 100e^{-0.10t}$ and has a useful life of 5 years. Market interest rates are at 8% compounded continuously. What would be a reasonable price for this mine discounting any risk involved?

65. Consumer Surplus Shafer and colleagues[20] created a mathematical model of a demand function for recreational boating in the Delaware River Area of Pennsylvania that is

[18] John A. Bishop, S. Chakraborti, and Paul D. Thistle. 1994. *Bull. Econ. Res.* 46(1):41–55.

[19] Denise J. Doiron and Garry F. Barrett. 1996. Inequality in male and female earnings: the role of hours and wages. *Rev. Econ. Stat.* 78(3):410–420.

[20] Elwood L. Shafer, Arun Upneja, Wonseok Seo, and Jihwan Yoon. 2000. Economic values of recreational power boating resources in Pennsylvania. *Environ. Manage.* 26(3):339–348.

given by the equation $p(q) = 279.6e^{-0.1433}$, where q is the number of visitor trips, that is, the number of individuals who participated in any one recreational power boating trip, and p is the cost (or price) per person per trip. The average number of visitor trips was 14.36. Find, as the researchers did, the consumer surplus.

Chapter 6 Cases

CASE 1 A Lorentz Curve

Recently, Chotikapanich[21] suggested the following function as a Lorentz curve.

$$y = f(x) = \frac{e^{kx} - 1}{e^k - 1}$$

Show that this curve is indeed a Lorentz curve by showing the following:

(a) $f(0) = 0$.
(b) $f(1) = 1$.
(c) $0 < f(x) < x < 1$.
(d) $f'(x) \geq 0$.
(e) $f''(x) > 0$

[21] Duangkamon Chotikapanich. 1993. A comparison of alternative functional forms for the Lorentz curve. *Econ. Lett.* 41:129–138.

CASE 2 Income Distribution in the United States

The following table shows the distribution of family income in the United States for recent selected years. The data was taken from Fischer[22] for the first 6 years listed and from the *1998 Wall Street Journal Almanac* for 1995.

[22] David Hackett Fischer. 1996. *The Great Wave (Price Revolutions and the Rhythm of History).* New York: Oxford University Press.

Year	Percent by Quintile				
1947	5.0	11.8	17.0	23.1	43.2
1959	4.9	12.3	17.9	23.8	41.1
1968	5.6	12.4	17.7	23.7	40.5
1979	5.2	11.6	17.5	24.1	41.7
1988	4.6	10.7	16.7	24.0	44.0
1992	4.4	10.5	16.5	24.0	44.6
1995	3.7	9.1	15.2	23.3	48.7
Taxable Wealth by Quintile					

Estimate the coefficient of inequality for each of the 7 years, graph, and explain what is happening.

CASE 3 Theory of Social Security

The figure shows graphs of the earnings $y = W(t)$ and consumption $y = C(t)$ functions for an individual, where R is the time of retirement and D the time of death. We assume a constant rate of interest r throughout this life cycle. (See Kotlikoff.[23])

[23] Laurence J. Kotlikoff. 1979. Testing the theory of social security and life cycle accumulation. *The Amer. Econ. Rev.* 69(3):396–400.

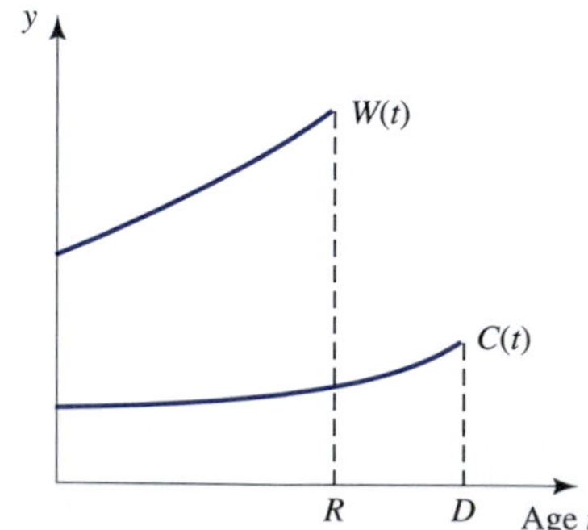

(a) Explain the significance of the graph of the consumption curve always being under the graph of the earnings curve. (Hint: What is $W(t) - C(t)$?)

(b) Show that the following budget constraint must hold:

$$\int_0^D C(t)e^{-rt}\,dt \le \int_0^R W(t)e^{-rt}\,dt$$

(Hint: Start by noticing that $\int_0^D = \int_0^R + \int_R^D$.)

CASE 4 **Density**

If $\delta(x)$ is a continuous function that gives the density in numbers of particles of a pollutant per unit length, then the total number in the interval $[a, b]$ is $P = \int_a^b \delta(x)\,dx$.

(a) Use Riemann sums to justify this formula.

(b) Use the formula to solve the following.

An industrial plant is located at the center of a town that extends 2 miles on either side of the plant. Let x be the distance from the plant. A straight highway goes through the town, and the concentration (density) of pollutants along this highway is given by

$$\delta(x) = 1000(24 - 3x^2)$$

in numbers of particles per mile per day at the point x. Find the total number of particles of pollutant per day in the town along this highway.

7 Additional Topics in Integration

This chapter contains a collection of important additional topics in integration. In the first section we will learn a very important technique of integration; in the second section we will see how to use a table of integrals; in the third section we will see how to evaluate definite integrals using numerical techniques. In the fourth section we extend the notion of integration to unbounded intervals.

CASE STUDY Continuous Money Flow on an Infinite Interval

In Chapter 6 we noted a variety of money flows. We may have a continuously compounding account in a bank. An electric utility's electricity is continuously used in varying amounts, resulting in a money flow for the company. An oil or gas well continually draws oil or gas from the ground, again resulting in a flow of money for the owner (assuming that the oil or gas has been contracted for sale).

To be specific, consider a producing oil well. Oil wells are regularly bought and sold. But how can a fair price be determined? This requires calculating a present value for the well. The present value is intuitively the amount of money needed at present interest rates to be able to generate the same income flow over the time interval under consideration as is generated by the given flow. Say the well is being offered at a price of p dollars. As difficult as it might be, you must estimate the amount of oil the well can produce over its lifetime and the price of oil. You must then compare the income stream from the well with an income stream you can obtain by investing the p dollars at current rates of return.

Usually, the life of the well is unknown, so we need to take the time period to be $(0, \infty)$. Suppose the rate of change of income in thousands of dollars per year from the oil well is projected to be $f(t) = 100e^{-0.1t}$, where t is measured in years. What is a fair price for this oil well as a function of the annual interest rate r, given as a decimal? What is the fair price when $r = 0.05$? When $r = 0.10$? For the answers to these questions, see Example 4 in Section 7.4.

7.1 Integration by Parts

APPLICATION **Application Sales of New Product**

The rate of sales of a new product will tend to increase rapidly initially and then decrease. Suppose the rate of sales of a new product is given by $S'(x) = 1000x^2e^{-x}$ items per week, where x is the number of weeks from the introduction of the product. How many items are sold in the first three weeks? The answer can be found in Example 4 on page 492.

Integration by Parts

If $f(x)$ and $g(x)$ are differentiable functions, then the product rule of differentiation says that

$$\frac{d}{dx}f(x)g(x) = f(x)\frac{d}{dx}g(x) + g(x)\frac{d}{dx}f(x)$$

Rearranging the terms yields

$$f(x)g'(x) = (f(x)g(x))' - g(x)f'(x)$$

Taking the indefinite integral gives

$$\int f(x)g'(x)\,dx = f(x)g(x) - \int g(x)f'(x)\,dx$$

In practice this is written in an alternative way. If we let $u = f(x)$ and $v = g(x)$, then $du = f'(x)\,dx$ and $dv = g'(x)\,dx$. The above formula then becomes $\int u\,dv = uv - \int v\,du$. This technique is called **integration by parts**.

Integration by Parts

$$\int u\,dv = uv - \int v\,du$$

At first glance it is difficult to believe that anything has been accomplished. However, the goal is to define u and v so that $\int v\,du$ is easier to compute than the original integral, $\int u\,dv$. The following examples will illustrate this. Keep in mind that your first choice of u and v might not work.

EXAMPLE 1 **Using Integration by Parts**

Find $\int xe^x\,dx$.

Solution

If we set $u = x$ and $dv = e^x\,dx$, then $du = dx$ and $v = \int dv = \int e^x\,dx = e^x$, and

$$\begin{aligned}
\int xe^x\,dx &= \int u\,dv \\
&= uv - \int v\,du \\
&= xe^x - \int e^x\,dx \\
&= xe^x - e^x + C = e^x(x-1) + C
\end{aligned}$$

■

Notice that $\int v\,du = \int e^x\,dx$ was simpler than the original integral. Had we made the substitution $u = e^x$ and $dv = x\,dx$, then we would have $\int v\,du = \int \frac{1}{2}x^2e^x\,dx$, which is *more* complicated than the original integral.

EXAMPLE 2 **Using Integration by Parts**

Find $\int x\sqrt{1+x}\,dx$.

Solution

We might first think of letting $u = x$, since du will be simple. Let us try this. Let

$$u = x \qquad \text{and} \qquad dv = \sqrt{1+x}\,dx$$

Of course, $du = dx$, and $v = \int \sqrt{1+x}\,dx$. This last integral is similar to others we have done before and can be readily done by substitution. If we let $w = 1+x$, then

$dw = dx$, and

$$v = \int \sqrt{1+x}\,dx = \int w^{1/2}\,dw = \frac{2}{3}w^{3/2} = \frac{2}{3}(1+x)^{3/2}$$

Then

$$\begin{aligned}\int x\sqrt{1+x}\,dx &= \textstyle\int u\,dv \\ &= uv - \int v\,du \\ &= x\frac{2}{3}(1+x)^{3/2} - \int \frac{2}{3}(1+x)^{3/2}\,dx\end{aligned}$$

Now we can evaluate the last integral in just the way we found v. Thus,

$$\begin{aligned}\int x\sqrt{1+x}\,dx &= x\frac{2}{3}(1+x)^{3/2} - \int \frac{2}{3}(1+x)^{3/2}\,dx \\ &= x\frac{2}{3}(1+x)^{3/2} - \frac{2}{3}\cdot\frac{2}{5}(1+x)^{5/2} + C \\ &= \frac{2}{3}x(1+x)^{3/2} - \frac{4}{15}(1+x)^{5/2} + C\end{aligned}$$

■

EXAMPLE 3 **Using Integration by Parts**

Find (a) $\int x^3\sqrt{1+x^2}\,dx$ and (b) $\int_0^{\sqrt{3}} x^3\sqrt{1+x^2}\,dx$.

Solution

(a) We might be tempted to set $u = x^3$ and $dv = \sqrt{1+x^2}$, but then we could *not* find v, since we cannot find $\int \sqrt{1+x^2}\,dx$. But we can integrate $\int x\sqrt{1+x^2}\,dx$. Thus, we set

$$u = x^2 \qquad \text{and} \qquad dv = x\sqrt{1+x^2}$$

Then

$$du = 2x\,dx \qquad \text{and} \qquad v = \frac{1}{3}(1+x^2)^{3/2}$$

(The reader should verify the correctness of v). Thus,

$$\begin{aligned}\int x^3\sqrt{1+x^2}\,dx &= \int u\,dv \\ &= uv - \int v\,du \\ &= x^2\frac{1}{3}(1+x^2)^{3/2} - \int \frac{1}{3}(1+x^2)^{3/2}2x\,dx \\ &= \frac{1}{3}x^2(1+x^2)^{3/2} - \frac{2}{15}(1+x^2)^{5/2} + C\end{aligned}$$

(b) To evaluate this definite integral, we use the antiderivative found in part (a).

$$\begin{aligned}\int_0^{\sqrt{3}} x^3\sqrt{1+x^2}\,dx &= \left(\frac{1}{3}x^2(1+x^2)^{3/2} - \frac{2}{15}(1+x^2)^{5/2}\right)\Big|_0^{\sqrt{3}}\\ &= \left[\frac{1}{3}\cdot 3(1+3)^{3/2} - \frac{2}{15}(1+3)^{5/2}\right] - \left[0 - \frac{2}{15}\right]\\ &= \left[4^{3/2} - \frac{2}{15}4^{5/2}\right] - \left[0 - \frac{2}{15}\right]\\ &= \frac{58}{15}\end{aligned}$$

■

Sometimes one must use integration by parts *more than once*. Sometimes $\int v\,du$ is definitely simpler but requires another integration by parts. The following is such an example.

Application

EXAMPLE 4 **Repeated Integration by Parts**

Suppose the rate of sales of a new product is given by $S'(x) = 1000x^2e^{-x}$ items per week, where x is the number of weeks from the introduction of the product. How many items are sold in the first three weeks?

Solution

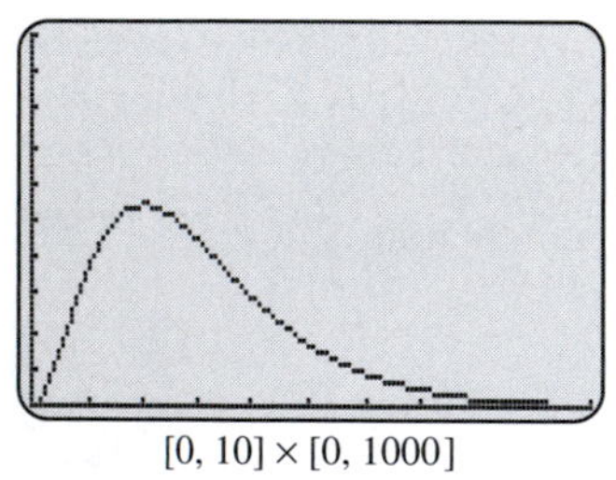

$[0, 10] \times [0, 1000]$

Screen 7.1
A graph of $y = 1000x^2e^{-x}$ giving the rate of sales of new product.

Screen 7.1 shows a graph of $y = 1000x^2e^{-x}$ using a window with dimensions [0, 10] by [0, 1000]. We are seeking $1000\int_0^3 x^2e^{-x}\,dx$. We first find $\int x^2e^{-x}\,dx$. We set

$$u = x^2 \quad \text{and} \quad dv = e^{-x}\,dx$$

Then

$$du = 2x\,dx \quad \text{and} \quad v = -e^{-x}$$

Then

$$\begin{aligned}\int x^2e^{-x}\,dx &= \int u\,dv\\ &= uv - \int v\,du\\ &= x^2\left(-e^{-x}\right) - \int -e^{-x}2x\,dx\\ &= -x^2e^{-x} + 2\int xe^{-x}\,dx\end{aligned}$$

The integral we are left with, $\int xe^{-x}\,dx$, is definitely simpler, but to evaluate it requires another integration by parts. To do this, we let

$$u = x \quad \text{and} \quad dv = e^{-x}\,dx$$

Then

$$du = dx \qquad \text{and} \qquad v = -e^{-x}$$

and

$$\begin{aligned}\int xe^{-x}\,dx &= \int u\,dv \\ &= uv - \int v\,du \\ &= x\left(-e^{-x}\right) - \int -e^{-x}\,dx \\ &= -xe^{-x} + \int e^{-x}\,dx \\ &= -xe^{-x} - e^{-x} + C^*\end{aligned}$$

Substituting this back in the above equation yields

$$\begin{aligned}\int x^2e^{-x}\,dx &= -x^2e^{-x} + 2\int xe^{-x}\,dx \\ &= -x^2e^{-x} + 2\left[-xe^{-x} - e^{-x} + C^*\right] \\ &= -e^{-x}\left[x^2 + 2x + 2\right] + C\end{aligned}$$

where we have set $C = 2C^*$ for convenience. Then

$$\begin{aligned}1000\int_0^3 x^2e^{-x}\,dx &= -1000e^{-x}\left[x^2 + 2x + 2\right]\Big|_0^3 \\ &= -1000e^{-3}[17] + 1000e^0[2] \\ &= 1000(2 - 17e^{-3}) \\ &\approx 1153.6\end{aligned}$$

■

Warm Up Exercise Set 7.1

1. Find (a) $\int \ln x\,dx$ and (b) $\int_1^e \ln x\,dx$. (Hint: Notice that $\ln x = 1 \cdot \ln x$.)

2. Find $\int \frac{xe^x}{(x+1)^2}\,dx$. (Hint: You might need to try several choices of u and dv before one works.)

Exercise Set 7.1

In Exercises 1 through 20, evaluate.

1. $\int 2xe^{2x}\,dx$

2. $\int xe^{-2x}\,dx$

3. $\int_1^{e^2} x\ln x\,dx$

4. $\int_e^{e^2} x^2\ln x\,dx$

5. $\int_1^e \ln x^2\,dx$

6. $\int_1^e \ln x^3\,dx$

7. $\int \frac{x}{\sqrt{1+x}}\,dx$

8. $\int x(1+x)^{3/2}\,dx$

9. $\int x(1+x)^{10}\,dx$

10. $\int x(1+x)^5\,dx$

11. $\int x(x+2)^{-2}\,dx$

12. $\int x(x+2)^{-3}\,dx$

13. $\int x^5\sqrt{1+x^3}\,dx$

14. $\int (x+1)^{2/3}x\,dx$

15. $\int x^3(1+x^2)^{10}\,dx$

16. $\int \frac{x^3}{\sqrt{1+x^2}}\,dx$

17. $\int x^2e^{2x}\,dx$

18. $\int x^2e^{-2x}\,dx$

19. $\int_0^2 x^3e^{x^2}\,dx$

20. $\int_1^2 (\ln x)^2\,dx$

Applications and Mathematical Modeling

21. Sales The rate of sales of a new product will tend to increase rapidly initially and then fall off. Suppose the rate of sales of a new product is given by $S'(t) = 1000te^{-3t}$ items per week, where t is the number of weeks from the introduction of the product. How many items are sold in the first four weeks? Assume that $S(0) = 0$.

22. Sales Suppose the rate of sales of a new product is given by $S'(t) = 1000t^2e^{-4t}$ items per month, where t is the number of months from the introduction of the product. How many items are sold in the first year? Assume that $S(0) = 0$.

23. Profits The marginal profits of an electric can opener manufacturer are given by $P'(x) = 0.1x\sqrt{x+2}$, where x is measured in thousands of can openers and $P(x)$ is measured in thousands of dollars. Find the total profits generated by increasing the number of can openers from 7000 to 14,000.

24. Consumers' Surplus Find the consumer's surplus if $p = D(x) = 25 - x^3\sqrt{5+x^2}$ is the demand equation with x measured in thousands of units and $p_0 = 1$ is the equilibrium price of the commodity. (Hint: $D(2) = 1$.)

25. Costs The cost function, in thousands of dollars, of a coat manufacturer is given by $x(1+x)^4$, where x is measured in thousands of coats. Find the average cost for the first 2000 coats.

26. Price The demand equation for a bath towel manufacturer is given by $p = p(x) = x(x+1)^{-2}$, where p is measured in dollars and x is measured in thousands of towels. Find the average price over the interval $[1, 8]$.

27. Population The population of a new town is given by $P(t) = 1000\ln t$, where $t \geq 1$ is measured in years. What is the average population over the time from $t = 1$ to $t = e$?

28. Income Distribution Find the coefficient of inequality for the Lorentz curve $y = xe^{x-1}$.

More Challenging Exercises

29. Establish the following reduction formula:

$$\int (\ln x)^n\,dx = x(\ln x)^n - n\int (\ln x)^{n-1}\,dx$$

30. If $f(0) = g(0) = 0$, show that

$$\int_0^a f(x)g''(x)\,dx = f(a)g'(a) - f'(a)g(a) + \int_0^a f''(x)g(x)\,dx$$

31. Use integration by parts to show that

$$\int f(x)\,dx = xf(x) - \int xf'(x)\,dx$$

Solutions to WARM UP EXERCISE SET 7.1

1. a. Our choices for u are very limited. Let

$$u = \ln x \quad \text{and} \quad dv = dx$$

Then

$$du = \frac{1}{x}\,dx \quad \text{and} \quad v = x$$

and

$$\begin{aligned}\int \ln x\,dx &= \int u\,dv\\ &= uv - \int v\,du\\ &= (\ln x)\,x - \int x\frac{1}{x}\,dx\\ &= x\ln x - \int dx\\ &= x\ln x - x + C\end{aligned}$$

b.

$$\int_1^e \ln x\,dx = (x\ln x - x)\big|_1^e$$
$$= (e\ln e - e) - (1\cdot\ln 1 - 1)$$
$$= (0) - (-1) = 1$$

2. Let

$$u = xe^x \qquad \text{and} \qquad dv = \frac{1}{(1+x)^2}\,dx$$

Then

$$du = e^x(x+1)\,dx \qquad \text{and} \qquad v = \frac{-1}{x+1}$$

$$\int \frac{xe^x}{(1+x)^2}\,dx = \int u\,dv$$
$$= uv - \int v\,du$$
$$= xe^x\left(\frac{-1}{x+1}\right) - \int\left(-\frac{1}{x+1}\right)e^x(x+1)\,dx$$
$$= -\frac{xe^x}{x+1} + \int e^x\,dx$$
$$= -\frac{xe^x}{x+1} + e^x + C$$
$$= \frac{e^x}{x+1} + C$$

7.2 Integration Using Tables

Using Tables of Integrals

Application

APPLICATION Application Sales

The rate of sales, in millions of dollars per week, of a toy is given by

$$S'(x) = \frac{e^x}{e^{2x}-1}$$

where x is the number of weeks after the end of an advertising campaign. Find the total dollar sales during the second and third weeks. See Example 4 on page 497 for the answer.

Using Tables of Integrals

In previous sections we saw how to integrate many expressions, including $\int x^n\,dx$, $\int e^{kx}\,dx$, and so on. We were also introduced to some techniques of integration, such as substitution and integration by parts. A substantial number of different techniques of integration exist; many require considerable skill and ingenuity. Unfortunately, no hard and fast rules determine when to use each technique. In the face of these difficulties the most convenient technique is to refer to a table of integrals. We can always check an answer in any table by merely differentiating the answer to see whether we obtain the integrand.

EXAMPLE 1 **Using a Table of Integrals**

Find $\int \frac{1}{x(2x+3)}\,dx$.

Solution

By looking through the table of integrals found in Appendix B.2, we notice that the integrand has the form of the formula in the sixth entry if we set $a = 2$ and $b = 3$. Thus, we have

$$\int \frac{1}{x(2x+3)}\,dx = \frac{1}{3}\ln\left|\frac{x}{2x+3}\right| + C \qquad \blacksquare$$

Remark Notice we have added the arbitrary constant C. To save space, the table does not include this constant.

EXAMPLE 2 **Using a Table of Integrals**

Find $\int \frac{1}{\sqrt{9x^2+1}}\,dx$.

Solution

If we look through the table in Appendix B.2, we do not actually find this integral, but it appears to look somewhat like entry 22. Suppose we factor out the 9 from the integrand. Then we obtain

$$\begin{aligned}\frac{1}{\sqrt{9x^2+1}} &= \frac{1}{\sqrt{9\left(x^2+\frac{1}{9}\right)}}\\ &= \frac{1}{3}\cdot\frac{1}{\sqrt{x^2+\frac{1}{9}}}\end{aligned}$$

So

$$\int \frac{1}{\sqrt{9x^2+1}}\,dx = \frac{1}{3}\int \frac{1}{\sqrt{x^2+\frac{1}{9}}}\,dx$$

This integral is of the form of entry 22 with $a = 1/3$. Thus,

$$\begin{aligned}\int \frac{1}{\sqrt{9x^2+1}}\,dx &= \frac{1}{3}\int \frac{1}{\sqrt{x^2+\frac{1}{9}}}\,dx\\ &= \frac{1}{3}\ln\left|x+\sqrt{x^2+\frac{1}{9}}\right| + C \qquad \blacksquare\end{aligned}$$

EXAMPLE 3 **Repeated Use of a Table of Integrals**

Find $\int x^2 e^{-x}\,dx$.

Solution

Although this integral can be done by using integration by parts, one can also do it by using entry 38 in the table in Appendix B.2 and taking $n = 2$ and $a = -1$. Doing this, we obtain

$$\begin{aligned}\int x^2e^{-x}\,dx &= \frac{1}{(-1)}x^2e^{-x} - \frac{2}{(-1)}\int xe^{-x}\,dx \\ &= -x^2e^{-x} + 2\int xe^{-x}\,dx\end{aligned}$$

Now we are left to evaluate $\int xe^{-x}\,dx$, which we can do by using the very same entry but this time taking $n = 1$ and $a = -1$. We then obtain for this latter integral

$$\begin{aligned}\int xe^{-x}\,dx &= \frac{1}{(-1)}xe^{-x} - \frac{1}{(-1)}\int x^0e^{-x}\,dx \\ &= -xe^{-x} + \int e^{-x}\,dx \\ &= -xe^{-x} - e^{-x} + C^*\end{aligned}$$

Now using this above yields

$$\begin{aligned}\int x^2e^{-x}\,dx &= -x^2e^{-x} + 2\int xe^{-x}\,dx \\ &= -x^2e^{-x} + 2\left[-xe^{-x} - e^{-x} + C^*\right] \\ &= -e^{-x}\left(x^2 + 2x + 2\right) + C\end{aligned}$$

where we have set $C = 2C^*$ for convenience. ■

Application

EXAMPLE 4 **Using Substitution in a Table of Integrals**

The rate of sales, in millions of dollars per week, of a toy is given by

$$S'(x) = \frac{e^x}{e^{2x} - 1}$$

where x is the number of weeks after the end of an advertising campaign. Find the total dollar sales during the second and third weeks.

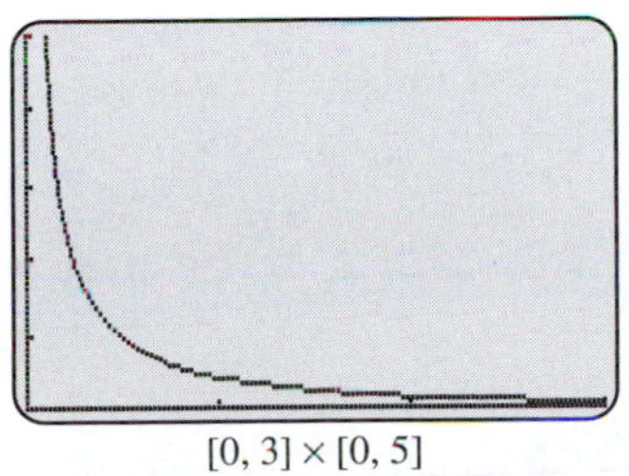

[0, 3] × [0, 5]

Screen 7.2
A graph of $y = \dfrac{e^x}{e^{2x} - 1}$ giving the rate of sales per week of a toy.

Solution

The graph of $y = \dfrac{e^x}{e^{2x} - 1}$ is shown in Screen 7.2 using a window with dimensions [0, 3] by [0, 5]. We are seeking $\displaystyle\int_1^3 \frac{e^x}{e^{2x} - 1}\,dx$. The integrand involves exponentials, yet no formula in the table under the exponential heading applies. If we make the substitution $u = e^x$, then $du = e^x\,dx$, and the integral becomes

$$\int \frac{e^x}{e^{2x} - 1}\,dx = -\int \frac{e^x}{1 - e^{2x}}\,dx = -\int \frac{1}{1 - u^2}\,du$$

This last integral can be found in entry 13 in the table of integrals with $a = 1$. Thus,

$$\begin{aligned}\int \frac{e^x}{e^{2x}-1}\,dx &= -\int \frac{1}{1-u^2}\,du \\ &= -\frac{1}{2}\ln\left|\frac{u+1}{u-1}\right| + C \\ &= \frac{1}{2}\ln\left|\frac{u-1}{u+1}\right| + C \\ &= \frac{1}{2}\ln\left|\frac{e^x-1}{e^x+1}\right| + C\end{aligned}$$

Then

$$\int_1^3 \frac{e^x}{e^{2x}-1}\,dx = \frac{1}{2}\ln\frac{e^x-1}{e^x+1}\bigg|_1^3 \approx 0.336$$

or about \$336,000. ■

Warm Up Exercise Set 7.2

1. Find $\int \frac{4x}{\sqrt{x^4+4}}\,dx$.

Exercise Set 7.2

In Exercises 1 through 26, evaluate, using the table of integrals in Appendix B.2.

1. $\int x\sqrt{3x+2}\,dx$

2. $\int \frac{1}{x\sqrt{4-x^2}}\,dx$

3. $\int \sqrt{x^2-9}\,dx$

4. $\int \frac{x}{2x-1}\,dx$

5. $\int \frac{1}{x\sqrt{1-x}}\,dx$

6. $\int \frac{\sqrt{4-x^2}}{x}\,dx$

7. $\int \frac{1}{2x^2+5x+2}\,dx$

8. $\int \frac{1}{x^2\sqrt{x^2+9}}\,dx$

9. $\int x\sqrt{x^2-9}\,dx$

10. $\int \frac{x^2}{(2x+3)^2}\,dx$

11. $\int \frac{1}{x^2(x-3)}\,dx$

12. $\int \frac{x}{(9-x^2)^{3/2}}\,dx$

13. $\int \frac{x^2}{(x^2+16)^{3/2}}\,dx$

14. $\int \frac{\sqrt{x^2+16}}{x}\,dx$

15. $\int \frac{1}{x(2x^3+1)}\,dx$

16. $\int \frac{1}{2+3e^{4x}}\,dx$

17. $\int \frac{1}{x^2\sqrt{1-4x^2}}\,dx$

18. $\int \frac{1}{(1-4x^2)^{3/2}}\,dx$

19. $\int \frac{x^2}{\sqrt{9x^2-1}}\,dx$

20. $\int \frac{1}{(4x^2+1)^{3/2}}\,dx$

21. $\int (\ln x)^3\,dx$

22. $\int x^3e^{2x}\,dx$

23. $\int \frac{e^{3x}}{e^x+1}\,dx$

24. $\int \frac{e^{2x}}{(e^x+1)^2}\,dx$

25. $\int x^2e^{-2x}\,dx$

26. $\int (\ln(x+1))^3\,dx$

Applications and Mathematical Modeling

27. Revenue Find the revenue function of a toaster manufacturer if marginal revenue, in hundreds of dollars, is given by $x\sqrt{x+1}$, where x is in hundreds of toasters.

28. Sales The rate of change of the number of sales of refrigerators by a firm is given by $\frac{t}{(t^2+9)^{3/2}}$, where t is in weeks. How many refrigerators are sold in the first four weeks?

29. Profit If the rate of change of profit in thousands of dollars per week is given by

$$P'(t) = \frac{10}{(t^2+9)^{3/2}}$$

where t is measured in weeks and $P(0) = 0$, find $P = P(t)$.

30. Costs The marginal cost, in thousands of dollars, of a brick manufacturer is given by $\frac{x}{\sqrt{x^2+9}}$, where x is in thousands of bricks. If fixed costs are \$10,000, find $C = C(x)$.

31. Consumers' Surplus Find the consumers' surplus if

$$p = D(x) = \frac{4e}{e + e^{0.10x}}$$

is the demand equation with x measured in thousands of units and $p_0 = 2$ is the equilibrium price of the commodity.

32. Producers' Surplus Find the producers' surplus if $p = S(x) = x\sqrt{x^2+5}$, is the supply equation with x measured in thousands of units and $p_0 = 6$ is the equilibrium price of the commodity. (Hint: $S(2) = 6$.)

33. Pollutants The concentration (density) of pollutants, measured in thousands of particles per mile per day, at a distance of x miles east of an industrial plant is given by

$$\delta(x) = \frac{1}{x^2+5x+4}$$

Find the amount of pollutants between $x = 0$ and $x = 5$ miles.

34. Biology The rate of change of a certain population of plant in a small region is given by

$$P'(t) = \frac{1000}{(9t^2+1)^{3/2}}$$

where t is in months. If the initial population is zero, find the population at any time. What happens as $t \to \infty$?

More Challenging Exercises

35. Often when using tables or other techniques, you can obtain two very different-looking answers. Consider the integral $\int \frac{x}{(x+1)^2}\,dx$. Integrate by parts, and obtain $-\frac{x}{x+1} + \ln|x+1| + C$. Integrate by substitution, and obtain $\frac{1}{x+1} + \ln|x+1| + C$. Does this mean that

$$-\frac{x}{x+1} + \ln|x+1| = \frac{1}{x+1} + \ln|x+1|$$

Let y_1 be the left-hand side, and let y_2 be the right-hand side. On your graphing calculator, graph y_1, y_2, and $y_3 = y_1 - y_2$. Explain what you have discovered.

36. Check the answer to Exercise 13. (Hint: Let $F(x)$ be the answer in the text.) Then show that

$$F'(x) = \frac{-16}{(x^2+16)^{3/2}} + \frac{\sqrt{x^2+16}+x}{x\sqrt{x^2+16}+x^2+16}$$

Denote the second fraction as $g(x)$. Since

$$F'(x) = \frac{x^2}{(x^2+16)^{3/2}}$$

we must have

$$g(x) = \frac{x^2+16}{(x^2+16)^{3/2}}$$

You probably cannot establish this algebraically. Establish it using your graphing calculator with a window of dimensions $[-20, 20]$ by $[0, 0.3]$. Use TRACE to move back and forth between the two functions.

Solutions to WARM UP EXERCISE SET 7.2

1. Use entry 22 in the table of integrals in Appendix B.2 after using the substitution $u = x^2$. Then $du = 2x\,dx$, and

$$\int \frac{4x}{\sqrt{x^4+4}}\,dx = 2\int \frac{2x}{\sqrt{x^4+4}}\,dx$$
$$= 2\int \frac{1}{\sqrt{u^2+4}}\,du$$
$$= 2\ln\left|u + \sqrt{u^2+4}\right| + C$$
$$= 2\ln\left|x^2 + \sqrt{x^4+4}\right| + C$$

7.3 Numerical Integration

APPLICATION Finding Approximations of Definite Integrals

Find approximations for $\int_0^2 \sqrt{4 - x^2}\, dx = \pi$ and $\int_1^2 \frac{1}{x}\, dx = \ln 2$. See Examples 3 and 5 on pages 506 and 507.

Introduction

In Chapter 6 we introduced the Riemann sum and used left- and right-hand sums to estimate definite integrals. The only other method we have for evaluating the definite integral $\int_a^b f(x)\, dx$ requires us to find an antiderivative $F(x)$ of $f(x)$, evaluate F at a and b, and use the formula

$$\int_a^b f(x)\, dx = F(b) - F(a)$$

If we cannot do this, our method fails, and we must return to numerical methods, such as those touched on in Chapter 6. This section introduces other methods of estimating definite integrals and develops ways to calculate the magnitude of errors associated with these estimates.

It turns out that the method for determining definite integrals by finding antiderivatives *does* fail for many important and rather simple-looking integrals, such as

$$\int_0^1 e^{-x^2}\, dx \qquad \text{and} \qquad \int_2^3 \frac{1}{\ln x}\, dx$$

The first of these integrals plays a vital role in continuous probability theory.

Of course, $\int_0^1 e^{-x^2}\, dx$ exists, since e^{-x^2} is continuous on [0, 1]. It has been *proven*, however, that e^{-x^2} has no elementary antiderivative, that is, one that can be expressed as finite combinations of algebraic, exponential, logarithmic, and trigonometric functions. In effect, we cannot write down the antiderivative in a form that would permit us to evaluate the antiderivative at the two points 0 and 1.

This difficulty is not as disastrous as might first appear, for two reasons. The first is the existence of *numerical* methods, which are the subject of this section, that will permit us to find definite integrals such as $\int_0^1 e^{-x^2}\, dx$ to (for all practical purposes) any desired accuracy. Thus, knowing this integral to an accuracy of, for example, five or ten decimal places will be sufficient for any practical purpose.

The second reason is that many of the functions that are encountered in practice come from data collected in various ways. In a typical situation we do not know what the function $f(x)$ is at *every* point x but rather only at the discrete points $x_1, x_2, \ldots, x_n$, where $f(x)$ has been measured (Figure 7.1). The numerical methods given in this section are well suited to give reasonable approximations to the definite integral $\int_a^b f(x)\,dx$ under such conditions.

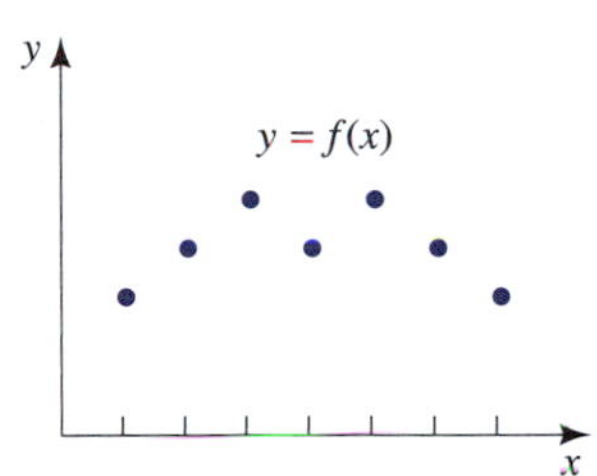

Figure 7.1
Typically, we know $f(x)$ only at discrete values of x.

If $f(x)$ is continuous on the interval $[a, b]$, we *defined* the definite integral $\int_a^b f(x)\,dx$ as the common limit of the left- and right-hand Riemann sums. That is, we divided the interval $[a, b]$ into n subintervals of equal length $\Delta x = \dfrac{b-a}{n}$, with endpoints at $a = x_0, x_1, \cdots, x_n = b$. If $f(x)$ is continuous, the limit of the left-hand sum $\lim_{n\to\infty} \sum_{k=0}^{n-1} f(x_k)\,\Delta x$ equals the limit of the right-hand sum $\lim_{n\to\infty} \sum_{k=1}^{n} f(x_k)\,\Delta x$, which equals the definite integral $\int_a^b f(x)\,dx$.

If n is large, we expect both the left- and right-hand Riemann sums $\sum_{k=0}^{n-1} f(x_k)\,\Delta x$ and $\sum_{k=1}^{n} f(x_k)\,\Delta x$ to approximate the definite integral $\int_a^b f(x)\,dx$. This represents a *numerical* method of approximating the definite integral by using certain thin rectangles. Refer to Figure 7.2, where the left-hand sum is indicated. We will not dwell on left- and right-hand Riemann sums here, since they were covered in Chapter 6.

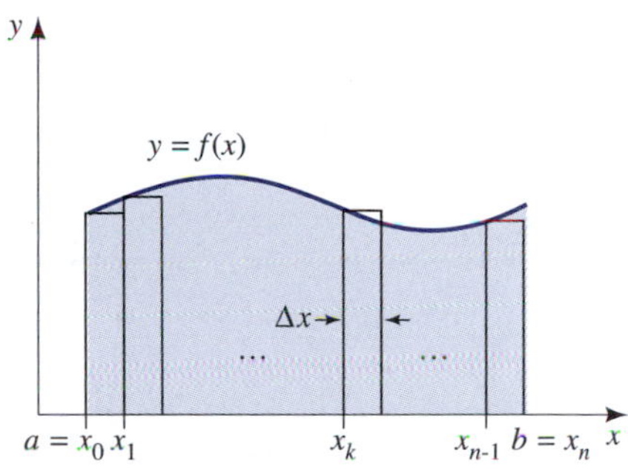

Figure 7.2

The sum of the areas of the rectangles is the left-hand sum $\sum_{k=0}^{n-1} f(x_k)\,\Delta x$.

The Trapezoidal Rule

The left- and right-hand Riemann sums approximate $\int_a^b f(x)\,dx$ by rectangles with heights given by the values of f evaluated at the endpoints of the subintervals. In general, a better approximation can be obtained by using trapezoids as shown in Figure 7.3. Once again the interval $[a, b]$ has been divided into n subintervals, each of length $\Delta x = \frac{b-a}{n}$. Also $x_0 = a$, $x_1 = a + \Delta x$, $x_2 = a + 2\Delta x, \ldots, x_n = a + n\Delta x$.

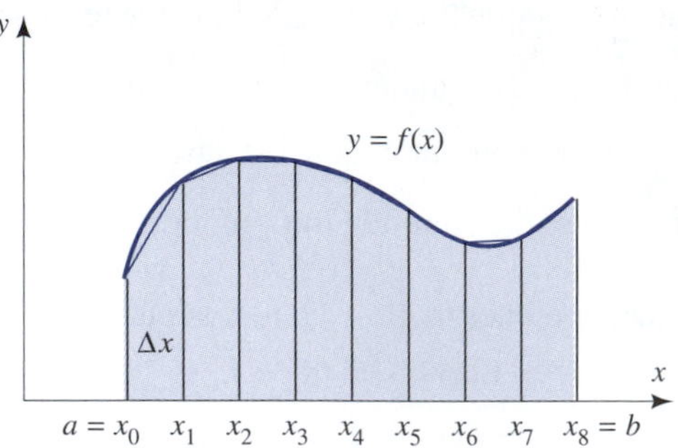

Figure 7.3
Approximate the area with trapezoids.

Recall that the area of a trapezoid is given by multiplying the length of the base times the average of the heights of the sides. Thus, the areas of the n trapezoids is given by

$$\begin{aligned} T_n &= \frac{f(x_0) + f(x_1)}{2}\Delta x + \frac{f(x_1) + f(x_2)}{2}\Delta x + \frac{f(x_2) + f(x_3)}{2}\Delta x \\ &\quad + \cdots + \frac{f(x_{n-2}) + f(x_{n-1})}{2}\Delta x + \frac{f(x_{n-1}) + f(x_n)}{2}\Delta x \\ &= \Delta x\left(\frac{1}{2}f(x_0) + f(x_1) + f(x_2) + \cdots + f(x_{n-1}) + \frac{1}{2}f(x_n)\right) \end{aligned}$$

We then have the following.

Trapezoidal Rule

For any integer n, divide the interval $[a, b]$ into n subintervals of equal length with endpoints $a = x_0, x_1, \ldots, x_n = b$. If

$$T_n = \frac{b-a}{n}\left[\frac{1}{2}f(x_0) + f(x_1) + f(x_2) + \ldots + f(x_{n-1}) + \frac{1}{2}f(x_n)\right]$$

then

$$\int_a^b f(x)\,dx \approx T_n$$

In general, the trapezoidal rule gives much closer approximations than using the left- or right-hand Riemann sums. We will return to this point later.

EXAMPLE 1 Using the Trapezoidal Rule

Approximate $\int_0^2 \sqrt{4-x^2}\,dx = \pi$ by using T_4.

Solution

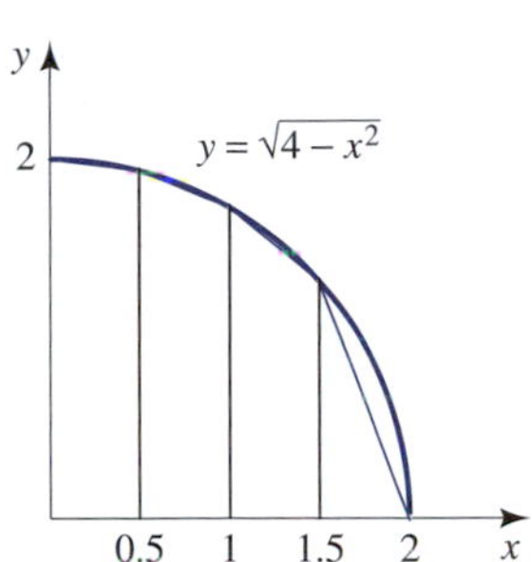

Figure 7.4
Using the trapezoid rule to approximate π.

See Figure 7.4, in which we see that T_4 must be smaller than the area, which is π. We have $\Delta x = \dfrac{2-0}{4} = 0.50$, and the points x_0, x_1, x_2, x_3, and x_4 are 0, 0.50, 1.00, 1.50, and 2, respectively. From the preceding formula, we have that

$$\begin{aligned} T_4 &= \frac{2-0}{4}\left[\frac{1}{2}f(0) + f(0.5) + f(1) + f(1.5) + \frac{1}{2}f(2)\right] \\ &= \frac{1}{2}\left[\frac{1}{2}\cdot 2 + \sqrt{4-.25} + \sqrt{4-1} + \sqrt{4-2.25} + \frac{1}{2}\cdot 0\right] \\ &\approx 2.996 \end{aligned}$$

■

Now notice that

$$\int_1^2 \frac{1}{x}\,dx = \ln x|_1^2 = \ln 2$$

Thus, we can approximate $\ln 2$ by approximating the integral $\int_1^2 \frac{1}{x}\,dx$.

EXAMPLE 2 Using the Trapezoidal Rule

Use T_4 to find an approximation to $\int_1^2 \frac{1}{x}\,dx = \ln 2$.

Solution

In Figure 7.5 we can see that T_4 must be larger than the area, that is, $T_4 > \ln 2$. We have $\Delta x = \frac{2-1}{4} = 0.25$, and the points x_0, x_1, x_2, x_3, and x_4 are 1, 1.25, 1.5, 1.75, and 2, respectively. Thus,

$$\begin{aligned} T_4 &= \frac{2-1}{4}\left[\frac{1}{2}f(1) + f(1.25) + f(1.5) + f(1.75) + \frac{1}{2}f(2)\right] \\ &= \frac{1}{4}\left[\frac{1}{2} + \frac{4}{5} + \frac{2}{3} + \frac{4}{7} + \frac{1}{4}\right] \\ &\approx 0.6970 \end{aligned}$$

Compare this to $\ln 2 = 0.6931$ to four decimal places.

■

You can explore this example further in Interactive Illustration 7.1.

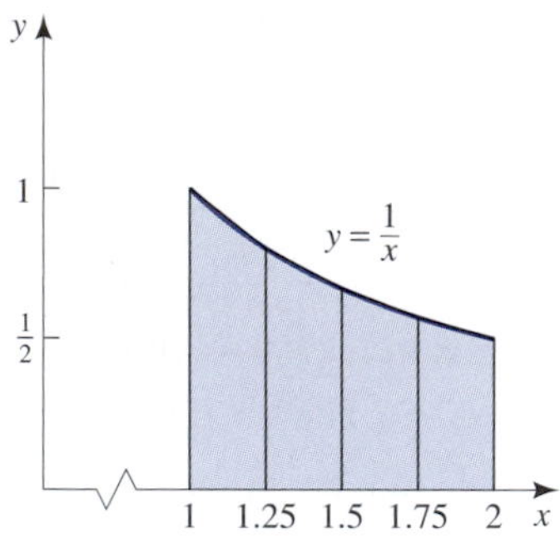

Figure 7.5
Using the trapezoidal rule to approximate ln 2.

Interactive Illustration 7.1

Numerical Integration

This interactive illustration allows you to explore Examples 1 and 2, other examples to follow in this text, and additional examples not in the text. For Example 1, click on the desired function, $n = 4$, and the trapezoidal rule.

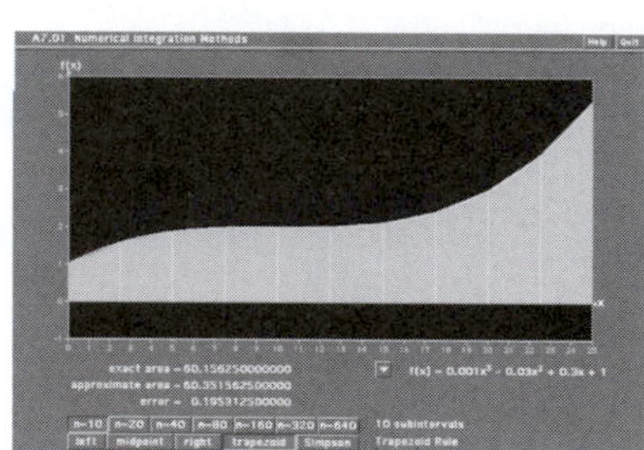

Simpson's Rule

When approximating the area under a curve using left- or right-hand Riemann sums, we are using "slices" consisting of thin rectangles, that is, a slice with the top a horizontal line (**polynomial of degree zero**). When approximating using the trapezoidal rule, we use thin slices with the top a line that connects the curve at two points (**polynomial of degree one**).

Intuitively, we might expect that trapezoids would be the better approximation, in general, since the straight line forming the top of a trapezoid hits the curve at *both*

The Granger Collection

James Gregory, 1638–1675

It was actually the English mathematician James Gregory who, in 1668, discovered Simpson's rule. Thomas Simpson (1710–1761), after whom Simpson's rule is named, wrote a very popular calculus textbook.

endpoints of the subinterval, whereas the top of a rectangle used in a left- or right-hand sum need only hit the curve at *one* endpoint. In general, it is true that trapezoids will yield better approximations than rectangles from left- or right-hand sums.

It is easy to criticize the use of trapezoids, however, since the typical curve actually "bends," whereas the top of a trapezoid is a straight line (polynomial of degree one). We might then think of using for the top a convenient curve that "bends." The obvious next choice would be a **polynomial of degree two**, that is, a parabola. For technical reasons we will need to divide that interval $[a, b]$ into an *even* number n of subintervals and, as usual, of equal length.

In Figure 7.6 the interval has been divided into four subintervals. We then wish to find a second-degree polynomial $Ax^2 + Bx + C$ that goes through the three points $(x_0, f(x_0))$, $(x_1, f(x_1))$, and $(x_2, f(x_2))$ and use this as a top over the interval $[x_0, x_2]$.

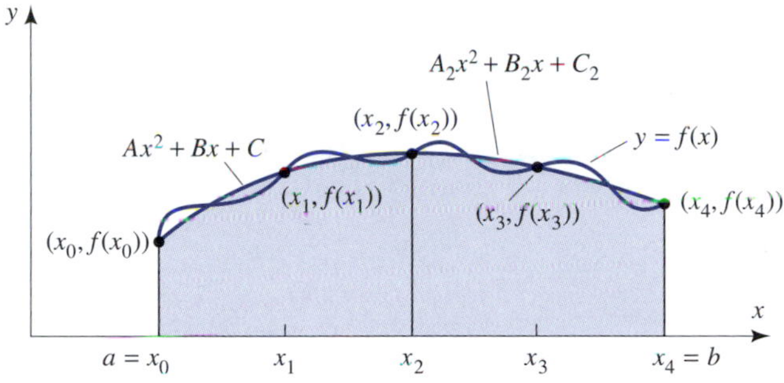

Figure 7.6
Simpon's rule uses different quadratics for the tops of pairs of subintervals.

In the same way we take the unique second-degree polynomial $A_2x^2 + B_2x + C_2$ that goes through the three points $(x_2, f(x_2))$, $(x_3, f(x_3))$, and $(x_4, f(x_4))$ as the top over the subinterval $[x_2, x_4]$. We refer to the sum of the areas under these two polynomials as S_4, the 4 referring to the original four subintervals.

It turns out (no proof given) that the area under the first parabola from x_0 to x_2 is

$$\frac{\Delta x}{3}[f(x_0) + 4f(x_1) + f(x_2)]$$

where, as usual, $\Delta x = (b - a)/n$. In the same way the area under the second parabola from x_2 to x_4 is

$$\frac{\Delta x}{3}[f(x_2) + 4f(x_3) + f(x_4)]$$

Thus, we have that the sum of these two is

$$S_4 = \frac{\Delta x}{3}[f(x_0) + 4f(x_1) + 2f(x_2) + 4f(x_3) + f(x_4)]$$

In general, we have the following.

Simpson's Rule

For any even integer n, divide the interval $[a, b]$ into n equal subintervals with endpoints $a = x_0, x_1, \ldots, x_n = b$, and let

$$S_n = \frac{b-a}{3n}[f(x_0) + 4f(x_1) + 2f(x_2) + 4f(x_3) + \cdots + 2f(x_{n-2}) + 4f(x_{n-1}) + f(x_n)]$$

Then

$$S_n \approx \int_a^b f(x)\,dx$$

Notice in this formula that with the exceptions of the first and last terms there is a 4 in front of a term $f(x_k)$ if k is *odd* and a 2 in front of a term $f(x_k)$ if k is *even*.

EXAMPLE 3 **Using Simpson's Rule**

Use Simpson's rule for S_4 to approximate $\int_0^2 \sqrt{4-x^2}\,dx = \pi$.

Solution

The interval $[0, 2]$ is divided into four subintervals with $x_0 = 0$, $x_1 = 0.50$, $x_2 = 1$, $x_3 = 1.50$, and $x_4 = 2$. We then have, using Simpson's rule, that

$$\begin{aligned} S_4 &= \frac{2-0}{3 \cdot 4}[f(0) + 4f(0.5) + 2f(1) + 4f(1.5) + f(2)] \\ &= \frac{1}{6}\left[2 + 4\sqrt{4-0.25} + 2\sqrt{4-1} + 4\sqrt{4-2.25} + 0\right] \\ &\approx 3.084 \end{aligned}$$

This compares to $T_4 = 2.996$ for the trapezoidal rule, and we see that S_4 is much closer to π than T_4 is. ■

EXAMPLE 4 **Comparing the Two Rules**

Find the trapezoidal and Simpson approximation for $\int_0^2 \sqrt{4-x^2}\,dx = \pi$ for $n =$ 100, 200, and 400. Compare the results.

Solution

The following table gives the results. The errors listed are $|T_n - \pi|$ and $|S_n - \pi|$.

n	T_n	Error	S_n	Error
100	3.14042	0.00117	3.14113	0.00046
200	3.14118	0.00041	3.14143	0.00016
400	3.14145	0.00014	3.14154	0.00005

Notice how the approximations improve as n becomes larger. Also notice that Simpson's rule is substantially better than the trapezoidal rule for each n. ■

EXAMPLE 5 **Using Simpson's Rule**

Use S_4 to find an approximation to $\int_1^2 \frac{1}{x}\,dx = \ln 2$.

Solution

Here $x_0 = 1$, $x_1 = 1.25$, $x_2 = 1.50$, $x_3 = 1.75$, and $x_4 = 2$, and

$$\begin{aligned} S_4 &= \frac{2-1}{3 \cdot 4}\left[f(1) + 4f(1.25) + 2f(1.5) + 4f(1.75) + f(2)\right] \\ &= \frac{1}{12}\left[1 + 4 \cdot \frac{4}{5} + 2 \cdot \frac{2}{3} + 4 \cdot \frac{4}{7} + \frac{1}{2}\right] \\ &\approx 0.6933 \end{aligned}$$

Compare this to $T_4 = 0.6970$ for the trapezoidal rule, and we see that S_4 is much closer to $\ln 2$ than T_4 is, since to four decimal places $\ln 2 = 0.6931$. ■

EXAMPLE 6 **Comparing the Two Rules**

Find the trapezoidal and Simpson's approximation for $\int_1^2 \frac{1}{x}\,dx = \ln 2$ for $n = 8$, 16, and 32. Compare the results.

Solution

The following table gives the results. The errors listed are $|T_n - \ln 2|$ and $|S_n - \ln 2|$.

n	T_n	Error	S_n	Error
8	0.69412185	0.00097467	0.69315453	0.00000735
16	0.69339120	0.00024402	0.69314765	0.00000047
32	0.69320821	0.00006103	0.69314721	0.00000003

Notice how the approximations improve as n becomes larger. Also notice that Simpson's rule is substantially better than the trapezoidal rule for each n. ■

Errors

An approximation is not worth much if we do not know how close it is to the correct answer. It is, of course, of vital importance to know just how accurate our approximations really are.

We now state two important results that give error estimates.

Error in the Trapezoidal Rule

The maximum possible error incurred in using the trapezoidal rule is

$$\frac{M(b-a)^3}{12n^2}$$

where $|f''(x)| \leq M$ on $[a, b]$.

Naturally, we need to assume that $f''(x)$ exists and is bounded on $[a, b]$.

Error in Simpson's Rule

The maximum possible error incurred in using Simpson's rule is

$$\frac{M(b-a)^5}{180n^4}$$

where $|f^{(4)}(x)| \leq M$ on $[a, b]$ where $f^{(4)}(x)$ is the fourth derivative.

Naturally, we need to assume that $f^{(4)}(x)$ exists and is bounded on $[a, b]$.

Remark Since these error formulas are correct, no matter what function f is used, they are extremely conservative error bounds. That is, the error in any particular problem will most likely be much less.

EXAMPLE 7 **Finding Error Bounds**

Use the preceding error bound formulas to find the largest possible errors incurred in the previous calculations of T_4 and S_4 for the definite integral $\int_1^2 \frac{1}{x}\,dx$.

Solution

(a) *Trapezoidal Rule Error.* We first need to find a bound of $f''(x)$ on the interval $[1, 2]$. We have $f(x) = x^{-1}$, $f'(x) = -x^{-2}$, and $f''(x) = 2x^{-3}$. So we need to find an upper bound on $2x^{-3}$ on $[1, 2]$. Since the function $2x^{-3}$ is decreasing on this interval, it attains its maximum at the left endpoint $x = 1$, where we see that $f''(1) = 2$. Thus, we take $M = 2$ in the error bound formula given above. Then

$$\frac{M(b-a)^3}{12n^2} = \frac{2(2-1)^3}{12(4)^2} \approx 0.01$$

As we noted above, the actual error was $T_4 - \ln 2 \approx 0.004$.

(b) *Simpson's Rule Error.* We first need to find a bound of $f^{(4)}(x)$ on the interval $[1, 2]$. We have $f''(x) = 2x^{-3}$, so $f'''(x) = -6x^{-4}$ and $f^{(4)}(x) = 24x^{-5}$. So we need to find an upper bound on $24x^{-5}$ on $[1, 2]$. Since the function $24x^{-5}$ is decreasing on this interval, it attains its maximum at the left endpoint $x = 1$, where we see that $f^{(4)}(1) = 24$. Thus, we take $M = 24$ in the error bound formula given above. Then

$$\frac{M(b-a)^5}{180n^4} = \frac{24(2-1)^5}{180(4)^4} \approx 0.0005$$

As we noted above, the actual error was $S_4 - \ln 2 \approx 0.0002$. ■

Exploration 1

Controlling the Error

Suppose you wish to find $\int_1^2 \frac{1}{x}\,dx = \ln 2$ with an error no greater than 0.0001.

(a) If you use the trapezoidal rule, how large must you take n?

(b) If you use Simpson's rule, how large must you take n?

If the original function $f(x)$ is a *first*-degree polynomial, that is, $f(x) = Ax + B$, then the error bound formula for the trapezoidal rule indicates that there is no error, since in this case $f''(x) = 0$ and therefore $M = 0$. This is naturally what we expect since the tops of the approximating trapezoids are themselves *first*-degree polynomials.

But very surprisingly, the Simpson's rule error bound formula indicates that Simpson's rule is exact not just for *second*-degree polynomials, as you would be inclined to suspect, but also for *third*-degree polynomials. This is true because the fourth derivative of a third-degree polynomial is identically zero, and therefore, $M = 0$ in the above formula. For this reason Simpson's rule gives much better approximations than one might otherwise suspect.

$f(x)$ on $[0, 2]$	x^2	x^3	$1/(x+1)$	$\sqrt{1+x^2}$
Trapezoidal rule ($n = 2$)	3.000	5.000	1.667	3.032
Simpson's rule ($n = 2$)	2.667	4.000	1.111	2.964
Exact (to three decimals)	2.667	4.000	1.099	2.958

Warm Up Exercise Set 7.3

1. Use T_8 to approximate $\int_1^2 \frac{1}{x}\,dx$.

2. Use S_8 to approximate $\int_1^2 \frac{1}{x}\,dx$.

3. Using the error bound formulas, find the maximum error in Exercises 1 and 2.

Exercise Set 7.3

In Exercises 1 through 8, find T_2 and S_2, and for Exercises 1 through 6, compare your approximations to the correct answer.

1. $\int_0^2 x^3\,dx$ **2.** $\int_0^2 x^2\,dx$

3. $\int_1^3 \frac{1}{x}\,dx$ **4.** $\int_0^2 x^4\,dx$

5. $\int_0^1 \sqrt{1-x^2}\,dx$ **6.** $\int_0^2 \frac{1}{x+1}\,dx$

7. $\int_0^1 e^{x^2}\,dx$ **8.** $\int_2^3 \frac{1}{\ln x}\,dx$

In Exercises 9 through 16, find T_4 and S_4, and for Exercises 9 through 14, compare your approximations to the correct answer and to the approximations obtained above.

9. $\int_0^2 x^3\,dx$ **10.** $\int_0^2 x^2\,dx$

11. $\int_1^3 \frac{1}{x}\,dx$ **12.** $\int_0^2 x^4\,dx$

13. $\int_0^1 \sqrt{1-x^2}\,dx$ **14.** $\int_0^2 \frac{1}{x+1}\,dx$

15. $\int_0^1 e^{x^2}\,dx$ **16.** $\int_2^3 \frac{1}{\ln x}\,dx$

In Exercises 17 through 24, find T_{100} and S_{100}, and for Exercises 17 through 22, compare your approximations to the correct answer and to the approximations obtained above.

17. $\int_0^2 x^3\,dx$ **18.** $\int_0^2 x^2\,dx$

19. $\int_1^3 \frac{1}{x}\,dx$ **20.** $\int_0^2 x^4\,dx$

21. $\int_0^1 \sqrt{1-x^2}\,dx$ **22.** $\int_0^2 \frac{1}{x+1}\,dx$

23. $\int_0^1 e^{x^2}\,dx$ **24.** $\int_2^3 \frac{1}{\ln x}\,dx$

Applications and Mathematical Modeling

25. Revenue In the following table the rate $R'(t)$ of revenue in millions of dollars per week is estimated in some manner for discrete equally spaced points $0 = t_0, t_1, \ldots, t_4 = 8$, where t is in weeks. Find (a) T_4 and (b) S_4. The answers will then be approximations to $\int_0^8 R'(t)\,dt = R(8) - R(0) = R(8)$, which is the total revenue for the eight-week period.

t_i	0	2	4	6	8
$R'(t_i)$	1.0	0.8	0.5	0.7	0.8

26. Sales In the following table the rate $R'(t)$ of sales for an entire industry in billions of dollars is given for discrete equally spaced points $0 = t_0, t_1, \ldots, t_{10} = 10$, where t is in weeks. Find (a) T_{10} and (b) S_{10}. The answers will then be approximations to $\int_0^{10} R'(t)\,dt = R(10) - R(0) = R(10)$, which is the total sales for the ten-week period.

t_i	0	1	2	3	4	5
$R'(t_i)$	10	12	15	16	20	18
t_i	6	7	8	9	10	
$R'(t_i)$	16	15	12	10	6	

27. Production In the following table the rate $P'(t)$ of production of steel in thousands of tons is estimated in some manner for discrete equally spaced points $0 = t_0, t_1, \ldots, t_4 = 16$, where t is in weeks. Find (a) T_4 and (b) S_4. The answers will then be approximations to $\int_0^{16} P'(t)\,dt = P(16) - P(0) = P(16)$, which is the total production of steel for the eight-week period.

t_i	0	4	8	12	16
$P'(t_i)$	1	2	3	5	4

28. Population In the following table the rate of growth $M'(t)$ of a bacteria has been estimated by measuring the rate of increase in mass at discrete, equally spaced points $0 = t_0, t_1, \ldots, t_{10} = 20$, where t is in hours. Find (a) T_{10} and

(b) S_{10}. The answers will then be approximations to $\int_0^{20} M'(t)\,dt = M(20) - M(0)$, which is the change in mass for the 20-hour period.

t_i	0	2	4	6	8	10
$M'(t_i)$	10	14	20	30	50	100
t_i	12	14	16	18	20	
$M'(t_i)$	100	80	50	20	0	

29. Velocity In the following table the velocity $v(t) = s'(t)$ is given for discrete, equally spaced points $0 = t_0, t_1, \ldots, t_6 = 3$. Find (a) T_6 and S_6. The answers will then be approximations to $\int_0^3 v(t)\,dt = s(3) - s(0)$, which is the total distance traveled.

t_i	0	0.5	1	1.5	2	2.5	3
$v(t_i)$	1.0	1.6	2.0	1.8	1.5	1.0	0.5

In Exercises 30 through 34, find the maximum error for T_{10} and S_{10}.

30. $\int_0^2 x^3\,dx$ **31.** $\int_0^2 x^2\,dx$ **32.** $\int_1^3 \frac{1}{x}\,dx$

33. $\int_0^2 x^4\,dx$ **34.** $\int_0^2 \frac{1}{x+1}\,dx$

35. A tract of city land bounded by four roads is shown in the figure, which shows measurements taken every 100 feet. If property in this location goes for \$10,000 an acre, approximately what is this tract worth? (Hint: Use Simpson's rule to estimate the area, and note that there are 43,560 square feet to an acre.)

36. If x is large, the logarithmic integral $\text{li}(x) = \int_2^x \frac{dt}{\ln t}$ approximates the number of prime numbers less than x. Find an approximation to $\text{li}(200)$. (Hint: There are 46 primes less than 200.)

More Challenging Exercises

37. In Example 1 the curve was concave down, and the trapezoid rule gave an underestimate. In Example 2 the curve was concave up, and the trapezoid rule gave an overestimate. Will the trapezoidal rule always give an underestimate in the case that the graph of the function is concave down and an overestimate in the case that the graph of the function is concave up? Explain.

38. Can you use the trapezoidal rule error bound formula to find an error bound for T_{10} for $\int_0^1 \sqrt{x}\,dx$? Why or why not. Explain using complete sentences.

39. For any $b > 0$, show directly (without using the Simpson's rule error bound formula) that S_2 gives the exact value for $\int_0^b x^3\,dx$.

40. In Examples 4 and 6 we used the trapezoidal rule and Simpson's rule to estimate $\int_1^2 \frac{1}{x}\,dx$ and $\int_0^2 \sqrt{4 - x^2}\,dx$. Recall that the approximations for the first integral were more accurate with $n = 32$ than the second with $n = 400$. By studying the error rule formulas and the integrands, give a reasonable explanation of how this could happen.

41. Show that you can obtain the trapezoidal rule by taking the average of the left- and right-hand sums.

42. Section 6.4 contains a discussion of the midpoint rule that uses the rectangles with heights equal to the values of f at the midpoint of the subintervals.

a. In Example 4 we used the trapezoidal rule to approximate $\int_0^2 \sqrt{4 - x^2}\,dx = \pi$ using $n = 100$, 200, and 400. Use the midpoint rule to approximate this same integral for the same values of n. Determine the errors. Compare the errors used in the midpoint rule with the errors in the trapezoidal rule.

b. In Example 6 we used the trapezoidal rule to approximate $\int_1^2 \frac{1}{x}\,dx = \ln 2$ using $n = 8$, 16, and 32. Use the midpoint rule to approximate this same integral for the same values of n. Determine the errors. Compare the errors used in the midpoint rule with the errors in the trapezoidal rule.

c. On the basis of your results in parts (a) and (b), speculate on how accurate the midpoint rule is in comparison to the trapezoidal rule.

Solutions to WARM UP EXERCISE SET 7.3

1. We have $\Delta x = \frac{2-1}{8} = 0.125$, and the points x_0, x_1, x_2, $\ldots, x_8$ are $1, 1.125, 1.25, \ldots, 1.875, 2$, respectively. Thus,

$$T_8 = \frac{1-0}{8}\left[\frac{1}{2}f(1) + f(1.125) + f(1.25) + f(1.375) + f(1.50) + f(1.625) + f(1.75) + f(1.875) + \frac{1}{2}f(2)\right]$$

$$= \frac{1}{8}\left[\frac{1}{2} + \frac{1}{1.125} + \frac{1}{1.25} + \frac{1}{1.375} + \frac{1}{1.5} + \frac{1}{1.625} + \frac{1}{1.75} + \frac{1}{1.875} + \frac{1}{4}\right]$$

$$\approx 0.6941$$

2. The points $x_0, x_1, \ldots, x_8$ are the same as in the previous exercise. Then

$$S_8 = \frac{2-1}{3\cdot 8}[f(1) + 4f(1.125) + 2f(1.25) + 4f(1.375) + 2f(1.5) + 4f(1.625) + 2f(1.75) + 4f(1.875) + f(2)]$$

$$= \frac{1}{24}\left[1 + 4\frac{1}{1.125} + 2\frac{1}{1.25} + 4\frac{1}{1.375} + 2\frac{1}{1.5} + 4\frac{1}{1.625} + 2\frac{1}{1.75} + 4\frac{1}{1.875} + \frac{1}{2}\right]$$

$$\approx 0.6931545$$

3. Just as in Example 5 of the text, $|f''(x)| \le 2$, $|f^{(4)}(x)| \le 24$, and $b - a = 2 - 1 = 1$. For this exercise $n = 8$. Thus, a bound on the trapezoidal error is

$$\frac{M(b-a)^3}{12n^2} = \frac{2(2-1)^3}{12(8)^2} \approx .0026$$

Compare this with the actual error of 0.001. A bound on the Simpson's error is

$$\frac{M(b-a)^5}{180n^4} = \frac{24(2-1)^5}{180(8)^4} \approx .00003$$

Compare this with the actual error of 0.000007.

7.4 Improper Integrals

Improper Integrals
Applications

APPLICATION Sales of an Engine Part over the Long Term

To sell a large number of its most recent engines to the aircraft manufacturers of the world, an aircraft engine manufacturer had to guarantee to supply all spare parts for this engine in perpetuity. The company plans to stop manufacturing this engine one year from now and wishes in the next year to make all the spare parts it will ever need. The company estimates that the rate at which engine manifolds will be needed per year is given by $r(t) = 200e^{-0.1t}$, where t is in years from now. How many manifolds will be needed? See Example 3 on page 515 for the answer.

Improper Integrals

Up to now we have considered definite integrals $\int_a^b f(x)\,dx$, where the interval $[a, b]$ of integration was *bounded*, that is, both a and b are finite numbers. But in some important applications one must consider *infinite* intervals of integration.

Consider, for example, $\int_1^\infty \frac{1}{x^2}\,dx$. We define this as follows. First calculate $\int_1^b \frac{1}{x^2}\,dx$, which is the area of the region between the graph of $y = \frac{1}{x^2}$ and the x-axis on the interval $[1, b]$ (Figure 7.7). We obtain

$$\int_1^b \frac{1}{x^2}\,dx = -\left.\frac{1}{x}\right|_1^b = -\frac{1}{b} + 1 = 1 - \frac{1}{b}$$

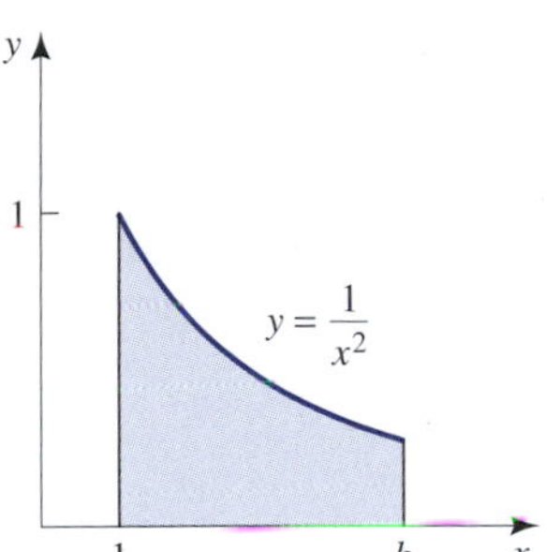

Figure 7.7
The area between the graph of the curve $y = \frac{1}{x^2}$ and the x-axis from 1 to b is $\int_1^b x^{-2}dx$.

We then take the limit of this expression as $b \to \infty$ (Figure 7.8). We obtain

$$\int_1^\infty \frac{1}{x^2}\,dx = \lim_{b\to\infty} \int_1^b \frac{1}{x^2}\,dx = \lim_{b\to\infty}\left(1 - \frac{1}{b}\right) = 1$$

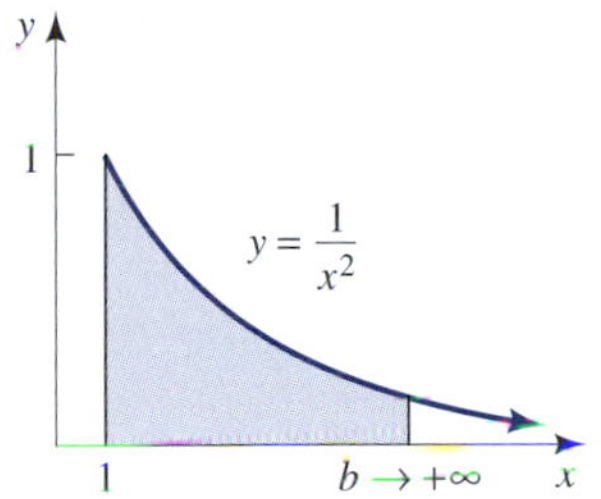

Figure 7.8
We define $\int_1^\infty x^{-2}dx$ as $\lim_{b\to\infty} \int_1^b x^{-2}dx$.

Thus, we declare that $\int_1^\infty \frac{1}{x^2}\,dx = 1$. We also declare that the area of the unbounded region between the graph of $y = \frac{1}{x^2}$ and the x-axis on the interval $[1, \infty)$ is 1.

Of course, there will be functions $y = f(x)$ for which the limit $\lim_{b\to\infty} \int_1^b f(x)\,dx$ does not exist. In such a case we say that the improper integral $\int_1^\infty f(x)\,dx$ *diverges*.

DEFINITION $\int_a^\infty f(x)\,dx$

If $f(x)$ is continuous on $[a, \infty)$, we define

$$\int_a^\infty f(x)\,dx = \lim_{b\to\infty} \int_a^b f(x)\,dx$$

if this limit exists as a number. Otherwise, we say that the integral **diverges**. If the limit exists as a number, we say that the integral **converges** to that number.

Remark When the integral is over an infinite interval, we call it an **improper integral**.

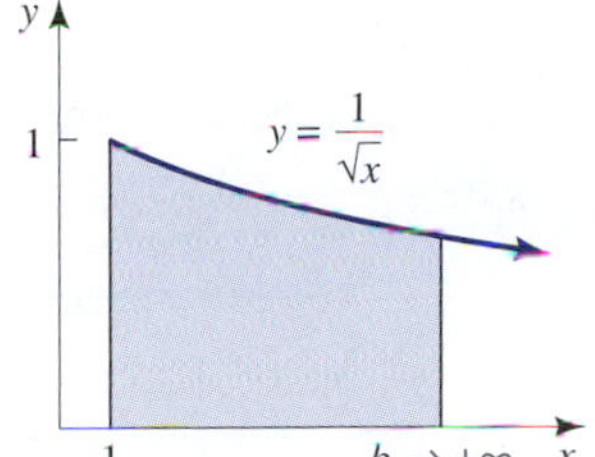

Figure 7.9
$\int_1^\infty \frac{1}{\sqrt{x}}dx = \lim_{b\to\infty} \int_1^b \frac{1}{\sqrt{x}}dx.$

EXAMPLE 1 Finding an Improper Integral

Find $\int_1^\infty \frac{1}{\sqrt{x}}\,dx$.

Solution

According to our definition, we first need to evaluate $\int_1^b \frac{1}{\sqrt{x}}\,dx$, for any $b > 1$. Of course, this just represents the area under the curve on the interval $[1, b]$ (Figure 7.9).

We have

$$\int_1^b x^{-1/2}\,dx = 2x^{1/2}\Big|_1^b$$
$$= 2(\sqrt{b} - 1)$$

Now we need to take the limit of this expression as $b \to \infty$ (Figure 7.9). Doing this, we obtain

$$\int_1^\infty x^{-1/2}\,dx = \lim_{b\to\infty} \int_1^b x^{-1/2}\,dx$$
$$= \lim_{b\to\infty} 2(\sqrt{b} - 1)$$

which is unbounded and therefore does not exist. This integral then diverges. ■

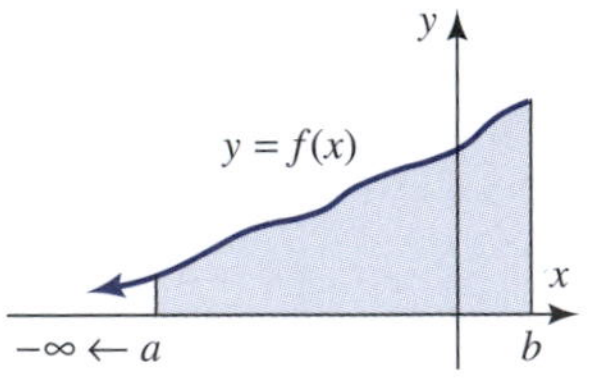

Figure 7.10

$\int_{-\infty}^{b} f(x)dx = \lim_{a\to-\infty} \int_a^b f(x)dx.$

Just as we defined an improper integral on the interval $[a, \infty)$, so also we can define an improper integral on the interval $(-\infty, b]$. We now state this definition (Figure 7.10).

DEFINITION $\int_{-\infty}^{b} f(x)\,dx$

If $f(x)$ is continuous on $(-\infty, b]$, we define

$$\int_{-\infty}^{b} f(x)\,dx = \lim_{a\to-\infty} \int_a^b f(x)\,dx$$

if this limit exists. If the limit does not exist, we say that the integral **diverges**.

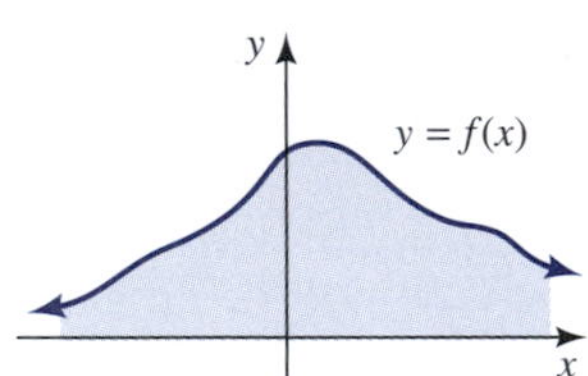

Figure 7.11

$\int_{-\infty}^{\infty} f(x)dx = \int_{-\infty}^{0} f(x)dx + \int_0^\infty f(x)dx$ if both of the latter two integrals converge.

Finally, we define an improper integral of the form $\int_{-\infty}^{\infty} f(x)\,dx$ (Figure 7.11).

DEFINITION $\int_{-\infty}^{\infty} f(x)\,dx$

If $f(x)$ is continuous on $(-\infty, \infty)$, we define

$$\int_{-\infty}^{\infty} f(x)\,dx = \int_{-\infty}^{0} f(x)\,dx + \int_0^\infty f(x)\,dx$$

if both integrals on the right converge. If one or both of the integrals on the right diverges, then we say that $\int_{-\infty}^{\infty} f(x)\,dx$ diverges.

EXAMPLE 2 **Finding an Improper Integral**

Find $\int_{-\infty}^{\infty} 2xe^{-x^2}\,dx$.

[-3, 3] × [-1, 1]

Screen 7.3
A graph of $y = 2xe^{-x^2}$.

Solution

A graph of the function is shown in Screen 7.3 using a window with dimensions $[-3, 3]$ by $[-1, 1]$. According to the previous definition, we need to find both $\int_{-\infty}^{0} 2xe^{-x^2}\,dx$ and $\int_{0}^{\infty} 2xe^{-x^2}\,dx$. To find the last integral, we need first to find $\int_{0}^{b} 2xe^{-x^2}\,dx$ for every $b > 0$. Then

$$\begin{aligned}\int_0^b 2xe^{-x^2}\,dx &= -e^{-x^2}\Big|_0^b \\ &= 1 - e^{-b^2} \\ &= 1 - \frac{1}{e^{b^2}}\end{aligned}$$

Now taking the limit of this last expression as $b \to \infty$, we obtain

$$\begin{aligned}\lim_{b\to\infty}\int_0^b 2xe^{-x^2}\,dx &= \lim_{b\to\infty}\left(1 - \frac{1}{e^{b^2}}\right) \\ &= 1\end{aligned}$$

For the other integral we have, for any $a < 0$,

$$\begin{aligned}\int_a^0 2xe^{-x^2}\,dx &= -e^{-x^2}\Big|_a^0 \\ &= -1 + e^{-a^2} \\ &= \frac{1}{e^{a^2}} - 1\end{aligned}$$

Now taking the limit of this last expression as $a \to -\infty$, we obtain

$$\begin{aligned}\lim_{a\to-\infty}\int_a^0 2xe^{-x^2}\,dx &= \lim_{a\to-\infty}\left(\frac{1}{e^{a^2}} - 1\right) \\ &= -1\end{aligned}$$

Thus,

$$\begin{aligned}\int_{-\infty}^{\infty} 2xe^{-x^2}\,dx &= \int_{-\infty}^{0} 2xe^{-x^2}\,dx + \int_{0}^{\infty} 2xe^{-x^2}\,dx \\ &= -1 + 1 = 0\end{aligned}$$

■

Applications

The large aircraft manufacturers of the world do not produce the engines for their aircraft. Rather they purchase them from several large engine manufacturers. In the United States, General Electric and United Technologies dominate this fiercely competitive industry.

EXAMPLE 3 **Spare Parts Production**

To sell a large number of its most recent engines to the aircraft manufacturers of the world, an aircraft engine manufacturer had to guarantee to supply all spare parts for this engine in perpetuity. The company plans to stop manufacturing this engine one year from now and wishes in the next year to make all the spare parts it will ever need. The company estimates that the rate at which engine manifolds will be needed per year is given by $r(t) = 200e^{-0.1t}$, where t is in years from now (Figure 7.12). How many manifolds will be needed?

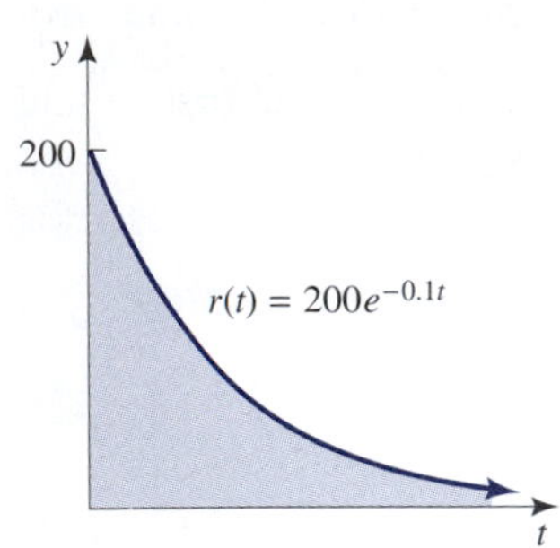

Figure 7.12
Graph of the rate of required engine manifolds.

Solution

Since the integral $\int_0^b r(t)\,dt$ is the amount needed from now to b years from now, $\int_0^\infty r(t)\,dt$ is the amount needed for all time. Then

$$\begin{aligned}\int_0^b r(t)\,dt &= 200\int_0^b e^{-0.10t}\,dt \\ &= -2000e^{-0.10t}\Big|_0^b \\ &= -2000(e^{-0.10b} - 1)\end{aligned}$$

$$\begin{aligned}\int_0^\infty r(t)\,dt &= \lim_{b\to\infty}\int_0^b r(t)\,dt \\ &= \lim_{b\to\infty} -2000(e^{-0.10b} - 1) \\ &= 2000\end{aligned}$$

Thus, the company needs to make 2000 engine manifolds. ■

We now assume that we have a continuous income flow of money and that the positive function $f(t)$ represents the rate of change of this flow. If the flow of money continues perpetually, then the function $f(t)$ must be given on the interval $[0, \infty)$ (Figure 7.13). The present value of all future income is denoted by $P_V(\infty)$. We have the following.

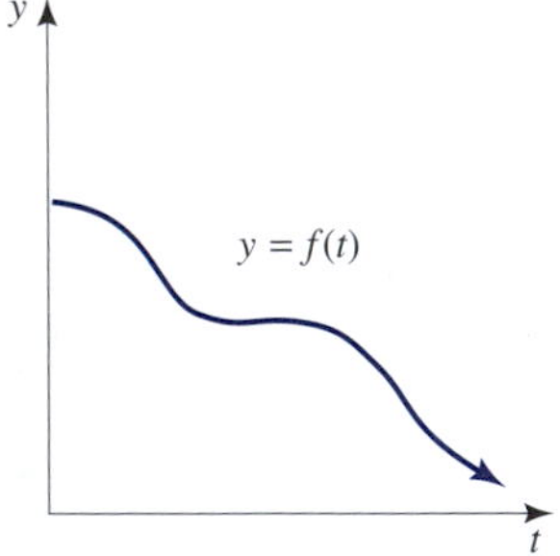

Figure 7.13
A graph of a typical rate of change of a continuous income flow.

Present Value of a Continuous Income Flow on $[0, \infty)$

Let $f(t)$ be the continuous rate of change of total income on the interval $[0, \infty)$, and suppose that one can invest current money at an interest rate of r in perpetuity, where r is a decimal, compounded continuously forever. Then the present value $P_V(\infty)$ of this continuous income flow over $[0, \infty)$ is given by

$$\begin{aligned}P_V(\infty) &= \int_0^\infty f(t)e^{-rt}\,dt \\ &= \lim_{T\to\infty}\int_0^T f(t)e^{-rt}\,dt\end{aligned}$$

EXAMPLE 4 **Present Value on an Infinite Interval**

The rate of change of income in thousands of dollars per year from an oil well is projected to be $f(t) = 100e^{-0.1t}$, where t is measured in years. What would be a fair initial price for the oil well as a function of r?

Solution

$$\begin{aligned}
P_V(\infty) &= \int_0^{\infty} f(t)e^{-rt}\,dt \\
&= \lim_{T\to\infty} \int_0^{T} f(t)e^{-rt}\,dt \\
&= \lim_{T\to\infty} \int_0^{T} 100e^{-0.1t}e^{-rt}\,dt \\
&= 100 \lim_{T\to\infty} \int_0^{T} e^{-(r+0.1)t}\,dt \\
&= 100 \lim_{T\to\infty} \frac{-1}{r+0.1} e^{-(r+0.1)t}\Big|_0^T \\
&= \frac{100}{r+0.1} \lim_{T\to\infty} (1 - e^{-(r+0.1)T}) \\
&= \frac{100}{r+0.1}
\end{aligned}$$

■

Remark Notice that as r increases, the value of the oil well goes down. Thus, if long-term interest rates go up, the value of the well will drop.

Warm Up Exercise Set 7.4

1. Repeat Example 3 if $r(t) = 1000(t+1)^{-3/2}$.

2. Find $\int_{-\infty}^{\infty} f(x)\,dx$, where $f(x) = 1$ if $|x| < 1$ and $f(x) = x^{-2}$ if $|x| \geq 1$.

Exercise Set 7.4

In Exercises 1 through 20, evaluate each improper integral whenever it is convergent.

1. $\int_1^{\infty} \frac{1}{x^3}\,dx$

2. $\int_1^{\infty} \frac{1}{\sqrt[3]{x}}\,dx$

3. $\int_1^{\infty} \frac{1}{\sqrt[4]{x}}\,dx$

4. $\int_1^{\infty} \frac{1}{x^4}\,dx$

5. $\int_1^{\infty} \frac{1}{x^{1.01}}\,dx$

6. $\int_1^{\infty} \frac{1}{x^{0.99}}\,dx$

7. $\int_1^{\infty} \frac{1}{x}\,dx$

8. $\int_{-\infty}^{-1} \frac{1}{x^2}\,dx$

9. $\int_{-\infty}^{0} e^x\,dx$

10. $\int_0^{\infty} e^{-x}\,dx$

11. $\int_{-\infty}^{-1} \frac{1}{x^3}\,dx$

12. $\int_{-\infty}^{-1} \frac{1}{\sqrt[3]{x}}\,dx$

13. $\int_0^{\infty} x^2e^{-x^3}\,dx$

14. $\int_{-\infty}^{0} x^2e^{x^3}\,dx$

15. $\int_{-\infty}^{0} \frac{x}{(x^2+1)^2}\,dx$

16. $\int_{-\infty}^{\infty} e^{-|x|}\,dx$

17. $\int_{-\infty}^{\infty} \frac{x}{(x^2+1)^4}\,dx$

18. $\int_{-\infty}^{\infty} \frac{x}{x^2+1}\,dx$

19. $\int_{-\infty}^{\infty} x^3e^{-x^4}\,dx$

20. $\int_{-\infty}^{\infty} \frac{x^3}{(x^4+3)^2}\,dx$

Applications and Mathematical Modeling

21. Production Repeat Example 3 with $r(t) = 1000(t+1)^{-5/2}$.

22. Mining It is estimated that platinum is being extracted from a certain mine at the rate of

$$r(t) = 10\frac{t}{(t^2+1)^2}$$

tons per year, where t is in years. How much will be extracted from this mine in the long term?

23. Sales A company estimates that the rate of increase in millions of dollars of sales for a new product is given by $r(t) = 40e^{-2t}$, where t is in years. If this rate continues forever, what will be the eventual sales?

24. Profits A company estimates that the rate of increase in millions of dollars of profits from a new product is given by $(t+1)^{-4/3}$, where t is in years. If this rate continues forever, what will be the eventual profits?

25. Radioactive Waste Suppose radioactive waste from a closed dump site is entering the atmosphere over an area at a rate given by $r(t) = 100e^{-kt}$ tons per year, where t is measured in years. Assuming that this rate continues forever and that $k = 0.05$, find the total amount of waste that will enter the atmosphere.

26. Radioactive Waste Repeat Exercise 25 if $k = 0.10$.

27. Disease A contagious disease is spreading into the world's human population at the rate of $p'(t) = 10e^{-0.05t}$, where $p(t)$ is measured in millions and t is measured in years. If this rate continues forever, what will be the total number of people who are infected with this disease?

28. Biology A population of mice in an area is growing at the rate of

$$p'(t) = \frac{1}{(t+1)^{5/2}}$$

where p is measured in thousands and t is measured in months. If this rate continues forever, what will be the total number of mice in this area?

29. Pollutants Pollutants are leaking from a closed dump site into a small lake nearby. The number of thousands of particles of pollutants that is accumulating in the lake is given by

$$N(t) = \frac{1}{(t+1)^2}$$

where t is measured in years. Assuming that this equation holds forever, how many particles of pollutants will eventually enter this lake from the dump?

30. By making the change of variable $x = -\ln u$, show that the improper integral $\int_0^{\infty} f(x)\,dx$ becomes $\int_0^1 u^{-1} f(-\ln u)\,du$. Use this and Simpson's rule for $n = 128$ to approximate $\int_0^{\infty} \frac{e^{-x}}{1+x^2}\,dx$.

In Exercises 31 through 34, $f(t)$ is the rate of change of total income per year, and r is the annual interest rate compounding continuously. Find $P_V(\infty)$.

31. $f(t) = 100$, $r = 8\%$

32. $f(t) = 1000$, $r = 12\%$

33. $f(t) = 100e^{-0.1t}$, $r = 8\%$

34. $f(t) = 1000e^{-0.2t}$, $r = 10\%$

35. Continuous Income Flow An investment with a continuous income flow of \$10,000 per year forever is being sold. Assume that the current annual interest rate of 10% compounded continuously will continue forever. Would you buy if the price was (a) \$110,000? (b) \$90,000?

36. Continuous Income Flow An investment with a continuous income flow of \$16,000 per year forever is being sold. Assume that the current annual interest rate of 8% compounded continuously will continue forever. Would you buy if the price was (a) \$250,000? (b) \$180,000?

37. Investment Decision An investment with a continuous income flow of $24{,}000e^{0.04t}$ in dollars per year forever is being sold. Assume that the current annual interest rate of 8% compounded continuously will continue forever. Would you buy if the price was (a) \$700,000? (b) \$550,000?

38. Investment Decision An investment with a continuous income flow of $10{,}000e^{0.05t}$ in dollars per year forever is being sold. Assume that the current annual interest rate of 10% compounded continuously will continue forever. Would you buy if the price was (a) \$300,000? (b) \$250,000?

More Challenging Exercises

39. A student argues that $\int_{-\infty}^{\infty} x\,dx = 0$, since $\int_{-a}^{a} x\,dx = 0$ for every a. Explain why this is or is not correct, using the definitions given in this section.

40. Suppose $0 \le f(x) \le \frac{1}{x^2}$ on the interval $[1, \infty)$. What can you say about $\int_1^{\infty} f(x)\,dx$? Explain carefully.

41. Suppose $f(x) \geq \dfrac{1}{\sqrt{x}}$ on the interval $[1, \infty)$. What can you say about $\displaystyle\int_1^\infty f(x)\,dx$? Explain carefully.

42. The integral $\displaystyle\int_0^1 \frac{1}{\sqrt{x}}\,dx$ is an improper integral (of a type not considered in this section), since the function $f(x) = \dfrac{1}{\sqrt{x}}$ is unbounded on the interval $(0, 1]$. Give a reasonable definition of this improper integral, and then evaluate using your definition.

43. If $f(x)$ is continuous, $\displaystyle\int_{-\infty}^\infty f(x)\,dx$ is convergent, and c is any real number, show that

$$\int_{-\infty}^\infty f(x)\,dx = \int_{-\infty}^c f(x)\,dx + \int_c^\infty f(x)\,dx$$

44. Law Enforcement Shavell[1] let $x(h)$ be the enforcement effort specific to apprehending those who commit acts causing harm h and considered the term $\displaystyle\int_0^\infty x(h)\,dh$. Explain what you think this integral stands for and explain why.

[1] Steven Shavell. 1991. Law enforcement: specific versus general enforcement of law. *J. of Political Econ.* 99(5):1088–1108.

Solutions to WARM UP EXERCISE SET 7.4

1. In the integral $\displaystyle\int_0^b (t+1)^{-3/2}\,dx$, make the change of variable $u = t + 1$. Then $du = dt$, and

$$\begin{aligned}1000\int_0^b (t+1)^{-3/2}\,dt &= 1000\int_1^{b+1} u^{-3/2}\,du\\ &= -2000u^{-1/2}\Big|_1^{b+1}\\ &= 2000\left(1 - \frac{1}{\sqrt{b+1}}\right)\end{aligned}$$

$$\begin{aligned}1000\int_0^\infty (t+1)^{-3/2}\,dt &= 1000\lim_{b\to\infty}\int_0^b (t+1)^{-3/2}\,dt\\ &= \lim_{b\to\infty} 2000\left(1 - \frac{1}{\sqrt{b+1}}\right)\\ &= 2000\end{aligned}$$

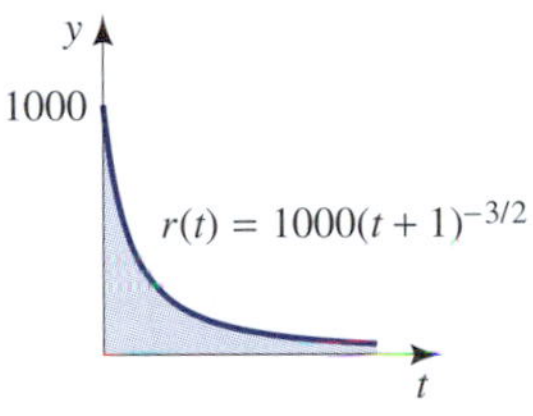

2. By symmetry, notice from the figure that $\displaystyle\int_{-\infty}^0 f(x)\,dx = \int_0^\infty f(x)\,dx$. Then

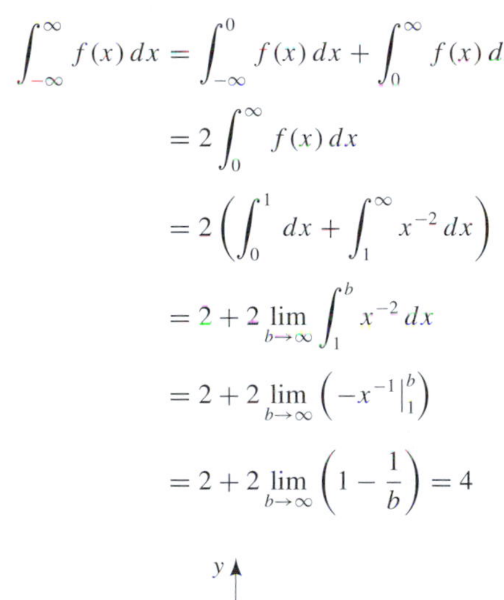

$$\begin{aligned}\int_{-\infty}^\infty f(x)\,dx &= \int_{-\infty}^0 f(x)\,dx + \int_0^\infty f(x)\,dx\\ &= 2\int_0^\infty f(x)\,dx\\ &= 2\left(\int_0^1 dx + \int_1^\infty x^{-2}\,dx\right)\\ &= 2 + 2\lim_{b\to\infty}\int_1^b x^{-2}\,dx\\ &= 2 + 2\lim_{b\to\infty}\left(-x^{-1}\Big|_1^b\right)\\ &= 2 + 2\lim_{b\to\infty}\left(1 - \frac{1}{b}\right) = 4\end{aligned}$$

Summary Outline

- **Integration by Parts** $\displaystyle\int u\,dv = uv - \int v\,du$ (p. 490)

- **Trapezoidal Rule** For any integer n, divide the interval $[a, b]$ into n subintervals of equal length with endpoints

$a = x_0, x_1, \ldots, x_n = b$. If

$$T_n = \frac{b-a}{n}\left[\frac{1}{2}f(x_0) + f(x_1) + f(x_2) + \cdots + f(x_{n-1}) + \frac{1}{2}f(x_n)\right]$$

then

$$\int_a^b f(x)\,dx \approx T_n. \qquad \text{(p. 502)}$$

- **Simpson's Rule** For any even integer n divide the interval $[a, b]$ into n equal subintervals with endpoints $a = x_0, x_1, \ldots, x_n = b$, and let

$$S_n = \frac{b-a}{3n}[f(x_0) + 4f(x_1) + 2f(x_2) + 4f(x_3) + \cdots + 2f(x_{n-2}) + 4f(x_{n-1}) + f(x_n)]$$

Then

$$S_n \approx \int_a^b f(x)\,dx. \qquad (p. 506)$$

- **Error in the Trapezoidal Rule** The maximum possible error incurred in using the trapezoidal rule is $\dfrac{M(b-a)^3}{12n^2}$, where $|f''(x)| \le M$ on $[a, b]$. (p. 508)
- **Error in Simpson's Rule** The maximum possible error incurred in using Simpson's rule is $\dfrac{M(b-a)^5}{180n^4}$, where $|f^{(4)}(x)| \le M$ on $[a, b]$. (p. 508)
- If $f(x)$ is continuous on $[a, \infty)$, define

$$\int_a^\infty f(x)\,dx = \lim_{b\to\infty}\int_a^b f(x)\,dx$$

if this limit exists. Otherwise, we say that the integral **diverges**. If the limit exists we say that the integral **converges** to that number. (p. 513)

- If $f(x)$ is continuous on $(-\infty, b]$, define

$$\int_{-\infty}^b f(x)\,dx = \lim_{a\to-\infty}\int_a^b f(x)\,dx$$

if this limit exists. If the limit does not exist, we say that the integral **diverges**. (p. 514)

- If $f(x)$ is continuous on $(-\infty, \infty)$, define

$$\int_{-\infty}^\infty f(x)\,dx = \int_{-\infty}^0 f(x)\,dx + \int_0^\infty f(x)\,dx$$

if both integrals on the right exist. If one or both of the integrals on the right diverges then we say that $\int_{-\infty}^\infty f(x)\,dx$ **diverges**. (p. 514)

- Let $f(t)$ be the continuous rate of change of income on the interval $[0, \infty)$ and suppose that one can invest current money at an interest rate of r in perpetuity, where r is a decimal, compounded continuously forever. Then the **present value** $P_V(\infty)$ of this continuous income flow over $[0, \infty)$ is given by

$$P_V(\infty) = \int_0^\infty f(t)e^{-rt}\,dt = \lim_{T\to\infty}\int_0^T f(t)e^{-rt}\,dt$$

(p. 516)

Review Exercises

In Exercises 1 through 6, integrate by parts.

1. $\int xe^{5x}\,dx$

2. $\int \frac{x}{\sqrt{5+3x}}\,dx$

3. $\int x^5 \ln x\,dx$

4. $\int \frac{x}{(x+1)^2}\,dx$

5. $\int x^3\sqrt{9-x^2}\,dx$

6. $\int \frac{\ln x}{\sqrt{x}}\,dx$

In Exercises 7 through 12, evaluate using the table of integrals.

7. $\int \frac{x}{2x+3}\,dx$

8. $\int \frac{1}{x(x+4)^2}\,dx$

9. $\int \frac{1}{x^2\sqrt{4-x^2}}\,dx$

10. $\int \frac{\sqrt{9-x^2}}{x}\,dx$

11. $\int \sqrt{x^2+1}\,dx$

12. $\int \frac{1}{1+e^x}\,dx$

In Exercises 13 through 16, evaluate the improper integrals.

13. $\int_1^\infty x^{-1.1}\,dx$

14. $\int_1^\infty x^{-0.9}\,dx$

15. $\int_{-\infty}^\infty x^2e^{-|x^3|}\,dx$

16. $\int_{-\infty}^\infty e^{-x}\,dx$

In Exercises 17 and 18, approximate using the trapezoidal rule and Simpson's rule for $n = 2$, and compare your answers to the correct answer.

17. $\int_0^2 5x^4\,dx$

18. $\int_1^4 \frac{1}{x}\,dx$

19. Repeat Exercise 17 for $n = 4$.

20. Repeat Exercise 18 for $n = 4$.

21. A mine produces a rate of change of total income in thousands of dollars per year given by $f(t) = 100e^{-0.10t}$ in perpetuity. Market interest rates are at 8% compounded continuously and will stay the same forever. What would be a reasonable price for this mine, discounting any risk involved?

Chapter 7 Cases

CASE 1 Coefficient of Inequality

The following table gives the accumulated U.S. income by population deciles for 1966. (See Suits[2])

Use the table to calculate the coefficient of inequality. Do this by using the trapezoidal rule to approximate the needed definite integral.

Population Decile	Cumulated Adjusted Family Income
1	1.21
2	3.88
3	8.13
4	13.92
5	21.16
6	30.22
7	40.02
8	52.29
9	67.45
10	100.00

[2] Daniel B. Suits. 1977. Measurement of tax progressivity. *Amer. Econ. Rev.* 67(4):747–752

CASE 2 Measuring Cardiac Output

An important medical procedure is to measure cardiac output. Insufficient output indicates a problem with the heart.

To measure a patient's cardiac output, a measured amount of dye is placed in a vein close to where the blood enters the right atrium of the heart. The dye then mixes with the blood in this chamber. Samples of the blood are taken regularly in a peripheral artery near the aorta where the blood leaves the heart. An instrument called a densitometer can make the withdrawals and dye concentration estimates as frequently as five times a second. The figure shows a model of the situation, where the pulsing of the heart acts to mix the dye and the blood.

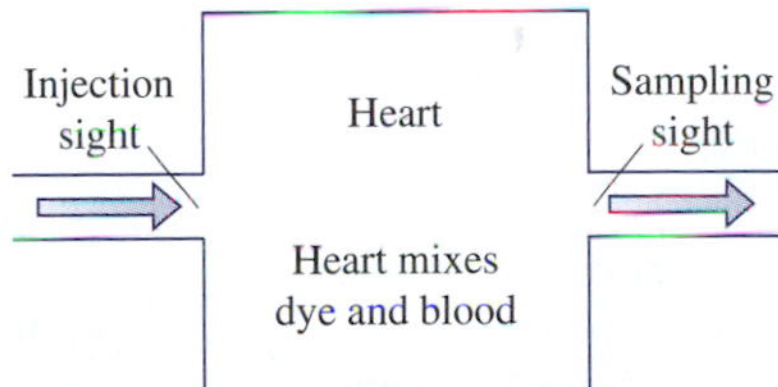

A dye must be selected that is removed rapidly from the circulatory system so as to minimize the effects of recirculation. One such dye is indocyanine, which is removed by the liver.

Let F measure cardiac output in liters per second, and let $c(t)$ be the concentration of dye in the blood in milligrams per liter (measured by the densitometer). Then the rate at which dye leaves the heart is given by $r(t) = Fc(t)$. If the recirculation problem has been minimized, then the amount A of dye injected is also the total amount of dye leaving the heart. Then $A = \int_0^{\infty} Fc(t)\,dt$. We then solve for F as follows:

$$A = \int_0^{\infty} Fc(t)\,dt$$

$$= F\int_0^{\infty} c(t)\,dt$$

$$F = \frac{A}{\int_0^{\infty} c(t)\,dt}$$

In a matter of only 10 or 15 seconds, the concentration $c(t)$ of dye will be essentially zero. Thus, the integral $\int_0^{\infty} c(t)\,dt$ can be approximated by $\int_0^{b} c(t)\,dt$, where b is about 10 or 15. This latter integral can then be approximated by using Simpson's rule. Do this using the data in the following table.

t (sec)	0	1	2	3	4	5	6
$c(t)$ (mg/l)	0	0	1.7	5.6	9.2	8.4	5.2
t (sec)	7	8	9	10	11	12	
$c(t)$ (mg/l)	3.8	2.1	1.0	0.5	0.2	0	

8 Functions of Several Variables

We now consider functions of more than one variable. In the first section we will see that the graph of a function $z = f(x,y)$ of two variables represents a surface in three dimensions. In the second section we will see that the partial derivatives measure the rates of change of a function when moving in certain directions, just as your rate of change of height when moving in different directions from a point on the side of a mountain will differ depending on the direction in which you move. In the third and fourth sections we develop ways of finding the relative and absolute extrema of functions of several variables. In the fifth and sixth sections we extend the notions of local linearity and the integral to functions of several variables.

CASE STUDY Competitive Demand Relations in the Breakfast Cereal Industry

We consider here some results of a detailed study by Cotterill and Haller[1] of the demand relationships among a number of brands of breakfast cereals. The results were in support of Cotterill's testimony as an expert economic witness for the State of New York in *State of New York* v. *Kraft General Foods et al.* It is the first and, as of January 1998, only full-scale attempt to present in a federal district court analysis of a merger's impact by using scanner-generated brand-level data and econometric techniques to estimate brand- and category-level elasticities of demand. Keep in mind that the data in this study was obtained from Kraft by court order as part of New York's challenge of the acquisition of Nabisco Shredded Wheat by Kraft General Foods. Otherwise, such data would be extremely difficult, and most likely impossible, to obtain.

In this introductory piece we restrict ourselves to only two of the cereals considered in the Cotterill and Haller study: Shredded Wheat and Grape Nuts. Let p and q be the prices of a 16-ounce box of Shredded Wheat and a 16-ounce box of Grape Nuts, respectively. Also let x and y be the number of pounds sold of Shredded Wheat and Grape Nuts, respectively.

Cotterill and Haller were able to quantify the relationships among x, y, p, and q defined above. They

[1] Ronald W. Cotterill and Lawrence E. Haller. 1997. An economic analysis of the demand for RTE cereal: product market definition and unilateral market power effects. Research Report No. 35. Food Marketing Policy Center. University of Connecticut.

found that the following equations held approximately:

$$x = Ap^{-1.7326}q^{0.7235}$$
$$y = Bp^{1.2280}q^{-2.0711}$$

where A and B are positive constants. What happens to the demand for these cereals if the price of one of them goes up? For the answers to these questions, see Example 3 of Section 8.2.

The point of the Cotterill and Haller study was to see whether the expectations mentioned in the previous paragraph relating the price to the demand existed between pairs of different brands of breakfast cereal. Suppose it were discovered that a price increase for one brand does not lead to a decrease in demand for this brand or to an increase in demand for another brand that is believed to be a competitor. Then a case could be made by the government that anticompetitive behavior existed, in possible violation of existing law. Thus, the exact nature of the price to demand relationship are a critical factor in the government's case.

8.1 Functions of Several Variables

Functions of Several Variables
The Cartesian Coordinate System
Level Curves

APPLICATION **Revenue That Depends On Two Variables**

In one of their later studies, Cobb and Douglas[2] found that if $f(L, K)$ was an index of physical volume of manufacturing, L was an index of the average number of employed wage earners, and K was an index of the value of plants, buildings, tools, and machinery in the U.S. manufacturing sector, then $f(L, K) = 0.84L^{0.63}K^{0.30}$. What is $f(10, 20)$? What is the meaning of the exponents? For the answers, see Example 4 on page 527.

[2] Paul Douglas. 1976. The Cobb-Douglas production function one again: its history, its testing, and some new empirical values. *J. Polit. Econ.* October 1976:903–915.

Functions of Several Variables

We have become familiar with functions of *one* variable. For example, the area A of a circle is a function of its radius r and can be written as $A = A(r) = \pi r^2$. But to describe the area of a rectangle of width w and length l will require the function of *two* variables given by $A = A(w, l) = wl$. The notation $A(2, 3)$ means the area of a rectangle with width $w = 2$ and length $l = 3$, which is obtained by replacing w with 2 and replacing l with 3 in the formula for area; thus, $A(2, 3) = 2 \cdot 3 = 6$. In the same way, $A(7, 4) = 7 \cdot 4 = 28$ is the area of a rectangle with width $w = 7$ and length $l = 4$.

If a company sells pens at \$2 each and sells x pens, then the revenue is a function of *one* variable and is given by $R = R(x) = 2x$ dollars. If, in addition, the company sells y pencils at 5 cents each, then the revenue, in dollars, now becomes a function of *two* variables and is given by $R = R(x, y) = 2x + 0.05y$. Thus, if 3000 pens and 4000 pencils are sold, the revenue is

$$R(3000, 4000) = 2(3000) + 0.05(4000) = 6200$$

dollars.

DEFINITION **Functions of Two Variables**

A **function of two variables**, denoted by $z = f(x, y)$, is a rule that associates to every point (x, y) in some set D, called the **domain**, a unique number denoted by $z = f(x, y)$.

The variables x and y are the **independent variables**, and z is the **dependent variable**.

If a domain is not specified, we assume that the domain is the largest possible set for which $f(x, y)$ makes sense.

EXAMPLE 1 **Evaluating Functions**

Let $f(x, y) = 2x - 3y$, $g(x, y) = \dfrac{1}{x + y}$, and $h(x, y) = \sqrt{9 - x^2 - y^2}$. Find $f(5, 1)$, $g(2, 3)$, and $h(1, 2)$.

Solution

$$f(5, 1) = 2(5) - 3(1) = 7$$

$$g(2, 3) = \frac{1}{2+3} = \frac{1}{5}$$

$$h(1, 2) = \sqrt{9 - (1)^2 - (2)^2} = 2$$

■

EXAMPLE 2 **Finding the Domain**

Find the domain of each of the functions (a) $f(x, y) = 2x - 3y$, (b) $g(x, y) = \dfrac{1}{x+y}$, and (c) $h(x, y) = \sqrt{9 - x^2 - y^2}$.

Solution

(a) One can calculate $2x - 3y$ for any x and any y, so the domain is the entire xy-plane.

(b) One can calculate $\dfrac{1}{x+y}$ only if the denominator is not zero. This happens when $y \neq -x$. Thus, the domain of g is all points in the xy-plane except for points on the line $y = -x$. See Figure 8.1a.

(c) One can form the square root only of a nonnegative number. Thus, the domain of h is all points (x, y) for which $9 - x^2 - y^2 \geq 0$, which is the same as $x^2 + y^2 \leq 9$. This is a disk of radius 3 centered at the origin. See Figure 8.1b.

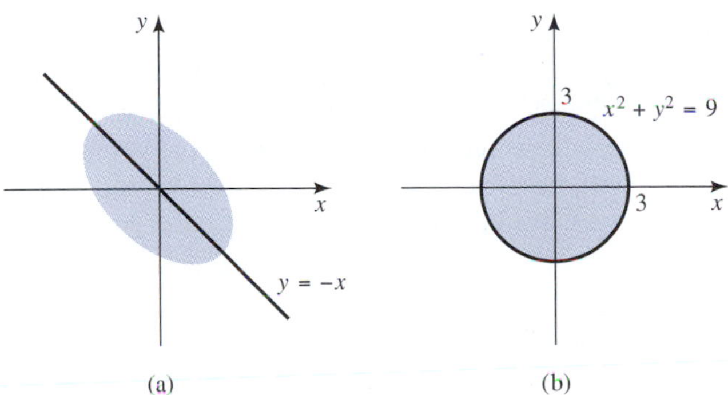

Figure 8.1
(a). The domain of $g(x, y) = 1/(x + y)$ excludes the line $y = -x$. (b). The domain of $h(x, y) = \sqrt{9 - x^2 - y^2}$ is the region inside and on the circle $x^2 + y^2 = 9$.

■

In Exercise 49 in Section 1.5 we pointed out that Cotterill and Haller[3] had discovered advertising "effectiveness" using data provided by a major breakfast cereal

[3] Ronald W. Cotterill and Lawrence E. Haller. 1997. An economic analysis of the demand for RTE cereal: product market definition and unilateral market power effects. Research Report No. 35. Food Marketing Policy Center. University of Connecticut.

producer under a court order. Naturally, a specific ad has its biggest impact immediately and a lessening impact as time goes on. An important problem is to quantify this relationship. We now consider another example.

EXAMPLE 3 How Advertising Affects Sales Of Crest Toothpaste

Schreiber[4] was able to obtain data on the annual sales and advertising budgets for Crest toothpaste. On the basis of this data, he estimated that Crest sales S in millions of dollars in the current year was approximated by the equation

$$S(x, y) = -51.78 + 5.45x + 7.27y$$

where x and y is the advertising budget in millions of dollars for the previous year and two years ago, respectively. In 1978 the advertising budget for Crest toothpaste was \$23,056,000, and in 1979 it was \$26,000,000. What does the model predict 1980 sales will be?

Solution

We have $x = 26.000$ and $y = 23.056$. Then

$$S(x, y) = S(26.000, 23.056) = -51.78 + 5.45(26.000) + 7.27(23.056) \approx 257.537$$

Thus, sales are predicted to be approximately \$257,537,000. ■

Remark From the formula in Example 3 we notice that the advertising of two years ago has a *greater* impact than that of one year ago. Where is the advertising decay mentioned in the Cotterill and Haller study on advertising decay in the paragraph preceding Example 3? The Schreiber study notes that there is normally a lead time between advertising and actual sales. For example, November and December advertisements may persuade consumers to purchase Crest but perhaps not until January or later. Schreiber concludes from his data and the formula given in Example 3 that for Crest the diminishing returns effect appears to set in after two years.

Next we consider production functions, which play an important role in business. The total output of a firm P, measured by the total units produced per year, certainly depends on labor costs L and capital investment K, with labor costs measured by dollars per year spent on wages and capital measured by dollars of capital investment per year. Assume that we explicitly know this relationship to be $P = f(K, L)$. Then we call $f(K, L)$ the **production function** of the firm.

One important class of production functions are the **Cobb-Douglas production functions**. They are

$$f(L, K) = AL^bK^c$$

where A, b, and c are positive constants.

[4] Max M. Schreiber. 1982. Forecasting sales on the basis of advertising budgets: the case of Crest toothpaste. *Bus. Econ.* 17(3):43–45.

EXAMPLE 4 **Evaluating a Cobb-Douglas Production Function**

In one of their later studies, Cobb and Douglas[5] found that if $f(L, K)$ was an index of physical volume of manufacturing, L was an index of the average number of employed wage earners, and K was an index of the value of plants, buildings, tools, and machinery in the U.S. manufacturing sector, then

$$f(L, K) = 0.84L^{0.63}K^{0.30}$$

Find $f(10, 20)$. Also explain in words the meaning of the exponents.

Solution

$$f(10, 20) = 0.84(10)^{0.63}(20)^{0.30} \approx 0.84(4.266)(2.456) \approx 8.80$$

The exponent 0.30 means that a 1% increase in capital input results in approximately a 0.30% increase in output. This is true because

$$\frac{f(L, 1.01K)}{f(L, K)} = \frac{0.84L^{0.63}K^{0.30}(1.01)^{0.30}}{0.84L^{0.63}K^{0.30}} = (1.01)^{0.30} \approx 1.003$$

In the same way the exponent 0.63 means that a 1% increase in labor input results in approximately a 0.63% increase in output. ■

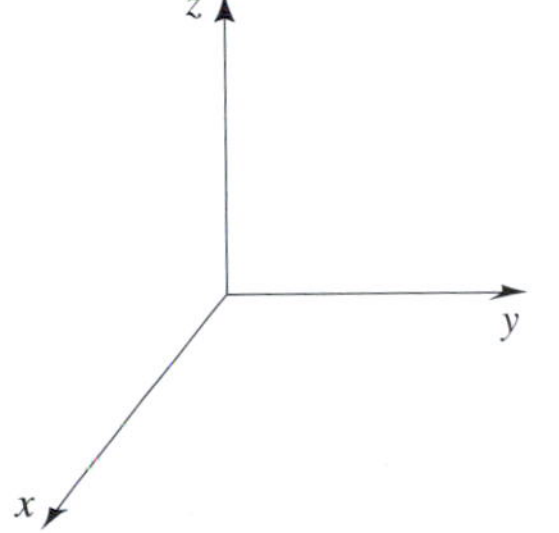

Figure 8.2
The three-dimensional coordinate system.

The Cartesian Coordinate System

We are already familiar with a two-dimensional (Cartesian) coordinate system. This is necessary if we are to graph a function $y = f(x)$ of *one* variable, since we need to plot the pair of points $(x, y) = (x, f(x))$ for all x in the domain of f. To graph a function $z = f(x, y)$ of *two* variables, we will need a three-dimensional (Cartesian) coordinate system.

A point in three-dimensional space can be uniquely represented by an ordered triple of numbers (x, y, z), and every ordered triple (x, y, z) can uniquely represent a point in three-dimensional space. The coordinate system is shown in Figure 8.2. The three axes shown in Figure 8.2—the x-axis, the y-axis, and the z-axis—are all perpendicular to each other.

EXAMPLE 5 **Using the Cartesian Coordinate System**

Locate $(1, -2, 5)$ in a three-dimensional Cartesian coordinate system.

Solution

To find the point $(1, -2, 5)$, start at the origin $(0, 0, 0)$ and move one unit in the positive x-direction along the x-axis, then move two units in the direction of the negative y-axis, then move five units vertically in the positive z-axis. See Figure 8.3. ■

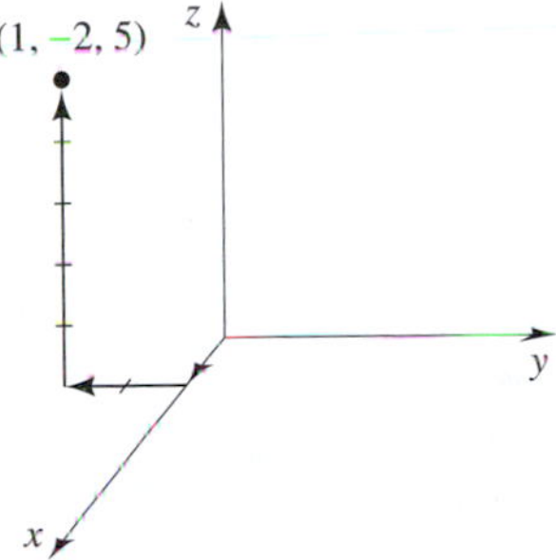

Figure 8.3
To get to $(1, -2, 5)$, go one unit in the x-direction, then two units in the negative y-direction, and then five units in the z-direction.

[5] Paul Douglas. 1976. The Cobb-Douglas production function once again: its history, its testing, and some new empirical values. *J. Polit. Econ.* October 1976:903–915.

We would like to find a formula for the distance d between the two points $P(x_1, y_1, z_1)$ and $Q(x_2, y_2, z_2)$ shown in Figure 8.4. We construct a rectangular box so that the line from P to Q forms its diagonal, as shown in Figure 8.4. The bottom of the box lies in a plane, and so the distance s can be found from the Pythagorean theorem to be

$$s = \sqrt{(x_2 - x_1)^2 + (y_2 - y_1)^2}$$

Now the triangle through the three points R, P, and Q forms a right triangle, and again using the Pythagorean theorem, we have that

$$\begin{aligned} d^2 &= s^2 + (z_2 - z_1)^2 \\ &= (x_2 - x_1)^2 + (y_2 - y_1)^2 + (z_2 - z_1)^2 \end{aligned}$$

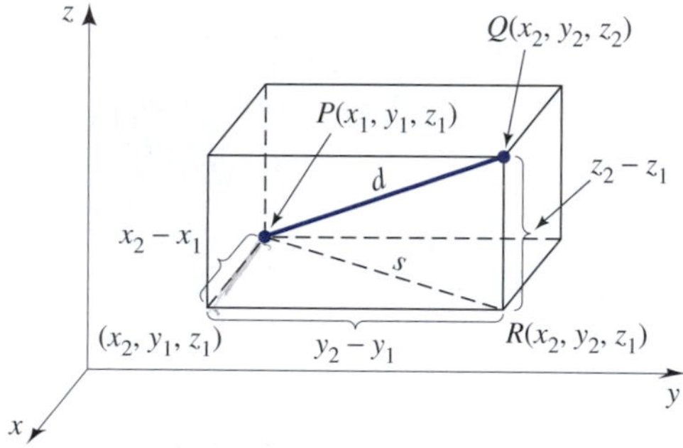

Figure 8.4
The line segment PQ is the diagonal of the box.

Thus, we have the following.

The Distance Formula

The distance between the two points (x_1, y_1, z_1) and (x_2, y_2, z_2) is

$$d = \sqrt{(x_2 - x_1)^2 + (y_2 - y_1)^2 + (z_2 - z_1)^2}$$

EXAMPLE 6 **Finding the Distance Between Two Points**

Find the distance between the two points $(2, 3, -2)$ and $(1, -4, -1)$ shown in Figure 8.5.

Solution

Setting $(2, 3, -2) = (x_1, y_1, z_1)$ and $(1, -4, -1) = (x_2, y_2, z_2)$ yields

$$\begin{aligned} d &= \sqrt{(x_2 - x_1)^2 + (y_2 - y_1)^2 + (z_2 - z_1)^2} \\ &= \sqrt{[(1) - (2)]^2 + [(-4) - (3)]^2 + [(-1) - (-2)]^2} \\ &= \sqrt{1 + 49 + 1} = \sqrt{51} \end{aligned}$$

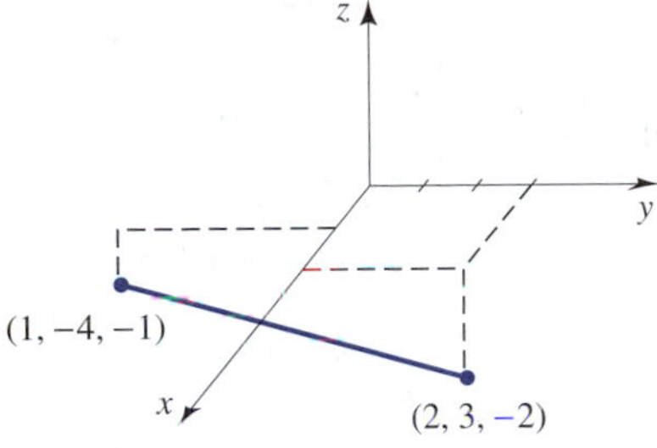

Figure 8.5
The distance from $(1, -4, -1)$ to $(2, 3, -2)$ is $\sqrt{51}$.

We now define the **graph** of a function $z = f(x, y)$. A graph of a function will enable us to see a picture of the function. See Figure 8.6.

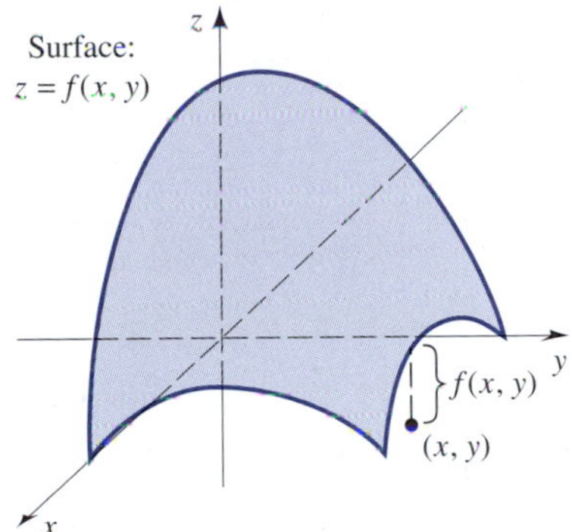

Figure 8.6
The graph of $z = f(x, y)$ is a suface in three dimensions.

DEFINITION **Graph Of a Function of Two Variables**

Given a function $z = f(x, y)$ with domain D, the **graph** of f is the set of points (x, y, z) with $z = f(x, y)$ and $(x, y) \in D$. Since this is described by *two* variables, the result is a two-dimensional object called a **surface**.

For example, if (x, y) represents a point in the state of Vermont and $z = f(x, y)$ is the height above sea level, then the graph of $z = f(x, y)$ would be the surface of Vermont, including the mountains, valleys, and plains.

EXAMPLE 7 **Graphing a Function**

Graph the function

$$z = f(x, y) = \sqrt{9 - x^2 - y^2}$$

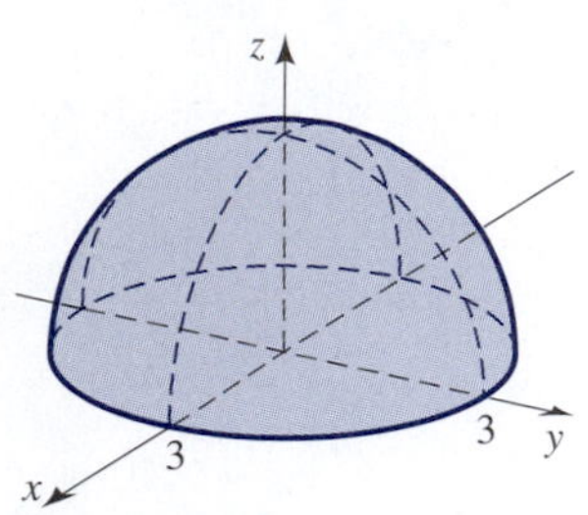

Figure 8.7
A graph of $z = f(x, y) = \sqrt{9 - x^2 - y^2}$.

Solution

In Example 2 the domain was found to be all points inside and on the circle in the xy-plane with center at (0, 0) and radius 3. To graph the function $f(x, y)$, square each side of $z = \sqrt{9 - x^2 - y^2}$ and obtain

$$z^2 = 9 - x^2 - y^2$$

which can be written as

$$x^2 + y^2 + z^2 = 9 \tag{1}$$

The left-hand side is the square of the distance from the point (x, y, z) to (0, 0, 0). Thus, a point (x, y, z) satisfies Equation (1) if and only if this point is on the sphere with center at (0, 0, 0) and radius 3. But since z must be *positive*, the graph of $f(x, y)$ is the surface given by the top half of this sphere. See Figure 8.7. ■

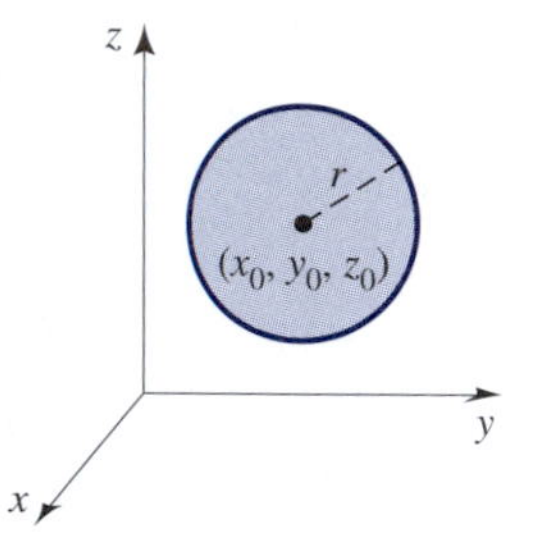

Figure 8.8
A sphere of radius r and center (x_0, y_0, z_0).

We can use the distance formula to find the equation of the sphere centered at the point (x_0, y_0, z_0) with radius r shown in Figure 8.8. The point (x, y, z) will be on this sphere if and only if the distance from the point (x, y, z) to the point (x_0, y_0, z_0) is r, that is,

$$\sqrt{(x - x_0)^2 + (y - y_0)^2 + (z - z_0)^2} = r$$

or

$$(x - x_0)^2 + (y - y_0)^2 + (z - z_0)^2 = r^2$$

Equation of Sphere

The equation of the sphere centered at (x_0, y_0, z_0) with radius r is given by

$$(x - x_0)^2 + (y - y_0)^2 + (z - z_0)^2 = r^2$$

The graph of an equation of the form $Ax + By + Cz = D$ is a plane. We do not show this here because this is beyond the scope of this text.

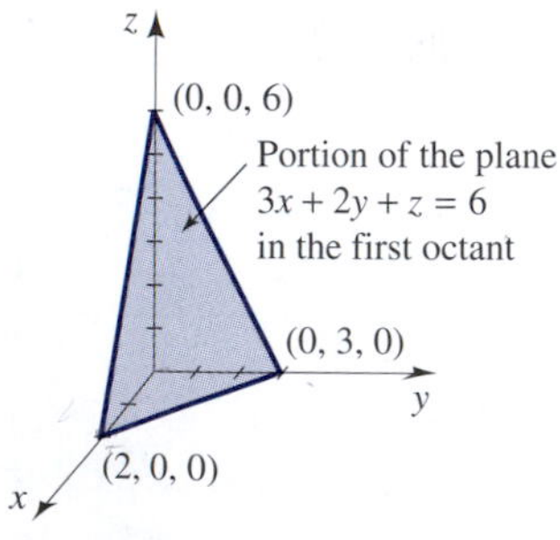

Figure 8.9
Portion of the plane $3x + 2y + z = 6$ in the first octant.

EXAMPLE 8 **Graphing a Plane**

Graph $3x + 2y + z = 6$.

Solution

Since we know the graph is a plane, we can determine the plane by locating the intercepts. Setting $y = z = 0$ yields $x = 2$. Thus, the point (2, 0, 0) is on the plane. Also setting $x = z = 0$ yields $y = 3$. Thus, (0, 3, 0) is on the plane. Finally, setting $x = y = 0$ yields $z = 6$. Thus, (0, 0, 6) is on the plane. From this we can graph the plane. See Figure 8.9. ■

Level Curves

A topographic map conveys a description of a *three*-dimensional portion of the surface of the earth by using a *two*-dimension map. This is done by using **level curves**. If we intersect a surface given by the graph of $z = f(x, y)$ with a plane given by $z = z_0$, we obtain the equation $z_0 = f(x, y)$. The graph of this equation in the xy-plane is called a level curve for f. See Figure 8.10.

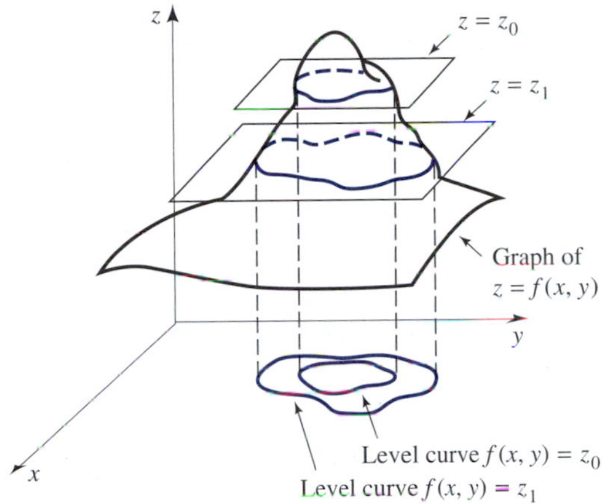

Figure 8.10
Some level curves for the given $z = f(x, y)$.

Thus, in Figure 8.11 the curve with a 3000 next to it means that the surface above this curve has the constant height 3000. The next curve shown inside this one has the number 3500 next to it. This conveys the information that the surface is *rising* as we move from the level curve $z_0 = 3000$ to the level curve $z_1 = 3500$. This indicates that we we are moving up a hill, as shown in Figure 8.11.

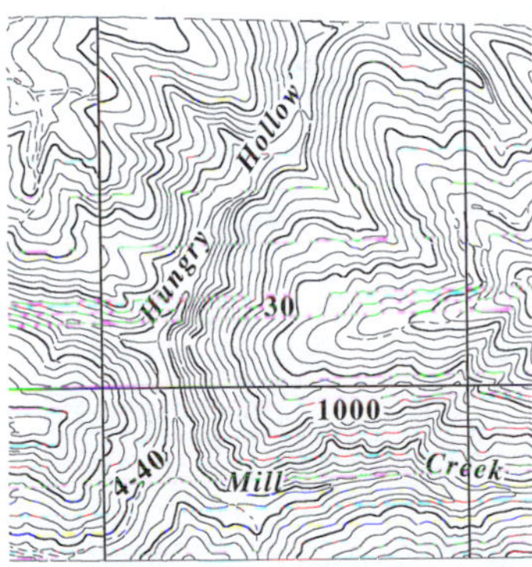

(© United States Geological Survey.)

Figure 8.12 gives a typical weather chart. Here the temperature is $T = T(x, y)$. A level curve is also referred to as an **isotherm**, that is, a curve on which the temperature is the same.

Figure 8.13 shows a typical chart of the barometric pressure. On each level curve, called an **isobar**, the barometric pressure is the same.

Example 9 shows how we can exploit level curves to determine how a surface looks.

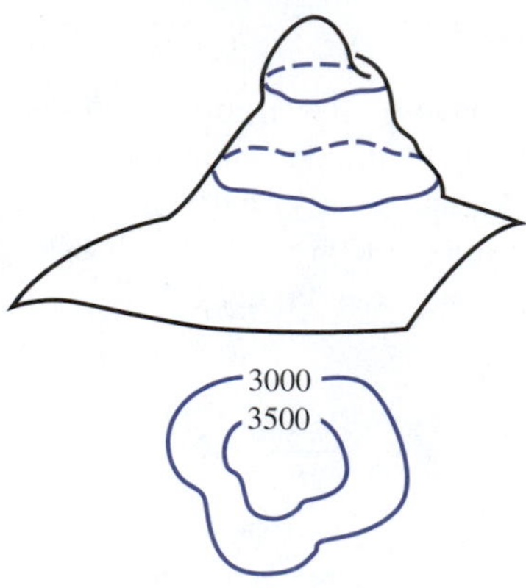

Figure 8.11
The curve with 3000 indicates the points at which the surface is 3000 units above the xy-plane.

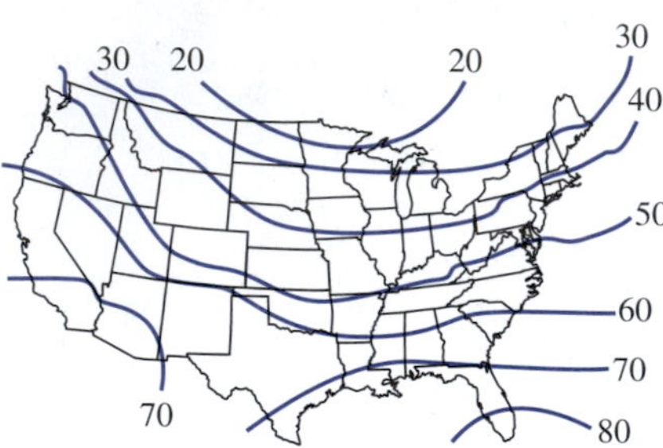

Figure 8.12
Isotherms give level curves on which the temperature is the same.

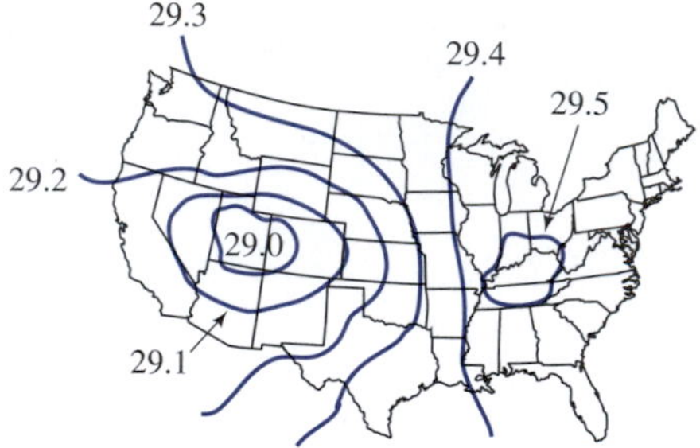

Figure 8.13
Level curves on which the barometric pressure is the same.

EXAMPLE 9 **Graphing Using Level Curves**

Graph $z = 4 - x^2 - y^2$ by using level curves.

Solution

To find the level curves, we find the intersection of each horizontal plane $z = z_0$ with the surface and obtain

$$
\begin{aligned}
z_0 &= 4 - x^2 - y^2 \\
x^2 + y^2 &= 4 - z_0
\end{aligned}
$$

Clearly, we obtain the empty set if $z_0 > 4$. This conveys the information that no part of the surface extends above the horizontal plane $z = 4$. If $z_0 = 4$, we obtain $x^2 + y^2 = 0$, which is true only if $x = 0$ and $y = 0$. Thus, the level curve for $z_0 = 4$ consists of the single point $(0, 0)$. If now $z_0 < 4$, we obtain a circle $x^2 + y^2 = 4 - z_0$ in the xy-plane with center at $(0, 0)$ and radius $\sqrt{4 - z_0}$. For example, if $z_0 = 1$, we have the level curve shown in Figure 8.14 with a "1" next to it. This indicates that along this level curve, the surface is always one unit above the xy-plane. The level curve is a circle centered at $(0, 0)$ with radius $\sqrt{3}$. The level curve for $z_0 = 0$ is also shown in Figure 8.14 and is a circle centered at $(0, 0)$ with radius 2. This conveys the information that moving outward from $(0, 0)$ in any direction results in the height of the surface decreasing. Figure 8.15 then gives a three-dimensional picture of the graph.

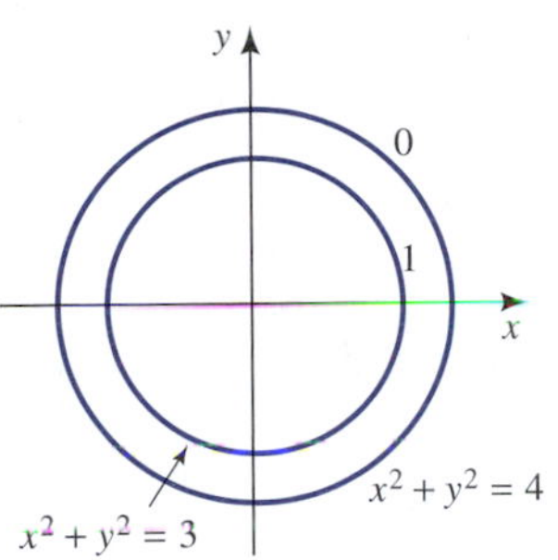

Figure 8.14
By setting $z = f(x, y) = 4 - x^2 - y^2 = 1$, we obtain the level curve $x^2 + y^2 = 3$. By setting $z = f(x, y) = 4 - x^2 - y^2 = 0$, we obtain the level curve $x^2 + y^2 = 4$.

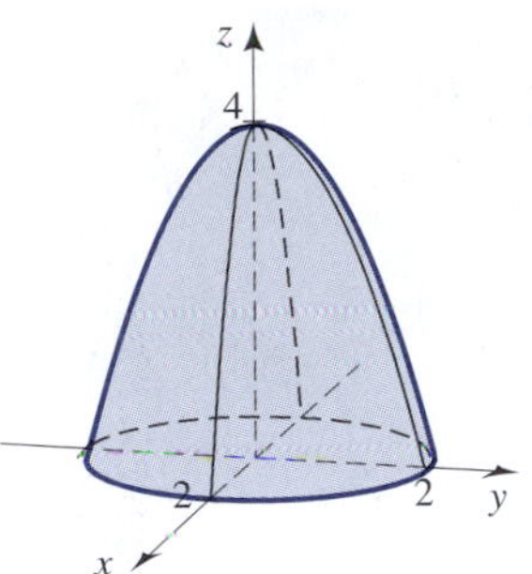

Figure 8.15
A graph of $z = 4 - x^2 - y^2$.

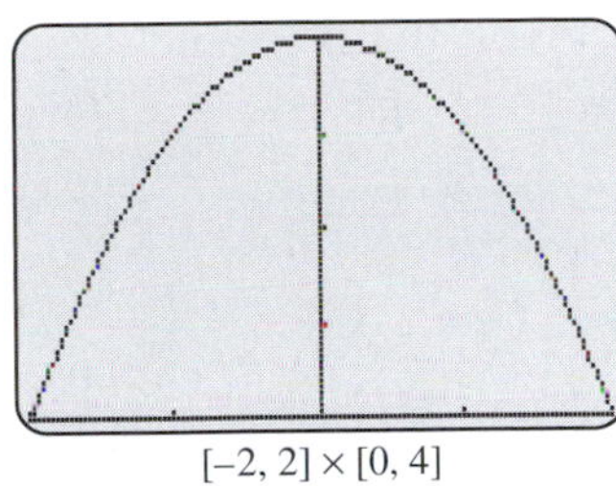

$[-2, 2] \times [0, 4]$

Screen 8.1
Shown is the graph of the level curve $y = 0$ for $z = 4 - x^2 - y^2$ given by $z = 4 - x^2$.

■

Remark We can, of course, use our graphers in a variety of ways. We can use them to find level curves, for example, or to find the graph in the various coordinate planes. For example, setting $y = 0$ in the equation of Example 9 gives $z = 4 - x^2$ and is the graph in the xz-plane. This is shown in Screen 8.1. Setting $x = 0$ gives $z = 4 - y^2$ and is the graph in the yz-plane, a graph of which is the same as shown in Screen 8.1.

Interactive Illustration 8.1, 8.2, 8.3

Level Curves

These interactive illustrations allow you to explore level curves and their relationship to the graph of a function of two variables. Each of these interactive illustrations presents a graph of a function $z = f(x, y)$. A slider is provided for z. Selecting z gives the corresponding level curve. (For now, ignore the two boxes on the right entitled linear constraint and elliptic constraint.)

1. Make z larger. Explain what happens to the level curves. Explain the connection between what is happening to the level curves and the 3-dimensional graph that you see.
2. Make z smaller. Explain what happens between the level curves. Explain the connection to what is happening to the level curves and the 3-dimensional graph that you see.

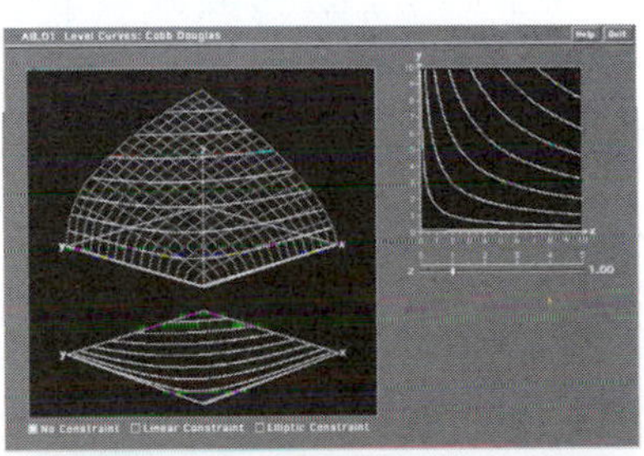

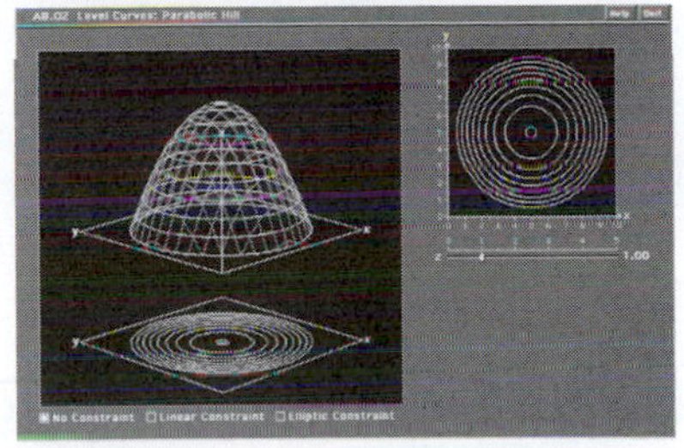

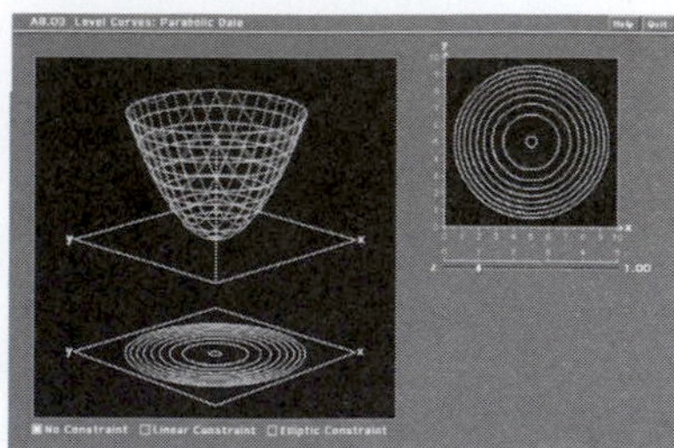

Warm Up Exercise Set 8.1

1. The box shown in the figure has a square base and open top. Find the surface area $S(x, y)$ as a function of x and y. Then find $S(5, 8)$.

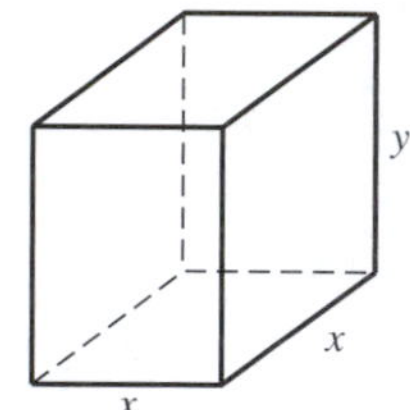

2. Plot the two points $(1, 2, -5)$ and $(0, 4, -3)$, and find the distance between them.

3. Use level curves to graph $z = x^2 + y^2 - 3$.

Exercise Set 8.1

In Exercises 1 through 10, evaluate the function at the indicated points.

1. $f(x, y) = 3x + 4y + 2$, $(1, 2)$, $(0, 2)$, $(-2, 4)$
2. $f(x, y) = 2x - 5y$, $(0, 0)$, $(-2, 4)$, $(2, -3)$
3. $f(x, y) = 2x^2 + y^2 - 4$, $(1, 2)$, $(2, -3)$, $(-1, -2)$
4. $f(x, y) = x^2 + x - y^2$, $(2, 3)$, $(-1, 4)$, $(-2, -3)$
5. $f(x, y) = x/(x + y)$, $(1, 2)$, $(-1, 2)$, $(-2, -4)$
6. $f(x, y) = y/(x + y)$, $(2, 1)$, $(-1, 2)$, $(-2, -1)$
7. $f(x, y) = \sqrt{8 - x^2 - y}$, $(1, 1)$, $(1, -2)$, $(-2, 0)$
8. $f(x, y) = \sqrt{9 + x^2 + 2y^2}$, $(0, 0)$, $(1, 2)$, $(-1, -3)$
9. $f(x, y) = 1/\sqrt{x - y}$, $(2, 1)$, $(5, 1)$, $(-1, -5)$
10. $f(x, y) = \sqrt[3]{xy}$, $(8, 1)$, $(9, 3)$, $(-1, -1)$

In Exercises 11 through 16, find the domain of the given function.

11. $f(x, y) = x + 3y$
12. $f(x, y) = x - 3y$
13. $f(x, y) = x/(x + y)$
14. $f(x, y) = y/(x - 2y)$
15. $f(x, y) = \sqrt{16 - x^2 - y^2}$
16. $f(x, y) = 1/\sqrt{16 - x^2 - y^2}$

In Exercises 17 and 18, find the distance between the given two points.

17. $(1, 2, 3)$ and $(0, 4, 5)$
18. $(3, -1, -2)$ and $(2, -3, -5)$

In Exercises 19 through 26, describe the surface.

19. $x + 2y + 3z = 12$
20. $3x - 2y + z = 6$
21. $2x + 5y + 10z = 20$
22. $y = 4$
23. $z = 3$
24. $x^2 + y^2 + z^2 = 25$
25. $x^2 + y^2 + z^2 = 36$
26. $(x - 1)^2 + (y - 2)^2 + z^2 = 4$

Exercises 27 through 32 give graphs of functions. Match the functions with the level curves shown in the figures labeled with letters from (a) to (f).

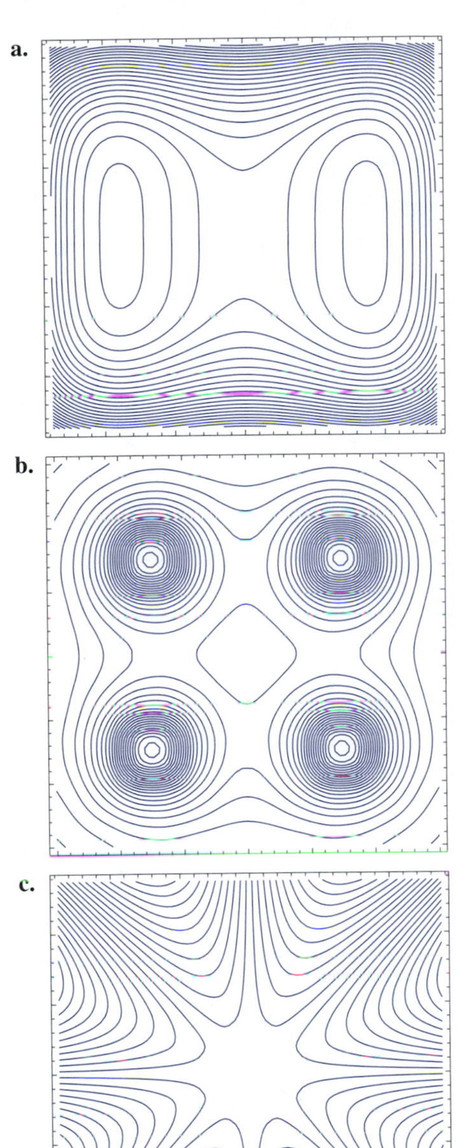
a.
b.
c.

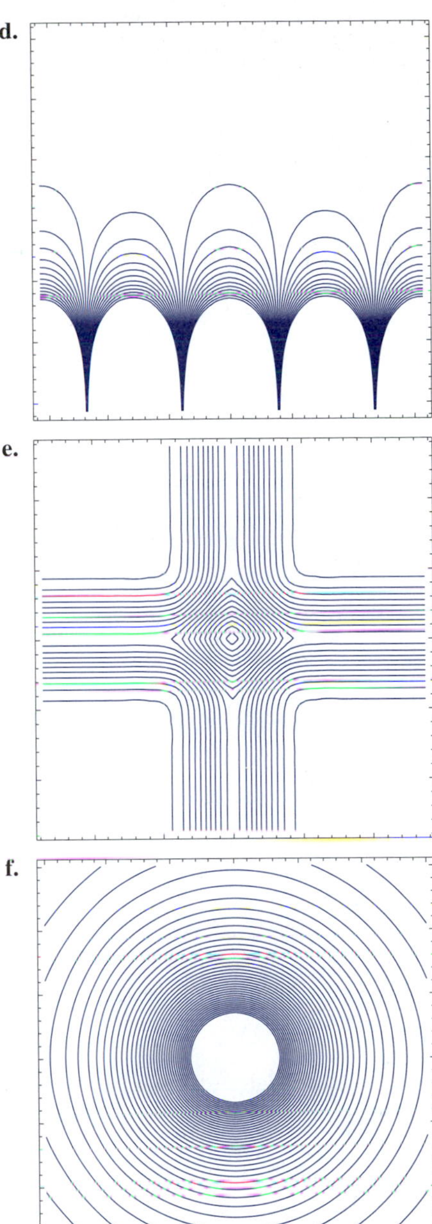
d.
e.
f.

27.

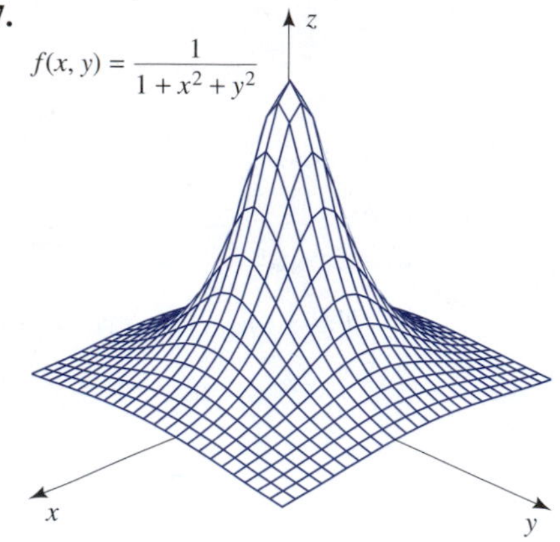

28.

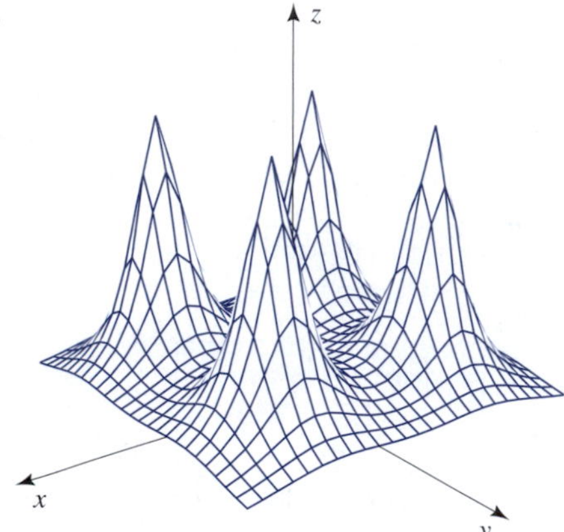

29.

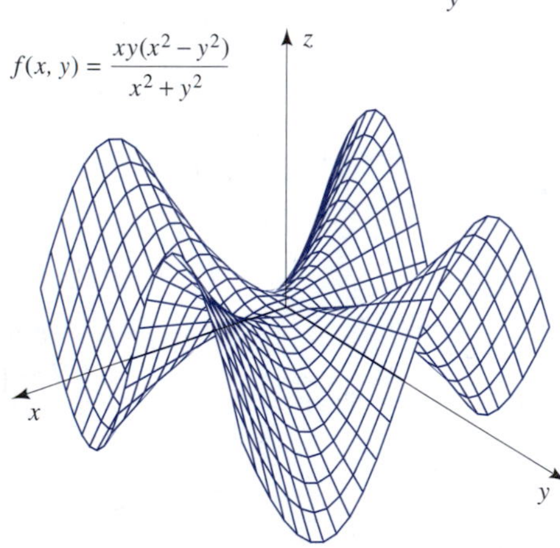

30.

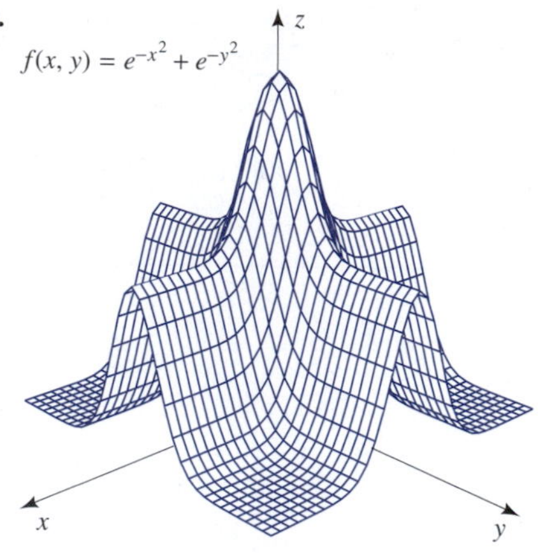

31.

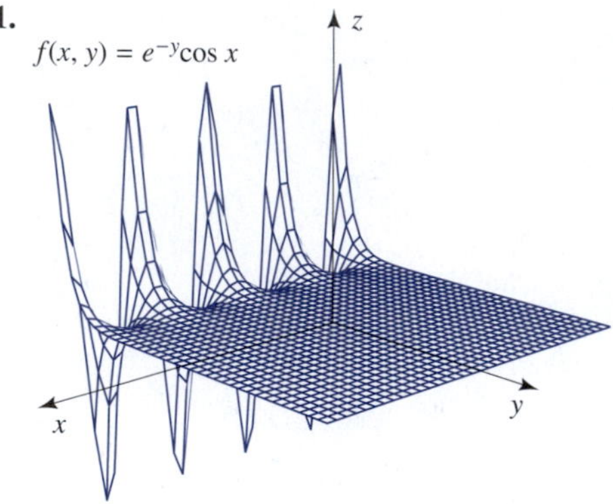

32.

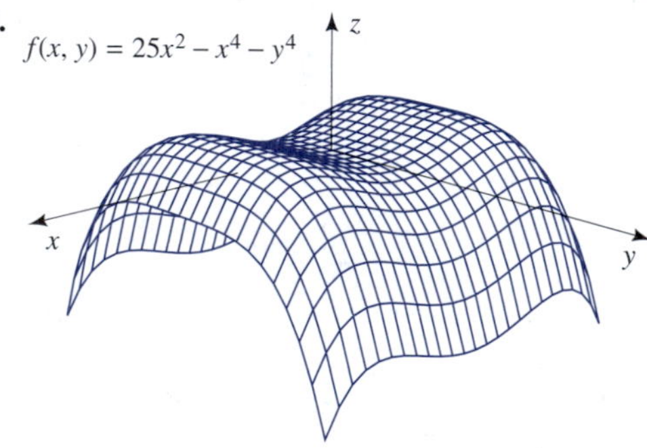

In Exercises 33 through 36, find the given level curves for the indicated functions and describe the surface.

33. $z = f(x, y) = x^2 + y^2$, $z_0 = 0$, $z_0 = 1$, $z_0 = 4$, $z_0 = 9$.

34. $z = f(x, y) = 1 + x^2 + y^2$, $z_0 = 1$, $z_0 = 2$, $z_0 = 5$, $z_0 = 10$.

35. $z = f(x, y) = 1 - x^2 - y^2$, $z_0 = 1$, $z_0 = 0$, $z_0 = -3$, $z_0 = -8$.

36. $z = f(x, y) = 2 - x^2 - y^2$, $z_0 = 2$, $z_0 = 1$, $z_0 = -2$, $z_0 = -7$.

37. Find the surface area $S(x, y)$ of the box with square base and lid as a function of x and y. Find $S(3, 5)$.

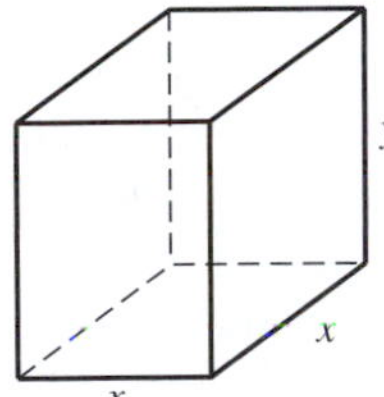

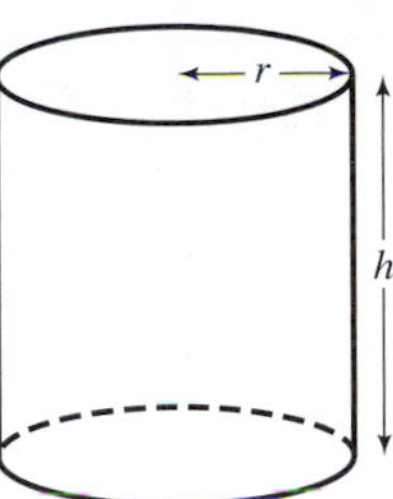

38. Find the surface area $S(r, h)$ of the cylinder as a function of r and h. Find $S(5, 10)$.

Applications and Mathematical Modeling

39. **Demand** A company sells cameras and film to be used in the camera. Let x and y be the number of cameras sold per day and the number of rolls of film sold per day, respectively, and let p and q be the respective prices. Suppose $p = 1400 - 12x - y$ and $q = 802 - 3x - 0.5y$. Find the revenue function $R(x, y)$ and $R(100, 200)$.

40. **Cost** Suppose the daily cost function $C(x, y)$ for the company in Exercise 39 is $C(x, y) = 15{,}000 + 50x + 0.5y$. Find $C(100, 200)$.

41. **Profits** Find the profit function $P(x, y)$ for the company discussed in Exercises 39 and 40. Find $P(100, 200)$.

42. **Interest** If \$1000 is compounded continuously at an annual rate of r and for t years, write the amount $A(r, t)$ as a function of r and t. Find $A(0.10, 5)$.

43. **Interest** If \$1000 is compounding m times a year at an annual rate of 8% for t years, write the amount $A(m, t)$ as a function of m and t. Find $A(12, 5)$.

44. **Cobb-Douglas Production Function** Cocks[6] determined a Cobb-Douglas production function for Eli Lilly and Company. He found that $f(L, K) = AL^{1.31939}K^{2.09583}$, where A is a positive constant, f is a measure of physical output, L is a measure of labor input, and K is a measure of physical capital input. Explain in words the meaning of the two exponents.

45. **Medicine** In the Fick method for directly measuring cardiac output,[7] the cardiac output C is given by $C = 100x/y$. Here x is the carbon dioxide, in cubic centimeters per minute, released by the lungs and y represents the change in carbon dioxide content of the blood leaving the lungs from when it entered, measured in cubic centimeters of carbon dioxide per 100 cm^3 of blood per minute.
a. What is the domain of this function?
b. Find $C(6, 3)$.
c. Find $C(1, 2)$.

46. **Biology** The transfer of energy by convection from an animal results from a temperature difference between the animal's surface temperature and the surrounding air temperature. The convection coefficient is given by

$$h(V, D) = \frac{kV^{1/3}}{D^{2/3}}$$

where k is a constant, V is the wind velocity in centimeters per second, and D is the diameter of the animal's body in centimeters.[8]
a. What is the domain of this function?
b. Find $h(8, 27)$.
c. Find $C(64, 8)$.

47. **Medicine** The flow, Q, in cubic centimeters per second of blood from a large vessel to a small capillary has been described by $Q(d, P) = 0.25C\pi d^2\sqrt{P}$, where C is a constant, d is the diameter of the capillary, and P is the difference in pressure from the (large) vessel from that in the capillary.
a. What is the domain of this function?
b. Find $Q(4, 9)$ if $C = 2$.
c. Find $Q(2, 16)$ if $C = 2$.

48. **Biology** The following formula[9] relates the surface area A in square meters of a human to the weight w in kilograms and height h in meters: $A(w, h) = 2.02w^{0.425}h^{0.725}$. What is the domain of this function? Estimate your surface area.

---

6 Douglas L. Cocks. 1981. Company total factor productivity: refinements, production functions, and certain effects of regulation. *Bus. Econ.* 16(3):5–14.

7 Duane J. Clow and N. Scott Urquhart. 1974. *Mathematics in Biology*. New York: Ardsley House, page 387.

8 Ibid., page 412.

9 Ibid., page 414.

More Challenging Exercises

49. Continuity We say that $f(x, y)$ is continuous at (a, b) if $f(x, y)$ approaches $f(a, b)$ no matter how (x, y) approaches (a, b). Let

$$f(x, y) = \frac{x^2 y}{x^4 + y^2}$$

if $(x, y) \neq (0, 0)$ and define $f(0, 0) = 0$. Let (x, y) approach $(0, 0)$ along the straight line $y = mx$ and see what happens to $f(x, y)$. In fact, show that $\lim_{x \to 0} f(x, mx) = 0$, no matter what m is. Can you conclude that f is continuous at $(0, 0)$? If you answered yes, be sure to do the next exercise.

50. Continuity Let f be the function in Exercise 49. Now let (x, y) approach $(0, 0)$ along the curves $y = mx^2$ and see what happens to $f(x, y)$. That is, find $\lim_{x \to 0} f(x, mx^2)$. Do you obtain the same limit for every value of m? What do you now conclude about the continuity of f at $(0, 0)$?

Solutions to WARM UP EXERCISE SET 8.1

1. The area of the base is x^2, while the area of each of the four sides is xy. Thus, the total surface area is $S(x, y) = x^2 + 4xy$. Then $S(5, 8) = 5^2 + 4(5)(8) = 185$.

2. The two points $(1, 2, -5)$ and $(0, 4, -3)$ are plotted in the figure. The distance between the two points is given by

$$\begin{aligned} d &= \sqrt{(x_2 - x_1)^2 + (y_2 - y_1)^2 + (z_2 - z_1)^2} \\ &= \sqrt{[0 - 1]^2 + [4 - 2]^2 + [(-3) - (-5)]^2} \\ &= \sqrt{1 + 4 + 4} = 3 \end{aligned}$$

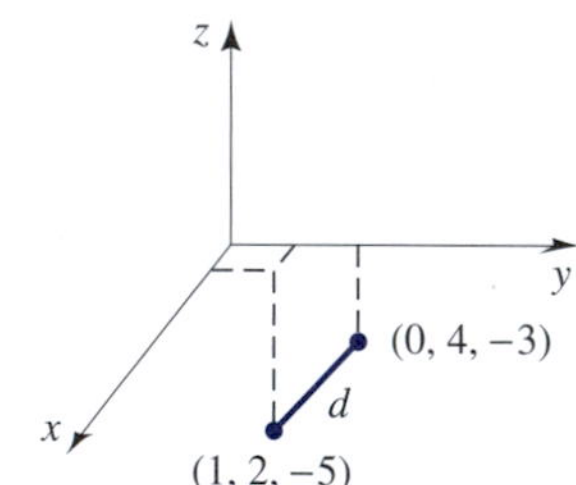

3. To find the level curves to $z = x^2 + y^2 - 3$, set $z = z_0$, $z_0 = x^2 + y^2 - 3$ or $x^2 + y^2 = z_0 + 3$. If $z_0 < -3$, this yields the empty set. If $z_0 = -3$, this gives one single point $(0, 0)$. If $z_0 > -3$, this gives a circle centered at $(0, 0)$ with radius $\sqrt{z_0 + 3}$. As z_0 increases, the radius increases. Thus, for example, if $z_0 = 1$, the level curve is a circle with radius 2 shown in the figure. If $z_0 = 6$, the level curve is a circle with radius 3 shown in the figure. Moving away from $(0, 0)$ results in z increasing.

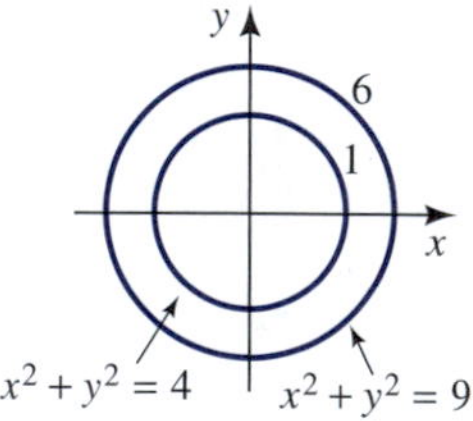

8.2 Partial Derivatives

Partial Derivatives
Competitive and Complementary Demand Relations
Second-Order Partial Derivatives

APPLICATION **Competing Commodities**

Let p and q be the price per pound of Shredded Wheat and Grape Nuts, respectively. Also let x and y be the number of pounds sold of Shredded Wheat and Grape Nuts, respectively. Cotterill and Haller[10] found that

$$x = Ap^{-1.7326}q^{0.7235}$$
$$y = Bp^{1.2280}q^{-2.0711}$$

where A and B are positive constants. Can you determine whether these two commodities compete with each other? What do the coefficients mean? For the answer, see Example 3 on page 546.

[10] Ronald W. Cotterill and Lawrence E. Haller. 1997. An economic analysis of the demand for RTE cereal: product market definition and unilateral market power effects. Research Report No. 35. Food Marketing Policy Center. University of Connecticut.

Partial Derivatives

Suppose we are given a function $z = f(x, y)$ with domain D. Given a point $(a, b) \in D$, the plane $x = a$ that passes through this point also intersects the surface as indicated in Figure 8.16. The intersection of the plane with the surface forms a curve $z = f(a, y) = g(y)$ shown in Figure 8.16. In Figure 8.17 the plane $x = a$ has been removed for better examination. Notice that with $x = a$ this curve is a function of *one* variable y. The slope of the tangent line T_y to this curve is then given simply by

$$g'(b) = \lim_{\Delta y \to 0} \frac{f(a, b + \Delta y) - f(a, b)}{\Delta y}$$

We refer to this as the **partial derivative of f with respect to y** and denote it by $\frac{\partial f}{\partial y}(a, b)$. The important practical point is that this derivative is calculated as an ordinary derivative with the x variable treated as a *constant*.

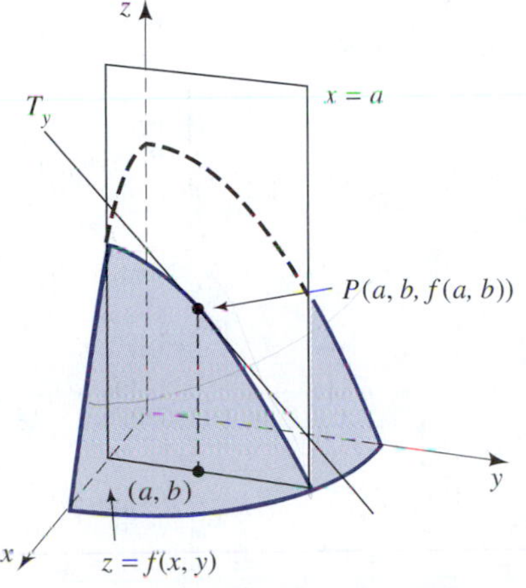

Figure 8.16

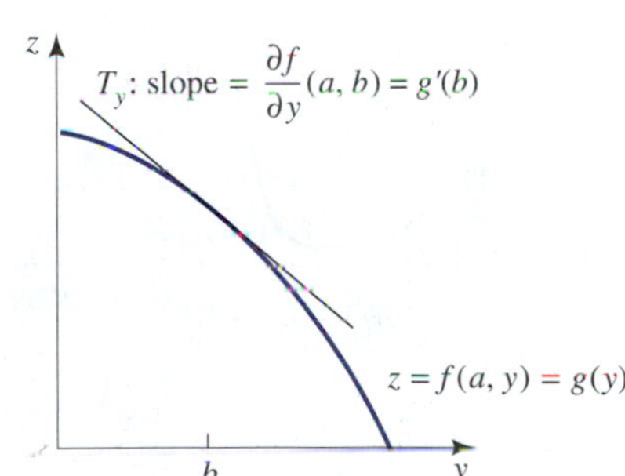

Figure 8.17
$\frac{\partial f}{\partial y}(a, b) = g'(b)$.

In the same way the plane $y = b$ intersects the surface in a curve shown in Figure 8.18. This curve is given by $z = f(x, b) = h(x)$, and the slope of the tangent line T_x to this curve is given by

$$h'(a) = \lim_{\Delta x \to 0} \frac{f(a + \Delta x, b) - f(a, b)}{\Delta x}$$

See Figure 8.19. We refer to this as the **partial derivative of f with respect to x** and denote it by $\frac{\partial f}{\partial x}(a, b)$. It is calculated as an ordinary derivative with the y variable treated as a *constant*.

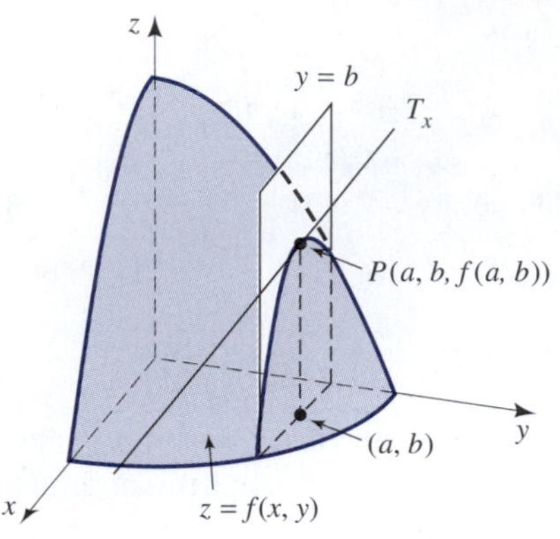

Figure 8.18

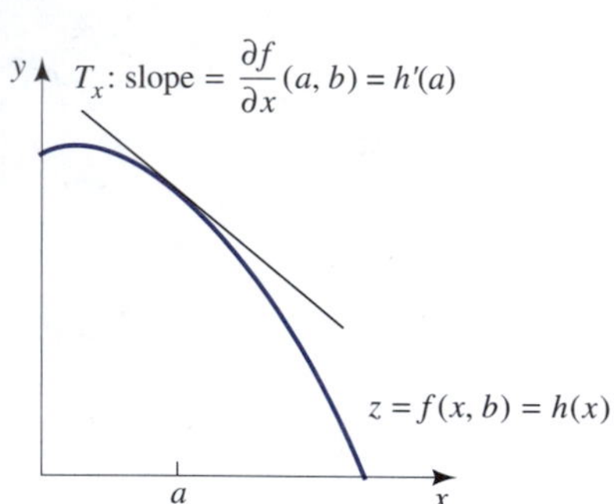

Figure 8.19
$\frac{\partial f}{\partial y}(a, b) = h'(a)$.

We then have the following definitions.

DEFINITION Partial Derivatives

If $z = f(x, y)$, then the **partial derivative of f with respect to x** is

$$\frac{\partial f}{\partial x}(x, y) = \lim_{\Delta x \to 0} \frac{f(x + \Delta x, y) - f(x, y)}{\Delta x}$$

and the **partial derivative of f with respect to y** is

$$\frac{\partial f}{\partial y}(x, y) = \lim_{\Delta y \to 0} \frac{f(x, y + \Delta y) - f(x, y)}{\Delta y}$$

if the limits exist.

Remark If $z = f(x, y)$, we also denote $\frac{\partial f}{\partial x}(x, y)$ by $\frac{\partial z}{\partial x}$, f_x, or $f_x(x, y)$ and $\frac{\partial f}{\partial y}(x, y)$ by $\frac{\partial z}{\partial y}$, f_y or $f_y(x, y)$.

If we move in the positive x-direction from the point (x, y), the partial derivative with respect to x, $\frac{\partial f}{\partial x}(x, y)$, is the instantaneous rate of change of f with respect to x. If we move in the positive y-direction from the point (x, y), the partial derivative with respect to y, $\frac{\partial f}{\partial y}(x, y)$, is the instantaneous rate of change of f with respect to y.

If we think of the surface of our function as the surface of a mountain and the point $(x, y, f(x, y))$ as a point on the surface that we are currently standing on, then as we move in the positive x-direction, $\frac{\partial f}{\partial x}(x, y)$ measures our rate of ascent if this partial derivative is positive or rate of descent if the partial derivative is negative (Figure 8.20). In the same way, as we move in the positive y-direction, $\frac{\partial f}{\partial y}(x, y)$ measures our rate of ascent if this partial derivative is positive or rate of descent if the partial derivative is negative (Figure 8.21).

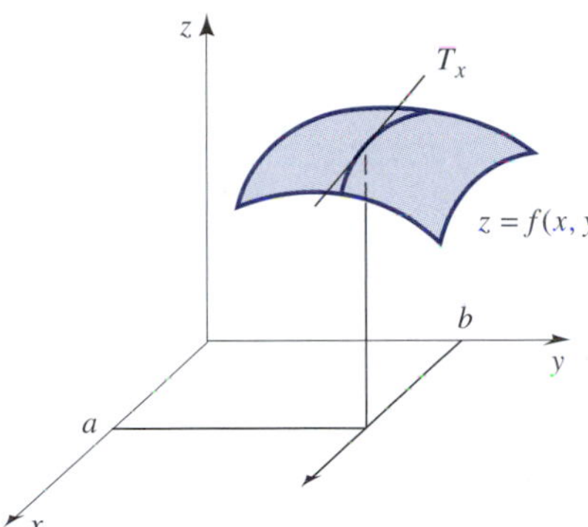

Figure 8.20
The slope of the tangent line T_x is $\frac{\partial f}{\partial x}(a, b)$.

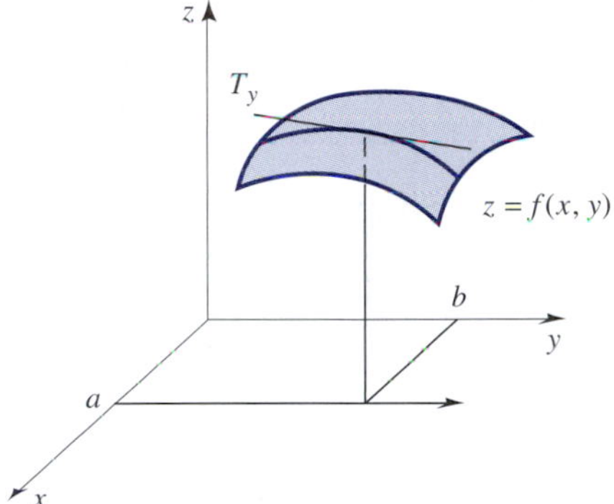

Figure 8.21
The slope of the tangent line T_y is $\frac{\partial f}{\partial y}(a, b)$.

Interactive Illustration 8.4

Partial Derivatives

This illustrative illustration allows you to explore the geometric meaning of the partial derivatives of a function of two variables. A Cobb-Douglas function $z = x^c y^{1-c}$ is graphed in the top box. In the bottom two boxes you see graphs of $z = f(a, y)$ and $z = f(x, b)$, where $(a, b) = (1, 1)$.

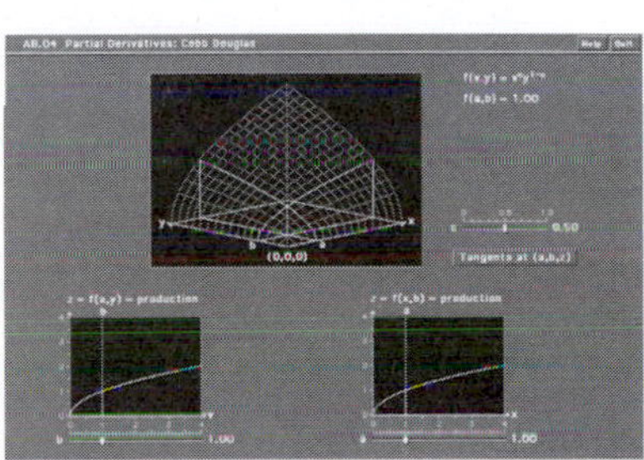

1. Set $c = 0.5$. From the figure estimate the slope of the tangent line to $z = f(a, y)$ and relate this to a partial derivative.
2. Set $c = 0.5$. From the figure estimate the slope of the tangent line to $z = f(x, b)$ and relate this to a partial derivative.
3. (a) Make c larger and then smaller and repeat parts (1) and (2).
   (b) When c increases does the slope of the tangent line to $z = f(x, b)$ increase or decrease? Verify algebraically by calculating the appropriate partial derivative.
   (c) When c increases does the slope of the tangent line to $z = f(a, y)$ increase or decrease? Verify algebraically by calculating the appropriate partial derivative.

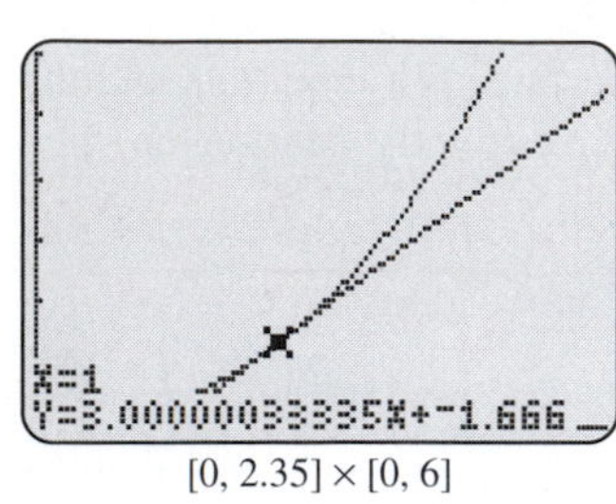

[0, 2.35] × [0, 6]

Screen 8.2
The slope of the tangent line to the curve $z = f(x, 0) = x^2 + \frac{1}{3}x^3$ gives $\frac{\partial f}{\partial x}(1, 0) \approx 3$.

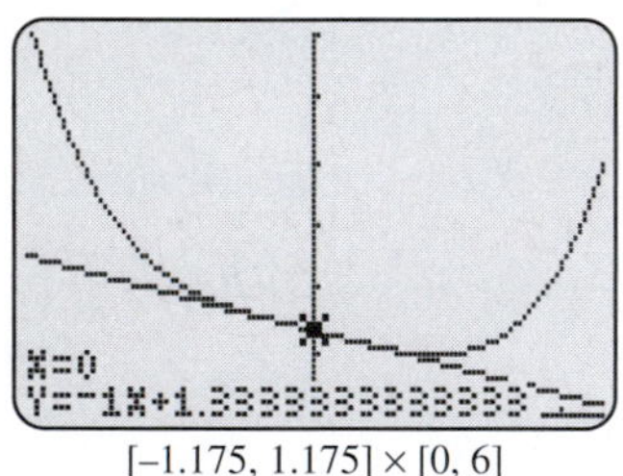

[−1.175, 1.175] × [0, 6]

Screen 8.3
The slope of the tangent line to the curve $z = f(1, y) = \frac{4}{3} + 2y^4 - y$ gives $\frac{\partial f}{\partial y}(1, 0) \approx -1$.

EXAMPLE 1 Finding Partial Derivatives

Let $f(x, y) = x^2 + 2x^3y^4 + \frac{1}{3}x^3 - y$. Find $\partial f/\partial x$ at $(1, 0)$ and $\partial f/\partial y$ at $(1, 0)$ in two ways.

(a) As the slopes of certain tangent lines.
(b) Analytically. Interpret these numbers.

Solution

(a) Refer to Figures 8.18 and 8.19. Suppose $(a, b) = (1, 0)$. Notice that

$$z = h(x) = f(x, b) = f(x, 0) = x^2 + \frac{1}{3}x^3$$

A graph is shown in Screen 8.2 together with a tangent line at $x = 1$. We see that the slope is approximately 3. Thus, $\frac{\partial f}{\partial x}(1, 0) \approx 3$.

Now refer to Figures 8.16 and 8.17. Suppose $(a, b) = (1, 0)$. Notice that

$$z = g(y) = f(a, y) = f(1, y) = \frac{4}{3} + 2y^4 - y$$

A graph is shown in Screen 8.3 together with a tangent line at $y = 0$. We see that the slope is approximately -1. Thus, $\frac{\partial f}{\partial y}(1, 0) \approx -1$.

(b) To find $\partial f/\partial x$, treat y as a *constant*, and differentiate with respect to x and obtain

$$\begin{aligned}\frac{\partial f}{\partial x}(x, y) &= \frac{\partial}{\partial x}(x^2) + \frac{\partial}{\partial x}(2y^4x^3) + \frac{\partial}{\partial x}\left(\frac{1}{3}x^3\right) - \frac{\partial}{\partial x}(y)\\ &= \frac{\partial}{\partial x}(x^2) + 2y^4\frac{\partial}{\partial x}(x^3) + \frac{1}{3}\frac{\partial}{\partial x}(x^3) - \frac{\partial}{\partial x}(y)\\ &= 2x + 2y^4 3x^2 + x^2 + 0\\ &= 2x + 6x^2y^4 + x^2\end{aligned}$$

Then

$$\frac{\partial f}{\partial x}(1, 0) = 2(1) + 6(1)^2(0)^4 + (1) = 3$$

With y kept at $y = 0$ the instantaneous rate of change of f with respect to x when $x = 1$ is 3. So as a point moves from $(1, 0)$ in the positive x-direction the function is increasing at three z-units per x-unit (Figure 8.22). To find $\partial f/\partial y$, treat x as a *constant*, and differentiate with respect to y to obtain

$$\begin{aligned}\frac{\partial f}{\partial y}(x, y) &= \frac{\partial}{\partial y}(x^2) + \frac{\partial}{\partial y}(2x^3y^4) + \frac{\partial}{\partial y}\left(\frac{1}{3}x^3\right) - \frac{\partial}{\partial y}(y)\\ &= \frac{\partial}{\partial y}(x^2) + 2x^3\frac{\partial}{\partial y}(y^4) + \frac{\partial}{\partial y}\left(\frac{1}{3}x^3\right) - \frac{\partial}{\partial y}(y)\\ &= 0 + 2x^3 4y^3 + 0 - 1\\ &= 8x^3y^3 - 1\end{aligned}$$

Then

$$\frac{\partial f}{\partial y}(1, 0) = -1$$

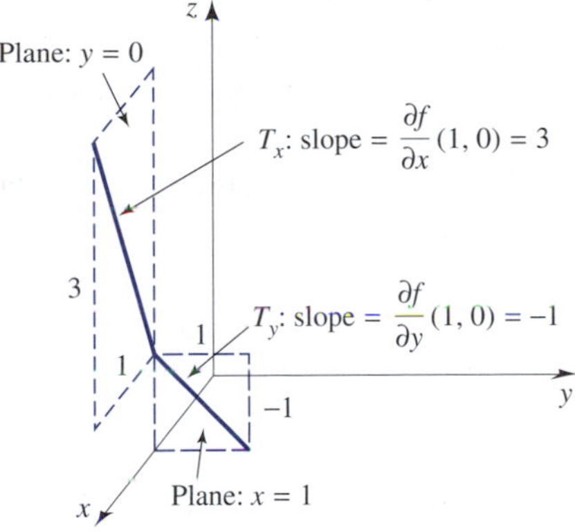

Figure 8.22
$\frac{\partial f}{\partial x}(1, 0) = 3$ means that the instantaneous rate of change of f with respect to x is 3 as we move in the positive x-direction from the point $(1, 0)$. $\frac{\partial f}{\partial y}(1, 0) = -1$ means that the instantaneous rate of change of f with respect to y is -1 as we move in the position y-direction from the point $(1, 0)$.

With x kept at $x = 1$, the instantaneous rate of change of f with respect to y when $y = 0$ is -1. So as a point moves from $(1, 0)$ in the positive y-direction, the function is decreasing at one z-unit per y-unit (Figure 8.22). ■

In Section 8.1 Cobb-Douglas production functions $f(L, K) = AL^bK^c$ were introduced. Recall that $P = f(L, K)$ is the total output of a firm, L is the cost of labor, and K is capital investment. The partial derivative $\partial f/\partial K$ is called the **marginal productivity of capital,** and the partial derivative $\partial f/\partial L$ is called the **marginal productivity of labor**.

EXAMPLE 2 Finding Marginal Productivities

In their original study, Cobb and Douglas[11] found that if $f(L, K)$ was an index of physical volume of manufacturing, L was an index of the average number of employed wage earners, and K was an index of the value of plants, buildings, tools, and machinery in the U.S. manufacturing sector from 1899 to 1922, then

$$f(L, K) = 1.01L^{0.75}K^{0.25}$$

Find the marginal productivities when $L = 16$ and $K = 81$, and interpret the result.

Solution

The marginal productivity of labor and the marginal productivity of capital are given by

$$\frac{\partial f}{\partial L}(L, K) = 0.7575L^{-0.25}K^{0.25}$$

[11] Paul Douglas. 1948. Are there laws of production? *Amer. Econ. Review* 38(1):1–41.

and

$$\frac{\partial f}{\partial K}(L, K) = 0.2525L^{0.75}K^{-0.75}$$

respectively. Substituting $L = 16$ and $K = 81$ yields

$$\frac{\partial f}{\partial L}(16, 81) = 0.7575(16)^{-0.25}(81)^{0.25} = 0.7575\left(\frac{1}{2}\right)(3) = 1.13625$$

$$\frac{\partial f}{\partial K}(16, 81) = 0.2525(16)^{0.75}(81)^{-0.75} = 0.2525(8)\left(\frac{1}{27}\right) = \frac{2.02}{27} \approx 0.07481$$

Thus, with $L = 16$ the rate of change of production with respect to L at $K = 81$ is 1.13625. With $K = 81$ the rate of change of production with respect to K at $L = 16$ is 0.07481. We notice that production increases much more rapidly with an increase in labor costs, as compared with an increase in capital spending. Putting all of a (small) sum of money into labor will then lead to a much greater increase in production than putting all of this money into increasing capital spending from this point. ■

Exploration 1

Partial Derivative at a Point

Verify the partial derivatives found in Example 2 by using the numerical derivative.

Competitive and Complementary Demand Relations

We are already familiar with a demand relation that relates the demand for a product to the price of that product. But the demand for a commodity such as home heating oil depends not only on the price of oil but also on the price of natural gas, and vice versa.

Suppose then that there are two commodities A and B such that the price of one affects the price of the other. Let the unit price of the first commodity be p, and let the unit price of the second commodity be q. We assume that their respective demands x and y are functions of p and q. Thus,

$$x = f(p, q) \qquad \text{and} \qquad y = g(p, q)$$

We then have the following definitions that generalize the idea of marginal demand from one dimension:

$$\frac{\partial x}{\partial p} = \text{marginal demand for } A \text{ with respect to } p$$

$$\frac{\partial x}{\partial q} = \text{marginal demand for } A \text{ with respect to } q$$

$$\frac{\partial y}{\partial p} = \text{marginal demand for } B \text{ with respect to } p$$

$$\frac{\partial y}{\partial q} = \text{marginal demand for } B \text{ with respect to } q$$

If the price of the second commodity is held constant, then the demand x should decrease if p increases. But this implies[12] that

$$\frac{\partial x}{\partial p} < 0$$

In the same way, if the price of the first commodity is held constant, then the demand y should decrease if q increases. But this implies that

$$\frac{\partial y}{\partial q} < 0$$

Now suppose that the two commodities A and B are home heating oil and natural gas, respectively. Then an increase in the price q of natural gas will naturally lead to an increase in demand x_A for home heating oil, and an increase in the price p of home heating oil will lead to an increase in demand x for natural gas. This implies that

$$\frac{\partial x}{\partial q} > 0 \qquad \text{and} \qquad \frac{\partial y}{\partial p} > 0$$

In this case we say that the two commodities are **competitive**.

On the other hand, suppose that the two commodities A and B are automobiles and gasoline. Then an increase in the price of one will lead to a decrease in the demand for the other. This implies that

$$\frac{\partial x}{\partial q} < 0 \qquad \text{and} \qquad \frac{\partial y}{\partial p} < 0$$

In this case we say that the two commodities are **complementary**.

Figure 8.23 shows the graphs of typical demand functions when the two commodities are competitive and when they are complementary.

DEFINITION **Competitive and Complementary Demand Relations**

Suppose we have two commodities A and B with respective unit prices p and q and respective demand equations $x = f(p, q)$ and $y = g(p, q)$. If

$$\frac{\partial x}{\partial q} > 0 \qquad \text{and} \qquad \frac{\partial y}{\partial p} > 0$$

then we say that the two commodities are **competitive**. If

$$\frac{\partial x}{\partial q} < 0 \qquad \text{and} \qquad \frac{\partial y}{\partial p} < 0$$

then we say that the two commodities are **complementary**.

[12] Mathematically, we could have the partial derivative equal to zero, but in practice this rarely happens.

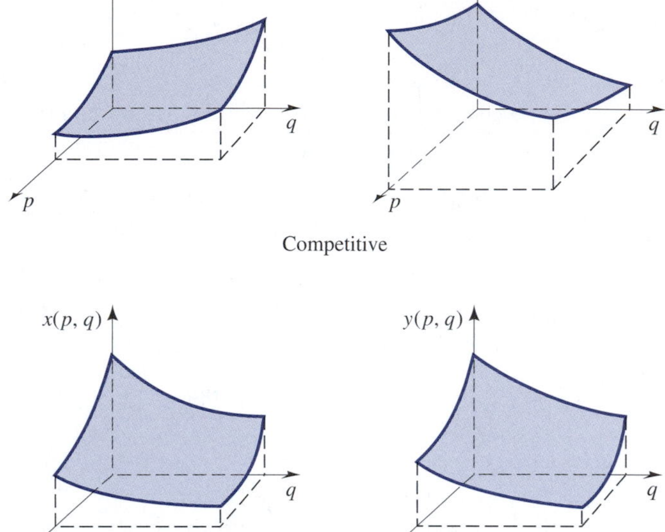

Competitive

Complementary

Figure 8.23

EXAMPLE 3 Competitive and Complementary Commodities

Let p and q be the price per pound of Shredded Wheat and Grape Nuts, respectively. Also let x and y be the number of pounds sold of Shredded Wheat and Grape Nuts, respectively. Cotterill and Haller[13] found that

$$x = Ap^{-1.7326}q^{0.7235}$$
$$y = Bp^{1.2280}q^{-2.0711}$$

where A and B are positive constants.

(a) Determine whether Shredded Wheat and Grape Nuts are competitive, complementary, or neither.

(b) Explain what all the coefficients mean.

Solution

(a) First notice that these functions are reasonable demand functions, since

$$\frac{\partial x}{\partial p} = -1.7326Ap^{-2.7326}q^{0.7235} < 0$$

and

$$\frac{\partial y}{\partial q} = -2.0711Bp^{1.2280}q^{-3.0711} < 0$$

---

[13] Ronald W. Cotterill and Lawrence E. Haller. 1997. An economic analysis of the demand for RTE cereal: product market definition and unilateral market power effects. Research Report No. 35. Food Marketing Policy Center. University of Connecticut.

Since

$$\frac{\partial x}{\partial q} = 0.7235Ap^{-1.7326}q^{-0.2765} > 0$$

and

$$\frac{\partial y}{\partial p} = 1.2280Bp^{0.2280}q^{-2.0711} > 0$$

we see that the two commodities are competitive.

(b) If the price p of Shredded Wheat is increased by 1%, then the demand for Shredded Wheat will decrease by about 1.7%, and the demand for Grape Nuts will increase by about 1.2%. If the price q of Grape Nuts is increased by 1%, then the demand for Grape Nuts will decrease by about 2.1%, and the demand for Shredded Wheat will increase by about 0.7%. ■

EXAMPLE 4 **Competitive and Complementary Commodities**

If

$$x = 50 - 2p^3 - q^2 \qquad \text{and} \qquad y = 75 - p^2 - 5q^3$$

determine whether A and B are competitive, complementary, or neither.

Solution

First notice that these functions are reasonable demand functions, since

$$\frac{\partial x}{\partial p} = -6p^2 < 0 \qquad \text{and} \qquad \frac{\partial y}{\partial q} = -15q^2 < 0$$

Since

$$\frac{\partial x}{\partial q} = -2q < 0 \qquad \text{and} \qquad \frac{\partial y}{\partial p} = -2p < 0$$

we see that the two commodities are complementary. ■

Second-Order Partial Derivatives

We know that the second derivative of a function of one variable is defined to be the derivative of the derivative. We do something very similar for partial derivatives.

DEFINITION **Second-Order Partial Derivatives**

Given a function $z = f(x, y)$, we define the following four second-order partial derivatives:

$$\frac{\partial^2 f}{\partial x^2} = \frac{\partial}{\partial x}\left(\frac{\partial f}{\partial x}\right) \qquad \frac{\partial^2 f}{\partial y^2} = \frac{\partial}{\partial y}\left(\frac{\partial f}{\partial y}\right)$$

$$\frac{\partial^2 f}{\partial y \partial x} = \frac{\partial}{\partial y}\left(\frac{\partial f}{\partial x}\right) \qquad \frac{\partial^2 f}{\partial x \partial y} = \frac{\partial}{\partial x}\left(\frac{\partial f}{\partial y}\right)$$

Remark We also use the notation

$$\frac{\partial^2 f}{\partial x^2} = f_{xx} \qquad \frac{\partial^2 f}{\partial y^2} = f_{yy}$$

$$\frac{\partial^2 f}{\partial y \partial x} = (f_x)_y = f_{xy} \qquad \frac{\partial^2 f}{\partial x \partial y} = (f_y)_x = f_{yx}$$

We will see in the next section that second partial derivatives play an important role in finding the maximum and minimum values of a function of more than one variable.

EXAMPLE 5 **Finding Second-Order Partial Derivatives**

Find all four second-order partial derivatives of $f(x, y) = e^{xy}$.

Solution

We first find the two first-order partial derivatives.

$$f_x = ye^{xy} \qquad \text{and} \qquad f_y = xe^{xy}$$

Then

$$f_{xx} = \frac{\partial}{\partial x}(f_x) = \frac{\partial}{\partial x}\left(ye^{xy}\right) = y^2 e^{xy}$$

$$\begin{aligned} f_{xy} &= \frac{\partial}{\partial y}(f_x) = \frac{\partial}{\partial y}\left(ye^{xy}\right) \\ &= y\frac{\partial}{\partial y}e^{xy} + e^{xy}\frac{\partial}{\partial y}y \\ &= xye^{xy} + e^{xy} \end{aligned}$$

$$f_{yy} = \frac{\partial}{\partial y}(f_y) = \frac{\partial}{\partial y}\left(xe^{xy}\right) = x^2 e^{xy}$$

$$\begin{aligned} f_{yx} &= \frac{\partial}{\partial x}(f_y) = \frac{\partial}{\partial x}\left(xe^{xy}\right) \\ &= x\frac{\partial}{\partial x}e^{xy} + e^{xy}\frac{\partial}{\partial x}x \\ &= xye^{xy} + e^{xy} \end{aligned}$$

■

Remark Notice that in the previous example $f_{xy} = f_{yx}$. This is no accident. *This will always happen if all second-order partial derivatives are continuous.*

So far in this chapter we have considered only functions of *two* variables. We can have functions of any number of variables. For example, $w = f(x, y, z)$ is a function of three variables. The partial of f with respect to one of the variables is obtained by taking the derivative with respect to that variable while treating the other variables as constants.

EXAMPLE 6 Finding Partial Derivatives of a Function of Three Variables

Keyser[14] studied large manufacturing corporations and found that sales S were approximated by $S(M, B, A) = 6.6273M^{0.1217}B^{0.0412}A^{0.7015}$, where M is cash and demand deposits, B is investment in financial assets, and A is other current assets. (Consult the paper for details as to why all of this should be so.) Find all three first-order partial derivatives of $S(M, B, A)$, and explain in words what one of them means.

Solution

We have

$$S_M = 6.6273(0.1217)M^{-0.8783}B^{0.0412}A^{0.7015}$$
$$S_B = 6.6273(0.0412)M^{0.1217}B^{-0.9588}A^{0.7015}$$
$$S_A = 6.6273(0.7015)M^{0.1217}B^{0.0412}A^{-0.2985}$$

For example, S_A is the rate of change of sales with respect to current assets. ■

Warm Up Exercise Set 8.2

1. If $f(x, y) = x^2y^3$, find the two first-order partial derivatives, evaluate them at (1, 1), and interpret this result.

2. Find all second-order partial derivatives of the function in Exercise 1.

Exercise Set 8.2

In Exercises 1 through 20, find both first-order partial derivatives. Then evaluate each partial derivative at the indicated point.

1. $f(x, y) = x^2 + y^2$, (1, 3)

2. $f(x, y) = x^2 - y^2$, (1, 2)

3. $f(x, y) = x^2y - x^3y^2 + 10$, (1, 2)

4. $f(x, y) = xy - x^2y^3 + y^4$, (1, −1)

5. $f(x, y) = \sqrt{xy}$, (1, 1)

6. $f(x, y) = \sqrt{1 + xy}$, (0, 1)

7. $f(x, y) = \sqrt{1 + x^2y^2}$, (1, 0)

8. $f(x, y) = \sqrt{1 + x^4y^3}$, (1, 1)

9. $f(x, y) = e^{2x+3y}$, (1, 1)

10. $f(x, y) = e^{x-y^2}$, (0, 3)

11. $f(x, y) = xye^{xy}$, (1, 1)

12. $f(x, y) = xye^{-xy}$, (1, 1)

13. $f(x, y) = \ln(x + 2y)$, (1, 0)

14. $f(x, y) = \ln(x^2 - y)$, (1, e)

15. $f(x, y) = e^{xy} \ln x$, (1, 0)

16. $f(x, y) = e^{xy} \ln y$, (0, 1)

17. $f(x, y) = 1/xy$, (1, 2)

18. $f(x, y) = y/x$, (2, 1)

19. $f(x, y) = (x - y)/(x^2 + y^2)$, (2, 1)

20. $f(x, y) = (x^2 - y^2)/(x^2 + y^2)$, (2, 3)

In Exercises 21 through 30, find all four of the second-order partial derivatives. In each case, check to see whether $f_{xy} = f_{yx}$.

21. $f(x, y) = x^2y^4$

22. $f(x, y) = x^2/y^3$

23. $f(x, y) = e^{2x-3y}$

24. $f(x, y) = e^{x-y}$

25. $f(x, y) = \sqrt{xy}$

26. $f(x, y) = \ln(x + 2y)$

27. $f(x, y) = x^2e^y$

28. $f(x, y) = x \ln y$

29. $f(x, y) = \sqrt{x^2 + y^2}$

[14] L. Richard Keyser. 1980. Corporate demand for cash balances. *Bus. Econ.* 15(4):59–63.

30. $f(x, y) = xye^{xy}$

In Exercises 31 through 38, find all three first-order partial derivatives.

31. $f(x, y, z) = xyz$

32. $f(x, y, z) = e^{xyz}$

33. $f(x, y, z) = \sqrt{x^2 + y^2 + z^2}$

34. $f(x, y, z) = x^2/(x^2 + y^2)$

35. $f(x, y, z) = e^{x+2y+3z}$

36. $f(x, y, z) = xe^y + ye^z$

37. $f(x, y, z) = \ln(x + 2y + 5z)$

38. $f(x, y, z) = xy\ln(y + 2z)$

In Exercises 39 through 44, determine whether the demand relations are competitive or complementary.

39. $x = 100 - p - q$, $y = 200 - 2p - 4q$

40. $x = 100 - p + 2q$, $y = 50 + 2p - 4q$

41. $x = 20 - p^3 + q^3$, $y = 10 + p - q^3$

42. $x = e^{-pq}$, $y = e^{-p-q}$

43. $x = 10 - \dfrac{p}{q}$, $y = 10 + p - q$

44. $x = 30 - \dfrac{p}{q}$, $y = 40 + \dfrac{p}{q}$

45. Let $z = f(x, y)$. Suppose $\dfrac{\partial f}{\partial x}(1, 2) = 3$. Interpret this partial derivative as an instantaneous rate of change.

46. Let $z = f(x, y)$. Suppose $\dfrac{\partial f}{\partial y}(1, 2) = -5$. Interpret this partial derivative as an instantaneous rate of change.

Applications and Mathematical Modeling

47. Compounding If \$1000 is invested at an annual rate of r and compounded monthly, then the amount after t years is given by

$$A(r, t) = 1000\left(1 + \frac{r}{12}\right)^{12t}$$

Find $\partial A(r, t)/\partial r$. Interpret your answer.

48. Continuous Compounding If \$1000 is invested at an annual rate of r and compounded continuously, the amount after t years is given by $A(r, t) = 1000e^{rt}$. Find $\partial A(r, t)/\partial r$. Interpret your answer.

49. Fishery Let N denote the number of prey in a school of fish, let r denote the detection range[15] of the school by a predator, and let c be a constant that defines the average spacing of fish within the school. The volume of the region within which a predator can detect the school, called the *visual volume*, has been given by biologists as

$$V(r, N) = \frac{4\pi r^3}{3}\left(1 + \frac{c}{r}N^{1/3}\right)^3$$

Find (a) $\partial V(r, N)/\partial r$. (b) $\partial V(r, N)/\partial N$. Interpret your answers.

50. Biology The transfer of energy by convection from an animal[16] results from a temperature difference between the animal's surface temperature and the surrounding air temperature. The convection coefficient is given by

$$h(V, D) = \frac{kV^{1/3}}{D^{2/3}}$$

where k is a constant, V is the wind velocity in centimeters per second, and D is the diameter of the animal's body in centimeters. Find the two first-order partial derivatives, and interpret your answers.

51. Biology If p represents the total density of leaves in a tree per unit of ground area, the leaves are uniformly distributed among n layers, and r is the circular radius of the leaf, then the amount L of light penetrating all n layers[17] is given approximately by

$$L(r, n, p) = \left(1 - \frac{p\pi r^2}{n}\right)^n$$

Find the first-order partial derivatives of L with respect to r and with respect to p, and state what each of these mean.

52. Biology The concentration C of oxygen at the surface of the roots of plants is given by

$$C(Q, r, R) = A + \frac{Qr^2}{2D}\ln\frac{r}{R}$$

where A and D are constants, Q is the oxygen consumption of the roots, r is the radius of the root, and R is the radius of the root plus the moisture film.[18] Find the three first-order partial derivatives, and interpret your answers.

53. Forestry Smith and colleagues[19] estimated that the leaf area of lodgepole pine is

$$A = A(S, D) = 0.079S^{1.43}D^{-0.73}$$

[15] Colin Clark. 1976. *Mathematical Bioeconomics*. New York: John Wiley and Sons.

[16] Duane J. Clow and N. Scott Urquhart. 1974. *Mathematics in Biology*. New York: Ardsley House.

[17] Ibid.

[18] Ibid.

[19] Frederick W. Smith, D. Arthur Sampson, and James N. Long. 1991. Comparison of leaf area index estimates from tree allometrics and measured light interception. *Forest Sci.* 37:1682–1688.

where A is the leaf area, S is the cross-sectional area at breast height, and D is the distance from breast height to the center of the live crown. Find the two first-order partial derivatives of $A(S, D)$.

54. Engineering Production Function An engineering production function in the natural gas transmission industry was given by Cullen[20] as

$$Q = Q(H, d, L) = 0.33\frac{H^{0.27}d^{1.8}}{L^{0.36}}$$

where Q is the output in cubic feet of natural gas, H is the station horsepower, d is the inside diameter of the transmission line in inches, and L is the length of the pipeline. Find the three first-order partial derivatives of $Q(H, d, L)$.

[20] Jeffrey L. Cullen. 1978. Production, efficiency, and welfare in the natural gas transmission industry. *Amer. Econ. Rev.* 68(3):311–323.

55. Temperature Keleher and Rahel[21] created a mathematical model relating temperature with the latitude and altitude and given by the equation $T(l, a) = -11.468 + 2.812l - 0.007a - 0.043l^2$, where T is the mean July air temperature in degrees Celsius in the Rocky Mountain region, l is latitude in decimal degrees, and a is altitude in meters. Find $\frac{\partial T}{\partial a}$ and $\frac{\partial T}{\partial l}$. What is the sign of $\frac{\partial T}{\partial a}$? Does this make sense?

[21] Christopher J. Keleher and Frank J. Rahel. 1996. Thermal limits to salmonid distribution in the Rocky Mountain region and potential habitat loss due to global warming: a geographic information system (GIS) approach. *Trans. Amer. Fish. Soc.* 125:1–13.

More Challenging Exercises

56. Product Liability The economists Viscusi and Moore[22] used the function $C(s, L)$ to describe unit liability cost, where s is a measure of product safety and L measures the percentage of the total liability to be paid. The standard assumption is that

$$\frac{\partial C}{\partial s} < 0 \qquad \text{and} \qquad \frac{\partial C}{\partial L} > 0$$

Explain why you think these two conditions are reasonable.

[22] Skip Viscusi and Michael J. Moore. 1993. Product liability, research and development and innovation. *J. Polit. Econ.* 101(1):161–184.

57. Advertising The economists Dorfman and Steiner[23] used the function $x = x(A, P)$ to describe the quantity demanded when advertising at a level of A and P is the price of a unit of advertising. The assumption is made that

$$\frac{\partial x}{\partial A} > 0 \qquad \text{and} \qquad \frac{\partial x}{\partial P} < 0$$

Explain why you think these two conditions are reasonable.

[23] R. Dorfman and P. O. Steiner. 1954. Optimal advertising and optimal quality. *Amer. Econ. Rev.* 44:826–836.

Solutions to WARM UP EXERCISE SET 8.2

1.

$$\frac{\partial f}{\partial x}(x, y) = 2xy^3 \qquad \frac{\partial f}{\partial y}(x, y) = 3x^2y^2$$

$$\frac{\partial f}{\partial x}(1, 1) = 2(1)(1)^3 = 2 \qquad \frac{\partial f}{\partial y}(1, 1) = 3(1)^2(1)^2 = 3$$

With y kept at $y = 1$, the instantaneous rate of change of f with respect to x when $x = 1$ is 2. Thus, as a point (x, y) moves in the positive x-direction, the function f is increasing at two z-units per x-unit. Also with x kept at $x = 1$, the instantaneous rate of change of f with respect to y when $y = 1$ is 3. Thus, as a point (x, y) moves in the positive y-direction, the function f is increasing at three z-units per y-unit.

2.

$$\frac{\partial^2 f}{\partial x^2} = \frac{\partial}{\partial x}\left(\frac{\partial f}{\partial x}\right) = \frac{\partial}{\partial x}(2xy^3) = 2y^3$$

$$\frac{\partial^2 f}{\partial y \partial x} = \frac{\partial}{\partial y}\left(\frac{\partial f}{\partial x}\right) = \frac{\partial}{\partial y}(2xy^3) = 6xy^2$$

$$\frac{\partial^2 f}{\partial y^2} = \frac{\partial}{\partial y}\left(\frac{\partial f}{\partial y}\right) = \frac{\partial}{\partial y}(3x^2y^2) = 6x^2y$$

$$\frac{\partial^2 f}{\partial x \partial y} = \frac{\partial}{\partial x}\left(\frac{\partial f}{\partial y}\right) = \frac{\partial}{\partial x}(3x^2y^2) = 6xy^2$$

8.3 Extrema of Functions of Two Variables

Relative Extrema
Absolute Extrema

APPLICATION Finding an Absolute Maximum

A package mailed in the United States must have length plus girth of no more than 108 inches. What are the dimensions of a rectangular package of greatest volume that can be mailed? See Example 4 on page 558 for the answer.

Relative Extrema

For a function of *one* variable we have already defined relative and absolute extrema. In the search for these extrema we noted the need to find all critical points and found both the first and second derivatives useful in determining the extrema. For a function of *two* variables we will in this section give the corresponding definitions of relative extrema, absolute extrema, and critical point and also give an appropriate second derivative test.

DEFINITION Relative Maximum And Relative Minimum

Suppose $z = f(x, y)$ is a function defined on some domain D.

We say that f has a **relative maximum** at $(a, b) \in D$ if there exists a circle centered at (a, b) and entirely in D such that

$$f(x, y) \leq f(a, b)$$

for all points (x, y) inside this circle.

We say that f has a **relative minimum** at $(a, b) \in D$ if there exists a circle centered at (a, b) and entirely in D such that

$$f(x, y) \geq f(a, b)$$

for all points (x, y) inside this circle.

We say that the function $z = f(x, y)$ has a **relative extremum** at $(a, b) \in D$ if f has either a relative maximum or a relative minimum at (a, b).

See Figure 8.24 for examples.

For a typical domain D shown in Figure 8.25 we see that a point $(a, b) \in D$ will have a circle centered at (a, b) and entirely in D if and only if the point (a, b) is not on the boundary of D. This is true in general and simply means that a relative extremum cannot occur at a boundary point of D. This is analogous to the one-variable

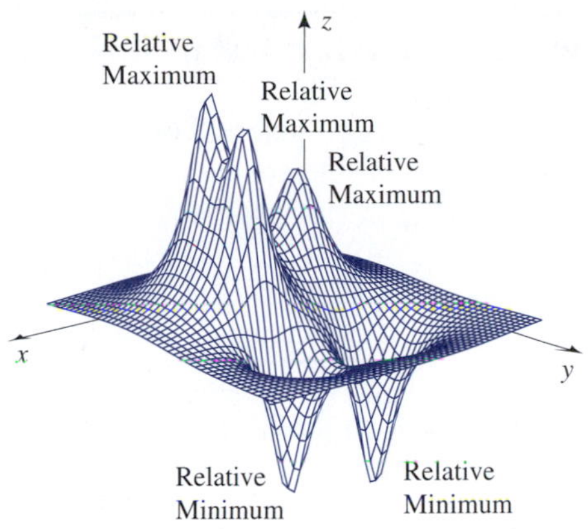

Figure 8.24
The relative extrema of $z = f(x, y)$.

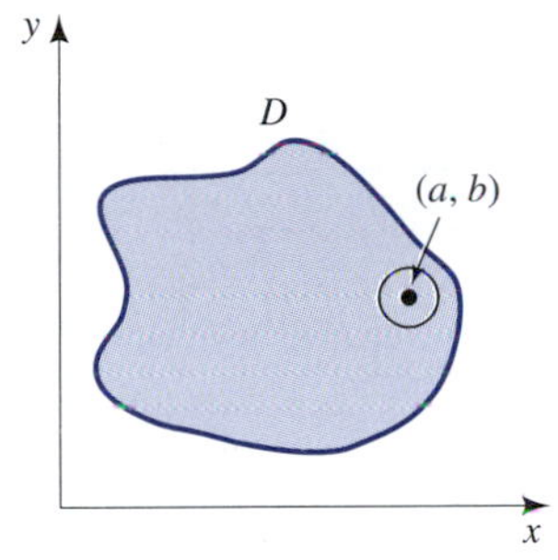

Figure 8.25
A point (a, b) is interior to D if there exists a circle centered at (a, b) and entirely inside D.

case, in which the relative extrema were by definition excluded from occurring at the endpoints (boundary) of the interval $[a, b]$.

For the remainder of this chapter *we always assume that the given function* $z = f(x, y)$ *has continuous first- and second-order partial derivatives in its domain of definition.* (We say that $f(x, y)$ is continuous at (a, b) if $f(x, y)$ approaches $f(a, b)$ no matter how the point (x, y) approaches (a, b). See Exercises 49 and 50 in Section 8.1. Geometrically, $z = f(x, y)$ will be a continuous function in a region D if the graph does not have any holes or gaps.)

Suppose we then have such a function $z = f(x, y)$ and that this function has a relative maximum at (a, b). If we intersect the surface given by the graph of $z = f(x, y)$ with the plane $y = b$ as in Figure 8.26, then $f(a, b)$ is also a relative maximum of the curve $z = g(x) = f(x, b)$ of intersection. This implies from our knowledge of *one* variable that the slope m_x of the tangent line at this point must be *zero*. But the slope of this tangent line is the partial derivative with respect to x. Thus, $f_x(a, b) = 0$. In the same way we see from Figure 8.26 that at (a, b) the slope of the tangent line to the curve of intersection of the surface with the plane $x = a$ is also zero and is just $f_y(a, b)$. Thus, $f_y(a, b) = 0$. Similar remarks apply for the case of a relative minimum. We thus have the following important result.

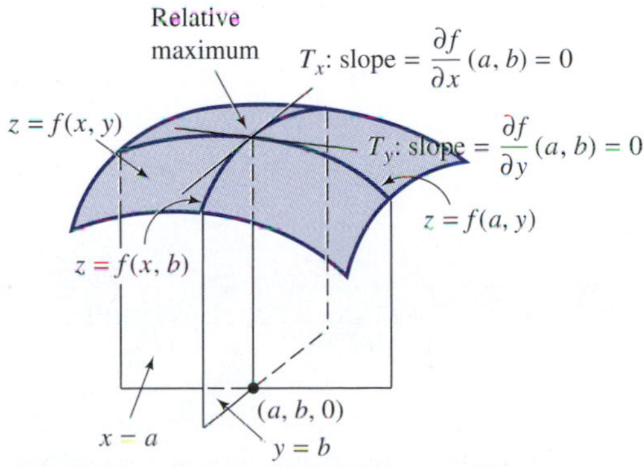

Figure 8.26
If $z = f(x, y)$ attains a relative maximum at (a, b), then $\dfrac{\partial f}{\partial x}(a, b) = 0$.

Necessary Condition for Relative Extrema

If $z = f(x, y)$ is defined, both first-order partial derivatives exist for all values of (x, y) inside some circle about (a, b), and f assumes a relative extremum at (a, b), then

$$\frac{\partial f}{\partial x}(a, b) = 0 \qquad \text{and} \qquad \frac{\partial f}{\partial y}(a, b) = 0$$

A point that is not on the boundary of the domain of f where *both* first-order partial derivatives are zero is called a **critical point**. (Recall that we are dealing only with functions whose first-order partial derivatives exist in their domains of definition.)

DEFINITION **A Critical Point**

Assume that both first-order partial derivatives of $z = f(x, y)$ exist in the domain of definition of f. **Critical points** are points (a, b), not on the boundary of the domain of f, for which

$$\frac{\partial f}{\partial x}(a, b) = 0 \qquad \text{and} \qquad \frac{\partial f}{\partial y}(a, b) = 0$$

In Figure 8.27 we see the three possibilities that can happen at a critical point. We can have a relative maximum, a relative minimum, or a **saddle point**. As Figure 8.27 indicates, at a saddle point both partial derivatives are zero, but the function does not attain a relative maximum or a relative minimum.

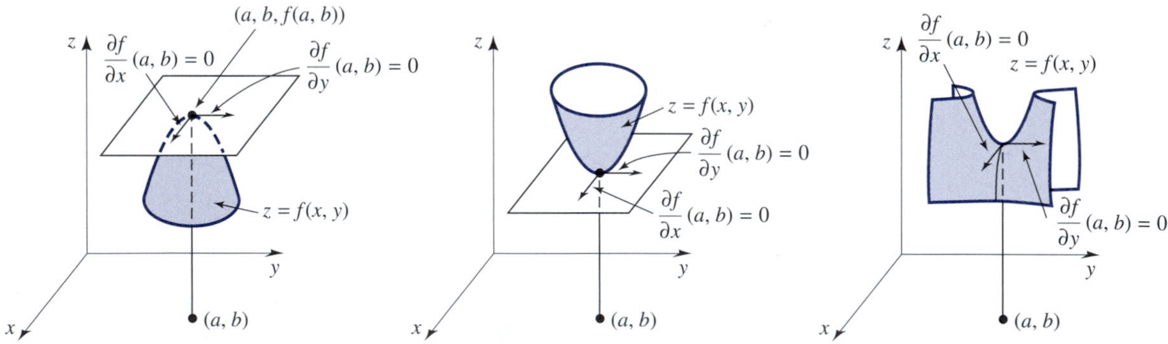

Figure 8.27
Each function has a critical point at (a, b). Function (a) has a relative maximum at (a, b), function (b) has a relative minimum, and function (c) has neither a relative maximum nor a relative minimum.

When dealing with a function of one variable, we know from the second derivative test that if the first derivative is zero at a point and the second derivative is positive at the same point, then the function has a relative minimum, with a corresponding result if the second derivative is negative. One might then suspect that if, at a critical point for a function of *two* variables, *all* the second-order partial derivatives were positive, the function would have a relative minimum. *This is not true, in general.* The following example illustrates this.

EXAMPLE 1 **Function with Positive Second Derivatives but No Extrema**

For the function $f(x, y) = x^2 + 3xy + y^2$, show that

$$f_x(0, 0) = 0, \qquad f_y(0, 0) = 0, \qquad f_{xx}(0, 0) > 0,$$
$$f_{yy}(0, 0) > 0, \qquad f_{xy}(0, 0) > 0$$

but f assumes no relative extrema at $(0, 0)$.

Solution

Since $f_x(x, y) = 2x + 3y$, $f_x(0, 0) = 0$, and since $f_y(x, y) = 3x + 2y$, $f_y(0, 0) = 0$. Also $f_{xx}(x, y) = 2 > 0$, $f_{yy}(x, y) = 2 > 0$, and $f_{xy}(x, y) = 3 > 0$. If now we move along the line $y = x$, we see that

$$f(x, x) = x^2 + 3xx + x^2 = 5x^2$$

so f increases as we move away from $(0, 0)$ along this line. But along the line $y = -x$,

$$f(x, -x) = x^2 - 3xx + x^2 = -x^2$$

so f decreases as we move away from $(0, 0)$ along this line.

In summary, moving away from the origin in one direction, we ascend, but moving away from the origin in a different direction, we descend. Thus, we have neither a minimum nor a maximum.

The second derivative test for functions of more than one variable will then be more complicated. We now state this test. ■

Second Derivative Test for Functions of Two Variables

For $z = f(x, y)$, assume that f_{xx}, f_{yy}, and f_{xy} are all continuous at every point inside a circle centered at (a, b) and that (a, b) is a critical point, that is,

$$f_x(a, b) = 0 \quad \text{and} \quad f_y(a, b) = 0$$

Define the number $\Delta(a, b)$ by

$$\Delta(a, b) = f_{xx}(a, b) f_{yy}(a, b) - \left[f_{xy}(a, b)\right]^2$$

Then

1. $\Delta(a, b) > 0$ and $f_{xx}(a, b) < 0$ implies that f has a relative maximum at (a, b).
2. $\Delta(a, b) > 0$ and $f_{xx}(a, b) > 0$ implies that f has a relative minimum at (a, b).
3. $\Delta(a, b) < 0$ implies that (a, b) is a saddle point. That is, f has neither a relative minimum nor a relative maximum at (a, b).
4. If $\Delta(a, b) = 0$, f can have a relative maximum, a relative minimum, or a saddle point.

Figure 8.28 summarizes this test.

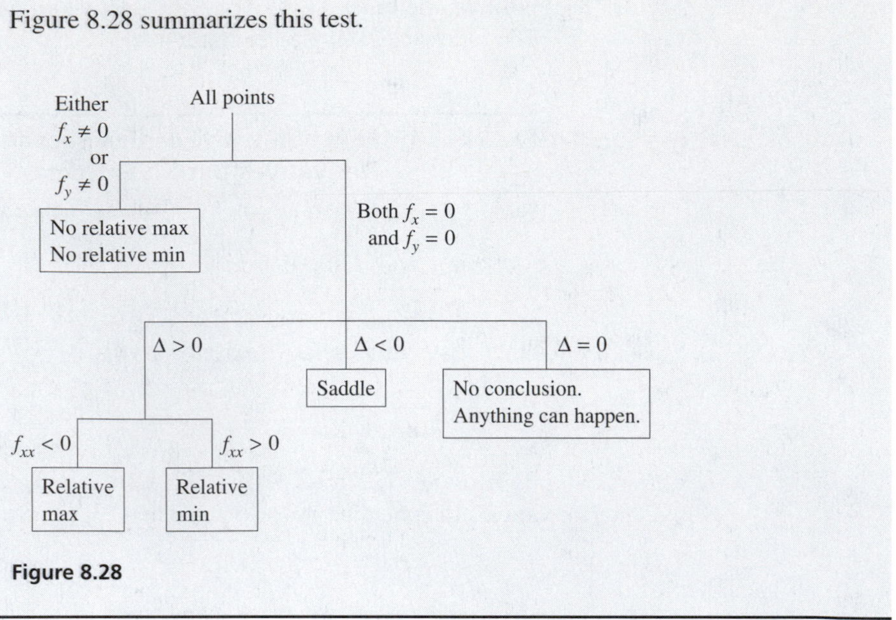

Figure 8.28

EXAMPLE 2 **Using the Second Derivative Test**

Find the relative extrema of

$$f(x, y) = -x^2 - y^2 + 2x + 4y + 5$$

Solution

We first locate all critical points by finding both first-order partial derivatives and setting them both equal to zero. Thus,

$$f_x = -2x + 2 = 0 \quad \text{and} \quad f_y = -2y + 4 = 0$$

The first equation implies that $x = 1$ and the second equation implies that $y = 2$; these are the only solutions; therefore, $(1, 2)$ is the only critical point. We note that

$$f_{xx} = -2 \qquad f_{yy} = -2 \qquad f_{xy} = 0$$

Thus,

$$\Delta(1, 2) = (-2)(-2) - (0)^2 = 4$$

Since $\Delta(1, 2) > 0$ and $f_{xx}(1, 2) < 0$, the second derivative test indicates that f has a relative maximum at $(1, 2)$. ■

EXAMPLE 3 **Using the Second Derivative Test**

Find the relative extrema of

$$f(x, y) = x^2 - 2xy + \frac{1}{3}y^3 - 3y$$

Solution

We first locate all critical points by finding both first-order partial derivatives and setting them both equal to zero. Thus,

$$f_x = 2x - 2y = 0 \qquad \text{and} \qquad f_y = -2x + y^2 - 3 = 0$$

The first equation implies that $y = x$. Substituting this into the second equation yields $-2x + x^2 - 3 = 0$, or

$$0 = x^2 - 2x - 3 = (x - 3)(x + 1)$$

This has two solutions, $x = -1$ and $x = 3$. Since $x = y$, this then indicates that the critical points are $(-1, -1)$ and $(3, 3)$. Notice that

$$f_{xx} = 2 \qquad f_{yy} = 2y \qquad f_{xy} = -2$$

Thus, for the first critical point $(-1, -1)$, $f_{yy}(-1, -1) = -2$, and

$$\Delta(-1, -1) = (2)(-2) - (-2)^2 = -8$$

Since $\Delta(-1, -1) < 0$, the second derivative test indicates that f has a saddle point at $(-1, -1)$.

For the second critical point $(3, 3)$, $f_{yy}(3, 3) = 6$, and

$$\Delta(3, 3) = (2)(6) - (-2)^2 = 8$$

Since $\Delta(3, 3) > 0$ and $f_{xx}(3, 3) > 0$, the second derivative test indicates that f has a relative minimum at $(3, 3)$. ■

Absolute Extrema

We have the following definitions for functions of two variables.

DEFINITION Absolute Maximum and Absolute Minimum

We say that f has an **absolute maximum** on D at $(a, b) \in D$ if for all $(x, y) \in D$,

$$f(x, y) \le f(a, b)$$

We say that f has an **absolute minimum** on D at $(a, b) \in D$ if for all $(x, y) \in D$,

$$f(x, y) \ge f(a, b)$$

Also we say that f has an **absolute extremum** on D at $(a, b) \in D$ if f has either an absolute maximum or an absolute minimum on D at (a, b).

When we have a differentiable function of one variable $y = f(x)$, we know that a point at which the function attains an absolute extremum on a *bounded and closed* interval $[a, b]$ can occur at the critical points on the open interval (a, b) or *can occur at the endpoints* where the derivative need not be zero. A similar situation exists for functions of two variables.

We call the interval $[a, b]$ *closed*, since it includes its endpoints. In the same way we say that a region D in the xy-plane is **closed** if it includes its boundary.

We know that a continuous function of one variable on a closed and bounded interval $[a, b]$ assumes its absolute maximum and its absolute minimum on the interval. We have a similar theorem for functions of two variables.

Existence of Absolute Extrema on a Closed and Bounded Region

If $z = f(x, y)$ is continuous on the closed and bounded region D, then f assumes its absolute extrema in D, that is, there exists two points $(a, b) \in D$ and $(A, B) \in D$ such that

$$f(a, b) \leq f(x, y) \leq f(A, B) \qquad \text{for all } (x, y) \in D$$

If (x_c, y_c) is a critical point of $f(x, y)$, then it is a candidate for a relative extremum and therefore a candidate for an absolute extremum. But also *every point on the boundary of D is a candidate for an absolute extremum.*

EXAMPLE 4 **Finding an Absolute Maximum**

A package mailed in the United States must have length plus girth of no more than 108 inches. Find the dimensions of a rectangular package of greatest volume that can be mailed.

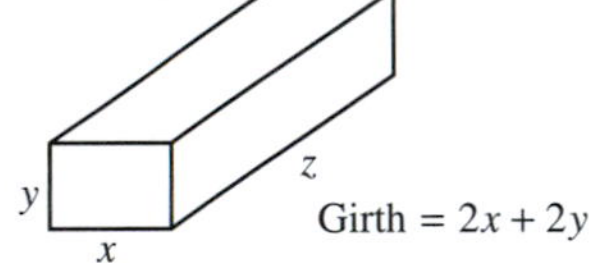

Figure 8.29

Solution

Let x, y, and z represent the dimensions of the package, with z denoting the length. See Figure 8.29. We wish to find the maximum volume which is given by $V = xyz$. As Figure 8.29 indicates, the girth is $2x + 2y$, so the condition that the girth plus the length be no more than 108 inches can be written as

$$\text{girth} + \text{length} = 2x + 2y + z \leq 108$$

Naturally, $x \geq 0$, $y \geq 0$, and $z \geq 0$. Since we wish the largest possible volume, we can assume that

$$2x + 2y + z = 108$$

Solving this equation for z and substituting into the equation for V yields

$$\begin{aligned} V &= xy(108 - 2x - 2y) \\ &= 108xy - 2x^2y - 2xy^2 \end{aligned}$$

Since $2x + 2y + z = 108$ and $z \geq 0$, we must have

$$2x + 2y \leq 108$$

Thus, we are searching for the absolute maximum of $V(x, y) = 108xy - 2x^2y - 2xy^2$ on the region D shown in Figure 8.30.

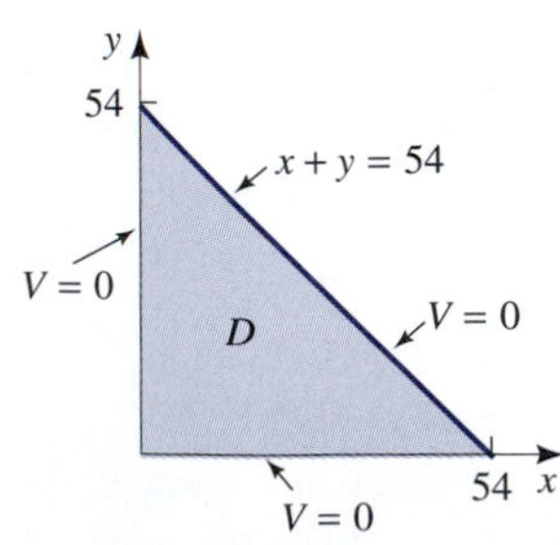

Figure 8.30

We first search for critical points inside D. Setting the two first-order partial derivatives equal to zero, we obtain

$$\begin{aligned} 0 &= V_x = 108y - 4xy - 2y^2 = y(108 - 4x - 2y) \\ 0 &= V_y = 108x - 2x^2 - 4xy = x(108 - 2x - 4y) \end{aligned}$$

The first equation implies that $y = 0$ or $108 - 4x - 2y = 0$. Since $y = 0$ is on the boundary of D, we discard this solution. The second equation implies that $x = 0$ or

$108 - 2x - 4y = 0$. Since $x = 0$ is on the boundary of D, we discard this solution. Thus, the only critical point in the interior of D is given by the solution to

$$108 - 4x - 2y = 0 \qquad \text{and} \qquad 108 - 2x - 4y = 0$$

Subtracting the second equation from twice the first yields $108 - 6x = 0$ or $x = 18$. Substituting this back into one of the two equations then yields $y = 18$. Thus, $(18, 18)$ is the only critical point of $V(x, y)$ inside the region D; therefore, it is certainly a candidate for the absolute maximum.

But all points of the boundary of D are also candidates for an absolute maximum. However, we notice that on the boundary of D either $x = 0$ or $y = 0$ or $z = 0$, and all of these give $V = 0$, that is, $V = 0$ on the boundary of D. Thus, V does not assume its absolute maximum on the boundary of D.

We conclude that V assumes its absolute maximum on D at $(18, 18)$ *without even checking to see whether $V(18, 18)$ is a relative maximum*, since the critical point $(18, 18)$ is now the only possible place in D where V could assume its absolute maximum. Finally, when $x = 18$ and $y = 18$ are substituted into the equation $2x + 2y + z = 108$, we obtain $z = 36$ for the final dimension of the box.

Let us double-check our work to see whether $V(18, 18)$ is indeed a relative maximum. We have

$$V_{xx} = -4y \qquad V_{yy} = -4x \qquad V_{xy} = 108 - 4x - 4y$$

and

$$V_{xx}(18, 18) = -72 \qquad V_{yy}(18, 18) = -72 \qquad V_{xy}(18, 18) = -36$$

Thus,

$$\Delta(18, 18) = (-72)(-72) - (-36)^2 > 0$$

and $V_{xx}(18, 18) < 0$, and so by the second derivative test V has a relative maximum at this point. ■

EXAMPLE 5 **Finding an Absolute Minimum**

Find the dimensions of the rectangular box of volume 8 ft^3 that minimizes the surface area.

Solution

If, as in Figure 8.31, the dimensions of the box are given by x, y, and z, the volume is $V = xyz$, and also, according to the problem, $V = 8$. Thus, $8 = xyz$. The surface area S is given by $S = 2xy + 2xz + 2yz$. Since $8 = xyz$, we have $z = 8/xy$, and

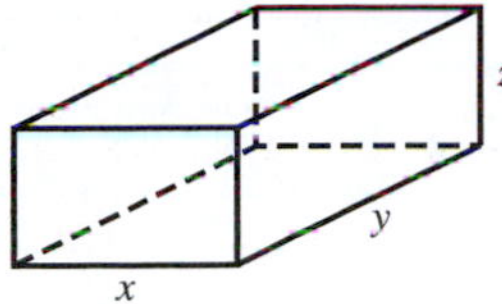

Figure 8.31

$$\begin{aligned} S &= 2xy + 2x\frac{8}{xy} + 2y\frac{8}{xy} \\ &= 2xy + \frac{16}{y} + \frac{16}{x} \end{aligned}$$

We thus wish to minimize S over the region D given by $x > 0$ and $y > 0$.

We find the critical points by taking both first-order partial derivatives and setting them equal to zero, obtaining

$$0 = S_x = 2y - \frac{16}{x^2} = \frac{2}{x^2}(x^2y - 8)$$
$$0 = S_y = 2x - \frac{16}{y^2} = \frac{2}{y^2}(xy^2 - 8)$$

The first of these equations gives $x^2y = 8$, and the second gives $xy^2 = 8$. Then $x^2y = xy^2$, and since $x > 0$ and $y > 0$, this implies that $x = y$. Then $8 = x^2y = x^3$, and so $x = 2$, and also $y = 2$. Thus, $(2, 2)$ is the only critical point of $S(x, y)$ in the region D.

If we now look carefully at the formula $S(x, y) = 2xy + 16/y + 16/x$, we see that since $x > 0$ and $y > 0$, $S(x, y)$ becomes unbounded if either x or y becomes large without bound or if either x or y goes to zero. Thus, we can visualize the surface $S(x, y)$ as extremely high near either axis and also for large x and y. Refer to Figure 8.32. We conclude from this that $S(x, y)$ cannot have a minimum for any small value of either x or y or that $S(x, y)$ cannot have a minimum for any large value of either x or y.

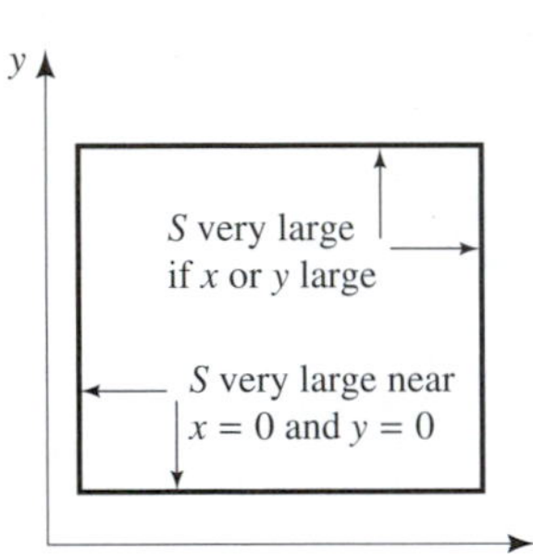

Figure 8.32

Thus, the single critical point $(2, 2)$ is the *only* candidate for a point at which S assumes an absolute minimum; therefore, S *must* assume its absolute minimum at this point. Substituting $x = 2$ and $y = 2$ into the equation $8 = xyz$ yields $z = 2$. Thus, the box must be a cube with each side equal to 2 feet.

Let us double-check our work to see whether $(2, 2)$ is indeed a relative minimum. We have

$$S_{xx} = \frac{32}{x^3} \qquad S_{yy} = \frac{32}{y^3} \qquad S_{xy} = 2$$

and

$$S_{xx}(2, 2) = 4 \qquad S_{yy}(2, 2) = 4 \qquad S_{xy}(2, 2) = 2$$

Thus,

$$\Delta(2, 2) = (4)(4) - (2)^2 = 12 > 0$$

and $S_{xx}(2, 2) > 0$, and so by the second derivative test, V has a relative minimum at this point. ■

Warm Up Exercise Set 8.3

1. Determine the relative extrema, if any, for the function $f(x) = -x^3 - y^3 - 3xy$.

2. Suppose a firm has two products X and Y that compete with each other. Let the unit prices for the products X and Y be p and q, respectively, and let the demand equations be

$$p(x, y) = 4 - x + y$$
$$q(x, y) = 8 + x - 2y$$

where x and y are the number of items of X and Y, respectively, produced and sold. Find where the revenue attains a relative maximum.

Exercise Set 8.3

In Exercises 1 through 20, find all critical points, and determine whether each point is a relative minimum, relative maximum, or a saddle point.

1. $f(x, y) = x^2 + xy + 2y^2 - 8x + 3y$
2. $f(x, y) = x^2 + 2y^2 - xy + 3y$
3. $f(x, y) = -x^2 + xy - y^2 + 3x + 8$
4. $f(x, y) = -5x^2 - xy - y^2 - 4x - 8y$
5. $f(x, y) = -x^2 + 2xy + 3y^2 - 8y$
6. $f(x, y) = 3x^2 - 2xy + y^2 + x$
7. $f(x, y) = -3x^2 + xy - y^2 - 4x - 3y$
8. $f(x, y) = -x^2 - xy - 3y^2 + 4x + 2y$
9. $f(x, y) = 2x^2 + xy + y^2 + 4$
10. $f(x, y) = 2x^2 - xy + y^2 - 8x + 7$
11. $f(x, y) = -x^2 + 2xy + 2y^2 - 4x - 8y$
12. $f(x, y) = -2x^2 + 2xy - y^2 + 2x + 4y$
13. $f(x, y) = x^3 + y^3 - 3xy - 1$
14. $f(x, y) = 2x^3 - 3x^2 - 12x + y^2 - 2y$
15. $f(x, y) = xy - x^3 - y^2$
16. $f(x, y) = -x^2 + y^3 + 6x - 12y$
17. $f(x, y) = 3/xy - 1/x^2y - 1/xy^2$
18. $f(x, y) = e^{xy}$
19. $f(x, y) = e^{x^2-y^2}$
20. $f(x, y) = xy + x/y^2$
21. Show that $f(x, y) = x^4y^4$ has a relative minimum at $(0, 0)$ and that $\Delta(0, 0) = 0$.
22. Show that $f(x, y) = -x^4y^4$ has a relative maximum at $(0, 0)$ and that $\Delta(0, 0) = 0$.
23. Show that $f(x, y) = x^3y^3$ has a saddle point at $(0, 0)$ and that $\Delta(0, 0) = 0$.
24. Show that if $\Delta(a, b) > 0$ and $f_{xx}(a, b) > 0$, then $f_{yy}(a, b) > 0$.

In Exercises 25 through 28, determine whether the given function is increasing or decreasing at the point (a) $(1, 0)$ as y increases, (b) $(0, 1)$ as x increases, (c) $(1, 5)$ as y increases, and (d) $(1, 5)$ as x increases.

25. $f(x, y) = x^2 + y^2 + xy + x + y$
26. $f(x, y) = -x^2 - y^2 + xy + x + y$
27. $f(x, y) = x^2 + y^2 - xy - x - y$
28. $f(x, y) = -x^2 - y^2 - xy - x - y$

Applications and Mathematical Modeling

In Exercises 29 and 30, do the following:

a. Find the single critical point in the interior of the first quadrant.

b. Show that the region over which the function is to be maximized is a triangular region and that the function is zero on the boundary of this triangle. Conclude from this alone that the function assumes its maximum at the critical point found in part (a).

c. Check your work by using the second derivative test to verify that the critical point found in part (a) is a point at which the function attains a relative maximum.

29. **Construction** An architect is attempting to fit a rectangular closet of maximal volume into a chopped-off corner of a building. She determines that the available space is as shown in the figure where the constricting plane is given by $4x + 2y + z = 12$ with the dimensions given in feet. What are the dimensions of the closet of maximum volume?

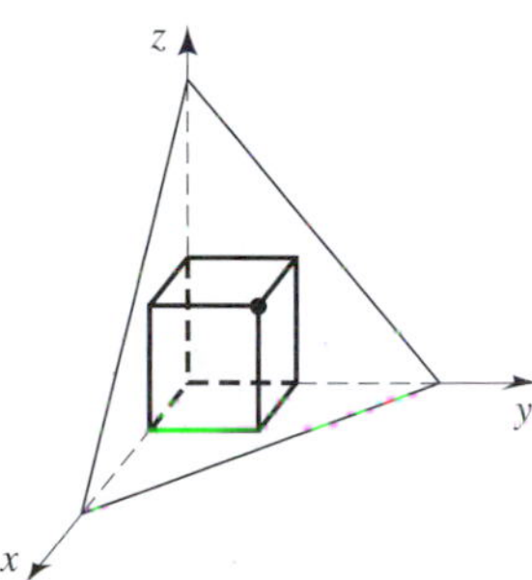

30. **Number Theory** Find three positive numbers whose sum is 48 and whose product is as large as possible.

In Exercises 31 and 32, do the following:

a. Find the single critical point in the first quadrant.

b. Show that the function to be minimized is large for large x and y and also for small x and y. Conclude from this alone that the function assumes its minimum at the critical point found in part (a).

c. Check your work by using the second derivative test to verify that the critical point found in part (a) is a point at which the function attains a relative minimum.

31. **Cost** Find the dimensions of the cheapest rectangular box with open top and volume of 96 cubic feet with the cost in square feet of the base per three times that of the cost of each side.

32. **Number Theory** Find three positive numbers whose product is 64 and whose sum is as large as possible.

In the remaining exercises (except Exercise 36) there is only one critical point in the domain under consideration. Locate this critical point, and use the second derivative test. Assume that the relative extrema are the absolute extrema.

33. **Cost** A firm has two separate plants producing the same item. Let x and y be the amounts produced in the respective plants with respective cost functions given by $C_1 = 100 - 12x + 3x^2/100$ and $C_2 = 100 - 10y + y^3/3000$. What allocation of production in the two plants will minimize the firm's cost?

34. **Profit** A firm manufactures and sells two products, X and Y, that sell for \$15 and \$10 each, respectively. The cost of producing x units of X and y units of Y is

$$C(x, y) = 400 + 7x + 4y + 0.01(3x^2 + xy + 3y^2)$$

Find the values of x and y that maximize the firm's profit.

35. **Competitive Pricing** If the demand equations are given by $p(x, y) = 4 - x + 3y$ and $q(x, y) = 8 + x - 5y$, find where the revenue attains an absolute maximum for $x \geq 0$ and $y \geq 0$.

36. **Competitive Pricing** If the demand equations are given by $p(x, y) = 4 - x + 2y$ and $q(x, y) = 8 + x - 2y$, show that the revenue does *not* attain an absolute maximum for $x \geq 0$ and $y \geq 0$. Do this by showing that along the line $y = \frac{3}{4}x$,

$$R(x, y) = R(x, \frac{3}{4}x) = 10x + \frac{1}{8}x^2$$

which is not bounded.

37. **Competitive Pricing** If the demand equations are the same as those in Exercise 35 and the cost function is given by $C(x, y) = 2x + 4y + 10{,}000$, find the point at which the profit attains a maximum for $x \geq 0$ and $y \geq 0$.

38. **Geometry** A rectangular box without a top is to have a surface area of 48 square feet. What dimensions will yield the maximum volume?

39. **Cost** A firm has three separate plants producing the same product and has a contract to sell 1000 units of its product. Each plant has a different cost function. Let x, y, and z be the number of items produced at the three plants, and assume that the respective cost functions are

$$C_1(x) = 100 + 0.03x^2 \qquad C_2(y) = 100 + 2y + 0.05y^2$$
$$C_3(z) = 100 + 12z$$

Find the allocation of production in the three plants that will minimize the firm's total cost.

40. **Facility Location** A company has three stores. The first is located 4 miles east and 2 miles north of the center of town, the second is located 2 miles east and 1 mile north of the center of town, and the third store is located 1 mile east and 6 miles south of the center of town. A warehouse is to be located convenient to these three stores. Find the location of the warehouse if the sum of the squares of the distances from each store to the warehouse is to be minimized.

More Challenging Exercises

41. **Profit** A firm sells its product in the United States and in Japan. The cost function given in dollars is $C(z) = 100{,}000 + 80z + 0.05z^2$, where z is the amount produced. Let x be the amount sold in the United States, and let y be the amount sold in Japan. Then $x \geq 0$ and $y \geq 0$. Considering the two different cultures, we assume the following two different demand equations (given in dollars):

$$p_U(x) = 500 - 0.05x \qquad \text{and} \qquad p_J(y) = 400 - 0.03y$$

Suppose that at most 3000 units of the product can be sold in each country. What are the values of x and y that will maximize the firm's profit $P(x, y)$?

42. **Profit** Suppose that in Exercise 41 the production facilities are located in the United States, and the transportation cost of moving the product to Japan is \$110 per item shipped to Japan. With this additional cost, now find the values of x and y that will maximize the firm's profit.

43. **Profit** Suppose that in Exercise 42 the transportation costs of moving the product to Japan is \$330 per item shipped to Japan. Now find the values of x and y that will maximize the firm's profit.

44. **Profit** Suppose that in Exercise 41 the production facilities limit the total number of items produced to no more than 1500. Now find the values of x and y that will maximize the firm's profit.

45. **Agriculture** Paris[24] showed that an approximate relationship between yield y of corn and the amounts of nitrogen

[45] Quirino Paris. 1992. The von Liebig hypothesis. *Amer. Agr. Econ.* 74:1019–1028.

N and phosphorus P fertilizer used is given by

$$y = -0.075 + 0.584N + 0.664P - 0.158N^2 - 0.18P^2 + 0.081PN$$

where y, N, and P are in appropriate units. Find the number of units of nitrogen and phosphorus that will maximize the corn yield according to this model. Use your grapher to find the critical point.

46. Agriculture The study mentioned in Exercise 45 indicated that another approximate relationship between yield y of corn and the amounts of nitrogen N and phosphorus P fertilizer used is also given by

$$y = -0.057 - 0.316N - 0.417P + 0.635N^{1/2} + 0.852P^{1/2} + 0.341(PN)^{1/2}$$

where y, N, and P are in appropriate units. Find the number of units of nitrogen and phosphorus that will maximize the corn yield according to this model. Use your grapher to find the critical point.

47. The Production Function Suppose a production function is in the form of a Cobb-Douglas production function $P = ax^b y^{1-b}$ and a cost function is $C = p_1 x + p_2 y$, where x is the number of items made at a cost of p_1 and y is the number of items made at a cost of p_2. Find the cost as a function of P, that is, find $C = f(P)$. Hint: Solve for x in the production function, substitute this x in the cost function. The cost function then becomes a function of y. Differentiate with respect to y, set equal to zero, and solve for y. Use this to find x, substitute into the cost equation, and finally obtain

$$C = \left(\frac{1}{b} - 1\right)^b \left(\frac{p_1}{p_2}\right)^b \left[\frac{b}{1-b} p_2 + 1\right] \frac{P}{a}$$

Solutions to WARM UP EXERCISE SET 8.3

1. Taking the first-order partial derivatives and setting them equal to zero gives

$$f_x = -3x^2 - 3y = 0 \qquad \text{and} \qquad f_y = -3y^2 - 3x = 0$$

From the first of these we obtain $y = -x^2$. Substituting this into the second then gives $3x^4 + 3x = 3x(x^3 + 1)$. Therefore, $x = 0$ or $x = -1$. The critical points are thus $(0, 0)$ and $(-1, -1)$. Also

$$f_{xx} = -6x \qquad f_{yy} = -6y \qquad f_{xy} = -3.$$

At $(0, 0)$,

$$\Delta(0, 0) = f_{xx}(0, 0) f_{yy}(0, 0) - (f_{xy}(0, 0))^2 = -9 < 0$$

and therefore, $(0, 0)$ is a saddle point.
At $(-1, -1)$,

$$\begin{aligned}\Delta(-1, -1) &= f_{xx}(-1, -1) f_{yy}(-1, -1) - (f_{xy}(-1, -1))^2 \\ &= (6)(6) - (-3)^2 = 27\end{aligned}$$

and $f_{xx}(-1, -1) = 6 > 0$. Therefore, $f(-1, -1)$ is a relative minimum.

2. The revenue $R(x, y)$ is

$$\begin{aligned}R(x, y) &= xp(x, y) + yq(x, y) \\ &= 4x + 8y + 2xy - x^2 - 2y^2\end{aligned}$$

Take

$$D = \{(x, y)\colon x \geq 0, y \geq 0\}$$

We then are seeking to maximize $R(x, y)$ over the region D.

We first look for critical points interior to D by setting the first-order partial derivatives equal to zero:

$$\begin{aligned}0 &= R_x = 4 + 2y - 2x \\ 0 &= R_y = 8 + 2x - 4y\end{aligned}$$

This yields the two linear equations

$$\begin{aligned}x - y &= 2 \\ x - 2y &= -4\end{aligned}$$

Subtracting the two equations yields $y = 6$. This gives $x = 8$.

Using the second derivative test to classify this critical point, we have

$$R_{xx} = -2 \qquad R_{yy} = -4 \qquad R_{xy} = 2$$

Thus,

$$\Delta(8, 6) = (-2)(-4) - (2)^2 = 4 > 0$$

and $R_{xx}(8, 6) < 0$, which implies that the $R(x, y)$ has a relative maximum at $(8, 6)$. (It can be shown that this is also an absolute maximum.)

8.4 Lagrange Multipliers

The Method of Lagrange Multipliers
Applications

Joseph Louis Lagrange, 1736–1813

After Leonhard Euler, Lagrange is considered the greatest mathematician of the eighteenth century. Lagrange had wide-ranging interests and had fundamental accomplishments in the theory of equations, differential equations, and number theory. He was the first great mathematician to recognize the unsatisfactory state of the foundations of analysis and calculus and to introduce more rigor into these subjects.

APPLICATION **Maximizing Production**

Suppose we have a Cobb-Douglass production function $f(x, y) = 15x^{1/3}y^{2/3}$, where x is the number of units of labor, y is the number of units of capital, and f is the number of units of a certain product that is produced. If each unit of labor costs \$200, each unit of capital costs \$100, and the total expense for both is limited to \$7,500,000, what is the number of units of labor and capital needed to maximize production? The answer can be found in Example 3 on page 568.

The Method of Lagrange Multipliers

There is no maximum volume of a rectangular box of dimensions x, y, and z, since $V(x, y, z) = xyz$ is obviously unbounded in the first octant. However, we were asked in Example 4 of Section 8.3 to maximize the volume of such a box subject to the **constraint** that the girth plus the length be 108, that is, subject to $g(x, y, z) = 2x + 2y + z - 108 = 0$. Subject to the given constraint, a maximum existed, and we found it. Consider a second example. The minimum surface area of a rectangular box is clearly zero. However, in Section 8.3 we were asked in Example 5 to minimize the surface area $S(x, y, z) = 2xy + 2xz + 2yz$ of a box subject to the *constraint* that the volume be 8, that is, $g(x, y, z) = xyz - 8 = 0$. Subject to the given constraint, a *positive* minimum existed, and we found it. Consider now a third example. The maximum of $f(x, y) = 4 - x^2 - y^2$ is clearly 4. However, the maximum of f subject to the constraint that $g(x, y) = y - 1 = 0$ is 3 and therefore different. See Figure 8.33. Notice that the maximum or minimum of a function can be very different from the maximum or minimum subject to a *constraint*.

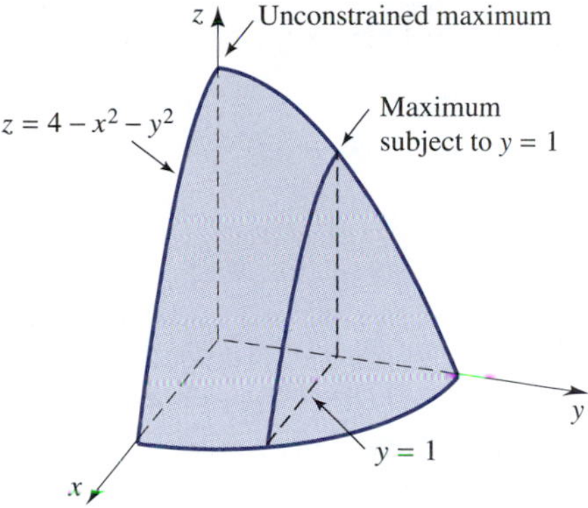

Figure 8.33
The maximum is different if subjected to the constraint $y = 1$.

These are examples of a situation that arises often. We wish to maximize or minimize some function $f(x, y, z)$ subject to a constraint $g(x, y, z) = 0$, or we wish to maximize or minimize some function of two variables $f(x, y)$ subject to a constraint $g(x, y) = 0$. When solving such problems in Section 8.3, we merely solved the constraint equation $g(x, y, z) = 0$ for z and substituted the result in $f(x, y, z)$, obtaining a function of only *two* variables, and then found the extrema. In general, this can be extremely difficult, if not impossible, to do. But even if it can be done, the resulting equation can sometimes be unnecessarily complicated. There is an alternative method called the *method of Lagrange multipliers* that avoids these difficulties. We state the method for three independent variables.

Method of Lagrange Multipliers

Candidates (x_c, y_c, z_c) for a relative extrema of $f(x, y, z)$ subject to the constraint $g(x, y, z) = 0$ can be found among the critical points (x_c, y_c, z_c, λ) of the auxiliary function

$$F(x, y, z, \lambda) = f(x, y, z) + \lambda g(x, y, z)$$

that is, among the solutions of the equations

$$F_x = 0 \qquad F_y = 0 \qquad F_z = 0 \qquad g = 0$$

It is important to realize that this method yields only candidates for the extrema. The test to determine whether the relative extrema are minima or maxima is too complicated to be given here. However, we noted in Section 8.3 that often there is only one candidate and that we can argue in some way that this candidate is the one we are seeking.

Applications

We first revisit a fence problem from an earlier chapter. We do this to gain some geometric insight into Lagrange multipliers.

EXAMPLE 1 **Using Lagrange Multipliers**

What are the dimensions of the rectangular pasture of the largest area that can be enclosed by a fence of fixed perimeter P?

Solution

Let x and y be the dimensions of the rectangular pasture. Then we are seeking to maximize the area $A(x, y) = xy$ subject to the constraint $2x + 2y = P$. Let us see whether we can solve this problem geometrically. Figure 8.34 shows several level curves for the function $A(x, y)$. Notice from the figure that the maximum cannot occur on the level curve $A(x, y) = A_1$, since there are other level curves (such as $A(x, y) = A_2$) on which $A(x, y)$ is larger and that also intersect the constraint equation given by $g(x, y) = 2x + 2y - P = 0$. Notice that *the maximum must occur at a point where a level curve is tangent to the constraint curve.* Because of the symmetry of the curves, this must be where $x = y$.

Now let us solve, using Lagrange multipliers. Form the function

$$F(x, y) = A(x, y) + \lambda g(x, y) = xy + \lambda(2x + 2y - P)$$

and solve the equations

$$\begin{aligned} F_x &= y + 2\lambda = 0 \\ F_y &= x + 2\lambda = 0 \\ g &= 2x + 2y - P = 0 \end{aligned}$$

The first two equations yield $x = y$. Substituting this into the constraint equation then yields $x = y = P/4$.

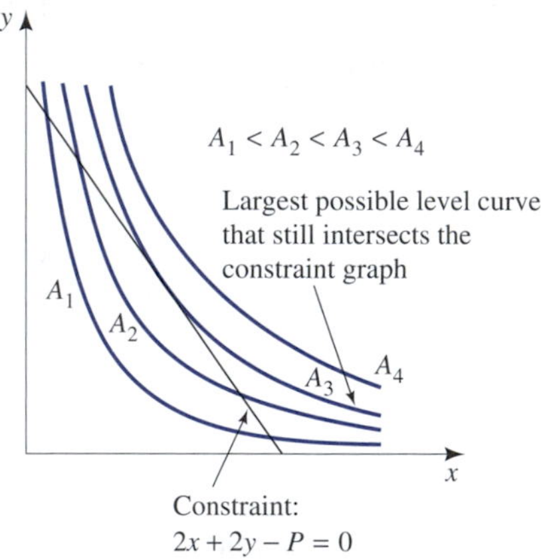

Figure 8.34
Level curves for $A(x, y) = xy$.

■

It turns out in general (though not shown here) that *the critical points of the auxiliary function F in the Lagrange multipliers method occur at points where a level curve of f is tangent to the constraint curve g.* This is much less useful when the dimension of the problem is higher because of the difficulty of visualizing the graphs involved.

Interactive Illustration 8.5

Constraints

This interactive illustration allows you to experimentally verify that the critical points of the auxiliary function F in the Lagrange multipliers method occur at points where a level curve of f is tangent to the constraint curve g. Go back to each of the Interaction Illustrations 8.1, 8.2, and 8.3.

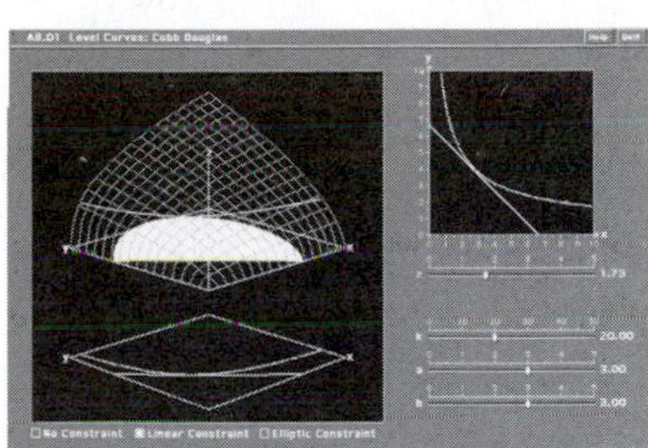

1. Click on the linear constraint box. (The linear constraint is $g(x, y) = ax + by - k$.) Now move the slider on z and observe the 3-dimensional graph. From this observation find the value of z that yields an extremum. Now check the bottom box and the top right box and note that the level cuve is tangent to the constraint curve. Use the sliders on a, b, and k to change these values and repeat.
2. Click on the elliptic constraint box. (The elliptic constraint is $g(x) = \left(\frac{x}{a}\right)^2 + \left(\frac{y}{b}\right)^2 - k$.) Now repeat (1).

Let us redo Example 4 of Section 8.3 using this new method.

EXAMPLE 2 Using Lagrange Multipliers

A package mailed in the United States must have length plus girth of no more than 108 inches. Find the dimensions of a rectangular package of greatest volume that can be mailed.

Solution

Let x, y, and z represent the dimensions of the package, with z denoting the length. See Figure 8.35. We wish to maximize the volume $V = xyz$. The condition that the girth plus the length be no more than 108 inches can be written as

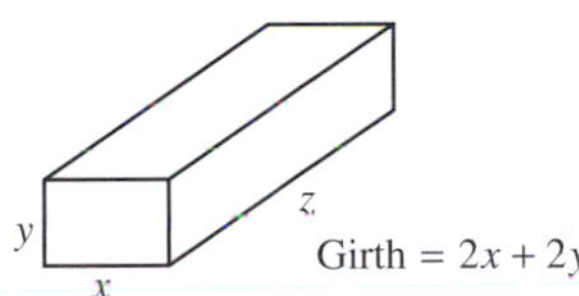

Figure 8.35

$$\text{girth} + \text{length} = 2x + 2y + z \leq 108$$

Naturally, $x \geq 0$, $y \geq 0$, and $z \geq 0$. Since we want to find the largest possible volume, we can assume that

$$2x + 2y + z = 108$$

Thus, our constraint equation is given by $g(x, y, z) = 2x + 2y + z - 108 = 0$.

We then form the auxiliary function $F(x, y, z, \lambda)$ given by

$$F(x, y, z, \lambda) = xyz + \lambda(2x + 2y + z - 108)$$

and solve the equations

$$\begin{aligned} F_x &= yz + 2\lambda = 0 \\ F_y &= xz + 2\lambda = 0 \\ F_z &= xy + \lambda = 0 \\ g(x, y, z) &= 2x + 2y + z - 108 = 0 \end{aligned}$$

Solving the first three equations for $-\lambda$ and setting them equal yields

$$-\lambda = \frac{yz}{2} = \frac{xz}{2} = xy$$

This immediately yields $x = y$, and then $z = 2y = 2x$. Substituting this into the constraint equation then yields

$$2x + 2x + 2x = 108$$

or $x = 18$, and then $y = x = 18$ and $z = 2x = 36$. We then must argue as we did in Section 8.3 that this point must be where the function attains its maximum volume.

■

EXAMPLE 3 **Using Lagrange Multipliers**

Suppose we have a Cobb-Douglass production function

$$f(x, y) = 15x^{1/3}y^{2/3}$$

where x is the number of units of labor, y is the number of units of capital, and f is the number of units of a certain product that is produced. If each unit of labor costs \$200, each unit of capital costs \$100, and the total expense for both is limited to \$7,500,000, find the number of units of labor and capital needed to maximize production.

Solution

We have that $200x + 100y = 7{,}500{,}000$. Thus, the constraint equation is

$$g(x, y) = 200x + 100y - 7{,}500{,}000 = 0$$

Form the auxiliary function $F(x, y, \lambda)$ given by

$$F(x, y, \lambda) = 15x^{1/3}y^{2/3} + \lambda(200x + 100y - 7{,}500{,}000)$$

and solve the equations

$$\begin{aligned} F_x &= 5x^{-2/3}y^{2/3} + 200\lambda = 0 \\ F_y &= 10x^{1/3}y^{-1/3} + 100\lambda = 0 \\ g(x, y) &= 200x + 100y - 7{,}500{,}000 = 0 \end{aligned}$$

Solving the first two equations for $-\lambda$ and setting them equal yields

$$-\lambda = \frac{x^{-2/3}y^{2/3}}{40} = \frac{x^{1/3}y^{-1/3}}{10}$$

Multiplying each side of this equation by $x^{2/3}y^{1/3}$ yields $y = 4x$. Substituting this into the constraint equation yields

$$\begin{aligned} 200x + 100(4x) &= 7{,}500{,}000 \\ x &= 12{,}500 \end{aligned}$$

and $y = 50{,}000$. The relative maximum production is then

$$f(12\,500, 50\,000) = 15(12\,500)^{1/3}(50\,000)^{2/3} \approx 472{,}500$$

units. It turns out that this is also the maximum production. ■

Every plane has an equation of the form $ax + by + cz = d$. There are a number of ways of finding the perpendicular distance from a point to a plane. The following way uses Lagrange multipliers.

EXAMPLE 4 **Using Lagrange Multipliers**

Find the shortest distance from the point (0, 0, 0) to the plane $2x + 2y + z = 9$.

Solution

We will minimize the square of the distance d to avoid working with square roots. Thus, we wish to minimize

$$d^2 = f(x, y, z) = x^2 + y^2 + z^2$$

subject to the constraint

$$g(x, y, z) = 2x + 2y + z - 9 = 0$$

We form the auxiliary function $F(x, y, z, \lambda)$ given by

$$F(x, y, z, \lambda) = x^2 + y^2 + z^2 + \lambda(2x + 2y + z - 9)$$

and solve the equations

$$\begin{aligned} F_x &= 2x + 2\lambda = 0 \\ F_y &= 2y + 2\lambda = 0 \\ F_z &= 2z + \lambda = 0 \\ g(x, y, z) &= 2x + 2y + z - 9 = 0 \end{aligned}$$

Solving the first three equations for $-\lambda$ and setting them equal yields

$$-\lambda = x = y = 2z$$

This yields $y = x$ and $z = \frac{1}{2}x$. Now substituting back into the constraint equation yields

$$\begin{aligned} 9 &= 2x + 2y + z \\ &= 2x + 2x + \frac{1}{2}x \\ &= \frac{9}{2}x \\ x &= 2 \end{aligned}$$

Then $y = 2$ and $z = 1$. Substituting this into f yields

$$\begin{aligned} d^2 &= 2^2 + 2^2 + 1^2 \\ &= 9 \end{aligned}$$

Thus, $d = 3$. From geometric considerations, we know this must be the absolute minimum. ■

Warm Up Exercise Set 8.4

1. Find the dimensions of the rectangular box with a lid of volume 8 ft^3 that minimizes the surface area using Lagrange multipliers. (This was done in Example 5 of Section 8.3 without using Lagrange multipliers.)

Exercise Set 8.4

In Exercises 1 through 32, solve using Lagrange multipliers.

1. Minimize $f(x, y) = 3x^2 + y^2$ subject to the constraint $x + y - 4 = 0$.
2. Minimize $f(x, y) = x^2 + 3y^2$ subject to the constraint $x - y + 4 = 0$.
3. Maximize $f(x, y) = -x^2 + xy - 4y^2$ subject to the constraint $x + y + 4 = 0$.
4. Maximize $f(x, y) = -5x^2 - xy - y^2$ subject to the constraint $x + y + 20 = 0$.
5. Minimize $f(x, y) = 2x^2 + xy + y^2 + x$ subject to the constraint $2x + y - 3 = 0$.
6. Minimize $f(x, y) = 2x^2 - xy + y^2 + 7y$ subject to the constraint $x - 2y + 1 = 0$.
7. Maximize $f(x, y) = -2x^2 + xy - y^2 + 3x + y$ subject to the constraint $2x + 3y + 11 = 0$.
8. Maximize $f(x, y) = -x^2 - xy - 3y^2 + x - y$ subject to the constraint $-x + y + 2 = 0$.
9. Maximize $f(x, y) = xy$ in the first quadrant subject to the constraint $x^2 + y^2 - 8 = 0$.
10. Minimize $f(x, y) = xy^2$ in the first quadrant subject to the constraint $x^2 + y^2 - 8 = 0$.
11. Maximize $f(x, y) = xy$ in the first quadrant subject to the constraint $x + y - 8 = 0$.
12. Maximize $f(x, y) = x^2y$ in the first quadrant subject to the constraint $x + y - 9 = 0$.
13. Minimize $f(x, y) = x^2 + 2y^2 + 4z^2$ subject to the constraint $x + y + z - 7 = 0$.
14. Minimize $f(x, y) = x + 2y - 3z$ subject to the constraint $x^2 + 4y^2 - z = 0$.
15. Maximize $f(x, y) = xyz$ in the first quadrant subject to the constraint $x^2 + y^2 + 4z^2 - 12 = 0$.
16. Maximize $f(x, y) = xy + yz$ subject to the constraint $x^2 + y^2 + z^2 - 16 = 0$.
17. Find two positive numbers whose sum is 20 and whose product is as large as possible.
18. Find two positive numbers whose sum is 40 such that the sum of their squares is as small as possible.
19. Find three positive numbers whose sum is 36 and whose product is as large as possible.
20. Find three positive numbers whose product is 64 and whose sum is as small as possible.

Applications and Mathematical Modeling

21. **Construction** Find the most economical proportions for a closed cylindrical can (soft drink can) that will hold 16π cubic inches if the costs of the top, bottom, and side are the same.
22. **Construction** A manufacturer wishes to construct an open rectangular box by removing squares from the corners of a square piece of cardboard and bending up sides as shown in the figure. What are the dimensions of the box of maximum volume?

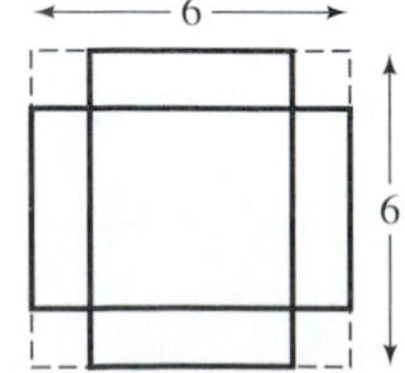

23. **Fencing** A person wishes to enclose a rectangular parking lot, using a building as one boundary and adding fence for the other boundaries. If 400 feet of fencing are available, find the dimensions of the largest parking lot that can be enclosed.
24. **Cost** What are the dimensions of the rectangular field of 20,000 ft^2 that will minimize the cost of fencing if one side cost three times as much as the other three.
25. **Cost** A closed rectangular box of volume 324 in^3 is to be made with a square base. If the material for the bottom costs twice as much per square foot as the material for the sides and top, find the dimensions of the box that minimize the cost of materials.
26. **Production** Suppose a Cobb-Douglass production function is given by $f(x, y) = 100x^{0.75}y^{0.25}$, where x is the

number of units of labor, y is the number of units of capital, and f is the number of units of a certain product that is produced. If each unit of labor costs \$100, each unit of capital costs \$200, and the total expense for both is limited to \$1,000,000, find the number of units of labor and capital needed to maximize production.

27. **Production** Suppose a Cobb-Douglass production function is given by $f(x, y) = 100x^{0.20}y^{0.80}$, where x is the number of units of labor, y is the number of units of capital, and f is the number of units of a certain product that is produced. If each unit of labor costs \$100, each unit of capital costs \$200, and the total expense for both is limited to \$1,000,000, find the number of units of labor and capital needed to maximize production.

28. **Geometry** A rectangular box without a top is to have a surface area of 48 ft^2. What dimensions will yield the maximum volume?

29. **Construction** An architect is attempting to fit a rectangular closet of maximal volume into a chopped-off corner of a building. She determines that the available space is as shown in the figure where the constricting plane is given by $4x + 2y + z = 12$ with the dimensions given in feet. What are the dimensions and volume of the closet of maximum volume?

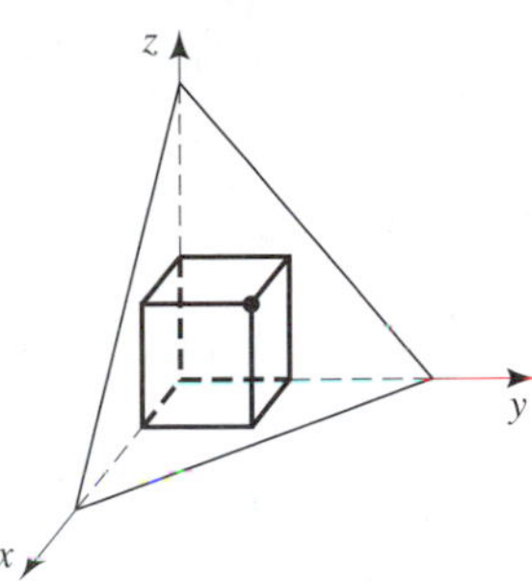

30. **Cost** Find the dimensions of the cheapest rectangular box with open top and volume of 96 ft^3 with the cost in square feet of the base per three times that of the cost of each side.

31. **Cost** A firm has three separate plants producing the same product and has a contract to sell 1000 units of its product. Each plant has a different cost function. Let x, y, and z be the number of items produced at the three plants, and assume that the respective cost functions are

$$C_1(x) = 100 + 0.03x^2 \qquad C_2(y) = 100 + 2y + 0.05y^2$$
$$C_3(z) = 100 + 12z$$

Find the allocation of production in the three plants that will minimize the firm's total cost.

More Challenging Exercises

32. Show that the shortest distance from the point (x_0, y_0, z_0) to the plane $ax + by + cz = E$ is

$$d = \frac{|E - ax_0 - by_0 - cz_0|}{\sqrt{a^2 + b^2 + c^2}}$$

33. Suppose that a production function is given by $P = f(x, y)$ and a cost function is given by $C = p_1x + p_2y$, where x is the number of items X made at a cost of p_1 and y is the number of items Y made at a cost of p_2. Use Lagrange multipliers to establish the important economic observation that if the firm wishes to produce P_1 items while minimizing cost, then

$$\frac{f_x}{p_1} = \frac{f_y}{p_2}$$

34. Suppose the production function in the previous exercise was a Cobb-Douglas production function of the form $P = ax^by^{1-b}$ and the cost function was the same. Use the result in that exercise to show that

$$p_2y = \frac{1-b}{b}p_1x$$

Solutions to WARM UP EXERCISE SET 8.4

1. Let the dimensions of a box be given by the three numbers x, y, and z. Then the volume is $V = xyz$, and also, according to the problem, $V = 8$. Thus, $8 = xyz$. From this we obtain the constraint equation $g(x, y, z) = xyz - 8 = 0$. The surface area S is given by $S(x, y, z) = 2xy + 2xz + 2yz$. It is this surface area that we wish to minimize.

   We then form the auxiliary function $F(x, y, z, \lambda)$ given by

$$F(x, y, z, \lambda) = 2xy + 2xz + 2yz + \lambda(xyz - 8)$$

and solve the equations

$$F_x = 2y + 2z + \lambda yz = 0$$
$$F_y = 2x + 2z + \lambda xz = 0$$
$$F_z = 2x + 2y + \lambda xy = 0$$
$$g(x, y, z) = xyz - 8 = 0$$

Solving the first three equations for $-\lambda$ and setting them equal yields

$$-\lambda = 2\frac{y+z}{yz} = 2\frac{x+z}{xz} = 2\frac{x+y}{xy}$$

Setting the first two fractions equal yields $xy + xz = xy + yz$, giving $x = y$, while the other equation yields $xy + yz = xz + yz$, which implies that $y = z$. Thus, $x = y = z$. Substituting this into the constraint equation yields $x^3 - 8 = 0$ or $x = 2$. Then $y = 2$, and $z = 2$. We then must argue as we did in Section 8.3 that this point must be where the function attains its minimum surface.

8.5 Tangent Plane Approximations

Tangent Plane Approximations

Applications

APPLICATION Finding Easy-to-Calculate Approximations

Suppose we have a Cobb-Douglass production function $f(x, y) = 81x^{1/3}y^{2/3}$, where x is the number of units of labor, y is the number of units of capital, and f is the number of units of a certain product that is produced. Find an approximation for the change in production given changes in labor Δx and changes in capital Δy as a *linear* function of Δx and Δy. For the answer, see Example 2 on page 574 and the discussion that follows.

Tangent Plane Approximations

Recall that for a function of *one* variable, $y = f(x)$, the actual change in y as x changes from a to $a + \Delta x$ is given by

$$\Delta y = f(a + \Delta x) - f(a)$$

Recall also that if Δx is small, this actual change in y can be approximated by the change along the tangent line, that is,

$$\Delta y \approx dy = f'(a) \cdot \Delta x$$

Thus, for Δx small the graph of $y = f(x)$ is approximately the same as the graph of the tangent line. See Figure 8.36. We will now extend these ideas to functions of *two* variables.

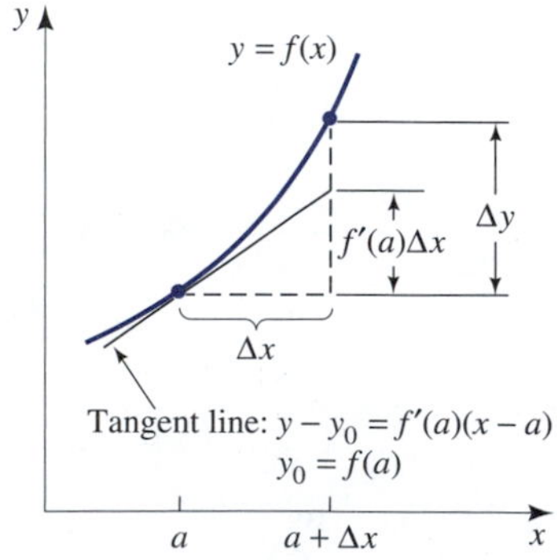

Figure 8.36
If $f'(a)$ exists, the graph of $y = f(x)$ is nearly linear near $x = a$.

We have already seen that $\dfrac{\partial f}{\partial x}(a, b)$ gives the slope of the line T_x tangent to the curve in which the plane $y = b$ intersects the surface $z = f(x, y)$ and that $\dfrac{\partial f}{\partial y}(a, b)$ gives the slope of the line T_y tangent to the curve in which the plane $x = a$ intersects the surface $z = f(x, y)$. The lines T_x and T_y determine a plane. This plane is called the **tangent plane** to the surface at the point $P = (a, b, z_0)$, where $z_0 = f(a, b)$. See Figure 8.37. It can be shown, though we do not do so here, that this tangent plane is given by the equation

$$z - z_0 = f_x(a, b)(x - a) + f_y(a, b)(y - b)$$

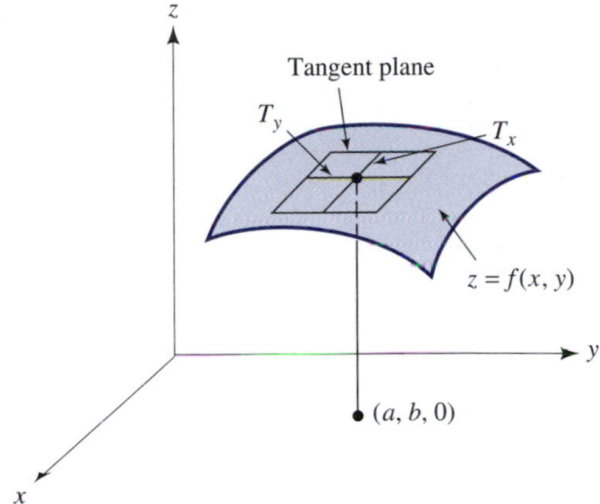

Figure 8.37
The two tangent lines T_x and T_y determine the tangent plane.

Notice carefully in the last equation that the point (a, b, z_0) is on the *surface*, whereas the point (x, y, z) is on the *tangent plane*. If we let

$$\Delta x = x - a \qquad \text{and} \qquad \Delta y = y - b$$

and denote the corresponding change in z as $\Delta z = z - z_0$, then the equation of the tangent plane can be written as

$$\Delta z = f_x(a, b)\Delta x + f_y(a, b)\Delta y$$

This indicates how the changes Δx and Δy produce the change Δz *on the tangent plane*. Since the tangent plane *is tangent* to the surface, we fully expect from the geometry of the situation (see Figure 8.37) that if the changes in Δx and Δy are small, the actual change in z, which is $f(a + \Delta x, b + \Delta y) - f(a, b)$, denoted by Δz, will be approximately the change on the tangent plane. Thus, we have the following.

Tangent Plane Approximations

Suppose $f_x(a, b)$ and $f_y(a, b)$ exist. If Δx and Δy are small, then

$$\Delta z = f(a + \Delta x, b + \Delta y) - f(a, b) \approx f_x(a, b) \cdot \Delta x + f_y(a, b) \cdot \Delta y$$

EXAMPLE 1 **Using Tangent Plane Approximations**

Use the tangent plane to approximate $\sqrt[3]{(2.06)^2 + (1.97)^2}$.

Solution

Since 2.06 and 1.97 are very close to 2 and we can easily take the cube root of $2^2 + 2^2 = 8$, we set $a = 2 = b$ and $\Delta x = 0.06$ and $\Delta y = -0.03$. We take

$$f(x, y) = \sqrt[3]{x^2 + y^2}$$

Then $f(a, b) = f(2, 2) = \sqrt[3]{8} = 2$, and

$$f_x(x, y) = \frac{2x}{3(x^2 + y^2)^{2/3}} \qquad f_y(x, y) = \frac{2y}{3(x^2 + y^2)^{2/3}}$$

and

$$f_x(2, 2) = \frac{1}{3} \qquad f_y(2, 2) = \frac{1}{3}$$

Thus,

$$\begin{aligned}
\sqrt[3]{(2.06)^2 + (1.97)^2} &= f(a + \Delta x, b + \Delta y) \\
&\approx f(a, b) + \Delta z \\
&= f(a, b) + f_x(a, b) \cdot \Delta x + f_y(a, b) \cdot \Delta y \\
&= 2 + \frac{1}{3}\Delta x + \frac{1}{3}\Delta y \\
&= 2 + \frac{1}{3}(0.06) + \frac{1}{3}(-0.03) \\
&= 2.01
\end{aligned}$$

The actual value of $\sqrt[3]{(2.06)^2 + (1.97)^2}$ is 2.0103 to four decimal places. Therefore, in this case the error using the tangent plane approximation is very small. ■

Remark Notice that for $x_0 = 2$ and $y_0 = 2$ we have the formula

$$\Delta z \approx \frac{1}{3} \cdot \Delta x + \frac{1}{3} \cdot \Delta y$$

which can be used to give quick estimates of Δz for a variety of values of Δx and Δy. More generally, for any (x, y) we have

$$\Delta z \approx \frac{2x}{3(x^2 + y^2)^{2/3}} \cdot \Delta x + \frac{2y}{3(x^2 + y^2)^{2/3}} \cdot \Delta y$$

where $\Delta z = f(x + \Delta x, y + \Delta y) - f(x, y)$.

Applications

We now give a number of applications.

EXAMPLE 2 **Using Tangent Plane Approximation**

Suppose we have a Cobb-Douglass production function

$$f(x, y) = 81x^{1/3}y^{2/3}$$

where x is the number of units of labor, y is the number of units of capital, and f is the number of units of a certain product that is produced. Using the tangent plane, find an approximation for the change in production as x changes from 27 to 29 and as y changes from 125 to 122.

Solution

We take $a = 27$ and $\Delta x = 2$. Also $b = 125$ and $\Delta y = -3$. We first need to calculate the first-order partial derivatives. Thus,

$$f_x(x, y) = 27x^{-2/3}y^{2/3} \qquad \text{and} \qquad f_y(x, y) = 54x^{1/3}y^{-1/3}$$

and

$$f_x(27, 125) = 27(27)^{-2/3}(125)^{2/3} = 27(9)^{-1}(25) = 75$$

and

$$f_y(27, 125) = 54(27)^{1/3}(125)^{-1/3} = 54(3)(5)^{-1} = 32.4$$

Then

$$\begin{aligned}\Delta z &= f(27+2, 125-3) - f(27, 125)\\ &= f(a + \Delta x, b + \Delta y) - f(a, b)\\ &\approx f_x(a, b) \cdot \Delta x + f_y(a, b) \cdot \Delta y\\ &= f_x(27, 125)(2) + f_y(27, 125)(-3)\\ &= (75)(2) + (32.4)(-3)\\ &= 52.8\end{aligned}$$

The actual change in z is

$$\Delta z = f(29, 122) - f(27, 125) \approx 46.5$$

■

Remark Notice that for $(a, b) = (27, 125)$ we have the formula

$$\Delta z \approx (75) \cdot \Delta x + (32.4) \cdot \Delta y$$

This formula can be used to give quick estimates of the change in production given various changes in x and y. More generally, for any (x, y) we have

$$\Delta z \approx 27x^{-2/3}y^{2/3} \cdot \Delta x + 54(x)^{1/3}(y)^{-1/3} \cdot \Delta y$$

where $\Delta z = f(x + \Delta x, y + \Delta y) - f(x, y)$.

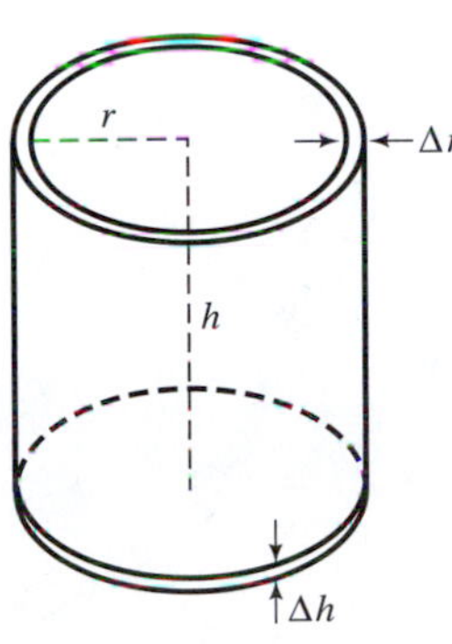

Figure 8.38
The side has thickness Δr and the top and bottom have thickness Δh.

EXAMPLE 3 **Using Tangent Line Approximation**

(a) Use the tangent line approximation to estimate the change in volume of a cylindrical can of inside radius r and height h for any small changes in r and h when $r = 2$ and $h = 5$. See Figure 8.38.

(b) Suppose the side of the can is 0.02 inch thick and the top and bottom are 0.03 inch thick. Estimate the volume of the metal in the can.

Solution

(a) The volume $V(r, h)$ is given by

$$V(r, h) = \pi r^2 h$$

The first-order partial derivatives are

$$V_r(r, h) = 2\pi r h \qquad V_h(r, h) = \pi r^2$$

Then

$$\begin{aligned} \Delta V &= V(r + \Delta r, h + \Delta h) - V(r, h) \\ &\approx V_r(r, h) \cdot \Delta r + V_h(r, h) \cdot \Delta h \\ &= 2\pi r h \cdot \Delta r + \pi r^2 \cdot \Delta h \\ &= 2\pi(2)(5) \cdot \Delta r + \pi(2)^2 \cdot \Delta h \\ &= 20\pi \cdot \Delta r + 4\pi \cdot \Delta h \end{aligned}$$

(b) Take $\Delta r = 0.02$. Since the top and bottom are each 0.03 inch thick, $\Delta h = 2(0.03)$. Then we have

$$\Delta V \approx 20\pi \cdot \Delta r + 4\pi \cdot \Delta h = 20\pi(0.02) + 4\pi(0.06) \approx 2.01$$

or 2.01 in^3. ■

Warm Up Exercise Set 8.5

1. A firm has a cost function given by

$$C(x, y) = x^3 + 3xy + y^2$$

Currently, $x = 20$ and $y = 100$. Use the tangent line approximation to approximate the change in costs ΔC as a function of the change in x and the change in y.

2. Take $\Delta x = \Delta y = 1$ in the answer in Exercise 1 to estimate the change in costs of producing one more of each item. Compare your approximation to the actual answer.

Exercise Set 8.5

In Exercises 1 through 16, use the tangent plane approximation to estimate

$$\Delta z = f(a + \Delta x, b + \Delta y) - f(a, b)$$

for the given function at the given point and for the given values of Δx and Δy.

1. $f(x, y) = x^2 + y^3$, $(a, b) = (2, 1)$, $\Delta x = 0.1$, $\Delta y = 0.2$

2. $f(x, y) = 2x^3 + xy$, $(a, b) = (1, 2)$, $\Delta x = 0.1$, $\Delta y = 0.3$

3. $f(x, y) = x^2y^3$, $(a, b) = (3, 2)$, $\Delta x = 0.1$, $\Delta y = -0.1$

4. $f(x, y) = x^3y^4$, $(a, b) = (-1, 1)$, $\Delta x = 0.1$, $\Delta y = -0.2$

5. $f(x, y) = \dfrac{x - y}{x + y}$, $(a, b) = (1, 2)$, $\Delta x = 0.2$, $\Delta y = 0.1$

6. $f(x, y) = \dfrac{x + y}{x - y}$, $(a, b) = (3, 4)$, $\Delta x = -0.1$, $\Delta y = 0.2$

7. $f(x, y) = \dfrac{x}{x^2 + y^2}$, $(a, b) = (3, 4)$, $\Delta x = -0.1$, $\Delta y = 0.2$

8. $f(x, y) = \dfrac{y}{x^2 + y^2}$, $(a, b) = (3, 4)$, $\Delta x = 0.02$, $\Delta y = -0.03$

9. $f(x, y) = \ln(x + 2y)$, $(a, b) = (2, 4)$, $\Delta x = 0.1$, $\Delta y = 0.2$

10. $f(x, y) = \ln(x^2 + y^2)$, $(a, b) = (1, 2)$, $\Delta x = 0.5$, $\Delta y = -0.5$

11. $f(x, y) = \sqrt{x + y}$, $(a, b) = (4, 5)$, $\Delta x = -0.1$, $\Delta y = -0.2$

12. $f(x, y) = \sqrt{x^2 + y}$, $(a, b) = (2, 5)$, $\Delta x = 1/30$, $\Delta y = 1/15$

13. $f(x, y) = e^{x+y}$, $(a, b) = (1, -1)$, $\Delta x = 0.02$, $\Delta y = 0.03$

14. $f(x, y) = e^{xy}$, $(a, b) = (2, 0.5)$, $\Delta x = 0.02$, $\Delta y = -0.01$

15. $f(x, y) = xe^y$, $(a, b) = (0, 1)$, $\Delta x = 0.1$, $\Delta y = 0.2$

16. $f(x, y) = xye^{xy}$, $(a, b) = (1, 2)$, $\Delta x = 0.02$, $\Delta y = 0.01$

In Exercises 17 through 28, use the tangent line approximation to estimate the given quantities.

17. $\sqrt{4.04}\sqrt{9.06}$

18. $\sqrt{4.08}\sqrt{8.94}$

19. $\sqrt{(3.02)^2 + (4.03)^2}$

20. $\sqrt{(3.01)^2 + (3.97)^2}$

21. $\sqrt{(2.06)^3 + (0.97)^2}$

22. $\sqrt{(1.94)^3 - (1.03)^2}$

23. $(1.02)^7 + (1.98)^3$

24. $(1.02)^7(1.98)^3$

25. $(0.99)^7(2.98)^3$

26. $(0.97)^7 + (1.98)^3$

27. $e^{(1.02)^2-(0.98)^2}$

28. $e^{(2.01)^3-7.97}$

Applications and Mathematical Modeling

29. Cost A cost function in thousands of dollars for a firm producing two products X and Y is given by $C(x, y) = 1000 + x^3 + xy + y^2$, where x and y are the respective number in thousands of X and Y produced. If 1000 of X and 2000 of Y are currently being produced, use the tangent plane approximation to estimate the change in cost if the firm produces an additional 100 of the X items and 300 of the y items.

30. Revenue Suppose the revenue function for the firm in Exercise 29 is given by $R(x, y) = \sqrt{xy}(10 - xy)$ and the firm is currently selling what they produce. Use the tangent plane approximation to estimate the change in revenue if the firm produces and sells an additional 100 of the X items and 300 of the Y items.

31. Production Suppose we have a Cobb-Douglass production function $f(x, y) = 100x^{1/4}y^{3/4}$, where x is the number of units of labor, y is the number of units of capital, and f is the number of units of a certain product that is produced. Using the tangent plane, find an approximation for the change in production as x changes from 16 to 18 and as y changes from 81 to 80.

32. Manufacturing Use the tangent plane approximation to estimate the volume of metal in a closed cylindrical can of inner radius 3 inches and inner height 6 inches if the metal is 0.03 inch thick.

33. Manufacturing Use the tangent plane approximation to estimate the volume of metal in a closed rectangular metal box with a square bottom and top with each side of inner length 1 foot and inner height 3 feet if the metal is 0.05 foot thick.

34. Biology In measuring the volume of a rod-shaped bacterium (right circular cylinder), suppose an error of 2% is made measuring its radius and a 1% error is made in measuring its length. Using the tangent line approximation, estimate the greatest relative error ($|\Delta V|/V$) in calculating its volume when $r = 1$ unit and $h = 10$ units.

35. Biology The following formula relates the surface area A in square meters of a human to the weight w in kilograms and height h in meters:[25] $A(w, h) = 2.02w^{0.425}h^{0.725}$. Use the tangent plane approximation to estimate the difference in surface area of two humans, one weighing 100 kg and 2 meters tall and the other weighing 101 kg and 2.1 meters tall.

36. Medicine Poiseville's law states that the resistance of a blood vessel of length L and radius r is given by

$$R = \frac{kL}{r^4}$$

where k is a constant. Use the tangent plane approximation to show how small changes in L and r affect the resistance R.

[25] Duane J. Clow and N. Scott Urquhart. 1974. *Mathematics in Biology*. New York: Ardsley House.

More Challenging Exercises

37. Find an approximation for $f(1.01, 0.02)$ where

$$f(x, y) = \frac{x^3 \ln(x + 1) + (x - 1)e^{yx^2 - 0.4y^2}}{\sqrt{x^3 + \ln(y + 1)}}$$

The function $f(x, y)$ is rather complicated. Calculating the first partial derivatives will be very time consuming and prone to error. Instead, use your calculator to find the numerical derivatives.

38. Find a square about the point (5, 7) in which the volume of the cylinder $V = \pi r^2 h$ given in Example 3 will not vary by more than approximately ± 0.1.

Solutions to WARM UP EXERCISE SET 8.5

1. We have $C_x(x, y) = 3x^2 + 3y$ and $C_y(x, y) = 3x + 2y$. Thus, $C_x(20, 100) = 1500$, $C_y(20, 100) = 260$, and

$$\Delta C \approx 1500\Delta x + 260\Delta y$$

2. Setting $\Delta x = \Delta y = 1$ in the previous formula yields

$$\Delta C \approx 1500(1) + 260(1) = 1760$$

The actual change in C is given by

$$C(21, 101) - C(20, 100) = 25{,}825 - 24{,}000 = 1825$$

8.6 Double Integrals

Riemann Sums and Integrals

Iterated Integrals

Applications

APPLICATION Concentration of Pollutants

An industrial plant is located at the precise center of a town shaped as a square with each side of length 4 miles. If the plant is placed at the point (0, 0) of the xy-plane, then certain pollutants are dispersed in such a manner that the concentration at any point (x, y) in town is given by $C(x, y) = 1000(24 - 3x^2 - 3y^2)$, where $C(x, y)$ is the number of particles of pollutants per square mile of surface per day at a point (x, y) in town. What is the average concentration of these pollutants in this town? For the answer, see Example 5 on page 588.

Riemann Sums and Integrals

In this section we will first be interested in finding the volume V between a surface given by $z = f(x, y)$ and a region D in the xy-plane. We will then see that this can be used for many applications. We will assume that $f(x, y)$ is nonnegative on D. See Figure 8.39.

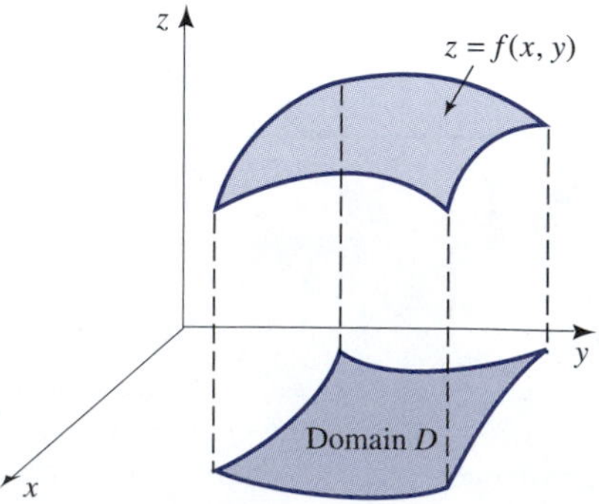

Figure 8.39

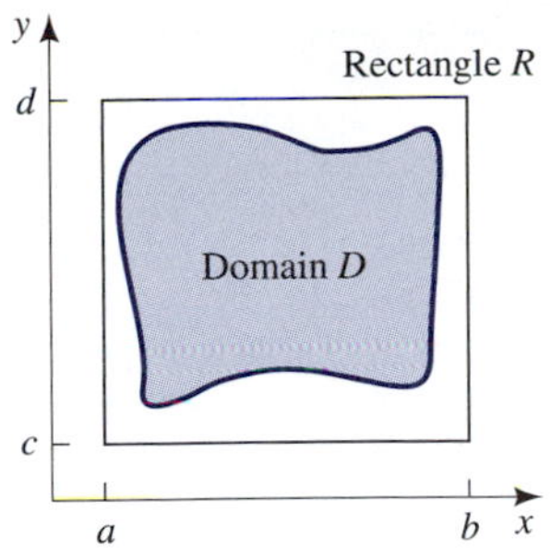

Figure 8.40
Encase the domain D in a rectangle.

The domain D is shown in the xy-plane in Figure 8.40. In this figure, the domain D has been encased with a rectangle R as shown, where

$$R(x, y) = \{(x, y): a \leq x \leq b, c \leq y \leq d\}$$

For any integer n we divide both intervals $[a, b]$ and $[c, d]$ in n equal subintervals and define

$$\Delta x = \frac{b-a}{n} \qquad \Delta y = \frac{d-c}{n}$$

This gives us the collection of subrectangles shown in Figure 8.41. We label all subrectangles that lie *entirely inside D* in some manner by going from left to right and top to bottom as indicated in Figure 8.41. Let the total number of these subrectangles be $N = N(n)$, and label them as $R_1, R_2, \ldots, R_N$. (If the region D were the rectangle R, then $N = N(n) = n^2$.) In each R_i, pick some point $(x_i, y_i) \in R_i$. Now for each $i = 1, 2, \ldots, N$, construct a rectangular box as in Figure 8.42 of height $f(x_i, y_i)$ and base R_i. If n is large, the volume

$$V_i = f(x_i, y_i)\Delta x \Delta y$$

of this rectangular box will approximate the volume under the surface $z = f(x, y)$ and over R_i. The sum of all such volumes will then approximate the volume we are seeking. This sum is called a **Riemann sum of f on D** and is denoted by S_n. We thus have

$$\begin{aligned} V \approx S_n &= V_1 + V_2 + \cdots + V_N \\ &= f(x_1, y_1)\Delta x \Delta y + f(x_2, y_2)\Delta x \Delta y + \cdots + f(x_N, y_N)\Delta x \Delta y \\ &= \sum_{i=1}^{N} f(x_i, y_i)\Delta x \Delta y \end{aligned}$$

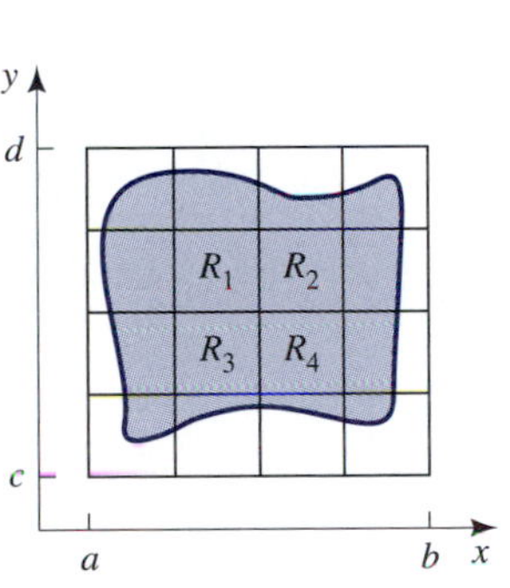

Figure 8.41
Divide the interval $[a, b]$ and $[c, d]$ into n equal subintervals giving n^2 subrectangles.

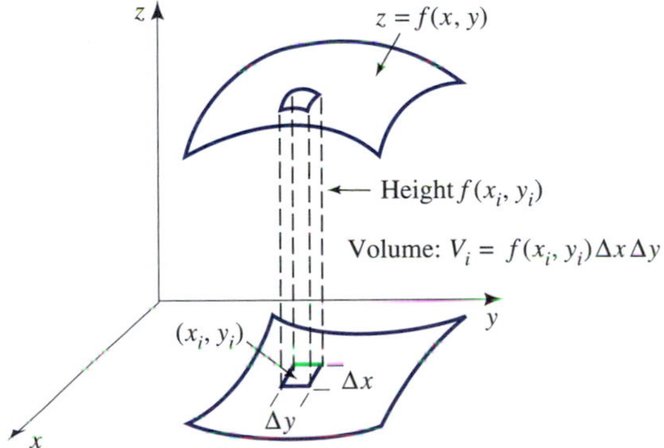

Figure 8.42
The Riemann sum is the sum of the volumes of boxes like the one shown.

EXAMPLE 1 **Finding a Riemann Sum**

Find the Riemann sum of $f(x, y) = 6xy^2$ on the rectangle

$$R = \{(x, y): 0 \leq x \leq 2, 0 \leq y \leq 1\}$$

for $n = 2$ by taking the points (x_i, y_i) at the top right of each subrectangle.

Solution

Each interval [0, 2] and [0, 1] is divided into two parts, and

$$\Delta x = \frac{2-0}{2} = 1 \qquad \Delta y = \frac{1-0}{2} = \frac{1}{2}$$

The four subrectangles are obtained as shown in Figure 8.43, where the four points (x_1, y_1), (x_2, y_2), (x_3, y_3), and (x_4, y_4) are $(1, 1)$, $(2, 1)$, $(1, 0.50)$, and $(2, 0.50)$, respectively. Note that the function increases as either x or y increases. Thus, on any subrectangle R_i the function is largest at the top right point (x_i, y_i). Thus, the top of each of the rectangular boxes will be above the surface, and the Riemann sum will be larger than the volume. Then

$$\begin{aligned} S_2 &= f(x_1, y_1)\Delta x\Delta y + f(x_2, y_2)\Delta x\Delta y + f(x_3, y_3)\Delta x\Delta y + f(x_4, y_4)\Delta x\Delta y \\ &= f(1, 1)\Delta x\Delta y + f(2, 1)\Delta x\Delta y + f(1, 0.50)\Delta x\Delta y + f(2, 0.50)\Delta x\Delta y \\ &= (6)\frac{1}{2} + (12)\frac{1}{2} + (1.50)\frac{1}{2} + (3)\frac{1}{2} \\ &= 11.25 \end{aligned}$$

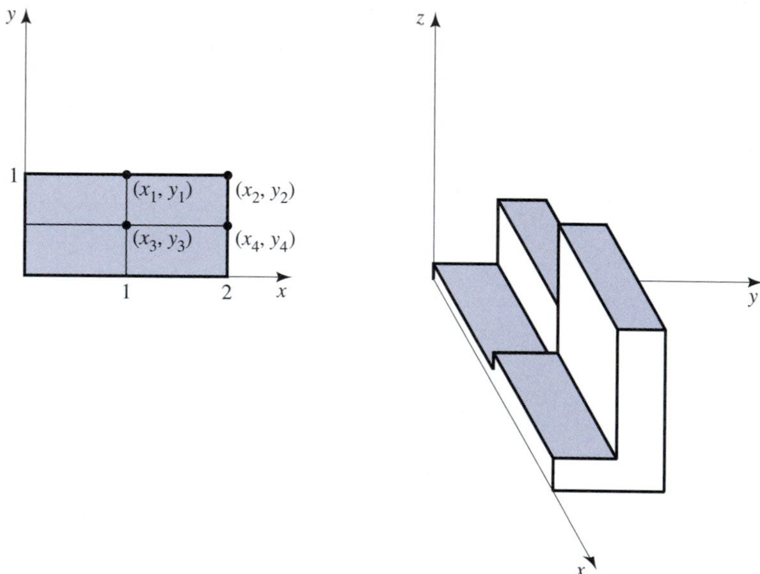

Figure 8.43
The Riemann sum S_2 is the sum of the volumes of the four boxes on the right.

■

If we take the points (x_i, y_i) at the lower left of R_i, then the tops of the rectangular boxes will be under the surface, and the corresponding Riemann sum S_n^* will be less than the volume.

The table shows the results of calculating the Riemann sums by taking (x_i, y_i) at the lower left and the upper right.

n	n^2	S_n^*	S_n
4	16	1.969	7.031
8	64	2.871	5.379
100	10,000	3.901	4.100
500	250,000	3.980	4.020

If, in the previous discussion, we then let n become larger and larger, we expect the Riemann sum to become a better and better approximation to the volume V. By letting $n \to \infty$, we intuitively see that we should obtain the volume V. According to the table, the volume appears to be 4. (In Example 2 we will see that this is the case.)

For a function $f(x, y)$ that is continuous on D, the limit $\lim_{n\to\infty} S_n$ will exist and will be the same limit no matter how the points (x_i, y_i) in R_i are chosen. The result also holds whether or not $f(x, y)$ is nonnegative. We have the following theorem.

DEFINITION **Definite Integral**

Let $f(x, y)$ be continuous on the bounded region D. Then the **definite integral of f over D** is

$$\int\int_D f(x, y)\, dA = \lim_{n\to\infty} \sum_{i=1}^{N} f(x_i, y_i)\Delta x \Delta y$$

where the limit exists and is the same no matter how the points (x_i, y_i) are chosen in R_i.

When $f(x, y) \geq 0$, then the indefinite integral is intuitively the volume under the surface and over the domain D. We then *define* the volume to be this Riemann integral.

DEFINITION **Volume as a Definite Integral**

Let $f(x, y)$ be continuous on the bounded region D and nonnegative there. Then the volume under the surface $z = f(x, y)$ and over the domain D is the definite integral of f over D, or

$$V = \int\int_D f(x, y)\, dA$$

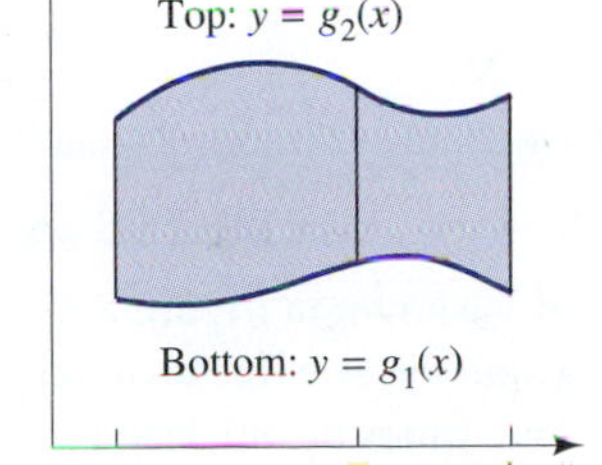

Figure 8.44
The domain has a top and bottom each given by one function.

Iterated Integrals

Suppose the domain D of $z = f(x, y)$ has a "top" and a "bottom" given by $y = g_2(x)$ and $y = g_1(x)$, respectively, as in Figure 8.44. For any fixed $\bar{x} \in [a, b]$, draw a straight line connecting the top and the bottom of the domain D, as shown in Figure 8.44. As shown in Figure 8.45, the plane $x = \bar{x}$ cuts out a plane region with top $z = f(\bar{x}, y)$. We can readily find the area $A(\bar{x})$ on the plane $x = \bar{x}$ under the curve $z = f(\bar{x}, y)$ as

$$A(\bar{x}) = \int_{g_1(\bar{x})}^{g_2(\bar{x})} f(\bar{x}, y)\, dy$$

This is the cross-sectional area of a slice of the volume.

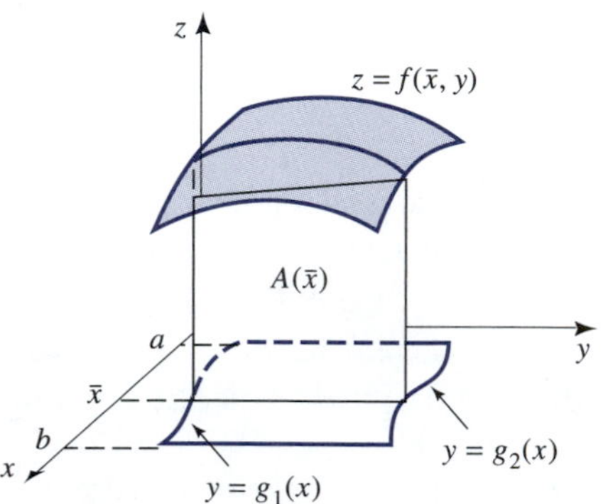

Figure 8.45
The slice at $\bar{x}$ cuts out an area between the curve $z = f(\bar{x}, y)$ and the xy-plane from $g_1(\bar{x})$ to $g_2(\bar{x})$.

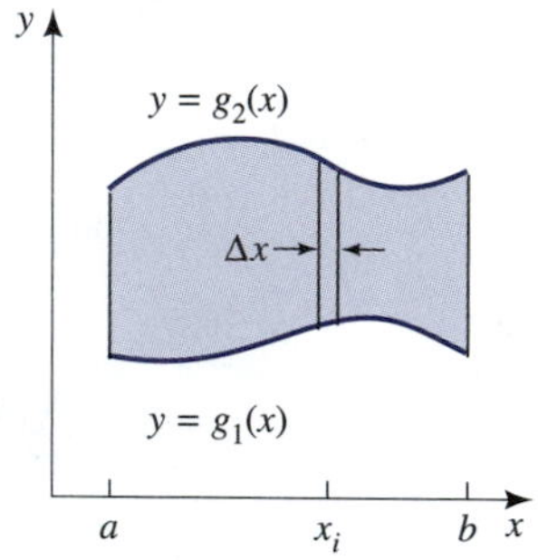

Figure 8.46
We take a vertical slice of thickness Δx.

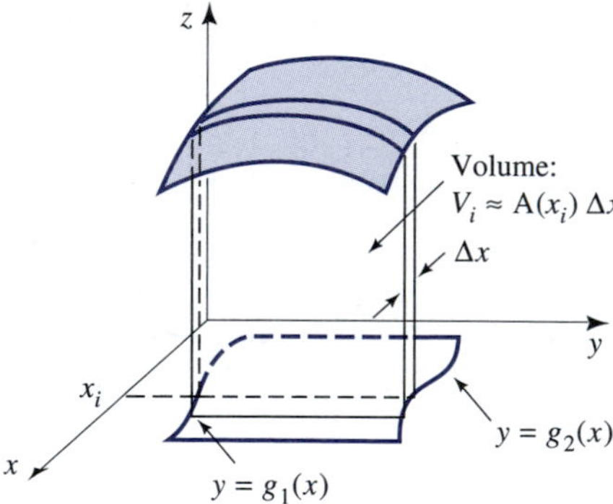

Figure 8.47
The slice in Figure 8.46 then cuts out the volume shown.

For any integer n we now divide the interval $[a, b]$ into n equal subintervals, each with base of length $\Delta x = \dfrac{b-a}{n}$. The ith such interval I_i is shown in Figure 8.46. Each interval then slices out a thin volume under the surface as shown in Figure 8.47. For each i we pick any point $x_i \in I_i$. By virtue of continuity and the smallness of Δx, the volume V_i of this slice is approximately the thickness, Δx, times the surface area of either side. The surface area of either side is approximately $A(x_i)$, given before. Thus,

$$V_i \approx A(x_i)\Delta x$$

If we sum all n of these up, we will approximately obtain the volume V under the surface $z = f(x, y)$ over the domain D. We have

$$\begin{aligned} V &= V_1 + V_2 + \cdots + V_n \\ &\approx A(x_1)\Delta x + A(x_2)\Delta x + \cdots + A(x_n)\Delta x \\ &= \sum_{i=1}^{n} A(x_i)\Delta x \end{aligned}$$

But we recognize this as a Riemann sum of $A(x)$ on the interval $[a, b]$. Thus, as $n \to \infty$,

$$\sum_{i=1}^{n} A(x_i)\Delta x \to \int_a^b A(x)\,dx$$

Thus, we have

$$V = \int_a^b A(x)\,dx$$

where

$$A(x) = \int_{g_1(x)}^{g_2(x)} f(x, y)\,dy$$

We must understand that in this last integral for $A(x)$, *x is held constant* during the integration. Putting this together, we have

$$V = \int_a^b \left(\int_{g_1(x)}^{g_2(x)} f(x, y)\,dy \right) dx$$

We thus have the following result.

Iterated Integral

If the domain D is given as in Figure 8.44 and $z = f(x, y) \geq 0$ is continuous on D, then the volume V between the surface $z = f(x, y)$ and the domain D is

$$\int\!\!\int_D f(x, y)\,dA = \int_a^b \left(\int_{g_1(x)}^{g_2(x)} f(x, y)\,dy \right) dx$$

Remarks The following two remarks are critical:

1. In the above, $\int_{g_1(x)}^{g_2(x)} f(x, y)\,dy$ is to be done *first*, that is, we integrate with respect to y *first*.
2. When calculating $\int_{g_1(x)}^{g_2(x)} f(x, y)\,dy$, treat x as a *constant*. Thus, we integrate first with respect to y while holding x constant.

Since there will be two basic types of regions to master, we must be able to distinguish between the two. Toward this end we should get in the habit, when taking a double integral over a region such as that shown in Figure 8.44, to take the following steps.

Procedure for Setting up the Iterated Integral

1. Clearly label the "top" and the "bottom" of the domain D.
2. Draw the slice shown in Figure 8.44 so that one realizes that x is being held constant while the y is varying from the "bottom" to the "top." This gives the limits of integration for the *inner* integral.
3. Now "sum" these slices "on x," that is, x goes from a to b. This gives the limits of integration on the *outer* integral.

EXAMPLE 2 An Iterated Integral over a Rectangle

Find the volume under the surface given by $z = 6xy^2$ over the domain given by the rectangle

$$D = \{(x, y) \colon 0 \leq x \leq 2, 0 \leq y \leq 1\}$$

that is, find

$$\int\int_D 6xy^2\, dA$$

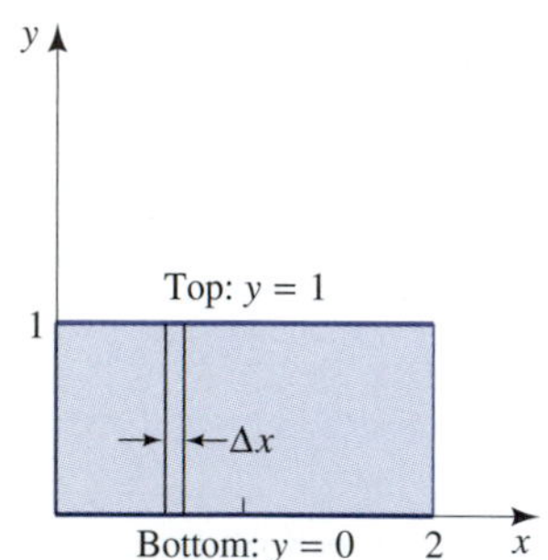

Figure 8.48
We take a vertical slice of thickness Δx.

Solution

The region D has a bottom given by $y = g_1(x) = 0$ and a top given by $y = g_2(x) = 1$. See Figure 8.48. The slice shown in Figure 8.48 or Figure 8.49 indicates that while holding x constant, we integrate with respect to y from the bottom to the top. Thus, the limits of integration on the inner integral go from $y = 0$ to $y = 1$. Now the slices are "summed" on x from 0 to 2. This gives the limits of integration on the outer integral. Thus,

$$\begin{aligned}
\int\int_D f(x, y)\, dA &= \int_a^b \left(\int_{g_1(x)}^{g_2(x)} f(x, y)\, dy \right) dx \\
&= \int_0^2 \left(\int_0^1 6xy^2\, dy \right) dx \\
&= \int_0^2 \left(2xy^3 \Big|_{y=0}^{y=1} \right) dx \\
&= \int_0^2 2x\, dx \\
&= x^2 \Big|_0^2 \\
&= 4
\end{aligned}$$

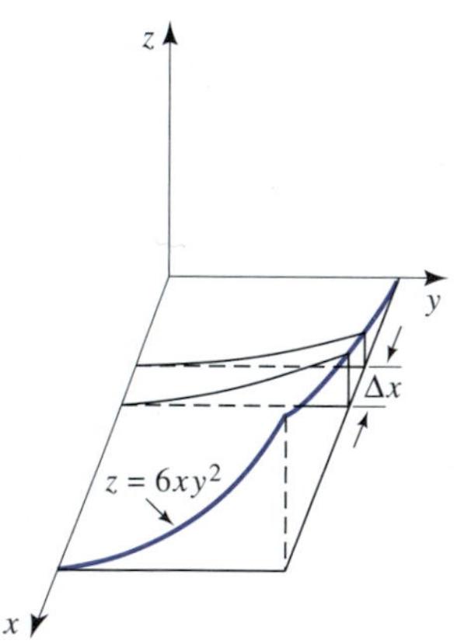

Figure 8.49
The slice shown in Figure 8.48 cuts out the volume shown.

EXAMPLE 3 Finding an Iterated Integral

Find the volume under the surface given by $z = 48xy$ over the domain given by the region D in the first quadrant of the xy-plane bounded by $y = x$ and $y = x^2$.

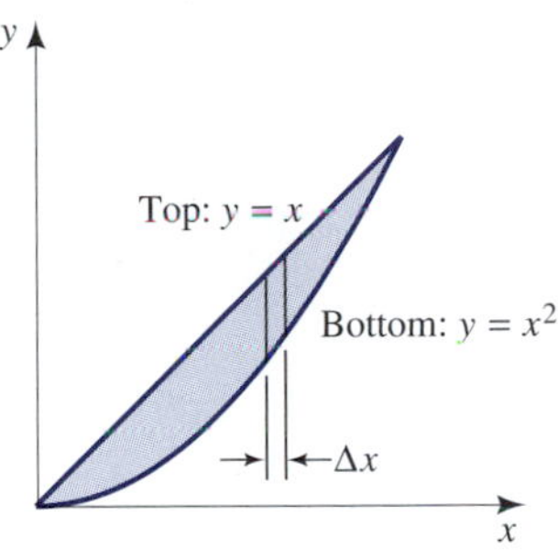

Figure 8.50
We take a vertical slice of thickness Δx.

Solution

The region D has a bottom given by $y = g_1(x) = x^2$ and a top given by $y = g_2(x) = x$. See Figure 8.50. The slice shown in Figures 8.50 and 8.51 indicates that while holding x constant, we integrate with respect to y from the bottom to the top. Thus, the limits of integration on the inner integral go from $y = x^2$ to $y = x$. Now the slices are "summed" on x from 0 to 1. This gives the limits of integration on the outer integral.

Thus,

$$\begin{aligned}\int\int_D f(x, y)\, dA &= \int_a^b \left(\int_{g_1(x)}^{g_2(x)} f(x, y)\, dy\right) dx \\ &= \int_0^1 \left(\int_{x^2}^{x} 48xy\, dy\right) dx \\ &= \int_0^1 \left(24xy^2\Big|_{y=x^2}^{y=x}\right) dx \\ &= \int_0^1 (24x^3 - 24x^5)\, dx \\ &= (6x^4 - 4x^6)\Big|_0^1 \\ &= 2\end{aligned}$$

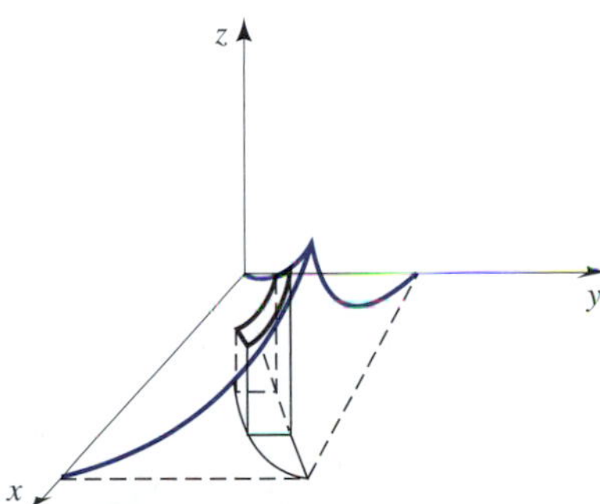

Figure 8.51
The slice shown in Figure 8.50 cuts out the volume shown.

■

Now let us consider the case in which the domain D of $z = f(x, y)$ has a "left" and a "right" given by $x = g_1(y)$ and $x = g_2(y)$, respectively, as in Figure 8.52. We proceed in a manner similar to what we did before, this time summing thin slices of volumes shown in Figure 8.53. This leads to the following iterated integral.

Iterated Integral

If the domain D is given as in Figure 8.52 and $z = f(x, y) \geq 0$ is continuous on D, then the volume V between the surface $z = f(x, y)$ and the domain D is

$$\int\int_D f(x, y)\, dA = \int_c^d \left(\int_{g_1(y)}^{g_2(y)} f(x, y)\, dx\right) dy$$

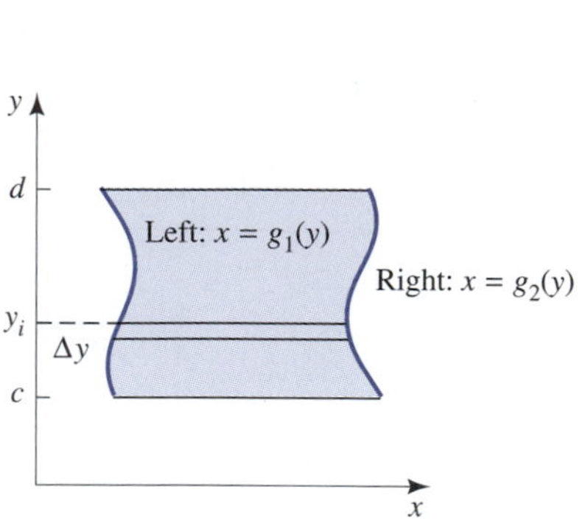

Figure 8.52
We take a horizontal slice of thickness Δy.

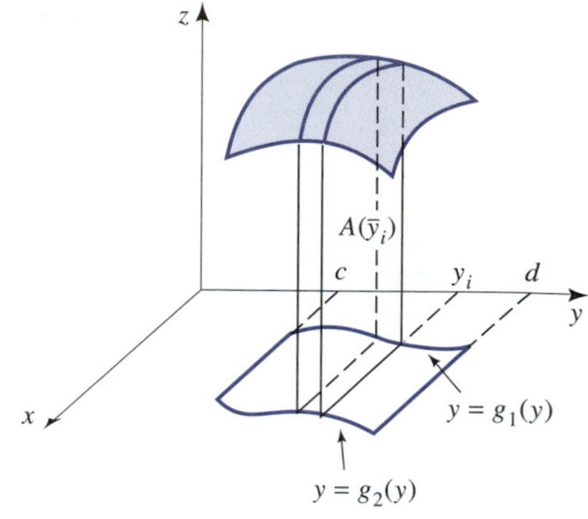

Figure 8.53
The slice shown in Figure 8.52 cuts out the thin slice of volume shown.

In this iterated integral we say that we are *integrating with respect to x first*, whereas in the previous iterated integral we say that we are *integrating with respect to y first.*

Remarks The following two remarks are critical:

1. Notice that $\int_{g_1(y)}^{g_2(y)} f(x, y)\,dx$ is to be done *first*.
2. When calculating $\int_{g_1(y)}^{g_2(y)} f(x, y)\,dx$, *treat y as a constant.*

One should get in the habit when taking a double integral over a region such as that shown in Figure 8.52 of following the procedures listed below.

Procedure for Setting up an Iterated Integral

1. Clearly label the "left" and the "right" of the domain D.
2. Draw the slice shown in Figure 8.52 so that one realizes that y is being held constant while x is varying from the "left" to the "right." This gives the limits of integration for the *inner* integral.
3. Now sum these slices "on y," that is, the slices go from c to d. This gives the limits of integration on the *outer* integral.

EXAMPLE 4 **Integrating with Respect to *x* First**

Redo Example 3 by integrating first with respect to x.

Solution

This region has a left side and a right side. The left side is given by $x = g_1(y) = y$, and the right side is given by $x = g_2(y) = \sqrt{y}$. See Figure 8.54. The slice shown in the figure indicates that we will integrate from the left to the right while holding

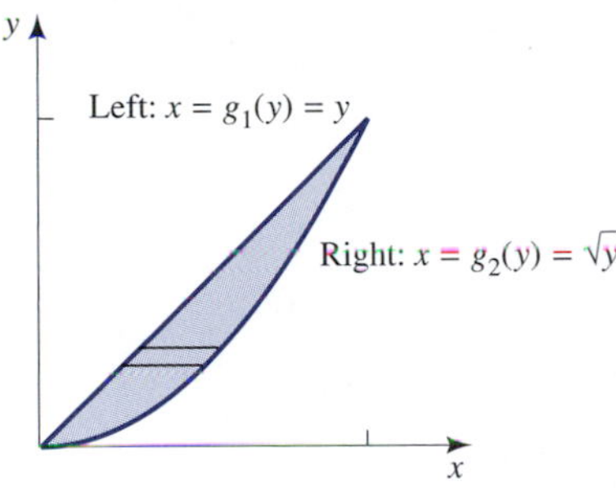

Figure 8.54
We take a horizontal slice of the thickness Δy.

y constant. Thus, the limits of integration on the inner integral go from $x = y$ to $x = \sqrt{y}$. Now the slices are summed on y from 0 to 1. This gives the limits of integration on the outer integral. Then

$$\begin{aligned}
\int\int_D f(x, y)\,dA &= \int_c^d \left(\int_{g_1(y)}^{g_2(y)} f(x, y)\,dx \right) dy \\
&= \int_0^1 \left(\int_y^{\sqrt{y}} 48xy\,dx \right) dy \\
&= \int_0^1 \left([24x^2y]\Big|_{x=y}^{x=\sqrt{y}} \right) dy \\
&= \int_0^1 (24y^2 - 24y^3)\,dy \\
&= (8y^3 - 6y^4)\Big|_0^1 \\
&= 2
\end{aligned}$$

■

Applications

We have already defined the average value of a continuous function $f(x)$ of *one* variable on the interval $[a, b]$ to be

$$\text{average value} = \frac{1}{b-a}\int_a^b f(x)\,dx$$

For a nonnegative function f this is the same as the area under the function $y = f(x)$ on the interval $[a, b]$ divided by the length of this interval.

There is a similar definition for the **average value** of a function $z = f(x, y)$ of *two* variables on a domain D in the xy-plane such as that shown in Figure 8.55.

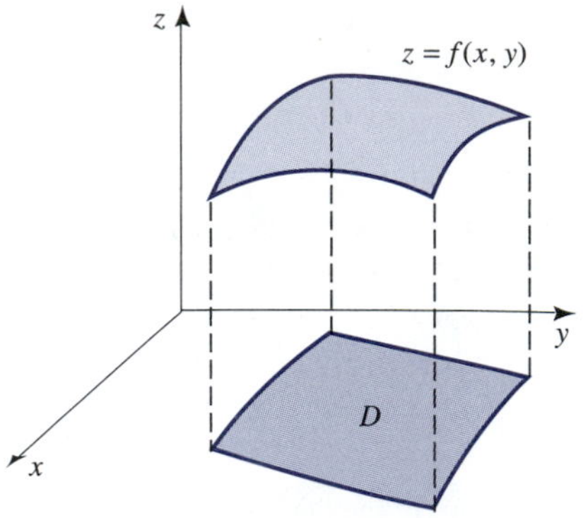

Figure 8.55
We want to define the average value of $z = f(x, y)$ over the region D.

DEFINITION Average Value

If $f(x, y)$ is a continuous function on a bounded domain D, the **average value** of f on D is

$$\frac{1}{A(D)} \int\int_D f(x, y)\, dA$$

where $A(D)$ is the area of the domain D.

Notice that for a nonnegative function f this is the same as the volume under the surface $z = f(x, y)$ and above the domain D divided by the area of the domain.

EXAMPLE 5 Average Concentration of Pollutants

An industrial plant is located at the precise center of a town shaped as a square with each side of length 4 miles. If the plant is placed at the point $(0, 0)$ of the xy-plane (see Figure 8.56), then certain pollutants are dispersed is such a manner that the concentration at any point (x, y) in town is given by

$$C(x, y) = 1000(24 - 3x^2 - 3y^2)$$

where $C(x, y)$ is the number of particles of pollutants per square mile of surface per day at a point (x, y) in town. What is the average concentration of these pollutants in this town?

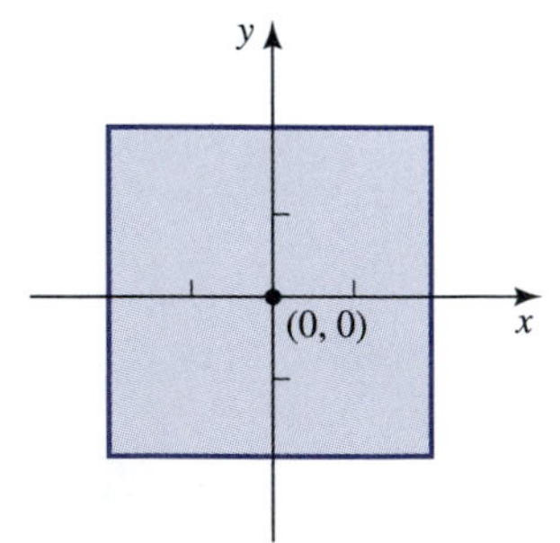

Figure 8.56
The industrial plant is located at the center of the town shaped as a square.

Solution

This concentration varies over the area of the town, being largest, naturally, at the center of town where the plant is located and smallest at the four corners of town. In fact, $C(0, 0) = 24{,}000$, while $C(2, 2) = C(2, -2) = C(-2, -2) = C(-2, 2) = 0$. The concentration function $C(x, y)$ could be obtained by taking samples at various points around the town and then "fitting a surface to the data" in a fashion analogous to the way we fitted curves to data points by the method of least squares in Chapter 2 when dealing with functions of one variable. Each sample could be a pollutant count over, say, a single square foot of surface. Multiplying this number by the number of square feet in a square mile then gives the number of particles per square mile. This last number is then $C(x, y)$ at the center of the 1 mile by 1 mile square.

The average value is the double integral $\int\int_D C(x, y)\,dA$ divided by the area of D, where D is the square shown in Figure 8.56. If the domain D were some irregularly shaped region, we would have to resort to Riemann sums to approximate this double integral. But the region D has a "bottom" and "top" and also a "left" and "right." Thus, we can evaluate this double integral as an iterated integral and also can then integrate first with respect to x or first with respect to y. We integrate first with respect to x, as indicated by the slice in Figure 8.57. The left is given by $x = -2$, and the right is given by $x = 2$. The slices are summed from $y = -2$ to $y = 2$. Thus,

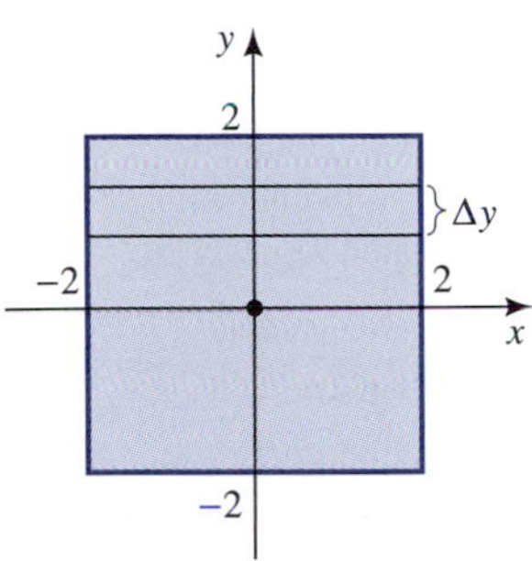

Figure 8.57
We take a horizontal slice of thickness Δy.

$$\begin{aligned}
\int\int_D C(x, y)\,dA &= 1000\int_{-2}^{2}\left(\int_{-2}^{2}\left[24 - 3x^2 - 3y^2\right]dx\right)dy \\
&= 1000\int_{-2}^{2}\left(\left[24x - x^3 - 3xy^2\right]\Big|_{x=-2}^{x=2}\right)dy \\
&= 1000\int_{-2}^{2}\left([48 - 8 - 6y^2] - [-48 + 8 + 6y^2]\right)dy \\
&= 1000\int_{-2}^{2}\left(80 - 12y^2\right)dy \\
&= 1000\left(80y - 4y^3\right)\Big|_{-2}^{2} \\
&= 1000[(160 - 32) - (-160 + 32)] \\
&= 256{,}000
\end{aligned}$$

The area of D is the area of the square and is 16. Thus, the average concentration of pollutants is $256{,}000/16 = 16{,}000$ particles per square mile, or about 0.0006 particle per square foot per day. ■

Warm Up Exercise Set 8.6

1. Find the volume under the surface given by $z = 10 - x^2 - y^2$ and over the domain given by the rectangle

$$D = \{(x, y): 0 \le x \le 1, 0 \le y \le 3\}$$

that is, find

$$\int\int_D (10 - x^2 - y^2)\,dA$$

2. Find the volume under the surface given by $z = 8xy$ over the domain given by the triangular region D in the xy-plane bounded by $y = x$, $y = 0$, and $x = 1$ by integrating first with respect to y.

3. Redo Exercise 2 by integrating first with respect to x.

Exercise Set 8.6

In Exercises 1 through 4, find the Riemann sum S_n for the given n for $f(x, y) = 6xy^2$ on the rectangle $R = \{(x, y): 0 \le x \le 2, 0 \le y \le 1\}$ by taking the points (x_i, y_i) at the indicated points of each subrectangle. (Compare with Example 1 of the text.)

1. S_2, (x_i, y_i) at the center of each subrectangle
2. S_3, (x_i, y_i) at the center of each subrectangle
3. S_2, (x_i, y_i) at the lower left of each subrectangle

4. S_4, (x_i, y_i) at the center of each subrectangle

5. Find S_2 for $z = x^2y$ for $D = \{(x, y) | 0 \le x \le 4, 0 \le y \le 2\}$, where (x_i, y_i) are taken at the lower left of each subrectangle.

6. Repeat Exercise 5 for S_4.

In Exercises 7 through 28, find the double integral over the indicated region D in two ways. (a) Integrate first with respect to x. (b) Integrate first with respect to y.

7. $\int\int_D (2x + 3y)\,dA$, $D = \{(x, y) : 0 \le x \le 1, 0 \le y \le 1\}$

8. $\int\int_D (x^2 + y)\,dA$, $D = \{(x, y) : 0 \le x \le 1, 0 \le y \le 1\}$

9. $\int\int_D xy\,dA$, $D = \{(x, y) : 0 \le x \le 2, 0 \le y \le 1\}$

10. $\int\int_D xy\,dA$, $D = \{(x, y) : 0 \le x \le 1, 0 \le y \le 3\}$

11. $\int\int_D 6x^2y\,dA$, $D = \{(x, y) : 1 \le x \le 2, 2 \le y \le 3\}$

12. $\int\int_D 6xy^2\,dA$, $D = \{(x, y) : 1 \le x \le 2, 2 \le y \le 3\}$

13. $\int\int_D y\sqrt{y^2 + x}\,dA$, $D = \{(x, y) : 0 \le x \le 1, 0 \le y \le 1\}$

14. $\int\int_D x\sqrt{x^2 + y}\,dA$, $D = \{(x, y) : 0 \le x \le 1, 0 \le y \le 4\}$

15. $\int\int_D \frac{1}{xy}\,dA$, $D = \{(x, y) : 1 \le x \le 2, 1 \le y \le 3\}$

16. $\int\int_D \left(1 + \frac{x}{y}\right)\,dA$, $D = \{(x, y) : 1 \le x \le 2, 1 \le y \le 3\}$

17. $\int\int_D e^{x+y}\,dA$, $D = \{(x, y) : 0 \le x \le 2, 0 \le y \le 3\}$

18. $\int\int_D xye^{0.5(x^2+y^2)}\,dA$, $D = \{(x, y) : 0 \le x \le 2, 0 \le y \le 3\}$

19. $\int\int_D 1/(1 + x + y)^3\,dA$, $D = \{(x, y) : 0 \le x \le 2, 0 \le y \le 3\}$

20. $\int\int_D 1/(1 + x + y)^2\,dA$, $D = \{(x, y) : 1 \le x \le 2, 1 \le y \le 3\}$

21. $\int\int_D xe^{x^2}\,dA$, $D = \{(x, y) : 0 \le x \le 2, 0 \le y \le 1\}$

22. $\int\int_D ye^{y^2}\,dA$, $D = \{(x, y) : 0 \le x \le 2, 0 \le y \le 1\}$

23. $\int\int_D (x + y)\,dA$, D is the triangular region with vertices at $(0, 0)$, $(1, 0)$, $(1, 1)$

24. $\int\int_D (x + y)\,dA$, D is the triangular region with vertices at $(0, 0)$, $(0, 1)$, $(1, 1)$

25. $\int\int_D (x^2 + y^2)\,dA$, D is the triangular region with vertices at $(0, 0)$, $(1, 0)$, $(1, 1)$

26. $\int\int_D (x^2 + y^2)\,dA$, D is the triangular region with vertices at $(0, 0)$, $(0, 1)$, $(1, 1)$

27. $\int\int_D xy\,dA$, D is the triangular region with vertices at $(0, 0)$, $(1, 0)$, $(1, 1)$

28. $\int\int_D x^2y\,dA$, D is the triangular region with vertices at $(1, 0)$, $(2, 0)$, $(2, 1)$

In Exercises 29 through 34, find the volume under the surface of the given function and over the indicated region.

29. $f(x, y) = 1$, D is the region in the first quadrant bounded by the curves $y = x$ and $y = x^3$.

30. $f(x, y) = 1$, D is the region in the first quadrant bounded by the curves $y^2 = x^3$ and $y = x$.

31. $f(x, y) = xy^2$, D is the region in the first quadrant bounded by the curves $y = x$ and $y = x^2$.

32. $f(x, y) = \frac{x^2}{y}$, D is the region in the first quadrant bounded by the curves $x = 0$, $x = y^{3/2}$, $y = 1$, and $y = 2$.

33. $f(x, y) = xy$, D is the region in the first quadrant bounded by the curves $y = x$ and $x = y^2$.

34. $f(x, y) = xe^y$, D is the region in the first quadrant bounded by the curves $y = 0$, $y = x^2$, $x = 0$, and $x = 1$.

In Exercises 35 through 42, integrating first with respect to one of x or y is difficult, while integrating first with respect to the other is not. Find $\int\int_D f(x, y)\,dA$ the easier way.

35. $f(x, y) = x$, D is the region bounded by the curves $y = x^2 + 1$, $y = x$, $x = -1$, and $x = 1$.

36. $f(x, y) = x$, D is the region in the fourth quadrant bounded by the curves $y = -x^2$, $y = -3x - 1$, $x = 0$, and $x = 2$.

37. $f(x, y) = 1$, D is the region in the first and fourth quadrants bounded by the curves $x = 1 - y$, $y = x^2 - 1$, and $x = 0$.

38. $f(x, y) = 1$, D is the region in the first quadrant bounded by the curves $x = y - 2$, $x = y^3$, $x = 0$, and $x = 1$.

39. $f(x, y) = 1$, D is the region in the first quadrant bounded by the curves $xy = 1$, $y = x$, and $y = 2$.

40. $f(x, y) = y$, D is the region in the first quadrant bounded by the curves $x = 1$ and $x = y^2$.

41. $f(x, y) = y$, D is the region in the first quadrant bounded by the curves $y = 2x$, $y = x$, and $x = 2$.

42. $f(x, y) = y$, D is the region in the first quadrant bounded by the curves $y = \frac{1}{2}x$, $y = x$, and $y = 2$.

Applications and Mathematical Modeling

43. Concentration of Pollutants In Example 5 of the text, find the average concentration of pollutants in the region

$$D_1 = \{(x, y) : 0 \le x \le 1, 0 \le y \le 2\}$$

44. Concentration of Pollutants If D is the region given in Example 5 and D_1 is the region given in Exercise 43, find the average concentration of pollutants over the region consisting of D minus D_1.

45. Cobb-Douglas Production Function Find the average value of the Cobb-Douglass production function $f(x, y) = x^{1/3}y^{2/3}$ for the range of x and y given by

$$D = \{(x, y) : 8 \le x \le 27, 1 \le y \le 8\}$$

46. Cobb-Douglas Production Function Find the average value of the Cobb-Douglass production function $f(x, y) = x^{2/3}y^{1/3}$ for the range of x and y given by

$$D = \{(x, y) : 8 \le x \le 27, 1 \le y \le 8\}$$

More Challenging Exercises

47. The iterated integral

$$\int_0^1 \left(\int_y^1 2e^{x^2}\, dx \right) dy$$

is impossible to do if the integration must be first with respect to x. Change the order of integration, and find the double integral.

48. Density The concentration $C(x, y)$ given in Example 5 is an example of a **density function** that we usually denote in general by $\delta(x, y)$. The density function gives the number per unit area. In the case of the pollutants we counted the number of particles per square mile. In general, if $\delta(x, y)$ is a continuous function that gives the density in number per unit of area, then the total number in D is given by

$$\int\int_D \delta(x, y)\, dA$$

Find the total number of particles of pollutant in town per day in Example 5.

49. Population The density function for a coastal town is given by

$$\delta(x, y) = \frac{120{,}000}{(2 + x + y)^3}$$

in people per square mile, where x and y are in miles and where the town is the square

$$D = \{(x, y) : 0 \le x \le 2, 0 \le y \le 2\}$$

(The shore runs along the x-axis.) Find the total population of this town.

50. Population Find the total population living in the town in Exercise 49 that are within one mile of shore.

51. For the integral in Exercise 9, use your computer or graphing calculator to find one or two Riemann sums for each $n = 20$, 40, and 100. For example, you could take (x_i, y_i) at the lower left of the subrectangles to obtain one Riemann sum and take (x_i, y_i) at the upper right of the subrectangles for another Riemann sum. By examining the integrand, show that the first must be a lower bound and the second must be an upper bound.

52. For the integral in Exercise 15, use your computer or graphing calculator to find one or two Riemann sums for each $n = 20$, 40, and 100. For example, you could take (x_i, y_i) at the lower left of the subrectangles to obtain one Riemann sum and take (x_i, y_i) at the upper right of the subrectangles for another Riemann sum. By examining the integrand, show that the first must be an upper bound and the second must be a lower bound.

Solutions to WARM UP EXERCISE SET 8.6

1. The region D has a bottom given by $y = g_1(x) = 0$ and a top given by $y = g_2(x) = 3$. See the figure. The slice shown in the figure indicates that while holding x constant, we integrate with respect to y from the bottom to the top. Thus, the limits of integration on the inner integral go from $y = 0$ to $y = 3$. Now the slices are summed on x from 0 to 1. This gives the limits of integration on the outer integral. Thus,

$$\int\int_D f(x, y)\,dA = \int_a^b \left(\int_{g_1(x)}^{g_2(x)} f(x, y)\,dy \right) dx$$
$$= \int_0^1 \left(\int_0^3 (10 - x^2 - y^2)\,dy \right) dx$$
$$= \int_0^1 \left(\left[10y - x^2y - \frac{1}{3}y^3 \right] \Big|_{y=0}^{y=3} \right) dx$$
$$= \int_0^1 (30 - 3x^2 - 9)\,dx$$
$$= \int_0^1 (21 - 3x^2)\,dx$$
$$= (21x - x^3)|_0^1 = 20$$

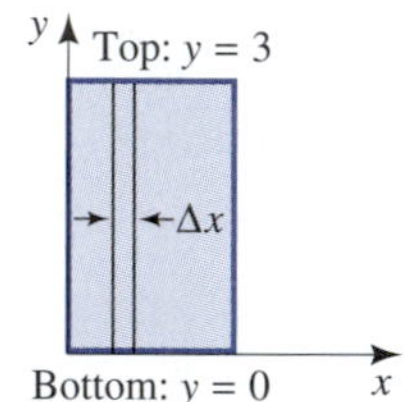

2. The region D has a bottom given by $y = g_1(x) = 0$ and a top given by $y = g_2(x) = x$. See the figure. The slice shown in the figure indicates that while holding x constant, we integrate with respect to y from the bottom to the top. Thus, the limits of integration on the inner integral go from $y = 0$ to $y = x$. Now the slices are summed on x from 0 to 1. This gives the limits of integration on the outer integral. Thus,

$$\int\int_D f(x, y)\,dA = \int_a^b \left(\int_{g_1(x)}^{g_2(x)} f(x, y)\,dy \right) dx$$
$$= \int_0^1 \left(\int_0^x 8xy\,dy \right) dx$$
$$= \int_0^1 \left([4xy^2]|_{y=0}^{y=x} \right) dx$$
$$= \int_0^1 4xx^2\,dx$$
$$= x^4|_0^1 = 1$$

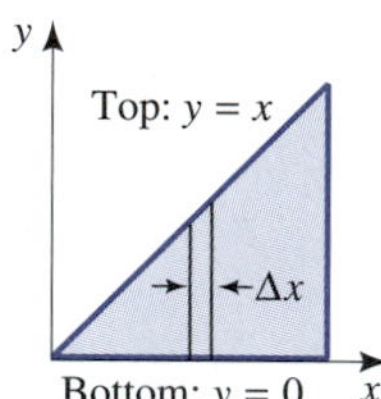

3. We notice that this region does have a left side and a right side. The left side is given by $x = g_1(y) = y$, and the right side is given by $x = g_2 = 1$. See the figure. The slice shown in the figure indicates that we will integrate from the left to the right while holding y constant. Thus, the limits of integration on the inner integral go from $x = y$ to $x = 1$. Now the slices are summed on y from 0 to 1. This gives the limits of integration on the outer integral. Then

$$\int\int_D f(x, y)\,dA = \int_c^d \left(\int_{g_1(y)}^{g_2(y)} f(x, y)\,dx \right) dy$$
$$= \int_0^1 \left(\int_y^1 8xy\,dx \right) dy$$
$$= \int_0^1 \left(4x^2y|_{x=y}^{x=1} \right) dy$$
$$= \int_0^1 (4y - 4y^3)\,dy$$
$$= (2y^2 - y^4)|_0^1 = 1$$

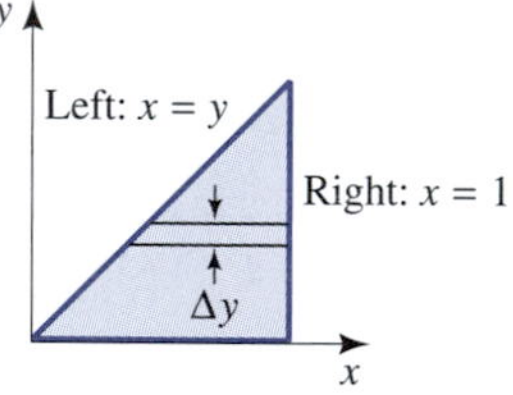

Summary Outline

- A **function of two variables**, denoted by $z = f(x, y)$, is a rule that associates to every point (x, y) in some set D (called the **domain**) a unique point denoted by $z = f(x, y)$. (p. 524)
- The **distance between the two points** (x_1, y_1, z_1) and (x_2, y_2, z_2) is

$$d = \sqrt{(x_2 - x_1)^2 + (y_2 - y_1)^2 + (z_2 - z_1)^2} \quad \text{(p. 528)}$$

- Given a function $z = f(x, y)$ with domain D, the **graph** is the set of points (x, y, z) with $z = f(x, y)$ and $(x, y) \in D$. Since this is described by *two* variables, the result is a two-dimensional object called a **surface**. (p. 529)
- The equation of the **sphere** centered at (x_0, y_0, z_0) with radius r is given by

$$(x - x_0)^2 + (y - y_0)^2 + (z - z_0)^2 = r^2 \quad \text{(p. 530)}$$

- The graph of an equation of the form $Ax + By + Cz = D$ is a **plane**. (p. 530)
- The graph of $f(x, y) = z_0$ in the xy-plane is called a **level curve**. (p. 531)
- We say that $f(x, y)$ is **continuous** at (a, b), if $f(x, y)$ approaches $f(a, b)$ no matter how (x, y) approaches (a, b). (p. 538)
- If $z = f(x, y)$, then the **partial derivative of f with respect to x** is

$$\frac{\partial f}{\partial x}(x, y) = \lim_{\Delta x \to 0} \frac{f(x + \Delta x, y) - f(x, y)}{\Delta x} \quad \text{(p. 540)}$$

and the **partial derivative of f with respect to y** is

$$\frac{\partial f}{\partial y}(x, y) = \lim_{\Delta y \to 0} \frac{f(x, y + \Delta y) - f(x, y)}{\Delta y}$$

if the limits exist. (p. 540)

- Suppose we have two commodities with unit prices p and q and respective demand equations $x = f(p, q)$ and $y = g(p, q)$. If

$$\frac{\partial x}{\partial q} > 0 \quad \text{and} \quad \frac{\partial x}{\partial p} > 0$$

then we say that the two commodities are **competitive**. If

$$\frac{\partial x}{\partial q} < 0 \quad \text{and} \quad \frac{\partial x}{\partial p} < 0$$

then we say that the two commodities are **complementary**. (p. 545)

- Given a function $z = f(x, y)$, we define the following four **second-order partial derivatives**:

$$\frac{\partial^2 f}{\partial x^2} = \frac{\partial}{\partial x}\left(\frac{\partial f}{\partial x}\right) \qquad \frac{\partial^2 f}{\partial y^2} = \frac{\partial}{\partial y}\left(\frac{\partial f}{\partial y}\right)$$

$$\frac{\partial^2 f}{\partial y \partial x} = \frac{\partial}{\partial y}\left(\frac{\partial f}{\partial x}\right) \qquad \frac{\partial^2 f}{\partial x \partial y} = \frac{\partial}{\partial x}\left(\frac{\partial f}{\partial y}\right) \quad \text{(p. 547)}$$

- We say that f has a **relative maximum** at $(a, b) \in D$ if there exists a circle centered at (a, b) and entirely in D such that

$$f(x, y) \le f(a, b)$$

for all points (x, y) inside this circle. (p. 552)
We say that f has a **relative minimum** at $(a, b) \in D$ if there exists a circle centered at (a, b) and entirely within D such that

$$f(x, y) \ge f(a, b)$$

for all points (x, y) inside this circle. (p. 552)
We say that the function $z = f(x, y)$ has a **relative extremum** at $(a, b) \in D$ if f has either a relative maximum or a relative minimum at (a, b). (p. 552)

- **Necessary Condition for Relative Extrema** If $z = f(x, y)$ is defined, both first partial derivatives exist for all values of (x, y) inside some circle about (a, b), and f assumes a relative extremum at (a, b), then

$$\frac{\partial f}{\partial x}(a, b) = 0 \quad \text{and} \quad \frac{\partial f}{\partial y}(a, b) = 0 \quad \text{(p. 554)}$$

- Suppose that both partial derivatives of $z = f(x, y)$ exist in its domain of definition. **Critical points** are points (a, b), not on the boundary of the domain of f, for which

$$\frac{\partial f}{\partial x}(a, b) = 0 \quad \text{and} \quad \frac{\partial f}{\partial y}(a, b) = 0 \quad \text{(p. 554)}$$

- **Second Derivative Test for Functions of Two Variables** For $z = f(x, y)$, assume that f_{xx}, f_{yy}, and f_{xy} all exist and are continuous at every point near (a, b). Suppose that (a, b) is a critical point, that is, assume that $f_x(a, b) = 0$ and $f_y(a, b) = 0$. Define

$$\Delta(a, b) = f_{xx}(a, b) f_{yy}(a, b) - \left[f_{xy}(a, b)\right]^2$$

Then:

1. $\Delta(a, b) > 0$ and $f_{xx}(a, b) < 0$ implies that $f(a, b)$ is a relative maximum.
2. $\Delta(a, b) > 0$ and $f_{xx}(a, b) > 0$ implies that $f(a, b)$ is a relative minimum.
3. $\Delta(a, b) < 0$ implies (a, b) is a saddle point, that is, (a, b) is neither a relative minimum nor a relative maximum.
4. $\Delta(a, b) = 0$ implies that the test fails, that is, when $\Delta(a, b) = 0$, there can be a relative maximum, a relative minimum, or a saddle. (p. 555)

- We say that f has an **absolute maximum** on D at $(a, b) \in D$ if for all $(x, y) \in D$, $f(x, y) \le f(a, b)$.
We say that f has an **absolute minimum** on D at $(a, b) \in D$ if for all $(x, y) \in D$, $f(x, y) \ge f(a, b)$.
Also we say that f has an **absolute extremum** on D at $(a, b) \in D$ if f has either an absolute maximum or an absolute minimum on D at (a, b). (p. 557)
- **Existence of Absolute Extrema on a Closed and Bounded Region** If $z = f(x, y)$ is continuous on the closed and bounded region D, then f assumes its absolute extrema in D, that is, there exist two points $(a, b) \in D$ and $(A, B) \in D$ such that

$$f(a, b) \le f(x, y) \le f(A, B) \qquad \text{for all } (x, y) \in D$$

(p. 558)

- **Method of Lagrange Multipliers** Candidates (x_c, y_c, z_c) for a relative extremum of $f(x, y, z)$, subject to the constraint $g(x, y, z) = 0$, can be found among the critical points (x_c, y_c, z_c, λ) of the auxiliary function

$$F(x, y, z, \lambda) = f(x, y, z) + \lambda g(x, y, z)$$

that is, among the solutions of the equations

$$F_x = 0 \qquad F_y = 0 \qquad F_z = 0 \qquad g = 0 \quad \text{(p. 565)}$$

- **Equation of Tangent Plane** The plane that is tangent to the surface $z = f(x, y)$ at the point $P = (a, b, z_0)$ is given by the equation

$$z - z_0 = f_x(a, b)(x - a) + f_y(a, b)(y - b)$$

where $z_0 = f(a, b)$. (p. 573)

- **Tangent Plane Approximations** If the first-order partial derivatives of $f(x, y)$ are continuous at (a, b) and if Δx and Δy are small, then

$$\begin{aligned}\Delta z &= f(a + \Delta x, b + \Delta y) - f(a, b)\\ &\approx f_x(a, b) \cdot \Delta x + f_y(a, b) \cdot \Delta y \quad \text{(p. 573)}\end{aligned}$$

- **Riemann Sum** For any integer n, divide both intervals $[a, b]$ and $[c, d]$ into n equal subintervals and define

$$\Delta x = \frac{b - a}{n} \qquad \text{and} \qquad \Delta y = \frac{d - c}{n}$$

This gives a collection of subrectangles. Label all subrectangles that lie *entirely inside* D (the domain of $f(x, y)$) in some manner. Let the total number of these subrectangles be N. In each of these subrectangle R_i, pick some point $(x_i, y_i) \in R_i$. The Riemann sum is defined to be

$$\sum_{i=1}^{N} f(x_i, y_i)\Delta x \Delta y \quad \text{(p. 579)}$$

- Let $f(x, y)$ be continuous on the bounded region D. Then the **definite integral of f over D** is

$$\int\int_D f(x, y)\,dA = \lim_{n\to\infty} \sum_{i=1}^{N} f(x_i, y_i)\Delta x \Delta y$$

where the limit exists and is the same no matter how the points (x_i, y_i) are chosen in R_i. (p. 581)

- Let $f(x, y)$ be continuous on the bounded region D and nonnegative there. Then the volume under the surface $z = f(x, y)$ and over the domain D is the definite integral of f over D or

$$V = \int\int_D f(x, y)\,dA \quad \text{(p. 581)}$$

- If the domain D is given as in the following figure and $z = f(x, y) \geq 0$ is continuous on D, then the volume V between the surface $z = f(x, y)$, and the domain D is

$$\int\int_D f(x, y)\,dA = \int_a^b \left(\int_{g_1(x)}^{g_2(x)} f(x, y)\,dy\right) dx \quad \text{(p. 583)}$$

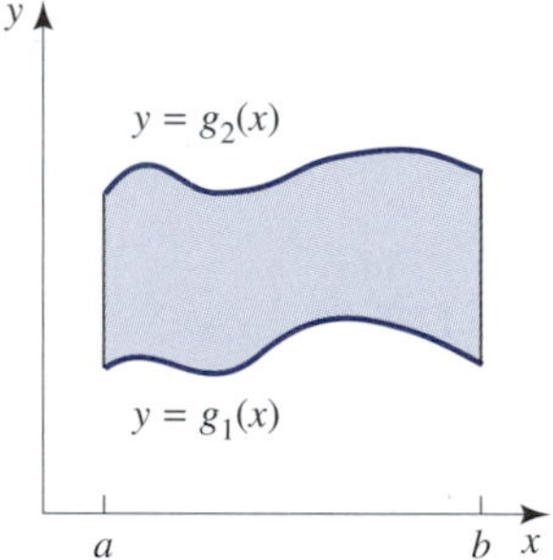

- If the domain D is given as in the figure and $z = f(x, y) \geq 0$ is continuous on D, then the volume V between the surface $z = f(x, y)$, and the domain D is

$$\int\int_D f(x, y)\,dA = \int_c^d \left(\int_{g_1(y)}^{g_2(y)} f(x, y)\,dx\right) dy \quad \text{(p. 585)}$$

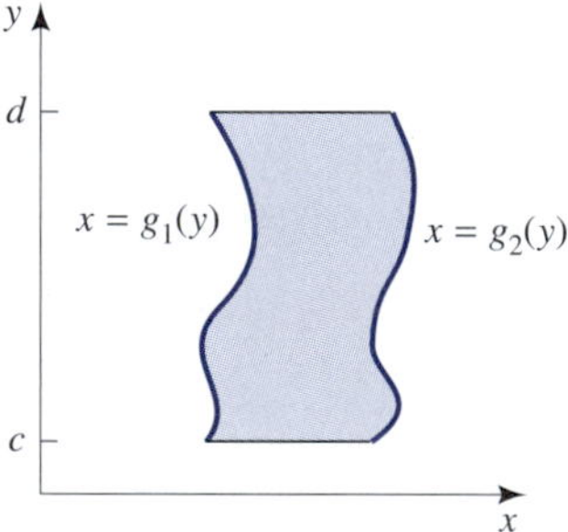

Review Exercises

1. Evaluate the function $z = f(x, y) = 2x^2 + 3y - 1$ at the points $(0, 0)$, $(1, 2)$, and $(2, -1)$.
2. Evaluate the function $z = f(x, y) = \dfrac{2x^2}{5y}$ at the points $(0, 1)$, $(1, 2)$, and $(2, -1)$.
3. Determine the domains of the functions in Exercises 1 and 2.
4. Find the distance between the two points $(1, -2, 5)$ and $(3, -4, -1)$.
5. For the function $z = f(x, y) = 16 - x^2 - y^2$, find the level surfaces corresponding to $z_0 = 16$, $z_0 = 12$, $z_0 = 7$. Describe the surface $z = f(x, y)$.
6. Find both first-order partial derivatives to the function $z = f(x, y) = x + x^2y^5 + 10$.
7. Evaluate the two first-order partial derivatives found in Exercise 6 at the point $(2, 1)$.
8. Find both first-order partial derivatives to the function $z = f(x, y) = e^{-xy} + \ln xy$.

9. Find all three second-order partial derivatives of the function in Exercise 6.

10. Find all three second-order partial derivatives of the function in Exercise 8.

11. Find all three first-order partial derivatives to the function $z = f(x, y, z) = e^{-xyz} + xy^2z^3$.

In Exercises 12 through 19, find all the critical points, and determine whether each critical point is a relative minimum, a relative maximum, or a saddle point.

12. $f(x, y) = x^2 - xy + 4y^2$

13. $f(x, y) = -x^2 + 2xy - 8y^2$

14. $f(x, y) = -x^2 + xy - y^2 - 3x$

15. $f(x, y) = (x - 2)^2 + (y + 3)^2$

16. $f(x, y) = x^3 + y^3 + 3xy$

17. $f(x, y) = 3x^2 + y^3 - 6xy - 9y$

18. $f(x, y) = y^3 - 3y^2 - x^2 - 9y + 10x$

19. $f(x, y) = 12xy - x^3 - 36y^2$

20. Let $z = f(x, y) = ax^2 + 2bxy + cy^2$, where a, b, and c are constants. Show that $(0, 0)$ is a critical point. Show that the point $(0, 0)$ is a relative minimum if $ac - b^2 > 0$ and $a > 0$ and a relative maximum if $ac - b^2 > 0$ and $a < 0$.

21. Using Lagrange multipliers, minimize $z = 4x^2 = 2xy - 3y^2$ subject to the constraint $x + y = 1$.

22. Using Lagrange multipliers, maximize $z = -2x^2 - xy - y^2 - x$ subject to the constraint $2x + y - 1 = 0$.

23. Use the tangent line approximation to estimate

$$\Delta z = f(x_0 + \Delta x, y_0 + \Delta y) - f(x_0, y_0)$$

if $f(x, y) = 2x^2 + 3y^2$, $(x_0, y_0) = (1, 2)$, $\Delta x = 0.1$, $\Delta y = 0.2$.

24. Repeat Exercise 23 if $f(x, y) = x^2y^4$, $(x_0, y_0) = (1, 2)$, $\Delta x = 0.2$, and $\Delta y = -0.1$.

25. Use the tangent line approximation to estimate $\sqrt{3.01^2 + 3.98^2}$.

In Exercises 26 through 31, find the double integral over the indicated region D in two ways. (a) Integrate first with respect to x; (b) integrate first with respect to y.

26. $\int\int_D (3x + 6y)\, dA$, $D = \{(x, y)|0 \leq x \leq 2, 1 \leq y \leq 2\}$.

27. $\int\int_D (12x^2y^3)\, dA$, $D = \{(x, y)|0 \leq x \leq 1, 1 \leq y \leq 3\}$.

28. $\int\int_D (x^2 + y)\, dA$, where D is the triangular region with vertices at $(0, 0)$, $(1, 0)$, and $(1, 1)$.

29. $\int\int_D (x + xy)\, dA$, where D is the triangular region with vertices at $(0, 0)$, $(0, 1)$, and $(1, 1)$.

30. $\int\int_D (x + y)\, dA$, where D is the region between the two curves $y = 1$ and $y = x^2$.

31. $\int\int_D (x + y)\, dA$, where D is the region between the two curves $y = x$ and $y = \sqrt{x}$.

In Exercises 32 through 35, change the order of integration, and then evaluate.

32. $\int_0^1 \int_0^x (x + y)\, dy\, dx$

33. $\int_0^1 \int_0^{\sqrt{x}} (x + y)\, dy\, dx$

34. $\int_0^1 \int_0^{\sqrt{y}} (x + y)\, dx\, dy$.

35. $\int_0^1 \int_y^{\sqrt{y}} dx\, dy$.

36. Find the average value of $f(x, y) = 8xy^3$ on the region $D = \{(x, y)|0 \leq x \leq 2, 1 \leq y \leq 3\}$.

37. **Cobb-Douglas Production Function** Synnott[26] reported that real GNP in the United States in billions of 1972 dollars was approximated by the equation $Q(L, E) = 2.014L^{0.99}E^{0.47}$, where L is employment in thousands and E is energy consumption in quadrillions of BTUs. Find the first partial derivatives of the production function, and state what each means.

38. **Supply Equation** To analyze the impact of U.S.-Mexico trade in the North American dry onion economy, Fuller and colleagues[27] estimated a supply equation for dry onions to be

$$Q = 5868X^{0.17}Y^{0.13}T^{0.02}$$

where Q is the total seasonal onion production in 1000 cwts, X and Y two indices of onion prices, and T the time in years, with $T = 1$ corresponding to the year 1969. Find the partial derivative of Q with respect to T and explain what this means.

39. **Forestry** When planning stand-density management regimes, foresters need an objective means of determining how much growing space must be allocated to individual

---

[26] Thomas W. Synnott III. 1982. Energy and productivity. *Bus. Econ.* 17(1):63–64.

[27] Stephen Fuller, Melanie Gillis, and Houshmand A. Ziari. 1996. Effect of liberalized U.S.-Mexico dry onion trade: A spacial and intertemporal equilibrium analysis. *J. Agr. Appl. Econ.* 28:135–147.

trees to ensure that they reach the desired size at a specific time in stand development. Seymour and Smith[28] determined that the crown projection area A for white pine was approximated by the equation $A = 1265D^{3.589}H^{-2.514}$, where D is diameter at breast height in inches and H is height in feet. The crown projection area is the area of ground that the tree takes up in a stand. If the goal is to harvest white pine when they are 80 feet tall and have a diameter at breast height of 16 inches, find the crown projection area and the number of such trees possible per acre. (One acre is 43,560 ft^2.)

40. **Cost** A cost function is given by $C(x, y) = 1000 + x^2 + y^2 - xy$. Find $C(10, 1)$.

41. **Cost** Find the first partial derivatives of the cost function given in the previous exercise.

42. **Cost** Find all the second partial derivatives of the cost function given in the previous exercise.

43. **Continuous Compounding** If $\$P$ is invested at an annual rate of r and compounded continuously, the amount after t years is $A(P, r, t) = Pe^{rt}$. Find the three first partial derivatives of this function.

44. **Packaging** Find the dimensions of the cheapest rectangular box of volume 16 ft^3 for which the cost per square foot of the top and bottom is twice that of the sides.

45. **Minimizing Production Costs** The daily cost function of a firm that produces x units of a product X and y units of another product Y is given by $C(x, y) = 244 + x^2 - 20x + y^2 - 24y$. Find the number of units of each of the products that will minimize the cost.

46. **Maximizing Profit** The firm in Exercise 45 can sell each unit of product X for \$2 and each unit of product Y for \$4. Find the levels of production that will maximize the profits.

47. **Competitive Pricing** A firm makes x thousand units of regular toothpaste and y thousand units of mint. Suppose the prices are p and q, respectively, and the demand equations are given by

$$p = 4 - x + y$$
$$q = 14 + x - 2y$$

Find the number of units of each that will maximize revenue.

48. **Fishery** A small lake is to be stocked with two types of fish. There are x thousand of the first fish and y thousand of the second, the average weight of the first is $(3 - 2x - y)$ pounds, and that of the second is $(4 - x - 3y)$. Find the numbers of each fish that maximize the total weight of these fish.

49. **Measurements** A rectangular cardboard box of dimensions 10 in. by 10 in. by 20 in. is being produced. Use the tangent plane approximation to approximate the volume of cardboard if the sides are 0.15 in. thick and the top and bottom are 0.25 in. thick.

50. **Fishery** Karpoff[29] used the function $h = h(e, N, X)$, where h = individual firm harvest rate, e = individual firm effort rate, N = number of participating firms (or vessels), and X = resource stock size. He then assumed that

$$\frac{\partial h}{\partial e} > 0 \qquad \frac{\partial h}{\partial N} < 0 \qquad \frac{\partial h}{\partial X} > 0$$

Explain why you think these three conditions are reasonable.

[28] Robert S. Seymour and David M. Smith. 1987. A new stocking guide formulation applied to Eastern white pine. *Forest Sci.* 33(2):469–484.

[29] Jonathan M. Karpoff. 1987. Suboptimal controls in common resource management: the case of the fishery. *J. Polit. Econ.* 95(1):179–194.

Chapter 8 Cases

CASE 1 The Theory of Economic Growth

This discussion is based on a paper by Robert M. Solow[30] (winner of the Nobel Prize in Economics).

Let the total rate of production for a country at time t be designated as $Y(t)$. Part of the output at any instant is consumed, and the rest is saved. We assume that the fraction s of output that is saved is a constant, so the rate of savings is $sY(t)$. We let $K(t)$ denote the stock of capital, and then net investment is the rate of increase of this capital stock, or $\frac{dK}{dt}$. Thus, we must have

$$\frac{dK}{dt} = sY$$

[30] Robert M. Solow. 1956. A contribution to the theory of economic growth. *Quart. J. Econ.* 70:65–94.

We assume a Cobb-Douglass production function of the form

$$Y = K^a L^{1-a}$$

where L is the rate of labor input. Thus,

$$\frac{dK}{dt} = sY = sK^a L^{1-a}$$

We now assume that the labor force increases at a constant rate n. Thus,

$$L(t) = L_0 e^{nt}$$

where L_0 is the initial labor force. Substituting this into the last equation then gives

$$\frac{dK}{dt} = sK^a (L_0 e^{nt})^{1-a}$$

(a) Integrate this last equation and find $K(t)$, assuming that K_0 is the initial capital stock. Now define $k = K/L$ as the level of output per unit of labor, and find an expression for $k(t)$.

(b) From part (a), determine approximately what $k(t)$ is for large t. (Note: The term $\left(\frac{s}{n}\right)^{1/b}$ is referred to as the equilibrium value of the capital-labor ratio.)

(c) What happens to the long-term value of $k(t)$ if the savings rate s is larger? If the rate of labor growth n is larger? Does this model indicate that a country with higher savings rate will be richer or poorer? What about a country with a higher labor growth rate?

By looking at a large set of economic data on many countries, Mankiw et al.[31] showed that the Solow model given above is correct to a first approximation. Their paper then modifies the Solow model and shows that the modifications are even better predictors of the direction and magnitudes of the effects of savings and labor. The improvement comes from including another term that accounts for human capital. Specifically, this model assumes a production function of the form

$$Y(t) = K(t)^a H(t)^b L(t)^{1-a-b}$$

where H is the stock of human capital and the other variables are the same as in the Solow model.

[31] N. Gregory Mankiw, David Romer, and David N. Weil. 1992. A contribution to the empirics of economic growth. *Quart. J. Econ.* 107(2):407–437.

A Review

A.1 Exponents and Roots

Exponents
Roots
Rational Exponents
Rationalizing

Exponents

The symbol x^2 denotes $x \cdot x$, the symbol x^3 denotes $x \cdot x \cdot x$, and, in general, we have the following.

DEFINITION Exponents

If n is a positive integer

$$x^n = \underbrace{x \cdot x \cdot \cdots \cdot x}_{n \text{ factors}}.$$

The number n is called the **exponent** and x is called the **base.**

Examples

$2^4 = 2 \cdot 2 \cdot 2 \cdot 2 = 16$

$$\left(-\frac{1}{2}\right)^3 = \left(-\frac{1}{2}\right)\left(-\frac{1}{2}\right)\left(-\frac{1}{2}\right) = -\frac{1}{8}.$$

DEFINITION **Zero Exponent**

If $x \neq 0$, then

$$x^0 = 1.$$

The expression 0^0 is not defined.

Examples

$2^0 = 1 \qquad \pi^0 = 1 \qquad (a+b)^0 = 1$

DEFINITION **Negative Exponent**

If n is a positive integer, then

$$x^{-n} = \frac{1}{x^n}.$$

Examples

$2^{-1} = \frac{1}{2} \qquad x^{-3} = \frac{1}{x^3} \qquad \left(\frac{2}{3}\right)^{-2} = \frac{1}{\left(\frac{2}{3}\right)^2} = \frac{1}{\frac{4}{9}} = \frac{9}{4}$

Properties of Exponents

If m and n are integers, the following properties are true.

	Property	Example
1.	$x^m x^n = x^{m+n}$	$x^2 \cdot x^3 = x^{2+3} = x^5$
2.	$(x^m)^n = x^{mn}$	$(x^2)^3 = x^{2 \cdot 3} = x^6$
3.	$(xy)^n = x^n y^n$	$(2 \cdot 4)^3 = 2^3 \cdot 4^3 = 8 \cdot 64 = 512$
4.	$\left(\frac{x}{y}\right)^n = \frac{x^n}{y^n}$ if $y \neq 0$	$\left(\frac{2}{3}\right)^2 = \frac{2^2}{3^2} = \frac{4}{9}$
5.	$\frac{x^m}{x^n} = x^{m-n}$ if $x \neq 0$	$\frac{x^5}{x^2} = x^{5-2} = x^3$
	$\frac{x^m}{x^n} = \frac{1}{x^{n-m}}$ if $x \neq 0$	$\frac{x^2}{x^5} = \frac{1}{x^{5-2}} = \frac{1}{x^3}$

Examples

$(2x^2y^3)(3x^3y^4) = (2)(3)(x^2x^3)(y^3y^4) = 6x^5y^7$

$(2x^2y^3z)^4 = 2^4(x^2)^4(y^3)^4(z)^4 = 16x^8y^{12}z^4$

$\left(\frac{2x^2}{3y^3}\right)^3 = \frac{(2x^2)^3}{(3y^3)^3} = \frac{8x^6}{27y^9}$

$$(a^4b^{-2})^{-3} = (a^4)^{-3}(b^{-2})^{-3} = (a^{-12})(b^6) = \frac{b^6}{a^{12}}$$

$$(a^2b^3)^2(a^4b)^{-1} = (a^2b^3)^2\frac{1}{a^4b} = \frac{a^4b^6}{a^4b} = b^5$$

We can also do the last example as follows.

$$(a^2b^3)^2(a^4b)^{-1} = (a^4b^6)(a^{-4}b^{-1}) = a^{4-4}b^{6-1} = b^5$$

Roots

DEFINITION Roots

If n is a positive integer and $y^n = x$, then we say that y is an nth root of x.
If n is odd, there is exactly one root, which is denoted by $y = \sqrt[n]{x} = x^{1/n}$.
If n is even and x is negative, there is no (real) root.
If n is even and x is positive, we use the symbols $y = \sqrt[n]{x} = x^{1/n}$ to denote the *positive* nth root.
If $n = 2$, we write $\sqrt[2]{x} = \sqrt{x}$.

The symbol $\sqrt{\ }$ is called a **radical sign**.

Examples

$\sqrt{9} = 3$ $\qquad \sqrt[3]{-64} = -4$ $\qquad \sqrt[4]{\frac{1}{16}} = \frac{1}{2}$

Properties of Roots

	Radical Formulation	Exponent Formulation	
1.	$\sqrt[n]{xy} = \sqrt[n]{x}\sqrt[n]{y}$	$(xy)^{1/n} = x^{1/n}y^{1/n}$	
2.	$\sqrt[n]{\frac{x}{y}} = \frac{\sqrt[n]{x}}{\sqrt[n]{y}}$	$\left(\frac{x}{y}\right)^{1/n} = \frac{x^{1/n}}{y^{1/n}}$,	$y \neq 0$
3.	$\sqrt[m]{\sqrt[n]{x}} = \sqrt[mn]{x}$	$(x^{1/n})^{1/m} = x^{1/mn}$	

Examples

$$\sqrt{40} = \sqrt{4 \cdot 10} = \sqrt{4}\sqrt{10} = 2\sqrt{10}$$

$$\sqrt[3]{-24} = \sqrt[3]{(-8)(3)} = \sqrt[3]{-8}\sqrt[3]{3} = -2\sqrt[3]{3}$$

$$\sqrt{\sqrt[3]{64}} = \sqrt[6]{64} = 2$$

$$\sqrt[3]{8x^6y^{10}} = \sqrt[3]{8}\sqrt[3]{x^6}\sqrt[3]{y^{10}} = 2\sqrt[3]{(x^2)^3}\sqrt[3]{(y^3)^3y} = 2x^2\sqrt[3]{(y^3)^3}\sqrt[3]{y} = 2x^2y^3\sqrt[3]{y}$$

Rational Exponents

DEFINITION Rational Exponents

If m and n are integers, $x \neq 0$, and $x^{1/n}$ exists, then

$$x^{m/n} = \left(x^{1/n}\right)^m = \left(\sqrt[n]{x}\right)^m$$

or, equivalently,

$$x^{m/n} = \left(x^m\right)^{1/n} = \sqrt[n]{x^m}.$$

Examples

$4^{3/2} = (\sqrt{4})^3 = 2^3 = 8$ or $4^{3/2} = \sqrt{4^3} = \sqrt{64} = 8$

$(-8)^{2/3} = (\sqrt[3]{-8})^2 = (-2)^2 = 4$ or $(-8)^{2/3} = \sqrt[3]{(-8)^2} = \sqrt[3]{64} = 4$

The properties of exponents are also true for rational exponents.

Example

$$\begin{aligned} 4x^{3/2}y^{2/3}\left(\sqrt[5]{4}x^{1/3}y^{1/2}\right)^{-5/2} &= 4x^{3/2}y^{2/3}(4^{1/5})^{-5/2}(x^{1/3})^{-5/2}(y^{1/2})^{-5/2} \\ &= 4x^{3/2}y^{2/3}4^{-1/2}x^{-5/6}y^{-5/4} \\ &= 4x^{3/2-5/6}\,\frac{1}{2}\,y^{2/3-5/4} \\ &= 2x^{4/6}y^{-7/12} = 2\frac{x^{2/3}}{y^{7/12}} \end{aligned}$$

Rationalizing

To remove a radical from the denominator of a quotient, multiply both the numerator and denominator of the quotient by an appropriate expression.

Examples

$$\frac{1}{\sqrt{3}} = \frac{1}{\sqrt{3}}\,\frac{\sqrt{3}}{\sqrt{3}} = \frac{\sqrt{3}}{(\sqrt{3})^2} = \frac{\sqrt{3}}{3}$$

$$\sqrt[5]{\frac{x}{y^2}} = \sqrt[5]{\frac{x}{y^2}\,\frac{y^3}{y^3}} = \frac{\sqrt[5]{xy^3}}{\sqrt[5]{y^5}} = \frac{\sqrt[5]{xy^3}}{y}$$

We can also rationalize the numerator.

Example

$$2 + \sqrt{3} = \frac{2+\sqrt{3}}{1}\,\frac{2-\sqrt{3}}{2-\sqrt{3}} = \frac{(2)^2 - (\sqrt{3})^2}{2-\sqrt{3}} = \frac{1}{2-\sqrt{3}}.$$

Exercise Set A.1

In Exercises 1 through 64, simplify. Write the answers without using negative exponents.

1. x^2x^8
2. y^3y^{11}
3. $(x^3)^5$
4. $(y^4)^3$
5. $(x+2)^4(x+2)$
6. $(y+3)^5(y+3)^2$
7. $\dfrac{x^9}{x^4}$
8. $\dfrac{y^3}{y^7}$
9. $\dfrac{(x+1)^5}{(x+1)^2}$
10. $\dfrac{(x^2y)^2}{(xy^3)^3}$
11. 8^0
12. $(2^0+2)^0$
13. 2^{-1}
14. 4^{-2}
15. $\left(\dfrac{1}{10}\right)^{-1}$
16. $\left(\dfrac{4}{9}\right)^{-2}$
17. $\left(\dfrac{xy^2}{z}\right)^{0}$
18. $(a^3bc^0)^{-2}$
19. $(a^2b)^2(ab^3)^{-1}$
20. $(ab^2c^{-1})^{-2}$
21. $\left(\dfrac{x^{-1}y^{-1}}{z}\right)^{-1}$
22. $\left(\dfrac{x^2y^{-2}z}{x^3yz^{-2}}\right)^{-2}$
23. $\left(\dfrac{x^2y^{-5}}{x^{-2}y^{-3}}\right)^{-3}$
24. $\dfrac{(x+1)^3(x+2)^4}{[(x+1)(x+2)]^3}$
25. $\sqrt[3]{-64}$
26. $\sqrt[5]{32}$
27. $\sqrt[5]{-32}$
28. $\sqrt[3]{-\dfrac{1}{9}}$
29. $\sqrt[4]{(-7)^4}$
30. $\sqrt[4]{7^4}$
31. $\sqrt[3]{(-3)^3}$
32. $\sqrt[3]{3^3}$
33. $\sqrt{\dfrac{25}{4}}$
34. $\sqrt[3]{\dfrac{8}{27}}$
35. $\sqrt[3]{-\dfrac{8}{125}}$
36. $\sqrt[5]{\dfrac{243}{32}}$
37. $\sqrt{\sqrt{16}}$
38. $\sqrt[3]{\sqrt{64}}$
39. $8^{2/3}$
40. $64^{2/3}$
41. $(0.001)^{1/3}$
42. $\left(-\dfrac{1}{125}\right)^{1/3}$
43. $36^{-3/2}$
44. $125^{2/3}$
45. $32^{4/5}$
46. $32^{-4/5}$
47. $64^{4/3}$
48. $\left(\dfrac{9}{16}\right)^{-5/2}$
49. $\left(\dfrac{1000}{27}\right)^{4/3}$
50. $(-1)^{3/5}$
51. $\left(\dfrac{8}{27}\right)^{-2/3}$
52. $100^{3/2}$
53. $\sqrt{4x^4}$
54. $\sqrt[4]{16y^8}$
55. $\sqrt[3]{ab^2}\sqrt[3]{a^2b}$
56. $\sqrt[3]{8x^6}$
57. $\sqrt{72a^4b^3c^5}$
58. $\sqrt[3]{8x^4b^8}$
59. $\sqrt[5]{32a^{10}b^{15}}$
60. $\sqrt{18a^4b^5}$
61. $\sqrt[3]{\dfrac{8a^9b^6}{c^{12}}}$
62. $\sqrt[5]{\dfrac{4x^6}{x^{-4}y^5}}$
63. $\sqrt{ab}\sqrt{a^3b^3}$
64. $\sqrt[3]{ab}\sqrt[3]{a^2b^2}$

In Exercises 65 through 72, write the expressions, using rational exponents rather than radicals.

65. $\dfrac{1}{\sqrt{x}}+\sqrt{x}$
66. $\sqrt{\sqrt{x}}$
67. $\sqrt[3]{(x+1)^2}$
68. $\left(\sqrt[3]{x}+1\right)^2$
69. $\sqrt[3]{x^2}$
70. $\sqrt{x^3}$
71. $\sqrt{\dfrac{x}{y}}$
72. $\sqrt[3]{\dfrac{x}{y}}$

In Exercises 73 through 84, simplify.

73. $(x^2+1)^{3/4}(x^2+1)^{2/3}$
74. $\dfrac{(x^2+1)^{1/3}}{(x^2+1)^{-3/4}}$
75. $\dfrac{x^{2/3}x^{-1/3}}{x^{2/3}}$
76. $\dfrac{y^{-1/4}y^{-1/2}}{y^{-3/4}}$
77. $\dfrac{a^{2/3}b^{3/4}}{a^{-1/3}b^{-1/4}}$
78. $\dfrac{a^{1/3}b^{-2/5}}{a^{-2/3}b^{1/5}}$
79. $\dfrac{a^{2/3}b^{4/3}}{a^{2/3}b^{1/3}}$
80. $\dfrac{a^{-1/3}b^{-3/4}}{a^{-2/3}b^{1/4}}$
81. $\dfrac{(x+1)^{2/3}(x+2)^{1/3}}{(x+1)^{-1/3}(x+2)^{-2/3}}$
82. $\dfrac{(x+1)^{-1/3}(y+1)^{-2/3}}{(x+1)^{2/3}(y+1)^{-5/3}}$
83. $[(x+1)^3]^{5/3}$
84. $[(y-2)^{1/3}]^6$

In Exercises 85 through 92, rationalize the denominator.

85. $\dfrac{1}{\sqrt{5}}$
86. $\dfrac{1}{\sqrt{7}}$
87. $\dfrac{1}{\sqrt{2}+1}$
88. $\dfrac{1}{3-\sqrt{5}}$
89. $\dfrac{1}{\sqrt[3]{5}}$
90. $\dfrac{3}{\sqrt[4]{3}}$
91. $\dfrac{1+\sqrt{3}}{1-\sqrt{3}}$
92. $\dfrac{2-\sqrt{5}}{2+\sqrt{5}}$

In Exercises 93 through 96, rationalize the numerator.

93. $\dfrac{\sqrt{5}-\sqrt{2}}{3}$
94. $\dfrac{\sqrt{7}-\sqrt{3}}{4}$
95. $\dfrac{\sqrt{10}-2}{6}$
96. $\sqrt{10}-3$

A.2 Polynomials and Rational Expressions

Polynomials
Factoring
Rational Expressions

Polynomials

We begin with the following definition.

DEFINITION Polynomial

A polynomial is an expression of the form

$$a_n x^n + a_{n-1} x^{n-1} + \cdots + a_1 x + a_0,$$

where n is a nonnegative integer and each coefficient a_i is a real number.

If $a_n \neq 0$, the polynomial is said to have **degree** $\boldsymbol{n}$.

Lower order polynomials have special names. A polynomial such as 3 has degree equal to *zero* and is called a **constant**. A polynomial such as $x - \frac{1}{2}$ has degree equal to *one* and is called **linear**. A polynomial such as $4x^2 - 2x + 4$ has degree equal to *two* and is called a **quadratic**. A polynomial such as $2x^3 - 5x^2 - \frac{3}{2}x - 6$ has degree equal to *three* and is called a **cubic**.

Since for any real number x, $a_n x^n + a_{n-1} x^{n-1} + \cdots + a_1 x + a_0$ is a real number, polynomials must follow the same rules as real numbers.

EXAMPLE 1

Simplify the following.
(a) $(x^3 + 2x^2 + 4x - 7) + (2x^3 - x^2 + 2x + 10)$
(b) $(x^4 - 2x^2 + 3) - (x^4 + x^3 - x - 4)$
(c) $(x^2 - 2)(3x^3 - x + 1)$.

Solution

(a) $(x^3 + 2x^2 + 4x - 7) + (2x^3 - x^2 + 2x + 10)$
$$= (1 + 2)x^3 + (2 - 1)x^2 + (4 + 2)x + (-7 + 10)$$
$$= 3x^3 + x^2 + 6x + 3.$$
(b) $(x^4 - 2x^2 + 3) - (x^4 + x^3 - x - 4)$
$$= (1 - 1)x^4 - x^3 - 2x^2 + x + (3 + 4)$$
$$= -x^3 - 2x^2 + x + 7$$
(c) $(x^2 - 2)(3x^3 - x + 1) = x^2(3x^3 - x + 1) - 2(3x^3 - x + 1)$
$$= (3x^5 - x^3 + x^2) + (-6x^3 + 2x - 2)$$
$$= 3x^5 + (-1 - 6)x^3 + x^2 + 2x - 2$$
$$= 3x^5 - 7x^3 + x^2 + 2x - 2$$
■

We now list some products that occur frequently. They can be checked by multiplying the terms on the left-hand side.

Special Products

1. $(a+b)(a-b)=a^2-b^2$
2. $(a+b)^2=a^2+2ab+b^2$
3. $(a-b)^2=a^2-2ab+b^2$
4. $(ax+b)(cx+d)=acx^2+(ad+bc)x+bd$
5. $(x+y)^3=x^3+3x^2y+3xy^2+y^3$
6. $(x-y)^3=x^3-3x^2y+3xy^2-y^3$
7. $(a-b)(a^2+ab+b^2)=a^3-b^3$
8. $(a+b)(a^2-ab+b^2)=a^3+b^3$

Factoring

We will often find it useful to **factor** a polynomial, that is, write the polynomial as a product of two or more nonconstant polynomials, each of which is as low an order as possible.

For example, using the first special product listed above we see that

$$x^2-1=(x+1)(x-1)$$

Although, in general, factoring a polynomial can be extremely difficult, many polynomials can be factored by using the other special products listed above. The following are some examples.

EXAMPLE 2

Factor (a) $4x^2-9$, (b) $8x^3-27$, (c) $81x^4-16$, (d) $4x^{-4}-9x^{-6}$.

Solution

(a) Use Formula 1 with $a=2x$ and $b=3$. Then

$$\begin{aligned}4x^2-9&=(2x)^2-3^2\\&=(2x+3)(2x-3)\end{aligned}$$

(b) Use Formula 7 with $a=2x$ and $b=3$:

$$\begin{aligned}8x^3-27&=(2x)^3-(3)^3\\&=(2x-3)[(2x)^2+(2x)(3)+(3)^2]\\&=(2x-3)(4x^2+6x+9)\end{aligned}$$

(c) Use Formula 1 twice as follows:

$$\begin{aligned}81x^4-16&=(9x^2)^2-(4)^2\\&=(9x^2+4)(9x^2-4)\\&=(9x^2+4)(3x+2)(3x-2)\end{aligned}$$

(d) There are two ways of factoring this expression. In calculus, one normally factors out the highest negative power of x first. Doing this gives

$$\begin{aligned} 4x^{-4} - 9x^{-6} &= x^{-6}(4x^2 - 9) \\ &= x^{-6}(2x + 3)(2x - 3) \end{aligned}$$

■

Rational Expressions

A **rational expression** is a sum, difference, product, or quotient of terms, each of which is the quotient of polynomials. Thus, the following are rational expressions.

$$\frac{x^2 + 1}{x^3 - x - 3} \qquad 2x^3 + \frac{5x + 1}{x^2 - 2}$$

Rational expressions must satisfy the same rules as real numbers.

Rules for Rational Expressions

Let $a(x)$, $b(x)$, $c(x)$, $d(x)$ be rational expressions with $b(x) \neq 0$ and $d(x) \neq 0$. Also $c(x) \neq 0$ in Rule 5. Then

Rule 1. $\dfrac{a(x)}{b(x)} + \dfrac{c(x)}{b(x)} = \dfrac{a(x) + c(x)}{b(x)}$

Rule 2. $\dfrac{a(x)}{b(x)} - \dfrac{c(x)}{b(x)} = \dfrac{a(x) - c(x)}{b(x)}$

Rule 3. $\dfrac{a(x)d(x)}{b(x)d(x)} = \dfrac{a(x)}{b(x)}$

Rule 4. $\dfrac{a(x)}{b(x)} \cdot \dfrac{c(x)}{d(x)} = \dfrac{a(x)c(x)}{b(x)d(x)}$

Rule 5. $\dfrac{a(x)}{b(x)} \div \dfrac{c(x)}{d(x)} = \dfrac{a(x)}{b(x)} \cdot \dfrac{d(x)}{c(x)}$

EXAMPLE 3 **Rule 2**

Simplify $\dfrac{x^2 + x + 1}{x^2 + 1} - \dfrac{x + 1}{x^2 + 1}$.

Solution

$$\frac{x^2 + x + 1}{x^2 + 1} - \frac{x + 1}{x^2 + 1} = \frac{(x^2 + x + 1) - (x + 1)}{x^2 + 1} = \frac{x^2}{x^2 + 1}$$

■

EXAMPLE 4 **Rule 3**

Simplify $\dfrac{x^2 - 1}{x^2 + 4x + 3}$.

Solution

$$\frac{x^2-1}{x^2+4x+3}=\frac{(x+1)(x-1)}{(x+1)(x+3)}=\frac{x-1}{x+3}$$ ■

EXAMPLE 5 **Rule 4**

Simplify $\dfrac{x^2+3x+2}{2x^2-x-1}\,\dfrac{x-1}{x+1}$.

Solution

$$\frac{x^2+3x+2}{2x^2-x-1}\cdot\frac{x-1}{x+1}=\frac{(x+1)(x+2)(x-1)}{(x-1)(2x+1)(x+1)}=\frac{x+2}{2x+1}$$ ■

EXAMPLE 6 **Rule 5**

Simplify $\dfrac{x^3-2x^2+x}{x^2+2x}\div\dfrac{x-1}{x^2-4}$.

Solution

$$\begin{aligned}\frac{x^3-2x^2+x}{x^2+2x}\div\frac{x-1}{x^2-4}&=\frac{x^3-2x^2+x}{x^2+2x}\cdot\frac{x^2-4}{x-1}\\&=\frac{x(x^2-2x+1)}{x(x+2)}\cdot\frac{x^2-4}{x-1}\\&=\frac{(x-1)^2(x-2)(x+2)}{(x+2)(x-1)}\\&=(x-1)(x-2)\end{aligned}$$ ■

Least Common Multiple

To combine the two fractions

$$\frac{1}{x^2-2x+1}-\frac{1}{x^3-x}$$

we must first factor, obtaining

$$\frac{1}{(x-1)^2}-\frac{1}{x(x-1)(x+1)}$$

We now take as the least common denominator the product of each of the factors, taking the power of each factor as the highest power that appears. Thus, the least common denominator for the above example is $x(x-1)^2(x+1)$. Thus,

$$\begin{aligned}\frac{1}{(x-1)^2}-\frac{1}{x(x-1)(x+1)}&=\frac{1}{(x-1)^2}\,\frac{x(x+1)}{x(x+1)}-\frac{1}{x(x-1)(x+1)}\,\frac{x-1}{x-1}\\&=\frac{x(x+1)-(x-1)}{x(x-1)^2(x+1)}\end{aligned}$$

$$= \frac{x^2 + x - x + 1}{x(x-1)^2(x+1)}$$
$$= \frac{x^2 + 1}{x(x-1)^2(x+1)}$$

The following is a problem that comes up in calculus.

EXAMPLE 7

Simplify $\dfrac{\dfrac{1}{x+h} - \dfrac{1}{x}}{h}$.

Solution

$$\begin{aligned}
\frac{\frac{1}{x+h} - \frac{1}{x}}{h} &= \frac{1}{h}\left(\frac{1}{x+h} - \frac{1}{x}\right) \\
&= \frac{1}{h}\left(\frac{1}{x+h}\,\frac{x}{x} - \frac{1}{x}\,\frac{x+h}{x+h}\right) \\
&= \frac{1}{h}\,\frac{x-(x+h)}{x(x+h)} \\
&= \frac{1}{h}\,\frac{-h}{x(x+h)} \\
&= \frac{-1}{x(x+h)}
\end{aligned}$$

■

Rationalizing the Numerator

In calculus, one often must rationalize the numerator.

EXAMPLE 8

Rationalize $\dfrac{\sqrt{x}-2}{x-4}$.

Solution

$$\begin{aligned}
\frac{\sqrt{x}-2}{x-4} &= \frac{\sqrt{x}-2}{x-4}\,\frac{\sqrt{x}+2}{\sqrt{x}+2} \\
&= \frac{(\sqrt{x})^2 - (2)^2}{(x-4)(\sqrt{x}+2)} \\
&= \frac{(x-4)}{(x-4)(\sqrt{x}+2)} \\
&= \frac{1}{\sqrt{x}+2}
\end{aligned}$$

■

Exercise Set A.2

In Exercises 1 through 14, simplify.

1. $(x^2 + 2x - 1) + (2x^2 + x + 3)$
2. $(x^2 - x + 2) + (-x^2 - x - 1)$
3. $(-2x^2 + x - 2) + (-x^2 - 2x + 1)$
4. $(-3x^2 - x + 1) + (x^2 + x - 1)$
5. $(x^4 + x^2 + x + 1) - (x^3 + 2x - 3)$
6. $(x^4 - 2x^3 - 1) - (2x^4 + x^2 + x - 1)$
7. $(2x^3 - 2x^2 + 1) - (x^3 - x^2 - x - 1)$
8. $(-x^3 - x + 1) - (x^3 - x^2 + 2)$
9. $(x + 1)(x^2 + 2x + 3)$
10. $(x + 2)(x^2 - x - 2)$
11. $(2x - 1)(x^2 - x + 3)$
12. $(2x - 3)(x^2 - 2)$
13. $(x^2 + 3)(x^2 + 1)$
14. $(x^2 + 2)(x^2 + x + 2)$

In Exercises 15 through 58, factor.

15. $x^3 - 27$
16. $x^3 - 8$
17. $x^3 - \frac{1}{8}$
18. $x^3 - \frac{1}{64}$
19. $27x^3 - 8$
20. $8x^3 - 125$
21. $\frac{1}{8}x^3 - 27$
22. $\frac{1}{64}x^3 - 27$
23. $\frac{1}{64}x^3 - \frac{1}{27}$
24. $\frac{1}{125}x^3 - \frac{1}{8}$
25. $a^3x^3 - b^3$
26. $a^3x^3 - \frac{1}{b^3}$
27. $x^2 - 16$
28. $x^2 - 81$
29. $16x^2 - 1$
30. $81x^2 - 1$
31. $x^4 - \frac{1}{16}$
32. $x^4 - \frac{1}{81}$
33. $\frac{1}{16}x^4 - 1$
34. $\frac{1}{81}x^4 - 1$
35. $x^4 - 4$
36. $x^4 - 9$
37. $\frac{1}{4}x^4 - 1$
38. $\frac{1}{9}x^4 - 1$
39. $\dfrac{x^2 - x - 2}{x^2 - 3x + 2}$
40. $\dfrac{x^2 + 3x + 2}{x^2 + 5x + 6}$
41. $\dfrac{x^3 + x^2}{x - x^3}$
42. $\dfrac{x^2 - 4}{x^2 + x - 6}$
43. $\dfrac{x^2 - 1}{x^2 - 4} \cdot \dfrac{x - 2}{x + 1}$
44. $\dfrac{x^2 - 2x + 1}{x^2 + 3x + 2} \cdot \dfrac{x + 1}{x - 1}$
45. $\dfrac{x^3 - 1}{x^2 - 16} \cdot \dfrac{x - 4}{x^2 + x + 1}$
46. $\dfrac{x^4 - 1}{x + 2} \cdot \dfrac{x}{x^2 - 1}$
47. $\dfrac{x^2 - 1}{x + 1} \div \dfrac{x - 1}{2x + 1}$
48. $\dfrac{2x^2 + x - 1}{x^2 - 2x + 1} \div \dfrac{2x - 1}{x - 1}$
49. $\dfrac{x^2 + 1}{x + 1} \div \dfrac{x}{x + 1}$
50. $\dfrac{a^3 - b^3}{a^2 - b^2} \div \dfrac{1}{a + b}$
51. $\dfrac{x}{x + 1} - \dfrac{1}{x + 1}$
52. $\dfrac{x + 1}{x^2 + 1} - \dfrac{2x + 3}{x^2 + 1}$
53. $\dfrac{x}{x^2 - 1} - \dfrac{1}{x^2 - 1}$
54. $\dfrac{x^2}{x^4 - 8} + \dfrac{4}{x^4 - 8}$
55. $\dfrac{x^2}{x^2 - 1} - \dfrac{x}{x - 1}$
56. $\dfrac{1}{x^2 + x - 2} - \dfrac{1}{x^2 - 4}$
57. $\dfrac{x^2}{x^4 - 16} - \dfrac{1}{x^2 + 4}$
58. $\dfrac{1}{x - 3} + \dfrac{1}{x^2 - 9} - \dfrac{1}{x^2 - 2x - 3}$

In Exercises 59 through 66, rationalize the denominator.

59. $\dfrac{1}{\sqrt{x} + 1}$
60. $\dfrac{1}{\sqrt{x} - 1}$
61. $\dfrac{1}{\sqrt{x} + \sqrt{y}}$
62. $\dfrac{1}{\sqrt{x} - \sqrt{3}}$
63. $\dfrac{y - 9}{\sqrt{y} - 3}$
64. $\dfrac{a - b}{\sqrt{a} - \sqrt{b}}$
65. $\dfrac{a^2 - 1}{\sqrt{a} - 1}$
66. $\dfrac{x^2 - y^2}{\sqrt{x} - \sqrt{y}}$

In Exercises 67 through 70, rationalize the numerator.

67. $\dfrac{\sqrt{x} - 1}{x - 1}$
68. $\dfrac{\sqrt{x} - \sqrt{y}}{x - y}$
69. $\dfrac{\sqrt{x + h} - \sqrt{x}}{h}$
70. $\dfrac{\sqrt{x} - \sqrt{x - h}}{h}$

A.3 Equations

Linear Equations
Quadratic Equations
Polynomial Equations
Linear Equations with Absolute Values

Linear Equations

The polynomial $ax + b$ is called a **linear** expression, and the equation $ax + b = 0$ is called a **linear equation**.

If $a \neq 0$, the linear equation $ax + b = 0$ can always be solved as follows:

$$\begin{aligned} ax + b &= 0 \\ ax &= -b \\ x &= -\frac{b}{a} \end{aligned}$$

EXAMPLE 1

Solve $\frac{2x}{3} - \frac{x-1}{3} = \frac{1}{4} - \frac{1}{2}\left(3x - \frac{3x-2}{3}\right)$.

Solution

First remove the parentheses.

$$\frac{2x}{3} - \frac{x-1}{3} = \frac{1}{4} - \frac{3}{2}x + \frac{3x-2}{6}$$

Remove the fractions by multiplying by 12:

$$\begin{aligned} 4(2x) - 4(x-1) &= 3 - 6(3x) + 2(3x-2) \\ 8x - 4x + 4 &= 3 - 18x + 6x - 4 \end{aligned}$$

Move all x-terms to the left and all constant terms to the right:

$$\begin{aligned} 8x - 4x + 18x - 6x &= 3 - 4 - 4 \\ 16x &= -5 \\ x &= -\frac{5}{16} \end{aligned}$$

■

Quadratic Equations

If $a \neq 0$, the polynomial $ax^2 + bx + c$ is called a **quadratic expression**. The equation

$$ax^2 + bx + c = 0 \qquad a \neq 0$$

is called a **quadratic equation**. We shall always assume that a, b, and c are *real* numbers and that x is a real-valued variable.

Some quadratic equations can be solved easily by factoring.

EXAMPLE 2

Solve (a) $x^2 - 4 = 0$, (b) $9x^2 - 16 = 0$ and (c) $3x^2 + 7x - 6 = 0$.

Solution

(a) To solve $x^2 - 4 = 0$, first recognize the expression $x^2 - 4$ as a difference of squares $a^2 - b^2$ with $a = x$ and $b = 2$. Then factoring gives

$$0 = x^2 - 4 = x^2 - 2^2 = (x + 2)(x - 2)$$

The only way the product of two expressions can be zero is if at least one of them is zero. Thus, either $x + 2 = 0$ or $x - 2 = 0$. This in turn implies that $x = -2$ or $x = 2$. Thus, the quadratic equation $x^2 - 4 = 0$ has two solutions, $x = -2$ and $x = 2$.

(b) To solve $9x^2 - 16 = 0$, first recognize the expression $9x^2 - 16$ as a difference of squares $a^2 - b^2$ with $a = 3x$ and $b = 4$. Factoring then gives

$$0 = 9x^2 - 16 = (3x)^2 - 4^2 = (3x + 4)(3x - 4)$$

implies that

$$3x + 4 = 0 \qquad \text{or} \qquad 3x - 4 = 0$$

$$x = -\frac{4}{3} \qquad \text{or} \qquad x = \frac{4}{3}$$

(c) To solve $3x^2 + 7x - 6 = 0$, first factor $3x^2 + 7x - 6$ into a product of two linear expressions. Since the first term is $3x^2$ and the last term is negative, we try $(3x + \quad)(x - \quad)$ or $(3x - \quad)(x + \quad)$. Now 6 can be written as $6 = 3 \cdot 2$ or $6 = 6 \cdot 1$. Using trial and error with the first way gives the following:

$$\begin{aligned}
(3x - 3)(x + 2) &= 3x^2 + 3x - 6 && \text{wrong} \\
(3x + 3)(x - 2) &= 3x^2 - 3x - 6 && \text{wrong} \\
(3x + 2)(x - 3) &= 3x^2 - 7x - 6 && \text{wrong} \\
(3x - 2)(x + 3) &= 3x^2 + 7x - 6 && \text{correct}
\end{aligned}$$

Thus, we have $3x^2 + 7x - 6 = (3x - 2)(x + 3)$. Therefore,

$$0 = 3x^2 + 7x - 6 = (3x - 2)(x + 3)$$

implies that

$$3x - 2 = 0 \qquad \text{or} \qquad x + 3 = 0$$

$$x = \frac{2}{3} \qquad \text{or} \qquad x = -3$$

■

The procedures used in the last example do not always work. A procedure that does always work is to use the quadratic formula.

Solutions of the Quadratic Equation

The solutions of the quadratic equation

$$ax^2 + bx + c = 0 \qquad a \neq 0$$

are given by the quadratic formula

$$x = \frac{-b \pm \sqrt{b^2 - 4ac}}{2a}$$

The term $b^2 - 4ac$ is called the **discriminant**.

The quadratic formula breaks down into three cases.

Case 1. If $b^2 - 4ac > 0$, there are the two solutions

$$\frac{-b \pm \sqrt{b^2 - 4ac}}{2a}$$

Case 2. If $b^2 - 4ac = 0$, there is one solution given by

$$-\frac{b}{2a}$$

Case 3. If $b^2 - 4ac < 0$, there are no solutions in the set of real numbers.

The term $b^2 - 4ac$ is called the **discriminant**.

EXAMPLE 3

Solve $2x^2 - 3x - 1 = 0$.

Solution

Use the quadratic formula with $a = 2$, $b = -3$, $c = -1$ and obtain

$$\begin{aligned} x &= \frac{-(-3) \pm \sqrt{(-3)^2 - 4(2)(-1)}}{2(2)} \\ &= \frac{3 \pm \sqrt{17}}{4} \end{aligned}$$

■

Polynomial Equations

We now consider solving polynomial equations of any order. We have just seen how any quadratic equation can be solved using the quadratic formula. It would be convenient if any polynomial equation such as $x^5 + 3x^3 - 2x^2 - 8 = 0$ could be solved by using some corresponding "polynomial formula." Unfortunately, no such formula exists for polynomials of degree five or higher except for very special cases. We can solve polynomial equations in the cases in which the polynomial can be completely factored. We now consider such cases.

EXAMPLE 4

Solve (a) $4x^4 - 36x^2 = 0$, (b) $2x^5 - 2x = 0$, and (c) $10x^3 - 5x^2 - 5x = 0$.

Solution

(a)
$$\begin{aligned} 0 = 4x^4 - 36x^2 &= 4x^2(x^2 - 9) && \text{common factor} \\ &= 4x^2(x + 3)(x - 3) && \text{difference of squares} \end{aligned}$$

From this factorization we can immediately obtain the solutions $x = 0$, $x = -3$, and $x = 3$.

(b)
$$\begin{aligned} 0 = 2x^5 - 2x &= 2x(x^4 - 1) && \text{common factor} \\ &= 2x(x^2 + 1)(x^2 - 1) && \text{difference of squares} \\ &= 2x(x^2 + 1)(x + 1)(x - 1) && \text{difference of squares} \end{aligned}$$

From this factorization we can immediately obtain the solutions $x = 0$, $x = -1$, and $x = 1$.

(c) $0 = 10x^3 - 5x^2 - 5x = 5x(2x^2 - x - 1)$ common factor

$= 5x(2x + 1)(x - 1)$ factoring a quadratic

From this factorization we can immediately obtain the solutions $x = 0$, $x = 1$, and $x = -1/2$. ■

Linear Equations with Absolute Value

We now give the definition of absolute value.

DEFINITION Absolute Value

$$|x| = \begin{cases} x & \text{if } x \geq 0 \\ -x & \text{if } x \leq 0 \end{cases}$$

Remark You may have learned to find the absolute value by "dropping the sign." Doing this for algebraic expressions can be disastrous. For example, $|-x|$ is x if x is positive but $-x$ if x is negative. Thus, $|-x|$ is sometimes equal to x and sometimes to $-x$.

It follows immediately that the absolute value has the following property.

A Property of Absolute Value

For any $a \geq 0$,

$$|x| = a \quad \text{if and only if} \quad x = a \quad \text{or} \quad x = -a.$$

EXAMPLE 5

Solve $|2x - 3| = 2$.

Solution

This will be true if $2x - 3 = 2$ or if $2x - 3 = -2$. Thus,

$$\begin{aligned} 2x - 3 &= 2 &\quad \text{or} \quad 2x - 3 &= -2 \\ 2x &= 5 &\quad \text{or} \quad 2x &= 1 \\ x &= \frac{5}{2} &\quad \text{or} \quad x &= \frac{1}{2} \end{aligned}$$

■

Exercise Set A.3

In Exercises 1 through 10, solve.

1. $x - 2(x - 3) = 4 - 2(x + 1)$
2. $3(x - 1) - 4(x - 2) - 3 = -(3x - 1)$
3. $-(2x - 1) + 3(x + 4) - 4 = 2(4x - 1)$
4. $(4x - 5) - 3(2x + 1) - 4 = 3 + 2(x - 1) - 3(2x + 1)$
5. $\frac{1}{2}x - \frac{x - 2}{4} - 1 = 2(x - 2)$

6. $\frac{x-1}{2} - \frac{2x-3}{4} - \frac{1}{2} = \frac{1}{2}(3x-1)$

7. $\frac{2x-3}{3} - \frac{2x+1}{2} - \frac{2}{3} = \frac{1}{3}(4x+3)$

8. $\frac{7}{12}x - \frac{2x-1}{3} + \frac{5}{6} = \frac{1}{2}(5x-2)$

9. $\frac{1}{2}\left(3x - \frac{1}{2}\frac{x-1}{2}\right) = 2x - \frac{3}{2}$

10. $\frac{1}{2}\left(\frac{3x-2}{3} - \frac{1}{3}\frac{2x-1}{2}\right) = \frac{1}{3}\left(\frac{2x-3}{2} + \frac{3}{2}\right)$

In Exercises 11 through 50, set the quadratic expression equal to zero and solve by factoring.

11. $y^2 - 4$
12. $x^2 - 9$
13. $4y^2 - 9$
14. $36x^2 - 25$
15. $\frac{1}{4}x^2 - 1$
16. $\frac{1}{9}y^2 - 1$
17. $x^2 - \frac{1}{4}$
18. $y^2 - \frac{1}{9}$
19. $25x^2 - 9$
20. $16x^2 - 25$
21. $\frac{1}{4}y^2 - 9$
22. $\frac{1}{16}y^2 - \frac{1}{9}$
23. $4y^2 - \frac{1}{9}$
24. $9y^2 - \frac{1}{4}$
25. $a^2x^2 - b^2$
26. $a^2y^2 - \frac{1}{b^2}$
27. $x^2 + 4x + 4$
28. $x^2 + 4x + 3$
29. $x^2 - 5x + 4$
30. $x^2 + 2x - 8$
31. $2x^2 + 3x - 2$
32. $2x^2 - 3x + 1$
33. $3x^2 + 7x + 2$
34. $6x^2 + 5x + 1$
35. $6x^2 + x - 1$
36. $6x^2 + x - 1$
37. $10x^2 - 12x + 2$
38. $10x^2 - 11x - 6$
39. $6x^2 + 13x + 6$
40. $6x^2 - 5x - 6$
41. $12x^2 - 11x + 2$
42. $12x^2 - x - 6$
43. $9x^2 - 6x + 1$
44. $x^2 - 6x - 9$
45. $9x^2 + 12x + 4$
46. $4x^2 + 20x + 25$
47. $x^2 - x + \frac{1}{4}$
48. $x^2 - \frac{1}{2}x + \frac{1}{16}$
49. $4x^2 + 2x + \frac{1}{4}$
50. $9x^2 - 3x + \frac{1}{4}$

In Exercises 51 through 58, set the polynomial expression equal to zero and solve by factoring.

51. $x^4 - 16$
52. $x^4 - 81$
53. $16x^4 - 81$
54. $81x^4 - 16$
55. $x^5 + x^4 - 2x^3$
56. $x^7 + x^6 - 6x^5$
57. $2x^6 - 162x^2$
58. $3x^7 - 48x^3$

In Exercises 59 through 72, set each of the quadratic expressions equal to zero. If there is a solution, solve by using the quadratic formula.

59. $2x^2 + 4x + 2$
60. $2x^2 - 5x + 3$
61. $5x^2 - 7x + \frac{1}{5}$
62. $4x^2 - 9x + 2$
63. $-x^2 + x + 1$
64. $-2x^2 - 3x + 2$
65. $-3x^2 - 4x + 5$
66. $-2x^2 - 3x + 4$
67. $-x^2 + 4x - 1$
68. $-2x^2 + 5x - 2$
69. $-3x^2 + 7x - \frac{2}{3}$
70. $-x^2 + 3x - 2$
71. $x^2 + x + 1$
72. $x^2 + 2x - 2$

In Exercises 73 through 84, solve.

73. $|2x - 1| = 1$
74. $|2x - 3| = 2$
75. $|3x - 1| = 5$
76. $|5x - 2| = 4$
77. $|2 - 3x| = 2$
78. $|3 - 2x| = 1$
79. $|4 - x| = 5$
80. $|3 - 5x| = 3$
81. $\left|\frac{1}{2} - x\right| = \frac{2}{3}$
82. $\left|x - \frac{1}{3}\right| = \frac{1}{4}$
83. $\left|\frac{1}{2}x - \frac{2}{3}\right| = \frac{1}{6}$
84. $\left|\frac{1}{3}x - \frac{2}{5}\right| = \frac{1}{2}$

A.4 Inequalities

Linear Inequalities
Inequalities with Absolute Values
Nonlinear Inequalities

Linear Inequalities

We begin this section with some fundamental properties of inequalities.

Properties of Inequalities

1. $a > b$ implies that $a + c > b + c$.
2. $a > b, c > 0$ implies that $ac > bc$.
3. $a > b, c < 0$ implies that $ac < bc$.

EXAMPLE 1

Solve the inequality $5x - 4 < 2x + 5$.

Solution

$$\begin{aligned} 5x - 4 &< 2x + 5 \\ 5x - 2x &< 5 + 4 \\ 3x &< 9 \\ x &< 3 \end{aligned}$$

Figure A.1

Thus, all numbers *less* than 3 is the solution set. This set can also be represented by using the interval notation $(-\infty, 3)$. See Figure A.1. ■

EXAMPLE 2

Solve the inequality $-2 \leq 6 - 2x < 8$.

Solution

$$\begin{aligned} -2 &\leq 6 - 2x & \text{and} \qquad 6 - 2x &< 8 \\ 2x &\leq 6 + 2 & -2x &< 8 - 6 = 2 \\ 2x &\leq 8 & 2x &> -2 \\ x &\leq 4 & x &> -1 \end{aligned}$$

Figure A.2

Thus, x must simultaneously satisfy $x \leq 4$ and $x > -1$. We can describe this solution set by the interval notation $(-1, 4]$. See Figure A.2. ■

Inequalities with Absolute Values

Properties of Inequalities with Absolute Values

For any $a > 0$,

1. $|x| < a$ if and only if $-a < x < a$.
2. $|x| > a$ if and only if either $x > a$ or $x < -a$.

EXAMPLE 3

Solve the inequality $|2x + 4| < 2$.

Solution

According to Property 1, this is equivalent to

$$-2 < 2x + 4 < 2$$

$$\begin{array}{rcr} -2 < 2x + 4 & \text{and} & 2x + 4 < 2 \\ -2x < 6 & & 2x < -2 \\ x > -3 & & x < -1 \end{array}$$

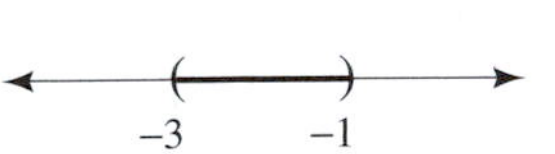

Figure A.3

In interval notation the solution set is $(-3, -1)$. See Figure A.3. ∎

EXAMPLE 4

Solve the inequality $|3 - 4x| > 5$.

Solution

According to Property 2, this is equivalent to

$$\begin{array}{rcr} 3 - 4x > 5 & \text{or} & 3 - 4x < -5 \\ -4x > 2 & & -4x < -8 \\ x < -\dfrac{1}{2} & & x > 2 \end{array}$$

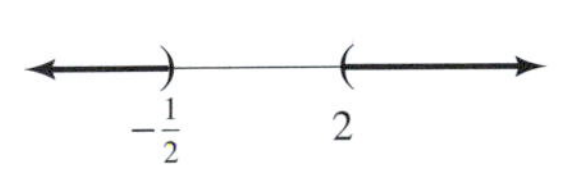

Figure A.4

In interval notation the solution set is $(-\infty, -\frac{1}{2})$ together with $(2, +\infty)$. See Figure A.4. ∎

Nonlinear Inequalities

EXAMPLE 5

Solve the inequality $x^2 - 3x < -2$.

Solution

This is equivalent to

$$x^2 - 3x + 2 < 0$$

Factoring the left side, this is equivalent to

$$(x - 1)(x - 2) < 0$$

The only way the product of the two terms $(x - 1)$ and $(x - 2)$ can be negative is for one to be positive and the other negative. Thus, we need to look carefully at the signs of these two factors. Clearly, $x - 1 < 0$ if $x < 1$ and $x - 1 > 0$ if $x > 1$. In the same way $x - 2 < 0$ if $x < 2$ and $x - 2 > 0$ if $x > 2$. We then indicate this with a **sign chart** of these two factors as shown in Figure A.5. The sign of $(x-1)(x-2)$ is then obtained by taking the product of the signs of the individual factors. Thus, on the interval $(-\infty, 1)$ the sign of both of the factors is negative, so the sign of the product is positive. On the interval $(1, 2)$ one factor is negative and the other factor is positive. Thus, the sign of the product is negative. On the interval $(2, +\infty)$ both factors are positive, so the sign of the product is also positive.

From this we see that the product $(x-1)(x-2)$ is negative only when $1 < x < 2$. Thus, the solution set is $(1, 2)$.

		1		2	
$(x-1)$	$- - -$		$+ + +$		$+ + +$
$(x-2)$	$- - -$		$- - -$		$+ + +$
$(x-1)(x-2)$	$+ + +$		$- - -$		$+ + +$

Figure A.5

■

EXAMPLE 6

Solve the inequality $2x^2 - x \geq 3$.

Solution

This is equivalent to

$$2x^2 - x - 3 \geq 0.$$

Factoring the left side, this is equivalent to

$$(x+1)(2x-3) \geq 0$$

Figure A.6 gives the sign chart of the two factors $(x+1)$ and $(2x-3)$. It is suggested that the first factor be the one with the smallest zeroes. This seems to be the least confusing and most systematic, but is not mandatory. In the last line is given the product of the signs in the previous two lines. In this way we find all places where the product $(x+1)(2x-3)$ is greater or equal to zero. This is the set $(-\infty, -1]$ together with the set $[\frac{3}{2}, +\infty)$.

		-1		$\frac{3}{2}$	
$(x+1)$	$- - -$		$+ + +$		$+ + +$
$(2x-3)$	$- - -$		$- - -$		$+ + +$
$(x+1)(2x-3)$	$+ + +$		$- - -$		$+ + +$

Figure A.6

■

EXAMPLE 7

Solve the inequality $\dfrac{x^2-2}{x} < 1$.

Solution

Notice that we cannot have $x = 0$.

We must reduce the inequality to one with zero on the right. Thus,

$$\frac{x^2-2}{x} < 1$$

$$\frac{x^2-2}{x} - 1 < 0$$

$$\frac{x^2-2}{x} - \frac{x}{x} < 0$$

$$\frac{x^2-x-2}{x} < 0$$

$$\frac{(x+1)(x-2)}{x} < 0$$

The sign of the quotient will be determined by the signs of the three factors $(x+1)$, x, and $(x-2)$. We note the signs of the three factors in the sign chart given in Figure A.7. The sign chart of the quotient given in the last line in Figure A.7 is obtained by multiplying the individual signs on the previous three lines. Thus, since the sign of all three factors is negative on $(-\infty, -1)$, the product is negative. On the interval $(-1, 0)$, two of the factors have negative signs and one has a positive sign, so the product is positive. We see from Figure A.7 that the sign of the quotient is negative on the interval $(-\infty, -1)$ together with $(0, 2)$. The "N" in the sign chart indicates that the quotient is not defined at $x = 0$.

	$(-\infty,-1)$	-1	$(-1,0)$	0	$(0,2)$	2	$(2,\infty)$
$(x+1)$	− − −		+ + +		+ + +		+ + +
x	− − −		− − −		+ + +		+ + +
$(x-2)$	− − −		− − −		− − −		+ + +
$\frac{(x+1)(x-2)}{x}$	− − −		+ + +	N	− − −		+ + +

Figure A.7

■

EXAMPLE 8

Solve the inequality $\dfrac{x^2-4}{x^2+1} \le 0$.

Solution

We factor and obtain

$$\frac{x^2-4}{x^2+1} \le 0$$

$$\frac{(x+2)(x-2)}{x^2+1} \le 0$$

We notice that the factor x^2+1 is *always* positive. Figure A.8 shows the sign chart of the three factors. The sign of the quotient is shown on the last line, where we see that the quotient is less than or equal to zero on $[-2, 2]$.

	-2		2	
$(x+2)$	$- - -$	$+ + +$	$+ + +$
$(x-2)$	$- - -$	$- - -$	$+ + +$
(x^2+1)	$+ + +$	$+ + +$	$+ + +$
$\frac{(x+2)(x-2)}{(x^2+1)}$	$+ + +$	$- - -$	$+ + +$

Figure A.8

Exercise Set A.4

Solve the following.

1. $4x - 1 < x + 3$

2. $9x + 3 > 5x + 7$

3. $3x - 5 > 5x + 7$

4. $3x - 2 < 7x - 10$

5. $5x - 3 \leq 2x - 6$

6. $2x - 7 \geq 4x + 3$

7. $3x + 7 \geq 5x - 7$

8. $3 - x \leq 2 - 3x$

9. $\frac{5x - 1}{3} < \frac{2x + 1}{6}$

10. $\frac{2x - 1}{2} > -\frac{x - 1}{4}$

11. $1 - x \geq \frac{x - 1}{2}$

12. $\frac{x}{2} - \frac{1}{3} \leq \frac{x + 3}{4}$

13. $|x + 1| < 3$

14. $|x + 2| \leq 4$

15. $|2x - 1| \leq 1$

16. $|3 - 2x| < 6$

17. $|x - 1| > 2$

18. $|x + 2| \geq 3$

19. $|3x - 2| \geq 6$

20. $|6 - x| > 3$

21. $|2x + 3| < \frac{1}{6}$

22. $\left|\frac{x}{2} - 3\right| \leq 4$

23. $\frac{|2x - 5|}{3} \geq 5$

24. $\left|\frac{1}{2} - 2x\right| > 4$

25. $(x - 1)(x - 3) < 0$

26. $(x + 1)(x - 2) > 0$

27. $(x + 2)(x - 3) \geq 0$

28. $(x + 2)(x + 1) \leq 0$

29. $x^2 - 4 < 0$

30. $x^2 - x - 6 > 0$

31. $x^2 + 2x - 3 \geq 0$

32. $x^2 + 2x \leq 0$

33. $2x^2 - 5x < -2$

34. $2x^2 + x > 3$

35. $3x^2 + 7x \geq 6$

36. $6x^2 + 3 \leq 11x$

37. $\frac{x - 1}{x + 2} < 0$

38. $\frac{x + 2}{x + 3} > 0$

39. $\frac{x - 1}{x - 2} \geq 0$

40. $\frac{x - 2}{x + 3} \leq 0$

41. $\frac{x}{x + 2} > \frac{1}{x + 2}$

42. $\frac{2x}{x + 3} < \frac{3}{x + 3}$

43. $\frac{1}{x - 1} \leq \frac{x}{1 - x}$

44. $\frac{2x}{x - 4} \geq \frac{1}{x - 4}$

45. $x \geq \frac{1}{x}$

46. $x \leq \frac{4}{x}$

47. $\frac{x^2 + 2}{x} > 3x$

48. $\frac{x^2 - 3}{x} < 2x$

49. $\frac{(x - 2)(x - 3)}{x - 1} < 0$

50. $\frac{x^2 - 1}{x - 3} > 0$

51. $\frac{x^2 - 1}{x - 2} \geq 0$

52. $\frac{x^2 - 2x + 1}{x} \leq 0$

53. $\frac{x^2 + 2x + 1}{x} \leq 0$

54. $\frac{x^2 + 1}{x - 2} \geq 0$

55. $\frac{x^2 - x - 2}{x^2} < 0$

56. $\frac{x^2 - 1}{x^2 + 1} > 0$

A.5 The Cartesian Coordinate System

The Cartesian Coordinate System
The Distance Between Two Points
Creating Equations
Graphs

APPLICATION **Getting to the Church on Time**

A prospective groom is located at point O in Figure A.9 and must drive his car to his wedding taking place in 25 minutes at the church located at point C. The church at C is 13 miles east and 4 miles north of O. The only highway goes through a town located at B, which is 10 miles due east of O. From B the highway goes straight to C. From O to B is a nice open stretch of highway, and he can average 50 miles per hour. The stretch from B to C has traffic lights and some congestion, and he can average only 25 miles per hour on this leg of the trip. Can he get to the church on time? See Example 2 on page 621 for the answer.

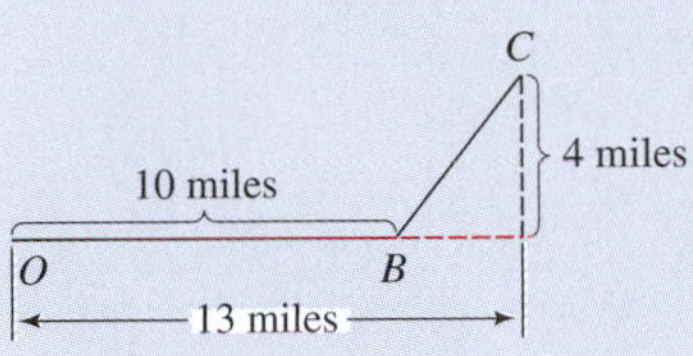

Figure A.9

The Cartesian Coordinate System

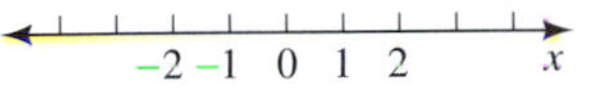

Figure A.10

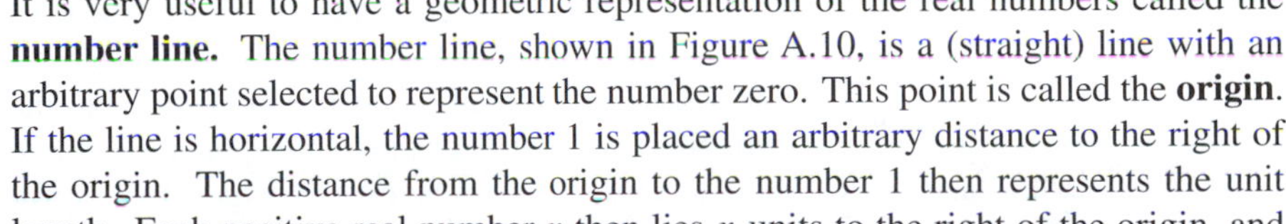

It is very useful to have a geometric representation of the real numbers called the **number line.** The number line, shown in Figure A.10, is a (straight) line with an arbitrary point selected to represent the number zero. This point is called the **origin**. If the line is horizontal, the number 1 is placed an arbitrary distance to the right of the origin. The distance from the origin to the number 1 then represents the unit length. Each positive real number x then lies x units to the right of the origin, and each negative real number x lies $-x$ units to the left of the origin. In this manner

North Wind Picture Archives

René Descartes, 1596–1650

It is sometimes said that Descartes' work "La Geometrie" marks the turning point between medieval and modern mathematics. He demonstrated the interplay between algebra and geometry, tying together these two branches of mathematics. Because of poor health as a child, he was always permitted to remain in bed as long as he wished. He maintained this habit throughout his life and did his most productive thinking while lying in bed in the morning. Descartes was a great philosopher as well as a mathematician and believed that mathematics should be a model for other branches of study. The following is a quote from his famous "Discours sur la Methode": "The long chain of simple and easy reasonings by means of which geometers are accustomed to reach conclusions of their most difficult demonstrations led me to imagine that all things, to the knowledge of which man is competent, are mutually connected in the same way, and that there is nothing so far removed from us as to be beyond our reach, or so hidden that we cannot discover it, provided only we abstain from accepting the false for the true, and always preserve in our thoughts the order necessary for the deduction of one truth from another."

each real number corresponds to exactly one point on the number line, and each point on the number line corresponds to exactly one real number.

We will use the following standard notation.

Interval Notation	Definition	Geometric Representation
(a, b)	$\{x\vert a < x < b\}$	
$[a, b]$	$\{x\vert a \le x \le b\}$	
$(a, b]$	$\{x\vert a < x \le b\}$	
$[a, b)$	$\{x\vert a \le x < b\}$	
$[a, \infty)$	$\{x\vert a \le x\}$	
(a, ∞)	$\{x\vert a < x\}$	
$(-\infty, b)$	$\{x\vert x < b\}$	
$(-\infty, b]$	$\{x\vert x \le b\}$	

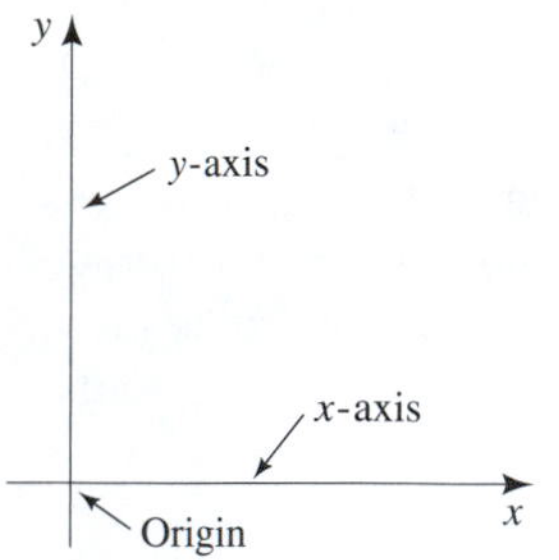

Figure A.11

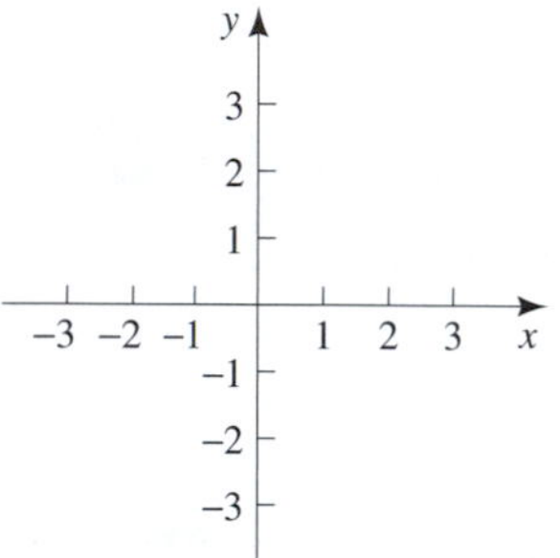

Figure A.12

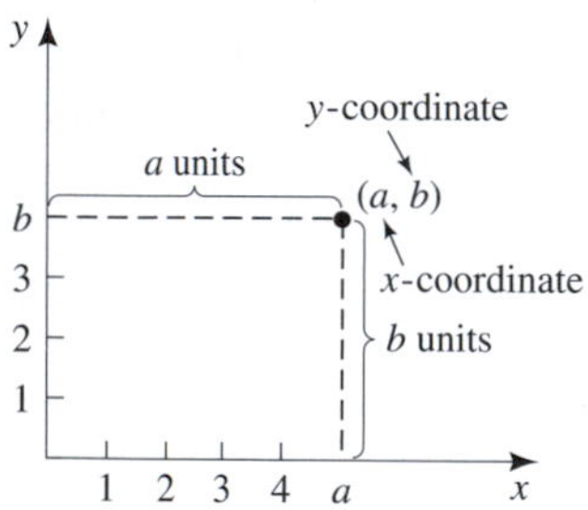

Figure A.13

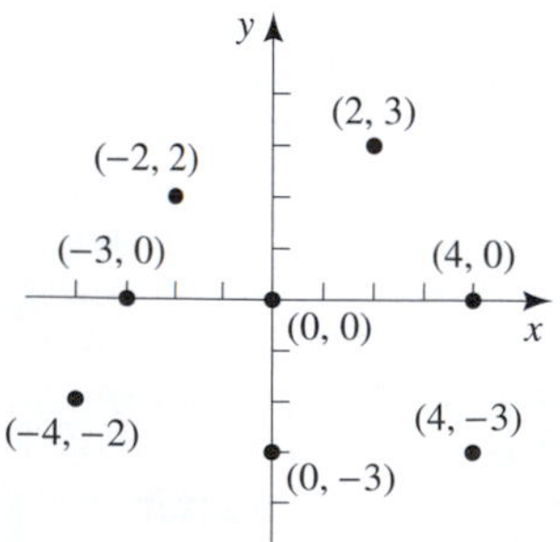

Figure A.14

The Cartesian coordinate system (named after René Descartes) permits us to express data or relationships between two variables as geometric pictures. Representing numbers and equations as geometric pictures using the Cartesian coordinate system yields deeper insight into the equations and data since we can see them graphically.

To describe a **Cartesian coordinate system**, we begin with a horizontal line called the **x-axis** and a vertical line called the **y-axis** both drawn in the plane. The point of intersection of these two lines is called the **origin**. The plane is then called the **xy-plane** or the **coordinate plane.** See Figure A.11.

For each axis, select a unit of length. The units of length for each axis need not be the same. Starting from the origin as zero, mark off the scales on each axis. For the x-axis, positive numbers are marked off to the right of the origin and negative numbers to the left. For the y-axis, positive numbers are marked off above the origin and negative numbers below the origin. See Figure A.12.

Each point P in the xy-plane is assigned a pair of numbers (a, b) as shown in Figure A.13. The first number a is the horizontal distance from the point P to the y-axis and is called the **x-coordinate,** and the second number b is the vertical distance to the x-axis and is called the **y-coordinate**.

Conversely, every pair of numbers (a, b) determines a point P in the xy-plane with x-coordinate equal to a and y-coordinate equal to b. Figure A.14 indicates some examples.

We use the standard convention that $P(x, y)$ is the point P in the plane with Cartesian coordinates (x, y).

Warning The symbol (a, b) is used to denote both a point in the xy-plane and an interval on the real line. The context will always indicate the way in which this symbol is being used.

The x-axis and the y-axis divide the plane into four **quadrants**. The **first quadrant** is all points (x, y) with both $x > 0$ and $y > 0$. The **second quadrant** is all points (x, y) with both $x < 0$ and $y > 0$. The **third quadrant** is all points (x, y) with both $x < 0$ and $y < 0$. The **fourth quadrant** is all points (x, y) with both $x > 0$ and $y < 0$. See Figure A.15.

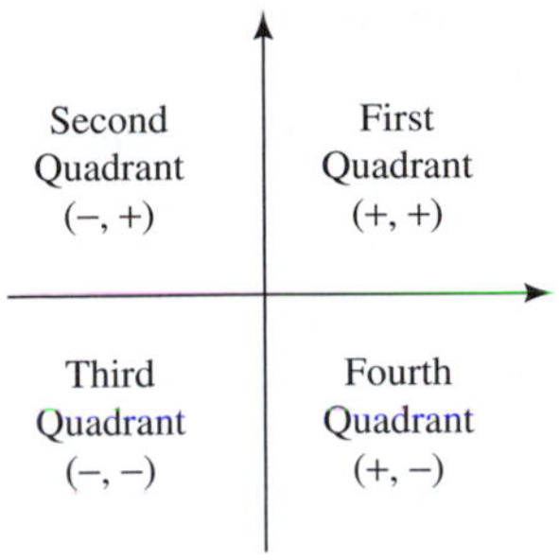

Second
Quadrant
(−, +)
First
Quadrant
(+, +)
Third
Quadrant
(−, −)
Fourth
Quadrant
(+, −)

Figure A.15

The Distance Between Two Points

We wish now to find the distance d between two given points $P_1(x_1, y_1)$ and $P_2(x_2, y_2)$ in the xy-plane shown in Figure A.16.

y
x
$P_2(x_2, y_2)$
d
$y_2 - y_1 = b$
$P_1(x_1, y_1)$
$x_2 - x_1 = a$

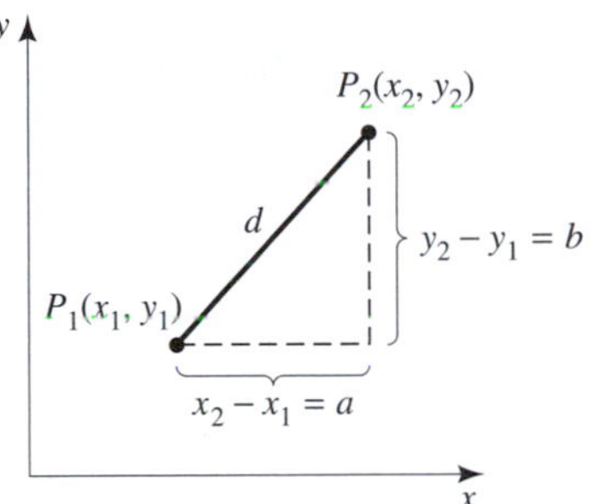

Figure A.16

Recall the Pythagorean theorem that says that $d^2 = a^2 + b^2$ in Figure A.16. Then

$$d^2 = |x_2 - x_1|^2 + |y_2 - y_1|^2 = (x_2 - x_1)^2 + (y_2 - y_1)^2$$

Taking square roots of each side gives the following.

Distance Formula

The distance $d = d(P_1, P_2)$ between the two points $P_1(x_1, y_1)$ and $P_2(x_2, y_2)$ in the xy-plane is given by

$$d(P_1, P_2) = \sqrt{(x_2 - x_1)^2 + (y_2 - y_1)^2}.$$

y
(−2, 4)
x
(4, −3)

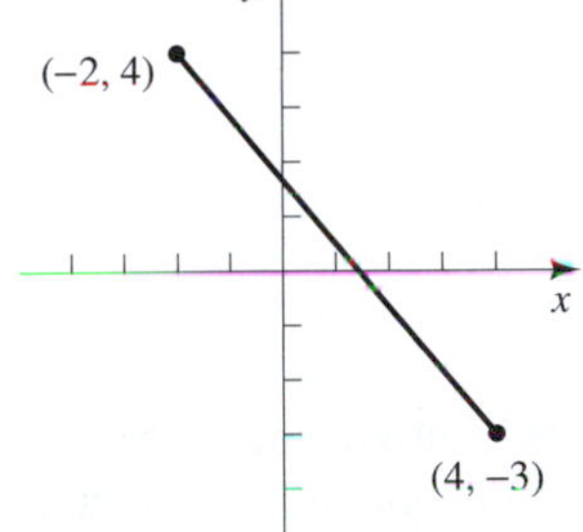

Figure A.17

EXAMPLE 1 **Finding the Distance Between Two Points**

Find the distance between the two points $(-2, 4)$ and $(4, -3)$ shown in Figure A.17.

Solution

We have $(x_1, y_1) = (-2, 4)$ and $(x_2, y_2) = (4, -3)$. Thus

$$\begin{aligned} d(P_1, P_2) &= \sqrt{(x_2 - x_1)^2 + (y_2 - y_1)^2} \\ &= \sqrt{[(4) - (-2)]^2 + [(-3) - (4)]^2} \\ &= \sqrt{(6)^2 + (-7)^2} = \sqrt{85} \end{aligned}$$

■

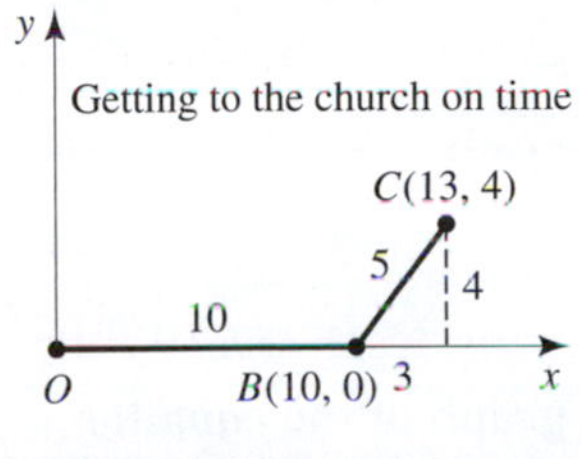

y
Getting to the church on time
C(13, 4)
5
4
10
O
B(10, 0)
3
x

Figure A.18

EXAMPLE 2 **Using the Distance Between Two Points**

A prospective groom is located at point O in Figure A.18 and must drive his car to his wedding taking place in 25 minutes at the church located at point C. The church at C is located 13 miles east and 4 miles north of O. The only highway goes through a town located at B, which is 10 miles due east of O. From B the highway goes straight to C. From O to B is a nice open stretch of highway, and he can average 50 miles per hour. The stretch from B to C has traffic lights and some congestion, and he can average only 25 miles per hour on this leg of the trip. Can he get to the church on time?

Solution

The distance from O to B is 10 miles, and the distance from B to C is $\sqrt{(13-10)^2+(4-0)^2}=5$ miles. Distance is the product of velocity with time, $d=v\cdot t$. Thus, $t=d/v$. Then the time to travel the given route is

$$\frac{10}{50}+\frac{5}{25}=\frac{2}{5}$$

of an hour, or 24 minutes. This is less than 25 minutes, so he can make his wedding on time. ■

Creating Equations

To solve any applied problem, we must first take the problem and translate it into mathematical terms. Doing this requires creating equations.

EXAMPLE 3 Creating an Equation

A small shop makes two styles of shirts. The first style costs \$7 to make, and the second costs \$9. The shop has \$252 a day to produce these shirts. If x is the number of the first style produced each day and y is the number of the second style, find an equation that x and y must satisfy.

Solution

Create a table that includes the needed information. See the following table.

Style	First	Second	Both
Cost of each shirt	\$7	\$9	
Number of each style produced	x	y	
Total cost	$7x$	$9y$	252

Since x are produced at a cost of \$7 each, the cost in dollars of producing x of these shirts is $7x$. Since y are produced at a cost of \$9 each, the cost in dollars of producing y of these shirts is $9y$. This is indicated in the table. Since the cost of producing the first style plus the cost of producing the second style must total \$252, the equation we are seeking is

$$7x+9y=252$$

■

Graphs

Given any equation in x and y (such as $y=2x-1$), the **graph of the equation** is the set of all points (x, y) such that x and y satisfy the equation.

DEFINITION Graph of an Equation

The **graph of an equation**, is the set

$$\{(x, y)|x \text{ and } y \text{ satisfy the equation}\}$$

For example, the point $(0, -1)$ is on the graph of the equation $y = 2x - 1$, because when x is replaced with 0 and y replaced with -1, the equation is satisfied. That is, $(-1) = 2(0) - 1$.

To find the graph of an equation, plot points until a pattern emerges. More complicated equations will require more points.

EXAMPLE 4 Determining a Graph of an Equation

Sketch the graph of $y = 2x - 1$, that is, find the set $\{(x, y)|y = 2x - 1\}$.

Solution

We take a number of values for x, put them into the equation, and solve for the corresponding values of y. For example, if we take $x = -3$, then $y = 2(-3) - 1 = -7$. The following table summarizes this work.

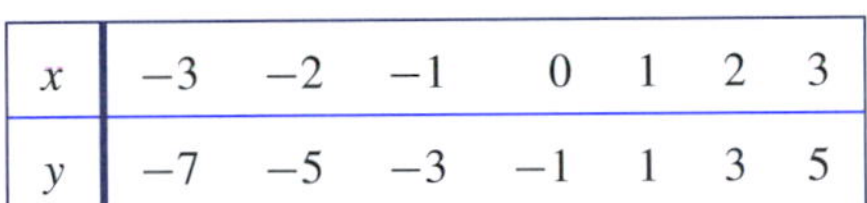

x	-3	-2	-1	0	1	2	3
y	-7	-5	-3	-1	1	3	5

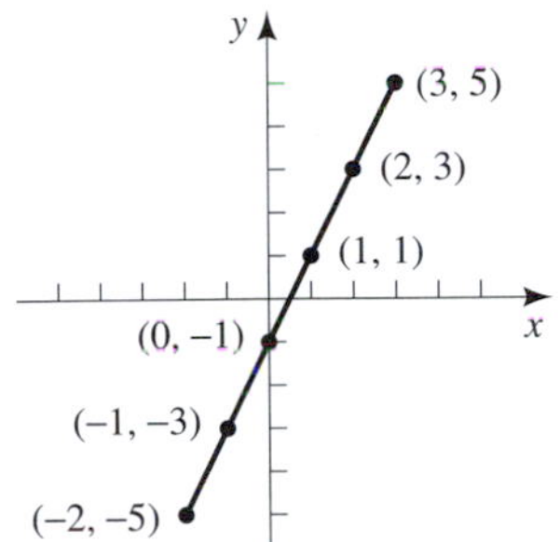

Figure A.19

The points (x, y) found in the table are graphed in Figure A.19. We see that the graph is a straight line. ■

EXAMPLE 5 Determining the Graph of an Equation

Sketch the graph of $x = 2$, that is, find the set $\{(x, y)|x = 2\}$.

Solution

Points such as $(2, -1)$, $(2, 1)$, and $(2, 2)$ are all on the graph. The graph is a vertical line, as shown in Figure A.20.

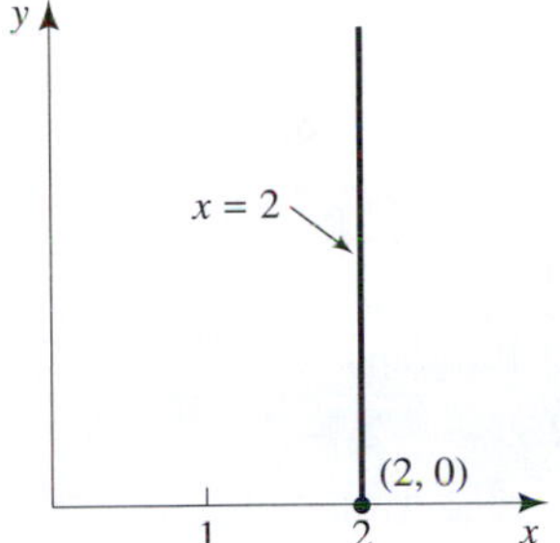

Figure A.20

EXAMPLE 6 Determining a Graph of an Equation

Sketch the graph of $y = x^2$, that is, find the set $\{(x, y)|y = x^2\}$.

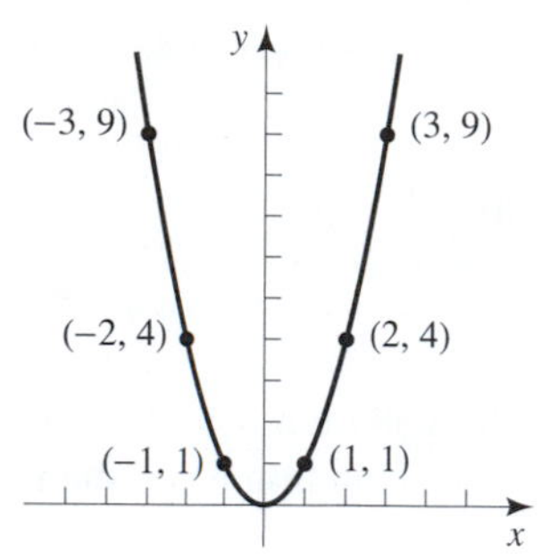

Figure A.21

Solution

We take a number of values for x, put them into the equation, and solve for the corresponding values of y. For example, if we take $x = -2$, then $y = (-2)^2 = 4$. The table summarizes this work.

x	−3	−2	−1	0	1	2	3
y	9	4	1	0	1	4	9

The points (x, y) found in the table are graphed in Figure A.21. ■

Warm Up Exercise Set A.5

1. Plot the points $(-2, 5)$, $(-2, -4)$, and $(3, -2)$.
2. Find the distance between the two points $(-4, -2)$ and $(3, -4)$.
3. Sketch the graph of the equation $7x + 9y = 252$ found in Example 3.

Exercise Set A.5

In Exercises 1 through 8, draw a number line and identify.

1. $(-1, 3)$
2. $[-2, 6]$
3. $(2, 5]$
4. $[-3, -1)$
5. $(-2, \infty)$
6. $[2, \infty)$
7. $(-\infty, 1.5]$
8. $(-\infty, -2)$

In Exercises 9 through 20, plot the given point in the xy-plane.

9. $(1, 2)$
10. $(-3, -4)$
11. $(-2, 1)$
12. $(2, -3)$
13. $(1, 0)$
14. $(-2, 0)$
15. $(0, -2)$
16. $(0, 4)$
17. $(-2, -1)$
18. $(2, 2)$
19. $(3, -4)$
20. $(-3, 0)$

In Exercises 21 through 32, find the set of points (x, y) in the xy-plane that satisfies the given equation or inequality.

21. $x \geq 0$
22. $y \geq 0$
23. $x < 0$
24. $y < 0$
25. $x = 2$
26. $x = -1$
27. $y = 3$
28. $y = -2$
29. $xy > 0$
30. $xy < 0$
31. $xy = 0$
32. $x^2 + y^2 = 0$

In Exercises 33 through 40, find the distance between the given points.

33. $(2, 4)$, $(3, 2)$
34. $(-2, 5)$, $(2, -3)$
35. $(-2, 5)$, $(-3, -1)$
36. $(1, 3)$, $(1, 6)$
37. $(-3, -2)$, $(4, -3)$
38. $(-5, -6)$, $(-3, -2)$
39. (a, b), (b, a)
40. (x, y), $(0, 0)$

In Exercises 41 through 50, sketch the graphs.

41. $y = -2x + 1$
42. $y = -2x - 2$
43. $y = 3x - 2$
44. $y = 2x + 3$
45. $y = 3$
46. $x = 4$
47. $y = 3x^2$
48. $y = -x^2$
49. $y = x^3$
50. $y = \sqrt{x}$

Applications and Mathematical Modeling

51. Transportation A truck travels a straight highway from O to A and then another straight highway from A to B. How far does the truck travel?

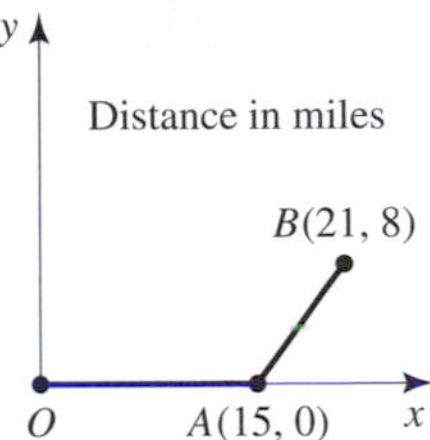

52. Transportation If the truck in the previous exercise travels 30 miles per hour from O to A and 60 miles per hour from A to B, how long does the trip take?

53. Navigation A ship leaves port and heads due east for 4 hours with a speed of 3 miles per hour, after which it turns due north and maintains this direction at a speed of 2.5 miles per hour. How far is the ship from port 6 hours after leaving?

54. Navigation A ship leaves port and heads due east for 1 hour at 5 miles per hour. It then turns north and heads in this direction at 4 miles per hour. How long will it take the ship to get 13 miles from port?

55. Oil Slick The shore line stretches along the x-axis as shown with the ocean consisting of the first and second quadrants. An oil spill occurs at the point $(2, 4)$, and a coastal town is located at $(5, 0)$, where the numbers are in miles. If the oil slick is approaching the town at a rate of 0.5 mile per day, how many days will it be until the oil slick reaches the coastal town?

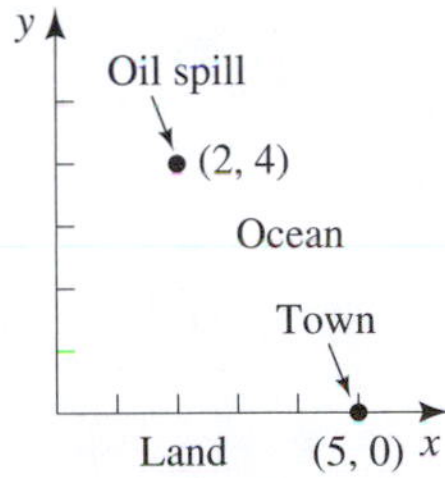

56. Transportation Charges A furniture store is located 5 miles east and 14 miles north of the center of town. You live 10 miles east and 2 miles north of the center of town. The store advertises free delivery within a 12-mile radius of the store. Do you qualify for free delivery?

57. Profits A furniture store sells chairs at a profit of \$100 each and sofas at a profit of \$150 each. Let x be the number of chairs sold each week, and let y be the number of sofas sold each week. If the profit in one week is \$3000, write an equation that x and y must satisfy. Graph this equation.

58. Revenue A restaurant has two specials: steak and chicken. The dinner with steak is \$13, and the dinner with chicken is \$10. Let x be the number of steak specials served, and let y be the number of chicken specials. If the restaurant made \$390 in sales on these specials, find the equation that x and y must satisfy. Graph this equation.

59. Nutrition An individual needs about 800 mg of calcium daily in his diet but is unable to eat any dairy products because he is allergic to them. However, this individual does enjoy eating canned sardines and steamed broccoli. Let x be the number of ounces of canned sardines consumed each day and let y be the number of cups of steamed broccoli. If there is 125 mg of calcium in each ounce of canned sardines and 190 mg in each cup of steamed broccoli, what equation must x and y satisfy if this person is to obtain his daily need of calcium from these two sources?

60. Costs A contractor builds ranch and split-level style homes. The ranch costs \$130,000 to build, and the split-level costs \$150,000. Let x be the number of ranch-style homes built, and let y be the number of split-level style homes. If the contractor has \$1,360,000 to build these homes, find the equation that x and y must satisfy. Graph this equation.

61. Biology A fisheries biologist decides to set a gill net on her next trip to a remote mountain lake. The net must stretch between points A and B shown in the figure. Since she must backpack to the lake, she wants to carry no more net than necessary. She walks off the distance a from C to B and the distance b from A to C, being careful that AC is perpendicular to AB. Find an equation that gives the distance c from A to B in terms of a and b.

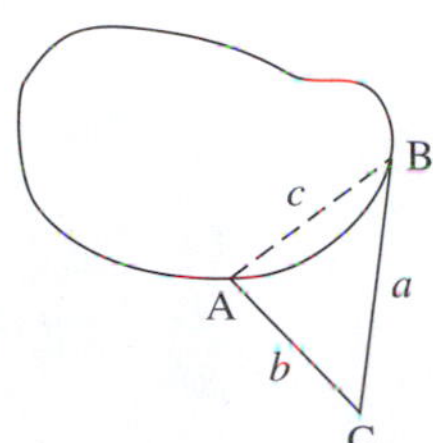

Solutions to WARM UP EXERCISE SET A.5

1. The points are plotted in the following figure.

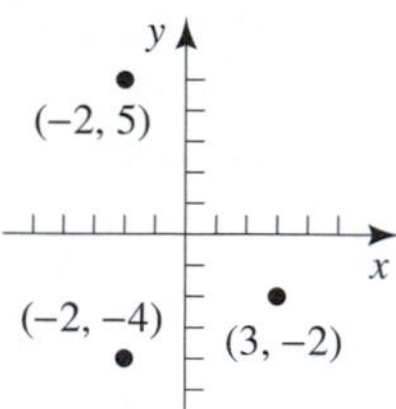

2. The distance between the two points $(-4, -2)$ and $(3, -4)$ is given by

$$\sqrt{[3-(-4)]^2 + [-4-(-2)]^2} = \sqrt{7^2 + (-2)^2} = \sqrt{53}$$

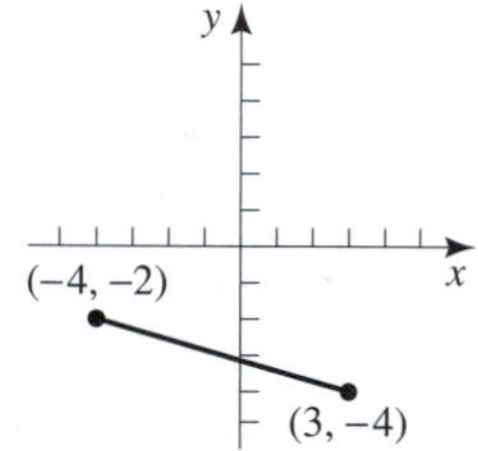

3. First solve for y, and obtain $y = -\frac{7}{9}x + 28$. Now take a number of values for x, put them into the equation, and solve for the corresponding values of y. For example, if we take $x = 9$, then $y = -\frac{7}{9}(9) + 28 = 21$. The following table summarizes this work.

x	0	9	18	27	36
y	28	21	14	7	0

The points (x, y) found in the table are graphed in the figure. We see that apparently the graph is a straight line.

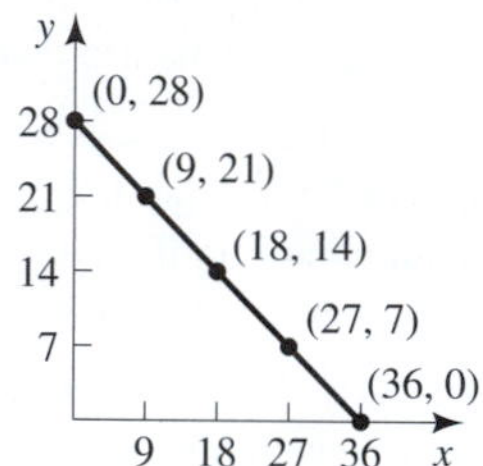

A.6 Lines

Slope
Equations of Lines
Applications

APPLICATION Tropical Rain Forest Destruction

According to data collected by Aiken,[1] the equation $p = -1.10t + 2229$ approximates the percentage rain forest cover in peninsular Malaysia, where t is the calendar year. What is the significance of the constant -1.10? For the answer, see Example 9 on page 635.

[1] S. R. Aiken and C. H. Leigh. 1984. A second national park for peninsular Malaysia? Biological Conservation 29:253–76.

Slope

To describe a straight line, one must first describe the "slant" or **slope** of the line. See Figure A.22.

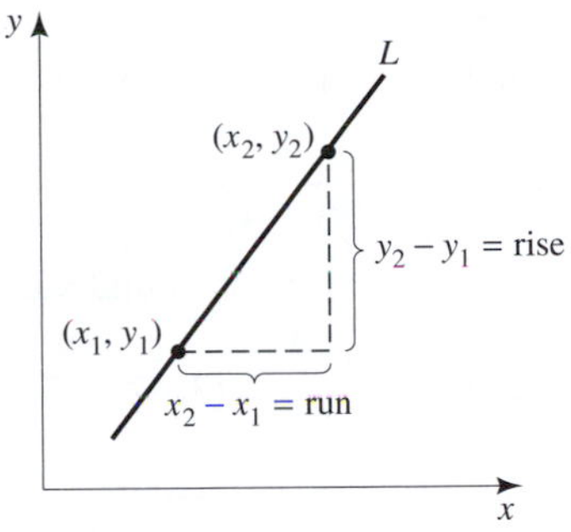

Figure A.22

DEFINITION **Slope of a Line**

Let (x_1, y_1) and (x_2, y_2) be two points on a line L. If the line is nonvertical $(x_1 \neq x_2)$, the **slope** m of the line L is defined to be

$$m = \frac{y_2 - y_1}{x_2 - x_1}$$

If L is vertical $(x_2 = x_1)$, then the slope is said to be **undefined**.

Remark The term $x_2 - x_1$ is called the **run**, and $y_2 - y_1$ is called the **rise**. See Figure A.22. Thus, the slope m can be thought of as

$$m = \frac{y_2 - y_1}{x_2 - x_1} = \frac{\text{rise}}{\text{run}}$$

If (x_1, y_1) and (x_2, y_2) are two points on a vertical line L, then we must have $x_1 = x_2$. (Why?) Using the formula for slope would yield

$$m = \frac{y_2 - y_1}{x_2 - x_1} = \frac{y_2 - y_1}{0}$$

which is undefined. This is why a vertical line is said to have an undefined slope.

The slope of the line does not depend on which two points on the line are chosen to use in the above formula. To see this, notice in Figure A.23 that the two right triangles are similar, since the corresponding angles are equal. Thus the ratio of the corresponding sides must be equal. That is,

$$\frac{b}{a} = \frac{d}{c}$$

where the ratio on the left is the slope using the points A and D and the ratio on the right is the slope using the points B and C.

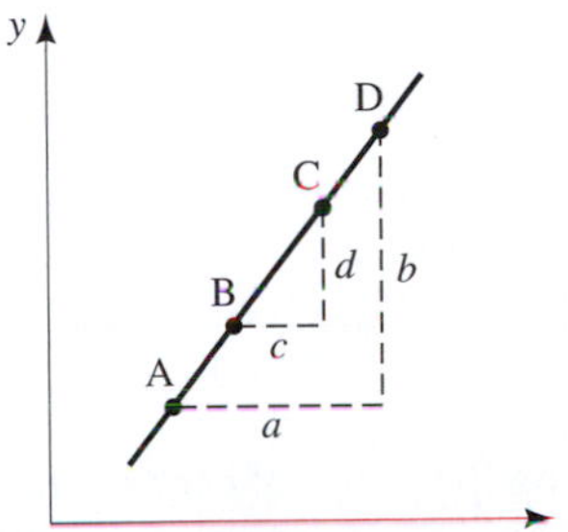

Figure A.23

EXAMPLE 1 **Finding the Slope of Lines**

Find the slope (if it exists) of the line through each pair of points. Sketch the line.

(a) (1, 2), (3, 6),

(b) (−3, 1), (3, −2),

(c) (−1, 3), (2, 3),

(d) (3, 4), (3, 1)

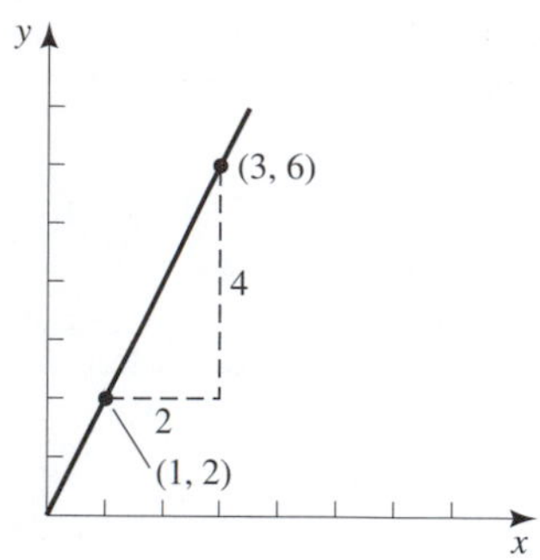

Figure A.24

Solution

(a) $(x_1, y_1) = (1, 2)$ and $(x_2, y_2) = (3, 6)$. So

$$m = \frac{y_2 - y_1}{x_2 - x_1} = \frac{(6) - (2)}{(3) - (1)} = \frac{4}{2} = 2$$

See Figure A.24. Notice that for each unit we move to the right, the line moves up $m = 2$ units.

(b) $(x_1, y_1) = (-3, 1)$ and $(x_2, y_2) = (3, -2)$. So

$$m = \frac{y_2 - y_1}{x_2 - x_1} = \frac{(-2) - (1)}{(3) - (-3)} = \frac{-3}{6} = -\frac{1}{2}$$

See Figure A.25. Notice that for each unit we move to the right, the line moves down $1/2$ unit.

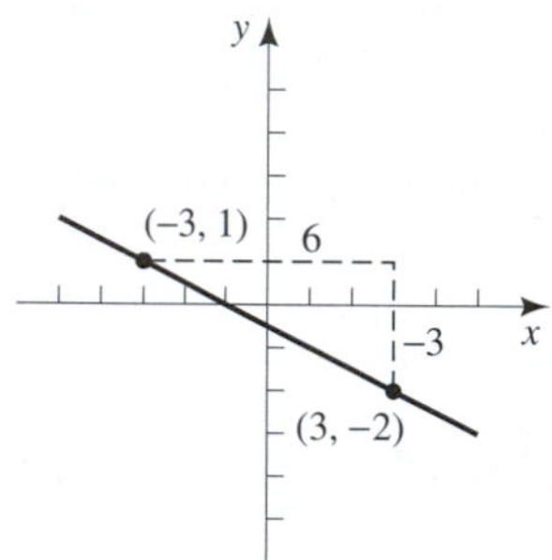

Figure A.25

(c) $(x_1, y_1) = (-1, 3)$ and $(x_2, y_2) = (2, 3)$. So

$$m = \frac{y_2 - y_1}{x_2 - x_1} = \frac{(3) - (3)}{(2) - (-1)} = \frac{0}{3} = 0$$

See Figure A.26. Notice that this line is horizontal.

(d) $(x_1, y_1) = (3, 4)$ and $(x_2, y_2) = (3, 1)$. So

$$m = \frac{y_2 - y_1}{x_2 - x_1} = \frac{(1) - (4)}{(3) - (3)} = \frac{-3}{0} = \text{undefined}$$

See Figure A.27. Notice that this line is vertical.

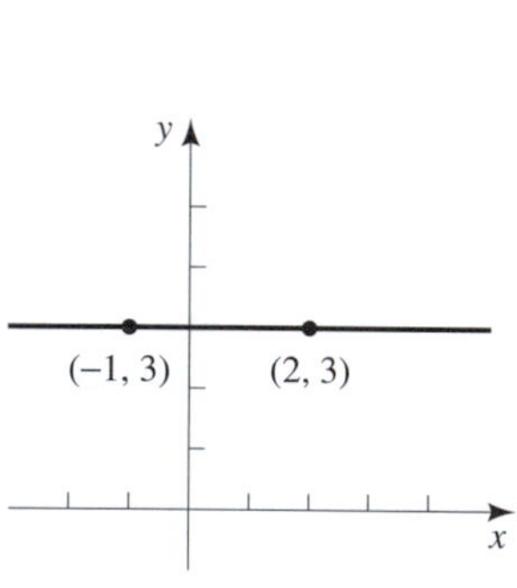

Figure A.26

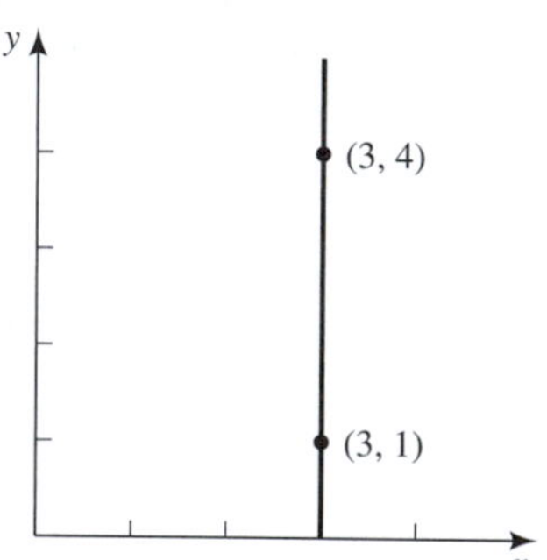

Figure A.27

■

Exploration 1

Slope

(a) **(For calculators with Grid On Mode)** Set the MODE of your graphing calculator to Grid On. Use a window with dimensions $[-3, 3]$ by $[0, 5]$, and set $x_{\text{scl}} = y_{\text{scl}} = 1$. Use a square window, if possible. Graph the following four equations on the same standard viewing window.

$$y = x + 3 \qquad y = 2x + 3$$
$$y = -x + 3 \qquad y = -2x + 3$$

Using the grid, estimate the slope of each line.

(b) Using TRACE move along the graph of each of the lines $y_1 = 2x + 3$ and $y_2 = -2x + 3$ in turn. Verify that as the cursor moves to the right, the y-coordinate on the first line changes twice as much as the x-coordinate, while the y-coordinate on the second line changes negatively twice as much as the x-coordinate.

Exploration 2

Equations of Lines

(For calculators with Grid On Mode) Set the MODE of your graphing calculator to Grid On. Use a square window if possible. Graph $y_1 = 2x$ and $y_2 = -0.5x - 0.5$ on a screen with dimensions $[-3, 3]$ by $[-2, 6]$ and with $x_{scl} = y_{scl} = 0.5$. Verify that the graphs are the line shown in Figures A.24 and A.25, and go through the points shown in these figures. By using the grid, notice the change in the y-coordinate when the x-coordinate is increased by 1. What relationship does this change in y-coordinate have with the slope? Establish using algebra.

Consider any line L that is not vertical, and let $P_1(x_1, y_1)$ be one point on the line. We noted earlier that in determining the slope of a line, *any* two points on the line can be used. We will now take a second point $P_2(x_2, y_2)$ on the line where P_2 is chosen with $x_2 = x_1 + 1$. Thus, we have run $= x_2 - x_1 = 1$, and therefore,

$$m = \frac{y_2 - y_1}{x_2 - x_1} = \frac{y_2 - y_1}{1} = y_2 - y_1 = \text{rise} \qquad \text{(A.1)}$$

This says that if the run is taken as 1, then the slope will equal the rise.

If now m is positive, as in Figure A.28a, then moving one unit to the right results in moving *up* m units. Thus the line *rises* (moving from left to right) and the larger the value of m the steeper the rise.

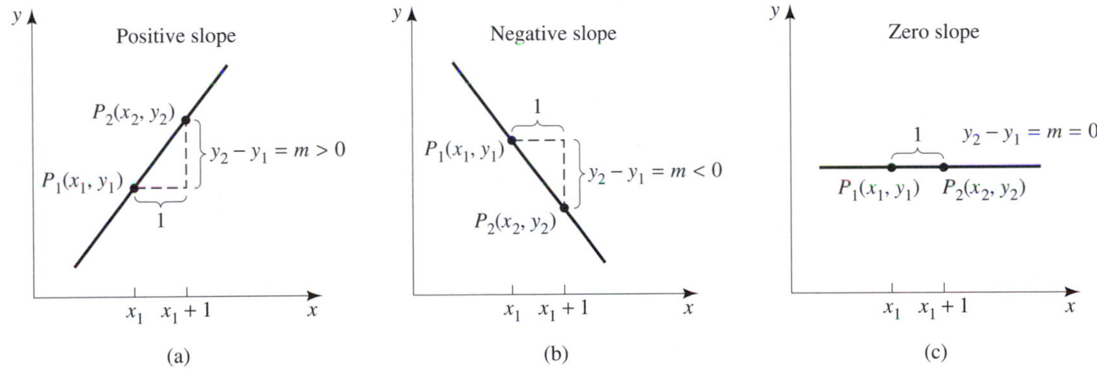

Figure A.28

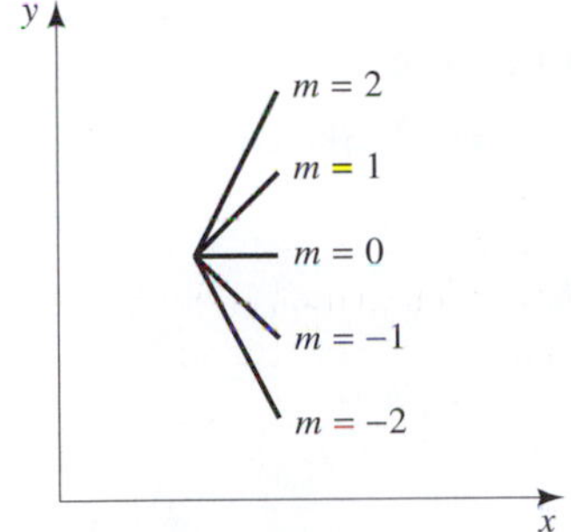

Figure A.29

If m is negative, as in Figure A.28b, then moving one unit to the right results in moving *down* $|m| = -m$ units. Thus, the line *falls*, and the more negative the value of m, the steeper the fall.

If m is zero, as in Figure A.28c, then moving over one unit results in moving up *no* units. Thus, the line is *horizontal*.

The first three parts of Example 1 give specific examples of these three general cases.

Figure A.29 shows several lines through the same point with different slopes.

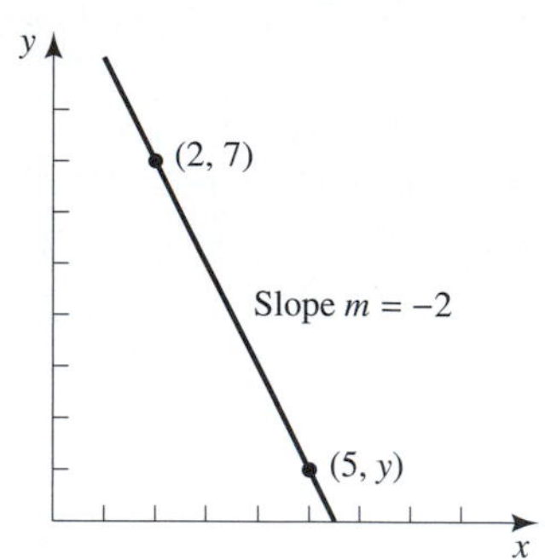

Figure A.30

EXAMPLE 2 Using the Slope of a Line

Find the y-coordinate of point P if P has x-coordinate 5 and P is on the line through (2, 7) with slope -2.

Solution

See Figure A.30. Using the definition of slope, we have

$$\begin{aligned} -2 &= \frac{y-7}{5-2} \\ &= \frac{y-7}{3} \\ -6 &= y-7 \\ y &= 1 \end{aligned}$$

■

Equations of Lines

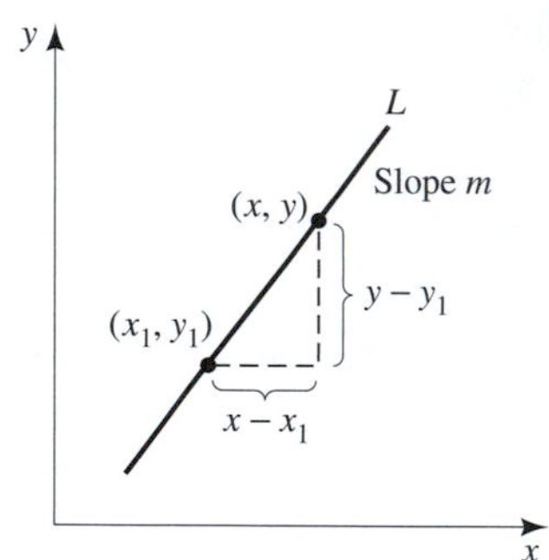

Figure A.31

Suppose we have a line through the point (x_1, y_1) with slope m as shown in Figure A.31. If (x, y) is any other point on the line, then we must have

$$m = \frac{y - y_1}{x - x_1}$$

or

$$y - y_1 = m(x - x_1)$$

This equation is the **point-slope equation.**

The Point-Slope Equation of a Line

The equation of the line through the point (x_1, y_1) with slope m is given by

$$y - y_1 = m(x - x_1)$$

EXAMPLE 3 Finding the Point-Slope Equation of a Line

Find an equation of the line through $(-2, 3)$ with slope $m = -2$. Sketch the graph.

Solution

Since $(x_1, y_1) = (-2, 3)$, the point-slope equation is

$$\begin{aligned} y - y_1 &= m(x - x_1) \\ y - 3 &= -2[x - (-2)] \\ y &= -2x - 1 \end{aligned}$$

One quick way of finding the graph is to notice that when $x = 0$, $y = -2(0) - 1 = -1$. Thus, $(0, -1)$ is a second point on the graph. See Figure A.32. ■

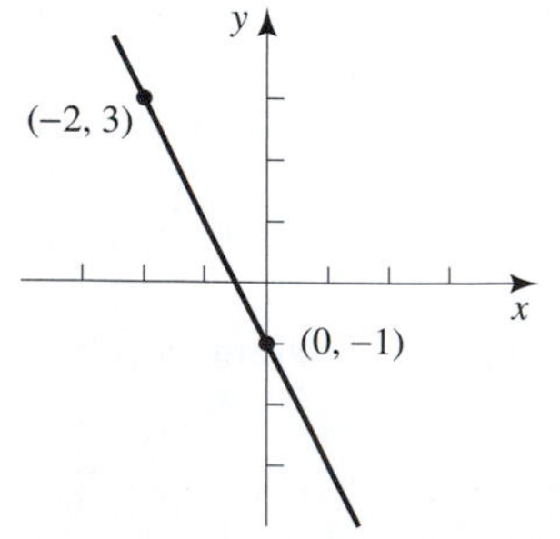

Figure A.32

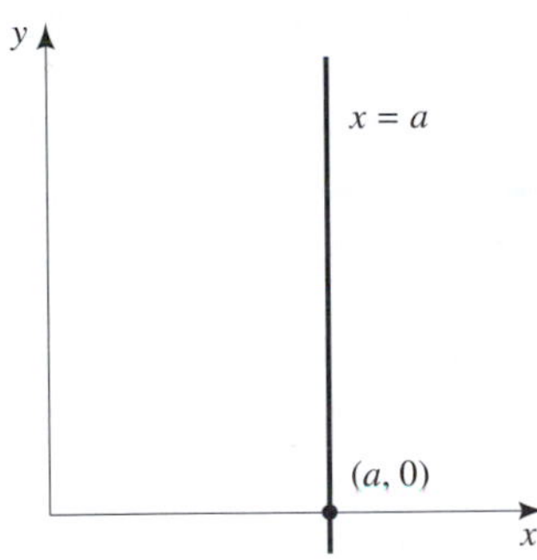

Figure A.33

A point (x, y) will be on the graph of a vertical line such as the one shown in Figure A.33 if and only if $x = a$, where $(a, 0)$ is the point where the vertical line crosses the x-axis. Thus, we have the following.

Vertical Lines

A vertical line has the equation

$$x = a$$

where $(a, 0)$ is the point at which the vertical line crosses the x-axis.

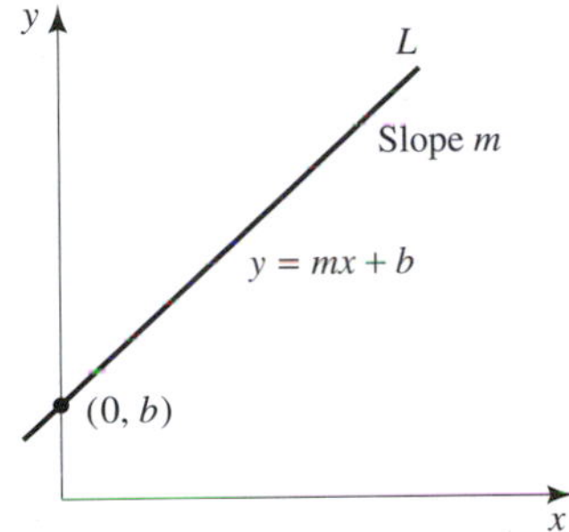

Figure A.34

Given any line that is not vertical, then the line has a slope m and, by Figure A.34, must cross the y-axis at some point $(0, b)$. The number b is called the **y-intercept**. Thus, we can use the point-slope equation where $(x_1, y_1) = (0, b)$ is a point on the line. This gives

$$y - b = m(x - 0)$$
$$y = mx + b$$

This is the **slope-intercept equation** of a nonvertical line.

The Slope-Intercept Equation of a Line

The equation of the line with slope m and y-intercept b is given by

$$y = mx + b$$

Slope	Graph	Equation
$m > 0$	rises (↗)	$y = mx + b$
$m < 0$	falls (↘)	$y = mx + b$
$m = 0$	horizontal (↔)	$y =$ constant
m undefined	vertical (↕)	$x =$ constant

EXAMPLE 4 **Finding the Slope-Intercept Equation of a Line**

Find the equation of a line with slope 3 and y-intercept -2. Draw a graph.

Solution

We have $m = 3$ and $b = -2$. Thus,

$$y = mx + b = 3x - 2$$

To draw a graph, we first notice that since the y-intercept is -2, the point $(0, -2)$ is on the graph (see Figure A.35). Since the slope is 3, if we move over one unit to the right, the line will move up three units, so the point $(1, 1)$ is also on the line. ■

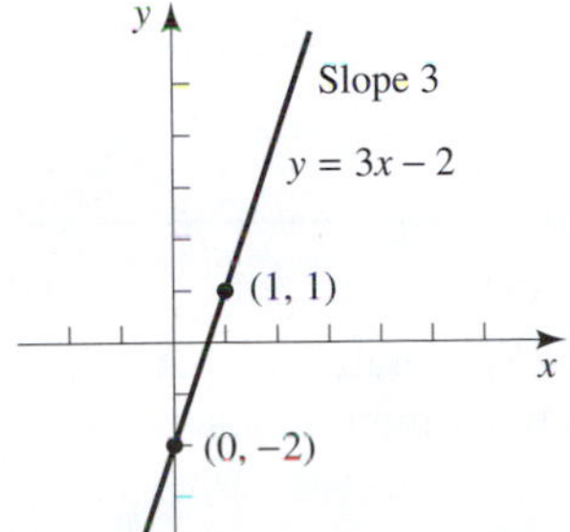

Figure A.35

If a and b are not both zero, the equation $ax + by = c$, is called a **linear equation**. The following can be proven. (See Exercise 65.)

The General Form of the Equation of a Line

If a and b are not both zero, then the graph of a linear equation $ax + by = c$ is a straight line, and, conversely, every line is the graph of a linear equation.

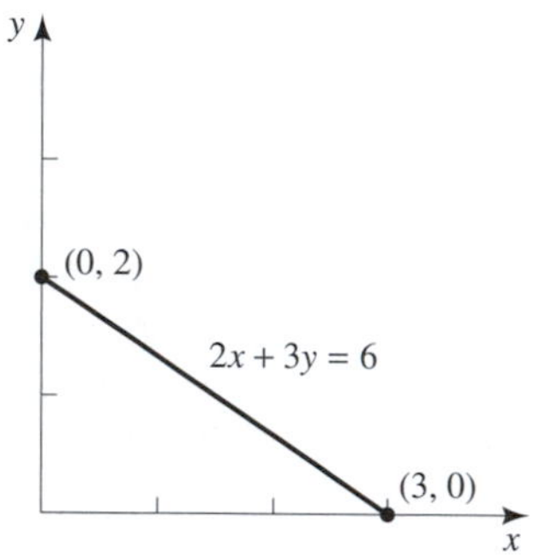

Figure A.36

We have already defined the y-intercept. We now need to define the x-intercept. Any line that is not horizontal must cross the x-axis at some point $(a, 0)$. The number a is called the x**-intercept**. To find the x-intercept of a line, simply set $y = 0$ in the equation of the line and solve for x. Similarly, to find the y-intercept of a line, set $x = 0$ in the equation of the line and solve for y. The intercepts are very useful in graphing, as the next example illustrates.

EXAMPLE 5 Graphing an Equation of a Line

Sketch a graph of $2x + 3y = 6$ by finding the x- and y-intercepts. Support with a grapher.

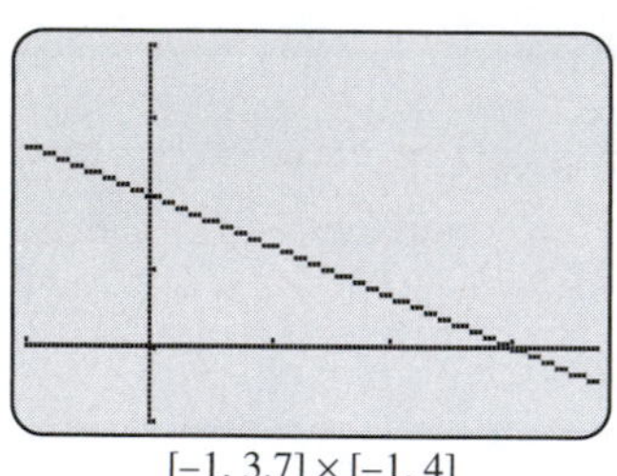

$[-1, 3.7] \times [-1, 4]$

Screen A.1

A graph of $y = -\frac{2}{3}x + 2$.

Solution

Since the equation is in the form $ax + by = c$, we know that the graph of this equation is a line. Setting $x = 0$, yields $3y = 6$ or $y = 2$ as the y-intercept. Setting $y = 0$, yields $2x = 6$ or $x = 3$ as the x-intercept. We then obtain Figure A.36.

To support with a grapher, we must first solve for y and obtain $y = -\frac{2}{3}x + 2$. Using a window with dimension $[-1, 3.7]$ by $[-1, 4]$, the graph is shown in Screen A.1. This agrees with Figure A.36, and we also verify the intercepts. ■

We will use technology in this text in two important ways. First, we will solve a problem mathematically and then support our answer graphically, as we did in Example 5. Second, we will solve a problem graphically and then confirm the solution mathematically, as we do in the next exploration.

Exploration 3

Finding an Intercept Graphically

Find the y-intercept of $2y - 4x = -15$ graphically. Confirm algebraically.

Exploration 4

Parallel Lines

Graph the following four lines on your graphing calculator. Change the grapher format to *simultaneous* if you can.

$$y = -2x + 10 \qquad y = -2x + 8$$
$$y = -2x + 6 \qquad y = -2x + 4$$

Do these line appear parallel? Speculate on what is common about the equations that might indicate that they are all parallel.

Since slope is a number that indicates the direction or slant of a line, the following theorem can be proven.

Slope and Parallel Lines

Two lines are parallel if and only if they have the same slope.

EXAMPLE 6 Finding the Equation of a Line

Let L_1 be the line that goes through the two points $(-3, 1)$ and $(-1, 7)$. Find the equation of the line L through the point $(2, 1)$ and parallel to L_1. See Figure A.37.

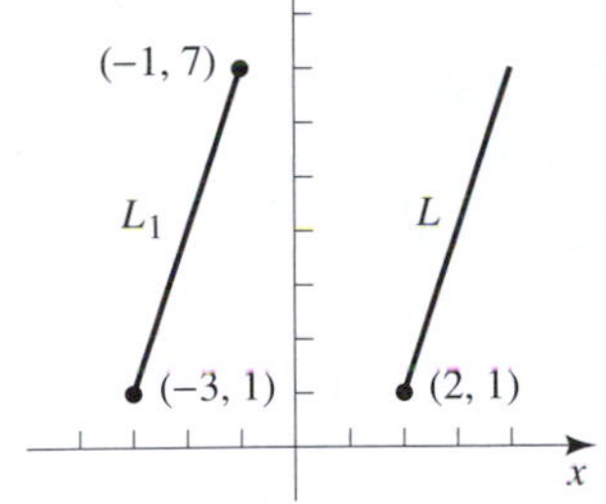

Figure A.37

Solution

We already have the point $(2, 1)$ on the line L. To use the point-slope equation, we need to find the slope m of L. But since L is parallel to L_1, m will equal the slope of L_1. The slope of L_1 is

$$m = \frac{y_2 - y_1}{x_2 - x_1} = \frac{(7) - (1)}{(-1) - (-3)} = \frac{6}{2} = 3$$

Now let us use the point-slope equation for L, where $(2, 1)$ is a point on the line L with slope $m = 3$. Then we obtain

$$\begin{aligned} y - y_1 &= m(x - x_1) \\ y - 1 &= 3(x - 2) \\ y &= 3x - 5 \end{aligned}$$

■

We also have the following theorem. (See Exercise 66.)

Slope and Perpendicular Lines

Two nonvertical lines with slope m_1 and m_2 are perpendicular if and only if

$$m_2 = -\frac{1}{m_1}$$

EXAMPLE 7 Finding the Equation of a Line

Find the equation of the line L through the point $(3, -1)$ and perpendicular to the line L_1 given by $2x + 3y = 6$.

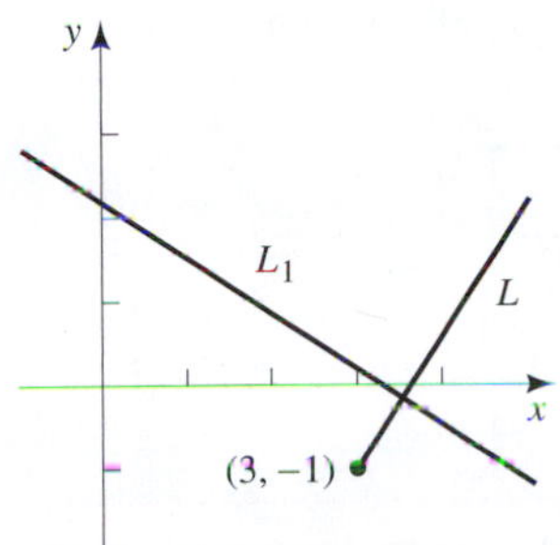

Figure A.38

Solution

See Figure A.38. We rewrite $2x + 3y = 6$ as

$$y = -\frac{2}{3}x + 2$$

This indicates that the slope of L_1 is $-2/3$. Thus, the slope m of L will be the negative reciprocal, or $m = 3/2$. Now use the point-slope equation for L to get

$$y - y_1 = m(x - x_1)$$

$$y - (-1) = \frac{3}{2}(x - 3)$$

$$y = \frac{3}{2}x - \frac{11}{2}$$

■

Interactive Illustration A.1

Graphs of Lines

This interactive illustration allows you to explore what happens to the graph of $y = mx + b$ as m and b change. Sliders are provided for m and b.

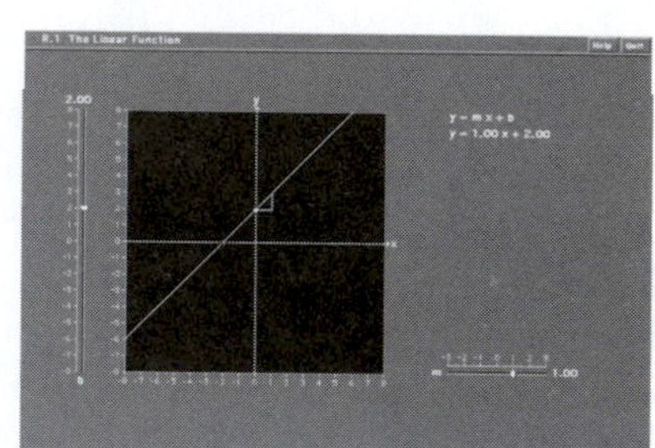

1. Increase m. What happens to the graph? Decrease m. What happens to the graph? What does the graph look like when m is positive? Zero? Negative?
2. Increase b. What happens to the graph? Decrease b. What happens to the graph? What does the graph look like when b is positive? Zero? Negative?

Applications

EXAMPLE 8 **Production of Vacuum Cleaners**

A firm produces two models of vacuum cleaners: regular and deluxe. The regular model requires 8 hours of labor to assemble, and the deluxe requires 10 hours of labor. Let x be the number of regular models produced, and let y be the number of deluxe models. If the firm wishes to use 500 hours of labor to assemble these cleaners, find an equation that the two quantities x and y must satisfy. If the number of the regular models that are assembled is increased by 5, how many fewer deluxe models will be assembled?

Solution

To find the required equation, construct a table with the given information, as follows.

Type	Regular	Deluxe	Total
Work-hours for each cleaner	8	10	
Number of each model produced	x	y	
Total work-hours for each model	$8x$	$10y$	500

The number of work-hours needed to assemble all of the regular models is the number of work-hours required to produce one of these models times the number produced. This gives $8x$. In a similar fashion we see that the number of work-hours needed to produce all of the deluxe models is $10y$. The number of hours allocated to assemble the regular model plus the number allocated to assemble the deluxe must equal 500. That is,

$$8x + 10y = 500$$

This is the required equation. Solving for y yields

$$y = -\frac{4}{5}x + 50$$

This is a line with slope $-\frac{4}{5}$. See Figure A.39. Thus, increasing the number x of regular models produced by 5 will decrease the number y of deluxe models produced by 4. ■

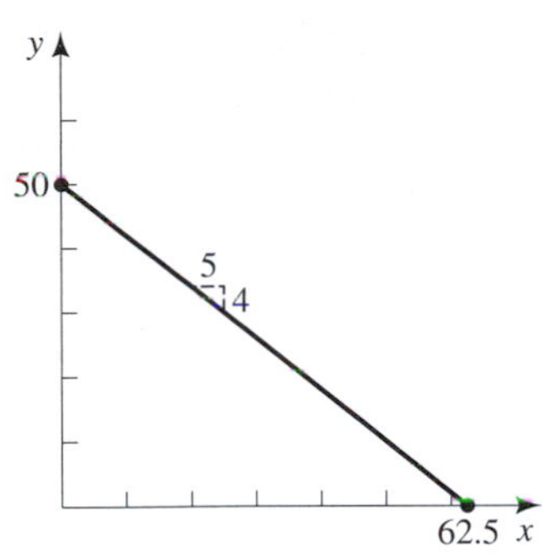

Figure A.39

Exploration 5

Slope

Graph the line given in Example 8 on your grapher. Use the TRACE feature to move the x coordinate to the right by five units and note the change in the y coordinate. Did you get 4?

EXAMPLE 9 **Tropical Rain Forest Destruction**

According to data collected by Aiken,[2] the equation $p = -1.10t + 2229$ approximates the percentage rain forest cover in peninsular Malaysia, where t is the calendar year. What is the significance of the constant -1.10? According to this equation what was the percentage of rain forest cover in 1990 in peninsular Malaysia?

Solution

The line $p = -1.10t + 2229$ has slope $m = -1.10$. This implies that each year brings a 1.10% loss of rain forest cover in peninsular Malaysia. Using this equation, we have that $-1.10(1990) + 2229 = 40$. Thus, according to this equation, the percent rain forest cover was 40 in peninsular Malaysia in 1990. Refer to Figure A.40.

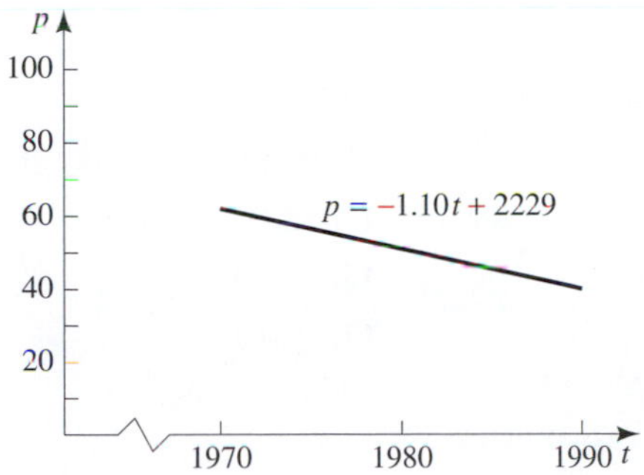

Figure A.40

■

[2] Ibid.

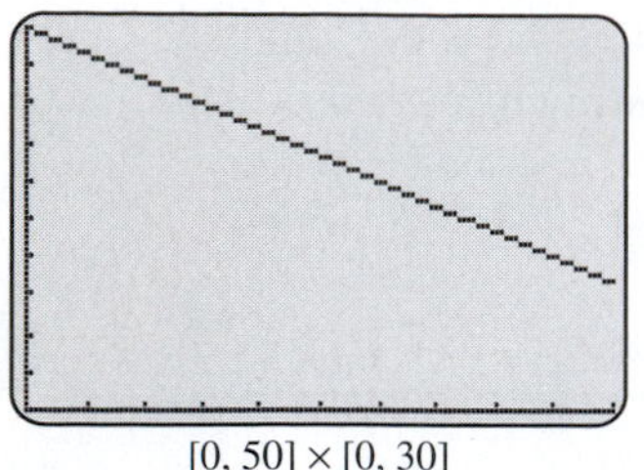

[0, 50] × [0, 30]

Screen A.2
A graph of $y = 30 - 0.4x$.

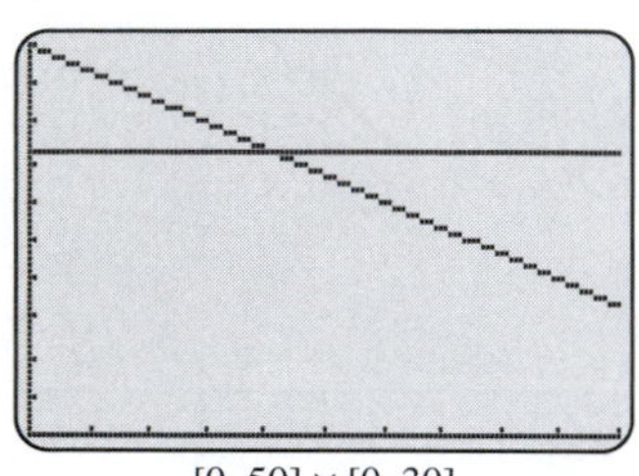

[0, 50] × [0, 30]

Screen A.3
A graph of $y_1 = 30 - 0.4x$ and $y_2 = 22$.

EXAMPLE 10 **Intersection of Graphs**

Chapman[3] estimated that the relationship between the price x of a barrel of oil and the number y of billions of barrels of oil used in the world per year is $y = 30 - 0.4x$. Graph on your graphing calculator. Determine the price of a barrel of oil if 22 billions of barrels of oil are used in the world per year. Confirm algebraically.

Solution

We graph $y_1 = 30 - 0.4x$ using a window with dimensions [0, 50] by [0, 30] and obtain Screen A.2. Now in the same window, also graph $y_2 = 22$ and obtain Screen A.3. Now use the "intersect" operation on your graphing calculator to find that the x-coordinate of the point of intersection is 20.

To confirm algebraically, we need to solve for x in the equation $22 = 30 - 0.4x$. We have

$$\begin{aligned} 22 &= 30 - 0.4x \\ 0.4x &= 30 - 22 = 8 \\ x &= 8/0.4 = 20 \end{aligned}$$

■

Warm Up Exercise Set A.6

1. Find the equation of the line through the two points (3, 1) and $(-2, -3)$ and then find the y-intercept. Sketch.
2. Find the equation of the line through the point $(2, -1)$ and perpendicular to the line in the previous exercise. Sketch.
3. A household has \$100 to purchase quarts of milk for \$0.80 a quart or candy bars at \$0.40 a bar. Let x be the number of quarts of milk, and let y be the number of candy bars purchased.
   a. Write a linear equation that x and y must satisfy. (This equation is called the **budget equation** in household economics.)
   b. Plot this line.
   c. What does the slope of this line mean?

Exercise Set A.6

In Exercises 1 through 8, find the slope (if it exists) of the line through the given pair of points. Sketch the line.

1. (1, 3), (2, 7)
2. $(3, -2)$, (2, 5)
3. $(-2, -2)$, $(-3, 1)$
4. $(-3, 2)$, (5, 4)
5. $(-1, 3)$, (2, 3)
6. $(-1, 2)$, $(-3, -4)$
7. $(-3, -1)$, $(-3, -7)$
8. $(-5, -2)$, $(-3, -4)$

In Exercises 9 through 20, find the slope (if it exists) of the line given by the each of the equations. Also find the x- and y-intercepts if they exist. Graph.

9. $y = 2x - 1$
10. $y = -3x + 2$
11. $x + y = 1$
12. $x - 2y - 3 = 0$
13. $-x + 2y - 4 = 0$
14. $x + y = 0$
15. $x = 3$
16. $x = 0$
17. $y = 4$
18. $y = 0$
19. $x = 2y + 1$
20. $x = -3y + 2$

[3] Duane J. Chapman. 1987. Computation techniques for intertemporal allocation of natural resources. *Amer. J. Agr. Econ.* 69:134–142.

In Exercises 21 through 24, find the x- and y-intercepts, and use these intercepts to sketch a graph of the line. Support your sketch with a graphing utility.

21. $3x + 5y = 15$

22. $5x + 3y = 15$

23. $4x - 5y = 20$

24. $-6x + 5y = 30$

In Exercises 25 through 47, find the equation of the line given the indicated information.

25. Through (2, 1) with slope -3

26. Through $(-3, 4)$ with slope 3

27. Through $(-1, -3)$ with slope 0

28. Through (2, 3) with slope undefined

29. Through $(-2, 4)$ and $(-4, 6)$

30. Through $(-1, 3)$ and $(-1, 6)$

31. Through $(-2, 4)$ and $(-2, 6)$

32. Through $(-1, 4)$ and $(-3, 2)$

33. Through $(a, 0)$ and $(0, b)$, $ab \neq 0$

34. Through (a, b) and (b, a), $a \neq b$

35. Through $(-2, 3)$ and parallel to the line $2x + 5y = 3$

36. Through $(2, -3)$ and parallel to the line $2x - 3y = 4$

37. Through (0, 0) and parallel to the line $y = 4$

38. Through (0, 0) and parallel to the line $x = 4$

39. Through (1, 2) and perpendicular to the line $y = 4$

40. Through (1, 2) and perpendicular to the x-axis

41. Through $(-1, 3)$ and perpendicular to the line $x + 2y - 1 = 0$

42. Through $(2, -4)$ and perpendicular to the line $x - 3y = 4$

43. Through $(3, -1)$ and with y-intercept 4

44. Through $(-3, -1)$ and with y-intercept 0

45. y-intercept 4 and parallel to $2x + 5y = 6$

46. y-intercept -2 and perpendicular to $x - 3y = 4$

47. Are the two lines $2x + 3y = 6$ and $3x + 2y = 6$ perpendicular?

48. A person has $2.35 in change consisting entirely of dimes and quarters. If x is the number of dimes and y is the number of quarters, write an equation that x and y must satisfy.

Applications and Mathematical Modeling

49. Investment A person invests some money in a bond that yields 8% a year and some money in a money market fund that yields 5% a year and obtains $576 of interest in the first year. If x is the amount invested in the bond and y is the amount invested in the money market fund, find an equation that x and y must satisfy. If the amount of money invested in bonds is increased by $100, by how much must the money invested in the money market decrease if the total interest is to stay the same?

50. Transportation Costs New cars are transported from Detroit to Boston and from Detroit to New York City. It costs $90 to transport each car to Boston and $80 to transport each car to New York City. Suppose $4000 is spent on transporting these cars. If x is the number transported to Boston and y is the number transported to New York City, find an equation that x and y must satisfy. If the number of cars transported to Boston is increased by 8, how many fewer cars must be transported to New York City if the total transportation costs remained the same?

51. Size of Deer Antlers Clow[4] showed that the size of a deer's antlers depends primarily on the age of the deer. Mule deer in northern Colorado have no antlers at age 10 months, but after that the weight of the antlers increases by 0.12 kg for every 10 months of additional age. Let x be the age of these mule deer in months, and let y be the weight of their antlers in kilograms. Assuming a linear relationship, write an equation that x and y must satisfy. Graph using a graphing calculator. Using the TRACE function, determine the weight in kilograms of antlers carried by a 40-month-old mule deer.

52. Scheduling An attorney has two types of documents to process. The first document takes 3 hours to process, and the second takes 4 hours. If x is the number of the first type that she processes and y is the number of the second type, find an equation that x and y must satisfy if she works 45 hours on these documents. If she processes 8 additional of the first type of document, how many fewer of the second type of document would she process?

53. Milk Production The amount of milk produced by a cow depends on the quality and therefore the cost of feed. According to a study in Wisconsin the annual feed cost c in dollars per cow per year is approximately related to the annual pounds x of milk per cow per year by $c = 201.38 + 0.03x$. What is the significance of the number 0.03 in this equation, and what will be the increase in cost for a 1000-pound increase in milk production per year?

[4] Duane J. Clow and N. Scott Urquhart. 1984. Mathematics in biology. New York: Ardsley House.

54. Biology In an extensive study of 82 species of desert lizards from three continents, Pianka[5] found that the body temperature T_b of the lizards was given approximated by $T_b = 38.8 + (1 - \beta)(T_a - 38.8)$, where T_a is the air temperature in degrees Celsius and β is a constant between 0 and 1 that depends on the specific species. Explain the significance of the number $1 - \beta$. What is the significance if $\beta = 1$?

55. Biology Simpson[6] reports that in females of the Atlantic Central American milk snake (*Lampropeltis triangulum polyzona*) the total length L is linear in the tail length l. Find the linear equation relating L to l if for one of these snakes $L = 140$ mm when $l = 60$ mm and for another $L = 1050$ mm when $l = 455$ mm. Find the slope and its significance.

56. Biology Clarke and McKenzie[7] report that weight w in milligrams of the sheep blowfly pupa is approximated by $w = 41.61 - 0.25T$, where T is the temperature in degrees Celsius. Explain the significance of the number -0.25. Graph.

57. Veterinary Entomology Hribar and coworkers[8] found that the number y of flies on a bull was approximated by $y = 10.95 + 2.27x$, where x is the number of flies on the front legs of the bull. What is the significance of the number 2.27? 10.95? Graph.

58. Environmental Quality Pierzynski and colleagues[9] found that the relative yield y of soybeans was approximated by $y = 1.54 - 0.00056x$, where x is the concentration (mg/kg) of zinc in the soybean tissue. What is the significance of the fact that the coefficient of x is *negative*? Graph for $0 \leq x \leq 1000$.

59. Profits A furniture store sells chairs at a profit of \$100 each and sofas at a profit of \$150 each. Let x be the number of chairs sold each week, and let y be the number of sofas sold each week.

a. If the profit in one week is \$3000, write an equation that x and y must satisfy. Graph using your graphing calculator.

b. Use the TRACE function to determine how much y changes when x changes by 1.

c. Use TRACE to find the point on the graph with x-coordinate equal to 3. What is the y-coordinate, and what does the point (x, y) mean in terms of chairs and sofas?

d. Find the x-coordinate of the point where the graph intersects the x-axis.

60. Revenue A restaurant has two specials: steak and chicken. The steak special costs \$13, and the chicken special costs \$10. Let x be the number of steak specials served, and let y be the number of chicken specials.

a. If the restaurant took in \$390 in total sales on these specials, find the equation that x and y must satisfy. Graph using your graphing calculator.

b. Use the TRACE function to determine how much y changes when x changes by 1.

c. Use TRACE to find the point on the graph with x-coordinate equal to 10. What is the y-coordinate, and what does the point (x, y) mean in terms of the steak and chicken specials?

d. Find the x-coordinate of the point where the graph intersects the x-axis.

61. Nutrition An individual needs about 820 mg of calcium daily in his diet but is unable to eat any dairy products because he is allergic to them. However, this individual does enjoy eating canned sardines and steamed broccoli. Let x be the number of ounces of canned sardines consumed each day, and let y be the number of cups of steamed broccoli.

a. If there is 125 mg of calcium in each ounce of canned sardines and 190 mg in each cup of steamed broccoli, what equation must x and y satisfy if this person is to satisfy his daily need for calcium from these two sources? Graph using your graphing calculator.

b. Use the TRACE function to determine how much y changes when x changes by 1.

c. Use TRACE to find the point on the graph with x-coordinate equal to 2. What is the y-coordinate, and what does the point (x, y) mean in terms of the amount of calcium and canned sardines?

d. Find the x-coordinate of the point where the graph intersects the x-axis.

62. Costs A contractor builds ranch and split-level style homes. The ranch costs \$130,000 to build, and the split-level costs \$150,000. Let x be the number of ranch-style homes built and y the number of split-level style homes.

a. If the contractor has \$1,360,000 to build these homes, find the equation that x and y must satisfy. Graph using your graphing calculator.

b. Use the TRACE function to determine how much y changes when x changes by 1.

c. Use TRACE to find the point on the graph with x-coordinate equal to 7. What is the y-coordinate, and what does the point (x, y) mean in terms of the cost of each style home?

---

[5] E. R. Pianka. 1986. Ecology and natural history of desert lizards. Princeton University Press.

[6] G. G. Simpson, A. Roe, and R. C. Lewontin. 1960. Quantitative zoology. Harcourt, Brace.

[7] Geoffrey Clarke and Leslie McKenzie. 1992. Fluctuating asymmetry as a quality control indicator for insect mass rearing process. *J. Econ. Entomol.* 85:2045–2050.

[8] Lawrence Hribar, Daniel LePrince, and Lane Foil. 1992. Feeding sites of some Louisiana tabanidae on fenvalevate-treated and control cattle. *J. Econ. Entomol.* 85:2279–2285.

[9] Gary Pierzynski and A. Paul Schwab. 1993. Bioavailability of zinc, cadmium, and lead in a metal contaminated alluvial soil. *J Environ. Qual.* 22:247–254.

d. Find the x-coordinate of the point where the graph intersects the x-axis.

63. Sociology Leigh[10] showed that the equation $y = -2011.44 + 1.06x$ was approximately true, where x is the calendar year (from 1958 to 1967) and y the percentage of whites who would not move if blacks came to live next door. Graph using your graphing calculator. Graphically find the year in which, according to the given equation, 69.34% of whites would not move if blacks came to live next door.

64. Forestry Ewel[11] showed that the equation $T_s = 8.71 + 0.533T_a$ approximates the relationship between the soil and ground temperatures in a Florida pine plantation, where T_s is soil temperature and T_a is air temperature in degrees Celsius. Graph using your graphing calculator. Graphically find the air temperature if the soil temperature is 19.37°.

[10] Wilhelmina A. Leigh. 1988. The social preference for fair housing during the civil rights movement and since. *Amer. Econ. Rev.* 78(2): 156–162.

[11] Katherine C. Ewel and Henry L. Gholz. 1991. A simulation model of the role of belowground dynamics in a Florida pine plantation. *Forest Sci.* 37(2):397–438.

Enrichment Exercises

65. Sketch a proof of the following. If a and b are not both zero, then the graph of a linear equation $ax + by = c$ is a straight line, and conversely, every line is the graph of a linear equation.

66. Establish the following theorem: Two lines with slope m_1 and m_2 are perpendicular if and only if

$$m_2 = -\frac{1}{m_1}$$

Give a proof in the special case shown in the figure in which the two lines intersect at the origin. (The general proof is very similar). (Hint: The two lines L_1 and L_2 are perpendicular if and only if the triangle AOB is a right triangle. By the Pythagorean theorem this will be the case if, and only if,

$$[d(A, B)]^2 = [d(A, O)]^2 + [d(B, O)]^2$$

Now use the distance formula for each of the three terms in the previous equation and simplify.)

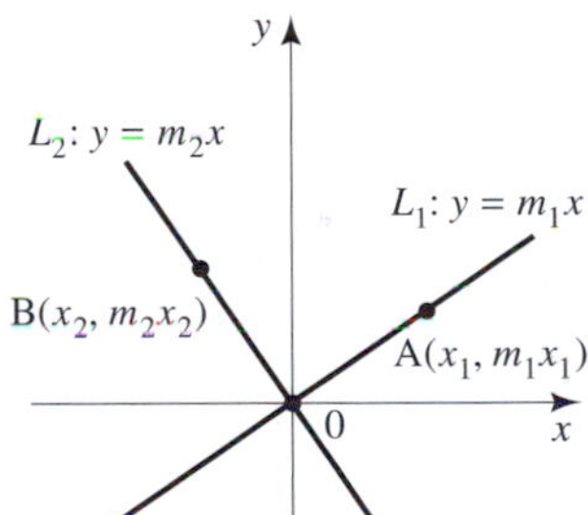

Solutions to WARM UP EXERCISE SET A.6

1. To find the equation of the line through the two points (3, 1) and (−2, −3), we first must find the slope of the line through the two points. This is

$$m = \frac{y_2 - y_1}{x_2 - x_1} = \frac{1 - (-3)}{3 - (-2)} = \frac{4}{5}$$

We now have the slope of the line and *two* points on the line. Pick one of the points, say, (3, 1), and use the point-slope form of the equation of the line and obtain

$$y - y_1 = m(x - x_1)$$

$$y - 1 = \frac{4}{5}(x - 3)$$

$$y = \frac{4}{5}x - \frac{7}{5}$$

This last equation is in the form of $y = mx + b$ with $b = -\frac{7}{5}$. Thus, the y-intercept is $-\frac{7}{5}$.

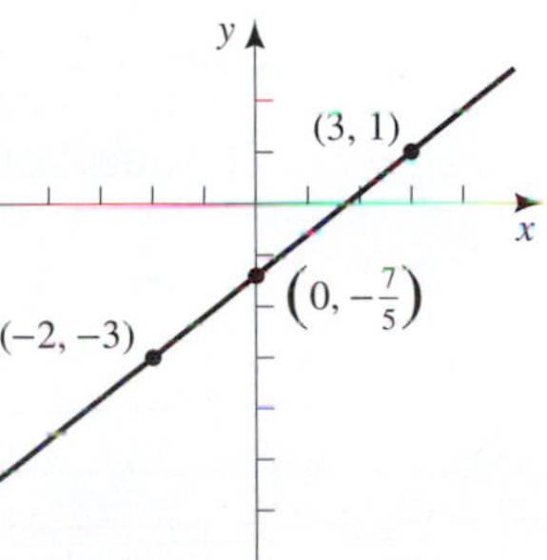

2. The equation through the point (2, −1) and perpendicular to the line in the previous exercise must have slope equal to the negative reciprocal of the slope of that line. Then the slope of the line whose equation we are seeking is $-\frac{5}{4}$. Using the point-slope form of the equation yields

$$y - y_1 = m(x - x_1)$$

$$y + 1 = -\frac{5}{4}(x - 2)$$

$$y = -\frac{5}{4}x + \frac{3}{2}$$

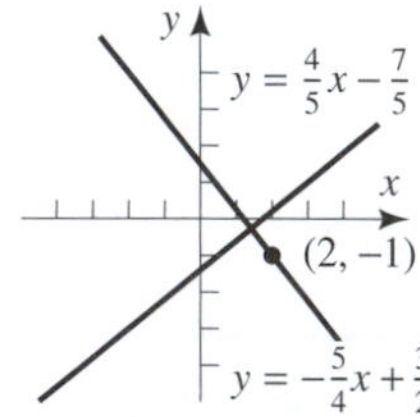

3. We organize the information in the following table:

Item	Quarts of Milk	Candy Bars	Total
Number of each item purchased	x	y	
Price of each item	0.80	0.40	
Total amount spent on each item	$0.80x$	$0.40y$	100

a. Since $0.80x$ is spent on milk and $0.40y$ on candy, the budget equation is given by

$$0.80x + 0.40y = 100$$

b. To find the y-intercept, we set $x = 0$ and obtain $100 = 0.40y$, or $y = 250$. To find the x-intercept, we set $y = 0$ and obtain $100 = 0.80x$, or $x = 125$. The graph is shown in the figure.

c. To find the slope of the line, we have

$$0.40y = -0.80x + 100$$

$$y = -2x + 250$$

We see that the slope is $m = -2$. This means that for each additional quart of milk purchased, two fewer candy bars can be purchased.

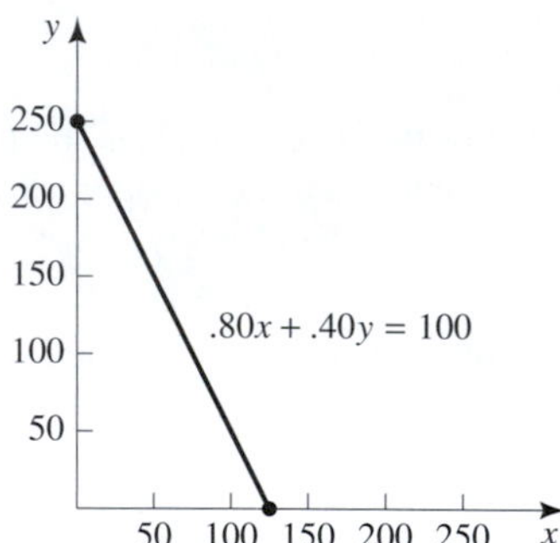

A.7 Quadratic Functions

Quadratic Functions

The function

$$f(x) = ax^2 + bx + c \qquad a \neq 0$$

is a **quadratic function**. In this section we graph quadratic functions, find the x-intercepts (zeros), and find where each function takes either a maximum or minimum.

We shall see that all quadratics have graphs that look basically like the graphs of ax^2. We begin with the following example.

EXAMPLE 1 **Graphs of Some Simple Quadratics**

Graph the quadratics $y = \frac{1}{2}x^2$, $y = x^2$, $y = 2x^2$, $y = -x^2$.

Solution

By using the values found in the following table the graphs are then given in Figure A.41.

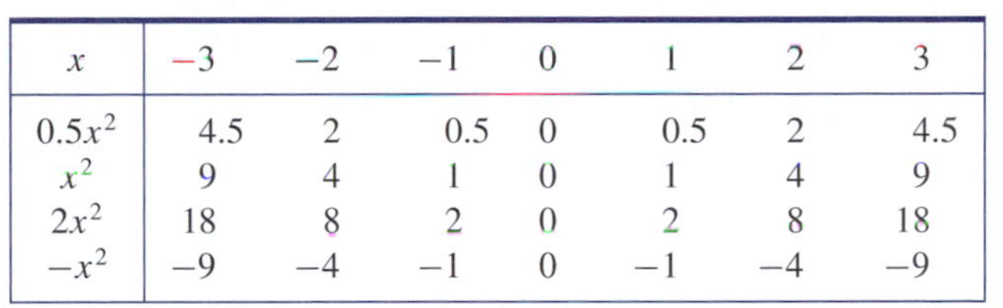

x	-3	-2	-1	0	1	2	3
$0.5x^2$	4.5	2	0.5	0	0.5	2	4.5
x^2	9	4	1	0	1	4	9
$2x^2$	18	8	2	0	2	8	18
$-x^2$	-9	-4	-1	0	-1	-4	-9

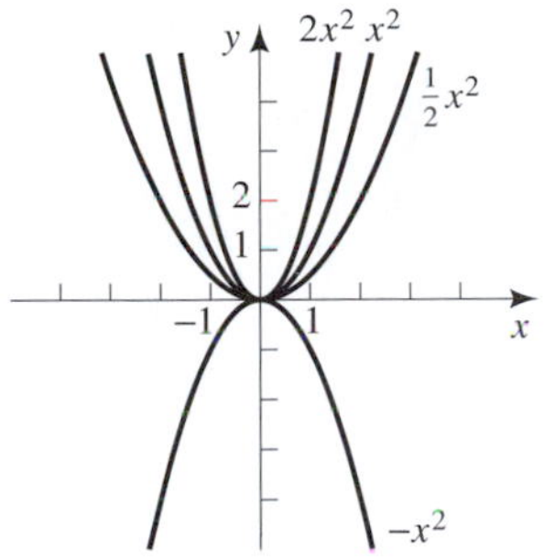

Figure A.41
The graphs of some simple quadratics.

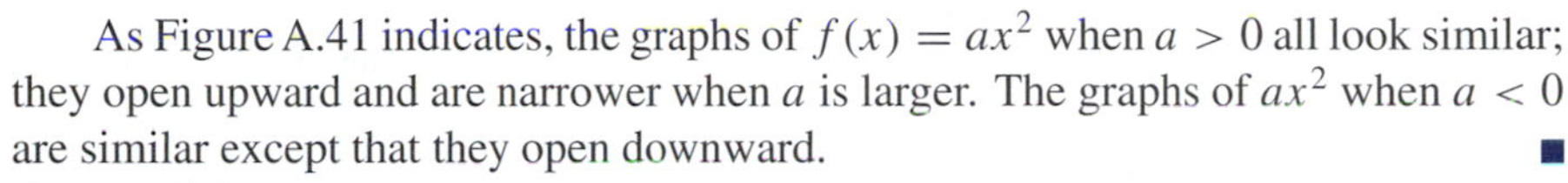

As Figure A.41 indicates, the graphs of $f(x) = ax^2$ when $a > 0$ all look similar; they open upward and are narrower when a is larger. The graphs of ax^2 when $a < 0$ are similar except that they open downward. ■

Interactive Illustration A.2

Graphs of Certain Quadratics

This interactive illustration allows you to experiment changing the parameters a in the quadratic ax^2, and see the result on a graph. A slider is provided for this parameter.

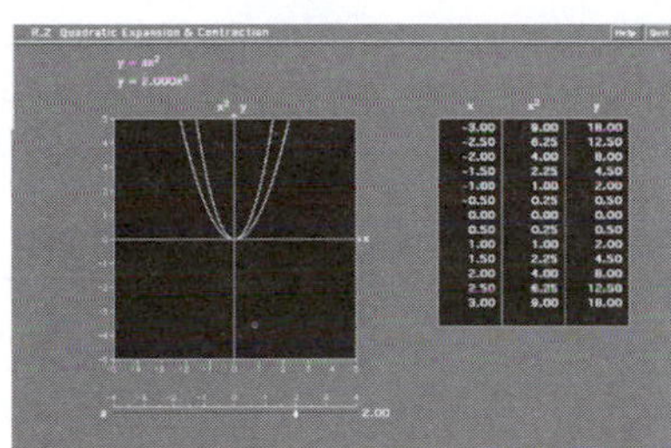

1. Make a larger. What happens?
2. Make a smaller. What happens?
3. Make a negative. What happens?

We now see how to effectively graph some quadratics by shifting the graph of $y = x^2$ up, down, left, or right.

EXAMPLE 2 **Shifting a Graph Up or Down**

Graph $y = x^2 + 3$ and $y = x^2 - 1$.

Solution

In Figure A.42a we show the graph of $y = x^2$ found in Figure A.41. To graph $y = x^2 + 3$ in Figure A.42b, we simply add 3 to every y value obtained in Figure A.42a. This results in the graph of $y = x^2$ shifted upward three units to obtain the graph of $y = x^2 + 3$. To obtain the graph of $y = x^2 - 1$ in Figure A.42c, we shift the graph of $y = x^2$ down one unit.

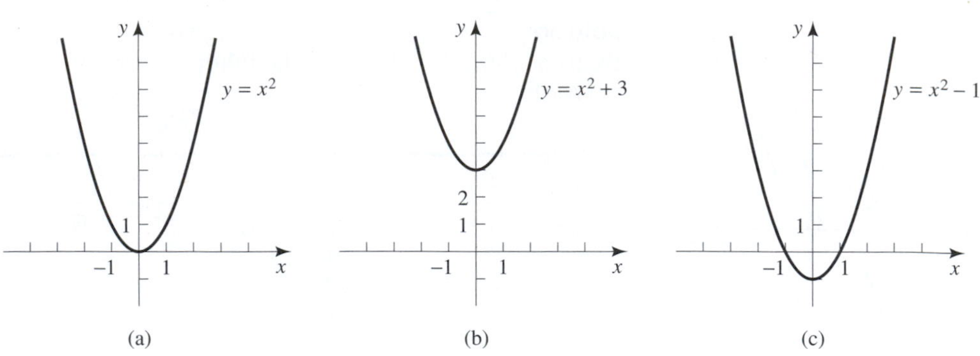

Figure A.42
Adding 3 shifts the graph up three units. Subtracting 1 shifts the graph down one unit.

Interactive Illustration A.3

Graphs of Certain Quadratics

This interactive illustration allows you to experiment with changing the parameter c in the quadratic $x^2 + c$ and see the result on a graph. A slider is provided for the this parameter.

1. Make c larger. What happens to the graph?
2. Make c smaller. What happens to the graph?

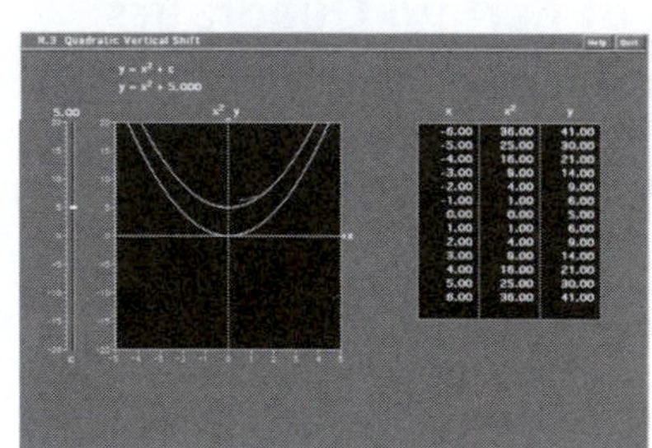

EXAMPLE 3 **Shifting a Graph Left or Right**

Graph $y = (x - 2)^2$ and $y = (x + 1)^2$.

Solution

First notice that these are indeed quadratics since

$$(x - 2)^2 = x^2 - 4x + 4 \qquad (x + 1)^2 = x^2 + 2x + 1$$

The first table compares the values of x^2 with $(x - 2)^2$. Notice that the y values of $y = (x - 2)^2$ are the same as the y values of $y = x^2$ except that they occur two units to the right on the x-axis. This is seen in Figure A.43a.

x	x^2	$(x-2)^2$
−5	25	49
−4	16	36
−3	9	25
−2	4	16
−1	1	9
0	0	4
1	1	1
2	4	0
3	9	1
4	16	4
5	25	9

x	x^2	$(x+1)^2$
−5	25	16
−4	16	9
−3	9	4
−2	4	1
−1	1	0
0	0	1
1	1	4
2	4	9
3	9	16
4	16	25
5	25	36

The second table compares the values of x^2 with $(x+1)^2$. Notice that the y values of $y=(x+1)^2$ are the same as the y values of $y=x^2$ except that they occur 1 unit to the left on the x-axis. This is seen in Figure A.43b.

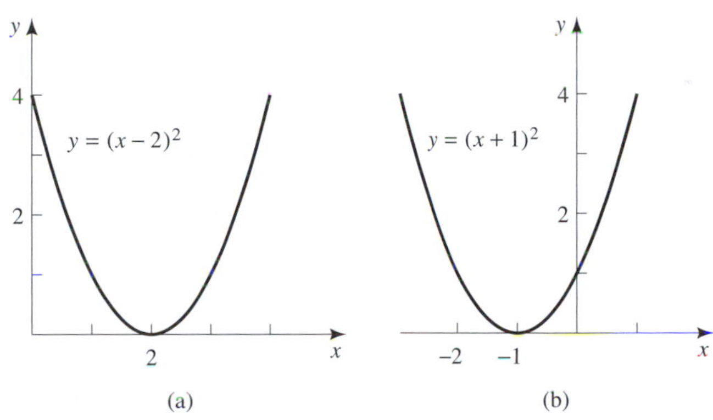

Figure A.43
Replacing x with $x-2$ shifts the graph two units to the right. Replacing x with $x+1$ shifts the graph one unit to the left.

■

Interactive Illustration A.4

Graphs of Certain Quadratics

This interactive illustration allows you to experiment changing with the parameters h in the quadratic $(x-h)^2$ and see the result on a graph. A slider is provided for the three parameters.

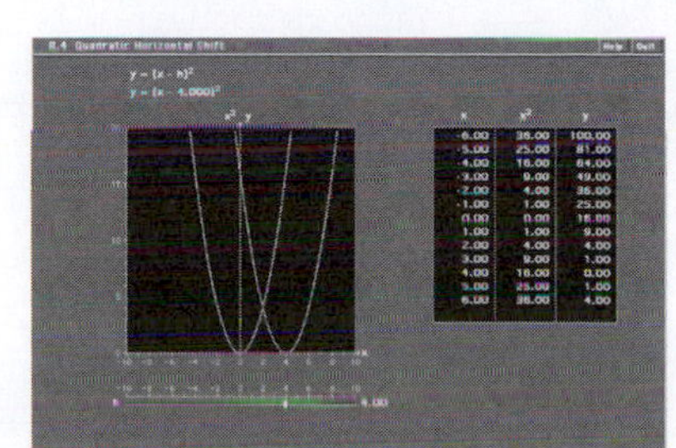

1. Make h larger. What happens to the graph?
2. Make h smaller. What happens to the graph?

We now see how to graph a general quadratic by shifting the graph of a simple quadratic of the form $y = ax^2$ first to the left or right and then up or down. Before we can do this, we must first do some algebraic manipulation to the quadratic (called completing the square) to put the quadratic expression into a convenient form.

EXAMPLE 4 **Graphing a Quadratic Function**

Sketch the graph of $y = 2x^2 - 16x + 14$.

Solution

We complete the square.

Step 1. Factor out the coefficient of x^2:

$$2x^2 - 16x + 14 = 2(x^2 - 8x + 7)$$

Step 2. Complete the square inside the bracket. Do this by adding and subtracting the square of one half of the coefficient of x:

$$\left(\frac{-8}{2}\right)^2 = (-4)^2 = 16$$

$$\begin{aligned} 2(x^2 - 8x + 7) &= 2(x^2 - 8x + 16 - 16 + 7) && \text{Add and subtract 16} \\ &= 2([x^2 - 8x + 16] - 9) && \text{Collect terms} \\ &= 2([x - 4]^2 - 9) && \text{Complete square} \\ &= 2(x - 4)^2 - 18 && \text{Multiply by 2} \end{aligned}$$

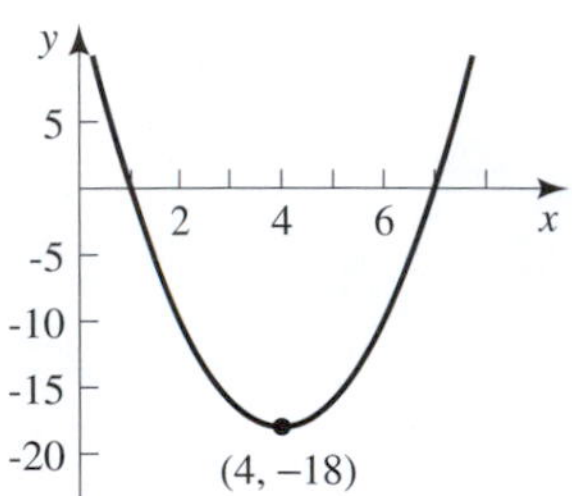

Figure A.44
The vertex of this quadratic is at $(4, -18)$.

In view of the previous examples the graph of $2(x - 4)^2 - 18$ is obtained by shifting the graph of $2x^2$ four units to the right and 18 units down, as shown in Figure A.44. ■

The same procedure can be used for any quadratic $ax^2 + bx + c$, with $a \neq 0$.

$$\begin{aligned} f(x) &= ax^2 + bx + c \\ &= a\left(x^2 + \frac{b}{a}x + \frac{c}{a}\right) && \text{Factor out } a \\ &= a\left(\left[x^2 + \frac{b}{a}x + \frac{b^2}{4a^2}\right] - \frac{b^2}{4a^2} + \frac{c}{a}\right) && \text{Add and subtract } \left(\frac{b}{2a}\right)^2 \\ &= a\left(\left[x + \frac{b}{2a}\right]^2 + \frac{c}{a} - \frac{b^2}{4a^2}\right) && \text{Complete the square} \\ &= a\left(x + \frac{b}{2a}\right)^2 + \left(c - \frac{b^2}{4a}\right) && \text{Multiply by } a \end{aligned}$$

If we set

$$h = -\frac{b}{2a} \quad \text{and} \quad k = c - \frac{b^2}{4a}$$

the quadratic can be written in **standard form**

$$f(x) = a(x - h)^2 + k$$

The point (h, k) is called the **vertex**. With the quadratic in the standard form $y = a(x - h)^2 + k$, and $a > 0$, notice that y is smallest when the term $a(x - h)^2 = 0$. This can happen only if $x = h$. Thus, when $a > 0$, $y = a(x - h)^2 + k$ takes a *minimum* value of y when $x = h$. The minimum is $y = k$.

With the quadratic in the standard form $y = a(x - h)^2 + k$, and $a < 0$, notice that y is largest when the term $a(x - h)^2 = 0$. This can happen only if $x = h$. Thus, when $a < 0$, $y = a(x - h)^2 + k$ takes a *maximum* value of y when $x = h$. The maximum is $y = k$.

The graphs, which are called **parabolas**, are given in Figure A.45. Notice that when $a > 0$, the parabola is concave up, and $y = k$ is the *minimum* value of y. When $a < 0$, the parabola is concave down, and $y = k$ is the *maximum* value of y.

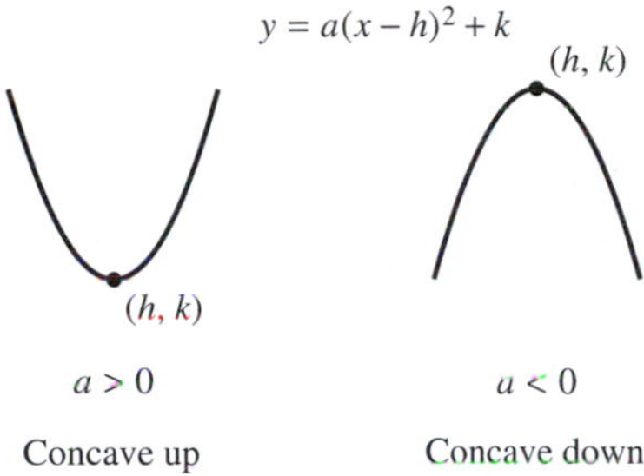

Figure A.45
The two types of quadratics.

The Quadratic $ax^2 + bx + c$

Every quadratic with $a \neq 0$ can be put into the standard form

$$ax^2 + bx + c = a(x - h)^2 + k$$

where

$$h = -\frac{b}{2a} \qquad \text{and} \qquad k = c - \frac{b^2}{4a}$$

The graphs of the quadratics are shown in Figure A.45.
When $a > 0$, the parabola opens upward and $y = k$ is the *minimum* value of y.
When $a < 0$, the parabola opens downward and $y = k$ is the *maximum* value of y.

Remark Refer to Figure A.45. As we see, if $a > 0$, the quadratic is decreasing on $(-\infty, h)$ and increasing on (h, ∞). If $a < 0$, the quadratic is increasing on $(-\infty, h)$ and decreasing on (h, ∞).

Interactive Illustration A.5

The Graph of $y = a(x - h)^2 + k$

This interactive illustration allows you to change the parameters a, h, and k to see the effect on the graph of $y = a(x - h)^2 + k$. Sliders are provided for the three parameters. Change each of these parameters in turn, and notice the effect on the graph.

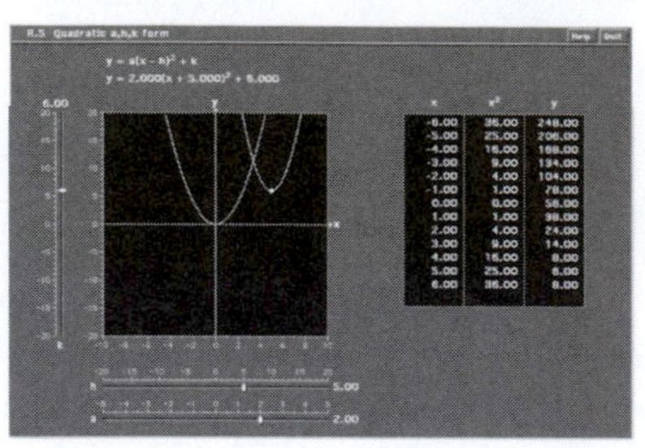

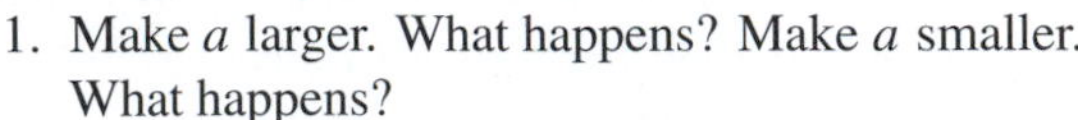

1. Make a larger. What happens? Make a smaller. What happens?
2. Make h larger. What happens? Make h smaller. What happens?
3. Make k larger. What happens? Make k smaller. What happens?

EXAMPLE 5 **Graphing a Quadratic Function**

Graph $y = -2x^2 + 10x - 5$.

Solution

We first note that $a = -2 < 0$, and therefore, the graph is concave down and the vertex will be at the maximum.

We can now proceed algebraically in either of two ways: complete the square or use the formulas listed earlier. We illustrate the latter.

$$h = -\frac{b}{2a} = -\frac{10}{2(-2)} = 2.5$$

$$k = c - \frac{b^2}{4a} = -5 - \frac{10^2}{-8} = 7.5$$

Thus, the vertex is at $(2.5, 7.5)$. A maximum of 7.5 occurs at $x = 2.5$. The graph is shown in Figure A.46. ■

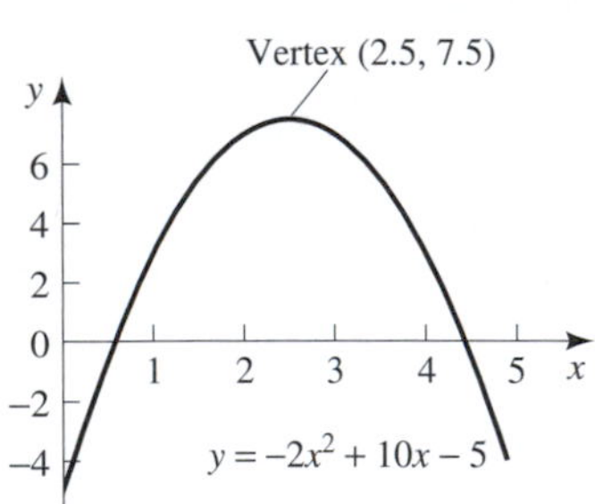

Figure A.46

Zeros of Quadratics

A quadratic $y = ax^2 + bx + c$ need not have zeros, that is, x-intercepts. For example, the quadratic $y = x^2 + 3$ does not have an x-intercept. This is the case since $x^2 + 3 \geq 3$ and therefore y is never zero. The graph is shown in Figure A.42b.

But other quadratics do have zeros. For example, the quadratic $y = x^2$ whose graph is shown in Figure A.42a has just one zero. The quadratic $y = x^2 - 1$ whose graph is shown in Figure A.42c has two zeros. One certain way of finding the zeros of quadratic (if they exist) is to use the quadratic formula.

The Quadratic Formula

The zeros of the quadratic $y = ax^2 + bx + c$ are given by

$$\frac{-b \pm \sqrt{b^2 - 4ac}}{2a}$$

If $b^2 - 4ac > 0$, there are two zeros.
If $b^2 - 4ac = 0$, there is one zero.
If $b^2 - 4ac < 0$, there are no zeros.

EXAMPLE 6 **Finding the Zeros of a Quadratic**

Find the zeros of the quadratic $y = x^2 - 2x - 1$ and graph.

Solution

Here we have $a = 1$, $b = -2$, and $c = -1$. Then using the quadratic formula yields

$$\begin{aligned} \frac{-b \pm \sqrt{b^2 - 4ac}}{2a} &= \frac{-(-2) \pm \sqrt{(-2)^2 - 4(1)(-1)}}{2(1)} \\ &= \frac{2 \pm \sqrt{4+4}}{2} \\ &= 1 \pm \sqrt{2} \end{aligned}$$

There are two zeros: $x = 1 + \sqrt{2}$ and $x = 1 - \sqrt{2}$.

To graph, we first note that $a = 1 > 0$, and so the graph is concave up and the vertex will be a minimum. To find the vertex, we have

$$h = -\frac{b}{2a} = -\frac{-2}{2} = 1$$

$$k = c - \frac{b^2}{4a} = -1 - \frac{4}{4} = -2$$

So the vertex is at $(1, -2)$. A graph is shown in Figure A.47.

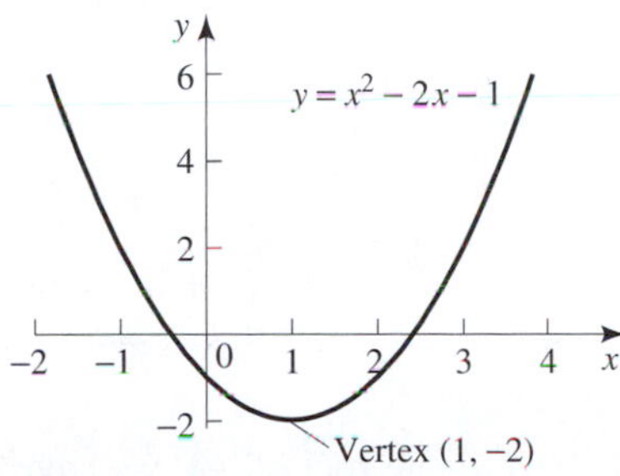

Figure A.47

■

EXAMPLE 7 **Finding the Zeros of a Quadratic**

Find the zeros of the quadratic $y = x^2 - 4x + 4$ and graph.

Solution

Here we have $a = 1$, $b = -4$, and $c = 4$. Then using the quadratic formula yields

$$\begin{aligned}\frac{-b \pm \sqrt{b^2 - 4ac}}{2a} &= \frac{-(-4) \pm \sqrt{(-4)^2 - 4(1)(4)}}{2(1)} \\ &= \frac{4 \pm \sqrt{16 - 16}}{2} \\ &= 2\end{aligned}$$

There is only one zero: $x = 2$.

To graph, we first note that $a = 1 > 0$, and so the graph is concave up and the vertex will be a minimum. To find the vertex, we have

$$h = -\frac{b}{2a} = -\frac{-4}{2} = 2$$

$$k = c - \frac{b^2}{4a} = 4 - \frac{16}{4} = 0$$

So the vertex is at $(2, 0)$. A graph is shown in Figure A.43a. (This is true since $(x - 2)^2 = x^2 - 4x + 4$.) ■

Remark We could have found the zero of the quadratic in Example 7 by factoring. We could have noted that

$$x^2 - 4x + 4 = (x - 2)^2$$

The only way this can be zero is for x to be 2. It is not realistic to think we could factor the quadratic in Example 6! Some quadratics are too difficult to factor, but the quadratic formula can always be used.

EXAMPLE 8 **Finding the Zeros of a Quadratic**

Find the zeros of the quadratic $y = -x^2 - 2x - 2$, and graph.

Solution

Here we have $a = -1$, $b = -2$, and $c = -2$. Then using the quadratic formula yields

$$\begin{aligned}\frac{-b \pm \sqrt{b^2 - 4ac}}{2a} &= \frac{-(-2) \pm \sqrt{(-2)^2 - 4(-1)(-2)}}{2(1)} \\ &= \frac{2 \pm \sqrt{4 - 8}}{2} \\ &= \frac{2 \pm \sqrt{-4}}{2}\end{aligned}$$

Since there is no (real) square root of -4, we have no zeros.

To graph, we first note that $a = -1 < 0$, and so the graph is concave down, and the vertex will be a maximum. To find the vertex, we have

$$h = -\frac{b}{2a} = -\frac{-2}{-2} = -1$$

$$k = c - \frac{b^2}{4a} = -2 - \frac{4}{-4} = -1$$

So the vertex is at $(-1, -1)$. A graph is shown in Figure A.48.

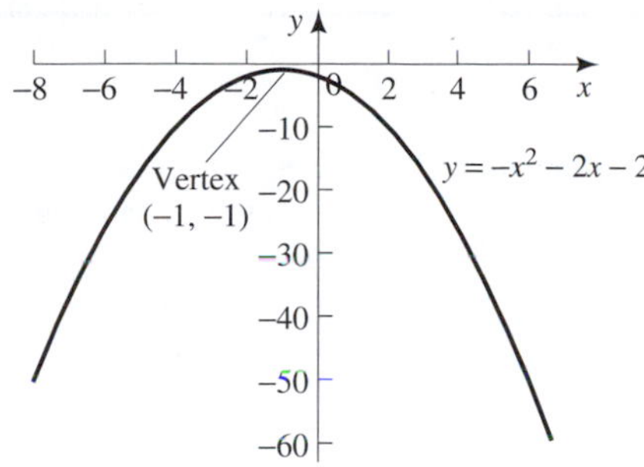

Figure A.48

■

Enrichment: Factoring and Zeros

We now work through an example that illustrates again that finding the zeros of a quadratic is equivalent to factoring the quadratic.

EXAMPLE 9 **Factoring a Quadratic Graphically**

Consider $y = -x^2 + 8.8x - 14.95$. Graph on your computer or graphing calculator and use your available operations to determine the zeros. Now use this information to factor the quadratic and then confirm your answer algebraically.

Solution

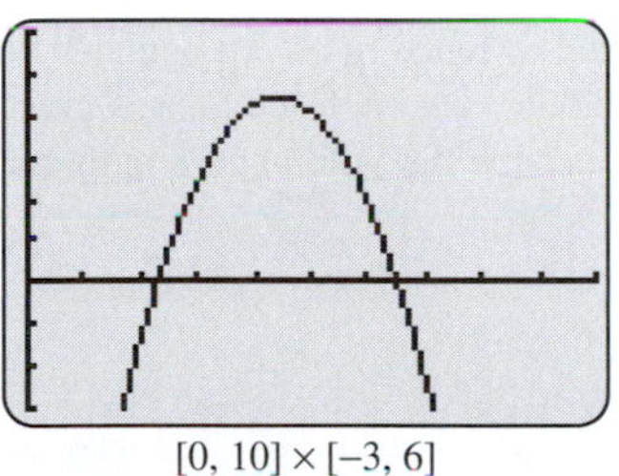

Screen A.4

A graph is shown in Screen A.4. Using the available operations, we find the x-intercepts (or the zeros of the quadratic) to be 2.3 and 6.5. If this is correct, then we must have

$$-x^2 + 8.8x - 14.95 = -(x^2 - 8.8x + 14.95) = -(x - 2.3)(x - 6.5)$$

Because this is indeed correct, we have confirmed our answer algebraically.

Of course, we could have used the quadratic formula to find the zeros of the quadratic. However, there is no corresponding formula for finding the zeros of many other types of important functions. In these instances where algebraic formulas do not exist, using our computer or graphing calculator is the only way of finding the zeros. ■

Exploration 1

Finding Zeros Using Our Technology

Find the zeros of $y = 1.1x^2 - 9.68x + 12.045$ on your computer or graphing calculator. Confirm your answer using the quadratic formula.

Warm Up Exercise Set A.7

1. Graph $y = (x + 3)^2 - 8$ and locate the vertex.
2. Complete the square for $y = -2x^2 + 8x - 6$ and graph. Also, check your work by using the formulas for h and k to find the vertex. Also find the x-intercepts.

Exercise Set A.7

In Exercises 1 through 8, graph the given quadratic function and locate the vertex. Also determine the interval on which each function is increasing and the interval on which it is decreasing.

1. $y = (x-1)^2 + 2$ **2.** $y = 2(x+2)^2 + 3$

3. $y = 2(x+2)^2 - 3$ **4.** $y = (x+1)^2 - 4$

5. $y = -(x-1)^2 + 3$ **6.** $y = -2(x-2)^2 + 1$

7. $y = -(x+1)^2 - 2$ **8.** $y = -3(x+2)^2 - 1$

In Exercises 9 through 12, complete the square and put in standard form. Locate the vertex and graph.

9. $y = x^2 - 4x + 4$ **10.** $y = x^2 - 8x + 16$

11. $y = -2x^2 + 12x - 20$ **12.** $y = -x^2 + 2x - 4$

In Exercises 13 through 24, locate the vertex, find the zeros, and graph.

13. $y = x^2 + 2x + 1$ **14.** $y = x^2 + 6x + 9$

15. $y = -4x^2 - 8x - 3$ **16.** $y = -5x^2 - 10x - 4$

17. $y = x^2 - 2x + 4$ **18.** $y = 2x^2 - 12x + 20$

19. $y = -x^2 - 2x - 2$ **20.** $y = -3x^2 - 6x - 4$

21. $y = x^2 - 2x$ **22.** $y = 2x^2 - 8x + 7$

23. $y = -x^2 + 4x - 2$ **24.** $y = -2x^2 + 4x + 1$

25. Let α and β be positive constants. Match the graphs of the following quadratics with those found in the figure.

a. $y = -\alpha(x+\alpha)^2 + \beta$
b. $y = \alpha(x+\alpha)^2 - \beta$
c. $y = -\alpha(x-\alpha)^2 - \beta$
d. $y = \alpha(x-\alpha)^2 + \beta$

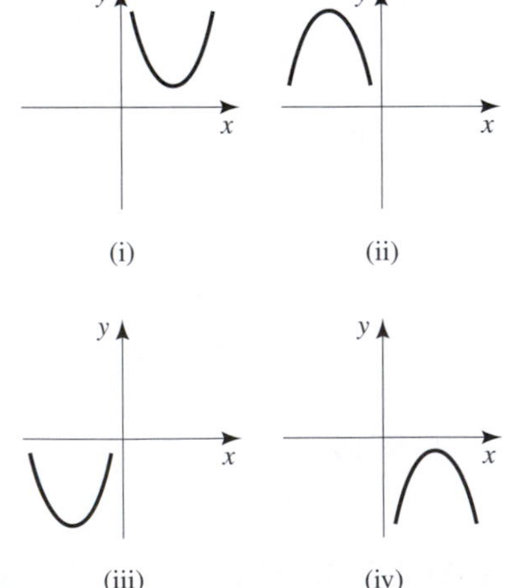

26. Let α and β be positive constants. Match the graphs of the following quadratics with those found in the figure.

a. $y = \alpha(x+\alpha)^2 + \beta$
b. $y = -\alpha(x+\alpha)^2 - \beta$
c. $y = \alpha(x-\alpha)^2 - \beta$
d. $y = -\alpha(x-\alpha)^2 + \beta$

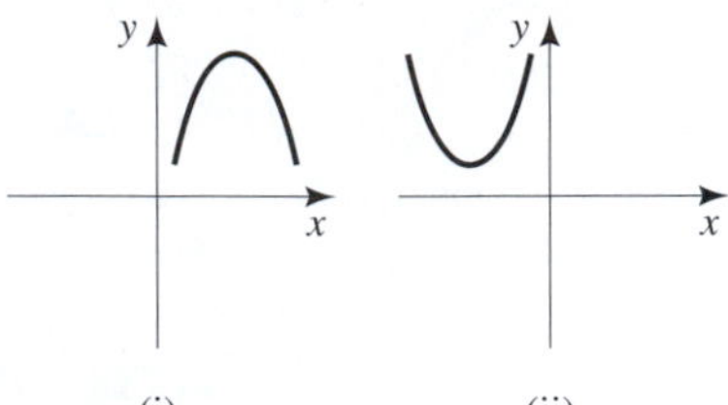

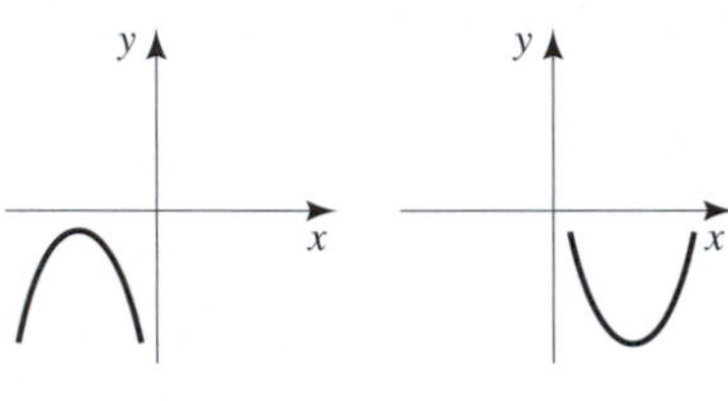

27. The sum of two numbers is 9. What is the largest possible value for their product?

28. If 200 feet of fence is used to fence in a rectangular field and x is the width of the field, the enclosed area is given by $A = x(100 - x)$. Determine the width and length that give the maximum area.

Enrichment Exercises

29. Is the sum of two quadratic functions always a quadratic function? Explain carefully.

30. Suppose you have two quadratic functions, $y = f(x)$ and $y = g(x)$, both of which have graphs that open downward. Consider the function $y = f(x) + g(x)$. Show that this function is quadratic and its graph opens downward.

31. Give an intuitive geometric argument (not an algebraic one) that shows that the quadratic $y = f(x) + g(x)$ in the previous exercise peaks at a value in between the values for which the first two quadratics peak.

32. Can the graph of a quadratic function cross the x-axis in three different places? Why or why not?

In Exercises 33 through 36, use your grapher to find zeros of the given quadratics.

33. $y = -x^2 + 3.1x - 1.98$

34. $y = -x^2 + 5.9x - 7.92$

35. $y = -4x^2 + 38.4x - 59.36$

36. $y = -2.4x^2 + 14.64x - 14.112$

Solutions to WARM UP EXERCISE SET A.7

1. The quadratic $y = (x + 3)^2 - 8$ is in standard form $y = a(x - h)^2 + k$ with $a = 1$, $h = -3$, and $k = -8$. Since $a > 0$, the parabola opens upward and the vertex $(h, k) = (-3, -8)$ is the point on the graph where the function attains its minimum.

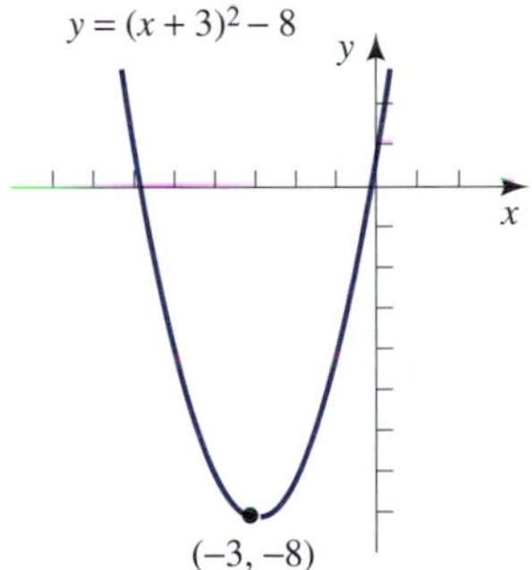

2. Complete the square.

$$\begin{aligned} y &= -2x^2 + 8x - 6 \\ &= -2(x^2 - 4x + 3) \\ &= -2([x^2 - 4x + 4] - 4 + 3) \qquad \left(\frac{-4}{2}\right)^2 = 4 \\ &= -2([x - 2]^2 - 1) \\ &= -2(x - 2)^2 + 2 \end{aligned}$$

The quadratic is now in standard form. The vertex is at $(2, 2)$. since $a = -2 < 0$, the graph is concave down.

We can also use the formulas

$$h = -\frac{b}{2a} = -\frac{8}{2(-2)} = 2$$

$$k = c - \frac{b^2}{4a} = -6 - \frac{64}{-8} = 2$$

Use the quadratic formula

$$\frac{-b \pm \sqrt{b^2 - 4ac}}{2a} = \frac{-(8) \pm \sqrt{(8)^2 - 4(-2)(-6)}}{2(-2)}$$

$$= \frac{-8 \pm \sqrt{64 - 48}}{-4}$$

$$= 2 \pm -1$$

so the x-intercepts are $x = 1$ and $x = 3$.

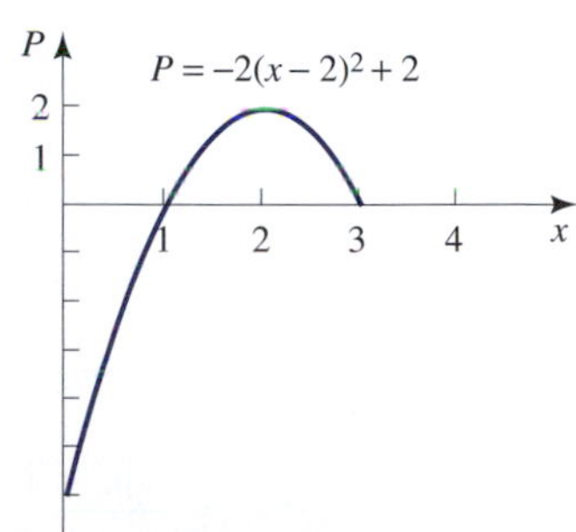

A.8 Some Special Functions and Graphing Techniques

Some Special Functions

Graphing Techniques

Some Special Functions

The area of a square with sides of length s is $A = f(s) = s^2$. The volume of a cube with sides of length s is $V = g(s) = s^3$. The volume of a sphere of radius r is $V = h(r) = \frac{4}{3}\pi r^3$. All of these functions are *power functions*. Functions such as $f(x) = x^{1/2}$ and $f(x) = \frac{1}{x^2} = x^{-2}$ are also **power functions**. In general we have the following definition.

DEFINITION **Power Functions**

A **power function** is a function of the form

$$f(x) = kx^r$$

where k and r are any real numbers.

Warning There is a substantial difference between the power function x^r and exponential function r^x. (See Exercises 59–62.)

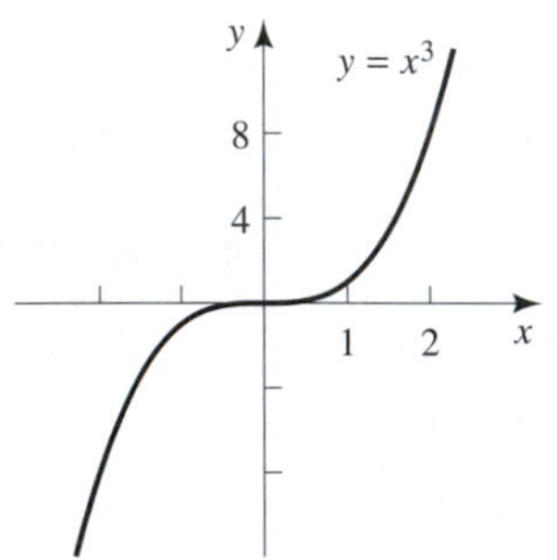

Figure A.49
A graph of $y = x^3$.

Let us consider the domains of some power functions. The domain of $y = x^3$ and $y = \sqrt[3]{x}$ is the entire real line while the domain of $y = \sqrt{x}$ is $[0, \infty)$. From Tables A.1, A.2, and A.3, which show a number of values for these functions, the graphs are constructed in Figures A.49, A.50, and A.51.

Table A.1

x	-3	-2	-1	0	1	2	3
x^3	-27	-8	-1	0	1	8	27

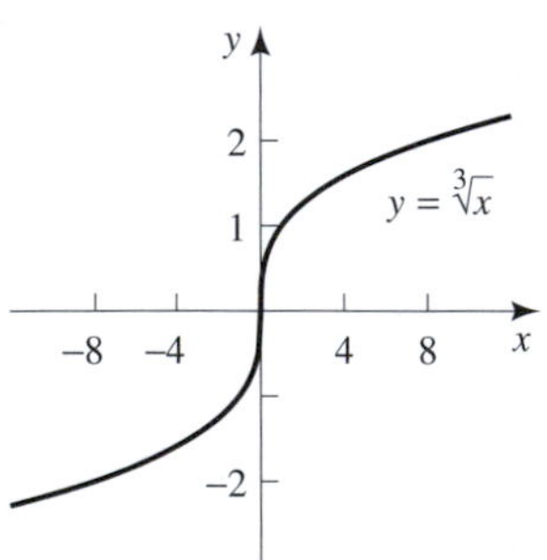

Figure A.50
A graph of $y = \sqrt[3]{x}$.

Table A.2

x	0	1	4	9	16	25	36
$\sqrt{x}$	0	1	2	3	4	5	6

Table A.3

x	-27	-8	-1	0	1	8	27
$\sqrt[3]{x}$	-3	-2	-1	0	1	2	3

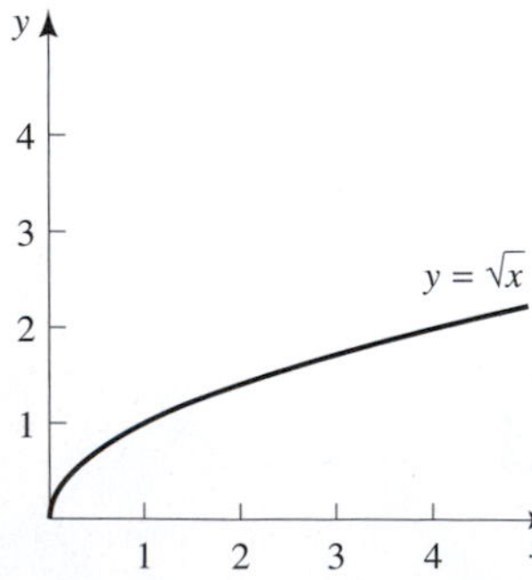

Figure A.51
A graph of $y = \sqrt{x}$.

If $r > 0$, the domains of power functions are either all real numbers or else all nonnegative numbers, depending on r. For example, the two rational functions $f(x) = x^{1/2}$ and $g(x) = x^{5/4}$ have $[0, \infty)$ as their domains, whereas $f(x) = x^{1/3}$ and $g(x) = x^{4/5}$ have all real numbers as their domains. If $r < 0$, then $x = 0$ cannot be in the domain. For example, if $f(x) = x^{-1/3} = \frac{1}{x^{1/3}}$, then the domain is the real line except for $x = 0$. If $x = x^{-1/2} = \frac{1}{x^{1/2}}$, then the domain is $(0, \infty)$.

Remark Be careful when using your grapher to graph $y = x^r$ when r is a rational number $r = m/n$. For example, the domain of $f(x) = x^{2/3}$ is the set of all real numbers. However, your grapher may graph this function incorrectly. If your grapher does give an incorrect graph rewrite $f(x)$ as $f(x) = (x^{1/3})^2$ or $(x^2)^{1/3}$.

We next consider polynomial functions.

DEFINITION **Polynomial Function**

A **polynomial function** of degree n is a function of the form

$$f(x) = a_n x^n + a_{n-1}x^{n-1} + \cdots + a_1 x + a_0 \qquad (a_n \neq 0)$$

where $a_n, a_{n-1}, \ldots, a_1, a_0$, are constants and n is a nonnegative integer. The domain is all real numbers. The coefficient a_n is called the **leading coefficient**.

A polynomial function of degree 1 ($n = 1$), $f(x) = a_1 x + a_0$, ($a_1 \neq 0$) is a linear function and has already been encountered earlier. A polynomial function of degree 2 ($n = 2$), $f(x) = a_2 x^2 + a_1 x + a_0$ ($a_2 \neq 0$), is a quadratic function. The functions

$$f(x) = -x^5 + 10x^3 - 9x - 10 \qquad \text{and} \qquad g(x) = x^6 - 9x^3 + 8x - 10$$

are polynomials of degree 5 and 6, respectively, with leading coefficients of -1 and 1, respectively. The graphs are shown in Figure A.52.

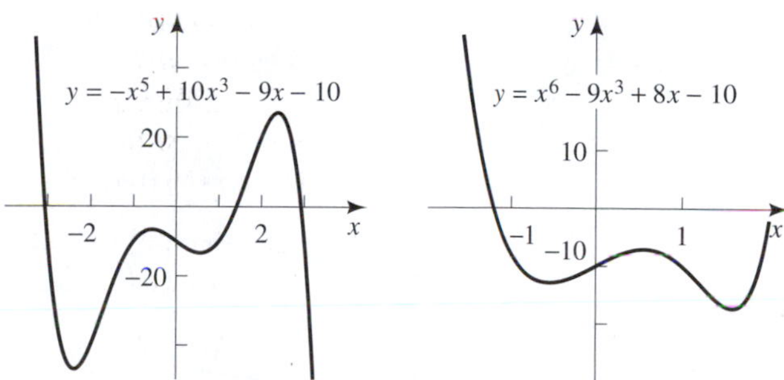

Figure A.52
Graphs of typical polynomial functions.

We continue with the definition of *rational function*.

DEFINITION **Rational Function**

A **rational function** $R(x)$ is the quotient of two polynomial functions and thus is of the form $R(x) = \frac{f(x)}{g(x)}$ where $f(x)$ and $g(x)$ are polynomial functions. The domain of $R(x)$ is all real numbers for which $g(x) \neq 0$.

Thus,

$$R(x) = \frac{x^5 + x - 2}{x - 3} \qquad \text{and} \qquad Q(x) = \frac{x + 1}{x^2 - 1}$$

are rational functions with the domain of $R(x)$ all real numbers other than $x = 3$ and the domain of $Q(x)$ is all real numbers other than $x = \pm 1$.

Graphing Techniques

We develop a number of techniques to aid in graphing.

In Section A.7 we noted that the graph of $y = x^2 + 3$ is the graph of $y = x^2$ shifted upward by 3 units, while the graph of $y = x^2 - 1$ is the graph of $y = x^2$ shifted downward by 1 unit. This is an example of the following general graphing principal.

Vertical Shift

For $k > 0$, the graph of $y = f(x) + k$ is the graph of $y = f(x)$ shifted upward by k units, while the graph of $y = f(x) - k$ is the graph of $y = f(x)$ shifted downward by k units.

EXAMPLE 1 **Vertical Shift**

Graph $y = \sqrt{x}$, $y = \sqrt{x} + 2$, and $y = \sqrt{x} - 1$ on the same graph.

Solution

The graph of $y = \sqrt{x}$ is shown in Figure A.53. The graph of $y = \sqrt{x} + 2$ is the graph of $y = \sqrt{x}$ shifted upward by two units, while the graph of $y = \sqrt{x} - 1$ is the graph of $y = \sqrt{x}$ shifted downward by one unit. See Figure A.53.

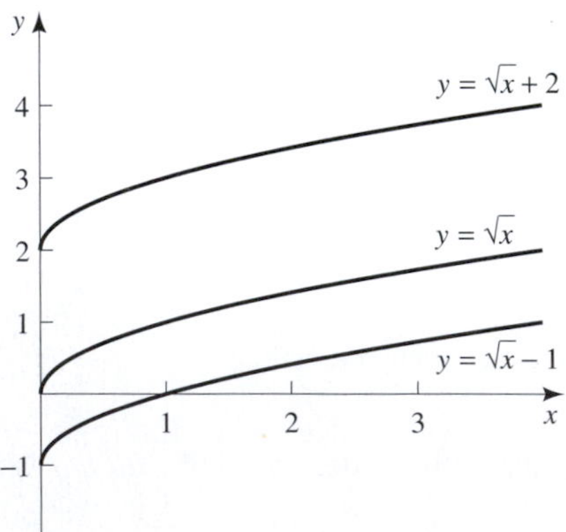

Figure A.53
Adding 2 shifts the graph up two units, whereas subtracting 1 shifts the graph down one unit.

■

We also noted in Section A.7 that the graph of $y = (x - 2)^2$ is the graph of $y = x^2$ shifted to the right by two units, while the graph of $y = (x + 1)^2$ is the graph of $y = x^2$ shifted to the left by one unit. This is an example of the following general graphing principle.

Horizontal Shift

For $c > 0$, the graph of $y = f(x - c)$ is the graph of $y = f(x)$ shifted to the right by c units, while the graph of $y = f(x + c)$ is the graph of $y = f(x)$ shifted to the left by c units.

EXAMPLE 2 **Horizontal Shift**

Graph $y = x^3$, $y = (x - 2)^3$, and $y = (x + 1)^3$ on the same graph.

Solution

The graph of $y = x^3$ is shown in Figure A.54. The graph of $y = (x - 2)^3$ is the graph of $y = x^3$ shifted to the right by two units, while the graph of $y = (x + 1)^3$ is the graph of $y = x^3$ shifted to the left by one unit. See Figure A.54. ∎

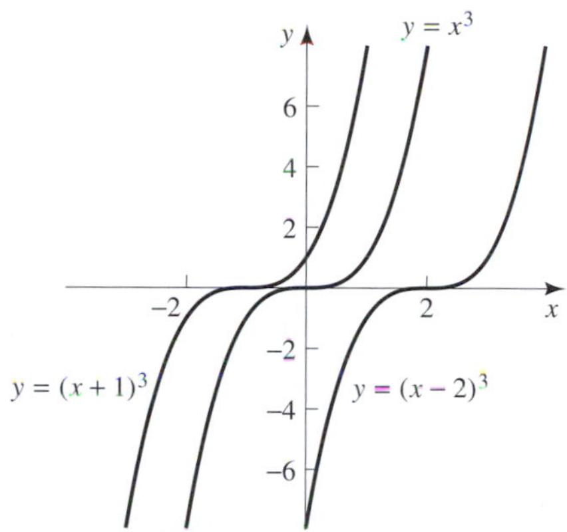

Figure A.54
Replacing x with $x + 1$ shifts the graph one unit to the left, whereas replacing x with $x - 2$ shifts the graph two units to the right.

Interactive Illustration A.6

Seeing Horizontal and Vertical Shifts

This interactive illustration allows you to see horizontal and vertical shifts. Click on one of the four functions. Each is in the form of $f(x - a) + b$. Sliders are provided on the parameters a and b.

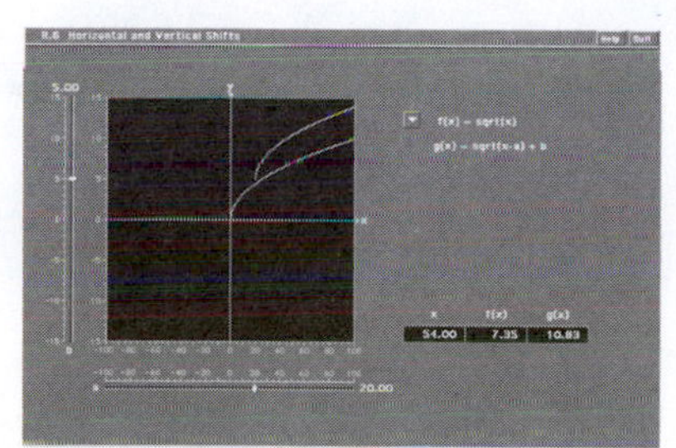

1. Increase a. What happens to the graph? Decrease a. What happens to the graph?
2. Increase b. What happens to the graph? Decrease b. What happens to the graph?

In Section A.7 we compared the graphs of $y = f(x) = x^2$ with that of $y = -f(x) = -x^2$. The graphs are shown in Figure A.55. As we see the graph of $y = -f(x) = -x^2$ is the reflection of the graph of $y = f(x) = x^2$ in the x-axis. This is an example of the following general graphing principal.

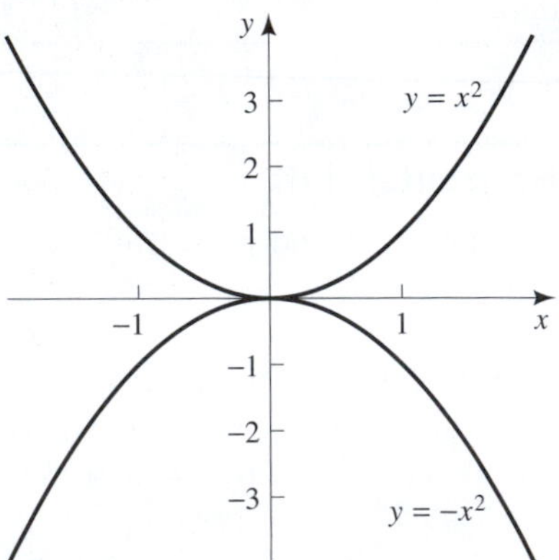

Figure A.55
The graph of $y = -x^2$ is a reflection across the x-axis of the graph of $y = x^2$.

Reflection

The graph of $y = -f(x)$ is the reflection of the graph of $y = f(x)$ in the x-axis.

EXAMPLE 3 **Reflection**

Graph $y = x^3$ and $y = -x^3$ on the same graph.

Solution

The graphs are shown in Figure A.56.

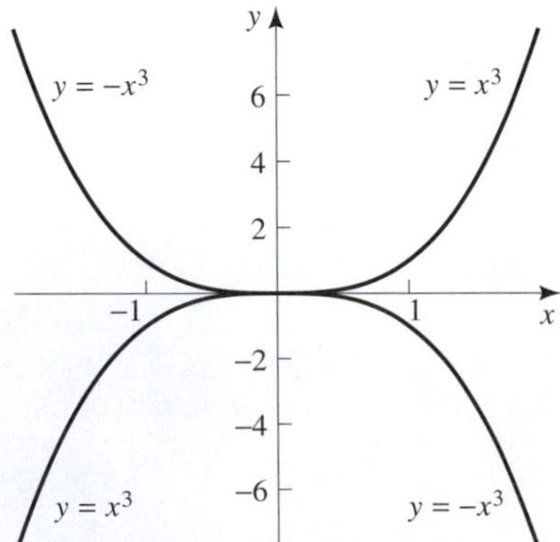

Figure A.56
The graph of $y = -x^3$ is a reflection across the x-axis of the graph of $y = x^3$. ■

Interactive Illustration A.7

Reflections

This interactive illustration allows you to explore reflections further. Three different functions are provided. Each is in the form of $f(x-a)+b$. Sliders are available for a and b. Click on one of the functions. What does the graph of the negative of this function look like?

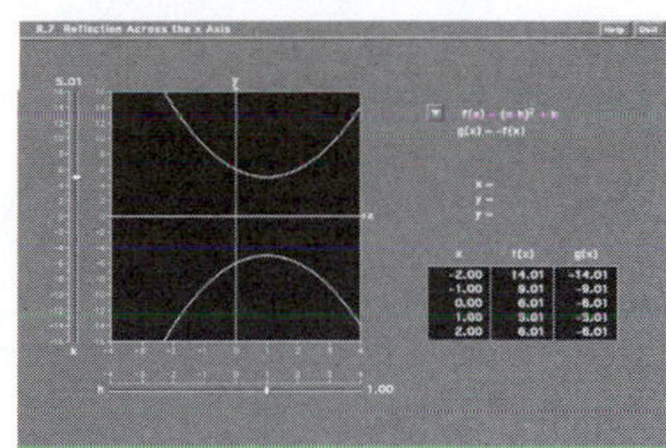

In Section A.7 we compared the graphs of $y = ax^2 = af(x)$ for various values of a with that of $y = x^2 = f(x)$. Figure A.57 shows some examples. We note that for $a = 2$ the graph of $y = 2x^2$ expands (vertically) the graph of $y = x^2$ by a factor of 2, while the graph of $y = 0.5x^2$ contracts (vertically) the graph of $y = x^2$ by a factor of 0.5. This is an example of the following general graphing principal.

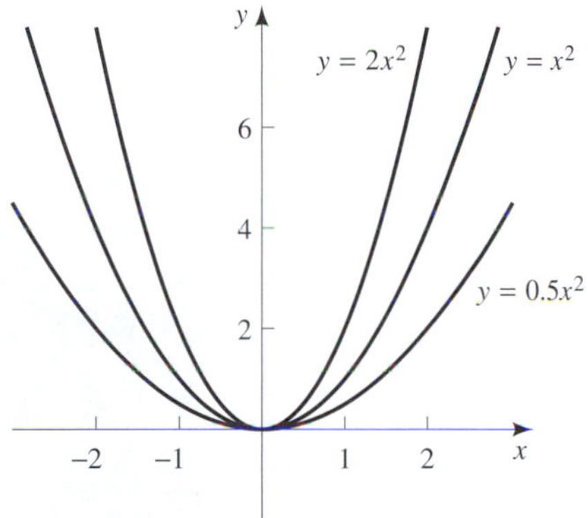

Figure A.57
The graph of $y = ax^2$ for $a > 0$ expands or contracts the graph of $y = x^2$ depending on whether $a > 1$ or $a < 1$.

Expansion and Contraction

For $a > 0$, the graph of $y = af(x)$ is an expansion (vertically) of the graph of $y = f(x)$ if $a > 1$, and a contraction (vertically) if $0 < a < 1$.

EXAMPLE 4 **Expansion and Contraction**

Draw the graphs of $y = |x|$, $y = 2|x|$, and $y = 0.5|x|$ on the same graph.

Solution
The graphs are shown in Figure A.58.

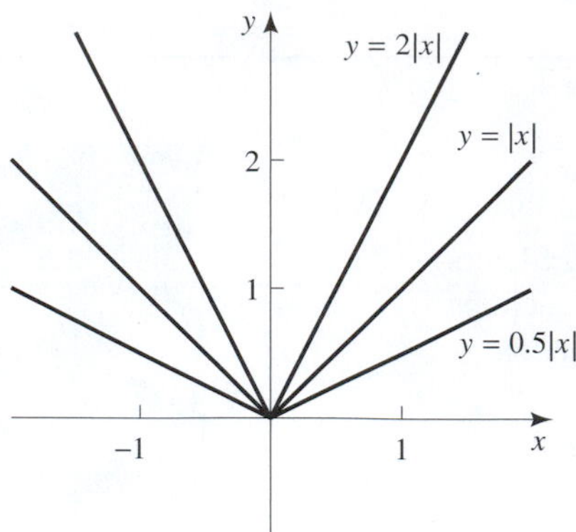

Figure A.58
The graph of $y = a|x|$ for $a > 0$ expands or contracts the graph of $y = |x|$ depending on whether $a > 1$ or $a < 1$.

Interactive Illustration A.8

Expansions and Contractions

This interactive illustration allows you to explore expansions and contractions further. Three functions are provided. Click on one function and denote it by $f(x)$. Notice that you can graph $y = af(x)$ with a slider provided for a.

1. Make a larger. What happens to the graph?
2. Make a smaller. What happens to the graph?

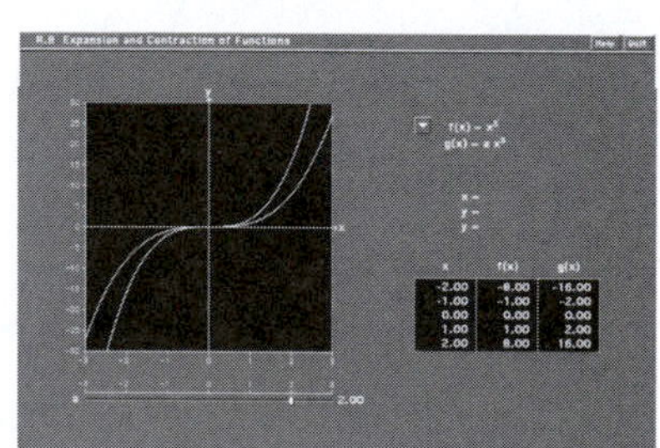

We can combine several graphing principals to graph a function.

EXAMPLE 5 **Combining Graphing Principals**

Graph $y = g(x) = \sqrt{x-1} + 2$.

Solution

Let $y = f(x) = \sqrt{x}$. The graph is shown in Figure A.59a. We know then that the graph of $y = \sqrt{x-1}$ is the graph of $y = f(x) = \sqrt{x}$ shifted to the right by one unit. (See Figure A.59b.) The graph of $y = g(x) = \sqrt{x-1} + 2$ is then the graph of $y = \sqrt{x-1}$ shifted upward by two units. See Figure A.59c.

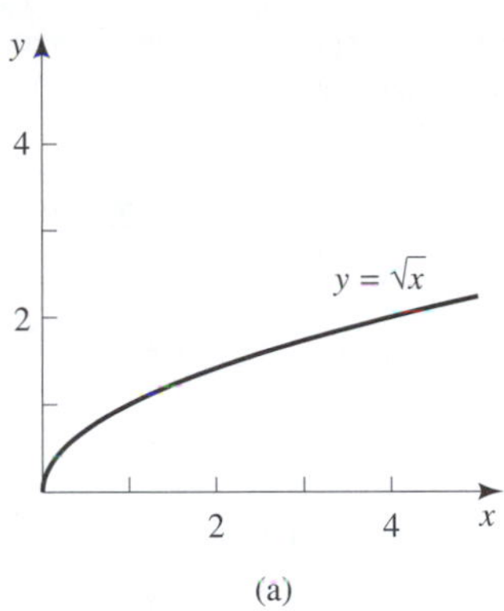

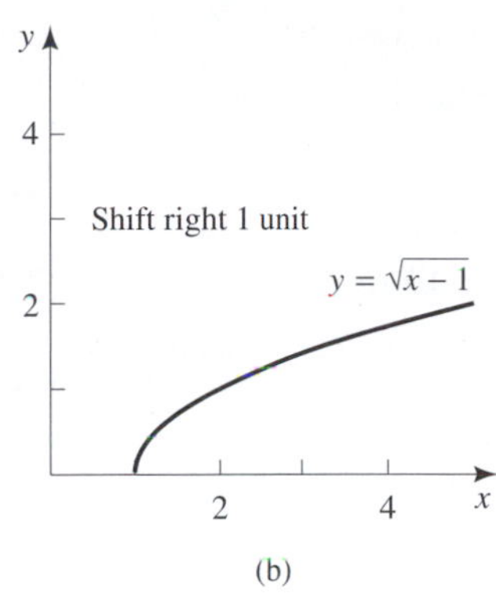

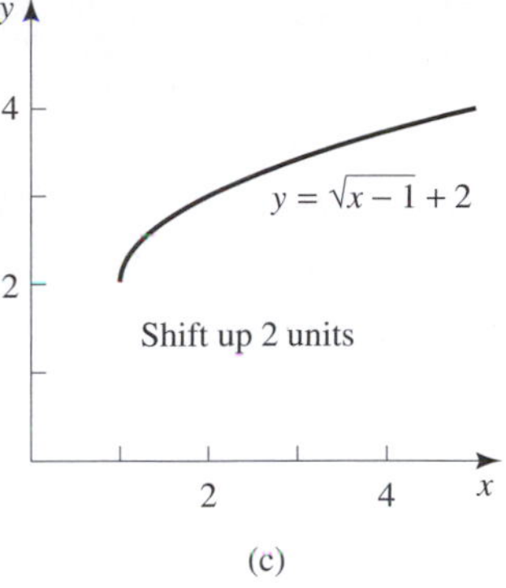

Figure A.59
Replacing x with $x - 1$ shifts the graph one unit to the right. Adding two then shifts the graph up by two units.

Warm Up Exercise Set A.8

1. Using the basic graphing principles, sketch graphs of
   a. $y = f(x) + 2$
   b. $y = f(x + 1)$
   c. $y = 2f(x)$

   where $y = f(x)$ is given in the following figure.

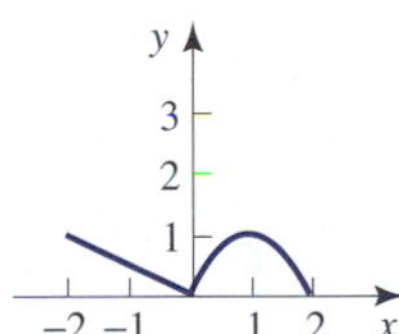

2. Graph $y = |x^3|$, $y = |(x - 1)^3|$, $y = 0.5|(x - 1)^3|$

Exercise Set A.8

In Exercises 1 through 4, determine the degree of each of the given polynomial functions, their domains, and the leading coefficients.

1. $p(x) = x^8 - x + 2$
2. $p(x) = 0.001x^6 + x^5 + x - 3$
3. $p(x) = 1000x^9 - 10x^7 + x^2$
4. $p(x) = 100x^{101} + x^{50} - x^9 + 2$

In Exercises 5 through 10, determine the domains of each of the given functions.

5. $\dfrac{x+1}{x-3}$
6. $\dfrac{x^3-2}{x^2-1}$
7. $\dfrac{x^3+1}{x^2+1}$
8. $x^{1/6}$
9. $x^{-1/7}$
10. $x^{-5/6}$

In Exercises 11 through 24, refer to the following two graphs.

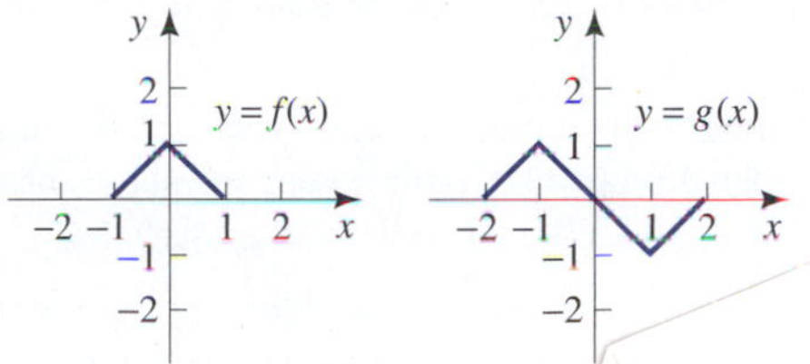

Use the graph of f or g and the basic g

graph each of the following.

11. $y = f(x) + 2$
12. $y =$
13. $y = g(x) - 1$
14. y

15. $y = f(x+2)$ **16.** $y = f(x-2)$

17. $y = g(x-1)$ **18.** $y = g(x+1)$

19. $y = -f(x)$ **20.** $y = -g(x)$

21. $y = 2f(x)$ **22.** $y = 2g(x)$

23. $y = 0.5f(x)$ **24.** $y = 0.5g(x)$

In Exercises 25 through 32, graph using the graphing principles.

25. $y = x^3$, $y = x^3 - 1$, $y = x^3 - 2$

26. $y = |x|$, $y = |x| - 1$, $y = |x| - 2$

27. $y = \sqrt[3]{x}$, $y = \sqrt[3]{x-1}$, $y = \sqrt[3]{x-2}$

28. $y = |x|$, $y = |x-1|$, $y = |x-2|$

29. $y = |x|$, $y = -|x|$

30. $y = \sqrt[3]{x}$, $y = -\sqrt[3]{x}$

31. $y = \sqrt{x}$, $y = \sqrt{x-1}$, $y = \sqrt{x-2}$

32. $y = x^3$, $y = 2x^3$, $y = 0.5x^3$

In Exercises 33 through 40, graph using the graphing principles.

33. $y = |x-1| + 2$ **34.** $y = (x-1)^3 - 1$

35. $y = \sqrt{x+1} - 2$ **36.** $y = \sqrt[3]{x+1} + 2$

37. $y = (x+2)^3 - 1$ **38.** $y = (x+2)^3 + 1$

39. $y = -|x-1| + 2$ **40.** $y = -(x-1)^3 - 1$

In Exercises 41 through 46, use the basic graphing principles to sketch the graphs. Then support your answers using your graphing calculator.

41. $y = 3^x$, $y = 3^{x+3}$, $y = 3^{x-3}$

42. $y = 2^x$, $y = 2 \cdot 2^x$, $y = 0.25 \cdot 2^x$

43. $y = 5^{-x}$, $y = 2 \cdot 5^{-x}$, $y = 0.25 \cdot 5^{-x}$

44. $y = 7^{-x}$, $y = 7^{-(x+3)}$, $y = 7^{-(x-3)}$

45. $y = 3^x$, $y = -3^x$

46. $y = 2^x$, $y = -2^{-x}$

Exercises 47 through 53 show given transformations in a specified order to be applied to the graph of the given function. Give an equation for the function associated with the transformed graph. Support your answer with a grapher.

47. $y = f(x) = |x|$, vertical shrink by 0.5, shift up by 3.

48. $y = f(x) = |x|$, vertical stretch by 2, shift down by 3.

49. $y = f(x) = |x^3|$, vertical stretch by 2, shift down by 3.

50. $y = f(x) = |x^3|$, vertical shrink by 0.5, shift up by 2.

51. $y = f(x) = |x|$, reflect through x-axis, shift up by 3, shift right by 2.

52. $y = f(x) = |x|$, reflect through x-axis, shift down by 3, shift left by 2.

53. Reverse the order of the transformations in Exercises 51. Do you obtain the same function?

Enrichment Exercises

54. On the same screen (with dimension [0, 1] by [0, 1]) graph $y = x^2$, $y = x^3$, $y = x^4$, and $y = x^5$. What can you say about how these graphs compare to each other on the interval [0, 1]. What do you think is true about the graphs of $y = x^n$ on [0, 1] for n an integer?

55. Refer to Exercise 54. Change the dimension of the screen to [0, 4] by [0, 64]. What can you say about how these graphs compare to each other on the interval [1, 4]? What do you think is true about the graphs of $y = x^n$ on $[1, \infty)$ for an integer n?

56. Speculate on how the graphs of $y = x^{1/n}$ for an integer n are related on [0.1]. Confirm your suspicions by graphing $y = x^{1/2}$, $y = x^{1/3}$, $y = x^{1/4}$, and $y = x^{1/5}$.

57. Speculate on how the graphs of $y = x^{1/n}$ for an integer n are related on $[1, \infty)$. Confirm your suspicions by graphing $y = x^{1/2}$, $y = x^{1/3}$, $y = x^{1/4}$, and $y = x^{1/5}$.

8. On your computer or graphing calculator, graph $y = x^n$ for some *even* powers n. Then graph $y = x^n$ for some *odd* owers n. What is a major distinguishing feature of the phs when n is even and when n is odd? Explain why this must be so.

59. There is a substantial difference between a^x and x^a. Let $y_1 = 2^x$ and $y_2 = x^2$.

a. Clearly, $y_1 = y_2$ when $x = 2$. Find two other values of x for which $2^x = x^2$.

b. Determine values of x for which $2^x < x^2$. Determine values of x for which $x^2 < 2^x$.

c. To obtain a further insight into the difference between a^x and x^a, compare the values for 2^x and x^2 for $x = -100$ and $x = 100$. Do the same for other large positive and negative values of x. What do you conclude?

60. Given any positive integer n, speculate on whether $y_1 = 2^x$ or $y_2 = x^n$ is larger for large x. Experiment on your graphing utility to decide.

61. Let $y_1 = (1.1)^x$ and $y_2 = x^3$.

a. Which is larger when $x = 2$?

b. On a screen of dimension [0, 200] by [0, 20,000,000] graph and determine which is larger for large values of x.

62. Given any positive integer n, speculate on whether $y_1 = (1.1)^x$ or $y_2 = x^n$ is larger for large x. Experiment on your graphing calculator to decide.

63. Two graphs are shown below. One is the graph of $y = x^{11}$, and one is the graph of $y = x^{12}$. Which is which? Explain why you think so.

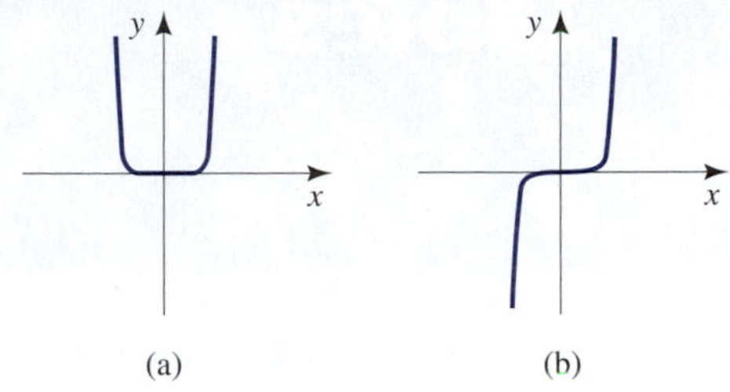

(a) (b)

Solutions to WARM UP EXERCISE SET A.8

1.

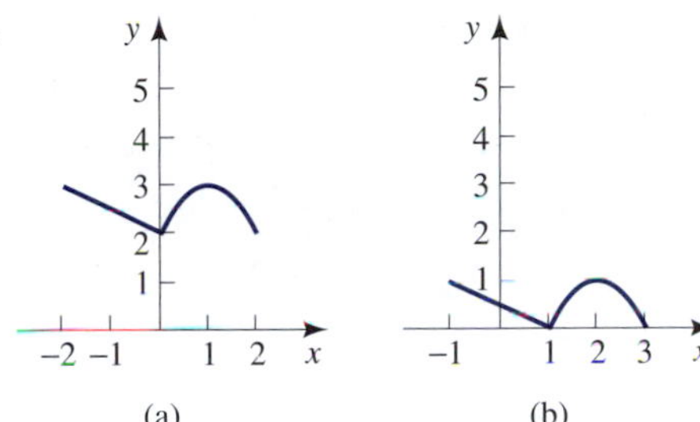

(a) (b)

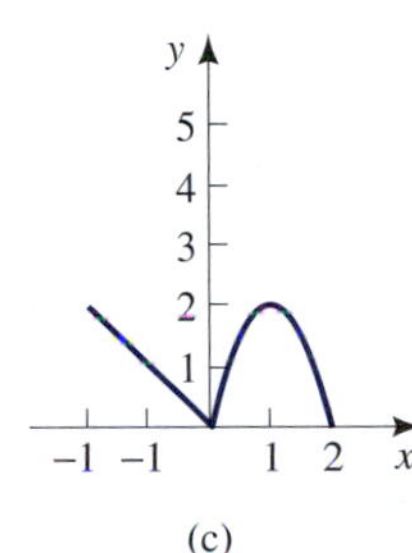

(c)

2.

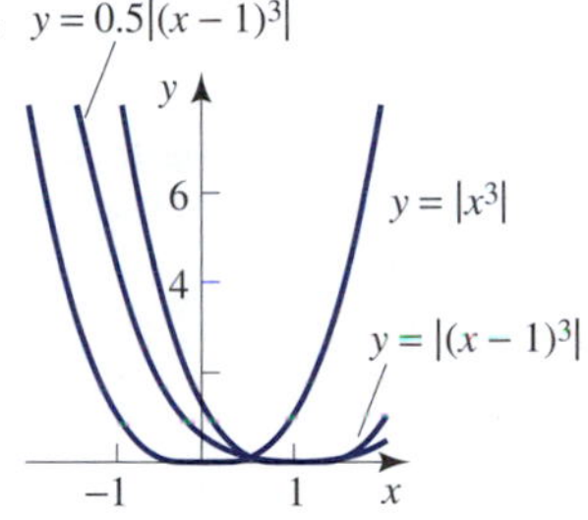

B Tables

B.1 Basic Geometric Formulas

- **Pythagorean Theorem**
  $c^2 = a^2 + b^2$

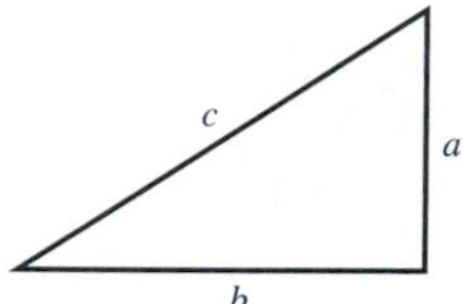

- **Rectangle**
  Area $= ab$
  Perimeter $= 2a + 2b$

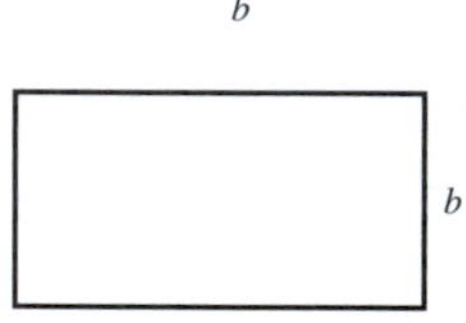

- **Parallelogram**
  Height $= h$
  Area $= ah$
  Perimeter $= 2a + 2b$

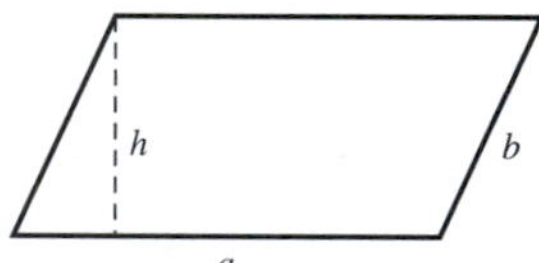

- **Triangle**
  Height $= h$
  Area $= \frac{1}{2}hc$
  Perimeter $= a + b + c$

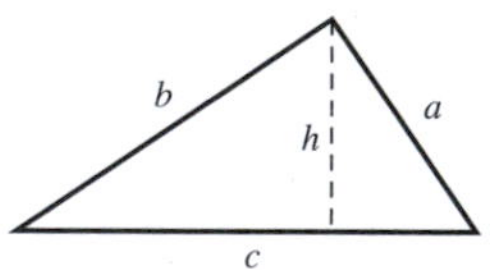

- **Trapezoid**
  Area $= \frac{1}{2}(a + b)h$

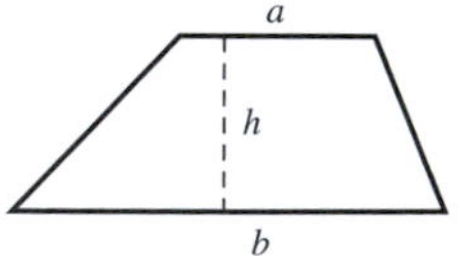

- **Circle**
  Radius $= r$
  Diameter $= d = 2r$
  Area $= \pi r^2$
  Circumference $= 2\pi r$
  $\pi = 3.14159$

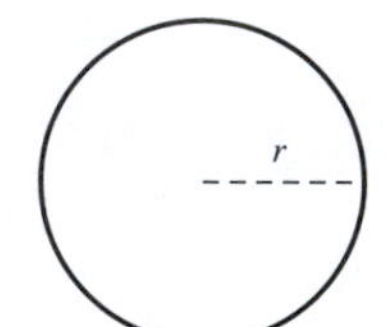

- **Rectangular Solid**
  Volume $= abc$
  Total surface area $= 2ab + 2ac + 2bc$

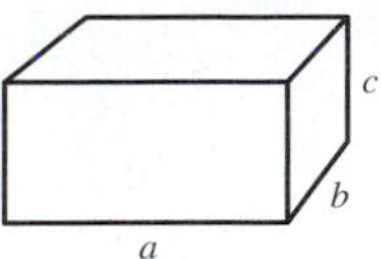

- **Right Circular Cylinder**
  Radius of base $= r$
  Height $= h$
  Volume $= \pi r^2 h$
  Total surface area $= 2\pi r^2 + 2\pi r h$

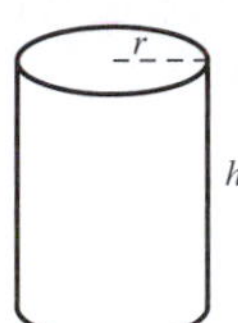

- **Sphere**
  Radius $= r$
  Diameter $= d = 2r$
  Volume $= \frac{4}{3}\pi r^3$
  Surface area $= 4\pi r^2$

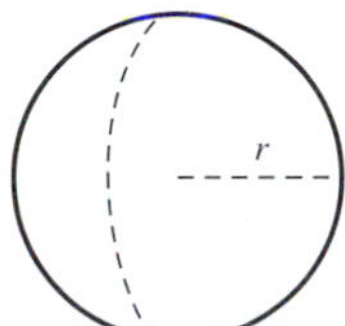

- **Right Circular Cone**
  Radius of base $= r$
  Height $= h$
  Volume $= \frac{1}{3}\pi r^2 h$
  Total surface area $= \pi r^2 + \pi r\sqrt{r^2 + h^2}$

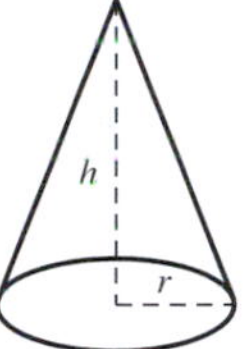

B.2 Tables of Integrals

Integrands Containing $(ax + b)$

1. $\displaystyle\int x(ax+b)^n\,dx = \frac{(ax+b)^{n+1}}{a^2}\left[\frac{ax+b}{n+2} - \frac{b}{n+1}\right] \qquad (n \neq -1, -2)$

2. $\displaystyle\int \frac{x}{ax+b}\,dx = \frac{x}{a} - \frac{b}{a^2}\ln|ax+b|$

3. $\displaystyle\int \frac{x}{(ax+b)^2}\,dx = \frac{1}{a^2}\left[\ln|ax+b| + \frac{b}{ax+b}\right]$

4. $\displaystyle\int \frac{x^2}{ax+b}\,dx = \frac{1}{a^3}\left[\frac{1}{2}(ax+b)^2 - 2b(ax+b) + b^2\ln|ax+b|\right]$

5. $\displaystyle\int \frac{x^2}{(ax+b)^2}\,dx = \frac{1}{a^3}\left[ax+b - \frac{b^2}{ax+b} - 2b\ln|ax+b|\right]$

6. $\displaystyle\int \frac{1}{x(ax+b)}\,dx = \frac{1}{b}\ln\left|\frac{x}{ax+b}\right| \qquad (b \neq 0)$

7. $\displaystyle\int \frac{1}{x^2(ax+b)}\,dx = -\frac{1}{bx} + \frac{a}{b^2}\ln\left|\frac{ax+b}{x}\right| \qquad (b \neq 0)$

8. $\displaystyle\int \frac{1}{x(ax+b)^2}\,dx = \frac{1}{b(ax+b)} - \frac{1}{b^2}\ln\left|\frac{ax+b}{x}\right| \qquad (b \neq 0)$

Integrands Containing $\sqrt{ax+b}$

9. $\displaystyle\int x\sqrt{ax+b}\,dx = \frac{2}{a^2}\left[\frac{(ax+b)^{5/2}}{5} - \frac{b(ax+b)^{3/2}}{3}\right]$

10. $\displaystyle\int x^2\sqrt{ax+b}\,dx = \frac{2}{a^3}\left[\frac{(ax+b)^{7/2}}{7} - \frac{2b(ax+b)^{5/2}}{5} + \frac{b^2(ax+b)^{3/2}}{3}\right]$

11. $\displaystyle\int \frac{x}{\sqrt{ax+b}}\,dx = \frac{2ax-4b}{3a^2}\sqrt{ax+b}$

12. $\displaystyle\int \frac{1}{x\sqrt{ax+b}}\,dx = \frac{1}{\sqrt{b}}\ln\left|\frac{\sqrt{ax+b}-\sqrt{b}}{\sqrt{ax+b}+\sqrt{b}}\right| \qquad (b>0)$

Integrands Containing $a^2 \pm x^2$

13. $\displaystyle\int \frac{1}{a^2-x^2}\,dx = \frac{1}{2a}\ln\left|\frac{x+a}{x-a}\right|$

14. $\displaystyle\int \frac{x}{a^2\pm x^2}\,dx = \pm\frac{1}{2}\ln|a^2\pm x^2|$

15. $\displaystyle\int \frac{1}{x(a^2\pm x^2)}\,dx = \frac{1}{2a^2}\ln\left|\frac{x^2}{a^2\pm x^2}\right|$

Integrands Containing $\sqrt{a^2-x^2}$

16. $\displaystyle\int \frac{x}{\sqrt{a^2-x^2}}\,dx = -\sqrt{a^2-x^2}$

17. $\displaystyle\int \frac{1}{x\sqrt{a^2-x^2}}\,dx = -\frac{1}{a}\ln\left|\frac{a+\sqrt{a^2-x^2}}{x}\right|$

18. $\displaystyle\int \frac{1}{x^2\sqrt{a^2-x^2}}\,dx = -\frac{\sqrt{a^2-x^2}}{a^2x}$

19. $\displaystyle\int \frac{1}{(a^2-x^2)^{3/2}}\,dx = \frac{1}{a^2}\frac{x}{\sqrt{a^2-x^2}}$

20. $\displaystyle\int \frac{x}{(a^2-x^2)^{3/2}}\,dx = \frac{1}{\sqrt{a^2-x^2}}$

21. $\displaystyle\int \frac{\sqrt{a^2-x^2}}{x}\,dx = \sqrt{a^2-x^2} - a\ln\left|\frac{a+\sqrt{a^2-x^2}}{x}\right|$

Integrands Containing $\sqrt{x^2\pm a^2}$

22. $\displaystyle\int \frac{1}{\sqrt{x^2\pm a^2}}\,dx = \ln|x+\sqrt{x^2\pm a^2}|$

23. $\displaystyle\int \frac{x}{\sqrt{x^2\pm a^2}}\,dx = \sqrt{x^2\pm a^2}$

24. $\displaystyle\int \frac{x^2}{\sqrt{x^2 \pm a^2}}\,dx = \frac{1}{2}x\sqrt{x^2 \pm a^2} \mp \frac{1}{2}a^2 \ln|x + \sqrt{x^2 \pm a^2}|$

25. $\displaystyle\int \frac{1}{x\sqrt{x^2 + a^2}}\,dx = -\frac{1}{a}\ln\left|\frac{a + \sqrt{x^2 + a^2}}{x}\right|$

26. $\displaystyle\int \frac{1}{x^2\sqrt{x^2 \pm a^2}}\,dx = \mp\frac{\sqrt{x^2 \pm a^2}}{a^2 x}$

27. $\displaystyle\int \frac{1}{(x^2 \pm a^2)^{3/2}}\,dx = \pm\frac{1}{a^2}\frac{x}{\sqrt{x^2 \pm a^2}}$

28. $\displaystyle\int \frac{x}{(x^2 \pm a^2)^{3/2}}\,dx = -\frac{1}{\sqrt{x^2 \pm a^2}}$

29. $\displaystyle\int \frac{x^2}{(x^2 \pm a^2)^{3/2}}\,dx = -\frac{x}{\sqrt{x^2 \pm a^2}} + \ln|x + \sqrt{x^2 \pm a^2}|$

30. $\displaystyle\int x\sqrt{x^2 \pm a^2}\,dx = \frac{1}{3}(x^2 \pm a^2)^{3/2}$

31. $\displaystyle\int \sqrt{x^2 \pm a^2}\,dx = \frac{1}{2}x\sqrt{x^2 \pm a^2} \pm \frac{1}{2}a^2 \ln|x + \sqrt{x^2 \pm a^2}|$

32. $\displaystyle\int \frac{\sqrt{x^2 + a^2}}{x}\,dx = \sqrt{x^2 + a^2} - a\ln\left|\frac{a + \sqrt{x^2 + a^2}}{x}\right|$

33. $\displaystyle\int \frac{\sqrt{x^2 \pm a^2}}{x^2}\,dx = -\frac{\sqrt{x^2 \pm a^2}}{x} + \ln|x + \sqrt{x^2 \pm a^2}|$

Integrands Containing $(ax^2 + bx + c)$

34. $\displaystyle\int \frac{1}{ax^2 + bx + c}\,dx = \frac{1}{\sqrt{b^2 - 4ac}}\ln\left|\frac{2ax + b - \sqrt{b^2 - 4ac}}{2ax + b + \sqrt{b^2 - 4ac}}\right|$
$(b^2 - 4ac > 0)$

Integrands Containing $(ax^n + b)$

35. $\displaystyle\int \frac{1}{x(ax^n + b)}\,dx = \frac{1}{nb}\ln|\frac{x^n}{ax^n + b}| \qquad (n \neq 0, b \neq 0)$

36. $\displaystyle\int \frac{1}{x\sqrt{ax^n + b}}\,dx = \frac{1}{n\sqrt{b}}\ln\left|\frac{\sqrt{ax^n + b} - \sqrt{b}}{\sqrt{ax^n + b} + \sqrt{b}}\right| \qquad (b > 0)$

Integrands Containing Exponentials and Logarithms

37. $\displaystyle\int xe^{ax}\,dx = \frac{1}{a^2}(ax - 1)e^{ax}$

38. $\displaystyle\int x^n e^{ax}\,dx = \frac{1}{a}x^n e^{ax} - \frac{n}{a}\int x^{n-1}e^{ax}\,dx$

39. $\displaystyle\int \frac{1}{b+ce^{ax}}\,dx = \frac{1}{ab}[ax - \ln(b+ce^{ax})] \qquad (ab \neq 0)$

40. $\displaystyle\int x^n \ln|x|\,dx = \frac{1}{n+1}x^{n+1}\left[\ln|x| - \frac{1}{n+1}\right] \qquad (n \neq -1)$

41. $\displaystyle\int (\ln x)^n\,dx = x\,(\ln x)^n - n\int (\ln x)^{n-1}\,dx$

Miscellaneous Integrals

42. $\displaystyle\int \sqrt{\frac{x+a}{x+b}}\,dx = \sqrt{(x+b)(x+a)} + (a-b)\ln|\sqrt{x+b} + \sqrt{x+a}|$

Answers to Selected Exercises

Chapter 1

Section 1.1

1. Yes **3.** No **5.** Yes

7. **a.** $-4, -3, -2, 2, 3$ **b.** $0, 4.5$
c. $[-4, -3], [-2, 2], [3, 5]$
d. $[-5, -4], [-3, -2], [2, 3]$

9. **a.** Fourth; **b.** third; **c.** second;
d. first. The remaining graph is the last: Profits have steadily decreased.

11. $x \neq 1$ **13.** $x \geq 0$ **15.** $x > 1$

17. $x \neq 2, 3$ **19.** All real numbers

21. $(-2, 0) \cup (0, \infty)$ **23.** All real numbers

25. Domain: $[0, 2]$ **27.** Domain: $[0, 2]$

29. Domain: $[0, 6]$

31. (25) increasing on $(0, 1)$, decreasing on $(1, 2)$, concave down on $(0, 2)$; (26) increasing on $(-1, 1)$, concave up on $(-1, 0)$, concave down on $(0, 1)$; (28) increasing on $(-1, 1)$ and $(1, 2)$, concave up on $(1, 2)$, concave down on $(-1, 1)$

33. $\frac{1}{2}, \frac{2}{3}, -1, \frac{1}{\sqrt{2}}, \frac{1}{x+3}, \frac{x+1}{x+2}$

35. $3, 0, \sqrt{2x^2-2}, \sqrt{\frac{2}{x}-1}, \sqrt{-2x-1}, -\sqrt{2x-1},$

37. $1, -1, |x|, x^2$ **39.** $1, 2, 2, 2$

41. $0, 1, 0, 1$ **43.** Yes

45. No **47.** No **49.** Yes

51. Continuous everywhere except $x = -1$.

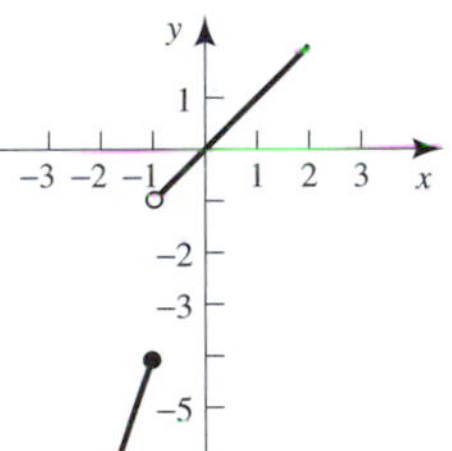

53. Continuous everywhere

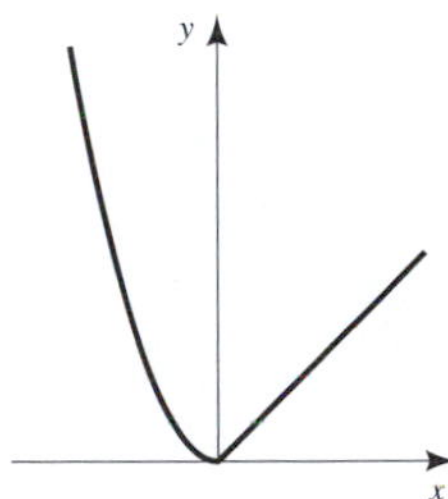

55. Continuous everywhere except $x = 1$

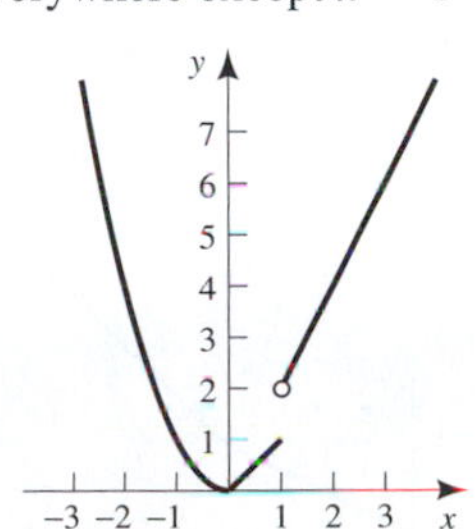

57. Continuous except at $x = 1$ and $x = 2$

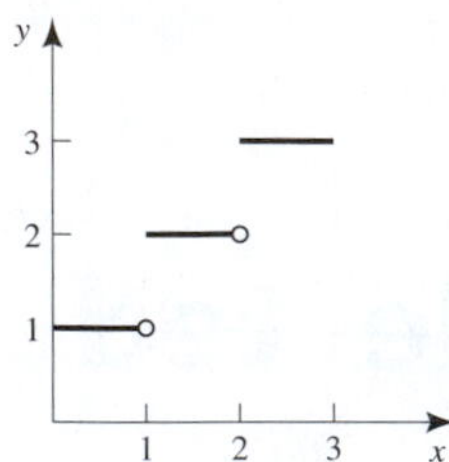

59. All real numbers

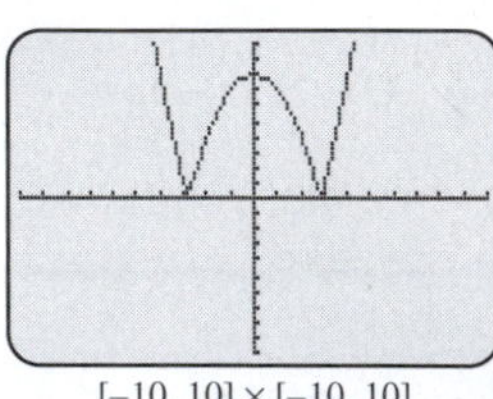

$[-10, 10] \times [-10, 10]$

61. $[-3, 3]$

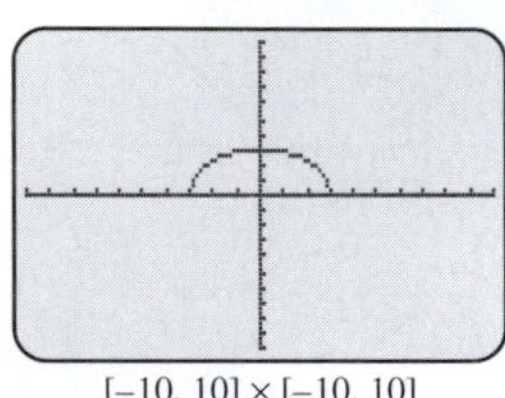

$[-10, 10] \times [-10, 10]$

63. $[0, 5) \cup (5, \infty)$

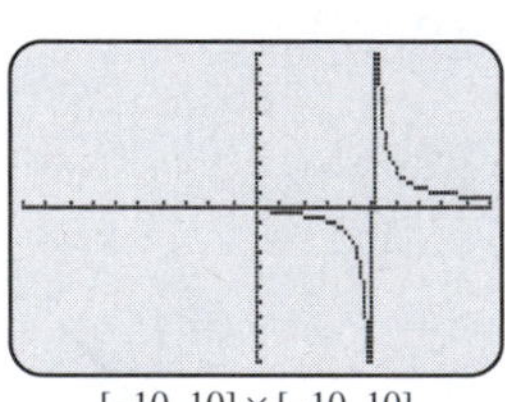

$[-10, 10] \times [-10, 10]$

65. 0, 2, 3, 4

67. $V(x) = 3x^3$

69. $d(t) = 60t$

71. $R(x) = 60x$

73. $C(x) = \begin{cases} x & \text{if } 0 \le x \le 500 \\ 500 + 2(x - 500) & \text{if } x > 500 \end{cases}$

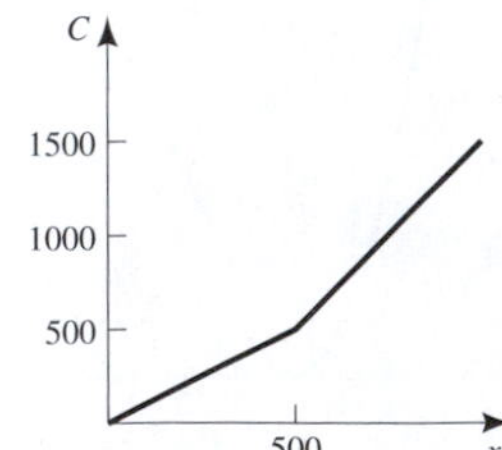

This function is continuous on $[0, \infty)$

75. $T(x) = \begin{cases} 0 & \text{if } 0 \le x \le 50 \\ 0.01(x - 50) & \text{if } x > 50 \end{cases}$

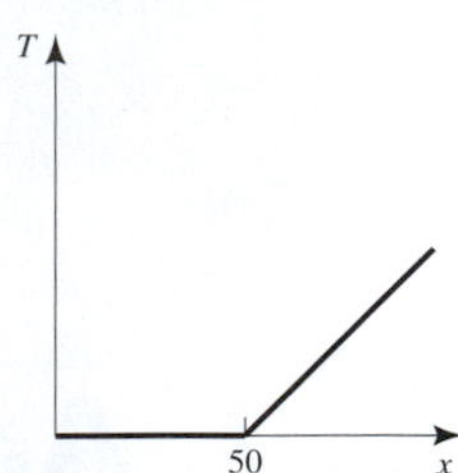

Continuous on $[0, \infty)$

77. $P(x) = \begin{cases} 0.37 & \text{if } 0 \le x \le 1 \\ 0.60 & \text{if } 1 < x \le 2 \\ 0.83 & \text{if } 2 < x \le 3 \end{cases}$

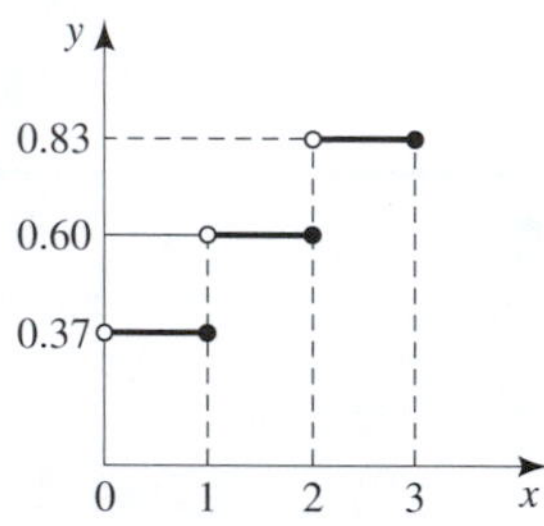

Continuous on $[0, 3]$ except at $x = 1$ and 2.

79. $A(10) = 0.80$. There are 0.80 fish on average in a cubic meter of water that is at 10°C. $A(13) = 1.15$. There are 1.15 fish on average in a cubic meter of water that is at 13°C. $A(19) = 0$. There are no fish on average in a cubic meter of water that's at 19°C.

81. $N(0) = 0$. There are no fish at the surface. $N(120) = 20$. Twenty fish were caught at a depth of 120 meters. $N(260) \approx 3$. There are about 3 fish at a depth of 260 meters.

83. 109.6, 70.5

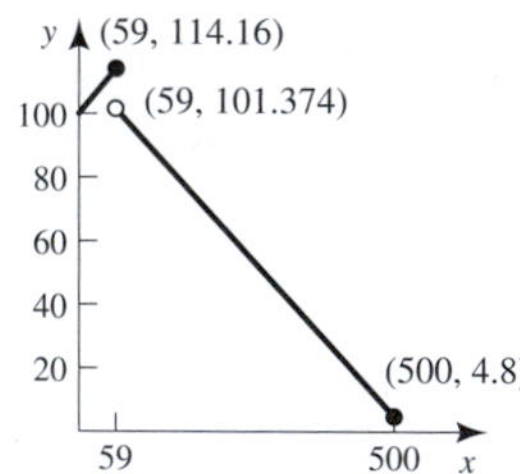

85.

$$T(x) = \begin{cases} 0.15x & \text{if } x \le 27,050 \\ 4057.50 + 0.275(x - 27,050) & \text{if } 27,050 < x \le 65,550 \\ 14,645 + 0.305(x - 65,550) & \text{if } 65,550 < x \le 136,750 \\ 36,361 + 0.355(x - 136,750) & \text{if } 136,750 < x \le 297,350 \\ 93,374 + 0.391(x - 297,350) & \text{if } x > 297,350 \end{cases}$$

87. $S(118) \cong 7$. This means that there will be about 7 plant species in 118 square meter area of woodland. S is reduced by a factor of $(0.5)^{0.248} \approx 0.82$

89. 3

91. $6x + 3h$

93. $f(x) = \begin{cases} 3x - 2 & \text{if } x \geq 2 \\ x + 2 & \text{if } x < 2 \end{cases}$

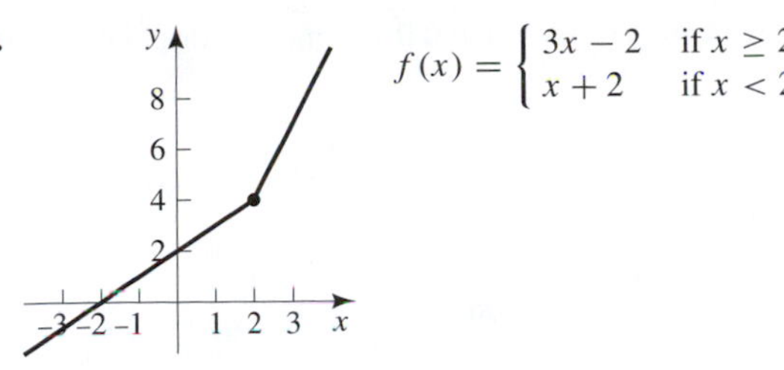

95. The two functions are the same.

97. $f(a + b) = m(a + b) = ma + mb = f(a) + f(b)$

99. $3f(x) = 3 \cdot 3^x = 3^{x+1} = f(x + 1)$, $f(a + b) = 3^{a+b} = 3^a \cdot 3^b = f(a) \cdot f(b)$

Section 1.2

1. $C = 3x + 10{,}000$

3. $R = 5x$

5. $P = 2x - 10{,}000$

7. $x = 2$

9. $x = 20$

11. $(1.5, 4.5)$

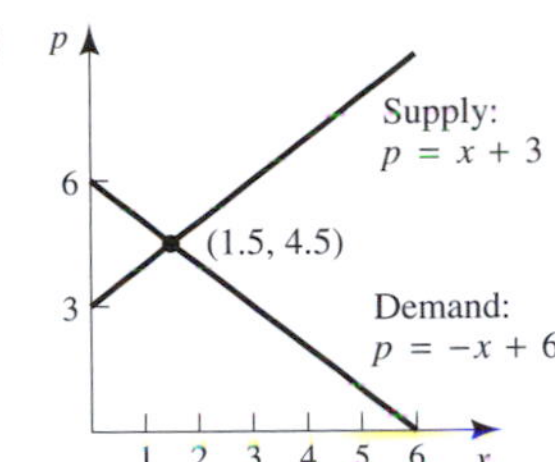

13. $(1, 15)$

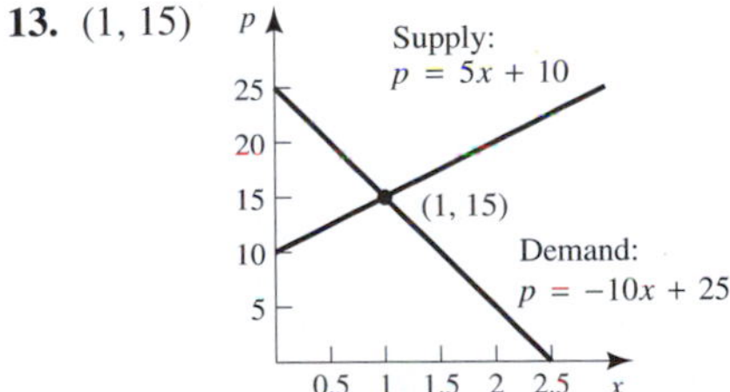

15. $x = 5$

17. $x = 1.5$

19.

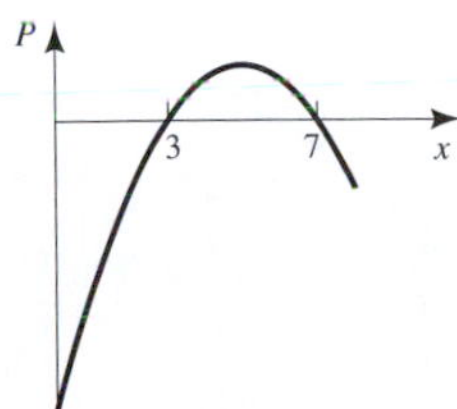

$P(x) = -2x^2 + 20x - 42$, maximum when $x = 5$, break-even quantities are $x = 3, 7$

21.

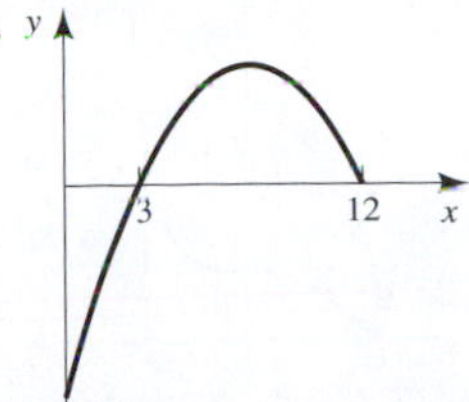

$P(x) = -x^2 + 15x - 36$, maximum when $x = \frac{15}{2}$, break-even quantities are $x = 3, 12$

23. Break-even quantities are $x = 0.9$ and 2.2. Maximum at $x = 1.55$.

25. Break-even quantities are approximately $x = 1.936436$, 7.663564. Maximum at $x = 4.8$.

27. $V = -5000t + 50{,}000$, \$45,000, \$25,000

29. $p = -0.005x + 22.5$

31. Let x be the number of pairs of fenders manufactured. Then the costs functions in dollars are as follows:

Steel: $C(x) = 260{,}000 + 5.26x$
Aluminum: $C(x) = 385{,}000 + 12.67x$
RMP: $C(x) = 95{,}000 + 13.19x$
NPN: $C(x) = 95{,}000 + 9.53x$
PPT: $C(x) = 95{,}000 + 12.555x$

33. 12,000

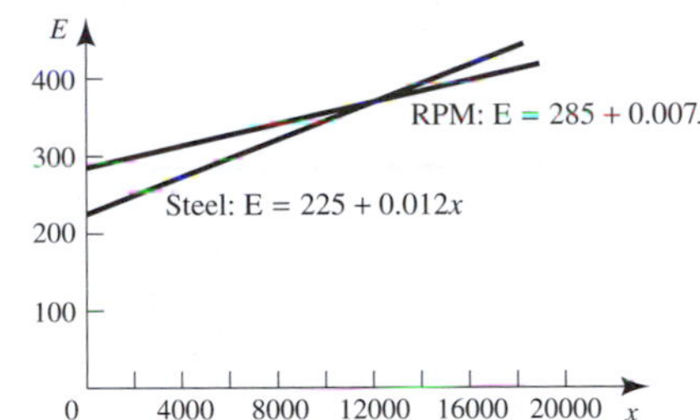

35. Let x be the number of cows, then the cost $C(x)$ in dollars is $C(x) = 13{,}386 + 393x$. The revenue function in dollars is $R(x) = 470x$. The profit function in dollars is $P(x) = 77x - 13{,}386$. The profit for an average of 97 cows is $P(97) = -5917$, that is, a loss of \$5917. For many such farms the property and buildings have already been paid off. Thus, the fixed costs for these farms are lower than stated in the table.

37. $C = 209.03x + 447{,}917$, $R = 266.67x$, $P = 57.64x - 447{,}917$

39. Outside

41. **a.** $f = 0.3056x$; **b.** $c = 3525 + 0.3056x$; **c.** \$4289

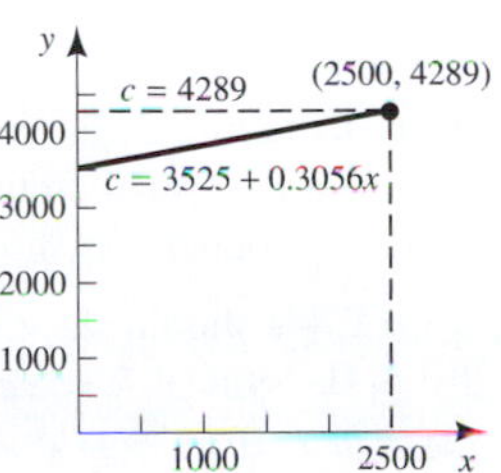

43. Let x be the number of copies. The cost in cents with no plastic card is $C(x) = 10x$. The cost with a plastic card is $C(x) = 50 + 7x$. 17.

45. Decreases by 0.108475 yen per ton. This number is the slope of the line.

47. 110

49. 5593 bales, 30,000 bales

51. 63.8, 2.95116

53. About 286 pounds, the relative yield of vegetables is maximized

55. \$11.667 billion, 58.46% **57.** 35

59. **a.** 30.74; **b.** 19.91; **c.** 41.57

61. $C = 5x + 1000$ **63.** $C = 3x + 2000$

65. $R = 11x$

67. **a.** Manual; **b.** automatic.

Volume	Total Costs Manual	Total Costs Automatic
1000	\$2600	\$8020
10,000	\$17,000	\$8200

69.

Volume	Costs per Unit Manual	Costs per Unit Automatic
1000	\$2.60	\$8.02
10,000	\$1.70	\$0.82
100,000	\$1.61	\$0.10

$\overline{C}_m = \dfrac{1000}{x} + 1.6$, $\overline{C}_a = \dfrac{8000}{x} + 0.02$. The cost per unit for a manual machine tends toward \$1.60 and becomes greater than the cost per unit for an automatic machine, which tends toward \$0.02.

71. $\alpha = 2$, $\beta = 8$

73. Since we expect profits to turn negative if too many products are made and sold (the price would have to be very low to unload so many products), a should be negative.

75. The graph of $p = mx + e$ should slope downward, requiring $m < 0$ and $e > 0$. Since $R = xp = mx^2 + ex$ and $R = ax^2 + bx$, $a = m < 0$ and $b = e > 0$.

77. Approximately 2%, 4%.

79. The fewer the number of boat trips, the more the owner needs to charge to make a living. But at some point the price is so high that nobody will be willing to pay such a high price. The industry refers to this as the "choke" price.

81. About 20. There are about 0.72 times the original, or a 28% drop

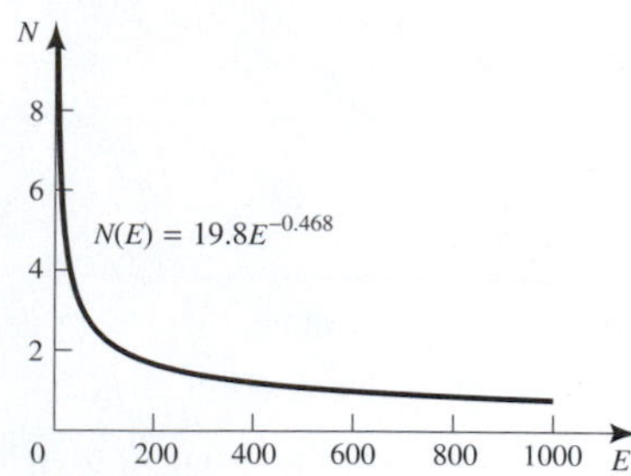

83. About 7. about 66%.

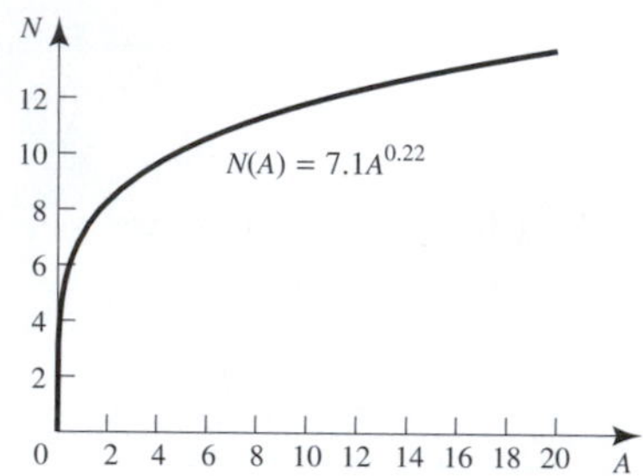

Section 1.3

1.

3.

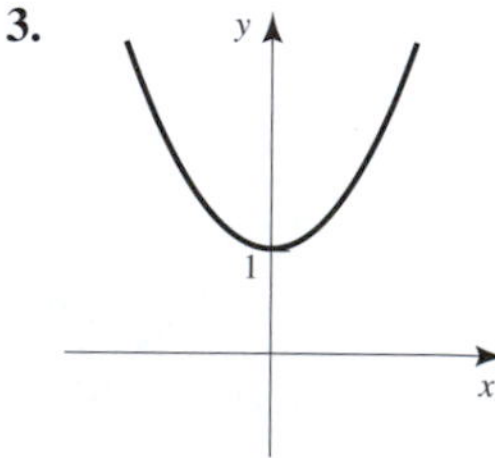

5.

7.

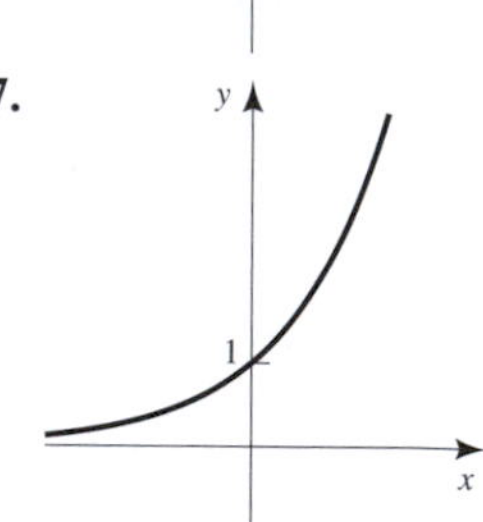

9. 2 **11.** $\dfrac{3}{4}$

13. $\dfrac{3}{4}$ **15.** $-\dfrac{45}{23}$

17.

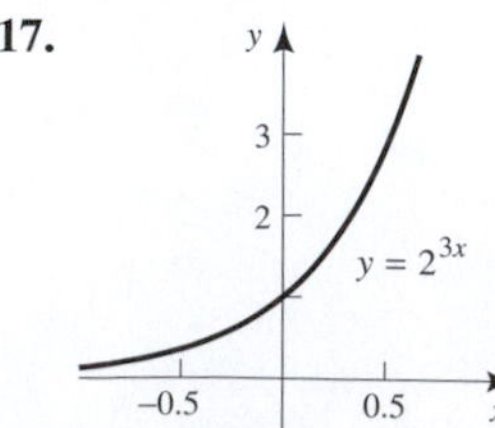

19.

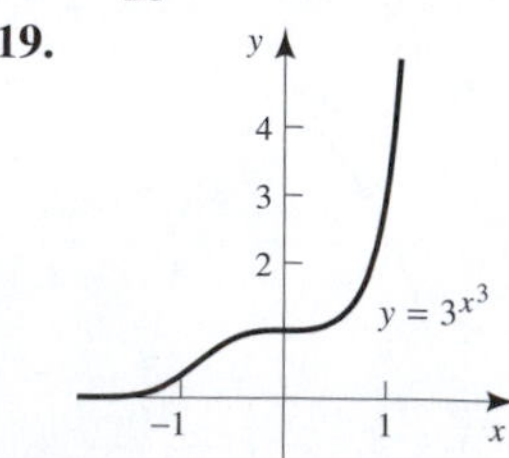

21.

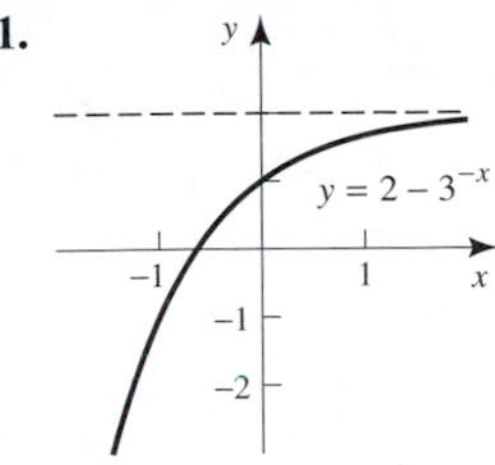

23.

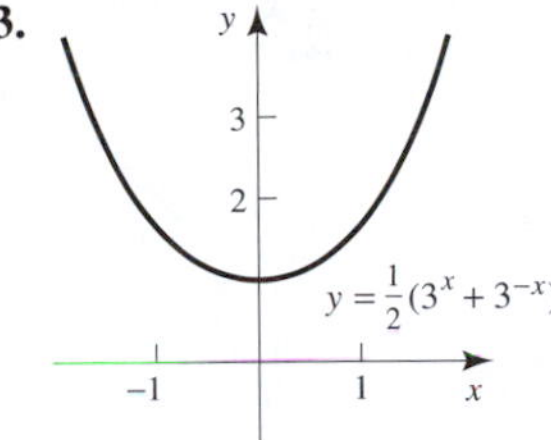

25. -1 **27.** -3

29. $-\frac{1}{2}$ **31.** $-\frac{1}{2}$

33. 1 **35.** -1

37. 0, 1

39. **a.** \$1080; **b.** \$1082.43; **c.** \$1083.00; **d.** \$1083.22; **e.** \$1083.28; **f.** \$1083.29

41. **a.** \$3262.04; **b.** \$7039.99; **c.** \$14,974.46; **d.** \$31,409.42; **e.** \$93,050.97; **f.** \$267,863.55

43. **a.** 8.16%; **b.** 8.243%; **c.** 8.300%; **d.** 8.322%; **e.** 8.328% **f.** 8.329%

45. **a.** \$1784.31; **b.** \$1664.13; **c.** \$1655.56; **d.** \$1652.99

47. \$31,837.58 **49.** \$26,897.08

51. \$5.5 trillion **53.** \$1104.79, no

55. 6.66%, 6.82%, the second bank

57. \$67,297.13 **59.** 11.58 years

61. 8.2%

63. The second bank, the second effective interest rate is greater than the first one.

65. About 3.7 years **67.** \$568.95

69. 6.8 billion **71.** 5%

73. 16.82 **75.** $(0, \infty)$, $(-\infty, 0)$

77. $P = 32(0.5)^x$

79. Increasing. 346 mm. 736 grams.

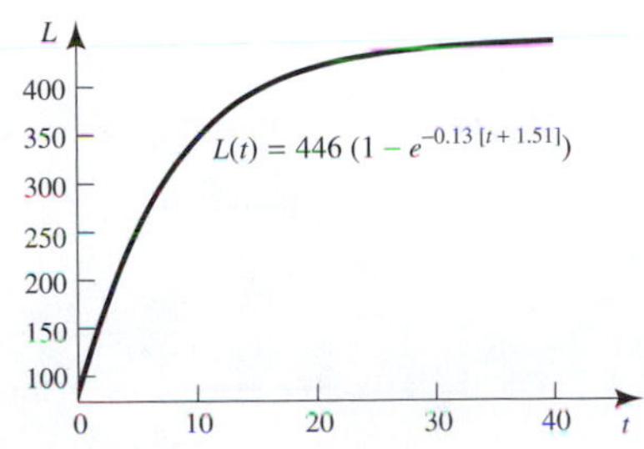

81. 2.1

Section 1.4

1. **a.** 11 **b.** 3 **c.** 28 **d.** $\frac{7}{4}$

3. $x^2 + 2x + 4$, $-x^2 + 2x + 2$, $(2x + 3)(x^2 + 1)$, $\dfrac{2x + 3}{x^2 + 1}$. All domains are $(-\infty, \infty)$.

5. $3x + 4$, $x + 2$, $2x^2 + 5x + 3$, $\dfrac{2x + 3}{x + 1}$. The domain is $(-\infty, \infty)$ in the first three cases and $(-\infty, -1) \cup (-1, \infty)$ in the last case.

7. $\sqrt{x + 1} + x + 2$, $\sqrt{x + 1} - x - 2$, $\sqrt{x + 1}(x + 2)$, $\dfrac{\sqrt{x + 1}}{x + 2}$. Domains are all $[-1, \infty)$.

9. $2x + 1 + \dfrac{1}{x}$, $2x + 1 - \dfrac{1}{x}$, $\dfrac{2x + 1}{x}$, $x(2x + 1)$. Domains are all $(-\infty, 0) \cup (0, \infty)$.

11. $[-1, \infty)$ **13.** $\left[\dfrac{5}{2}, \infty\right)$

15. **a.** 125; **b.** -1

17. $6x - 3$, $6x + 1$. Domains are all $(-\infty, \infty)$.

19. x, x, Domains: $(-\infty, \infty)$, $(-\infty, \infty)$

21. **a.** 5; **b.** 54; **c.** 7; **d.** 16

23. $f(x) = x^5$, $g(x) = x + 5$

25. $f(x) = \sqrt[3]{x}$, $g(x) = x + 1$

27. $f(x) = |x|$, $g(x) = x^2 - 1$

29. $f(x) = \dfrac{1}{x}$, $g(x) = x^2 + 1$

31. **a.** 0; **b.** -1; **c.** 0; **d.** 1

33. $9t + 10$. The firm has a daily start-up cost of \$10,000 and costs of \$9000 per hour.

35. $R(x) - C(x)$, which is the profit $P(x)$.

37. $V(t) = \dfrac{4\pi(30 - 2t)^3}{3}$

39. $g(x) = 40x$, $f(r) = 5.5r$, $y = f(g(x)) = (f \circ g)(x) = 40(5.5)x = 220x$. There are 220 yards in a furlong.

41. Increasing. 686 mm. 3178 grams. $W(L(t)) = 1.30 \times 10^{-6} \cdot (960(1 - e^{-0.12[t+0.45]}))^{3.31}$

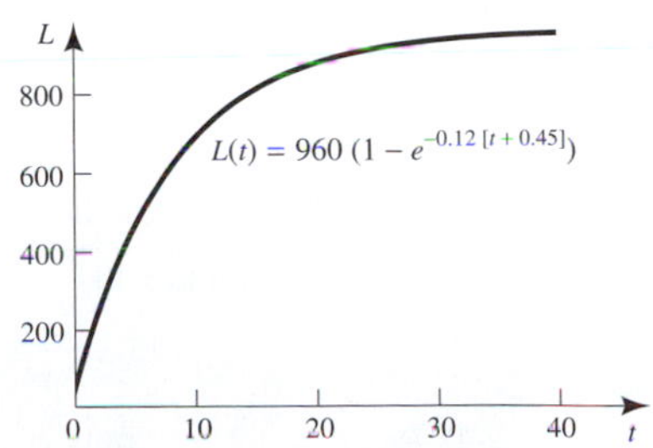

43. $(-\infty, \infty)$ **45.** $[-5, \infty)$ **47.** $(-\infty, 5]$

Section 1.5

1. 100 **3.** $e^{-\pi}$

5. $3^{5/3}$ **7.** 4

9. -1

11. 2π

13. 8

15. 81

17. $2\log x + \frac{1}{2}\log y + \log z$

19. $\frac{1}{2}\log x + \frac{1}{2}\log y - \log z$

21. $\frac{1}{2}(\log x + \log y - \log z)$

23. $\log x^2 y$

25. $\log \frac{\sqrt{x}}{\sqrt[3]{y}}$

27. $\ln \frac{x^3 y}{\sqrt[3]{z}}$

29. $0.2\log(0.6)$

31. $-0.5\log 3$

33. $\frac{1 - \log 2}{3}$

35. $\pm\sqrt{\ln 4}$

37. $10^{-1.5} - 7$

39. 2.5

41.

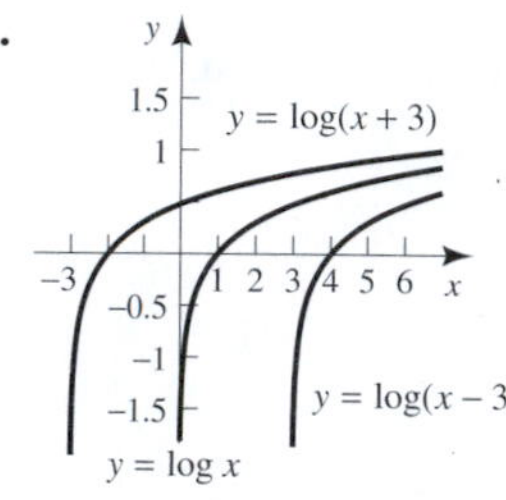

43.

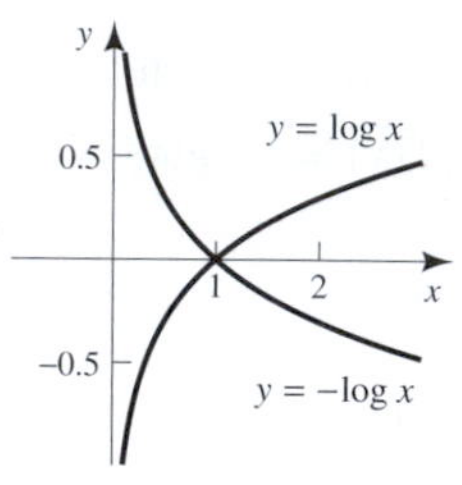

45. $x \approx 0.01$ and $x \approx 1.4724$

47. $10^x = e^{(\ln 10)x} \approx e^{2.3026x}$

49. About 6.5 weeks

51. About 7.3 years

53. About 0.0084

55. 24.8 years

57.

$$\begin{aligned} y &= P_0 e^{-kt} \\ P_0 e^{-kT} &= \frac{1}{2} P_0 \\ e^{-kT} &= \frac{1}{2} \\ -kT &= \ln\frac{1}{2} = -\ln 2 \\ kT &= \ln 2 \end{aligned}$$

59. $\ln 3 \approx 1.0986$ years

61. About 5.8 hours

63. 24.94%

65. About \$40

67. $\log x = \frac{\ln x}{\ln 10}$

69. No. Since for $y > 0$, $\log_1 x = y$ if and only if $x = 1^y = 1$ has no solution if $x \neq 1$.

71. y_2

73. 5.25 years

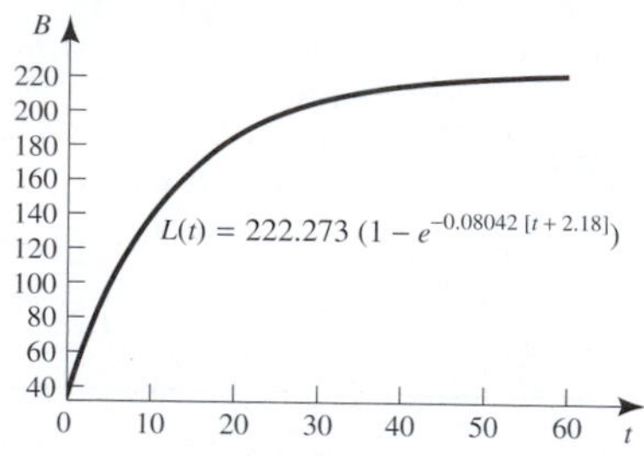

Chapter 1 Review Exercises

1. $5, 11, 2x^2 + 4x + 5, 2x^4 + 3$

2. $0, -\frac{3}{5}, \frac{x+1}{x^2 + 4x + 5}, \frac{x^3 - 1}{x^6 + 1}$

3. Yes

4. No

5. Yes

6. No

7. No

8. Yes

9. $\{x | x \neq 3\}$

10. $\{x | x > 5\}$

11. All x except for $x = \pm 1$

12. $|x| > 1$

13. $x \neq 0$

14. $x > 0$

15. $x \leq 2$

16. All x

17. $1, \frac{1}{3}, \frac{1}{a^2 - 2a + 2}, \frac{1}{x^2 + 2x + 2}$

18. $1, \frac{1}{\sqrt{2} + 1}, \frac{1}{a}, \frac{x+1}{x+2}$

19.

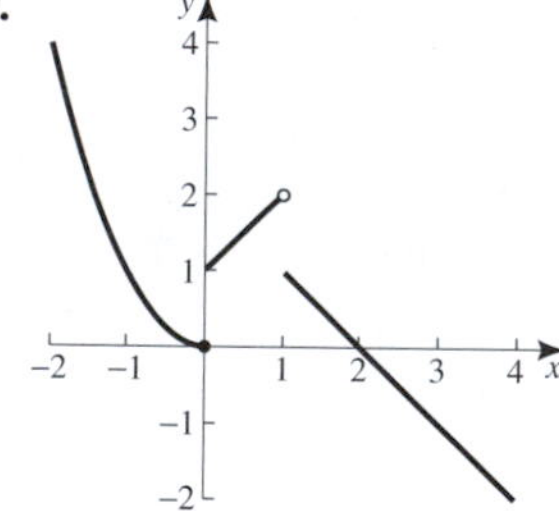

20. If your bill is \$50.00 you pay no tax. If your bill is \$50.01, you pay $.01(50.01) = 0.50$ dollars. A graph is shown.

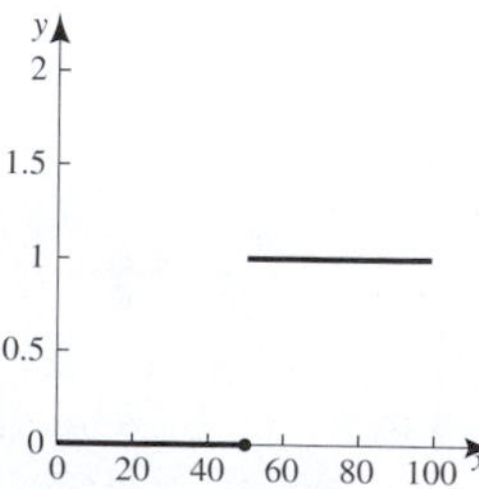

The break in the graph at 50.00 means this is a point of discontinuity.

So a one cent increase in your bill, when your bill is \$50.00, results in an increase of \$0.50 in tax.

21. The account is at \$100,000 until the end of the first quartrer, when it abruptly becomes \$102,000. It stays at this amount until the end of the second quarter when it abruptly becomes $1.02(102,000) = 104,040$ dollars. It stays at this amount until the end of the third quarter when it abruptly becomes $1.02(104,000) = 106,120.80$ dollars. It stays at this azmount until the end of the foursth quarter when it abruptly becomes $1.02(106,120.80) = 108,243.22$ dollars.

The graph is shown. Breaks occur in the graph at the quarter points, so these are points of discontinuity. Notice that if you withdrew your money 1 second before the end of the first quarter, you would obtain \$100,000. But if you withdrew your money 1 second into the second quarter, you would obtain \$102,000. Such a point of discontinuity would certainly affect when you withdraw your money.

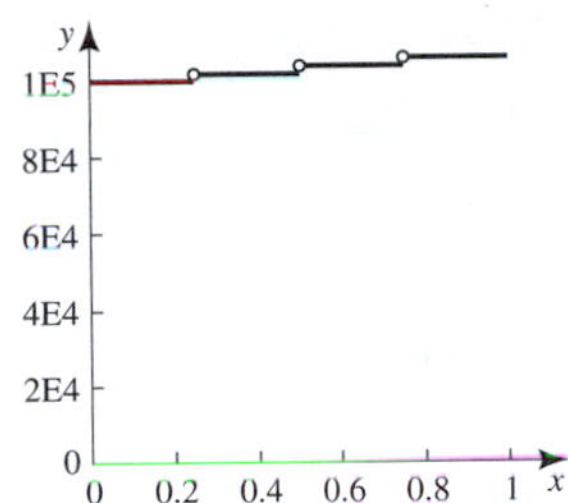

22. A graph is shown.

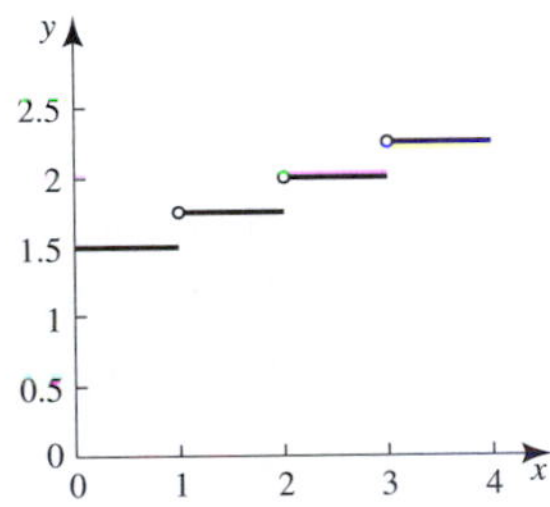

Discontinuities occur at the minute marks. For example, a 59.9 second call costs \$1.50, but a 60.1 second call costs \$1.75.

23. A graph is shown.

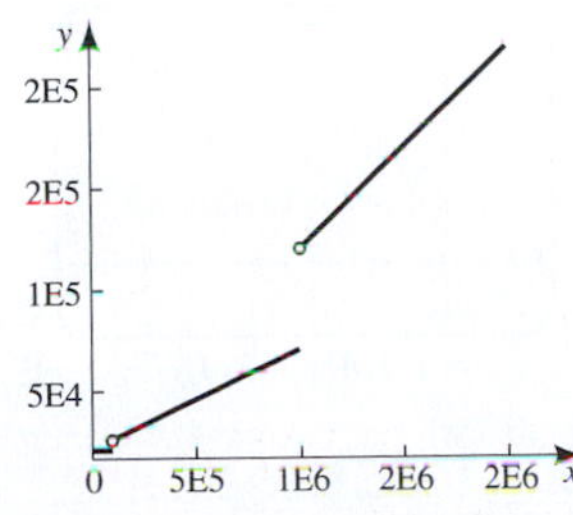

Discontinuities occur at 100,000 and 1,000,000. If the salesman sales \$100,000, he earns \$20,000. But if he sales \$100,000.01, he earns $20,000 + .05(100,000.01) = 25,000$ dollars. So a 1 cent increase in sales results in a \$5000 increase in salary.

If his sales are \$1,000,000, he earns $20,000 + 0.05(1,000,000 - 100,000) = 65,000$ dollars. But if sales are at \$1,000,000.01, he earns $20,000 + 0.10(1,000,000.01) = 120,000$ dollars. So a 1 cent increase in sales results in a $120,000 - 65,000 = 55,000$ dollar increase in earnings.

24. $x^3 + 2x + 3$. Domain: $(-\infty, \infty)$

25. $x^3 - 2x + 5$. Domain: $(-\infty, \infty)$

26. $(x^3 + 4)(2x - 1)$. Domain: $(-\infty, \infty)$

27. $\dfrac{x^3 + 4}{2x - 1}$. Domain: $(-\infty, 0.50) \cup (0.50, \infty)$

28. $9x^2 + 6x + 3,\ 3x^2 + 7$. Domains: $(-\infty, \infty), (-\infty, \infty)$

29. $\sqrt{x},\ \sqrt{x - 1} + 1$. Domains: $[0, \infty), [1, \infty)$

30.

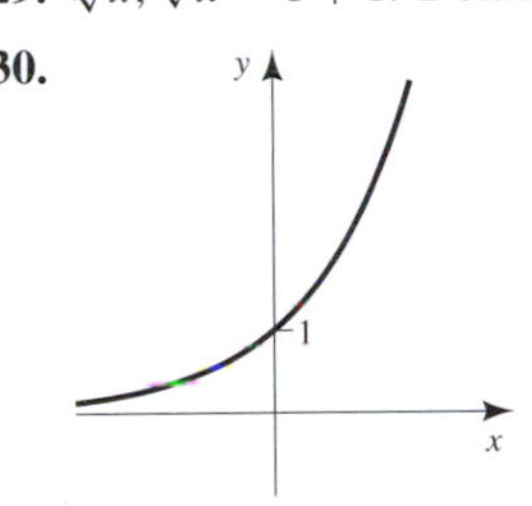

31.

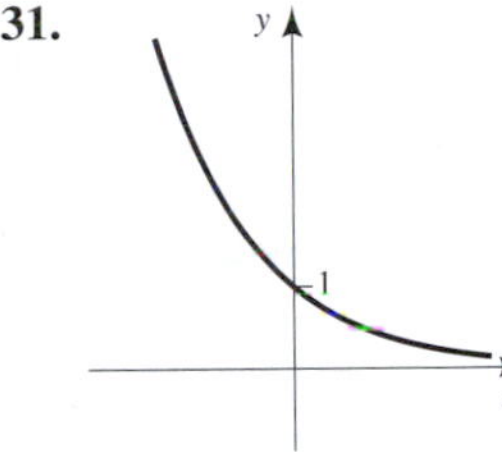

32.

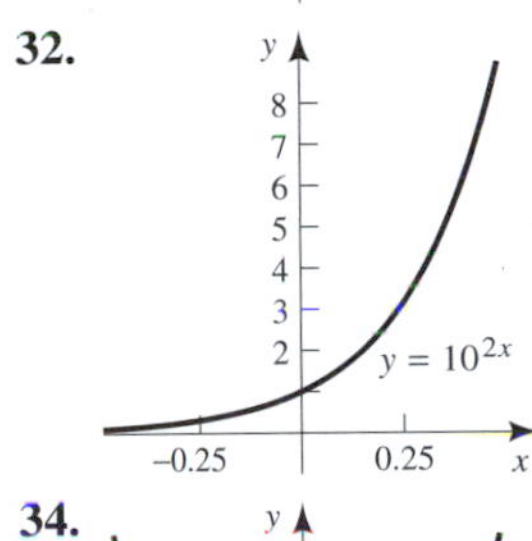

33.

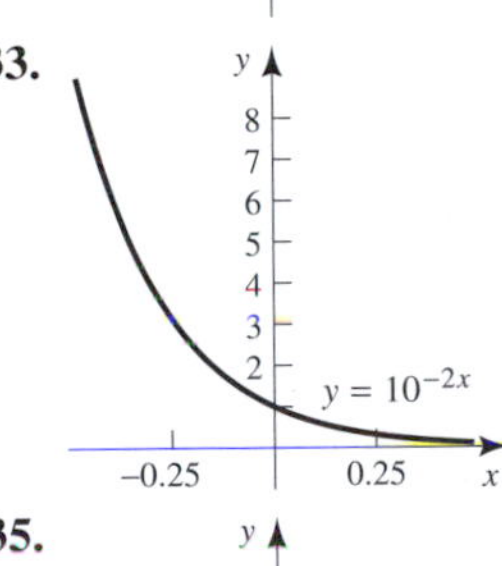

34.

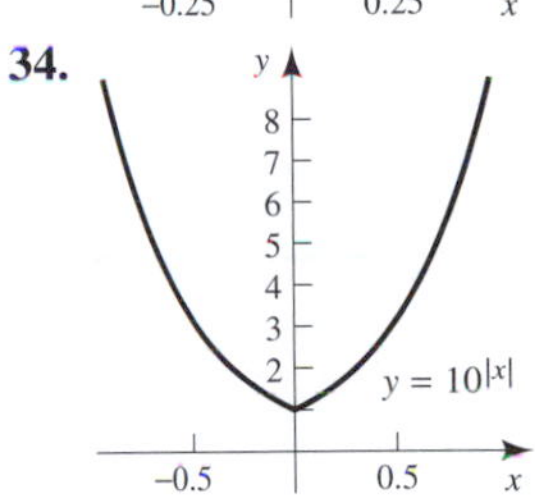

35.

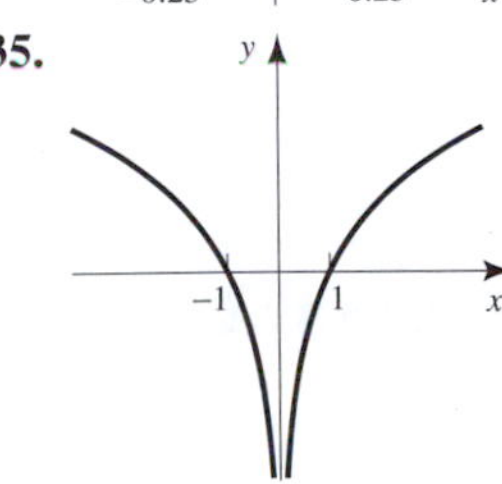

36. $\dfrac{3}{2}$

37. 3

38. e

39. 0.50

40. 4

41. 5

42. Approximately day 227

43. $x \approx 1.5953$

44. $C = 6x + 2000$

45. $R = 10x$

46. $P = 4x - 2000$, 500

47. $x = 150$

48. $x = 1000$, $b = 2000$

49. 150 miles

50. 42 minutes

51. $p - 1 = -\dfrac{1}{200,000}(x - 100,000)$

52. \$29.37

53. 12,000

54. $4.5x + 5.5y = 15$. $m = -\frac{9}{11}$. For every additional 11 cups of kidney beans, she must decrease the cups of soybeans by 9.

55. If x is tons, then $C(x) = 447,917 + 209.03x$, $R(x) = 266.67x$, $P(x) = 57.64x - 447,917$

56. **a.** Boston **b.** Houston

57.

Quadruped	l	h	$l : h^{2/3}$
Ermine	12 cm	4 cm	4.762
Dachshund	35 cm	12 cm	6.677
Indian tiger	90 cm	45 cm	7.114
Llama	122 cm	73 cm	6.985
Indian elephant	153 cm	135 cm	5.814

58. 10, 20, 30

59. For each unit distance away from the pocket margin, the proportion of dead roots decrease by 0.023 unit.

60. For 10,000 more units of shoot length, there will be 275 more buds.

61. $c = -6x + 12$

62. 6.84 years, 508.29 pounds

63. 103.41 pounds, 4294.39 pounds

64. 149.31, 313.32, 657.48

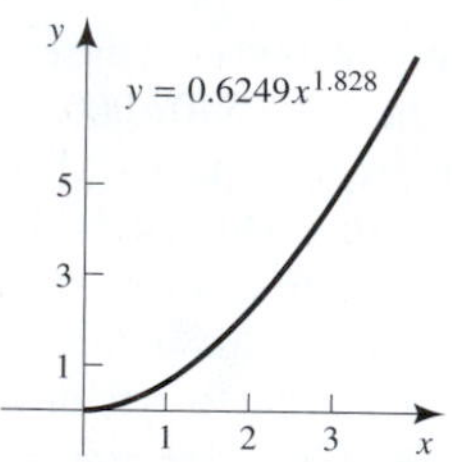

65. 11.18

66. $10x^2$

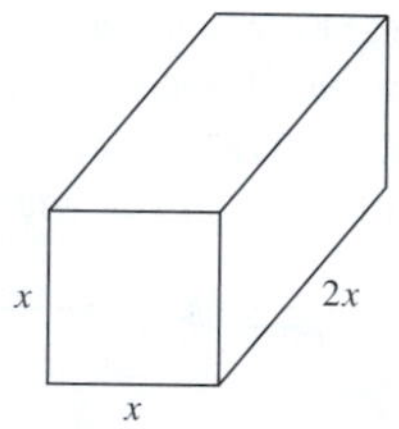

67.

68. $4\pi t^2$

69. 13.25 years

70. $k = 0.02$

71. In the year 2022

72. $\ln 10/k$

73. Day 253

74. 0.47

75. 231 years

Chapter 2

Section 2.1

1. $y = 0.50x + 0.50$, 0.5000

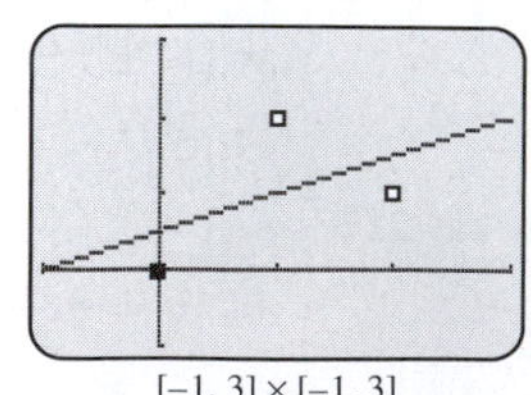

$[-1, 3] \times [-1, 3]$

3. $= 1.1x + 0.1$, 0.9467

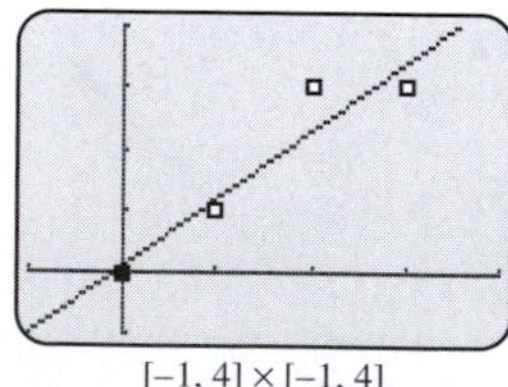

$[-1, 4] \times [-1, 4]$

5. $y = -0.9x + 4.5$, -0.9234

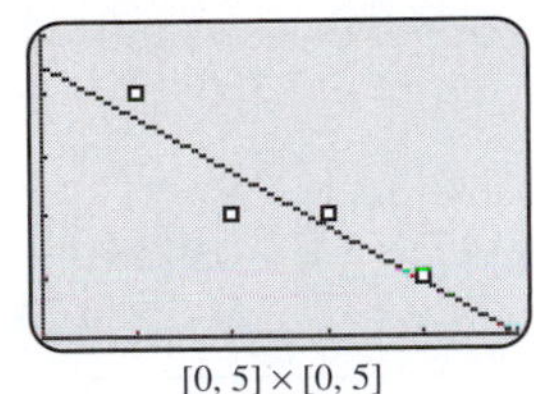

[0, 5] × [0, 5]

7. $y = -0.7x + 3.4$, -0.9037

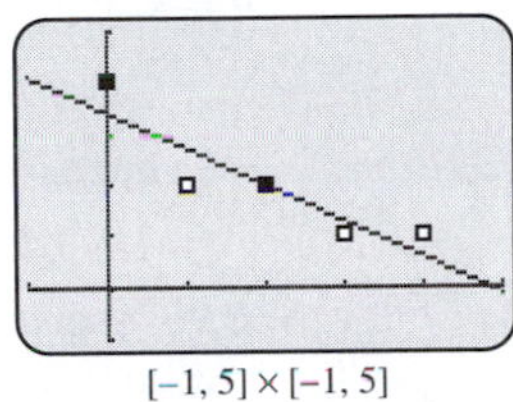

[−1, 5] × [−1, 5]

9. If x is thousands of bales and y is thousands of dollars, then $y = 37.32x + 82.448$, $r = 0.9994$

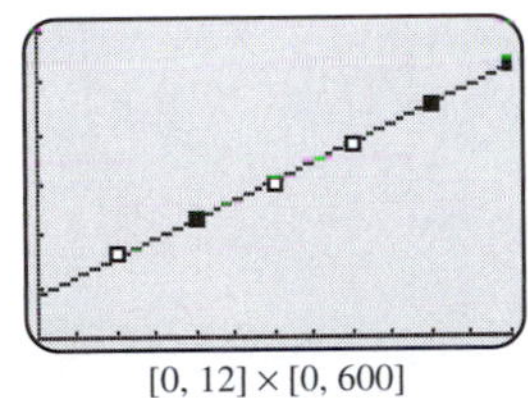

[0, 12] × [0, 600]

(b) 3180; (c) \$4659 loss; \$21,271 profit

11. **a.** $y = -0.0055x + 1.8244$, $r = -0.9262$.
b. Since the slope is negative, the larger the machine size, the fewer employee-hours are needed per ton.
c. 1.27
d. 149

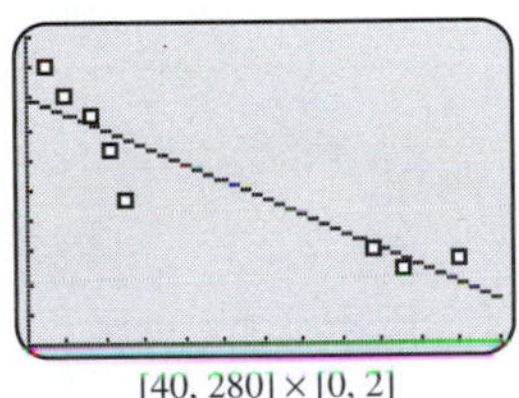

[40, 280] × [0, 2]

13. $y = -0.0005892x + 0.07298$, $r = -0.9653$. The line slopes down. Thus, an increase in price leads to a decrease in demand.

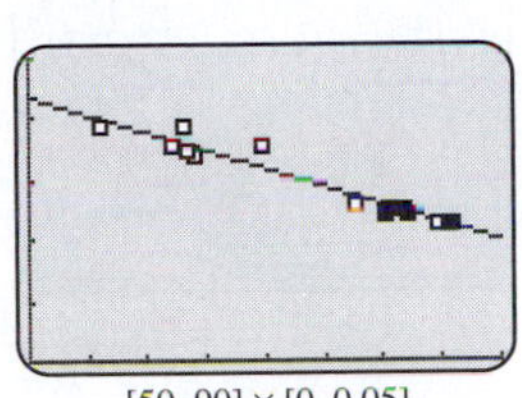

[50, 90] × [0, 0.05]

15. **a.** $y = 2.0768x + 1.4086$, $r = 0.9961$.
b. \$209,000
c. 59,511 dozen

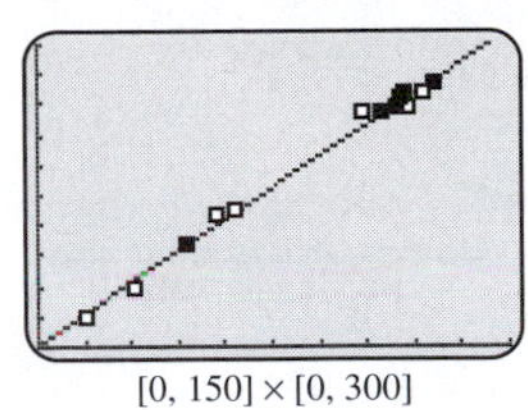

[0, 150] × [0, 300]

17. **a.** $y = 0.8605x + 0.5638$, $r = 0.9703$.
b. 6.6%
c. 7.5%

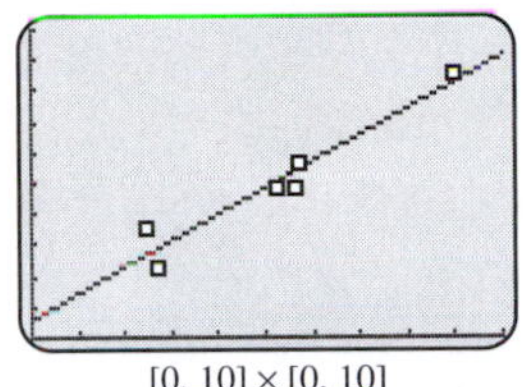

[0, 10] × [0, 10]

19. **a.** $y = 0.3601x - 2.5162$, $r = 0.9070$.
b. 8.3%
c. 26.4%

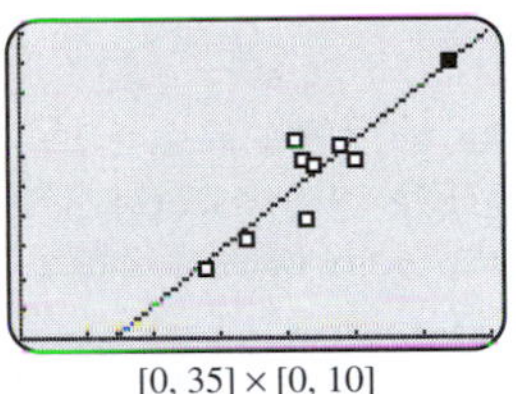

[0, 35] × [0, 10]

21. $y = -87.16x + 1343$, $r = -0.8216$.

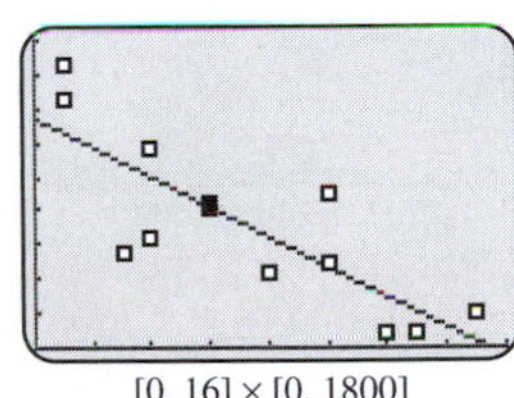

[0, 16] × [0, 1800]

23. **a.** $y = 3.4505x + 14.4455$, $r = 0.9706$, $z = 2.0446x + 1.0990$, $r = 0.9413$;
b. For each additional aphid per plant the percentage of times the virus is transmitted to the fruit increases by about 3.5 for the brown aphid and about 2 for the melon aphid;
c. The melon aphid is more destructive, since the melon aphid transmits the virus more often than the brown aphid.

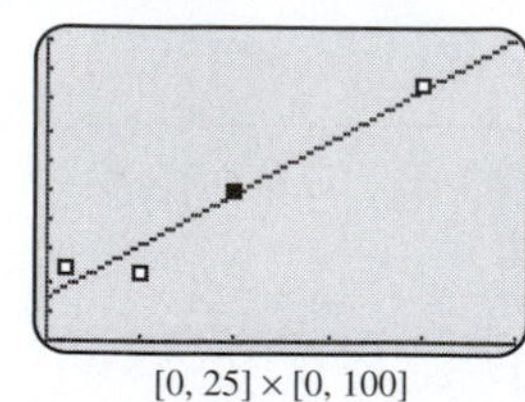
[0, 25] × [0, 100]

25. **a.** $y = 0.1264x - 19.0979$;
b. $r = 0.9141$;
c. for each additional trichome per 6.25 mm^2, there is an increase 0.1264% of damaged pods.

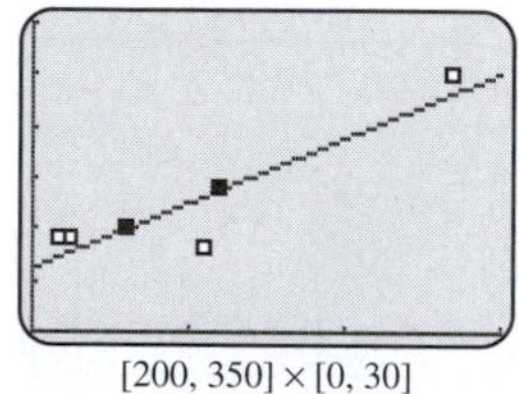
[200, 350] × [0, 30]

Section 2.2

1. **a.** $y = 0.1848x + 0.6758$, $r^2 = 0.9800$.
b. $y = 0.0006357x^2 + 0.0807x + 3.5705$, $r^2 = 0.9992$. Quadratic regression is better because of its higher correlation coefficient.

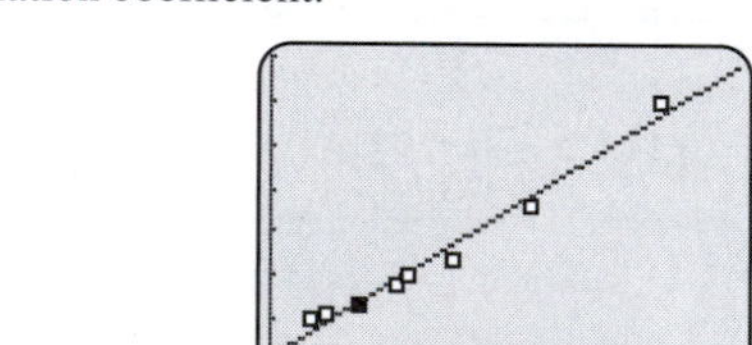
[0, 180] × [0, 35]

3. **a.** $y = 0.0011x^2 + 0.0943x + 2.1498$, $r^2 = 0.9978$.
b. \$5932;
c. 16,544;
d. approximately 45.1

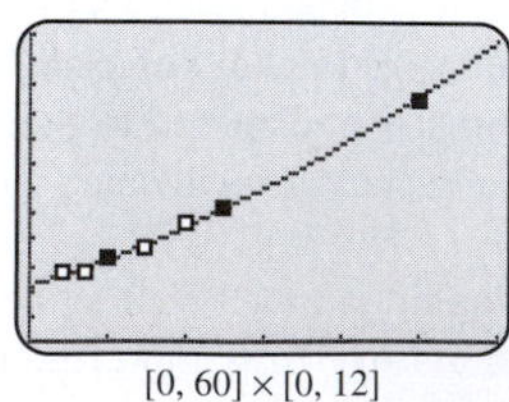
[0, 60] × [0, 12]

5. **a.** $y = 0.000083x^2 - 0.02396x + 2.80925$; about 144

7. $y = 0.0031x^2 - 0.04x + 0.494$. About 6 years.

9. **a.** Linear gives $y = 28.545x + 155.237$, $r^2 = 0.9995$, quadratic gives $y = 0.1150x^2 + 26.0465x + 165.0653$, $r^2 = 0.99999997$. Quadratic is better.
b. 4499 bales. **c.** 20,000 bales

11. $y = -0.0044x^2 + 0.2851x - 3.0462$, about 32.45

13. $-0.00349x^2 + 0.39551x + 21.8891$. About 57.

Section 2.3

1. **a.** $y = 80.55x^{-0.3145}$, $r = -0.9758$.
b. \$20.72; **c.** about 23

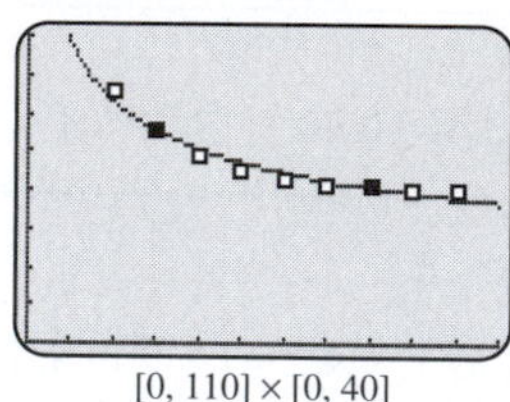
[0, 110] × [0, 40]

3. **a.** $y = 136.04x^{0.5263}$, $r^2 = 0.9678$, quadratic is best;
b. 5037 bales; **c.** 20,000 bales

5. $y = -0.00000217x^3 + 0.000425x^2 - 0.021x + 1.034$, $r^2 = 0.5356$.
Minimum at $x = 33.218567$

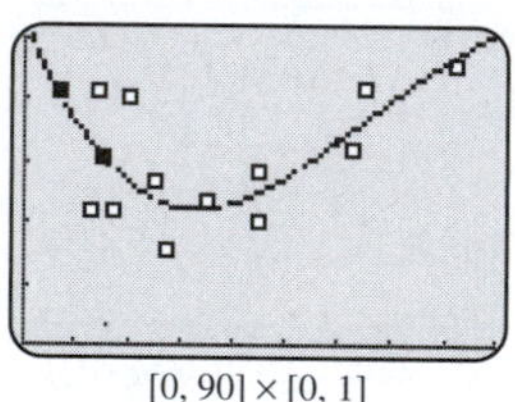
[0, 90] × [0, 1]

7. Cubic regression is $y = -0.00031x^3 + 0.0523x^2 - 2.182x + 51.17$, $r^2 = 0.9838$.
Linear and quadratic regressions are $y = 0.548x + 7.958$, $r^2 = 0.9548$, $y = -0.00354x^2 + 0.588x + 6.968$, $r^2 = 0.9550$. Cubic is better.

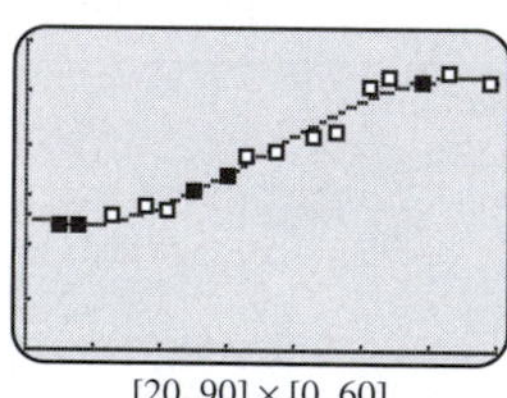
[20, 90] × [0, 60]

9. $y = 0.0056x^4 - 0.581x^3 + 22.61x^2 - 388x + 2484$, $r^2 = 0.9393$.

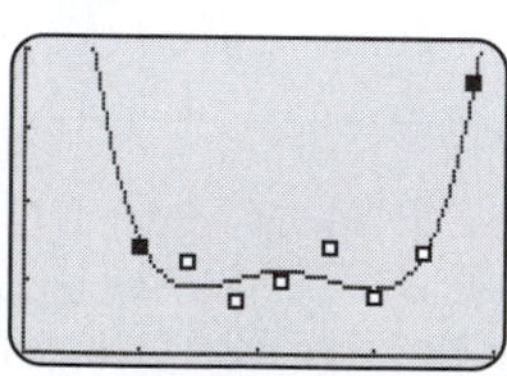
[15, 35] × [0, 20]

11. $y = 36.23x^{0.167}$, $r = 0.9738$

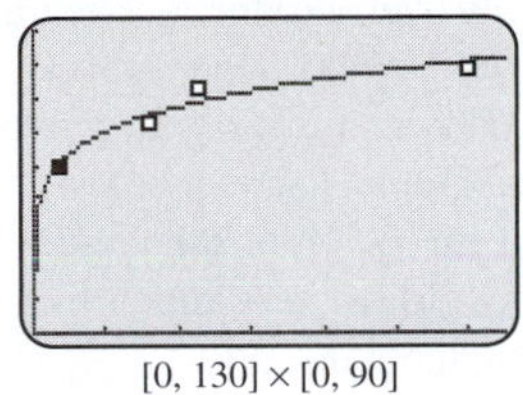

[0, 130] × [0, 90]

Section 2.4

1. $y = 8.027(1.1424)^x$, $r = 0.9932$.
Payments in 1997 are about $292 billion.

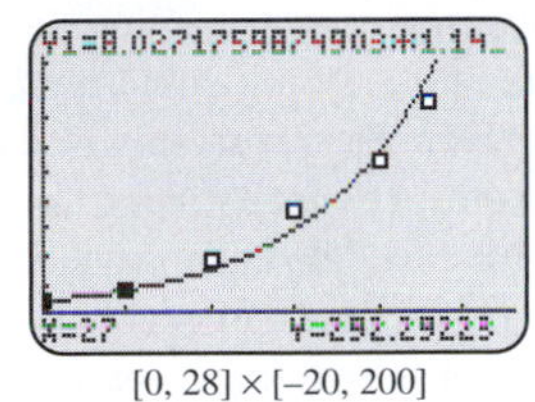

[0, 28] × [−20, 200]

3. $y = 0.0646(2.5489)^x$, $r = 0.9809$
The number in 1997 is about 115 million.

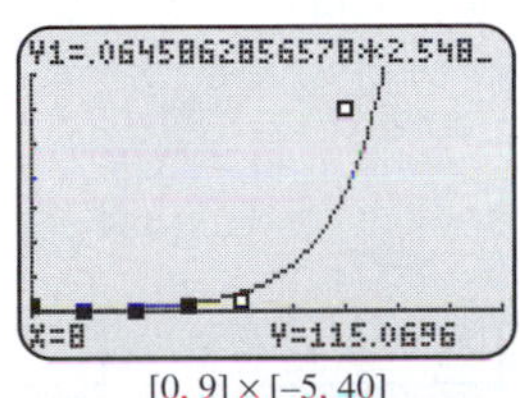

[0, 9] × [−5, 40]

5. a. $y = 0.2989(1.5861)^x$, $r = 0.9889$.
The number in 1997 is 120 million.
b. 1995

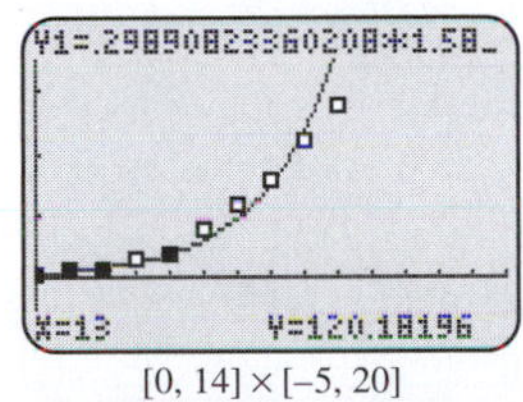

[0, 14] × [−5, 20]

7. a. $y = 3.016(1.4362)^x$, $r = 0.9985$
The number in 1997 is about 115 million.
b. According to the model, the number reached 20 million about one and a quarter years after 1993.

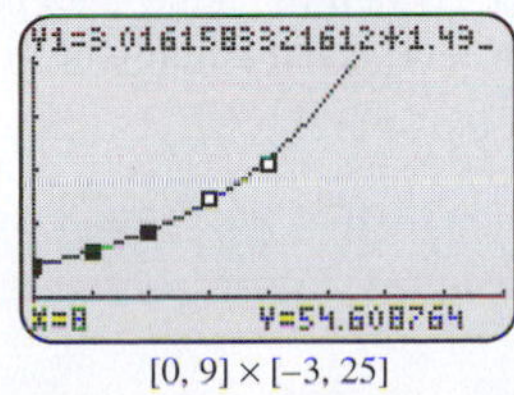

[0, 9] × [−3, 25]

9. a. $y = 925(0.9546)^x$, $r = -0.9975$.
The model predicts about 230 ordinations in 1997.
b. According to the model, the number reached 150 in 2006.

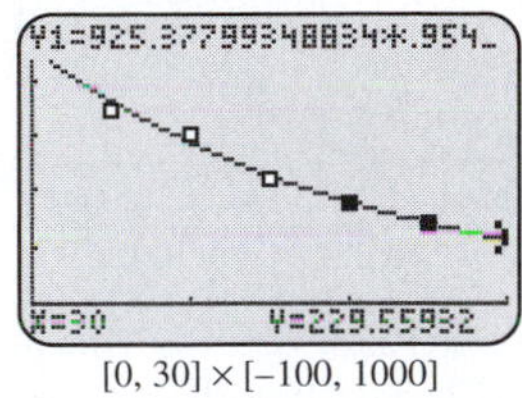

[0, 30] × [−100, 1000]

11. $y = 123 - 4.599 \ln x$, $r = -0.9986$.
The model predicts about 109 days with 20%.

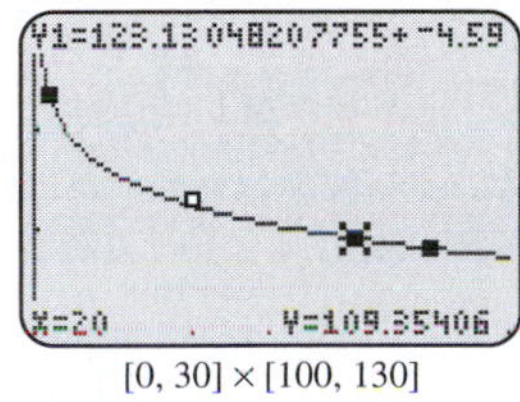

[0, 30] × [100, 130]

13. $y = 1457 - 212 \ln x$, $r = -0.7525$
According to the model, there would be a 142% increase with a 1988 population of 500.

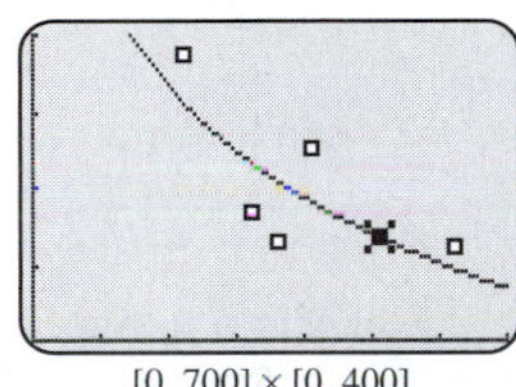

[0, 700] × [0, 400]

Section 2.5

1. a. $y = 2.198(1.0217)^x$, $r = 0.9961$. This model estimates the population in 1990 to be 161.3 million.

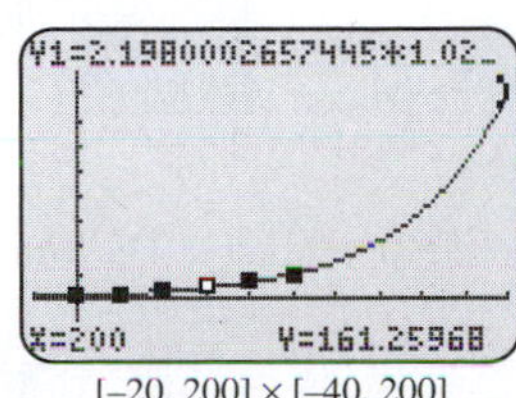

[−20, 200] × [−40, 200]

b. $y = 41.2/(1 + 18.33e^{-0.259})$. This model estimates the population in 1990 to be 37.3 million, a much better approximation.

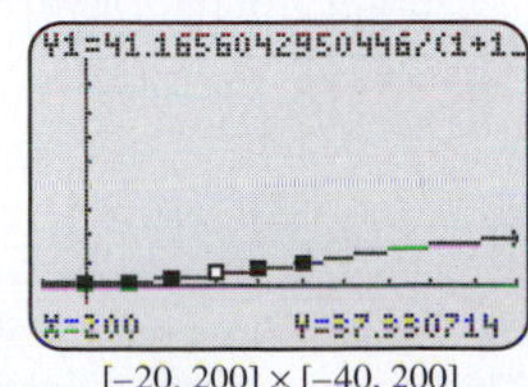

[−20, 200] × [−40, 200]

3. a. $y = 0.248(1.766)^x$, $r = 0.9938$. This model estimates the sales in 1992 to be 713 million.

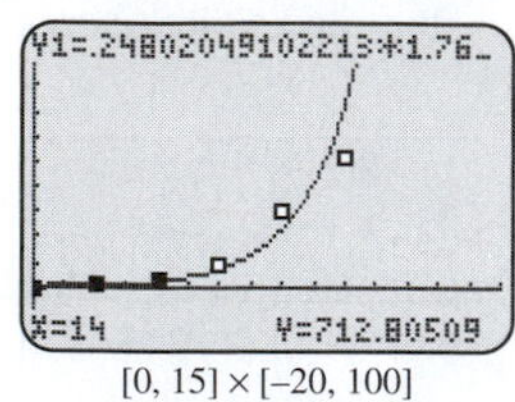

[0, 15] × [−20, 100]

b. $y = 59.9/(1 + 1038e^{-0.8752})$. This model estimates the sales in 1992 to be 60 million, a much better approximation.

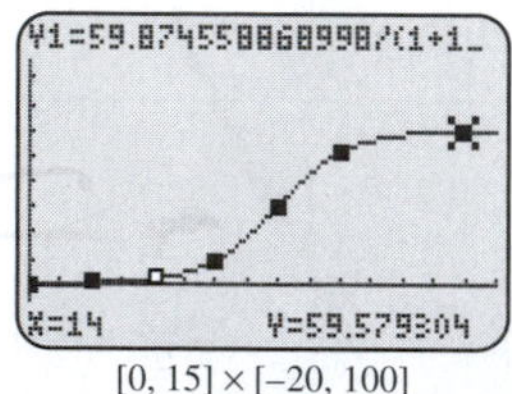

[0, 15] × [−20, 100]

5. a. $y = 33.11(1.0322)^x$, $r = 0.9939$. This model estimates the sales in 1992 to be 125.6 quadrillion Btu.

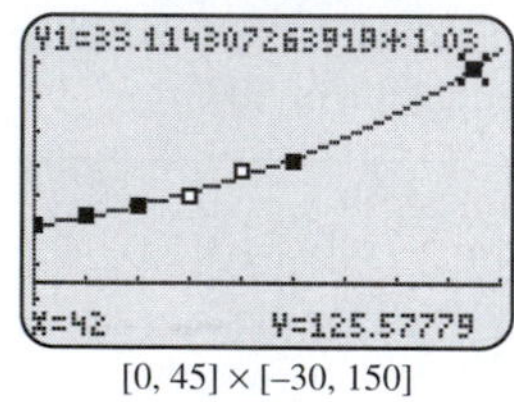

[0, 45] × [−30, 150]

b. $y = 155/(1 + 3.800e^{-0.0477})$. This model estimates the sales in 1992 to be 102.5 quadrillion Btu, a much better approximation.

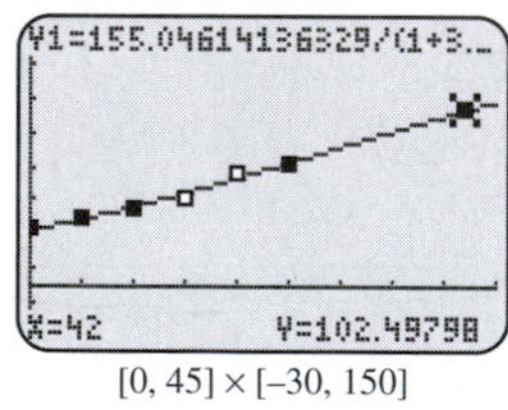

[0, 45] × [−30, 150]

7. $y = 422.5/(1 + 55.03e^{-0.0218})$. The limiting value is 422.5 million.

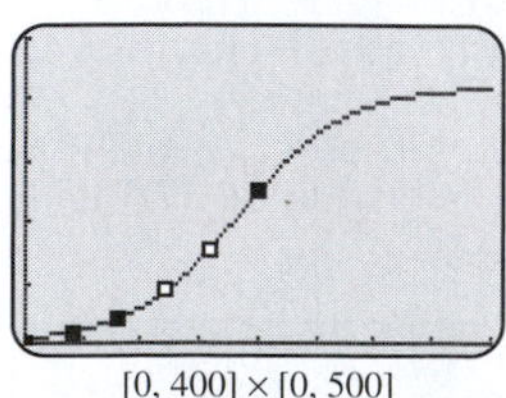
[0, 400] × [0, 500]

Section 2.6

1. Exponential regression gives $y = 85.59(0.9859)^x$ with $r^2 = 0.5209$. This is not an outstanding fit. Power regression gives $y = 156.75x^{-0.3225}$ with $r^2 = 0.7570$. This is an improvement over exponential regression. Quadratic regression gives $y = 0.0645x^2 - 4.325x + 116.05$ with $r^2 = 0.8787$. This is a distinct improvement over power regression. The graph is shown below. Notice the good fit and notice that the graph of the quadratic turns up at the right. The graph looks very much like we would expect based on our business insights.

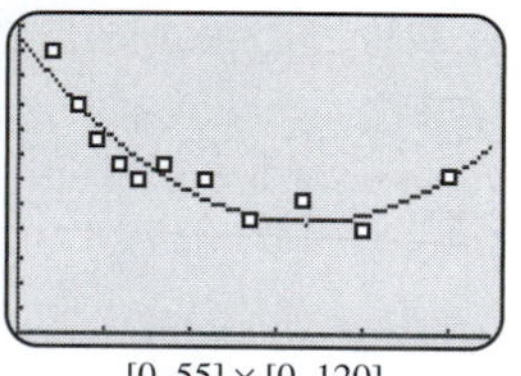
[0, 55] × [0, 120]

Naturally, cubic regression will lead to a better fit than quadratic regression. Doing cubic regression on a TI-83 gives $r^2 = 0.8946$. However, this is probably not a big enough improvement to justify using the more complicated cubic model. We conclude that the quadratic model is best. The quadratic model gives an excellent fit, is simple, and gives a graph that is reasonable based on basic business considerations.

3. Cubic is best with $r^2 = 0.7751$. (Quartic gives only a very slight improvement with $r^2 = 0.7856$.)

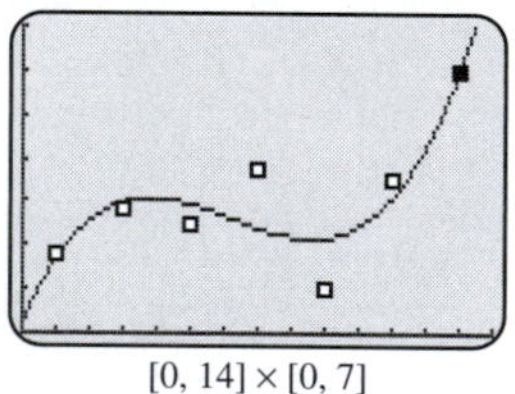
[0, 14] × [0, 7]

5. Exponential ($r^2 = 0.9973$) is best, since quadratic ($r^2 = 0.9982$) turns up just to the left of the our prescribed interval.

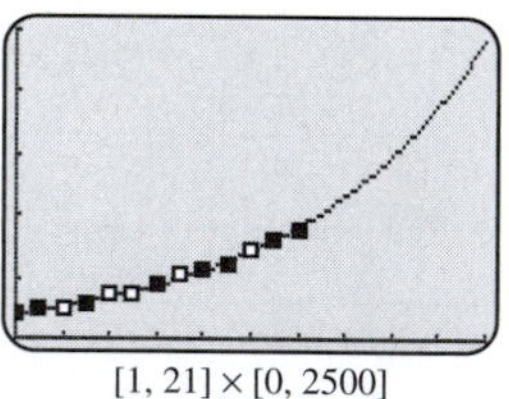
[1, 21] × [0, 2500]

7. Graphs clearly show the logistic curve fits well. The cubic ($r^2 = 0.9824$) is rejected, since the graph turns down at the end of the interval. The quartic ($r^2 = 0.9981$) is also rejected, since the graph dips on the latter part of the interval, and also because the graph initially decreases.

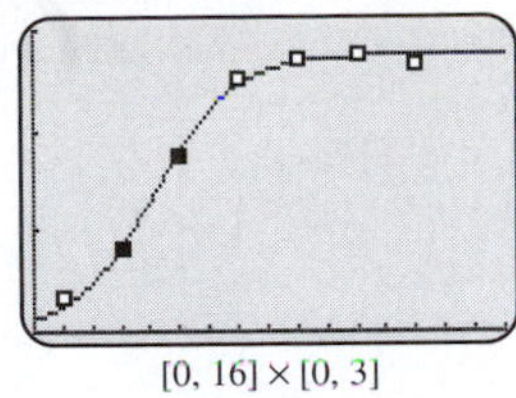

[0, 16] × [0, 3]

9. Linear ($r^2 = 0.5684$). Quadratic ($r^2 = 0.6048$) probably does not give a sufficient improvement for a more complicated model.

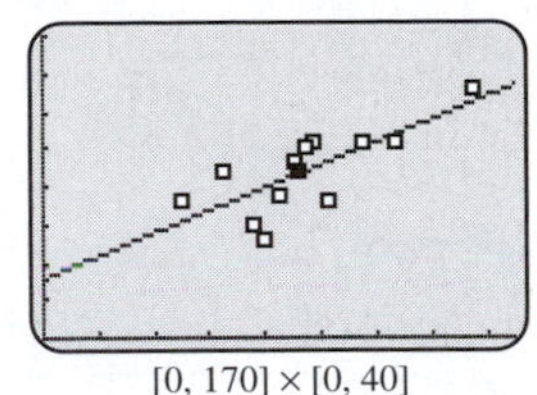

[0, 170] × [0, 40]

Chapter 3

Section 3.1

1. 2, 2, 2

3. 2, 2, 2

5. 1, 2, does not exist

7.

x	0.9	0.99	0.999	→	1	←	1.001	1.01	1.1
$f(x)$	4.7	4.97	4.997	→	5	←	5.003	5.03	5.3

$$\lim_{x\to 1} f(x) = 5$$

9.

x	1.9	1.99	1.999	→	2	←	2.001	2.01	2.1
$f(x)$	2.859	3.881	3.988	→	4	←	4.012	4.121	5.261

$$\lim_{x\to 2} f(x) = 4$$

11.

x	1.9	1.99	1.999	→	2	←	2.001	2.01	2.1
$f(x)$	−10	−100	−1000	→	undefined	←	1000	100	10

$\lim_{x\to 2} f(x)$ does not exist

13.

x	3.9	3.99	3.999	→	4	←	4.001	4.01	4.1
$f(x)$	3.9748	3.9975	3.9998	→	4	←	4.0002	4.0025	4.0248

$$\lim_{x\to 4} f(x) = 4$$

15. 1, 0, does not exist

17. 1, 1, 1

19. 0, does not exist, does not exist

21. −1, 1, does not exist

23. None of the limits exist.

25. 4, does not exist, does not exist

27. 13

29. 13

31. −3

33. $-\frac{3}{4}$

35. 1

37. 4

39. 2

41. −2

43. Does not exist

45. 4

47. **a.** 0.37, 0.60, does not exist; **b.** 0.60, 0.60, 0.60

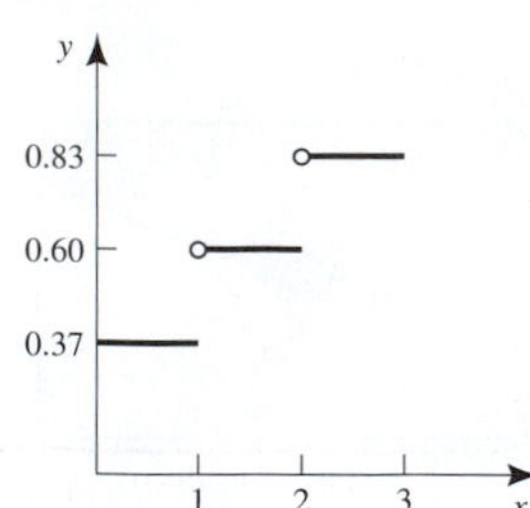

49. 1, 10, 100, 1000. $C(x)$ becomes large without bound

51. a.

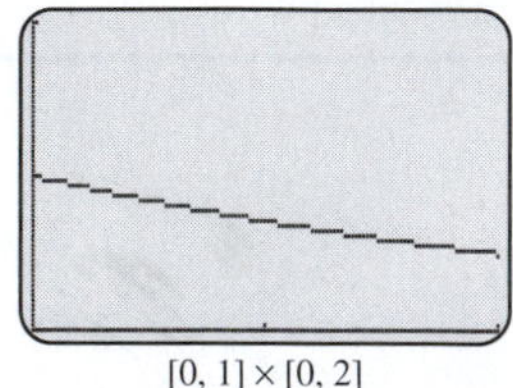
[0, 1] × [0, 2]

x	−0.1	−0.01	−0.001	→	0	←	0.001	0.01	0.1
$f(x)$	0.9330	0.9931	0.9993	→	1	←	0.9993	0.9931	0.9330

$$\lim_{x\to 0} f(x) = 1$$

b.

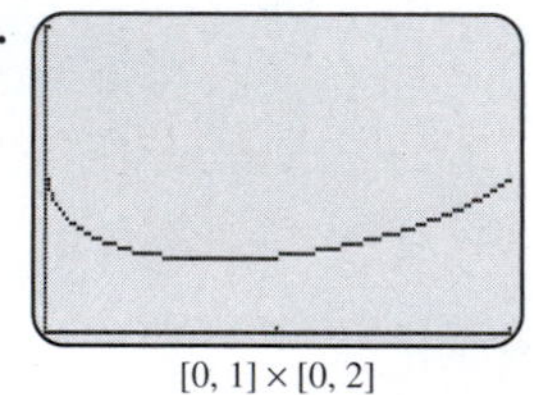
[0, 1] × [0, 2]

x	0	←	0.0001	0.001	0.01	0.1
$f(x)$	1	←	0.9982	0.9863	0.9120	0.6310

$$\lim_{x\to 0^+} f(x) = 1$$

c.

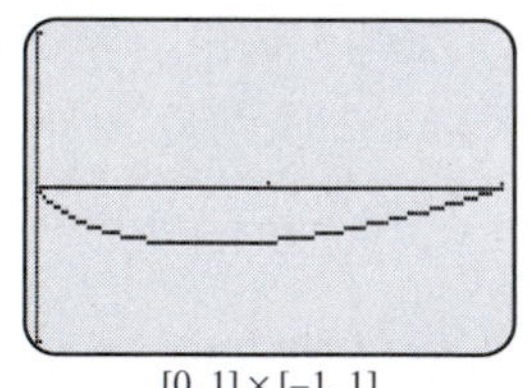
[0, 1] × [−1, 1]

x	0	←	0.0001	0.001	0.01	0.1
$f(x)$	0	←	−0.0009	−0.0069	−0.0461	−0.2303

$$\lim_{x\to 0^+} f(x) = 0$$

d.

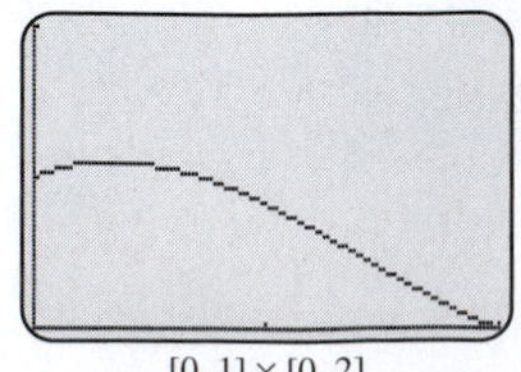
[0, 1] × [0, 2]

x	0	←	0.0001	0.001	0.01	0.1
$f(x)$	1	←	1.0002	1.0019	1.0154	1.0870

$$\lim_{x\to 0^+} f(x) = 1$$

e. 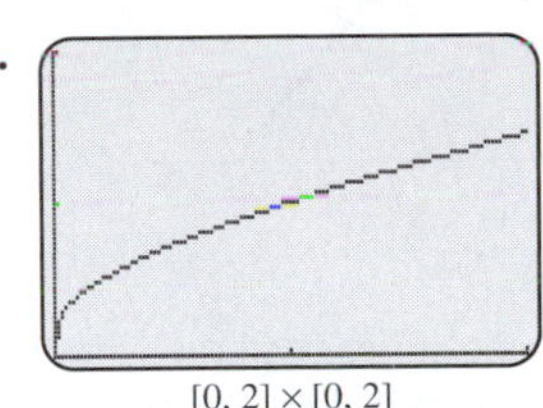

$[0, 2] \times [0, 2]$

x	0.9	0.99	.999	→	1	←	1.001	1.01	1.1
$f(x)$	0.9491	0.9950	0.9995	→	1	←	1.0005	1.0050	1.0492

$$\lim_{x\to 1} f(x) = 1$$

53. Since $\lim_{x\to 0^-} f(x) = -1 \neq 1 = \lim_{x\to 0^+} f(x)$, it is impossible to define $f(0)$ so that $\lim_{x\to 0} f(x)$ exists.

55.

$$\frac{\sqrt{x}-1}{x-1} = \frac{\sqrt{x}-1}{x-1}\cdot\frac{\sqrt{x}+1}{\sqrt{x}+1} = \frac{(x-1)}{(x-1)(\sqrt{x}+1)} = \frac{1}{\sqrt{x}+1},$$

As $x \to 1$, this goes to $1/2$.

57. No. For example, $f(x) = \frac{1}{x}$, $g(x) = 1 - \frac{1}{x}$. Both $\lim_{x\to 0} f(x)$ and $\lim_{x\to 0} g(x)$ do not exist, but $\lim_{x\to 0}(f(x)+g(x)) = \lim_{x\to 0}\left(\frac{1}{x} + 1 - \frac{1}{x}\right) = 1$.

59. For very small farms of, say, less than an acre, the owner has the additional time to give additional nurture to the plants and to protect them from disease, insects, and animals.

61. Everywhere

63. Everywhere except $x = 0$.

65. Everywhere except $x = 1$.

67. Everywhere except $x = 0$.

69. Everywhere

71. Everywhere except $x = \pm 1$

73. Continuous everywhere.

75. Everywhere

77. Continuous on $[-8, \infty)$, where it is defined

79. $k = -1$

81. Every polynomial is continuous at every point. The indicated function in the graph is not continuous at the point where the graph "blows up" and therefore cannot be the graph of any polynomial.

Section 3.2

1. $-2, 2$ **3.** $-0.25, -0.05$ **5.** $-6, -2$

7. $-4, -9$ **9.** $2, -2$ **11.** 96

13. 96 **15.** 0 **17.** -64

19. $y - 48 = 32(t - 1)$

21. 2 **23.** 2 **25.** -8

27. 0 **29.** -16 **31.** 3

33. BC, AB and DE, CD

35. Positive at A and D, negative at B, zero at C

37. 0.5 **39.** $y - 6 = 7(x - 2)$

41. Peoria: -627. From 1980 to 1997, Peoria lost on average 627 people each year.
Springfield: 315. From 1980 to 1997, Springfield gained on average 315 people each year.

43. -1.2 percent inflation per percent unemployment. For each 1% increase in the percentage of unemployment, the inflation rate as measured by percent will decrease by 1.2.

45. 0.79 mm per day. On average they grow 0.79 mm each day.

47. Virginia's warbler: about -2.6 birds per centimeter of precipitation. For each centimeter increase in precipitation, there were on average about 2.6 fewer of these warblers.
Orange-crowned warbler: about 1.8 birds per centimeter of precipitation. For each centimeter increase in precipitation, there were on average about 1.8 more of these warblers.

49. About 0.25 kg of birth mass per yearly age. Each year the females birthed calves that were on average 0.25 kg heavier.

51. About \$0.40 per week worked. About \$6.00 per week worked. On average, hourly wages are rising much faster on the larger interval of weeks worked.

53. **a.** An increase of about 200 private-equity investments per quarter.

b. A drop of about 100 private-equity investments per quarter. Privite-equity investments increased during the first period and then decreased during the second period.

55. Increases about 0.4% per mile. Decreases about 1.3% per mile. The price of an average home first increases as you move away from the city and then sharply decreases.

57. **a.** Increases about \$300 per year.

b. Decreases about \$250 per year. Income increases when younger and then decreases as one gets older.

59. **a.** 0.12 degree body temperature per degree ambient temperature.

b. 0.32 degree body temperature per degree ambient temperature. The bird's body temperature increases at a more rapid rate for higher ambient temperatures.

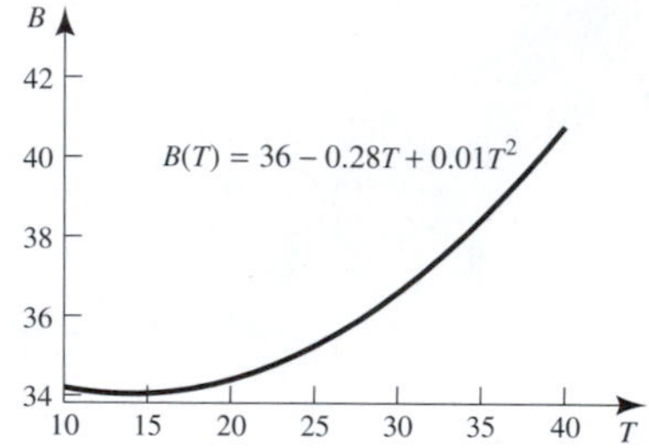

61. **a.** 0.2 bushel of soybeans per pound. At 100 pounds of fertilizer per acre the yield is increasing by about 0.2 bushel per pound of fertilizer.

b. −0.4 bushel of soybeans per pound. At 200 pounds of fertilizer per acre the yield is decreasing by about 0.4 bushel per pound of fertilizer. Initially, increasing the amount of fertilizer increases the yield, but at some point it begins to decrease the yield. Clearly, some fertilizer is helpful, but too much is actually counterproductive.

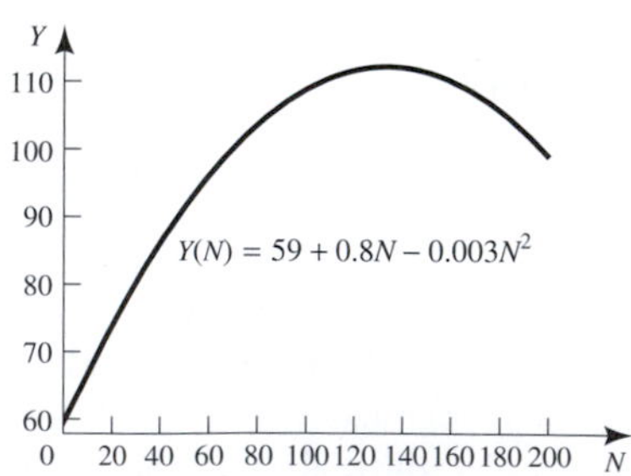

63. −0.5 day per degree. −7.1 days per degree. There is a more rapid increase in the early start in the growing season as the mean July temperature increases.

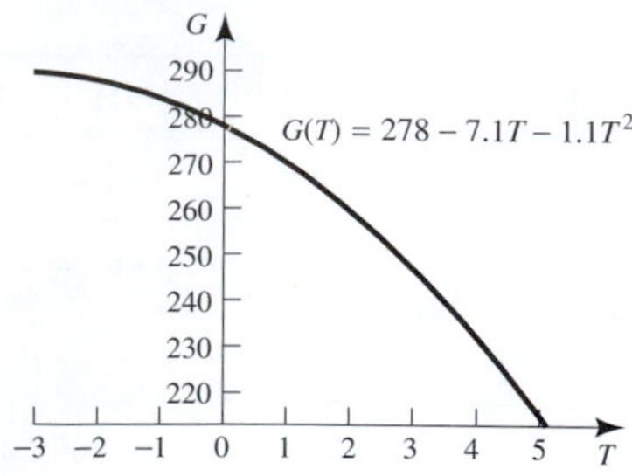

65. **a.** 1.7% per day.

b. −0.1% per day. On the given lactation interval, the percentage of lipid in the milk first increased and then decreased.

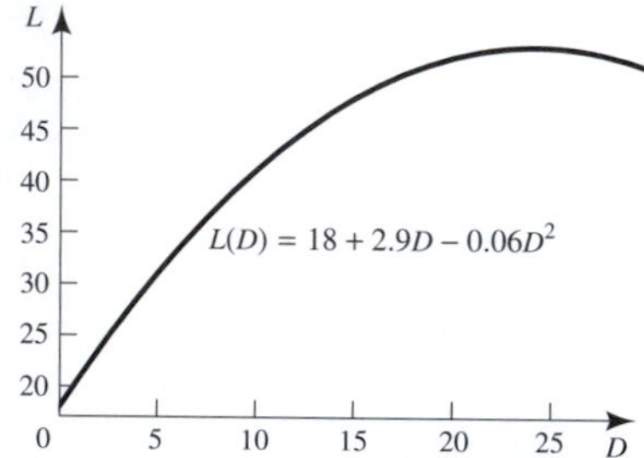

67. When 150 tons of steel is being produced, the cost is increasing at \$300 per ton. If cost were to increase at this constant rate, then the cost of the next ton would be \$300.

69. When the tax is set at 3% of taxable income, tax revenue is increasing at \$3 billion per percent increase in tax rate. If the tax revenue were to increase at this constant rate than increasing the tax rate by an additional 1% would increase tax revenue by \$3 billion.

71.

[0, 75.2] × [0, 120]

x	12	20	32	40	52
Slope	−2.57	−1.58	−0.09	0.90	2.39

For example, when $x = 12$, the slope of the tangent line and the rate of change are both −2.57. When $x = 12$, the average costs were decreasing at the rate of \$2.57 per one thousand pairs of shoes increase in output. Average costs initially decrease but at some point begin to increase.

73. 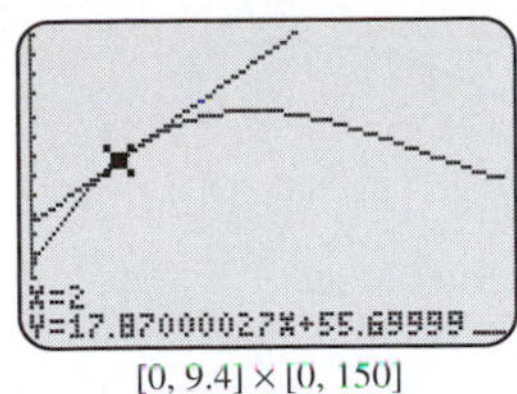

[0, 9.4] × [0, 150]

x	2	3	5	6
Slope	17.87	9.30	−2.98	−6.69

For example, when $x = 2$, the slope of the tangent line and the rate of change are both 17.87. When $x = 2$, the units of smoke in cities were increasing at the rate of 17.87 units per one thousand dollar increase in GDP per capita. The units of smoke in cities initially increase with increasing GDP per capita income but at some point begin to decrease.

77. a. $y = 0.057x^2 - 4.07x + 114.55$, $r^2 = 0.8702$

b. 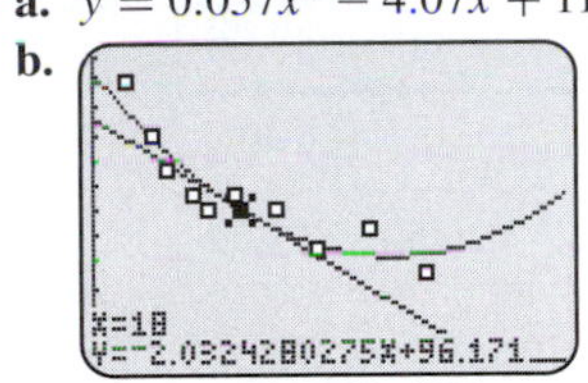

[0, 56.4] × [0, 120]

x	18	27	36	42	48
Slope	−2.03	−1.01	0.01	0.69	1.37

For example, when $x = 18$, the slope of the tangent line and the rate of change are both −2.03. When $x = 18$, the average costs were decreasing at the rate of \$2.03 per one thousand pairs of shoes increase in output. Average costs initially decrease but at some point begin to increase.

79. a. $y = 0.0174x^3 - 0.324x^2 + 1.72x + 0.274$, $r^2 = 0.7751$

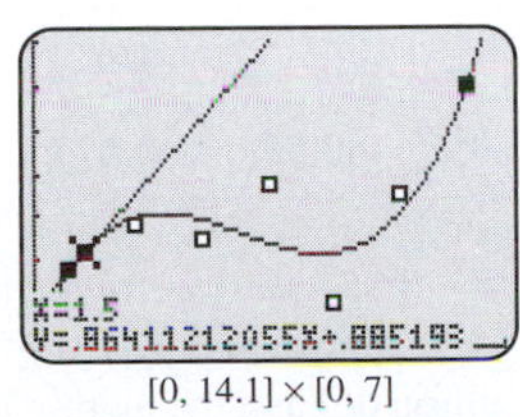

[0, 14.1] × [0, 7]

x	1.5	2.7	3.9	6	9.6	12
Slope	0.86	0.35	−0.02	−0.29	0.30	1.45

For example, when $x = 1.5$, the slope of the tangent line and the rate of change are both 0.86. When $x = 1.5$, the units of coliform in waters were increasing at the rate of 0.86 units per one thousand dollar increase in GDP per capita. The units of coliform in waters initially increase with increasing GDP per capita income, then decrease, and then increase again.

Section 3.3

1. 5 **3.** $2x$ **5.** $6x + 3$

7. $-6x^2 + 1$ **9.** $\dfrac{-1}{(x+2)^2}$ **11.** $\dfrac{-2}{(2x-1)^2}$

13. (1) $y - 2 = 5(x - 1)$; (3) $y - 5 = 2(x - 1)$; (5) $y - 5 = 9(x - 1)$

15. $y = x - 1$

17. Everywhere except $x = 0$

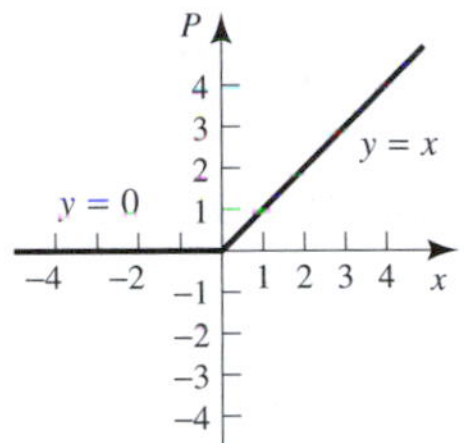

19. Everywhere except $x = 1$

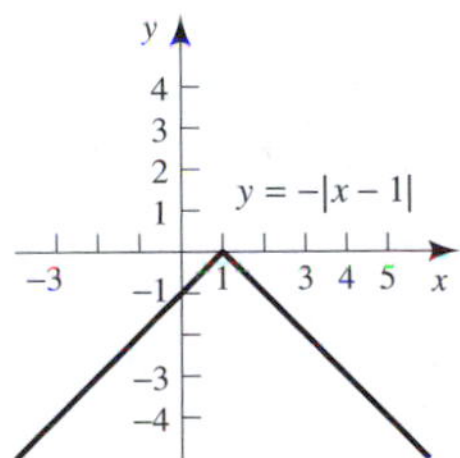

21. Everywhere except $x = 1$

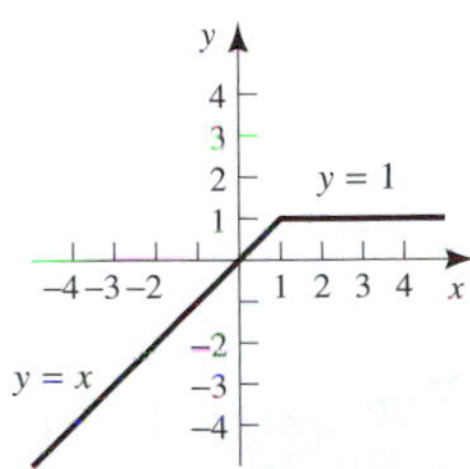

23. 1, 2, 4

25. a. $(0, \infty)$
b. $(-\infty, 0)$
c. $f' \to 0$. f' becomes large. f' becomes a large negative number.
d. $(-\infty, -1)$, $(-1, 0)$ and $(0, \infty)$

27. (b) **29.** (c) **31.** (e)

33. 3, 1, 0, −1, 0, 1 **35.** Does not exist

37.

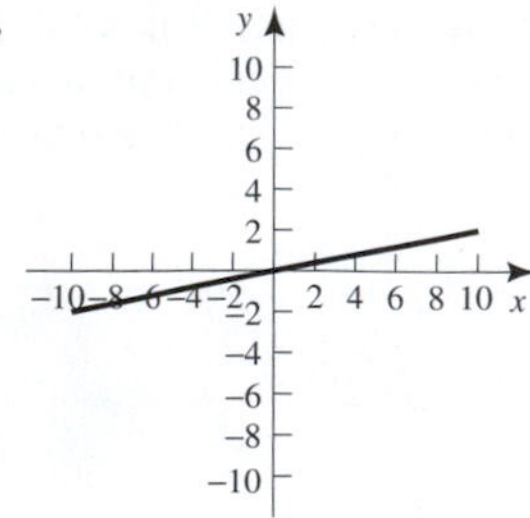

39.

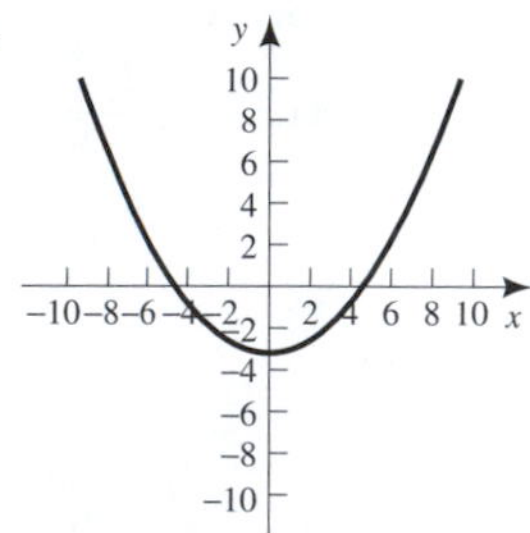

41. $B'(T) = -0.28 + 0.02T$ degree body temperature per degree ambient temperature. The bird's body temperature increases at a more rapid rate for higher ambient temperatures.

43. $Y'(N) = 0.8 - 0.006N$ bushels of soybeans per pound. Initially, increasing the amount of fertilizer increases the yield, but at some point it begins to decrease the yield. Clearly, some fertilizer is helpful, but too much is actually counterproductive.

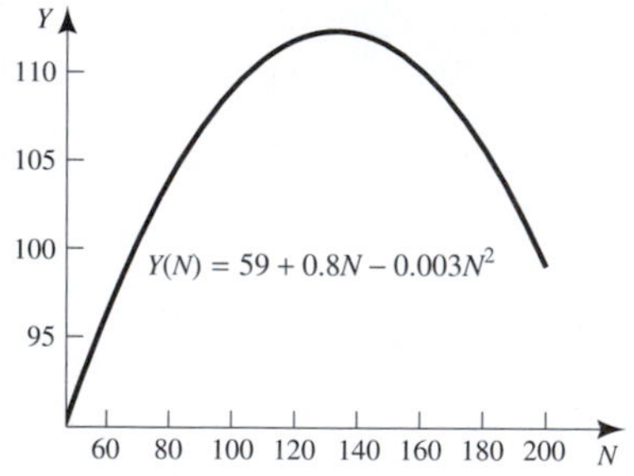

45. $G'(T) = -7.1 - 2.2T$ days per degree. There is a more rapid increase in the early start in the growing season as the mean July temperature increases.

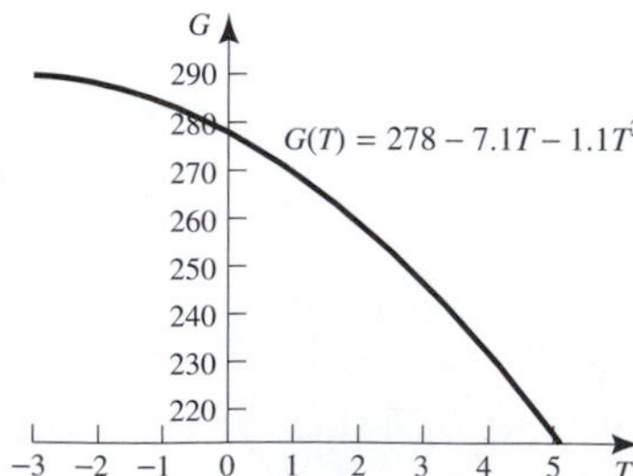

47. $L'(D) = 2.9 - 0.12D$ % per day. On the given lactation interval, the percentage of lipid in the milk first increased, and then decreased.

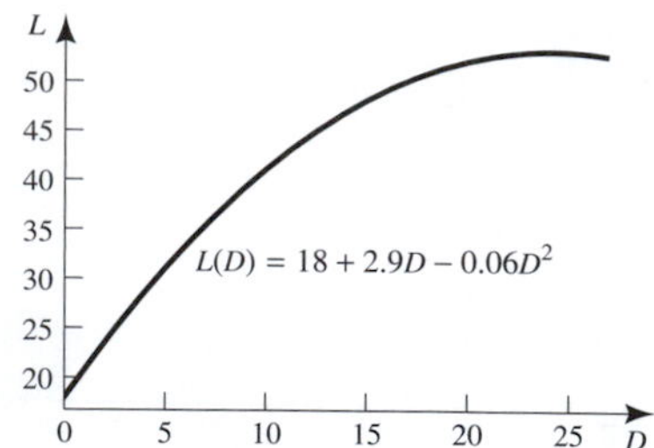

49.

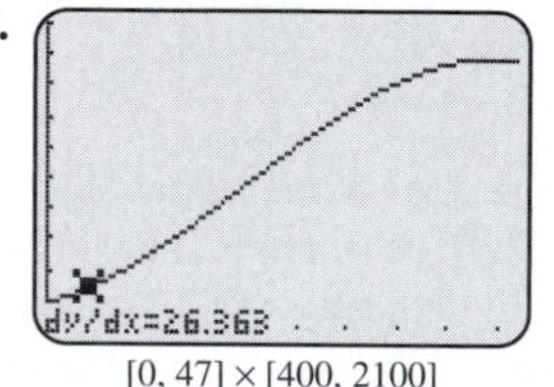

[0, 47] × [400, 2100]

x	5	10	15	20
$f'(x)$	26.36	35.63	41.19	43.06
x	25	30	35	40
$f'(x)$	41.24	35.72	26.50	13.59

Since $f'(5) \approx 26.36$, the slope of the tangent line to the curve at $x = 5$ is approximately 26.36. Thus, when output was \$5000, yard labor costs were increasing at \$26.36 per \$1000 increase in output. Notice that the rate of increase of labor costs itself first increases, and then at larger values of output it decreases.

51.

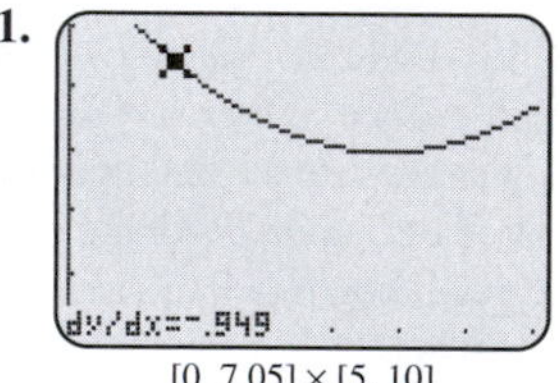

[0, 7.05] × [5, 10]

x	1.5	3	4.5	6	6.6
$f'(x)$	−0.95	−0.51	−0.073	0.365	0.54

Since $f'(1.5) \approx -0.95$, the slope of the tangent line to the curve at $x = 1.5$ is approximately -0.95. Thus, when assets were 1.5 million dollars, the margin ratio was decreasing at 0.95 units per 1 million dollar increase in assets. The rate of change of the margin ratio initially decreases then at higher values of assets begins to increase.

53. $\dfrac{1}{2\sqrt{t+1}}$ **55.** $\dfrac{1}{\sqrt{2t+5}}$

57. $f'(0) = \lim_{h\to 0} \frac{f(h) - f(0)}{h}$

$= \lim_{h\to 0} \frac{\sqrt[3]{h^2} - 0}{h}$

$= \lim_{h\to 0} h^{2/3-1} = \lim_{h\to 0} h^{-1/3}$

$= \lim_{h\to 0} \frac{1}{\sqrt[3]{h}}$

does not exist

59. a.

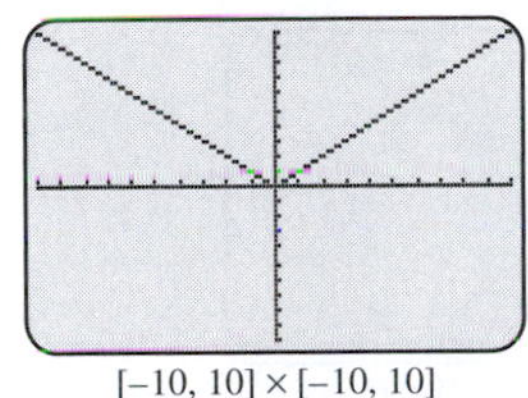

[−10, 10] × [−10, 10]

$f'(0)$ does not appear to exist since the graph appears to have a corner when $x = 0$,

b.

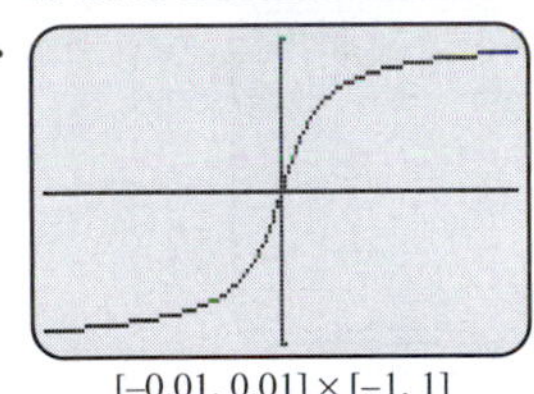

[−0.01, 0.01] × [−1, 1]

From the graph it appears that as $h \to 0$, $\frac{\sqrt{h^2 + 0.000001} - 0.001}{h} \to 0$. This implies that $f'(0) = 0$.

h	0.0001	0.00001
$\frac{\sqrt{h^2 + 0.000001} - 0.001}{h}$	0.0499	0.0050

c. The following screen was obtained after using the ZOOM about a half dozen times. Notice that the "corner" observed in part (a) now appears to have rounded out and the tangent to the curve at $x = 0$ now appears to be 0. This supports the conclusion found in part (b).

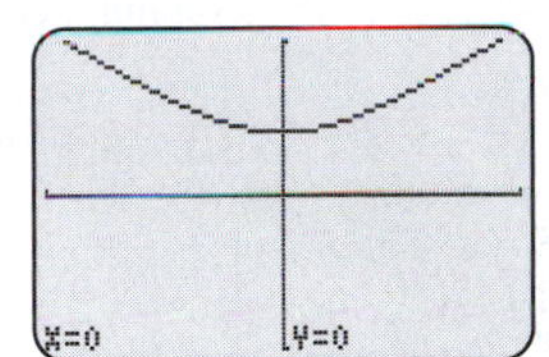

[−0.00244, 0.00244] × [−0.00244, 0.00244]

61. From the figure it is difficult to tell whether the curve has a corner when $x = 0$.

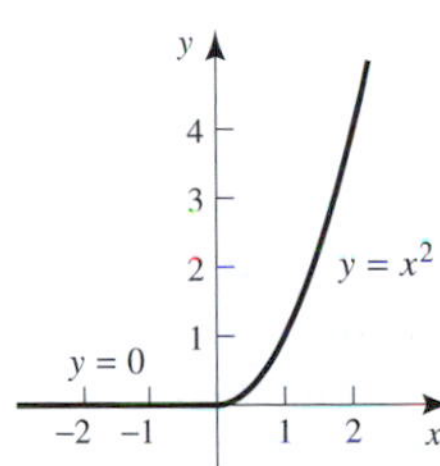

h	−0.1	−0.01	−0.001	→	0	←	0.001	0.01	0.1
$\frac{f(0+h) - f(0)}{h}$	0	0	0	→	0	←	0.001	0.01	0.1

From the numerical work in the table it appears that $f'(0) = 0$. Since $\lim_{x\to 0^-} \frac{f(0+h) - f(0)}{h} = \lim_{x\to 0^-} \frac{0 - 0}{h} = 0$, and $\lim_{x\to 0^+} \frac{f(0+h) - f(0)}{h} = \lim_{x\to 0^+} \frac{h^2 - 0}{h} = \lim_{x\to 0^+} h = 0$, then $\lim_{x\to 0} \frac{f(0+h) - f(0)}{h} = 0$

63. $f'(0) = 1$

65. $f'(1.57) \approx -1$

67. The same, yes

69. a. $y = 29.434x^{-0.5335}$, $r = -0.9245$

b.

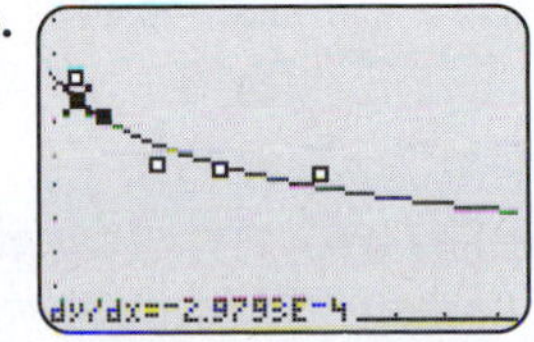

[1000, 4760] × [0, 0.9]

x	1200	1600	2000
$f'(x)$	−0.000298	−0.000192	0.000136
x	2400	2800	3200
$f'(x)$	−0.000103	−0.000081	−0.000066

For example, since $f'(1200) \approx -0.000298$, the slope of the tangent line to the curve at $x = 1200$ is approximately -0.000298. Thus, when the total number of open accounts was 1200, salary cost per account was decreasing at \$0.000298 per account. Notice that the rate of decrease of salary cost per account lessens as the total number of open account increases. (Salary cost efficiencies lessen as the total number of open accounts increase.)

71. a. $y = 0.939x^2 - 7.237x + 30.785$, $r^2 = 0.2033$

b.

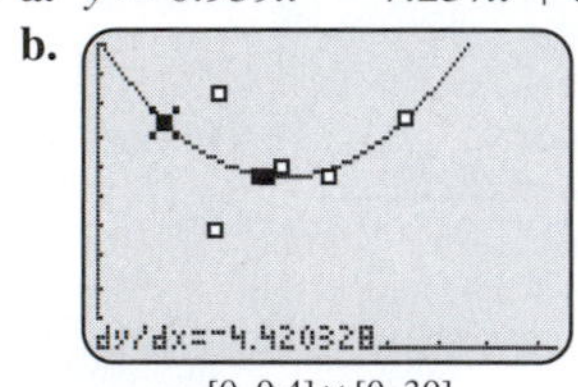

[0, 9.4] × [0, 30]

x	1.5	3	4.5	6	6.6
$f'(x)$	−4.42	−1.60	1.21	4.03	5.16

For example, since $f'(1.5) \approx -4.42$, the slope of the tangent line to the curve at $x = 1.5$ is approximately -4.42. Thus, when the assets were \$1.5 million, the variable cost ratio was decreasing at 4.42 per \$1 million increase in assets. Notice that the rate of change of the variable cost ratio increased from negative to positive as assets increased.

[0, 47] × [0, 0.5]

39. 0.02

41. Δ area $\approx 2\pi l \cdot \Delta r$

43. a. $y = 1.5x + 19.817$, $r = 0.9741$, $r^2 = 0.9488$

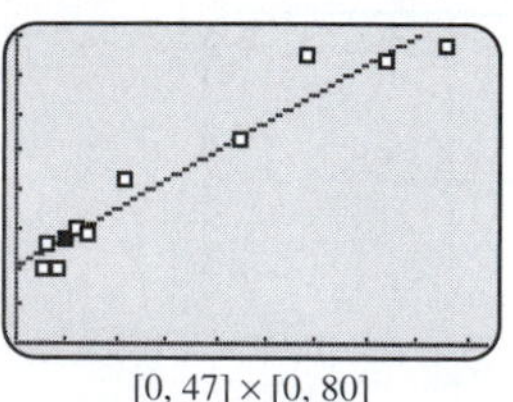

[0, 47] × [0, 80]

b. $y = -0.0254x^2 + 2.5787x + 14.292$, $r^2 = 0.9720$

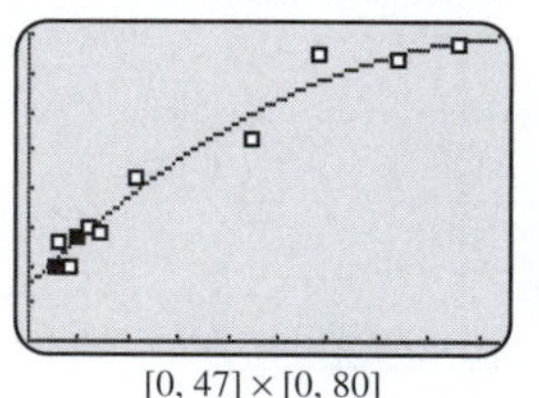

[0, 47] × [0, 80]

c. The quadratic with a larger square of the correlation coefficient.

d. marginal cost decreases

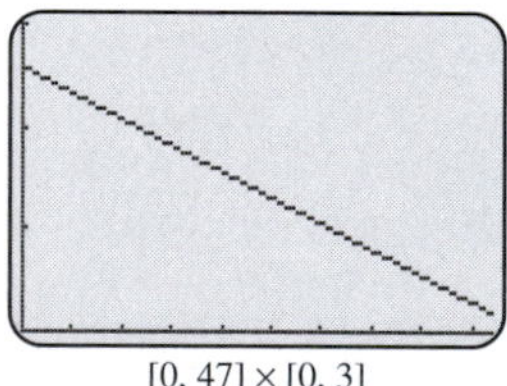

[0, 47] × [0, 3]

Section 3.4

1. 4.03 **3.** 4.97 **5.** −0.05

7. $\ln 2 + 0.01 \approx 0.70$

9. 0.95 **11.** 0.51

13. $f'(1) \approx \frac{1}{3}$ **15.** $f'(1) \approx 2.72$ **17.** $f'(1) \approx 1$

19. 1.0333 **21.** −0.272 **23.** −0.40

25. −1.6 **27.** 4 **29.** 2

31. −0.015 **33.** 288,000 **35.** 5.03

37. $C'(x) = 0.4490 - 0.03126x + 0.000555x^2$, $C'(10) \approx 0.19$, $C'(30) \approx 0.01$, $C'(40) \approx 0.09$. Marginal cost decreases and then increases.

Chapter 3 Review Exercises

1. $x = -1$: 1, 1, 1; $x = 0$: 2, 1, does not exist; $x = 1$: 2, does not exist, does not exist; $x = 2$: 1, 1, 1

2. −1, 0, 1 **3.** −1, 0, 1, 2 **4.** 12.7

5. π^3 **6.** 0 **7.** −3

8. −4 **9.** 1 **10.** −7

11. 3 **12.** 7 **13.** Does not exist

14. Does not exist **15.** 2 **16.** 3

17. 1.5 **18.** Does not exist **19.** −1

20. 0 **21.** 0.5 **22.** 2

23. 0 **24.** $4x$ **25.** $-2x$

26. $4x + 1$ **27.** $2x + 5$ **28.** $\frac{-2}{x^3}$

29. $\dfrac{-2}{(2x+5)^2}$ **30.** $\dfrac{1}{2\sqrt{x+7}}$ **31.** $\dfrac{1}{\sqrt{2x-1}}$

32. $y - 9 = 12(x - 2)$ **33.** $y - 10 = 3(x - 4)$

34. $y - 6 = 11(x - 1)$ **35.** $y - \dfrac{1}{3} = -\dfrac{1}{54}(x - 9)$

36. $f'(1) \approx 0.20$ **37.** $f'(8) \approx 0.33$

38. $f'(2) \approx -0.19$ **39.** $f'(4) \approx 0.19$

40. 1.01 **41.** 3.9

42. 0.11 **43.** 2.503

44. **a.** iii **b.** iv **c.** i **d.** ii

45. **a.** v **b.** i **c.** vi **d.** iii **e.** iv **f.** ii

46.

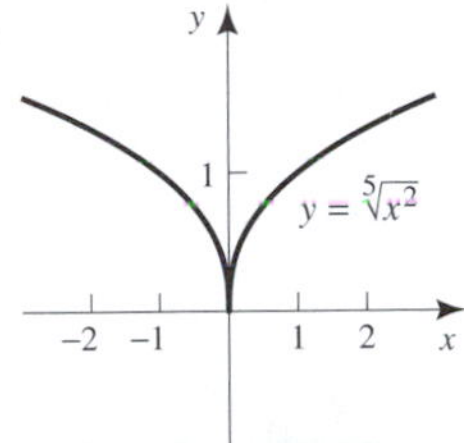

From the graph, the tangent at $x = 0$ appears vertical. Thus, $f'(0)$ does not exist.

h	−0.1	−0.01	−0.001	→	0	←	0.001	0.01	0.1
$\dfrac{f(0+h)-f(0)}{h}$	4.0	−15.9	−63.1	→	No	←	63.1	15.9	4.0

From the data in the table, $f'(0)$ does not appear to exist.

47. $$\lim_{x\to 0} \frac{f(0+h)-f(0)}{h} = \lim_{x\to 0} \frac{\sqrt[5]{x^2}-0}{h}$$
$$= \lim_{x\to 0} h^{2/5-1} = \lim_{x\to 0} h^{-3/5}$$

does not exist, so $f'(0)$ does not exist.

48. Yes, since $\lim_{x\to 2} f(x) = f(2)$

49. Since

$$\lim_{h\to 0^-} \frac{f(h)-f(0)}{h} = \lim_{h\to 0^-} \frac{0-0}{h} = 0$$

and

$$\lim_{h\to 0^+} \frac{f(h)-f(0)}{h} = \lim_{h\to 0^+} \frac{h^3-0}{h} = \lim_{h\to 0^+} h^2 = 0,$$

then

$$\lim_{h\to 0^-} \frac{f(h)-f(0)}{h} = \lim_{h\to 0^+} \frac{f(h)-f(0)}{h}$$
$$= \lim_{h\to 0} \frac{f(h)-f(0)}{h}$$
$$= f'(0) = 0$$

50. \$50

51. Does not exist. (i becomes large without bound.)

52. 0

53. Everywhere except the integer points of minute.

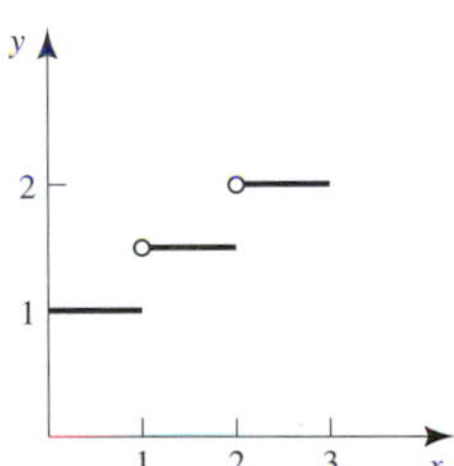

54. 200, 0, −200

55. −2, the distance s is decreasing at the rate of 2 ft/sec when $t = 3$. 0, the distance s stops decreasing when $y = 4$. 2, the distance s is increasing at the rate of 2 ft/sec when $t = 5$.

56. $\Delta \approx 12.57$

57. About 22,580 per year. During this time there was on average an increase of about 22,580 Hispanic businesses each year.

58. About 0.05 pounds per year. During this time there was for each person on average an increase of about 0.05 pound of waste per year.

59. −80 farms per year. During this time there was a decrease of about 80 farms during each year.

60. \$0.0032 per hour worked during a year. Male workers who worked between 1000 and 3400 hours a year saw their average hourly wage increase on average by \$0.0032 for each additional hour they worked that year.

61. 0.7387% per hour. For each additional hour of light exposure, whole milk lost 0.7387% of its vitamin A.

62. −0.09% per day. For each additional day postpartum the percentage of protein in the milk dropped by 0.09.

63. 2.4722% per hour. For each additional hour of light exposure, nonfat milk lost 2.4722% of its vitamin A.

64. 6558 per mm. A female bass had 6558 more eggs for each millimeter increase in length.

65. 0.042 per mm. For each mm increase in carapice length, the clutch size increased on average by 0.042.

66. 0.19 mm of length of prey for each millimeter length of bass. The length of the prey increased by 0.19 mm for each millimeter increase in the length of the bass.

67. About 0.00023 species per hectare. For each hectare increase in the size of a lake the number of species increased on average by 0.00023.

68. about 5 degrees per month, −4.5 degrees per month

69.

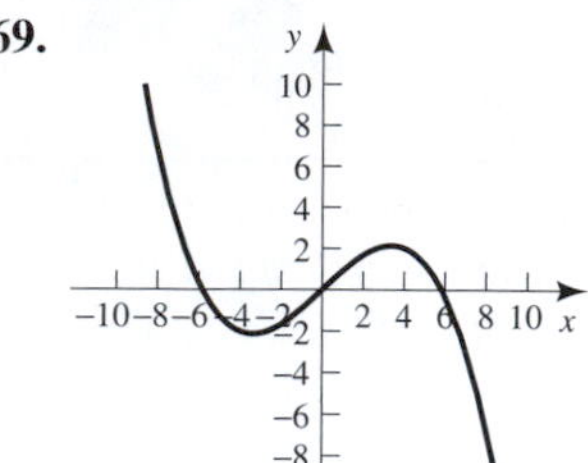

Chapter 4

Section 4.1

1. 0

3. 0

5. $23x^{22}$

7. $1.4x^{0.4}$

9. ex^{e-1}

11. $\frac{4}{3}x^{1/3}$

13. $-5.1x^{-2.7}$

15. $0.12x^{-1/2}$

17. $2e^x$

19. $\pi + \frac{2}{x}$

21. $-2x - 1$

23. $\frac{2}{\pi}x + \pi$

25. $-3x^2 - 2x - 1 - \frac{10}{x}$

27. $5x^4 - 5e^x$

29. $0.7x - 0.3e^x$

31. $-\frac{1}{x} - 0.02x - 0.09x^2$

33. $\frac{1}{6}(e^x + 2x - 2)$

35. $9(x^2 - x^{-4})$

37. $2x + \frac{1}{x^2}$

39. $4x^{-1} + 9x^{-4} + 8x$

41. $-81t^2 - e^{t-3}$

43. $-\frac{1}{t^2} - \frac{2}{t^3} - \frac{3}{t^4}$

45. $\frac{1}{3}\left(\frac{1}{\sqrt[3]{t^2}} + \frac{1}{\sqrt[3]{t^4}}\right)$

47. $5u^{2/3} - 2u^{-1/3}$

49. $4u + 3$

51. $\frac{1}{2\sqrt{u}} - \frac{9}{2\sqrt{u^5}}$

53. $\frac{1}{\sqrt{x}} + \frac{1}{6\sqrt{x^3}}$

55. $\pi^2 x^{\pi-1} + \frac{2}{x}$

57. $\sqrt{2}(x^{\sqrt{2}-1} + x^{-\sqrt{2}-1})$

59. $-\frac{3}{2}x^{-5/2} - \frac{4}{5}x^{-9/5} + \frac{1}{2x}$

61. $\frac{2}{3}x^{-1/3}$

63. $-\frac{2}{5}x^{-7/5}$

65. $y = x + 2$

67. $x = \pm\sqrt{3}$

69. 10

71. 83

73. $B'(T) = -0.28 + 0.02T$ degree body temperature per degree ambient temperature. $B'(25) = 0.22$. At an ambient temperature of 25 degrees, the body temperature is increasing at 0.22 degree for each degree of ambient temperature.

75. $Y'(N) = 0.824 - 0.00638N$ bushels of soybeans per pound. $Y'(150) = -0.133$. When 150 pounds of fertilizer per acre is used, the yield of soybeans is decreasing at 0.133 bushel per acre for each additional pound of fertilizer per acre. Clearly, too much fertilizer is counterproductive.

77. $G'(T) = -7.1 - 2.2T$ days per degree. $G'(-1) = -4.9$. When the mean July temperature is -1^0C, the beginning of the growing season is 4.9 days earlier for each additional degree Celsius increase in the mean July temperature.

79. $0.053378x + 21.7397$, 63.25, $-0.053378x + 41.5103$

81. $g'(r) = -0.066r + 0.024r^2$ and is the rate of change of the growth rate with respect to the real interest rate; $g'(-2) \approx 0.228$ and means that when the real interest rate is −2%, growth is increasing at the rate 0.228% per each percentage increase in real interest rates; $g'(2) \approx -0.036$ and means that when the real interest rate is 2%, growth is decreasing at the rate 0.036% per each percentage increase in real interest rates;

83. $5.62L$

85. $0.24(L + 0.0039L^2)$

87. $\frac{292.6T^{0.75}}{(273)^{1.75}P} \approx \frac{0.016}{P}T^{0.75}$

89. **a.** −1.36 and 0.92.
b. −1.36 and 0.92.
c. Solutions in part (b) are equal to values in part (a).
d. Exact solutions cannot be found.

91. L_1: $y = -4x - 3$, L_2: $y = 4x - 3$. Solve $4x - 3 = -4x - 3$ and obtain $x = 0$.

93. No. For example,

$$f(x) = \begin{cases} 1 & x \leq 0 \\ 0 & x > 0 \end{cases} \quad \text{and} \quad g(x) = \begin{cases} 0 & x \leq 0 \\ 1 & x > 0 \end{cases}$$

Notice that $f(x) + g(x) = 1$

95. $3x^2 - 4\sin x - 3\cos x$

97. a. $0.03x^2 - 2x + 50$;

b. 22, the cost of the 21st item is approximately \$22; 18, the cost of the 41st item is approximately \$18; 38, the cost of the 41st item is approximately \$38.

c.

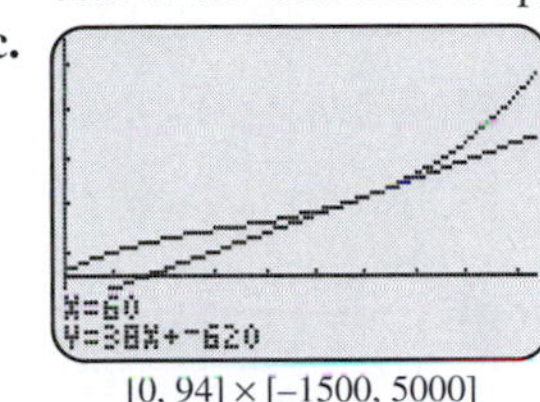

[0, 94] × [–1500, 5000]

Marginal cost first decreases, and then at some point, begins to increase.

99. a. $y = -0.0023x^2 + 1.1x + 19.437$, $r^2 = 0.9893$

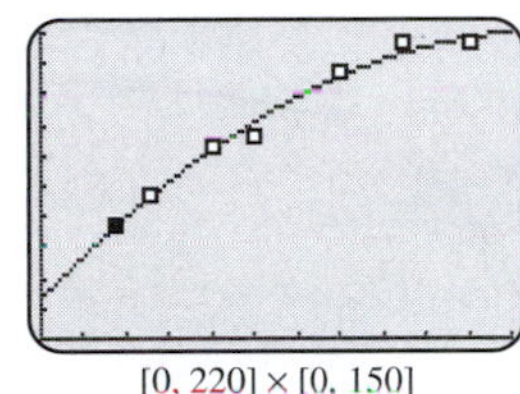

[0, 220] × [0, 150]

b.

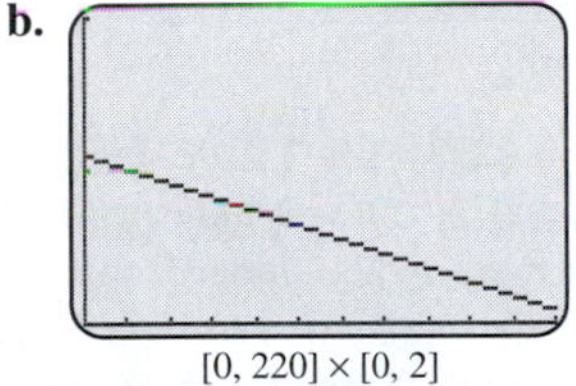

[0, 220] × [0, 2]

Marginal cost decreases.

Exercise 4.2

1. $3x^2e^x + x^3e^x$

3. $\dfrac{\ln x + 2}{2\sqrt{x}}$

5. $e^x(x^2 + 2x - 3)$

7. $(4e^x - 5x^4)\ln x + (4e^x - x^5)/x$

9. $e^x(\sqrt{x} + 1) + \dfrac{(e^x + 1)}{2\sqrt{x}}$

11. $(2x - 3\ln x)(1 - x^{-2}) + \left(2 - \dfrac{3}{x}\right)(x + x^{-1})$

13. $\dfrac{1/x - \ln x}{e^x}$

15. $-\dfrac{1}{x(\ln x)^2}$

17. $\dfrac{2}{(x+2)^2}$

19. $\dfrac{8}{(x+5)^2}$

21. $\dfrac{e^x(x-3)}{(x-2)^2}$

23. $\dfrac{-1}{(2x-1)^2}$

25. $\dfrac{3e^u(u-1)^2}{(u^2+1)^2}$

27. $\dfrac{-u^2 - 2u + 2}{(u^2+2)^2}$

29. $\dfrac{u^4 + 9u^2 + 2u}{(u^2+3)^2}$

31. $\dfrac{-8u^3 + (3u-1)e^u - 1}{3\sqrt[3]{u^2}(u^3 - e^u - 1)^2}$

33. $\dfrac{1 - x^2}{(1+x^2)^2}$

35. $\dfrac{1}{10(10 - 0.1x)^2}$

37. $p'(x) = -145499(138570 + x)^{-2} < 0$. This indicates that the demand curve slopes down, as it should.

39. $\dfrac{-6t^2 + 15}{(2t^2+5)^2}$

41. $\dfrac{2(I^2 + 100{,}000I)}{5(I + 50{,}000)^2}$

43. $\dfrac{7.12(2x - 13.74)}{(13.74x - x^2)^2}$

45. a. $2e^{2x}$; **b.** $3e^{3x}$; **c.** ne^{nx}

47. $f'(x) = (x-a)[2g(x) + (x-a)g'(x)]$ implies $f'(a) = 0$. The equation of the tangent line to $y = f(x)$ at $x = a$ is $y - f(a) = f'(a)(x - a)$ or $y = 0$, which is the x-axis.

49. No. Letting $f(x) = |x|$ and $g(x) = x$ and $h(x) = f(x) \cdot g(x)$, we recall that $f'(0)$ does not exist. So the product rule cannot be used.

51. $\sin x + x\cos x$

53. $e^x(\cos x - \sin x)$

55. $\dfrac{x\cos x - \sin x}{x^2}$

57. $1 + \tan^2 x$

59. a. $2.05x^2 - 36.51x + 415.14$. $r^2 = 0.9996$.

b. −11.91 dollars per ton. When 6000 tons are produced, the average total cost is dropping by a rate of \$11.91 per ton.

c. $C'(x) = 1000[q(x) + x \cdot q'(x)]$. $C'(6) = 198,600$ is the rate of change of total cost when 6000 tons are produced.

d. $Q(x) = 79.365x^2 + 201,214x + 408771.1$. $Q'(6) = 202,167$.

61. a. $428.29x^{-0.25536}$. The square of the correlation coefficient is 0.9948.

b. −11.53 dollars per ton. When 6000 tons are produced, the average total cost is dropping by a rate of \$11.53 per ton.

c. $C'(x) = 1000[q(x) + x \cdot q'(x)]$. $C'(6) = 201,860$ is the rate of change of total cost when 6000 tons are produced.

d. $Q(x) = 428,290x^{0.744639}$. $Q'(6) = 201,825$.

Section 4.3

1. $14(2x+1)^6$

3. $8x(x^2+1)^3$

5. $3e^x\sqrt{2e^x - 3}$

7. $\dfrac{1}{2}\sqrt{e^x}$

9. $\dfrac{1}{\sqrt{2x+1}}$

11. $4x(x^2+1)^{-2/3}$

13. $-\dfrac{e^x}{(e^x+1)^2}$

15. $\dfrac{6x^2}{(x^3+1)^2}$

17. $\dfrac{x^2(1-\ln x)+1}{x(x^2+1)^{3/2}}$

19. $(3x-2)^4(36x+11)$

21. $\dfrac{4(\ln x)^3}{x}$

23. $e^x(x^2+1)^7(x^2+16x+1)$

25. $(1-x)^3(2x-1)^4(14-18x)$

27. $\dfrac{(4x+3)^2}{2\sqrt{x}}(28x+3)$

29. $e^x \ln x\left(\ln x+\dfrac{2}{x}\right)$

31. $-12(x+1)^{-5}$

33. $\dfrac{2xe^x-3e^x-6}{(x+3)^4}$

35. $\dfrac{1}{4\sqrt{x(\sqrt{x}+1)}}$

37. $-x(1+x^2)^{-3/2}$

39. $g(x)>0$ for $0<x<3.2$. $g(x)=0$ if $x=3.2$. The biomass will be greater next year if the current biomass is less than 3.2 million tons and will be less if the current biomass is greater than 3.2 million tons. Apparently, too many fish result in overuse of the finite resources, causing the total mass to decrease the next year.

$$g'(x)=0.30355\left(1-\frac{x}{3.2}\right)^{-0.6135}\left[1-\frac{1.35865}{3.2}x\right]$$

$g'(x)>0$ when $0<x<\frac{3.2}{1.35865}\approx 2.355$ and $g'(x)<0$ when $x>2.355$. The rate of increase of the additional mass of fish next year increases until $x=2.355$ and then begins to decrease.

41. $\dfrac{4\pi r^2 c}{3n^{2/3}}\left(1+\dfrac{c}{r}n^{1/3}\right)^2$

43. $\dfrac{a(3n-2b)}{2\sqrt{n-b}}$

45. $\dfrac{1}{8\sqrt{\sqrt{\sqrt{x}+1}+1}\cdot\sqrt{\sqrt{x}+1}\cdot\sqrt{x}}$

47. Write as a product $(e^x+1)(x+3)^{-3}$

49. $3\sin^2 x\cos x$

51. $-\dfrac{\sin x}{2\sqrt{\cos x}}$

53. $r'(t)=\dfrac{0.0001}{\sqrt{36-t}}\left[-1008+186t-5t^2\right]$

Section 4.4

1. $4e^{4x}$

3. $-3e^{-3x}$

5. $4xe^{2x^2+1}$

7. $\dfrac{e^{\sqrt{x}}}{2\sqrt{x}}$

9. $2xe^x+x^2e^x$

11. $e^x x^4(5+x)$

13. $(1-x)e^{-x}$

15. $2xe^{x^2}(1+x^2)$

17. $\dfrac{e^x}{2\sqrt{e^x+1}}$

19. $\dfrac{1}{2\sqrt{x}}e^{\sqrt{x}}+\dfrac{1}{2}e^{\sqrt{x}}$

21. $\dfrac{1+e^x-xe^x}{(1+e^x)^2}$

23. $\dfrac{8}{(e^{2x}+e^{-2x})^2}$

25. $\dfrac{3e^{3x}}{2\sqrt{e^{3x}+2}}$

27. $5^x\ln 5$

29. $\dfrac{3^{\sqrt{x}}\ln 3}{2\sqrt{x}}$

31. $3^x+x3^x\ln 3$

33. $\dfrac{3x^2+2x}{x^3+x^2+1}$

35. $4x\ln|x|+2x$

37. $\dfrac{2x+1}{2(x^2+x+1)\sqrt{\ln|x^2+x+1|}}$

39. $\dfrac{1}{x(x+1)}$

41. $-4xe^{-x^2}\ln|x|+\dfrac{2}{x}e^{-x^2}$

43. $2x\ln x+x$

45. $\dfrac{1}{x(x^2+1)}-\dfrac{2x\ln x}{(x^2+1)^2}$

47. $\dfrac{11}{x}(\ln x)^{10}$

49. $\dfrac{3}{x}$

51. $e^{-3x}(1-3x),\ \dfrac{1}{3}$

53. $-2xe^{-x^2},0$

55. $xe^x(2+x),0,-2$

57. $xe^{-x}(2-x),0,2$

59. $\ln x^2+2,\ \dfrac{1}{e}$

61. $\dfrac{1-2\ln x}{x^3},\ \sqrt{e}$

63. $e^{-2x}(1-2x),\ \dfrac{1}{2}$

65. $\dfrac{2x}{x^2+5},\ (0,\infty)$

67. $L'(t)=(0.13)(446)e^{-0.13[t+1.51]}>0$. The fact that $L'(t)>0$ just says that the fish get bigger with age.

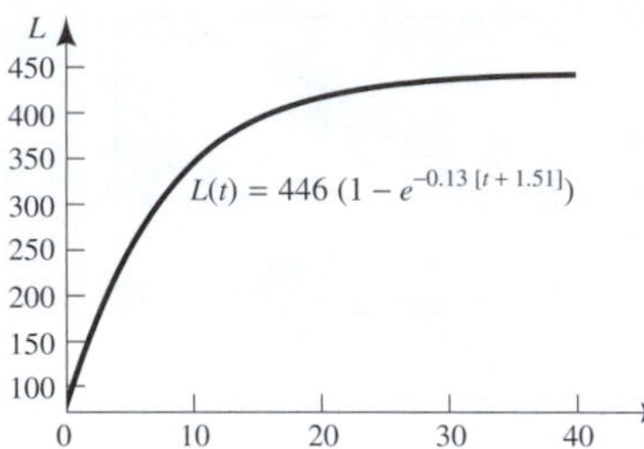

69. $N'(t)=(0.172)(2.979)(0.0223)(1+2.979e^{-0.0223t})^{-2}e^{-0.0223t}>0$. The fact that $N'(t)>0$ indicates that the nitrogen is increasing in these fields. Indeed, that is why fields are rested. As t becomes large, we see from the graph that the tangent becomes small. Therefore, $N'(t)$ must become small. This indicates that the nitrogen restoration rate slows in time.

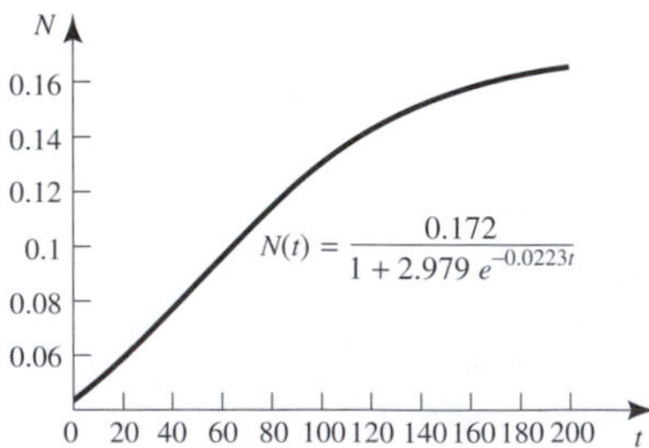

71. $\dfrac{3e^{3x}+2x}{(e^{3x}+x^2)\cdot\ln 2}$

73. $e^{e^{e^x}}\cdot e^{e^x}\cdot e^x$

75. **a.** 500;
b. $1488e^{-0.024t}(1+124e^{-0.024t})^{-2}$;

c. $p'(t) > 0$;
d.

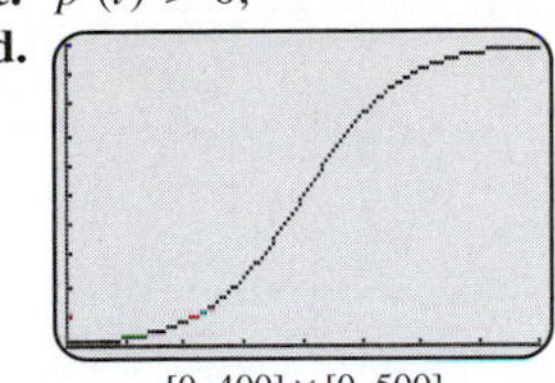

[0, 400] × [0, 500]

The graph increases and is asymptotic to $y = 500$;
e. The two graphs are about the same for the first 50 years.

77. Let $u = f(x)$; then

$$\begin{aligned}\frac{d}{dx}(\sin f(x)) &= \frac{d}{dx}(\sin u)\\ &= \frac{d}{du}(\sin u)\cdot\frac{du}{dx}\\ &= \cos u\cdot\frac{du}{dx}\\ &= (\cos f(x))\, f'(x)\end{aligned}$$

$$\begin{aligned}\frac{d}{dx}(\cos f(x)) &= \frac{d}{dx}(\cos u)\\ &= \frac{d}{du}(\cos u)\cdot\frac{du}{dx}\\ &= -\sin u\cdot\frac{du}{dx}\\ &= -(\sin f(x))\, f'(x)\end{aligned}$$

79. $-3x^2 \sin x^3$ **81.** $\dfrac{\cos(\ln x)}{x}$

83. $-\sin(\sin x)\cos x$

85. $g'(x) = re^{-bx}[1 - bx]$. $g'(x) > 0$ when $0 < x < 1/b$. $g'(x) < 0$ when $x > 1/b$. This indicates that when the population is small, it will increase next year. However, when the population is large, it will be less next year. In this case the population is so large that the finite resources available cannot hold the large population, and this population must decrease.

87.

$$\begin{aligned}\frac{d}{dt}W(L(t)) &= \frac{dW}{dL}\cdot\frac{dL}{dt}\\ &\approx 3839(1 - e^{-0.12[t+0.45]})^{2.31}e^{-0.12[t+0.45]}\end{aligned}$$

89. $P'(D) = (0.06D - 0.4)10^{0.03D^2 - 0.4D + 1}\ln 10$.

91.

$$\begin{aligned}f(t) &= (1-t)^t\\ \ln f(t) &= \ln(1-t)^t\\ &= t\cdot\ln(1-t)\\ \frac{f'(t)}{f(t)} &= \ln(1-t) - \frac{t}{1-t}\end{aligned}$$

$$\begin{aligned}f'(t) &= f(t)\left[\ln(1-t) - \frac{t}{1-t}\right]\\ &= (1-t)^t\left[\ln(1-t) - \frac{t}{1-t}\right]\end{aligned}$$

93. $e(t) = 3927.29(1.0330)^t$. Square of correlation coefficient is 0.9998. $e'(3.5) \approx 143$. This means that half way into the year 1793, the United States was growing at a rate of 143,000 per year.

Section 4.5

1. The demand drops by 7.7%

3. The demand drops by 5.0%

5. The demand drops by 3.1%

7. The demand drops by 13.2%

9. The demand drops by 5.6%

11. 0.5

13. **a.** $\frac{1}{4}$ inelastic; **b.** 1 unit; **c.** 9 elastic

15. **a.** $\frac{1}{7}$ inelastic; **b.** 1 unit; **c.** 3 elastic

17. **a.** $\frac{1}{3}$ inelastic; **b.** $\frac{1}{2}$ inelastic; **c.** $\frac{10}{11}$ inelastic

19. **a.** $\frac{2}{1.001}$ elastic; **b.** $\frac{2}{11}$ inelastic

21. **a.** 2 elastic; **b.** 2 elastic;

23. **a.** 15; **b.** 15

25. **a.** $\dfrac{1800p}{126.5 - 1800p}$
b. 0.74 inelastic, 2.47 elastic

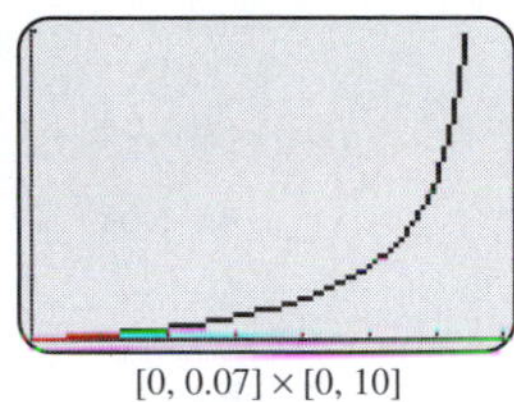

[0, 0.07] × [0, 10]

27. 2.0711, for every one percentage increase in the price of Grape Nuts, the demand decreases by 2.0711%.

29. $E = -\dfrac{a}{x}\dfrac{dx}{da}$, 0.0229, for every one percent increase in the amount of coupons the demand for Shredded Wheat increases by 0.0229%.

Section 4.6

1. 1.34×10^8 kilograms. 0.67×10^8

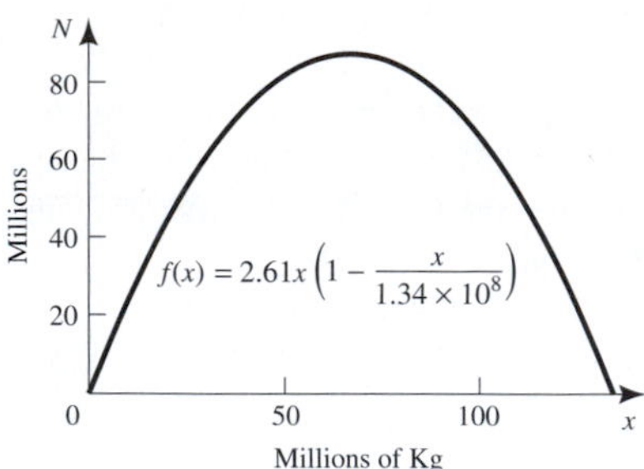

3. $f(x) = 0$ implies $x = 0$ or $x = K$.
$f'(x) = r\left(1 - \dfrac{2x}{K}\right) = 0$ implies $x = \dfrac{K}{2}$.

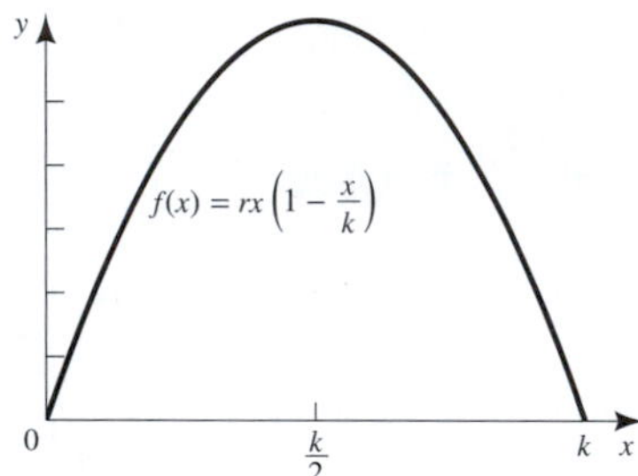

5. 16,000 pounds, 6,912,000 pounds

7. msy = 243 million pounds, mvp $\approx$ 2.09 million pounds, cc $\approx$ 12.91 million pounds

Chapter 4 Review Exercises

1. $6x^5 - 6x$

2. $\dfrac{24}{5}x^{1/5} + 2x^{-3/5}$

3. $-6x^{-4} - 2x^{-5/3}$

4. $2.6x^{0.3} - 7.2x^{-3.4}$

5. $2\pi^2 x + 2$

6. $3\pi^3 x^2 + 2e^2 x$

7. $\dfrac{2}{3}x^{-1/3}$

8. $-\dfrac{5}{3}x^{-8/3}$

9. $(x^3 + x + 1) + (x + 3)(3x^2 + 1)$

10. $2x(x^5 - 3x + 2) + (x^2 + 5)(5x^4 - 3)$

11. $\dfrac{x^4 - 2x^2 + 3}{2\sqrt{x}} + \sqrt{x}(4x^3 - 4x)$

12. $\dfrac{x^3 - 3x + 5}{3\sqrt[3]{x^2}} + \sqrt[3]{x}(3x^2 - 3)$

13. $\dfrac{3}{2}x^{1/2}(x^2 + 4x + 7) + (x^{3/2})(2x + 4)$

14. $-\dfrac{7}{2}x^{-9/2}(x^3 + 2x + 1) + (x^{-7/2})(3x^2 + 2)$

15. $\dfrac{2}{(x + 1)^2}$

16. $\dfrac{1 - 2x}{2\sqrt{x}(2x + 1)^2}$

17. $\dfrac{x^4 + 3x^2}{(x^2 + 1)^2}$

18. $\dfrac{3x^2 - 4x - 5}{(3x - 2)^2}$

19. $\dfrac{-2x^5 + 4x^3 + 2x}{(x^4 + 1)^2}$

20. $30(3x + 1)^9$

21. $40x(x^2 + 4)^{19}$

22. $\dfrac{5(\sqrt{x} - 5)^{3/2}}{4\sqrt{x}}$

23. $\dfrac{2x + 1}{2\sqrt{x^2 + x + 1}}$

24. $x^2(x^2 + x + 2)^4(13x^2 + 8x + 6)$

25. $(x^4 + x + 1)^4(22x^5 + 60x^3 + 7x^2 + 2x + 15)$

26. $\dfrac{-14x^5 - 16x^3 - 2x^2 + 2x - 4}{(x^4 + x + 1)^5}$

27. $\dfrac{3\sqrt{x}}{4\sqrt{x^{3/2} + 1}}$

28. $\dfrac{1}{6\sqrt{x}\sqrt[3]{(\sqrt{x} + 1)^2}}$

29. $-7e^{-7x}$

30. $6x^2 e^{2x^3 + 1}$

31. $e^x(x^2 + 3x + 3)$

32. $\dfrac{2x}{x^2 + 1}$

33. $2xe^{x^2}\ln x + \dfrac{1}{x}e^{x^2}$

34. $\dfrac{2x + e^x}{x^2 + e^x}$

35. $\dfrac{1}{5}x^{-4/5}$

36. $\dfrac{2}{3}x^{-1/3}$

37. $-3x^{-4}$

38. $\dfrac{x - 1}{2x^{3/2}}$

39.

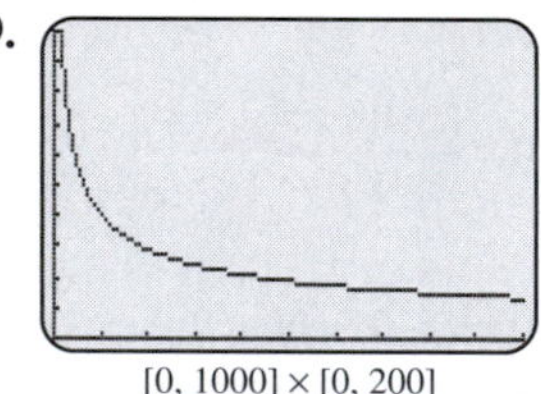

[0, 1000] × [0, 200]

x	100	200	300	400
$h'(x)$	−0.38	−0.14	−0.08	−0.05

$h'(100) \approx -0.38$ means that when the 100th camera was produced, direct labor hours were decreasing at the rate of 0.38 hour per camera. The rate is decreasing less, indicating that fewer additional hours are needed to make each additional camera.

40. a.

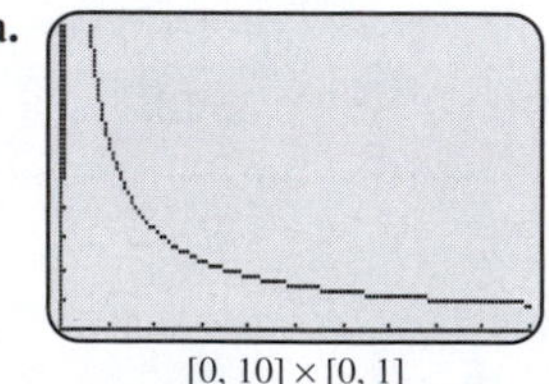

[0, 10] × [0, 1]

n	2	4	6	8
$C'(n)$	−0.1504	−0.0376	−0.0167	−0.0094

$C'(2) \approx -0.1504$ means that when the number of hours worked per day was 2, the unit cost was decreasing at a rate of \$0.15 for each additional hour per day worked. The data indicates that the more hours worked per day, the less the unit costs, that is, the greater the efficiency.

b. 1.05 hours

41.

[0, 1000] × [0, 2]

Q	100	200	400	800
$C'(Q)$	−0.00064	−0.00030	−0.00014	−0.00007

$C'(100) \approx -0.00064$ means that when 100,000 kg of milk was processed, the average cost was decreasing at the rate of \$0.00064 per 1000 kg increase in production. Average cost decreases less as the size of the plant increases, indicating economies of scale.

42.

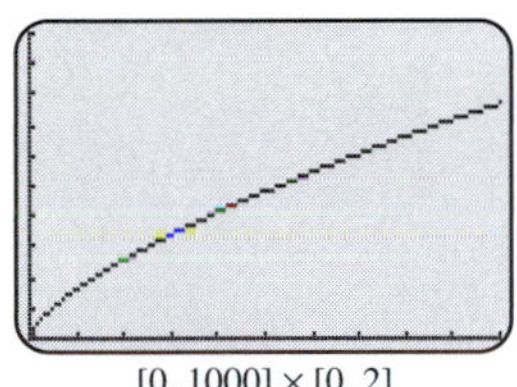

[0, 1000] × [0, 2]

D	100	200	400	800
$R'(D)$	0.0022	0.0018	0.0014	0.0011

$R'(100) \approx 0.0022$ means that when the distance was 100 miles, the rail transportation cost per ton was increasing at a rate of \$0.0022 per 1 mile increase in distance. This rate becomes less as the distance increases, indicating economies of scale.

43. a. 606.11, at $x = 2$ items sold, one more item sold results in a \$606.11 profit;

b. 329.14, at $x = 3$ items sold, one more item sold results in a \$329.14 profit;

c. −225.98, at $x = 4$ items sold, one more item sold results in a \$225.98 loss;

d. −1434.13, at $x = 5$ items sold, one more item sold results in a \$1434.13 loss;

44. 56, at $t = 1$, the particle is moving at 56 ft/sec in the positive direction; 0, the particle is at rest at the instant $t = 2$; −152, at $t = 3$, the particle is moving at 152 ft/sec in the negative direction.

45. $3.2572t^{1.395}$, the instantaneous rate of change of the number of wheat aphids at time t is given by $3.2572t^{1.395}$.

46. $7.08x^2 - 41.66x + 63.77$, at the age of x, the instantaneous rate of change of the number of the eggs is given by $7.08x^2 - 41.66x + 63.77$. Since the quadratic formula indicates that $f'(x)$ is never zero and since $f''(0) > 0$, $f'(x) > 0$ for all x. This means that *Ostrinia furnacalis* lays more eggs on older plants.

47. $f'(x) = -8.094\ln(1.053)\cdot(1.053)^{-x} \approx -0.418(1.053)^{-x}$. $f'(x) < 0$ for all x means that more immature sweetpotato whiteflies will cause smaller tomato leaves.

48. 25

49. a. $\frac{2}{3}$ inelastic; **b.** 1 unit; **c.** $\frac{3}{2}$ elastic

50. $-\dfrac{900}{(3x+1)^2}$

51. For smaller outputs, average cost decreases as output increases (there are economies of scale), but for larger outputs, average costs increase as output increases.

52. For smaller percentages of industry capacity, average cost decreases (there are economies of scale) and then levels off as the percent of industry capacity increases. But for large percentages of industry capacity, average cost increases for increases of percentage of industry capacity.

53. For each 1% increase in household disposable income, dollars spent overseas increase by 0.7407%.

54. 2.174

55. The first, since $\dfrac{dx}{dp}$ is the larger in absolute value and $\dfrac{p_0}{x_0}$ is the same for both.

56. $E = n$

Chapter 5

Section 5.1

1. Relative maximum on about May 30 and June 12; relative minimum on about June 5.

3. Relative maximum at about $x = 1300$, $x = 1850$, and $x = 1970$; relative minimum at about $x = 1400$ and $x = 1950$.

5. Relative maximum at about $x = 2500$ and $x = 3300$; relative minimum at about $x = 2800$.

7. Critical values are $x = 2$ and $x = 4$
increasing on $(-\infty, 2)$, $(4, \infty)$
decreasing on $(2, 4)$
relative maximum at $x = 2$
relative minimum at $x = 4$

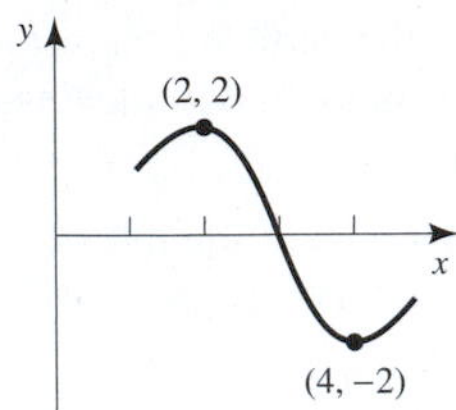

9. Critical values are $x = 1, 3, 5$
increasing on $(-\infty, 1)$, $(3, \infty)$
decreasing on $(1, 3)$.
relative maximum at $x = 1$
relative minimum at $x = 3$

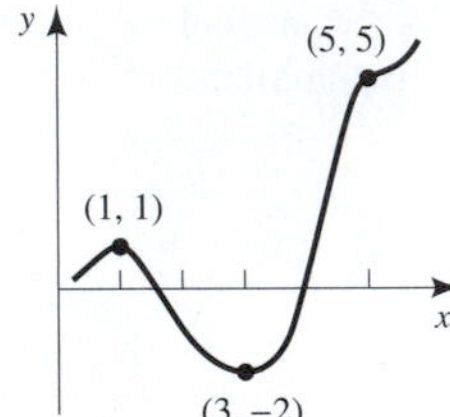

11. Critical values are $x = -1, 1, 2$
increasing on $(-1, 1)$, $(2, \infty)$
decreasing on $(-\infty, -1)$, $(1, 2)$.
relative maximum at $x = 1$
relative minima at $x = -1$ and $x = 2$

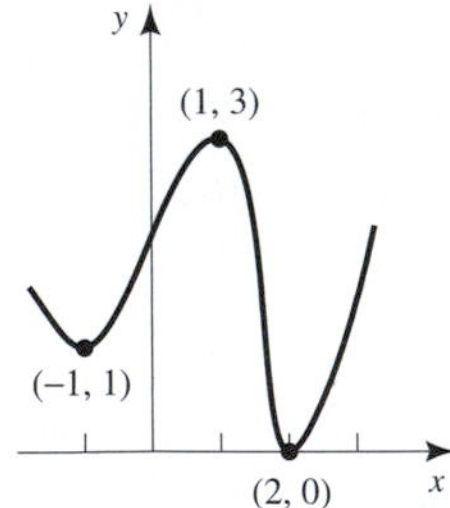

13. Critical value at $x = 2$
increasing on $(2, \infty)$
decreasing on $(-\infty, 2)$
no relative maxima
relative minimum at $x = 2$

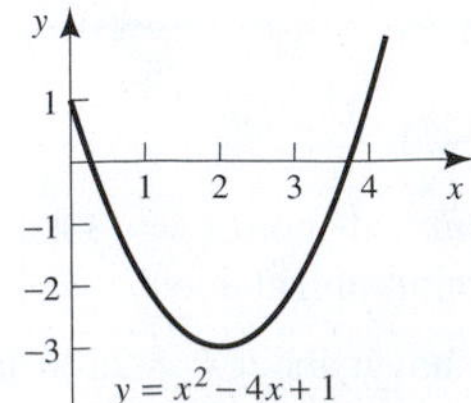

15. No critical values
increasing on $(-\infty, \infty)$
never decreasing
no relative maximum
no relative minimum

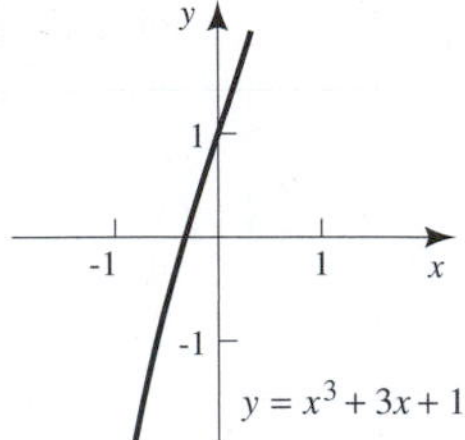

17. Critical value at $x = 0$
increasing on $(0, \infty)$
decreasing on $(-\infty, 0)$
no relative maximum
relative minimum at $x = 0$

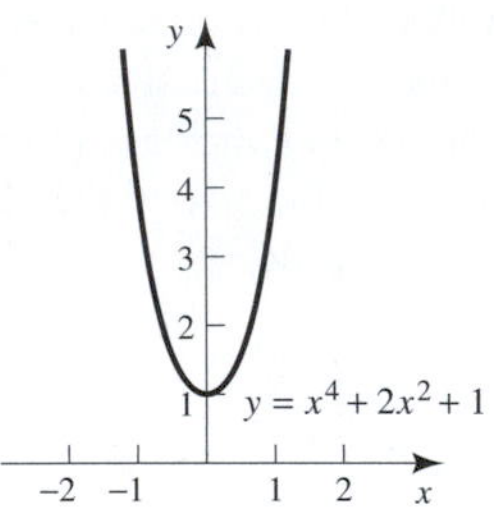

19. Critical value at $x = 1$
increasing on $(-\infty, 1)$
decreasing on $(1, \infty)$
relative maximum at $x = 1$
no relative minima

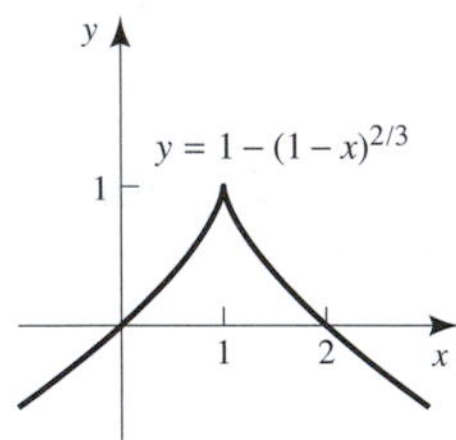

21. Critical values at $x = -1$ and $x = 1$
increasing on $(-\infty, -1)$, $(1, \infty)$
decreasing on $(-1, 1)$
relative maximum at $x = -1$
relative minimum at $x = 1$

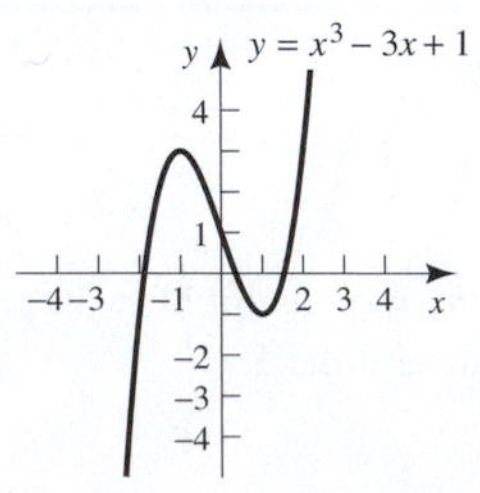

23. Critical values at $x = 1/2$ and $x = 3/2$
increasing on $(-\infty, 1/2)$, $(3/2, \infty)$
decreasing on $(1/2, 3/2)$
relative maximum at $x = 1/2$
relative minimum at $x = 3/2$

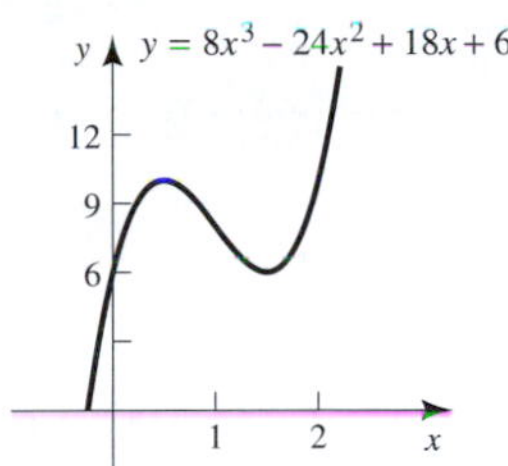

25. Critical values at $x = -2, 0, 2$
increasing on $(-2, 0)$, $(2, \infty)$
decreasing on $(-\infty, -2)$, $(0, 2)$
relative maximum at $x = 0$
relative minima at $x = -2, 2$

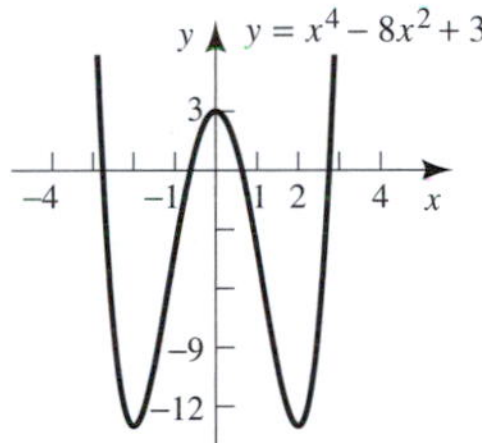

27. Critical values at $x = \pm 1$
increasing on $(-1, 1)$
decreasing on $(-\infty, -1)$, $(1, \infty)$
relative maximum at $x = 1$
relative minimum at $x = -1$

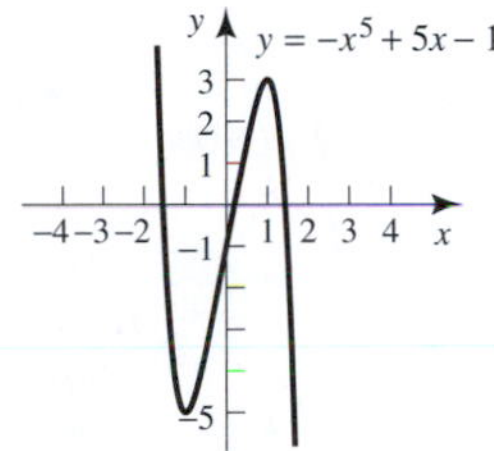

29. Critical values at $x = 0, \pm 2$
increasing on $(-\infty, -2)$, $(2, \infty)$
decreasing on $(-2, 2)$
relative maximum at $x = -2$
relative minimum at $x = 2$

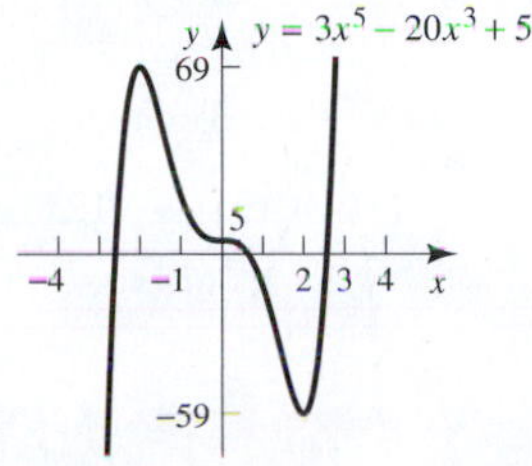

31. Critical value at $x = 1$
increasing on $(1, \infty)$
decreasing on $(0, 1)$
no relative maxima
relative minimum at $x = 1$

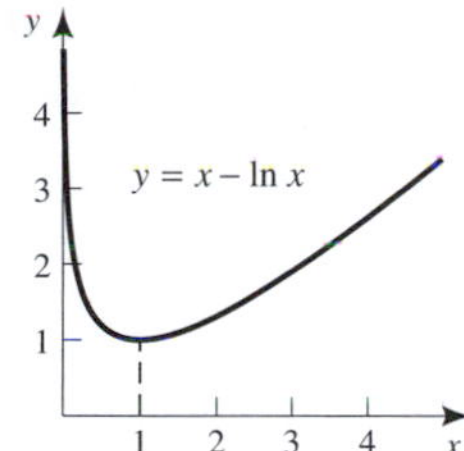

33. Critical value at $x = 0$
increasing on $(-\infty, 0)$
decreasing on $(0, \infty)$
relative maximum at $x = 0$
no relative minima

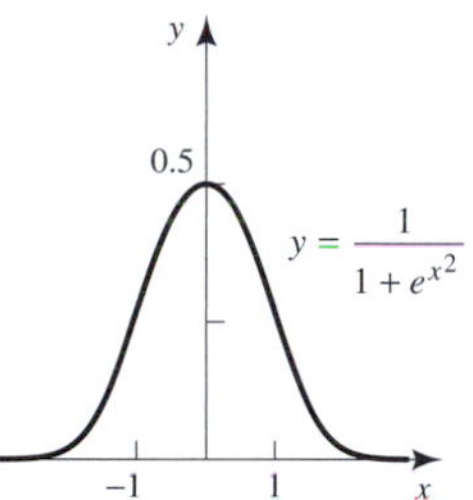

35. Critical values at $x = 0, 2$
increasing on $(0, 2)$
decreasing on $(-\infty, 0)$, $(2, \infty)$
relative maximum at $x = 2$
relative minimum at $x = 0$

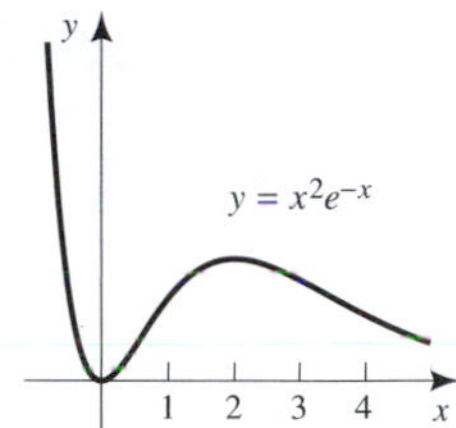

37. Critical value at $x = 2b$
increasing on $(2b, \infty)$
decreasing on $(-\infty, 2b)$
no relative maxima
relative minimum at $x = 2b$

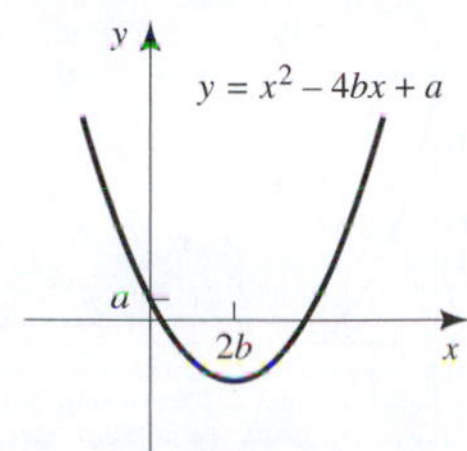

39. Critical value at $x = 0$
increasing on $(0, \infty)$
decreasing on $(-\infty, 0)$
no relative maxima
relative minimum at $x = 0$

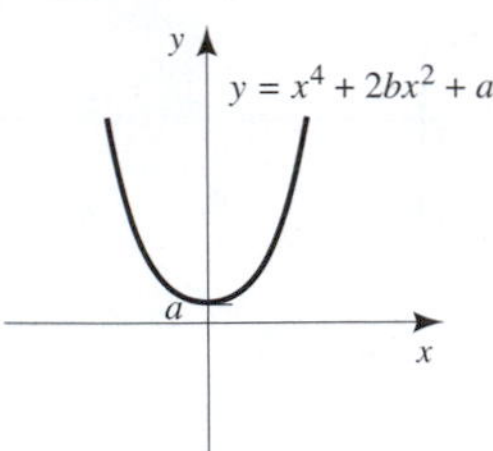

41. Critical values at $x = \pm\sqrt{b}$
increasing on $(-\infty, -\sqrt{b})$, $(\sqrt{b}, \infty)$
decreasing on $(-\sqrt{b}, \sqrt{b})$
relative maxima at $x = -\sqrt{b}$
relative minimum at $x = \sqrt{b}$

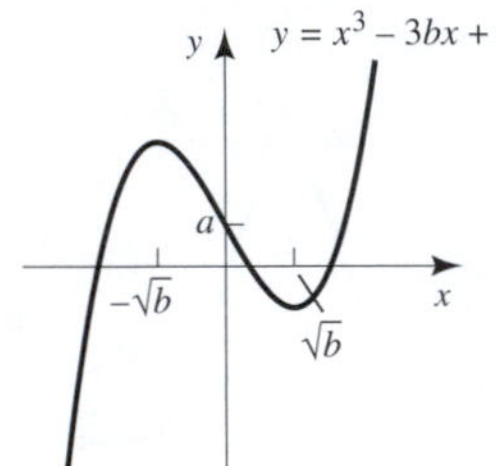

43. $f'(x) = 3ax^2 + b$ is positive if both $a > 0$ and $b > 0$ are positive and negative if both $a < 0$ and $b < 0$ are negative.

49. 500 **51.** $x = 20$ **53.** $x = 3$

55. $(0, \infty)$

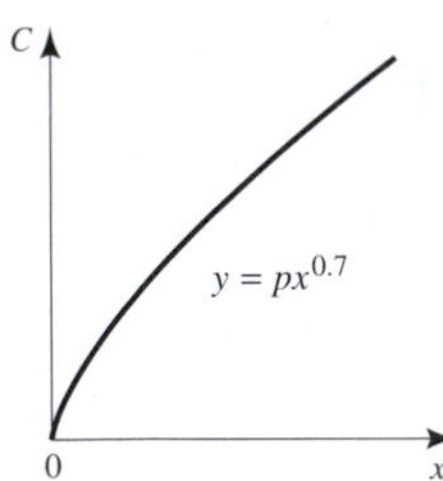

57. $x = 10.309052$ **59.** 33.95

61. 1.044 **63.** 30. 24,400

65. $I/2$

67. $f'(x) = -55.232e^{-0.64x} < 0$ for all real x, $y \to 9.1$

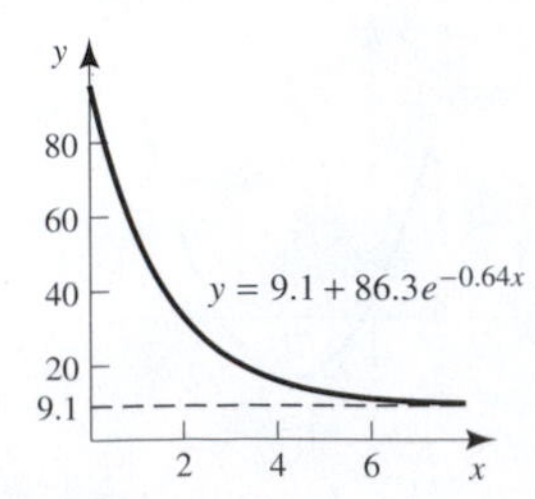

69. $x = 36.1$

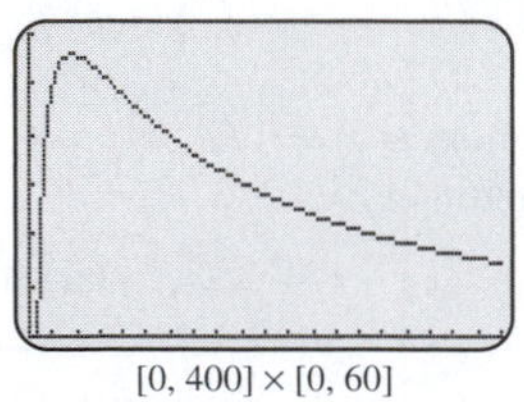

71. $f'(x) = \dfrac{-0.0899 \ln 10 \cdot 10^{3.8811-0.1798\sqrt{x}}}{\sqrt{x}} < 0$ for all $x > 0$, $y \to 0$

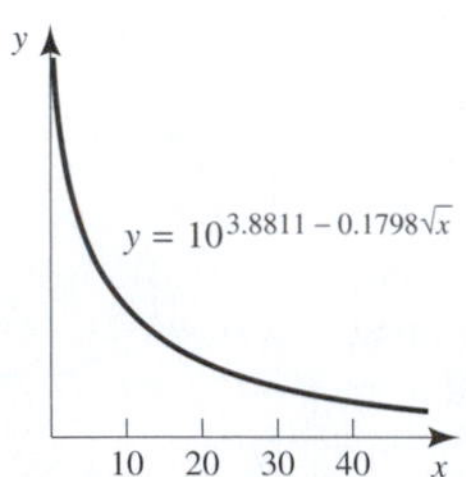

73. Neither. Consider the interval $(0, 10)$. If $f'(x) \le 0$ on $(0, 10)$, then f is decreasing there and $f(10) \le f(0) = 0$. But, $f(10) = 100$. From this contradiction, we conclude that f' is positive somewhere on $(0, 10)$. If f' were negative somewhere on $(0, \infty)$, then f' would change sign, indicating a relative extrema. Since this is not the case, we must have $f'(x) \ge 0$ on $(0, \infty)$. A similar argument shows that $f'(x) \ge 0$ on $(-\infty, 0)$. So we have that $f'(x) \ge 0$ on $(-\infty, \infty)$. Therefore, $f(0)$ cannot be a relative extrema.

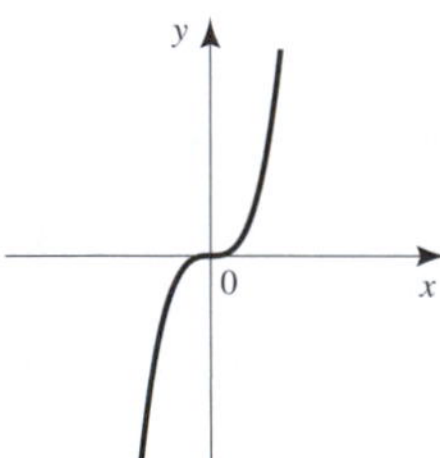

75. Relative maximum

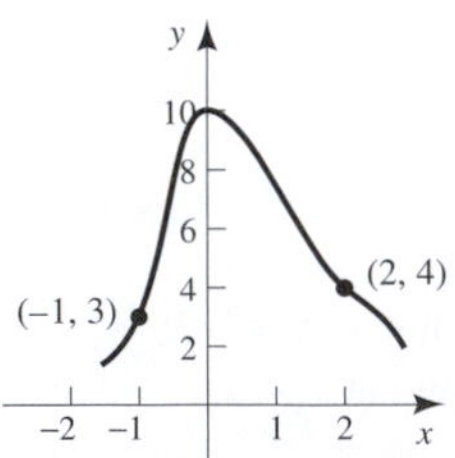

77. Let $f(x) = x^{13} - x^{12} - 10$. Then,

$$\begin{aligned} f'(x) &= 13x^{12} - 12x^{11} \\ &= x^{11}(13x - 12) \end{aligned}$$

Thus, there are critical points at $x = 0$ and $x = 12/13$. Notice that $f'(x) > 0$ and f increasing on $(-\infty, 0)$ and $(12/13, \infty)$ and that $f'(x) < 0$ and f decreasing on $(0, 12/13)$. Note $f(0) = -10$.

We conclude that $f(x) < 0$ on $(-\infty, 12/13]$ and increasing on $[12/13, \infty)$. Therefore, the graph crosses the x-axis exactly once.

79. Let $g(x) = e^x - (1 + x)$, then $g'(x) = e^x - 1$. Then $g'(x) < 0$ on $(-\infty, 0)$. So $g(x) > g(0) = 0$ for any $x < 0$. That is, $g(x) > 0$ on $(-\infty, 0)$. Since $e^x > 1$ on $(0, \infty)$, $g'(x) > 0$ on $(0, \infty)$. So $g(x) > g(0) = 0$ for any $x > 0$. That is, $g(x) > 0$ on $(0, \infty)$. Therefore, $g(x) \geq 0$ on $(-\infty, \infty)$, that is, $e^x \geq 1 + x$ for all x.

81. $h'(x) = f'(x)g(x) + f(x)g'(x) < 0$ since $f' > 0$, $g < 0$, $f < 0$, and $g' > 0$

83. $b^2 < 3c$

85. **a.** $f(x)$ has a relative minimum at $x = 1$, since f decreases to the left of 1 ($f'(x) < 0$ there) and increases to the right of 1 ($f'(x) > 0$ there).

b. $g(x)$ has a relative maximum at $x = 1$, since g increases to the left of 1 ($g'(x) > 0$ there) and decreases to the right of 1 ($g'(x) < 0$ there).

87. Positive

89. Negative

91. Let

$$f(m) = 1 - \left(\frac{m}{b}\right)^{x-1} - \left(1 - \frac{k}{m}\right)(x - 1)$$

$$f'(m) = -(x - 1)\left(\frac{m}{b}\right)^{x-2} - (x - 1)\left(\frac{k}{m^2}\right)$$

$$= -(x - 1)\left[\left(\frac{m}{b}\right)^{x-2} + \frac{k}{m^2}\right] < 0$$

since $x > 1$. Thus, $f(m)$ is a decreasing function. We have

$$f(k) = 1 - \left(\frac{k}{b}\right)^{x-1} > 0$$

and $f(b) = -\left(1 - \frac{k}{b}\right)(x - 1) < 0$

since $\frac{k}{b} < 1$ and $x - 1 > 0$. Therefore, by the constant sign theorem, $f(m)$ has at least one root in (k, b). Since f is also monotonic, f has exactly one root in (k, b), that is, the equation has a unique solution for m in (k, b).

93. 0.32

95. 5.982

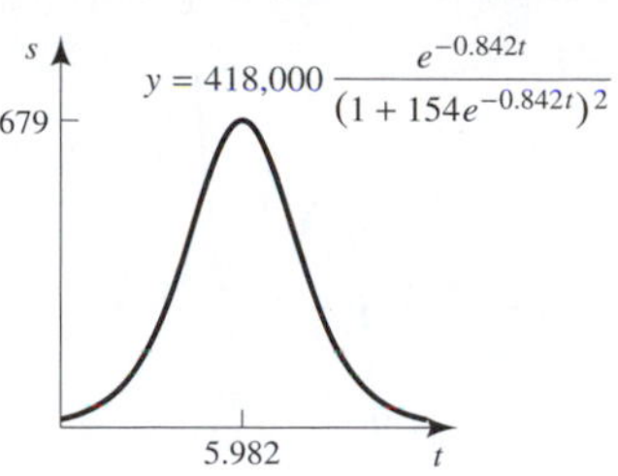

97. $y = -0.00000217x^3 + 0.000425x^2 - 0.021x + 1.034$, $r^2 = 0.5356$.
Decreasing on $(0, 33.218567)$.
Increasing on $(33.218567, 90)$.

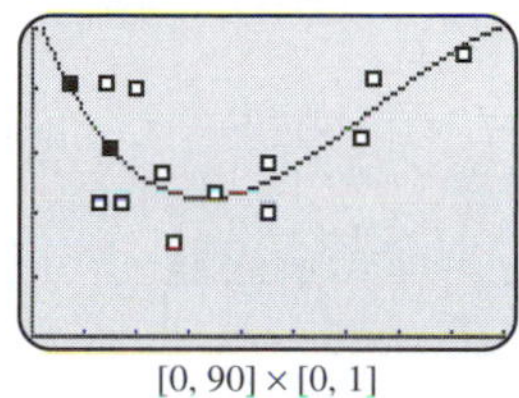

$[0, 90] \times [0, 1]$

99. $-1.10964x^2 + 44.71013x - 361.72713$. Approximately 20.146.

Section 5.2

1. $80x^3 + 2$

3. $2(x + 1)^{-3}$

5. $-(2x + 1)^{-3/2}$

7. $240x^2, 480x$

9. $-6x^{-4}, 24x^{-5}$

11. $\frac{8}{27}x^{-7/3}, -\frac{56}{81}x^{-10/3}$

13. No inflection value; concave down on $(0, 4)$

15. Inflection value at $x = 2$; concave up on $(2, 4)$; concave down on $(0, 2)$

17. Inflection value at $x = 2$; concave up on $(2, 4)$; concave down on $(0, 2)$

19. No inflection values; concave up on $(2, 4)$; concave down on $(0, 2)$;

21. 1, 3, 5. Concave up on $(1, 3)$ and $(5, \infty)$, concave down on $(-\infty, 1)$ and $(3, 5)$

23. -2. Concave up on $(-2, \infty)$, concave down on $(-\infty, -2)$

25. Critical values at $x = 0, 2$
relative maximum at $x = 0$
relative minimum at $x = 2$
inflection value at $x = 1$
increasing on $(-\infty, 0)$ and $(2, \infty)$
decreasing on $(0, 2)$
concave up on $(1, \infty)$
concave down on $(-\infty, 1)$

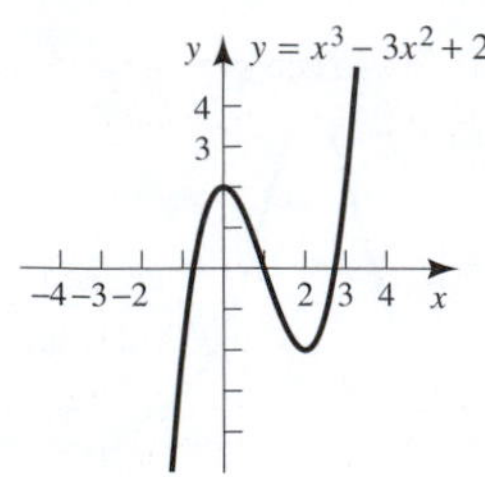

27. Critical values at $x = 0, 1$
relative maximum at $x = 0$
a relative minimum at $x = 1$
inflection value at $x = 0.75$
increasing on $(-\infty, 0)$ and $(1, \infty)$
decreasing on $(0, 1)$
concave up on $(0.75, \infty)$
concave down on $(-\infty, 0.75)$

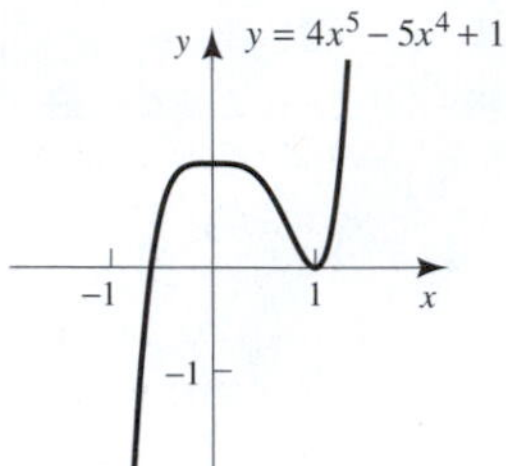

29. Critical value at $x = 0$
relative minimum at $x = 0$
no inflection values
increasing on $(0, \infty)$
decreasing on $(-\infty, 0)$
concave up on $(-\infty, \infty)$

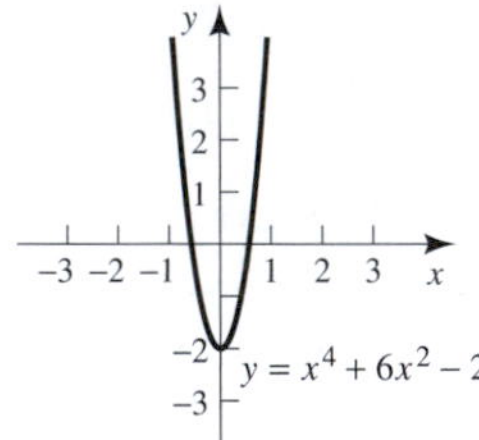

31. Critical values at $x = 0, 2$
relative maximum at $x = 2$
inflection values at $x = 0, \dfrac{4}{3}$
increasing on $(-\infty, 2)$
decreasing on $(2, \infty)$
concave up on $(0, 4/3)$
concave down on $(-\infty, 0)$, $(4/3, \infty)$

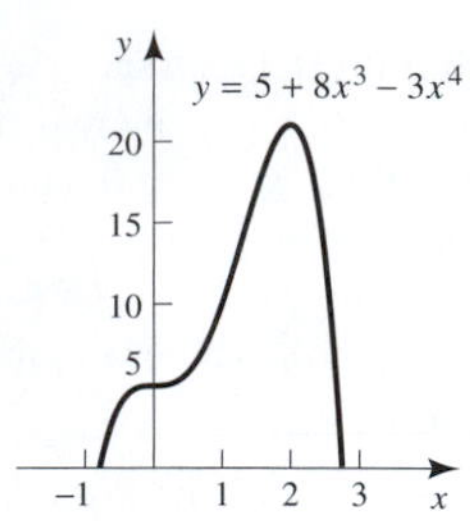

33. Critical values at $x = -2, 0$
relative maximum at $x = -2$
relative minimum at $x = 0$
inflection values at $x = -2 \pm \sqrt{2}$
increasing on $(-\infty, -2)$, $(0, \infty)$
decreasing on $(-2, 0)$
concave up on $(-\infty, -2-\sqrt{2})$, $(-2+\sqrt{2}, \infty)$
concave down on $(-2-\sqrt{2}, -2+\sqrt{2})$

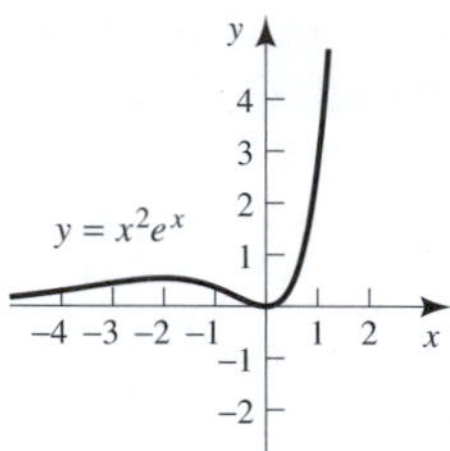

35. Critical value at $x = 0$
relative maximum at $x = 0$
no relative minimum
inflection values at $x = \pm\sqrt[4]{0.75}$
increasing on $(-\infty, 0)$
decreasing on $(0, \infty)$
concave up on $(-\infty, -\sqrt[4]{0.75})$, $(\sqrt[4]{0.75}, \infty)$
concave down on $(-\sqrt[4]{0.75}, \sqrt[4]{0.75})$
asymptote: $y = 0$

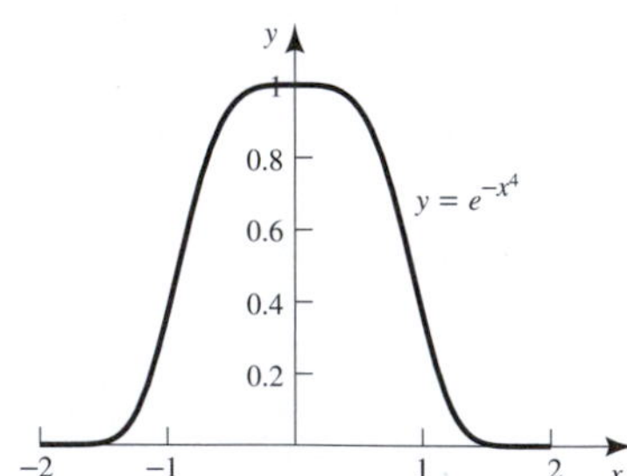

37. Since $f''(x) = 2a$, the sign of $f''(x)$ and a are the same.

39. $f''(x) = 6ax + 2b = 0$ only if $x = -\dfrac{b}{3a}$, where f'' changes sign.

41. (i) c; (ii) a; (iii) d; (iv) b

43. -1, relative minimum

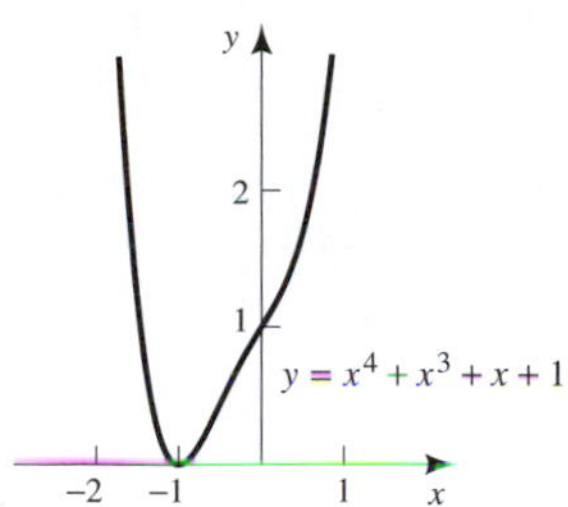

45. $t = 3$. Marginal profit is increasing on $(0, 3)$ and decreasing on $(3, 6)$

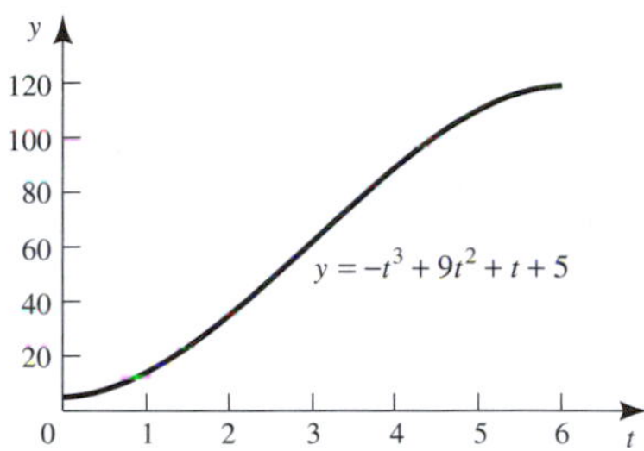

47. $t = 7.5$. Marginal profit is increasing on $(0, 7.5)$ and decreasing on $(7.5, 10)$

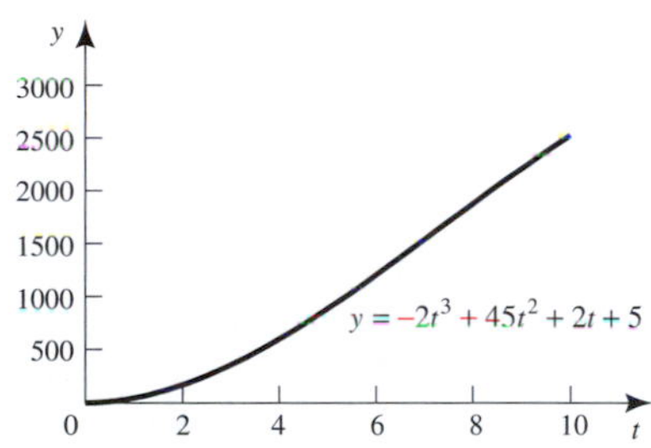

49. Yes. The graph is concave down, indicating that marginal costs are decreasing.

51. No. $C''(x) = 0.2 > 0$ so marginal cost $C'(x)$ is increasing.

53. Marginal sales decrease on $(0, 10)$ and increase for $x > 10$. Sales drop off ever more rapidly during the first 10 months, but afterward the drop off slows.

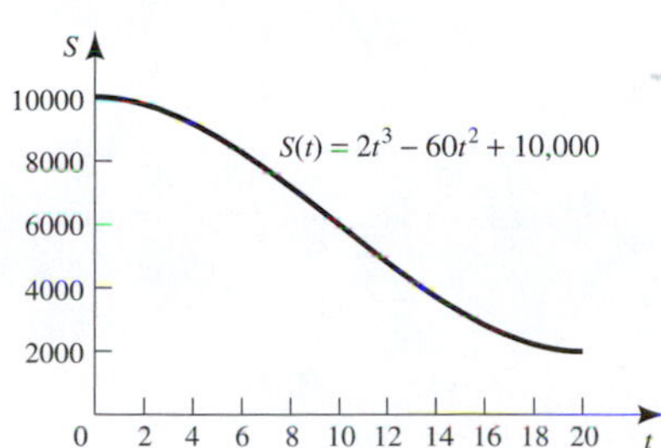

55. Decreasing. Concave up. The rate of decrease is slowing to zero. The economies of size become almost nil for larger schools.

57. Concave down. Since the graph is concave down, the rate of change of yield loss is decreasing.

59. 2

61. a. $y' = -0.175ax^{-1.175}$, $y'' \approx 0.206ax^{-2.175}$.

b. The average amount of needed labor decreases by a factor of 88.6%. Less labor is needed for production since workers become more efficient in performing the repeated manual tasks.

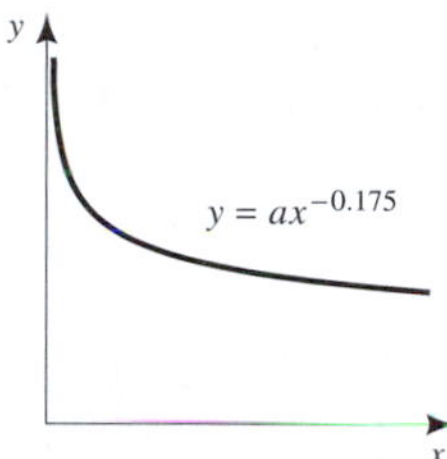

63. In parts (a), (b), and (c), $f''(0) = 0$. In part (d), $f''(0)$ does not exist. Thus, the second derivative test cannot be applied in any of these examples. In part (a), $f'(x) = 4x^3$ and is negative on $(-\infty, 0)$ and positive on $(0, \infty)$. Thus, $x = 0$ is a relative minimum. In part (b), $f'(x) = -4x^3$ and is positive on $(-\infty, 0)$ and negative on $(0, \infty)$. Thus, $x = 0$ is a relative maximum. In part (c), $f'(x) = 3x^2$ and is positive on $(-\infty, 0)$ and positive on $(0, \infty)$. Thus, there is no relative extrema at $x = 0$. In part (d), $f'(x) = 12x^{1/3}$ and is negative on $(-\infty, 0)$ and positive on $(0, \infty)$. Thus, $x = 0$ is a relative minimum.

65. Critical values at $x = 0, b$
relative maximum at $x = 0$
relative minimum at $x = b$
inflection value at $x = 3b/4$
increasing on $(-\infty, 0)$, (b, ∞)
decreasing on $(0, b)$
concave up on $(3b/4, \infty)$
concave down on $(-\infty, 3b/4)$

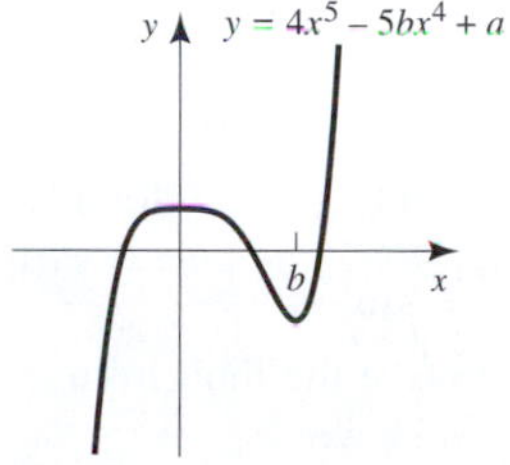

67. Critical values at $x = 0, 2b$
relative maxima at $x = 2b$
no relative minima
inflection value at $x = 0, 4b/3$
increasing on $(-\infty, 2b)$
decreasing on $(2b, -\infty)$
concave up on $(0, 4b/3)$
concave down on $(-\infty, 0)$, $(4b/3, \infty)$

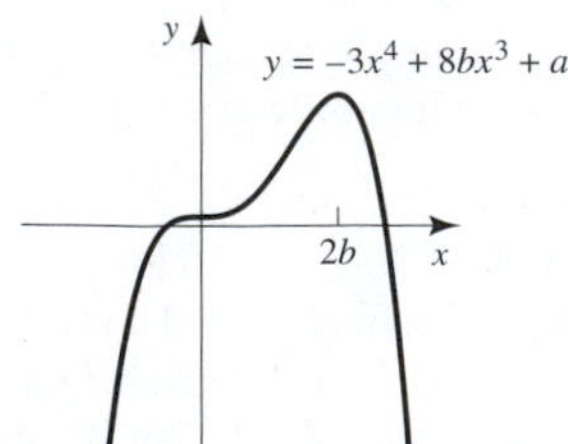

69. Recall that if $g'(a)$ exists, then $g(x)$ is continuous at $x = a$. Since $f''(a)$ exists, $f'(x)$ is continuous at $x = a$. This certainly means that $f'(a)$ exists. This in turn, implies that $f(x)$ is continuous at $x = a$.

71. a. **b.** 4 **c.** 2

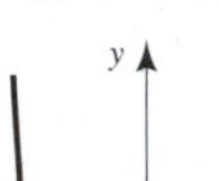

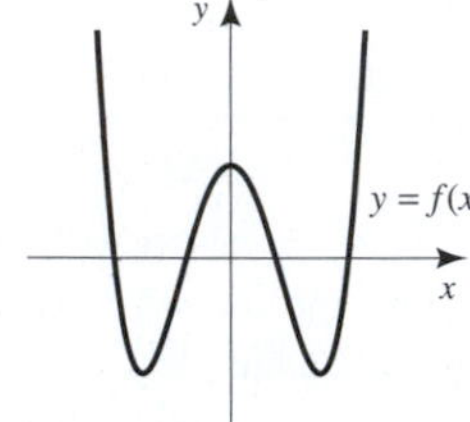

73. a.

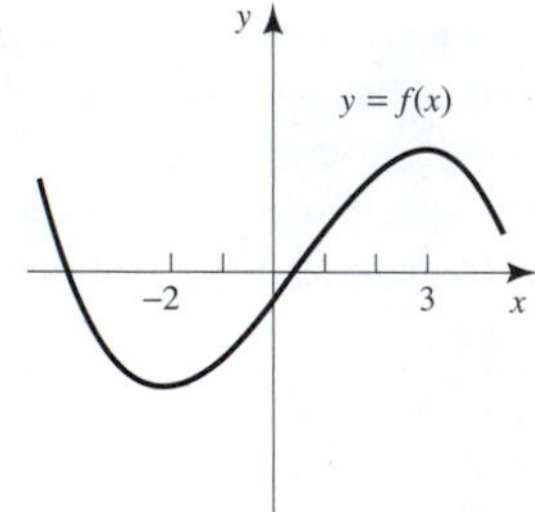

b. Yes. Since $f(x)$ is a polynomial, $f''(x)$ is also a polynomial. As a polynomial, $f''(x)$ is continuous everywhere. Since $f''(x)$ changes sign going from -2 to 3, $f''(x)$ must be zero on the interval $(-2, 3)$. So there must be an inflection point on this interval, since $f''(x)$ changes sign.

75. $g''(x) = \dfrac{-f''(x)[f(x)]^2 + 2f(x)[f'(x)]^2}{[f(x)]^4} < 0,$

since $f''(x) > 0$ and $f(x) < 0$ for all x.

77. We have $f'(0) = \lim\limits_{h\to 0} \dfrac{f(0+h) - f(0)}{h}$. To evaluate this limit we need to take the limit from the left and the limit from the right. We have

$$\lim_{h\to 0^-} \frac{f(0+h) - f(0)}{h} = \lim_{h\to 0^-} \frac{0-0}{h}$$

$$\lim_{h\to 0^+} \frac{f(0+h) - f(0)}{h} = \lim_{h\to 0^+} \frac{h^3-0}{h}$$

$$= \lim_{h\to 0^+} h^2$$

$$= 0$$

This shows that $f'(0) = 0$. So we now have that $f'(x) = 0$ when $x \le 0$ and $f'(x) = 3x^2$ when $x > 0$.

We have $f''(0) = \lim\limits_{h\to 0} \dfrac{f'(0+h) - f'(0)}{h}$. To evaluate this limit we need to take the limit from the left and the limit from the right. We have

$$\lim_{h\to 0^-} \frac{f'(0+h) - f'(0)}{h} = \lim_{h\to 0^-} \frac{0-0}{h}$$

$$\lim_{h\to 0^+} \frac{f'(0+h) - f'(0)}{h} = \lim_{h\to 0^+} \frac{3h^2-0}{h}$$

$$= \lim_{h\to 0^+} 3h$$

$$= 0$$

This shows that $f''(0) = 0$.

79. $-\sin x, -\cos x$

81. Critical values at $x = 1, 3$;
increasing on $(1, 3)$;
decreasing on $(-\infty, 1), (3, \infty)$;
concave up on $(-\infty, 2)$;
concave down on $(2, \infty)$;
inflection value at $x = 2$

83. a. $y = -0.0254x^2 + 2.5787x + 14.292,\ r^2 = 0.9720$
b. Yes, since the derivative, marginal cost, is decreasing.

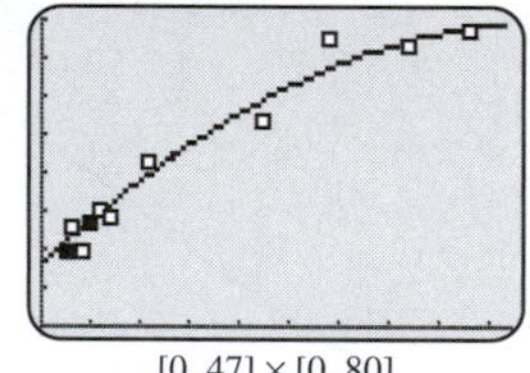

[0, 47] × [0, 80]

85. Logistic: $y = \dfrac{1085}{1 + 203e^{-0.0716t}}$. The graph is concave up at first, then turns concave down. As $t \to \infty$, $y \to 1085$.

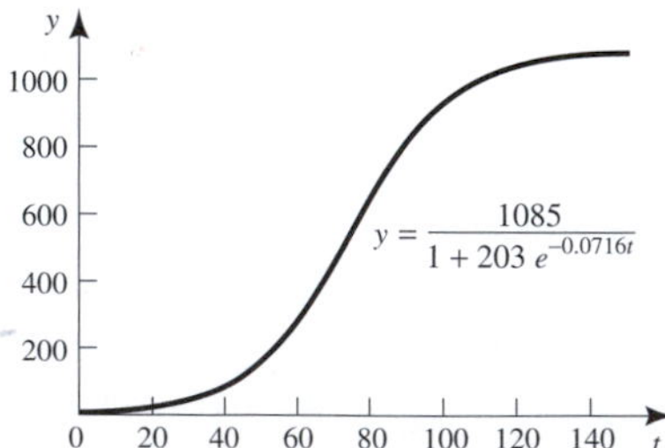

Section 5.3

1. $\dfrac{2}{3}$ **3.** 0

5. ∞ **7.** $-\infty$

9. 1 **11.** 5

13. 1 **15.** 0

17. 2 **19.** $r \to c$

21. 100%

23. Since $f'(x) > 0$ and $f''(x) < 0$ for positive x, f increases and is concave down. As $x \to \infty$, $f(x) \to 0.259$.

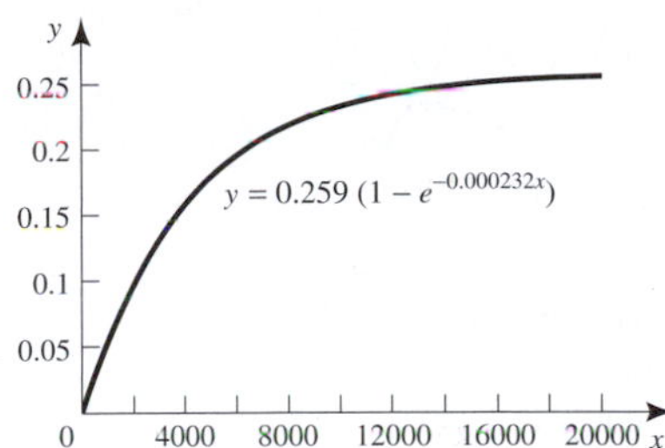

25. Since $L'(t) > 0$, $L(t)$ is always increasing. As $t \to \infty$, $L(t) \to 0.26e^{2.876}$, so this number is a bound for the length of the larva.

27. **a.** As $t \to \infty$, $T(t) \to 70$. Thus, the temperature of the milk returns ultimately to the temperature of the room.
b. About 1.01 hours.

29. As $x \to \infty$, $y \to 1.27$.

31. For large t, $P(t) \approx 4$, so the profit is approximately \$4 million.

33. 0

35. **a.** ii
b. iii
c. iv
d. i

37. As x becomes large without bound, any polynomial (other than a constant) must become either positively or negatively large without bound. Thus, a nonconstant polynomial cannot be asymptotic to any horizontal line. Obviously, the indicated polynomial is not a constant.

39. **a.** $w = 0.10293(0.91227)^x$
b. As $t \to \infty$, $w \to 0$, as you would expect.

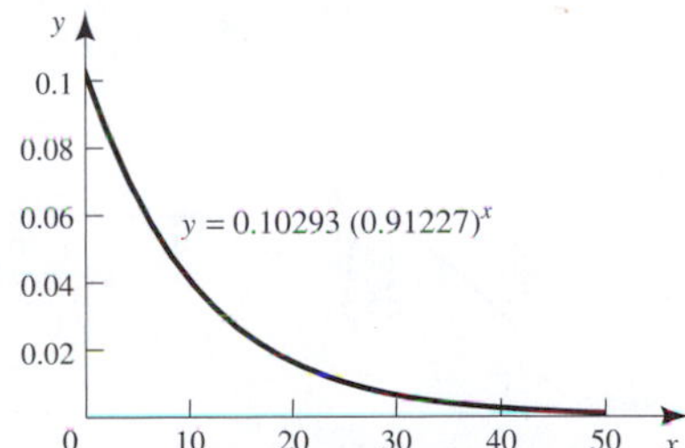

Section 5.4

1. Critical values at $x = 0, -3$
no relative maximum
relative minimum at $x = -3$
inflection values at $x = 0, -2$
increasing on $(-3, \infty)$
decreasing on $(-\infty, -3)$
concave up on $(-\infty, -2)$, $(0, \infty)$
concave down on $(-2, 0)$

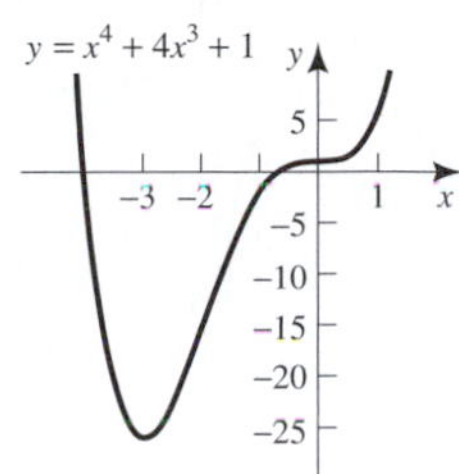

3. Critical values at $x = -4, 0, 4$
relative maximum at $x = 0$
relative minimum at $x = -4, 4$
inflection values at $x = \pm 4/\sqrt{3}$
increasing on $(-4, 0)$, $(4, \infty)$
decreasing on $(-\infty, -4)$, $(0, 4)$
concave up on $(-\infty, -4/\sqrt{3})$, $(4/\sqrt{3}, \infty)$
concave down on $(-4/\sqrt{3}, 4/\sqrt{3})$

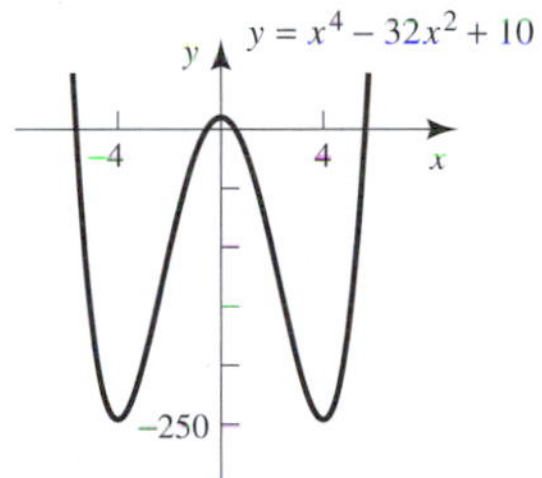

5. Critical values at $x = 0, \pm\sqrt{2}$
relative maximum at $x = \sqrt{2}$
relative minimum at $x = -\sqrt{2}$
inflection values at $x = 0, \pm 1$
increasing on $(-\sqrt{2}, \sqrt{2})$
decreasing on $(-\infty, -\sqrt{2})$ and $(\sqrt{2}, \infty)$
concave up on $(-\infty, -1)$ and $(0, 1)$
concave down on $(-1, 0)$ and $(1, \infty)$

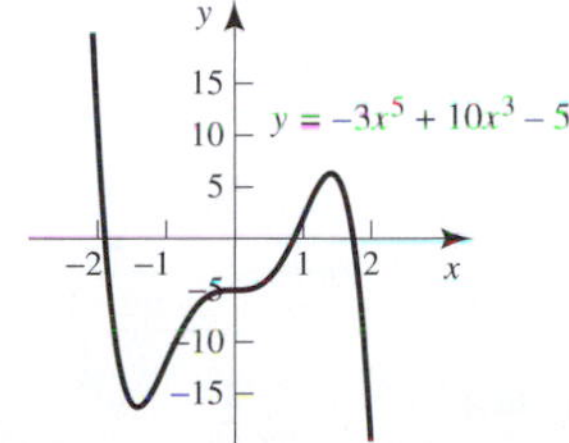

7. No critical values
no relative maximum
no relative minimum
no inflection values
nowhere increasing
decreasing on $(-\infty, 1)$, $(1, \infty)$
concave up on $(1, \infty)$
concave down on $(-\infty, 1)$
asymptotes: $x = 1$, $y = 0$

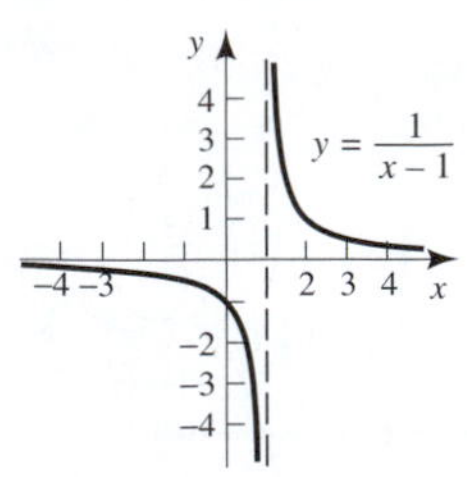

9. No critical or inflection values
no maxima or minima
increasing nowhere
decreasing on $(-\infty, 1)$, $(1, \infty)$
concave up on $(1, \infty)$
concave down on $(-\infty, 1)$
asymptotes: $x = 1$, $y = 1$

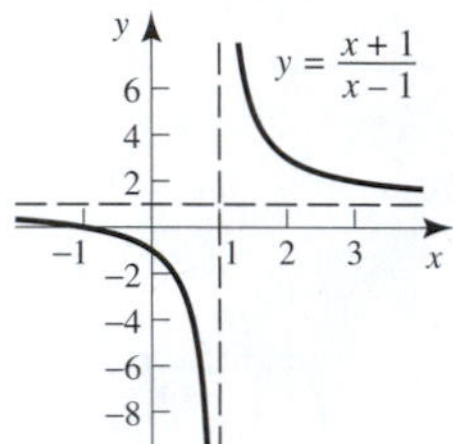

11. Critical value at $x = 0$
relative maximum at $x = 0$
no relative minimum
inflection values at $x = \pm\dfrac{1}{\sqrt{3}}$
increasing on $(-\infty, 0)$
decreasing on $(0, \infty)$
concave up on $(-\infty, -1/\sqrt{3})$, $(1/\sqrt{3}, \infty)$
concave down on $(-1/\sqrt{3}, /\sqrt{3})$
asymptote: $y = 0$

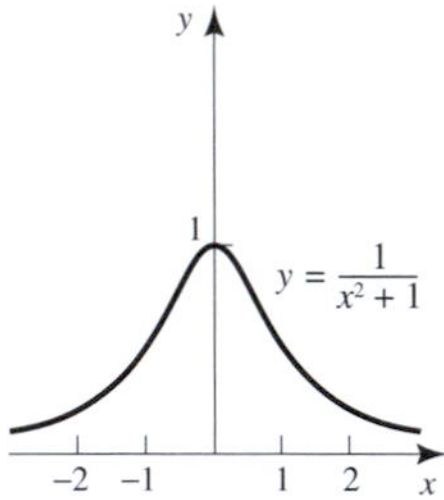

13. Critical values at $x = \pm 3$
relative maximum at $x = -3$
relative minimum at $x = 3$
no inflection values
increasing on $(-\infty, -3)$, $(3, \infty)$
decreasing on $(-3, 0)$, $(0, 3)$
concave up on $(0, \infty)$
concave down on $(-\infty, 0)$
asymptote: $x = 0$

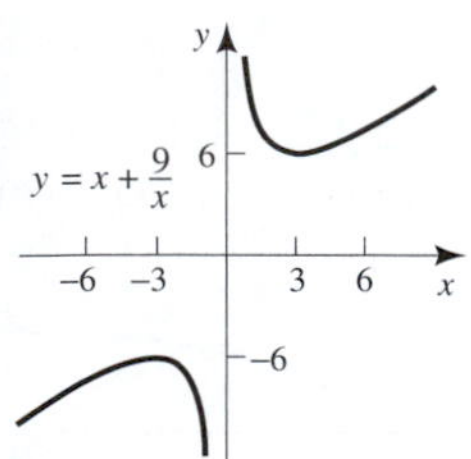

15. Critical values at $x = \pm 1$
relative maximum at $x = -1$
relative minimum at $x = 1$
no inflection values
increasing on $(-\infty, -1)$, $(1, \infty)$
decreasing on $(-1, 0)$, $(0, 1)$
concave up on $(0, \infty)$
concave down on $(-\infty, 0)$
asymptote: $x = 0$

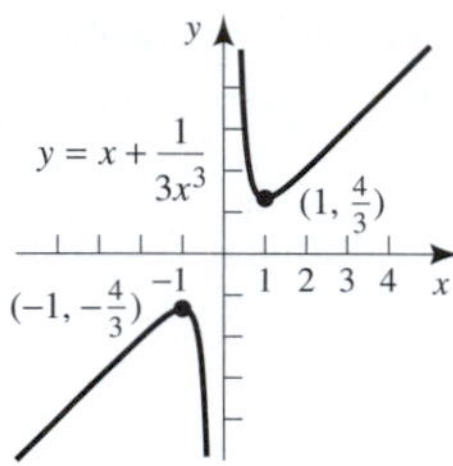

17. Critical values at $x = 0, 2$
relative maximum at $x = 0$
relative minimum at $x = 2$
no inflection values
increasing on $(-\infty, 0)$, $(2, \infty)$
decreasing on $(0, 1)$, $(1, 2)$
concave up on $(1, \infty)$
concave down on $(-\infty, 1)$
asymptote: $x = 1$

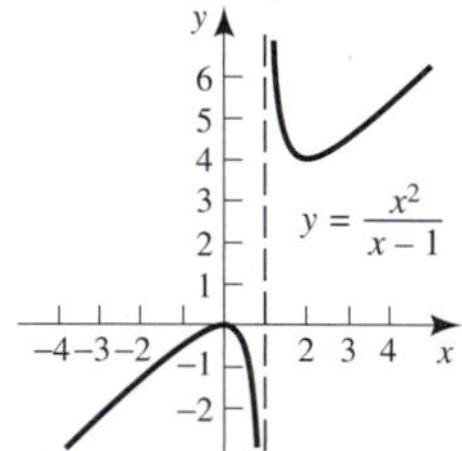

19. Critical values at $x = \pm 1$
relative maximum at $x = 1$
relative minimum at $x = -1$
inflection values at $x = 0, \pm\sqrt{3}$
increasing on $(-1, 1)$
decreasing on $(-\infty, -1)$, $(1, \infty)$
concave up on $(-\sqrt{3}, 0)$, $(\sqrt{3}, \infty)$
concave down on $(-\infty, -\sqrt{3})$, $(0, \sqrt{3})$
asymptote: $y = 0$

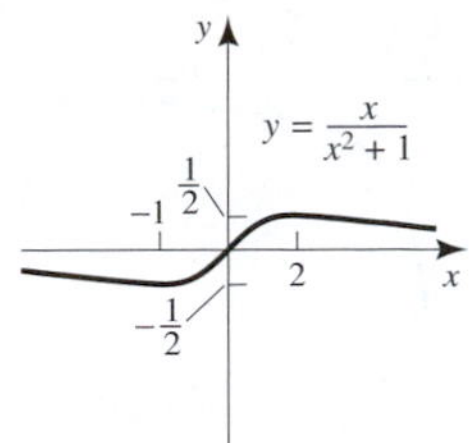

21. Critical value at $x = 2$
relative maximum at $x = 2$
no relative minima
inflection value at $x = 3$
increasing on $(0, 2)$
decreasing on $(-\infty, 0)$, $(2, \infty)$
concave up on $(3, \infty)$
concave down on $(-\infty, 0)$, $(0, 3)$
asymptotes: $x = 0$, $y = 0$

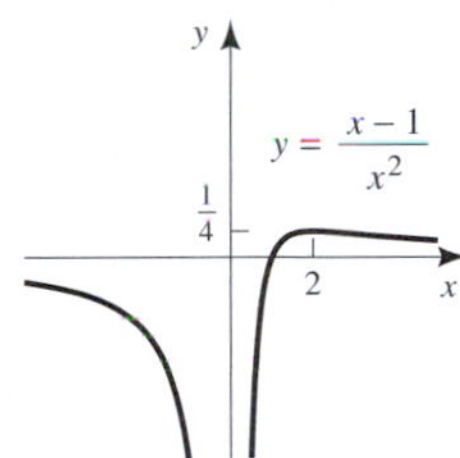

23. Critical value at $x = 4$
relative maximum at $x = 4$
no relative minima
inflection value at $x = 6$
increasing on $(0, 4)$
decreasing on $(-\infty, 0)$, $(4, \infty)$
concave up on $(6, \infty)$
concave down on $(-\infty, 0)$, $(0, 6)$
asymptotes: $x = 0$, $y = 1$

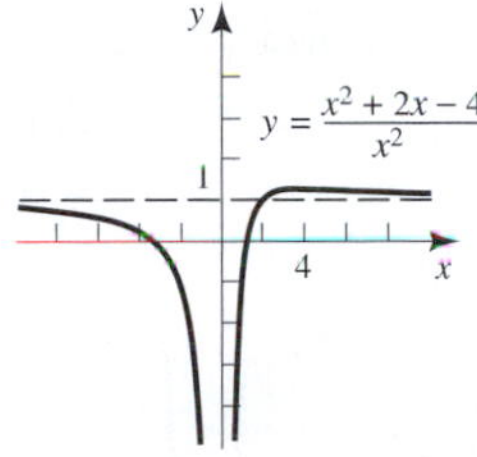

25. Critical values at $x = 0, 1$
relative maximum at $x = 0$
relative minima at $x = 1$
inflection value at $x = -\dfrac{1}{2}$
increasing on $(-\infty, 0)$, $1, \infty)$
decreasing on $(0, 1)$
concave up on $\left(-\frac{1}{2}, 0\right)$, $(0, \infty)$
concave down on $(-\infty, -1/2)$
no asymptotes

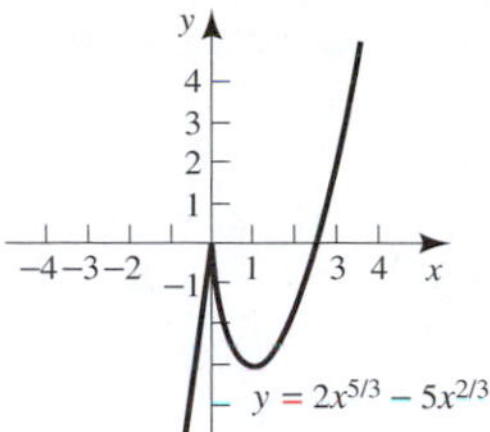

27. Critical values at $x = 0, 1$
relative maximum at $x = 1$
relative minimum at $x = 0$
no inflection value
increasing on $(0, 1)$
decreasing on $(-\infty, 0)$, $(1, \infty)$
concave up nowhere
concave down on $(-\infty, 0)$, $(0, \infty)$

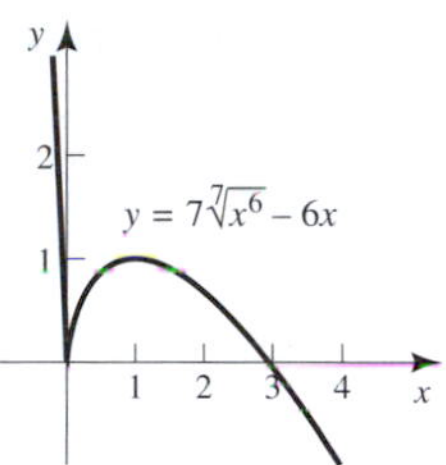

29. Critical values at $x = 0, \pm\sqrt{2}$
relative maxima at $x = \pm\sqrt{2}$
relative minimum at $x = 0$
inflection values at $x = \pm\sqrt{\dfrac{5 \pm \sqrt{17}}{2}}$
increasing on $(-\infty, -\sqrt{2})$, $(0, \sqrt{2})$
decreasing on $(-\sqrt{2}, 0)$, $(\sqrt{2}, \infty)$
concave up on $\left(-\infty, -\sqrt{\frac{5+\sqrt{17}}{2}}\right)$, $\left(-\sqrt{\frac{5-\sqrt{17}}{2}}, \sqrt{\frac{5-\sqrt{17}}{2}}\right)$
and $\left(\sqrt{\frac{5+\sqrt{17}}{2}}, \infty\right)$
concave down on $\left(-\sqrt{\frac{5+\sqrt{17}}{2}}, -\sqrt{\frac{5-\sqrt{17}}{2}}\right)$ and
$\left(\sqrt{\frac{5-\sqrt{17}}{2}}, \sqrt{\frac{5+\sqrt{17}}{2}}\right)$
asymptote: $y = 0$

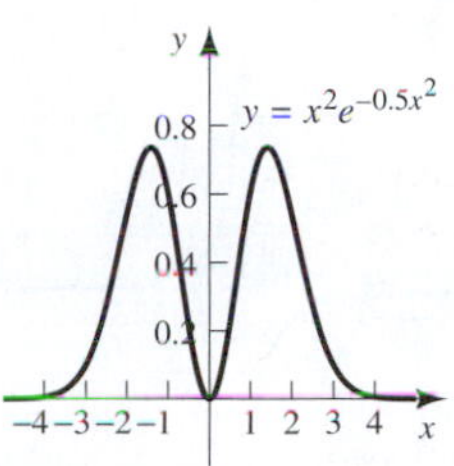

31. Critical value at $x = 0$
relative maximum at $x = 0$
no relative minimum

inflection values at $x = \ln(\sqrt{2} \pm 1)$
increasing on $(-\infty, 0)$
decreasing on $(0, \infty)$
concave up on $(-\infty, \ln(\sqrt{2} - 1)), (\ln(\sqrt{2} + 1), \infty)$
concave down on $(\ln(\sqrt{2} - 1), \ln(\sqrt{2} + 1))$
asymptote: $y = 0$

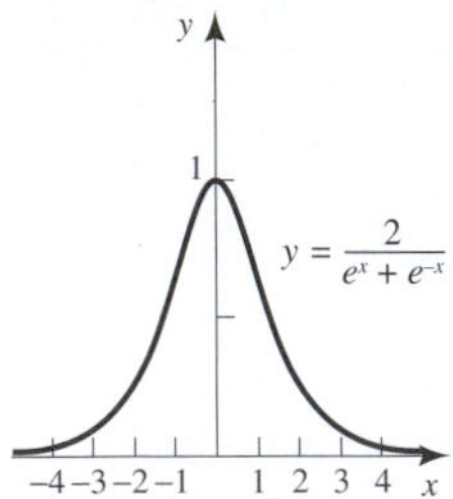

33. Critical value at $x = \sqrt{e}$
relative maximum at $x = \sqrt{e}$
no relative minimum
inflection values at $x = e^{5/6}$
increasing on $(0, \sqrt{e})$
decreasing on $(\sqrt{e}, \infty)$
concave up on $(e^{5/6}, \infty)$
concave down on $(0, e^{5/6})$
asymptote: $x = 0$, $y = 0$

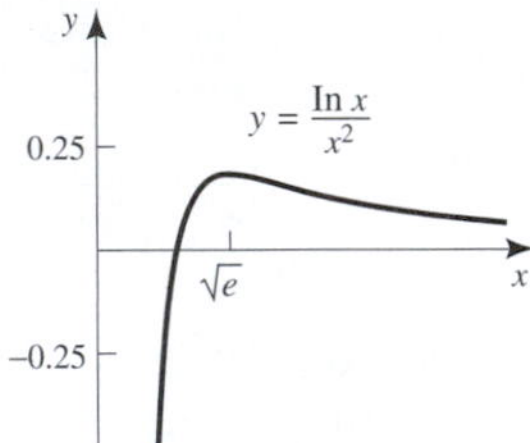

35. No critical value
no relative maximum
no relative minimum
no inflection values
nowhere increasing
decreasing on $(-\infty, 0), (0, \infty)$
concave up on $(0, \infty)$
concave down on $(-\infty, 0)$
asymptote: $x = 0$, $y = 0$

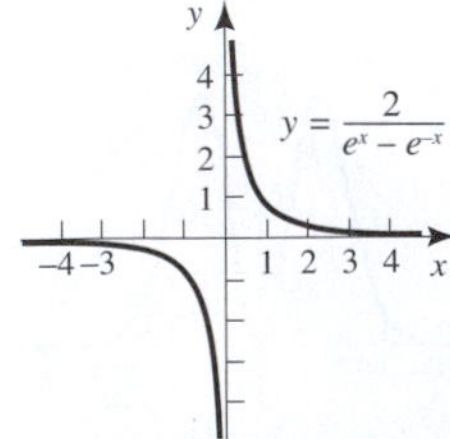

37. $n(t)$ is increasing and concave down on $(0, \infty)$ and $n(t) \to 9$, as $t \to \infty$

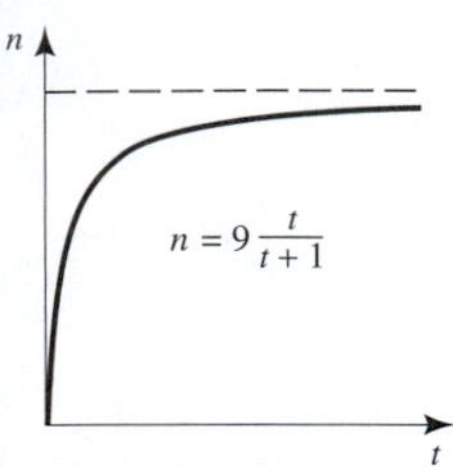

39. Critical value at $x = 0$
inflection values at $x = \pm b/\sqrt{3}$
asymptote: $y = 0$

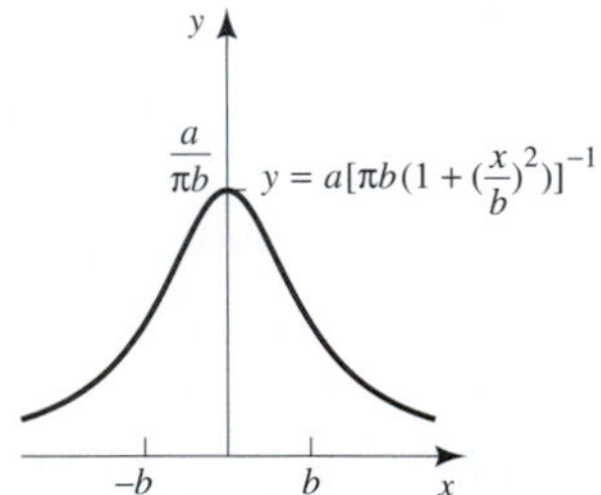

41. $\sqrt[3]{0.5}$

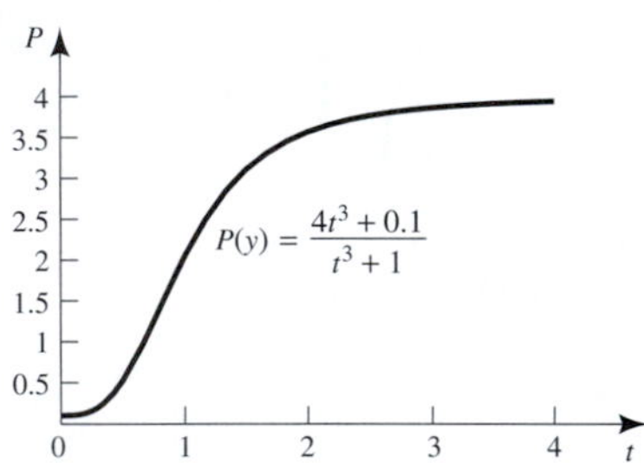

43. Critical values at $x = 0, \pm 4\sqrt{b}$
relative maximum at $x = 0$
relative minimum at $x = \pm 4\sqrt{b}$
inflection values at $\pm 4\sqrt{b/3}$
increasing on $(-4\sqrt{b}, 0), (4\sqrt{b}, \infty)$
decreasing on $(-\infty, -4\sqrt{b}), (0, 4\sqrt{b})$
concave up on $(-\infty, -4\sqrt{b/3}), (4\sqrt{b/3}, \infty)$
concave down on $(-4\sqrt{b/3}, 4\sqrt{b/3})$

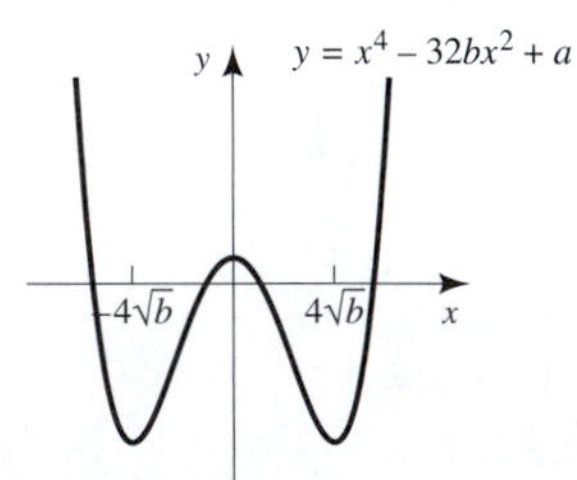

45. Critical values at $x = \pm\sqrt{b}$
relative maximum at $x = -\sqrt{b}$
relative minimum at $x = \sqrt{b}$
no inflection values
increasing on $(-\infty, -\sqrt{b}), (\sqrt{b}, \infty)$

decreasing on $(-\sqrt{b}, 0)$, $(0, \sqrt{b})$
concave up on $(0, \infty)$
concave down on $(-\infty, 0)$
asymptote: $x = 0$

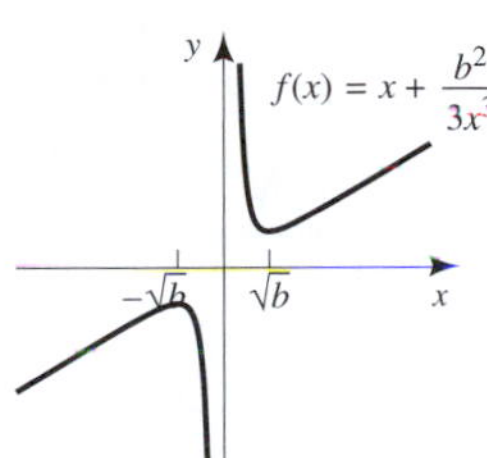

47.

49.

51. $C(t) = k(e^{-at} - e^{-bt}) = ke^{-at}[1 - e^{(a-b)t}]$
Since $a - b < 0$, $1 - e^{(a-b)t}$ is approximately 1 for large t.
Therefore, $C(t) \approx ke^{-at}$.

Section 5.5

1. Absolute minimum in June.
absolute maximum in November.

3. Absolute minimum in September.
absolute maximum in April.

5. Absolute minimum in 1934,
absolute maximum in 1910.

7. **a.** Absolute minimum at $x = 2$,
absolute maximum at $x = 4$;
b. absolute minimum at $x = 1$,
absolute maximum at $x = 2$;
c. absolute minimum at $x = -2, 1$,
absolute maximum at $x = -1, 2$.

9. **a.** Absolute minimum at $x = 0$,
absolute maximum at $x = -2$;
b. absolute minimum at $x = 2$,
absolute maximum at $x = 1$;
c. absolute minimum at $x = 2$,
absolute maximum at $x = -2$.

11. Absolute minimum at $x = 1$,
absolute maximum at $x = -1$.

13. Absolute minimum at $x = 0$,
absolute maximum at $x = 2$.

15. Absolute minimum at $x = 1$,
absolute maximum at $x = 3$.

17. Absolute minimum at $x = -3, 0$,
absolute maximum at $x = 2$.

19. Absolute minimum at $x = -2, 2$,
absolute maximum at $x = -3, 3$.

21. Absolute minimum at $x = 2$,
absolute maximum at $x = 8$.

23. Absolute minimum at $x = \frac{1}{2}$,
no absolute maximum.

25. Absolute minimum at $x = 0$,
no absolute maximum.

27. Absolute minimum at $x = 0$,
no absolute maximum.

29. **a.** Absolute minimum at $x = 1$,
absolute maximum at $x = \frac{1}{2}$;
b. no absolute minimum, no absolute maximum;
c. absolute minimum at $x = 1$, no absolute maximum;
d. no absolute minimum, absolute maximum at $x = -1$.

31. 8, 8

33. 10, 10

35. 5, 20

37. 10, -10

39. $2a^2$

41. $\left(\frac{c}{2a}, \frac{c}{2b}\right)$

43. $x = 5$

45. $r/2$

47. 200 pounds. 145 bushels.

49. b/c. $D = 200$ corresponds to about July 21.

51. $D'(t) = 0.00007(36 - t)^{-0.6}[-2.4t^2 + 90.2t - 468]$
Maximum when $D \approx 31.4$.

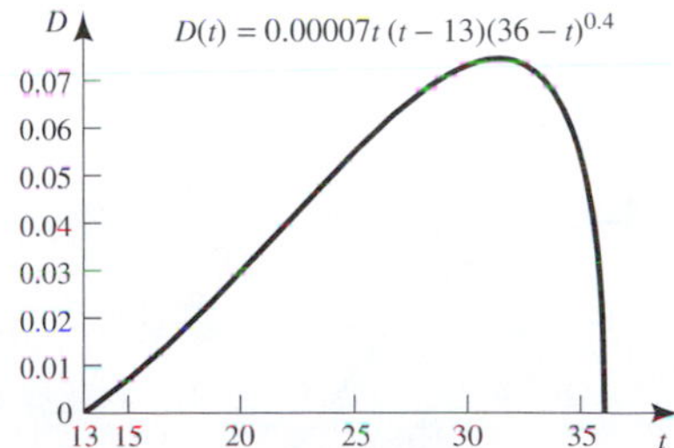

53. **a.** $0.0008227x^2 - 0.1981x + 21.04$
b. 120

55. $0.03321x^2 - 0.39969x + 1.01396$. 6.02.

57. $-0.0057x^2 + 0.297x - 2.8402$. About 26 degrees.

Section 5.6

1. 200×200 3. $x = 1.5$ 5. 10×20

7. 100×200 9. 40 11. \$450

13. $4 \times 4 \times 2$, 48 15. $6 \times 6 \times 9$

17. About 185 pounds 19. Four runs of 4000 each

21. C is $10\sqrt{3}$ feet to the right of D

23. Cut at $\dfrac{48}{3\sqrt{3}+4} \approx 5.22$, and use the shorter piece for the square.

25. $\dfrac{500}{9+\sqrt{3}} \approx 46.6$ yards

27. Absolute minimum at $x = -1$, absolute maximum at $x = 4$

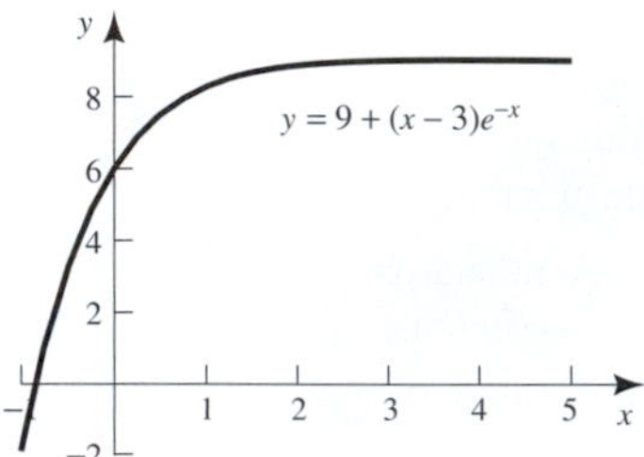

Section 5.7

1. 100%

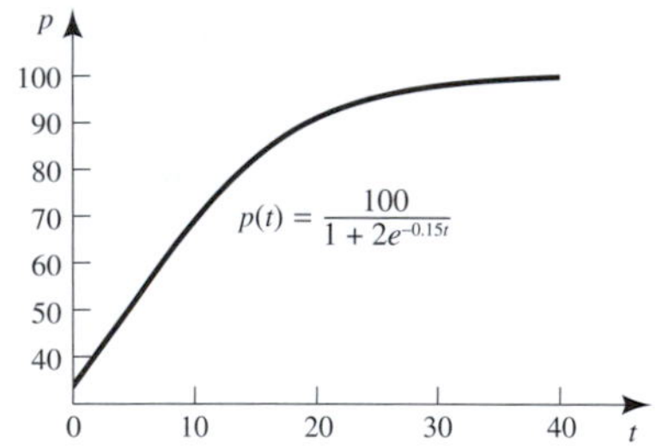

3. About 61.

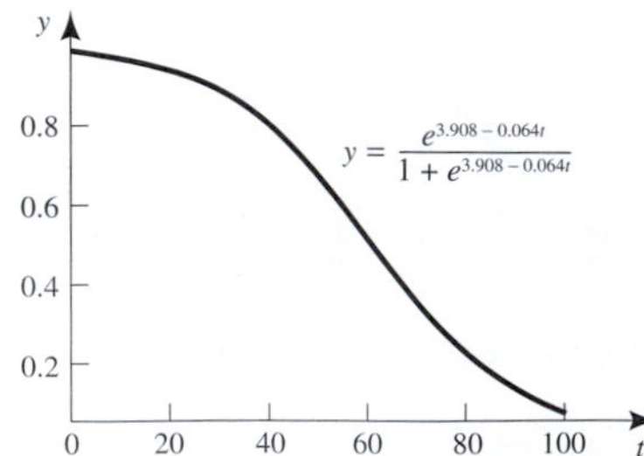

5. The graph is concave up to about $x = 49$ and concave down afterward. The rate of nitrogen accumulation increases until about 49 years later and then begins to decrease to zero. The level of nitrogen approaches 0.172.

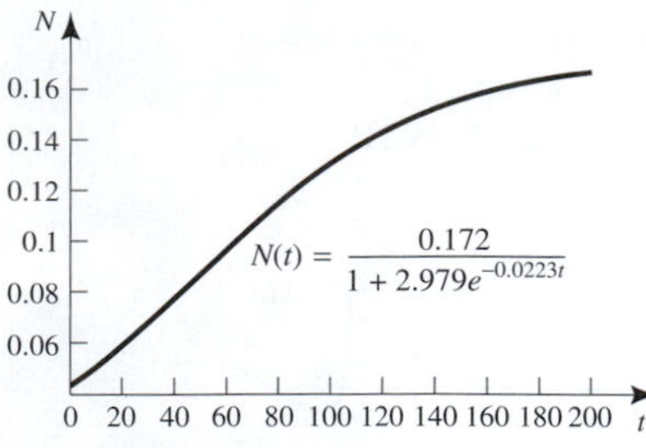

7. The graph is concave up to about $t = 190.5$ and concave down afterward. The rate of citrus growth increases until about 190.5 days and then begins to decrease to zero. The fruit surface area approaches 146.3346 cm^2.

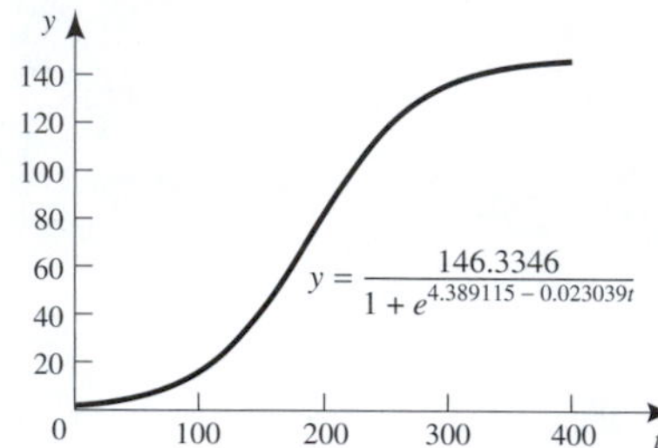

9. The graph is concave up to about $x = 4$ and concave down afterward. The rate of head capsule width growth increases until about $x = 4$ (the fourth instar) and then begins to decrease to zero. The head capsule width approaches 2.75 mm.

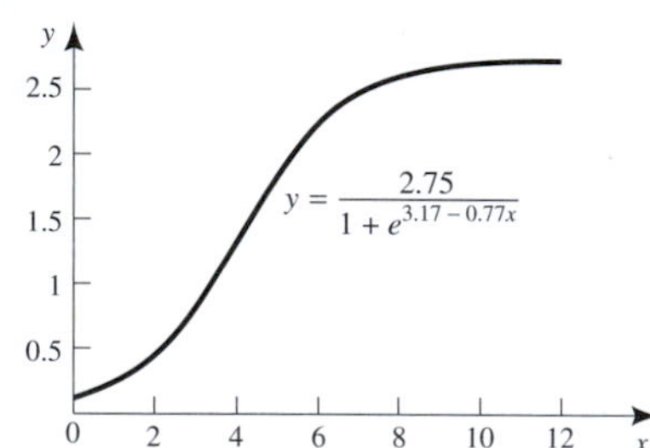

11. $5 \ln 20 \approx 14.98$, 10

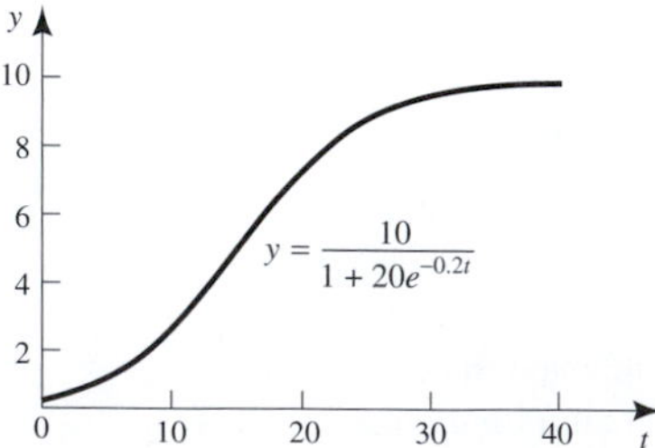

13. 5.4 years

15. $P'(t) = \dfrac{akLe^{-kt}}{(1+ae^{-kt})^2} > 0$ for all $t \geq 0$

17. $P''(t) = ak^2L\dfrac{e^{-kt}(ae^{-kt}-1)}{(1+ae^{-kt})^3} = 0$ if and only if $ae^{-kt} - 1 = 0$. This implies $t = \dfrac{1}{k}\ln a$.

19. Since $e^{-kc} = e^{-\ln a} = \dfrac{1}{a}$,
$$P(c) = \frac{L}{1+ae^{-kc}} = \frac{L}{1+a/a} = \frac{L}{2}$$

21. Since $P_0 = \dfrac{100}{1+a}$, $a = \dfrac{100}{P_0} - 1$. Since
$$Q = P(T) = \frac{100}{1+ae^{-kT}} = \frac{100}{1+(100/P_0-1)e^{-kT}},$$
$e^{-kT} = \dfrac{P_0(100-Q)}{Q(100-P_0)}$. This implies that
$$k = -\frac{1}{T}\ln\frac{P_0(100-Q)}{Q(100-P_0)} = \frac{1}{T}\ln\frac{Q(100-P_0)}{P_0(100-Q)}$$

Section 5.8

1. -2 **3.** $-\dfrac{3}{5}$

5. $\dfrac{9}{2}$ **7.** $-\dfrac{2}{7}$

9. -2 **11.** $-\dfrac{1}{2}$

13. $\dfrac{16}{31}$ **15.** $-\dfrac{1}{2}$

17. -3 **19.** $-\dfrac{15}{7}$

21. $-\dfrac{3}{4}$ **23.** $-\dfrac{5}{4}$

25. $-\dfrac{\sqrt[4]{al}}{4}$ **27.** 6

29. 6 **31.** 16

33. 12 **35.** -1

37. -1 **39.** -2

41. -200 **43.** 400

45. 1.6 miles per hour **47.** -3 feet per second

49. $\dfrac{125}{16\pi}$ yards per minute

51. 40 square feet per second

53. Since $f(g(x)) = x$, $\dfrac{df(g(x))}{dx} = 1$ and $f'(g(x))\cdot g'(x) = 1$. When $x = c$, $g(c) = y_0$, we have $f'(g(c))g'(c) = 1$. Then $f'(y_0) = \dfrac{1}{g'(c)}$, since $g'(c) \neq 0$.

55. $f(g(x)) = \sqrt[3]{g(x)} = \sqrt[3]{x^3} = x$, $c = 2$, $y_0 = g(c) = c^3 = 8$. Then $f'(y_0) = \dfrac{1}{3}y_0^{-2/3} = \dfrac{1}{12}$ and $g'(c) = 3c^2 = 12$. Therefore, $f'(y_0) = \dfrac{1}{g'(c)}$.

Chapter 5 Review Exercises

1. Critical values at $x = 3, 5, 7, 11, 13, 15$; increasing on $(3, 7)$, $(11, 13)$, $(15, 18)$; decreasing on $(1, 3)$, $(7, 11)$, $(13, 15)$

2. Critical values at $x = -2, 1$; increasing on $(-\infty, -2)$, $(1, \infty)$; decreasing on $(-2, 1)$; relative maximum at $x = -2$; relative minimum at $x = 1$

3. Critical values at $x = -4, 2, 5$; increasing on $(-\infty, -4)$, $(5, \infty)$; decreasing on $(-4, 5)$; relative maximum at $x = -4$; relative minimum at $x = 5$

4. Critical value at $x = -\dfrac{3}{4}$
relative minimum at $x = -\dfrac{3}{4}$
increasing on $(-3/4, \infty)$
decreasing on $(-\infty, -3/4)$
concave up on $(-\infty, \infty)$
concave down nowhere

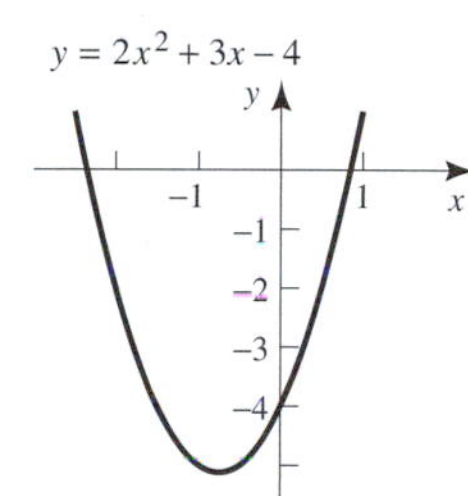

5. Critical value at $x = 1$
relative maximum at $x = 1$
increasing on $(-\infty, 1)$
decreasing on $(1, \infty)$
concave up nowhere
concave down on $(-\infty, \infty)$

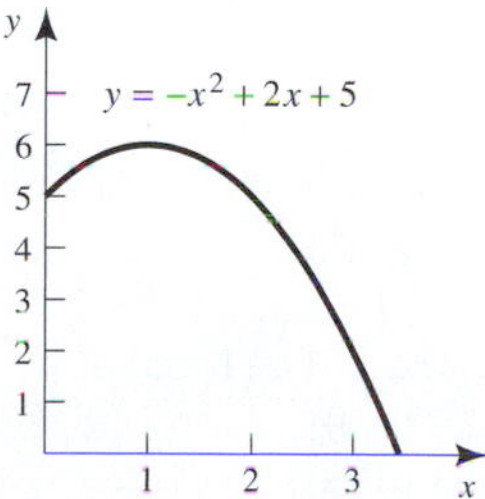

6. Critical value at $x = -1$
relative maximum at $x = -1$
increasing on $(-\infty, -1)$
decreasing on $(-1, \infty)$
concave up nowhere
concave down on $(-\infty, \infty)$

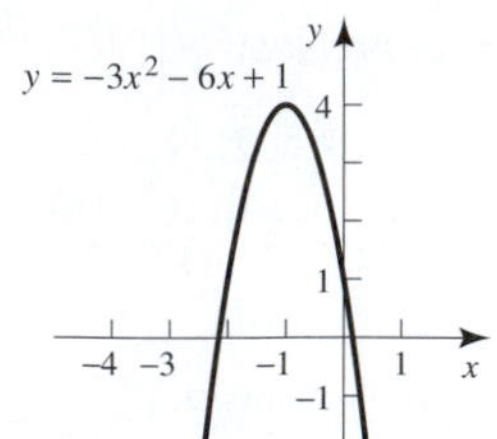

7. Critical value at $x = -2$
relative minimum at $x = -2$
increasing on $(-2, \infty)$
decreasing on $(-\infty, -2)$
concave up on $(-\infty, \infty)$
concave down nowhere

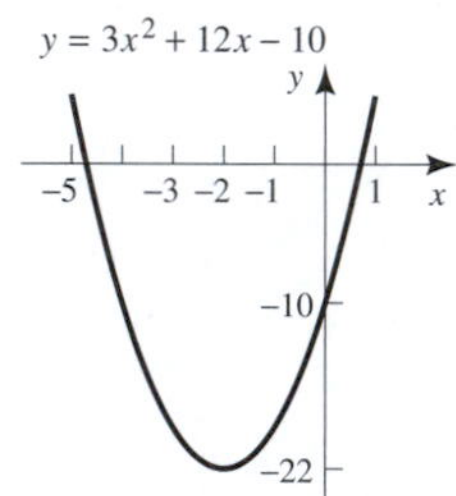

8. Critical values at $x = -1, 1$
relative maximum at $x = -1$
relative minimum at $x = 1$
increasing on $(-\infty, -1), (1, \infty)$
decreasing on $(-1, 1)$
concave up on $(0, \infty)$
concave down on $(-\infty, 0)$
inflection value at $x = 0$

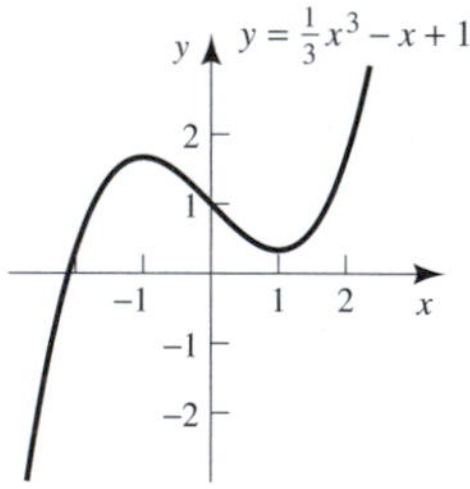

9. Critical values at $x = -1, 1$
relative maximum at $x = 1$
relative minimum at $x = -1$
increasing on $(-1, 1)$
decreasing on $(-\infty, -1), (1, \infty)$
concave up on $(-\infty, 0)$
concave down on $(0, \infty)$
inflection value at $x = 0$

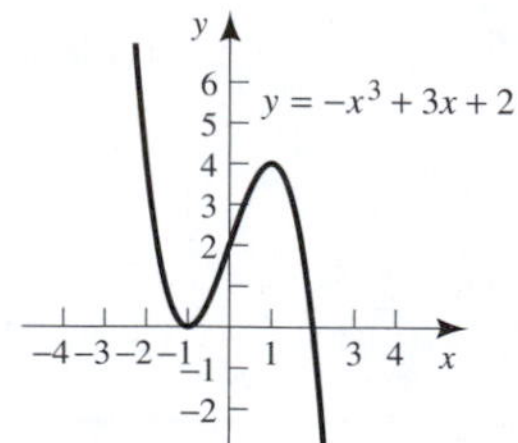

10. Critical values at $x = 0, 3$
relative minimum at $x = 3$
increasing on $(3, \infty)$
decreasing on $(-\infty, 3)$
concave up on $(-\infty, 0), (2, \infty)$
concave down on $(0, 2)$
inflection values at $x = 0, 2$

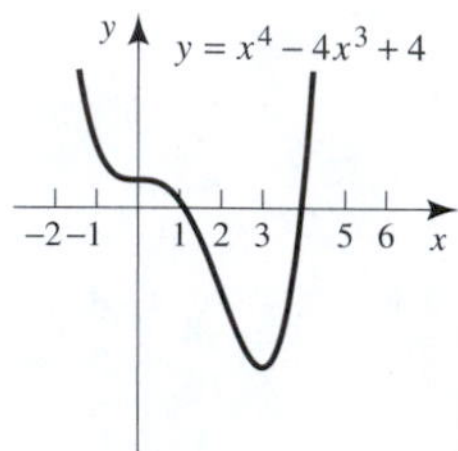

11. Critical values at $x = 0, 12$
relative maximum at $x = 0$
relative minimum at $x = 12$
increasing on $(-\infty, 0), (12, \infty)$
decreasing on $(0, 12)$
concave up on $(9, \infty)$
concave down on $(-\infty, 9)$
inflection value at $x = 9$

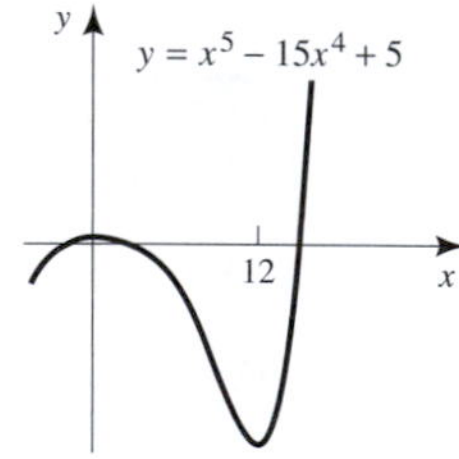

12. Critical values at $x = -3, 0, 3$
relative maximum at $x = 0$
relative minima at $x = -3, 3$
increasing on $(-3, 0), (3, \infty)$
decreasing on $(-\infty, -3), (0, 3)$
concave up on $(-\infty, -\sqrt{3}), (\sqrt{3}, \infty)$
concave down on $(-\sqrt{3}, \sqrt{3})$
inflection value at $x = \pm\sqrt{3}$

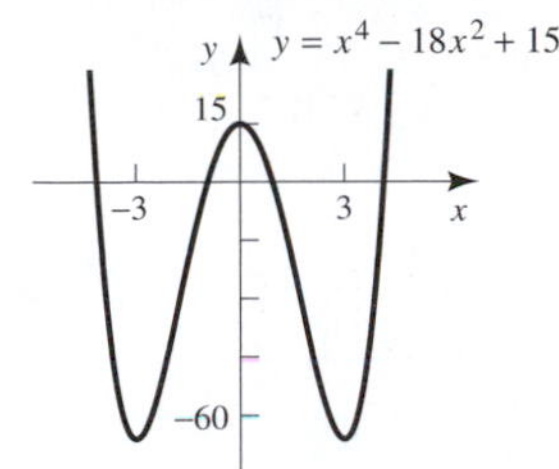

13. Critical values at $x = -3, 0, 3$
relative maximum at $x = 3$
relative minimum at $x = -3$
increasing on $(-3, 3)$
decreasing on $(-\infty, -3)$, $(3, \infty)$
concave up on $(-\infty, -3/\sqrt{2})$, $(0, 3/\sqrt{2})$
concave down on $(-3/\sqrt{2}, 0)$, $(3/\sqrt{2}, \infty)$
inflection value at $x = 0, \pm 3/\sqrt{2}$

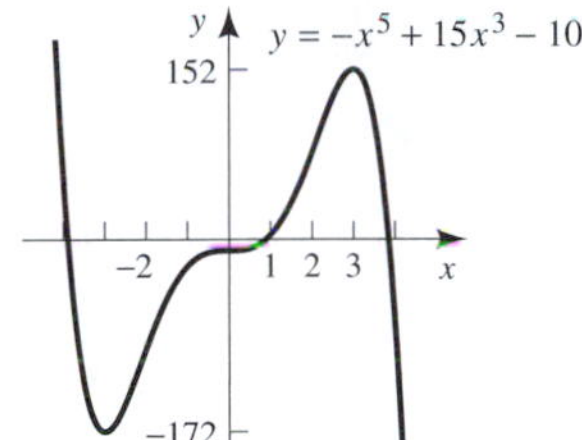

14. Critical value at $x = 2$
relative minimum at $x = 2$
increasing on $(-\infty, 0)$, $(2, \infty)$
decreasing on $(0, 2)$
concave up on $(-\infty, 0)$, $(0, \infty)$
concave down nowhere
no inflection values
asymptote at $x = 0$

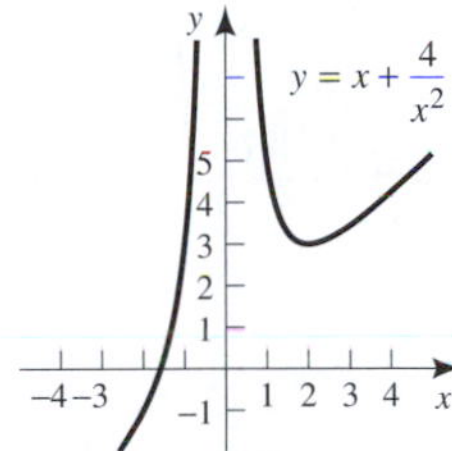

15. Critical values at $x = -1, 1$
relative maximum at $x = -1, 1$
increasing on $(-1, 0)$, $(1, \infty)$
decreasing on $(-\infty, -1)$, $(0, 1)$
concave up on $(-\infty, 0)$, $(0, \infty)$
concave down nowhere
no inflection values
asymptote at $x = 0$

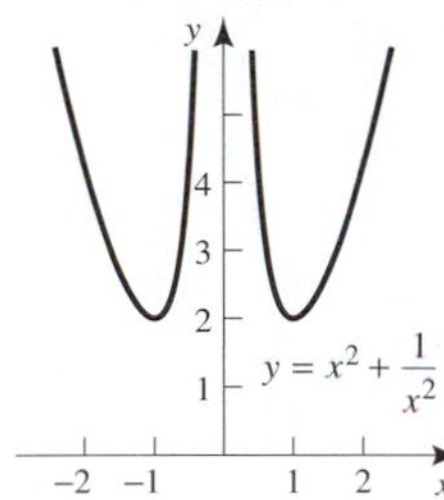

16. Critical value at $x = -\sqrt[3]{2}$
relative maximum at $x = -\sqrt[3]{2}$
increasing on $(-\infty, -\sqrt[3]{2})$, $(0, \infty)$
decreasing on $(-\sqrt[3]{2}, 0)$
concave up nowhere
concave down on $(-\infty, 0)$, $(0, \infty)$
no inflection values
asymptote at $x = 0$

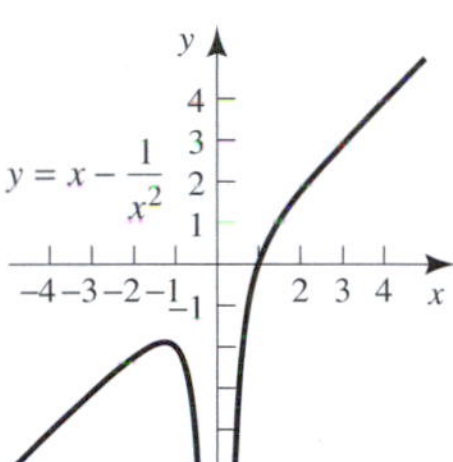

17. No critical values
increasing on $(0, \infty)$
decreasing on $(-\infty, 0)$
concave up on $(-\infty, -3)$, $(3, \infty)$
concave down on $(-3, 0)$, $(0, 3)$
inflection values at $x = \pm 3$
asymptote at $x = 0$

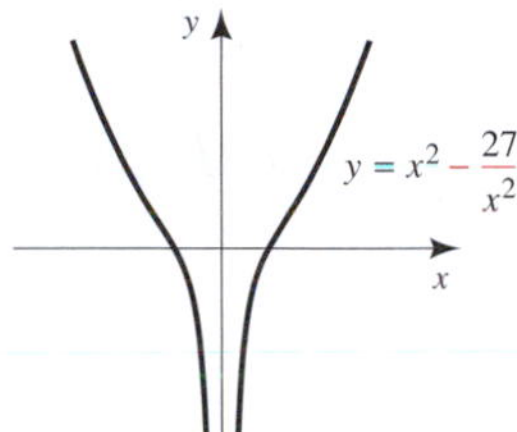

18. No critical values
increasing nowhere
decreasing on $(-\infty, 5)$, $(5, \infty)$
concave up on $(5, \infty)$
concave down on $(-\infty, 5)$
no inflection values
asymptotes at $x = 5$, $y = 1$

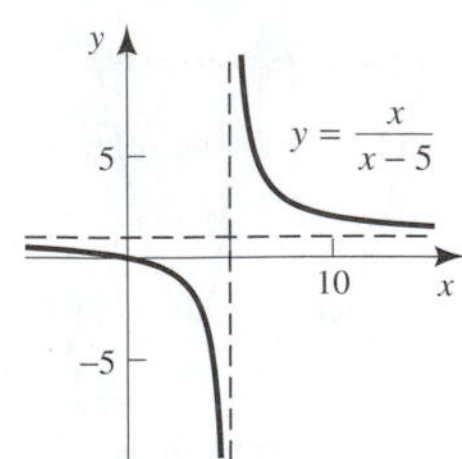

19. No critical values
increasing nowhere
decreasing on $(-\infty, 5)$, $(5, \infty)$
concave up on $(5, \infty)$
concave down on $(-\infty, 5)$
no inflection values
asymptotes at $x = 5$, $y = 1$

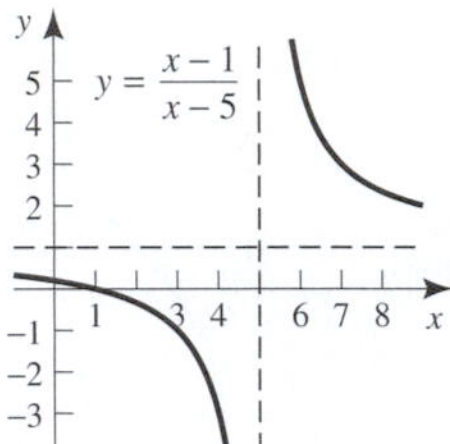

20. Critical values at $x = 0$
relative minimum at $x = 0$
increasing on $(0, \infty)$
decreasing on $(-\infty, 0)$
concave up on $(-\infty, \infty)$
concave down nowhere
no inflection values

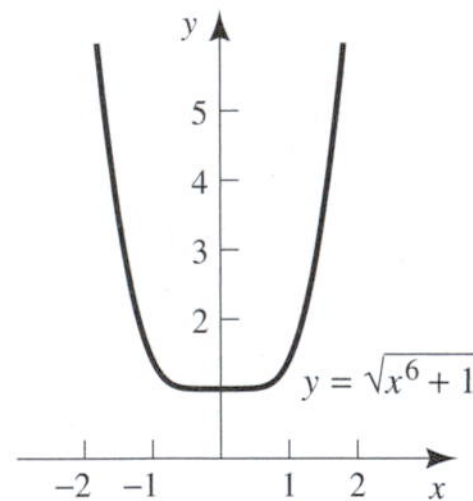

21. Critical value at $x = 9$
increasing nowhere
decreasing on $(-\infty, \infty)$
concave up on $(9, \infty)$
concave down $(-\infty, 9)$
inflection value at $x = 9$

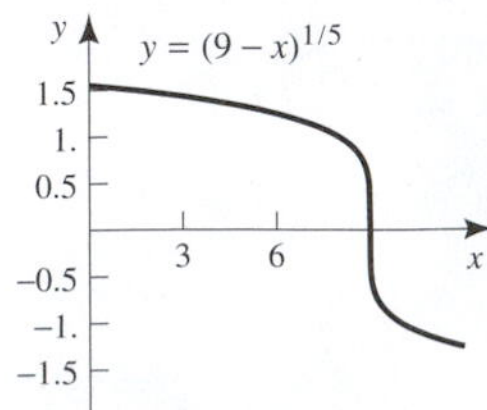

22. Critical value at $x = 0$
relative maximum at $x = 0$
increasing on $(-\infty, 0)$
decreasing on $(0, \infty)$
concave up on $(-\infty, -1/\sqrt{2})$, $(1/\sqrt{2}, \infty)$
concave down $(-1/\sqrt{2}, 1/\sqrt{2})$
inflection values at $x = \pm 1/\sqrt{2}$
asymptote at $y = 0$

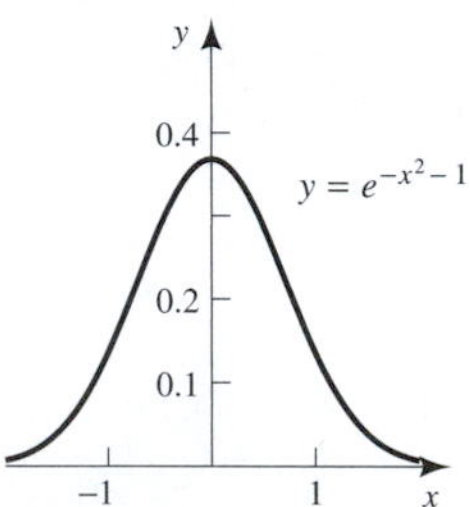

23. Critical value at $x = 0$
relative minimum at $x = 0$
increasing on $(0, \infty)$
decreasing on $(-\infty, 0)$
concave up on $(-1/\sqrt{2}, 1/\sqrt{2})$
concave down $(-\infty, -1/\sqrt{2})$, $(1/\sqrt{2}, \infty)$
inflection values at $x = \pm 1/\sqrt{2}$
asymptote at $y = 1$

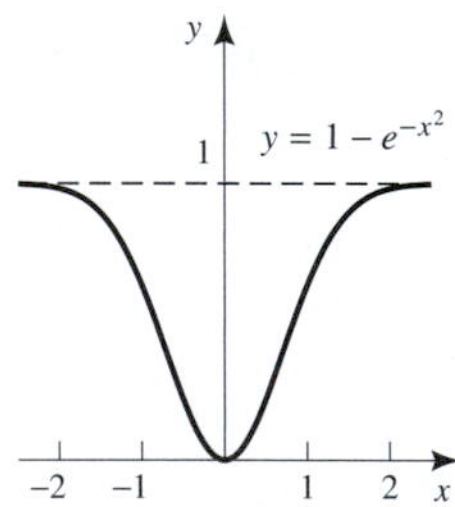

24. Critical value at $x = 0$
relative minimum at $x = 0$
increasing on $(0, \infty)$
decreasing on $(-\infty, 0)$
concave up on $(-1, 1)$
concave down $(-\infty, -1)$, $(1, \infty)$
inflection values at $x = \pm 1$

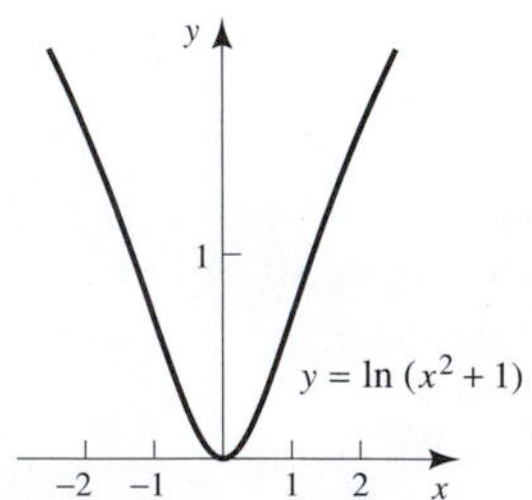

25. No critical values
increasing on $(-\infty, \infty)$
decreasing nowhere

concave up on $(-\infty, \infty)$
concave down nowhere
no inflection values
asymptotes at $y = 0$

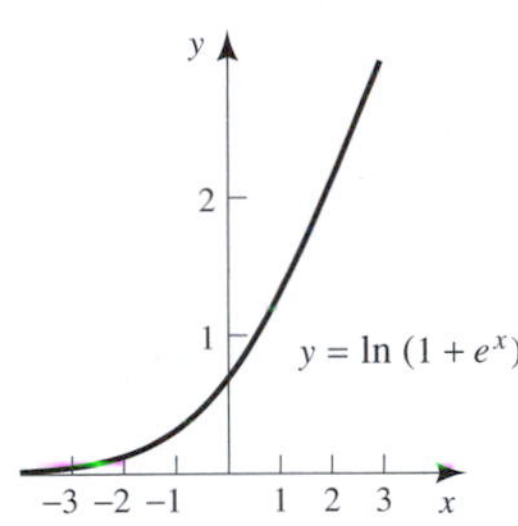

26. Absolute maximum at $x = 1$; absolute minimum at $x = -1$

27. Absolute maximum at $x = -2$; absolute minimum at $x = 2$

28. No absolute maximum; absolute minimum at $x = 1$

29. Absolute maximum at $x = 1$; no absolute minimum

30. Absolute maximum at $x = 2$; absolute minimum at $x = -1$

31. Absolute maximum at $x = 3$; absolute minimum at $x = 0$

32. Absolute maximum at $x = 0$; absolute minimum at $x = 2$

33. Absolute maximum at $x = 1$; absolute minimum at $x = 2$

34. Absolute maximum at $x = 1$; absolute minimum at $x = 2$

35. absolute maximum at $x = 4$; absolute minimum at $x = 1$

36. $-\dfrac{11}{30}$ **37.** $-\dfrac{5}{16}$

38. -1 **39.** 0

40. $f''(x) = 12ax^2 + 2b$. $f'' > 0$ if $a, b > 0$ and $f'' < 0$ if $a, b < 0$.

41. Relative minimum at $x = a$; relative maximum at $x = b$

42. $g''(x) = e^{f(x)}\left[(f'(x))^2 + f''(x)\right] \geq 0$, since $e^{f(x)} > 0$ and $f''(x) > 0$ for all x.

43. $g''(x) = \dfrac{f''(x)f(x) - [f'(x)]^2}{[f(x)]^2} < 0$, since $f > 0$ and $f'' < 0$ for all x.

44. Increasing on $(\sqrt{a/b}, \infty)$; decreasing on $(0, \sqrt{a/b})$

45. 10

46. Solving $\bar{C} = C'$ for x, we obtain $x = 10$, which is the value where the average cost is minimized.

47. $\bar{C}$ is minimized at the value where $\bar{C}' = 0$, that is, $\bar{C}'(x) = \dfrac{xC'(x) - C(x)}{x^2} = 0$. This implies that $C'(x) = \dfrac{C(x)}{x} = \bar{C}(x)$

48. $\dfrac{N}{k}\dfrac{1-a}{2-a}$ **49.** $\sqrt[3]{\dfrac{a}{2b}}$

50. Build only a square enclosure.

51. \$70 **52.** \$72

53. $r = \dfrac{3}{\sqrt[3]{\pi}}$, $h = \dfrac{12}{\sqrt[3]{\pi}}$ **54.** One run of 1000 cases each

55. When $P'(x) = 0$, $x = \dfrac{1}{|p'(x)|}(p(x) - a)$. Thus, increasing a decreases x.

56. 180 **57.** $\dfrac{\sqrt{6}}{3} \approx 0.82$ ft/sec

58. 0, 70.13, everywhere

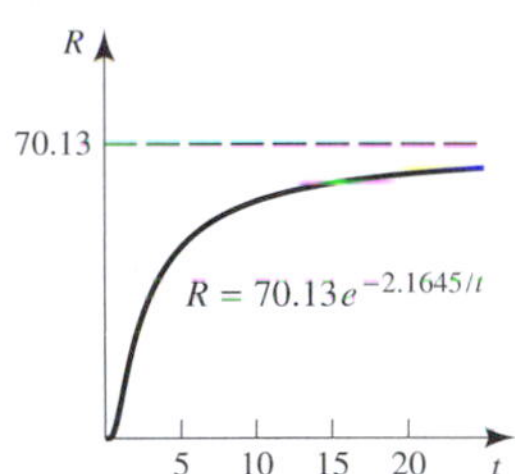

59. Everywhere **60.** $bk \leq a$ **61.** 0.82

62. Concave down. As advertising increases, sales increase less and less, indicating diminishing returns.

63. Concave up. As assets increase, average cost increases less and less, indicating economies of scale.

64. Concave down on $(0, 20000/36)$, concave up on $(20000/36, \infty)$ On the interval $(0, 20000/36)$ the rate of change of the earnings of the greatest earner with respect to the secondary worker is decreasing. On the interval $(20000/36, \infty)$ the rate of change of the earnings of the greatest earner with respect to the secondary worker is increasing.

65. **a.** (ii) **b.** (vi) **c.** (i) **d.** (v) **e.** (iv) **f.** (iii)

66. 2 **67.** 0

68. Does not exist (∞) **69.** 0

70. 1 **71.** 0

72. Concave up on $(0, 3)$
concave down on $(3, \infty)$
inflection point at $t = 3$
point of diminishing returns at $t = 3$

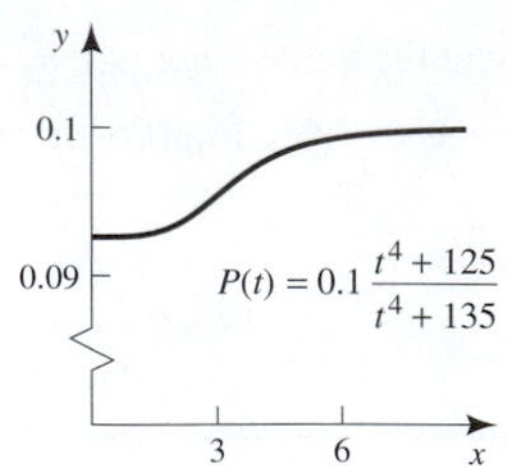

Chapter 6

Section 6.1

1. $\frac{1}{100}x^{100} + C$ **3.** $-\frac{1}{98}x^{-98} + C$

5. $5x + C$ **7.** $-\frac{5}{2}y^{-2} + C$

9. $\frac{\sqrt{2}}{5}y^{5/2} + C$ **11.** $\frac{3}{5}u^{5/3} + C$

13. $2x^3 + 2x^2 + C$ **15.** $\frac{1}{3}x^3 + \frac{1}{2}x^2 + x + C$

17. $\frac{10\sqrt{2}}{11}u^{1.1} - \frac{1}{21}u^{2.1} + C$

19. $\frac{1}{3}t^3 - t + C$ **21.** $\frac{2}{3}t^{3/2} - \frac{3}{8}t^{8/3} + C$

23. $t^6 - t^4 + t + C$ **25.** $x + 3\ln|x| + C$

27. $\pi x + \ln|x| + C$ **29.** $\frac{2}{3}t^{3/2} + 2t^{1/2} + C$

31. $e^x - \frac{3}{2}x^2 + C$ **33.** $5e^x + C$

35. $5e^x - 4x + C$ **37.** $x + \ln|x| + C$

39. $R(x) = 30x - 0.25x^2$

41. $p(x) = 2x^{-1/2} + 1$

43. $C(x) = 100x - 0.1e^x + 1000.1$

45. $C(x) = 20x^{3/2} - 2x^3$

47. $s(t) = -16t^2 + 30t + 15$

49. $s(t) = -16t^2 + 10t + 6$

51. Approximately \$3,200,000

53. $C(x) = x - 0.25x^2 + 0.2$

55. $L(x) = 0.28x + .025x^2 - \frac{1}{30000}x^3$

57. $\frac{1}{2}e^{2x} + C$ **59.** $\frac{1}{3}\sin 3x + C$

Section 6.2

1. $\frac{2}{11}(3x + 1)^{11} + C$ **3.** $-\frac{1}{16}(3 - x^2)^8 + C$

5. $\frac{4}{5}(2x^2 + 4x - 1)^{5/2} + C$

7. $\frac{1}{48}(3x^4 + 4x^3 + 6x^2 + 12x + 1)^4 + C$

9. $\frac{4}{3}(x + 1)^{3/2} + C$ **11.** $\frac{1}{3}(x^2 + 1)^{3/2} + C$

13. $\frac{3}{4}(x^2 + 1)^{2/3} + C$ **15.** $2(x^{1/3} + 1)^{3/2} + C$

17. $\frac{2}{3}(\ln x)^{3/2} + C$ **19.** $-e^{1-x} + C$

21. $-\frac{1}{2}e^{1-x^2} + C$ **23.** $2e^{\sqrt{x}} + C$

25. $\frac{1}{3}\ln|3x + 5| + C$ **27.** $-\frac{1}{6}(x^2 + 3)^{-3} + C$

29. $-\ln(e^{-x} + 1) + C$ **31.** $\frac{1}{2}\ln(e^{2x} + e^{-2x}) + C$

33. $\frac{1}{2}\ln|\ln|x|| + C$

35. $p(x) = 4(3x^2 + 1)^{-1} + 9$

37. $R(x) = 20x^2 - x^4$

39. $C(x) = 2\ln(x^2 + 1) + 1000$

41. 6 tons

43. $20{,}000e^{2.2} + 80{,}000 \approx 260{,}500$

45. $\frac{1}{2}\ln 7 \approx 0.97$ **47.** \$16,703.20

49. Let $u = x^5 + x^4 + 1$; then $du = x(5x^3 + 4x^2)\,dx$, and

$$\int (x^5 + x^4 + 1)^2(5x^3 + 4x^2)\,dx = \int u^2 \cdot \frac{du}{x}$$

There is no way to eliminate the factor $\frac{1}{x}$; therefore, the method fails.

51. $-\cos x^2 + C$ **53.** $\frac{1}{3}\sin x^3 + C$

Section 6.3

1. a.

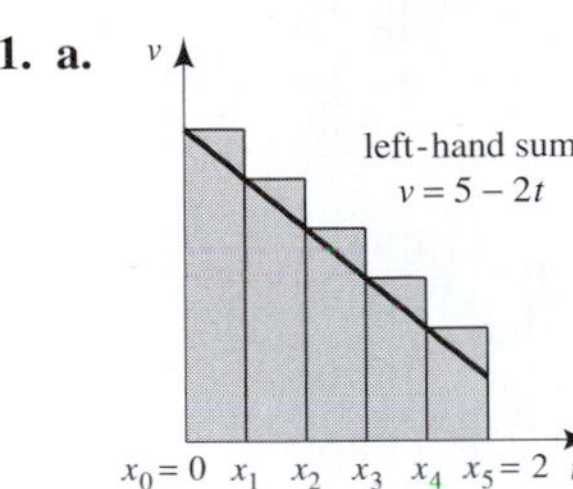

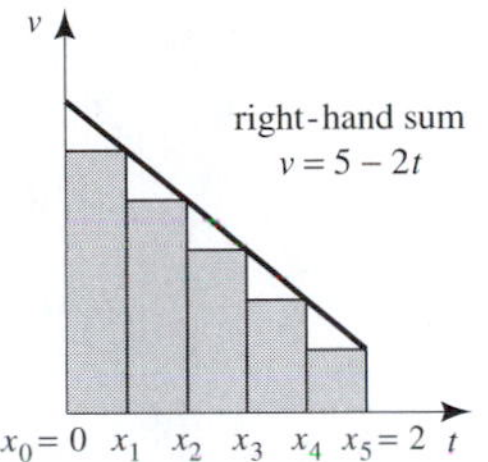

b. 6.8, 5.2, 1.6, 6 **c.** 6.4, 5.6, 0.8, 6

3. a.

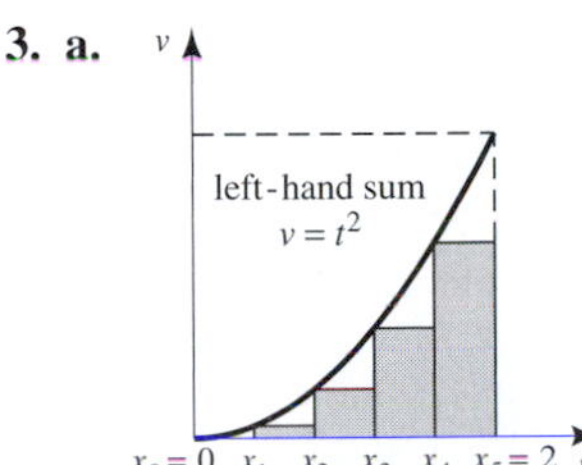

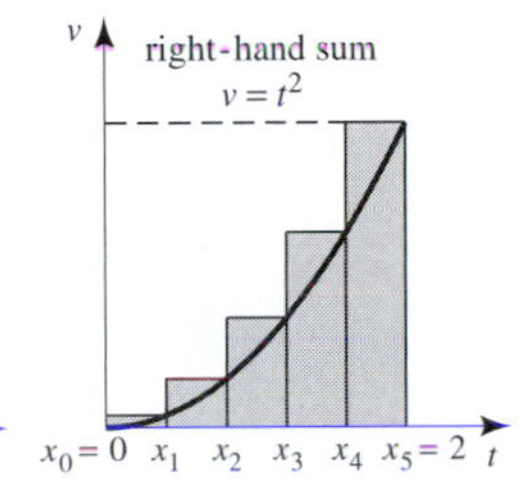

b. 1.92, 3.52, 1.6, 2.72 **c.** 2.28, 3.08, 0.8, 2.68

5. a.

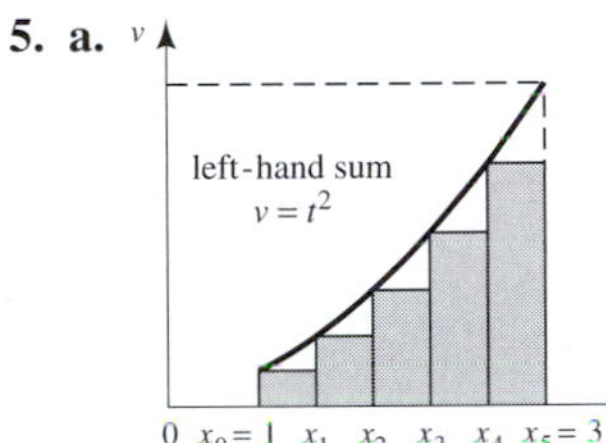

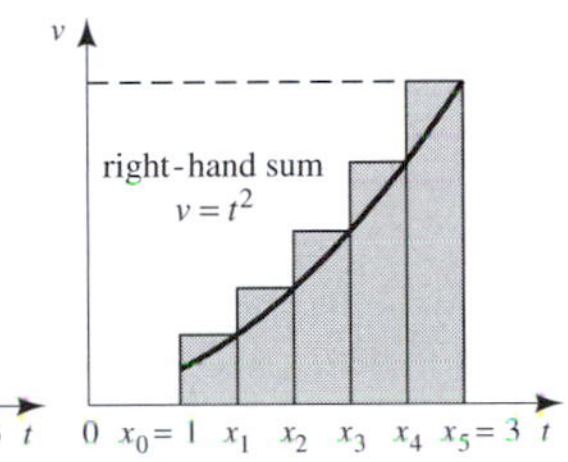

b. 7.12, 10.32, 3.2, 8.72 **c.** 7.88, 9.48, 1.6, 8.68

7. a.

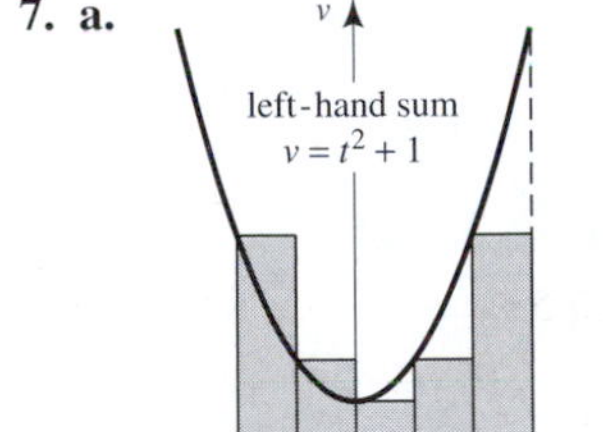

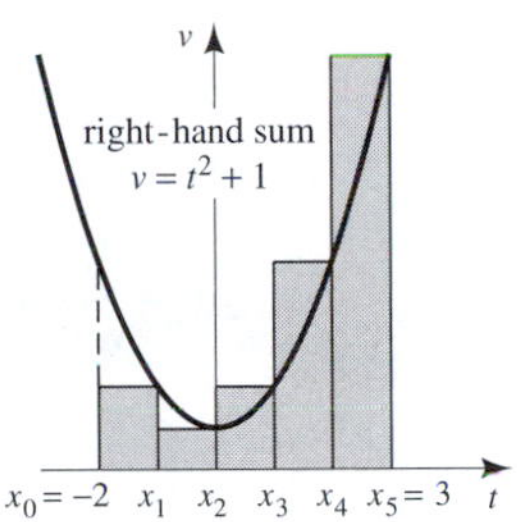

b. 15, 20, 5, 17.5 **c.** 15.625, 18.125, 2.5, 16.875

9. a.

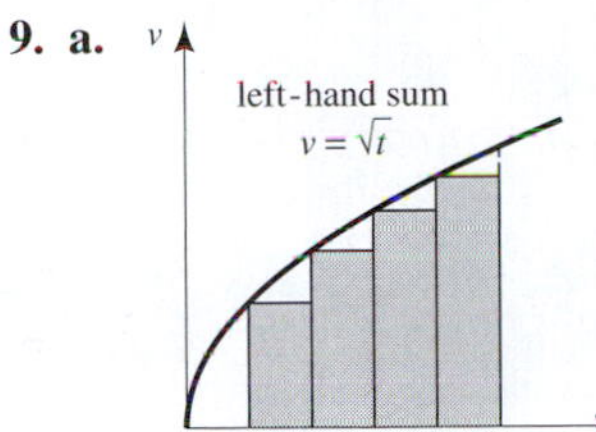

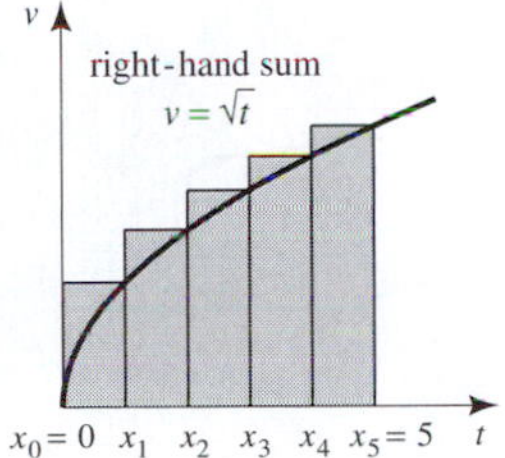

b. 6.15, 8.38, 2.23, 7.265 **c.** 6.83, 7.94, 1.11, 7.385

11.

n	Left-Hand Sum	Right-Hand Sum	Average of Sums
4	0.646003	0.350120	0.498062
8	0.572467	0.424526	0.498497
40	0.513438	0.483850	0.498644

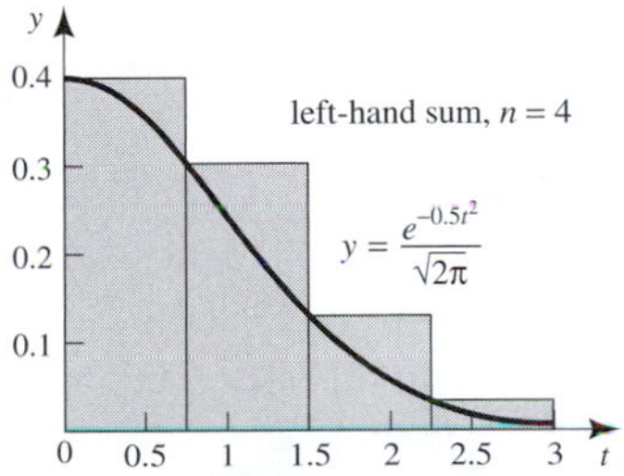

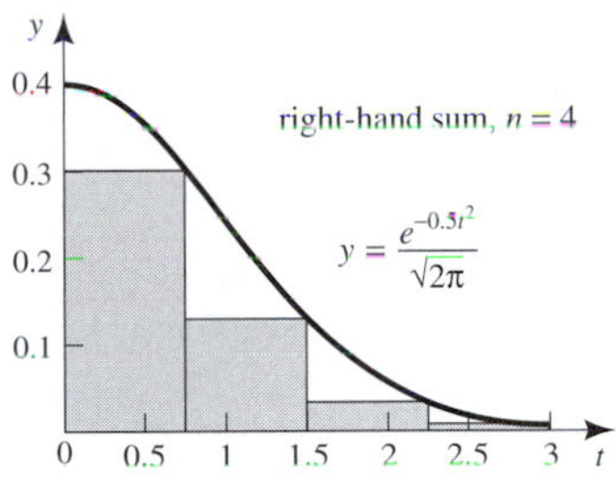

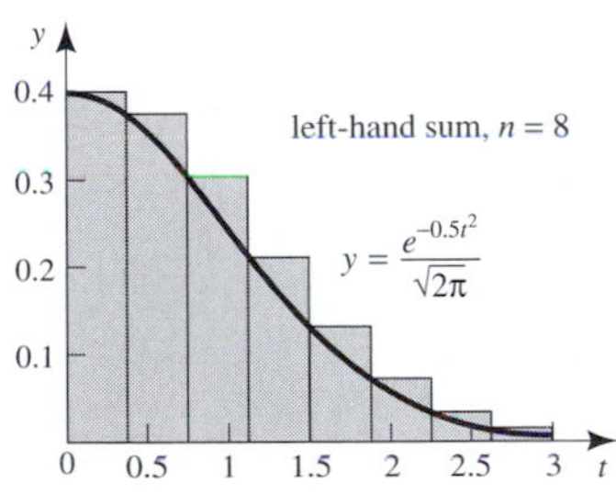

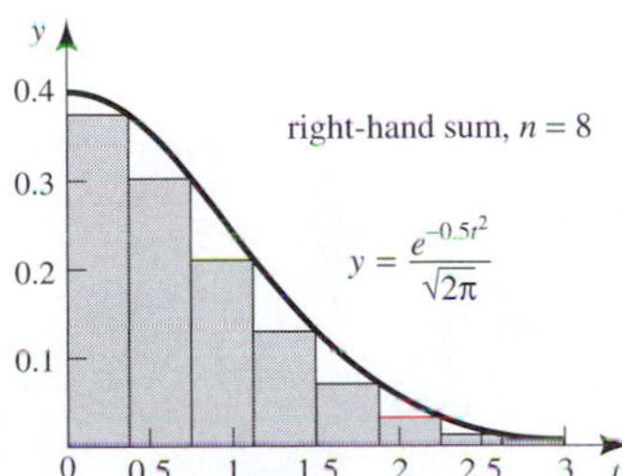

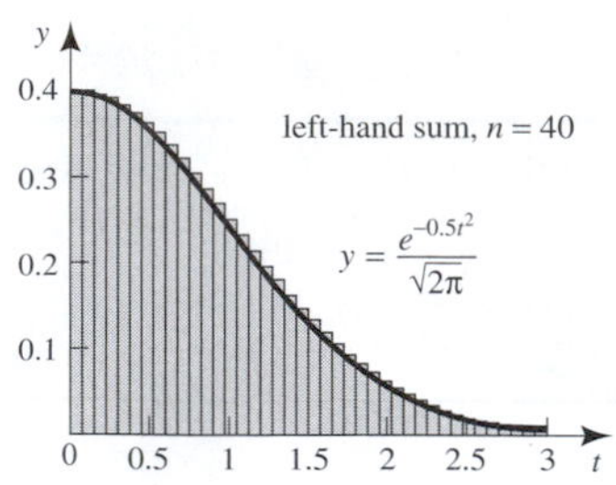

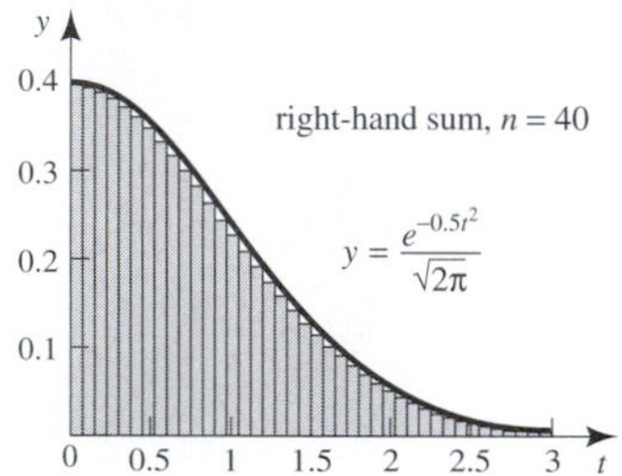

13.

n	Left-Hand Sum	Right-Hand Sum	Average of Sums
4	7.946161	9.179429	8.562795
8	8.302919	8.919553	8.611236
40	8.564839	8.688166	8.626503

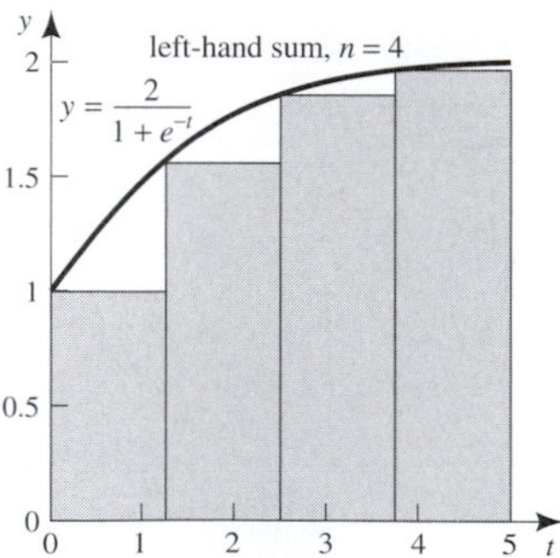

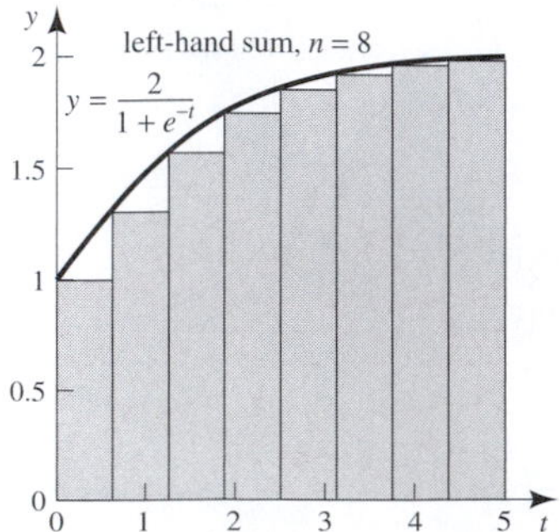

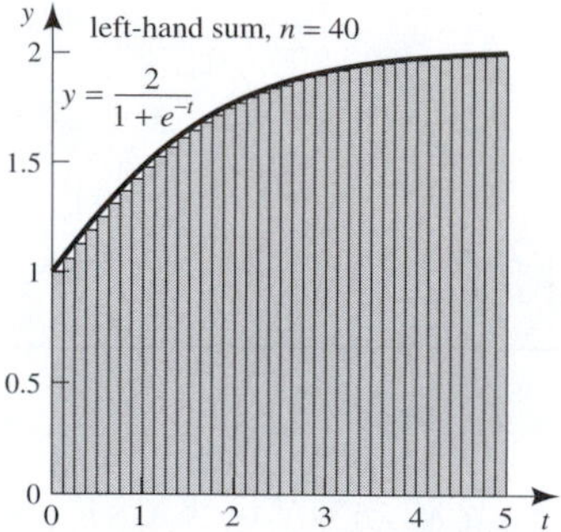

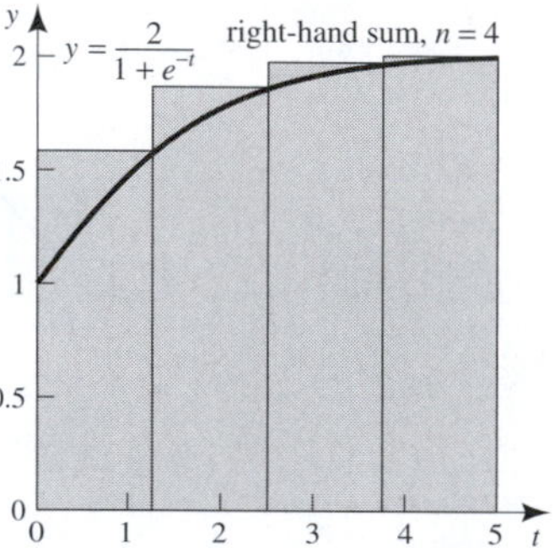

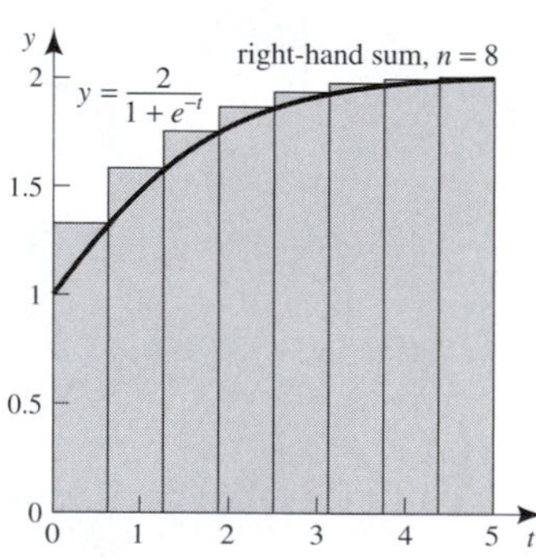

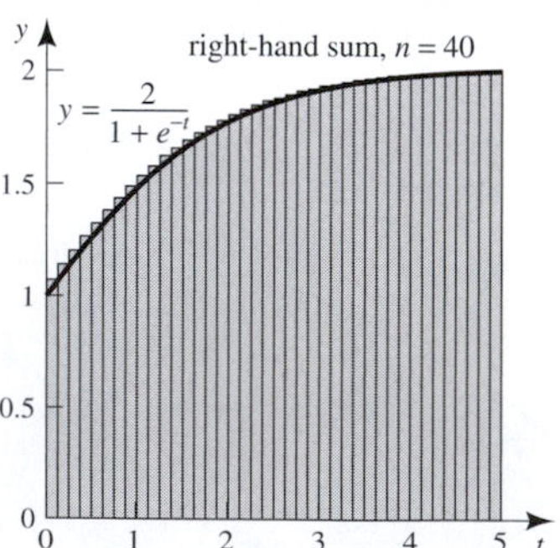

15.

n	Left-Hand Sum	Right-Hand Sum	Average of Sums
4	2.000000	6.000000	4.000000
8	3.000000	5.000000	4.000000
16	3.800000	4.200000	4.000000

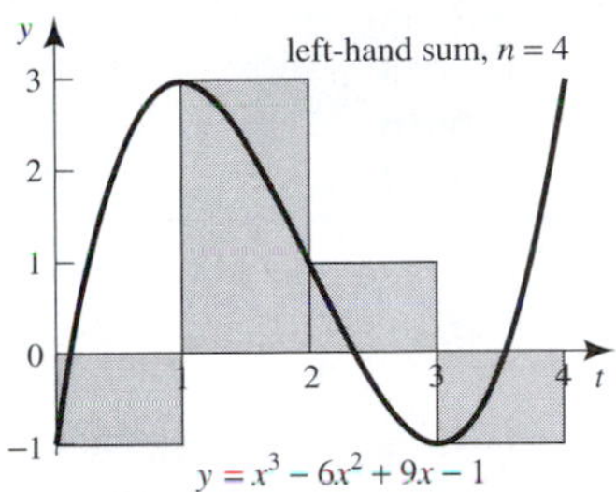

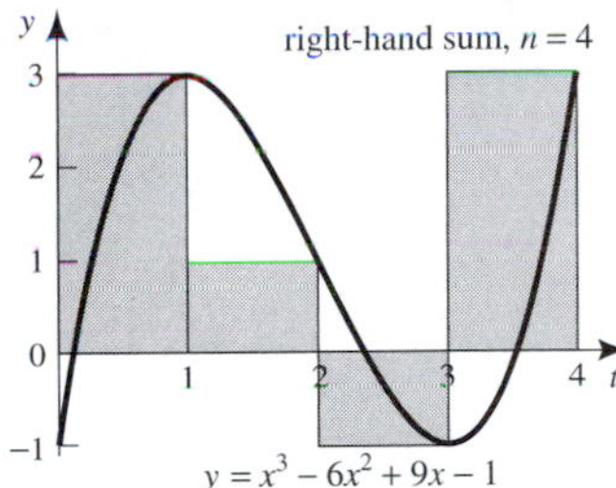

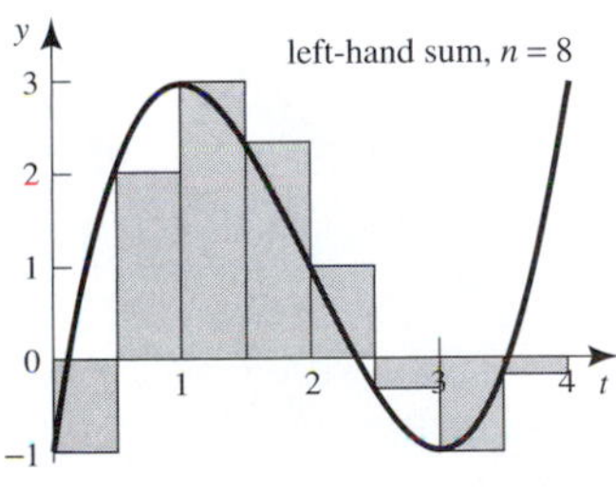

right-hand sum, $n = 8$

$y = x^3 - 6x^2 + 9x - 1$

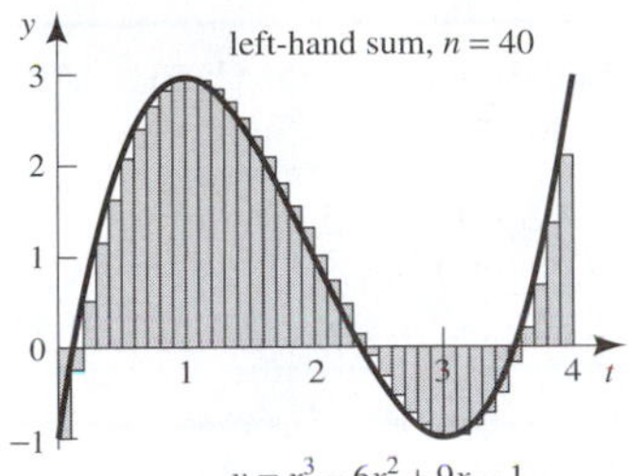

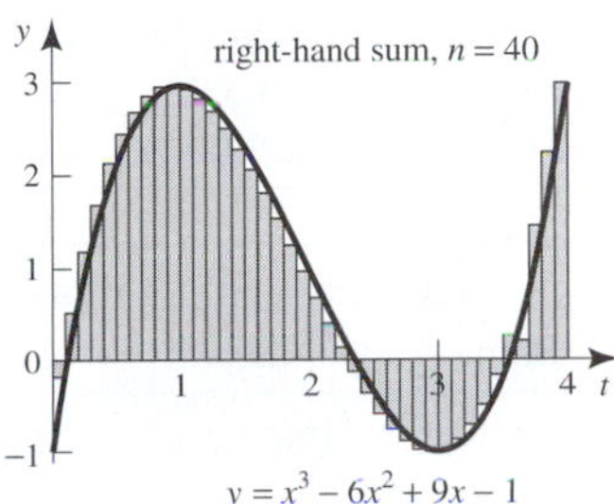

17. 44. Taking $n = 10$, the left-hand sum is 43, and the right-hand sum is 45. The average is 44.

19. Use rectangles with base of length 1 unit and obtain a lower bound of 15 and an upper bound of 24.

21. 135, 95

23. 0.01, 0.001

25. We know that right-hand sum − left-hand sum $= [f(b) - f(a)]\Delta x$. If $f(a) = f(b)$, then $f(b) - f(a) = 0$ and the difference of the two sums is zero.

27. 10, 100

Exercises 6.4

1.

n	Left-Hand Sum	Right-Hand Sum	Average of Sums
10	0.610509	0.710509	0.660509
100	0.661463	0.671463	0.666463
1000	0.666160	0.667160	0.666667

3.

n	Left-Hand Sum	Right-Hand Sum	Average of Sums
10	0.202500	0.302500	0.252500
100	0.245025	0.255025	0.250025
1000	0.249500	0.250500	0.250000

5.

n	Left-Hand Sum	Right-Hand Sum	Average of Sums
10	4.920000	5.720000	5.320000
100	5.293200	5.373200	5.333200
1000	5.329332	5.337332	5.333332

7.

n	Right-Hand Sum	Left-Hand Sum	Average of Sums
10	1.034896	1.168229	1.101563
100	1.091975	1.105309	1.098642
1000	1.097946	1.099279	1.098613

9.

n	Left-Hand Sum	Right-Hand Sum	Average of Sums
10	1.633799	1.805628	1.719714
100	1.709705	1.726888	1.718297
1000	1.717423	1.719141	1.718282

11.

n	Left-Hand Sum	Right-Hand Sum	Average of Sums
10	0.675611	0.710268	0.692939
100	0.691412	0.694878	0.693145
1000	0.692974	0.693321	0.693147

13.

n	Left-Hand Sum	Right-Hand Sum	Average of Sums
80	0.506046	0.491251	0.498649
160	0.502348	0.494951	0.498650
320	0.500499	0.496801	0.498650
640	0.499575	0.497725	0.498650

15.

n	Left-Hand Sum	Right-Hand Sum	Average of Sums
80	8.596146	8.657810	8.626978
160	8.611681	8.642513	8.627097
320	8.619419	8.634834	8.627127
640	8.623280	8.630988	8.627134

17.

n	Left-Hand Sum	Right-Hand Sum	Average of Sums
80	3.900000	4.100000	4.000000
160	3.950000	4.050000	4.000000
320	3.975000	4.025000	4.000000
640	3.987500	4.012500	4.000000

19.

n	Left-Hand Sum	Right-Hand Sum
10	10.560000	10.560000
100	10.665600	10.665600
1000	10.666656	10.666656

21.

n	Left-Hand Sum	Right-Hand Sum
10	2.622043	2.700019
100	2.636816	2.644614
1000	2.639529	2.640309

23. 20 **25.** 6 **27.** $\frac{9}{2}$

29. $\frac{\pi}{2}$ **31.** $\frac{\pi}{2}$ **33.** 16

35. 0.5

37. 6

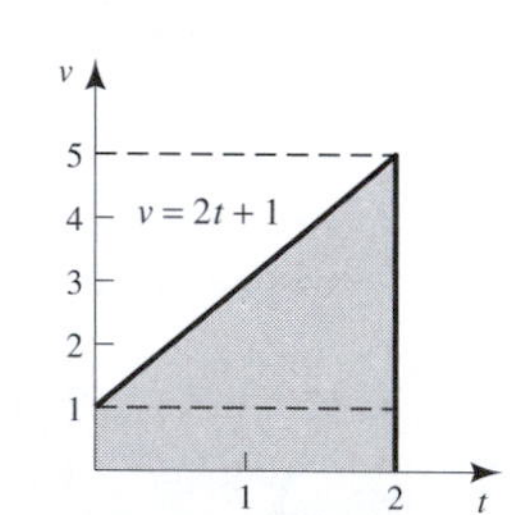

39. 4

41. 1

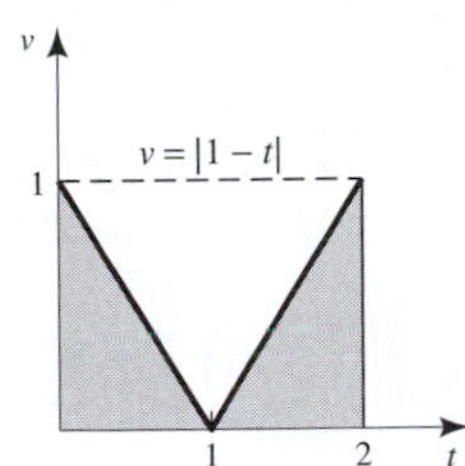

43. On $[1, 4]$, $1 \le \sqrt{x} \le 2$. Take $m = 1$, $M = 2$. Then $\int_1^4 \sqrt{x}\,dx \le M(b-a) = 2(4-1) = 6$. $\int_1^4 \sqrt{x}\,dx \ge m(b-a) = 1(4-1) = 6$.

45. On $[1, 2]$, $0.5 \le 1/x \le 1$. Take $m = 0.5$, $M = 1$. Then $\int_1^2 \frac{1}{x}\,dx \le M(b-a) = 1(2-1) = 1$. $\int_1^2 \frac{1}{x}\,dx \ge m(b-a) = 0.5(2-1) = 0.5$.

47. On $[0, 1]$, $x^3 \le x$. Then $\int_0^1 x^3\,dx \le \int_0^1 x\,dx = 0.5$

49. On $[-1, 1]$, $\sqrt{3} \le \sqrt{3+x^2} \le 2$. Take $m = \sqrt{3}$ and $M = 2$. Then

$$\int_{-1}^{1} \sqrt{1+x^2}\,dx \le M(b-a) = 2[1-(-1)] = 4$$

$$\int_{-1}^{1} \sqrt{1+x^2}\,dx \ge m(b-a) = \sqrt{3}[1-(-1)] = 2\sqrt{3}$$

51. Left-hand sum is 0.747140; right-hand sum is 0.746508; left-hand sum must be greater since the function is decreasing on $[0, 1]$.

53. t^2

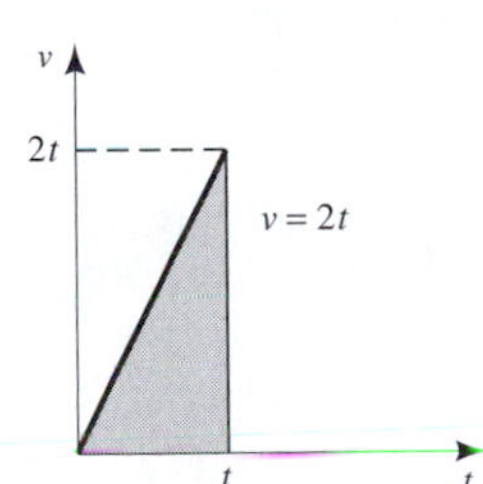

55. $n = 10$, left-hand sum gives a lower estimate of 0.687405, right-hand sum gives an upper estimate of 0.787405

57. $n = 160$, left-hand sum gives a lower estimate of 3.950156, right-hand sum gives an upper estimate of 4.050156

59. $n = 80$, left-hand sum gives a lower estimate of 5.283125, right-hand sum gives an upper estimate of 5.383125

61. No difference

63. 1.09861214. Correct answer to eight decimal places is 1.09861229. The left-hand sum is 1.09927925. The right-hand sum is 1.09794592.

65. $N = 22$

67. **a.** The first runner is ahead after the first minute, since the first runner has a larger velocity during every moment of the first minute.

b. The second runner is ahead after the first 5 minutes. The second runner gets behind the first runner at the 1-minute mark a distance equal to the area A_1 between the two curves on $[0, 1]$. From the 1-minute mark to the end of the fifth minute the second runner gains a distance equal to the area A_2 between the two curves on $[1, 5]$. Notice that $A_2 > A_1$. Thus, at the end of the fifth minute the second runner is ahead of the first runner by a distance equal to $A_2 - A_1$.

Section 6.5

1. 15

3. 0

5. 2

7. 14

9. $\frac{11}{8}$

11. $\frac{1}{2}(e^{-2} - e^{-4})$

13. $\frac{1}{2}\ln 3$

15. 0

17. $-\frac{1}{16}$

19. $\frac{26}{3}$

21. 0

23. $\frac{1}{2}\ln\frac{5}{3}$

25. 6

27. 5

29. 0

31. 0

33. $\frac{1}{3}$

35. $\int_1^5 f(x)\,dx$

37. $\int_8^{10} f(x)\,dx$

39. **a.** 5; **b.** 7

41. The lower estimate is 13; the upper estimate is 21. Take $\Delta t = 1$.

43. 948, 928

45. $2(e^2 - e) \approx 9.342$ million barrels.

47. 71

49. About 11 years

51. About 8535 gal

53. **a.** $V(0) = I$. Thus,

$$\begin{aligned} V(T) &= I + \int_0^T -k(T-t)\,dt \\ &= I - kT\int_0^T 1\,dt + k\int_0^T t\,dt \\ &= I - kT \cdot t\big|_0^T + \frac{k}{2}t^2\bigg|_0^T \\ &= I - kT^2 + \frac{k}{2}T^2 \\ &= I - \frac{1}{2}kT^2 \end{aligned}$$

b. Suppose the equipment is worth $\frac{1}{2}I$ after T^* years. Then

$$V(T^*) - V(0) = \int_0^{T^*} -k(T-t)\,dt$$
$$\frac{1}{2}I - I = -kT\int_0^{T^*} 1\,dt + k\int_0^{T^*} t\,dt$$
$$-\frac{1}{2}I = -kT \cdot t\Big|_0^{T^*} + \frac{1}{2}kt^2\Big|_0^{T^*}$$
$$= -kTT^* + \frac{1}{2}kT^{*2}$$
$$\frac{1}{2}k(T^*)^2 - kT(T^*) + \frac{1}{2}I = 0$$

$$T^* = \frac{kT \pm \sqrt{k^2T^2 - 4 \cdot \frac{1}{2}k \cdot \frac{1}{2}I}}{2 \cdot \frac{1}{2}k}$$
$$= T \pm \sqrt{T^2 - \frac{I}{k}}$$

Since $T^* \le T$, $T^* = T - \sqrt{T^2 - \frac{I}{k}}$.

55. $\frac{74}{3}$ thousand **57.** \$3194.53

59. \$8 million **61.** About 10.23 **63.** 129,744

65. a. First store. Revenue from the first store during the first year is

$$R(1) - R(0) = \int_0^1 R_1'(t)\,dt,$$

which is also the area between the graph of the curve $y = R_1'(t)$ and the x-axis on the interval [0, 1]. This is greater than the area under the graph of $y = R_2$ and the x-axis on [0, 1].

b. Second store. Sales for the second store less sales of the first store during the three-week period is $[R_2(3) - R_2(0)] - [R_1(3) - R_1(0)]$ equals

$\int_0^3 R_2'(t)\,dt - \int_0^3 R_1'(t)\,dt$ equals

$$\int_1^3 [R_2'(t) - R_1'(t)]\,dt - \int_1^3 [R_1'(t) - R_2'(t)]\,dt$$

The first integral on the right-hand side of the preceding equation is the area under the graph of $y = R_2'(t)$ and above the graph of $y = R_1'(t)$ on [1, 3] and is greater than the second integral, which is the area under the graph of $y = R_1'(t)$ and above the graph of $y = R_2'(t)$ on [0, 1].

67. $\int_a^b f(x)\,dx$ is a real number. $\int f(x)\,dx$ is a function of x. For example, $\int_0^1 x\,dx = 0.5$ and $\int x\,dx = \frac{1}{2}x^2 + C$.

69. $\frac{a}{t}\left(\frac{1}{m} - \frac{1}{m+tb}\right)$

71.

$$\int_a^b cf(x)\,dx = \lim_{n\to\infty}\sum_{k=1}^n cf(x_i)\Delta x$$
$$= \lim_{n\to\infty} c\sum_{k=1}^n f(x_i)\Delta x$$
$$= c\lim_{n\to\infty}\sum_{k=1}^n f(x_i)\Delta x$$
$$= c\int_a^b f(x)\,dx$$

Section 6.6

1. 3.75 **3.** $10e - 20$

5. $\frac{1}{6}$ **7.** $\frac{1}{2}(e^4 - e^2) - 1$

9. $\frac{9}{2}$ **11.** $\frac{1}{2}$

13. $\frac{16}{3}$ **15.** $\frac{8}{3}$

17. $\frac{125}{6}$ **19.** $\frac{125}{6}$

21. $e^2 + e - 2$ **23.** 16

25. 1.5 **27.** $\frac{1}{2}$

29. 1 **31.** 8

33. $\frac{23}{2}$ **35.** 20.75

37. x-coordinates of intercepts are $x = -1.53$, $x = 0$, $x = 1.28$. The upper estimate is 5.5821; the lower estimate is 5.5815.

39. x-coordinates of intercepts are $x = -1.80$, $x = 0$, $x = 1.19$. The lower estimate is 5.7052; the upper estimate is 5.7069.

41. a. 10 **b.** −6
c. 4 **d.** 6

43. $\frac{1}{3}$ **45.** $\frac{1}{12}$

47. 0.1 **49.** $\frac{1}{300}$

51. \$4.8 billion **53.** $\frac{26}{3}$

55.

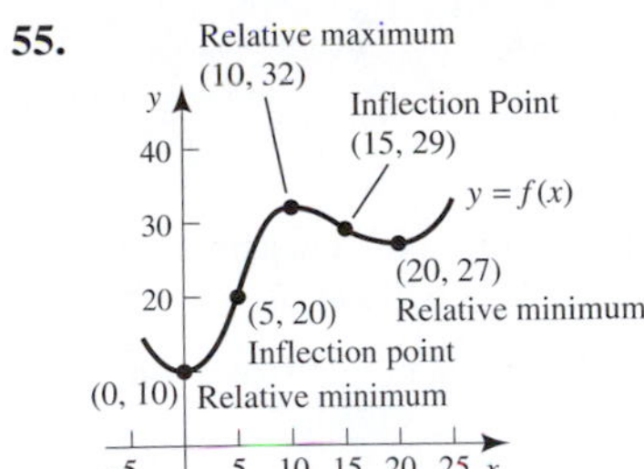

57. The coefficient of inequality has become larger.

59. 15

61. 22, 15

63. 0.382 **65.** 0.361

Section 6.7

1. a. \$100;
b. $\$200(1 - e^{-0.50}) = \$78.69$;
c. $\$200(e^{0.50} - 1) = \$129.74$

3. a. $\$200(e - 1) = \$3436.56$;
b. \$2000;
c. $\$2000e = \$5436.56$

5. a. $\$400(1 - e^{-0.50}) = \$157.39$;
b. $\$2000\dfrac{1 - e^{-1.5}}{15} = \$103.58$;
c. $\$2000\dfrac{e - e^{-0.5}}{15} = \$281.57$

7. a. \$50; **b.** \$26.42; **c.** \$71.83

9. a. \$333.33; **b.** \$160.60; **c.** \$436.56

11. a. Yes, since $P_V(10) = \$948{,}180.84$;
b. no, since $P_V(10) = \$776{,}869.84$

13. a. Yes, since $P_V(5) \approx \$316{,}000$;
b. no, since $P_V(5) \approx \$303{,}000$

15. $\dfrac{128}{3}$ **17.** $0.9 - 0.1 \ln 10$

19. $\dfrac{1}{2}$ **21.** $\dfrac{2}{9}$

23. 8, 16, where equilibrium point is (4, 8)

25. $\dfrac{128}{3}$, 8, where equilibrium point is (4, 4)

27. \$1416.40

Chapter 6 Review Exercises

1. $\dfrac{1}{10}x^{10} + C$ **2.** $\dfrac{2}{3}x^{1.5} + C$

3. $-2y^{-1} + C$ **4.** $\dfrac{\sqrt{3}}{3}y^3 + C$

5. $2x^3 + 4x^2 + C$ **6.** $x^2 - 3\ln|x| + C$

7. $\dfrac{2}{5}t^{5/2} + 2t^{1/2} + C$ **8.** $\dfrac{1}{3}e^{3x} + C$

9. $-e^{-x} - \dfrac{1}{5}e^{-5x} + C$ **10.** $\dfrac{2}{21}(2x + 1)^{21} + C$

11. $(2x^3 + 1)^{10} + C$ **12.** $\dfrac{2}{3}\sqrt{x^3 + 1} + C$

13. $\dfrac{1}{2}\ln|x^2 + 2x| + C$ **14.** $\dfrac{1}{2}e^{x^2+5} + C$

15. $-\dfrac{5}{7}(x^2 + 1)^{-7} + C$ **16.** $-2e^{-\sqrt{x}} + C$

17. $\ln(e^x + 1) + C$ **18.** $\dfrac{1}{4}(\ln x)^4 + C$

19. $n = 10$: 1.77, 1.57; $n = 100$: 1.6767, 1.6567; $n = 1000$: 1.667667, 1.665667;

$$\int_0^1 (2x^2 + 1)\,dx$$

20.

n	Left-Hand Sum	Right-Hand Sum
10	1.2025	1.3025
100	1.2450	1.2550
1000	1.2495	1.2505

$$\int_0^1 (x^3 + 1)\,dx$$

21. $\dfrac{9\pi}{4}$

22. 1.283. The area of the rectangles is the left-hand sum and is greater than the area under the curve.

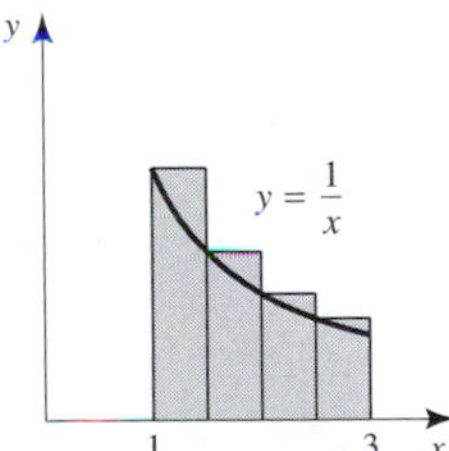

23. 0.95. The area of the rectangles is the right-hand sum and is less than the area under the curve.

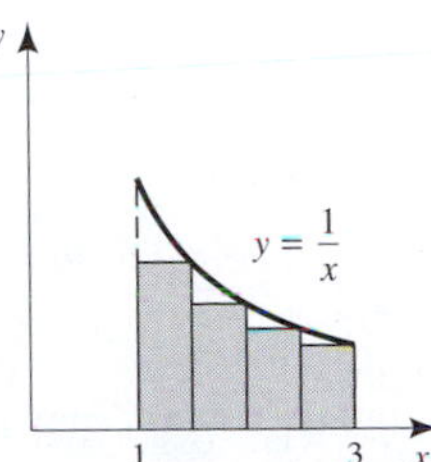

24. 52 **25.** −8 **26.** 8

27. $\dfrac{1}{2}e^4 - \dfrac{1}{2}e^2 + \ln 2 \approx 24.3$

28. $\dfrac{1}{5}(2^{10} - 1) = 204.6$ **29.** $\dfrac{1}{3}(e^9 - e^1) \approx 2700.1$

30. $\dfrac{1}{2}\ln 5 \approx 0.8$ **31.** $3 - \sqrt{3} \approx 1.26795$

32. $\frac{1}{2}\int_0^2 e^{t^4}\,dt$ **33.** 0 **34.** $\frac{1}{2}\ln 3$

35. 9 **36.** $\frac{8}{3}$ **37.** $\frac{8}{3}$

38. 1 **39.** $\frac{20}{3}$ **40.** 72

41. $\frac{32}{3}$ **42.** 9 **43.** $\frac{2}{3}$

44. 13 **45.** $4\ln 2 - 2\ln 3$

46. $\frac{38}{3}$ **47.** $x^2 - 10x$

48. $x^3 + x + 1000$ **49.** \$98,419.70

50. 1.96 thousand if $S(0) = 0$

51. 34, 24 million **52.** 6906

53. 10.82 **54.** 17,183

55. $x_0 = 2$, $p_0 = 6$ **56.** $\frac{16}{3}$

57. 6 **58.** $L = \frac{3}{5}$

59. There is a greater inequality of income distribution in the United Kingdom than there is in Norway.

60. There is a greater inequality of income distribution for females than there is for males.

61. $\$250(1 - e^{-0.4}) = \$82.42$

62. $\frac{\$250}{3}(1 - e^{-1.2}) = \$58.23$

63. $\frac{\$250}{3}(e^{0.8} - e^{-0.4}) = \$129.59$

64. $\frac{\$5000}{9}(1 - e^{-0.9}) = \$329.68$ thousand

65. \$1189.12

Chapter 7

Section 7.1

1. $\left(x - \frac{1}{2}\right)e^{2x} + C$

3. $\frac{3}{4}e^4 + \frac{1}{4}$ **5.** 2

7. $2x\sqrt{1+x} - \frac{4}{3}(1+x)^{3/2} + C$

9. $\frac{1}{11}x(1+x)^{11} - \frac{1}{132}(1+x)^{12} + C$

11. $-\frac{x}{x+2} + \ln|x+2| + C$

13. $\frac{2}{9}x^3(1+x^3)^{3/2} - \frac{4}{45}(1+x^3)^{5/2} + C$

15. $\frac{x^2}{22}\left(1+x^2\right)^{11} - \frac{1}{264}\left(1+x^2\right)^{12} + C$

17. $\frac{1}{2}x^2e^{2x} - \frac{1}{2}xe^{2x} + \frac{1}{2}e^{2x} + C$

19. $\frac{1}{2}(3e^4 + 1)$ **21.** 111 **23.** \$26,306.67

25. \$36.47 **27.** $\frac{1000}{e-1} \approx 582$

29. Let $u = (\ln x)^n$ and $dv = dx$. Then $du = \frac{n}{x}(\ln x)^{n-1}\,dx$, $v = x$, and

$$\begin{aligned}\int (\ln x)^n\,dx &= \int u\,dv \\ &= uv - \int v\,du \\ &= x(\ln x)^n - \int nx(\ln x)^{n-1}\,dx\end{aligned}$$

31. Let $u = f(x)$ and $dv = dx$. Then $du = f'(x)\,dx$, $v = x$, and

$$\begin{aligned}\int f(x)\,dx &= \int u\,dv \\ &= uv - \int v\,du \\ &= xf(x) - \int xf'(x)\,dx\end{aligned}$$

Section 7.2

1. $\frac{2}{9}\left[\frac{1}{5}(3x+2)^{5/2} - \frac{2}{3}(3x+2)^{3/2}\right] + C$

3. $\frac{1}{2}x\sqrt{x^2-9} - \frac{9}{2}\ln\left|x + \sqrt{x^2-9}\right| + C$

5. $\ln\left|\frac{\sqrt{1-x}-1}{\sqrt{1-x}+1}\right| + C$ **7.** $\frac{1}{3}\ln\left|\frac{2x+1}{2x+4}\right| + C$

9. $\frac{1}{3}(x^2-9)^{3/2} + C$ **11.** $\frac{1}{3x} + \frac{1}{9}\ln\left|\frac{x-3}{x}\right| + C$

13. $-\frac{x}{\sqrt{x^2+16}} + \ln\left|x + \sqrt{x^2+16}\right| + C$

15. $\frac{1}{3}\ln\left|\frac{x^3}{2x^3+1}\right| + C$ **17.** $-\frac{\sqrt{1-4x^2}}{x} + C$

19. $\frac{1}{18}x\sqrt{9x^2-1} + \frac{1}{54}\ln\left|x + \frac{1}{3}\sqrt{9x^2-1}\right| + C$

21. $x(\ln x)^3 - 3x(\ln x)^2 + 6x\ln x - 6x + C$

23. $\frac{1}{2}(e^x+1)^2 - 2(e^x+1) + \ln|e^x+1| + C$

25. $-\frac{1}{2}e^{-2x}\left(x^2+x+\frac{1}{2}\right)+C$

27. $R(x)=\frac{2}{5}(x+1)^{5/2}-\frac{2}{3}(x+1)^{3/2}+\frac{4}{15}$

29. $\frac{10t}{9\sqrt{t^2+9}}$ **31.** 4.8 **33.** 327

35. $y_3(x)=-1$. Thus, $y_1(x)=y_2(x)-1$.

Section 7.3

1. $T_2=5$, $S_2=4$, exact $=4$

3. To four decimal places, $T_2=1.1667$, $S_2=1.1111$, exact $=1.0986$

5. To four decimal places, $T_2=0.6830$, $S_2=0.7440$, exact $=0.7854$

7. To six decimal places, $T_2=1.571583$, $S_2=1.475731$, exact $=1.462652$

9. $T_4=4.25$, $S_2=S_4=4$, exact $=4$

11. To four decimal places, $T_4=1.1167$, $S_4=1.1$, exact $=1.0986$

13. To four decimal places, $T_4=0.7489$, $S_4=0.7709$, exact $=0.7854$

15. To six decimal places, $T_4=1.490679$, $S_4=1.463711$, exact $=1.462652$

17. $T_{100}=4.0004$, $S_{100}=4$, exact $=4$

19. $T_{100}=1.098642$, $S_{100}=1.098612294$, exact to nine decimal places $=1.098612289$

21. $T_{100}=0.785104$, $S_{100}=0.785283$, exact to six decimal places $=0.785398$

23. $T_{100}=1.462697$, $S_{100}=1.462652$, exact to six decimal places $=1.462652$

25. a. $T_4=5.8$; **b.** $S_4\approx 5.87$

27. a. $T_4=50$; **b.** $S_4=52$

29. a. $T_6=4.325$; **b.** $S_6=4.35$

31. a. $\frac{1}{75}$; **b.** 0

33. a. 0.32; **b.** 0.00043

35. \$78,818

37. Yes. When the graph of the function is concave down, the line segment connecting the two endpoints of any two subintervals lies below the curve. Thus, the trapezoid rule always gives an underestimate. When the graph of the function is concave up, the line segment connecting the two endpoints of any subinterval lies above the curve. Thus, the trapezoid rule always gives an overestimate.

39. We have $\Delta x=\frac{b-0}{2}=\frac{b}{2}$. Then

$$\begin{aligned}S_2&=\frac{b}{6}\left[f(0)+4f\left(\frac{b}{2}\right)+f(b)\right]\\&=\frac{b}{6}\left[0+4\cdot\frac{b^3}{8}+b^3\right]\\&=\frac{b^4}{4}\end{aligned}$$

And

$$\int_0^b x^3\,dx=\frac{1}{4}x^4\Big|_0^b=\frac{1}{4}(b^4-0)=\frac{b^4}{4}$$

Thus, $S_2=\int_0^b x^3\,dx$.

41. Left-hand sum is $f(x_0)\Delta x+f(x_1)\Delta x+\cdots+f(x_{n-1})\Delta x$. Right-hand sum is $f(x_1)\Delta x+f(x_2)\Delta x+\cdots+f(x_n)\Delta x$. $\frac{(\text{left-hand sum}+\text{right-hand sum})}{2}$ is $\frac{f(x_0)+f(x_1)}{2}\Delta x+\frac{f(x_1)+f(x_2)}{2}\Delta x+\cdots+\frac{f(x_{n-1})+f(x_n)}{2}\Delta x=T_n$

Section 7.4

1. $\frac{1}{2}$ **3.** Diverges

5. 100 **7.** Diverges

9. 1 **11.** $-\frac{1}{2}$

13. $\frac{1}{3}$ **15.** $-\frac{1}{2}$

17. 0 **19.** 0

21. Approximately 667 engine manifolds

23. \$20 million **25.** 200 million

27. 2000 tons **29.** 1000

31. \$1250 **33.** \$555.56

35. a. No, since $P_V(\infty)=\$100{,}000$; **b.** yes, since $P_V(\infty)=\$100{,}000$

37. a. No, since $P_V(\infty)=\$600{,}000$; **b.** yes, since $P_V(\infty)=\$600{,}000$

39. $\int_{-\infty}^{\infty}x\,dx=\int_{-\infty}^{0}x\,dx\int_0^{+\infty}x\,dx$. Since $\int_{-\infty}^{0}x\,dx$ diverges, $\int_{-\infty}^{\infty}x\,dx$ diverges. Notice that

$$\begin{aligned}&\lim_{a\to-\infty}\int_a^0 f(x)\,dx+\lim_{b\to\infty}\int_0^b f(x)\,dx\\&\neq\lim_{a\to\infty}\int_{-a}^{a}f(x)\,dx\end{aligned}$$

41. Since

$$\int_1^\infty \frac{1}{\sqrt{x}}\,dx = \lim_{b\to\infty}\int_1^b x^{-1/2}\,dx = \lim_{b\to\infty}(2\sqrt{b}-2)$$

diverges, and $f(x) \geq \frac{1}{\sqrt{x}}$ on $[1, \infty)$, we have

$$\int_1^\infty f(x)\,dx \geq \int_1^\infty \frac{1}{\sqrt{x}}\,dx$$

Thus, $\int_1^\infty f(x)\,dx$ diverges.

43.

$$\begin{aligned}&\int_{-\infty}^c f(x)\,dx + \int_c^\infty f(x)\,dx\\ &= \lim_{a\to-\infty}\int_a^c f(x)\,dx + \lim_{b\to\infty}\int_c^b f(x)\,dx\\ &= \lim_{a\to-\infty}\int_a^0 f(x)\,dx + \int_0^c f(x)\,dx\\ &\quad + \int_c^0 f(x)\,dx + \lim_{b\to\infty}\int_0^b f(x)\,dx\\ &= \lim_{a\to-\infty}\int_a^0 f(x)\,dx + \lim_{b\to\infty}\int_0^b f(x)\,dx\\ &= \int_{-\infty}^\infty f(x)\,dx\end{aligned}$$

Chapter 7 Review Exercises

1. $\frac{1}{25}e^{5x}(5x-1)+C$

2. $\frac{2}{3}x\sqrt{5+3x} - \frac{4}{27}(5+3x)^{3/2}+C$

3. $\frac{1}{6}x^6\ln x - \frac{1}{36}x^6 + C$

4. $-\frac{x}{x+1} + \ln|x+1| + C$

5. $-\frac{1}{5}(9-x^2)^{3/2}(x^2+6)+C$

6. $2\sqrt{x}\ln x - 4\sqrt{x}+C$

7. $\frac{x}{2} - \frac{3}{4}\ln|2x+3|+C$

8. $\frac{1}{4(x+4)} - \frac{1}{16}\ln\left|\frac{x+4}{x}\right|+C$

9. $-\frac{\sqrt{4-x^2}}{4x}+C$

10. $\sqrt{9-x^2} - 3\ln\left|\frac{3+\sqrt{9-x^2}}{x}\right|+C$

11. $\frac{1}{2}x\sqrt{x^2+1}+\frac{1}{2}\ln|x+\sqrt{x^2+1}|+C$

12. $x-\ln(1+e^x)+C$

13. 10 **14.** Diverges

15. $\frac{2}{3}$ **16.** Diverges

17. **a.** 45; **b.** $\frac{100}{3}$; **c.** correct = 32

18. To four decimal places:
a. 1.5375; **b.** 1.425;
c. correct = 1.3863

19. To four decimal places:
a. 35.3125; **b.** 32.0833; **c.** correct = 32

20. To four decimal places:
a. 1.4281; **b.** 1.3916;
c. correct = 1.3863

21. $555,600

Chapter 8

Section 8.1

1. **a.** 13; **b.** 10; **c.** 12

3. **a.** 2; **b.** 13; **c.** 2

5. **a.** $\frac{1}{3}$; **b.** -1; **c.** $\frac{1}{3}$

7. **a.** $\sqrt{6}$; **b.** 3; **c.** 2

9. **a.** 1; **b.** $\frac{1}{2}$; **c.** $\frac{1}{2}$

11. All x and y **13.** $\{(x, y) | x \neq -y\}$

15. $\{(x, y) | x^2 + y^2 \leq 16\}$ **17.** 3

19. Plane with intercepts (0, 0, 4), (0, 6, 0), (12, 0, 0)

21. Plane with intercepts (0, 0, 2), (0, 4, 0), (10, 0, 0)

23. Horizontal plane three units above xy-plane

25. Sphere of radius 6 centered at (0, 0, 0)

27. (f) **29.** (c) **31.** (d)

33. **a.** $z_0 = 0$: $x = 0 = y$;
b. $z_0 = 1$: $x^2 + y^2 = 1$;
c. $z_0 = 4$: $x^2 + y^2 = 4$;
d. $z_0 = 9$: $x^2 + y^2 = 9$;
e. surface is a bowl

35. a. $z_0 = 1$: $x = 0 = y$;
b. $z_0 = 0$: $x^2 + y^2 = 1$;
c. $z_0 = -3$: $x^2 + y^2 = 4$;
d. $z_0 = -8$: $x^2 + y^2 = 9$;
e. surface is a mountain

37. $S(x, y) = 2x^2 + 4xy$, $S(3, 5) = 78$

39. $R(x, y) = 1400x + 802y - 12x^2 - 4xy - 0.5y^2$, 80,400

41. $P(x, y) = -15000 + 1350x + 801.5y - 12x^2 - 4xy - 0.5y^2$, 60,300

43. $A(m, t) = 1000\left(1 + \dfrac{0.08}{m}\right)^{mt}$, \$1489.85

45. a. $\{(x, y) | y \neq 0\}$; **b.** 200; **c.** 50

47. a. $\{(d, P) | P \geq 0\}$; **b.** 24π; **c.** 8π

49. When $y = mx$,

$$\lim_{x\to 0} f(x, mx) = \lim_{x\to 0} \frac{mx^3}{x^4 + m^2x^2} = \lim_{x\to 0} \frac{mx}{x^2 + m^2} = \frac{0}{0 + m^2} = 0$$

for all m. But this does not imply that f is continuous at $(0, 0)$.

Section 8.2

1. $f_x = 2x$, $f_y = 2y$, 2, 6

3. $f_x = 2xy - 3x^2y^2$, $f_y = x^2 - 2x^3y$, -8, -3

5. $f_x = \dfrac{\sqrt{y}}{2\sqrt{x}}$, $f_y = \dfrac{\sqrt{x}}{2\sqrt{y}}$, $\dfrac{1}{2}$, $\dfrac{1}{2}$

7. $f_x = \dfrac{xy^2}{\sqrt{1 + x^2y^2}}$, $f_y = \dfrac{x^2y}{\sqrt{1 + x^2y^2}}$, 0, 0

9. $f_x = 2e^{2x+3y}$, $f_y = 3e^{2x+3y}$, $2e^5$, $3e^5$

11. $f_x = ye^{xy} + xy^2e^{xy}$, $f_y = xe^{xy} + x^2ye^{xy}$, $2e$, $2e$

13. $f_x = \dfrac{1}{x + 2y}$, $f_y = \dfrac{2}{x + 2y}$, 1, 2

15. $f_x = ye^{xy} \ln x + \dfrac{1}{x}e^{xy}$, $f_y = xe^{xy} \ln x$, 1, 0

17. $f_x = \dfrac{-1}{x^2y}$, $f_y = \dfrac{-1}{xy^2}$, $-\dfrac{1}{2}$, $-\dfrac{1}{4}$

19. $f_x = \dfrac{y^2 - x^2 + 2xy}{(x^2 + y^2)^2}$, $f_y = \dfrac{y^2 - x^2 - 2xy}{(x^2 + y^2)^2}$, $\dfrac{1}{25}$, $-\dfrac{7}{25}$

21. $f_{xx} = 2y^4$, $f_{xy} = f_{yx} = 8xy^3$, $f_{yy} = 12x^2y^2$

23. $f_{xx} = 4e^{2x-3y}$, $f_{xy} = f_{yx} = -6e^{2x-3y}$, $f_{yy} = 9e^{2x-3y}$

25. $f_{xx} = -\dfrac{1}{4}\dfrac{\sqrt{y}}{x^{3/2}}$, $f_{xy} = f_{yx} = \dfrac{1}{4\sqrt{xy}}$, $f_{yy} = -\dfrac{1}{4}\dfrac{\sqrt{x}}{y^{3/2}}$

27. $f_{xx} = 2e^y$, $f_{xy} = f_{yx} = 2xe^y$, $f_{yy} = x^2e^y$

29. $f_{xx} = \dfrac{y^2}{(x^2 + y^2)^{3/2}}$, $f_{xy} = f_{yx} = \dfrac{-xy}{(x^2 + y^2)^{3/2}}$, $f_{yy} = \dfrac{x^2}{(x^2 + y^2)^{3/2}}$

31. $f_x = yz$, $f_y = xz$, $f_z = xy$

33. $f_x = \dfrac{x}{\sqrt{x^2 + y^2 + z^2}}$, $f_y = \dfrac{y}{\sqrt{x^2 + y^2 + z^2}}$, $f_z = \dfrac{z}{\sqrt{x^2 + y^2 + z^2}}$

35. $f_x = e^{x+2y+3z}$, $f_y = 2e^{x+2y+3z}$, $f_z = 3e^{z+2y+3z}$

37. $f_x = \dfrac{1}{x + 2y + 5z}$, $f_y = \dfrac{2}{x + 2y + 5z}$, $f_z = \dfrac{5}{x + 2y + 5z}$

39. Complementary **41.** Competitive

43. Competitive

45. $A_r = 1000t(1 + r/12)^{12t-1}$. This gives the rate of change of the amount with respect to changes in the interest rate r.

47. $\dfrac{\partial V}{\partial r} = 4\pi r^2\left(1 + \dfrac{c}{r}N^{1/3}\right)^2$. This gives the rate of change of the detection volume with respect to changes in the detection radius r. $\dfrac{\partial V}{\partial N} = \dfrac{4\pi cr^2}{3N^{2/3}}\left(1 + \dfrac{c}{r}N^{1/3}\right)^2$. This gives the rate of change of the detection volume with respect to changes in the number of prey N.

49. $\dfrac{\partial L}{\partial r} = -2p\pi r\left(1 - \dfrac{p\pi r^2}{n}\right)^{n-1}$. This gives the rate of change of light with respect to changes in the radius r of the leaves. $\dfrac{\partial L}{\partial p} = -\pi r^2\left(1 - \dfrac{p\pi r^2}{n}\right)^{n-1}$. This gives the rate of change of light with respect to changes in the total density p of the leaves.

51. $\dfrac{\partial A}{\partial S}(S, D) = 0.11297S^{0.43}D^{-0.73}$,
$\dfrac{\partial A}{\partial D}(S, D) = -0.05767S^{1.43}D^{-1.73}$

53. $\dfrac{\partial x}{\partial A} > 0$ means that there is more demand when advertising is at a higher level. $\dfrac{\partial x}{\partial P} < 0$ means that there is less demand when advertising is at a higher price.

55. a. The instantaneous rate of change of f with respect to x at the point $x = 1$ and $y = 2$ is 3.
b. The instantaneous rate of change of f with respect to y at the point $x = 1$ and $y = 2$ is -5.

Section 8.3

1. Relative minimum at $(5, -2)$

3. Relative maximum at $(2, 1)$

5. Saddle at (1, 1)

7. Relative maximum at $(-1, -2)$

9. Relative minimum at (0, 0)

11. Saddle at (0, 2)

13. Saddle at (0, 0), relative minimum at (1, 1)

15. Saddle at (0, 0), relative maximum at (1/6, 1/12)

17. Relative maximum at (1, 1)

19. Saddle at (0, 0)

21. Since $f(x, y) = x^4y^4 \geq 0 = f(0, 0)$ for all x and all y, f has a relative minimum at (0, 0). Since $f_{xx} = 12x^2y^4$, $f_{yy} = 12x^4y^3$, and $f_{xy} = 16x^3y^3$, $\Delta(0, 0) = 0$.

23. If $y = x$, then $f(x, y) = x^3x^3 = x^6$, and $f(x, x)$ increases as x increases. But if $y = -x$, $f(x, -x) = -x^6$, and $f(x, -x)$ decreases as x increases. Since $f_{xx} = 6xy^3$, $f_{yy} = 6x^3y$, and $f_{xy} = 9x^2y^2$, $\Delta(0, 0) = 0$.

25. Increasing in all cases.

27. Increasing in the third case and decreasing in all the others.

29. $x = 1, y = 2, z = 4$ **31.** Base 4×4 and height 6

33. $x = 200, y = 100$ **35.** $x = 18, y = 8$

37. $x = 9, y = 4$

39. $x = 200, y = 100, z = 700$

41. $x = 1600, y = 1000$ **43.** $x = 2100, y = 0$

45. $N \approx 2.463$, $P \approx 2.399$

47. (1) $x = (P/a)^{1/b}y^{1-1/b}$; (2) $C = C(y) = p_1(P/a)^{1/b}y^{1-1/b} + p_2y$; (3) $C'(y) = 0$ implies that $y = \left(\frac{p_1}{p_2}\right)^b\left(\frac{1}{b} - 1\right)^b\frac{P}{a}$; (4) $x = \left(\frac{p_1}{p_2}\right)^{b-1}\left(\frac{1}{b} - 1\right)^{b-1}\frac{P}{a}$; (5) $C = C(P) = \left(\frac{1}{b} - 1\right)^b\left(\frac{p_1}{p_2}\right)^b\left(\frac{p_2}{1-b}\right)\frac{P}{a}$

Section 8.4

1. Minimum at $x = 1, y = 3$

3. Maximum at $x = -3, y = -1$

5. Minimum at $x = 1, y = 1$

7. Maximum at $x = -1, y = -3$

9. Minimum at $x = 2, y = 2$

11. Maximum at $x = 4, y = 4$

13. Minimum at $x = 4, y = 2, z = 1$

15. Maximum at $x = 2, y = 2, z = 1$

17. 10 and 10 **19.** 12, 12, and 12

21. $r = 2, h = 4$ **23.** $x = 200, y = 100$

25. $6 \times 6 \times 9$ **27.** $x = 2000, y = 4000$

29. $x = 1, y = 2, z = 4$

31. $x = 200, y = 100, z = 700$

33. Let $F(x, y, \lambda) = p_1x + p_2y + \lambda f(x, y) - \lambda P_1$; then

$$\begin{cases} 0 = F_x = p_1 + \lambda f_x \\ 0 = F_y = p_2 + \lambda f_y \end{cases}$$

which implies that

$$\lambda = -\frac{p_1}{f_x} = -\frac{p_2}{f_y}$$

That is, $\dfrac{f_x}{p_1} = \dfrac{f_y}{p_2}$.

Section 8.5

1. 1 **3.** -6 **5.** $\frac{1}{15}$

7. $\frac{-11}{1250}$ **9.** $\frac{1}{20}$ **11.** $\frac{-1}{20}$

13. 0.05 **15.** $\frac{e}{10} \approx 0.2718$

17. 6.05 **19.** 5.036 **21.** 3.11

23. 8.9 **25.** 24.57 **27.** 1.08

29. \$2000 **31.** 118.75 **33.** 0.70 ft^3

35. 0.9573 m^2 **37.** 0.7116

Section 8.6

1. $\frac{15}{4}$ **3.** $\frac{3}{4}$ **5.** 8

7. 2.5 **9.** 1 **11.** 35

13. $\frac{4}{15}(2\sqrt{2} - 1)$ **15.** $\ln 2 \ln 3$

17. $(e^2 - 1)(e^3 - 1)$ **19.** $\frac{7}{24}$

21. $\frac{e^4 - 1}{2}$ **23.** $\frac{1}{2}$ **25.** $\frac{1}{3}$

27. $\frac{1}{8}$ **29.** $\frac{1}{4}$ **31.** $\frac{1}{40}$

33. $\frac{1}{24}$ **35.** $-\frac{2}{3}$ **37.** $\frac{7}{6}$

39. $\frac{3}{2} - \ln 2$ **41.** 4 **43.** 19,000

45. $\frac{3627}{532}$ **47.** $e - 1$ **49.** 10,000

51.

n	LLSUM	URSUM
20	0.9025	1.1025
40	0.950625	1.050625
100	0.9801	1.0201

Chapter 8 Review Exercises

1. $-1, 7, 4$ **2.** $0, \frac{1}{5}, -\frac{8}{5}$

3. (1) All x and y; (2) $\{(x, y)|y \neq 0\}$

4. $2\sqrt{11}$

5. $z_0 = 16$: $x^2 + y^2 = 0$ or the single point $(0, 0)$; $z_0 = 12$: $x^2 + y^2 = 4$, circle of radius 2 centered at $(0, 0)$; $z_0 = 7$: $x^2 + y^2 = 9$, circle of radius 3 centered at $(0, 0)$.

6. $f_x(x, y) = 1 + 2xy^5$, $f_y(x, y) = 5x^2y^4$

7. $f_x(2, 1) = 5$, $f_y(2, 1) = 20$

8. $f_x = -ye^{-xy} + \dfrac{1}{x}$, $f_y = -xe^{-xy} + \dfrac{1}{y}$

9. $f_{xx} = 2y^5$, $f_{yy} = 20x^2y^3$, $f_{xy} = 10xy^4$

10. $f_{xx} = y^2e^{-xy} - \dfrac{1}{x^2}$, $f_{yy} = x^2e^{-xy} - \dfrac{1}{y^2}$, $f_{xy} = -e^{-xy} + xye^{-xy}$

11. $f_x = -yze^{-xyz} + y^2z^3$, $f_y = -xze^{-xyz} + 2xyz^3$, $f_z = -xye^{-xyz} + 3xy^2z^2$

12. Relative minimum at $(0, 0)$

13. Relative maximum at $(0, 0)$

14. Relative maximum at $(-2, -1)$

15. Relative minimum at $(2, -3)$

16. Saddle at $(0, 0)$, relative maximum at $(-1, -1)$

17. Saddle at $(-1, -1)$, relative minimum at $(3, 3)$

18. Saddle at $(5, 3)$, relative maximum at $(5, -1)$

19. Saddle at $(0, 0)$, relative maximum at $(2/3, 1/9)$

20. Since $f_{xx}(0, 0) = 2a$ and $\Delta(0, 0) = 4(ac - b^2)$, the result follows.

21. Minimum at $(4, -3)$ 22. Maximum at $(1/4, 1/2)$

23. 2.8 24. 3.2 25. 4.99

26. 24 27. 80 28. $\dfrac{5}{12}$

29. $\dfrac{7}{24}$ 30. $\dfrac{4}{5}$ 31. $\dfrac{3}{20}$

32. $\dfrac{1}{2}$ 33. $\dfrac{13}{20}$ 34. $\dfrac{13}{20}$

35. $\dfrac{1}{6}$ 36. 80

37. $Q_L = 2.014(0.99)L^{-0.01}E^{0.47}$, rate of change of GNP with respect to employment. $Q_E = 2.014(0.47)L^{0.99}E^{-0.53}$, rate of change of GNP with respect to energy consumption

38. $\dfrac{\partial Q}{\partial T} = (0.02)(5868)X^{0.17}Y^{0.13}T^{-0.98}$
The instantaneous rate of change of onion production with respect to time. Units are 1000 cwt per year.

39. Approximately 436 sq. ft. and 100 trees

40. 1091

41. $C_x = 2x - y$, $C_y = 2y - x$

42. $C_{xx} = 2$, $C_{yy} = 2$, $C_{xy} = -1$

43. $A_P = e^{rt}$, $A_r = tPe^{rt}$, $A_t = Pre^{rt}$

44. $2 \times 2 \times 4$ 45. $x = 10$, $y = 12$

46. $x = 11$, $y = 14$

47. 11,000 of regular and 9000 of mint

48. 500 of each 49. 170 cubic inches

50. $\dfrac{\partial h}{\partial e} > 0$ means that the individual firm has a higher harvest rate with a higher effort rate. $\dfrac{\partial h}{\partial N} < 0$ means that more participating firms result in less of a harvest. $\dfrac{\partial h}{\partial X} > 0$ means that larger resource stock size results in more of a harvest.

Appendix

Section A.1

1. x^{10} 3. x^{15} 5. $(x + 2)^5$

7. x^5 9. $(x + 1)^3$ 11. 1

13. 1/2 15. 10 17. 1

19. a^3/b 21. xyz 23. y^6/x^{12}

25. −4 27. −2 29. 7

31. −3 33. 5/2 35. −2/5

37. 2 39. 4 41. 0.1

43. 1/216 45. 16 47. 256

49. 10,000/81 51. 9/4 53. $2x^2$

55. ab 57. $6\sqrt{2}a^2b^{3/2}c^{3/2}$

59. $2a^2b^3$ 61. $2a^3b^2/c^4$ 63. a^2b^2

65. $x^{-1/2} + x^{1/2}$ 67. $(x + 1)^{2/3}$

69. $x^{2/3}$ 71. $(x/y)^{1/2}$ 73. $(x^2 + 1)^{17/12}$

75. $1/x^{1/3}$ 77. ab 79. b

81. $(x + 1)(x + 2)$ 83. $(x + 1)^5$

85. $\sqrt{5}/5$ 87. $\sqrt{2} - 1$

89. $\sqrt[3]{25}/5$ 91. $-(1 + \sqrt{3})^2/2$

93. $1/(\sqrt{5} + \sqrt{2})$ 95. $1/(\sqrt{10} + 2)$

Section A.2

1. $3x^2 + 3x + 2$ 3. $-3x^2 - x - 1$

5. $x^4 - x^3 + x^2 - x + 4$ 7. $x^3 - x^2 + x + 2$

9. x^3+3x^2+5x+3 **11.** $2x^3-3x^2+7x-3$
13. x^4+4x^2+3 **15.** $(x-3)(x^2+3x+9)$
17. $\left(x-\frac{1}{2}\right)\left(x^2+\frac{1}{2}x+\frac{1}{4}\right)$
19. $(3x-2)(9x^2+6x+4)$
21. $\left(\frac{1}{2}x-3\right)\left(\frac{1}{4}x^2+\frac{3}{2}x+9\right)$
23. $\left(\frac{1}{4}x-\frac{1}{3}\right)\left(\frac{1}{16}x^2+\frac{1}{12}x+\frac{1}{9}\right)$
25. $(ax-b)(a^2x^2+abx+b^2)$
27. $(x-4)(x+4)$ **29.** $(4x-1)(4x+1)$
31. $\left(x^2+\frac{1}{4}\right)\left(x+\frac{1}{2}\right)\left(x-\frac{1}{2}\right)$
33. $\left(\frac{1}{4}x^2+1\right)\left(\frac{1}{2}x+1\right)\left(\frac{1}{2}x-1\right)$
35. $(x^2+2)(x+\sqrt{2})(x-\sqrt{2})$
37. $\left(\frac{1}{2}x^2+1\right)\left(\frac{1}{\sqrt{2}}x+1\right)\left(\frac{1}{\sqrt{2}}x-1\right)$
39. $\frac{x+1}{x-1}$ **41.** $\frac{x}{1-x}$
43. $\frac{x-1}{x+2}$ **45.** $\frac{x-1}{x+4}$
47. $2x+1$ **49.** $\frac{x^2+1}{x}$
51. $\frac{x-1}{x+1}$ **53.** $\frac{1}{x+1}$
55. $-\frac{x}{(x-1)(x+1)}$ **57.** $\frac{4}{(x^2+4)(x+2)(x-2)}$
59. $\frac{\sqrt{x}-1}{x-1}$ **61.** $\frac{\sqrt{x}-\sqrt{y}}{x-y}$
63. $\sqrt{y}+3$ **65.** $(a+1)(\sqrt{a}+1)$
67. $\frac{1}{\sqrt{x}+1}$ **69.** $\frac{1}{\sqrt{x+h}+\sqrt{x}}$

Section A.3

1. -4 **3.** $11/7$ **5.** 2
7. $-19/10$ **9.** $13/5$ **11.** ± 2
13. $\pm 3/2$ **15.** ± 2 **17.** $\pm 1/2$
19. $\pm 3/5$ **21.** ± 6 **23.** $\pm 1/6$
25. $\pm b/a$ **27.** -2 **29.** $1, 4$
31. $1/2, -2$ **33.** $-1/3, -2$ **35.** $1/3, -1/2$
37. $1/5, 1$ **39.** $-2/3, -3/2$ **41.** $1/4, 2/3$
43. $1/3$ **45.** $-2/3$ **47.** $1/2$
49. $-1/4$ **51.** ± 2 **53.** $\pm 3/2$
55. $0, 1, -2$ **57.** $x = 0, \pm 3$ **59.** -1
61. $\frac{7}{10} \pm \frac{3}{10}\sqrt{5}$ **63.** $\frac{1}{2} \pm \frac{1}{2}\sqrt{5}$
65. $-\frac{2}{3} \pm \frac{\sqrt{19}}{3}$ **67.** $2 \pm \sqrt{3}$
69. $\frac{7}{6} \pm \frac{\sqrt{41}}{6}$ **71.** No real solution
73. $1, 0$ **75.** $2, -4/3$ **77.** $0, 4/3$
79. $-1, 9$ **81.** $-1/6, 7/6$ **83.** $5/3, 1$

Section A.4

1. $x < \frac{4}{3}$ **3.** $x < -6$ **5.** $x \le -1$
7. $x \le 7$ **9.** $x < \frac{3}{8}$ **11.** $x \le 1$
13. $-4 < x < 2$ **15.** $0 \le x \le 1$
17. $x < -1$ or $x > 3$ **19.** $x \le -\frac{4}{3}$ or $x \ge \frac{8}{3}$
21. $-\frac{19}{12} < x < -\frac{17}{12}$ **23.** $x \le -5$ or $x \ge 10$
25. $1 < x < 3$ **27.** $x \le -2$ or $x \ge 3$
29. $-2 < x < 2$
31. $x \le -3$ or $x \ge 1$ **33.** $\frac{1}{2} < x < 2$
35. $x \le -3$ or $x \ge \frac{2}{3}$ **37.** $-2 < x < 1$
39. $x \le 1$ or $x > 2$ **41.** $x < -2$ or $x > 1$
43. $-1 \le x < 1$ **45.** $-1 \le x < 0$ or $x \ge 1$
47. $x < -1$ or $0 < x < 1$ **49.** $x < 1$ or $2 < x < 3$
51. $-1 \le x \le 1$ or $x > 2$ **53.** $x < 0$
55. $-1 < x < 2$ with $x \ne 0$

Section A.5

1. (———) −1 3
3. (———] 2 5
5. (———→ −2
7. ←———] 1.5

9–19.

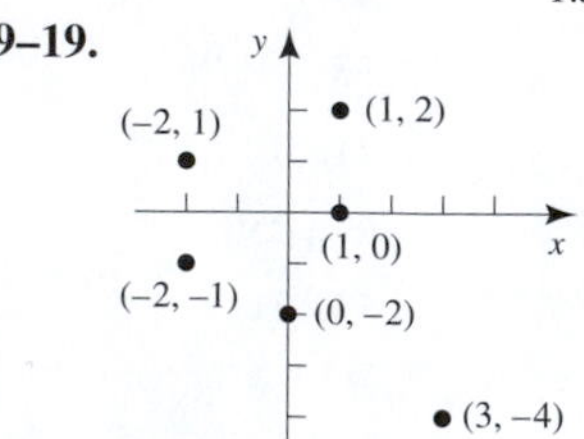

21.

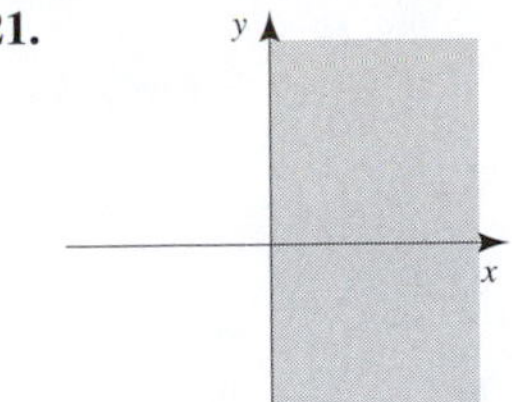

23.

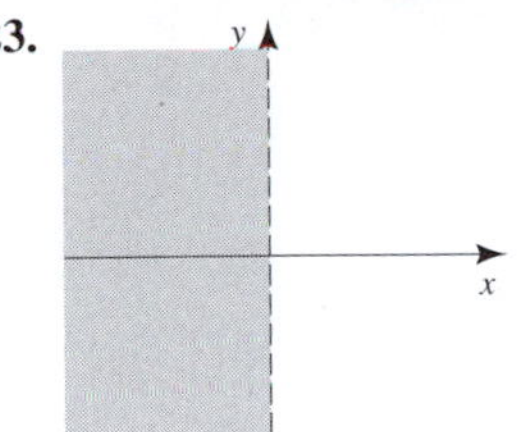

25.

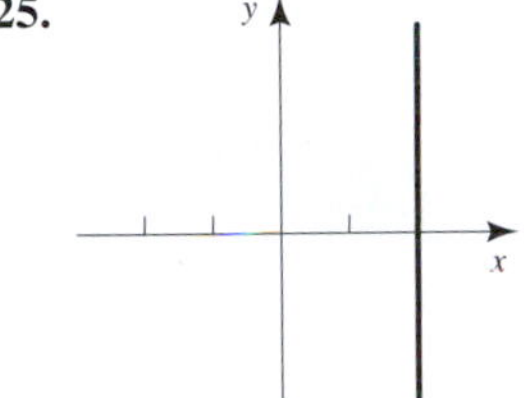

27.

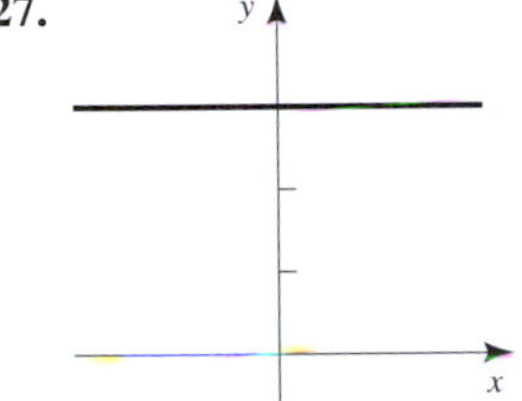

29.

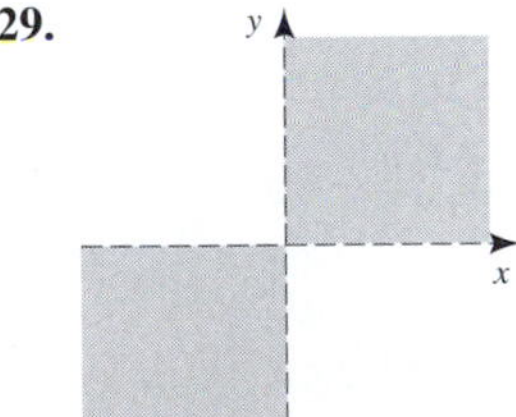

31.

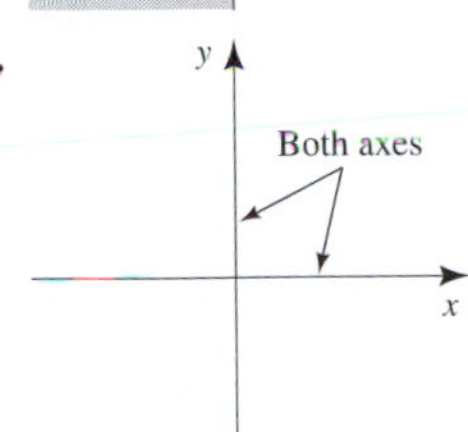

33. $\sqrt{5}$ **35.** $\sqrt{37}$

37. $5\sqrt{2}$ **39.** $\sqrt{2}|b - a|$

41.

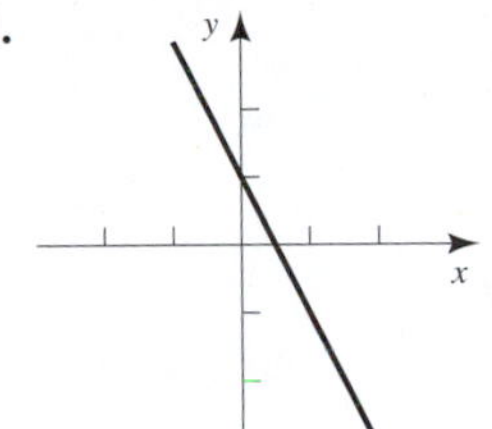

43.

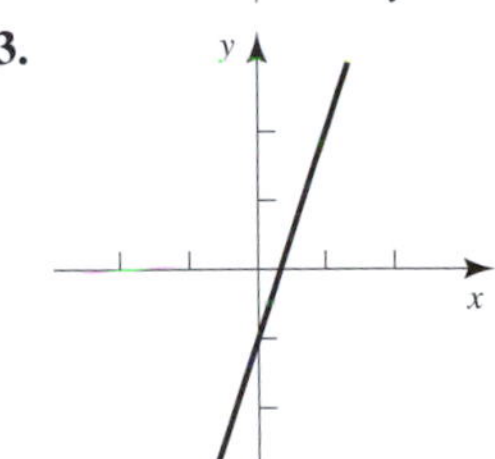

45.

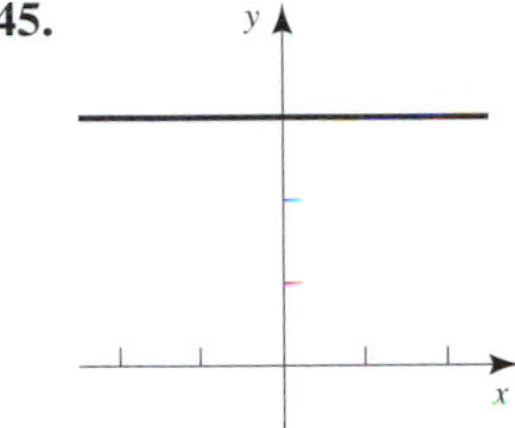

47.

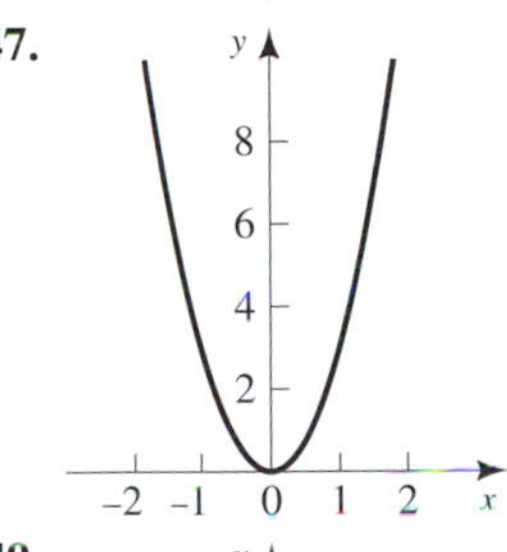

49.

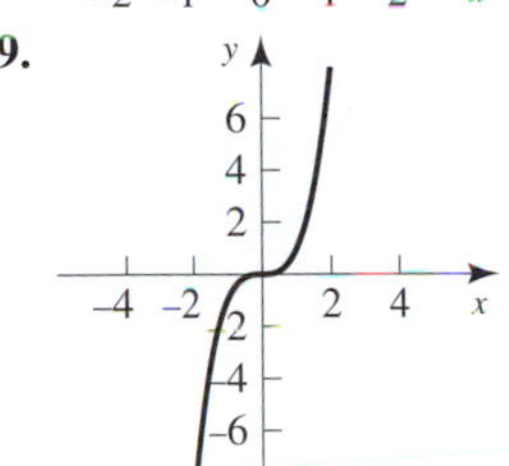

51. 25 miles **53.** 13 miles

55. 10 days **57.** $100x + 150y = 3000$

59. $125x + 190y = 800$ **61.** $c = \sqrt{a^2 - b^2}$

Section A.6

1. 4

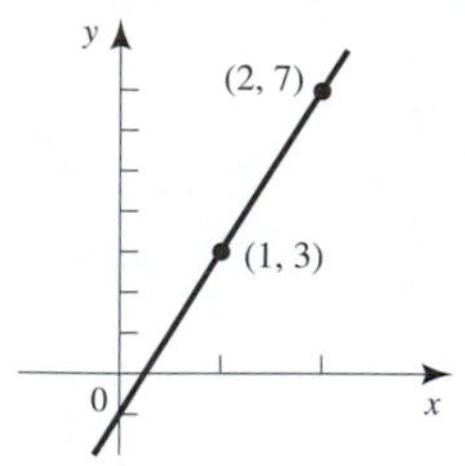

3. -3

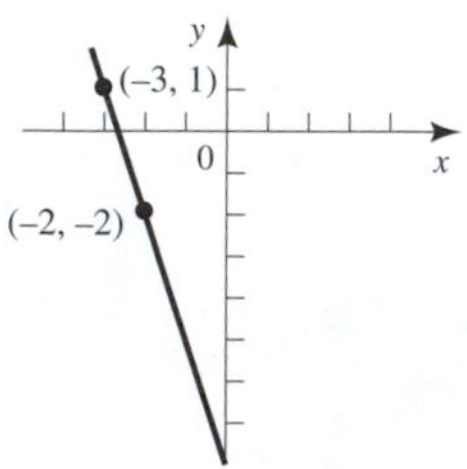

5. 0

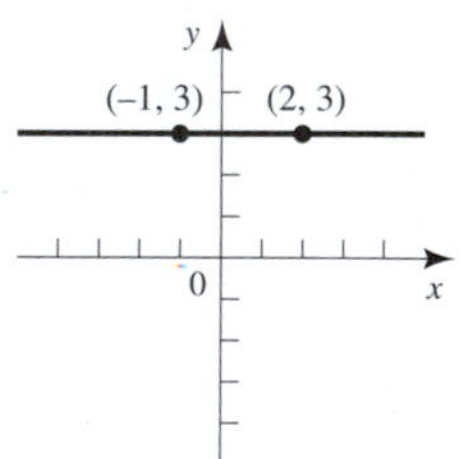

7. Undefined

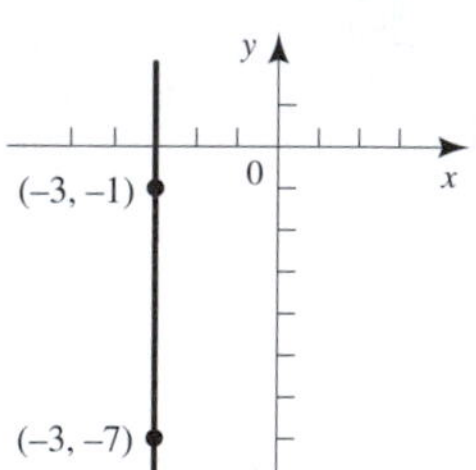

9. $2, \frac{1}{2}, -1$

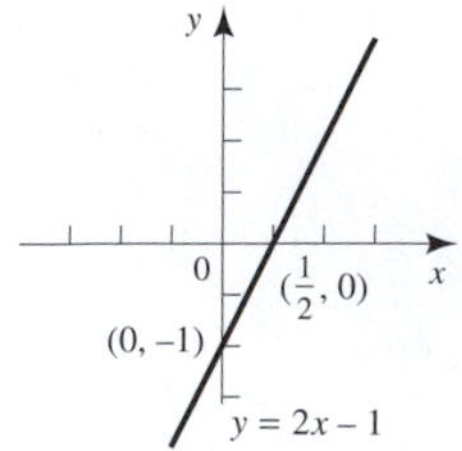

11. $-1, 1, 1$

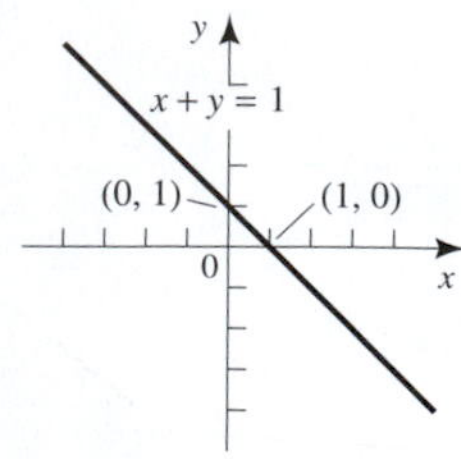

13. $\frac{1}{2}, -4, 2$

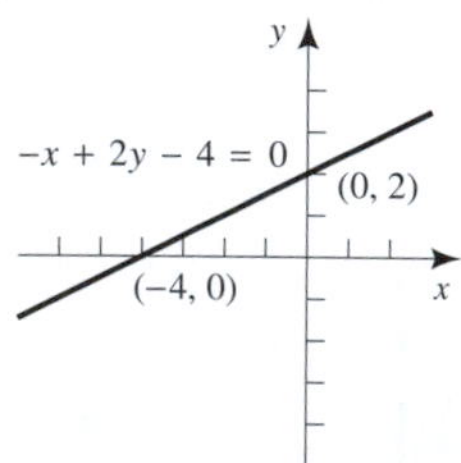

15. Undefined, 3, no y-intercept

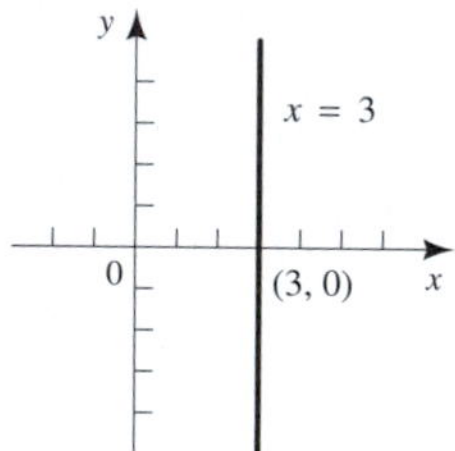

17. 0, no x-intercept, 4

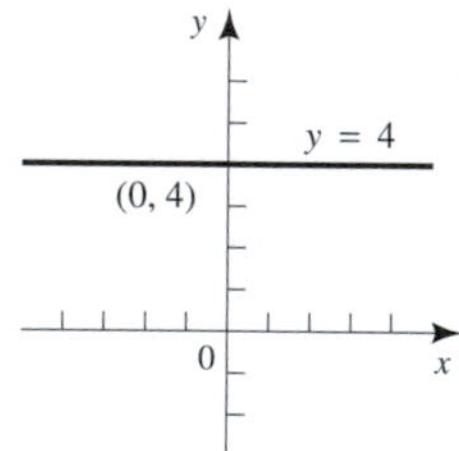

19. $\frac{1}{2}, 1, -\frac{1}{2}$

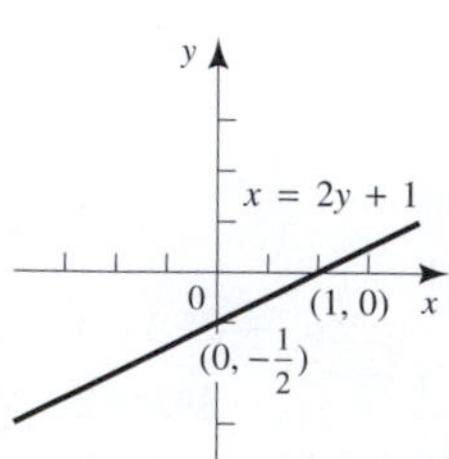

21. 5, 3

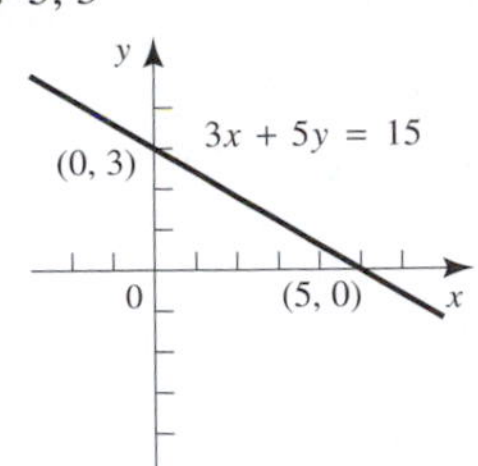

23. 5, −4

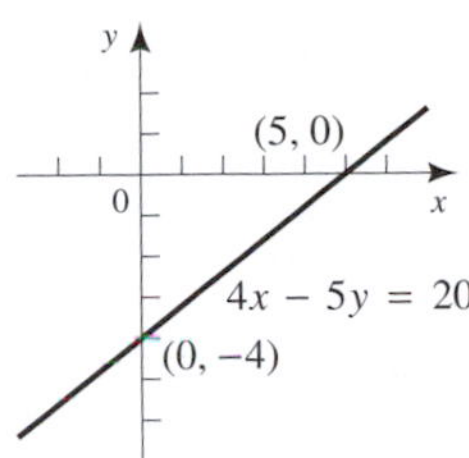

25. $y = -3x + 7$

27. $y = -3$

29. $y = -x + 2$

31. $x = -2$

33. $y = -\frac{b}{a}x + b$

35. $y = -\frac{2}{5}x + \frac{11}{5}$

37. $y = 0$

39. $x = 1$

41. $y = 2x + 5$

43. $y = -\frac{5}{3}x + 4$

45. $y = -\frac{2}{5}x + 4$

47. No

49. $0.08x + 0.05y = 576$, \$160

51. $y = 0.012x - 0.12$, 0.36 kilograms

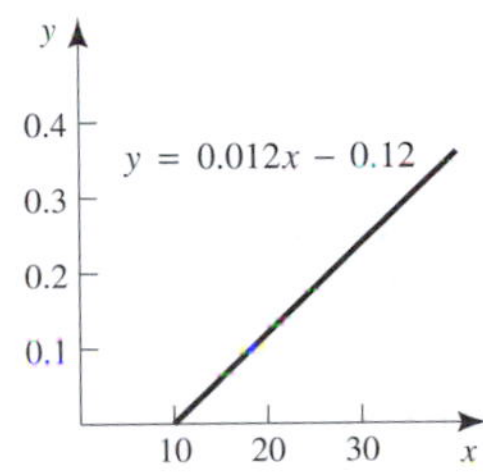

53. Each additional pound of milk per cow per year requires an additional 3 cents in cost per cow per year. \$30.

55. $L - 140 = \frac{182}{79}(l - 60)$, $\frac{182}{9}$, each 79-mm of increase in the tail brings a 182-mm increase in the total length.

57. For each additional fly on the front legs of the bull, there are (on average) 2.27 additional flies on the bull: no flies on the front legs, implies 10.95 flies (on average) on the bull.

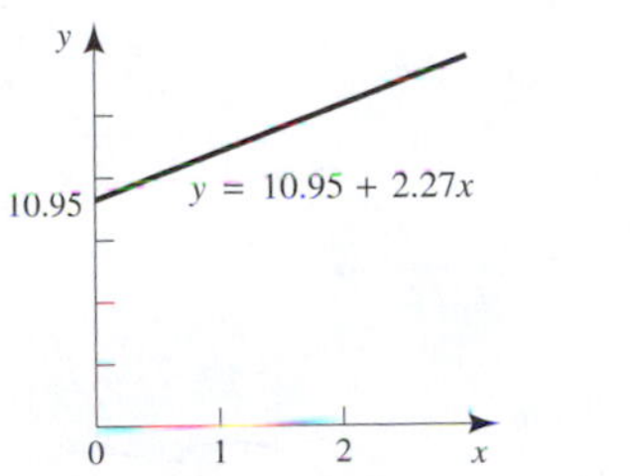

59. a. $100x + 150y = 3000$

b. $-\frac{2}{3}$

c. 18. If three chairs are sold, then 18 sofas need to be sold for the profit to be \$3000.

d. 30

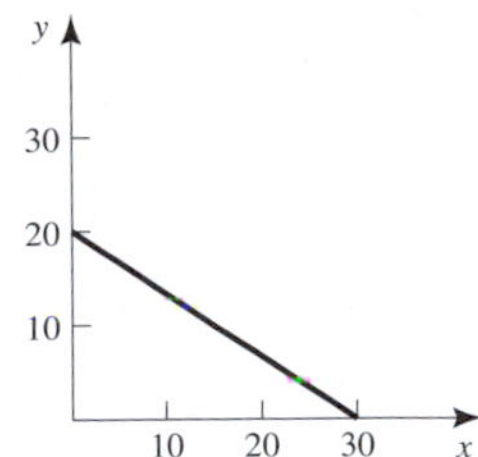

61. a. $125x + 190y = 820$

b. $-125/190$

c. 3. If 2 ounces of sardines are consumed, then 3 cups of broccoli need to be eaten to obtain 820 mg of calcium.

d. 6.56

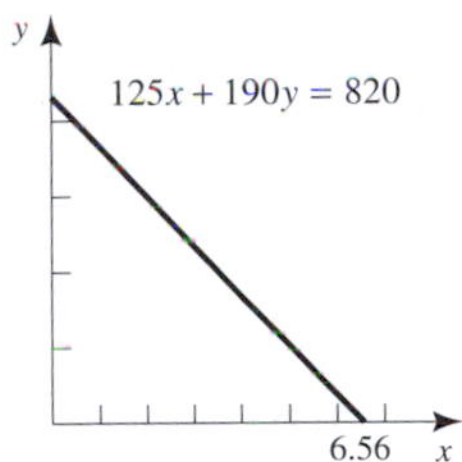

63. 1963

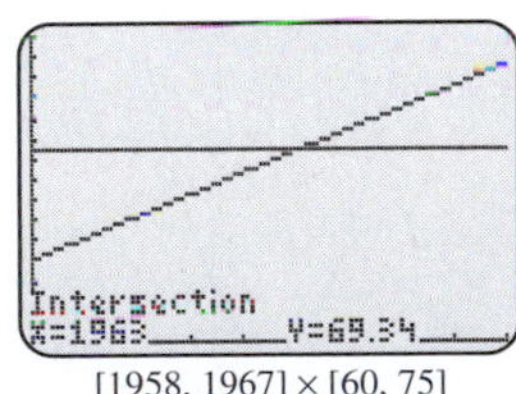

[1958, 1967] × [60, 75]

65. If $b \neq 0$, then $ax + by = c$ if and only if $y = -\frac{a}{b}x + \frac{c}{b}$. This is the straight line with slope $-\frac{a}{b}$ and y-intercept $\frac{c}{b}$. If $b = 0$, then by assumption $a \neq 0$, and we have $x = \frac{c}{a}$ which is a vertical line. Therefore, $ax + by = c$ is a straight line if a and b are not both zero. Conversely, every line is the graph of either the equation $y - y_1 = m(x - x_1)$ if and only if $-mx + y = y_1 - mx_1$ with $a = -m$, $b = 1$, and $c = y_1 - mx_1$, or the equation $x = x_0$ if m fails to exist, in which case $a = 1$, $b = 0$, and $c = x_0$. Therefore, every line is the graph of a linear equation.

Section A.7

1.

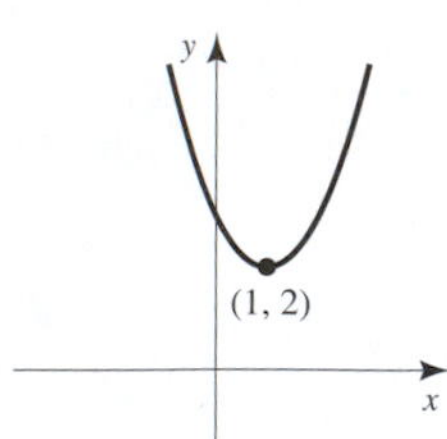

3.

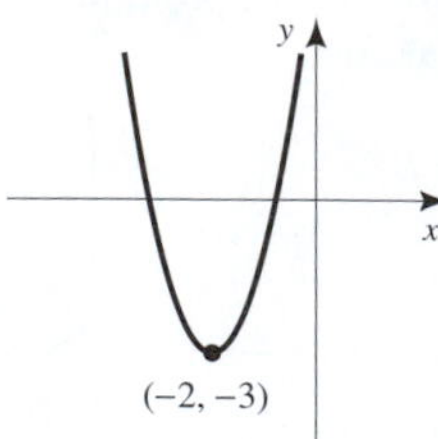

5.

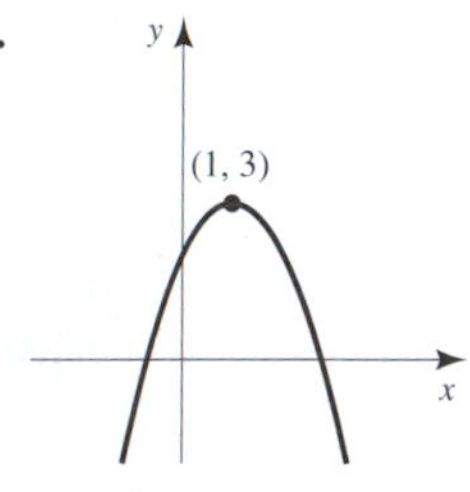

7.

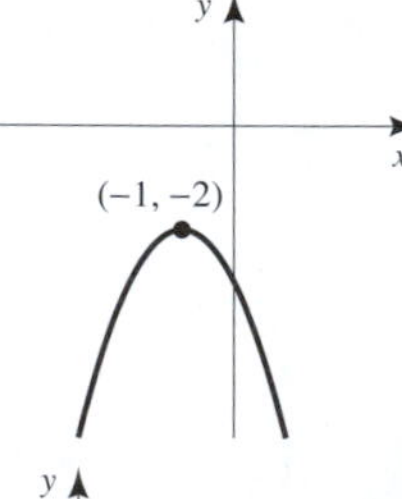

9.

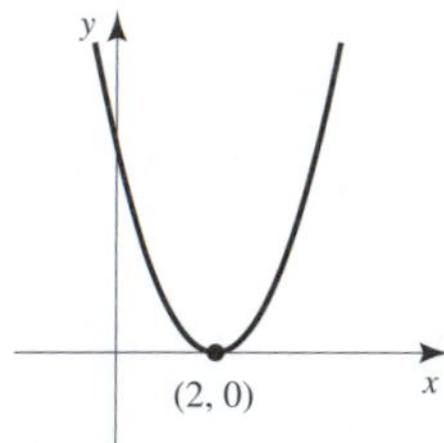

11.

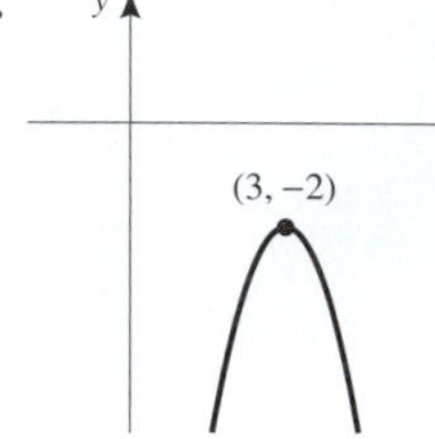

13. Zero at $x = -1$

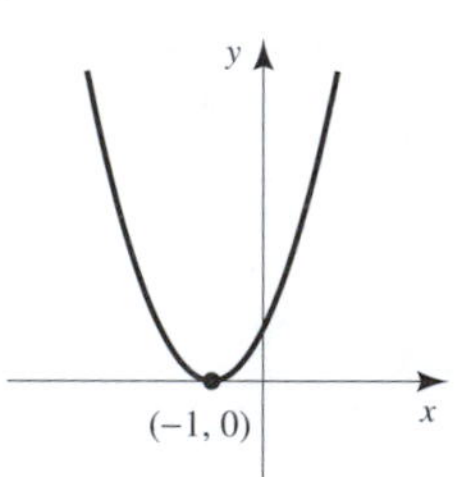

15. Zeros at $x = -1/2$, $x = -3/2$

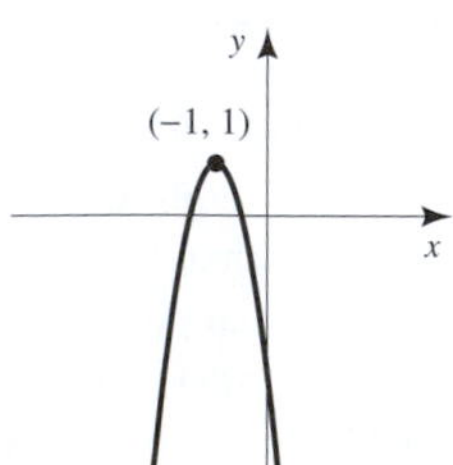

17. No zeros

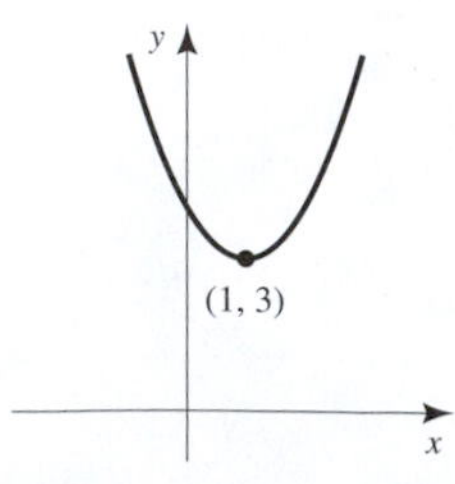

19. No zeros

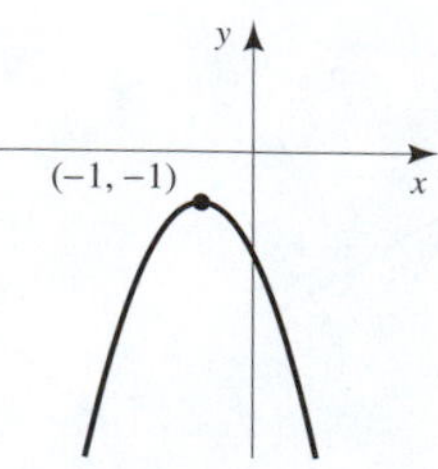

21. $(1, -1)$ **23.** $(2, 2)$

25. a. ii; **b.** iii; **c.** iv; **d.** i

27. 20.25

29. Both $p(x) = x^2$ and $q(x) = -x^2$ are quadratics, but $p(x) + q(x) = 0$ and is not a quadratic. Thus, the sum of two quadratics need not be a quadratic.

31. Both $f(x)$ and $g(x)$ are increasing for $x < x_1$; thus, the peak of $f + g$ must occur after x_1. Both $f(x)$ and $g(x)$ are decreasing for $x > x_2$; thus, the peak of $f + g$ must occur before x_2.

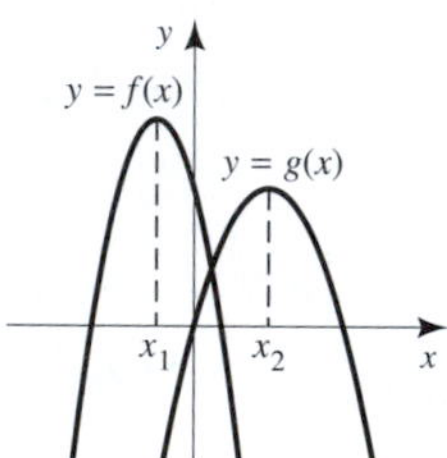

33. 0.9, 2.2

35. Approximately 1.9364358 and 7.6635642

Section A.8

1. 8, $(-\infty, \infty)$, 1 **3.** 9, $(-\infty, \infty)$, 1000

5. $x \neq 3$ **7.** All x **9.** $x \neq 0$

11.

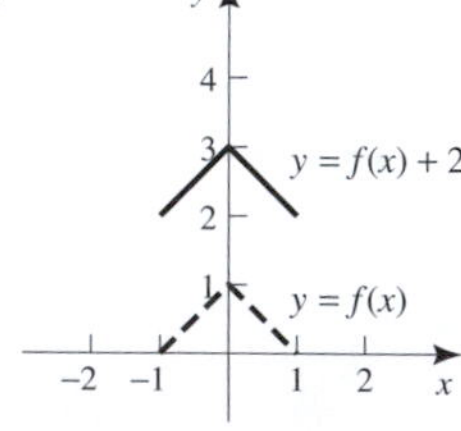

13.

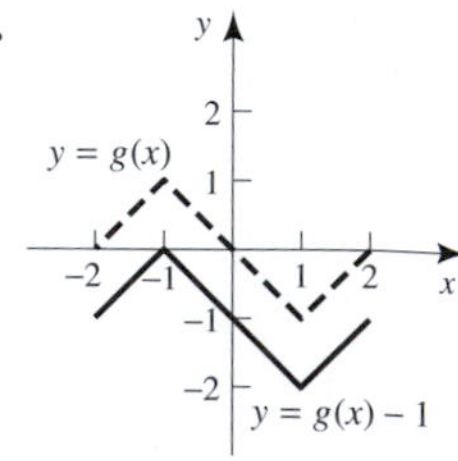

15.

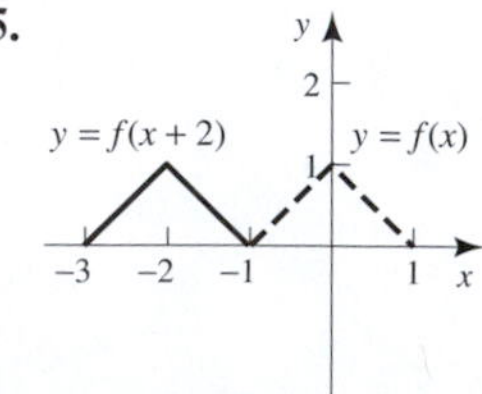

17.

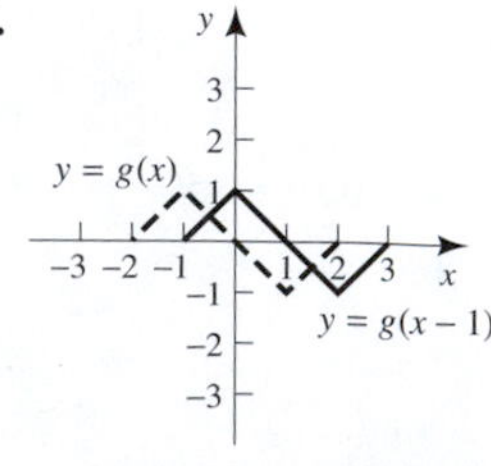

19.

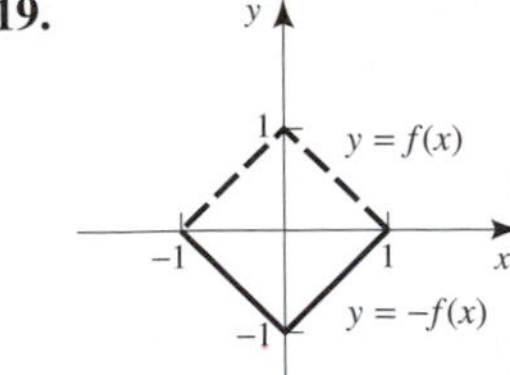

21.

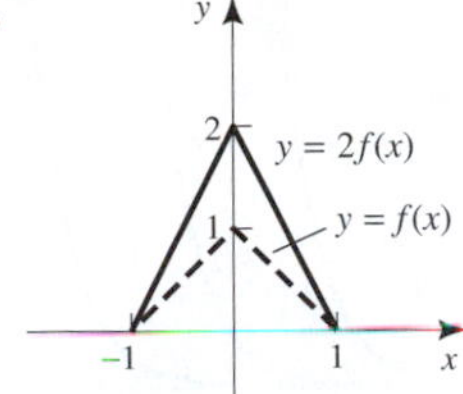

33.

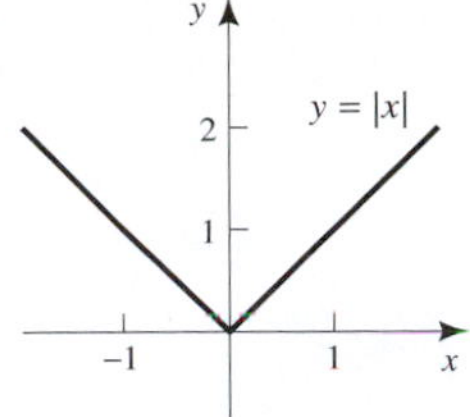

23.

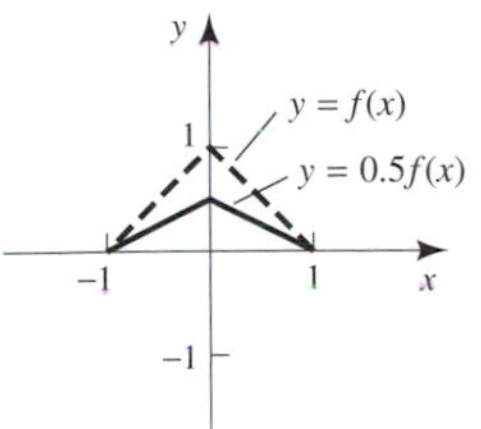

25.

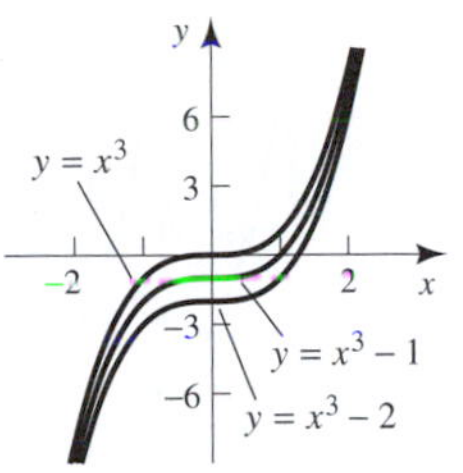

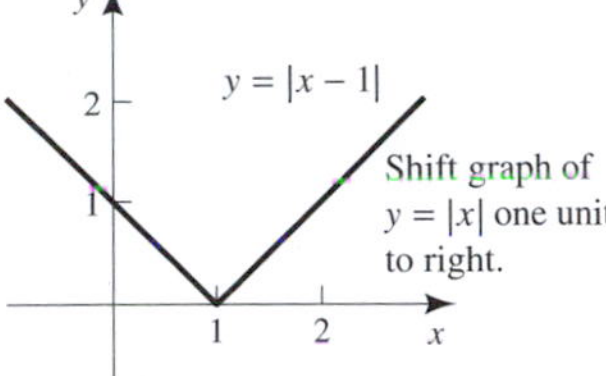

27.

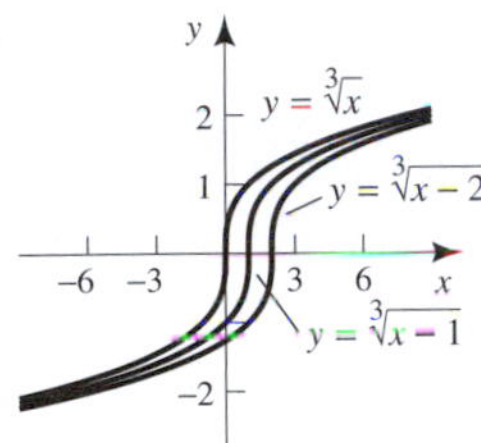

29.

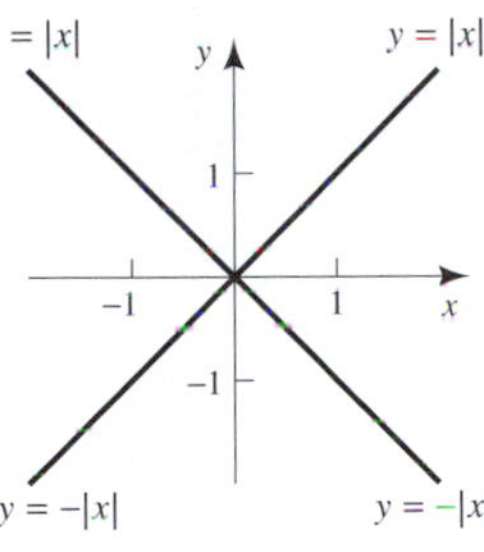

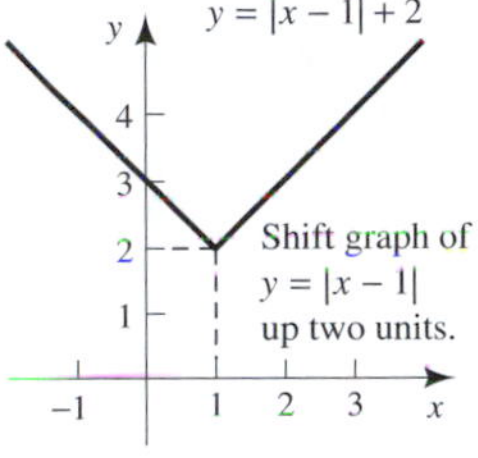

31.

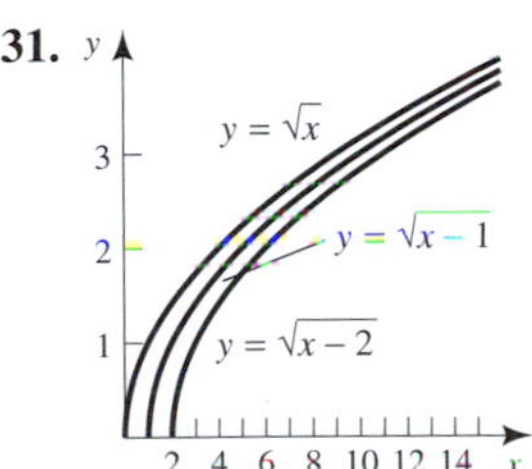

35.

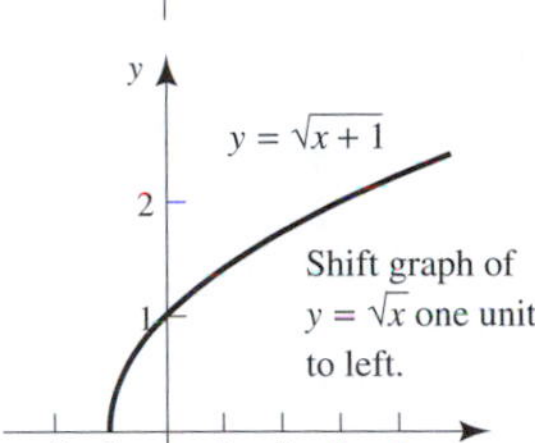

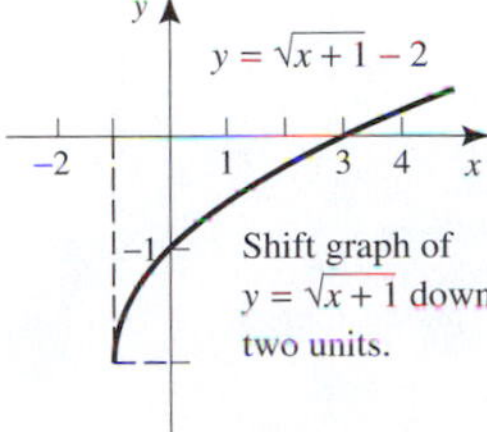

37.

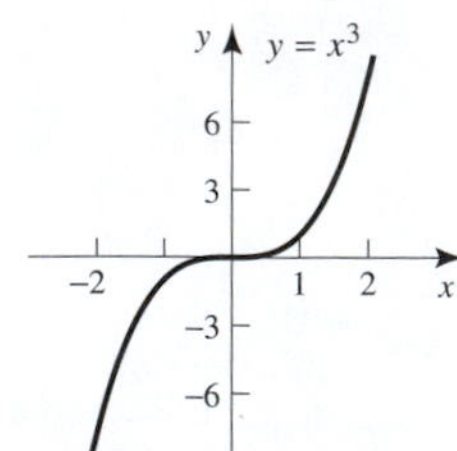

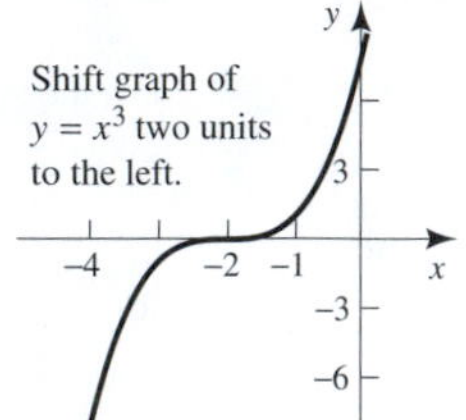

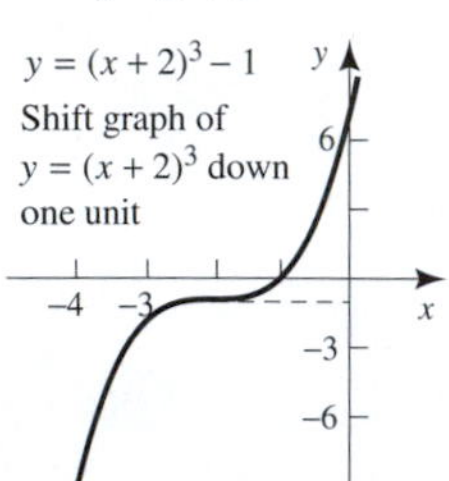

39.

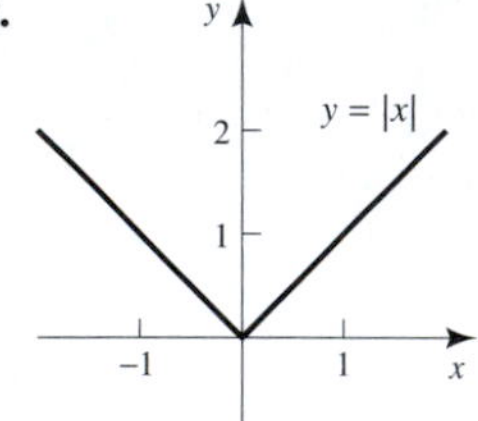

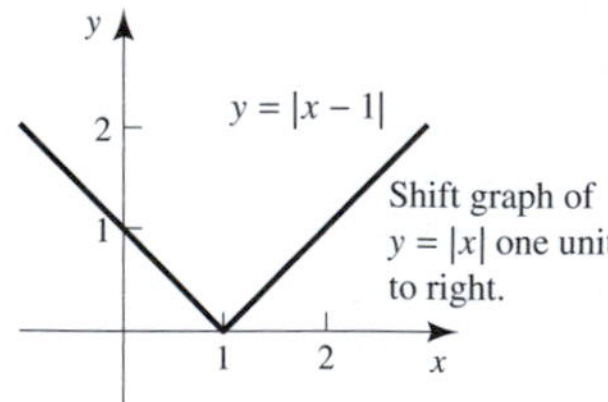

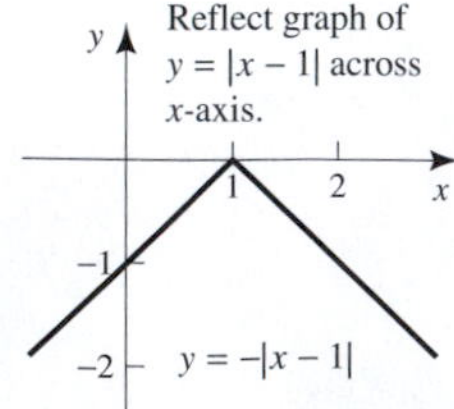

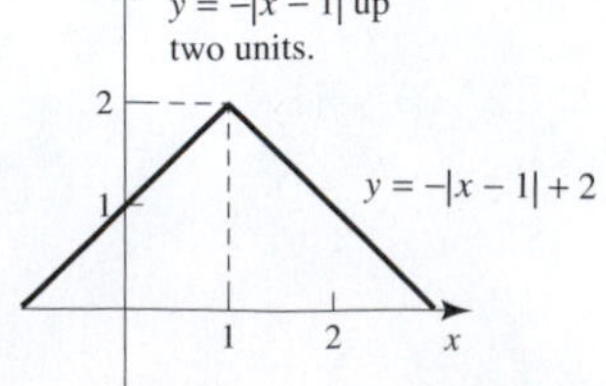

41.

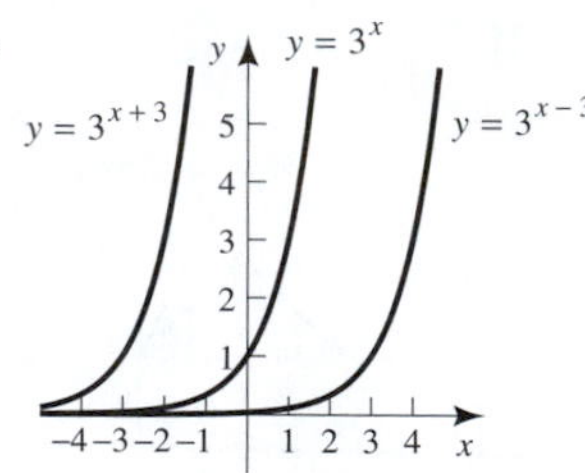

43.

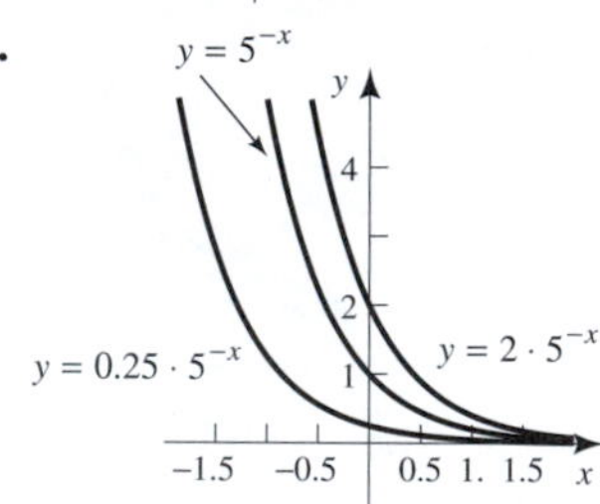

45.

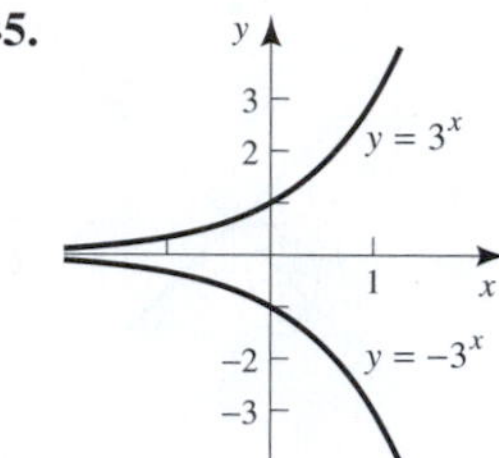

47. $y = 0.5|x| + 3$

49. $y = 2|x^3| - 3$

51. $y = -|x - 2| + 3$

53. No

55. $x^2 < x^3 < x^4 < x^5$, $x^m < x^n$ if $m < n$

57. $x^{1/2} > x^{1/3} > x^{1/4} > x^{1/5}$, $x^{1/m} > x^{1/n}$ if $m < n$

59. **a.** $x = 4$ and $x \approx -0.7666647$;
b. $2^x < x^2$ on $(-\infty, -0.7666647)$ and $(2, 4)$ and $x^2 < 2^x$ on $(-0.7666647, 2)$ and $(4, \infty)$;
c. When x is a large negative number, 2^x is very close to zero, whereas x^2 is very large. For large values of x, 2^x is much larger than x^2.

61. **a.** y_2;
b. $(1.1)^x$ eventually becomes larger than x^3.

63. **a.** is the graph of $y = x^{12}$ since $(-x)^{12} = x^{12}$. (The graph is symmetrical about the y-axis.)
b. is the graph of $y = x^{11}$.

Index